CHEMISTRY
Structure and Dynamics

CHEMISTRY
Structure and Dynamics
Second Edition

James N. Spencer
Franklin and Marshall College

George M. Bodner
Purdue University

Lyman H. Rickard
Millersville University

JOHN WILEY & SONS

ACQUISITIONS EDITOR D. Brennan
MARKETING MANAGER B. Smith
SENIOR PRODUCTION EDITOR Patricia McFadden
SENIOR DESIGNER Karin Gerdes Kincheloe
PRODUCTION MANAGEMENT SERVICES Ingrao Associates

Cover photo Adri Berger/Stone

This book was set in 10.5/12.5 Times Roman by TechBooks and printed and bound by Courier/Westford. The cover was printed by Lehigh Press.

This book is printed on acid-free paper. ∞

ISBN: 0-471-20125-1

Printed in the United States of America

10 9 8 7 6 5 4 3 2 1

To the Instructor:

The annotated version of the text was prepared to describe the philosophy behind the text and to provide a rationale for the use of new methods of presenting traditional chemistry concepts. The new methodologies were implemented because we believe that providing students with experimental data before they are exposed to theoretical models allows them to discover concepts by themselves. For example, in the text, students are led to develop electron configurations from experimental photoelectron spectroscopy data rather than from the Schrödinger model.

Annotations are provided whenever the new methods are introduced to show how these approaches enhance a student's understanding and to acquaint instructors with material that they may not have encountered. Previous editions of the text have included these discussions in the Instructor's Manual; this information now appears in annotated form in the text for more convenient reference. Because these innovative presentations may be unfamiliar to instructors, we have included additional explanation and references for further understanding and to assist the instructor in using these approaches. An example of one new approach is found in the treatment of thermodynamics. The student begins with a thorough background in the origin of heat-producing or -absorbing processes involved in chemical reactions—that is, the making and breaking of bonds—before being introduced to traditional formation parameters. The goal is to have students develop a richer and deeper understanding of what takes place in chemical reactions beyond the usual algorithm of products minus reactants.

Annotations also refer to the ChemActivities in *Chemistry: A Guided Inquiry,* by R. S. Moog and J. J. Farrell, appropriate for that chapter. Each activity introduces experimental data or a model to the student. This is followed by a series of critical thinking questions that require the student to analyze the experimental data in order to discover chemical principles and relationships. We encourage the use of these activities with the text. The ChemActivities may be used to teach the course using cooperative group learning, as a workbook, in recitations, or in class discussions.

We believe these new approaches in content and in process are beneficial to student understanding and offer new and fresh approaches to this course that will enliven the classroom for students and instructors.

JAMES N. SPENCER
GEORGE M. BODNER
LYMAN H. RICKARD

Preface

The second edition of Spencer/Bodner/Rickard, *Chemistry: Structure and Dynamics,* retains its hallmark features—brevity, flexibility, and currency—while increasing its ease of use in the classroom. As in the first edition,

- The text contains 15 core chapters followed by optional modules that the instructor may cover at his or her discretion. Thus considerable flexibility is provided in structuring the course.
- Moog/Farrell's *Chemistry: A Guided Inquiry* will also be available in a new edition to correspond with and extend the text instruction. Moog and Farrell provide a cooperative teaching pedagogy which can be used for recitation sections, as an in-class workbook or as a take-home study guide.
- The text and related materials are based on a hands-on approach to learning chemistry. Students are given models or data where possible and are guided by leading questions to develop the concepts by themselves.

We solicited reviews and instructor comments which led to the following changes in the second edition.

- **Sixty Percent More Problems.** The end-of-chapter problem sets have been increased by some 60%. Each section of the text now has a set of questions designed to take the student through the section following a guided inquiry approach. A student can read through the section and then go to the problem set for questions that check the student's understanding of that section. Additional integrated problems have been added to the problem sets.
- **New Thermodynamics Section.** New sections on thermodynamics have been added, and several other chapters have been restructured.
- **New 2-Color Illustrations.** New 2-color figures have been added, and old figures have been revised.
- **Now in Hardcover.** The text is now available in hardcover, yet remains substantially lower priced than other first-year texts. Instructors have indicated that price is important but that durability is also necessary, so we have moved to a more durable format but have kept the price at a reasonable level.
- **New Annotated Instructor's Edition.** An annotated edition of the text is now available for instructors to explain why nontraditional approaches to some traditional topics have been adopted and to give instructors ideas on how to implement these approaches.

In 1989 the Division of Chemical Education of the American Chemical Society recognized the need to foster the development of alternative introductory chemistry curricula. The Task Force on the General Chemistry Curriculum was created to meet the need. We were members of the Task Force and our book, *Chemistry: Structure and Dynamics,* is one of the products of the work of the Task Force.

This text encourages innovation in the teaching of general chemistry by providing flexible materials that allow instructors to build a custom curriculum appropriate for their students. We achieved this flexibility through the use of a core/modular structure. The text contains the core and chapter appendices called Special Topics. The modules are furnished as a separate shrink wrapped packet that accompanies the text. The core provides the fundamentals that all students need to prepare them for the degree programs in which an introductory chemistry course is a prerequisite. By selecting topics from the chapter appendices and modules instructors can customize their course to fit the needs of their particular students.

The text is written for students taking introductory chemistry for science and math majors. Its flexibility, however, makes it appropriate for most introductory chemistry courses. We decided which concepts to include in the core based on two criteria. First, the concepts should be the fundamental building blocks for understanding chemistry—concepts that provide the basis for both the core and the modules. Second, these concepts should be perceived by *the students* as being directly applicable to their majors or careers.

Chemistry: Structure and Dynamics is characterized by the following key features:

- **Brevity.** The core contains approximately 600 pages with 75 pages of chapter appendices (Special Topics).
- **Flexibility.** Modules and chapter appendices can be added to the core to provide a curriculum that meets the needs of the students at a particular institution.
- **Currency.** New models and current methods of understanding chemical concepts not yet found in traditional texts are used to introduce material.
- **Unifying Themes.** Major themes are used to link the core material into a unified whole.
- **Balanced Coverage.** Organic and biochemical examples are used throughout the text to provide a more balanced coverage of all areas of chemistry.
- **Conceptual Nature.** There is an increased emphasis on conceptual questions and problems at the end of the chapters. A large selection of traditional problems are provided, but these are supplemented with conceptual problems and discussion questions that ask students to explain, describe, or suggest experiments.
- **Case Studies.** A chapter on chemical analysis that is unique to introductory texts has been included in the core. This chapter uses case studies to introduce students to the methods and tools that chemists use to solve real-world problems.

THE CORE

Traditional texts present concepts and principles in isolation, with little, if any, connection between each concept or principle and the rest of the material. In *Chemistry: Structure and Dynamics* unifying themes are used to integrate the core topics.

- The *process of science* theme incorporates experimental data in discussions that traditional texts present as the *product* of scientific process. Our goal is to provide students with the evidence that will allow them to understand why chemists believe what they do.

- A second theme is the interrelationships between chemistry on the macroscopic and microscopic scales. This theme is developed to help students understand both how and why chemists make observations on the macroscopic scale so as to comprehend the microscopic world of atoms and molecules, and then use the resulting understanding of microscopic structure and properties to explain, predict, and—most importantly—control the macroscopic properties of matter.
- Atomic and molecular structure are developed early in the text and then repeatedly used to help students understand the physical and chemical properties of matter.

The final chapter of the core introduces students to the types of real problems that chemists can help solve and to the instruments they use to do so. Although most of the students in the typical introductory chemistry course are not chemistry majors, many of them will have careers that will require them to interact with chemists. Most of these interactions will involve analyses of samples. This chapter introduces information on how samples are analyzed and some of the major tools that chemists use. Case studies representing problems from a variety of fields have been used to present real-world problems. This chapter is not meant to emphasize instrumental analysis or spectral interpretation. Instead it is meant to make instruments seem less like mysterious black boxes and the interpretation of spectra seem less like magic.

THE CHAPTER APPENDICES (SPECIAL TOPICS) AND MODULES

The chapter appendices extend the core material. The core chapter on bonding, for example, covers Lewis structures, molecular geometry, and the concept of polarity. A chapter appendix is then available that includes hybridization, valence bond theory, and molecular orbital theory. The modules are designed to introduce new topics, such as biochemistry, polymer chemistry, nuclear chemistry, and coordination chemistry.

The core/modular approach has advantages over both the traditional 1000-page texts and new shorter texts. When traditional texts are used, sections or even whole chapters are skipped. This can be frustrating to those students who depend heavily on the text for understanding new material because later topics in the text are often explained using concepts that have been omitted. *Chemistry: Structure and Dynamics* avoids this problem by building the modules on the concepts presented in the core. The core covers all concepts that are prerequisite to the modules.

NEW MODELS

The new models used in this text can be divided into three types: (1) data-driven models, (2) models that reflect current understanding of chemical theories, and (3) models that make it easier for students to understand traditional concepts. The development of electron configurations from experimental photoelectron spectral (PES) data is an example of a data-driven model that supports the unifying theme of the process of science by demonstrating to students how experimental data can be used to construct models. This approach gives students a more concrete, and still scientifically correct, foundation on which to base their understanding of electron configuration than does use of the more abstract quantum numbers. The use of experimental data and the graphical representation of that data to develop the gas laws is another example of this type of model. Once the kinetic molecular theory is developed, it is used throughout the text to provide a consistent background for understanding temperature, heat, and equilibrium processes.

Models that reflect current understanding of chemical theories include the replacement of the valence shell electron pair repulsion (VSEPR) theory for predicting molecular geometry with Gillespie's more recent electron domain (ED) model and the use of bond-type triangles to explain the interrelationship of covalent, ionic, and metallic bonding. Bond-type triangles are then used in the text to help explain physical and chemical properties of compounds and to predict properties of new materials.

New models in the text are also used to present familiar concepts in innovative ways that make it easier for students to understand. For example, enthalpies of atom combination replace enthalpies of formation. Because chemistry is concerned with the making and breaking of bonds, thermochemical calculations are done using the energetics of breaking reactant bonds to produce atoms in the gas phase and then allowing these atoms to combine to form product molecules. In the enthalpy of formation approach found in most traditional texts, the standard states are the elements in their stable states of aggregation. This construct places another concept between the student and the idea that chemical reactions liberate or absorb heat through the making and breaking of bonds. The major strength of the atom combination method is the use of the powerful visual model of reactants being broken down into gaseous atoms and then recombining to form products. The advantage to this method is that bond breaking always requires an input of energy, whereas bond making produces energy. Instructors accustomed to thinking of enthalpy changes in terms of enthalpies of formation may at first find this method more awkward. However, our experience indicates that students find this approach much easier to visualize than enthalpies of formation.

A major advantage of the atom combination approach is that the same standard states and reaction diagrams are used for all thermodynamic parameters. Traditionally, third-law entropies are used in conjunction with enthalpies based on elemental standard states to introduce free energy. The use of third-law entropies that are based on yet a different concept and standard state further confuses students. The atom combination approach clearly shows students the origin of the entropies of substances and, because of the direct relationship with enthalpies and free energies, makes all these concepts more accessible to students.

Another example of the use of models that reflect current understanding of chemical theories is the introduction of average valence electron energies (AVEE), calculated from PES data, which describe how tightly an atom holds on to its electrons. AVEE values are used to predict which elements will form cations and which will form anions and to explain the diagonal line in the periodic table that separates metals from nonmetals. AVEE is then used to develop the concept of electronegativity, thus giving electronegativity a more concrete meaning for students. Not only do the students develop the electronegativity concept from the same data used to determine the electronic structure of atoms, but they also see how electronegativities relate to oxidation numbers, partial charges, and formal charges. This theme is carried through the text and provides a unifying thread for oxidation–reduction, resonance structures, and polarity.

USING THE CORE TEXT AND MODULES

The core text consists of 15 chapters. With the exception of Chapter 2, the core chapters are designed to be covered in order. Chapter 2 covers the mole, chemical equations, molarity, and stoichiometry. This information is provided early in the text for those instructors who traditionally cover these topics at the beginning of the course. However, the mole concept is used only qualitatively in Chapters 3–5. Mole calculations and stoichiometry are not required until Chapter 6. Molarity and solution stoichiometry are not required until Chapter 8. Instructors can choose to cover portions of Chapter 2 as fits their curriculum.

The core may be supplemented using the chapter appendices, called Special Topics (found at the end of Chapters 4, 6, 8, 10, 11, 12, and 14), or modules. The chapter appendices extend the core topics found in the chapters they accompany. The material in the chapter appendices may be skipped at the discretion of the instructor because subsequent core chapters depend only on the concepts covered in the core chapters. Topics covered in the core chapters and their accompanying appendices are listed in the table of contents.

The modules may be inserted at various points in the text depending on the core material prerequisite to the modules. The table below lists the core chapters that are prerequisite for each module.

Prerequisites for Modules

Module	Prerequisites
Biochemistry	Chapter 11 and Organic Chemistry: Functional Groups Module
Complex Ion Equilibria	Chapter 11
Materials Science	Chapter 9
Chemistry of the Nonmetals	Chapter 12
Nuclear Chemistry	Chapter 14
Organic Chemistry: Structure and Nomenclature of Hydrocarbons	Chapter 7, Chapters 3 and 4 Appendices
Organic Chemistry: Functional Groups	Structure and Nomenclature Module
Organic Chemistry: Reaction Mechanisms	Functional Groups Module
Polymer Chemistry	Chapter 7, and Organic Chemistry: Functional Groups Module, Chapters 3 and 4 Appendices
Solubility Equilibria	Chapter 11
Transition-Metal Chemistry	Chapter 5, Chapters 3 and 4 Appendices

TO THE STUDENT

This text, *Chemistry: Structure and Dynamics,* differs from most texts currently available for introductory chemistry. The text contains a core of material that provides a basis for understanding the fundamental concepts of chemistry, and presents new models to teach familiar concepts in ways that should make the concepts easier to grasp. Consistent themes throughout the text tie together many seemingly isolated topics.

The chapters in this text are designed to be read linearly, from the beginning to the end, rather than reading individual topics within the chapter. It is important to answer the Checkpoints and to work through the Exercises as you encounter them in your reading. Important new concepts are sometimes introduced in the Checkpoints and Exercises. In addition, answering the Checkpoints and working the Exercises will change your reading of the chapter from passive to active reading. The checkpoints will assist you in determining if you have understood what you have just read. Answers to the checkpoints can be found in Appendix D. The exercises are problems similar to the problems found at the end of the chapter. Solutions to selected end-of-chapter problems are found in Appendix C. Work the assigned end-of-chapter problems and then check your answer using Appendix C.

SUPPLEMENTS

Chemistry; A Guided Inquiry, written by Richard S. Moog and John J. Farrell, of Franklin and Marshall College. This supplement facilitates implementation of cooperative learning in general chemistry and can be used as part of a recitation section, as a workbook, or as a means of interacting with students in the classroom.

This supporting book uses all of the new approaches featured in *Chemistry: Structure and Dynamics,* and features problem assignments from the Core text. The book uses guided inquiry in which data, written descriptions, models, and figures are used to develop chemical concepts.

Instructor's Manual, prepared by James Spencer, George Bodner, Lyman Rickard, and Alex Grushow. The Instructor's Manual highlights material in the Core text, which differs from more traditional texts, focusing on how the text's method of presentation can best be utilized in the course. The manual contains complete solutions to all text problems as well as additional case studies. Further information is given on PES, atom combination parameters, bond-type triangles, AVEE, and partial charge calculations.

Annotated Instructor's Edition, prepared by James Spencer, George Bodner, and Lyman Rickard. The Annotated Instructor's Edition explains why nontraditional approaches to some traditional topics have been adopted and provides information on how to implement these approaches.

Student Solutions Manual, prepared by Alex Grushow. The Solutions Manual provides complete solutions to all text problems.

Computerized Test Bank, prepared by George Bodner. This Test Bank contains multiple-choice and short-answer questions.

Test Bank, prepared by George Bodner. This Test Bank contains multiple-choice and short-answer questions designed for an introductory chemistry course.

ACKNOWLEDGMENTS

There are many people who contributed to this textbook. First are all the members of the Task Force on the General Chemistry Curriculum, without whose discussions and ideas this project would never have been initiated. We are also indebted to the many reviewers:

Janice Alexander
Flathead Valley Community College

Linda Allen
Louisiana State University

Dennis M. Anjo
Cal State, Long Beach

Chris Bailey
Wells College

David W. Ball
Cleveland State University

Jay Bardole
Vincennes University

Elisabeth Bell-Loncella
University of Pittsburgh, Johnstown

Sheila Cancella
Raritan Valley Community College

David A. Cleary
Gonzaga University

Martin Cowie
University of Alberta

Michael Doyle
Trinity University

Robert Eierman
University of Wisconsin—Eau Claire

William Evans
University of California, Irvine

Daniel Freedman
SUNY-New Paltz

Keith Hansen
Lamar University

Lee Hansen
Brigham Young University

Tom Gilbert
Northeastern University

Thomas Grover
Gustavus Adolphus College

L. Peter Gold
Pennsylvania State University

Stan Grenda
University of Nevada, Las Vegas

Eugene Grimley III
Elon College

Mark Iannone
Millersville University

Keith Kester
Colorado College

Leslie Kinsland
University of Southwestern Louisiana

Nancy Konigsberg-Kerner
University of Michigan, Ann Arbor

David Lewis
Colgate University

Robert Loeschen
California State University, Long Beach

Baird Lloyd
Miami University of Ohio

David MacInnes Jr.
Guilford College

Doug Martin
Sonoma State University

Claude Mertzenich
Luther College

Patricia Metz
Texas Tech University

Susan Morante
Mont Royal College

Edward Paul
Stockton College

E. B. Robertson
University of Calgary

Carey Rosenthal
Drexel University

Doug Rustad
Sonoma State University

Patricia Schroeder
Johnson County Community College

Karl Sohlberg
Drexel University

William Stanclift
Northern Virginia Community College, Annandale

Wayne E. Steinmetz
Pomona College

Robert Stewart, Jr.
Miami (OH) University

Robert L. Swofford
Wake Forest University

Sandra Turchi
Millersville University

John Woolcock
Indiana University of Pennsylvania

Andrew Zanella
Claremont College

Particular contributions were made by John Farrell and Rick Moog of Franklin and Marshall College; Ron Gillespie of McMaster University; Dudley Herschbach of Harvard University; Lee Allen of Princeton University; Gordon Sproul of the University of South Carolina at Beaufort; and Alex Grushow at Rider University.

It is a particular pleasure to acknowledge the support and assistance of the staff at John Wiley and Sons, Inc. David Harris, our editor, guided and encouraged us during the writing of this text. We appreciate his support for a new and different text for introductory chemistry. Jennifer Yee, Debbie Brennan, Patricia McFadden, Cathy Donovan, and Suzanne Ingrao also attended to countless details.

Finally, to Kathy and Lynette and Killer, we owe a debt for their patience and encouragement.

JAMES N. SPENCER
GEORGE M. BODNER
LYMAN H. RICKARD

Contents

Chapter 6
Gases 211

SPECIAL TOPICS

Chapter 7
Making and Breaking of Bonds 252

Chapter 8
Liquids and Solutions 298

SPECIAL TOPICS

Chapter 12 Oxidation-Reduction Reactions 495

SPECIAL TOPICS

Chapter 13 Chemical Thermodynamics 551

Chapter 14
Kinetics 593

SPECIAL TOPICS

Chapter 15
Chemical Analysis 640

Appendix A

Appendix B

CHEMISTRY
Structure and Dynamics

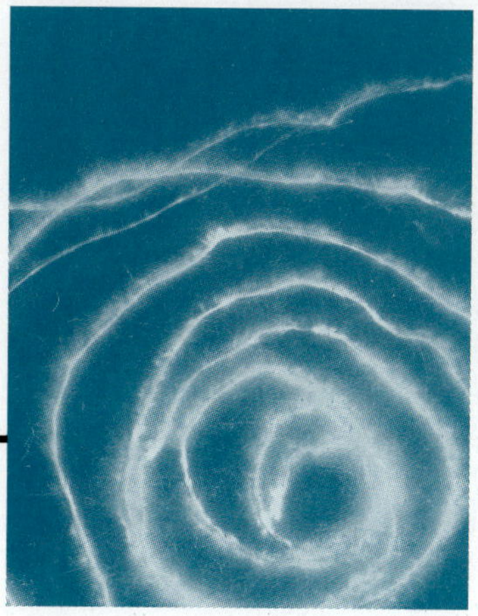

Chapter One

ELEMENTS AND COMPOUNDS

1.1 Chemistry: A Definition

Chapter 1 begins immediately with chemistry rather than the traditional unit conversions, scientific notation, and significant figures. This is not to imply that these are not important topics. However, our philosophy is that a chemistry course should begin with chemistry. The first chapter sets the tone of the book and the course.

ChemActivity 1 in Chemistry: A Guided Inquiry[1] *is designed to accompany this chapter.*

[1]Richard S. Moog and John J. Farrell, *Chemistry: A Guided Inquiry*, John Wiley & Sons, Inc. New York, NY, 2002.

It seems logical to start a book of this nature with the question: What is chemistry? Most dictionaries define chemistry as *the science that deals with the composition, structure, and properties of substances and the reactions by which one substance is converted into another*. Knowing the definition of chemistry, however, is not the same as understanding what it means.

Perhaps the best way to understand the nature of chemistry is to look at examples of what it isn't. In 1921, a group from the American Museum of Natural History began excavations at an archaeological site on Dragon-Bone Hill, near the town of Chou-k'outien, 34 miles southwest of Beijing, China. Fossils found at this site were assigned to a new species, *Homo erectus pekinensis,* commonly known as Peking man. The excavations suggest that for at least 500,000 years, people have known enough about the properties of stone to make tools, and they have been able to take advantage of the chemical reactions involved in combustion in order to cook food. But even the most liberal interpretation would not allow us to call this chemistry because of the absence of any evidence of control over these reactions or processes.

The ability to control the transformation of one substance into another can be traced back to the origin of two different technologies: brewing and metallurgy. People have been brewing beer for at least 12,000 years, since the time when the first cereal grains were cultivated, and the process of extracting metals from ores has been practiced for at least 6000 years, since copper was first produced by heating the ore malachite.

But brewing beer by burying barley until it germinates and then allowing the barley sprouts to ferment in the open air wasn't chemistry. Neither was extracting copper metal from one of its ores because this process was carried out without any understanding of what was happening, or why. Even the discovery around 3500 B.C. that copper mixed with 10% to 12% tin gave a new metal that was harder than copper, and yet easier to melt and cast, was not chemistry. The preparation of bronze was a major breakthrough in metallurgy, but it didn't provide us with an understanding of how to make other metals.

Between the sixth and the third centuries B.C., the Greek philosophers tried to build a theoretical model for the behavior of the natural world. They argued that the world was made up of four primary, or *elementary,* substances: fire, air, earth, and water. These substances differed in two properties: hot versus cold, and dry versus wet. Fire was hot and dry; air was hot and wet; earth was cold and dry; water was cold and wet.

This model was the first step toward the goal of understanding the properties and compositions of different substances and the reactions that convert one substance to another. But some elements of modern chemistry were still missing. This model could explain certain observations of how the natural world behaved, but it couldn't predict new observations or behaviors. It was also based on pure speculation. In fact, its proponents weren't interested in using the results of experiments to test the model:

Modern chemistry is based on certain general principles.

- **One of the goals of chemistry is to recognize patterns in the behaviors of different substances.** An example of this might be the discovery in 1794 by the French chemist Antoine Lavoisier that many substances which burn in air gain weight.
- **Once a pattern is recognized, it should be possible to develop a model that explains these observations.** Lavoisier concluded that substances that

burn in air combine with the oxygen in the air to form products that weigh more than the starting material.

- **These models should allow us to predict the behavior of other substances.** In 1869, Dmitri Mendeléeff used his model for the behavior of the known elements to predict the properties of elements that had not yet been discovered.
- **When possible, the models should be quantitative.** They should not only predict what happens, but by how much.
- **The models should be able to make predictions that can be tested experimentally.** Mendeléeff's periodic table was accepted by other chemists because of the agreement between his predictions and the results of experiments based on the predictions.

In essence, *chemistry is an experimental science*. Experiment serves two important roles. It forms the basis of observations that define the problems which theories must explain, and it provides a way of checking the validity of new theories. This text emphasizes an experimental approach to chemistry. As often as possible, it presents the experimental basis of chemistry before the theoretical explanations of these observations.

1.2 Physical and Chemical Properties

At the end of each chapter in the text there is a set of problems that corresponds to each chapter section. After reading a section, students may be assigned to do the corresponding problems. The problems range from essay questions to numerical calculations and are designed to take the student through the major concepts in each section.

Chemists differentiate between the composition and structure of a substance. The **composition** is described in terms of the number and types of atoms in the substance. Vitamin B_{12}, for example, is described in terms of the formula $C_{63}H_{88}N_{14}O_{14}PCo$, which suggests that each molecule contains 63 carbon atoms, 88 hydrogen atoms, 14 nitrogen atoms, 14 oxygen atoms, and one atom each of phosphorus and cobalt. The **structure** of a substance describes the arrangement of these atoms in three-dimensional space. The structure of Vitamin B_{12} is shown in Figure 1.1.

Every substance has a unique set of chemical and physical properties that can be used to identify that substance. **Physical properties** are characteristic properties of the substance, by itself. Physical properties include color, hardness, density, vapor pressure, melting point, and boiling point. Although the appearance of a substance changes when it melts or boils, the composition of the substance stays the same. Butane (C_4H_{10}), for example, has the same number of atoms of carbon and hydrogen and the same three-dimensional structure when the liquid butane in a disposable cigarette lighter escapes to form a gas at room temperature and atmospheric pressure.

The **chemical properties** of a substance are observed when it undergoes a chemical reaction. The fact that butane burns in the presence of oxygen and a spark, for example, is one of the most important chemical properties of this substance. In the course of the reaction, the butane is transformed into two new substances, carbon dioxide and water. Table 1.1 gives examples of several common chemical and physical properties.

1.3 Elements, Compounds, and Mixtures

Matter is defined as anything that has mass and occupies space. All substances that we encounter—whether natural or synthetic—are matter. Matter can be divided

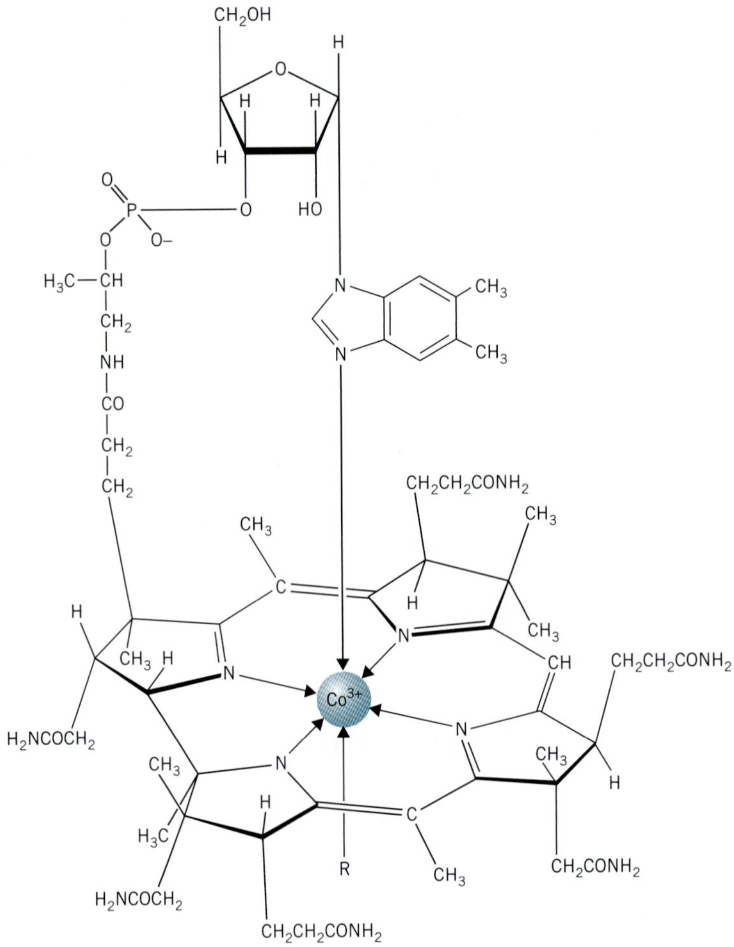

Fig. 1.1 The structure of vitamin B_{12}.

into pure substances and mixtures, as shown in Figure 1.2. Pure substances can be further divided into elements and compounds.

Pure substances have a constant composition. Water, for example, is always 88.81% oxygen by weight, regardless of where it is found. When pure, the salt used to flavor food has exactly the same composition regardless of whether it was dug from mines beneath the surface of the earth or obtained by evaporating seawater. No matter where it comes from, salt always contains 1.54 times as much chlorine by weight as sodium. Pure substances also have constant chemical and physical properties. Pure water always freezes at 0°C and boils at 100°C at atmospheric pressure. Since it always has the same composition, salt that is pure always has the same chemical and physical properties.

Table 1.1
Examples of Physical and Chemical Properties

Lead has a density of 11.34 g/cm^3	Physical
Milk sours	Chemical
Sulfur is yellow	Physical
Gasoline is flammable	Chemical
Iron rusts	Chemical
Sugar melts at 185°C	Physical
Diamonds are hard	Physical

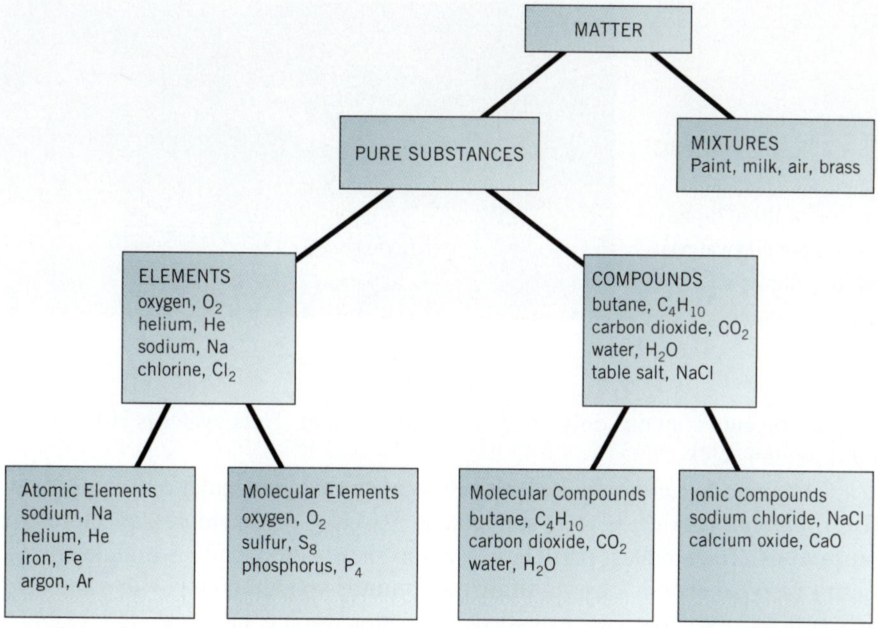

Fig. 1.2 The classification of matter.

Mixtures, such as a cup of coffee, have different compositions from sample to sample, and therefore varying properties. If you are a coffee drinker you will have noted that cups of coffee from your home, the college cafeteria, and a gourmet coffeehouse aren't the same. They vary in appearance, aroma, and flavor because of differences in the composition of this mixture.

Pure substances can be divided into elements and compounds. **Elements** are substances that contain only one kind of atom. To date, 114 elements have been discovered. They include a number of substances with which you are familiar, such as the oxygen in the atmosphere, the aluminum in aluminum foil, the iron in nails, the copper in electrical wires, and the mercury in thermometers. Elements are the fundamental building blocks from which all other substances are made.

Imagine cutting a piece of gold metal in half and then repeating this process again and again and again. In theory, we should eventually end up with a single gold atom. If we tried to split this atom in half we would end up with something that no longer retains any of the properties of gold. An **atom,** in other words, is the smallest particle of an element that has any of the properties of that element.

Compounds (such as water or table salt) contain more than one element combined in fixed proportions. Water is composed of the elements hydrogen and oxygen in the ratio of two atoms of hydrogen to one atom of oxygen. Table salt is composed of the elements sodium and chlorine in the ratio of one atom of sodium per atom of chlorine.

If we tried to divide a compound, such as water, into infinitesimally small portions we would eventually end up with a single molecule of water containing two hydrogen atoms and one oxygen atom. If we tried to break this molecule into its individual atoms we would no longer have water. A **molecule** is therefore the smallest particle that has any of the properties of a compound.

The composition of a compound can be represented by a **chemical formula** as shown in Figure 1.3. The subscripts in a chemical formula represent the relative numbers of atoms present in the compound. By convention, no subscript is written

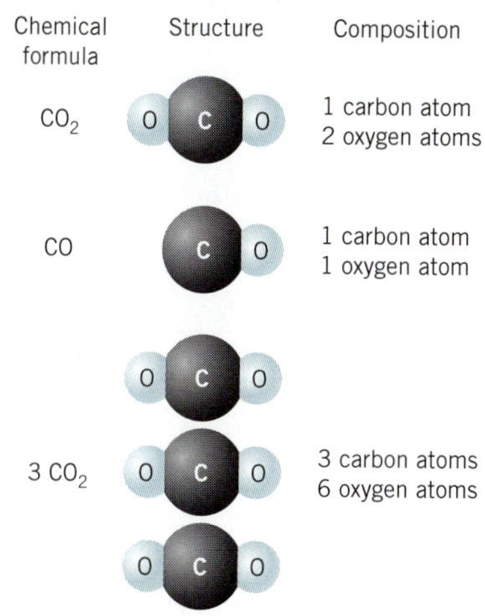

Fig. 1.3 The formula CO_2 describes a molecule that contains one carbon atom and two oxygen atoms. The formula CO tells us that this molecule consists of one carbon and one oxygen atom. A collection of three CO_2 molecules would be described by writing "3 CO_2."

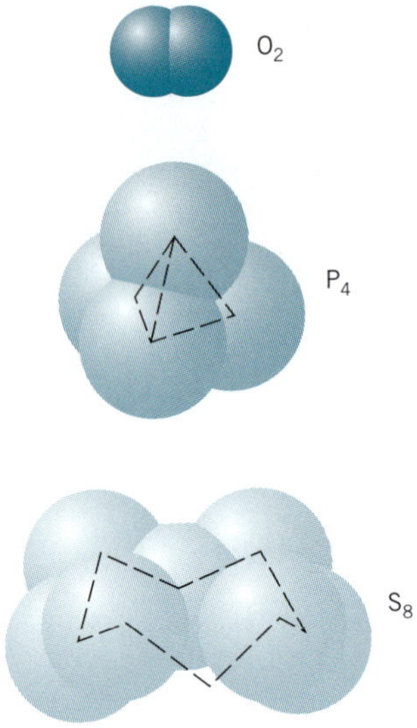

Fig. 1.4 At room temperature, oxygen exists as O_2 molecules, phosphorus forms P_4 molecules, and sulfur forms cyclic S_8 molecules.

Fig. 1.5 It isn't obvious from their appearance that one of these liquids is a compound (water) and the other is an element (mercury).

when a molecule contains only one atom of an element. Thus, water is H_2O and carbon dioxide is CO_2.

Compounds can be divided into two categories: molecular and ionic. Water (H_2O), carbon dioxide (CO_2), and butane (C_4H_{10}) are examples of **molecular compounds.** The smallest particle in each of these compounds is a molecule that doesn't carry an electric charge. **Ionic compounds** contain both positive and negative particles that form an extended three-dimensional network. The chemical formula of an ionic compound describes the ratio of positive and negative particles in this network. Sodium chloride (NaCl) is the best-known example of an ionic compound.

Elements can also exist in the form of molecules, but these molecules are composed of identical atoms. The oxygen we breathe, for example, consists of molecules that contain two oxygen atoms, O_2 (Figure 1.4). Elemental phosphorus molecules are composed of four phosphorus atoms (P_4), and elemental sulfur contains molecules composed of eight sulfur atoms (S_8).

Elements and compounds are classified as pure substances because they have constant properties and composition. As a result, a substance can't be classified as an element or compound on the basis of its physical properties (Figure 1.5).

The only way to determine whether a substance is an element or a compound is to try to break it down into simpler substances. Elements can be broken down into only one kind of atom.

If a substance can be decomposed into more than one kind of atom, it is a compound. Water, for example, can be decomposed into hydrogen and oxygen by passing an electric current through the liquid, as shown in Figure 1.6.

In a similar fashion, salt can be decomposed into its elements—sodium and chlorine—by passing an electric current through a molten sample. Table 1.2 provides examples of common elements, compounds, and mixtures.

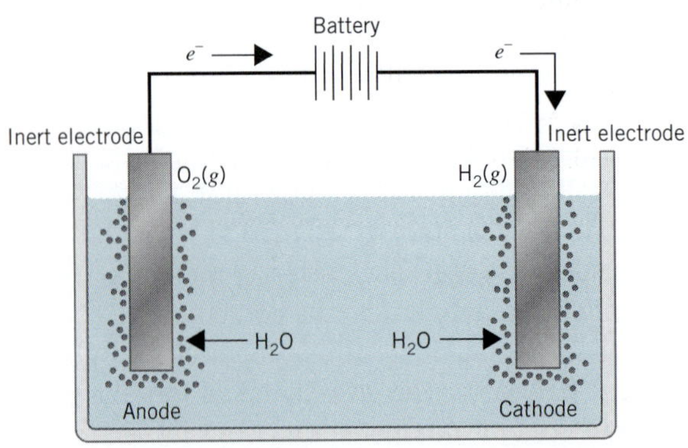

Fig. 1.6 Electrolysis of water results in the production of oxygen gas and hydrogen gas.

Exercise 1.1

At breakfast one morning, one of the authors was faced with a choice between two cereals. One was raisin bran and the other was "Crispix," which contains flakes of rice fused with flakes of corn. Describe the characteristic properties of these cereals that makes one an analogy for a mixture of elements and the other an analogy for a chemical compound.

Solution

Raisin bran has the following characteristic properties of a *mixture*.

- The cereal does not have a constant composition; the ratio of raisins to bran flakes changes from sample to sample.
- It is easy to physically separate the two "elements," to pick out the raisins, for example, and eat them separately.

Crispix has some of the characteristic properties of a *compound*.

- The ratio of rice flakes to corn flakes is constant; it is 1:1 in every sample.
- There is no way to separate the "elements" without breaking the bonds that hold them together.

Table 1.2
Examples of Elements, Compounds, and Mixtures

Oxygen gas	Element
Carbon dioxide gas	Compound
Gasoline	Mixture
Distilled water	Compound
Tap water	Mixture
Sugar	Compound
Air	Mixture

In addition to problem sets designed to guide students through a section and reemphasize the major concepts, checkpoints appear frequently throughout the text. Checkpoints allow students to test their understanding of the material they have read. Students should be encouraged to do the checkpoints as they read. The checkpoints are designed to be an integral and important part of the chapter. Doing the checkpoints can help change the student's reading of the text from passive to active reading. Answers to the checkpoints are given at the end of the text.

1.4 Atomic Symbols

As we have seen, chemists use a shorthand notation to save both time and space when describing atoms. Each element is represented by a unique symbol. Most of these symbols make sense because they are derived from the name of the element.

H = hydrogen	B = boron
C = carbon	N = nitrogen
O = oxygen	P = phosphorus
Se = selenium	Si = silicon
Mg = magnesium	Br = bromine
Al = aluminum	Ca = calcium
Cr = chromium	Zn = zinc

Symbols that don't seem to make sense come from the Latin or German names of the elements. Fortunately, there are only a handful of elements in this category.

Ag = silver	Na = sodium
Au = gold	Pb = lead
Cu = copper	Sb = antimony
Fe = iron	Sn = tin
Hg = mercury	W = tungsten
K = potassium	

Additional symbols and names of the elements can be found at the end of the text.

➤ CHECKPOINT

Describe the difference between the symbols 8 S and S_8.

1.5 Evidence for the Existence of Atoms

Most students in a beginning chemistry course already believe in atoms. If asked to describe the evidence on which they base this belief, however, they hesitate. Our senses argue against the existence of atoms. The atmosphere in which we live feels like a continuous fluid. We don't feel bombarded by collisions with individual particles in the air. The water we drink looks like a continuous fluid. We can take a glass of water, pour out half, divide the remaining water in half, and repeat this process again and again, without ever appearing to reach the point at which it is impossible to divide it one more time. Because our senses suggest that matter is continuous, it isn't surprising that the debate about the existence of atoms goes back to the ancient Greeks and continued well into the twentieth century.

Experiments with gases that first became possible at the turn of the nineteenth century led John Dalton in 1803 to propose a model for the atom based on the following assumptions:

- Matter is made up of atoms that are indivisible and indestructible.
- All atoms of an element are identical.
- Atoms of different elements have different weights and different chemical properties.
- Atoms of different elements combine in simple whole-number ratios to form compounds.
- Atoms cannot be created or destroyed. When a compound is decomposed, the atoms are recovered unchanged.

Dalton's assumptions form the basis of the modern atomic theory. However, modern experiments have shown that not all atoms of an element are exactly the same and that atoms can be broken down into subatomic particles. It is only recently that direct evidence for the existence of atoms has become available. Using the scanning tunneling microscope developed in the 1980s, scientists have finally been able to observe and even manipulate individual atoms.

Research for the New Millennium

SCANNING TUNNELING MICROSCOPY

Many chapters contain a discussion of recent research related to one of the concepts introduced in the chapter. These sections, Research for the new Millennium, were designed to help students understand the role of chemistry in today's world.

In 1982 a paper describing a new technique known as *scanning tunneling microscopy* (STM) was published by a group of scientists from the IBM Research Laboratory in Zurich [G. Binnig, H. Rohrer, Ch. Geber, and E. Weibel, *Physical Review Letters,* **49,** 57 (1982)].[1] This paper was built on prior work in the same laboratory that showed that the current that flows between the tip of a sharp piece of tungsten metal and the surface of platinum metal over which it is moved is very sensitive to the distance between the tip and the metal surface. In their 1982 paper, the IBM researchers showed how this information could be used to study the surface of the metal. As the tungsten tip moves over the surface being studied, it is raised or lowered as needed to give a constant current. Measurements of the motion of the tip are then recorded and analyzed to yield an image of individual atoms or molecules.

[1]Binnig and Rohrer received the Nobel Prize in physics in 1986 for the development of STM.

STM allows scientists to study the physical nature and chemical composition of the surface of solids on the atomic level. Of particular commercial importance to the microelectronics industry is the ability of STM to observe crystal defects on the surface of silicon. With the ever-decreasing size of integrated circuit components these defects become very important.

Experiments have demonstrated that the STM probe can be used to move individual atoms or molecules. The "molecular man" shown in Figure 1.7 was formed by moving 28 CO molecules into position on a platinum surface. The ability to manipulate individual atoms or molecules has the potential to allow scientists to control reactions of single atoms and molecules. This could lead to the production of new chemical substances that are not possible using normal chemical methods.

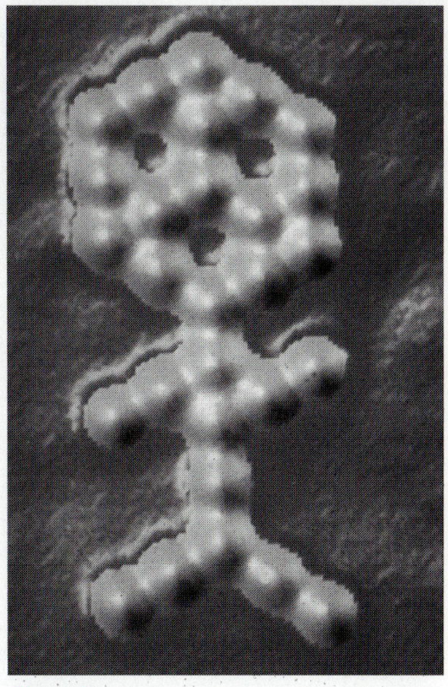

Fig. 1.7 This "molecular man" was formed by moving carbon monoxide molecules into position on a platinum surface.

1.6 The Structure of Atoms

Contrary to Dalton's model of the atom, atoms are not indivisible. Today, we recognize that atoms are composed of the three fundamental subatomic particles listed in Table 1.3: **electrons, protons,** and **neutrons.** Chemists normally refer to these particles as fundamental because they are the building blocks of all atoms.[2] Although gold atoms and oxygen atoms are quite different from one another, the electrons, protons, and neutrons found within gold atoms are indistinguishable from the electrons, protons, and neutrons found within oxygen atoms.

The electric charges on an electron and a proton are given in the third column of Table 1.3 in units of coulombs (C), the fundamental unit of electric charge. We are often more interested in the relationship between two measurements than we are in their **absolute** value. In such a case we calculate the ratio between the two measurements. Because the magnitude of the charge on an electron and a proton is the same—they differ only in the sign of the charge—the **relative** charges on these particles are -1 and $+1$, as shown in the fourth column in this table.

The data in Table 1.3 suggest that the charge on one proton exactly balances the charge on an electron, and vice versa. Thus, atoms are electrically neutral when they contain the same number of electrons and protons.

The absolute masses of the three subatomic particles are given in the fifth column of Table 1.3 in units of grams, the fundamental unit for measurements of mass. The last column gives the relative mass of these particles. Because the mass of a proton is almost the same as that of a neutron, both particles are assigned a relative mass of 1. Because the ratio of the mass of an electron to that of a proton is very small, the electron is assigned a relative mass of zero.

Table 1.3

Fundamental Subatomic Particles

Particle	Symbol	Absolute Charge (C)	Relative Charge	Absolute Mass (g)	Relative Mass
Electron	e^-	-1.60×10^{-19}	-1	9.11×10^{-28}	0
Proton	p^+	1.60×10^{-19}	$+1$	1.673×10^{-24}	1
Neutron	n^0	0	0	1.675×10^{-24}	1

[2]Protons and neutrons are not fundamental particles in the strictest sense because they are composed of still smaller particles, the so-called up and down quarks.

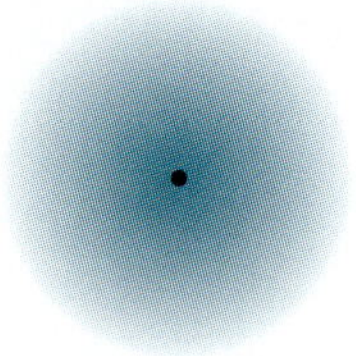

Fig. 1.8 The exact position of an electron in an atom cannot be determined. Often electrons are described as a cloud of negative charge spread out in the space surrounding the nucleus. The boundary of the atom is not a physical boundary but instead is a volume that contains most of the electron density of the atom.

The protons and neutrons in an atom are concentrated in the **nucleus,** which contains most of the mass of the atom. For example, 99.97% of the mass of a carbon atom can be found in the nucleus of that atom. The term *nucleus* comes from the Latin word meaning "little nut." This term was chosen to convey the image that the nucleus of an atom occupies an infinitesimally small fraction of the volume of an atom. The radius of an atom is approximately 10,000 times larger than its nucleus. To appreciate the relative size of an atom and its nucleus, imagine that we expand an atom until it is the size of the Superdome. The nucleus would be the size of a small pea suspended above the 50-yard line with electrons moving throughout the arena. Thus, most of the volume of an atom is empty space through which the electrons move.

It is impossible to determine the exact position or path of an electron. Because of this chemists often visualize electrons in two ways. Electrons can be described as very small particles or as a cloud of negative charge spread out through the volume of space surrounding the nucleus that corresponds to the size of the atom (Figure 1.8).

1.7 Atomic Number and Mass Number

The number of protons in the nucleus of an atom determines the identity of the atom. Every carbon atom ($Z = 6$) has 6 protons in the nucleus of the atom, whereas sodium atoms ($Z = 11$) have 11. Each element has therefore been assigned an **atomic number** (Z) between 1 and 114 that describes the number of protons in the nucleus of an atom of that element. Neutral atoms contain enough electrons to balance the charge on the nucleus. The nucleus of a neutral carbon atom would be surrounded by 6 electrons; a neutral sodium atom would contain 11 electrons.

The nucleus of an atom is also described by a **mass number** (A), which is the sum of the number of protons and neutrons in the nucleus. The difference between the mass number and the atomic number of an atom is therefore equal to the number of neutrons in the nucleus of that atom. A carbon atom with a mass number of 12 would contain 6 protons and 6 neutrons. A sodium atom with a mass number of 23 would contain 11 protons and 12 neutrons.

A shorthand notation has been developed to describe the number of neutrons and protons in the nucleus of an atom. The atomic number is written in the bottom left corner of the symbol for the element and the mass number is written in the top left corner: ${}^{A}_{Z}X$. The diagrams in Figure 1.9 can be used to represent atoms that would be given the symbols ${}^{12}_{6}C$ and ${}^{23}_{11}Na$. Each element, however, has a unique atomic number and a unique symbol. It is therefore redundant to give both the symbol for the element and its atomic number. The atoms in Figure 1.9 are therefore usually written as ${}^{12}C$ and ${}^{23}Na$.

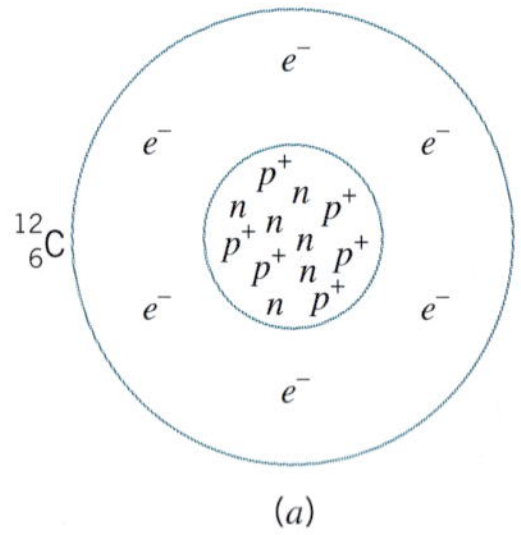

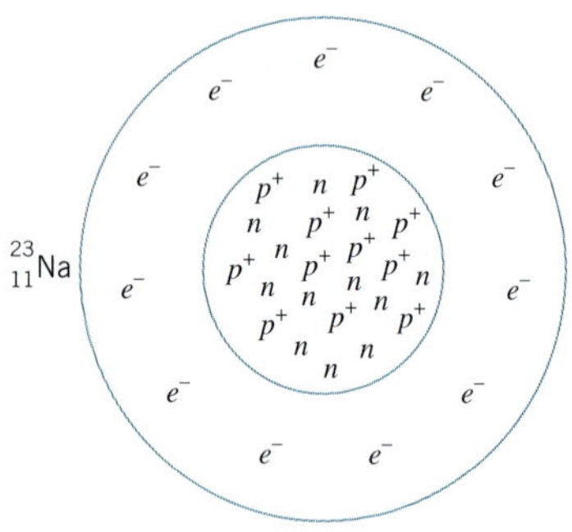

Fig. 1.9 (*a*) A neutral ${}^{12}_{6}C$ atom consists of a nucleus of 6 protons and 6 neutrons surrounded by 6 electrons to balance the positive charge on the nucleus. (*b*) A neutral ${}^{23}_{11}Na$ atom contains 11 protons, 12 neutrons, and 11 electrons.

1.8 Isotopes

The number of protons in the nucleus of an atom determines the identity of the atom. As a result, all atoms of an element must have the same number of protons. But they don't have to contain the same number of neutrons.

Atoms with the same atomic number but different numbers of neutrons are called **isotopes.** Carbon, for example, has three naturally occurring isotopes: ${}^{12}C$, ${}^{13}C$, and ${}^{14}C$. ${}^{12}C$ has 6 protons and 6 neutrons; ${}^{13}C$ has 6 protons and 7 neutrons; ${}^{14}C$ has 6 protons and 8 neutrons.

Table 1.4

Common Isotopes of Some of the Lighter Elements

Isotope	Natural Abundance (%)	Mass (g)	Mass (amu)
^{1}H	99.985	1.6735×10^{-24}	1.007825
^{2}H	0.015	3.3443×10^{-24}	2.01410
^{6}Li	7.42	9.9883×10^{-24}	6.01512
^{7}Li	92.58	1.1650×10^{-23}	7.01600
^{10}B	19.7	1.6627×10^{-23}	10.0129
^{11}B	80.3	1.8281×10^{-23}	11.00931
^{12}C	98.892	1.9926×10^{-23}	12.00000
^{13}C	1.108	2.1592×10^{-23}	13.003
^{16}O	99.76	2.6560×10^{-23}	15.99491
^{17}O	0.04	2.8228×10^{-23}	16.99913
^{18}O	0.20	2.9888×10^{-23}	17.99916
^{20}Ne	90.51	3.3198×10^{-23}	19.99244
^{21}Ne	0.27	3.4861×10^{-23}	20.99384
^{22}Ne	9.22	3.6518×10^{-23}	21.99138

Each element occurs in nature as a mixture of its isotopes. Consider a "lead" pencil, for example. These pencils don't contain the element lead, which is fortunate because many people chew on pencils and lead can be very toxic. They contain a substance once known as "black lead" and now known as graphite.[3] The graphite in a pencil contains a mixture of ^{12}C, ^{13}C, and ^{14}C atoms. The three isotopes, however, do not occur to the same extent. Most of the atoms (98.892%) are ^{12}C, a small percentage (1.108%) are ^{13}C, and only 1 in about 10^{12} is the radioactive isotope of carbon, ^{14}C. The percentage of atoms occurring as a given isotope is referred to as the **natural abundance** of that isotope. Some elements, such as fluorine, have only one naturally occurring isotope, ^{19}F, whereas other elements have several, as shown in Table 1.4.

The mass of each isotope in Table 1.4 is given in terms of both an absolute and a relative measurement. The absolute measurement is given in units of grams (g), the fundamental unit for measurements of mass. Because the mass of an atom is so very small, it is often more useful to know the relative mass of the atom in units of **atomic mass units (amu)**. The unit of amu is defined such that the mass of an atom of ^{12}C is exactly 12 amu.

➤ CHECKPOINT

There are two naturally occurring isotopes of lithium, ^{6}Li and ^{7}Li. If you selected 10,000 lithium atoms at random, how many would be ^{6}Li? How many would be ^{7}Li?

Exercise 1.2

Calculate the ratio of the mass of an ^{1}H atom to that of a ^{12}C atom when the masses are measured in units of grams. Use this ratio to calculate the mass of an ^{1}H atom in units of amu.

Solution

The relative mass of ^{1}H and ^{12}C atoms can be calculated from their absolute masses in grams.

$$\frac{^{1}H}{^{12}C}\ 5\ \frac{1.6735\ 3\ 10^{2\ 24}}{1.9926\ 3\ 10^{2\ 23}}\ 5\ 0.083986$$

[3]The graphite in most pencils is mixed with clay; the more clay, the harder the pencil.

If the mass of a ^{12}C atom is exactly 12 amu, then the mass of an ^{1}H atom to five significant figures must be 1.0078 amu.

$$12 \text{ amu} \times 0.083986 = 1.0078 \text{ amu}$$

Most students have been introduced to unit conversions, scientific notation, and significant figures in high school and therefore only require a review to understand these concepts. As each of these topics is first encountered in Chapter 1, students are referred to Appendix A for a brief review if needed. In addition, there is a worksheet in Appendix A that can be assigned for reviewing significant figures and unit conversions.

The calculation in Exercise 1.2 used exponential notation and the concept of significant figures. For a review of these concepts, see Appendix A at the end of the text.

1.9 The Difference Between Atoms and Ions

Imagine that you had a piece of aluminum foil that had been used to wrap a sandwich. Would it be easier to change the number of electrons on some of the aluminum atoms in the foil or to change the number of protons?

It is much easier to change the number of electrons on an atom than the number of protons in the nucleus. The best evidence for this is the fact that aluminum can conduct an electric current. Since the number of protons determines the identity of an atom, and the aluminum in the foil doesn't change to some other element when it conducts electricity, the charged particles that move through the foil must be electrons, not protons.

The electrically charged particles formed when electrons are added to or removed from a neutral atom are called **ions.**[4] Neutral atoms are turned into positively charged ions by removing one or more electrons (Figure 1.10). The positively charged ions are often called **cations.** An Na^{+} ion or cation that has 10 electrons and 11 protons is produced by removing one electron from a neutral sodium atom that contains 11 electrons and 11 protons. Ions with larger positive charges can be produced by removing more electrons. A neutral aluminum atom, for example, has 13 electrons and 13 protons. If we remove 3 electrons from this atom, we get a positively charged Al^{3+} ion that has 10 electrons and 13 protons, for a net charge of +3.

Atoms that gain extra electrons become negatively charged ions, or **anions,** as shown in Figure 1.11. A neutral chlorine atom, for example, has 17 protons and 17 electrons. By adding 1 more electron to this atom, a Cl^{-} ion is produced that has 18 electrons and 17 protons, for a net charge of −1.

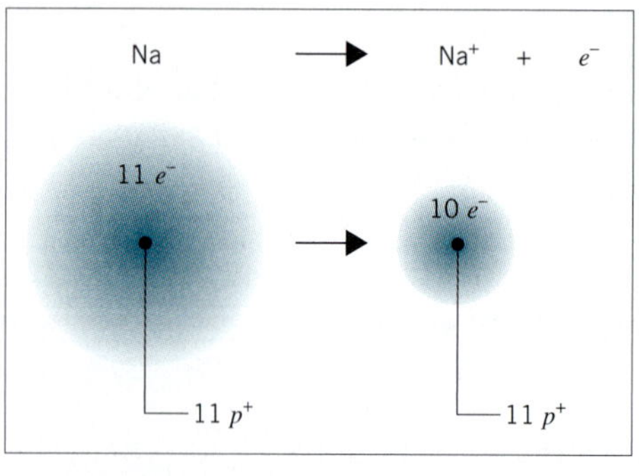

Fig. 1.10 Removing an electron from a neutral sodium atom produces an Na^{+} ion that has a net charge of +1.

[4]The force of attraction that holds protons within the nucleus of an atom is at least 10^6 times as large as the force of attraction between the nucleus and the electrons that surround it. As a result, ions can only be formed by the gain or loss of electrons.

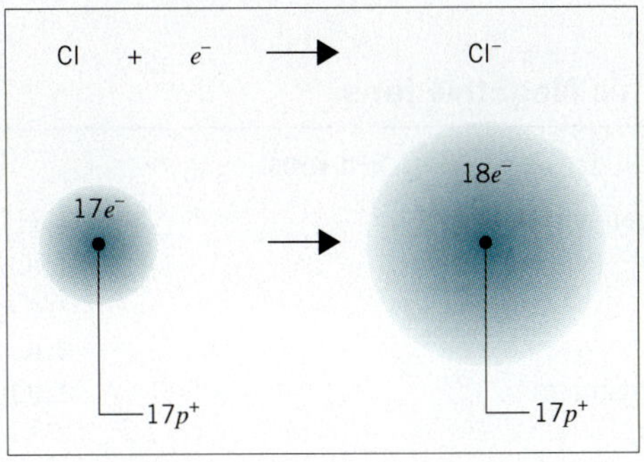

Fig. 1.11 Adding an extra electron to a neutral chlorine atom produces an Cl^- ion that has a net charge of -1.

The gain or loss of electrons by an atom to form negative or positive ions has an enormous impact on the chemical and physical properties of the atom. Sodium metal, which consists of neutral sodium atoms, produces a flame when it comes in contact with water. But positively charged Na^+ ions are so unreactive with water they are essentially inert.[5] Neutral chlorine atoms instantly combine to form Cl_2 molecules, which are so reactive that entire communities are evacuated when trains carrying chlorine gas derail. Negatively charged Cl^- ions are essentially inert to chemical reactions.

The enormous difference between the chemistry of neutral atoms and their ions means that you will have to pay close attention to the symbols you read, to make sure that you never confuse one with the other.

Exercise 1.3

Find the number of protons, electrons, and neutrons in each of the following atoms and ions.

(a) ^{7}Li

(b) $^{24}Mg^{2+}$

(c) $^{79}Br^-$

Solution

(a) Lithium has an atomic number of 3, which means that the nucleus of any atom of this element must contain 3 protons. Because the symbol for ^{7}Li describes a neutral atom, the number of protons and electrons in the atom must be the same. This atom therefore contains 3 electrons. The mass number is equal to the sum of the neutrons and protons. This atom has a mass number of 7 and the nucleus contains 3 protons. It therefore contains 4 neutrons.

(b) The atomic number of magnesium is 12, which means that this ion contains 12 protons. Because the ion carries a charge of $+2$, there must be two more protons (positive charges) than electrons (negative charges). This ion therefore contains 10 electrons. The mass number is 24 and there are 12 protons, which means there must also be 12 neutrons.

(c) Bromine has an atomic number of 35 and therefore has 35 protons. Because the ion has a -1 charge the ion must have 1 more electron than a neutral atom. This ion therefore contains 36 electrons. Because the mass number of the ion is 79 and it contains 35 protons, there must be 44 neutrons as well.

[5]A substance that seldom undergoes a chemical reaction is said to be inert.

Table 1.5
Common Polyatomic Negative Ions

	−1 ions		
HCO_3^-	hydrogen carbonate	OH^-	hydroxide
$CH_3CO_2^-$	acetate	ClO_4^-	perchlorate
NO_3^-	nitrate	ClO_3^-	chlorate
NO_2^-	nitrite	ClO_2^-	chlorite
MnO_4^-	permanganate	ClO^-	hypochlorite
CN^-	cyanide		
	−2 ions		
CO_3^{2-}	carbonate	O_2^{2-}	peroxide
SO_4^{2-}	sulfate	CrO_4^{2-}	chromate
SO_3^{2-}	sulfite	$Cr_2O_7^{2-}$	dichromate
$S_2O_3^{2-}$	thiosulfate		
	−3 ions		
PO_4^{3-}	phosphate	AsO_4^{3-}	arsenate
BO_3^{3-}	borate		

The material in Sections 1.10 and 1.11 can be postponed until Chapters 4 and 5, respectively. This material is not used again until then. It is included here for instructors who wish to cover polyatomic ions early in the semester in order to prepare students for the laboratory portion of the course.

1.10 Polyatomic Ions

Simple ions, such as Mg^{2+} and N^{3-} ions, are formed by adding or subtracting electrons from neutral atoms. **Polyatomic ions** are electrically charged substances composed of more than one atom. There are only two polyatomic cations that you will commonly encounter. These are the ammonium and hydronium ions, NH_4^+ and H_3O^+. A few of the more common anions are listed in Table 1.5.

1.11 Predicting the Formulas of Ionic Compounds

A famous Sherlock Holmes story that revolves around a dog that didn't bark reminds us that we often forget to notice things that don't happen. As a chemical equivalent of the dog that doesn't bark, consider what happens when you pour a few crystals of table salt into the palm of your hand. Even though table salt is an **ionic compound** that is composed of electrically charged ions, you don't feel the same thing that you would feel if you made the mistake of touching a bare wire that had been plugged into an electric socket. In other words, ionic compounds carry no net electric charge. This means that they must contain just as many positive charges as negative charges.

The formulas of compounds that contain ions can be predicted by assuming that the total charge on the positive ions must balance the total charge on the negative ions. The charges on the ions, therefore, dictate the formula of the ionic compound that is formed. Chemical formulas for ionic compounds are represented by the simplest whole number ratio of positive to negative ions that will yield an electric charge of zero on the compound.

> **➤ CHECKPOINT**
> What ions can be found in each of the following ionic compounds: NaOH, K_2SO_4, $BaSO_4$, and $Be_3(PO_4)_2$?

Exercise 1.4

(a) Calcium carbonate is the major component in chalk, limestone, and marble. It is composed of calcium ions, Ca^{2+}, and the polyatomic carbonate anion, CO_3^{2-}. Predict the formula of this compound.

(b) Calcium fluoride, which contains Ca^{2+} and F^- ions, can be found in the mineral known as fluorite. Predict the formula of this compound.

(c) Rubies are composed of aluminum oxide with traces of chromium(III) oxide, which contains Cr^{3+} and O^{2-} ions, that gives them their red color. Predict the formula of chromium(III) oxide.

(d) Strontium nitrate is a component of fireworks that produces a red color. This compound is composed of Sr^{2+} and NO_3^- ions. Predict the formula of this compound.

Solution

(a) The total charge from the positive ions must balance the charge from the negative ions. Since the calcium ion has a charge of +2 and the carbonate ion has a charge of −2, the charge on one calcium ion cancels the charge on one carbonate ion. The formula of calcium carbonate is $CaCO_3$.

(b) Once again, the total charge from the positive ions must balance the charge from the negative ions. The charge on each calcium ion is +2, whereas the charge on each fluoride ion is −1. In order for the charges to cancel there must be two fluoride ions for every calcium ion. The formula for this compound is therefore CaF_2.

(c) The formula for chromium (III) oxide can be found by looking for the smallest common multiple of the charges on the two ions. Two Cr^{3+} ions would have a total charge of +6, which can be balanced by three O^{2-} ions. The compound therefore has the formula Cr_2O_3.

(d) The +2 charge on the strontium requires two negative nitrate ions in order to balance the charge. The chemical formula is written as $Sr(NO_3)_2$. Parentheses are used in this formula to indicate that the subscript 2 represents two nitrate ions.

1.12 The Periodic Table

While trying to organize a discussion of the properties of the elements for a chemistry course at the Technological Institute in St. Petersburg, Dmitri Ivanovitch Mendeléeff[6] listed the properties of each element on a different card. As he arranged the cards in different orders, he noticed that the properties of the elements repeat in a periodic fashion when the elements are listed more or less in order of increasing atomic weight. In 1869 Mendeléeff published the first of a series of papers outlining a table of the elements in which the properties of the elements repeated in a periodic fashion. A copy of the table Mendeléeff published in 1871 is shown in Figure 1.12. This table lists elements with similar chemical properties in the same column.

More than 700 versions of the **periodic table** were proposed in the first 100 years after the publication of Mendeléeff's table. A modern version of the table is shown in Figure 1.13. In this version the elements are arranged in order of increasing atomic number. The atomic number is given above the atomic symbol. The vertical columns are known as **groups,** or families. Traditionally these groups have been distinguished by a **group number** consisting of a Roman numeral followed by either an A or a B. In the United States, the elements in the first column on the

[6]There are at least a half dozen ways of spelling Mendeléeff's name because of disagreements about translations from the Cyrillic alphabet. The version used here is the spelling that Mendeléeff himself used when he visited England in 1887.

Reihen	Gruppe I. — R^2O	Gruppe II. — RO	Gruppe III. — R^2O^3	Gruppe IV. RH^4 RO^3	Gruppe V. RH^3 R^2O^5	Gruppe VI. RH^1 RO^3	Gruppe VII. RH R^2O^7	Gruppe VII. — RO^4
1	H = 1							
2	Li = 7	Be = 9.4	B = 11	C = 12	N = 14	O = 16	F = 19	
3	Na = 23	Mg = 24	Al = 27.3	Si = 28	P = 31	S = 32	Cl = 35.5	
4	K = 39	Ca = 40	– = 44	Tl = 48	V = 51	Cr = 52	Mn = 55	Fe = 56, Co = 59, Ni = 59, Cu = 63.
5	(Cu = 63)	Zn = 65	– = 68	– = 72	As = 75	Se = 78	Br = 80	
6	Rb = 85	Sr = 87	?Yt = 88	Zr = 90	Nb = 94	Mo = 96	– = 100	Ru = 104, Rh = 104, Pd = 106, Ag = 108.
7	(Ag = 108)	Cd = 112	In = 113	Sa = 118	Sb = 122	Te = 125	J = 127	
8	Cs = 133	Ba = 137	?Di = 138	?Ce = 140	–	–	–	– – – –
9	(–)	–	–	–	–	–	–	
10	–	–	?Er = 178	?La = 180	Ta = 182	W = 184	–	Os = 195, Ir = 197, Pt = 198, Au = 199.
11	(Au = 199)	Hg = 200	Tl = 204	Pb = 207	Bi = 208	–	–	
12	–	–	–	Th = 231	–	U = 240	–	– – – –

Fig. 1.12 This version of Mendeléeff's periodic table was published in the journal *Annalen der Chemie* in 1871.

> **➤ CHECKPOINT**
>
> What are the atomic numbers of F and Pb? What are the atomic symbols of the elements with atomic numbers 24 and 74?

left-hand side of the table are Group IA. The next column is IIA, then IIIB, and so on across the periodic table to VIIIA.

Unfortunately the same notation isn't used in all countries. The elements known as Group VIA in the United States are Group VIB in Europe. A new convention for the periodic table has been proposed that numbers the columns from 1 to 18, reading from left to right. This convention has obvious advantages. It is perfectly regular and therefore unambiguous. The advantages of the old format are less obvious, but they are equally real. This book therefore introduces the new convention but retains the old.

The elements in a column of the periodic table have similar chemical properties. Elements in the first column, for example, combine in similar ways with chlorine to form compounds with similar chemical formulas: HCl, LiCl, NaCl, KCl, and so on.

The horizontal rows in the periodic table are called **periods.** The first period contains only two elements: hydrogen (H) and helium (He). The second period contains eight elements (Li, Be, B, C, N, O, F, and Ne). Although there are nine horizontal rows

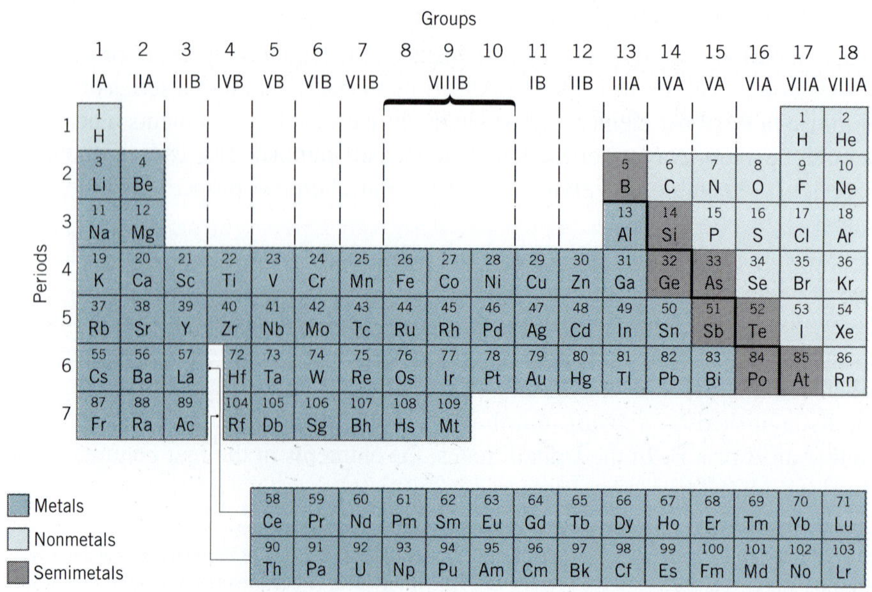

Fig. 1.13 A modern version of the periodic table.

in the periodic table in Figure 1.13, there are only seven periods. The two rows at the bottom of the table belong in the sixth and seventh periods. These rows are listed at the bottom to prevent the table from becoming so large it is unwieldy.

The elements in the periodic table can be divided into three categories: **metals, nonmetals,** and **semimetals.** The dividing line between the metals and the nonmetals in Figure 1.13 is marked with a heavy stairstep line. As you can see from Figure 1.13, more than 75% of the elements are metals. These elements are found toward the bottom-left side of the table.

Only 17 elements are nonmetals. With only one exception—hydrogen, which appears on both sides of the table—these elements are clustered in the upper-right corner of the periodic table. A cluster of elements that are neither metals nor nonmetals can be found between the metals and nonmetals in Figure 1.13. These elements are called the semimetals, or metalloids.

Exercise 1.5

Classify each element in Group IVA as either a metal, a nonmetal, or a semimetal.

Solution

Group IVA contains five elements: carbon, silicon, germanium, tin, and lead. According to Figure 1.13, these elements fall into the following categories.

Nonmetal:	C
Semimetal:	Si and Ge
Metal:	Sn and Pb

1.13 Metals, Nonmetals, and Semimetals

The tendency to divide elements into metals, nonmetals, and semimetals is based on differences between the chemical and physical properties of these three categories of elements. *Metals* have the following characteristic physical properties:

- They have a metallic shine or luster. (They look like metals!)
- They are usually solids at room temperature.
- They are *malleable*. They can be hammered, pounded, or pressed into different shapes without breaking.
- They are *ductile*. They can be drawn into thin sheets or wires without breaking.
- They conduct heat and electricity.

Nonmetals have the opposite physical properties:

- They seldom have a metallic luster. They tend to be colorless, like the oxygen and nitrogen in the atmosphere, or brilliantly colored, like bromine.
- They are often gases at room temperature.
- The nonmetallic elements that are solids at room temperature (such as carbon, phosphorus, sulfur, and iodine) are neither malleable nor ductile. They cannot be shaped with a hammer or drawn into sheets or wires.

- They are poor conductors of either heat or electricity. Nonmetals tend to be insulators, not conductors.

The *semimetals* have properties that lie between these extremes. They often look like metals but are brittle like nonmetals. They are neither conductors nor insulators but make excellent semiconductors.

DENSITY

Density, *d*, is a physical property of matter that chemists find particularly useful. Density is defined to be the mass of a substance divided by its volume.

$$d = m/V$$

The mass, *m*, is usually expressed in grams and the volume, *V*, in milliliters, mL, or cubic centimeters, cm^3. Because one milliliter is defined to be the same as one cubic centimeter, densities may be reported in units of g/mL or g/cm^3.

Densities are useful because pure substances have characteristic densities. Therefore densities can be used to differentiate between substances. Lead has a density of 11.3 g/cm^3 and the density of iron is 7.9 g/cm^3. Thus two unidentified metals, one of which is known to be iron and the other lead, could be identified by weighing out 1.0 g of each and immersing them in water. The volume of 1.0 g of lead is less than the volume of 1.0 g of iron, and the volume of water displaced by the metal would be greater for the iron.[7]

The density of a substance may be calculated by weighing out a known mass of the substance and determining its volume.

➤ CHECKPOINT

The density of diamond is 3.5 g/cm^3 and the density of anthracite coal is 1.8 g/cm^3. Which is heavier, 1.0 cm^3 of diamond or coal? Which is heavier, 2.0 cm^3 of coal or 1.0 cm^3 of diamond?

Exercise 1.6

(a) A gold brick of volume 10.6 cm^3 weighs 205 g. What is the density of gold?

(b) What is the weight in grams of 27.6 cm^3 of gold?

Solution

(a) Density is defined as mass divided by volume:

$$d = m/V$$

The two pieces of data needed to calculate the density are thus known.

$$d = 205\ \text{g}/10.6\ \text{cm}^3 = 19.3\ \text{g/cm}^3 = 19.3\ \text{g/mL}$$

(b) The density is known from part (a) then.

$$d = m/27.6\ \text{cm}^3 = 19.3\ \text{g/cm}^3$$

Solving for *m* gives

$$m = 27.6\ \text{cm}^3 \times 19.3\ \text{g/cm}^3 = 533\ \text{g}$$

[7]See Appendix A for a discussion of unit prefixes.

Key Terms

Absolute measurement
Anion
Atom
Atomic mass unit (amu)
Atomic number
Cation
Chemical formula
Chemical properties
Composition
Compound
Density
Electron
Element
Group
Group number
Ion
Ionic compound
Isotope
Mass number
Matter
Metal
Mixture
Molecular compound
Molecule
Natural abundance
Neutron
Nonmetal
Nucleus
Period
Periodic table
Physical properties
Polyatomic ions
Proton
Pure substance
Relative measurement
Semimetal
Structure

Problems

Chemistry: A Definition

1. How would you describe the goals of modern chemistry?
2. It was known by the eleventh century that addition of alum, prepared from a mineral, to animal skins aided in the tanning process. Could the practitioners of this tanning procedure be considered to be chemists?
3. The early Greek philosophers debated the idea of whether matter is continuous or consists of small indivisible particles. They performed no experiments. What role does experimentation play in chemistry?

Physical and Chemical Properties

4. Classify each of the following as an example of a physical or chemical property.

   Gold is heavy.
   Statues disintegrate over time.
   Water boils at 100°C.
   Natural gas burns.
   Mercury is a liquid.
   Sugar ferments.

5. What is meant by the composition of a substance?
6. What is the difference between the composition of a substance and the structure of the substance?

Elements, Compounds, and Mixtures

7. Define the following terms: *element, compound,* and *mixture*. Give an example of each.
8. Describe the difference between elements and compounds on the macroscopic (objects are visible to the naked eye) scale and on the atomic scale.
9. Classify the following substances into the categories of elements, compounds, and mixtures. Use as many labels as necessary to classify each substance. Use whatever reference books you need to identify each substance.
   (a) diamond (b) brass (c) soil
   (d) glass (e) cotton (f) milk of magnesia
   (g) salt (h) iron (i) steel
10. Granite consists primarily of three minerals in varying composition: feldspar, plagioclase, and quartz. Is granite an element, a compound, or a mixture?
11. Describe what the formula P_4S_3 tells us about this compound.
12. What information does the formula SO_3 give us about this compound?

Atomic Symbols

13. List the symbols for the following elements.
    (a) antimony (b) gold (c) iron
    (d) mercury (e) potassium (f) silver
    (g) tin (h) tungsten
14. Name the elements with the following symbols.
    (a) Na (b) Mg (c) Al (d) Si
    (e) P (f) Cl (g) Ar
15. Name the elements with the following symbols.
    (a) Ti (b) V (c) Cr (d) Mn (e) Fe
    (f) Co (g) Ni (h) Cu (i) Zn
16. Name the elements with the following symbols.
    (a) Mo (b) W (c) Rh (d) Ir (e) Pd
    (f) Pt (g) Ag (h) Au (i) Hg
17. Describe the difference between the following pairs of symbols.
    (a) Co and CO (b) Cs and CS_2
    (c) Ho and H_2O (d) 4 P and P_4

Evidence for the Existence of Atoms

18. Describe some of the evidence for the existence of atoms and some of the evidence from our senses that seems to deny the existence of atoms.
19. Choose one of Dalton's assumptions and design an experiment that would support or refute the assumption.
20. Why is the atomic theory so widely accepted?
21. Did any of Dalton's assumptions give any clues as to the structure of the atom?
22. According to Dalton, how do atoms of different elements differ?
23. One of Dalton's assumptions was that atoms cannot be created or destroyed. Does this mean that the number of atoms in the universe has remained unchanged?

The Structure of Atoms

24. Describe the differences between a proton, a neutron, and an electron.
25. One of Dalton's assumptions is now known to be in error. Which one is it?
26. What similarities are there between an atom of iron and an atom of mercury?
27. What are three fundamental subatomic particles that make up an atom? Give the relative charge on each of these particles.
28. What is a neutral atom?
29. Which of the particles that make up an atom is lightest?
30. Where is the weight of the atom concentrated?
31. How does the radius of an atom compare to the size of the nucleus?

Atomic Number and Mass Number

32. Describe the relationship between the atomic number, mass number, number of protons, number of neutrons, and number of electrons in a calcium atom, ^{40}Ca.
33. Write the symbol for the atom that contains 24 protons, 24 electrons, and 28 neutrons.
34. Calculate the number of protons and neutrons in the nucleus and the number of electrons surrounding the nucleus of a ^{39}K atom. What are the atomic number and the mass number of this atom?
35. Calculate the number of protons and neutrons in the nucleus and the number of electrons surrounding the nucleus of an ^{127}I atom. What are the atomic number and the mass number of this atom?
36. Identify the element that has atoms with mass numbers of 20 that contain 11 neutrons.
37. Give the symbol for the atom that has 34 protons, 45 neutrons, and 34 electrons.
38. Calculate the number of electrons in a ^{134}Ba atom.
39. Complete the following table.

Isotope	Atomic Number (Z)	Mass Number (A)	Number of Electrons
^{31}P	15	—	—
^{18}O	—	—	8
—	19	39	19
^{58}Ni	—	58	—

Isotopes

40. What is the ratio of the mass of a ^{12}C atom to a ^{13}C atom?
41. How many times heavier is an ^{6}Li atom than an ^{1}H atom?
42. If you were to select one oxygen atom at random, what would its mass in grams most likely be?
43. The ratio of the mass of a ^{12}C atom to that of an unknown atom is 0.750239. Identify the unknown atom.
44. Divide the mass of an ^{1}H atom in atomic mass units by the mass of the atom in grams. Do the same for ^{2}H and ^{12}C. Does this suggest a relationship between the atomic mass in grams and amu?
45. Complete the following table. Table 1.4 may be useful.

Mass (grams)	Z	A	Number of Neutrons	Mass (amu)
1.6627×10^{-23}	—	—	—	10.0129
—	12	—	12	23.9850
—	8	18	—	—
1.7752×10^{-22}	—	107	60	—

46. Without referring to Table 1.4, which is heavier, an atom of ^{11}B or ^{12}C? Justify your answer.
47. How many isotopes of oxygen occur naturally on earth?
48. What do all isotopes of oxygen have in common? What is different?
49. If you select one carbon atom at random, what is the mass of that atom likely to be (in grams and in amu)?
50. What is the mass (in amu) of 100 ^{12}C atoms? Of 100 ^{13}C atoms?
51. If you select 100 carbon atoms at random, will the total mass be
    (a) 1200.00 amu?
    (b) slightly more than 1200.00 amu?
    (c) slightly less than 1200.00 amu?
    (d) 1300.3 amu?
    (e) slightly less than 1300.3 amu?
    Explain your reasoning.

The Difference Between Atoms and Ions

52. Describe the difference between the following pairs of symbols.
    (a) H and H^+ (b) H and H^-
    (c) 2 H and H_2 (d) H^+ and H^-

53. Explain the difference between H^+ ions, H atoms, and H_2 molecules on the atomic scale.
54. How many electrons are in a $^{134}Ba^{2+}$ ion? How many protons? How many neutrons?
55. Write the symbol for the atom or ion that contains 24 protons, 21 electrons, and 28 neutrons.
56. How many protons, neutrons, and electrons are in the $^{127}I^-$ ion?
57. Give the symbol for the atom or ion that has 34 protons, 45 neutrons, and 36 electrons.
58. Complete the following table.

Isotope	Atomic Number (Z)	Mass Number (A)	Number of Electrons
$^{31}P^{3-}$	—	—	—
$^{18}O^{2-}$	—	—	—
$^{58}Ni^{2+}$	—	—	—
—	12	24	10
—	13	27	10
—	35	80	36

Polyatomic Ions

59. What are polyatomic ions?
60. List three polyatomic ions by name and chemical formula for which the charges are −1, −2, and −3.
61. Give two common polyatomic ions that have positive charges.

Predicting the Formulas of Ionic Compounds

62. Fluoride toothpastes convert the mineral apatite in tooth enamel into fluorapatite, $Ca_5(PO_4)_3F$. If fluorapatite contains Ca^{2+} and PO_4^{3-} ions, what is the charge on the fluoride ion in this compound?
63. Verdigris is a green pigment used in paint. The simplest formula for the compound is $Cu_3(OH)_2(CH_3CO_2)_4$. What is the charge on the copper ions in this compound, if the other polyatomic ions both carry a charge of −1?
64. Predict the formulas for neutral compounds containing the following pairs of ions.
(a) Mg^{2+} and NO_3^- (b) Fe^{3+} and SO_4^{2-}
(c) Na^+ and CO_3^{2-}
65. Predict the formulas for neutral compounds containing the following pairs of ions.
(a) Na^+ and O_2^{2-} (b) Zn^{2+} and PO_4^{3-}
(c) K^+ and $PtCl_6^{2-}$
66. Predict the formulas for sodium nitride and aluminum nitride if the formula for magnesium nitride is Mg_3N_2.
67. Compounds that contain the O^{2-} ion are called oxides. Those that contain the O_2^{2-} ion are called peroxides. If the formula for potassium oxide is K_2O, what is the formula for potassium peroxide?
68. What is the value of x in the $[Co(NO_2)_x]^{3-}$ ion if this complex ion contains Co^{3+} and NO_2^- ions?
69. Magnetic iron oxide has the formula Fe_3O_4. Explain this formula by assuming that Fe_3O_4 contains both Fe^{2+} and Fe^{3+} ions combined with O^{2-} ions.

The Periodic Table

70. Describe the differences between periods and groups of elements in the periodic table.
71. Mendeléeff placed both silver and copper in the same group as lithium and sodium. Look up the chemistry of these four elements in the *CRC Handbook of Chemistry and Physics*.[8] Describe some of the similarities that allow these elements to be classified in a single group on the basis of their chemical properties.
72. Which of the following are nonmetals?
(a) Li (b) Be (c) B (d) C
(e) N (f) O
73. Place each of the following elements in the correct group on the periodic table.
(a) K (b) Si (c) Ca (d) S
(e) Mg (f) He (g) I
74. Of the following sets of elements, which are in the same period of the periodic table? The same group?
(a) Be, B, C (b) Be, Mg, Ca (c) P, S, Al
(d) As, N, P (e) Sb, Te, Xe (f) K, Rb, Sr
75. How many elements are in Group IA?
76. How many elements are in the second period? The third period? The fourth period?
77. In which of the following sets of elements should all elements have similar chemical properties?
(a) O, S, Se (b) F, Cl, Te (c) Al, Si, P
(d) Ca, Sr, Ba (e) K, Ca, Sc (f) N, O, F

Metals, Nonmetals, and Semimetals

78. Classify each of the following elements as a metal, nonmetal, or semimetal.
(a) Na (b) Mg (c) Al (d) Si
(e) P (f) S (g) Cl (h) Ar
79. Classify each of the following elements as a metal, nonmetal, or semimetal.
(a) N (b) Sb (c) Sc (d) Se
(e) Ge (f) Sm (g) Sn (h) Sr
80. List three molecular elements and three atomic elements.
81. Classify the elements in Group VA as either metals, nonmetals, or semimetals. Describe what happens to the properties of the elements as we go down this column of elements.

[8]CRC Press, Boca Raton, Florida.

82. Classify the elements in the third period as either metals, nonmetals, or semimetals. Describe what happens to the properties of the elements as we go from left to right across this period.
83. Which weighs the most, 10.0 cm^3 of iron or 5.0 cm^3 of silver? The density of iron is 7.9 g/cm^3 and that of silver is 10.5 g/cm^3.
84. Two strips of metal each weighing 100 g are placed into a cylinder containing water. The volume of water displaced is 8.8 mL in one case and 37.0 mL in the other case. Identify the two metal strips.

Metal	Au	Fe	Al	Pb	Ag
Density (g/mL)	19.3	7.9	2.7	11.3	10.5

85. What is the density of a metal that has a volume of 20.0 mL and weighs 271 g?
86. Mercury is a liquid metal at room temperature with a density of 13.6 g/mL. What is the weight in grams of 5.6 mL of mercury?

Integrated Problems

87. The mass number of the atom *X* in Group IIA from which the ion is formed is 40. The formula of the ionic compound formed with the carbonate ion is XCO_3. How many electrons, protons, and neutrons does the ion *X* have? What is the chemical symbol for *X*?
88. Element *X* is a metal whose chemical properties are similar to potassium. There is only one isotope of atom *X*. The mass of *X* in amu is 22.98976. Use the trends in masses in Table 1.4 to identify element *X*.
89. Complete the following table for uncharged atoms.

Classification	Group	Period	Number of Electrons	Element Symbol
Metal	—	—	11	—
—	IVA	—	—	Ge
—	—	—	—	B
Semimetal	—	3	—	—
—	VIIA	4	—	—

Chapter Two

THE MOLE: THE LINK BETWEEN THE MACROSCOPIC AND THE ATOMIC WORLD OF CHEMISTRY

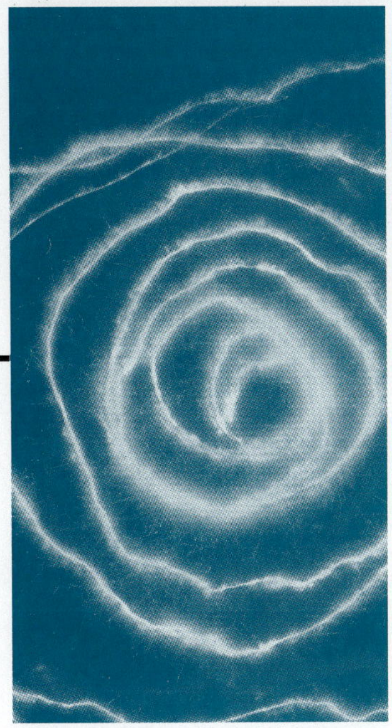

2.1 The Macroscopic, Atomic, and Symbolic Worlds of Chemistry

Chapter 2 is a traditional development of the mole, mole-gram conversions, chemical formulas, chemical equations, stoichiometry, solution concentration, and limiting reagents. This material is provided early in the text for those instructors who traditionally cover these topics at the beginning of the course.

ChemActivities 2 and 27–30 in Chemistry: A Guided Inquiry *are designed to accompany this chapter.*

Much of the material in this chapter is not prerequisite for the initial core chapters. It is therefore possible for instructors to delay coverage of portions of Chapter 2 until later in the course. The following sections of Chapter 2 are prerequisite to the indicated chapters.

For Chapter 3, Sections 2.1–2.4 are required.

For Chapter 5, Sections 2.10–2.13 are required.

For Chapter 6, Sections 2.5–2.9 and 2.14–2.16 are required.

Sections 2.17–2.21 may be covered with Chapter 8.

Chemists work in three very different worlds represented by Figure 2.1. Most measurements are done in the **macroscopic world**—with objects visible to the naked eye. Water, for example, on the macroscopic scale is a liquid that freezes at 0°C and boils at 100°C. When you walk into a chemical laboratory, you'll find a variety of bottles, tubes, flasks, and beakers designed to study samples of liquids and solids large enough to be seen. You may also find sophisticated instruments that can be used to analyze very small quantities of materials, but even these samples are visible to the naked eye.

Although they perform experiments on the macroscopic scale, chemists think about the behavior of matter in terms of a world of atoms and molecules. In this **atomic world,** water is no longer a liquid that freezes at 0°C and boils at 100°C, but individual molecules that contain two hydrogen atoms and an oxygen atom.

One of the challenges students face when they encounter chemistry for the first time is understanding the process by which chemists perform experiments on the macroscopic scale that can be interpreted in terms of the structure of matter on the atomic scale. The task of bridging the gap between the atomic and macroscopic worlds is made more difficult by the fact that chemists also work in a **symbolic world,** in which they represent water as H_2O and write equations such as the following to represent what happens when hydrogen and oxygen react to form water.

$$2\,H_2 + O_2 \longrightarrow 2\,H_2O$$

Chemists use the same symbols to describe what happens on both the macroscopic and the atomic scales. The symbol "H_2O," for example, is used to represent both a single water molecule and a beaker full of water.

It is easy to forget the link between the symbols chemists use to represent reactions and the particles involved in these reactions. Figure 2.2 provides an example of how you might envision the reaction described in the chemical equation written above on the atomic scale. The reaction starts with a mixture of H_2 and O_2 molecules, each containing a pair of atoms. It produces water molecules that contain two hydrogen atoms and an oxygen atom.

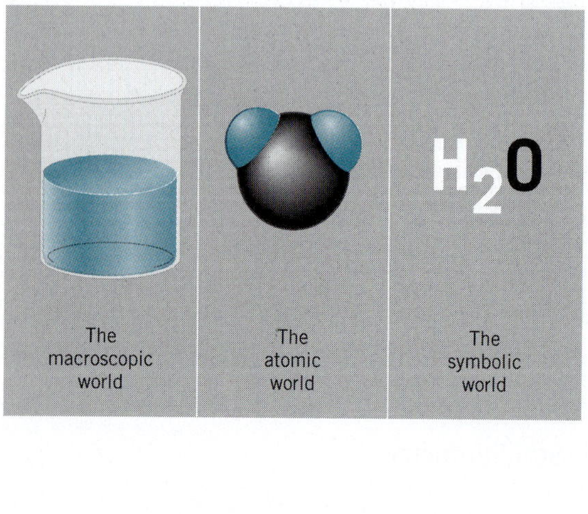

Fig. 2.1 Water on the scale of the macroscopic, atomic, and symbolic worlds.

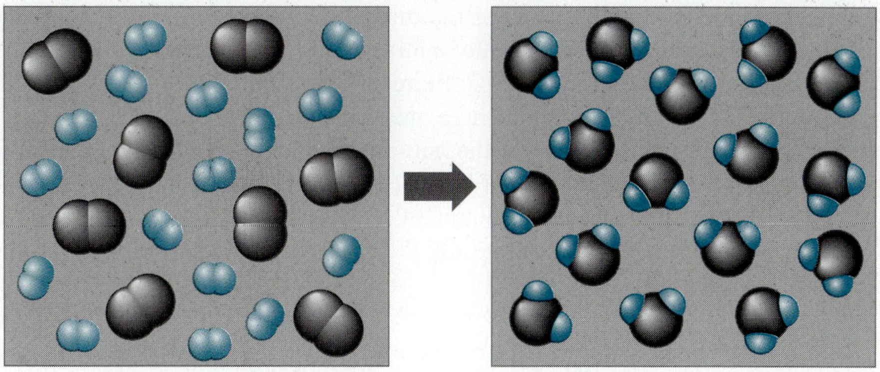

Fig. 2.2 A mechanical model for the reaction between H_2 and O_2 on the atomic scale to form water molecules.

2.2 The Mass of an Atom

Atoms are so small that a sliver of copper metal just big enough to be detected on a good analytical balance contains about 1×10^{17} atoms. As a result, it is impossible to measure the absolute mass of a single atom. We can, however, measure the relative masses of different atoms.

Figure 2.3 shows a diagram of a **mass spectrometer** that can be used to determine the relative mass of an atom or molecule. The sample is injected into an evacuated chamber. The particles in the sample flow past a filament, where they collide with high-energy electrons. As a result of these collisions, the neutral atoms or molecules in the sample lose electrons to form positively charged ions. As they pass between the poles of a magnet, the ions interact with the magnetic field. The interaction between the magnetic field and the charges on the ions bends the path along which the ions travel. The larger the mass of the ion, the smaller the angle through which its path is bent before it enters the detector.

Because the mass spectrometer can tell us only the relative mass of an atom, we need a standard with which our measurement can be compared. The standard used to calibrate these measurements is the ^{12}C isotope of carbon. The unit in which atomic mass measurements are reported is the **atomic mass unit (amu)** (see Section 1.8). By definition, the mass of a single atom of the ^{12}C isotope is exactly 12 atomic mass units or 12 amu.

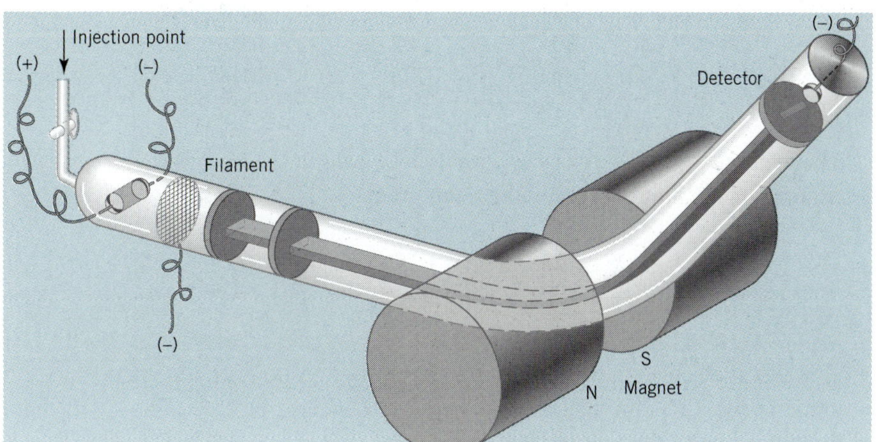

Fig. 2.3 Diagram of a mass spectrometer.

Most elements exist in nature as mixtures of isotopes. As we have seen, the graphite in a lead pencil is composed of a mixture of ^{12}C (98.892%, 12.000 amu), ^{13}C (1.108%, 13.003 amu), and an infinitesimally small amount of ^{14}C. It is therefore useful to calculate the average mass of a sample of carbon atoms. Because there is a large difference in the natural abundance of these isotopes, the average mass of a carbon atom must be a weighted average of the masses of the different isotopes. Because the amount of ^{14}C is so small, the average mass of a carbon atom is calculated using only the two most abundant isotopes of the element.

$$\left(12.000 \text{ amu} \times \frac{98.892}{100}\right) + \left(13.003 \text{ amu} \times \frac{1.108}{100}\right) = 12.011 \text{ amu}$$

The average mass of a carbon atom is much closer to the mass of a ^{12}C atom than a ^{13}C atom because the vast majority of the atoms in a sample of carbon are ^{12}C. This weighted average of all the naturally occurring isotopes of an atom is known as the **atomic weight** of the element. It is this value that is reported beneath the symbol of the element in the periodic table. It is important to recognize that the atomic weight of carbon is 12.011 amu even though no individual carbon atom actually has a mass of 12.011 amu.

Exercise 2.1

Assume that the following grades were obtained when a test was administered to 10 students: 84, 87, 84, 87, 92, 96, 92, 87, 84, 87. Calculate the average score on this exam as a weighted average of the results obtained by the individual students.

Solution

The first step in determining the average score involves calculating the percentage of the students who received each score.

Grade	Number of Grades	Percent of Grades
96	1	10%
92	2	20%
87	4	40%
84	3	30%

The average score on this exam can then be calculated from the grades and the percentage of the students receiving each grade.

$$\left(96 \times \frac{10}{100}\right) + \left(92 \times \frac{20}{100}\right) + \left(87 \times \frac{40}{100}\right) + \left(84 \times \frac{30}{100}\right) = 88$$

The average score on this exam is 88 even though no individual student received a grade of 88. In a similar fashion, the atomic weight of carbon is 12.011 amu even though no individual atom has this mass.

Exercise 2.2

Calculate the atomic weight of chlorine if 75.77% of the atoms have a mass of 34.97 amu and 24.23% have a mass of 36.97 amu.

Solution

Percent literally means "per hundred." Chlorine is therefore a mixture of atoms for which 75.77 parts per hundred have a mass of 34.97 amu and 24.23 parts per hundred have a mass of 36.97 amu. The atomic weight of chlorine is therefore 35.45 amu.

$$\left(34.97 \text{ amu} \times \frac{75.77}{100}\right) + \left(36.97 \text{ amu} \times \frac{24.23}{100}\right) = 35.45 \text{ amu}$$

2.3 The Mole as the Bridge Between the Macroscopic and Atomic Scales

Imagine that you pick the following items off the shelves of a grocery store: a dozen eggs, a 2-lb bag of sugar, a 5-lb bag of flour, and a quart of milk. When you open the egg carton, you know exactly how many eggs it should contain—a dozen. But the same can't be said about the sugar, flour, and milk. A recipe may call for 1 egg, but it never calls for 1 grain of sugar because a grain of sugar is too small to be useful. By the time you get to adding the flour or milk, the problem becomes even more serious; it is physically impossible to add a single grain of flour. Recipes therefore call for half a cup of sugar or two cups of flour or a cup of milk.

Chemists face a similar problem because it takes an enormous number of atoms to give a sample large enough to be seen with the naked eye. (A dot of graphite from a pencil just large enough to be weighed on an analytical balance contains approximately 5×10^{19} atoms.) Chemists therefore created a unit known as the **mole** (from Latin, meaning "a huge pile") that can serve as the bridge between chemistry on the macroscopic and atomic scales. The mole is defined as follows.

> **A mole of any substance contains the same number of particles as the number of atoms in exactly 12 grams of the ^{12}C isotope of carbon.**

Note that a single ^{12}C atom has a mass of exactly 12 amu and a mole of these atoms has a mass of exactly 12 grams.

$$1\ ^{12}\text{C atom} = 12.0000\ldots \text{ amu}$$
$$1 \text{ mole of } ^{12}\text{C atoms} = 12.000\ldots \text{ g}$$

As noted in the previous section, the average mass of naturally occurring carbon atoms is 12.011 amu. Thus, one mole of carbon atoms would have a mass of 12.011 grams.

The mole is the most fundamental unit of chemistry because it allows us to determine the number of elementary particles in a sample of a pure substance by simply determining the mass of the sample. Assume, for example, that we want a sample of aluminum metal that contains the same number of atoms as a mole of carbon atoms. We can start by looking up the atomic weight of aluminum in the periodic table.

$$1 \text{ Al atom} = 26.982 \text{ amu}$$

The atomic weight of this element is known to five significant figures (see the end of the text). When the atomic weight is used in a calculation, however, one only needs to retain the number of digits that corresponds to the appropriate number of significant figures (see Appendix A.3).[1]

The atomic weight of aluminum is a little more than twice that of a ^{12}C atom. As a result, aluminum atoms are a little more than twice as heavy as a ^{12}C atom.

$$\frac{1 \text{ Al atom}}{1\ ^{12}\text{C atom}} = \frac{26.982 \text{ amu}}{12 \text{ amu}} = 2.2485$$

(Note that the ratio of the mass of an aluminum atom to a ^{12}C atom in this calculation can be given to five significant figures because the mass of ^{12}C is based on a definition, not a measurement.)

If a mole of aluminum contains exactly the same number of atoms as a mole of ^{12}C, then a mole of aluminum must have a mass that is 2.2485 times the mass of a mole of ^{12}C atoms.

$$\frac{1 \text{ mol Al}}{1 \text{ mol } ^{12}\text{C}} = \frac{26.982 \text{ g}}{12 \text{ g}} = 2.2485$$

Thus, the mass of a mole of aluminum is 26.982 g.

Figure 2.4 shows two beakers. The beaker on the left contains 12.011 g of carbon that contains a naturally occurring mixture of the isotopes of this element. The beaker on the right contains 26.982 g of aluminum. Aluminum has only one naturally occurring isotope, therefore the weight that appears on the periodic chart is the mass in amu of an aluminum atom. Both beakers contain 1 mole of the respective elements. There are therefore the same number of atoms in both beakers.

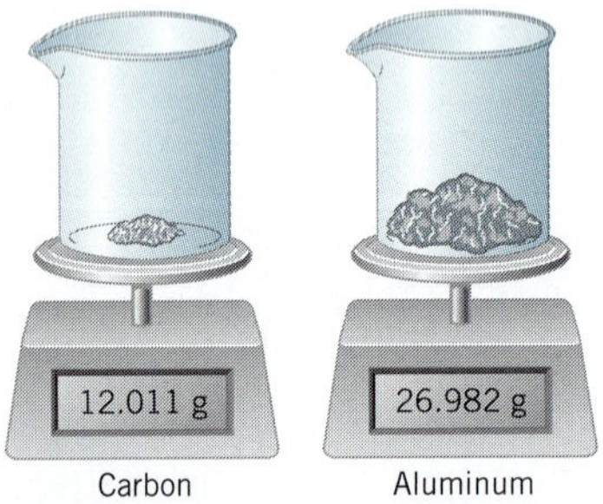

Fig. 2.4 Because each beaker contains 1 mole of the element, the two beakers contain the same number of atoms.

The results of this discussion of the relative masses of carbon and aluminum on the atomic and macroscopic world can be generalized as follows:

A mole of atoms of any element has a mass in grams equal to the atomic weight of the element.

The mass of a mole of a substance is often called the **molar mass.** The molar mass of ^{12}C, for example, is 12 g per mole (abbreviated "mol"). The molar mass of a sample of carbon that contains both ^{12}C and ^{13}C atoms in their natural abundances would be 12.011 g/mol. The relationship between atomic weight and molar mass is illustrated by the following examples.

Element	Atomic Weight	Molar Mass
Carbon	12.011 amu	12.011 g
Aluminum	26.982 amu	26.982 g
Iron	55.847 amu	55.847 g

[1]It is sometimes useful to use an extra significant figure during a calculation, and round off to the correct number of significant figures at the end of the calculation. This minimizes the probability of introducing systematic errors into the calculation.

The key to understanding the concept of the mole is recognizing that 12.011 grams of carbon contain the same number of atoms as 26.982 grams of aluminum or 55.847 grams of iron.

> **CHECKPOINT**
>
> What is the atomic weight and molar mass of potassium? Of uranium?

2.4 The Mole as a Collection of Atoms

For many years, chemists used the concept of a mole without knowing exactly how many particles there were in a mole of elemental carbon or aluminum metal. What matters is the fact that there are the same number of particles in a mole of each of these elements.

The only way to determine the number of particles in a mole is to measure the same quantity on both the atomic and the macroscopic scales. In 1910 Robert Millikan measured the charge in coulombs on a single electron: 1.6×10^{-19} C. Because the charge on a mole of electrons, 96,485 C, was already known, it was possible to estimate the number of electrons in a mole for the first time. Using more recent data, we get the following results.

$$\frac{96{,}485 \text{ C}}{1 \text{ mol}} \times \frac{1 \text{ electron}}{1.60217733 \times 10^{-19} \text{ C}} = 6.0221 \times 10^{23} \frac{\text{electrons}}{\text{mol}}$$

This number of electrons is the number of particles in a mole and is known as **Avogadro's number** or **Avogadro's constant.**

Avogadro's number is so large it is difficult to comprehend. It would take 6 million million galaxies the size of the Milky Way to yield 6.02×10^{23} stars. At the speed of light, it would take 102 billion years to travel 6.02×10^{23} miles. There are only about 40 times this number of drops of water in all the oceans on earth.

In everyday life units such as dozen (12) and gross (144) are used to describe a collection of items. The mole is sometimes referred to as the "chemist's dozen." The concept of the mole can be applied to any particle. In addition to talking about a mole of ^{12}C atoms, we can talk about a mole of Mg atoms, a mole of Na^+ ions, a mole of electrons, or a mole of glucose molecules ($C_6H_{12}O_6$). Each time we use the term, we refer to Avogadro's number of items.

1 mole of ^{12}C atoms contains 6.022×10^{23} atoms of ^{12}C.

1 mole of Mg atoms contains 6.022×10^{23} atoms of Mg.

1 mole of Na^+ ions contains 6.022×10^{23} ions of Na^+.

1 mole of electrons contains 6.022×10^{23} electrons.

1 mole of $C_6H_{12}O_6$ molecules contains 6.022×10^{23} $C_6H_{12}O_6$ molecules.

1 mole of photons contains 6.022×10^{23} photons.

Once we know the number of elementary particles in a mole, we can determine the number of particles in a sample of a pure substance by weighing the sample. To see how this is done, let's consider a process by which objects of known mass are counted by weighing a sample.

The relationship between counting and weighing can be demonstrated with the following example. A dozen balls are placed on a balance, as shown in Figure 2.5. The mass of the dozen balls is found to be 107 grams. Now assume that an unknown number of balls has a mass of 178 grams. How many balls are in the unknown sample?

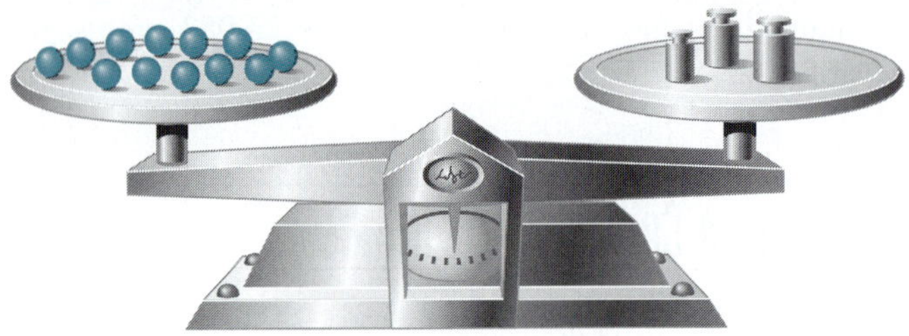

Fig. 2.5 A dozen balls that weigh 107 g.

This is the first use of the unit factor method for performing calculations. Students who are not familiar with this method can be referred to Appendix A at the end of the text.

We can build two unit factors from our knowledge of the mass of a dozen balls.

$$\frac{107\text{ g}}{1\text{ dozen balls}} \quad \text{or} \quad \frac{1\text{ dozen balls}}{107\text{ g}}$$

The question is: Which unit factor should we use? A technique known as *dimensional analysis* can guide us to the correct conversion factor. All we have to do is keep track of what happens to the units during the calculation. If the units cancel as expected, the calculation has been set up properly.

In this case, we know the mass of the unknown sample and the mass of a dozen balls. We therefore set up the calculation as follows:

$$178\text{ g} \times \frac{1\text{ dozen balls}}{107\text{ g}} = 1.66\text{ dozen balls}$$

We can now calculate the number of balls in the sample from the fact that there are 12 balls in a dozen.

$$1.66\text{ dozen} \times \frac{12\text{ balls}}{1\text{ dozen}} = 20\text{ balls}$$

In this example a dozen is analogous to a mole and 107 g/dozen is analogous to the molar mass of an element. You may well be asking yourself, wouldn't it be easier to just count the 20 balls? In the case of the balls it would be easier just to count them rather than weighing them and doing the above calculation. However, what if the balls represent atoms, which are too small and too numerous to count? Counting atoms would be an impossible task. The only method to determine the number of atoms in a pure sample is to weigh the sample and calculate the number of atoms.

Students who are unfamiliar with the use of prefixes in the metric system should be referred to Appendix A.

We can use the logic developed in the example shown above to calculate the number of carbon atoms in a one-carat diamond. All we need to know is that a diamond can be thought of as a single crystal that contains only carbon atoms and that the mass of a carat is defined as 200.0 milligrams. Because 1 gram is equal to 1000 milligrams, a one-carat diamond has a mass of 0.2000 grams.[2]

$$200.0\text{ mg} \times \frac{1\text{ g}}{1000\text{ mg}} = 0.2000\text{ g}$$

The atomic weight of carbon is 12.011 amu, which means that the molar mass of carbon is 12.011 g/mol.

$$1\text{ mol C} = 12.011\text{ g}$$

[2]See Appendix A for a discussion of unit prefixes and unit conversions.

We therefore start the calculation by determining the number of moles of carbon in the diamond.

$$0.2000 \text{ g C} \times \frac{1 \text{ mol C}}{12.011 \text{ g C}} = 0.01665 \text{ mol C}$$

We then use Avogadro's number to calculate the number of carbon atoms in the diamond.

$$0.01665 \text{ mol C} \times \frac{6.022 \times 10^{23} \text{ atoms}}{1 \text{ mol C}} = 1.003 \times 10^{22} \text{ C atoms}$$

➤ CHECKPOINT

How many aluminum atoms are there in 1.0 grams of pure aluminum?

2.5 Converting Grams Into Moles and Number of Atoms

The mole is the bridge between chemistry on the macroscopic scale, where we do experiments, and the atomic scale, where we think about the implications of these experiments. As a result, one of the most common calculations in chemistry involves converting measurements of the mass of a sample into the number of moles of the substance it contains. To show how this is done, consider the following question: How many moles of sulfur atoms does 45.5 grams of sulfur contain?

What information do we need to convert grams of a substance into moles?

$$\text{grams} \longrightarrow \text{moles}$$

It seems reasonable to try to find out the number of grams per mole of sulfur. According to the periodic table, the atomic weight of sulfur is 32.066 amu. This means that a mole of sulfur atoms would have a mass of 32.066 grams.

$$1 \text{ mol S} = 32.066 \text{ g}$$

We don't need five significant figures, however, because we only know the mass of the sample to three significant figures. Let's therefore carry four significant figures as we transform this equality into two unit factors.

$$\frac{1 \text{ mol S}}{32.07 \text{ g S}} \quad \text{and} \quad \frac{32.07 \text{ g S}}{1 \text{ mol S}}$$

Multiplying the size of the sample in grams by the unit factor on the left gives us the number of moles of sulfur atoms in the sample.

$$45.5 \text{ g S} \times \frac{1 \text{ mol S}}{32.07 \text{ g S}} = 1.42 \text{ mol S}$$

Once we know the number of moles of sulfur atoms, we can use Avogadro's number to calculate the number of atoms in the sample.

$$1.42 \text{ mol S} \times \frac{6.022 \times 10^{23} \text{ S atoms}}{1 \text{ mol S}} = 8.55 \times 10^{23} \text{ S atoms}$$

In general, you need two pieces of information to do calculations of this nature. You need to know the mass of a mole of the substance and the number of particles in a mole.

$$\underset{\textit{Macroscopic Scale}}{\text{Mass}} \xleftrightarrow{\textit{Molar mass}} \text{Moles} \xleftrightarrow{\textit{Avogadro's number}} \underset{\textit{Atomic Scale}}{\text{Atoms}}$$

Exercise 2.3

Calculate the mass of a sample of iron metal that contains 0.250 mol of iron atoms.

Solution

According to the periodic table, the molar mass of iron is 55.85 g/mol. This can be represented in terms of either of the following units factors.

$$\frac{1 \text{ mol Fe}}{55.85 \text{ g Fe}} \quad \text{or} \quad \frac{55.85 \text{ g Fe}}{1 \text{ mol Fe}}$$

In order to convert from moles to grams, we need the unit factor that tells us the number of grams of iron in one mole of this metal.

$$0.250 \text{ mol Fe} \times \frac{55.85 \text{ g FeS}}{1 \text{ mol Fe}} = 14.0 \text{ g Fe}$$

Exercise 2.4

Calculate the number of atoms in a 0.123-gram sample of aluminum foil.

Solution

Before we can do anything else, we need to know the number of moles of aluminum metal in the sample. This can be calculated from the mass of the sample and the molar mass of aluminum, which is 26.98 g/mol, to four significant figures.

$$0.123 \text{ g Al} \times \frac{1 \text{ mol Al}}{26.98 \text{ g Al}} = 4.56 \times 10^{-3} \text{ mol Al}$$

We can then use Avogadro's number to calculate the number of atoms in the sample.

$$4.56 \times 10^{-3} \text{ mol Al} \times \frac{6.022 \times 10^{23} \text{ Al atoms}}{1 \text{ mol Al}} = 2.75 \times 10^{21} \text{ Al atoms}$$

2.6 The Mole as a Collection of Molecules

Before we can apply the concept of the mole to compounds such as carbon dioxide (CO_2) or the sugar known as glucose ($C_6H_{12}O_6$), we have to be able to calculate the molecular weight of these compounds. As might be expected, the molecular weight of a compound is the sum of the atomic weights of the atoms in the formula of the compound.

Exercise 2.5

Calculate both the average mass of a single molecule of carbon dioxide and glucose and the molecular weight of these compounds.

Solution

The average mass of a *molecule of carbon dioxide* is the sum of the atomic weights of the three atoms in a CO_2 molecule.

Mass of a single CO_2 molecule:

$$\begin{aligned} 1\text{ C atom} &= 1(12.011\text{ amu}) = 12.011\text{ amu} \\ 2\text{ O atoms} &= 2(15.999\text{ amu}) = \underline{31.998\text{ amu}} \\ & \qquad\qquad\qquad\qquad\quad 44.009\text{ amu} \end{aligned}$$

The mass of a *mole of carbon dioxide* therefore would be 44.009 grams.

The average mass of a molecule of glucose is the sum of the atomic weights of the 24 atoms in a $C_6H_{12}O_6$ molecule.

Mass of a single $C_6H_{12}O_6$ molecule:

$$\begin{aligned} 6\text{ C atoms} &= 6(12.011\text{ amu}) = 72.066\text{ amu} \\ 12\text{ H atoms} &= 12(1.0079\text{ amu}) = 12.095\text{ amu} \\ 6\text{ O atoms} &= 6(15.999\text{ amu}) = \underline{95.994\text{ amu}} \\ & \qquad\qquad\qquad\qquad\quad 180.155\text{ amu} \end{aligned}$$

The molecular weight of the compound is therefore 180.155 g/mol.

For many years, chemists referred to the results of the calculations in the previous exercise as the **molecular weight** of the compound. This term is somewhat misleading for several reasons. First, no $C_6H_{12}O_6$ molecule ever has a mass equal to 180.155 amu. This is the *average mass* of the sugar molecules, which contain ^{12}C and ^{13}C atoms as well as a mixture of hydrogen and oxygen isotopes. Second, some compounds, as we shall see, don't exist as molecules, so it is misleading to talk about their "molecular" weight. Some chemists therefore recommend that we describe the results of these calculations as the *mass of a mole* or the *molar mass* of a compound. Because the term *molecular weight* has been so extensively used by chemists and is widely found in the chemical literature, we will use the terms *molar mass* and *molecular weight* interchangeably.

Exercise 2.6

Describe the difference between the mass of a mole of oxygen atoms (O) and the mass of a mole of oxygen molecules (O_2).

Solution

Because the atomic weight of oxygen is 15.999 amu, a mole of oxygen atoms has a mass of 15.999 grams. Each O_2 molecule has two atoms, however, so the molecular weight of O_2 molecules is twice as large as the atomic weight of the atom.

$$1 \text{ mol O} = 15.999 \text{ g} \qquad 1 \text{ mol } O_2 = 31.998 \text{ g}$$

The diagram we used to summarize mass-mole conversions for elements can be used for chemical compounds. In this case, however, we can take the calculation one step further by using the formula of the compound to calculate the number of atoms of a given element in the sample.

$$\text{Mass} \xleftrightarrow{\textit{Molar mass}} \text{Moles} \xleftrightarrow{\textit{Avogadro's number}} \text{Molecules} \xleftrightarrow{\textit{Chemical formula}} \text{Atoms}$$

Macroscopic Scale — *Atomic Scale*

To illustrate the power of the mole concept, consider the following question: What is the formula of carbon dioxide if 2.73 grams of carbon combine with 7.27 grams of oxygen molecules (O_2) when the carbon burns?

The first step toward answering an unfamiliar problem often involves drawing a diagram that helps us organize the information in the problem and visualize the process taking place. We could start, for example, with the diagram in Figure 2.6, which summarizes the relationship between the mass of carbon and the oxygen gas consumed in this reaction.

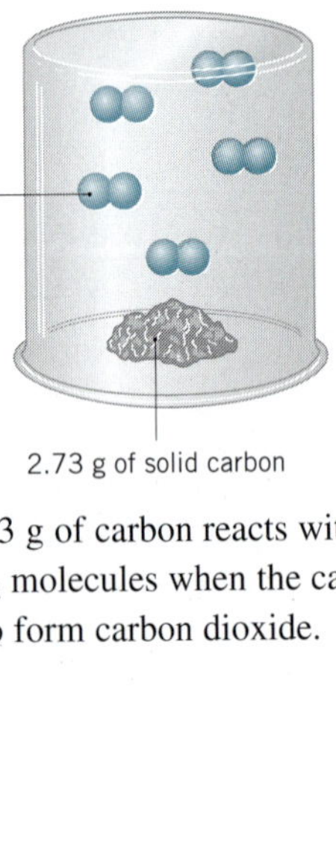

Fig. 2.6 2.73 g of carbon reacts with 7.27 g of O_2 molecules when the carbon burns to form carbon dioxide.

The next step in any problem of this kind is to convert grams into moles. To do this, we need to know the relationship between the number of grams and the number of moles of the substance. It doesn't matter which element we start with because we eventually have to work with both, so let's arbitrarily start with carbon.

The atomic weight of carbon is 12.011 amu, which means that a mole of carbon has a mass of 12.011 grams. We can use this information to construct two unit factors.

$$\frac{1 \text{ mol C}}{12.011 \text{ g C}} \quad \text{and} \quad \frac{12.011 \text{ g C}}{1 \text{ mol C}}$$

Converting grams of carbon into moles requires a unit factor that has units of moles in the numerator and grams in the denominator. Dimensional analysis therefore suggests setting up the problem as follows, in which the information in the unit factor is given to four significant figures.

$$2.73 \text{ g C} \times \frac{1 \text{ mol C}}{12.01 \text{ g C}} = 0.227 \text{ mol C}$$

The same format can be used to convert grams of oxygen into moles of oxygen molecules

$$7.27 \text{ g } O_2 \times \frac{1 \text{ mol } O_2}{32.00 \text{ g } O_2} = 0.227 \text{ mol } O_2 \text{ molecules}$$

and then moles of oxygen atoms.

$$0.227 \text{ mol } O_2 \text{ molecules} \times \frac{2 \text{ O atoms}}{1 \ O_2 \text{ molecule}} = 0.454 \text{ mol O atoms}$$

So far we have found that reacting 2.73 g of carbon and 7.27 g of oxygen corresponds to 0.227 mol of carbon reacting with 0.454 mol of oxygen atoms. Because atoms are neither created nor destroyed in a chemical reaction, the same number of atoms of each element must be found on both sides of the equation used to describe the reaction. The product of this reaction must therefore have a mass of 10.00 g (2.73 g + 7.27 g), and must contain 0.227 mol of carbon atoms and 0.454 mol of oxygen atoms.

We can now reread the question and ask: "Have we made any progress toward the answer?" In this case, we are trying to find the chemical formula for carbon dioxide, which gives the ratio of carbon atoms to oxygen atoms. The next step in the problem might therefore involve determining the relationship between the number of moles of carbon atoms and moles of oxygen atoms in the sample of carbon dioxide.

$$\frac{0.454 \text{ mol O}}{0.227 \text{ mol C}} = 2.00$$

There are twice as many moles of oxygen atoms as there are moles of carbon atoms in the sample. Because a mole of atoms always contains the same number of atoms, the only possible conclusion is that there are twice as many oxygen atoms as carbon atoms in the compound. In other words, the formula for carbon dioxide must be CO_2.

This calculation shows how the mole concept can be used as the bridge between macroscopic measurements (the mass of carbon and oxygen) and the microscopic atomic world (the number of carbon and oxygen atoms in a carbon dioxide molecule).

Exercise 2.7

Determine the number of carbon atoms in 0.500 grams of carbon dioxide, CO_2.

Solution

The first step in this calculation involves converting the mass of the sample into the number of moles of CO_2 using the molecular weight of CO_2 calculated in Exercise 2.5.

$$0.500 \text{ g } CO_2 \times \frac{1 \text{ mol } CO_2}{44.01 \text{ g } CO_2} = 1.14 \times 10^{-2} \text{ moles of } CO_2$$

Once we know the number of moles of CO_2 in the sample, we can use Avogadro's number to calculate the number of CO_2 molecules.

$$1.14 \times 10^{-2} \text{ mol } CO_2 \times \frac{6.022 \times 10^{23} \text{ } CO_2 \text{ molecules}}{1 \text{ mol } CO_2} = 6.86 \times 10^{21} \text{ } CO_2 \text{ molecules}$$

We can now use the chemical formula for carbon dioxide to determine the number of carbon atoms in the sample.

$$6.86 \times 10^{21} \text{ } CO_2 \text{ molecules} \times \frac{1 \text{ C atom}}{1 \text{ } CO_2 \text{ molecule}} = 6.86 \times 10^{21} \text{ C atoms}$$

➤ CHECKPOINT

How many atoms of carbon are there in one molecule of C_2H_2? How many moles of carbon atoms are in one mole of C_2H_2? How many atoms of carbon are in one mole of C_2H_2?

2.7 Percent Mass

It is often useful to know the **percent by mass** of the different elements in a compound. Percent mass can be determined experimentally or from the chemical formula. When the formula of a compound is known, the first step in this calculation involves determining the molecular weight of the compound. The molecular weight of ethanol (CH_3CH_2OH), for example, would be calculated as follows.

Mass of a single CH_3CH_2OH molecule:

$$\begin{aligned} 2 \text{ C atoms} &= 2(12.011 \text{ amu}) = 24.022 \text{ amu} \\ 6 \text{ H atoms} &= 6(1.0079 \text{ amu}) = 6.0474 \text{ amu} \\ 1 \text{ O atom} &= 15.999 \text{ amu} = \underline{15.999 \text{ amu}} \\ & \qquad\qquad\qquad\quad 46.068 \text{ amu} \end{aligned}$$

A mole of ethanol therefore weighs 46.068 grams.

Percent literally means "parts per hundred." The percent by mass of carbon in ethanol is therefore the mass of carbon in a mole of ethanol divided by the mass of a mole of ethanol, times 100.

$$\frac{24.022 \text{ g C}}{46.068 \text{ g } CH_3CH_2OH} \times 100 = 52.145\% \text{ C}$$

The percent by mass of hydrogen and oxygen in ethanol can be calculated in a similar fashion.

$$\frac{6.0474 \text{ g H}}{46.068 \text{ g } CH_3CH_2OH} \times 100 = 13.127\% \text{ H}$$

$$\frac{15.999 \text{ g O}}{46.068 \text{ g } CH_3CH_2OH} \times 100 = 34.729\% \text{ O}$$

2.8 Determining the Formula of a Compound

Section 2.6 showed one way to determine the formula of a compound. By carefully measuring the amount of carbon and oxygen that combined to form carbon dioxide, it was possible to show that the formula of this compound is CO_2. Let's look at another way to approach this problem using percent by mass data. This time we will examine the compound methane, once known as "marsh gas" because it was first collected above certain swamps, or marshes, in Britain.

Methane is 74.9% carbon and 25.1% hydrogen by mass. A 100-gram sample of the gas therefore contains 74.9 g of carbon and 25.1 g of hydrogen.

$$100 \text{ g methane} \times \frac{74.9 \text{ g C}}{100 \text{ g methane}} = 74.9 \text{ g C}$$

$$100 \text{ g methane} \times \frac{25.1 \text{ g H}}{100 \text{ g methane}} = 25.1 \text{ g H}$$

This is useful information because we can use the molar mass of these elements to convert the grams of carbon and hydrogen in this sample into moles of each element.

$$74.9 \text{ g C} \times \frac{1 \text{ mol C}}{12.01 \text{ g C}} = 6.24 \text{ mol C}$$

$$25.1 \text{ g H} \times \frac{1 \text{ mol H}}{1.008 \text{ g H}} = 24.9 \text{ mol H}$$

We now know the number of moles of carbon atoms and the number of moles of hydrogen atoms in a sample. We also know that there are always the same number of atoms in a mole of atoms of any element. It therefore might be useful to look at the ratio of the moles of these elements in the sample.

$$\frac{24.9 \text{ mol H}}{6.24 \text{ mol C}} = 3.99$$

The 100-gram sample of methane contains about four times as many moles of hydrogen atoms as moles of carbon atoms, within experimental error. This means that there are four times as many hydrogen atoms as carbon atoms in this sample.

This experiment tells us the simplest or **empirical formula** of the compound, but not necessarily the **molecular formula.** These results are consistent with molecules that contain one carbon atom and four hydrogen atoms: CH_4. But they are also consistent with formulas such as C_2H_8, C_3H_{12}, C_4H_{16}, and so on. All we know at this point is that the molecular formula for the molecule is some multiple of the empirical formula, CH_4. Using other experimental techniques it is possible to show that the molecular weight of methane is 16 g/mol, which is consistent with a molecular formula of CH_4.

Exercise 2.8

Vitamin C is found in citrus fruits, or it can be obtained from dietary supplements such as vitamin C tablets. Calculate the empirical formula for vitamin C, which is 40.9% C, 54.5% O, and 4.58% H by mass. The molar mass of vitamin C is 176 g/mol. What is the molecular formula for vitamin C?

Natural and synthetic sources of vitamin C.

Solution

We start by calculating the number of grams of each element in a 100-gram sample of vitamin C.

$$100 \text{ g} \times 40.9\% \text{ C} = 40.9 \text{ g C}$$

$$100 \text{ g} \times 54.5\% \text{ O} = 54.5 \text{ g O}$$

$$100 \text{ g} \times 4.58\% \text{ H} = 4.58 \text{ g H}$$

We then convert the number of grams of each element into the number of moles of atoms of that element.

$$40.9 \text{ g C} \times \frac{1 \text{ mol C}}{12.01 \text{ g C}} = 3.41 \text{ mol C}$$

$$54.5 \text{ g O} \times \frac{1 \text{ mol O}}{16.00 \text{ g O}} = 3.41 \text{ mol O}$$

$$4.58 \text{ g H} \times \frac{1 \text{ mol H}}{1.008 \text{ g H}} = 4.54 \text{ mol H}$$

Because we are interested in the simplest whole-number ratio of these elements, we now divide through by the element with the smallest number of moles of atoms.

$$\frac{3.41 \text{ mol O}}{3.41 \text{ mol C}} = 1.00 \qquad \text{and} \qquad \frac{4.54 \text{ mol H}}{3.41 \text{ mol C}} = 1.33$$

The ratio of C to H to O atoms in vitamin C is therefore 1: 1⅓: 1. It doesn't make sense to write the ratio of atoms as $CH_{1^1/_3}O$, because there is no such thing as one-third of a hydrogen atom. We therefore multiply this ratio by small whole numbers until we get a formula in which all of the coefficients are integers.

$$2(CH_{1^1/_3}O) = C_2H_{2^2/_3}O_2$$
$$3(CH_{1^1/_3}O) = C_3H_4O_3$$

The empirical formula of vitamin C is therefore $C_3H_4O_3$.

The mass of a single $C_3H_4O_3$ molecule would be 88.06 amu, to four significant figures.

$$\begin{aligned} 3 \text{ C atoms} &= 3(12.01 \text{ amu}) = 36.03 \text{ amu} \\ 4 \text{ H atoms} &= 4(1.008 \text{ amu}) = 4.032 \text{ amu} \\ 3 \text{ O atoms} &= 3(16.00 \text{ amu}) = \underline{48.00 \text{ amu}} \\ & \qquad\qquad\qquad\quad\; 88.06 \text{ amu} \end{aligned}$$

The mass of a mole of $C_3H_4O_3$ molecules is therefore 88.06 g.

This is as far as we can go with percent by mass data. Once we know the molar mass of vitamin C, however, we can compare the mass of a mole of this compound with the mass of a mole of $C_3H_4O_3$ molecules.

> **CHECKPOINT**
>
> Calcium carbide was once used in miner's lamps. A sample of calcium carbide large enough to contain one mole of calcium atoms also contains 24 grams of carbon. What is the empirical formula of the compound? What additional information is necessary to determine the molecular formula?

$$\frac{176 \text{ g vitamin C/mol}}{88.06 \text{ g } C_3H_4O_3\text{/mol}} = 2.00$$

The only possible conclusion is that a molecule of vitamin C is twice as large as the empirical formula for this compound. In other words, the molecular formula of vitamin C is $C_6H_8O_6$.

2.9 Elemental Analysis

Percent by mass data for a compound are obtained by a process known as **elemental analysis.** The analysis is carried out by burning a small sample in a microanalysis apparatus, as shown in Figure 2.7. A few milligrams of the compound are added to a tiny platinum boat, which is placed in a furnace heated to about 850°C, and a stream of oxygen (O_2) gas is passed over the sample.

Compounds that contain only carbon and hydrogen burn to form a mixture of CO_2 and H_2O. (If elements besides carbon and hydrogen are present, other gases

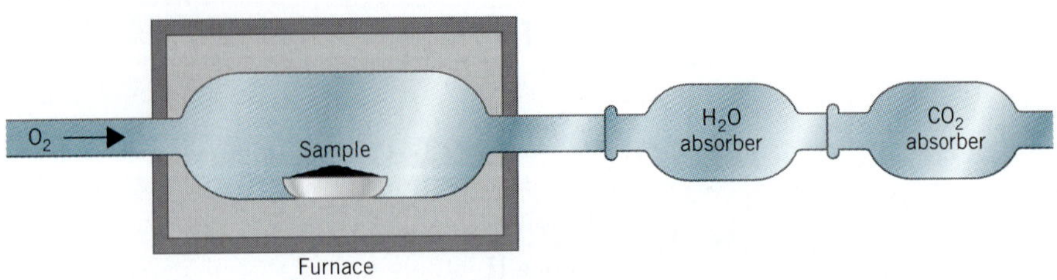

Fig. 2.7 Diagram of the microanalysis apparatus used to determine the percent by mass of carbon and hydrogen in a compound.

may be formed as well.) The CO_2 and H_2O produced in the combustion reaction are swept out of the furnace by the stream of oxygen gas and trapped in a pair of absorbers. The water vapor is absorbed onto a sample of magnesium perchlorate [$Mg(ClO_4)_2$] of known mass. The carbon dioxide is absorbed onto a known mass of the mineral ascharite ($Mg_2Br_2O_4 \cdot 2H_2O$).

A 3.00-mg sample of aspirin was analyzed in the apparatus shown in Figure 2.7. Aspirin is known to contain three elements: carbon, hydrogen, and oxygen. The results of the analysis found that 6.60 mg of CO_2 and 1.20 mg of H_2O were formed by burning the aspirin.

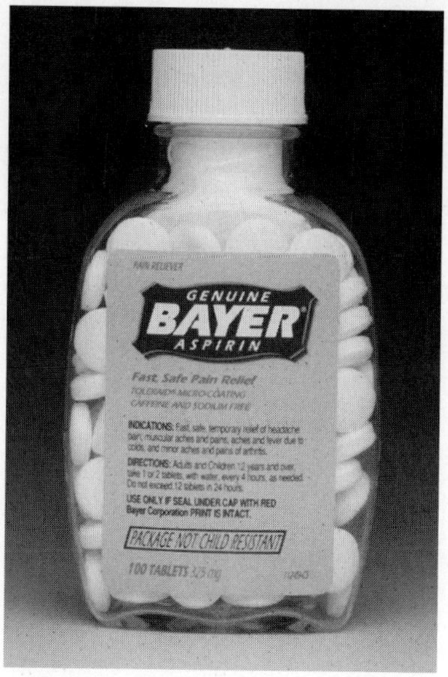

Elemental analysis can be used to determine the empirical formula for compounds such as aspirin.

Our goal is to convert this information into the percent by mass of carbon and hydrogen in aspirin. The logical place to start is by converting the number of grams of CO_2 and H_2O given off in this reaction into moles of these compounds.

$$0.00660 \text{ g } CO_2 \times \frac{1 \text{ mol } CO_2}{44.01 \text{ g } CO_2} = 1.50 \times 10^{-4} \text{ mol } CO_2$$

$$0.00120 \text{ g } H_2O \times \frac{1 \text{ mol } H_2O}{18.02 \text{ g } H_2O} = 6.66 \times 10^{-5} \text{ mol } H_2O$$

It would now be useful to know how many moles of carbon atoms and hydrogen atoms are present in the gases given off in this reaction. We therefore start by noting that there is one carbon atom in each CO_2 molecule, which means there is a mole of carbon atoms in each mole of CO_2 trapped in the CO_2 absorber. Because all of the carbon came from the aspirin, the aspirin must have contained 1.50×10^{-4} mol of carbon atoms.

$$1.50 \times 10^{-4} \text{ mol } CO_2 \times \frac{1 \text{ mol C}}{1 \text{ mol } CO_2} = 1.50 \times 10^{-4} \text{ mol C}$$

There are two hydrogen atoms in each H_2O molecule and therefore 2 moles of hydrogen atoms in each mole of H_2O molecules. The 3.00-mg sample of aspirin therefore must have contained 1.33×10^{-4} mol of hydrogen atoms.

$$6.66 \times 10^{-5} \text{ mol } H_2O \times \frac{2 \text{ mol H}}{1 \text{ mol } H_2O} = 1.33 \times 10^{-4} \text{ mol H}$$

We now know the number of moles of carbon atoms and hydrogen atoms in the original sample, so we can calculate the number of grams of each element in the sample.

$$1.50 \times 10^{-4} \text{ mol C} \times \frac{12.01 \text{ g C}}{1 \text{ mol C}} = 1.80 \times 10^{-3} \text{ g C}$$

$$1.33 \times 10^{-4} \text{ mol H} \times \frac{1.008 \text{ g H}}{1 \text{ mol H}} = 1.34 \times 10^{-4} \text{ g H}$$

According to this calculation, the 3.00-mg sample of aspirin contains 1.80 mg of carbon and 0.134 mg of hydrogen. Aspirin is therefore 60.0% C and 4.47% H by mass.

$$\frac{1.80 \text{ mg C}}{3.00 \text{ mg aspirin}} \times 100 = 60.0\% \text{ C}$$

$$\frac{0.134 \text{ mg H}}{3.00 \text{ mg aspirin}} \times 100 = 4.47\% \text{ H}$$

Carbon and hydrogen add up to only 64.5% of the total mass of the aspirin. The remaining mass (35.5%) must be due to the third element: oxygen.

Microanalysis therefore suggests that aspirin is 60.0% C, 4.47% H, and 35.5% O by mass. Once we know the percent by mass of each element in aspirin we can use the techniques shown in Exercise 2.8 to determine that the empirical formula for aspirin is $C_9H_8O_4$.

2.10 Chemical Reactions and the Law of Conservation of Atoms

We have focused so far on individual compounds such as carbon dioxide (CO_2) and glucose ($C_6H_{12}O_6$). Much of the fascination of chemistry, however, revolves around chemical reactions. The first breakthrough in the study of chemical reactions resulted from the work of the French chemist Antoine Lavoisier between 1772 and 1794. Lavoisier noted that the mass of the products of a chemical reaction is always the same as the mass of the starting materials consumed in the reaction. His results led to one of the fundamental laws of chemical behavior: The **law of conservation of mass,** which states that matter is conserved in a chemical reaction.

We now understand why matter is conserved—atoms are neither created nor destroyed in a chemical reaction. The hydrogen atoms in an H_2 molecule can combine with oxygen atoms in an O_2 molecule to form H_2O, as shown in Figure 2.8. But the number of hydrogen and oxygen atoms before and after the reaction is the same. The total mass of the products of a reaction therefore must be the same as the total mass of the reactants that undergo reaction.

2.11 Chemical Equations as a Representation of Chemical Reactions

It is possible to describe a chemical reaction in words, but it is much easier to describe it with a **chemical equation.** The formulas of the starting materials, or **reactants,** are written on the left side of the equation, and the formulas of the **products** are written on the right. Instead of an equal sign, the reactants and products are separated by an arrow. The reaction between hydrogen and oxygen to form water shown in Figure 2.8 is represented by the following equation.

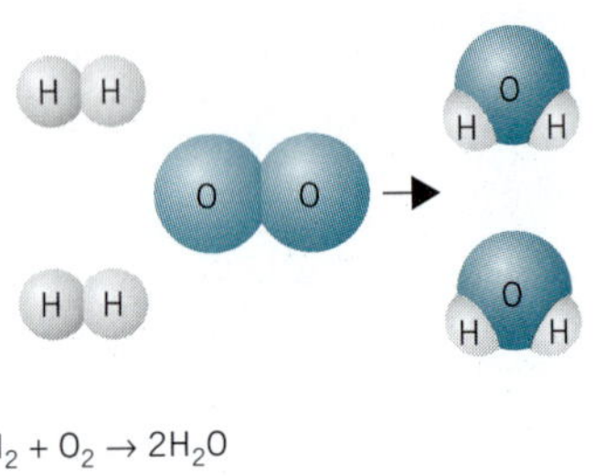

Fig. 2.8 Mass is conserved in chemical reactions such as the reaction between hydrogen, H_2, and oxygen, O_2, to form water because atoms are neither created nor destroyed.

$$2\ H_2 + O_2 \longrightarrow 2\ H_2O$$

It is often useful to indicate whether the reactants or products are solids, liquids, or gases by writing as *s*, *l*, or *g* in parentheses after the symbol for the reactants or products.

$$2\ H_2(g) + O_2(g) \longrightarrow 2\ H_2O(g)$$

Many of the reactions you will encounter in this course will occur when solutions of two substances dissolved in water are mixed. These **aqueous** solutions (from the Latin word *aqua,* "water") are so important we use the special symbol *aq* to describe them. This way we can distinguish between glucose as a solid, $C_6H_{12}O_6(s)$, and solutions of this sugar dissolved in water, $C_6H_{12}O_6(aq)$, or between salt as an ionic solid, NaCl(s), and solutions of salt dissolved in water, NaCl(aq). The process by which a sample dissolves in water will be indicated by equations such as the following.

$$C_6H_{12}O_6(s) \xrightarrow{H_2O} C_6H_{12}O_6(aq)$$

Ionic compounds and some molecular compounds break up into their component ions when they dissolve in water. Therefore, the aqueous forms of these compounds may be written as aqueous ions. Since salt is an ionic compound, the chemical equation describing the dissolution of salt can be written as an **ionic equation.**

Ionic equations are briefly introduced in this section because they are an important type of chemical equation that students will encounter often in the text. The concepts of an ionic equation and net ionic equation will be expanded in later chapters after students have a better understanding of ionic compounds.

$$NaCl(s) \xrightarrow{H_2O} Na^+(aq) + Cl^-(aq)$$

Chemical equations are such a powerful shorthand for describing chemical reactions that we tend to think about reactions in terms of these equations. It is important to remember that a chemical equation is a statement of *what can happen,* not necessarily *what will happen.* The following equation, for example, does not guarantee that hydrogen will react with oxygen to form water.

$$2\,H_2(g) + O_2(g) \longrightarrow 2\,H_2O(g)$$

It is possible to fill a balloon with a mixture of hydrogen and oxygen and find that no reaction occurs until the balloon is touched with a flame. All the equation tells us is what would happen if, or when, the reaction occurs.

➤ CHECKPOINT

The reaction between $HCl(aq)$ and $NaOH(aq)$ can be described by the following equation.

$$HCl(aq) + NaOH(aq) \longrightarrow NaCl(aq) + H_2O(l)$$

Assume that HCl, NaOH, and NaCl break up into their respective ions in aqueous solution. Write the ionic equation for this reaction.

2.12 Two Views of Chemical Equations: Molecules Versus Moles

Chemical equations such as the following can be used to represent reactions on either the atomic or macroscopic scale.

$$2\,H_2(g) + O_2(g) \longrightarrow 2\,H_2O(g)$$

Thus, this equation can be read in either of the following ways.

- When hydrogen reacts with oxygen, 2 molecules of hydrogen and 1 molecule of oxygen are consumed for every 2 molecules of water produced.
- When hydrogen reacts with oxygen, 2 moles of hydrogen and 1 mole of oxygen are consumed for every 2 moles of water produced.

Regardless of whether we think of the reaction in terms of molecules or moles, chemical equations must be balanced—there must be the same number of atoms of each element on both sides of the equation. As a result, the total mass of the reactants will be equal to the total mass of the products of the reaction. On the atomic scale, the following equation is balanced because the total mass of the reactants in atomic mass units is equal to the mass of the products.

$$\begin{array}{ccc} 2\,H_2(g) + O_2(g) & \longrightarrow & 2\,H_2O(g) \\ 2 \times 2 \text{ amu} + 32 \text{ amu} & & 2 \times 18 \text{ amu} \\ 36 \text{ amu} & & 36 \text{ amu} \end{array}$$

On the macroscopic scale, it is balanced because the mass of 2 moles of hydrogen and 1 mole of oxygen is equal to the mass of 2 moles of water.

$$\begin{array}{ccc} 2\,H_2(g) + O_2(g) & \longrightarrow & 2\,H_2O(g) \\ 2 \times 2\text{g} + 32\text{g} & & 2 \times 18\text{g} \\ 36\text{g} & & 36\text{g} \end{array}$$

Reaction diagrams show the reactants being broken apart into individual atoms and the recombination of these atoms to form the products or reactants.

Students can visualize that the same atoms are present in the products and reactants and that a chemical reaction is just a re-arrangement of atoms. The diagrams are a very powerful model for teaching thermodynamics and are used extensively in Chapters 7 and 13.

The following diagram is a useful way of visualizing the relationship between the mass of the starting materials and products of the reaction. The box on the left shows the reactants, and the box on the right shows the products. The box centered above the arrow for the reaction represents all of the atoms found in either the products or the reactants.

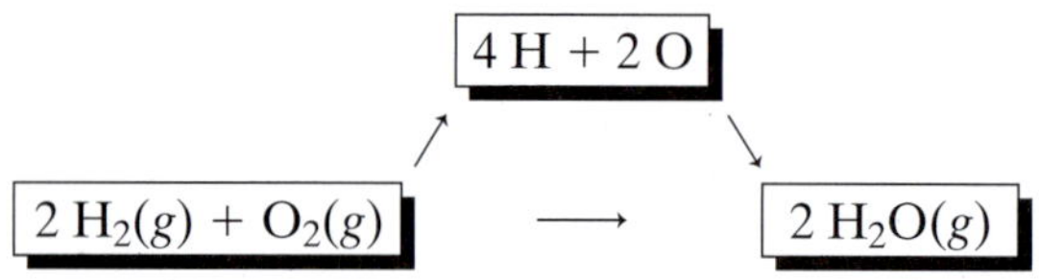

If we think about this reaction in terms of H_2 and O_2 molecules combining to form H_2O molecules, the equation is balanced because we have 4 hydrogen atoms and 2 oxygen atoms on both sides of the equation. If we think about the reaction in terms of moles of starting materials and products, the equation must be balanced because we have 4 moles of hydrogen atoms and 2 moles of oxygen atoms on both sides of the arrow.

It is important to recognize that reactions seldom occur by passing through an intermediate stage in which they form isolated atoms. But this approach can be a useful way to emphasize the fact that atoms are conserved in a chemical reaction. Each and every atom among the starting materials must be found in one of the products of the reaction.

2.13 Balancing Chemical Equations

There is no sequence of rules that can be blindly followed to get a balanced chemical equation. All we can do is manipulate the coefficients written in front of the formulas of the reactants and products until the number of atoms of each element on both sides of the equation is the same. The subscripts in the chemical formulas cannot be changed when balancing an equation because that would change the identity of the products and reactants. Persistence is required to balance chemical equations; the equation must be explored until the number of atoms of each element is the same on both sides of the equation.

While doing this it is usually a good idea to tackle the easiest part of a problem first. Consider, for example, the equation for the combustion of glucose ($C_6H_{12}O_6$). Everything that we digest, at one point or another, gets turned into a sugar that is oxidized to give the energy that fuels our bodies. Although there are a variety of sugars that can be used as fuels, the primary source of energy that drives our bodies is glucose, or *blood sugar* as it is also known. The bloodstream delivers both glucose and oxygen to tissues, where they react to give a mixture of carbon dioxide and water.

$$C_6H_{12}O_6(aq) + O_2(g) \longrightarrow CO_2(g) + H_2O(l)$$

If you look at this equation carefully, you might conclude that it is easier to balance the carbon and hydrogen atoms than the oxygen atoms in this reaction. All of the carbon atoms in glucose end up in CO_2 and all of the hydrogen atoms end up in H_2O, but there are two sources of oxygen among the starting materials and two compounds that contain oxygen among the products. This means that there is no way to predict the number of O_2 molecules consumed in this reaction until we know how many CO_2 and H_2O molecules are produced.

We can start the process of balancing this equation by noting that there are 6 carbon atoms in each $C_6H_{12}O_6$ molecule. Thus, 6 CO_2 molecules are formed for every $C_6H_{12}O_6$ molecule consumed.

$$1\ C_6H_{12}O_6 + \underline{\qquad}\ O_2 \longrightarrow \boldsymbol{6\ CO_2} + \underline{\quad}\ H_2O$$

There are 12 hydrogen atoms in each $C_6H_{12}O_6$ molecule, which means there must be 12 hydrogen atoms, or 6 H_2O molecules, on the right-hand side of the equation.

$$1\ C_6H_{12}O_6 + \underline{\quad\quad}\ O_2 \longrightarrow 6\ CO_2 + \boldsymbol{6\ H_2O}$$

Now that the carbon and hydrogen atoms are balanced, we can try to balance the oxygen atoms. There are 12 oxygen atoms in 6 CO_2 molecules and 6 oxygen atoms in $6H_2O$ molecules. To balance the 18 oxygen atoms in the products of this reaction we need a total of 18 oxygen atoms in the starting materials. But each $C_6H_{12}O_6$ molecule already contains 6 oxygen atoms. We therefore need 6 O_2 molecules among the reactants.

$$6\ C + 12\ H + 18\ O$$

$$1\ C_6H_{12}O_6 + 6\ O_2 \longrightarrow 6\ CO_2 + 6\ H_2O$$

The balanced equation for this reaction is therefore written as follows.

$$C_6H_{12}O_6(aq) + 6\ O_2(g) \longrightarrow 6\ CO_2(g) + 6\ H_2O(l)$$

There are now 6 carbon atoms, 12 hydrogen atoms, and 18 oxygen atoms on each side of the equation, as shown in Figure 2.9.

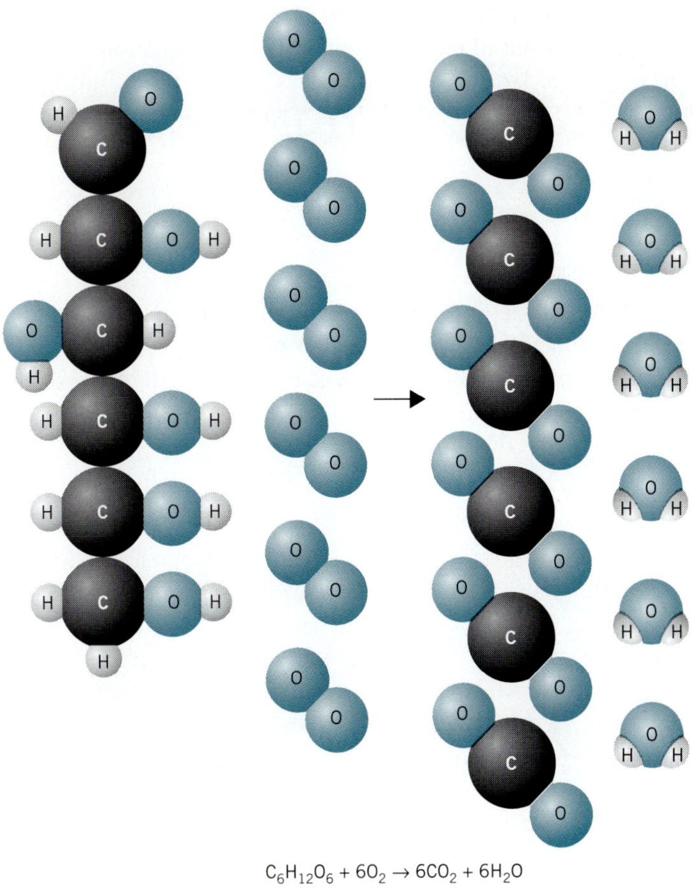

Fig. 2.9 A mechanical model on the atomic scale for the reaction between glucose ($C_6H_{12}O_6$) and O_2 to form CO_2 and H_2O. Note that the number of carbon, hydrogen, and oxygen atoms is the same in both the reactants and the products of the reaction.

Exercise 2.9

Write a balanced equation for the reaction that occurs when ammonia (NH_3) burns in air to form nitrogen oxide (NO) and water.

$$NH_3 + \_\_\_\_ O_2 \longrightarrow \_\_\_\_ NO + \_\_\_\_ H_2O$$

Solution

We might start by balancing the nitrogen atoms because all of the nitrogen atoms in ammonia end up in nitrogen oxide. If we start with 1 molecule of ammonia and form 1 molecule of NO, the nitrogen atoms are balanced.

$$\mathit{1\ NH_3} + \_\_\_\_ O_2 \longrightarrow \mathit{1\ NO} + \_\_\_\_ H_2O$$

We can then turn to the hydrogen atoms. We have 3 hydrogen atoms on the left and 2 hydrogen atoms on the right in this equation. One way of balancing the hydrogen atoms is to look for the lowest common multiple: $2 \times 3 = 6$. We therefore set up the equation so that there are 6 hydrogen atoms on both sides. Doing this doubles the amount of NH_3 consumed in the reaction, so we have to double the amount of NO produced.

$$\mathit{2\ NH_3} + \_\_\_\_ O_2 \longrightarrow \mathit{2\ NO + 3\ H_2O}$$

Because the nitrogen and hydrogen atoms are both balanced, the only task left is to balance the oxygen atoms. There are 5 oxygen atoms on the right side of this equation, so we need 5 oxygen atoms on the left. This could be accomplished by using a coefficient of 2½ in front of oxygen in the following equation:

$$2\ NH_3 + 2\tfrac{1}{2}\ O_2 \longrightarrow 2\ NO + 3\ H_2O$$

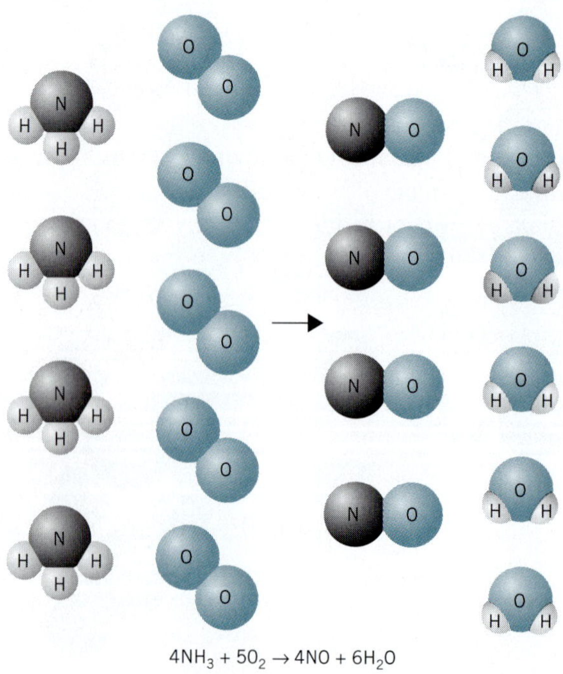

Fig. 2.10 The reaction between ammonia and oxygen to form NO and water on the atomic scale. All atoms in the reactants must be accounted for in the products.

This equation works on the macroscopic scale, where 2 moles of ammonia would react with 2.5 moles of oxygen. If we insist that chemical equations work on both the atomic and macroscopic scales, we must multiply the equation by 2. The balanced equation for this reaction is therefore written as follows.

$$4\ NH_3(g) + 5\ O_2(g) \longrightarrow 4\ NO(g) + 6\ H_2O(g)$$

All of the atoms in the reactants are now accounted for in the products. The 12 hydrogen atoms in 4 NH_3, for example, are found in the 6 water molecules, as shown in Figure 2.10.

➤ CHECKPOINT

What is the value of x in the following equation when it is balanced?

$$x\ SO_3(g) \longrightarrow 2\ SO_2(g) + O_2(g)$$

2.14 Mole Ratios and Chemical Equations

Science has two fundamental goals: (1) explaining observations about the world around us, and (2) predicting what will happen under a particular set of conditions. Any chemical equation explains something about the world, but a balanced chemical equation has the added advantage of allowing us to predict what happens when the reaction takes place.

We describe a chemical reaction with a balanced chemical equation just as we might describe the preparation of oatmeal cookies with a recipe. A recipe for 4 dozen oatmeal cookies might call for 1 egg, 1 cup of flour, ¼ cup of water, 3 cups of oatmeal, and 1 cup of brown sugar. However, you aren't restricted to preparing 4 dozen cookies. By reducing all the ingredients by a factor of two you could prepare only 2 dozen; by doubling the amounts of the ingredients you could prepare 8 dozen. In the same way, we aren't restricted to working with the number of moles specified in the balanced equation for a reaction. Using the ratio of moles described by the balanced equation, we can predict the amount of product that would be formed from a given amount of reactant, or the amount of reactants needed to form a given amount of product.

Consider the reaction that occurs when the rocket fuel known as hydrazine (N_2H_4) burns in air to form N_2 gas and water vapor, as shown in Figure 2.11.

$$N_2H_4(l) + O_2(g) \longrightarrow N_2(g) + 2\ H_2O(g)$$

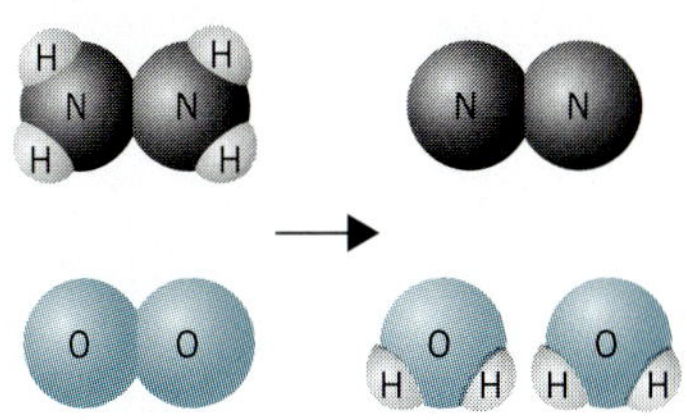

Fig. 2.11 The reaction between N_2H_4 and oxygen. Note that 1 mole of hydrazine reacts with 1 mole of O_2.

Let's start with a sample of 5.20 mol of hydrazine. How many moles of oxygen are required to consume this sample of hydrazine?

The coefficients in front of hydrazine and oxygen in this equation imply that 1 mole of hydrazine is consumed in this reaction for each mole of oxygen. We can express this information in terms of the following **mole ratios.**

$$\frac{1\ \text{mol}\ O_2}{1\ \text{mol}\ N_2H_4} \quad \text{or} \quad \frac{1\ \text{mol}\ N_2H_4}{1\ \text{mol}\ O_2}$$

To determine the moles of oxygen required to react with 5.20 mol of N_2H_4, we must decide which of these mole ratios to use. Dimensional analysis suggests that we should use the mole ratio on the left because it allows us to convert moles of hydrazine to moles of oxygen.

$$5.20\ \text{mol}\ N_2H_4 \times \frac{1\ \text{mol}\ O_2}{1\ \text{mol}\ N_2H_4} = 5.20\ \text{mol}\ O_2$$

Exercise 2.10

How many moles of water are formed when 5.20 mol of hydrazine react with excess oxygen?

Solution

Since oxygen is in excess, this reaction will occur until all of the hydrazine is consumed and then stop. We now need the mole ratio of water to hydrazine. According to the balanced equation, 2 moles of water are formed for every mole of hydrazine consumed in this reaction.

$$\frac{2 \text{ mol } H_2O}{1 \text{ mol } N_2H_4} \quad \text{or} \quad \frac{1 \text{ mol } N_2H_4}{2 \text{ mol } H_2O}$$

We can therefore calculate the number of moles of water produced in this reaction from the following equation.

$$5.20 \text{ mol } N_2H_4 \times \frac{2 \text{ mol } H_2O}{1 \text{ mol } N_2H_4} = 10.4 \text{ mol } H_2O$$

Let's now use what we have learned to see how a balanced equation can be used to predict the amount of O_2 we have to breathe to digest 10.00 g of sugar. We start with the balanced equation for the reaction.

$$C_6H_{12}O_6(aq) + 6\ O_2(g) \longrightarrow 6\ CO_2(g) + 6\ H_2O(l)$$

We then ask the fundamental question: How many moles of $C_6H_{12}O_6$ molecules does 10.00 g of this compound contain?

The only way we can convert grams of a substance into moles is to know something about the ratio of grams per mole in a sample of this substance. In other words, we need to know the molecular weight of the substance. The molecular weight of glucose calculated in Exercise 2.5 can be used to construct a pair of unit factors.

$$\frac{1 \text{ mol } C_6H_{12}O_6}{180.16 \text{ g } C_6H_{12}O_6} \quad \text{or} \quad \frac{180.16 \text{ g } C_6H_{12}O_6}{1 \text{ mol } C_6H_{12}O_6}$$

By paying attention to the units during the calculation, we can choose the correct unit factor to convert grams of sugar into moles.

$$10.00 \text{ g } C_6H_{12}O_6 \times \frac{1 \text{ mol } C_6H_{12}O_6}{180.16 \text{g } C_6H_{12}O_6} = 0.05551 \text{ mol } C_6H_{12}O_6$$

We now turn to the balanced equation for the reaction.

$$C_6H_{12}O_6(aq) + 6\ O_2(g) \longrightarrow 6\ CO_2(g) + 6\ H_2O(g)$$

This equation can be used to construct two mole ratios that describe the relationship between the moles of sugar and moles of oxygen consumed in the reaction.

$$\frac{6 \text{ mol } O_2}{1 \text{ mol } C_6H_{12}O_6} \quad \text{or} \quad \frac{1 \text{ mol } C_6H_{12}O_6}{6 \text{ mol } O_2}$$

By focusing on the units of this problem, we can select the correct mole ratio to convert moles of sugar into an equivalent number of moles of oxygen.

$$0.05551 \text{ mol } C_6H_{12}O_6 \times \frac{6 \text{ mol } O_2}{1 \text{ mol } C_6H_{12}O_6} = 0.3331 \text{ mol } O_2$$

We now need only one more step to complete our calculation—we need to convert the number of moles of O_2 consumed in the reaction into grams of oxygen. In Exercise 2.6 we concluded that the molecular weight of O_2 is exactly twice the atomic weight of the element. The next step in the calculation therefore involves multiplying the number of moles of O_2 consumed in the reaction by the molecular weight of oxygen.

$$0.3331 \text{ mol } O_2 \times \frac{32.00 \text{ g } O_2}{1 \text{ mol } O_2} = 10.66 \text{ g } O_2$$

We now have the answer to our original question. We need to breathe 10.66 g of oxygen to digest 10.00 g of the glucose that we carry through our bloodstream as the source of the energy needed to fuel our bodies.

2.15 Stoichiometry

By now, you have encountered all the steps necessary to do calculations of the sort that are grouped under the heading **stoichiometry.** The goal of these calculations is to use a balanced equation to predict the relationships between the amounts of the reactants and products of a reaction. There are three steps in these calculations.

- Find the starting material or product of the reaction for which you know both the mass of the sample and the formula of the substance. Use the molecular weight of this substance to convert the number of grams in the sample into an equivalent number of moles.
- Use the balanced equation for the reaction to create a mole ratio that can convert the number of moles of this substance into moles of another component of the reaction.
- Use the molecular weight of the other component of the reaction to convert the number of moles involved in the reaction into grams of that substance.

Exercise 2.11

Calculate the number of grams of ammonia (NH_3) needed to prepare 3.00 grams of nitrogen oxide (NO).

$$4\, NH_3(g) + 5\, O_2(g) \longrightarrow 4\, NO(g) + 6\, H_2O(g)$$

➤ CHECKPOINT

How many moles of H_2 are required to react completely with 2 moles of CO in the following reaction?

$$CO(g) + 2\,H_2(g) \longrightarrow CH_3OH(g)$$

How many molecules of hydrogen, H_2, would this be? How many atoms of hydrogen, H, would this be?

➤ CHECKPOINT

How many grams of H_2 are required to consume 2 grams of CO in the following reaction?

$$CO(g) + 2\,H_2(g) \longrightarrow CH_3OH(g)$$

Solution

The only component of this reaction about which we know both the formula of the compound and the mass of the sample is nitrogen oxide. We therefore start by converting 3.00 g of NO into an equivalent number of moles of the compound. To do this, we need to calculate the molecular weight of NO, which is 30.01 g/mol, to four significant figures. The number of moles of NO formed in this reaction can therefore be calculated as follows.

$$3.00 \text{ g NO} \times \frac{1 \text{ mol NO}}{30.01 \text{ g NO}} = 0.100 \text{ mol NO}$$

We now use the balanced equation for the reaction to determine the mole ratio that allows us to calculate the number of moles of NH_3 needed to produce 0.100 mole of NO.

$$0.100 \text{ mol NO} \times \frac{4 \text{ mol NH}_3}{4 \text{ mol NO}} = 0.100 \text{ mol NH}_3$$

We then use the molecular weight of NH_3 to calculate the mass of ammonia consumed in the reaction.

$$0.100 \text{ mol NH}_3 \times \frac{17.03 \text{ g NH}_3}{1 \text{ mol NH}_3} = 1.70 \text{ g NH}_3$$

According to this calculation, we need to start with 1.70 grams of ammonia to obtain 3.00 grams of nitrogen oxide.

Research for the New Millennium

THE STOICHIOMETRY OF THE BREATHALYZER

A patent was issued to R. F. Borkenstein in 1958 for the Breathalyzer, which is one method for determining whether an individual is DUI—driving under the influence—or DWI—driving while intoxicated. The chemistry behind the Breathalyzer is described by the following equation.

$$3\,CH_3CH_2OH(g) + 2\,Cr_2O_7^{2-}(aq) + 16\,H^+(aq) \longrightarrow 3\,CH_3CO_2H(aq) + 4\,Cr^{3+}(aq) + 11\,H_2O(l)$$

The instrument contains two ampules that each hold 0.75 mg potassium dichromate ($K_2Cr_2O_7$) dissolved in sulfuric acid (H_2SO_4). One of the ampules is used as a reference. The other is opened and the breath sample to be analyzed is added. If alcohol is present in the breath, it reacts with the yellow-orange $Cr_2O_7^{2-}$ ion to form a green Cr^{3+} ion. There is enough potassium dichromate in the ampule to consume the maximum amount of ethanol that might be expected to be present in someone's breath. The extent to which the color balance between the two ampules is disturbed is therefore a direct measure of the amount of alcohol in the breath sample.

Measurements of the alcohol on the breath are then converted into estimates of the concentration of alcohol in the blood. The link between these quantities is the assumption that 2100 mL of air exhaled from the lungs contains the same amount of alcohol as 1 mL of blood.

Measurements taken with the Breathalyzer are reported in units of percent blood-alcohol concentration (BAC) from 0 to 0.40%.[3] In most U.S. states, a BAC of 0.10% is sufficient for a DWI conviction. (This corresponds to a blood-alcohol concentration of 0.10 g of alcohol per 100 mL of blood.)

Between January 1989 and December 1990, almost 50 papers were published that described research related to the measurement of blood-alcohol concentration. Several studies probed the implications of the fact that the ratio of alcohol in the breath to the blood-alcohol concentration varies from one individual to another. However, research has shown that Breathalyzers can be used within limits to estimate BAC.[4] Other research has shown that there is no significant difference in the rate at which an individual metabolizes alcohol with age.[5] This research has also led to the development of more accurate methods for determining blood-alcohol concentration, particularly during autopsies that are done on those who drink and drive.[6]

Work with general chemistry students has revealed an interesting misconcept. Some students believed they can "cheat" on a Breathalyzer test by placing a copper penny in their mouth. (Modern folklore apparently suggests that this decreases the amount of alcohol on the breath.) Copper metal will, in fact, catalyze the following reaction, in which ethyl alcohol is oxidized to acetaldehyde.

$$CH_3CH_2OH \xrightarrow{Cu} CH_3CHO + H_2$$

There is only one minor problem—*the copper penny has to be heated until it glows red-hot before it will do this!*

2.16 The Nuts and Bolts of Limiting Reagents

According to Exercise 2.11, we need 1.70 g of ammonia to make 3.00 g of nitrogen oxide by the following reaction.

$$4\,NH_3(g) + 5\,O_2(g) \longrightarrow 4\,NO(g) + 6\,H_2O(g)$$

But we also need something else—we need enough oxygen for the reaction to take place.

Figure 2.12 shows the amount of NO produced when 1.70 g of ammonia is allowed to react with different amounts of oxygen. At first, the amount of NO produced is directly proportional to the amount of O_2 present when the reaction begins. At some point, however, the yield of the reaction reaches a maximum. No matter how much O_2 we add to the system, no more NO is produced.

We eventually reach a point at which the reaction runs out of NH_3 before all the O_2 is consumed. When this happens, the reaction must stop. No matter how much O_2 is added to the system, we can't get more than 3.00 g of NO from 1.70 g of NH_3.

When there isn't enough NH_3 to consume all the O_2 in the reaction, the amount of NH_3 limits the amount of NO that can be produced. Ammonia is therefore the **limiting reagent** in this reaction. Because there is more O_2 than we need, it is the **excess reagent.**

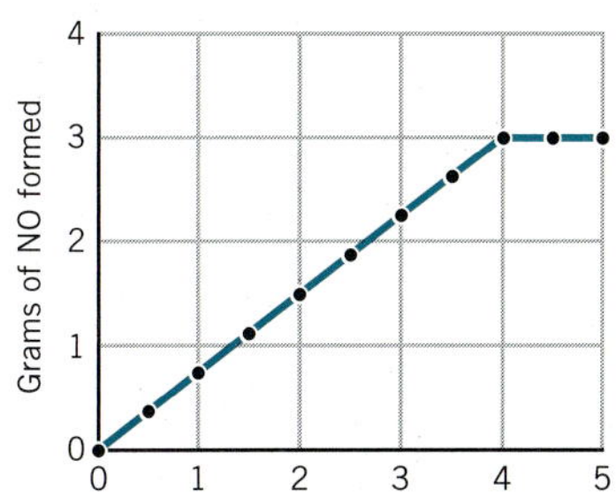

Fig. 2.12 A graph of the amount of NO that can be produced by adding different amounts of O_2 to 1.70 g of NH_3. Addition of oxygen up to 4.0 g produces more NO, but from this point on no matter how much O_2 is added no more NO is produced.

[3] D. A. Labianca, *Journal of Chemical Education*, **67**(3), 259 (1990).

[4] G. Simpson, *Journal of Analytical Toxicology*, **13**(2), 120 (1989).

[5] P. M. Hem and R. Volk, *Blutalkohol*, **26**(4), 276 (1990).

[6] J. V. Maracini, T. Carroll, S. Grant, S. Halleran, and J. A. Benz, *Journal of Forensic Science*, **41**, 181 (1989).

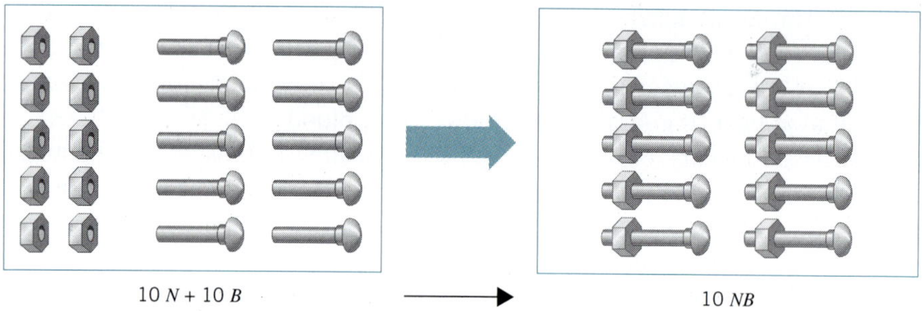

Fig. 2.13 Starting with 10 nuts (N) and 10 bolts (B), we can make 10 NB molecules, with no nuts or bolts left over.

The concept of limiting reagent is important because chemists frequently run reactions in which only a limited amount of one of the reactants is present. An analogy might help clarify what goes on in limiting reagent problems.

Let's start with exactly 10 nuts and 10 bolts, as shown in Figure 2.13. How many NB "molecules" can be made by screwing 1 nut (N) onto each bolt (B)? The answer is obvious: 10. After that, we run out of both nuts and bolts. Because we run out of both nuts and bolts at the same time, neither is a limiting reagent.

Now let's assemble N_2B molecules by screwing 2 nuts onto each bolt. Starting with 10 nuts and 10 bolts, we can make only 5 N_2B molecules, as shown in Figure 2.14. Because we run out of nuts, they must be the limiting reagent. Because 5 bolts are left over, they are the excess reagent.

Let's now extend the analogy to a slightly more difficult problem in which we assemble as many N_2B molecules as possible from a collection of 30 nuts and 20 bolts. There are three possibilities: (1) we have too many nuts and not enough bolts, (2) we have too many bolts and not enough nuts, or (3) we have just the right number of both nuts and bolts.

One way to approach the problem is to pick one of these alternatives and test it. Because there are more nuts (30) than bolts (20), let's assume that we have too many nuts and not enough bolts. In other words, let's assume that bolts are the limiting reagent in this problem. Now let's test that assumption. According to the formula, N_2B, we need 2 nuts for every bolt. Thus, we need 40 nuts to consume 20 bolts.

$$20 \text{ bolts} \times \frac{2 \text{ nuts}}{1 \text{ bolt}} = 40 \text{ nuts}$$

According to this calculation, we need more nuts (40) than we have (30). Thus, our original assumption is wrong. We don't run out of bolts; we run out of nuts.

Because our original assumption is wrong, let's turn it around and try again. Now let's assume that nuts are the limiting reagent and calculate the number of bolts we need.

$$30 \text{ nuts} \times \frac{1 \text{ bolt}}{2 \text{ nuts}} = 15 \text{ bolts}$$

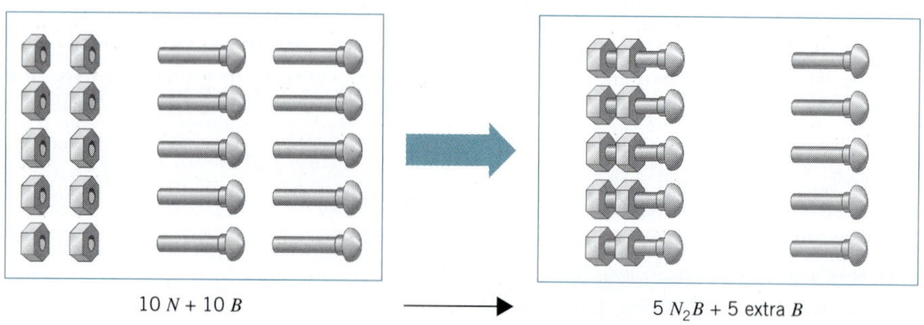

Fig. 2.14 Starting with 10 nuts and 10 bolts we can make only 5 N_2B molecules, and we will have 5 bolts left over. Because the number of N_2B molecules is limited by the number of nuts, the nuts are the limiting reagent in this analogy.

Do we have enough bolts to use up all the nuts? Yes, we need only 15 bolts and we have 20 bolts to choose from.

Our second assumption is correct. The limiting reagent in this case is nuts and the excess reagent is bolts. We can now calculate the number of N_2B molecules that can be assembled from 30 nuts and 20 bolts. Because the limiting reagent is nuts, the number of nuts limits the number of N_2B molecules we can make. Because we get one N_2B molecule for every 2 nuts, we can make a total of 15 of the N_2B molecules.

$$30 \text{ nuts} \times \frac{1\ N_2B \text{ molecule}}{2 \text{ nuts}} = 15\ N_2B \text{ molecules}$$

The following sequence of steps is helpful in working limiting reagent problems.

- Recognize that you have a limiting reagent problem, or at least consider the possibility that there might be a limiting amount of one of the reactants.
- Assume that one of the reactants is the limiting reagent.
- See if you have enough of the other reactant to consume the material you have assumed to be the limiting reagent.
- If you do, your original assumption was correct.
- If you don't, assume that another reagent is the limiting reagent and test this assumption.
- Once you have identified the limiting reagent, calculate the amount of product formed.

Exercise 2.12

Magnesium metal burns rapidly in air to form magnesium oxide. This reaction gives off an enormous amount of energy in the form of light and is used in both flares and fireworks. What mass of magnesium oxide (MgO) is formed when 10.0 g of magnesium reacts with 10.0 g of O_2?

Solution

It might be useful to start with a simple diagram, such as Figure 2.15, that summarizes the relevant information in the problem.

The next step toward solving the problem involves writing a balanced equation for the reaction.

$$2\ \text{Mg}(s) + \text{O}_2(g) \longrightarrow 2\ \text{MgO}(s)$$

The white light emitted during firework displays is produced by burning magnesium metal.

We then pick one of the reactants and assume it is the limiting reagent. For the sake of argument, let's assume that magnesium is the limiting reagent and O_2 is present in excess. Our immediate goal is to test the validity of this assumption. If it is correct, we will have more O_2 than we need to burn 10.0 g of magnesium. If it is wrong, O_2 is the limiting reagent.

We start by converting grams of magnesium into moles of magnesium.

$$10.0 \text{ g Mg} \times \frac{1 \text{ mol Mg}}{24.31 \text{ g Mg}} = 0.411 \text{ mol Mg}$$

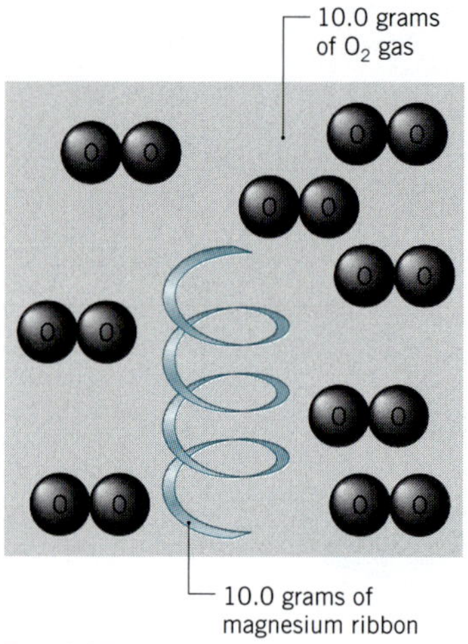

Fig. 2.15 A summary of the relevant information for Exercise 2.12.

We then use the balanced equation to predict the number of moles of O_2 needed to burn this much magnesium. According to the equation for the reaction, it takes 1 mole of O_2 to burn 2 moles of magnesium. We therefore need 0.206 mol of O_2 to consume all of the magnesium.

$$0.411 \text{ mol Mg} \times \frac{1 \text{ mol } O_2}{2 \text{ mol Mg}} = 0.206 \text{ mol } O_2$$

We now calculate the mass of the O_2.

$$0.206 \text{ mol } O_2 \times \frac{32.00 \text{ g } O_2}{1 \text{ mol } O_2} = 6.59 \text{ g } O_2$$

According to this calculation, we need 6.59 grams of O_2 to burn all the magnesium. Because we have 10.0 grams of O_2, our original assumption was correct. We have more than enough O_2 and only a limited amount of magnesium.

We can now calculate the amount of magnesium oxide formed when all of the limiting reagent is consumed. The balanced equation suggests that 2 moles of MgO are produced for every 2 moles of magnesium consumed. Thus, 0.411 mol of MgO can be formed in this reaction.

$$0.411 \text{ mol Mg} \times \frac{2 \text{ mol MgO}}{2 \text{ mol Mg}} = 0.411 \text{ mol MgO}$$

We can use the molar mass of MgO to calculate the number of grams of MgO that can be formed.

$$0.411 \text{ mol MgO} \times \frac{40.30 \text{ g MgO}}{1 \text{ mol MgO}} = 16.6 \text{ g MgO}$$

We can check the result of our calculations by noting that 10.0 grams of magnesium combine with 6.59 grams of O_2 to form 16.6 grams of MgO (3.4 g of O_2 remains unused). Mass is therefore conserved and we can feel confident that our calculations are correct.

➤ CHECKPOINT

If 4 grams of $I_2(s)$ are combined with 4 grams of $Mg(s)$, what will be the limiting reagent for the following chemical reaction?

$$Mg(s) + I_2(s) \longrightarrow MgI_2(s)$$

Sections 2.17–2.21 introduce solutions, molarity, dilutions, and solution stoichiometry. These sections can be postponed until Chapter 8. They are introduced here for instructors who need to cover this material early for the laboratory portion of the course.

2.17 Solute, Solvent, and Solution

Hydrogen chloride (HCl) and ammonia (NH_3) are gases at room temperature that are notoriously difficult to work with. Chemists have therefore traditionally found it easier to handle these compounds by dissolving them in water to form hydrochloric acid and aqueous ammonia, respectively.

$$HCl(g) \xrightarrow{H_2O} HCl(aq)$$

$$NH_3(g) \xrightarrow{H_2O} NH_3(aq)$$

The result of dissolving one of these gases in water is a **solution.** Solutions are uniform mixtures. The composition is the same throughout the mixture.

Solutions contain two components: a solute and a solvent. The substance that dissolves is the **solute.** The substance in which the solute dissolves is called the

solvent. Two general rules can be used to decide which component of a solution is the solute and which is the solvent.

- Any reagent that undergoes a change in state when it forms a solution is the *solute*. Thus, when gaseous HCl dissolves in water to form an aqueous solution, HCl is the solute.
- If neither component of the solution undergoes a change in state, the component present in the smallest quantity is the *solute*.

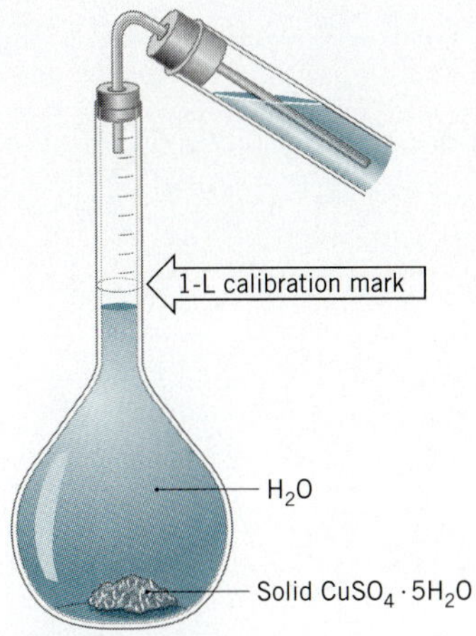

Fig. 2.16 A solution of $CuSO_4$ in water is made by dissolving $CuSO_4 \cdot 5\ H_2O$ in a solvent (H_2O).

Figure 2.16 shows how a solid such as copper(II) sulfate pentahydrate [$Cu(SO_4) \cdot 5\ H_2O$] can be dissolved in water to form an aqueous solution. $Cu(SO_4) \cdot 5\ H_2O$ is the solute because it dissolves in water, which is the solvent, to form an aqueous solution.

Table 2.1 gives examples of different kinds of solutions. $CuSO_4 \cdot 5\ H_2O$ is a dark-blue solid which forms a blue solution when dissolved in water. When H_2 gas dissolves in platinum metal to form a solid solution, H_2 is the solute. When liquid mercury dissolves in sodium metal, mercury is the solute. Wine that is 12% ethanol (CH_3CH_2OH) by volume is a solution of a small quantity of liquid ethanol (the solute) in a larger volume of liquid water (the solvent). In a 50:50 mixture either component of the mixture can be thought of as the solute.

Table 2.1
Examples of Solutions

Solute	Solvent	Solution
$CuSO_4 \cdot 5\ H_2O(s)$	$H_2O(l)$	$CuSO_4 \cdot 5\ H_2O(aq)$
$H_2(g)$	$Pt(s)$	$H_2/Pt(s)$
$Hg(l)$	$Na(s)$	$Na/Hg(s)$
$CH_3CH_2OH(l)$	$H_2O(l)$	Wine

2.18 Concentration

The ratio of the amount of solute to the amount of solvent or solution is known as the **concentration** of the solution.

$$\text{Concentration} = \frac{\text{amount of solute}}{\text{amount of solvent or solution}}$$

The concept of concentration is a common one. We talk about concentrated orange juice, which must be *diluted* with water. We even describe certain laundry products as *concentrated,* which means that we don't have to use as much of them.

2.19 Molarity as a Way of Counting Particles in Solution

Chemists use one concentration unit more than any other: **molarity (*M*).** The molarity of a solution is defined as the number of moles of solute per liter of solution.

Molarity is calculated by dividing the number of moles of solute in the solution by the volume of the solution in liters.

$$\text{Molarity} = \frac{\text{moles of solute}}{\text{liters of solution}}$$

Exercise 2.13

Copper sulfate is available as blue crystals that contain water molecules coordinated to the Cu^{2+} ions in the crystal. Because the crystals contain five water molecules per Cu^{2+} ion, the compound is called a *pentahydrate,* and the formula is written as $CuSO_4 \cdot 5\,H_2O$. Calculate the molarity of a solution prepared by dissolving 1.25 g of this compound in enough water to give 50.0 mL of solution.

Solution

A useful strategy for solving problems involves looking at the goal of the problem and asking: What information do we need to answer the question? The molarity of a solution is calculated by dividing the number of moles of solute by the volume of the solution. We therefore need two pieces of information: the number of moles of solute and the volume of the solution in liters.

The volume of the solution needs to be expressed in liters.

$$50.0 \text{ mL} \times \frac{1 \text{ L}}{1000 \text{ mL}} = 0.0500 \text{ L}$$

The number of moles of solute can be calculated from the mass of solute used to prepare the solution and the mass of a mole of this compound.

$$1.25 \text{ g } CuSO_4 \cdot 5\,H_2O \times \frac{1 \text{ mol } CuSO_4 \cdot 5\,H_2O}{249.7 \text{ g } CuSO_4 \cdot 5\,H_2O} = 0.00501 \text{ mol } CuSO_4 \cdot 5\,H_2O$$

The molarity of the solution is then calculated by dividing the number of moles of solute in the solution by the volume of the solution.

$$\frac{0.00501 \text{ mol } CuSO_4 \cdot 5\,H_2O}{0.0500 \text{ L}} = 0.100\,M\; CuSO_4 \cdot 5\,H_2O$$

Sections 2.5 and 2.6 described how to use measurements of the mass of a pure substance to count the number of moles that are present in a sample. Molarity can be used to count the number of moles of solute in a given volume of a solution. Note that molarity has units of moles per liter. The product of the molarity of a solution times its volume in liters is therefore equal to the number of moles of solute dissolved in the solution.

$$\frac{\text{mol}}{\text{L}} \times \text{L} = \text{mol}$$

We can write this relationship in terms of the following generic equation, where M is molarity, V is volume in liters, and n is number of moles.

$$M \times V = n$$

Exercise 2.14

How many moles of sodium sulfate, Na_2SO_4, are present in 250 mL of a 0.150 *M* solution of sodium sulfate?

Solution

We start with the relationship between the molarity of the solution (M), the volume of the solution being studied (V), and the number of moles of solute in the sample (n).

$$M \times V = n$$

We then substitute the known values of the concentration and the volume of solution into this equation.

$$(0.150\ M) \times (0.250\ \text{L}) = n$$

Solving for the value of n gives us the number of moles of Na_2SO_4 in this volume of this solution.

$$n = 0.0375\ \text{mol of } Na_2SO_4$$

2.20 Dilution Calculations

Anyone who has ever made a pitcher of orange juice by adding water to a can of frozen orange juice should appreciate that adding more solvent to a solution to decrease the solute concentration is known as **dilution.** Starting with a known volume of a solution of known molarity, we can prepare a more dilute solution of any desired concentration, as shown in Figure 2.17.

Fig. 2.17 The number of moles of solute is conserved during a dilution. The moles of solute removed from the initial solution with the pipet are the same moles of solute found in the final solution.

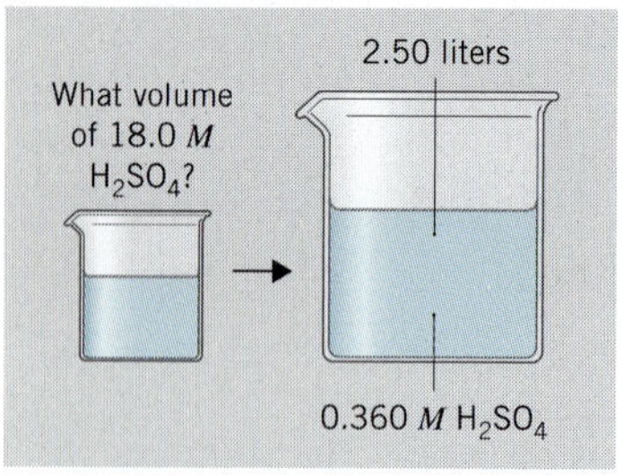

Fig. 2.18 The volume and concentration of the dilute H_2SO_4 solution and the concentration of the concentrated solution in Exercise 2.15.

Exercise 2.15

Describe how you would prepare 2.50 L of a 0.360 *M* solution of sulfuric acid (H_2SO_4) starting with concentrated sulfuric acid that is 18.0 *M*.

Solution

We have one piece of information about the concentrated H_2SO_4 solution (the concentration is 18.0 mol per liter) and two pieces of information about the dilute solution (the volume is 2.50 liters and the concentration is 0.360 mol per liter), as shown in Figure 2.18.

It seems reasonable to start with the solution about which we know the most. If we know the concentration (0.360 *M*) and the volume (2.50 liters) of the dilute sulfuric acid solution we are trying to prepare, we can calculate the number of moles of H_2SO_4 it must contain.

$$\frac{0.360 \text{ mol sulfuric acid}}{1 \text{ L}} \times 2.50 \text{ L} = 0.900 \text{ mol sulfuric acid}$$

What volume of concentrated H_2SO_4 contains the same number of moles of H_2SO_4? We can start with the equation that describes the relationship between the molarity (M) of a solution, the volume of the solution (V), and the number of moles of solute in the solution (n).

$$M \times V = n$$

We then substitute into this equation the molarity of the concentrated sulfuric acid solution and the number of moles of sulfuric acid molecules needed to prepare the dilute solution.

$$\frac{18.0 \text{ mol sulfuric acid}}{1 \text{ L}} \times V = 0.900 \text{ mol sulfuric acid}$$

We then solve this equation for the volume of the solution.

$$V = 0.0500 \text{ L}$$

According to this calculation, we can prepare 2.50 L of 0.360 *M* H_2SO_4 solution by adding 50.0 mL of concentrated sulfuric acid to enough water to give a total volume of 2.50 L.

2.21 Solution Stoichiometry

In Section 2.15 we saw how measurements of mass could be used to understand the relationship between the amounts of starting materials consumed and products generated in a chemical reaction between pure substances. In this section we will see how measurements of molarity can be used to obtain a similar goal for reactions that occur when solutions are mixed.

Exercise 2.16

The reaction between oxalic acid ($H_2C_2O_4$) and sodium hydroxide (NaOH) can be described by the following equation.

$$H_2C_2O_4(aq) + 2\ NaOH(aq) \longrightarrow 2\ Na^+(aq) + C_2O_4{}^{2-}(aq) + 2\ H_2O(l)$$

Calculate the concentration of an oxalic acid solution if it takes 34.0 mL of a 0.200 *M* NaOH solution to consume the acid in a 25.0-mL sample of the oxalic acid solution.

Solution

The above chemical equation can be rewritten to simplify it. We start by noting that NaOH dissociates into Na^+ and OH^- ions when it dissolves in water.

$$H_2C_2O_4(aq) + 2\ Na^+(aq) + 2\ OH^-(aq) \longrightarrow 2\ Na^+(aq) + C_2O_4{}^{2-}(aq) + 2\ H_2O(l)$$

Because 2 $Na^+(aq)$ appears on both sides of this equation, we can simplify it by writing the **net ionic equation.**

$$H_2C_2O_4(aq) + 2\ OH^-(aq) \longrightarrow C_2O_4{}^{2-}(aq) + 2\ H_2O(l)$$

We know only the volume of the oxalic acid solution, but we know both the volume and the concentration of the NaOH solution, as shown in Figure 2.19.

It therefore seems reasonable to start by calculating the number of moles of NaOH in this solution.

$$\frac{0.200 \text{ mol NaOH}}{1 \text{ L}} \times 0.0340 \text{ L} = 6.80 \times 10^{-3} \text{ mol NaOH}$$

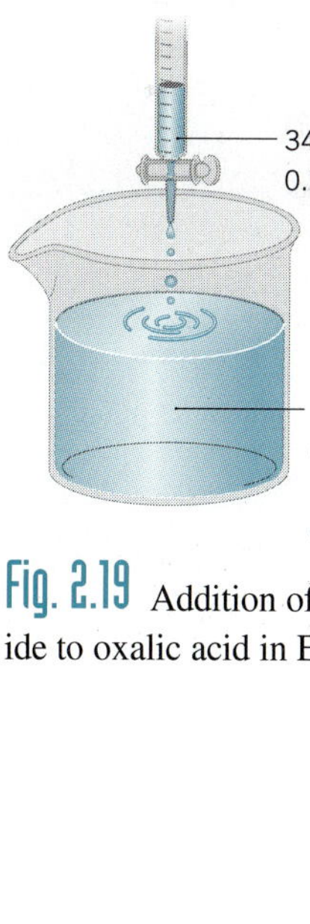

Fig. 2.19 Addition of sodium hydroxide to oxalic acid in Exercise 2.16.

We now know the number of moles of NaOH consumed in the reaction, and we have a balanced chemical equation for the reaction that occurs when the two solutions are mixed. We can therefore calculate the number of moles of $H_2C_2O_4$ needed to consume this much NaOH.

$$6.80 \times 10^{-3} \text{ mol NaOH} \times \frac{1 \text{ mol } H_2C_2O_4}{2 \text{ mol NaOH}} = 3.40 \times 10^{-3} \text{ mol } H_2C_2O_4$$

We now know the number of moles of $H_2C_2O_4$ in the original oxalic acid solution. Using the volume of this solution given in the statement of the problem, we can calculate the number of moles of oxalic acid per liter of this solution, which is the molarity of the solution.

$$\frac{3.40 \times 10^{-3} \text{ mol } H_2C_2O_4}{0.0250 \text{ L}} = 0.136\ M\ H_2C_2O_4$$

The oxalic acid solution therefore has a concentration of 0.136 mol per liter.

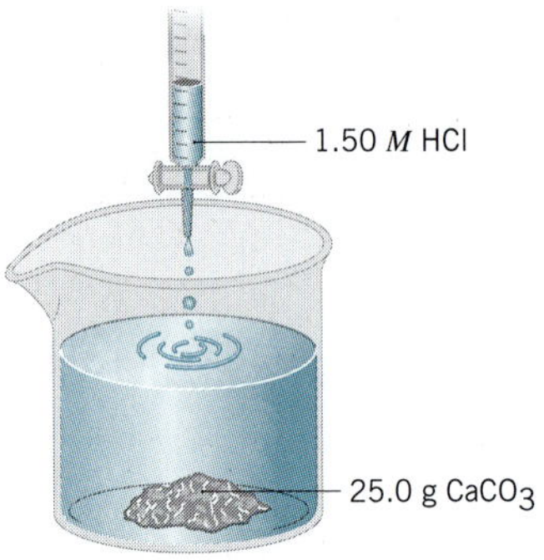

Fig. 2.20 The reaction between hydrochloric acid and calcium carbonate described in Exercise 2.17.

Exercise 2.17

Calculate the volume of 1.50 *M* HCl that would consume 25.0 g of $CaCO_3$ according to the following balanced equation.

$$CaCO_3(s) + 2\,HCl(aq) \longrightarrow CaCl_2(aq) + CO_2(g) + H_2O(l)$$

Solution

We have only one piece of information about the HCl solution (its concentration is 1.50 *M*) and only one piece of information about the $CaCO_3$ (it has a mass of 25.0 g), as shown in Figure 2.20.

Which of these numbers do we start with? There is nothing we can do with the concentration of the HCl solution unless we know either the number of moles of HCl consumed in the reaction or the volume of the solution. We can work with the mass of $CaCO_3$ consumed in the reaction, however. We can start by converting grams of $CaCO_3$ into moles of $CaCO_3$.

$$25.0 \text{ g } CaCO_3 \times \frac{1 \text{ mol } CaCO_3}{100.1 \text{ g } CaCO_3} = 0.250 \text{ mol } CaCO_3$$

We then ask: Does this information get us any closer to our goal of calculating the volume of HCl consumed in this reaction? We know the number of moles of $CaCO_3$ present initially, and we have a balanced equation that states that 2 moles of HCl is consumed for every mole of $CaCO_3$. We can therefore calculate the number of moles of HCl consumed in the reaction.

$$0.250 \text{ mol } CaCO_3 \times \frac{2 \text{ mol HCl}}{1 \text{ mol } CaCO_3} = 0.500 \text{ mol HCl}$$

We now know the number of moles of HCl (0.500 mol) and the molarity of the solution (1.50 *M*). The number of moles and the molarity are related to volume by the following equation.

$$M \times V = n$$

Substituting the known values of the concentration of the HCl solution and the number of moles of HCl into this equation gives the following result.

$$\frac{0.150 \text{ mol HCl}}{1 \text{ L}} \times V = 0.500 \text{ mol HCl}$$

We then solve the equation for the volume of the solution that would contain this amount of HCl.

$$V = 0.333 \text{ L}$$

According to this calculation, we need 333 mL of 1.50 *M* HCl to consume 25.0 g of $CaCO_3$.

With a little imagination, the concept of concentration can be used to do far more interesting calculations.

Exercise 2.18

Assume that a metal, M, reacts with hydrochloric acid according to the following balanced equation.

$$M(s) + 2\ HCl(aq) \longrightarrow M^{2+}(aq) + 2\ Cl^{-}(aq) + H_2(g)$$

Calculate the molar mass of the metal if 125 mL of 0.200 *M* HCl reacts with 0.304 g of the metal.

Solution

At first glance, it seems that we don't have enough information to solve this problem. The only way to proceed with a question like this is to start by identifying what you know, do what can be done, and see where this leads you. In other words, start by exploring the problem. What do we know?

- The metal reacts with hydrochloric acid according to the balanced equation given above.
- We start with 0.304 g of the metal.
- It takes 125 mL of 0.200 *M* HCl to consume the metal.

What can we do? We know the volume (125 mL) and the concentration (0.200 *M*) of a solution. We might therefore start by calculating the number of moles of solute in this solution.

$$\frac{0.200 \text{ mol HCl}}{1 \text{ L}} \times 0.125 \text{ L} = 0.0250 \text{ mol HCl}$$

Now what? We know the number of moles of HCl and we have a balanced equation. Furthermore, we are interested in one of the properties of the metal. It seems reasonable to convert moles of HCl consumed in this reaction into moles of metal consumed.

$$0.0250 \text{ mol HCl} \times \frac{1 \text{ mol M}}{2 \text{ mol HCl}} = 0.0125 \text{ mol M}$$

It is important never to lose sight of the goal of the problem. In this case, the problem asks for the molar mass of the metal. It might be useful to go to the end of the problem and work backward. Molar mass has units of grams per mole. If we knew both the number of grams and the number of moles of metal in a sample, we could calculate the molar mass of the metal.

But we already have that information. We know the number of moles of metal (0.0125 mol) in a sample of known mass (0.304 g). The ratio of these numbers is the molar mass of the metal.

$$\frac{0.304 \text{ g M}}{0.0125 \text{ mol M}} = 24.3 \text{ g/mol}$$

By looking at a table of atomic weights we can deduce that the metal is magnesium.

Key Terms

Aqueous
Atomic mass unit (amu)
Atomic weight
Atomic world
Avogadro's number
Chemical equation
Concentration
Dilution
Elemental analysis
Empirical formula
Excess reagent
Ionic equation
Law of conservation of mass
Limiting reagent
Macroscopic world
Mass spectrometer
Molar mass
Molarity (M)
Mole
Mole ratio
Molecular formula
Molecular weight
Net ionic equation
Percent by mass
Product
Reactant
Solute
Solution
Solvent
Stoichiometry
Symbolic world

Problems

The Macroscopic, Atomic, and Symbolic Worlds of Chemistry

1. Which of the following belong on the macroscopic scale?
   (a) an atom of gold (b) a gold ring
   (c) gold in an ore (d) gold dust
2. How would a chemist symbolize gold in the atomic world?
3. (a) How would a chemist symbolize gold in the macroscopic world?
   (b) Give a symbolic representation that chemists would use for a bar of iron. What symbolic representation would the chemist use for an atom of iron?

The Mass of an Atom

4. Calculate the atomic weight of bromine if naturally occurring bromine is 50.69% ^{79}Br atoms with a mass of 78.9183 amu and 49.31% ^{81}Br atoms with a mass of 80.9163 amu.
5. Naturally occurring zinc is 48.6% ^{64}Zn atoms (63.9291 amu), 27.9% ^{66}Zn atoms (65.9260 amu), 4.1% ^{67}Zn atoms (66.9721 amu), 18.8% ^{68}Zn atoms (67.9249 amu), and 0.6% ^{70}Zn atoms (69.9253 amu). Calculate the atomic weight of zinc.
6. What is the total mass in amu of a sample of 100,000 carbon atoms selected at random? What is the average mass of a carbon atom? Does any carbon atom have this mass?
7. What is the average mass of an Mg atom in amu for a large collection of magnesium atoms?
8. What is the average mass in amu of an I atom?
9. Identify the element that contains atoms that have an average mass of 28.086 amu.
10. There are two naturally occurring isotopes of element X. One of these isotopes has a natural abundance of 80.3% and a relative mass of 11.00931 amu. The second isotope is lighter. Identify element X and state your reasoning.
    Give your best estimate of the number of neutrons, protons, and electrons in each isotope.
11. Element X has only two naturally occurring isotopes. The most abundant or these two isotopes has a mass of 7.01600 amu and accounts for more than 90% of the isotopic atoms.
    (a) Identify element X. Explain your reasoning.
    (b) Give the mass number and the number of protons, electrons, and neutrons for each of the two isotopes.
    (c) The element X combines with various polyatomic anions to produce several compounds. The formulas of the compounds formed are XBr, X_2SO_4, and X_3PO_4. What is the charge on the ion formed by X? How many electrons does this ion have?
12. (a) There are two naturally occurring isotopes of silver, ^{107}Ag (106.90509 amu) is 51.84% and ^{109}Ag (108.90476 amu) is 48.16% abundant. Calculate the average atomic mass of silver. How will you know if your answer is correct?
    (b) How many protons does ^{107}Ag have? How many protons does ^{109}Ag have?
    (c) Give the number of neutrons and electrons in ^{107}Ag and ^{109}Ag.
13. Element X has only two naturally occurring isotopes. One has a relative mass of 78.9183 amu and the other has a relative mass of 80.9163 amu.
    (a) Which element is this most likely to be? Explain.
    (b) Without doing a calculation, estimate the percent abundance of these two isotopes. Explain how you arrived at your answer.
14. Complete the following table:

Isotope	Atomic Number	Mass Number	Number of Electrons	Number of Neutrons	% Abundance
^{6}Li	—	—	—	—	7.42
—	3	7	—	—	92.58
^{20}Ne	—	—	—	—	90.51
—	10	21	—	—	0.27
^{-}Ne	—	22	—	—	9.22

15. 100 Li atoms are selected at random. Answer true or false to each of the following. Refer to Problem 14. The total mass will be:
(a) more than 600 amu
(b) less than 600 amu
(c) 694.1 amu
(d) 700.0 amu
Explain your answer.
If 10,000 Ne atoms are selected at random, how many will have a mass number of 20?
16. When calculating the average atomic mass from percent abundance, you can always quickly check your answer. How?
17. There is only one naturally occuring isotope of this element. If that isotope has a mass of 26.982 amu, identify the element.

The Mole as a Bridge Between the Macroscopic and Atomic Scales

18. If a new scale of atomic weights was defined based on the assumption that the mass of a ^{12}C atom was exactly 1 amu, what would be the atomic weight of neon?
19. Identify the element that has an atomic weight 4.33 times as large as carbon.
20. Give the atomic weight and the molar mass of the following atoms.
(a) Li (b) C (c) Mg (d) Cu
21. Which of the following pairs of elements contains the same number of atoms?
(a) 12.011 g C, 12.011 g Na
(b) 22.99 g Na, 12.011 g C
(c) 39.01 g K, 9.012 g Be
(d) 85.47 g Rb, 6.941 g Li
22. Which would weigh the most, 1000 atoms of Al or 1000 atoms of Si?
23. Which contains the most atoms, a mole of Fe or a mole of Cu?
24. How many grams of each of the following would be required to make one mole?
(a) Ca (b) Sr (c) Se (d) Ge
25. Of each set of the following atoms, pick the heavier atom.
(a) Ni, Co (b) Zn, Al (c) Ga, Ge (d) Pb, Sn
26. If the average mass of a chromium atom is 51.996 amu, what is the mass of a mole of chromium atoms?
27. If the average sulfur atom is approximately twice as heavy as the average oxygen atom, what is the ratio of the mass of a mole of sulfur atoms to the mass of a mole of oxygen atoms?
28. Calculate the mass in grams of a mole of atoms of the following elements.
(a) C (b) Ni (c) Hg
29. If eggs sell for $0.90 a dozen, what does it cost to buy 2.5 dozen eggs? If the molar mass of carbon is 12.011 g, what is the mass of 2.5 moles of carbon atoms?

The Mole as a Collection of Atoms

30. What would be the value of Avogadro's number if a mole were defined as the number of ^{12}C atoms in 12 pounds of ^{12}C?
31. Calculate the number of atoms in 16 g of O_2, 31 g of P_4, and 32 g of S_2.
32. Benzaldehyde has the pleasant, distinctive odor of almonds. What is the weight of a mole of benzaldehyde if a single molecule has a mass of 1.762×10^{-22} gram?
33. What is the mass in grams of one ^1H atom? Of one ^{12}C atom?
34. What is the mass in grams of 4.35×10^6 atoms of ^{12}C?
35. What is the mass in grams of 6.022×10^{23} atoms of ^{12}C?
36. What is the mass in grams of a molecule of carbon dioxide that has one ^{12}C atom and two ^{16}O atoms?

Converting Grams into Moles and Number of Atoms

37. How many atoms are there in 25.0 g of Sn?
38. Calculate the mass in grams of a sample of copper metal that contains 1.65 mol of copper atoms.
39. How many grams of lithium contain 4.56×10^{23} atoms of lithium?
40. 2.0 mol of silver metal contains how many silver atoms?
41. How many atoms are there in
(a) one mole of Si? (b) two moles of Si?
(c) one-half mole of Si (d) 0.10 mol Si?
42. What is the weight of an atom of iron in grams?
43. If one atom of ^1H weighs 1.6735×10^{-24} g, what is the weight of 6.022×10^{23} atoms of ^1H?

The Mole as a Collection of Molecules

44. How many carbon and hydrogen atoms could be found in a sample of one dozen methane, CH_4, molecules? In one mole of methane molecules?
45. What is the average mass in amu of one methane molecule? What is the mass in grams of one mole of methane?
46. How many hydrogen atoms are present in 1.00 mol of hydrogen gas, H_2? How many H_2 molecules? What is the mass of the sample?
47. Indicate whether each of the following statements is true or false, and explain your reasoning.
(a) One mole of NH_3 weighs more than 1 mole of H_2O.
(b) There are more carbon atoms in 48 g of CO_2 than in 12 g of diamond (a pure form of carbon).
(c) There are equal numbers of nitrogen atoms in 1 mole of NH_3 and 1 mole of N_2.
(d) The number of Cu atoms in 100 g of Cu(*s*) is the same as the number of Cu atoms in 100 g of copper(II) oxide, CuO.

(e) The number of Ni atoms in 1 mol of Ni(s) is the same as the number of Ni atoms in 1 mole of nickel(II) chloride, $NiCl_2$.

48. Which pair of samples contains the same number of hydrogen atoms?
(a) 1 mol of NH_3 and 1 mol of N_2H_4
(b) 2 mol of NH_3 and 1 mol of N_2H_4
(c) 2 mol of NH_3 and 3 mol of N_2H_4
(d) 4 mol of NH_3 and 3 mol of N_2H_4

49. Which of the following contains the largest number of carbon atoms?
(a) 0.10 mol of acetic acid, CH_3CO_2H
(b) 0.25 mol of carbon dioxide, CO_2
(c) 0.050 mol of glucose, $C_6H_{12}O_6$
(d) 0.0010 mol of sucrose, $C_{12}H_{22}O_{11}$

50. Calculate the molecular weights of formic acid, HCO_2H, and formaldehyde, H_2CO.

51. Calculate the molecular weight of the following compounds.
(a) methane, CH_4
(b) glucose, $C_6H_{12}O_6$
(c) diethyl ether, $(CH_3CH_2)_2O$
(d) thioacetamide, CH_3CSNH_2

52. Calculate the molecular weight of the following compounds.
(a) tetraphosphorus decasulfide, P_4S_{10}
(b) nitrogen dioxide, NO_2
(c) zinc sulfide, ZnS
(d) potassium permanganate, $KMnO_4$

53. Calculate the molecular weight of the following compounds.
(a) chromium hexacarbonyl, $Cr(CO)_6$
(b) iron(III) nitrate, $Fe(NO_3)_3$
(c) potassium dichromate, $K_2Cr_2O_7$
(d) calcium phosphate, $Ca_3(PO_4)_2$

54. Root beer hasn't tasted the same since the FDA outlawed the use of sassafras oil as a food additive because sassafras oil is 80% safrole, which has been shown to cause cancer in rats and mice. Calculate the molecular weight of safrole, $C_{10}H_{10}O_2$.

55. MSG ($C_5H_8NNaO_4$) is a spice used in Chinese cooking that causes some people to feel light-headed (a disorder known as *Chinese restaurant syndrome*). Calculate the molecular weight of MSG.

56. Calculate the molecular weight of the active ingredients in the following prescription drugs.
(a) Darvon, $C_{22}H_{30}ClNO_2$
(b) Valium, $C_{16}H_{13}ClN_2O$
(c) tetracycline, $C_{22}H_{24}N_2O_8$

57. Calculate the atomic weight of platinum if 0.8170 mol of the metal has a mass of 159.4 g.

58. Calculate the mass of 0.0582 mol of carbon tetrachloride, CCl_4.

59. Calculate the moles of carbon tetrachloride, CCl_4, in 100 grams of CCl_4.

60. Calculate the number of moles in 5.72 g of Al.

Percent Mass

61. Calculate the percent by mass of chromium in each of the following oxides.
(a) CrO (b) Cr_2O_3 (c) CrO_3

62. Calculate the percent by mass of nitrogen in the following fertilizers.
(a) $(NH_4)_2SO_4$ (b) KNO_3
(c) $NaNO_3$ (d) $(H_2N)_2CO$

63. Calculate the percent by mass of carbon, hydrogen, and chlorine in DDT, $C_{14}H_9Cl_5$.

64. Emeralds are gem-quality forms of the mineral beryl, $Be_3Al_2(SiO_3)_6$. Calculate the percent by mass of silicon in beryl.

65. Osteoporosis is a disease common in older women who have not had enough calcium in their diets. Calcium can be added to the diet by tablets that contain either calcium carbonate ($CaCO_3$), calcium sulfate ($CaSO_4$), or calcium phosphate [$Ca_3(PO_4)_2$]. On a per-gram basis, which is the most efficient way of getting Ca^{2+} ions into the body?

Determining the Formula of a Compound

66. Calculate the number of moles of carbon atoms in 0.244 g of calcium carbide, CaC_2.

67. Calculate the number of moles of phosphorus in 15.95 g of tetraphosphorus decaoxide, P_4O_{10}.

68. Calculate the number of chlorine atoms in 0.756 g of K_2PtCl_6.

69. Calculate the number of oxygen atoms in the following samples.
(a) 0.100 mol of potassium permanganate, $KMnO_4$
(b) 0.25 mol of dinitrogen pentoxide, N_2O_5
(c) 0.45 mol of penicillin, $C_{16}H_{17}N_2O_5SK$

70. A molecule containing only nitrogen and oxygen contains 36.8% N by mass.
(a) How many grams of N would be found in a 100-g sample of the compound? How many grams of O would be found in the same sample?
(b) How many moles of N would be found in a 100-g sample of the compound? How many moles of O would be found in the same sample?
(c) What is the ratio of the number of moles of O to the number of moles of N?
(d) What is the empirical formula of the compound?

71. Stannous fluoride, or "Fluoristan," is added to toothpaste to help prevent tooth decay. What is the empirical formula for stannous fluoride if the compound is 24.25% F and 75.75% Sn by mass?

72. Iron reacts with oxygen to form three compounds: FeO, Fe_2O_3, and Fe_3O_4. One of these compounds,

known as magnetite, is 72.36% Fe and 27.64% O by mass. What is the formula of magnetite?

73. The most abundant ore of manganese is an oxide known as pyrolusite, which is 36.8% O and 63.2% Mn by mass. Which of the following oxides of manganese is pyrolusite?
(a) MnO (b) MnO_2 (c) Mn_2O_3
(d) MnO_3 (e) Mn_2O_7

74. Nitrogen combines with oxygen to form a variety of compounds, including N_2O, NO, NO_2, N_2O_3, N_2O_4, and N_2O_5. One of these compounds is called nitrous oxide, or "laughing gas." What is the formula of nitrous oxide if this compound is 63.65% N and 36.35% O by mass?

75. Chalcopyrite is a bronze-colored mineral that is 34.59% Cu, 30.45% Fe, and 34.96% S by mass. Calculate the empirical formula for the mineral.

76. A compound of xenon and fluorine is found to be 53.5% xenon by mass. What is the empirical formula of the compound?

77. In 1914, E. Merck and Company synthesized and patented a compound known as MDMA as an appetite suppressant. Although it was never marketed, it has reappeared in recent years as a street drug known as ecstasy. What is the empirical formula of the compound if it contains 68.4% C, 7.8% H, 7.2% N, and 16.6% O by mass?

78. What is the empirical formula of the compound that contains 0.483 g of nitrogen and 1.104 g of oxygen?
(a) N_2O (b) NO (c) NO_2
(d) N_2O_3 (e) N_2O_4

79. What is the empirical formula of the compound formed when 9.33 g of copper metal reacts with excess chlorine to give 14.54 g of the compound?

80. Is it possible to determine the molecular formula of a compound solely from its percent composition? Why or why not?

81. β-Carotene is the protovitamin from which nature builds vitamin A. It is widely distributed in the plant and animal kingdoms, always occurring in plants together with chlorophyll. Calculate the molecular formula for β-carotene if the compound is 89.49% C and 10.51% H by mass and its molecular weight is 536.89 g/mol.

82. The phenolphthalein used as an indicator in acid-base titrations has also been used as the active ingredient in laxatives such as ExLax. Calculate the molecular formula for phenolphthalein if the compound is 75.46% C, 4.43% H, and 20.10% O by mass and has a molecular weight of 318.31 grams per mole.

83. Caffeine is a central nervous system stimulant found in coffee, tea, and cola nuts. Calculate the molecular formula of caffeine if the compound is 49.48% C, 5.19% H, 28.85% N, and 16.48% O by mass and has a molecular weight of 194.2 grams per mole.

84. Aspartame, also known as NutraSweet, is 160 times sweeter than sugar when dissolved in water. The true name for this artificial sweetener is *N*-L-α-aspartyl-L-phenylalanine methyl ester. Calculate the molecular formula of aspartame if the compound is 57.14% C, 6.16% H, 9.52% N, and 27.18% O by mass and has a molecular weight of 294.30 grams per mole.

Elemental Analysis

85. How accurately would you have to measure the percent by mass of carbon and hydrogen to tell the difference between diazepam (Valium), with the formula $C_{16}H_{13}ClN_2O$, and chlordiazepoxide (Librium), with the formula $C_{16}H_{14}ClN_3O$?

86. The methane in natural gas, the propane used in camping stoves, and the butane used in butane lighters are all members of a family of compounds known as the alkanes, which have the generic formula C_nH_{2n+2}. Calculate the value of n for butane if 3.15 mg of butane burns in air to form 9.54 mg of CO_2 and 4.88 mg of H_2O.

87. In small quantities, the nicotine in tobacco is addictive. In large quantities, it is a deadly poison. Calculate the molecular formula of nicotine, $C_xH_yN_z$, if the molecular weight of nicotine is 162.2 g/mol and 4.38 mg of the compound burns to form 11.9 mg of CO_2 and 3.41 mg of water.

Chemical Reactions and the Law of Conservation of Atoms

88. If a candle is burned in a closed container filled with oxygen, will the mass of the container and contents be the same as, more than, or less than the original mass of the container, oxygen, and candle? Explain.

89. When gasoline is burned in air, are there more atoms, fewer atoms, or the same number as before burning? Explain.

90. Give an interpretation on a microscopic scale for why mass is conserved in a reaction.

91. What observation did Lavoisier make that led him to formulate the law of conservation?

92. What does the conservation of atoms in a chemical reaction tell us about what must happen to the atoms during the reaction?

Chemical Equations as a Representation of Chemical Reactions

93. State in a complete, grammatically correct sentence what the following symbolic equation represents.

$$2\ H_2(g) + O_2(g) \longrightarrow 2\ H_2O(g)$$

Do the same for this reaction.

$$2\ H_2(g) + O_2(g) \longrightarrow 2\ H_2O(l)$$

94. State in a complete, grammatically correct sentence what the following symbolic equation represents.

$$KI(s) \longrightarrow K^+(aq) + I^-(aq)$$

95. State in words what the following symbolic equation means.

$$CO_2(g) + H_2O(l) \longrightarrow H_2CO_3(aq)$$

96. For the reaction in Problem 95, is the number of C atoms conserved? Is the number of H atoms conserved? Is the number of molecules conserved?

Two Views of Chemical Equations: Molecules Versus Moles

97. Give two ways of interpreting the following equation.

$$H_2(g) + Cl_2(g) \longrightarrow 2\ HCl(g)$$

98. Show that the following equation is balanced by calculating the masses of the products and reactants in both amu and grams.

$$3\ Ca(s) + N_2(g) \longrightarrow Ca_3N_2(s)$$

99. 2.0 mol of $H_2(g)$ is mixed with 1.0 mol of $O_2(g)$ and is allowed to react as shown in Problem 93. How many atoms of H are initially present? How many atoms of O are initially present? How many atoms of H and O will there be in the product?
   (a) How many moles of H_2O will be formed if all the H_2 and O_2 react?
   (b) How many molecules of H_2 and O_2 were initially present?
   (c) How many molecules of H_2O were formed?

Balancing Chemical Equations

100. Balance the following chemical equations.
   (a) $Cr(s) + O_2(g) \longrightarrow Cr_2O_3(s)$
   (b) $SiH_4(g) \longrightarrow Si(s) + H_2(g)$
   (c) $SO_3(g) \longrightarrow SO_2(g) + O_2(g)$

101. Balance the following chemical equations.
   (a) $Pb(NO_3)_2(s) \longrightarrow PbO(s) + NO_2(g) + O_2(g)$
   (b) $NH_4NO_2(s) \longrightarrow N_2(g) + H_2O(g)$
   (c) $(NH_4)_2Cr_2O_7(s) \longrightarrow N_2(g) + Cr_2O_3(s) + H_2O(g)$

102. Balance the following chemical equations.
   (a) $CH_4(g) + O_2(g) \longrightarrow CO_2(g) + H_2O(g)$
   (b) $H_2S(g) + O_2(g) \longrightarrow H_2O(g) + SO_2(g)$
   (c) $B_5H_9(g) + O_2(g) \longrightarrow B_2O_3(s) + H_2O(g)$

103. Balance the following chemical equations.
   (a) $PF_3(g) + H_2O(l) \longrightarrow H_3PO_3(aq) + HF(aq)$
   (b) $P_4O_{10}(s) + H_2O(l) \longrightarrow H_3PO_4(aq)$

104. Balance the following chemical equations.
   (a) $C_3H_8(g) + O_2(g) \longrightarrow CO_2(g) + H_2O(g)$
   (b) $C_2H_5OH(l) + O_2(g) \longrightarrow CO_2(g) + H_2O(g)$
   (c) $C_6H_{12}O_6(s) + O_2(g) \longrightarrow CO_2(g) + H_2O(l)$

Mole Ratios and Chemical Equations

105. How many moles of CO_2 are produced when 5 moles of O_2 are consumed in the following reaction? $2\ CO(g) + O_2(g) \rightarrow 2\ CO_2(g)$

106. Does the total number of moles of gas present increase, decrease, or remain the same when the following reaction occurs? $2\ CO(g) + O_2(g) \longrightarrow 2\ CO_2(g)$

107. How many moles of CuO would be required to produce 12 mol of copper metal in the following reaction? $CuO(s) + H_2(g) \longrightarrow Cu(s) + H_2O(g)$

108. Carbon disulfide burns in oxygen to form carbon dioxide and sulfur dioxide.

$$CS_2(l) + 3\ O_2(g) \longrightarrow CO_2(g) + 2\ SO_2(g)$$

Calculate the number of O_2 molecules it would take to consume 500 molecules of CS_2. Calculate the number of moles of O_2 it would take to consume 5.00 moles of CS_2.

109. Calculate the number of moles of oxygen produced when 6.75 moles of manganese dioxide decomposes to form Mn_3O_4 and O_2.

$$3\ MnO_2(s) \longrightarrow Mn_3O_4(s) + O_2(g)$$

110. Calculate the number of moles of carbon monoxide needed to reduce 3.00 mol of iron(III) oxide to iron metal.

$$Fe_2O_3(s) + 3\ CO(g) \longrightarrow 2\ Fe(s) + 3\ CO_2(g)$$

Stoichiometry

111. Describe the steps needed to calculate the number of grams of CO_2 produced in the following reaction if x grams of CO are consumed.

$$2\ CO(g) + O_2(g) \longrightarrow 2\ CO_2(g)$$

112. Calculate the mass of oxygen released when enough mercury(II) oxide decomposes to give 25 g of liquid mercury.

$$2\ HgO(s) \longrightarrow 2\ Hg(l) + O_2(g)$$

113. Calculate the mass of CO_2 produced and the mass of oxygen consumed when 10.0 g of methane (CH_4) is burned in oxygen to produce CO_2 and H_2O.

114. How many pounds of sulfur react with 10.0 pounds of zinc to form zinc sulfide, ZnS?

115. Calculate the mass of oxygen that can be prepared by decomposing 25.0 g of potassium chlorate.

$$2\ KClO_3(s) \longrightarrow 2\ KCl(s) + 3\ O_2(g)$$

116. Predict the formula of the compound produced when 1.00 g of chromium metal reacts with 0.923 g of oxygen, O_2.

117. Ethanol, or ethyl alcohol, is produced by the fermentation of sugars such as glucose.

$$C_6H_{12}O_6(aq) \longrightarrow 2\ C_2H_5OH(aq) + 2\ CO_2(g)$$

Calculate the number of kilograms of alcohol that can be produced from 1.00 kilogram of glucose.

118. Calculate the number of pounds of aluminum metal that can be obtained from 1.000 ton of bauxite, $Al_2O_3 \cdot 2\ H_2O$.

119. Calculate the mass of phosphine, PH_3, that can be prepared when 10.0 g of calcium phosphide, Ca_3P_2, reacts with excess water.

$$Ca_3P_2(s) + 6\ H_2O(l) \longrightarrow 3\ Ca(OH)_2(aq) + 2\ PH_3(g)$$

120. Hydrogen chloride can be obtained by reacting phosphorus trichloride with excess water and then boiling the HCl gas out of the solution.

$$PCl_3(g) + 3\ H_2O(l) \longrightarrow 3\ HCl(aq) + H_3PO_3(aq)$$

Calculate the mass of HCl gas that can be prepared from 15.0 g of PCl_3.

121. Nitrogen reacts with hydrogen to form ammonia,

$$N_2(g) + 3\ H_2(g) \longrightarrow 2\ NH_3(g)$$

which burns in the presence of oxygen to form nitrogen oxide,

$$4\ NH_3(g) + 5\ O_2(g) \longrightarrow 4\ NO(g) + 6\ H_2O(l)$$

which reacts with excess oxygen to form nitrogen dioxide,

$$2\ NO(g) + O_2(g) \longrightarrow 2\ NO_2(g)$$

which dissolves in water to give nitric acid,

$$3\ NO_2(g) + H_2O(l) \longrightarrow 2\ HNO_3(aq) + NO(g)$$

Calculate the mass of nitrogen needed to make 150 g of nitric acid, assuming an excess of all other reactants.

The Nuts and Bolts of Limiting Reagents

122. Calculate the number of water molecules that can be prepared from 500 H_2 molecules and 500 O_2 molecules.

$$2\ H_2(g) + O_2(g) \longrightarrow 2\ H_2O(l)$$

What would happen to the potential yield of water molecules if the amount of O_2 were doubled? What if the amount of H_2 were doubled?

123. Calculate the number of moles of P_4S_{10} that can be produced from 0.500 mol of P_4 and 0.500 mol of S_3.

$$4\ P_4(s) + 5\ S_8(s) \longrightarrow 4\ P_4S_{10}(s)$$

What would happen to the potential yield of P_4S_{10} if the amount of P_4 were doubled? What if the amount of S_3 were doubled?

124. Calculate the number of moles of nitrogen dioxide, NO_2, that could be prepared from 0.35 mol of nitrogen oxide and 0.25 mol of oxygen.

$$2\ NO(g) + O_2(g) \longrightarrow 2\ NO_2(g)$$

Identify the limiting reagent and the excess reagent in the reaction. What would happen to the potential yield of NO_2 if the amount of NO were increased? What if the amount of O_2 were increased?

125. Calculate the mass of hydrogen chloride that can be produced from 10.0 g of hydrogen and 10.0 g of chlorine.

$$H_2(g) + Cl_2(g) \longrightarrow 2\ HCl(g)$$

What would have to be done to increase the amount of hydrogen chloride produced in the reaction?

126. Calculate the mass of calcium nitride, Ca_3N_2, that can be prepared from 54.9 g of calcium and 43.2 g of nitrogen.

$$3\ Ca(s) + N_2(g) \longrightarrow Ca_3N_2(s)$$

127. PF_3 reacts with XeF_4 to give PF_5.

$$2\ PF_3(g) + XeF_4(s) \longrightarrow 2\ PF_5(g) + Xe(g)$$

How many moles of PF_5 can be produced from 100.0 g of PF_3 and 50.0 g of XeF_4?

128. Trimethyl aluminum, $Al(CH_3)_3$, must be handled in an apparatus from which oxygen has been rigorously excluded because the compound bursts into flame in the presence of oxygen. Calculate the mass of trimethyl aluminum that can be prepared from 5.00 g of aluminum metal and 25.0 g of dimethyl mercury.

$$2\ Al(s) + 3\ Hg(CH_3)_2(l) \longrightarrow 2\ Al(CH_3)_3(l) + 3\ Hg(l)$$

129. The thermite reaction, used to weld rails together in the building of railroads, is described by the following equation.

$$Fe_2O_3(s) + 2\ Al(s) \longrightarrow Al_2O_3(s) + 2\ Fe(l)$$

Calculate the mass of iron metal that can be prepared from 150 grams of aluminum and 250 grams of iron(III) oxide.

Solute, Solvent, and Solution

130. Define the terms *solution, solvent,* and *solute* and give an example of each.
131. Which of the following are solutions?
    (a) chicken noodle soup (b) air
    (c) wine (d) table salt (NaCl)

Concentration and Molarity as a Way of Counting Particles in Solution

132. Describe in detail the steps you would take to prepare 125 mL of 0.745 *M* oxalic acid, starting with solid oxalic acid dihydrate ($H_2C_2O_4 \cdot 2\ H_2O$). Describe the glassware you would need, the chemicals, the amounts of each chemical, and the sequence of steps you would take.
133. Hydrochloric acid was once known as "muriatic acid" because it was the "marine" acid—it was made from seawater. Muriatic acid is still sold in many hardware stores for cleaning bricks and tile. What is the molarity of this solution if 125 mL contains 27.3 g of HCl?
134. Silver chloride is only marginally soluble in water, only 0.00019 g of AgCl dissolves in 100 mL of water. Calculate the molarity of this solution.
135. Ammonia (NH_3) is relatively soluble in water. Calculate the molarity of a solution that contains 252 g of NH_3 per liter.
136. During a physical exam, one of the authors was found to have a cholesterol level of 1.60 milligrams per deciliter (0.100 L). If the molecular weight of cholesterol is 386.67 grams per mole, what is the cholesterol level in his blood in units of moles per liter?
137. At 25°C, 5.77 g of chlorine gas dissolves in 1.00 liter of water. Calculate the molarity of Cl_2 in this solution.
138. Calculate the mass of Na_2SO_4 needed to prepare 0.500 L of a 0.150 *M* solution.
139. When asked to prepare a liter of 1.00 *M* K_2CrO_4, a student weighed out exactly 1.00 mole of K_2CrO_4 and added this solid to 1.00 L of water in a volumetric flask. What did the student do wrong? Did the student get a solution that was more concentrated than 1.00 *M* or less concentrated than 1.00 *M*? How would you prepare the solution?
140. You can make the chromic acid bath commonly used to clean glassware in the lab by dissolving 92 g of sodium dichromate ($Na_2Cr_2O_7 \cdot 2\ H_2O$) in enough water to give 458 mL of solution and then adding 800 mL of concentrated sulfuric acid. Calculate the molarity of the $Cr_2O_7^{2-}$ ion in the solution.
141. People who smoke marijuana can be detected by looking for the tetrahydrocannabinols (THC) that are the active ingredient in marijuana. The present limit on detection of THC in urine is 20 nanograms of THC per milliliter of urine (20 ng/mL). Calculate the molarity of the solution at the limit of detection if the molecular weight of THC is 315 g/mol.
142. What is the molarity of a solution formed by the dissolution of 1.25 g of KCl in 500 mL of solution? How could you make a solution that is twice as concentrated?
143. If 2.75 g of $AgNO_3$ are dissolved in 250 mL of solution, what is the molarity of the solution? How could you prepare a solution that is half as concentrated?
144. How many grams of NaOH would need to be dissolved in 250.0 mL of solution to produce a 1.25 *M* solution?
145. 500 mL of 0.50 *M* solution of NaOH contain how many moles of NaOH?
146. How many grams of NaOH are in 500 mL of a 0.50 *M* solution?
147. Which is more concentrated, 500 mL of a 0.20 *M* solution of NaCl or 250 mL of a 0.25 *M* solution of NaCl?
148. What is the molarity of a solution containing 0.25 g of $CuSO_4$ in 125 mL of solution? Describe how you would prepare a solution of this molarity.

Dilution Calculations

149. 0.275 g of $AgNO_3$ is dissolved in 500 mL of solution.
    (a) What is the molarity of this solution?
    (b) If 10.0 mL of this solution are transferred to a flask and diluted to 500 mL, what is the concentration of the resulting solution?
    (c) If 10.0 mL of the solution in (b) are transferred to a flask and diluted to 250 mL, what is the concentration of the resulting solution?
150. Describe how you would prepare 500 mL of a 0.10 *M* solution of HCl from a 12.0 *M* solution of HCl.
151. A 0.050 *M* solution of $CuSO_4$ is diluted to double the volume. What is the concentration of the new solution?
152. 100.0 ml of an 18.0 *M* solution of H_2SO_4 is transferred to a flask and diluted to 500.0 mL. What is the concentration of this new solution?
153. It is desired to prepare 250 mL of a 0.10 *M* solution of HCl from a 6.0 M solution of HCl in water. Describe how to do this.
154. 100.0 mL of a 0.050 *M* solution of NaCl in water is diluted to 250.0 ml. What is the final concentration of the solution?

155. To what final volume must 100 mL of a 1.20 *M* solution of KF be diluted to produce a 0.45 *M* solution?

156. 1.00 L of a 1.0 *M* solution of a sugar is diluted to 1.75 L. What is the final concentration?

157. Calculate the volume of 17.4 *M* acetic acid needed to prepare 1.00 L of 3.00 *M* acetic acid.

158. Calculate the concentration of the solution formed when 15.0 mL of 6.00 *M* HCl is diluted with 25.0 mL of water.

159. Describe how you would prepare 0.200 liter of 1.25 *M* nitric acid from a solution that is 5.94 *M* HNO_3.

Solution Stoichiometry

160. Calculate the concentration of an aqueous KCl solution if 25.00 mL of this solution gives 0.430 g of AgCl when treated with excess $AgNO_3$.

$$KCl(aq) + AgNO_3(aq) \longrightarrow AgCl(s) + KNO_3(aq)$$

161. Calculate the volume of 0.25 *M* NaI that would be needed to precipitate all of the Hg^{2+} ion from 45 mL of a 0.10 *M* $Hg(NO_3)_2$ solution.

$$2\,NaI(aq) + Hg(NO_3)_2(aq) \longrightarrow HgI_2(s) + 2\,NaNO_3(aq)$$

162. Calculate the molarity of an acetic acid (CH_3CO_2H) solution if 34.57 mL of the solution is needed to neutralize 25.19 mL of 0.1025 *M* sodium hydroxide.

$$CH_3CO_2H(aq) + NaOH(aq) \longrightarrow Na^+(aq) + CH_3CO_2^-(aq) + H_2O(l)$$

163. Calculate the molarity of a sodium hydroxide solution if 10.42 mL of this solution is needed to neutralize 25.00 mL of 0.2043 *M* oxalic acid ($H_2C_2O_4$).

$$H_2C_2O_4(aq) + 2\,NaOH(aq) \longrightarrow Na_2C_2O_4\,(aq) + 2\,H_2O(l)$$

164. Calculate the volume of 0.0985 *M* sulfuric acid (H_2SO_4) that would be needed to neutralize 10.89 mL of a 0.01043 *M* aqueous ammonia (NH_3) solution.

$$H_2SO_4(aq) + 2\,NH_3(aq) \longrightarrow (NH_4)_2SO_4(aq)$$

165. α-D-Glucopyranose reacts with the periodate ion (IO_4^-) as follows.

$$C_6H_{12}O_6(aq) + 5\,IO_4^-(aq) \longrightarrow 5\,IO_3^-(aq) + 5\,HCO_2H(aq) + H_2CO(aq)$$

Calculate the molarity of the glucopyranose solution if 25.0 mL of 0.750 *M* IO_4^- is required to consume 10.0 mL of the sugar solution.

166. Oxalic acid reacts with the chromate ion in acidic solution as follows.

$$3\,H_2C_2O_4(aq) + 2\,CrO_4^{2-}(aq) + 10\,H^+(aq) \longrightarrow 6\,CO_2(g) + 2\,Cr^{3+}(aq) + 8\,H_2O(l)$$

Calculate the molarity of the oxalic acid ($H_2C_2O_4$) solution if 10.0 mL of the solution consumes 40.0 mL of 0.0250 *M* CrO_4^{2-}.

Integrated Problems

167. Calculate the atomic weight of the metal, M, that forms a compound with the formula MCl_2 that is 74.5% Cl by mass.

168. Halothane is an anesthetic that is 12.17% C, 0.51% H, 40.48% Br, 17.96% Cl, and 28.87% F by mass. What is the molecular formula of the compound if each molecule contains one hydrogen atom?

169. A compound that is 31.9% K and 28.9% Cl by mass decomposes when heated to give O_2 and a compound that is 52.4% K and 47.6% Cl by mass. Write a balanced chemical equation for the reaction.

170. A compound that combines in fixed amounts with one or more molecules of water is known as a hydrate. In the lab, a 5.00-gram sample of the hydrate of barium chloride, $BaCl_2 \cdot xH_2O$, is heated to drive off the water. After heating, 4.26 g of anhydrous barium chloride, $BaCl_2$, remains. What is the value of x in the formula of the hydrate, $BaCl_2 \cdot xH_2O$?

171. Predict the formula of the compound produced when 1.00 g of chromium metal reacts with 0.923 g of oxygen atoms, O.

172. A 3.500-g sample of an oxide of manganese contains 1.288 grams of oxygen. What is the empirical formula of the compound?

173. Cocaine is a naturally occurring substance that can be extracted from the leaves of the coca plant, which grows in South America (and is not to be confused with chocolate, or cocoa, which is extracted from the seeds of another South American plant). If the chemical formula for cocaine is $C_{17}H_{21}O_4N$, what is the percentage by mass of carbon, hydrogen, oxygen, and nitrogen in the compound? Comment on the ease with which elemental analysis of the carbon and hydrogen in a compound can be used to distinguish between the white, crystalline powder known as aspirin ($C_9H_8O_4$), which is used to cure headaches, and the white, crystalline powder known as cocaine, which is more likely to cause headaches.

174. The oxygen-carrying protein known as hemoglobin is 0.335% Fe by mass and contains four Fe atoms per hemoglobin molecule. Calculate the molecular weight of this protein.

175. Metal carbonates decompose when they are heated to form metal oxides and carbon dioxide.

$$MCO_3(s) \longrightarrow MO(s) + CO_2(g)$$

Which of the following metal carbonates would lose 35.1% of its mass when it decomposes?

(a) Li_2CO_3 (b) $MgCO_3$ (c) $CaCO_3$
(d) $ZnCO_3$ (e) $BaCO_3$

176. A crucible and sample of $CaCO_3$ weighing 42.670 g were heated until the compound decomposed to form CaO and CO_2

$$CaCO_3(s) \longrightarrow CaO(s) + CO_2(g)$$

The crucible had a mass of 35.351 g. What is the theoretical mass of the crucible and residue after the decomposition is complete?

177. Nitrogen reacts with red-hot magnesium to form magnesium nitride,

$$3\,Mg(s) + N_2(g) \longrightarrow Mg_3N_2(s)$$

which reacts with water to form magnesium hydroxide and ammonia.

$$Mg_3N_2(s) + 6\,H_2O(l) \longrightarrow 3Mg(OH)_2(aq) + 2\,NH_3(aq)$$

Calculate the number of grams of magnesium that would be needed to prepare 15.0 g of ammonia.

178. A sealed bottle contains oxygen gas (O_2) and liquid butyl alcohol ($C_4H_{10}O$). There is enough oxygen in the bottle to react completely with the butyl alcohol to produce carbon dioxide (CO_2) and water (H_2O) gas. Write a chemical equation to describe this reaction. Assume that the bottle remains sealed during the reaction. Compare the number of molecules in the bottle before the reaction occurs ($C_4H_{10}O$ and O_2) with the number of molecules present in the bottle after the reaction (CO_2 and H_2O). Will the number of molecules in the bottle increase, decrease, or remain the same as the reaction takes place?

179. A 2.50-g sample of bronze was dissolved in sulfuric acid. The copper in the alloy reacted with sulfuric acid as follows.

$$Cu(s) + 2\,H_2SO_4(aq) \longrightarrow CuSO_4(aq) + SO_2(g) + 2\,H_2O(l)$$

The $CuSO_4$ formed in the reaction was mixed with KI to form CuI.

$$2\,CuSO_4(aq) + 5\,I^-(aq) \longrightarrow 2\,CuI(s) + I_3^-(aq) + 2\,SO_4^{2-}(aq)$$

The I_3^- ion formed in this reaction was then titrated with $S_2O_3^{2-}$.

$$I_3^-(aq) + 2\,S_2O_3^{2-}(aq) \longrightarrow 3\,I^-(aq) + S_4O_6^{2-}(aq)$$

Calculate the percentage by mass of copper in the original sample if 31.5 mL of 1.00 *M* $S_2O_3^{2-}$ was consumed in the titration.

180. Assume that you start with a glass of water, a glass of methanol, and a teaspoon. Exactly one teaspoon of water is removed from the glass of water and added to the glass of methanol. The resulting mixture of methanol is stirred until the two liquids are thoroughly mixed. Exactly one teaspoon of this mixture is then transferred back to the water. Which of the following statements is true?

(a) The volume of water that ends up in the methanol is larger than the volume of methanol that ends up in the water.

(b) The net volume of water transferred to the methanol is smaller than the net volume of methanol transferred to the water.

(c) The net volume of water added to the methanol is exactly the same as the net volume of methanol added to the water.

181. Iron can react with O_2 to produce two different oxides, $Fe_2O_3(s)$ or $Fe_3O_4(s)$. Write the chemical equations that describe both reactions. If 167.6 g of Fe reacts completely with excess $O_2(g)$ to produce 231.6 g of product, which oxide was formed?

182. Draw and label a diagram like Figures 2.9, 2.10, and 2.11 to represent the following chemical reaction.

$$CS_2(g) + 3Cl_2(g) \longrightarrow S_2Cl_2(g) + CCl_4(g)$$

183. Assume that two experiments are performed on the chemical reaction given below.

$2\,Br^-(aq)$ +	$Cl_2(aq)$ $\longrightarrow$	$Br_2(aq)$ +	$2\,Cl^-(aq)$
colorless	colorless	red	colorless

- **Experiment 1:** 100 mL of a 0.0100 *M* solution of Br^- is added to 100 mL of a 0.0200 *M* solution of Cl_2.
- **Experiment 2:** 100 mL of a 0.0100 *M* solution of Br^- is added to 100 mL of a 0.0500 *M* solution of Cl_2.

If the reaction between aqueous solutions of the Br^- ion and Cl_2 go to completion, which of the following would you expect to observe after mixing the two solutions? Explain your answer.

(a) The solution formed in experiment 1 will be a darker red.

(b) The solution formed in experiment 2 will be a darker red.

(c) The solutions formed in both experiments will be the same shade of red.

Chapter Three

THE STRUCTURE OF THE ATOM

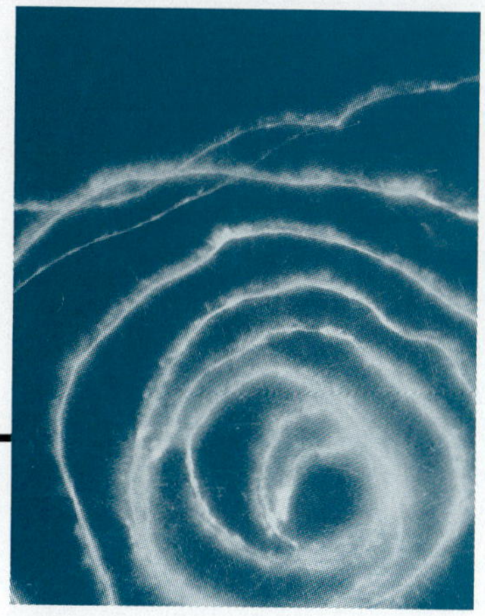

3.1 Rutherford's Model of the Atom

Often a historical development of models of the atom is used. However, the way in which scientists developed models of the atom may not be the best way for students to learn and understand new concepts. We do introduce Rutherford's and Bohr's experiments but then deviate from the use of historical experiments for our development of a model of the atom. ChemActivities 35, 3–11, and 19 in Chemistry: A Guided Inquiry *are designed to accompany this chapter.*

Shortly after radioactivity was discovered at the turn of the twentieth century, Ernest Rutherford became interested in α particles (positively charged helium nuclei) emitted by uranium metal and its compounds. Rutherford found that α particles were absorbed by a thin sheet of metal, but they could pass through metal foil if it was thin enough.

Rutherford observed that a narrow beam of α particles was broadened as it passed through the metal foil. Working with his assistant Hans Geiger, Rutherford measured the angle through which the α particles were scattered by a thin piece of metal foil. Because it is unusually ductile, gold can be made into a foil that is only 0.00004 cm—or about 1400 atoms—thick. When the foil was bombarded with α particles, Rutherford and Geiger found that the angle of scattering was small, on the order of 1°.

These results were consistent with Rutherford's expectations. He knew that the α particle had a considerable mass (for a subatomic particle) and that it moved quite rapidly. Although the α particles would be scattered slightly by collisions with the atoms through which they passed, Rutherford expected the α particles to pass through the metal foil much the same way a rifle bullet would pass through a bag of sand.

One day, Geiger suggested that a research project should be given to Ernest Marsden, who was working in Rutherford's laboratory. Rutherford responded, "Why not let him see whether any α particles can be scattered through a large angle?" When this experiment was done, Marsden found that a small fraction (perhaps 1 in 20,000) of the α particles were scattered through angles larger than 90°, as shown in Figure 3.1*a*. Many years later, reflecting on his reaction to these results, Rutherford said, "It was quite the most incredible event that has ever happened to me in my life. It was almost as incredible as if you fired a 15-inch shell at a piece of tissue paper and it came back and hit you."

Rutherford found that he could explain Marsden's results by assuming that the positive charge and most of the mass of an atom are concentrated in a small fraction of the total volume, which he called the **nucleus.** When he derived mathematical equations for the scattering that would occur, his equations predicted that the number of α particles scattered through a given angle should be proportional to the thickness of the foil and the square of the charge on the nucleus, and inversely proportional to the velocity with which the α particles moved raised to the fourth power. In a series of experiments, Geiger and Marsden verified each of these predictions.

J. J. Thomson (left), who showed that electrons were subatomic particles, and his student, Ernest Rutherford (right), who proposed that atoms were composed of negatively charged electrons orbiting an infinitesimally small positively charged nucleus.

Most of the α particles were able to pass through the gold foil without encountering anything large enough to significantly deflect their path. A small fraction of the α particles came close to the nucleus of a gold atom as they passed through the foil. When this happened, the force of repulsion between the positively charged α particle and the nucleus deflected the α particle by a small angle, as shown in Figure 3.1*b*. Occasionally, an α particle traveled along a path that would eventually lead to a direct collision with the nucleus of one of the 2000 or so atoms it had to pass through. When this happened, repulsion between the nucleus and the α particle deflected the α particle through an angle of 90° or more.

By carefully measuring the fraction of the α particles deflected through large angles, Rutherford was able to estimate the size of the nucleus. According to his calculations, the radius of the nucleus is at least 10,000 times smaller than

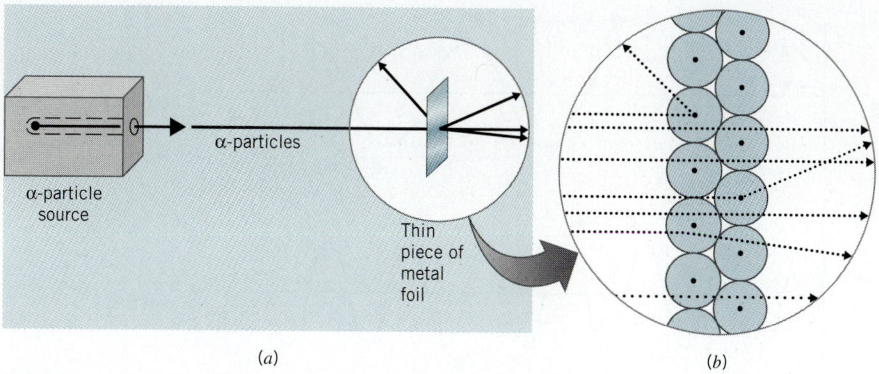

Fig. 3.1 (*a*) A block diagram of the Rutherford–Marsden–Geiger experiment. (*b*) Most α particles pass through empty space between the nuclei. A few come close enough to be repelled by the nucleus and are deflected through small angles. Occasionally, an α particle travels along a path that would lead to a direct hit with the nucleus. These particles are deflected through large angles by the force of repulsion between the α particle and the positively charged nucleus of the atom.

the radius of the atom. The vast majority of the volume of an atom is therefore empty space.

> **CHECKPOINT**
>
> Assume that a circle 1 cm in diameter is used to represent the nucleus of an atom. Calculate the size of the circle that would have to be used to represent the diameter of the atom.

3.2 Particles and Waves

Rutherford's model assumes that most of the mass and all of the positive charge of an atom are concentrated in an infinitesimally small nucleus surrounded by a sea of lightweight, negatively charged electrons. Our next goal is to develop a picture of how these electrons are distributed around the nucleus. Much of what we know about the arrangement of electrons in an atom has been obtained by studying the interaction between matter and different forms of **electromagnetic radiation.** We therefore need to understand what we mean when we say that electromagnetic radiation has some of the properties of both a particle and a wave.

Scientists divide matter into two categories: particles and waves. **Particles** are easy to understand because they have a definite mass and they occupy space. **Waves** are more challenging. They have no mass and yet they carry energy as they travel through space. The best way to demonstrate that waves carry energy is to watch what happens when a pebble is tossed into a lake. As the waves produced by this action travel across the lake, they set in motion any leaves that lie on the surface.

In addition to their ability to carry energy, waves have four other characteristic properties: speed, frequency, wavelength, and amplitude. As we watch waves travel across the surface of a lake, we can see that they move at a certain **speed** (v). By watching waves strike a pier at the edge of the lake, we can see that they are also characterized by a **frequency** (ν), which is the number of wave cycles that hit the pier per unit of time. The frequency of a wave is therefore reported in units of cycles per second (s^{-1}) or hertz (Hz).

The idealized drawing of a wave in Figure 3.2 illustrates the definitions of amplitude and wavelength. The **wavelength** (λ) is the distance between repeating

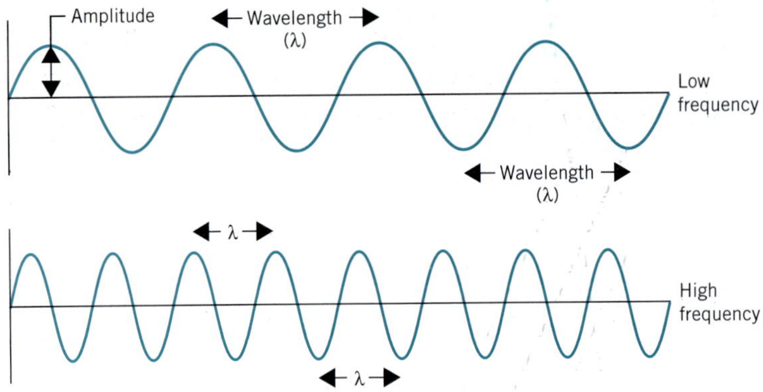

Fig. 3.2 The wavelength (λ) is the distance between repeating points on a wave. The *amplitude* is the difference between the height of the highest or lowest points on the wave and its midline. The *frequency* (ν) is the number of waves (or cycles) that pass a fixed point per unit time. The product of frequency times wavelength is the speed at which the wave moves through space.

points on the wave. The **amplitude** is the difference between highest or lowest points on the wave and the middle of the wave.

If we measure the frequency (ν) of any type of wave in cycles per second and the wavelength (λ) in meters, the product of the two numbers has the units of meters per second. The product of the frequency (ν) times the wavelength (λ) of a wave is therefore the speed (v) at which the wave travels through space.

$$\nu\lambda = \text{v}$$

To understand the relationship among the speed, frequency, and wavelength of a wave, it may be useful to consider a concrete example. Imagine that you are at a railroad crossing, watching a train that consists of 45-foot-long boxcars go by. Assume that it takes 3.0 seconds for each boxcar to pass in front of your car. The "frequency" of the train is 1 car every 3.0 seconds. How fast is the train moving? The speed at which the train moves through space is the product of the "frequency" of this phenomenon times its "wavelength."

$$\frac{1}{3.0\ \text{s}} \times 45\ \text{ft} = 15\frac{\text{ft}}{\text{s}} \approx 10\frac{\text{mi}}{\text{hr}}$$

Exercise 3.1

Orchestras in the United States tune their instruments to an "A" that has a frequency of 440 cycles per second, or 440 Hz. If the speed of sound is 1116 feet per second, what is the wavelength of this note?

Solution

The product of the frequency times the wavelength of any wave is equal to the speed with which the wave travels through space: $\nu\lambda = \text{v}$. Substituting the speed with which the note travels through space and the frequency of the note gives the following:

$$(440\ \text{s}^{-1})(\lambda) = 1116\ \text{ft/s}$$

Solving for λ gives a wavelength of 2.54 feet for the note.

3.3 Light and Other Forms of Electromagnetic Radiation

In 1865 James Clerk Maxwell proposed that light is a wave with both *electric* and *magnetic* components. Light is therefore a form of *electromagnetic radiation.* Because it is a wave, light is bent when it enters a glass prism. When white light, which is electromagnetic radiation made up of many different wavelengths, is focused on a prism, the light rays of different wavelengths are bent by differing amounts and the light is transformed into a band of colors. Radiation separated into its different wavelength components is called a **spectrum.** Starting from the side of the spectrum where the light is bent by the smallest angle, the colors are red, orange, yellow, green, blue, and violet.

Visible light contains the narrow band of frequencies and wavelengths in the small portion of the electromagnetic spectrum that our eyes can see. It includes radiation with wavelengths between about 400 nm (violet) and 700 nm (red). Because the wavelength of electromagnetic radiation can be as long as 40 m or as short as 10^{-5} nm, the visible spectrum is only a tiny portion of the total range of electromagnetic radiation. (See Appendix A.1, Table A.3, for the definition of metric prefixes.)

The electromagnetic spectrum includes radio and TV waves, microwaves, infrared radiation, visible light, ultraviolet radiation, X rays, γ rays, and cosmic rays, as shown in Figure 3.3. These different forms of radiation all travel at the speed of light in a vacuum (c). They differ, however, in their frequencies and wavelengths. The product of the frequency times the wavelength of electromagnetic radiation is always equal to the speed of light.

$$\nu\lambda = c$$

As a result, electromagnetic radiation that has a long wavelength has a low frequency, and radiation with a high frequency has a short wavelength.

> **➤ CHECKPOINT**
> What is the range of frequencies of the waves used in a microwave oven?

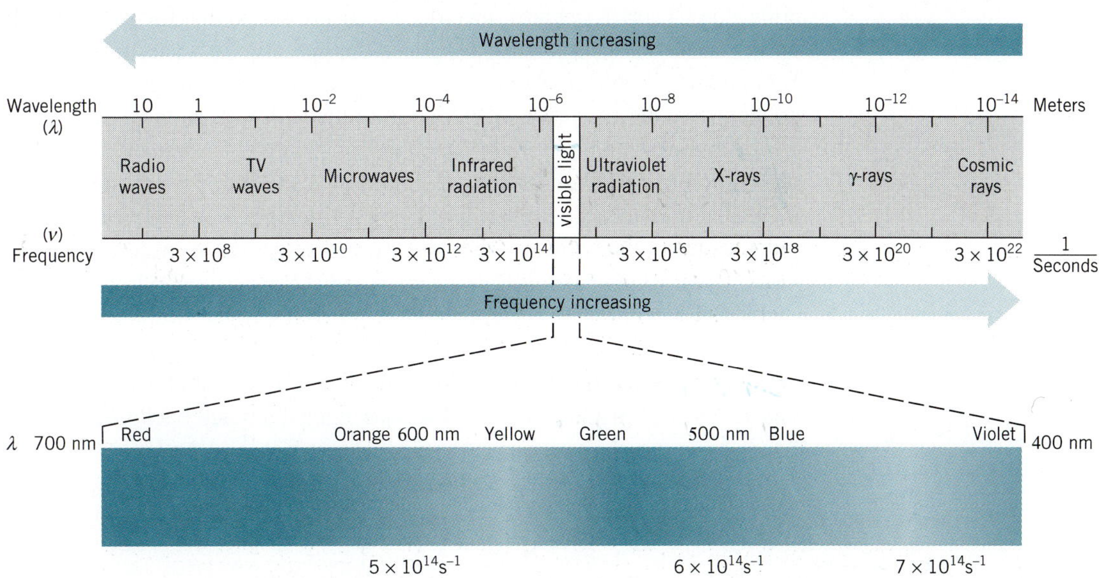

Fig. 3.3 The visible spectrum is the small portion of the total electromagnetic spectrum that our eyes can see. Other forms of electromagnetic radiation include radio and TV waves, microwaves, infrared and ultraviolet radiation, as well as X rays, γ rays, and cosmic rays.

Exercise 3.2

Calculate the frequency of red light that has a wavelength of 700.0 nm if the speed of light is 2.998×10^8 m/s.

Solution

The product of the frequency times the wavelength of any wave is equal to the speed with which the wave travels through space. In this case, the wave travels at the speed of light: 2.998×10^8 m/s. Before we can calculate the frequency of the radiation we have to convert the wavelength into units of meters.

$$700.0 \text{ nm} \times \frac{1 \text{ m}}{10^9 \text{ nm}} = 7.000 \times 10^{-7} \text{ m}$$

We can then substitute this wavelength and the speed of light into the following equation: $\nu\lambda = c$.

$$\nu(7.000 \times 10^{-7} \text{ m}) = 2.998 \times 10^8 \text{ m/s}$$

We then solve for the frequency of the light in units of cycles per second.

$$\nu = 4.283 \times 10^{14} \text{ s}^{-1}$$

3.4 Atomic Spectra

For more than 200 years chemists have known that sodium salts produce a yellow color when added to a flame. Robert Bunsen, however, was the first to systematically study this phenomenon. (Bunsen went so far as to design a new burner that would produce a colorless flame for this work.) Between 1855 and 1860, Bunsen and his colleague Gustav Kirchhoff developed a spectroscope that focused the light from the burner flame onto a prism that separated the light into its spectrum. Using this device, Bunsen and Kirchhoff were able to show that the **emission spectrum** of sodium salts contains two narrow bands of radiation in the yellow portion of the spectrum.

Chemists and physicists soon began using the spectroscope to catalog the wavelengths of light emitted or absorbed by a variety of compounds. The data were then used to detect the presence of certain elements in everything from mineral water to sunlight. No obvious patterns were discovered in these data, however, until 1885, when Johann Jacob Balmer analyzed the spectrum of hydrogen.

When an electric current is passed through a glass tube that contains hydrogen gas at low pressure, hydrogen atoms are produced and the tube gives off a pink-violet light. When the light is passed through a prism (as shown in Figure 3.4), four narrow lines of bright light are observed against a black background. These narrow bands have the characteristic wavelengths and colors shown in Table 3.1. Balmer found that these data fit the following equation to within ±0.02%.

$$\frac{1}{\lambda} = R_H\left(\frac{1}{2^2} - \frac{1}{n^2}\right)$$

In this equation, R_H is a constant known as the Rydberg constant, which is equal to 1.09737×10^{-2} nm^{-1} and n is an integer between 3 and 6.

Table 3.1
Characteristic Lines in the Visible Spectrum of Hydrogen

Color	Wavelength (nm)
Red	656.3
Blue-green	486.1
Blue-violet	434.0
Violet	410.2

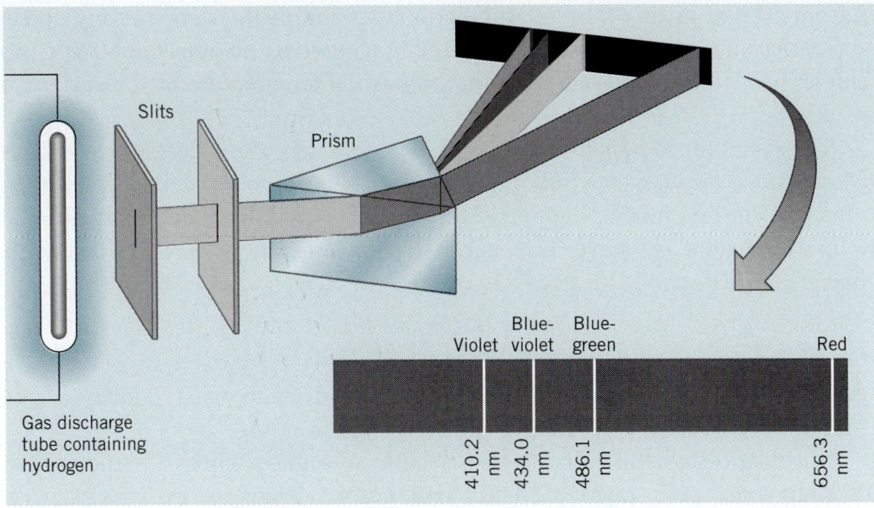

Fig. 3.4 The light given off when a tube filled with H_2 gas is excited with an electric discharge can be separated into four narrow lines of visible light when it is passed through a prism.

Between 1906 and 1924, four more series of lines were discovered in the emission spectrum of hydrogen by searching the infrared spectrum at longer wavelengths and the ultraviolet spectrum at shorter wavelengths. Lyman discovered one series of lines in the UV portion of the spectrum and Paschen, Brackett, and Pfund found three other series of lines in the IR spectrum. Each of these lines fits the same general equation, where n_1 and n_2 are integers and R_H is the Rydberg constant.

$$\frac{1}{\lambda} = R_H\left(\frac{1}{n_1^2} - \frac{1}{n_2^2}\right)$$

3.5 Quantization of Energy

The emission spectrum of the hydrogen atom raises several important questions.

- Why do hydrogen atoms give off only a handful of narrow lines of radiation when they emit light?
- Why do hydrogen atoms emit light when excited that has the same wavelengths as the light they absorb?
- Why do the lines in the hydrogen spectrum depend on the integer relationship observed by Balmer and others?

Answers to these questions came from work on a related topic.

It is common knowledge that objects give off light when heated. Examples range from the gentle red glow of an electric burner on a stove to the bright light emitted when the tungsten wire in a light bulb is heated by passing an electric current through the wire. What is less well known is a phenomenon discovered by Thomas Wedgwood in 1792. Wedgwood, whose father started the famous porcelain factory, noticed that many objects give off a red glow when heated to the same temperature. The bottom of a crucible and the iron triangle on which the crucible rests, for example, both glow red when heated with a Bunsen burner.

Wedgwood also noticed that the color of the light emitted by an object changes as it is heated to higher temperatures until the object glows white-hot, but the spectrum of light given off at a particular temperature is the same for any object. The fact that sunlight is equivalent to the light emitted by an object at 6000°C, for example, has led to the assumption that this is the temperature of the surface of the sun.

The spectrum of radiation that an object emits when it is heated was eventually understood through the work of Max Planck and Albert Einstein. According to the Planck–Einstein model, light and other forms of electromagnetic radiation behave to some extent as a wave. The energy (E) of the radiation is proportional to the frequency of this wave.

$$E = h\nu$$

In this equation, h is Planck's constant, which is equal to 6.626×10^{-34} joule · second (J · s).

But electromagnetic radiation also behaves to some extent as a particle. Light isn't continuous; it is a stream of small bundles or packets of energy. These "energy packets" were given the name **photons.** When an atom or molecule absorbs a photon it therefore gains a finite amount of energy equal to the product of Planck's constant times the frequency of the photon.

Exercise 3.3

Calculate both the energy of a single photon of red light with a wavelength of 700.0 nm and the energy of a mole of such photons.

Solution

In Exercise 3.2 we found that red light with a wavelength of 700.0 nm has a frequency of $4.283 \times 10^{14}\ s^{-1}$. Substituting the frequency into the Planck–Einstein equation gives the following result, in units of joules (J).

$$E = (6.626 \times 10^{-34}\ \text{J} \cdot \text{s})(4.283 \times 10^{14}\ \text{s}^{-1}) = 2.838 \times 10^{-19}\ \text{J}$$

A single photon of red light carries an insignificant amount of energy. But a mole of the photons has 170.9 kJ of energy.

$$\frac{2.838 \times 10^{-19}\ \text{J}}{1\ \text{photon}} \times \frac{6.022 \times 10^{23}\ \text{photons}}{1\ \text{mol}} = 1.709 \times 10^{5}\ \text{J/mol} = 170.9\ \text{kJ/mol}$$

This is enough energy to raise the temperature of 1 liter of water by more than 40°C.

As we have seen, atoms emit or absorb radiation at discrete frequencies or wavelengths. If the energy of the radiation is directly proportional to its frequency, this means that the atoms emit or absorb radiation with only discrete energies. This suggests that there are only specific stable *energy states* within an atom. If this is true, these *energy levels* are countable. In other words, the energy levels of an atom are **quantized.**

This explains why an object gives off red light when heated until it just starts to glow. Red light has the longest wavelength, and therefore the smallest frequency, of any form of visible radiation. Because the energy of electromagnetic radiation is proportional to its frequency, red light carries the smallest amount of energy of any form of visible radiation. The light given off when an object just starts to glow, therefore, has the color characteristic of the lowest energy form of electromagnetic radiation our eyes can see. As the object becomes hotter it also emits light at higher frequencies, until it eventually appears to give off white light.

> **➤ CHECKPOINT**
>
> Explain how the color of a heated metal bar could be used to determine its temperature.

3.6 The Bohr Model of the Atom

Although Rutherford was never able to incorporate electrons into his model of the atom, one of his students, Niels Bohr, proposed a model for the hydrogen atom that accounted for its spectrum. The **Bohr model** assumed that the negatively charged electron and the positively charged nucleus of a hydrogen atom were held together by the force of attraction between oppositely charged particles. According to Coulomb's law (shown below) this force is directly proportional to the charge on the electron (q_e) and the charge on the nucleus (q_p) and inversely proportional to the square of the distance between the particles (r^2).

Coulomb's law is sometimes introduced in texts or introductory chemistry courses when discussing Bohr's model of the atom but is not given much emphasis. In this text Coulomb's law will be very significant in the discussion of both ionization energies and photoelectron spectroscopy. Instructors are encouraged to emphasize Coulomb's law.

$$F = \frac{q_e \times q_p}{r^2}$$

Bohr found that he could derive the equation that fit the experimental data obtained by Balmer, Lyman, Paschen, Brackett, and Pfund if the following assumptions were made.

- The electron in a hydrogen atom travels around the nucleus of the atom in an orbit.
- The energy of the electron in a given orbit is proportional to its distance from the nucleus. (It takes energy to move an electron from a region close to the nucleus to one that is farther away.)
- Only orbits with certain energies are allowed. In other words, the energy of the electron in a hydrogen atom is quantized. The energy of the electron depends on the orbit it occupies.
- Light is absorbed when an electron moves from a lower energy orbit into one that has a higher energy. The energy of this radiation is equal to the difference between the energies of the two orbits.
- Light is emitted when an electron falls from a higher-energy orbit into a lower-energy orbit. Once again, the energy of this radiation is equal to the difference between the energies of the two orbits.

Although it was remarkably successful for the hydrogen atom, the Bohr model was unable to explain the properties of atoms more complex than hydrogen and was eventually abandoned for a **quantum mechanical model** of the atom. The quantum mechanical model has the advantage that it is more powerful, but it achieves this power at a significant cost in terms of the ease with which it can be visualized.

We will retain four ideas from the Bohr model as we try to visualize the quantum mechanical model of the atom, although these ideas will be modified slightly.

- Electrons are attracted to the nucleus of an atom by the force of attraction between oppositely charged objects.
- Electrons reside in regions in space that are at different distances from the nucleus.

- There are only certain regions in space in which an electron can reside. As a result, the energy of an electron in an atom is quantized.
- Atoms emit or absorb radiation when an electron moves from one of these regions of space to another.

3.7 The Energy States of the Hydrogen Atom

The only information that was given in Table 3.1 was the wavelength and the color of the radiation for the four lines in the visible spectrum of the hydrogen atom. Table 3.2 includes both the frequency and the energy of this radiation as well.

Table 3.2
Characteristic Lines in the Visible Spectrum of the Hydrogen Atom

Color	Wavelength (nm)	Frequency (s^{-1})	Energy (kJ/mol)
Red	656.3	4.568×10^{14}	182.2
Blue-green	486.1	6.167×10^{14}	246.0
Blue-violet	434.0	6.908×10^{14}	275.5
Violet	410.2	7.309×10^{14}	291.6

The data in Table 3.2 describe only the characteristic lines in the visible spectrum of the hydrogen atom analyzed by Balmer. Figure 3.5 also contains data for the lines discovered by Lyman, Paschen, Brackett, and Pfund. Each of these lines corresponds to a transition between a pair of energy levels labeled $n = 1, 2, 3, 4,$ and so on. By convention, the lowest energy level for the hydrogen atom is the $n = 1$ state. The highest energy level would be the $n = \infty$ state. According to this diagram, it would take a photon with a wavelength of 121.57 nm to excite an electron from the $n = 1$ to the $n = 2$ state. A photon with a wavelength of 121.57 nm would be emitted when the electron fell from the $n = 2$ to $n = 1$ state.

The energy for each level in Figure 3.5 is given on the left side of this diagram. These values are reported as negative numbers to convey that energy is given off when an electron falls from one of the energy levels toward the top of this diagram into an energy level that lies farther down on the diagram. Thus, moving an electron from the $n = 1$ to the $n = 2$ state would correspond to the absorption of a photon with an energy of 984 kJ/mol.

> **➤ CHECKPOINT**
>
> If electromagnetic radiation with an energy of 1166 kJ/mol were absorbed by a mole of hydrogen atoms in their lowest energy state, what would be the resulting energy of the hydrogen atoms? What would be the energy level (value of n)?

$$E = E_2 - E_1 = (-328 \text{ kJ/mol}) - (-1312 \text{ kJ/mol}) = 984 \text{ kJ/mol}$$

If the electron then falls back into the $n = 1$ state, a photon with an energy of 984 kJ/mol would be given off.

Exercise 3.4

Identify the transitions between different energy states that give rise to the four lines in the spectrum of the hydrogen atom given in Table 3.2.

Solution

Table 3.2 contains the following values for the energies of the four lines in the visible spectrum of the hydrogen atom: 182.2, 246.0, 275.5, and 291.6 kJ/mol.

Figure 3.5 suggests that each of these lines corresponds to a transition between the second energy state and a higher energy state.

182.2 kJ/mol	$3 \leftrightarrow 2$
246.0 kJ/mol	$4 \leftrightarrow 2$
275.5 kJ/mol	$5 \leftrightarrow 2$
291.6 kJ/mol	$6 \leftrightarrow 2$

Because of interactions between the electrons of an atom, the spectra for atoms that contain more than one electron are more complex than the spectrum for the hydrogen atom. As we'll see, however, the same basic postulates about the process by which radiation is emitted or absorbed by the hydrogen atom can be applied to investigate the energy levels of other atoms.

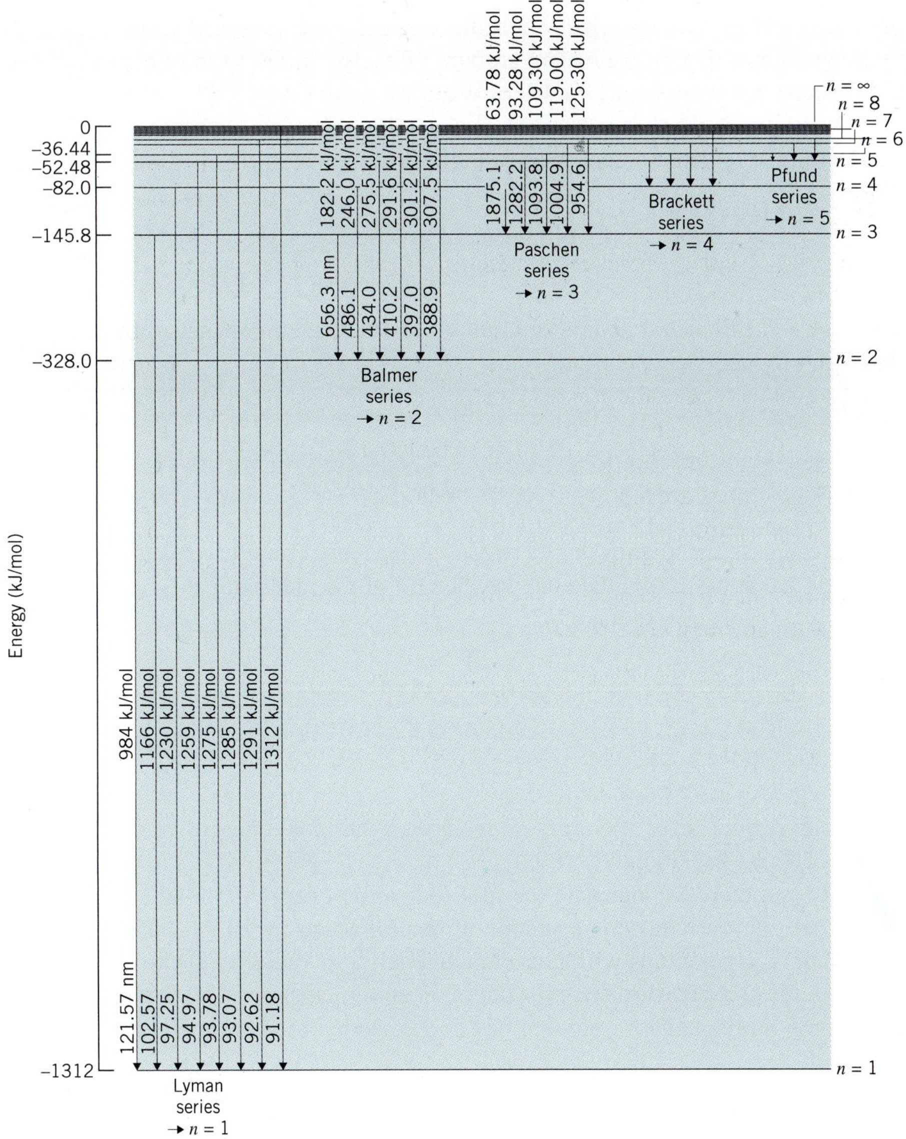

Fig. 3.5 According to the Bohr model, the wavelengths and corresponding energies in the ultraviolet spectrum of the hydrogen atom discovered by Lyman were the result of electrons dropping from higher energy states into the $n = 1$ state. The Balmer, Paschen, Brackett, and Pfund series result from electrons falling from high energy states into the $n = 2$, $n = 3$, $n = 4$, and $n = 5$ states, respectively.

Traditionally, a discussion of first ionization energies has followed the development of the concept of electron configuration. This order has been reversed to allow students to use experimental data to build their understanding of the electronic structure of the atom. First the shell model of the atom is developed and then continually refined until the quantum mechanical model is introduced.

The ionization energies of the most loosely held electron are presented as experimental data. If students recognize that an electron's potential energy is directly related to both the charge on the electron and the nucleus, and inversely related to the distance between them, they will be able to use first ionization energies to establish a simple model of the atom.

3.8 The First Ionization Energy

According to the data in Figure 3.5, the difference in energy between the $n = 1$ and $n = \infty$ levels is 1312 kJ/mol. This suggests that it should take 1312 kJ of energy to remove the electrons from a mole of isolated hydrogen atoms in the gas phase.

$$\mathrm{H}(g) + h\nu \longrightarrow \mathrm{H}^+(g) + e^-$$

This quantity is the first ionization energy of hydrogen, which is the energy needed to remove the electron from the lowest energy state of a neutral hydrogen atom in the gas phase to form a positive ion.

We can bring the magnitude of the first ionization energy for hydrogen into perspective by noting that it is more than one and a half times as large as the energy released when we ignite one mole of the methane that fuels the Bunsen burners in a chemistry laboratory.

$$\mathrm{CH_4}(g) + 2\,\mathrm{O_2}(g) \longrightarrow \mathrm{CO_2}(g) + 2\,\mathrm{H_2O}(g)$$

So much energy is consumed when the electrons are removed from a mole of hydrogen atoms that an equivalent amount of energy in the form of heat would be able to raise the temperature of 4 L of water by more than 75°C!

If we divide the first ionization energy of hydrogen by Avogadro's number we can obtain the ionization energy of a single hydrogen atom.

$$\frac{1312\,\mathrm{kJ}}{1\,\mathrm{mol}} \times \frac{1\,\mathrm{mol}}{6.022 \times 10^{23}\,\mathrm{atoms}} \times \frac{1000\,\mathrm{J}}{1\,\mathrm{kJ}} = 2.179 \times 10^{-18}\,\mathrm{J}$$

We can use the Planck–Einstein equation to calculate the frequency of the radiation that has this energy.

$$E = h\nu$$
$$2.179 \times 10^{-18}\,\mathrm{J} = (6.626 \times 10^{-34}\,\mathrm{J \cdot s})(\nu)$$
$$\nu = 3.288 \times 10^{15}\,\mathrm{s}^{-1}$$

We can then use the relationship between the frequency and wavelength of electromagnetic radiation to calculate the wavelength of the radiation.

$$\nu\lambda = c$$
$$(3.288 \times 10^{15}\,\mathrm{s}^{-1})(\lambda) = 2.998 \times 10^{8}\,\mathrm{m/s}$$
$$\lambda = 9.118 \times 10^{-8}\,\mathrm{m} = 91.18\,\mathrm{nm}$$

This frequency and wavelength corresponds to electromagnetic radiation in the ultraviolet region of the spectrum. If we need even more energy, we can turn to a source of X-ray radiation.

We can therefore measure the first ionization energy of an atom by shining ultraviolet light or X rays on a sample of neutral atoms in the gas phase. Some of the UV or X-ray photons will have enough energy to knock an electron out of the atom. The **first ionization energy (IE)** of an atom is measured by determining the radiation with the smallest amount of energy needed to remove an electron from the atom.

Experimental values of the first ionization energies for the neutral atoms in the gas phase for the first 20 elements are given in Table 3.3.

Table 3.3

First Ionization Energies for Gas Phase Atoms of the First 20 Elements

Symbol	Z	IE (kJ/mol)	Symbol	Z	IE (kJ/mol)
H	1	1312.0	Na	11	495.8
He	2	2372.3	Mg	12	737.7
Li	3	520.2	Al	13	577.6
Be	4	899.4	Si	14	786.4
B	5	800.6	P	15	1011.7
C	6	1086.4	S	16	999.6
N	7	1402.3	Cl	17	1251.1
O	8	1313.9	Ar	18	1520.5
F	9	1681.0	K	19	418.8
Ne	10	2080.6	Ca	20	589.8

A plot of the first ionization energy versus the atomic number of these elements is shown in Figure 3.6.

There are several clear patterns in these data.

- It doesn't matter whether we compare H and He, Li and Ne, or Na and Ar. The first ionization energy increases by a factor of between 2 and 4 as we go from left to right across a row of the periodic table.
- There is a dramatic drop in the first ionization energy as we go from the end of one row to the beginning of the next. It doesn't matter whether we compare helium and lithium, or neon and sodium: The first ionization energy decreases by more than a factor of 4.
- There is a gradual decrease in the first ionization energy as we go down a column of the periodic table. Consider going from He (IE = 2372 kJ/mol) to Ne (IE = 2081 kJ/mol) to Ar (IE = 1521 kJ/mol), for example.
- There are minor exceptions to the gradual increase in the first ionization energy across a row of the periodic table. The increase from Li (IE = 520 kJ/mol) to Be (IE = 899 kJ/mol), for example, is followed by a small decrease as we continue to B (IE = 801 kJ/mol).

➤ CHECKPOINT

Which would you expect to be larger, the first ionization energy of Rb or Sr? Explain why. Which would you expect to be larger, the first ionization energy of Cl or Br? Explain why.

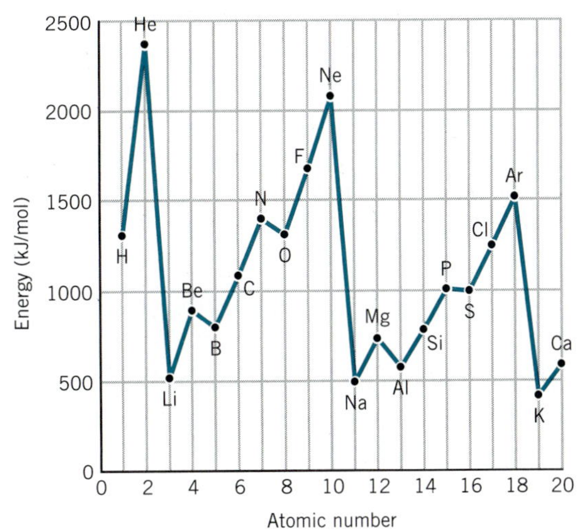

Fig. 3.6 Plot of the first ionization energies of the first 20 elements in the periodic table. There is a gradual increase in the first ionization energy across a row of the periodic table from H to He, from Li to Ne, and from Na to Ar. K and Ca appear to start a similar trend. There is a gradual decrease in the first ionization energy down a column of the periodic table, as can be seen by comparing the values for He, Ne, and Ar.

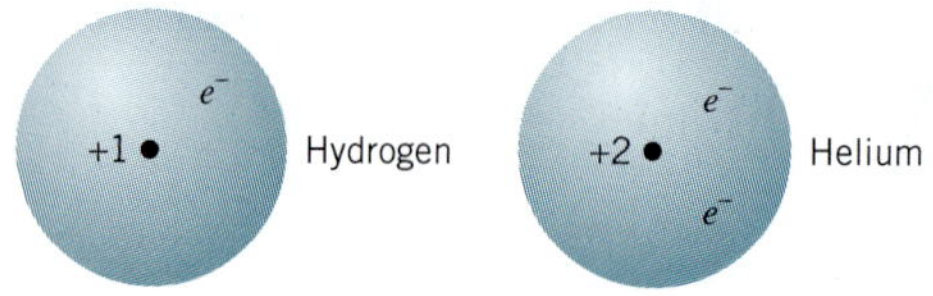

Fig. 3.7 The shell model assumes that the energy of an electron depends on the charge on the nucleus and the distance from the nucleus. The nucleus of a helium atom has twice the positive charge of the nucleus of a hydrogen atom. The first ionization energy of He is therefore about twice that of H.

Experimental data are used to develop electron configurations, because the exact solution of the Schrödinger equation is not possible for more than one-electron atoms and because the solution for the hydrogen atom is too difficult to introduce in general chemistry.

Students should understand the link between experimental data and the models that scientists develop to describe a concept. For this reason ionization energies and photoelectron spectroscopy are used to present students with experimental data and have them use these data to develop a model of the atom and electron configuration.

3.9 The Shell Model

The data in Table 3.3 are consistent with the idea that electrons are attracted to the nucleus more strongly as the charge on the nucleus increases and less strongly as the distance from the nucleus increases. Consider the first and second elements in the periodic table, for example. The first ionization energy increases by a factor of about 2 as we go from H (IE = 1312 kJ/mol) to He (IE = 2372 kJ/mol). This is consistent with the hypothesis that the helium electrons feel roughly twice the force of attraction for the nucleus because the positive charge on the helium nucleus is twice as large as the charge on the hydrogen nucleus.

If we extend this argument we would predict that the first ionization energy for a lithium atom would be 1.5 times as large as the first ionization energy of helium and 3 times as large as the first ionization energy of hydrogen, because the charge on the nucleus is now +3. Nothing could be further from the truth. The first ionization energy of lithium (IE = 520 kJ/mol) is only about 20% as large as that of helium (IE = 2372 kJ/mol), and it is less than half as large as that of hydrogen (IE = 1312 kJ/mol).

The only way to explain these data is to assume that the electron removed when the first ionization energy of lithium is measured is farther away from the nucleus than the two electrons on a helium atom. Because this electron is farther from the nucleus, it is much easier to remove from the atom.

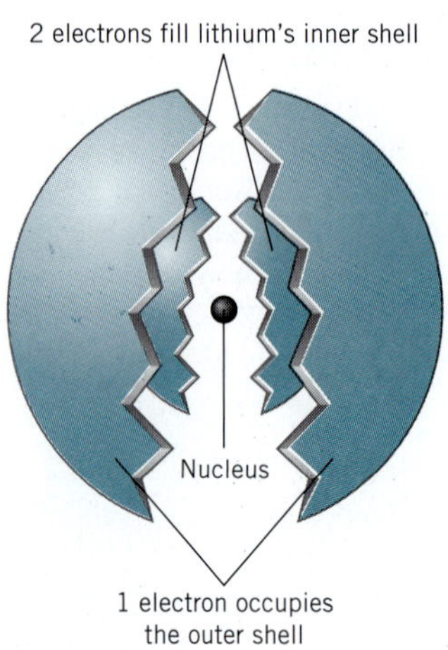

Fig. 3.8 The lithium atom consists of a nucleus with a positive charge of +3, two electrons in a shell close to the nucleus, and one electron in a shell farther from the nucleus. Reprinted from Carl Snyder, *The Extraordinary Chemistry of Ordinary Things,* John Wiley & Sons, Inc., New York, 1992, p. 24.

The data in Table 3.3 therefore suggest that the electrons in an atom are arranged in shells. The **shell model** of the atom assumes that the hydrogen and helium atoms consist of a nucleus surrounded by either one or two electrons in a single shell, relatively close to the nucleus as shown in Figure 3.7. Lithium contains two electrons that lie close to the nucleus of the atom, in the same shell as the electrons in H and He, and one electron in a shell that is farther from the nucleus, as shown in Figure 3.8.

The nucleus and the two inner electrons constitute the core of the lithium atom. The outermost electron in lithium doesn't experience the full +3 nuclear charge but rather a charge reduced by the underlying electrons. It is convenient to define a **core charge,** which, although it doesn't give the actual charge felt by an outer shell electron, is useful for organizing atomic properties. The core charge for lithium is the sum of the positive charge on the nucleus and the negative charge on the two inner shell core electrons:

$$\text{Li:}\quad (+3 \text{ nuclear charge}) + (-2 \text{ inner electron charge}) = +1 \text{ core charge}$$

Because the outer electron in lithium is at a greater distance from the nucleus and experiences a smaller attraction for the nucleus than the electrons in an He atom, it takes less energy to remove this electron from the atom. As a result, Li has a significantly smaller first ionization energy than helium.

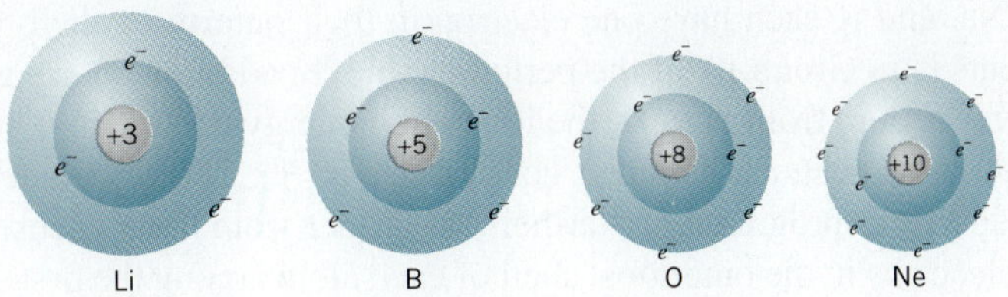

Fig. 3.9 The shell model for four atoms from the second period of the periodic table. The core charge can be found from the shell model. The core charges are +1 for Li, +3 for B, +6 for O, and +8 for Ne.

Lithium marks the start of a series of eight elements characterized by a general increase in the first ionization energy with atomic number. Although there are minor variations in this trend, it is consistent with the idea that eight electrons are added to the atom in the second shell of electrons.

The shell that lies closest to the nucleus is described as the $n = 1$ shell of electrons. The data in Table 3.3 suggest that hydrogen has a single electron in the $n = 1$ shell and that helium has two electrons in this shell. Lithium has two electrons in the $n = 1$ shell and a third electron in the $n = 2$ shell. The next element, Be, has a first ionization energy (899 kJ/mol) that is larger than that for Li, which reflects the increased charge on the core:

Be: (+4 nuclear charge) + (−2 inner electron charge) = +2 core charge

Note that for the elements in the first row of the periodic table the core charge is the same as the atomic number Z, whereas the core charge is Z − 2 for elements in the second row. Beryllium has two electrons in both the $n = 1$ and $n = 2$ shells, and neon has two electrons in the $n = 1$ shell and eight electrons in the $n = 2$ shell, as shown in Figure 3.9.

Once the eighth electron has been added to the $n = 2$ shell, there is a precipitous drop in the first ionization energy as we go from neon (*IE* = 2081 kJ/mol) to sodium (*IE* = 496 kJ/mol). This is similar to the drop observed from helium to lithium. It even has a similar magnitude: The first ionization energy drops by a factor of about 4 from neon to sodium. This suggests that the eleventh electron in the sodium atom is placed in a third shell ($n = 3$), at an even greater distance from the nucleus than the second shell of electrons used for the elements between Li and Ne, as shown in Figure 3.10.

The data in Table 3.3 for the elements between Na (Z = 11) and Ar (Z = 18) mirror the pattern observed between Li (Z = 3) and Ne (Z = 10). With slight variations, there is a gradual increase in the first ionization energy of these elements. This increase is then followed by another precipitous drop in the first ionization energy as we go from Ar (IE = 1521 kJ/mol) to potassium (IE = 419 kJ/mol). These data suggest that we leave the third shell at this point and enter a fourth shell ($n = 4$), as shown in Figure 3.11.

> **CHECKPOINT**
> What are the core charges for carbon and fluorine?

> **CHECKPOINT**
> Why aren't the first ionization data (Table 3.3) consistent with having nine electrons in the second shell of sodium?

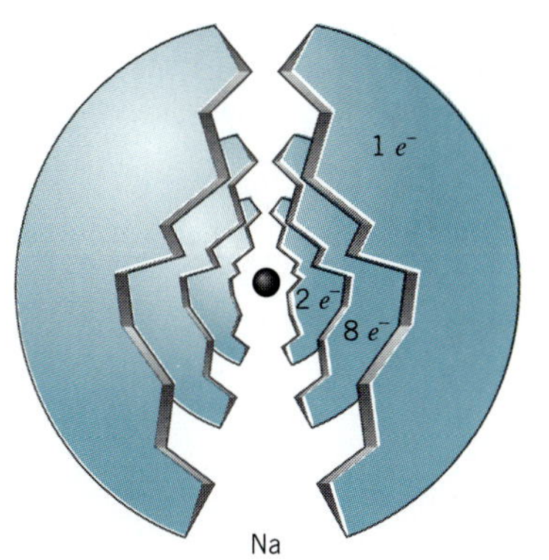

Fig. 3.10 The shell model of the sodium atom. There are three shells and a core charge of +1. Adapted from Carl Snyder, *The Extraordinary Chemistry of Ordinary Things,* John Wiley & Sons, Inc., New York, 1992.

3.10 The Shell Model and the Periodic Table

When the data in Table 3.3 are arranged in terms of the periodic table, many of the trends we observe are reinforced. There are two elements in the first row of the periodic table, and two electrons occupy the first shell in our shell model. There are eight electrons in the second and third shells, and eight elements in the second and third rows of the periodic table.

Furthermore, the number of electrons in each of the outermost shells on an atom is consistent with the group in which the element is found. We have concluded

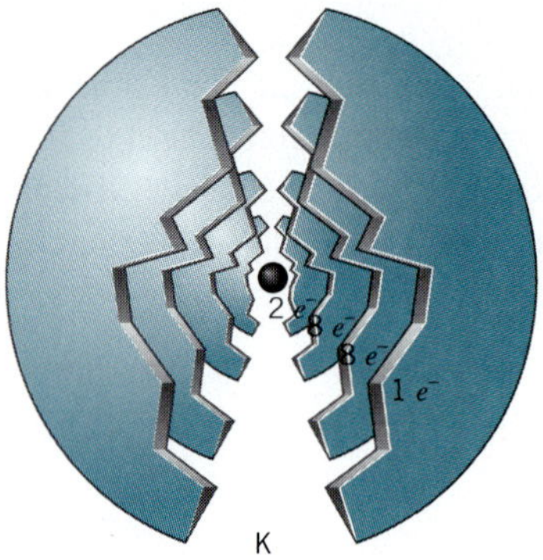

Fig. 3.11 The shell model of the potassium atom. Now there are four shells. The first shell is complete with two electrons, the $n = 2$ and $n = 3$ shells hold eight electrons, and the fourth shell has only one electron. Adapted from Carl Snyder, *The Extraordinary Chemistry of Ordinary Things,* John Wiley & Sons, Inc., New York, 1992.

that H, Li, Na, and K each have one electron in their outermost shell, and each of these elements is in Group IA of the periodic table. In addition, as we go down the column in this group, from H to K, the ionization energy decreases. This is consistent with all of these atoms having a core charge of +1, with the outermost electron in a shell that is progressively farther and farther from the nucleus. In fact, the number of electrons in the outermost shell of each atom among the first 20 elements in the periodic table corresponds exactly with the group number for that element.

As we will see in the next chapter, the electrons in the outermost shell of an atom are involved in the formation of bonds between atoms. Because these bonds are relatively strong, the electrons that form them are often called the **valence electrons** (from the Latin stem *valens,* "to be strong"). A more complete discussion of valence electrons is found in the next chapter.

The periodic table was originally created to group elements that had similar chemical and physical properties. The arrangement of the electrons in the atom deduced from the shell model is reflected in the arrangement of elements in the periodic table. This suggests that many of the chemical and physical properties of the elements are related to the number of electrons in the outermost shell, which are the valence electrons in these atoms. Also note that the core charge and the number of electrons in the outermost shell correspond to the group numbers IA through VIIIA.

➤ CHECKPOINT

Is the radius of the valence shell of Na larger, smaller, or the same as the radius of the valence shell of Li?

3.11 Photoelectron Spectroscopy and the Structure of Atoms

The shell model of the atom that we have deduced from first ionization energy data assumes that the electrons in an atom are arranged in shells about the nucleus, with successive shells being farther and farther from the nucleus of the atom. The data we used to derive this model represented the *minimum* energy needed to remove an electron from the atom. Thus, in each case, the first ionization energy reflects the ease with which the outermost electron can be removed from the atom.

For atoms that have many electrons, we would expect that it would take more energy to remove an electron from an inner shell than it does to remove the electron from the valence shell. It takes more energy to remove an electron from the $n = 2$ shell than from the $n = 3$ shell, for example, and even more energy to remove an electron from the $n = 1$ shell for a given atom.

This raises an interesting question: Do all of the electrons in a given shell have the same energy? This question can be answered by using a slightly different technique to measure the energy required to remove an electron from a neutral atom in the gas phase to form a positively charged ion.

This time, we will shine radiation on the sample that has enough energy to excite an atom to the point that one of its electrons from any shell is ejected from the atom to form a positively charged ion. The experiment, which is diagrammed in Figure 3.12, is known as **photoelectron spectroscopy (PES).**

The PES experiment begins with the absorption of a high-energy UV or X-ray photon. The energy of this photon is large enough to remove an electron from the atom (*IE*). The excess energy is carried off by the electron ejected from the atom in the form of kinetic energy (KE). We know the energy of the radiation ($h\nu$) used to excite the atom. If we measure the kinetic energy of the photoelectron ejected (KE) when this radiation is absorbed, we can calculate the energy required to remove this electron from the atom (*IE*).

$$IE = h\nu - \text{KE}$$

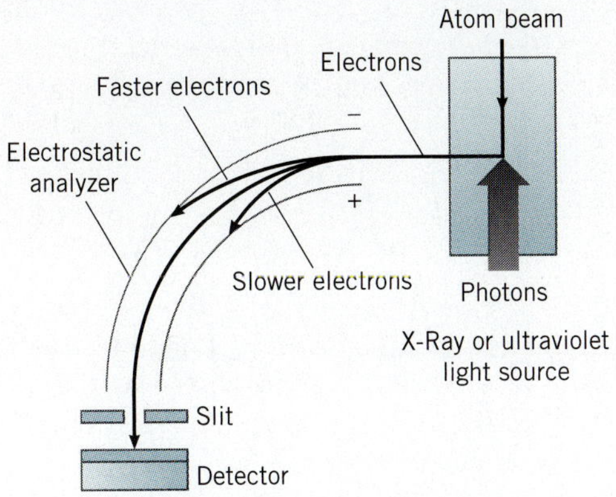

Fig. 3.12 A block diagram of photoelectron spectroscopy. Absorption of a high-energy electron leads to the ejection of an electron from the atom. The kinetic energy (KE) of the ejected electron is measured, and the energy required to remove the electron from the atom is calculated from the difference between the energy of the photon ($h\nu$) and the kinetic energy of the photoelectron. Because electrons can be removed from any shell in the atom, this experiment is different from measurements of first ionization energies, which remove electrons from only the outermost shell. Reprinted from R. J. Gillespie et al., *Atoms, Molecules and Reactions,* Prentice-Hall, Englewood Cliffs, New Jersey, 1994, p. 198.

PES differs from the experiment used to obtain the first ionization energies given in Table 3.3 by its ability to remove electrons *from any shell in the atom,* as shown in Figure 3.13. Not only can an electron from the outermost shell be removed, but an electron from one of the shells deep within the core of electrons that surround the nucleus can be ejected. Only a single electron is removed from a given atom, but that electron can come from any energy level. As a result, PES allows us to measure the energy needed to remove any electron on an atom.

PES data include not only the ionization energies of the most loosely held electrons but also the ionization energies of all electrons in the atom. These new data present a problem to the student, who finds that the simple shell model cannot rationalize these ionization energies. Thus, a rethinking of the model is necessary.

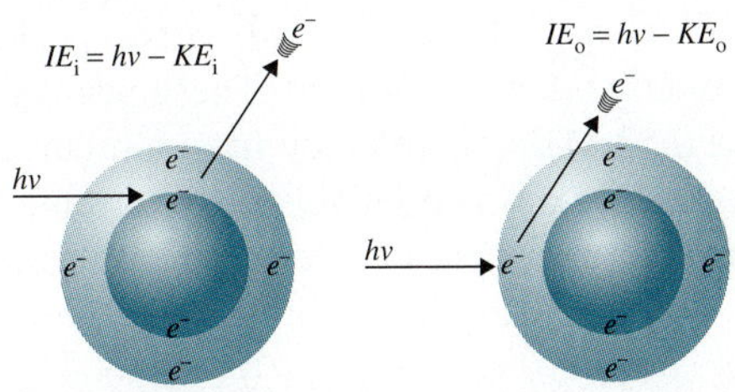

Fig. 3.13 Absorption of a high-energy photon can result in the ejection of an electron from any shell of an atom. The ionization energy of the inner shell electrons, IE_i, will be larger than that of the outer shell electrons, IE_o.

Data from PES experiments are obtained as peaks in a spectrum that plots the energy needed to eject an electron (IE) on the x axis versus the intensity of the observed signal on the y axis, as shown in Figure 3.14. The spectrum is plotted so that energy *increases* toward the *left* on the x axis. The intensity or the height of the peak is proportional to the number of electrons of equivalent energy ejected during the experiment. If we see two peaks, for example, that have a relative height of 2:1, we can conclude that one of the energy levels from which electrons are removed in this experiment contains twice as many electrons as the other.

As we examine PES data in the next section, it is important to remember that the electromagnetic radiation energy supplied may remove an electron from the outermost shell or remove an electron from one of the shells that are deep within the core of the atom.

There is a clear pattern in the first ionization energy data, which suggest that there are shells of electrons around an atom. Careful consideration of the PES data suggests that these shells not only can be but should be divided into subshells. PES ionization energies therefore provide a basis for developing electron configurations without the use of either quantum numbers or orbital shapes.

3.12 Electron Configurations from Photoelectron Spectroscopy

Hydrogen has one peak in the photoelectron spectrum in Figure 3.14 because it contains only a single electron. As expected, this peak comes at an energy of 1312 kJ/mol, or, as we'll express it from now on in megajoules, 1.312 MJ/mol. This is the energy required to eject the electrons from a mole of hydrogen atoms.

➤ CHECKPOINT

What factor determines the number of peaks in the photoelectron spectrum of an atom?

Helium also has only one peak in the PES experiment, which occurs at an energy of 2372 kJ/mol or 2.372 MJ/mol. Note that the peak for helium is shifted to

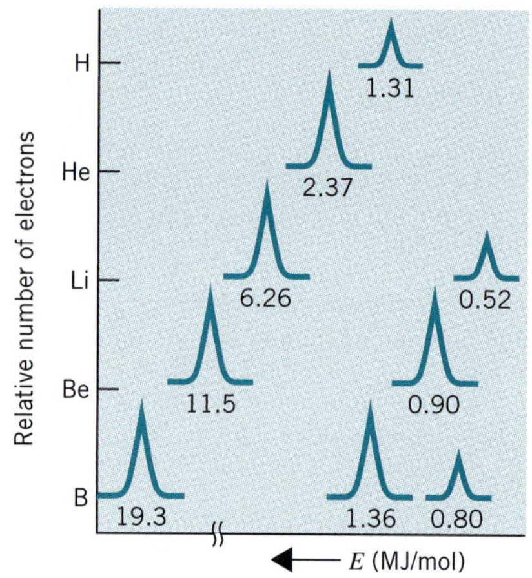

Fig. 3.14 Simulated photoelectron spectra of the first five elements in the periodic table. The energy needed to remove an electron from an atom increases from right to left. The energy required to remove an electron from the hydrogen atom is 1.312 MJ/mol and that for the helium atom is 2.372 MJ/mol. The height of the He spectrum is twice that of the H spectrum because He has twice as many electrons as H. The spectra have been adjusted so that the peak heights of spectra of different atoms are directly comparable.

> **➤ CHECKPOINT**
>
> Why is there only one peak in the photoelectron spectrum of He even though there are two electrons in an He atom?

the left in Figure 3.14 when compared to the peak in the spectrum for hydrogen. This corresponds to a shift toward a larger ionization energy, as expected. It takes more energy to remove an electron from a helium atom than the energy needed to remove an electron from hydrogen. The height of the peak for helium is twice that for the peak in the spectrum for hydrogen. This is consistent with our hypothesis that the two electrons in a helium atom both occupy the $n = 1$ shell.

Our shell model of the atom leads us to expect two peaks in the PES spectrum for lithium, which is exactly what is observed. These peaks occur at ionization energies (*IE*) of 6.26 and 0.52 MJ/mol, and they have a relative intensity of 2:1. The outermost electron in the Li atom is relatively easy to remove because it is in the $n = 2$ shell. But it takes a great deal of energy to reach into the $n = 1$ shell, as shown in Figure 3.14, because the electrons in that shell lie close to a nucleus that carries a charge of +3.

Two peaks are also observed in the PES spectrum of beryllium, with a relative intensity of 1:1. In this case, it takes an enormous amount of energy to reach into the $n = 1$ shell ($IE = 11.5$ MJ/mol) to remove one of the electrons that lie close to the nucleus with its charge of +4. It takes quite a bit less energy ($IE = 0.90$ MJ/mol) to remove one of the electrons in the $n = 2$ shell, as shown in Figure 3.14.

Ejected electrons may be separated out according to their kinetic energies by an electrostatic analyzer, which has two oppositely charged plates. As an ejected electron enters the electric field between the plates, it is deflected depending on its speed (kinetic energy). A detector is arranged so that electrons with the proper deflection may be recorded as "counts." Electrons of different kinetic energies may be counted by varying the voltage difference between the plates.

Students sometimes do not realize that in a photoelectron spectroscopy experiment only one electron is removed from an atom and that this could be any electron in the atom—not just electrons from the valence shell. It is, therefore, important to assign homework from this section.

An interesting phenomenon occurs when we compare the photoelectron spectrum of boron with the spectra for the first four elements shown in Figure 3.14. There are now three distinct peaks in the spectrum, at energies of 19.3, 1.36, and 0.80 MJ/mol. There is a peak in the PES spectrum that corresponds to removing one of the electrons from the $n = 1$ shell ($IE = 19.3$ MJ/mol). But there are also two additional peaks, with a relative intensity of 2:1, that correspond to removing an electron from the $n = 2$ shell ($IE = 1.36$ and 0.80 MJ/mol).

The same phenomenon occurs in the PES spectra for carbon, nitrogen, oxygen, fluorine, and neon, as shown in Table 3.4. In each case, we see three peaks. As the charge on the nucleus increases, it takes more and more energy to remove an electron from the $n = 1$ shell, until, by the time we get to neon, it takes 84.0 MJ/mol to remove an electron from this shell. (This is more than 100 times the energy given off in a typical chemical reaction.)

The PES spectra for B, C, N, O, F, and Ne contain a second peak, of gradually increasing energy because of the increasing nuclear charge, that has the same intensity as the peak for the $n = 1$ shell. And, in each case, we get a third peak, of gradually increasing energy, that corresponds to the electrons which are the easiest to remove from the atoms. The intensity of the third peak increases from element

Table 3.4

Ionization Energies for Gas Phase Atoms of the First 10 Elements Obtained from Photoelectron Spectra

IE (MJ/mol) Element	First Peak	Second Peak	Third Peak
H	1.31		
He	2.37		
Li	6.26	0.52	
Be	11.5	0.90	
B	19.3	1.36	0.80
C	28.6	1.72	1.09
N	39.6	2.45	1.40
O	52.6	3.12	1.31
F	67.2	3.88	1.68
Ne	84.0	4.68	2.08

Source: D. A. Shirley et al., *Physical Review B* (15), 544–552 (1977).

Low-resolution PES spectra are used so as not to have to introduce complicated coupling interactions between electron spins and angular momenta. A high-resolution photoelectron spectrum of the 2p *atomic state of argon shows two peaks lying very close to one another. A low-resolution photoelectron spectrum will not differentiate between the two peaks, and hence a single peak is seen. See the* Instructor's Manual.

The photoelectron spectra shown as figures in this text are computer generated. Normally photoelectron spectra of different elements are not directly comparable. The areas under the peaks of these spectra have been adjusted so that the areas under the peaks due to the 1s electrons are equivalent for all elements containing two 1s electrons.

to element, representing a single electron for boron up to a total of six electrons for neon, as shown in Figures 3.14 and 3.15.

The PES data for the first 10 elements reinforce our belief in the shell model of the atom. But they suggest that we have to refine the model to explain the fact that the electrons in the $n = 2$ shell seem to occupy different energy levels. In other words, we have to introduce the concept of **subshells** within the shells of electrons.

When the first evidence for subshells was discovered shortly after the turn of the twentieth century, a shorthand notation was introduced in which these subshells were described as either *s, p, d,* or *f*. Within any shell of electrons, it always takes the largest amount of energy to remove an electron from the *s* subshell.

The PES data in Table 3.4 suggest that there is only one subshell in the $n = 1$ shell. Chemists usually represent this by writing 1*s*, where the number represents the shell and the letter represents the subshell. These data also suggest that the 1*s* subshell can hold a maximum of two electrons. The third and fourth electrons on an atom seem to be added to a 2*s* subshell. Once we have four electrons, however, the 2*s* subshell seems to be filled, and we have to add the fifth electron to the next subshell: 2*p*.

By convention, the information we have deduced from the PES data is referred to as the atom's **electron configuration** and is written as follows.

H (Z = 1)	$1s^1$
He (Z = 2)	$1s^2$
Li (Z = 3)	$1s^2\ 2s^1$
Be (Z = 4)	$1s^2\ 2s^2$
B (Z = 5)	$1s^2\ 2s^2\ 2p^1$

The superscripts in the electron configurations designate the number of electrons in each subshell. The existence of two subshells within the $n = 2$ shell explains the minor inversion in the first ionization energies of boron and beryllium shown in Figure 3.6. It is slightly easier to remove the outermost electron from B (*IE* = 0.80 MJ/mol) than from Be (*IE* = 0.90 MJ/mol), in spite of the greater charge on the nucleus of the boron atom, because the outermost electron of

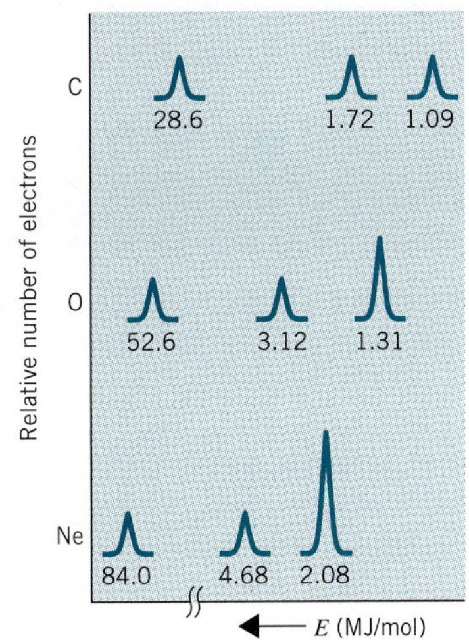

Fig. 3.15 Simulated photoelectron spectra of C, O, and Ne. The nuclear charge increases from carbon to oxygen to neon, and the spectra shift to the left to higher energies. The peak farthest to the right shows the electrons most easily removed, and the peak heights (relative number of electrons) increase in a ratio of 2:4:6. The spectra have been adjusted so that the peak heights for different atoms are directly comparable.

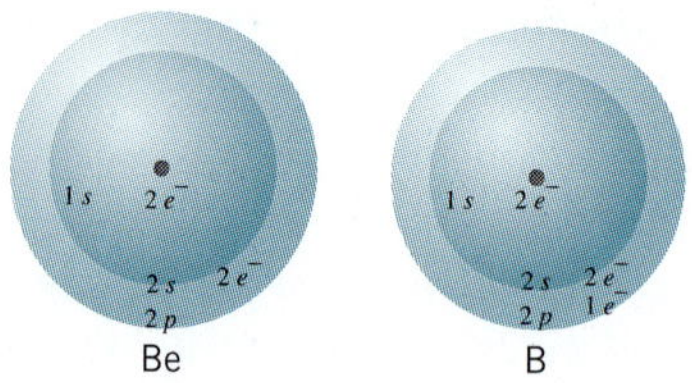

Fig. 3.16 The outermost electron in B is easier to remove than the outermost electron in Be because the electron is in the 2*p* subshell. The easiest electron to remove from Be is in the 2*s* subshell.

B is in the 2*p* subshell, whereas the outermost electron in Be has to be removed from the 2*s* subshell shown in Figure 3.16.

As we continue across the second row of the periodic table, from B to Ne, the number of electrons in the 2*p* subshell gradually increases to six. The best evidence for this is the growth in the intensity of the peak corresponding to the 2*p* subshell. In boron, this peak has half the height of the 1*s* or 2*s* peaks. When we get to neon, we find it has three times the height of the 1*s* or 2*s* peaks. This suggests that the 2*p* subshell can hold a maximum of six electrons. We can therefore continue the process of translating PES data into the electron configurations for the atoms as follows.

C (Z = 6) $1s^2\ 2s^2\ 2p^2$
N (Z = 7) $1s^2\ 2s^2\ 2p^3$
O (Z = 8) $1s^2\ 2s^2\ 2p^4$
F (Z = 9) $1s^2\ 2s^2\ 2p^5$
Ne (Z = 10) $1s^2\ 2s^2\ 2p^6$

➤ CHECKPOINT

Explain the relative intensity of the three peaks in the PES spectrum of Ne (Figure 3.17), which is 2:2:6. Assign each of these peaks to a subshell.

PES data for the next 11 elements in the periodic table are given in Table 3.5, in which the columns are labeled in terms of the representations for the subshells: 1*s*, 2*s*, 2*p*, 3*s*, 3*p*, 3*d*, and 4*s*.

As we might expect, sodium has four peaks in the photoelectron spectrum, corresponding to the loss of electrons from the 1*s*, 2*s*, 2*p*, and 3*s* subshells, as shown in Figure 3.17. It is easier to remove the electron in the $n = 3$ shell in sodium (IE = 0.50 MJ/mol) than the electrons in the $n = 2$ shell (IE = 6.84 and 3.67 MJ/mol), which in turn are easier to remove than the electrons in the $n = 1$ shell (IE = 104 MJ/mol).

Na (Z = 11) $1s^2\ 2s^2\ 2p^6\ 3s^1$

This approach illustrates the process of science because it uses experimental data to generate the electronic structure of the atom. Because the PES experiment is done on the macroscopic scale, and then interpreted in terms of the structure of matter on the microscopic scale, it provides an example of the interrelationship between chemistry on the two scales of matter. The net result is a concept of electron configuration that is built on a concrete basis.

Magnesium also gives four peaks in the PES experiment, which is consistent with the following electron configuration:

Mg (Z = 12) $1s^2\ 2s^2\ 2p^6\ 3s^2$

Aluminum and each of the subsequent elements up to argon have five peaks in the PES spectrum. The new peak corresponds to the 3*p* subshell, and these elements

When students are presented with experimental data from PES measurements of the ionization energies of the elements, they are able to develop many of these rules for themselves and thus obtain a fuller understanding of electron configurations—particularly with respect to the energies of the various subshells.

Table 3.5

Ionization Energies for Gas Phase Atoms of Elements 11 Through 21 Obtained from Photoelectron Spectra

IE (MJ/mol) Element	1*s*	2*s*	2*p*	3*s*	3*p*	3*d*	4*s*
Na	104	6.84	3.67	0.50			
Mg	126	9.07	5.31	0.74			
Al	151	12.1	7.79	1.09	0.58		
Si	178	15.1	10.3	1.46	0.79		
P	208	18.7	13.5	1.95	1.01		
S	239	22.7	16.5	2.05	1.00		
Cl	273	26.8	20.2	2.44	1.25		
Ar	309	31.5	24.1	2.82	1.52		
K	347	37.1	29.1	3.93	2.38		0.42
Ca	390	42.7	34.0	4.65	2.90		0.59
Sc	433	48.5	39.2	5.44	3.24	0.77	0.63

Source: D. A. Shirley et al., *Physical Review B* (15), 544–552 (1977).

have the following electron configurations.

Al (Z = 13)	$1s^2\ 2s^2\ 2p^6\ 3s^2\ 3p^1$
Si (Z = 14)	$1s^2\ 2s^2\ 2p^6\ 3s^2\ 3p^2$
P (Z = 15)	$1s^2\ 2s^2\ 2p^6\ 3s^2\ 3p^3$
S (Z = 16)	$1s^2\ 2s^2\ 2p^6\ 3s^2\ 3p^4$
Cl (Z = 17)	$1s^2\ 2s^2\ 2p^6\ 3s^2\ 3p^5$
Ar (Z = 18)	$1s^2\ 2s^2\ 2p^6\ 3s^2\ 3p^6$

The electron configurations of the elements in the third row of the periodic table therefore follow the same pattern as the corresponding elements in the second row.

By the time we get to potassium and calcium, we find six peaks in the PES spectrum, with the smallest ionization energy comparable to, but slightly less than, the ionization energies of the 3*s* electrons on sodium and magnesium. We therefore write the electron configurations of potassium and calcium as follows.

K (Z = 19)	$1s^2\ 2s^2\ 2p^6\ 3s^2\ 3p^6\ 4s^1$
Ca (Z = 20)	$1s^2\ 2s^2\ 2p^6\ 3s^2\ 3p^6\ 4s^2$

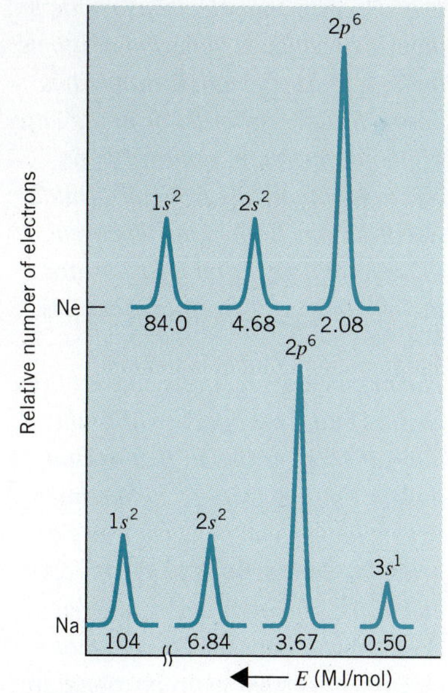

Fig. 3.17 Simulated photoelectron spectra of Ne and Na. Sodium has four peaks corresponding to loss of electrons from the 1*s*, 2*s*, 2*p*, and 3*s* subshells. The heights of the peaks correspond to the number of electrons in a subshell. The spectra have been adjusted so that the peak heights for different atoms are directly comparable.

Figure 3.18 shows the relative energies of the shells and subshells in the first six elements in the periodic table: H, He, Li, Be, B, and C. The same labels (1*s*, 2*s*, and 2*p*) are used to the describe the different energy states in these atoms. The energy associated with a given label, however, changes significantly from one element to another. Once again, the energies are given as negative numbers to indicate that the atom must absorb energy to remove an electron from one of these subshells. The energy of the 1*s* subshell gradually decreases from −1.31 MJ/mol in H to −2.37 MJ/mol in He, −6.26 MJ/mol in Li, and so on, until we reach an energy of −28.6 MJ/mol in C. This is explained by the shell model of the atom, which assumes that these electrons are close to a nucleus with a charge that increases from +1 to +6. The energy of the 2*s* orbital also decreases as the charge on the nucleus

In the traditional approach to introducing electron configurations, students are given rules for generating quantum numbers that can describe atomic orbitals and are then taught an algorithm for filling these orbitals.

This development of electron configurations from experimental PES data gives students a more concrete, while still scientifically correct, foundation on which to base their understanding of electron configuration than the more abstract quantum numbers and orbital shapes.

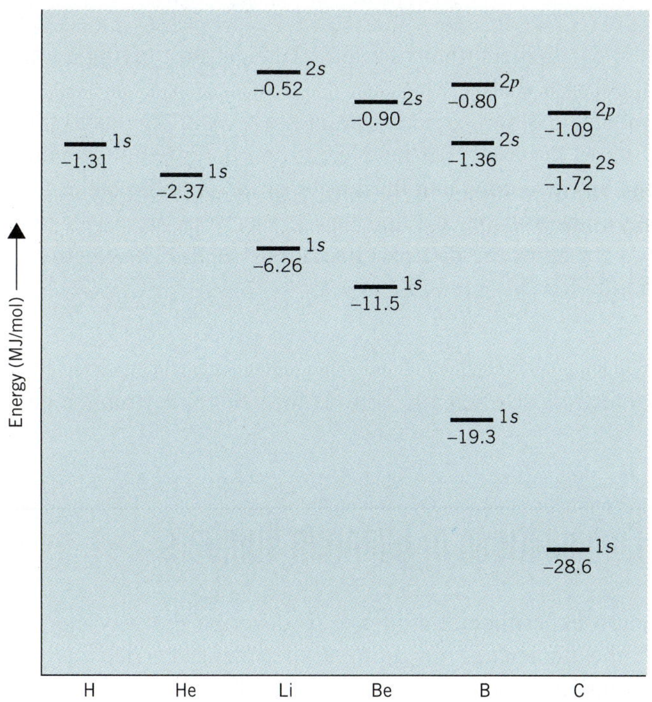

Fig. 3.18 Relative energies of the electrons within the shells and subshells in H, He, Li, Be, B, and C atoms in units of megajoules per mole (MJ/mol). The 1*s* electrons are more tightly held as the nuclear charge increases from H through C. The labels 1*s*, 2*s*, and 2*p* are used to describe the energy states.

It is assumed that electrons reside in atomic orbitals. Irradiation of atoms by light causes electron transitions between these orbitals, or in the case of photoionization, removal of an electron completely from an orbital. According to Koopmans' Theorem, these photoionization energies are the negative of the orbital energies.

The number of electrons ejected from each orbital is proportional to the number of electrons in that orbital; thus in the spectrum of Ne the intensity or height will be three times as great for 2p *electrons as for the* 2s *electrons. The area under a given peak (or the intensity) of the photoelectron spectrum is proportional to the number of electrons in that particular energy state.*

If an atom is bombarded with photons of sufficient energies (UV and X ray), electrons may be removed from all the energy states (orbitals) of an atom. Koopmans' theorem is usually interpreted as meaning that for multielectron atoms the one-electron energy as obtained from SCF is a good approximation of IE *and thus is assumed to be valid for the interpretation of PES data.*

> **CHECKPOINT**
>
> Using data from Table 3.5, sketch the photoelectron spectrum of scandium, Sc. Give the relative intensities of the peaks and assign energies to each peak. Use the photoelectron spectrum to list the order in which electrons are removed from subshells of scandium. Use the electron configuration of Sc to list the order in which its subshells are filled with electrons. Is there a difference in the orders in which electrons are added to and taken away from Sc?

increases, as we would expect. But the energy of the $1s$ orbital is always lower than the $2s$, which is lower than the $2p$.

An interesting phenomenon occurs in the next element in the periodic table, scandium (Sc). Our shell model predicts that the subshells used to hold electrons in calcium are all filled. The twenty-first electron therefore has to go into a new subshell. But the ionization energy for the new peak that appears in the PES spectrum doesn't occur at a lower energy than the subshells used previously, which has been observed for every other subshell as it has appeared. The new peak appears at a higher energy than the $4s$ subshell on scandium.

Evidence from other forms of spectroscopy suggest that the twenty-first electron on the scandium atom goes into the $n = 3$ shell, not the $n = 4$ shell. The subshell used to hold this electron is the $3d$ subshell. As we can see from the data in Table 3.5, the $3d$ subshell is very close in energy to the $4s$ subshell. For scandium, it is easier to remove a $4s$ electron than a $3d$ electron. The electron configuration of scandium may be written as follows:

$$\text{Sc } (Z = 21) \qquad 1s^2\,2s^2\,2p^6\,3s^2\,3p^6\,4s^2\,3d^1$$

This way of writing electron configurations is convenient because it follows the arrangement of the elements in the periodic table. Another way of writing electron configurations is to list the subshells in order of decreasing ionization energy.

$$\text{Sc } (Z = 21) \qquad 1s^2\,2s^2\,2p^6\,3s^2\,3p^6\,3d^1\,4s^2$$

In this text the first method is employed.

Scandium is followed by nine elements—known as the **transition metals**—that have no analogs in the second and third rows of the periodic table. In these elements, electrons continue to fill the $3d$ subshell until we reach zinc.

$$\begin{array}{ll} \text{Ti} & 1s^2\,2s^2\,2p^6\,3s^2\,3p^6\,4s^2\,3d^2 \\ \text{V} & 1s^2\,2s^2\,2p^6\,3s^2\,3p^6\,4s^2\,3d^3 \\ \vdots & \\ \text{Zn} & 1s^2\,2s^2\,2p^6\,3s^2\,3p^6\,4s^2\,3d^{10} \end{array}$$

Thus, the $3d$ subshell can contain a maximum of 10 electrons, giving a total of up to 18 electrons in the $n = 3$ shell.

The next element is gallium, at which point the $4p$ level begins to fill, giving six elements (gallium to krypton) that have electron configurations analogous to those of the elements in the second and third rows of the periodic table.

When forming a positive ion, we find that $4s$ electrons are easier to remove than the $3d$ electrons for all of the first-row transition metals. The electron configuration of the vanadium ion, V^{2+}, for example, is

$$V^{2+} \qquad 1s^2\,2s^2\,2p^6\,3s^2\,3p^6\,3d^3$$

The two electrons that are lost when a V^{2+} ion is formed come from the outermost shell, $4s$.

3.13 Allowed Combinations of Quantum Numbers

We have used modern experimental evidence to describe the way electrons are distributed around the nucleus of an atom. Experiments carried out near the beginning of the twentieth century were responsible for the development of the theory which remains the underpinning of our understanding of the atomic world. Scientists became more and more perplexed as they attempted to explain results of these early experiments. It was apparent that particles as small as atoms and electrons did not conform to the same physical laws as could be applied to everyday objects.

In 1926, the Austrian physicist Erwin Schrödinger worked out a mathematical way of dealing with the problem. This new study was known as wave mechanics or quantum mechanics. Schrödinger's formulation of the way electrons interacted with the nucleus led, of course, to the same results just derived from our study of photoelectron spectroscopy, but his method relied on a mathematical treatment. He assumed that the same mathematics which applied to a light wave also could be applied to explain the interaction of an electron with the electric charge field of a nucleus. His work showed that electrons of different energies were found in different regions around the nucleus. These different regions of stability resulted from a complex interplay of the electrostatic fields generated by charged electrons and nuclei. An electron could exist only in certain parts of this wave field, and its energy was therefore quantized. Schrödinger further showed that these stable states could be characterized by (1) the distance of the electron from the nucleus, (2) the momentum of the electron, and (3) the location of these regions of stability in space.

There were certain complications introduced by the new "wave mechanics" that were beyond what we experience in ordinary life. The idea of quantization, that there is a minimum amount of energy which every energy packet must have, places restrictions on our ability to make precise measurements at the atomic level. In our everyday world we can describe the motion of a baseball or an airplane by giving their successive positions and velocities. The motion of atomic particles, however, cannot be so described.

To measure the position of an object, the object must be illuminated with light so that we can see it. The photons of illuminating light are scattered by the object and strike our eye, allowing us to measure the position of the object. However, in the atomic world the energy of a photon of light can cause electrons to change energy states. This means that the electron is now disturbed from its original state, so either its velocity or its position is no longer the same. Therefore, if we know precisely the electron's position we cannot know its velocity, and vice versa. We cannot simultaneously be certain of an electron's position and velocity.

The German physicist Werner Heisenberg formulated this uncertainty in his famous principle, which states that the better the position of an electron is known, the less well its velocity is known. For an electron *not* to escape from an atom it must have a certain minimum velocity. This minimum velocity corresponds to an uncertainty in its position in the atom, which is as large as the atom itself. This means that an electron is simultaneously all around the nucleus. For this reason we often describe an electron as a cloud of electron density within the atom instead of describing it as a single particle.

Schrödinger's model tried to take into account the results of experiments that showed that electrons didn't always behave as if they were particles. They sometimes behaved as if they had a significant amount of the character of a wave. The Schrödinger model describes the regions in space, or **orbitals,** where electrons are most likely to be found. Instead of trying to tell us where the electron is at any time, this model gives the probability that an electron can be found in a given region of space. The model no longer tells us where the electron is; it only tells us where it might be.

The Schrödinger model uses three coordinates, or three *quantum numbers,* to describe the orbitals in which electrons can be found. These coordinates are known as the principal (n), angular (l), and magnetic (m_l) quantum numbers. The

For a ground state hydrogen atom, $1s^1$, the energy required to ionize the single electron is 1312 kJ/mol; the reverse of this, −1312 kJ/mol, is the 1s orbital energy in hydrogen. The ionization energy of He is 2372 kJ/mol, and therefore the 1s orbital energy is −2732 kJ/mol.

Instructors who do not wish to cover quantum numbers and orbital shapes can omit Sections 3.13, 3.14, and the last two paragraphs of 3.15. The concepts introduced here are not necessary for understanding the remaining material in the text.

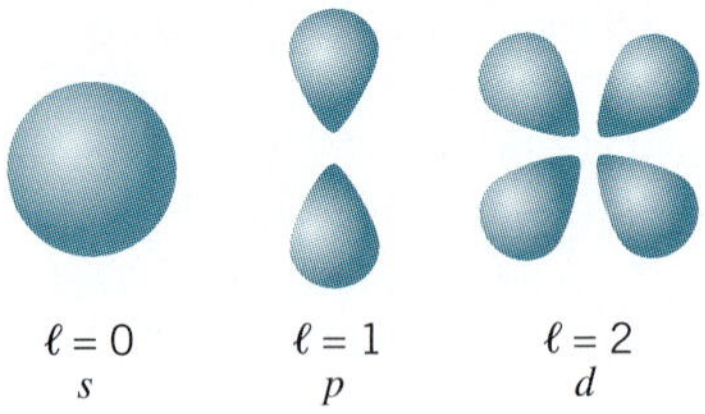

Fig. 3.19 The angular quantum number specifies the shape of the orbital. When $l = 0$, the orbital is spherical. When $l = 1$, it is polar. When $l = 2$, the orbital typically has the shape of a cloverleaf.

quantum numbers describe the size, shape, and orientation in space of the orbitals on an atom.

The **principal quantum number** (n) describes the size of the orbital. Orbitals for which $n = 2$ are larger than those for which $n = 1$, for example. As we have seen, it takes energy to excite an electron from an orbital in which the electron is close to the nucleus ($n = 1$) into an orbital in which it is farther from the nucleus ($n = 2$ or higher). The principal quantum number therefore indirectly describes the energy of an orbital.

The **angular quantum number** (l) describes the shape of the orbital. Orbitals have shapes that are best described as spherical ($l = 0$), polar ($l = 1$), or cloverleaf ($l = 2$), as shown in Figure 3.19. They can take on even more complex shapes as the value of the angular quantum number becomes larger.

There is only one way in which a sphere ($l = 0$) can be oriented in space. Orbitals that have polar ($l = 1$) or cloverleaf ($l = 2$) shapes, however, can point in different directions. We therefore need a third quantum number, known as the **magnetic quantum number** (m_l), to describe the orientation in space of a particular orbital. (It is called the *magnetic* quantum number because the effect of different orientations of orbitals was first observed in the presence of a magnetic field.)

Atomic spectra have shown that electrons in atoms occupy discrete energy states. The quantum mechanical model allows only certain combinations of quantum numbers to describe these states. Therefore a set of rules applies to the way quantum numbers can be assigned to describe an orbital.

> **CHECKPOINT**
>
> Explain why three quantum numbers are required to describe an orbital. What feature of the three-dimensional structure of an orbital is specified by each of these quantum numbers?

Selection Rules Governing Allowed Combinations of Quantum Numbers

- The three quantum numbers (n, l, and m_l) are integers: 0, 1, 2, 3, . . . so on.
- The principal quantum number (n) cannot be zero. The allowed values of n are therefore 1, 2, 3, 4, . . .
- The angular quantum number (l) can be any integer between 0 and $n - 1$. If $n = 3$, for example, l can be either 0, 1, or 2.
- The magnetic quantum number (m_l) can be any integer between $-l$ and $+l$. If $l = 2$, m_l can be either -2, -1, 0, 1, or 2.

3.14 Shells and Subshells of Orbitals

Orbitals that have the same value of the principal quantum number form a **shell.** Orbitals within a shell are divided into *subshells* that are labeled with the same value of the angular quantum number. As we have seen, chemists describe the shell and subshell in which an orbital belongs with a two-character code such as $2p$ or $4f$. The first character indicates the shell ($n = 2$ or $n = 4$). The second character identifies the subshell. By convention, the following lowercase letters are used to indicate different subshells.

$$s \quad l = 0$$
$$p \quad l = 1$$
$$d \quad l = 2$$
$$f \quad l = 3$$

Although there is no pattern in the first four letters (*s*, *p*, *d*, *f*), the letters progress alphabetically from that point (*g*, *h*, and so on). Some of the allowed combinations of the *n* and *l* quantum numbers are shown in Figure 3.20.

The third rule limiting allowed combinations of the *n*, *l*, and m_l quantum numbers has an important consequence. It says that the number of subshells in a shell will be equal to the principal quantum number for the shell. The $n = 3$ shell, for example, contains three subshells: the 3*s*, 3*p*, and 3*d* orbitals.

Let's look at some of the possible combinations of the *n*, *l*, and m_l quantum numbers. Start with the first shell, for which $n = 1$. According to the third rule, the angular quantum number (l) can be any integer between 0 and $n - 1$. Thus, if $n = 1$, l can only be 0. The fourth rule limits the magnetic quantum number (m_l) to integers between $-l$ and $+l$. Thus, when $l = 0$, m_l must be 0. There is only one orbital in the $n = 1$ shell because there is only one way in which a sphere can be oriented in space. The only allowed combination of quantum numbers for which $n = 1$ is the following.

n	l	m_l	
1	0	0	1*s*

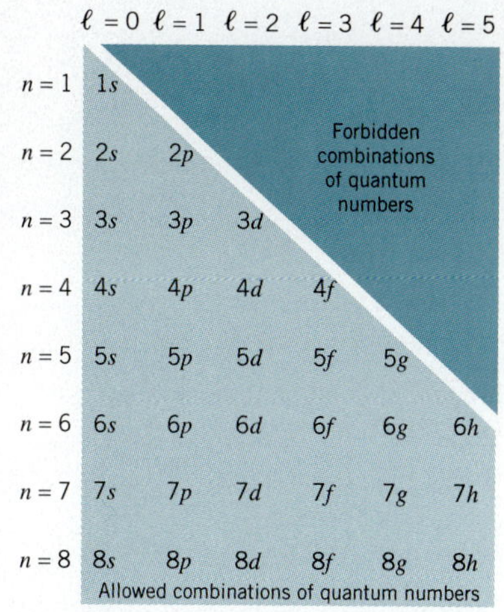

Fig. 3.20 Some allowed combinations of the principal (n) and angular (l) quantum numbers.

Let's now look at the orbitals in the second shell. When $n = 2$, l can be either 0 or 1. When $l = 0$, m_l must be zero. When $l = 1$, m_l can be either -1, 0, or 1. Thus, there are four orbitals in the $n = 2$ shell with the following combination of quantum numbers.

n	l	m_l	
2	0	0	2*s*
2	1	−1	2*p*
2	1	0	2*p*
2	1	1	2*p*

There is only one orbital in the 2*s* subshell. But there are three orbitals in the 2*p* subshell because there are three directions in which a *p* orbital can point. One of these orbitals is oriented along the *x* axis, another along the *y* axis, and the third along the *z* axis of a coordinate system, as shown in Figure 3.21. These orbitals are therefore known as the $2p_x$, $2p_y$ and $2p_z$ orbitals.

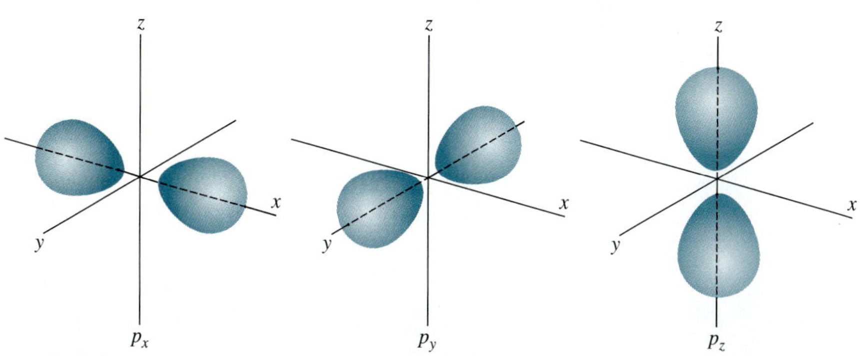

Fig. 3.21 There are three possible values of the magnetic quantum number (m_l) when the angular quantum number (l) is 1. The three values of m_l correspond to the three different orientations of the orbitals in space.

There are nine orbitals in the $n = 3$ shell.

n	l	m_l	
3	0	0	3s
3	1	−1	
3	1	0	3p
3	1	1	
3	2	−2	
3	2	−1	
3	2	0	3d
3	2	1	
3	2	2	

There is one orbital in the 3s subshell and there are three orbitals in the 3p subshell. The $n = 3$ shell, however, also includes five 3d orbitals.

The five different orientations of orbitals in the 3d subshell are shown in Figure 3.22. One of the orbitals lies in the xy plane of an xyz coordinate system and is called the $3d_{xy}$ orbital. The $3d_{xz}$ and $3d_{yz}$ orbitals have the same shape, but they lie between the axes of the coordinate system in the xz and yz planes. The fourth orbital in this subshell lies along the x and y axes and is called the $3d_{x^2-y^2}$ orbital. Most of the space occupied by the fifth orbital lies along the z axis, and this orbital is called the $3d_{z^2}$ orbital.

There are one orbital in the $n = 1$ shell, four orbitals in the $n = 2$ shell, and nine orbitals in the $n = 3$ shell. Thus, in general the number of orbitals in a shell is equal to the square of the principal quantum number: $1^2 = 1$, $2^2 = 4$, $3^2 = 9$, $4^2 = 16$, and so on. There are one orbital in an s subshell ($l = 0$), three orbitals in a p subshell ($l = 1$), and five orbitals in a d subshell ($l = 2$). The number of orbitals in a subshell is therefore $2(l) + 1$.

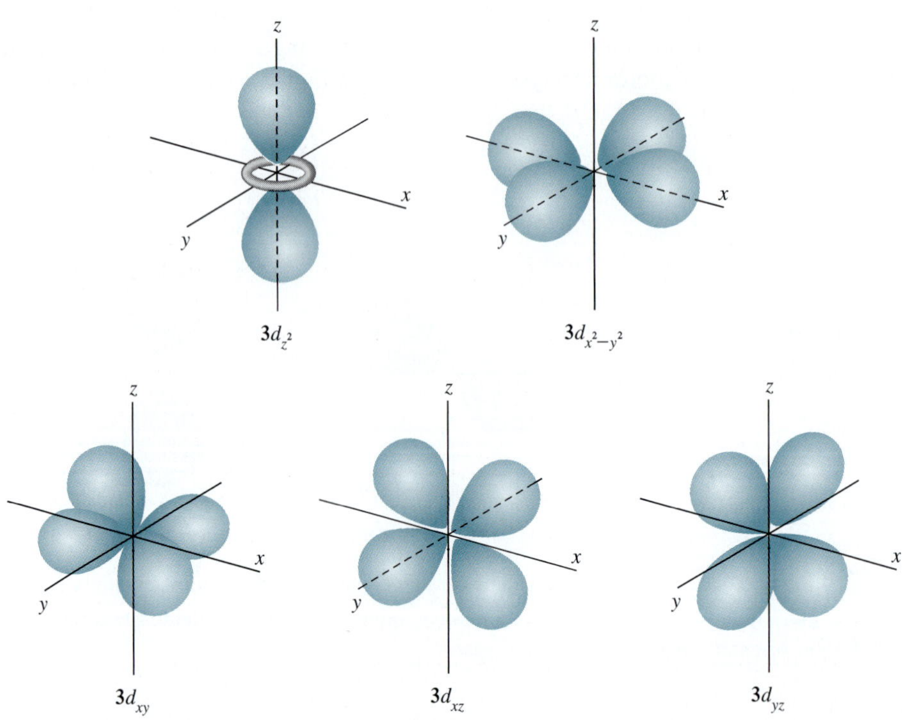

Fig. 3.22 There are five possible values of the magnetic quantum number (m_l) when the angular quantum number (l) is 2. The five values of m_l correspond to the five different orientations of the orbitals in space.

3.15 Orbitals and the Pauli Exclusion Principle

In Chapter 1, we discussed the characteristic charge and mass of an electron. The Schrödinger model is based on the fact that electrons exhibit some of the behavior of a wave as well. In 1920, the German physicists Otto Stern and Walter Gerlach discovered a fourth property of the electron.

The Stern–Gerlach experiment began by heating a piece of silver metal until silver atoms evaporated from the surface. These atoms were passed through slits into an evacuated chamber, where they were allowed to pass through a magnetic field. Because the magnetic field was inhomogeneous, the atoms felt a net force pushing them in a direction perpendicular to their path as they passed through the magnetic field. As a result, the path of the silver atoms was bent or deflected by the magnetic field.

Once they passed through the magnetic field, the atoms struck a glass plate, where they were deposited as silver metal. When the experiment was done at very low residual pressures and long exposure times, Stern and Gerlach found that the silver metal was deposited in two closely spaced areas, suggesting that silver atoms ($Z = 47$) interact with a magnetic field in two distinctly different ways. Similar results were obtained with lithium ($Z = 3$), copper ($Z = 29$), and gold ($Z = 79$), but not with zinc ($Z = 30$), cadmium ($Z = 48$), or mercury ($Z = 80$).

This experiment suggests that there is a fundamental difference between atoms that contain an odd number of electrons and at least some of those that contain an even number of electrons. Atoms that have an odd number of electrons seem to behave as if there is at least one electron that spins in either a clockwise or counterclockwise direction on its axis. A negatively charged electron that spins on its axis would produce a tiny magnetic field that would interact with the magnetic field through which the beam of atoms passes. Atoms in which all of the electrons were *paired* with an electron of opposite **spin** wouldn't have magnetic properties and would pass through the magnetic field without being deflected in either direction.

Silver has an odd number of electrons ($Z = 47$). Therefore, even if most of the electrons pair so that the spins on these electrons cancel, there must be at least one odd electron on each silver atom that is unpaired. This unpaired electron interacts with an external magnetic field to cause the paths of moving silver atoms to be deflected in one direction if the electron spins clockwise on its axis and in the other direction if it spins counterclockwise.

> **➤ CHECKPOINT**
>
> Predict the results of a Stern–Gerlach experiment on a beam of:
>
> (a) Li atoms (b) Be atoms
>
> (c) B atoms (d) N atoms

A further consequence of an electron having a spin is that electrons having the same spin (both electrons spinning in a clockwise direction, for example) have a low probability of being close together and a high probability of being far apart. There is no restriction on electrons of opposite spin being close together.

Because electrons of the same spin keep apart, an electron tends to exclude all other electrons of the same spin from the space that it occupies. However, an electron of opposite spin may enter this space. This space from which other electrons of the same spin tend to be excluded is called the *orbital* of the electron.

Because electrons have a negative charge, an electrostatic repulsion between electrons will also be present. The charge and spin reinforce one another to keep electrons of the same spin in separate orbitals. Electrons of opposite spin also undergo electrostatic repulsions, but exclusion of an electron having one spin from the space occupied by an electron of another spin does not come into play. Repulsion of like charges also tends to prevent the orbitals of electrons of opposite spins from coinciding. Thus, unless some other factor is present, all electrons tend to avoid one another.

In 1924, the Austrian physicist Wolfgang Pauli proposed the following hypothesis, which has become known as the Pauli exclusion principle.

Table 3.6

Allowed Combinations of Quantum Numbers for the n = 1, 2, 3 or 4 Shells

n	l	m_l	Subshell Notation	Number of Orbitals in Subshell	Number of Electrons Needed to Fill Subshell
1	0	0	1s	1	2 total in 1st shell = 2
2	0	0	2s	1	2
2	1	1, 0, −1	2p	3	6 total in 2nd shell = 8
3	0	0	3s	1	2
3	1	1, 0, −1	3p	3	6
3	2	2, 1, 0, −1, −2	3d	5	10 total in 3rd shell = 18
4	0	0	4s	1	2
4	1	1, 0, −1	4p	3	6
4	2	2, 1, 0, −1, −2	4d	5	10
4	3	3, 2, 1, 0, −1, −2, −3	4f	7	14 total in 4th shell = 32

Instructors who do not cover quantum numbers may omit the discussion of m_s in the last two paragraphs of this section.

Filling order is included here because it is the way chemists think and talk about electron configuration. Instructors using this should realize that there is an inconsistency between filling order and ionization energies obtained from photoelectron spectroscopy.

No more than two electrons can occupy an orbital, and, if there are two electrons in an orbital, the spins of the electrons must be paired.

The concept of the spin of an electron and the Pauli exclusion principle provide the last step needed to complete our model of the structure of the atom.

The electrons in a shell or a subshell occupy three-dimensional orbitals in space. An orbital can contain either one or two electrons. If there are two electrons in the orbital, they must have different spins. The different spins are described in terms of a **spin quantum number** (m_s). The allowed values for the spin quantum number are +1/2 and −1/2.

The allowed combinations of n, l, and m_l quantum numbers for the orbitals in the first four shells are given in Table 3.6. For each of these orbitals, there are two allowed values of the spin quantum number m_s. As a result, 2 electrons can occupy the 1s, 2s, 3s, or 4s orbitals; 6 electrons can occupy the 2p, 3p, or 4p orbitals; 10 electrons can occupy the 3d or 4d orbitals; and 14 electrons can occupy the 4f orbitals.

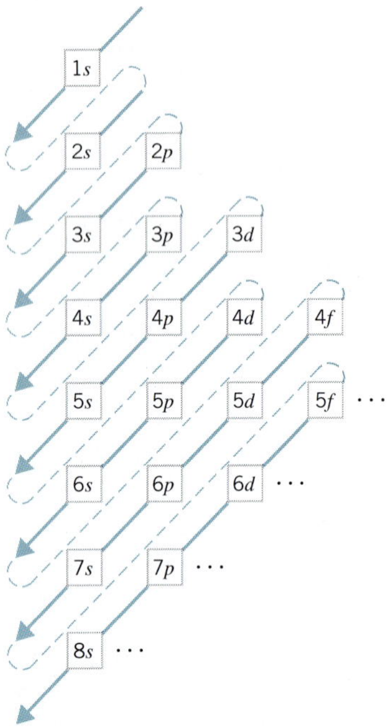

Fig. 3.23 The sequence of atomic orbitals can be predicted by following the arrows in this diagram.

3.16 Predicting Electron Configurations

It is clear from the electron configurations of the first 36 elements in the periodic table that they follow a regular pattern. These configurations can be predicted from the diagram in Figure 3.23. The diagram is read by following the arrows, starting at the top of the first line and proceeding on to the second, third, and fourth lines, and so on.

Exercise 3.5

Predict the electron configuration for a neutral tin atom (Sn, Z = 50).

Solution

We start by predicting the order of atomic orbitals from the diagram in Figure 3.23. We then add electrons to these orbitals, starting with the 1s orbital, until all 50 electrons have been included.

The electron configuration for an atom can be written by remembering that each orbital can hold two electrons. Because an *s* subshell contains only one orbital, only two electrons can be added to this subshell. There are three orbitals in a *p* subshell, however, so a set of *p* orbitals can hold up to six electrons. There are five orbitals in a *d* subshell, so a set of *d* orbitals can hold up to ten electrons. The following is the complete electron configuration for tin:

$$\text{Sn } (Z = 50) \qquad 1s^2\, 2s^2\, 2p^6\, 3s^2\, 3p^6\, 4s^2\, 3d^{10}\, 4p^6\, 5s^2\, 4d^{10}\, 5p^2$$

A list of electron configurations of the elements is found in Appendix B, Table B.15.

The order-of-filling diagram is just a device that enables students to write electron configurations. It should not be interpreted as giving the correct sequencing of energy levels.

There are many mnemonic devices for the order of filling of atomic orbitals. We have included one such diagram (Figure 3.23). It is important to note that the energy of the levels as determined from PES is not the same as the order of filling given by this or other similar diagrams.

*The order of filling for Ti is 1*s *2*s *2*p *3*s *3*p *4*s *3*d*. This order of filling often suggests to students that the 3*d *energy state is higher in energy than the 4*s *state, resulting in confusion when told that the 4*s *electron ionizes first. See* Instructor's Manual.

3.17 Electron Configurations and the Periodic Table

Atoms in Group VIIIA, such as He, Ne, and Ar, have electron configurations with filled shells of orbitals. By convention, we therefore write abbreviated electron configurations in terms of the number of electrons beyond the previous element with a filled *s* and *p* subshell electron configuration. The electron configuration of lithium, for example, could be written as follows.

$$\text{Li } (Z = 3) \qquad [\text{He}]\, 2s^1$$

When the electron configurations of the elements are arranged so that we can compare elements in the horizontal rows of the periodic table, we find that these rows correspond to the filling of shells or subshells. The second row, for example, contains elements in which the 2*s* and 2*p* subshells in the $n = 2$ shell are filled.

Li (Z = 3)	[He] $2s^1$
Be (Z = 4)	[He] $2s^2$
B (Z = 5)	[He] $2s^2\, 2p^1$
C (Z = 6)	[He] $2s^2\, 2p^2$
N (Z = 7)	[He] $2s^2\, 2p^3$
O (Z = 8)	[He] $2s^2\, 2p^4$
F (Z = 9)	[He] $2s^2\, 2p^5$
Ne (Z = 10)	[He] $2s^2\, 2p^6$

There is an obvious pattern within the vertical columns, or groups, of the periodic table as well. The elements in a group often have similar configurations for their outermost electrons. This relationship can be seen by looking at the electron configurations of elements in columns on either side of the periodic table.

Group IA		*Group VIIA*	
H	$1s^1$		
Li	[He] $2s^1$	F	[He] $2s^2\, 2p^5$
Na	[Ne] $3s^1$	Cl	[Ne] $3s^2\, 3p^5$
K	[Ar] $4s^1$	Br	[Ar] $4s^2\, 3d^{10}\, 4p^5$
Rb	[Kr] $5s^1$	I	[Kr] $5s^2\, 4d^{10}\, 5p^5$
Cs	[Xe] $6s^1$	At	[Xe] $6s^2\, 4f^{14}\, 5d^{10}\, 6p^5$

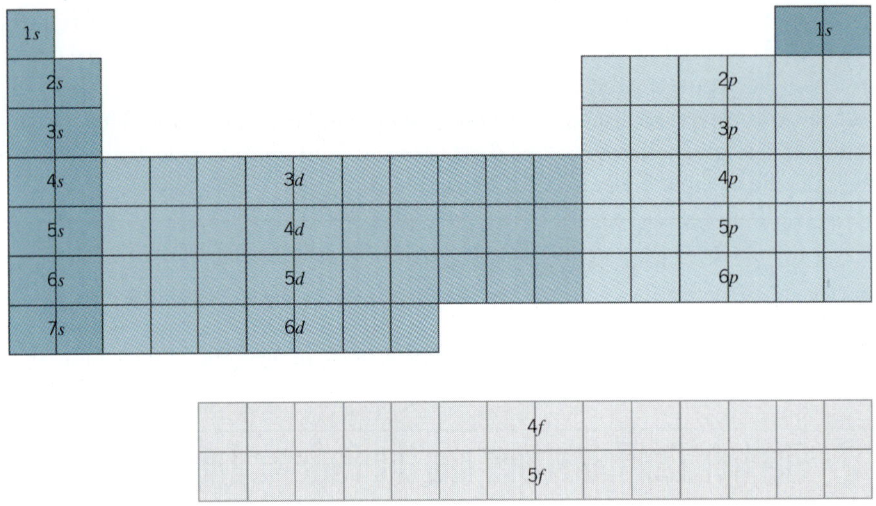

Fig. 3.24 The periodic table reflects the order in which atomic orbitals are filled. The *s* orbitals are filled in the two columns on the far left, and the *p* orbitals are filled in the six columns on the right. The *d* orbitals are filled along the transition between the *s* and *p* orbitals. The *f* orbitals are filled in the two long rows of elements at the bottom of the table.

Figure 3.24 shows the relationship between the periodic table and the orbitals being filled during the process by which the electron configuration is generated. The 2 columns on the left side of the periodic table correspond to the filling of an *s* orbital. The next 10 columns include elements in which the five orbitals in a *d* subshell are filled. The 6 columns on the right represent the filling of the three orbitals in a *p* subshell (except for He, which has only one occupied *s* orbital). Finally, the 14 columns at the bottom of the table correspond to the filling of the seven orbitals in an *f* subshell.

Elements are arranged by groups in the periodic table according to their chemical and physical properties. As shown above, elements in the same group have similar electron configurations for their outermost electrons. The group properties are a consequence of the outer electron configurations. Indeed, many of the properties of any atom are closely related to its electron configuration.

Exercise 3.6

Predict the electron configurations for calcium (Z = 20) and zinc (Z = 30) from their positions in the periodic table.

Solution

Calcium is in the second column and the fourth row of the table. The second column corresponds to the filling of an *s* orbital. The 1*s* orbital is filled in the first row, the 2*s* orbital in the second row, and so on. By the time we get to the fourth row, we are filling the 4*s* orbital. Calcium therefore has all of the electrons of argon, plus a filled 4*s* orbital.

$$\text{Ca } (Z = 20) \qquad [\text{Ar}]\ 4s^2$$

Zinc is the tenth element in the region of the periodic table where *d* orbitals are filled. Zinc therefore has a filled subshell of *d* orbitals. The only question is: Which set of *d* orbitals is filled? Although zinc is in the fourth row of the periodic table, the first time *d* orbitals occur is in the $n = 3$ shell. The following is therefore the abbreviated electron configuration for zinc.

$$\text{Zn } (Z = 30) \qquad [\text{Ar}]\ 4s^2\ 3d^{10}$$

3.18 Electron Configurations and Hund's Rules

It is sometimes useful to illustrate the electron configuration of an atom in terms of an **orbital diagram** that shows the spins of the electrons. By convention, electron spins are represented by arrows pointing up or down. Each orbital is represented by a single line or box.

Consider the electrons in the $n = 2$ shell of a boron atom, for example.

$$\text{B } (Z = 5) \qquad [\text{He}]\ 2s^2\ 2p^1 \qquad \underset{2s}{\underline{\uparrow\downarrow}}\ \ \underset{}{\underline{\uparrow}}\ \underset{2p}{\underline{\ \ }}\ \underline{\ \ }$$

The $2s$ orbital is designated by a single line, whereas three lines are needed to represent the three $2p$ orbitals. There is only one electron in the $2p$ subshell, so we add a single electron to one of the three orbitals in this subshell, with the spin of that electron shown in an arbitrarily chosen direction.

A problem arises when we try to adapt the diagram to the next element, carbon. Where do we put the second electron? Do we put it in the same orbital, to form a pair of electrons of opposite spin? Do we put it in a different orbital in the subshell with the same spin as the first electron? Or do we put it in a different orbital, but with the spins of the two electrons paired? The German physicist Friedrich Hund found that the most stable arrangement of electrons can be predicted from the following rules.

- One electron is added to each orbital in a subshell before two electrons are added to any orbital in the subshell.
- Electrons are added to a subshell with the same spin until each orbital in the subshell has at least one electron.

Hund's rules are a consequence of the spin and repulsion of electrons discussed earlier. Electrons tend to avoid one another and can exist in the same orbital only if their spins are paired.

According to **Hund's rules,** the electrons in the $2p$ subshell on a carbon atom occupy two different orbitals and have the same spin. These electrons can be represented as follows:

$$\text{C } (Z = 6) \qquad [\text{He}]\ 2s^2\ 2p^2 \qquad \underline{\uparrow}\ \underset{2p}{\underline{\uparrow}}\ \underline{\ \ }$$

When we get to N (Z = 7), we have to put one electron into each of the three orbitals in the $2p$ subshell, all with the same spins.

$$\text{N } (Z = 7) \qquad [\text{He}]\ 2s^2\ 2p^3 \qquad \underline{\uparrow}\ \underset{2p}{\underline{\uparrow}}\ \underline{\uparrow}$$

Because each orbital in the $2p$ subshell now contains one electron, the next electron added to the subshell must have the opposite spin, thereby filling one of the $2p$ orbitals.

$$\text{O } (Z = 8) \qquad [\text{He}]\ 2s^2\ 2p^4 \qquad \underline{\uparrow\downarrow}\ \underset{2p}{\underline{\uparrow}}\ \underline{\uparrow}$$

> **CHECKPOINT**
>
> How many unpaired electrons are there in each of the following atoms? Which of these atoms would have magnetic properties? (See Section 3.15.)
>
> (a) carbon (b) nitrogen
>
> (c) neon (d) fluorine

The ninth electron fills a second orbital in the subshell.

$$\text{F } (Z = 9) \qquad [\text{He}]\, 2s^2\, 2p^5 \qquad \underset{2p}{\underline{\uparrow\downarrow}\ \underline{\uparrow\downarrow}\ \underline{\uparrow\ }}$$

The tenth electron completes the $2p$ subshell.

$$\text{Ne } (Z = 10) \qquad [\text{He}]\, 2s^2\, 2p^6 \qquad \underset{2p}{\underline{\uparrow\downarrow}\ \underline{\uparrow\downarrow}\ \underline{\uparrow\downarrow}}$$

3.19 The Sizes of Atoms: Metallic Radii

The size of an atom influences many of the chemical and physical properties of the atom. It therefore would be useful to have an accurate measurement of the size of an isolated atom. Unfortunately, the size of an isolated atom can't be measured because we can't determine the location of the electrons that surround the nucleus. We therefore estimate the size of an atom by assuming that the radius of the atom is equal to one-half the distance between equivalent nuclei of adjacent atoms in a solid. This technique is best suited to elements that are metals, which form solids composed of extended planes of atoms of that element. The results of these measurements are known as **metallic radii.**

Because more than 75% of the elements are metals, metallic radii are available for most elements in the periodic table. Figure 3.25 shows the relationship between the metallic radii for elements in Groups IA and IIA. There are two general trends in the data.

- Atoms become *larger* as we go down a column of the periodic table.
- Atoms become *smaller* as we go from left to right across a row of the periodic table.

The first trend can be explained by looking at the electron configurations of the atoms. As we go down the periodic table, electrons are placed in larger and larger subshells but the core charge remains the same. When this happens, the size of the atom should increase.

The second trend is a bit surprising. We might expect atoms to become larger as we go across a row of the periodic table because each element has one more electron than the preceding element. But the additional electrons are added to the same shell. Because the number of protons in the nucleus increases as we

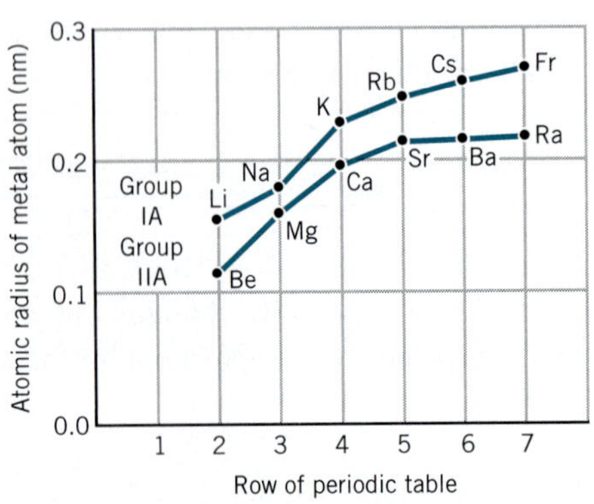

Fig. 3.25 Metallic radii increase as we go down a column of the periodic table. With rare exceptions, they decrease from left to right across a row of the table.

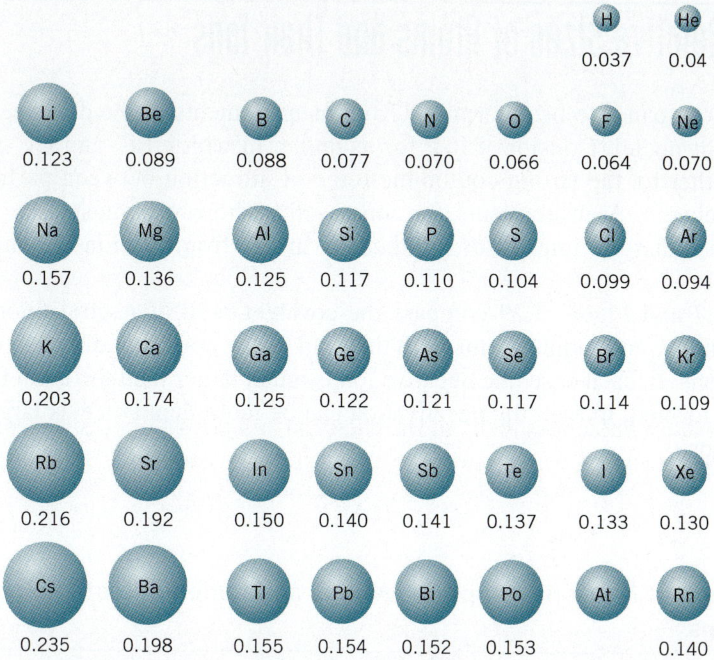

Fig. 3.26 Covalent radii for the main-group elements. Atoms become larger as we go down a column of the periodic table and smaller as we go from left to right across a row of the table.

go across a row of the table, the core charge increases and the force of attraction between the nucleus and the electrons that surround it also increases. The nucleus therefore tends to pull each electron in closer, and the atoms become smaller.

3.20 The Sizes of Atoms: Covalent Radii

The size of an atom can also be estimated by measuring the distance between adjacent atoms in compounds that are neither ionic or metallic. These compounds are called covalent compounds. The **covalent radius** of a chlorine atom, for example, is assumed to be equal to one-half the distance between the nuclei of the atoms in a Cl_2 molecule.

The covalent radii of the main-group elements are given in Figure 3.26. These data confirm the trends observed for metallic radii. Atoms *become larger* as we go down a column of the periodic table, and they *become smaller* as we go across a row of the table.

Table B.4 in the appendix contains covalent and metallic radii for a number of elements. The covalent radius for an element is usually a little smaller than the metallic radius. This can be explained by noting that covalent bonds tend to squeeze the atoms together, as shown in Figure 3.27.

➤ CHECKPOINT

What is the difference between covalent and metallic radii?

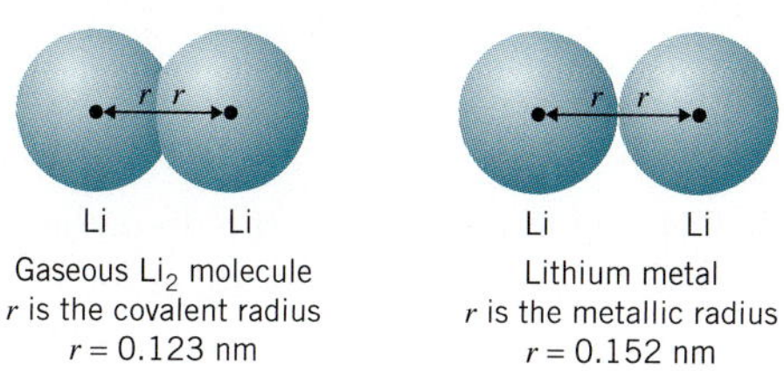

Fig. 3.27 The radii of atoms and ions can be determined by measuring the distance between adjacent nuclei. The covalent radius of lithium can be determined by measuring the internuclear distance in the gaseous Li_2 molecule, which exists only at very high temperatures. Li_2 is covalently bonded, and the Li atoms are pulled tightly together. The metallic radius of lithium can be determined from measurements of solid lithium metal. In lithium metal each lithium atom is surrounded by other lithium atoms, all immersed in a sea of electrons.

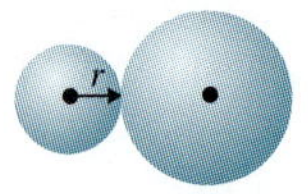

Fig. 3.28 The ionic radius of lithium can be determined from measurements of an ionic compound such as lithium chloride.

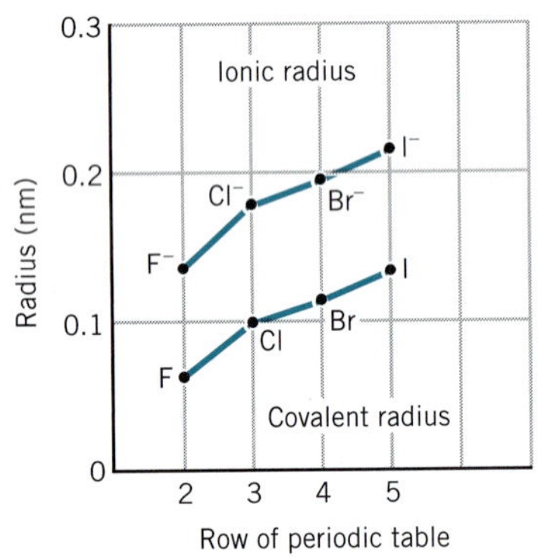

Fig. 3.29 A comparison of the radii of the F^-, Cl^-, Br^-, and I^- ions with the covalent radii of the corresponding neutral atoms. In each case, the negatively charged ion is larger than the neutral atom.

3.21 The Relative Sizes of Atoms and Their Ions

The radii of ions can also be determined from measurements of the distance between adjacent nuclei. Consider LiCl, for example, in which Li^+ and Cl^- ions are held together by the strong coulombic force of attraction between particles of opposite charge. Measurements for ionic solids, however, must take into account the fact that Li^+ ions differ significantly in size from Cl^- ions, as shown in Figure 3.28.

Table 3.7 and Figure 3.29 compare the covalent radii of neutral fluorine, chlorine, bromine, and iodine atoms with the radii of the corresponding F^-, Cl^-, Br^-, and I^- ions. In each case, the negative ion is much larger than the atom from which it was formed. In fact, the negative ion can be more than twice as large as the neutral atom.

Table 3.7

Covalent Radii of Neutral Group VIIA Atoms and Ionic Radii of Their Negative Ions

Element	Covalent Radius (nm)	Ion	Ionic Radius (nm)
F	0.064	F^-	0.136
Cl	0.099	Cl^-	0.181
Br	0.1142	Br^-	0.196
I	0.1333	I^-	0.216

The only difference between an atom and its ions is the number of electrons that surround the nucleus. A neutral chlorine atom, for example, contains 17 electrons, while a Cl^- ion contains 18 electrons. There is no room in the shells closest to the nucleus for an additional electron. The electron is, therefore, found in the outermost shell.

$$\text{Cl} \quad [\text{Ne}]\ 3s^2\ 3p^5 \qquad \text{Cl}^- \quad [\text{Ne}]\ 3s^2\ 3p^6$$

Because the nucleus can't hold the 18 electrons in the Cl^- ion as tightly as the 17 electrons in the neutral atom, the negative ion is significantly larger than the atom from which it forms.

Extending this line of reasoning suggests that positive ions should be smaller than the atoms from which they are formed. The electrons removed from an atom to form a positive ion are the electrons which are most loosely held, the electrons in the outer shell. The 11 protons in the nucleus of an Na^+ ion, for example, should be able to hold the 10 electrons on the ion more tightly than the 11 electrons on a neutral sodium atom. Removing an electron from the Na atom gives a sodium ion with electrons in only two shells. The Na^+ ion therefore should be much smaller than a sodium atom.

$$\text{Na} \quad 1s^2\ 2s^2\ 2p^6\ 3s^1 \qquad \text{Na}^+ \quad 1s^2\ 2s^2\ 2p^6$$

Table 3.8 and Figure 3.30 provide data to test this hypothesis. Here the covalent radii for neutral atoms of the Group IA elements are compared with the ionic radii for the corresponding positive ions. In each case, the positive ion is much smaller than the atom from which it forms.

Table 3.8

Covalent Radii of Neutral Group IA Atoms and Ionic Radii of Their Positive Ions

Element	Covalent Radius (nm)	Ion	Ionic Radius (nm)
Li	0.123	Li^+	0.068
Na	0.157	Na^+	0.095
K	0.2025	K^+	0.133
Rb	0.216	Rb^+	0.148
Cs	0.235	Cs^+	0.169

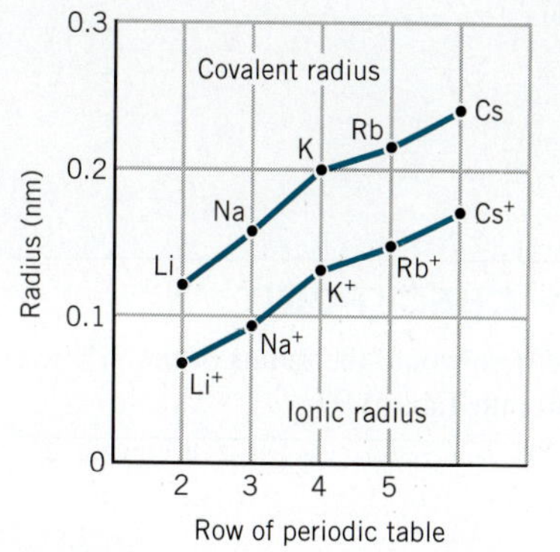

Fig. 3.30 A comparison of the radii of the Li^+, Na^+, K^+, Rb^+, and Cs^+ ions with the covalent radii of the corresponding neutral atoms. In each case, the positively charged ion is smaller than the neutral atom.

Exercise 3.7

Compare the sizes of neutral sodium and chlorine atoms and their Na^+ and Cl^- ions.

Solution

A neutral sodium atom is significantly larger than a neutral chlorine atom because chlorine has a larger core charge than sodium and their outer electrons are in the same shell.

$$\underset{(0.157\text{ nm})}{\text{Na}} \gg \underset{(0.099\text{ nm})}{\text{Cl}}$$

But an Na^+ ion is only one-half the size of a Cl^- ion because removal of an electron from sodium gives an ion with electrons in only two shells.

$$\underset{(0.095\text{ nm})}{\text{Na}^+} \ll \underset{(0.181\text{ nm})}{\text{Cl}^-}$$

These particles therefore increase in size in the following order:

$$\text{Na}^+ \approx \text{Cl} \ll \text{Na} \ll \text{Cl}^-$$

Tabulated values of ionic and covalent radii can be used to estimate the distance between atoms and ions in compounds. These distances are important because the properties of a compound are often dictated by how far one atom or ion is from another. For example, from the ionic radii in Tables 3.7 and 3.8 we can estimate the distance between Na^+ and Cl^- in sodium chloride to be 0.276 nm. For the distance between Na^+ and Br^- in sodium bromide we find 0.291 nm. It is no coincidence that NaCl melts at a higher temperature than NaBr. More detailed discussion of how properties depend on the sizes of the particles composing compounds is the subject of subsequent chapters.

3.22 Patterns in Ionic Radii

The ionic radii in Tables 3.7 and 3.8 confirm one of the patterns observed for both metallic and covalent radii: Atoms become larger as we go down a column of the periodic table. We can examine trends in ionic radii across a row of the periodic table by comparing data for atoms and ions that are **isoelectronic.** By definition, isoelectronic atoms or ions have the same number of electrons. Table 3.9 summarizes data on the radii of a series of isoelectronic ions and atoms of second- and third-row elements.

➤ **CHECKPOINT**

Where would the radius of the Si^{4+} ion fit into Table 3.9?

The data in Table 3.9 can be explained by noting that these atoms or ions all have 10 electrons, but the number of protons in the nucleus increases from 6 in the C^{4-} ion to 13 in the Al^{3+} ion. As the charge on the nucleus becomes larger, the nucleus can hold the same number of electrons more tightly. As a result, the atoms or ions become significantly smaller—in this series, by a factor of 5 from C^{4-} to Al^{3+}.

Table 3.9

Radii for Isoelectronic Second-Row and Third-Row Atoms or Ions

Atom or Ion	Radius (nm)	Electron Configuration
C^{4-}	0.260	$1s^2\ 2s^2\ 2p^6$
N^{3-}	0.171	$1s^2\ 2s^2\ 2p^6$
O^{2-}	0.140	$1s^2\ 2s^2\ 2p^6$
F^-	0.136	$1s^2\ 2s^2\ 2p^6$
Ne	0.112	$1s^2\ 2s^2\ 2p^6$
Na^+	0.095	$1s^2\ 2s^2\ 2p^6$
Mg^{2+}	0.065	$1s^2\ 2s^2\ 2p^6$
Al^{3+}	0.050	$1s^2\ 2s^2\ 2p^6$

3.23 Second, Third, Fourth, and Higher Ionization Energies

Sodium forms Na^+ ions, magnesium forms Mg^{2+} ions, and aluminum forms Al^{3+} ions. Why doesn't sodium form Na^{2+} ions, or even Na^{3+} ions? The answer can be obtained from data for the second, third, and higher ionization energies of the element.

As we have seen, more than one ionization energy can be measured for sodium, depending on whether the electron is removed from the $1s$, $2s$, $2p$, or $3s$ orbitals on the atom. The first ionization energy represents the energy it takes to remove the *outermost* electron from a neutral atom in the gas phase.

$$\mathrm{Na}(g) + \text{energy} \longrightarrow \mathrm{Na}^+(g) + e^-$$

For sodium this is the energy required to remove the electron from the $3s$ orbital.

$$\begin{array}{ll} \mathrm{Na} & 1s^2\ 2s^2\ 2p^6\ 3s^1 \\ \mathrm{Na}^+ & 1s^2\ 2s^2\ 2p^6 \end{array}$$

The *second ionization energy* of sodium is the energy it would take to remove another electron to form an Na^{2+} ion in the gas phase.

$$\mathrm{Na}^+(g) + \text{energy} \longrightarrow \mathrm{Na}^{2+}(g) + e^-$$

$$\mathrm{Na}^{2+} \quad 1s^2\ 2s^2\ 2p^5$$

The *third ionization energy* of sodium represents the process by which the Na^{2+} ion is converted to an Na^{3+} ion.

$$\mathrm{Na}^{2+}(g) + \text{energy} \longrightarrow \mathrm{Na}^{3+}(g) + e^-$$

$$\mathrm{Na}^{3+} \quad 1s^2\ 2s^2\ 2p^4$$

Table 3.10

First, Second, Third, and Fourth Ionization Energies (kJ/mol) of Gas Phase Atoms of Sodium, Magnesium, and Aluminum

	First *IE*	Second *IE*	Third *IE*	Fourth *IE*
Na	495.8	4562.4	6912	9,543
Mg	737.7	1450.6	7732.6	10,540
Al	577.6	1816.6	2744.7	11,577

➤ CHECKPOINT

How do the second, third, and higher ionization energies differ from ionization energies obtained from PES?

The energy required to form an Na^{3+} ion from an Na atom in the gas phase is the sum of the first, second, and third ionization energies of the element.

A complete set of data for the ionization energies of the elements is given in Table B.5 in the appendix. For the moment, let's look at the first, second, third, and fourth ionization energies of sodium, magnesium, and aluminum listed in Table 3.10.

It doesn't take much energy to remove one electron from a sodium atom to form an Na^{+} ion with a filled-shell electron configuration. Once this is done, it takes almost 10 times as much energy to remove a second electron because the next available electrons are in the $n = 2$ shell, which is closer to the nucleus. Because it takes so much energy to remove the second electron, sodium generally forms compounds that contain Na^{+} ions rather than Na^{2+} or Na^{3+} ions.

A similar pattern is observed when the ionization energies of magnesium are analyzed. The first ionization energy of magnesium is larger than that of sodium because magnesium has one more proton in its nucleus to hold onto the electrons in the 3*s* orbital.

$$\text{Mg} \qquad [\text{Ne}]\ 3s^2$$

The second ionization energy of Mg is larger than the first because it always takes more energy to remove an electron from a positively charged ion than from a neutral atom. The third ionization energy of magnesium is enormous, however, because an electron would have to be removed from the $n = 2$ shell to form an Mg^{3+} ion.

The same pattern can be seen in the ionization energies of aluminum. The first ionization energy of aluminum is smaller than that of magnesium because it involves removing an electron from a 3*p* rather than a 3*s* orbital. The second ionization energy of aluminum is larger than the first, and the third ionization energy is even larger. Although it takes a considerable amount of energy to remove three electrons from an aluminum atom to form an Al^{3+} ion, the energy needed to remove a fourth electron is astronomical. Thus, it would be a mistake to look for an Al^{4+} ion as the product of a chemical reaction.

Exercise 3.8

Predict the group in the periodic table in which an element with the following ionization energies would most likely be found.

1st *IE* = 786 kJ/mol

2nd *IE* = 1,577

3rd *IE* = 3,232

4th *IE* = 4,355

5th *IE* = 16,091

6th *IE* = 19,784

AVEE is the average one-electron energy of the valence electrons in ground state free atoms. As Z increases across a row the energy difference between subshells increases, the electrons are held more tightly, and AVEE increases.

Down a group the spacing between the subshells decreases, electrons are less strongly held, and AVEE decreases. Thus AVEE measures the attraction of an atom for electrons and the spacing of an atom's energy levels.

➤ CHECKPOINT

What is the maximum charge under normal conditions for an atom from Group IVA?

AVEEs derived from PES data have a number of useful properties. In addition to being a generalized electronegativity, AVEEs provide a rationale for many atomic chemical properties which are derivable from the periodic table.[1,2]

A large AVEE means strong attraction for electrons and large energy-level spacings. AVEE is a fundamental property of an atom that can be derived from data already used to relate electron configurations of atoms to their positions in the periodic table. Thus AVEE is fundamentally and intimately tied to the trends in the properties of the elements.

Be has a small AVEE, and its electrons are free to be delocalized onto neighboring atoms. The difference between the energies of the 2s and 2p orbitals is small enough to allow the formation of metallic bonds with neighboring atoms. This is why Be ($2s^2$) and H ($1s^2$) have such different chemistries despite both having filled subshells.

[1]L. C. Allen, *J. Am. Chem. Soc.* **114,** 1510 (1992).
[2]L. C. Allen, *J. Am. Chem. Soc.* **111,** 9003 (1989).

Solution

The gradual increase in the energy needed to remove the first, second, third, and fourth electrons from this element is followed by an abrupt increase in the energy required to remove one more electron. This is consistent with an element that has four electrons in the outermost shell, and we might expect the element to be in Group IVA of the periodic table. These data are in fact the ionization energies of silicon.

$$\text{Si} \quad [\text{Ne}]\, 3s^2\, 3p^2$$

The trends in the ionization energies of the elements can be used to explain why elements on the left side of the periodic table are more likely than those on the right to form positive ions. The ionization energies of elements on the left side of the table are much smaller than those of elements on the right. Consider the first ionization energies for sodium and chlorine, for example.

$$\text{Na} \quad \text{1st } IE = 495.8 \text{ kJ/mol}$$
$$\text{Cl} \quad \text{1st } IE = 1251.1 \text{ kJ/mol}$$

Elements on the left side of the periodic table are therefore more likely to form positive ions.

Trends in ionization energies can also be used to explain why the maximum positive charge found on atoms under normal conditions for the main-group elements is equal to the group number of the element. The number of valence electrons is equal to the group number, so the maximum positive charge on an ion is also equal to the group number. Because aluminum is in Group IIIA, for example, it can lose only three electrons before an inner subshell is reached. Thus, the maximum positive charge on an aluminum ion is +3.

3.24 Average Valence Electron Energy (AVEE)[3]

Ionization energies provide a measure of how tightly the electrons are held in an isolated atom. It would be useful to have a single quantity that reflects the *average* ionization energy of the valence electrons on an atom. This quantity, which is known as the **average valence electron energy (AVEE),** can be calculated from the ionization energies for the valence electrons obtained by photoelectron spectroscopy. Because there are different numbers of electrons in the various subshells, the AVEE is calculated as a weighted average.

Consider carbon, for example, which has two electrons in the valence 2*s* subshell and two electrons in the valence 2*p* subshell. The ionization energies of electrons from the 2*s* subshell (IE_s = 1.72 MJ/mol) and the 2*p* subshell (IE_p = 1.09 MJ/mol) are given in Table 3.4. The average valence electron energy for a carbon atom can therefore be calculated as follows.

$$\text{AVEE}_\text{C} = \left[\frac{(2 \times IE_s) + (2 \times IE_p)}{2+2}\right] = \left[\frac{(2 \times 1.72 \text{ MJ/mol}) + (2 \times 1.09 \text{ MJ/mol})}{4}\right]$$
$$= 1.41 \text{ MJ/mol}$$

[3]L. C. Allen, *Journal of the American Chemical Society,* **111,** 9003 (1989).

H							He
1.31							2.37
Li	Be		B	C	N	O	F	Ne
0.52	0.90		1.17	1.41	1.82	1.91	2.31	2.73
Na	Mg		Al	Si	P	S	Cl	Ar
0.50	0.74		0.92	1.13	1.39	1.35	1.59	1.85
K	Ca	Sc	Ga	Ge	As	Se	Br	Kr
0.42	0.59	0.68	1.00	1.07	1.26	1.3	1.53	1.69
Rb	Sr	Y	In	Sn	Sb	Te	I	Xe
0.40	0.55	0.57	0.94	1.04	1.13	1.2	1.35	1.47

Fig. 3.31 Average valence electron energies (AVEE) for the main-group elements, in megajoules per mole (MJ/mol) [L. C. Allen, *Journal of the American Chemical Society,* **111,** 9003 (1989)].

As the size of the atoms increases, the AVEE decreases, the electrons are held less tightly, and the spacing between energy levels decreases. As the s, p, and d levels become closer in energy, they mix in many different combinations to give nondirectional metallic bonding. Thus carbon is a nonmetal and lead is a metal.

Fluorine, on the other hand, would have an AVEE that is significantly larger.

$$\text{AVEE}_\text{F} = \left[\frac{(2 \times IE_s) + (5 \times IE_p)}{2 + 5}\right] = \left[\frac{(2 \times 3.88 \text{ MJ/mol}) + (5 \times 1.68 \text{ MJ/mol})}{7}\right]$$
$$= 2.31 \text{ MJ/mol}$$

The AVEE values for the main-group elements calculated from photoelectron spectroscopy data are given in Figure 3.31.

With the exception of phosphorus, there is a systematic increase in AVEE from left to right across each row of the periodic table. Because the charge on the nucleus steadily increases as we go across each row, and the size of the atom gradually decreases, the valence electrons on each atom are held more tightly as we go across the row.

In addition, as the atomic number increases across a row, the energy difference between the valence subshells increases. As a general rule, as we proceed down a group electrons are less strongly held and the energy difference between valence subshells decreases and AVEE decreases. Thus, AVEE measures two important quantities: the attraction of an atom for its electrons and the spacing of its valence energy levels. These trends are shown in Figure 3.32, which shows the energies of the 2*s* and 2*p* electrons in B, C, N, O, and F. The difference in energy between the 2*s* and 2*p* electrons becomes larger, as shown by the increased separation of the lines in Figure 3.32.

The metals on the left side of the periodic table have AVEE values that are relatively small compared with the nonmetals on the right side of the table. Sodium and magnesium, for example, have AVEE values of 0.50 and 0.74 MJ/mol, respectively, whereas the AVEE values for O and Cl are 1.91 and 1.59 MJ/mol. Thus sodium and magnesium are more likely to lose electrons to form positive ions than are O and Cl, and O and Cl are more likely to add electrons to form negative ions than are Na and Mg.

➤ CHECKPOINT

Using Table 3.4, calculate the difference in energy between the 2*s* and 2*p* subshells of oxygen. Using Table 3.5, calculate the difference in energy between the 3*s* and 3*p* subshells of sulfur. Predict whether S or Se would have the greater difference in energy between their outermost *s* and *p* subshells.

➤ CHECKPOINT

Why do AVEE values increase from left to right across a period and from bottom to top of a group? Is there another periodic property that follows those trends?

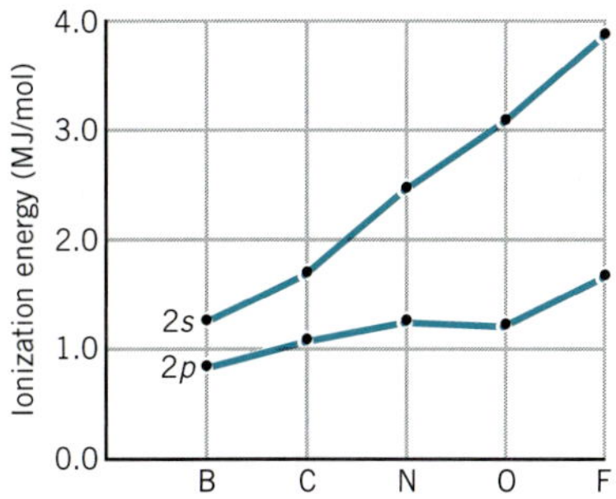

Fig. 3.32 Ionization energies of the electrons in the 2*s* and 2*p* subshells of B, C, N, O, and F. Two trends can be observed: The ionization energies of the subshells increase (the electrons are more tightly held) as we move across the period, and the difference in energy between the *s* and the *p* electrons also increases as we go across the period.

3.25 AVEE and Metallicity

Because the average valence electron energy provides a measure of how tightly an atom holds onto its valence electrons and the energy gap between valence electron subshells, it can be used to explore the dividing line between the metals and nonmetals in the periodic table. The semimetals (metalloids) that lie along the stair-step

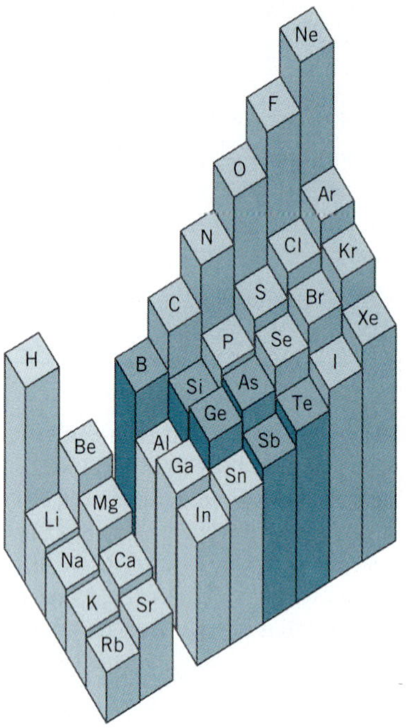

Fig. 3.33 Three-dimensional plot of the average valence electron energies (AVEE) of the main-group elements versus position in the periodic table. The AVEE is a measure of how tightly atoms hold onto their valence electrons and the energy gap between valence subshells.

line separating the metals from the nonmetals in the periodic table all have similar AVEE values. B, Si, Ge, As, Sb, and Te all have AVEE values that lie between 1.07 and 1.26 MJ/mol, and they are the only elements in the periodic table that have values within this range as shown in Figure 3.33.

The AVEE of an atom of a given element therefore can be used to decide whether the element is a metal or nonmetal. Atoms with an AVEE value below 1.07 MJ/mol are metals, whereas those with an AVEE greater than 1.26 MJ/mol are nonmetals. Elements with AVEE values in the range of 1.07 to 1.26 MJ/mol have properties intermediate between those of metals and nonmetals and are known as the *semimetals,* or *metalloids.*

The power of AVEE data can be illustrated by considering the difference between the second and third versus the fourth and fifth rows of the periodic table. In the second and third rows, there is only one semimetal each (B and Si). In the fourth and fifth rows, there are two (Ge and As, and Sb and Te). If we tried to extend the pattern seen in the fourth and fifth rows of the periodic table to the second and third rows, we would predict that beryllium and aluminum would be semimetals. But the AVEE values for beryllium (0.90) and aluminum (0.92) clearly indicate that these elements should be metals. This is consistent with the observation that both elements exhibit all of the characteristic properties of a metal. They have a metallic luster; they are good conductors of heat and electricity; and they are both malleable and ductile.

In general, there are two factors that contribute to metallic behavior. Valence electrons must be easy to remove, and the energy separation between valence subshells must be small. AVEE is a convenient measure of both effects. Both are necessary to observe metallic behavior. The increasing trend for metallicity down a group also follows the AVEE values. Carbon is a nonmetal; but as we go down Group IVA through Si, Ge, Sn, and Pb, metallic behavior increases. This reflects two things: The electrons are becoming easier to remove, and the valence subshells are becoming closer in energy.

Key Terms

Amplitude
Angular quantum number
Atomic spectra
Average valence electron energy (AVEE)
Bohr model
Core charge
Covalent radii
Electromagnetic radiation
Electron configuration
Emission spectrum
First ionization energy
Frequency
Hund's rules
Ionic radius
Isoelectronic
Magnetic quantum number
Metallic radii
Nucleus
Orbital diagram
Orbitals
Particles
Photoelectron spectroscopy (PES)
Photons
Principle quantum number
Quantized
Quantum mechanical model
Shell
Shell model
Spectrum
Speed
Spin
Spin quantum number
Subshell
Transition metals
Valence electrons
Wavelength
Waves

Problems

Rutherford's Model of the Atom

1. According to Rutherford's model of the atom, where is the positive charge concentrated?
2. What does Rutherford's model predict about the relative size of the nucleus and the radius of the atom?
3. What is the nucleus composed of according to Rutherford's model?
4. A chemistry text published in 1922 proposed a model of the atom based on only two subatomic particles: electrons and protons. All of the protons and some of the

electrons were concentrated in the nucleus of the atom. The other electrons revolved around the nucleus. The number of protons in the nucleus was equal to the mass of the atom. The charge on the nucleus was equal to the number of protons minus the number of electrons in the nucleus. Enough electrons were then added to the atom to neutralize the charge on the nucleus. Use this model to calculate the number of protons and electrons in a neutral fluorine atom, F, and a fluoride ion, F^-. Compare this calculation with results obtained by assuming that the nucleus is composed of protons and neutrons.

Particles and Waves

5. Describe the difference between a particle and a wave.
6. To examine the relationship among the frequency, wavelength, and speed of a wave, imagine that you are sitting at a railroad crossing, watching a train that consists of 45-foot boxcars go by. Furthermore assume it takes 1.0 second for each boxcar to pass in front of your car. Calculate the "frequency" of a boxcar and the product of the 45-foot "wavelength" times this frequency. Convert the final answer into units of miles per hour.
7. An octave in a musical scale corresponds to a change in the frequency of a note by a factor of 2. If a note with a frequency of 440 Hz is an A, then the A one octave above this note has a frequency of 880 Hz. What happens to the wavelength of the sound as the frequency increases by a factor of 2? What happens to the speed at which the sound travels to your ear?
8. The human ear is capable of hearing sound waves with frequencies between about 20 and 20,000 hertz. If the speed of sound is 340.3 meters per second at sea level, what is the wavelength in meters of the longest wave the human ear can hear?

Light and Other Forms of Electromagnetic Radiation

9. Calculate the wavelength in meters of blue-green light that has a frequency of 5.0×10^{14} cycles per second.
10. Calculate the frequency of red light that has a wavelength of 700 nanometers.
11. Which has the longer wavelength, red light or blue light?
12. Which has the larger frequency, radio waves or microwaves?
13. In a magnetic field of 2.35 tesla, ^{13}C nuclei absorb electromagnetic radiation that has a frequency of 25.147 MHz. Calculate the wavelength of this radiation. In which region of the electromagnetic spectrum does the radiation fall?
14. The meter has been defined as 1,650,763.73 wavelengths of the orange-red line of the emission spectrum of ^{86}Kr. Calculate the frequency of this radiation. In what portion of the electromagnetic spectrum does the radiation fall?
15. Soap bubbles pick up colors because they reflect light with wavelengths equal to the thickness of the walls of the bubble. What frequency of light is reflected by a soap bubble 6 nanometers thick?
16. Methylene blue, $C_{16}H_{18}ClN_3S$, absorbs light most intensely at wavelengths of 668 and 609 nanometers. What color is the light absorbed by the dye?

Atomic Spectra

17. A pink-violet light is observed when current is passed through a glass tube containing hydrogen gas. When this light emerges from a prism, four narrow lines of visible light are seen (Table 3.1). What are the frequencies of these four lines?
18. Sodium salts give off a characteristic yellow-orange light when added to the flame of a Bunsen burner. The yellow-orange color is due to two narrow bands of radiation with wavelengths of 588.9953 and 589.5923 nm. Calculate the frequencies of these emission lines.

Quantization of Energy

19. When sodium salts are added to a flame, the flame burns yellow. Strontium salts give a red flame, and copper salts impart a blue-green flame. What does this suggest about the difference in these atoms? Could the flame color be used to identify these atoms?
20. Which has more energy, ultraviolet or infrared radiation?
21. Which has more energy, yellow light with a wavelength of 580 nm or green light with a wavelength of 660 nm?
22. List the four lines in the visible region of the emission spectrum of hydrogen in order of increasing energy. See Table 3.1.
23. Calculate the energy in joules of the radiation in the emission spectrum of the hydrogen atom that has a wavelength of 656.3 nm.
24. Calculate the energy in joules of a single particle of radiation broadcast by an amateur radio operator who transmits at a wavelength of 10 m.
25. Cl_2 molecules can dissociate to form chlorine atoms by absorbing electromagnetic radiation. It takes 243.4 kJ of energy to break the bonds in a mole of Cl_2 molecules. What is the wavelength of the radiation that has just enough energy to decompose a Cl_2 molecule to chlorine atoms? In what portion of the spectrum is this wavelength found?
26. What does the statement "light is quantized" mean?

The Bohr Model of the Atom

27. Why do hydrogen atoms emit or absorb discrete amounts of energy?
28. How does Planck's idea of quantization explain how light is emitted by an atom?
29. How does the Bohr model of the atom differ from Rutherford's model?
30. In the Bohr atom, what holds the electron and nucleus together?

31. As the distance between the electron and the nucleus increases, what happens to the force of attraction between them according to Bohr's model?

32. For two charged particles separated by a distance r, the force between the particles is given by Coulomb's law:

$$F = \frac{q_1 \times q_2}{r^2}$$

(a) If the particles are separated by an infinite distance, what is the magnitude of the force of attraction between them?

(b) If the signs of the charges on the particles are the same, is the sign of the force between them positive or negative?

(c) If the signs of the charges on the particles are not the same, is the sign of the force between them positive or negative?

(d) Is the sign of the force between the particles in the hydrogen atom positive or negative?

(e) Which of the following has the greatest force between the particles?

(i) An electron 1000 pm from another particle of charge +1 or an electron 500 pm from another particle of charge +1?

(ii) An electron 1000 pm from another particle of charge +1 or an electron 1000 pm from another particle of charge +2?

33. Describe the hydrogen atom according to the Bohr model.

34. (a) Which of the following arrangements has the greatest force of attraction between the charged species?

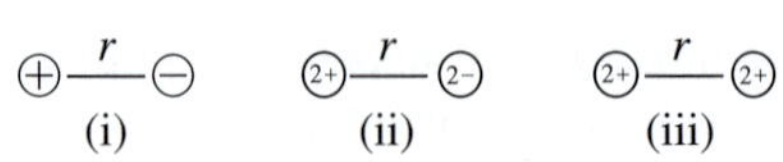

(b) For the following, which has the greatest force of attraction between charged species?

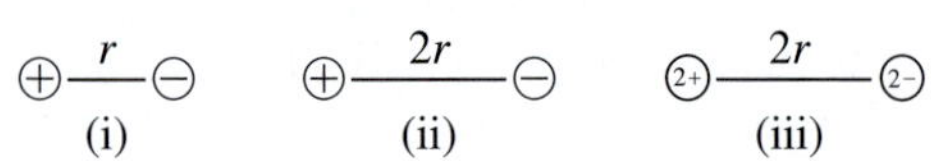

(c) For which of the following is the smallest amount of energy required to remove the negatively charged particle an infinite distance away from the positively charged particle?

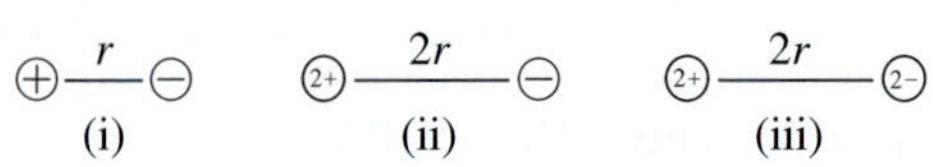

The Energy States of the Hydrogen Atom

35. Refer to Figure 3.5. How much energy is required to remove the electron completely from the hydrogen atom?

36. If the H-atom electron in the $n = 1$ state absorbs energy equivalent to 2.178×10^{-18} J, what energy state will it occupy?

37. How much energy in joules is required to change the energy state of the H-atom electron from $n = 2$ to $n = 3$? From $n = 2$ to $n = 4$? From $n = 2$ to infinity?

38. If the energy state of the H-atom electron is changed from $n = 3$ to $n = 2$, what is the energy change in joules and the frequency of the photon emitted?

39. From which energy state, $n = 1$ or $n = 2$, is it easier to remove the electron from the hydrogen atom? In which state is the electron most stable?

40. Suppose that 2.091×10^{-18} J is absorbed by the electron of the hydrogen atom in the $n = 1$ energy state. Describe the final energy state of the atom.

41. Propose a simple model for the hydrogen atom which accounts for the quantization of energy.

The First Ionization Energy

42. Estimate the first ionization energy for Br.

43. Explain why it takes energy to remove an electron from an isolated atom in the gas phase.

44. List the following elements in order of increasing first ionization energy.
(a) Li (b) Be (c) F (d) Na

The Shell Model

45. Use the data in Table 3.3 to construct a shell model of the atom similar to that in Figure 3.9 for Na through Ca.

46. Calculate the core charge for the atoms Na through Ca.

47. Estimate the first ionization energy of I according to the shell model.

48. How would you expect the ionization energy of Na to compare to that of He? Explain.

49. How would you expect the ionization energy of Cl^- to compare with that of Ar? Explain.

50. Predict the order of the first ionization energies for I, Xe, and Cs. Explain your reasoning.

51. Which electron is harder to remove from an Li atom, the one in the outermost shell or the one in the innermost shell? Explain.

52. Static electricity is due to the buildup of charge on a material. If wool is rubbed on a piece of rubber, the rubber becomes negatively charged and the wool becomes positively charged. Use the shell model for the structure of the atom to explain static electricity.

53. According to the shell model, why is the first ionization energy of Cl less than that of F?

54. If a single electron is removed from an Li atom, the resulting Li^+ cation has only two electrons, both in the $n = 1$ shell. In this respect it is very similar to an He atom. How would you expect the ionization

energy of Li^+ to compare with that of an He atom? Explain your reasoning.

55. If a single electron is added to an F atom, the resulting F^- negative ion has a total of eight valence electrons in the $n = 2$ shell. In this respect it is very similar to an Ne atom. How would you expect the ionization energy of F^- to compare with that of an Ne atom? Explain your reasoning.
56. Predict the order of the first ionization energies for the atoms Br, Kr, and Rb. Explain your reasoning.
57. Which has the lowest first ionization energy, He or Be? Explain.
58. Describe the general trend in first ionization energies from left to right across the second row of the periodic table.
59. Describe the general trend in first ionization energies from top to bottom of a column of the periodic table.
60. Explain why the first ionization energy of N is smaller than that of F.
61. Explain why the first ionization energy of hydrogen is so much larger than the first ionization energy of sodium.

The Shell Model and the Periodic Table

62. According to the shell model, why do first ionization energies increase across a row (period) of the periodic table?
63. According to the shell model, why do first ionization energies decrease down a column (group) of the periodic table?
64. Which of the following elements should have the largest first ionization energy?
(a) B (b) C (c) N (d) Mg (e) Al
65. Which of the following elements should have the smallest first ionization energy?
(a) Mg (b) Ca (c) Si (d) S (e) Se
66. How many valence electrons do each of the atoms H through Ne have?
67. What is the relationship among the core charge, number of valence electrons, and group number?
68. Why are valence electrons easier to remove from an atom than core electrons?

Photoelectron Spectroscopy and the Structure of Atoms

69. Do all of the electrons in a given shell have the same energy? Explain.
70. Refer to Table 3.3. If radiation of energy 520.2 kJ/mol is impinged upon a lithium atom, can an electron be removed?
71. If radiation of energy of 483.6 kJ/mol strikes a sodium atom, could an electron be removed? See Table 3.3.
72. In a single PES experiment, how many electrons are removed from an atom?
73. Is it possible to remove only a core electron in a PES experiment? Explain.
74. A photon of energy 920.6 kJ/mol strikes a B atom. If the kinetic energy of the electron ejected by the atom is 120.0 kJ/mol, what was the ionization energy of the ejected electron?
75. What determines the height of each peak in a photoelectron spectrum?
76. Why are the number of peaks in the PES for H and He the same?
77. Why does the PES of Li have two peaks? Why are the peaks of different heights?
78. In Figure 3.14, the energy associated with the peak representing the largest ionization energy increases from H to B. Why?
79. Why is the lowest energy peak in Figure 3.14 assigned to Li?
80. Why are there three peaks in the PES of B?
81. Refer to Figure 3.14. If photons of energy 1.40 MJ/mol bombard a B atom, what electrons could be removed? What electrons in a Be atom could be removed by photons of this energy?

Electron Configurations from Photoelectron Spectroscopy

Questions 82–86 should be answered in sequence

82. Why are two of the three peaks in the PES spectrum of neon assigned to the $n = 2$ shell rather than to the $n = 1$ shell? What is the rationale for assuming that the peak at 84.0 MJ/mol corresponds to electrons in the $n = 1$ shell?
83. Roughly sketch the photoelectron spectra for Al and S. Give the relative intensities of the peaks.
84. What element do you think should give rise to the photoelectron spectrum shown in Figure 3P.1? Explain your reasoning.

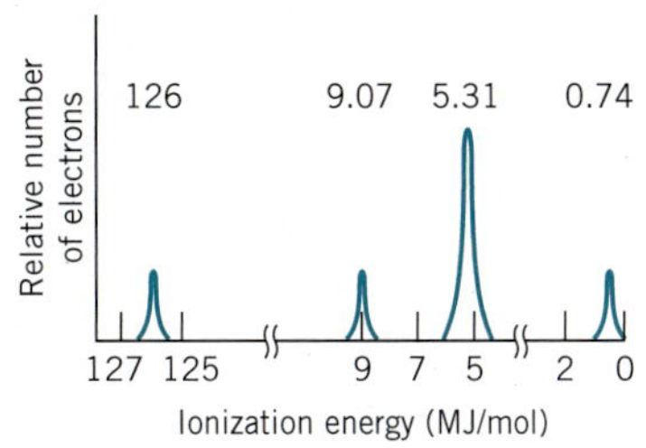

Fig. 3P.1

85. Use the ionization energies given below for Li, Na, and Ar to predict the photoelectron spectrum of K.

	Ionization Energy (MJ/mol)				
Element	1*s*	2*s*	2*p*	3*s*	3*p*
Li	6.26	0.52			
Na	104	6.84	3.67	0.50	
Ar	309	31.5	24.1	2.82	1.52

(a) First consider the first three shells (18 electrons) of K. For these 18 electrons, indicate the relative energies of the peaks and their relative intensities.

(b) If the nineteenth electron of K is found in the $n = 4$ shell, would the ionization energy be closest to 0.42, 1.4, or 2.0 MJ/mol? Explain. (Hint: Compare to Na and Li.) Show a predicted photoelectron spectrum based on this assumption.

(c) If the nineteenth electron of K is found in the third subshell of the $n = 3$ shell, would the ionization energy be closest to 0.42, 1.4, or 2.0 MJ/mol? Explain. (Hint: Compare to Ar.) Show a predicted photoelectron spectrum based on this assumption.

(d) Given the correct photoelectron spectrum of K (Figure 3P.2), predict whether the nineteenth electron of K is found in the $n = 4$ or $n = 3$ shell. Explain your reasoning.

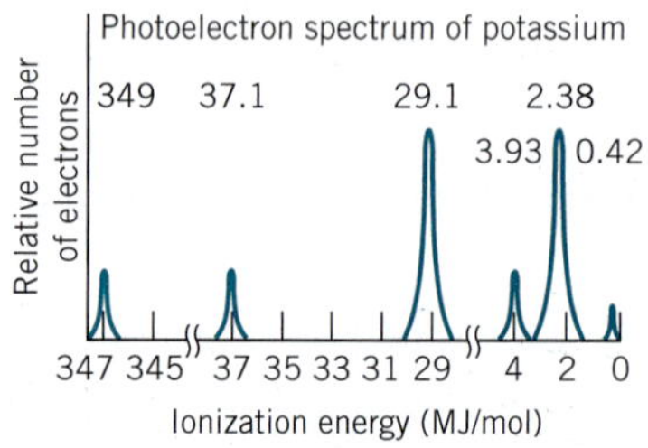

Fig. 3P.2

86. Identify the element whose photoelectron spectrum is shown in Figure 3P.3. (Note: In Figure 3P.3, the peak which arises from the 1*s* electrons has been omitted.)

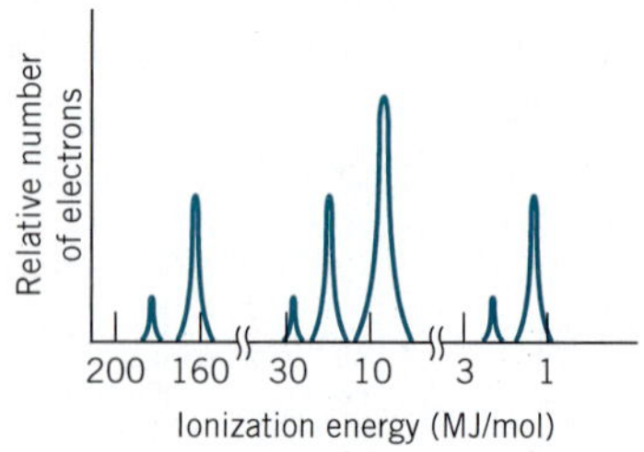

Fig. 3P.3

87. The PES for element *X* is given below:

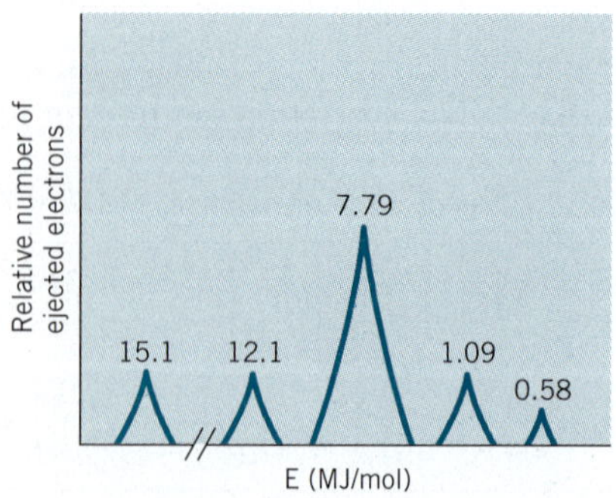

(a) What is the group number of this element?

(b) What is the maximum positive charge that this atom is likely to have under normal chemical conditions? Explain.

(c) What is this atom's core charge?

(d) Which element is this? Explain your reasoning.

(e) This atom is irradiated with light of 206 nm wavelength. Which, if any, of the atom's electrons could be removed by the photons of this light? 10^{-9} m = 1 nm.

88. The following PES spectrum is known for the atom *X*.

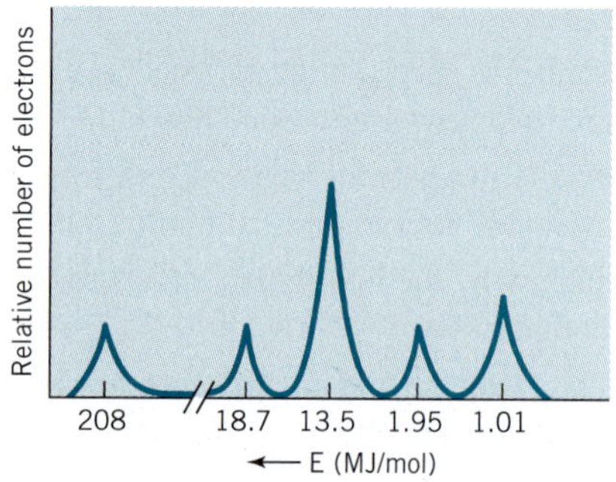

(a) Identify atom *X* and explain how you arrive at your conclusion.

(b) Sketch the shell model for atom *X* that corresponds to the PES spectrum.

(c) Sketch the PES spectrum for the atom having eight fewer protons than atom *X*. Label the axis clearly and show the approximate energies.

(d) If photons of wavelength 1.20×10^{-8} m are used to bombard atom *X*, which, if any, of the electrons of atom *X* could be completely removed from the atom?

Allowed Combinations of Quantum Numbers

89. Roughly sketch the shapes of the orbitals for which $l = 0$, $l = 1$, and $l = 2$.

90. If $l = 1$, what values can m_l have? What values can l have if $n = 2$?

91. Describe the rules used to determine what values of n, l, and m_l are allowed.

92. Identify the quantum number that specifies each of the following.
(a) The size of the orbital
(b) The shape of the orbital
(c) The way the orbital is oriented in space

Shells and Subshells of Orbitals

93. Determine the allowed values of the angular quantum number, l, when the principal quantum number is 4. Describe the difference between orbitals that have the same principal quantum number and different angular quantum numbers.

94. Determine the maximum value for the angular quantum number, l, when the principal quantum number is 5.

95. Write the combination of quantum numbers for every electron in the $n = 1$ and $n = 2$ shells using the selection rules outlined in this chapter.
96. Identify the symbols used to describe orbitals for which $l = 0, 1, 2,$ and 3.
97. Determine the number of orbitals in the $n = 3$, $n = 4$, and $n = 5$ shells.
98. Which of the following orbitals cannot exist?
(a) 6*s* (b) 3*p* (c) 2*d*
(d) 4*f* (e) 3*f*

Orbitals and the Pauli Exclusion Principle

99. What is meant by paired electrons?
100. What causes an atom to be magnetic?
101. Under what circumstances can an atom have no magnetic properties?
102. What would be the results of a Stern–Gerlach experiment on a beam of
(a) H atoms (b) He atoms
(c) Be atoms (d) F atoms?
103. In addition to magnetic properties, what is a further consequence of unpaired electrons?
104. Which of the following is a legitimate set of n, l, m_l, and m_s quantum numbers?
(a) 4, −2, −1, ½ (b) 4, 2, 3, ½
(c) 4, 3, 0, 1 (d) 4, 0, 0, −½
105. Which of the following is a legitimate set of n, l, m_l, and m_s quantum numbers?
(a) 0, 0, 0, ½ (b) 8, 4, −3, −½
(c) 3, 3, 2, +½ (d) 2, 1, −2, −½
(e) 5, 3, 3, −1
106. Determine the allowed values of the magnetic quantum number, m_l when the angular quantum number is 2. Describe the difference between orbitals that have the same angular quantum number and different magnetic quantum numbers.
107. Determine the number of allowed values of the magnetic quantum number when $n = 5$ and $l = 3$.
108. Calculate the maximum number of unpaired electrons that can be placed in a 5*d* subshell.
109. Explain why the difference between the atomic numbers of pairs of elements in a vertical column, or group, of the periodic table is either 8, 18, or 32.
110. Write the combination of quantum numbers for every electron in the $n = 1$ and $n = 2$ shells assuming that the selection rules for assigning quantum numbers are changed to the following.
(a) The principal quantum number can be any integer greater than or equal to 1.
(b) The angular quantum number can have any value between 0 and n.
(c) The magnetic quantum number can have any value between 0 and 1.
(d) The spin quantum number can have a value of either +1 or −1.
111. Which of the following sets of n, l, m_l, and m_s quantum numbers can be used to describe an electron in a 2*p* orbital?
(a) 2, 1, 0, −½ (b) 2, 0, 0, ½
(c) 2, 2, 1, ½ (d) 3, 2, 1, −½
(e) 3, 1, 0, ½
112. Calculate the maximum number of electrons that can have the quantum numbers $n = 4$ and $l = 3$.
113. Calculate the number of electrons in an atom that can simultaneously possess the quantum numbers $n = 4$ and $m_s = +½$.
114. Determine the allowed values of the spin quantum number, m_s, when $n = 5$, $l = 2$, and $m_l = -1$.
115. What happens to the number of subshells as the value of n becomes larger?
116. Determine the number of subshells in the $n = 3$ and $n = 4$ shells.
117. Calculate the maximum number of electrons in the $n = 1$, $n = 2$, $n = 3$, and $n = 4$ shells.
118. Calculate the maximum number of electrons that can fit into a 4*d* subshell.
119. Which of the following is a possible set of n, l, m_l, and m_s quantum numbers for the last electron added to form a gallium atom ($Z = 31$)?
(a) 3, 1, 0, −½ (b) 3, 2, 1, ½
(c) 4, 0, 0, ½ (d) 4, 1, 1, ½
(e) 4, 2, 2, ½
120. Which of the following is a possible set of n, l, m_l, and m_s quantum numbers for the last electron added to form an As^{3+} ion?
(a) 3, 1, −1, ½
(b) 4, 0, 0, −½
(c) 3, 2, 0, ½
(d) 4, 1, −1, ½
(e) 5, 0, 0, ½
121. What is an orbital?
122. Give two reason why electrons tend to avoid each other.
123. What requirement must two electrons meet if they are to occupy the same orbital?

Predicting Electron Configurations

124. Which of the following sets of subshells for yttrium is arranged in the correct order of filling?
(a) 3*d*, 4*s*, 4*p*, 5*s*, 4*d*
(b) 3*d*, 4*s*, 4*p*, 4*d*, 5*s*
(c) 4*s*, 3*d*, 4*p*, 5*s*, 4*d*
(d) 4*s*, 3*d*, 4*p*, 4*d*, 5*s*
125. Which of the following orders of filling of orbitals is incorrect?
(a) 3*s*, 4*s*, 5*s* (b) 5*s*, 5*p*, 5*d* (c) 5*s*, 4*d*, 5*p*
(d) 6*s*, 4*f*, 5*d* (e) 6*s*, 5*f*, 6*p*

126. As atomic orbitals are filled, the $6p$ orbitals are filled immediately after which of the following orbitals?
(a) $4f$ (b) $5d$ (c) $6s$ (d) $7s$

Electron Configurations and the Periodic Table

127. Describe some of the evidence that could be used to justify the argument that the modern periodic table is based on similarities in the chemical properties of the elements.
128. Describe some of the evidence that could be used to justify the argument that the modern periodic table groups elements with similar electron configurations.
129. Determine the row and column of the periodic table in which you would expect to find the first element to have $3d$ electrons in its electron configuration.
130. Determine the row and column of the periodic table in which you would expect to find the element that has five more electrons than the rare gas krypton.
131. Determine the group of the periodic table in which an element with the following electron configuration belongs.

$$[X] \quad 1s^2\,2s^2\,2p^6\,3s^2\,3p^6\,4s^2\,3d^{10}\,4p^6\,5s^2\,4d^{10}\,5p^3$$

132. In which group of the periodic table should element 119 belong if and when it is discovered?
133. Describe the groups of the periodic table in which the s, p, d, and f subshells are filled.
134. Write the electron configurations for the elements in the third row of the periodic table.
135. Give the electron configuration of the nitrogen atom.
136. Give the electron configuration for nickel.
137. The electron configuration of Si is $1s^2\,2s^2\,2p^6\,3s^2\,3p^x$, where x is which of the following?
(a) 1 (b) 2 (c) 3 (d) 4 (e) 6
138. Which of the following is the correct electron configuration for the P^{3-} ion?
(a) [Ne] (b) [Ne] $3s^2$
(c) [Ne] $3s^2\,3p^3$ (d) [Ne] $3s^2\,3p^6$
139. Which of the following is the correct electron configuration for the bromide ion, Br^-?
(a) [Ar] $4s^2\,4p^5$
(b) [Ar] $4s^2\,3d^{10}\,4p^5$
(c) [Ar] $4s^2\,3d^{10}\,4p^6$
(d) [Ar] $4s^2\,3d^{10}\,4p^6\,5s^1$
140. Determine the number of electrons in the third shell of a vanadium atom.
141. Determine the number of electrons in s orbitals in the Ti^{2+} ion.
142. Theoreticians predict that the element with atomic number 114 will be more stable than the elements with atomic numbers between 103 and 114. On the basis of its electron configuration, in which group of the periodic table should element 114 be placed?
143. Which is the first element to have $4d$ electrons in its electron configuration?
(a) Ca (b) Sc (c) Rb
(d) Y (e) La
144. Which of the following contains sets of atoms or ions that have equivalent electron configurations?
(a) B^{3+}, C^{4+}, H^+, He
(b) Na^+, Ne, N^{3+}, O^{2-}
(c) Mg^{2+}, F^-, Na^+, O^{2-}
(d) Ne, Ar, Xe, Kr
(e) O^{2-}, S^{2-}, Se^{2-}, Te^{2-}

Electron Configurations and Hund's Rules

145. Which of the following electron configurations for carbon satisfies Hund's rules?
(a) $1s^2\,2s^2\,2p^2\ \underbrace{\underline{\uparrow}\ \underline{\downarrow}\ \underline{\quad}}_{2p}$
(b) $1s^2\,2s^2\,2p^2\ \underbrace{\underline{\uparrow}\ \underline{\uparrow}\ \underline{\quad}}_{2p}$
(c) $1s^2\,2s^2\,2p^2\ \underbrace{\underline{\uparrow}\ \underline{\quad}\ \underline{\downarrow}}_{2p}$
(d) $1s^2\,2s^2\,2p^2\ \underbrace{\underline{\uparrow\downarrow}\ \underline{\quad}\ \underline{\quad}}_{2p}$
146. Draw orbital diagrams for:
(a) Si (b) V (c) Ga (d) Cl (e) Na
147. Which of the following orbital diagrams are incorrect for all electrons in the lowest energy levels of an atom?
(a) $\underset{s}{\underline{\uparrow\downarrow}}\ \ \underset{p}{\underline{\uparrow\downarrow}\ \underline{\uparrow}\ \underline{\quad}}$ (b) $\underset{s}{\underline{\uparrow}}\ \ \underset{p}{\underline{\uparrow\downarrow}\ \underline{\uparrow}\ \underline{\uparrow}}$
(c) $\underset{s}{\underline{\uparrow\downarrow}}\ \ \underset{p}{\underline{\uparrow}\ \underline{\uparrow}\ \underline{\uparrow}}$ (d) $\underset{s}{\underline{\quad}}\ \ \underset{p}{\underline{\uparrow\downarrow}\ \underline{\uparrow\downarrow}\ \underline{\downarrow}}$
148. Which of the following neutral atoms has the largest number of unpaired electrons?
(a) Na (b) Al (c) Si (d) P (e) S
149. Which of the following ions has five unpaired electrons?
(a) Ti^{4+} (b) Co^{2+} (c) V^{3+}
(d) Fe^{3+} (e) Zn^{2+}
150. Atoms that have unpaired electrons are magnetic. Those that have no unpaired electrons are not magnetic. Which of the following atoms or ions are magnetic? Show your work.

H, He, F^-, Na, Mg, Si, Cr^{3+}

151. Determine the maximum number of unpaired electrons that can be placed in a $5d$ subshell.

The Sizes of Atoms: Metallic Radii

152. Describe what happens to the sizes of the atoms as we go down a column of the periodic table. Explain.

153. Describe what happens to the sizes of the atoms as we go across a row of the periodic table from left to right. Explain.

154. At one time, the size of an atom was given in units of angstroms because the radius of a typical atom was about 1 angstrom (Å). Now they are given in a variety of units. If the radius of a gold atom is 1.442 Å, and 1 Å is equal to 10^{-8} cm, what is the radius of this atom in nanometers and in picometers?

155. Which of the following atoms has the smallest radius?
(a) Na (b) Mg (c) Al
(d) K (e) Ca

The Sizes of Atoms: Covalent Radii

156. Explain why the covalent radius of an atom is smaller than the metallic radius of the atom.

157. Which of the following atoms has the largest covalent radius?
(a) N (b) O (c) F (d) P (e) S

158. What happens to the covalent radii as the periodic table is traversed from left to right? From top to bottom? Explain.

159. Why does the covalent radius undergo a dramatic change from Xe to Cs?

The Relative Sizes of Atoms and Their Ions

160. Look up the covalent radii for magnesium and sulfur atoms and the ionic radii of Mg^{2+} and S^{2-} ions in Appendix B.4. Explain why Mg^{2+} ions are smaller than S^{2-} ions even though magnesium atoms are larger than sulfur atoms.

161. Explain why the radius of a Pb^{2+} ion (0.120 nm) is very much larger than that of a Pb^{4+} ion (0.084 nm).

162. Predict the relative sizes of the Fe^{2+} and Fe^{3+} ions, which can be found in a variety of proteins, including hemoglobin, myoglobin, and the cytochromes.

163. Explain how values of ionic radii can be used and why this information is important.

164. Describe what happens to the radius of an atom when electrons are removed to form a positive ion. Describe what happens to the radius of the atom when electrons are added to form a negative ion.

165. Predict the order of increasing ionic radius for the following ions: H^-, F^-, Cl^-, Br^-, and I^-. Compare your predictions with the data for the ions in Appendix B.4. Explain any differences between your predictions and the experiment.

Patterns in Ionic Radii

166. Predict whether the Al^{3+} or the Mg^{2+} ion is the smaller. Explain.

167. Which of the following ions has the largest radius? Explain.
(a) Na^+ (b) Mg^{2+} (c) S^{2-}
(d) Cl^- (e) Se^{2-}

168. Which of the following atoms or ions is the smallest? Explain.
(a) Na (b) Mg (c) Na^+
(d) Mg^{2+} (e) O^{2-}

169. Which of the following ions has the smallest radius? Explain.
(a) K^+ (b) Li^+ (c) Be^{2+}
(d) O^{2-} (e) F^-

170. Sort the following atoms or ions into isoelectronic groups.
(a) N^{3-} (b) Ar (c) F^- (d) Ne
(e) P^{3-} (f) Ca^{2+} (g) Al^{3+} (h) Si^{4+}
(i) Na^+ (j) S^{2-} (k) Cl^- (l) O^{2-}
(m) K^+ (n) Mg^{2+}

171. Which of the following isoelectronic ions is the largest? Explain.
(a) Mn^{7+} (b) P^{3-} (c) S^{2-}
(d) Sc^{3+} (e) Ti^{4+}

172. Which of the following ions is smallest? Explain.
(a) P^{3-} (b) S^{2-} (c) Cl^-

173. Which of the following ions is largest? Explain.
(a) Rb^+ (b) Sr^{2+} (c) In^{3+}

174. Arrange the following ions in order of increasing ionic radius.
(a) I^- (b) Cs^+ (c) Ba^{2+}

Second, Third, Fourth, and Higher Ionization Energies

175. Which of the following atoms or ions has the largest ionization energy?
(a) P (b) P^+ (c) P^{2+}
(d) P^{3+} (e) P^{4+}

176. Explain why the second ionization energy of sodium is so much larger than the first ionization energy of the element.

177. What is the most probable electron configuration for the element that has the following ionization energies?

1st *IE* = 578 kJ/mol
2nd *IE* = 1,817
3rd *IE* = 2,745
4th *IE* = 11,577
5th *IE* = 14,831

(a) [Ne] (b) [Ne] $3s^1$
(c) [Ne] $3s^2$ (d) [Ne] $3s^2\,3p^1$
(e) [Ne] $3s^2\,3p^2$ (f) [Ne] $3s^2\,3p^3$

178. Which of the following ionization energies is the largest?
(a) 1st *IE* of Ba (b) 1st *IE* of Mg
(c) 2nd *IE* of Ba (d) 2nd *IE* of Mg
(e) 3rd *IE* of Al (f) 3rd *IE* of Mg

179. Which of the following elements should have the largest second ionization energy? Explain.
(a) Na (b) Mg (c) Al
(d) Si (e) P

180. Which of the following elements should have the largest third ionization energy? Explain.
(a) B (b) C (c) N
(d) Mg (e) Al

181. List the following elements in order of increasing second ionization energy.
(a) Li (b) Be (c) Na (d) Mg (e) Ne

182. Some elements, such as tin and lead, have more than one common ion. Use the electron configurations to predict the most likely ions of Sn and Pb.

183. Use the electron configurations to rationalize why iron forms the Fe^{2+} and Fe^{3+} ions.

Average Valence Electron Energy (AVEE)

184. Why is Be more likely to form Be^{2+} than O is to form O^{2+}?

185. Use Table 3.4 to calculate the AVEE of B and F. Compare your results to the values given in Figure 3.31.

186. Without reference to Figure 3.31, arrange the following in order of increasing AVEE:
(a) P, Mg, Cl (b) S, O, Se, F (c) K, P, O

187. What two quantities does the AVEE measure?

188. How does the AVEE change from left to right across the periodic chart? Explain.

189. How does the AVEE change from top to bottom down a group on the periodic table?

190. Describe what happens to the difference between the energies of subshells as the value of *n* becomes larger.

AVEE and Metallicity

191. Refer to Figure 1.13 and explain why the elements N, P, As, Sb, and Bi range from nonmetals to semimetals to a metal using AVEE.

192. Are large values of AVEE associated with metals or nonmetals? Explain.

193. Use AVEE to explain why in passing from the left-hand side of the periodic chart to the right-hand side, metallic character decreases.

194. Explain why C is a nonmetal, Si is a semimetal, and Sn is a metal.

195. Arrange the following in order of increasing metallicity: Pb, Bi, Au, Ba.

196. Arrange the following in order of increasing nonmetallic behavior: B, Al, Ga, Tl.

197. Refer to Figure 3.31 and predict which is more metallic:
(a) P or S (b) As or Se
(c) Sn or Sb (d) Ga or Ge

198. As an atom gets larger, what happens to the energy required to remove valence electrons and to the distance in energy between shells?

Integrated Problems

199. The most recent estimates give values of about 10^{-10} meters for the radius of an atom and 10^{-14} meters for the radius of the nucleus of the atom. Calculate the fraction of the total volume of an atom that is essentially empty space.

200. What is the relationship between the core charges of the atoms in the third period and their atomic radii?

201. Consider the following ions/atoms: O^{2-}, F^{-}, Ne, Na^{+}, and Mg^{2+}. Order them in terms of increasing ionization energy. Also order them in terms of increasing radius. Now consider the following atoms: O, F, Ne, Na, Mg. Order them in terms of increasing ionization energy. Also order them in terms of increasing radius.

202. Chemists often describe substances from three perspectives: macroscopic, microscopic (atomic), and symbolic. Describe aluminum from these three perspectives.

203. Compare and contrast (what is the same and what is different) for the shell model of an oxygen atom obtained from first ionization energies (Table 3.3) and the quantum mechanical model developed from PES data (Table 3.4).

204. Two *hypothetical* shell models of a lithium atom are shown in Figure 3P.4.

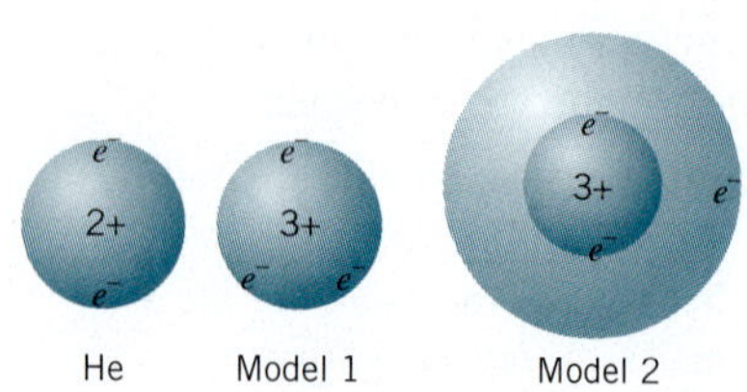

Fig. 3P.4

(a) Use Model 1 to predict the relationship (larger, smaller, or equal) of the first ionization energy of Li compared to the first ionization energy of He. Explain your reasoning.
(b) Use Model 2 to predict the relationship (larger, smaller, or equal) of the first ionization energy of

Li compared to the first ionization energy of He. Explain your reasoning.

(c) Use Table 3.3 to decide which model is most consistent with the observed ionization energy of Li. Explain your reasoning.

205. The following atoms and ions all have the same electronic structure (they are isoelectronic):

Ar S^{2-} K^+

Arrange them in order of increasing first ionization energy. Arrange them in order of increasing atomic radii. Explain your reasoning.

Questions 206–211 relate to the following data.

An atom with an equal number of spin-up and spin-down electrons is known as *diamagnetic,* and the atom is repelled by a magnetic field. In this case we say that all of the electrons are "paired." If this is not the case—that is, if there are unpaired electrons—the atom is attracted to a magnetic field, and it is known as *paramagnetic*. The strength of the attraction is an experimentally measurable quantity known as the magnetic moment. The magnitude of the *magnetic moment* (measured in magnetons) is related to (but not proportional to) the number of unpaired electrons present. In other words, the larger the number of unpaired electrons, the larger the magnetic moment. Here are some experimental data collected by an investigator of this phenomenon:

Magnetic Moments of Several Elements

Element	Type	Magnetic Moment (magnetons)
H	Paramagnetic	1.7
He	Diamagnetic	0
B	Paramagnetic	1.7
C	Paramagnetic	2.8
N	Paramagnetic	3.9
O	Paramagnetic	2.8
Ne	Diamagnetic	0

206. Why is the situation of equal numbers of spin-up and spin-down electrons referred to as all electrons being "paired"?

207. How many unpaired electrons are in the following atoms?
(a) C (b) N (c) O (d) Ne (e) F

208. How many "pairs" of electrons are there in a "filled" *p* shell?

209. On the basis of information provided in the table below, predict the results of a Stern–Gerlach experiment on each of the atoms listed.

210. An ion, X^{2+}, is known to be from the first transition metal series. The ion is paramagnetic with four unpaired electrons. What two possible elements could X be?

211. Element Z is diamagnetic. Its most common ion is Z^{2+}. Z has the next to the lowest first ionization energy in its group. The energy required to remove an electron from Z^{2+} is extremely high. Identify element Z. Give the chemical formula of the ionic compounds formed from Z^{2+} and the negative ions O^{2-} and Cl^-.

212. (a) Give complete electron configurations for

Ca, In, Si

(b) The table below gives the first four ionization energies for one of these atoms. Identify the atom. Explain how the four ionization energies match with the electron configuration.

	Energy (kJ/mol)
First Ionization Energy $A \longrightarrow A^+ + e$	558
Second Ionization Energy $A^+ \longrightarrow A^{2+} + e$	1820
Third Ionization Energy $A^{2+} \longrightarrow A^{3+} + e$	2704
Fourth Ionization Energy $A^{3+} \longrightarrow A^{4+} + e$	5200

213. Three atoms—X, Y, and Z—have the following relation to each other. Atom X has one less proton than atom Y and atom Z has one more proton than atom Y. The table below lists additional characteristics.

(a) Identify atoms X, Y, and Z. Explain how you arrived at your identification. Enter the identity in the table under Chemical Symbol.

(b) Fill in the missing underlined entries in the table below. In some cases your best estimates are acceptable.

Atom	Chemical Symbol	Behavior in a Magnetic Field	First Ionization Energy (*IE*) MJ/mol	Radius of Covalent Atom (nm)	Number of PES Peaks	Core Charge	AVEE	Number of Valence Electrons
X	__	not deflected	2.08	0.070	__	__	__	__
Y	__	__	__	0.16	4	__	0.50	__
Z		not deflected	0.74	__	__	__	__	__

214. The following data are known for the oxygen atom.

Ionization Energies (MJ/mol)

52.6 3.12 1.31

(a) Sketch the shell model for the oxygen atom.
(b) If photons of wavelength 2.39×10^{-8} m are used to bombard an O atom, which, if any, of the electrons could be completely removed from the atom?
(c) What is the first ionization energy, *IE*, for O?
(d) Calculate the AVEE for O.
(e) Roughly sketch the PES spectrum for O and F. Show clearly how the two spectra differ.

215. The following data are known for element *X*.

Number of Peaks in the PES	Covalent Radius of Neutral Atom	First *IE*	AVEE	Core Charge
5	0.099 nm	1.251 MJ/mol	1.59	+7

(a) Identify element *X*. Explain your reasoning.
(b) Give the complete electron configuration for element *X*.
(c) Give the electron configuration for X^{-1}. How would the radius of X^{-1} compare to the radius for *X*? Explain.
(d) How would *IE* for *X* compare to the energy required to remove the most loosely held electron from X^{-}? Explain.
(e) How would the AVEE for *X* compare to the AVEE for the element with one more proton? Explain.
(f) Qualitatively compare the covalent radius of *X* to that of the element with a core charge of +7 and a first *IE* of 1.681 MJ/mol.
(g) How many valence electrons does X have?

216. Three elements, *X*, *Y*, and *Z*, have the electron configurations:

$1s^2\,2s^2\,2p^6\,3s^2\,3p^6$

$1s^2\,2s^2\,2p^6\,3s^2$

$1s^2\,2s^2\,2p^6\,3s^2\,3p^6\,4s^1$

The first ionization energies are known to be (not in any order):

0.4188 MJ/mol

0.4958 MJ/mol

1.5205 MJ/mol

and the covalent radii are (not in any order):

0.157 nm

0.094 nm

0.202 nm

(a) Identify each element and match the appropriate values of ionization energy and atomic radius to each configuration.
(b) Which of *X*, *Y*, and *Z* has the smallest AVEE?
(c) Which is(are) paramagnetic?

217. (a) Which of the following is the most probable electron configuration for the atom that has these ionization energies? Explain.

	MJ/mol
1st *IE*	0.899
2nd *IE*	1.757
3rd *IE*	14.848
4th *IE*	21.006

(i) [He] $2s^1$
(ii) [He] $2s^2$
(iii) [He] $2s^2\,2p^1$
(iv) [He] $2s^2\,2p^2$
(v) [Ne] $3s^1$

(b) How many valence electrons does this atom have? Explain.
(c) What is the core charge of this atom?
(d) Why is the second electron more difficult to remove than the first electron?

Chapter Four

THE COVALENT BOND

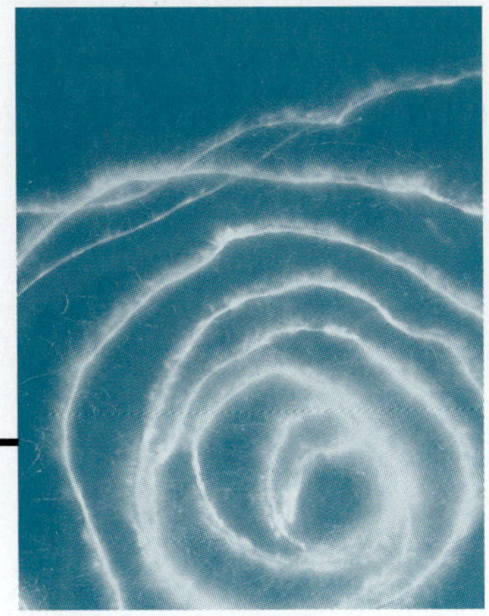

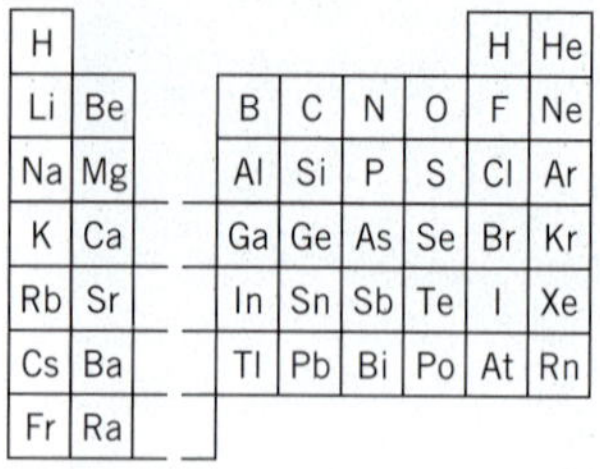

Fig. 4.1 The model developed by G. N. Lewis was first applied to the atoms of the main-group elements, which are found on either side of the periodic table.

This chapter develops the concept of covalent bonding and molecular geometry using Gillespie's electron domain (ED) model. ChemActivities 12–18 and 20–22 in Chemistry: A Guided Inquiry *are designed to accompany this chapter.*

Ever since Dalton introduced his atomic theory in 1803, chemists have tried to understand the forces that hold atoms together in chemical compounds. The goal of this chapter is to build a model for the bonding in molecules. The way in which atoms bond is important to chemists because the properties of compounds depend on their structure.

4.1 Valence Electrons

In 1902, while trying to find a way to explain the periodic table to a beginning chemistry class, G. N. Lewis discovered that the chemistry of the main-group elements shown in Figure 4.1 could be explained by assuming that atoms of these elements gain or lose electrons until they have eight electrons in the outermost shell of electrons of the atom. Eight electrons make an octet, and hence Lewis' discovery is often called the **octet rule.** The magnitude of this achievement can be appreciated by noting that this model was generated only 5 years after J. J. Thomson's discovery of the electron and 9 years *before* Ernest Rutherford proposed that the atom consisted of an infinitesimally small nucleus surrounded by a sea of electrons.

The electrons in the outermost shell eventually became known as the **valence electrons.** This name reflects the fact that the number of bonds an element can form is called its *valence.* Because the number of electrons in the outermost shells in the Lewis theory controls the number of bonds the atom can form, these outermost electrons are the valence electrons.

Figure 4.2, which utilizes the PES data from Chapter 3, shows how much easier it is to remove valence electrons than core electrons as atomic number increases. When $Z = 3$ (lithium) we already begin to see a large difference in the energy to remove the $1s$ core electrons as compared to the $2s$. The gap continues to widen, and by $Z = 10$ (neon) the $1s$ electron is extremely difficult to remove. At $Z = 18$ (argon) we see that the $2s$ and $2p$ core electrons are much more tightly held than the $3s$ and $3p$ valence electrons. There is not much difference, however, in the energy required to remove an electron from the $2p$ as compared to the $2s$ or the $3p$ compared to the $3s$. As Z increases the electrons in shells or subshells closer to the nucleus become increasingly buried in the atom and increasingly more difficult to remove. Figure 4.2 gives the ionization energies of both the inner shell and outer shell electrons and shows trends similar to those in Figure 3.32. Both figures show that AVEE is a measure of the energy required to remove valence electrons as well as the energy separation of the valence shells.

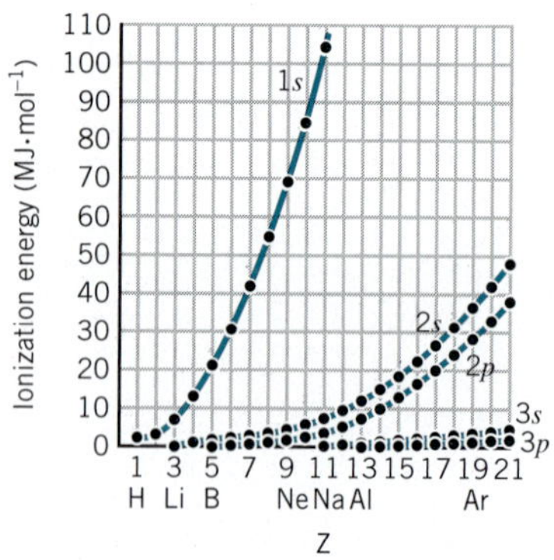

Fig. 4.2 Photoelectron ionization energies of the atoms of the first 21 elements. The energy required to remove a $1s$ electron increases rapidly with increasing nuclear charge, Z. The difference in the energy required to remove an s electron and a p electron from the same shell is never large. This difference gets smaller in subsequent shells. Reprinted from R. J. Gillespie et al., *Atoms, Molecules and Reactions,* Prentice-Hall, Inc., Englewood Cliffs, New Jersey, 1994, p. 200.

The valence electrons are those electrons of an atom that can be gained or lost in a chemical reaction. The number of valence electrons of an atom can be counted by totaling up all the electrons outside the core electrons. There is an exception to this method of counting the valence electrons. Filled d or f subshells are seldom involved in a chemical reaction, and consequently the electrons in filled d or f subshells aren't considered valence electrons. Consider, for example, gallium, which has the following electron configuration.

$$\text{Ga} \qquad [\text{Ar}]\ \mathbf{4s^2}\ 3d^{10}\ \mathbf{4p^1}$$

The [Ar] symbol represents all the core electrons, and since these electrons don't participate in chemical reactions they aren't valence electrons. Of the 13 electrons outside the core the $3d$ subshell is filled, and so the ten $3d$ electrons are not counted as valence electrons. This leaves two $4s$ electrons and one $4p$ electron. Gallium therefore has three valence electrons. Vanadium has the following electron configuration:

$$\text{V} \qquad [\text{Ar}]\ \mathbf{4s^2\ 3d^3}$$

The core electrons represented by [Ar] don't count as valence electrons, but the 4*s* and 3*d* electrons do. Thus vanadium has five valence-shell electrons.

Difficulty determining the number of valence electrons occurs only with the transition or rare earth elements. For most atoms dealt with in this course, the counting of valence electrons is not complicated. For fluorine, whose electron configuration is

$$\text{F} \quad [\text{He}]\ \mathbf{2s^2\ 2p^5}$$

all the electrons outside the [He] core are valence electrons, and fluorine thus has seven valence electrons. Figure 4.2 shows the energy gap separating the 2*s* and 2*p* electrons from the core 1*s* electrons in fluorine ($Z = 9$). Because of the energy difference between the first and second shells, only the second-shell electrons participate in bonding and are thus valence electrons. For main-group elements the number of valence electrons is equal to the group number. Fluorine, in Group VIIA, has seven valence electrons.

> **➤ CHECKPOINT**
>
> How many valence electrons does an atom in Group IVA have?

Exercise 4.1

Determine the number of valence electrons in neutral atoms of the following elements.

(a) Si

(b) Mn

(c) Sb

(d) Pb

Solution

We start by writing the electron configuration for each element.

Si	[Ne] $\mathbf{3s^2\ 3p^2}$
Mn	[Ar] $\mathbf{4s^2\ 3d^5}$
Sb	[Kr] $\mathbf{5s^2}\ 4d^{10}\ \mathbf{5p^3}$
Pb	[Xe] $\mathbf{6s^2}\ 4f^{14}\ 5d^{10}\ \mathbf{6p^2}$

Ignoring filled *d* and *f* subshells, we conclude that neutral atoms of these elements contain the following numbers of valence electrons.

(a) Si = **4**

(b) Mn = **7**

(c) Sb = **5**

(d) Pb = **4**

4.2 The Covalent Bond

By 1916 Lewis realized that there is another way atoms can combine to achieve an octet of valence electrons: They can share electrons with other atoms until each of their respective valence shells contains eight electrons. Two fluorine atoms, for example, can form a stable F_2 molecule in which each atom is surrounded by eight valence electrons by sharing a pair of electrons. A pair of oxygen atoms can form an O_2 molecule in which each atom has a total of eight valence electrons by sharing two pairs of electrons.

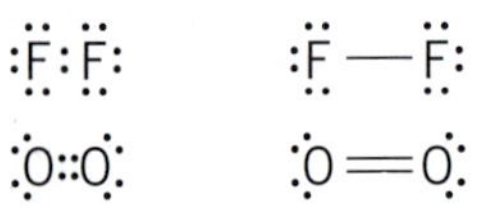

Fig. 4.3 Lewis structures of F_2 and O_2.

Whenever he applied this model, Lewis noted that the atoms seem to share pairs of electrons. He also noted that most molecules contain an even number of electrons, which suggests that the electrons exist in pairs. He therefore introduced a system of notation known as **Lewis structures** in which each atom is surrounded by up to four pairs of dots corresponding to the eight possible valence electrons. The Lewis structures of F_2 and O_2 are therefore written as shown in Figure 4.3. This symbolism is still in use today. The only significant change is the use of lines to indicate bonds between atoms formed by the sharing of a pair of electrons.

The prefix *co-* is used to indicate when things are joined or equal (for example, *coexist, cooperate,* and *coordinate*). It is therefore appropriate that the term **covalent bond** is used to describe the bonds in molecules that result from the sharing of one or more pairs of electrons. As might be expected, molecules held together by covalent bonds are called **covalent molecules.**

4.3 How Does the Sharing of Electrons Bond Atoms?

To understand how atoms can be held together by sharing a pair of electrons, let's look at the simplest covalent bond, the bond that forms when two isolated hydrogen atoms come together to form an H_2 molecule. In the Lewis model a single line is used to represent a pair of electrons in a bond as shown below.

$$\mathrm{H\cdot + \cdot H \longrightarrow H{-}H}$$

Each hydrogen atom contains a proton surrounded by a spherical cloud of electron density that corresponds to a single electron. When a pair of hydrogen atoms approach one another, the electron of each atom is attracted to the proton in the nucleus of the other atom, as shown in Figure 4.4. The magnitude of the force of attraction between the particles is equal to the charge on the electron (q_e) times the charge on the proton (q_p) divided by the square of the distance between particles (r^2).

$$F = \frac{q_e \times q_p}{r^2}$$

Two forces of repulsion are also created, however, because the two negatively charged electrons repel one another, as do the two positively charged nuclei, as shown in Figure 4.4.

At first glance, it might seem that the two new repulsive forces would balance the two new attractive forces. If this happened, the H_2 molecule would be no more stable than a pair of isolated hydrogen atoms. There must be a way in which the forces of attraction can be maximized, while the forces of repulsion are minimized.

The force of repulsion between the protons can be minimized if the pair of electrons are placed between the two nuclei. The distance between the electron in one atom and the nucleus of the other is now smaller than the distance between the two nuclei, as shown in Figure 4.5. As a result, the force of attraction between each electron and the nucleus of the other atom is larger than the force of repulsion between the two nuclei.

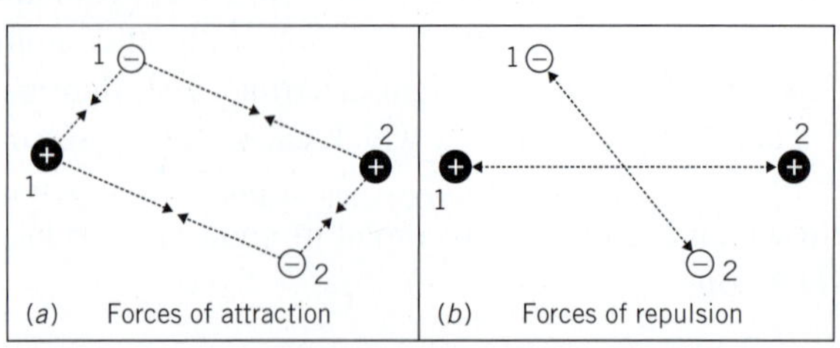

Fig. 4.4 (*a*) Two forces of attraction act to bring a pair of hydrogen atoms together: the forces of attraction between the electron of each atom and the proton of the other atom. (*b*) Two forces of repulsion drive a pair of hydrogen atoms apart: the repulsion between the two protons and the repulsion between the two electrons.

This can only occur if the hydrogen atoms have electrons of *opposite spin* so that each electron may enter the domain of the other electron. The net result of pairing the electrons is to create a region in space, in which both electrons can reside, that extends over both nuclei but concentrates the electron density between the two nuclei. These two electrons are said to occupy the same *bonding domain.* Each hydrogen atom is considered to share two electrons and so to have two electrons in its domain.

Pairing the electrons and placing them between the two nuclei creates a system that is more stable than a pair of isolated atoms if the nuclei are close enough to share the pair of electrons, but not so close that repulsion between the nuclei becomes too large. The hydrogen atoms in an H_2 molecule are therefore held together (or bonded) by the sharing of a pair of electrons, and this bond is the strongest when the distance between the two nuclei is 0.074 nm.

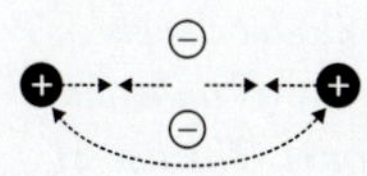

Fig. 4.5 If the electrons are paired and restricted to the region directly between the two nuclei, the attractive forces are larger than the repulsive forces in the hydrogen molecule. As a result, the molecule is more stable than a pair of isolated atoms.

4.4 Using Lewis Structures to Understand the Formation of Bonds

It is so easy to get caught up in the process of generating Lewis structures that we lose sight of the role they played in the development of chemistry. For at least a century before the theory was proposed, chemists had evidence that hydrogen and oxygen existed as diatomic H_2 and O_2 molecules. Lewis provided the first model that could explain why hydrogen atoms come together to form H_2 and why oxygen atoms spontaneously combine to form O_2. Although the model works well for many molecules, there are other bonding models that are superior in certain cases. The Lewis model has the advantage of simplicity and is often the first model used by chemists to describe the bonding in a molecule.

> **➤ CHECKPOINT**
>
> Explain why the distance between the nuclei of the two hydrogens in H_2^+ is 0.106 nm and that between the hydrogens in H_2 is 0.074 nm.

Lewis structures could also be used to understand why hydrogen and oxygen atoms combine to form a molecule with the formula H_2O. We start by noting that the hydrogen atom has only one valence electron and that the oxygen atom has six valence electrons.

$$\begin{array}{ll} \text{H} & \mathbf{1s^1} \\ \text{O} & 1s^2\ \mathbf{2s^2\ 2p^4} \end{array}$$

We then represent neutral atoms of these elements with the following symbols.

$$\text{H}\cdot \qquad \cdot\ddot{\underset{\cdot\cdot}{\text{O}}}\cdot$$

As we bring the atoms together, the electron on each hydrogen atom combines with an electron on the oxygen atom of opposite spin to occupy a region of space between the two nuclei. The hydrogen atoms can attract at most two electrons that must be of opposite spin into the region of space between the hydrogen and oxygen nuclei. These two electrons occupy the same bonding domain.

$$\text{H}\cdot + \cdot\ddot{\underset{\cdot\cdot}{\text{O}}}\cdot + \cdot\text{H} \longrightarrow \text{H}:\ddot{\underset{\cdot\cdot}{\text{O}}}:\text{H}$$

The result is an H_2O molecule in which each hydrogen atom shares a pair of electrons with the oxygen atom and the oxygen atom has access to an octet of valence electrons.

The Lewis structure of water is often written with a line representing each pair of electrons that is shared by two atoms.

$$\text{H}-\ddot{\underset{\cdot\cdot}{\text{O}}}-\text{H}$$

The electron domain model developed by Gillespie serves as an extension of his VSEPR model. The use of domains is particularly helpful later when explaining how double and triple bonds affect geometry.

There are four regions in space, or **domains,** where electrons can be found around the oxygen atom in the water molecule. Two of the domains contain **bonding electrons,** which are used to form the covalent bonds that hold the molecule together. These domains are called **bonding domains** because they hold the pairs of electrons used for bonding. Each hydrogen atom shares a bonding domain with oxygen. The other two domains around the oxygen atom contain pairs of electrons that are described as **nonbonding electrons.** Those domains are called **nonbonding domains.** The nonbonding electrons are in the valence shell of oxygen and are considered to belong exclusively to the oxygen atom in the molecule. Thus there are eight total electrons in the nonbonding and bonding domains of oxygen, and there are two electrons in the bonding domain of each hydrogen atom. The distance between the oxygen and hydrogen nuclei in each bond is experimentally found to be 0.096 nm.

Let's apply our technique for generating Lewis structures to carbon dioxide (CO_2). We start by determining the number of valence electrons on each atom. Carbon has four valence electrons, and each oxygen has six.

C [He] $2s^2\ 2p^2$
O [He] $2s^2\ 2p^4$

We can represent this information by the following symbols.

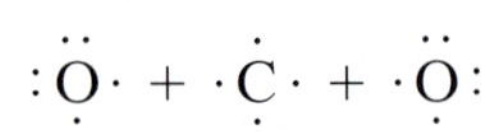

We now combine one electron from each atom to form covalent bonds between the atoms.

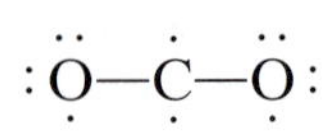

When this is done, each oxygen atom has a total of seven valence electrons, and the carbon atom has a total of six valence electrons. Because none of the atoms have a filled valence shell, we combine another electron on each atom to form two more bonds. The result is a Lewis structure in which each atom has eight electrons in its valence shell.

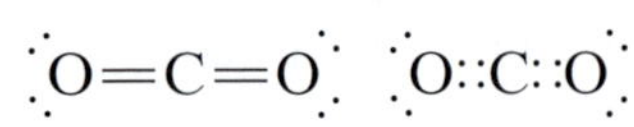

> **CHECKPOINT**
> In the CO_2 molecule, where are the nonbonding electrons? How many electrons are in bonding domains?

Drawing skeleton structures is difficult for students because they do not have sufficient knowledge to decide which atoms are bonded to one another and which atom is the central atom. The rules given here help, and it should be pointed out that many molecules have a symmetric arrangement of atoms.

4.5 Drawing Skeleton Structures

The most difficult step in generating the Lewis structure is the step in which the skeleton structure of the molecule is written. As a general rule, the element with the smallest AVEE (see Figure 3.31) is at the center of the structure. The first step in drawing the skeleton structure is the selection of the central atom. When there is only one atom of an element in a formula, that atom is often the central atom. Thus, the formulas of thionyl chloride ($SOCl_2$) and sulfuryl chloride (SO_2Cl_2) can be translated into the following skeleton structures. Skeleton structures are often symmetric.

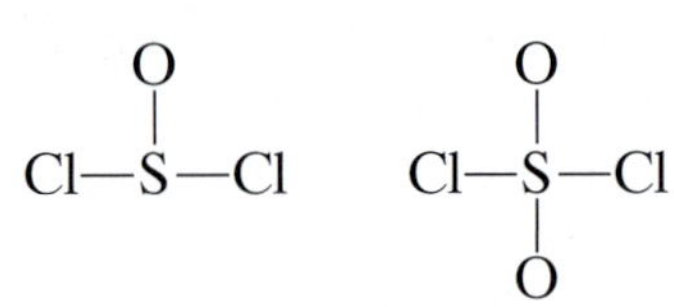

It is also useful to recognize that the formulas for complex molecules often provide hints about the skeleton structure of the molecule. Dimethyl ether, for example, is often written as CH_3OCH_3, which translates into the following skeleton structure.

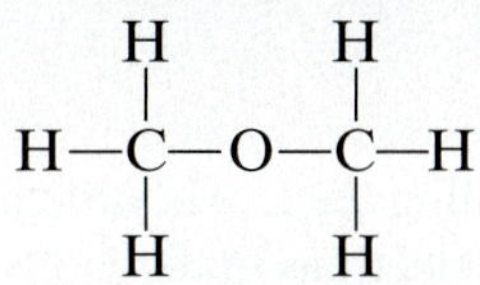

The atoms of many chemical formulas can be arranged in more than one skeleton structure. When there is more than one carbon atom in a formula, they often bond to one another. For example, the atoms in dimethyl ether shown above, C_2H_6O, can also be arranged into the following skeleton structure.

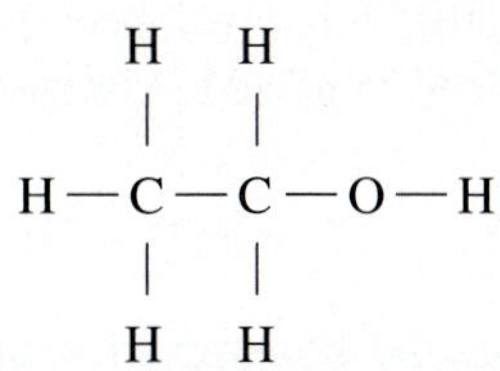

This is the skeleton structure for ethanol. As in this case, two or more possible structures may be found to be equally satisfactory. Section 4.6 details how to make certain a given structure represents a correct Lewis structure, and Section 4.12 provides another way to assist in determining which of two arrangements of atoms may be better. When two structures differ only in the arrangement of atoms, the structures are called isomers. Dimethyl ether and ethanol, shown above, are therefore isomers.

> **➤ CHECKPOINT**
>
> Why can H never serve as the central atom in a Lewis structure?

Finally, it is useful to recognize that structures containing 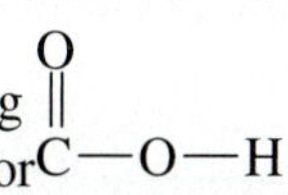 are part of a group of compounds known as acids. The formula for acetic acid, for example, is often written as CH_3CO_2H, because this molecule contains the following skeleton structure.

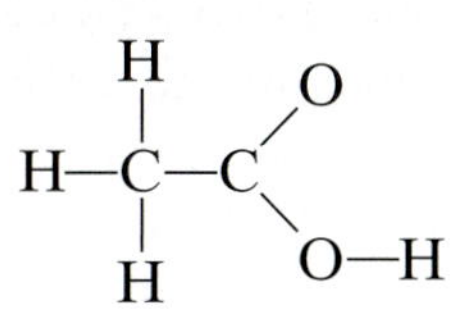

> **➤ CHECKPOINT**
>
> What is the difference between a skeleton structure and a Lewis structure?

4.6 A Step-by-Step Approach to Writing Lewis Structures

The method for writing Lewis structures used in Section 4.4 can be time consuming. For all but the simplest molecules, the following step-by-step process is faster.

- Write the skeleton structure of the molecule.
- Determine the number of valence electrons of the molecule.
- Use two valence electrons to form each bond in the skeleton structure.
- Try to place eight electrons in the valence shells of the atoms by distributing the remaining valence electrons as nonbonding electrons.

The first step in the process involves deciding which atoms in the molecules are connected by covalent bonds. As we have seen, the formula of the compound often provides a hint as to the skeleton structure. Consider PCl_3, for example. The formula for the molecule suggests the following skeleton structure.

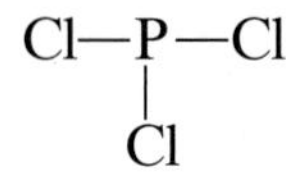

The second step in generating the Lewis structure of a molecule involves calculating the number of valence electrons in the molecule or ion. For a neutral molecule, this is the sum of the valence electrons on each atom. If the species has a charge, we add one electron for each negative charge and subtract an electron for each positive charge.

Chlorine is in Group VIIA of the periodic table, which means it contains 7 valence electrons. Therefore 21 valence electrons are contributed by the three chlorine atoms. A phosphorus atom has 5 valence electrons. Because PCl_3 has no charge, no additional electrons need to be added or subtracted. Thus, PCl_3 has a total of 26 valence electrons.

$$PCl_3 \qquad 5 + 3(7) = 26$$

The third step assumes that the skeleton structure of the molecule is held together by covalent bonds. The valence electrons are therefore divided into two categories: *bonding electrons* and *nonbonding electrons.* Because it takes two electrons to form a covalent bond, we can calculate the number of nonbonding electrons in the molecule by subtracting two electrons for each bond in the skeleton structure from the total number of valence electrons.

There are three covalent bonds in the skeleton structure for PCl_3. As a result, 6 of the 26 valence electrons must be used as bonding electrons. This leaves 20 nonbonding electrons in the valence shell.

$$\begin{array}{l} 26 \text{ valance electrons} \\ \underline{-6 \text{ bonding electrons}} \\ 20 \text{ nonbonding electrons} \end{array}$$

The last step in the process by which Lewis structures are generated involves using the nonbonding valence electrons to satisfy the octets of the atoms in the molecule. Each chlorine atom in PCl_3 already has two electrons—the electrons in the P—Cl covalent bond. Because each chlorine atom needs 6 nonbonding electrons to satisfy its octet, it takes 18 nonbonding electrons to satisfy the 3 chlorine atoms. This leaves one pair of nonbonding electrons, which can be used to fill the valence shell of the central atom.

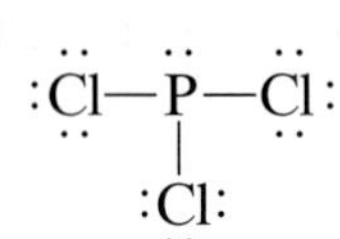

Lewis structures can also be used to describe polyatomic ions such as the ammonium ion, NH_4^+. We begin by drawing a skeleton structure.

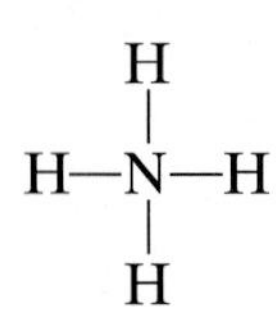

Nitrogen is in Group VA of the periodic table, which means that it contains five valence electrons. Each hydrogen atom has one valence electron. Thus the four hydrogens can contribute four electrons. This gives a total of nine valence electrons from the nitrogen and four hydrogens. Since this is a polyatomic cation, however, one electron is removed to give a positive charge, leaving a total of eight valence electrons.

Each of the four covalent bonds contains two electrons. Therefore, all eight valence electrons are used in the bonding, and no electrons remain to be distributed as nonbonding electrons.

$$\begin{array}{r} 8 \text{ valence electrons} \\ -8 \text{ bonding electrons} \\ \hline 0 \text{ nonbonding electrons} \end{array}$$

This leaves us with eight electrons surrounding the nitrogen and two electrons shared with each hydrogen. Thus, each atom is surrounded by the expected number of electrons. The final structure is drawn in brackets and shows the charge on the polyatomic ion at the upper-right-hand corner.

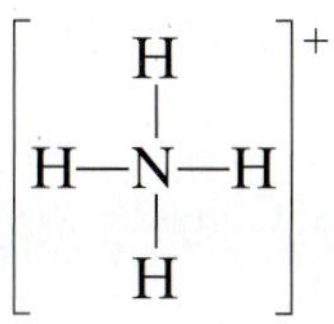

The brackets are used to remind us that the charge doesn't reside on any particular atom in this polyatomic ion. It is spread over the five atoms that contribute to the skeleton structure.

When calculating the number of valence electrons on a polyatomic negative ion, sufficient electrons must be added to account for the overall negative charge. Thus, the NO_3^- ion contains a total of 24 valence electrons.

$$NO_3^- \qquad 5 + 3(6) + 1 = 24$$

➤ CHECKPOINT

Draw the Lewis structures of carbon tetrachloride, CCl_4, and the hydroxide anion, OH^-.

4.7 Molecules That Don't Seem to Satisfy the Octet Rule

NOT ENOUGH ELECTRONS

Occasionally we encounter a molecule that doesn't seem to have enough valence electrons. When this happens, we have to remember why atoms share electrons in the first place. If we can't get a satisfactory Lewis structure by sharing a single pair of electrons, it may be possible to achieve that goal by sharing two or even three pairs of electrons.

Consider formaldehyde (H_2CO), for example, which contains 12 valence electrons.

$$H_2CO \qquad 2(1) + 4 + 6 = 12$$

The formula of the molecule suggests the following skeleton structure.

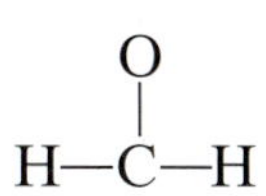

There are three covalent bonds in the skeleton structure, which means that six valence electrons must be used as bonding electrons. This leaves six nonbonding electrons. It is impossible, however, to satisfy the octets of the atoms in this molecule with only six nonbonding electrons. When the nonbonding electrons are used to satisfy the octet of the oxygen atom, the carbon atom has a total of only six valence electrons.

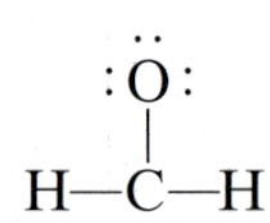

The carbon and oxygen atoms can share one of the non-bonding electron pairs to make a double bond, as shown below. There are now four bonds in the skeleton structure, which leaves only four nonbonding electrons. This is enough, however, to satisfy the octets of the carbon and oxygen atoms. Note that there are two single bond domains between the hydrogens and carbon and that there is one double bond domain between the carbon and oxygen atoms. Surrounding the oxygen atom there are one double bond domain and two nonbonding domains.

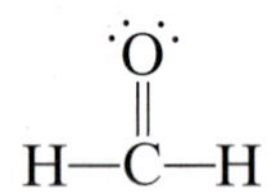

Every once in a while, we encounter a molecule for which it is impossible to write a satisfactory Lewis structure. Consider boron trifluoride (BF_3), for example, which contains 24 valence electrons.

$$BF_3 \qquad 3 + 3(7) = 24$$

There are three covalent bonds in the skeleton structure for the molecule. Because it takes 6 electrons to form the skeleton structure, there are 18 nonbonding valence electrons. But each fluorine atom needs 6 additional electrons to satisfy its octet. Thus, all of the nonbonding electrons are used by the three fluorine atoms. As a result, we run out of electrons while the boron atom has only 6 valence electrons.

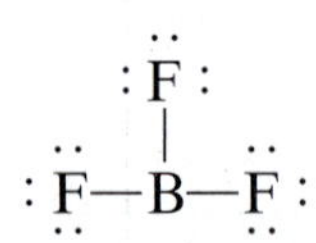

➤ CHECKPOINT

Could the following structure be written for BF_3?

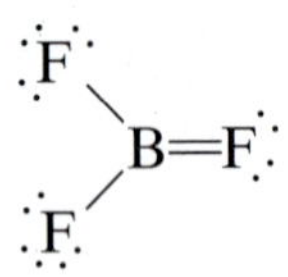

The next step would be to look for the possibility of forming a double or triple bond. However, chemists have learned that the atoms that form strong double or triple bonds are C, N, O, P, and S. Because neither boron nor fluorine belongs in that category, we have to stop with what appears to be an unsatisfactory Lewis structure.

TOO MANY ELECTRONS

It is also possible to encounter a molecule that seems to have too many valence electrons. When that happens, we expand the valence shell of the central atom. Consider the Lewis structure for sulfur tetrafluoride (SF_4), for example, which contains 34 valence electrons.

$$SF_4 \qquad 6 + 4(7) = 34$$

There are four covalent bonds in the skeleton structure for SF_4.

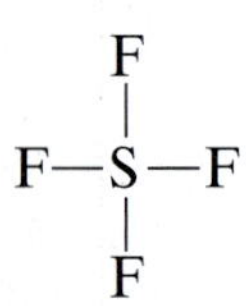

Because this requires using 8 valence electrons to form the covalent bonds that hold the molecule together, there are 26 nonbonding valence electrons.

Each fluorine atom needs 6 additional electrons to satisfy its octet. Because there are four F atoms, we need 24 nonbonding electrons for this purpose. But there are 26 nonbonding electrons in the molecule. We have satisfied the octets for all five atoms, and we still have one more pair of valence electrons. We therefore expand the valence shell of the central atom to hold more than 8 electrons. Usually, extra electron pairs go in nonbonding domains on the central atom.

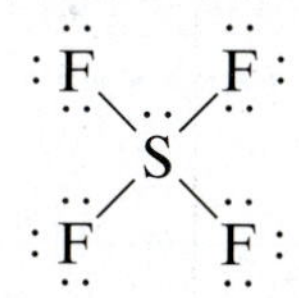

This raises an interesting question: How does the sulfur atom in SF_4 hold 10 electrons in its valence shell? The electron configuration for a neutral sulfur atom seems to suggest that only 8 electrons will fit in the valence shell of the atom because it takes 8 electrons to fill the 3*s* and 3*p* orbitals. But sulfur also has valence-shell 3*d* orbitals.

$$\text{S} \qquad [\text{Ne}]\ 3s^2\ 3p^4\ 3d^0$$

The traditional answer is that because the 3*d* orbitals on a neutral sulfur atom are all empty, one of them can be used to hold the extra pair of electrons on the sulfur atom in SF_4.

An enormous number of molecules follow the octet rule because this rule is obeyed by elements such as C, N, and O that form so many molecules. There are exceptions to the octet rule, however. The most common exception is elements in the first period (H and He), which fill their valence shell with only one pair of electrons. We've already seen two other exceptions: BF_3 and SF_4. BF_3 doesn't have enough valence electrons to obey the octet rule and SF_4 has too many. Elements in the third period or higher can exceed an octet but rarely are found to be electron deficient. Elements in the second period never exceed an octet but are sometimes electron deficient.

➤ CHECKPOINT

Elements in the first and second rows of the periodic table do not have valence-shell *d* orbitals. Use the fact that nitrogen and oxygen do not contain 2*d* orbitals to explain why these elements cannot expand their valence shell.

Students will find it helpful if it is emphasized that second-period atoms never exceed an octet but that they are sometimes electron deficient. Third-period and higher atoms can exceed an octet, but they are seldom electron deficient.

Exercise 4.2

Write the Lewis structure for xenon tetrafluoride (XeF_4).

Solution

Xenon (Group VIIIA) has 8 valence electrons, and fluorine (Group VIIA) has 7. Thus, there are 36 valence electrons in the molecule.

$$XeF_4 \qquad 8 + 4(7) = 36$$

The skeleton structure for the molecule contains four covalent bonds.

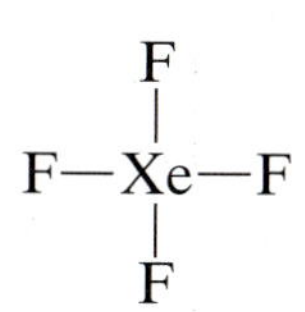

Because 8 electrons are used to form the skeleton structure, there are 28 nonbonding valence electrons. If each fluorine atom needs 6 nonbonding electrons, a total of 24 nonbonding electrons are used to complete the octets of the F atoms. This leaves 4 extra nonbonding electrons. Because the octet of each atom appears to be satisfied and we have electrons left over, we expand the valence shell of the central atom until it contains a total of 12 electrons.

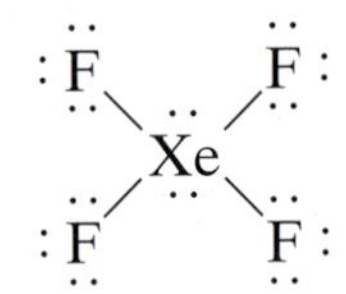

4.8 Bond Lengths

The **bond length** is defined as the distance between two atoms that are bonded together. Bond lengths are usually given in units of nanometers (nm), which is 10^{-9} meters. Carbon–carbon and carbon–hydrogen bond lengths are given for a number of compounds in Table 4.1

Table 4.1

Carbon–Carbon and Carbon–Hydrogen Bond Lengths in Selected Molecules[a]

Molecule	Carbon–Carbon Bond Length (nm)	Carbon–Hydrogen Bond Length (nm)
Ethane	0.154	0.110
Graphite	0.142	
Benzene	0.139	0.110[b]
Ethylene	0.133	0.109
Acetylene	0.120	0.106[b]

[a]From Emil J. Margolis, *Bonding and Structure,* Appleton, Century, Crofts, New York, 1968, with permission from Plenum Publishing Corp.
[b]David R. Lide, Ed., *CRC Handbook of Chemistry and Physics,* 75 ed., CRC Press, Boca Raton, FL, 1994.

The C—C and C—H bond lengths are illustrated for three of these compounds in Figure 4.6. The triple bond in acetylene is significantly shorter than the double bond in ethylene, which is shorter than the single bond in ethane. This suggests that the atoms in a triple bond are pulled closer to one another than in a double bond. The atoms in a double bond, however, are pulled closer together than those in a single bond.

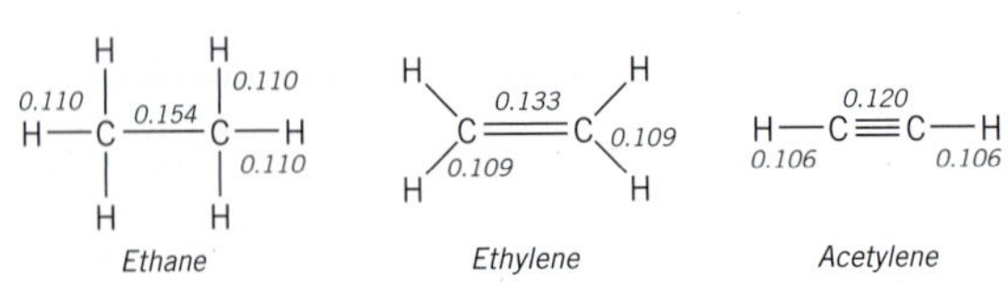

Fig. 4.6 Bond lengths (in nm) in ethane, ethylene, and acetylene. The carbon–carbon bond length gets shorter as the bond changes from a single to a double to a triple bond. The carbon–hydrogen bond length is about the same in all three compounds.

Because each C—H bond is formed by sharing a single pair of electrons, the C—H bond length is approximately the same in the three compounds in Figure 4.6. These C—H bonds, however, are shorter than any of the carbon–carbon bonds. This can be explained using the relative sizes of atoms discussed in Chapter 3. In covalently bonded compounds such as those shown here, we can estimate the length of a C—C bond by noting that the covalent radius of a carbon atom is 0.077 nm (see Figure 3.26). Thus, two covalently bonded carbon atoms should have their nuclei separated by 0.077 + 0.077 = 0.154 nm. Similarly, the carbon–hydrogen bond length should be about 0.077 + 0.037 = 0.114 nm.

When comparing bond lengths of single, double, and triple bonds, it is necessary to compare bonds between the same pair of atoms. For example, N≡O triple bonds are shorter than N═O double bonds, which are shorter than N—O single bonds. However, an N═O double bond isn't shorter than an N—H single bond.

Exercise 4.3

Use the covalent radii given in Chapter 3 to estimate the bond lengths for all bonds in the following compounds.

(a) H_2O
(b) CH_3OH
(c) CH_3OCH_3
(d) CH_3SCH_3

Solution

Before any bond lengths can be estimated, we need to write the Lewis structures of these molecules. The Lewis structures of the molecules are

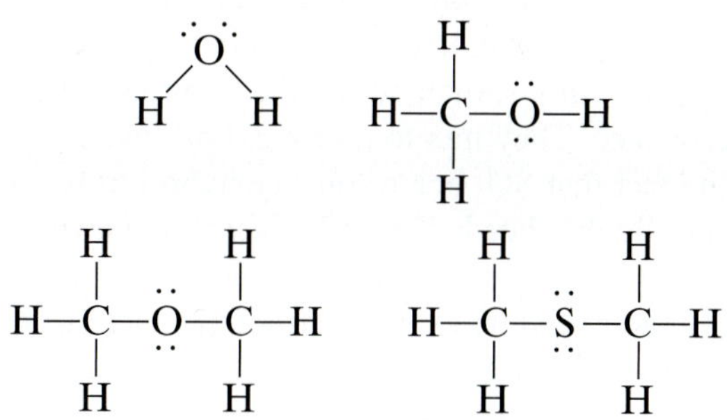

From Figure 3.26 the appropriate covalent radii are

H	0.037 nm
O	0.066 nm
C	0.077 nm
S	0.104 nm

The best estimates for the bond lengths in these molecules are

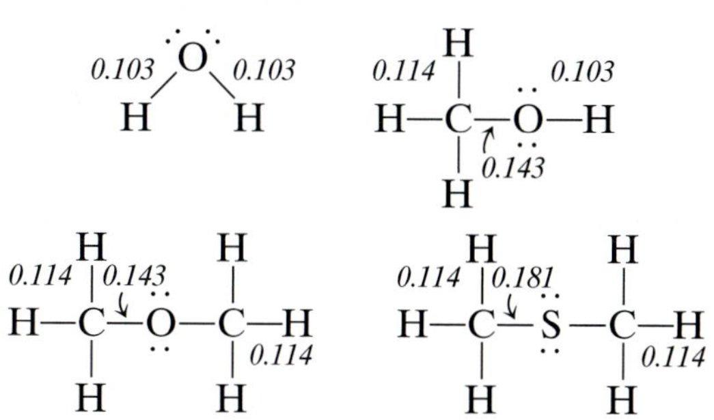

> **CHECKPOINT**
> In which compound is the nitrogen–nitrogen bond length the longest, $H_2N{-}NH_2$ or $HN{=}NH$? Explain.

The data given in Figure 3.26 are average values taken from a variety of compounds. Therefore, the bond lengths obtained by adding these radii are only approximations, but they are quite close to the actual values. It should also be noted that the covalent radii given in Figure 3.26 apply only to single bonds. If a compound contains a double or triple bond between pairs of atoms, these radii do not apply.

4.9 Resonance Hybrids

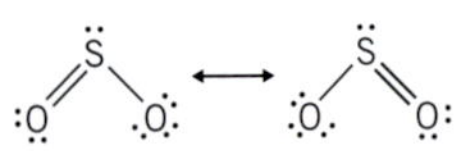

Fig. 4.7 Two equally satisfactory Lewis structures can be written for SO_2.

Two equivalent Lewis structures can be written for sulfur dioxide, as shown in Figure 4.7. The only difference between the two structures is the identity of the oxygen atom to which the double bond is formed. As a result, they must be equally satisfactory representations of the molecule. This raises an important question: Which of the Lewis structures for SO_2 is correct?

Interestingly enough, neither of the structures is correct. The two Lewis structures suggest that one of the sulfur–oxygen bond lengths is shorter than the other. Every experiment that is done to probe the structure of this molecule, however, suggests that the two sulfur–oxygen bonds have identical bond lengths. Moreover, the actual bond length is intermediate between that of a single and double bond.

Often when writing Lewis structures for molecules containing double or triple bonds, it is found that multiple bonds may be drawn satisfactorily in several different ways. When this occurs, as with SO_2, the best description of the structure of the molecule is a **resonance hybrid** of all the possible structures. The meaning of the term *resonance* can be best understood by an analogy. In music, the notes in a chord are often said to resonate, that is, they mix to give something that is more than the sum of its parts. In a similar sense, the two Lewis structures for the SO_2 molecule are in resonance. They mix to give a hybrid that is more than the sum of its components. The fact that SO_2 is a resonance hybrid of two Lewis structures is indicated by writing a double-headed arrow between the Lewis structures, as shown in Figure 4.7.

Isomeric structures and resonance structures are different. Isomers are structures with the same chemical formula but a different arrangement of the atoms. Resonance structures have the same arrangement of atoms but a different arrangement of electrons, leading to multiple bonds in more than one position in the structures.

The relationship between the SO_2 molecule and its Lewis structures can be illustrated by an analogy. Suppose a knight of the round table returns to Camelot to describe a wondrous beast encountered during his search for the Holy Grail. He suggests that the animal looked something like a unicorn because it had a large horn in the center of its forehead. But it also looked something like a dragon because it was huge, ugly, and thick skinned. The beast is, in fact, a rhinoceros. The animal is real, but it was described as a hybrid of two mythical animals. The SO_2 molecule is also real, but we have to use two mythical Lewis structures to describe its bonding.

Exercise 4.4

Acetic acid dissociates to some extent in water to give the acetate ion, $CH_3CO_2^-$. Write two alternative Lewis structures for the acetate ion.

Solution

We can start by noting that the acetate ion contains two carbon atoms (Group IVA), three hydrogen atoms (Group 1A), and two oxygen atoms (Group VIA). It also carries a negative charge, which means that $CH_3CO_2^-$ contains a total of 24 valence electrons.

$$CH_3CO_2^- \qquad 2(4) + 3(1) + 2(6) + 1 = 24$$

The skeleton structure contains six covalent bonds, which leaves 12 nonbonding electrons. Unfortunately, it takes all 12 nonbonding electrons to satisfy the octets of the oxygen atoms, which leaves no nonbonding electrons for the carbon atom bonded to the oxygens.

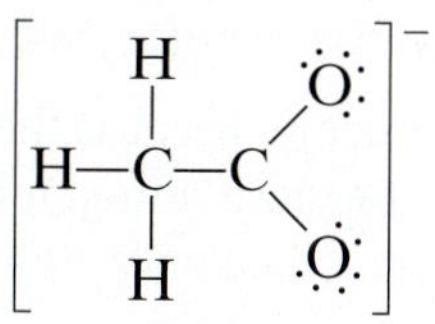

Because there aren't enough electrons to satisfy the octets of the atoms, we assume that there is at least one C=O double bond. There are now seven covalent bonds in the skeleton structure, which leaves only 10 nonbonding electrons. Fortunately, this is enough.

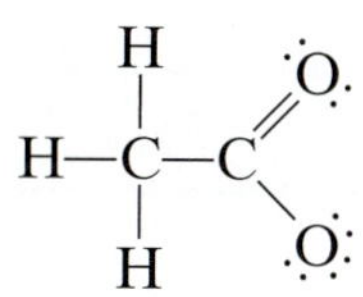

This process gives us one satisfactory Lewis structure for the acetate ion. We can get another by changing the location of the C=O double bond. As a result, the acetate ion is a resonance hybrid of the following Lewis structures.

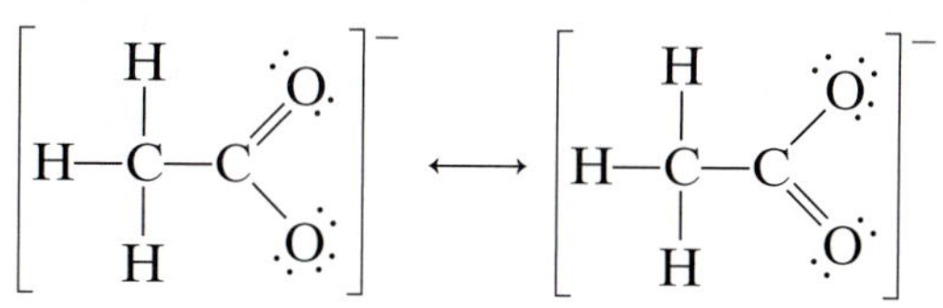

➤ CHECKPOINT

Why don't the two structures of the acetate ion at the left represent two isomers?

Exercise 4.5

The molecule N_2O has the skeletal structure NNO. Which of the Lewis structures given below are acceptable? For structures that are not acceptable explain why.

(a) :N̈—N—Ö:

(b) :N̈—N=Ȯ:

(c) :Ṅ=N=Ȯ:

(d) :N≡N—Ö:

(e) :Ṅ=N—Ö:

(f) :N≡N̈—Ȯ:

(g) :Ṅ=N̈—Ö:

(h) :N̈—N≡O:

➤ CHECKPOINT

Benzene, C_6H_6, has the best-known resonance structure in chemistry. Arrange the six carbon atoms in a ring and draw two resonance Lewis structures for benzene. All of the carbon–carbon bonds in benzene have the same bond length. Referring to Table 4.1, explain why the carbon–carbon bond length is less than that of ethane but greater than that of ethylene.

Solution

The N_2O molecule has 16 valence electrons: 5 from each nitrogen and 6 from the oxygen atom.

(a) This is not an acceptable Lewis structure because the valence shell of the central nitrogen is not filled. There are only four electrons around the central nitrogen atom.

(b) This is not an acceptable structure because there are only six electrons around the central nitrogen atom.

(c) This is an acceptable structure.

(d) This is an acceptable structure.

(e) This is not an acceptable structure because there are only six electrons around the central nitrogen atom.

(f) This is not an acceptable structure because the central nitrogen atom has more than a filled valence shell. Because nitrogen is from the second period in the periodic table, it can accommodate only eight electrons.

(g) The structure drawn has the correct filled valence shells for all of the atoms, but it includes 18 electrons. Only 16 electrons are available in the valence shells of the atoms.

(h) This is an acceptable Lewis structure.

Structures (c), (d), and (h) are resonance structures of one another.

Most electronegativity scales cannot be satisfactorily derived at the introductory level. AVEE is a measure of how tightly the valence electrons are held by atoms and is a generalized electronegativity. Thus the concept of electronegativity is easier for the student to grasp. AVEEs further show that electronegativity is a fundamental periodic property.

AVEE's and electronegativities can be calculated by students from the low-resolution PES data given in Chapter 3. Lee Allen has refined this data by using National Bureau of Standards atomic energy level tables to include multiplet-averaged calculations. Appendix B.7 and Figure 4.8 provide these more accurate values.

Electronegativities calculated by the PES data given in the text are quite close to Allen's values and for an introductory treatment are just as valid. The AVEEs given in Figure 3.31, when multiplied by 1.8 to adjust fluorine to a value of around 4, give electronegativities in good agreement with Allen's more refined values.

Allen's electronegativities have been compared with those of other scales. Sproul[1] has shown that Allen's electronegativities and those of Nagle are better for characterizing ionic, covalent, and metallic bonding than are the other 13 scales tested. Meek[2] points out that Allen's electronegativities provide values for the rare gases that are reasonable and require no further assumptions, as do other scales.

4.10 Electronegativity

A covalent bond involves the sharing of a pair of valence electrons by two atoms. When the atoms are identical, they must share the electrons equally. There is no difference between the electron density on the two oxygen atoms in an O_2 molecule, for example. The O_2 molecule is an example of a purely covalent compound.

The same can't be said about molecules that contain different atoms. Consider the HCl molecule. If there is any difference between the relative ability of hydrogen and chlorine to draw electrons toward itself, the electrons in the H—Cl bond won't be shared equally. The electrons in the bond will be drawn closer to one atom or the other.

The relative ability of an atom to draw electrons in a bond toward itself is called the **electronegativity** (*EN*) of the atom. For many years, chemists have recognized that some atoms attract electrons in a bond better than other atoms. Thus F and O are more electronegative than Na and Mg. But there is no direct way to measure the electronegativity of an atom. Instead, properties that are assumed to depend on electronegativity are measured and then compared to one another in order to determine a relative scale of electronegativity. There are currently at least 15 electronegativity scales in use. When there is close agreement between electronegativity values of an atom on different scales, our confidence in the value is increased.

The first scale of electronegativities was created by Linus Pauling. He assigned fluorine an electronegativity of 4.0 and determined the electronegativities of the atoms of the other elements relative to fluorine. Every electronegativity scale that has been proposed since then has been adjusted so that the electronegativity of fluorine is about 4.

[1]Gordon Sproul, *J. Phys. Chem.* **98,** 6699 (1994).

[2]Terry L. Meek, *J. Chem. Educ.* **72,** 17 (1995).

H 2.30																H 2.30	He 4.16
Li 0.91	Be 1.58											B 2.05	C 2.54	N 3.07	O 3.61	F 4.19	Ne 4.79
Na 0.87	Mg 1.29											Al 1.61	Si 1.92	P 2.25	S 2.59	Cl 2.87	Ar 3.24
K 0.73	Ca 1.03	Sc 1.2	Ti 1.3	V 1.4	Cr 1.5	Mn 1.6	Fe 1.7	Co 1.8	Ni 1.9	Cu 1.8	Zn 1.6	Ga 1.76	Ge 1.99	As 2.21	Se 2.42	Br 2.69	Kr 2.97
Rb 0.71	Sr 0.96	Y 1.0	Zr 1.1	Nb 1.3	Mo 1.4	Tc 1.5	Ru 1.7	Rh 1.8	Pd 1.9	Ag 2.0	Cd 1.5	In 1.66	Sn 1.82	Sb 1.98	Te 2.16	I 2.36	Xe 2.58
Cs 0.66	Ba 0.88										Hg 1.76						

(*a*) AVEE Scale

H 2.1																H 2.1	He –
Li 1.0	Be 1.5											B 2.0	C 2.5	N 3.0	O 3.5	F 4.0	Ne –
Na 0.9	Mg 1.2											Al 1.5	Si 1.8	P 2.1	S 2.5	Cl 3.0	Ar –
K 0.8	Ca 1.0	Sc 1.3	Ti 1.5	V 1.6	Cr 1.6	Mn 1.5	Fe 1.8	Co 1.9	Ni 1.9	Cu 1.9	Zn 1.6	Ga 1.6	Ge 1.8	As 2.0	Se 2.4	Br 2.8	Kr –
Rb 0.8	Sr 1.0	Y 1.2	Zr 1.4	Nb 1.6	Mo 1.8	Tc 1.9	Ru 2.2	Rh 2.2	Pd 2.2	Ag 1.9	Cd 1.7	In 1.7	Sn 1.8	Sb 1.9	Te 2.1	I 2.5	Xe –
Cs 0.7	Ba 0.9										Hg 1.9						

(*b*) Pauling Scale

Fig. 4.8 (*a*) Electronegativities of the elements calculated from photoelectron spectroscopy and refined AVEE. The AVEE data from Chapter 3 have been adjusted to give fluorine an electronegativity value close to 4. Reprinted from L. C. Allen and E. T. Knight, *Journal of Molecular Structure,* **261,** 313 (1992). (*b*) Electronegativities based on the Pauling scale.

A new scale was proposed in 1989 that links the electronegativity of an atom to the average valence electron energy (AVEE) obtained from photoelectron spectroscopy.[1] This scale assumes that atoms which best resist the loss of valence electrons are the atoms which are most likely to draw electrons in a bond toward themselves. Thus AVEE is, in fact, a generalized electronegativity, and the two terms can be used interchangeably. The AVEE data given in Figure 3.31 have been refined and used to calculate the electronegativities shown in Figure 4.8*a* by multiplying by a factor to give fluorine an electronegativity value of about 4. (These data also can be found in Table B.7 in the appendix and in the back of the text.) Figure 4.8 shows two electronegativity scales. Figure 4.8*a* shows the scale based on AVEE, and Figure 4.8*b* shows the original Pauling scale. Because the AVEE scale is based on PES experimental measurements discussed in Chapter 3, we will use the AVEE electronegativity scale exclusively in this text. This scale offers the additional advantage of providing electronegativities for the noble gases. These values will be useful when discussing the compounds formed by the noble gases.

When the magnitude of the electronegativities of the main-group elements is added to the periodic table as a third axis, we get the results shown in Figure 4.9. There are clear patterns in the data in Figures 4.8 and 4.9.

- Electronegativity increases in a regular fashion from left to right across a row of the periodic table.
- Electronegativity decreases down a column of the periodic table.

When atoms with large differences in electronegativities combine, the electron density in the resulting bond is pulled to the more electronegative atom.

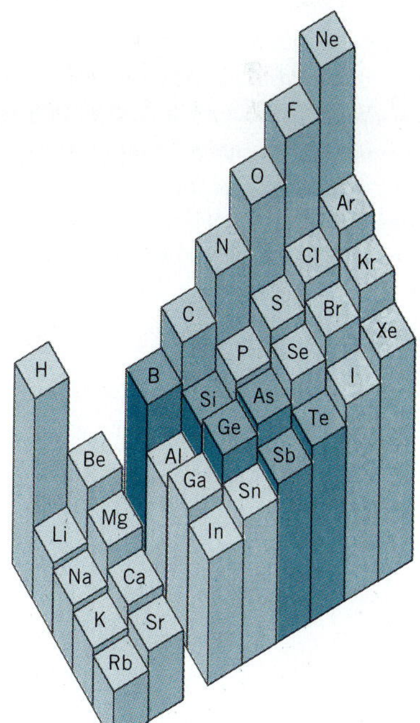

Fig. 4.9 Three-dimensional plot of the electronegativities of the main-group elements versus position in the periodic table. Reprinted from L. C. Allen, *Journal of the American Chemical Society,* **111,** 9003 (1989).

[1]L. C. Allen, *Journal of the American Chemical Society,* **111,** 9003 (1989).

Such compounds are called ionic compounds and are discussed in Chapter 5. NaCl is an example of an ionic compound. In NaCl most of the electron density in the bond is transferred from the sodium to the more electronegative atom chlorine.

In covalent molecules, such as those of this chapter, electrons will be shared between bonding atoms. If these atoms have the same electronegativity, the electrons will be shared equally and the resulting bond is referred to as a *pure covalent bond.* If the covalently bonded atoms have different electronegativities, the more electronegative atom will attract more of the electron density in the bond. This type of bond is referred to as a **polar covalent bond.** One end of the bond has a partial positive charge ($\delta+$), and the other end has a partial negative charge ($\delta-$).

> **CHECKPOINT**
>
> Classify the bonds in NO and O_2 as pure covalent or polar covalent.

A simple way to calculate partial charge on diatomic molecules is given that enables students to get a feeling for a partial charge. This calculation, when extended to polyatomic systems and systems containing multiple bonds, gives only an approximation to the partial charge, owing to nearest-neighbor effects. See the Instructor's Manual for sample calculations.

This approach gives reasonable estimates of partial charge but ignores the effects that neighboring atoms have on electron distribution. It is possible to correct for this, but doing so is beyond the scope of an introductory course. The idea that the model must be refined to take into account all neighbors can be an instructive insight on the part of the student.

4.11 Partial Charge

It would be useful to have a quantitative measure of the extent to which the charge in an individual covalent bond is located on one of the atoms that form the bond. This can be achieved by calculating the **partial charge** on the atom. The partial charge on an atom is determined by the difference between the electronegativities of the atoms that form the bond. When the difference is relatively small, the electrons are shared more or less equally. When it is relatively large, there is a significant charge separation in the bond.

To illustrate how the partial charge on an atom can be calculated, let's look at the HCl molecule.

$$\text{H—}\ddot{\underset{\cdot\cdot}{\text{Cl}}}\text{:}$$

In Figure 4.8*a* the electronegativity of chlorine is given as 2.87 and that of hydrogen as 2.30. Thus the electrons of the H—Cl bond should be attracted to the chlorine slightly more than to the hydrogen. This would result in a partial charge on both atoms, with an excess of electron density on the chlorine atom. The symbol δ is used to indicate the partial charge on an atom, resulting from unequal sharing of electrons. Thus $\delta-$ above an element's symbol means that the element has a greater share of the electron density in the bond, and $\delta+$ indicates a lesser share of electron density.

$$^{\delta^+}\text{H—Cl}^{\delta^-}$$

The partial charge of an atom in a molecule is determined by comparing the electron density associated with the free atom (when the atom is not involved in a bond) to the electron density associated with the atom when bonded in a molecule. Since only the outer shell valence electrons are involved in bonding, we consider only those electrons. We begin the calculation by determining the number of valence electrons (V) in a free atom. This can be determined from the group number in the periodic table. The valence electron density associated with an atom in a molecule is due to both the electrons in the covalent bond (B) and the nonbonding electrons (N) around the atom. The difference in the electronegativities of two bonded atoms has no effect on the number of nonbonding electrons. But the difference in electronegativity does affect the number of bonding electrons that are assigned to each atom. The calculation of partial charge therefore involves the electronegativities of the two atoms (EN_a and EN_b) that form the bond.

The partial charge on an atom engaged in a single bond is calculated by multiplying the number of bonding electrons, B_a, by the fraction of the total electronegativity of the two atoms that can be assigned to that atom.[2]

$$\delta_a = V_a - N_a - B_a\left(\frac{EN_a}{EN_a + EN_b}\right)$$

To illustrate how the formula is used, let's calculate the partial charge on the atoms in HCl.

$$\delta_{Cl} = 7 - 6 - 2\left(\frac{2.87}{2.87 + 2.30}\right) = -0.11$$

The partial charge on the chlorine is -0.11. The magnitude of the partial charge on the hydrogen atom is the same as the magnitude of the partial charge on the chlorine atom, but the sign of the charge is different. The partial charge on the hydrogen is $+0.11$, as shown below.

$$\delta_H = 1 - 0 - 2\left(\frac{2.30}{2.30 + 2.87}\right) = 0.11$$

The chlorine atom, being more electronegative than the hydrogen atom, attracts the electrons of the bond to itself so that the partial charge on chlorine is -0.11. This means that the chlorine atom has 11% more electron density associated with it in the HCl molecule when compared to a free chlorine atom. This molecule is therefore slightly polarized. HCl is a neutral molecule, so the sum of the partial charges on the two atoms must be zero. The electron density that is lost by the hydrogen atom must be gained by the chlorine atom. Table 4.2 lists additional partial charges on compounds formed between hydrogen and other Group VIIA atoms.

Table 4.2

Partial Charge on Group VIIA Atoms in Combination with Hydrogen

Molecule	δ
HF	−0.29
HCl	−0.11
HBr	−0.08

➤ CHECKPOINT

In a hypothetical molecule *AB*, if the electronegativity of *B* is much greater than that of *A*, which atom, *A* or *B*, would have the negative partial charge?

4.12 Formal Charge

The actual charge on an atom in a molecule is best represented by the partial charge. But partial charges are difficult to calculate by hand for complex molecules. Chemists therefore find it useful to calculate **formal charge,** which helps us identify the atoms that are most likely to carry a significant amount of positive or negative charge. Formal charges do not represent the real charges on atoms.

The first step in the calculation of formal charge involves dividing the electrons in each covalent bond equally between the atoms that form the bond. The number of valence electrons formally assigned to each atom is then compared with the number of valence electrons on a neutral atom of the element. If the atom has more valence electrons than a neutral atom, it is assigned a formal negative charge. If it has fewer valence electrons it carries a formal positive charge.

Formal charge is used as a tool to help decide which of several possible Lewis structures might be the best representation of the actual bonding in a compound. Many current introductory texts use minimization of formal charge in a compound or polyatomic ion by expanding the central atom's octet with the addition of multiple bonds as a criterion for the best Lewis structure.

Consider the amino acid known as glycine, which is often written as $H_3N^+CH_2CO_2^-$. We can use the concept of formal charge to explain the meaning of the positive and negative signs in the Lewis structure of this molecule.

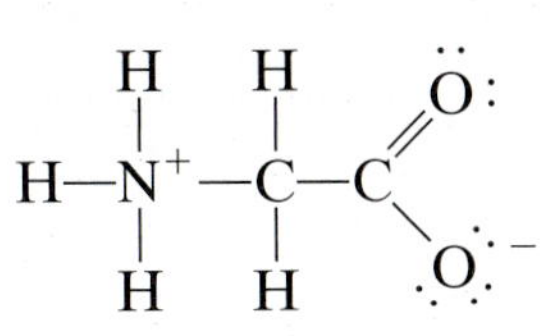

[2]L. C. Allen, *Journal of the American Chemical Society,* **111,** 9115 (1989).

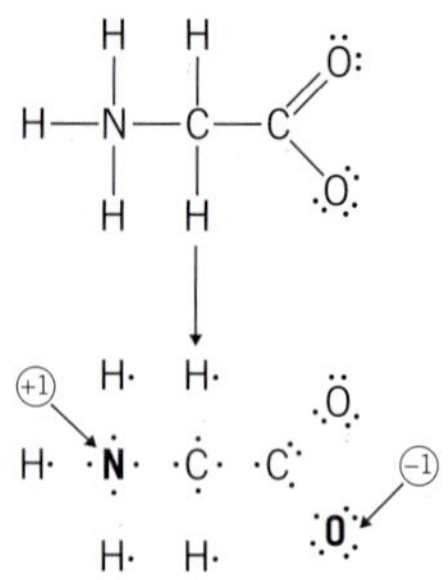

Fig. 4.10 The first step in calculating the formal charge on the atoms in glycine involves dividing pairs of bonding electrons between the atoms in each bond. Formal charges are shown in circles to distinguish them from actual charges.

We start by arbitrarily dividing pairs of bonding electrons so that each atom in a bond is formally assigned half of the electrons, as shown in Figure 4.10. Once this is done, the nitrogen has four valence electrons, one fewer than a neutral nitrogen atom. The nitrogen therefore carries a formal charge of +1 in this Lewis structure. Both of the carbon atoms formally have four electrons, which is equal to the number on a neutral carbon atom. As a result, neither carbon atom has a formal charge. The oxygen atom in the C═O double bond formally has six electrons, which means it has no formal charge. The other oxygen atom, however, has seven electrons, which means that it formally has a charge of −1.

Although the glycine molecule has no net charge, one end of the molecule does have a positive formal charge, and the other end has a negative formal charge. As a result, it isn't surprising that chemists write the formula for this amino acid as $H_3N^+CH_2CO_2^-$.

We can summarize the process by which the formal charge on an atom is calculated by noting that we start with the number of valence electrons (V) on a neutral atom of the element. We then subtract all of the nonbonding electrons on the atom (N) and half of the bonding electrons on the atom (B). The formal charge (FC) on an atom, a, is therefore given by the following equation.

$$FC_a = V_a - N_a - \frac{B_a}{2}$$

To illustrate how this equation is used, let's calculate the formal charge on the nitrogen atom in glycine. Nitrogen has five valence electrons, there are no nonbonding electrons on the atom, and there are eight bonding electrons. As we have seen, the formal charge on the nitrogen atom is therefore +1.

$$FC_N = 5 - 0 - \frac{8}{2} = +1$$

The following example shows why chemists find formal charge useful. Three equally valid Lewis structures can be drawn for dinitrogen oxide, N_2O, a gas known as laughing gas and used as an anesthetic.

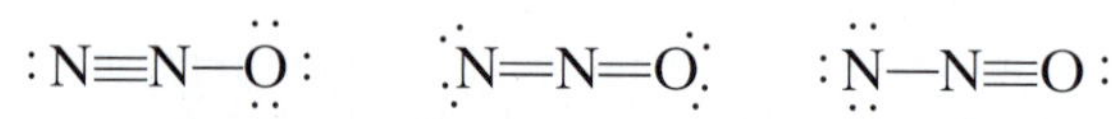

I *II* *III*

Which structure best represents the actual bonding in the molecule? Are the two nitrogen atoms bonded by a single, a double, or a triple bond? We begin by using what has come to be accepted as a general rule for deciding which is the best Lewis representation of a molecule. The best structure is considered to be the one in which the atoms have the smallest formal charges and the negative formal charges are on the more electronegative atoms. The formal charge on each atom in the three Lewis structures of N_2O are shown below.

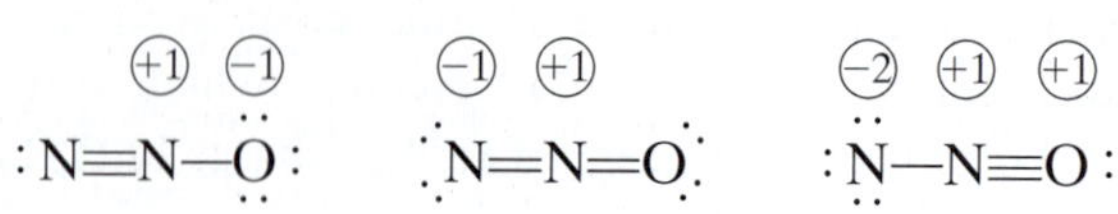

I *II* *III*

Note that the sum of the formal charges must always total to the charge on the species, which in this case is zero because N_2O is an electrically neutral molecule.

Structure III has two serious flaws. First, it puts a negative formal charge on the *less* electronegative nitrogen atom, rather than the *more* electronegative oxygen atom. It also has the largest formal charges. Structure II also puts a negative formal charge on nitrogen, instead of oxygen, and therefore isn't as good as structure I.

Are there experimental data that could be used to support structure I? Because nitrogen forms single, double, and triple bonds in the proposed structures, bond length data should be helpful. We expect the N—N bond length to be longer than N═N, which should be longer than the N≡N bond length. Typical nitrogen–nitrogen bond lengths are:

N—N	0.146 nm
N═N	0.125 nm
N≡N	0.110 nm

An experimental determination of the nitrogen–nitrogen bond length in N_2O gives a result of 0.113 nm, which indicates the bond must be a triple bond. The experimental data support the formal charge calculation that suggests structure I is the best Lewis structure for N_2O.

There are many instances in which the use of formal charge serves to rationalize chemical information. Nitric acid, HNO_3, is a very strong acid whereas nitrous acid, HNO_2, is not. The Lewis structures and formal charges for these compounds are shown below. Atoms with a formal charge of zero have no formal charge shown in the structures.

➤ CHECKPOINT

Calculate the formal charge on each atom in SO_2.

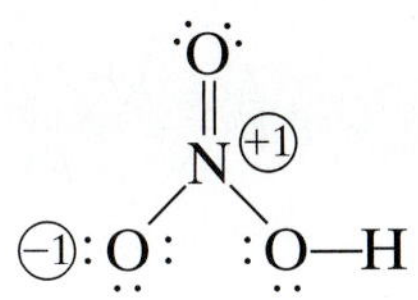

Nitric acid

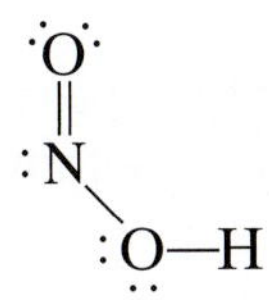

Nitrous acid

The formal charge on the nitrogen in nitric acid is $+1$, allowing the hydrogen atom with its positive partial charge to be easily removed. The nitrogen in nitrous acid has no formal charge and does not lose its hydrogen as easily as nitric acid. Thus nitric acid is a stronger acid than nitrous acid.

There are several schools of thought on what constitutes the best Lewis structures for molecules. The best evidence, of course, is experimental data. In the absence of conclusive data, some chemists prefer to arrange the electrons in a Lewis structure in order to minimize the formal charge on each atom. For example, the structure of SO_2 can be drawn with either one or two double bonds.

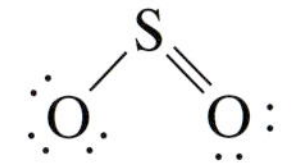

I

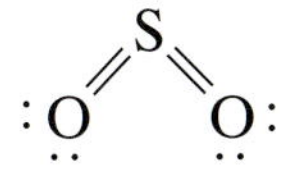

II

In structure I sulfur has eight valence electrons and a formal charge of $+1$. In structure II sulfur has ten valence electrons and a formal charge of zero. Without additional information, it is not possible to feel confident of the correct structure. One study of Lewis structures suggests that using additional multiple bonds and therefore expanding the octet around the central atom may not be the best representation of the structures of many molecules which have traditionally been drawn this way.[3] In this text we use expanded octets to explain the bonding in

Experimentally measured bond lengths have been used as evidence for expanded octets and multiple bonding. Recent work suggests that the best Lewis structural representation may not have an expanded octet. We have not used an expanded octet. In the Instructor's Manual, answers that require Lewis structures are shown using both ways of representing the structure.

[3]L. Suidan, J. K. Badenhoop, E. D. Glendenins, and F. Weinhold, *Journal of Chemical Education,* **72,** 583 (1995).

➤ CHECKPOINT

The acid strengths of the series of chloro-oxy acids are as follows:

$HClO_4$. $HClO_3$. $HClO_2$. HClO

Use formal charge to rationalize this trend.

structures such as SF_4 and XeF_4 discussed in Section 4.7. We also use formal charge as a tool to select between several possible Lewis structures, as discussed in this section. However, we do not expand the octet of an atom for the purpose of minimizing the formal charges in the structure. Therefore, until conclusive evidence is available, we'll use structure I to describe SO_2 rather than structure II.

The relationship used to calculate partial charge

$$\delta_a = V_a - N_a - B_a\left(\frac{EN_a}{EN_a + EN_b}\right)$$

is related to the expression used to determine formal charge on an atom

$$FC_a = V_a - N_a - \frac{B_a}{2}$$

The difference between the two expressions is that when formal charge is calculated, the electrons are assumed to be equally shared ($EN_a = EN_b$). If this relationship is used in the partial charge equation, we obtain the following:

$$\delta_a = V_a - N_a - B_a\left(\frac{EN_a}{EN_a + EN_b}\right)$$

$$\delta_a = V_a - N_a - \frac{B_a}{2} = FC_a$$

The calculated partial charge values are a better description of the actual distribution of electron density in the bond than is formal charge. Formal charge is a useful concept but is not intended to show how electrons are distributed between bonded atoms. Formal charges don't represent actual charges on atoms. The actual charge on an atom is determined in part by electronegativity differences between atoms. Note, for example, that the formal charge on both hydrogen and chlorine in HCl is zero. The electronegativity of chlorine is greater than that of hydrogen, so chlorine will attract electron density in the bond at the expense of hydrogen. In the calculation of formal charge, the electronegativities of hydrogen and chlorine are treated as if they were equal. Consequently, in formal charge calculations the bonding electrons are treated as if they were equally shared.

4.13 The Shapes of Molecules

The shape of a molecule can play an important role in its chemistry. The changes in the three-dimensional structure of proteins that occur when an egg is heated, for example, are the primary source of the differences between a raw egg and a cooked one.

An illustration of how sensitive biomolecules are to changes in their three-dimensional structure is provided by the chemistry of hemoglobin, a protein with a molecular weight of 65,000 amu that carries oxygen through the body. The protein contains four chains of amino acids, two α chains and two β chains.

The structure of hemoglobin is shown in Figure 4.11. Sickle-cell anemia occurs when the identity of a single amino acid among the 146 amino acids on a β chain is changed. The substitution of valine for glutamic acid at the sixth position on the chain produces a subtle change in the structure of the hemoglobin, which interferes with its ability to pick up oxygen at low pressures. The result is so severe that children who inherit this disorder from both parents seldom live past the age of 2 years.

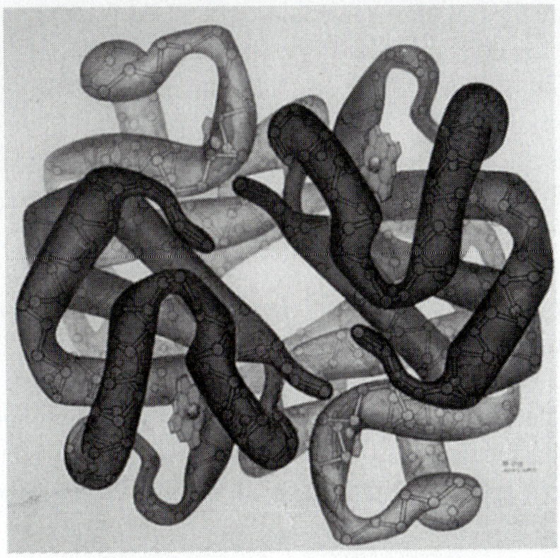

Fig. 4.11 Change in the identity of one of the 146 amino acids on one of the chains in hemoglobin causes a large enough change in the structure of hemoglobin to interfere with its ability to carry oxygen through the blood. Reprinted from R. E. Dickerson and I. E. Geis, *The Structure and Action of Proteins,* W. A. Benjamin, Inc., Menlow Park, CA, 1969.

Research for the New Millennium

THE SHAPES OF MOLECULES

Research in recent years has greatly expanded our understanding of how the shape of a molecule affects the way it binds to certain receptors on the membrane of a cell. It has long been known, for example, that cholera toxin binds to the cells that line the intestine. Once this happens, the toxin can slip through the cell membrane without destroying the cell. This initiates a series of reactions that induce the intestinal cells to secrete so much water—up to 9 gallons a day—that cholera victims die of the debilitating effects of diarrhea.

Until relatively recently, the mechanism by which cholera toxin achieved these results was unknown. In 1991 Rongguang Zhang and Edwin Westbrook at Argonne National Laboratory reported the structure of the cholera toxin [*Science,* **253,** 382–383 (1991)]. The protein contains two subunits. The B subunit, which has a molecular weight of about 55,000 amu, forms the doughnut structure shown in Figure 4.12.

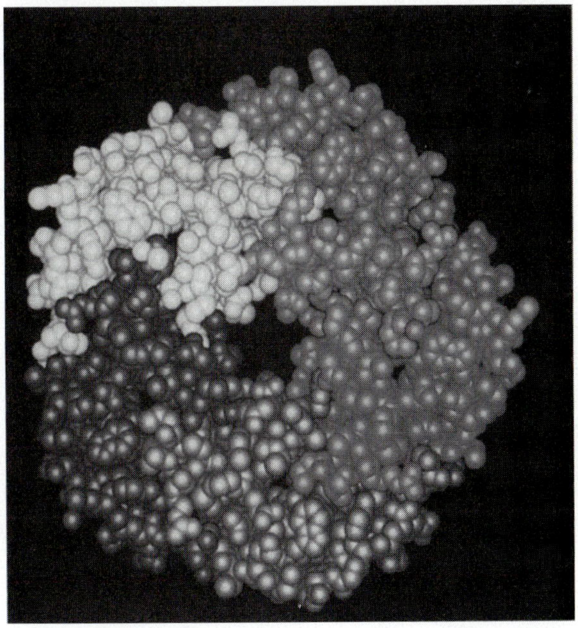

Fig. 4.12 Poison doughnut, the B subunit of cholera toxin.

The A subunit is smaller, with a molecular weight of 29,000 amu. The B subunit apparently binds to the cell and then pushes the smaller A subunit through the membrane of the cell. The A subunit then acts as an enzyme, initiating the reactions that lead to the secretion of water.

In the short term, knowledge of the structure of cholera toxin will drive research toward the discovery of vaccines that can prevent cholera. In the long term, it may provide the basis for the design of a system to deliver drugs across the cell membrane to destroy cancer cells.

Another example of how the structure of a molecule affects the way it binds to a receptor site is provided by progesterone and the synthetic analog of this steroid shown in Figure 4.13. Progesterone is a hormone that is secreted in the second half of the menstrual cycle. It binds to receptors in the endometrium, preparing the uterus for implantation of a fertilized egg. If pregnancy occurs, continued production of progesterone is essential for maintenance of the embryo, and eventually the fetus.

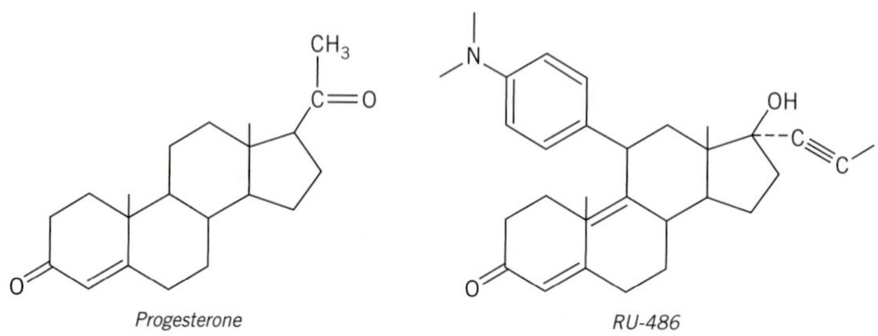

Fig. 4.13 Structures of progesterone and a synthetic analog that is the basis of the drug known as RU-486.

In 1975 Georges Teutsch initiated a project to study how small changes in the structure of progesterone might affect the ability of the molecule to bind to the five kinds of steroid receptors in a cell. Synthetic steroid hormones are divided into two categories. The *agonists* produce the same effect as the natural hormone when they bind. The *antagonists* bind to the receptor but don't switch on the activity induced when the natural hormone binds.

Teutsch and co-workers were searching for an antagonist for glucocorticoid receptors, which might increase the rate at which burns and other wounds heal. When they tested a compound originally known as RU-38486, they found that it was a powerful glucocorticoid antagonist. But it did more than that: It also bound very tightly to the progesterone receptor. As a progesterone antagonist, RU-486, as it became known, was a candidate for testing as a drug for fertility control.

In September 1988, RU-486 became available in France for the termination of pregnancy. The protocol for its use involves delivering 600 mg of the drug in a single dose, followed 36 to 48 hours later with a dose of a prostaglandin to induce the contractions necessary to expel the embryo from the uterus. When this protocol is followed, the rate of success is similar to that of surgical procedures [*New England Journal of Medicine,* **322**(10), 645–648 (1990)].

Because it binds so tightly to several kinds of hormone receptors, RU-486 has the potential to serve other functions. It could be administered with oxytocin to induce labor at the end of pregnancy, therefore reducing the number of cesarean deliveries that have to be done. It might be used to treat cancers that involve progesterone receptors, such as certain forms of breast cancer. It might also control the growth of certain noncancerous tumors that affect the brain. It might even be used for its original purpose—as a glucocorticoid antagonist. Research is sure to continue on this controversial drug and its potential uses.

4.14 Predicting the Shapes of Molecules [The Electron Domain Model]

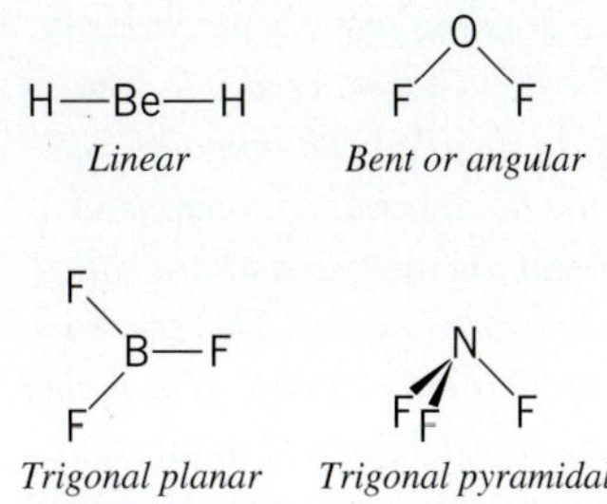

Fig. 4.14 There is no obvious relationship between the chemical formula and the shape of a molecule. BeH_2 is linear, whereas OF_2 is bent. BF_3 is a planar molecule, whereas NF_3 is pyramidal.

There is no direct relationship between the formula of a compound and the shape of its molecules, as shown by the examples in Figure 4.14. The three-dimensional shapes of these molecules can be predicted from their Lewis structures, however, with a model developed in the 1960s.[6,7] When it was introduced, it was called the *valence-shell electron pair repulsion (VSEPR) model.* The model has been modified and is now referred to as the **electron domain (ED) model.**[8]

It is surprising that the spatial arrangement of atoms in molecules conforms to a relatively small number of types. Angles between selected atoms tend generally to be 90°, 109.5°, 120°, or 180°. The limited number of ways of arranging atoms can be understood from the electron domain model.

The electron domain model provides the basic knowledge that makes the subsequent introduction of valence bond theory and hybrid orbitals in the Chapter 4 Special Topics or in future courses more understandable. In essence the electron domain model integrates a simple bonding picture with the VSEPR model.

The electrons in the valence shell of an atom in a molecule form pairs with opposite spins. Each pair occupies its own domain and is attracted to the central atom. The domains get as close to the central atom as they can but keep other domains as far away as possible. The ED model assumes that the geometry around each atom in a molecule can be predicted by keeping the domains of electron pairs separated. Thus two domains will be separated by 180°. For three domains the best arrangement is 120°, whereas four domains are 109.5° apart. Figure 4.15 shows how domains, treated as spheres, are arranged about a central atom in accordance with the ED model.

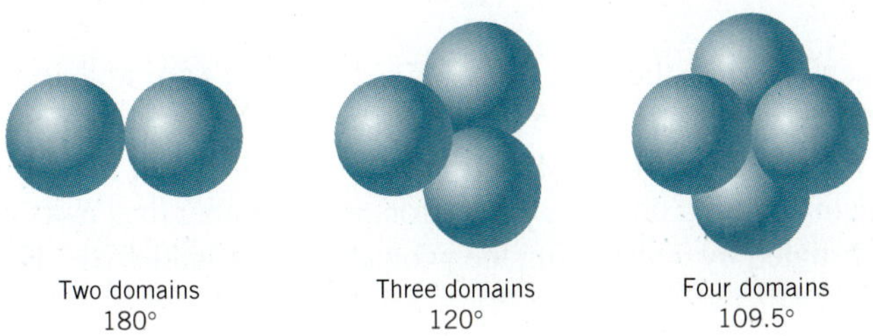

Fig. 4.15 Arrangement of domains about a central atom. Two domains take up a position on opposite sides of a central atom, three are arranged at an angle of 120°, and four at 109°.

A bonding domain is shared by two atoms while a nonbonding domain belongs entirely to the valence shell of a particular atom. Therefore, a nonbonding domain tends to spread out and occupy a larger space than does a bonding domain.

A domain is a region of space that can accommodate a pair of electrons. The electrons occupy either bonding or nonbonding domains. These domains are distributed about the central atom according to geometric considerations, as has long been done in VSEPR. In the ED model, multiple bonding is treated by combining two or more electron domains into a larger domain.

We can see how the ED model is used by applying it to BeH_2. The Lewis structure of the molecule suggests that there are two pairs of bonding electrons, or two bonding domains, in the valence shell of the central atom. Note that this molecule is an exception to the octet rule because only two pairs of electrons surround the central atom. However, this Lewis structure is supported by experimental data.

H—Be—H

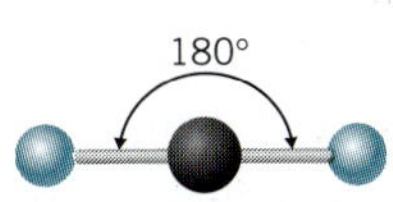

We can keep the two bonding domains as far apart as possible by arranging them on either side of the beryllium atom. The ED model therefore predicts that BeH_2 should be a **linear** molecule, with a 180° angle between the two Be—H bonds.

[6] R. J. Gillespie and R. S. Nyholm, *Quarterly Review of the Chemical Society,* **11,** 339 (1957).

[7] R. J. Gillespie, *Journal of Chemical Education,* **40,** 295 (1963).

[8] R. J. Gillespie, *Journal of Chemical Education,* **69,** 116 (1992).

A more complicated example shows how the ED model can be used for molecules with multiple bonds. Consider the Lewis structure of CO_2.

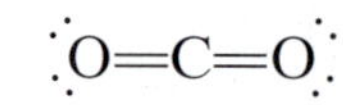

There are four pairs of bonding electrons in the valence shell of the carbon atom but only two domains in which those electrons can be found. (There are two pairs of electrons in the C=O double bond on the left and two pairs in the double bond on the right.) As a result, the ED model predicts that this molecule will also have a linear geometry, with a 180° angle between the two double bonds.

It is assumed that the two electron pairs which make up the double bond occupy more space than a single bond domain. The three electron pairs of a triple bond likewise require more space than a double bond domain. The relative sizes of the spaces occupied by single, double, and triple bonds will determine how these bonding regions are distributed around the central atoms. The space occupied by a double bond, which consists of two electron pairs, is called a double bond domain. Correspondingly, the three electron pairs that form a triple bond are collectively called a triple bond domain.

There are three domains in the valence shell of the central atom in boron trifluoride (BF_3) where electrons can be found. The domains correspond to the three pairs of bonding electrons in the B—F bonds.

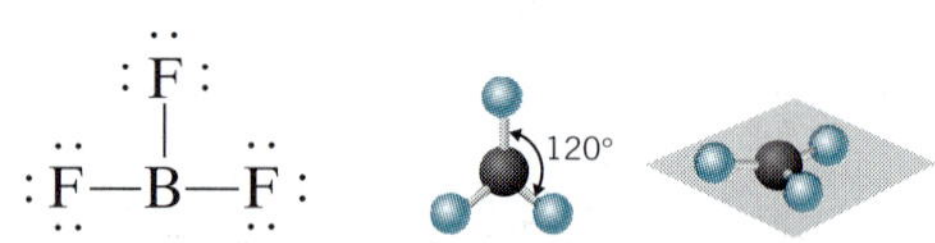

The optimum geometry to keep the three domains as far apart as possible is an equilateral triangle. The ED model therefore predicts a **trigonal planar** geometry for the BF_3 molecule, with a F—B—F bond angle of 120°, as shown in Figure 4.16.

The advantage of counting domains of electron density rather than pairs of electrons can also be illustrated by considering the geometry of formaldehyde, H_2CO.

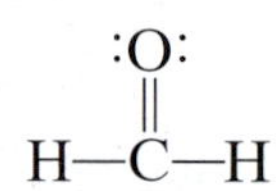

The Lewis structure of the molecule suggests that there are four pairs of electrons in the valence shell of the central atom. Two pairs form a double bond domain, and there are two single bond domains. The ED model therefore predicts that this molecule would have a trigonal planar geometry, just like BF_3.

At first glance, we might expect that the H—C—H bond angle in formaldehyde would be 120°, the internal angle in an equilateral triangle. Experiment suggests that it is a little smaller than this, only 118°. This can be explained by recalling that according to the ED model the C=O double bond domain occupies more space than the domains that contain the electrons in the C—H bonds. As a result, the H—C—O bond angles are slightly larger than 120°, and the H—C—H bond angle is slightly smaller than 120°, as shown in Figure 4.16.

BeH_2, CO_2, BF_3, and H_2O are all two-dimensional molecules, in which the atoms lie in the same plane. If we place the same restriction on methane (CH_4), we would get a square planar geometry, with an H—C—H angle of 90°.

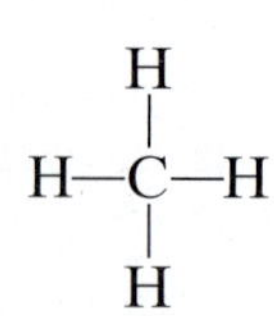

The two bonding pairs of electrons in a double bond are merged into what is termed a "double bond domain." Similarly, the three bonding domains of a triple bond are merged and are called a "triple bond domain." This terminology simplifies the counting of domains about a central atom and allows students to find correct geometries with a minimum of confusion.

The domain model provides a more understandable and less abstract way of obtaining the geometry of molecules. No use is made of hydrogen-like orbitals or hybrid orbitals, neither of which can be given adequate justification at this level. Because chemists use hybrid orbital terminology (sp^3, sp^2, etc.) it may be desirable to relate this terminology to the various geometries found through the ED model.

The use of the domain model and the influence of the double and triple bonds on geometry can be pointed out here with the example of Figure 4.16.

➤ CHECKPOINT

The structure of benzene was discussed in the checkpoint at the end of Section 4.9. Determine the number of electron pair domains around each carbon atom.

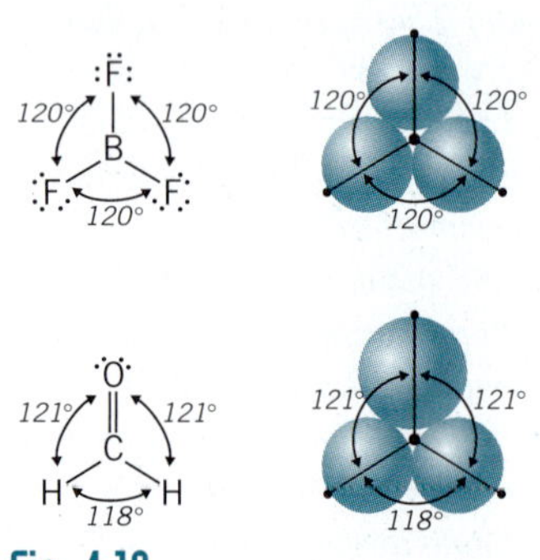

Fig. 4.16 Three equal bonding domains surround the central B atom in BF_3, giving an angle of 120°. The three bonding domains around the carbon atom in formaldehyde are not the same size, and the H—C—O angle increases to 121°.

There is a more efficient way of arranging the four domains in the valence shell of the central atom, however. If the four domains are arranged toward the corners of a **tetrahedron,** the H—C—H angle increases to 109.5°, as shown in Figure 4.17.

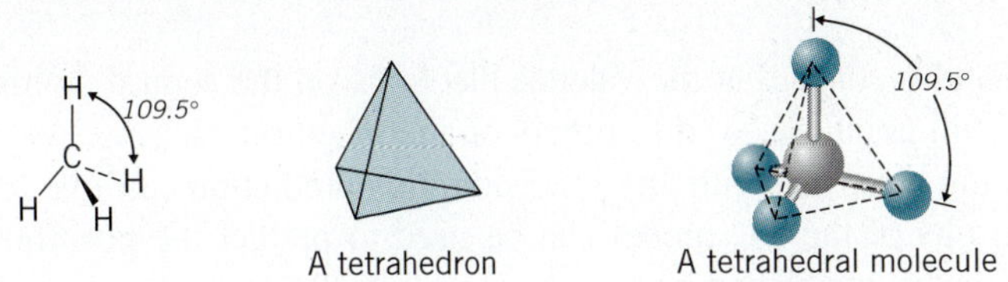

Fig. 4.17 Four equal bonding domains around the central carbon atom in methane give bond angles of 109.5°.

PF_5 has five single bond domains about the phosphorus atom.

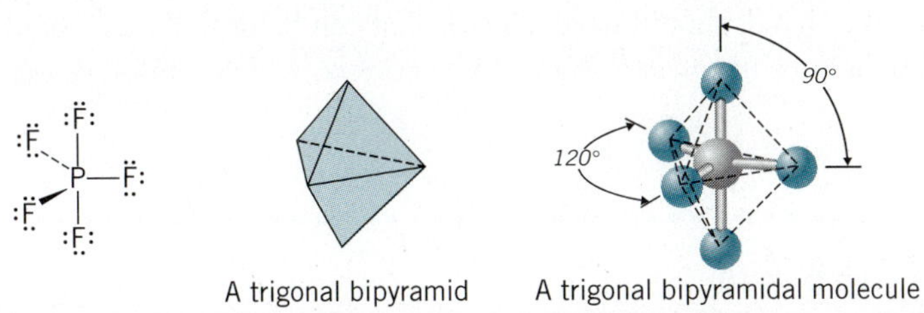

The best arrangement of the domains is toward the corners of a **trigonal bipyramid.** Three of the positions in a trigonal bipyramid are labeled equatorial because they lie along the equator of the molecule. The other two are axial because they lie along an axis perpendicular to the equatorial plane. The angle between the three equatorial positions is 120°, whereas the angle between an axial and an equatorial position is 90°.

There are six single bond domains on the central atom in SF_6.

➤ CHECKPOINT

Predict the shape of a molecule with five electron domains around the central atom. Is your answer dependent on how many of the five domains are bonding or nonbonding domains?

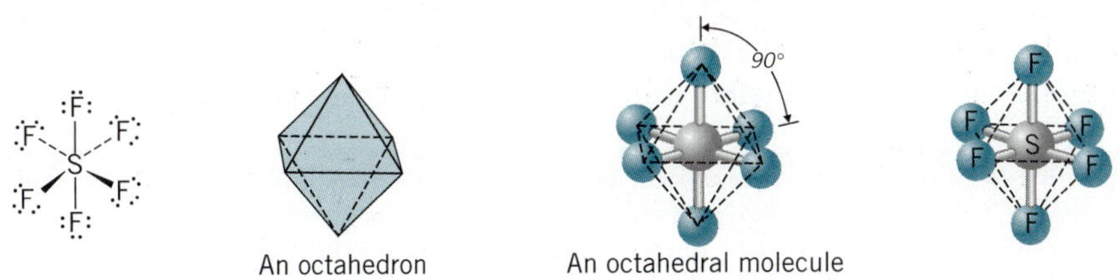

The optimum geometry for the molecule would involve arranging the six domains toward the corners of an **octahedron.** The term *octahedron* literally means "eight sides," but it is the six corners, or vertices, that interest us. To envision the geometry of an SF_6 molecule, imagine four fluorine atoms lying in a plane around the sulfur atom with one fluorine atom above the plane and another below. All F—S—F angles are 90°.

4.15 The Role of Nonbonding Electrons in the ED Model

All of our examples, so far, have contained only bonding electrons in the valence shell of the central atom. What happens when we apply the ED model to atoms that also contain nonbonding electrons? Consider ammonia (NH_3) and water (H_2O), for example.

$$\mathrm{H{-}\ddot{N}{-}H} \quad (\text{with H below N}) \qquad \mathrm{H{-}\ddot{\underset{\cdot\cdot}{O}}{-}H}$$

In each case, there are four domains in the valence shell of the central atom where electrons can be found. The valence electrons on the central atom in both NH_3 and H_2O therefore should be distributed toward the corners of a tetrahedron, as shown in Figure 4.18. Our goal, however, isn't the prediction of the distribution of valence

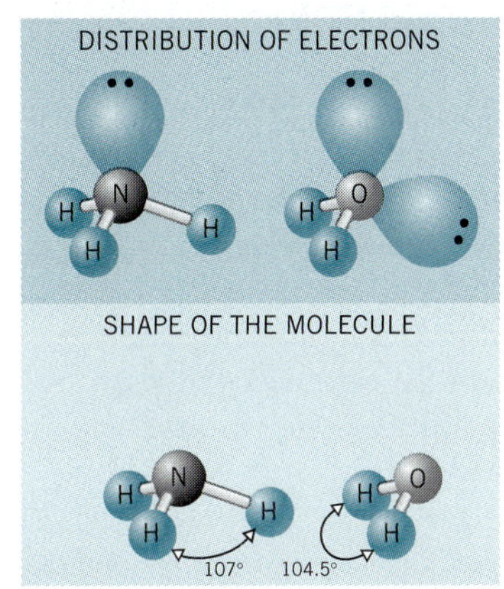

Fig. 4.18 The ED model predicts that the valence electrons on the central atom in NH_3 and H_2O will be oriented toward the corners of a tetrahedron. The shape of the molecules, however, is determined by the positions of the atoms. Ammonia is therefore described as trigonal pyramidal, and water is described as bent, or angular.

One of the advantages of the electron domain model it that is emphasizes that the most stable arrangement of electron-pair domains is for the domains to be as far apart from one another as possible due to the volume of space that the domains occupy. Students are sometimes confused by the term repulsion *in the VSEPR theory, believing that the repulsion is electrostatic repulsion due to the like charges on the electrons.*

electrons. It is to use the distribution of electrons to predict the geometry of the molecule, which describes how the atoms are distributed in space. In previous examples, the way the valence electrons were distributed and the geometry of the molecule have been the same. Once we include nonbonding electrons, this is no longer true.

The ED model predicts that the valence electrons on the central atoms in ammonia and water will point toward the corners of a tetrahedron. Because we can't locate the nonbonding electrons with any precision, this prediction can't be tested directly. But the results of the ED model can be used to predict the positions of the atoms in the molecules, which can be tested experimentally. If we focus on the positions of the atoms in ammonia, we predict that the NH_3 molecule should have a shape best described as **trigonal pyramidal,** with the nitrogen at the top of the pyramid. Water, on the other hand, should have a shape that can be described as **bent,** or **angular.** Both predictions have been shown to be correct, which reinforces our faith in the ED model.

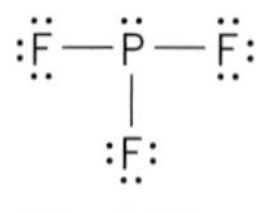

Fig. 4.19 Lewis structure of the PF_3 molecule. There are one nonbonding and three bonding domains.

Exercise 4.6

Use the Lewis structure of the PF_3 molecule shown in Figure 4.19 to predict the shape of this molecule.

Solution

The Lewis structure in Figure 4.19 suggests that there are four domains in the valence shell of the phosphorus atom where electrons can be found. There are electron pairs in the three P—F single bonds and a pair of nonbonding electrons. The geometry for the molecule is based on arranging these domains toward the corners of a tetrahedron. The shape of the molecule is therefore *trigonal pyramidal,* like the geometry of ammonia.

Bartell and Plato pointed out that ". . . the mutual avoidance (Pauli avoidance) of occupied, localized molecular orbitals is . . . the proper basis for the valence shell electron pair repulsion theory. . . . Discussion in many papers regard valence shell electron pair repulsions as if they were strictly electrostatic. . . . Frequenly this arises from a misconception of the VSEPR model. . ."[3]

Gillespie and Hargittai have stressed that electrons do undergo coulombic repulsion with each other but that ligands provide the attractive force necessary to overcome the electronic repulsions and cause two electrons to occupy the same region of space from which all other electrons are excluded.[4]

It should be emphasized here that the domain model assumes that nonbonding domains take up more space than bonding domains.

When we extend the ED model to molecules in which the domains are distributed toward the corners of a trigonal bipyramid, we run into the question of whether nonbonding electrons should be placed in equatorial or axial positions. Experimentally we find that nonbonding electrons usually occupy equatorial positions in a trigonal bipyramid.

To understand why, we have to recognize that nonbonding electron domains take up more space than bonding electron domains. Nonbonding domains are close to only one nucleus, and there is a considerable amount of space in which nonbonding electrons can reside and still be near the nucleus of the atom. Bonding electrons, however, must be simultaneously close to two nuclei, and only a small region of space between the nuclei satisfies this restriction.

Figures 4.20 and 4.21 can help us understand why nonbonding electrons are placed in equatorial positions in a trigonal bipyramid. If the nonbonding electron domain in SF_4 is placed in an axial position, it will be relatively close (90°) to *three* bonding-pair domains. But if the nonbonding domain is placed in an equatorial position, it will be 90° away from only *two* bonding domains. As a result, an axial position is more crowded than an equatorial position in a trigonal bipyramid arrangement. Larger nonbonding domains therefore occupy the equatorial positions.

The results of applying the ED model to SF_4, ClF_3, and the I_3^- ion are shown in Figure 4.21. When the nonbonding domain on the sulfur atom in SF_4 is placed in an equatorial position, the molecule can be best described as having a **seesaw,** or ***teeter-totter,*** shape. The bonding and nonbonding domains in the valence shell of

[3] L. S. Bartell and V. Plato, *J. Am. Chem. Soc.* **95,** 3097 (1973).

[4] R. J. Gillespie and L. Hargittai, *The VSEPR Model of Moleclar Geometry,* Allyn & Bacon, Boston (1991).

chlorine in ClF_3 can best be accommodated by placing both nonbonding domains in equatorial positions in a trigonal bipyramid. When this is done, we get a geometry that can be described as **T-shaped.** The Lewis structure of the triiodide (I_3^-) ion suggests a trigonal bipyramidal distribution of valence electrons on the central atom. When the three nonbonding electron domains on the central I atom are placed in equatorial positions, we get a linear molecule.

Molecular geometries based on an octahedral distribution of electron domains are easier to predict because the corners of an octahedron are all identical.

The ED model is summarized in Table 4.3. The number of bonding and nonbonding domains around the central atom determines the molecular geometry. For example, SO_2 has three electron domains around the central atom (two bonding domains and one nonbonding domain). We can determine the geometry around the central atom by locating the number of electron domains in column one. A central atom surrounded by three electron domains will have one of the three possible molecular geometries listed for that entry in column 5. Reading across the table we find that the electrons are distributed about the central atom in a trigonal planar

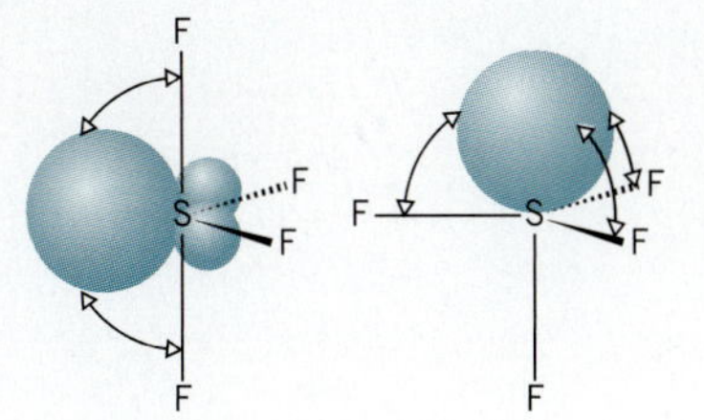

Fig. 4.20 Crowding of bonding and nonbonding domains in SF_4 is minimized if the nonbonding domain is placed in an equatorial position, as shown in the structure on the left.

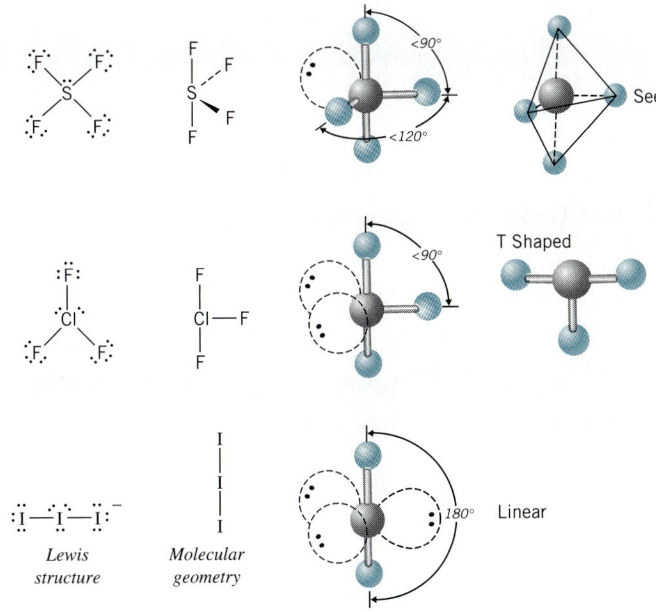

Fig. 4.21 Structures of SF_4, ClF_3, and the I_3^- ion.

Exercise 4.7

Use the Lewis structures of BrF_5 and XeF_4 shown in Figure 4.22 to predict the shapes of the molecules.

Solution

The valence-shell electrons on the bromine atom in BrF_5 are distributed toward the corners of an octahedron to form a molecule with a **square pyramidal** geometry.

For XeF_4 two of the six domains in the valence shell of the central atom are nonbonding domains and four are bonding domains. Because they occupy considerably more space, the nonbonding domains are kept as far apart as possible. The ED model therefore predicts a **square planar** structure.

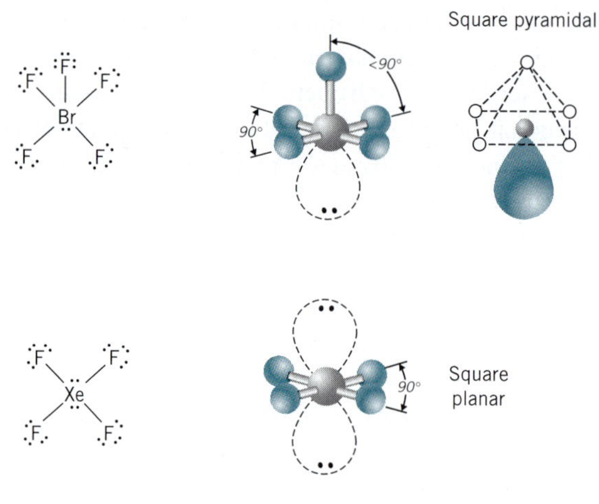

Fig. 4.22 Lewis structures and geometry of BrF_5 and XeF_4.

fashion. The two bonding and one nonbonding domains allow us to determine that the molecular geometry of SO2 is bent.

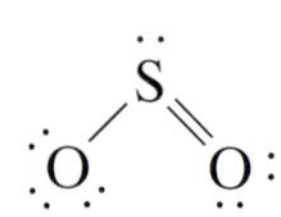

Table 4.3

Relationship Between Number of Electron Domains and Geometry Around an Atom

Electron Domains	Bonding Domains	Nonbonding Domains	Distribution of Electron Domains	Molecular Geometry	Examples
2 (*sp*)	2	0	Linear	Linear	BeH_2, CO_2
	1	1		Linear	CO, N_2
3 (sp^2)	3	0	Trigonal planar	Trigonal planar	BF_3, CO_3^{2-}
	2	1		Bent	O_3, SO_2
	1	2		Linear	O_2
4 (sp^3)	4	0	Tetrahedral	Tetrahedral	CH_4, SO_4^{2-}
	3	1		Trigonal pyramidal	NH_3, H_3O^+
	2	2		Bent	H_2O, ICl_2^+
	1	3		Linear	HF, OH^-
5 (sp^3d)	5	0	Trigonal bipyramidal	Trigonal bipyramidal	PF_5
	4	1		Seesaw	SF_4, IF_4^+
	3	2		T shaped	ClF_3
	2	3		Linear	I_3^-, XeF_2
6 (sp^3d^2)	6	0	Octahedral	Octahedral	SF_6, PF_6^-
	5	1		Square pyramidal	BrF_5, $SbCl_5^{2-}$
	4	2		Square planar	XeF_4, ICl_4^-

➤ CHECKPOINT

What is the shape of a molecule with three bonding domains and one non-bonding domain around the central atom?

Chemists frequently use special symbols to describe the geometry of the electron domain distribution. Two domains arranged linearly, are called an *sp* geometry. Three electron domains are called sp^2, and four electron domains are designated sp^3. These designations are also given in Table 4.3. The SO_2 molecule is thus described as having three electron domains or as sp^2. A more detailed description of the use of these labels is given in the Special Topics section at the end of this chapter.

4.16 Bond Angles

Hybrid orbitals are covered in more depth in the Special Topics section at the end of the chapter. Here only the designations sp, sp^2, sp^3, etc. are given. Because chemists use hybrid orbital terminology (sp^3, sp^2 etc), it may be desirable to relate this terminology to the various geometries found through the ED model.

Students have no more difficulty accepting sp^3 as meaning a tetrahedral arrangement than they do in accepting s, p or d designations for subshells. Hybrid orbitals, although known to be an incomplete representation of bonding, just as is the Lewis model, are useful and chemists use the symbology regularly.

Students frequently do not understand that three atoms are necessary to determine a bond angle. Thus some emphasis is placed on this point here.

Just as bond lengths are important to the structure and hence the properties of a compound, so are bond angles. To define a bond angle, at least three atoms must be located. Consider, for example, propyl alcohol, $CH_3CH_2CH_2OH$.

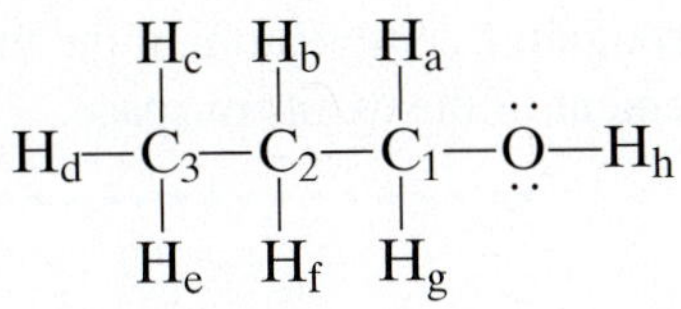

It is difficult to imagine the three-dimensional structure of a molecule from the flat, two-dimensional representations on the printed page of a text. Consider the geometry around the carbon atom labeled C_3. The best way of arranging the four bonding domains on this atom is toward the corners of a tetrahedron. Each of the bond angles on this carbon is therefore 109.5°, regardless of whether we look at the $H_c—C_3—H_d$, $H_d—C_3—H_e$, or $H_c—C_3—H_e$ angles. The bond angles on C_2 and C_1 are also 109.5° because there are four bonding domains on these atoms. The $C_1—O—H_h$ angle, however, is slightly smaller than 109.5° because the oxygen atom is surrounded by two single bond domains and two nonbonding domains. Because the nonbonding domains occupy more space than the bonding domains, the bonding pairs are somewhat squeezed to form an angle slightly less than 109°.

The ammonia molecule contains three bonding domains and one nonbonding domain, while the water molecule contains two bonding domains and two nonbonding domains. The ED model predicts that the four electron domains in water and ammonia will be arranged in a tetrahedral geometry around the central atom. On the basis of the tetrahedral geometry shown in Figure 4.17, we might expect the bond angles in ammonia and water to be 109.5° as found in methane, CH_4. Although this is a good approximation of the bond angle, the experimentally determined values for the bond angles in ammonia and water are 107° and 104.5°, respectively. The differences can be explained by nonbonding domains taking more space than bonding domains, thus pushing the bonding domains closer together and giving a smaller than expected bond angle.

➤ CHECKPOINT

Give the following bond angles for propyl alcohol:

$C_1—C_2—C_3$
$C_2—C_1—H_g$
$H_g—C_1—O$

Exercise 4.8

Predict the following bond angles for acetic acid.

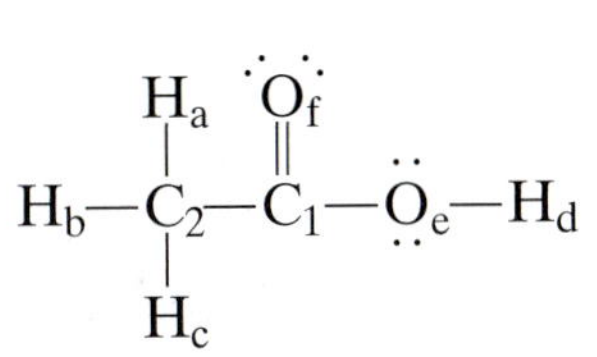

$H_a—C_2—H_b$	$C_1—O_e—H_d$
$C_2—C_1—O_f$	$O_f—C_1—O_e$
$C_2—C_1—O_e$	$H_a—C_2—C_1$

Solution

The Lewis structure of acetic acid shows that there are four bonding domains around C_2, three bonding domains around C_1, and a total of four domains (two

➤ CHECKPOINT

The structure of benzene was discussed in the checkpoint at the end of Section 4.9. Determine the bond angles around each carbon atom.

bonding and two nonbonding) around O_e. The resulting bond angles are as follows:

$$H_a—C_2—H_b = 109.5° \qquad C_1—O_e—H_d < 109°$$
$$C_2—C_1—O_f = 120° \qquad O_f—C_1—O_e = 120°$$
$$C_2—C_1—O_e = 120° \qquad H_a—C_2—C_1 = 109.5°$$

The number of domains around the central atom of the three atoms forming the angle determines the arrangement of the atoms in space.

4.17 The Difference Between Polar Bonds and Polar Molecules

The difference between the electronegativities (ΔEN) of chlorine ($EN = 2.87$) and hydrogen ($EN = 2.30$) is sufficient that the bond in HCl is polar. The partial charge on the hydrogen is positive and that on the chlorine is negative. Such a bond is said to be *polar.*

$$\overset{\delta+}{H}—\overset{\delta-}{Cl}$$
$$\longmapsto$$

Because it contains only this one bond, the HCl molecule can also be described as polar. The polarity of a bond in a structure is represented with an arrow having a plus sign at the positive end (the $\delta+$ and $\delta-$ shown in the above structure could be replaced by the ⟼). The arrow points in the direction of the more electronegative atom, and the plus sign is next to the least electronegative atom.

The polarity of a molecule can be predicted by considering the polarities of the individual bonds, the location of nonbonding pairs, and the three-dimensional geometric shape of the molecule. The magnitude of the polarity of a bond or of a molecule is measured using a quantity known as the **dipole moment,** μ. The dipole moment for a molecule depends on two factors: (1) the magnitude of the charge and (2) the distance between the negative and positive poles of the molecule. Dipole moments are reported in units of *debye* (D). The dipole moment for HCl is small, 1.08 D (see Table 4.4). This can be understood by noting that the difference in charge in the HCl bond is relatively small (the difference in electronegativity between the two atoms, $\Delta EN = 0.57$) and that the H—Cl bond is relatively short.

C—Cl bonds ($\Delta EN = 0.33$) are not as polar as H—Cl bonds ($\Delta EN = 0.57$), but they are significantly longer. As a result, the dipole moment for CH_3Cl is about the same as that for HCl, namely, 1.01 D. At first glance, we might expect a similar dipole moment for carbon tetrachloride (CCl_4), which contains four polar C—Cl bonds. The dipole moment of CCl_4, however, is zero. This can be rationalized by considering the structure of CCl_4 shown in Figure 4.23. The individual C—Cl bonds in this molecule are polar. However, since the dipole moments of the four C—Cl bonds are equivalent and the bonds are symmetrically arranged at the corners of a tetrahedron, the four C—Cl dipoles cancel each other. Carbon tetrachloride therefore illustrates an important point: Not all molecules that contain polar bonds have dipole moments.

Table 4.4
Dipole Moments of Common Compounds

Compound	Dipole Moment (D)
CH_4, methane	0
CH_3OH, methanol	1.70
CO, carbon monoxide	0.112
CO_2, carbon dioxide	0
HCl, hydrogen chloride	1.08
HI, hydrogen iodide	0.44
H_2O, water	1.85
H_2S, hydrogen sulfide	0.97

➤ CHECKPOINT

Does chloroform have a dipole moment?

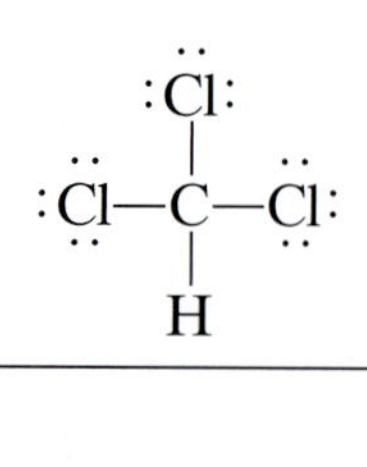

Exercise 4.9

The two-dimensional Lewis structure for dichloromethane is given below. On first inspection it would appear that the polarities of the two C—H bonds and the two C—Cl bonds should cancel one another, resulting in a nonpolar molecule with a dipole moment of zero. However, the measured dipole moment of dichloromethane is 1.60 D. Explain this discrepancy.

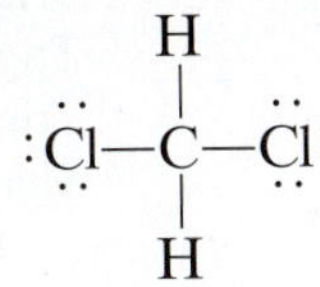

Solution

Dichloromethane is not the flat two-dimensional molecule shown in the Lewis structure but instead is tetrahedral as shown below. The dipole moments of the C—H bonds point toward the carbon atom. The dipole moments of the C—Cl bonds point away from the carbon atom. Therefore, the dipole moments of the bonds are additive to give a net dipole moment to the molecule as shown below.

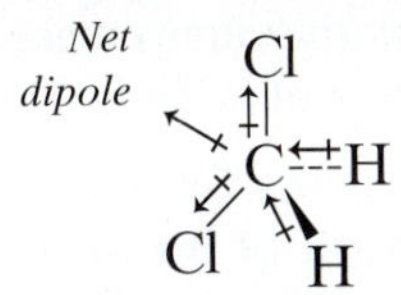

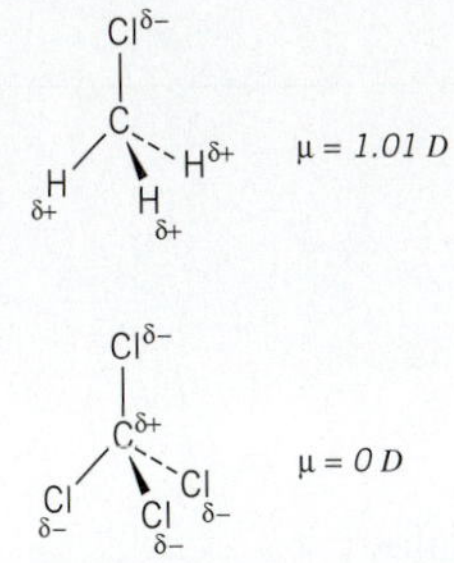

Fig. 4.23 Although the C—Cl bonds in both CH_3Cl and CCl_4 are polar, CCl_4 has no net dipole moment.

Exercise 4.10

Determine whether each of the following molecules is polar.

(a) NH_3

(b) CO_2

Solution

(a) The structure for ammonia is shown below. The dipoles of the N—H bonds are additive to give an overall dipole to the molecule. In addition, the nonbonding pair on the nitrogen adds to the overall dipole moment.

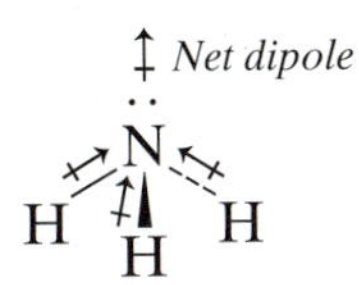

(b) The structure for carbon dioxide is shown below. Carbon dioxide is a linear molecule with an oxygen on each side of the carbon. Both C=O double bonds are polar, but their dipoles point in opposite directions, causing them to cancel one another and leading to a nonpolar molecule.

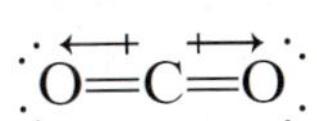

Key Terms

Angular
Bent
Bond length
Bonding domains
Bonding electrons
Covalent bond
Covalent molecules
Dipole moment
Domains
Electron domain (ED) model
Electronegativity
Formal charge
Ionic
Lewis structures
Linear
Nonbonding domains
Nonbonding electrons
Octahedron
Octet rule
Partial charge
Polar covalent bond
Resonance hybrid
Seesaw
Square planar
Square pyramidal
T-shaped
Tetrahedron
Trigonal bipyramid
Trigonal planar
Trigonal pyramidal
Valence electrons

Problems

Valence Electrons

1. Define the term *valence electrons.*
2. Determine the number of valence electrons in neutral atoms of the following elements.
(a) Li (b) C (c) Mg (d) Ar
3. Determine the number of valence electrons in neutral atoms of the following elements.
(a) Fe (b) Cu (c) Bi (d) I
4. Determine the number of valence electrons in the following negative ions and describe any general trends.
(a) C^{4-} (b) N^{3-} (c) S^{2-} (d) I^-
5. Determine the number of valence electrons in the following positive ions and describe any general trends.
(a) Na^+ (b) Mg^{2+} (c) Al^{3+} (d) Sc^{3+}
6. Explain why filled *d* and *f* subshells are ignored when the valence electrons on an atom are counted.
7. What is meant by the "octet rule"?
8. What is the difference between valence electrons and core electrons?
9. What is the relationship between the number of valence electrons and the group number?

The Covalent Bond

10. Write Lewis structures for the elements in the third row of the periodic table.
11. Describe the difference between the covalent bonds in O_2 and F_2. How many valence electrons surround an F atom in F_2? How many are around an O atom in O_2?
12. Draw the Lewis structure for N_2. What is a covalent molecule?

How Does the Sharing of Electrons Bond Atoms?

13. Why does the sharing of electrons between atoms offset the repulsive forces of the positively charged nuclei?
14. What must happen to the spin of two electrons if the electrons are to occupy a domain of space between two bonding nuclei?
15. What is a bonding domain?
16. If a stable bond is formed, what must be the relationship between attractive and repulsive forces between two nuclei?
17. Describe how the sharing of a pair of electrons by two chlorine atoms makes a Cl_2 molecule more stable than a pair of isolated Cl atoms.

Using Lewis Structures to Understand the Formation of Bonds

18. How many valence electrons does the hydrogen atom have? How many valence electrons does the chlorine atom have? Give a representation of these two neutral atoms using dots for electrons. Write an equation using dot representations that shows how the HCl molecule is formed from H and Cl atoms.
19. Give the electron configurations for N and H. Use symbols with dots as electrons to represent the distribution of valence electrons around each atom. Combine the N and H atoms to form the NH_3 molecule. Identify the bonding and nonbonding domains in NH_3.
20. For the molecule O_2, how many bonding electrons are there? How many nonbonding electrons?
21. In forming Lewis structures, how are the number of valence electrons that must be assigned to each atom determined?
22. How many bonding domains must there be in the following molecules?
(a) SO_2 (b) BeF_2 (c) CH_4 (d) Br_2
23. One of the following molecules has two bonding domains. Which one is it?
(a) BF_3 (b) NH_3 (c) O_3

Drawing Skeleton Structures

24. Draw skeleton structures of the following molecules.
(a) CH_4 (b) CH_3Cl (c) H_2CO
25. Which of the following skeletal structures is best for SO_2? Explain.
(a) O-S-O (b) O-O-S

26. Which of the following skeletal structures is best for NO_2? Explain.
(a) O-N-O (b) N-O-O

27. Draw skeleton structures for:
(a) $\underline{S}O_3$ (b) $\underline{S}O_2$ (c) O_3 (d) $\underline{N}H_3$ (e) $\underline{C}Cl_3H$
The central atom is underlined.

A Step-by-Step Approach to Writing Lewis Structures

28. Determine the total number of valence electrons in the following molecules or ions.
(a) BF_3 (b) CH_4 (c) NH_4^+ (d) H_2SO_4

29. Determine the total number of valence electrons in the following molecules or ions.
(a) KrF_2 (b) SF_4 (c) SiF_6^{2-} (d) ZrF_7^{3-}

30. Which of the following molecules or ions contain the same number of valence electrons?
(a) CO_2 (b) N_2O (c) CNO^- (d) NO_2^+
(e) SO_2 (f) O_3 (g) NO_2^-

31. Draw skeleton structures of the following ions.
(a) NH_4^+ (b) NO_3^- (c) SO_4^{2-}

32. Write Lewis structures for the following ions or molecules.
(a) NH_3 (b) CH_3^+ (c) H_3O^+ (d) BH_4^-

Molecules That Don't Seem to Satisfy the Octet Rule

33. Write Lewis structures for the following ions.
(a) NO_3^- (b) SO_3^{2-} (c) CO_3^{2-} (d) NO_2^+

34. Write Lewis structures for the following ions or molecules.
(a) C_2H_6 (b) C_2H_4 (c) C_2H_2 (d) C_2^{2-}

35. Write Lewis structures for the following nitrogen-containing molecules.
(a) N_2O (b) N_2O_3 ($ON—NO_2$)

36. Write Lewis structures for the following nitrogen-containing molecules.
(a) ClNO (b) $ClNO_2$ (c) NO^+
(d) NO_2^- (e) ONF_3

37. Explain why a satisfactory Lewis structure for N_2O_5 cannot be based on a skeleton structure that contains an N—N bond: $O_2N—NO_3$. Show how a satisfactory Lewis structure can be written if we assume that the skeleton structure is O_2NONO_2.

38. Write Lewis structures for the following ions or molecules.
(a) O_2 (b) O_3 (c) O_2^{2-} (d) O^{2-}

39. Write Lewis structures for the following ions or molecules.
(a) SO_2 (b) SO_3 (c) SO_3^{2-} (d) SO_4^{2-}

40. Write Lewis structures for the following ions or molecules.
(a) XeF_2 (b) XeF_4 (c) XeF_3^+ (d) $OXeF_4$

41. Which of the following ions or molecules have the same electron configurations as the N_2 molecule?
(a) CO (b) NO (c) CN^- (d) NO^+ (e) NO^-

42. Which of the following are exceptions to the Lewis octet rule?
(a) CO_2 (b) BeF_2 (c) SF_4 (d) SO_3

43. Which of the following are exceptions to the Lewis octet rule?
(a) BF_3 (b) H_2CO (c) XeF_4 (d) IF_3

44. A pair of NO_2 molecules can combine to form N_2O_4.

$$2\ NO_2(g) \longrightarrow N_2O_4(g)$$

Use the Lewis structures of these molecules to explain why.

45. Draw Lewis structures for:
(a) PF_5 (b) SF_6 (c) IF_4^+ (d) ICl_4^-
Do any of these molecules or ions obey the octet rule?

Bond Lengths

46. Define *bond length.*

47. The nitrogen-to-nitrogen bond lengths in the following compounds are: N_2, 0.110 nm; HNNH, 0.125 nm; H_2NNH_2, 0.146 nm. In which compound is the nitrogen-to-nitrogen bond most likely to be a single bond? A double bond? A triple bond?

48. Use covalent radii from Figure 3.26 to estimate bond lengths for all the bonds in:
(a) H_2S (b) OF_2 (c) NH_3 (d) BF_3

49. If the sulfur-to-oxygen bond lengths are about the same in SO_2 as in SO_3, what can be said about the bonding between sulfur and oxygen in the two molecules?

Resonance Hybrids

50. Draw all of the possible Lewis structures for the CO_3^{2-} ion.

51. Draw the three Lewis structures that may be used to describe the SCN^- ion.

52. Which of the following molecules or ions have possible resonance hybrids?
(a) HCO_2^- (b) PH_3 (c) HCN (d) C_2H_4

53. Which of the following molecules do not have a possible resonance hybrid?
(a) CO_2 (b) C_2H_2 (c) $CHCl_3$ (d) SO_3

Electronegativity

54. Define *electronegativity* and *AVEE.* Why can electronegativity and AVEE be used interchangeably?

55. Use Figure 3.31 and trends in covalent and metallic radii to explain the trend in electronegativity to increase from left to right across a row in the periodic table.

56. Which of the following atoms is the most electronegative?
(a) S (b) As (c) P (d) Se (e) Cl (f) Br

57. Which of the following series of atoms are arranged in order of decreasing electronegativity?
    (a) C > Si > P > As > Se
    (b) O > P > Al > Mg > K
    (c) Na > Li > B > N > F
    (d) K > Mg > Be > O > N
    (e) Li > Be > B > C > N
58. What is the trend in electronegativity from top to bottom of the periodic chart? Explain.

Partial Charge

59. Calculate the partial charge on the fluorine atom in the following molecules:
    (a) F_2 (b) HF (c) ClF
60. Calculate the partial charge on both atoms in the following molecules: IBr, ICl.
61. Calculate the partial charges on the two atoms in BF.
62. If ordinary purified table salt, NaCl, is heated to 801°C, the solid melts to produce Na^+ and Cl^-. Continued heating causes evaporation of the molten salt. Some of the species that are found in the gas phase are sodium chloride molecules, NaCl(*g*). Similar behavior is encountered for other solid salts when heated to sufficiently high temperatures. Calculate the partial charges on both atoms in the following gas phase molecules: LiCl, NaCl, KCl, RbCl, CsCl. Is the trend in partial charge what you would have expected on the basis of electronegativities?
63. Assign relative partial charges, δ, to the oxygen and hydrogen atoms in the molecule

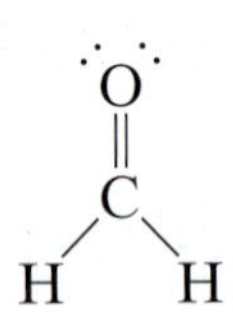

64. For the following molecules, assign relative partial charges, δ, to each atom.
    (a) SO_2 (b) ClO_2 (c) H_2O (d) OCS
    (e) CH_3OH
65. Which atom in each of the following has the most positive partial charge?
    (a) CH_4 (b) CCl_4 (c) BH_3 (d) NO_2 (e) ClNO

Formal Charge

66. Calculate the formal charge on the sulfur atom in the following molecules or ions.
    (a) SO_2 (b) SO_3 (c) SO_3^{2-}
67. Calculate the formal charge on the bromine atom in the following molecules.
    (a) HBr (b) Br_2 (c) HOBr (d) BrF_5
68. Calculate the formal charge on the nitrogen atom in the following ions, molecules, or compounds.
    (a) NH_3 (b) NH_4^+ (c) N_2H_4 (d) NH_2^-
69. Calculate the formal charge on the nitrogen atom in the following molecules.
    (a) N_2O (b) N_2O_3 (c) N_2O_5
70. Calculate the formal charge on the boron and nitrogen atoms in BF_3, NH_3, and the $F_3B—NH_3$ molecule formed when the two gases combine.
71. Calculate the formal charge on the two different sulfur atoms in the thiosulfate ion, $S_2O_3^{2-}$. Assume a skeleton structure that could be described as $S—SO_3$.
72. Draw two correct Lewis structures for ClNO and use formal charge to decide which is the best representation. What experimental evidence could be used to support your choice?
73. Draw two correct Lewis structures for PO_4^{3-} and decide on the basis of formal charge which is the best. What experimental evidence would help support your choice of structure?
74. Which would you expect to be more stable: ozone (O_3) or oxygen (O_2)? Use formal charge to support your answer.
75. Below are two skeleton structures for nitrous acid, HNO_2. Which is the best arrangement of atoms, on the basis of formal charge?

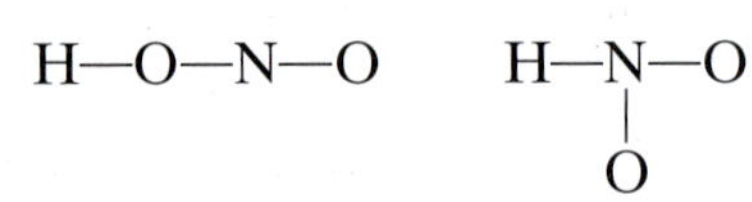

76. Draw all three Lewis structures for the SCN^- ion. Use formal charge to decide which is best.
77. Draw all three Lewis structures for the NCO^- ion. Use formal charge to decide which is best. Explain.

The Shapes of Molecules

78. If four pairs of electrons are around an atom, how will these four domains arrange themselves? How will three domains take position around a central atom? Two domains? Explain.
79. Predict the arrangement of atoms around the underlined central atom in the following molecules or ions.
    (a) $\underline{C}O_3^{2-}$ (b) $\underline{S}O_4^{2-}$ (c) $\underline{P}F_6^-$ (d) $\underline{As}F_5$
    Give the angle the central atom makes with the atoms around it.
80. Complete the table for molecules for which the cental atom has only bonding electrons.

Bonding Domains	Molecular Geometry	Example
—	linear	CO_2
3	—	—
—	tetrahedral	—
5	—	—
—	octrahedral	—

81. How many bonding domains are around the underlined central atom in each of the following?
(a) $\underline{N}H_4^+$ (b) $H\underline{C}Cl_3$ (c) $\underline{Be}H_2$
(d) $O\underline{C}Cl_2$ (e) $O\underline{C}S$
What is the bond angle around the central atom in each case?

The Role of Nonbonding Electrons in the ED Model

82. Determine the number of nonbonding pairs of electrons on the iodine atom in the following molecules or ions.
(a) I_2 (b) I_3^- (c) IF_3 (d) ICl_4^-

83. Which of the following molecules or ions have one or more pairs of nonbonding electrons?
(a) OH^- (b) O_2 (c) CO_3^{2-} (d) Br^-
(e) NH_3

84. Which of the following molecules or ions have one or more pairs of nonbonding electrons?
(a) H_2S (b) NH_4^+ (c) AlH_3 (d) CH_3^- (e) NH_2^-

85. Complete the following table.

Bonding Domains	Nonbonding Domains	Molecular Geometry
2	0	—
—	0	tetrahedral
2	—	bent
2	—	linear
5	0	—
—	—	T shaped
5	—	square pyramidal
—	2	square planar

Distinguish between the geometry associated with the arrangement of the electron domains around the central atom and the molecular geometry.

86. Predict the geometry around the central atom in the following molecules or ions.
(a) PH_3 (b) GaH_3 (c) ICl_3 (d) XeF_3^+

87. Predict the geometry around the central atom in the following molecules or ions.
(a) PO_4^{3-} (b) SO_4^{2-} (c) XeO_4 (d) MnO_4^-

88. The same elements often form compounds with very different shapes. Predict the geometry around the central atom in the following molecules or ions.
(a) SnF_2 (b) SnF_3^- (c) SnF_4 (d) SnF_6^{2-}

89. Sulfur reacts with fluorine to form a pair of neutral molecules—SF_4 and SF_6—that in turn form positive and negative ions. Predict the geometry around the sulfur atom in each of the following.
(a) SF_3^+ (b) SF_4 (c) SF_5^- (d) SF_6

90. Iodine and fluorine combine to form interhalogen compounds that can exist as either neutral molecules, positive ions, or negative ions. Predict the geometry around the iodine atom in each of the following.
(a) IF_2^- (b) IF_3 (c) IF_4^+ (d) IF_4^-

91. Predict the geometry around the central atom in each of the following oxides of nitrogen.
(a) N_2O (b) NO_2^- (c) NO_3^-

92. Predict the geometry around the central atom in the following molecule.
$Hg(CH_3)_2$

93. Which of the following compounds is best described as T shaped?
(a) XeF_3^+ (b) NO_3^- (c) NH_3 (d) ClO_3^- (e) SF_4

94. Which of the following molecules or ions have the same shape or geometry?
(a) NH_2^- and H_2O (b) NH_2^- and BeH_2
(c) H_2O and BeH_2 (d) NH_2^-, H_2O, and BeH_2

95. Which of the following molecules or ions have the same shape or geometry?
(a) SF_4 and CH_4 (b) CO_2 and H_2O
(c) CO_2 and BeH_2 (d) N_2O and NO_2
(e) PCl_4^+ and PCl_4^-

96. Which of the following molecules are best described as bent, or angular?
(a) H_2S (b) CO_2 (c) ClNO (d) NH_2^- (e) O_3

97. Which of the following molecules are planar?
(a) SO_3 (b) SO_3^{2-} (c) NO_3^- (d) PF_3 (e) BH_3

98. Which of the following molecules are tetrahedral?
(a) SiF_4 (b) CH_4 (c) NF_4^+ (d) BF_4^-
(e) TeF_4

99. Which of the following molecules are linear?
(a) C_2H_2 (b) CO_2 (c) NO_2^- (d) NO_2^+
(e) H_2O

100. Which of the following elements would form a linear compound with the formula XO_2?
(a) Se (b) P (c) C (d) O (e) F

101. Explain why the nonbonding electrons occupy equatorial positions in ClF_3, not axial positions.

Bond Angles

102. How many atoms are needed to define a bond angle?

103. Complete the following table.

Bonding Domains	Nonbonding Domains	Bond Angle Central Atom
2	0	—
5	0	—
3	—	120°
2	1	—
—	3	180°
—	0	109°
3	—	109°
2	2	—
2	3	—
3	2	—
4	2	—

104. Give the bond angle around the central atom in the following:
(a) CO_3^{2-} (b) SO_2 (c) H_3O^+ (d) IF_4^+
(e) BrF_5 (f) ICl_4^-

105. Give the bond angles around each underlined atom in the compounds:
(a) $H_3\underline{C}\underline{O}\underline{C}H_3$ (b) $\underline{C}H_2Cl_2$ (c) $H\underline{O}\underline{N}O$
(d) $H_3\underline{C}\overset{\overset{O}{\|}}{\underline{C}}OH$ (e) $\underline{S}O_3$

106. The Cl—B—Cl bond angle in BCl_3 is 120° and the H—C—H bond angle in H_2CCH_2 is 121°. Rationalize the difference using the electron domain model. The H—O—H bond angle in H_2O is 104.5°. Rationalize this angle using the electron domain model.

107. Give an estimate of all bond angles in:
(a) SF_4 (b) ClF_3 (c) I_3^- (d) XeF_4

108. Estimate the H—C—H bond angles in:
(a) C_2H_2 (b) C_2H_4 (c) C_2H_6

The Difference Between Polar Bonds and Polar Molecules

109. Explain why OCS is a polar molecule but CS_2 is not.

110. Use the Lewis structure of CO_2 to explain why the molecule has no dipole moment.

111. Which of the following molecules should be polar?
(a) CH_3OH (b) H_2O (c) CH_3OCH_3
(d) CH_3CO_2H

112. Explain why formaldehyde (H_2CO) is a polar molecule.

113. Explain why both thionyl chloride ($SOCl_2$) and sulfuryl chloride (SO_2Cl_2) are polar molecules.

114. Draw Lewis structures and identify the polar species in the following:
(a) BF_3 (b) C_2H_6 (c) PH_3 (d) N_2O

Integrated Problems

115. Polymerization is the combining together of individual units to produce a long chain. Phosphate detergent additives and many biologically important molecules consist of chains of individual PO_3^- units.

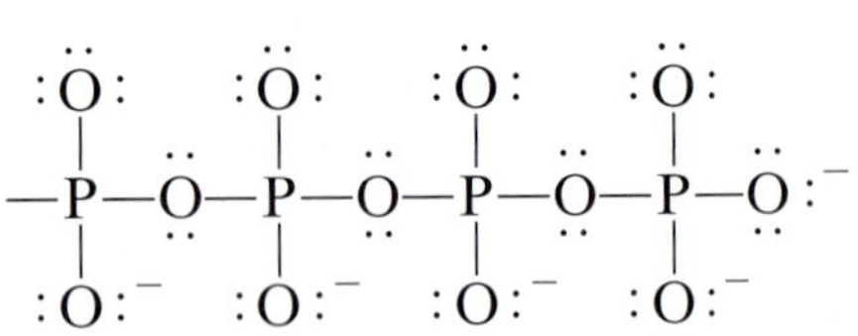

Draw the Lewis structure of PO_3^-. Use formal charges to explain why individual PO_3^- units aggregate with themselves to produce long chains. Would you expect SO_3 to polymerize?

116. (a) The partial charge in molecules with single bonds containing more than two atoms can be estimated if each bond is taken into account, although the procedure is not strictly correct. For example, in H_2O the partial charge on O is estimated by considering the two O—H bonds:

$$\delta_O = 6 - 4 - 2\left(\frac{EN_O}{EN_H + EN_O}\right) - 2\left(\frac{EN_O}{EN_H + EN_O}\right) = -0.44$$

Thus, for water the partial charges are

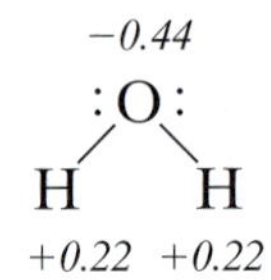

A more exact calculation gives a charge on oxygen that is less negative. Can you suggest why the simple calculation might have been anticipated to be in error? (b) For small molecules, however, these estimates of partial charge are acceptable. Calculate the partial charges on all the atoms in HOF and F_2O. Do the values seem reasonable? Why?

117. Chemical formulas usually give a good idea of the skeleton structure of a compound, but some formulas are misleading. The formulas of common acids are often written as follows: HNO_3, H_2SO_4, H_3PO_4, $HClO_4$. Write Lewis structures for those four acids, assuming that their skeleton structures are more appropriately described by the formulas $HONO_2$, $(HO)_2SO_2$, $(HO)_3PO$, and $HOClO_3$.

118. Write Lewis structures for the following compounds and then describe why the compounds are unusual.
(a) NO (b) NO_2 (c) ClO_2 (d) ClO_3

119. Which element forms a compound with the following Lewis structure?

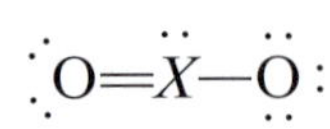

(a) Al (b) Si (c) P (d) S (e) Cl

120. Which element forms a compound with the following Lewis structure?

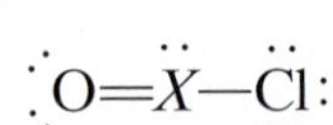

(a) Be (b) B (c) C (d) N (e) O (f) F

121. Which element forms a polyatomic ion with the formula XF_6^{2-} that has no nonbonding electrons in the valence shell of the central atom?
(a) N (b) C (c) Si (d) S (e) P

122. In which group of the periodic table does element X belong if there are two pairs of nonbonding electrons in the valence shell of the central atom in the XF_4^- ion?

123. In which group of the periodic table does element X belong if the distribution of electrons in the valence shell of the central atom in the XCl_4^- ion is octahedral?

124. Which of the following molecules does not contain a double bond?
(a) N_2 (b) CO_2 (c) C_2H_4 (d) NO_2 (e) SO_3

125. Which of the following contain only single bonds?
(a) CN^- (b) NO^+ (c) CO
(d) O_2^{2-} (e) Cl_2CO

126. Ingestion of oxalic acid ($H_2C_2O_4$), which can be found in a variety of vegetables and other plants, can produce nausea, vomiting, and diarrhea. When taken in excess, oxalic acid can be toxic. Draw the Lewis structure for the molecule. (Assume that the skeleton structure can be described as HO_2CCO_2H.)

127. In which group does element X belong if the shape of the XF_2^- ion is linear?

128. The charged species OCN^- is well known in chemistry.
(a) Write all possible Lewis structures for the species.
(b) Which of the structures is best? Explain.
(c) Typical experimental carbon-to-oxygen bond lengths are as follows:

C—O	C=O	C≡O
150 pm	*133 pm*	*120 pm*

What type of carbon–oxygen bond (single, double, triple, or somewhere in between) do you expect for OCN^-? Predict the carbon–oxygen bond length for OCN^-. Explain your answers. In addition to bond length, what other piece of experimental evidence would support your conclusion concerning the type of carbon–oxygen bond?
(d) What is the O—C—N bond angle? Explain your answer.
(e) Would you expect OCN^- to have a dipole moment? Explain.

129. Draw the Lewis structures and select the one species of each of the following pairs that has the largest dipole moment. Show your work and explain your answer. The central atom is underlined.
(a) $\underline{C}S_2$, $\underline{S}O_2$ (b) $H\underline{C}N$, $N\underline{C}CN$ (C_2N_2)
(c) $\underline{Ge}F_4$, $\underline{Se}F_4$

130. The molecule thymine, one of four DNA bases, is known to have the following structure:

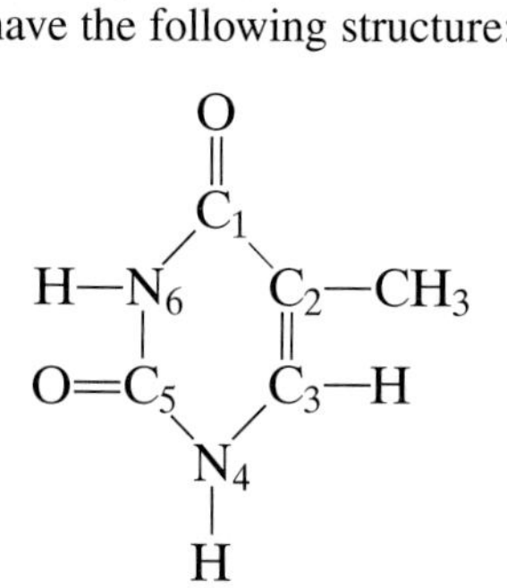

(a) Add the missing electrons to the structure.
(b) Give the bond angles around each numbered atom.

131. In Section 4.12 we discussed two possible structures for SO_2 and stated that without experimental evidence we could not distinguish between the two. Structure I contains one single bond and one double bond while structure II contains two double bonds. Explain why measurements of bond length cannot be used to distinguish between the two structures.

132. Each of the following Lewis structures is incorrect. Explain what is wrong with each one and give the correct Lewis structure.

(a) 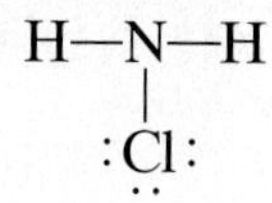
(b) 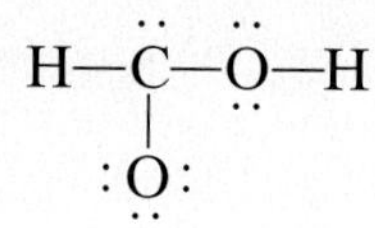
(c) 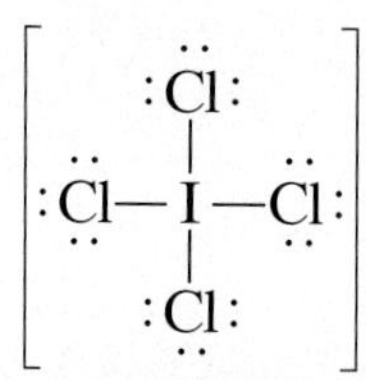
(d) 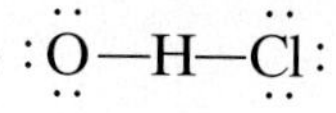
(e) 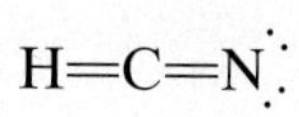

133. Which of the following Lewis structures can contribute to the description of the electron structure of HNO_3? Use formal charge to rationalize your choices.

(A) 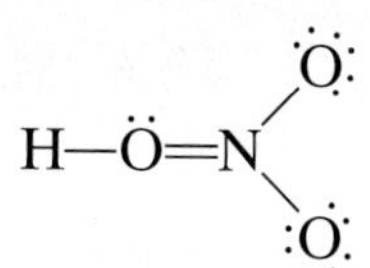
(B) 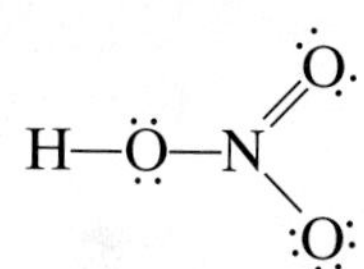

Chapter Four

SPECIAL TOPICS

4A.1 Valence Bond Theory

Molecular orbital theory was not considered to be a core subject because once the examples that illustrate how the approach is used are given, there is little further use a student can make of the theory at this level.

Because hybrid orbital theory is still widely used in organic chemistry, there is reason to consider including it in the course. However, much of organic chemistry can be done with only the electron domain model and the introduction to hybrid terminology.

The octet rule and Lewis structures give us a simple view of covalent bonding, but they don't tell us why covalent bonds are formed or how electrons are shared between atoms. There are two models that provide a deeper understanding of the covalent bond: molecular orbital theory (see Section 4A.4) and valence bond theory.

Valence bond theory assumes that valence atomic orbitals on adjacent atoms can overlap to form a bond that lies between the atoms. The covalent bond is formed by a pair of electrons with opposite spins. Consider, for example, the formation of an H_2 molecule from a pair of hydrogen atoms that each has an unpaired electron in an $1s$ orbital.

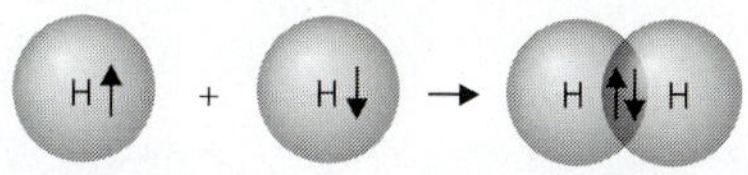

A similar approach can be used to explain the covalent bond in HF. Each atom has an orbital that has an unpaired electron.

H ↑ ($1s^1$)

F ↑↓ ($1s^2$) ↑↓ ($2s^2$) ↑↓ ↑↓ ↑ ($2p^5$)

But now the bond is formed by the interaction between a $1s$ orbital on one atom and a $2p$ orbital on the other.

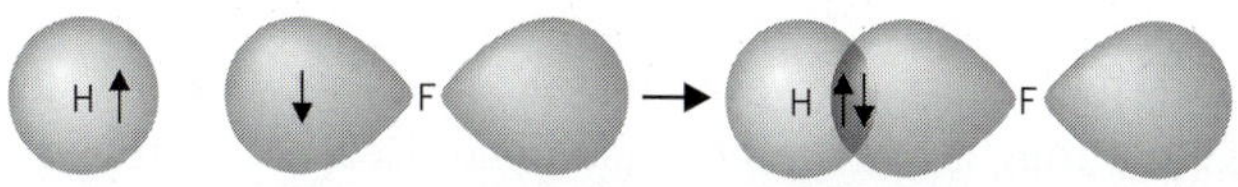

The valence bond model for the F_2 molecule is based on the head-to-head overlap between half-filled $2p$ orbitals on adjacent atoms.

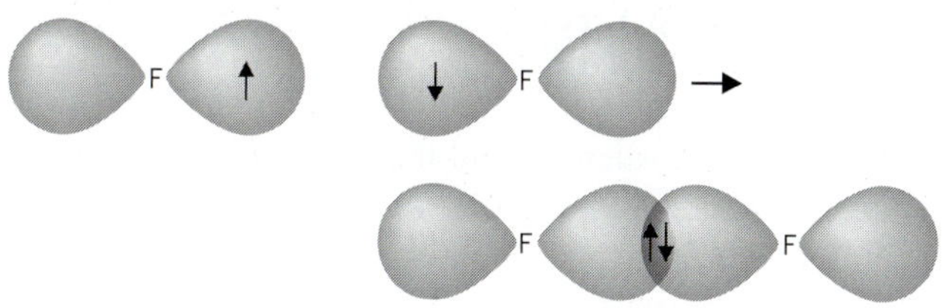

The covalent bonds in H_2, HF, and F_2 are called **sigma (σ) bonds** because they look like an s orbital when viewed along the bond. **Pi (π) bonds** are formed when orbitals overlap to form a bond that looks like a p orbital when viewed along the bond.

A way to differentiate between sigma and pi bonds is to consider the bonding in the O_2 molecule. The electron configuration of the oxygen atom tells us that there are two unpaired electrons on each atom.

O ↑↓ ($1s^2$) ↑↓ ($2s^2$) ↑↓ ↑ ↑ ($2p^4$)

Let's assume that one of the unpaired electrons is in the $2p_z$ orbital that points toward the neighboring atom. The interaction between the $2p_z$ orbitals on adjacent atoms leads to a sigma bond, just as it did in the F_2 molecule.

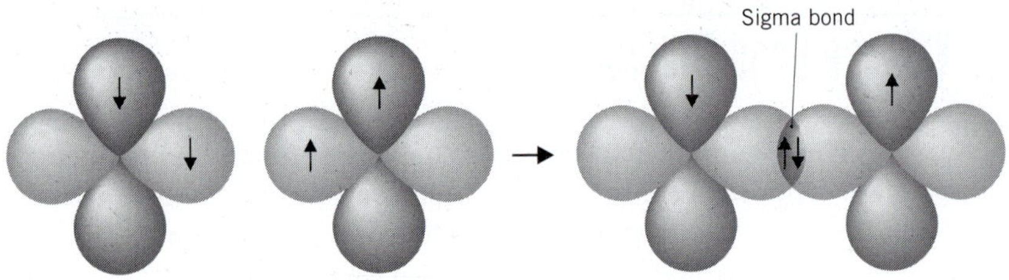

But this leaves us with an unpaired electron on each atom in one of the $2p$ orbitals perpendicular to the axis along which the sigma bond forms. The edge-on interaction between the half-filled orbitals leads to the formation of a bond that looks like a p orbital when viewed along the bond axis. In other words, it leads to a *pi bond.*

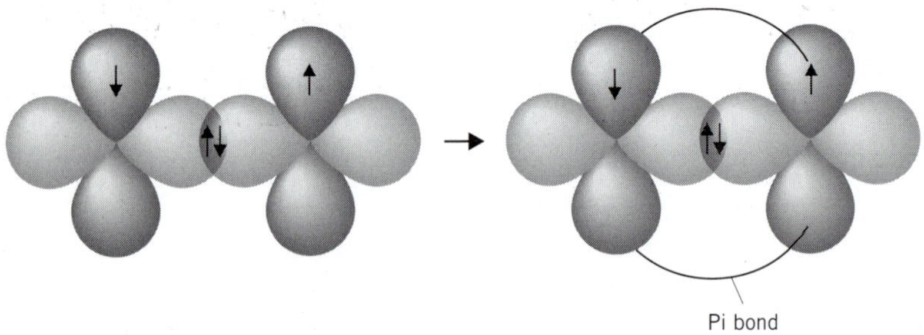

The combination of the sigma and pi bonds between the oxygen atoms suggests that the O_2 molecule is held together by a double bond, which is consistent with the Lewis structure for oxygen.

$$\ddot{\underset{\cdot\cdot}{O}}::\ddot{\underset{\cdot\cdot}{O}}$$

➤ CHECKPOINT

Draw the Lewis structure and the valence bond description of the bonding in N_2. Comment on any similarities or differences in the two descriptions.

The analysis also suggests, however, that the bond between the atoms is not necessarily twice as strong as an O—O single bond. Because there is greater overlap of orbitals for the sigma bond than for the pi bond, the sigma bond is generally stronger than the pi bond.

4A.2 Hybrid Atomic Orbitals

The simple version of the valence bond theory introduced in the previous section gives a satisfactory picture of the bonding in simple diatomic molecules such as H_2, HF, F_2, O_2, and N_2. But it gives less satisfactory results when applied to slightly more complex molecules, such as CH_4 and H_2O.

The electron configuration of carbon has only two unpaired electrons, and therefore carbon should form only two bonds.

$$\text{C} \quad \underset{1s^2}{\uparrow\downarrow} \quad \underset{2s^2}{\uparrow\downarrow} \quad \underset{2p^2}{\uparrow \ \ \uparrow \ \ \_\_}$$

The fact that carbon atoms form four bonds can be explained by noting that it takes relatively little energy to excite one of the electrons from the filled $2s$ orbital into one of the empty $2p$ orbitals to give four orbitals that each contain an unpaired electron.

$$\text{C*} \quad \underset{1s^2}{\uparrow\downarrow} \quad \underset{2s^1}{\uparrow} \quad \underset{2p^3}{\uparrow \ \ \uparrow \ \ \uparrow}$$

The energy invested in moving the electron is more than repaid by the energy released when two more bonds are formed.

A more serious problem arises when the valence bond theory is used to predict the geometry around an atom in a molecule. Consider water, for example. Combining a pair of hydrogen atoms with unpaired electrons in a 1*s* orbital with the two unpaired electrons on an oxygen atom (2*p* orbitals) gives the correct number of bonds.

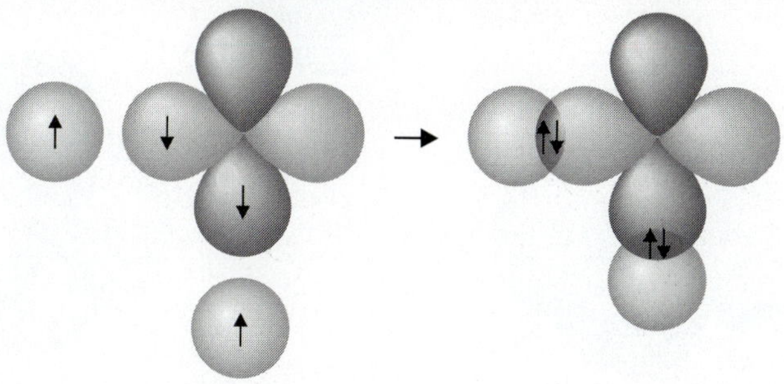

In this case, however, we predict that the H—O—H bond angle would be 90° because that is the angle between the 2*p* orbitals on the oxygen atom. The experimental angle is 105°.

The problem was solved by assuming that the valence atomic orbitals on an individual atom can be combined to form **hybrid atomic orbitals.** The geometry of a linear BeH_2 molecule can be explained, for example, by mixing the 2*s* orbital on the beryllium atom with one of the 2*p* orbitals to form a set of *sp* hybrid orbitals that point in opposite directions, as shown in Figure 4A.1. One of the valence electrons on the beryllium atom is then placed in each of the orbitals, and the orbitals are allowed to overlap with half-filled 1*s* orbitals on a pair of hydrogen atoms to form a linear BeH_2 molecule.

The geometry of trigonal planar molecules such as BF_3 can be explained by mixing a 2*s* orbital with a pair of 2*p* orbitals on the central atom to form three sp^2 hybrid orbitals that point toward the corners of an equilateral triangle. Molecules whose geometries are based on a tetrahedron, such as CH_4 and H_2O, can be understood by mixing a 2*s* orbital with all three 2*p* orbitals to obtain a set of four sp^3 orbitals that are oriented toward the corners of a tetrahedron.

The hybrid atomic orbital model can be extended to molecules whose shapes are based on trigonal bipyramidal or octahedral distributions of electrons by including valence-shell *d* orbitals. When one of the 3*d* orbitals is mixed with the 3*s* and the three 3*p* orbitals on an atom, the resulting sp^3d hybrid orbitals point toward the corners of a trigonal bipyramid. When two of the 3*d* orbitals are mixed with the 3*s* and 3*p* orbitals, the result is a set of six sp^3d^2 hybrid orbitals that point toward the corners of an octahedron.

The geometries of the five different sets of hybrid atomic orbitals (*sp*, sp^2, sp^3, sp^3d, and sp^3d^2) are shown in Figure 4A.2.

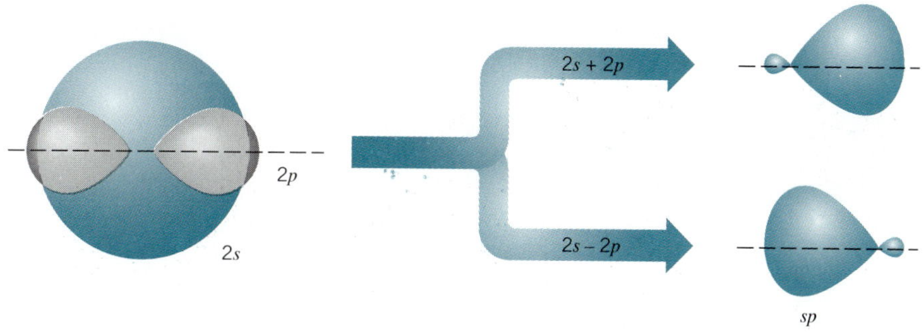

Fig. 4A.1 The *sp* hybrid orbitals used by the beryllium atom in the linear BeH_2 molecule are formed by combining the 2*s* and 2*p* orbitals on that atom. When the orbitals are added, we get an *sp* orbital that points in one direction. When one of the orbitals is subtracted from the other, we get an *sp* orbital that points in the opposite direction.

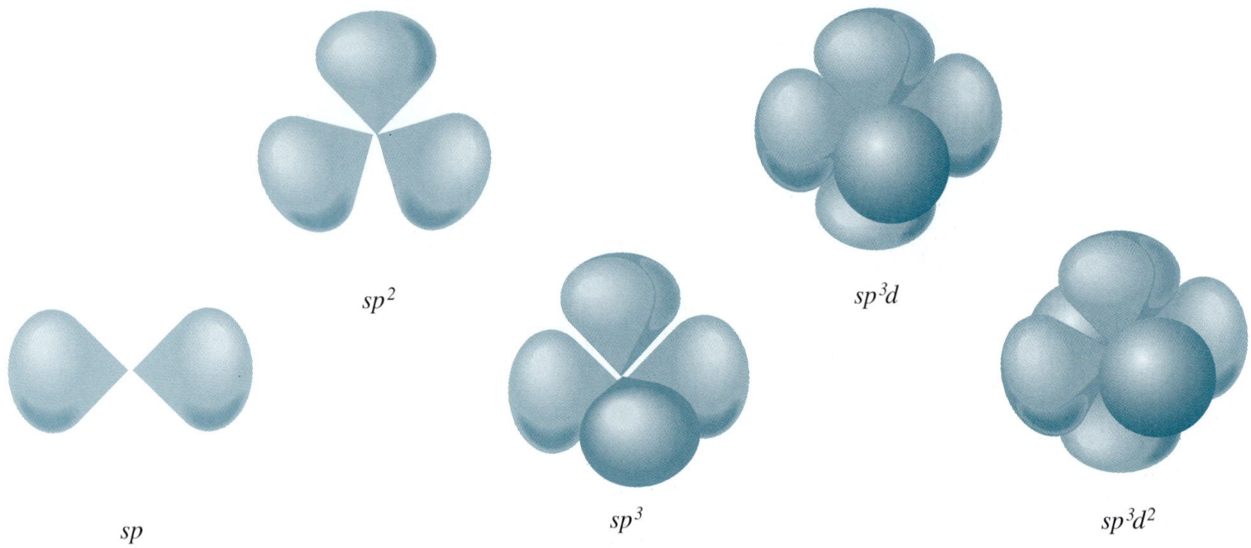

Fig. 4A.2 Shapes of the *sp*, sp^2, sp^3 sp^3d, and sp^3d^2 hybrid orbitals.

The relationship between hybridization and the distribution of electrons in the valence shell of an atom is summarized in Tables 4A.1 and 4.3.

Table 4A.1

Relationship Between the Distribution of Electron Domains on an Atom and the Hybridization of an Atom

Electron Domains	Distribution of Electron Domains	Hybridization	Examples
2	Linear	*sp*	BeH_2, CO_2, N_2
3	Trigonal planar	sp^2	BF_3, CO_3^{2-}, O_3
4	Tetrahedral	sp^3	CH_4, SO_4^{2-}, NH_3, H_2O
5	Trigonal bipyramidal	sp^3d	PF_5, ClF_3
6	Octahedral	sp^3d^2	SF_6, XeF_4

Exercise 4A.1

Determine the hybridization of the central atom in O_2, H_3O^+, $TeCl_4$, and ICl_4^- from the following Lewis structures

(a) O_2

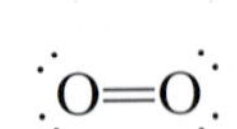

(b) H_3O^+

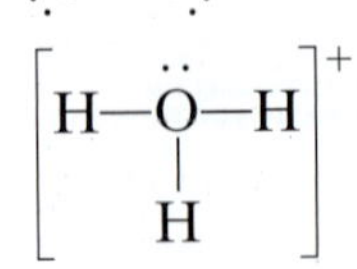

(c) $TeCl_4$

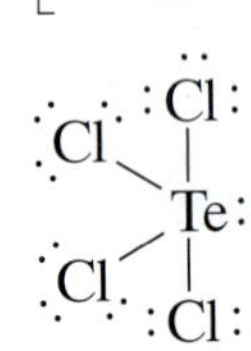

(d) ICl_4^-

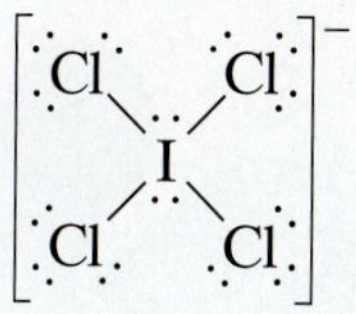

Solution

(a) The Lewis structure for the O_2 molecule suggests that the distribution of electron domains around each oxygen atom should be trigonal planar. The hybridization of the oxygen atoms in the O_2 molecule is therefore assumed to be sp^2.

(b) The Lewis structure suggests a tetrahedral distribution of electron domains around the oxygen atom in the ion. The oxygen atom in H_3O^+ is therefore assumed to be sp^3 hybridized.

(c) The Lewis structure suggests a trigonal bipyramidal distribution of domains in the valence shell of the tellurium atom. The tellurium atom therefore forms an sp^3d hybrid.

(d) The Lewis structure suggests an octahedral distribution of domains in the valence shell of the iodine atom. The iodine atom is therefore sp^3d^2 hybridized.

4A.3 Molecules with Double and Triple Bonds

The hybrid atomic orbital model can also be used to explain the formation of double and triple bonds. Let's consider the bonding in formaldehyde (H_2CO), for example, which has the following Lewis structure.

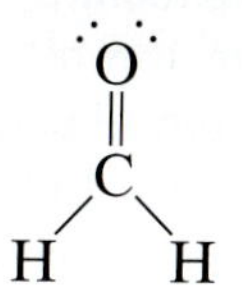

There are three places where electrons can be found in the valence shell of both the carbon and oxygen atoms in the molecule. As a result, the electron domain theory predicts that the valence electrons on the atoms will be oriented toward the corners of an equilateral triangle. The carbon and oxygen atoms are therefore assumed to be sp^2 hybridized.

When we create a set of sp^2 hybrid orbitals, we combine the 2*s* and two of the 2*p* orbitals on the atom. One of the valence electrons on the carbon atom is placed in each of the three sp^2 hybrid orbitals. The fourth valence electron is placed in the 2*p* orbital that wasn't used during hybridization.

There are six valence electrons on a neutral oxygen atom. A pair of electrons is placed in each of two of the sp^2 hybrid orbitals. One electron is then placed in the sp^2 hybrid orbital that points toward the carbon atom, and another is placed in the unhybridized 2*p* orbital.

The two C—H bonds are formed when the unpaired electron in one of the sp^2 hybrid orbitals on carbon interacts with a 1*s* electron on a hydrogen atom, as shown in Figure 4A.3. A C—O bond is formed when the electron in the third sp^2 hybrid orbital on carbon interacts with the unpaired electron in the sp^2 hybrid orbital on the oxygen atom. These bonds are sigma bonds.

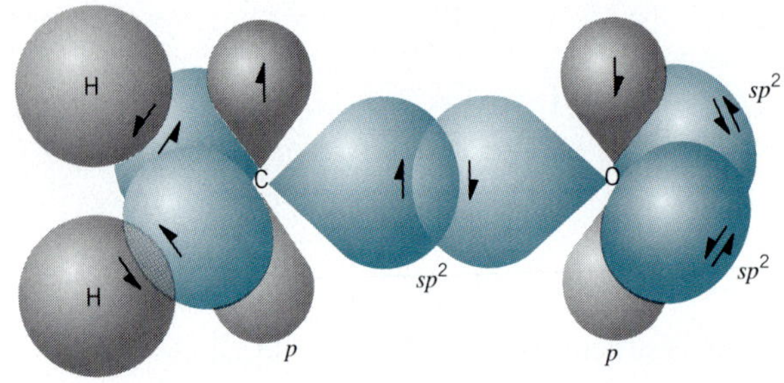

Fig. 4A.3 Hybrid atomic orbital picture of bonding in formaldehyde.

The electron in the unhybridized $2p$ orbital on the carbon atom then interacts with the electron in the unhybridized $2p$ orbital on the oxygen atom to form a second covalent bond between these atoms. This is called a pi bond.

When double bonds were first introduced, we noted that they occur most often in compounds that contain C, N, O, P, or S. There are two reasons for this. First, double bonds by their very nature are covalent bonds. They are therefore most likely to be found among the elements that form covalent compounds. Second, the interaction between p orbitals to form a π bond requires that the atoms come relatively close together, and thus π bonds tend to be the strongest for atoms that are relatively small.

4A.4 Molecular Orbital Theory

Bonding theory based on atomic orbitals focuses on the bonds formed between valence electrons on atoms and is called *valence bond* theory.

The valence bond model of SO_2 can't adequately explain the fact that the molecule contains two equivalent bonds with a bond order between that of an S—O single bond and an S=O double bond. The best it can do is suggest that SO_2 is a mixture, or hybrid, of the two Lewis structures that can be written for the molecule.

This problem, and many others, can be overcome by using a more sophisticated model of bonding based on **molecular orbitals.** Molecular orbital theory is more powerful than valence bond theory. But this power carries a significant cost in terms of the ease with which the model can be visualized. We can understand the basics of molecular orbital theory by constructing the molecular orbitals for the simplest possible molecules: homonuclear diatomic molecules such as H_2, O_2, and N_2.

Molecular orbitals are obtained by combining the atomic orbitals on the atoms to give orbitals that are characteristic of the molecule, not the individual atoms. Consider the H_2 molecule, for example. One of the molecular orbitals in the H_2 molecule is constructed by adding the two $1s$ atomic orbitals that come together to form the molecule. Another orbital is formed by subtracting one of these functions from the other, as shown in Figure 4A.4.

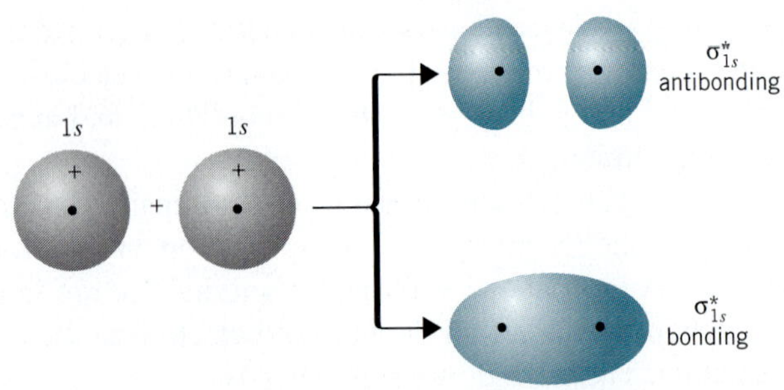

Fig. 4A.4 The interaction between a pair of $1s$ atomic orbitals on neighboring hydrogen atoms leads to the formation of bonding (σ) and antibonding (σ^*) molecular orbitals.

One of the orbitals is called a **bonding molecular orbital** because electrons in the orbital spend most of their time in the region directly between the two nuclei. It is called a *sigma* (σ) molecular orbital because it looks like an *s* orbital when viewed along the H—H bond. Electrons placed in the other orbital spend most of their time away from the region between the two nuclei. That orbital is therefore an **antibonding,** or *sigma star* (σ^*), molecular orbital.

The σ bonding molecular orbital concentrates electrons in the region directly between the two nuclei. Placing an electron in that orbital therefore stabilizes the H_2 molecule. Since the σ^* antibonding molecular orbital forces the electron to spend most of its time away from the area between the nuclei, placing an electron in that orbital makes the molecule less stable.

Electrons are added to molecular orbitals, one at a time, starting with the lowest-energy molecular orbital. The two electrons associated with a pair of hydrogen atoms are placed in the lowest-energy, or σ bonding, molecular orbital shown in Figure 4A.5. When this is done, the energy of an H_2 molecule is lower than that of a pair of isolated atoms. The H_2 molecule is therefore more stable than a pair of isolated atoms.

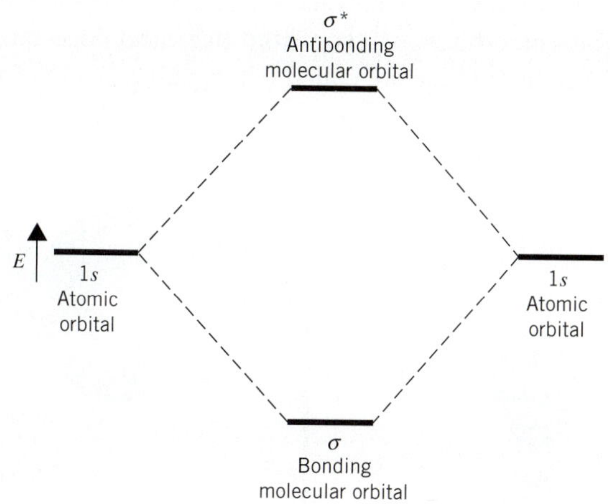

Fig. 4A.5 Relative energies of the 1*s* atomic orbitals on a pair of isolated hydrogen atoms and the σ and σ^* molecular orbitals they form when they interact.

The molecular orbital model can be used to explain why He_2 molecules don't exist. Combining a pair of helium atoms with $1s^2$ electron configurations would produce a molecule with pairs of electrons in both the σ bonding and the σ^* antibonding molecular orbitals. The total energy of an He_2 molecule would be essentially the same as the energy of a pair of isolated helium atoms, and there would be nothing to hold the He_2 molecule together.

As we have seen, the core electrons on an atom do not enter into bonding. Thus only the valence-shell atomic orbitals contribute to the description of the molecular orbitals of a molecule. The molecular orbital diagram for an O_2 molecule would therefore ignore the 1*s* electrons on both oxygen atoms and concentrate on the interactions between the 2*s* and 2*p* valence atomic orbitals.

The 2*s* orbitals on one oxygen atom combine with the 2*s* orbitals on another to form a σ_{2s} bonding and a σ_{2s}^* antibonding molecular orbital, just like the σ_{1s} and σ_{1s}^* orbitals formed from the 1*s* atomic orbitals. If we arbitrarily define the *z* axis of our coordinate system as the axis between the two atoms in an O_2 molecule, the $2p_z$ orbitals on the atoms in the molecule point directly toward each other. When those orbitals interact, they meet head-on to form a σ_{2p} bonding and a σ_{2p}^* antibonding molecular orbital, as shown in Figure 4A.6.

The $2p_x$ and $2p_y$ orbitals on one oxygen atom interact with the $2p_x$ and $2p_y$ orbitals on the other atom to form molecular orbitals that have a different shape, as

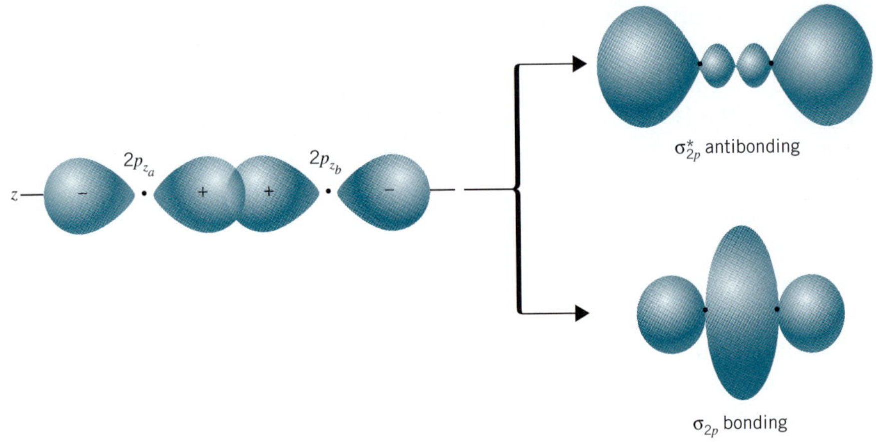

Fig. 4A.6 When a pair of $2p$ orbitals are combined so that they meet head-on, bonding (σ_{2p}) and antibonding (σ_{2p}*) molecular orbitals are formed. These are called σ orbitals because they concentrate the electrons along the axis between the two nuclei, just like the σ_{1s} and σ_{2s} orbitals.

shown in Figure 4A.7. These molecular orbitals are called *pi* (π) orbitals. Whereas σ and σ* orbitals concentrate the electrons along the axis on which the nuclei of the atoms lie, π and π* orbitals concentrate the electrons above and below the axis. Because there are two $2p$ atomic orbitals on each atom that meet edge-on, we get two π orbitals and two π* orbitals. These are called the π_x and π_y and the π_x* and π_y* molecular orbitals.

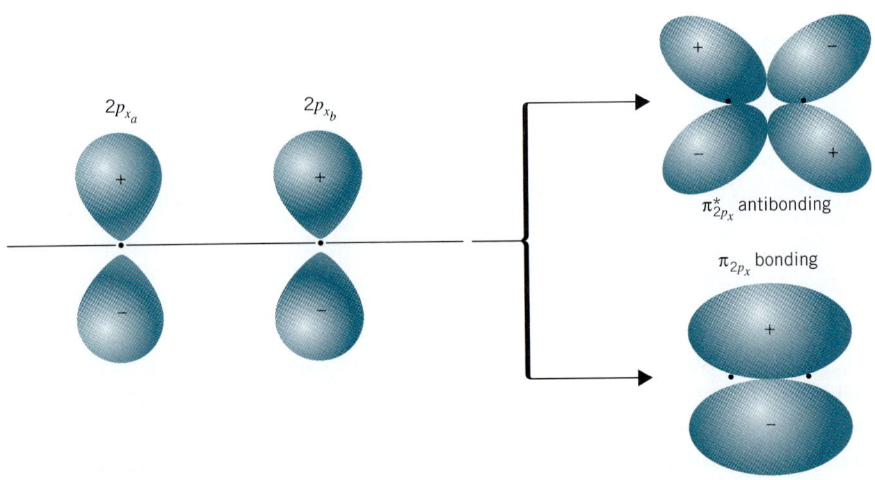

Fig. 4A.7 When a pair of $2p_x$ orbitals are combined so that they meet edge-on, bonding (π_x) and antibonding (π_x*) molecular orbitals are formed. A similar set of bonding (π_y) and antibonding (π_y*) orbitals is formed when a pair of $2p_y$ atomic orbitals are combined.

The interaction of four valence atomic orbitals on one atom with a set of four atomic orbitals on another atom therefore leads to the formation of a total of eight molecular orbitals: σ_{2s}, σ_{2s}*, σ_{2p}, σ_{2p}*, π_x, π_y, π_x*, and π_y*.

There is a difference between the energies of the $2s$ and $2p$ orbitals on an atom. As a result, the σ_{2s} and σ_{2s}* orbitals both lie at lower energies than the σ_{2p}, σ_{2p}*, π_x, π_y, π_x*, and π_y*. To sort out the relative energies of the six molecular orbitals formed when the $2p$ atomic orbitals on a pair of atoms are combined, we need to understand the relationship between the strength of the interaction between a pair of orbitals and the relative energies of the molecular orbitals they form.

Because the $2p_z$ orbitals meet head-on, the interaction between them is stronger than the interactions between the $2p_x$ or $2p_y$ orbitals, which meet edge-on. As a result, the σ_{2p} orbital lies at a lower energy than the π_x and π_y orbitals, and the σ_{2p}* orbital lies at a higher energy than the π_x* and π_y* orbitals, as shown in Fig. 4A.8.

Unfortunately an interaction is missing from the model. It is possible for the $2s$ orbital on one atom to interact with the $2p_z$ orbital on the other. This interaction

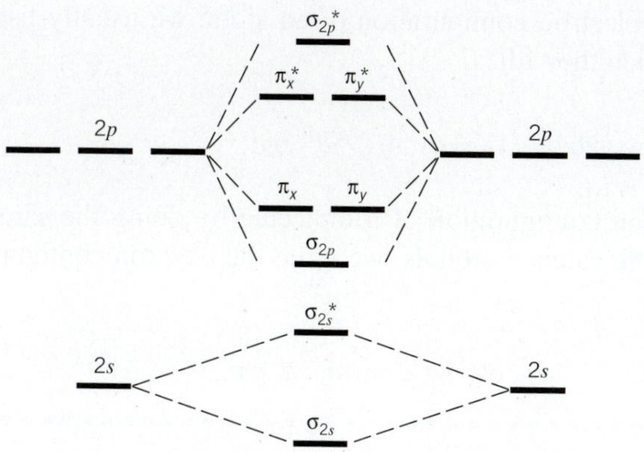

Fig. 4A.8 The interaction between $2p_z$ atomic orbitals that meet head-on is stronger than the interaction between $2p_x$ or $2p_y$ orbitals that meet edge-on. As a result, the difference between the energies of the σ_{2p} and σ_{2p}* orbitals is larger than the difference between the π_x and π_x* or π_y and π_y* orbitals.

introduces a slight change in the relative energies of the molecular orbitals, to give the diagram shown in Figure 4A.9. Experiments have shown that O_2 and F_2 are best described by the model in Figure 4A.8, but B_2, C_2, and N_2 are best described by the model shown in Figure 4A.9.

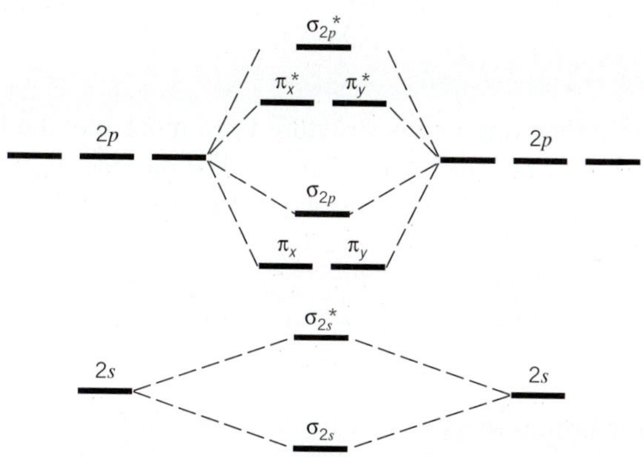

Fig. 4A.9 When the *s-p* orbital interaction is added to the molecular orbital theory, there is a slight change in the energies of the molecular orbitals formed by overlap of the atomic orbitals on neighboring atoms.

Exercise 4A.2

Construct a molecular orbital diagram for the O_2 molecule.

Solution

There are 6 valence electrons on a neutral oxygen atom and therefore 12 valence electrons in an O_2 molecule. These electrons are added to the diagram in Figure 4A.8, one at a time, starting with the lowest energy molecular orbital.

Because Hund's rules apply to the filling of molecular orbitals, molecular orbital theory predicts that there should be 2 unpaired electrons on this molecule—1 electron each in the π_x* and π_y* orbitals.

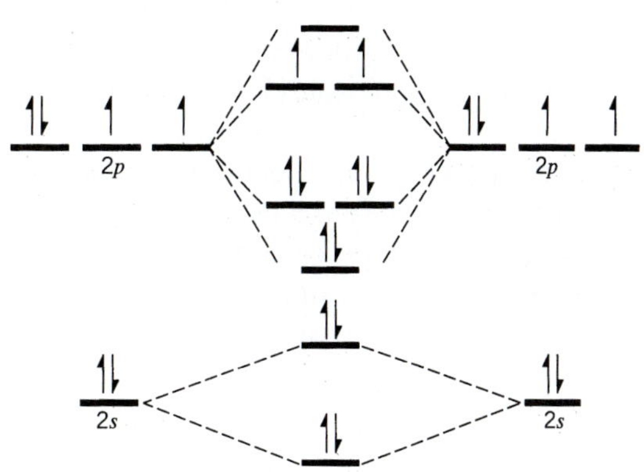

When writing the electron configuration of an atom, we usually list the orbitals in the order in which they fill.

$$\text{Pb} \qquad [\text{Xe}]\, 6s^2\, 4f^{14}\, 5d^{10}\, 6p^2$$

We can write the electron configuration of a molecule by doing the same thing. Concentrating only on the valence orbitals, we write the electron configuration of O_2 as follows.

$$O_2 \qquad \sigma_{2s}^{\ 2}\, \sigma_{2s}^{\ *2}\, \sigma_{2p}^{\ 2}\, \pi_x^{\ 2}\, \pi_y^{\ 2}\, \pi_x^{\ *1}\, \pi_y^{\ *1}$$

The number of bonds between a pair of atoms is called the **bond order.** Bond orders can be calculated from Lewis structures, which are the heart of the valence bond model. Oxygen, for example, has a bond order of 2.

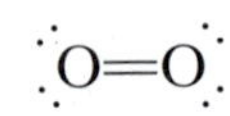

When there is more than one Lewis structure for a molecule, the bond order is an average of these structures. The bond order in sulfur dioxide, for example, is 1.5—the average of an S—O single bond in one Lewis structure and an S=O double bond in the other.

In molecular orbital theory, we calculate bond orders by assuming that two electrons in a bonding molecular orbital contribute one net bond and that two electrons in an antibonding molecular orbital cancel the effect of one bond. We can calculate the bond order in the O_2 molecule by noting that there are eight valence electrons in bonding molecular orbitals and four valence electrons in antibonding molecular orbitals in the electron configuration of the molecule given in Exercise 4A.2. Thus, the bond order is 2.

$$\text{Bond order} = \frac{\text{bonding electrons} - \text{antibonding electrons}}{2} = \frac{8 - 4}{2} = 2$$

Although the Lewis structure and molecular orbital models of oxygen yield the same bond order, there is an important difference between the models. The electrons in the Lewis structure are all paired, but there are two unpaired electrons in the molecular orbital description of the molecule. As a result, we can test the predictions of the theories by studying the effect of a magnetic field on oxygen.

Atoms or molecules in which the electrons are paired are **diamagnetic,** that is, they are repelled by both poles of a magnet. Those that have one or more unpaired electrons are **paramagnetic,** or attracted to a magnetic field. Liquid oxygen is attracted to a magnetic field and can actually bridge the gap between the poles of a horseshoe magnet. The molecular orbital model of O_2 is therefore superior to the valence bond model, which cannot explain the paramagnetism of oxygen.

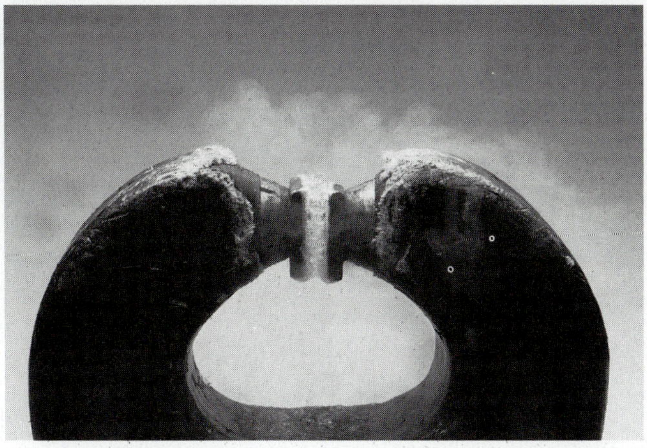

Liquid O_2 is so strongly attracted to a magnetic field that it will bridge the gap between the poles of a horseshoe magnet.

Exercise 4A.3

Use the molecular orbital diagram in Figure 4A.9 to calculate the bond order in nitrogen oxide (NO). Compare the results of this calculation with the bond order obtained from the Lewis structure.

Solution

There are 11 valence electrons in the NO molecule, and there are two possible Lewis structures, depending on whether we locate the unpaired electron on the nitrogen atom or on the oxygen atom.

$$\cdot\ddot{\text{N}}=\ddot{\text{O}}\colon \longleftrightarrow \colon\dot{\text{N}}=\ddot{\text{O}}\cdot$$

Both Lewis structures contain an N=O double bond, however, so the bond order in the valence bond model for NO is 2.

To calculate the bond order in the molecular orbital model of NO, we have to predict the electron configuration of the molecule by adding electrons, one at a time, to the molecular orbitals in Figure 4A.9.

$$NO_2 \quad \sigma_{2s}^{2}\, \sigma_{2s}^{*2}\, \pi_x^{2}\, \pi_y^{2}\, \sigma_{2p}^{2}\, \pi_x^{*1}$$

There are eight valence electrons in bonding molecular orbitals and three valence electrons in antibonding molecular orbitals in the compound. The bond order in the molecular orbital model of NO is therefore 2.5.

$$\text{Bond order} = \frac{\text{bonding electrons} - \text{antibonding electrons}}{2} = \frac{8-3}{2} = 2.5$$

The strength of the covalent bond in NO has been shown by experiment to be significantly stronger than a typical N=O double bond, in agreement with the predictions of molecular orbital theory for the molecule.

Problems

Valence Bond Theory

4A-1. What is meant by orbital overlap?

4A-2. Which of the following orbitals could overlap?
(a) $2s + 2s$ (b) $2p_x + 2p_y$ (c) $2p_y + 2p_y$
(d) $2p_z + 2p_z$ (e) $2s + 2p_z$

4A-3. Draw valence bond pictures showing the overlap of atomic orbitals in each of the following molecules: NH_3, H_2S, NO_2, C_2.

Hybrid Atomic Orbitals

4A-4. Determine the hybridization of the central atom in the following molecules or ions.
(a) NH_3 (b) NH_4^+ (c) NO
(d) NO_2 (e) NO_2^+

4A-5. Determine the hybridization of the central atom in the following molecules or ions.
(a) CH_4 (b) H_2CO (c) HCO_2^-

4A-6. Determine the hybridization of the central atom in the following molecules or ions.
(a) SF_4 (b) BrO_3^- (c) XeF_3^+ (d) Cl_2CO

Molecules with Double and Triple Bonds

4A-7. Write the Lewis structures for molecules of ethylene, C_2H_4, and acetylene, C_2H_2. Use hybrid atomic orbitals to describe the bonding in the compounds.

4A-8. Write the Lewis structures for carbon monoxide and carbon dioxide. Use hybrid atomic orbitals to describe the bonding in the compounds.

Molecular Orbital Theory

4A-9. Describe the difference between σ and π molecular orbitals.

4A-10. Describe the difference between bonding and antibonding molecular orbitals.

4A-11. Explain why the difference between the energies of the σ_{2p} bonding and σ_{2p}* antibonding orbitals is larger than the difference between the π_x or π_y bonding and π_x* or π_y* antibonding orbitals.

4A-12. Describe the molecular orbitals formed by the overlap of the following atomic orbitals. (Assume that the bond lies along the z axis of the coordinate system.)
(a) $2s + 2s$ (b) $2p_x + 2p_x$ (c) $2p_y + 2p_y$
(d) $2p_z + 2p_z$ (e) $2s + 2p_z$

4A-13. Write the electron configuration for the following diatomic molecules and calculate the bond order in each molecule.
(a) H_2 (b) C_2 (c) N_2 (d) O_2 (e) F_2

4A-14. Use molecular orbital theory to predict whether the H_2^+, H_2^-, and H_2^{2-} ions should be more stable or less stable than a neutral H_2 molecule.

4A-15. Use molecular orbital theory to explain why the oxygen–oxygen bond is stronger in the O_2 molecule than in the O_2^{2-} ion.

4A-16. Use molecular orbital theory to predict whether the bond order in the superoxide ion, O_2^-, should be higher or lower than the bond order in a neutral O_2 molecule.

4A-17. Use molecular orbital theory to predict whether the peroxide ion, O_2^{2-}, should be paramagnetic.

4A-18. Write the electron configuration for the following diatomic molecules. Calculate the bond order in each molecule.
(a) HF (b) CO (c) CN^-
(d) ClO^- (e) NO^+

4A-19. Classify the following molecules as paramagnetic or diamagnetic.
(a) HF (b) CO (c) CN^- (d) NO (e) NO^+

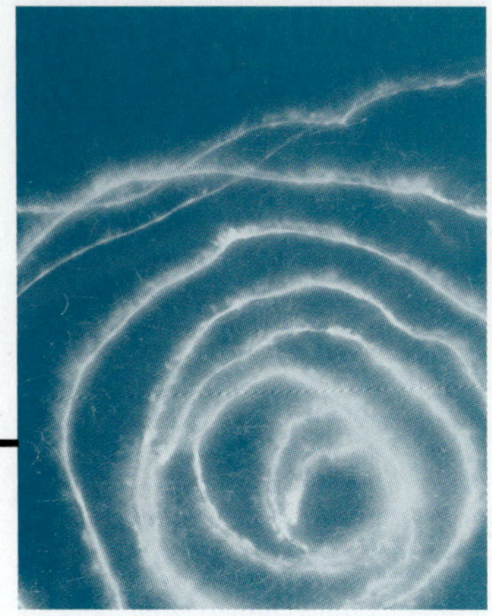

Chapter Five

IONIC AND METALLIC BONDS

5.1 The Active Metals
5.2 Main-Group Metals and Their Ions
5.3 Main-Group Nonmetals and Their Ions
5.4 Transition Metals and Their Ions
5.5 Predicting the Products of Reactions That Produce Ionic Compounds
5.6 Oxides, Peroxides, and Superoxides
5.7 The Ionic Bond
5.8 Structures of Ionic Compounds
5.9 Metallic Bonds
5.10 The Relationship Among Ionic, Covalent, and Metallic Bonds
5.11 Bond-Type Triangles
5.12 Properties of Metallic, Covalent, and Ionic Compounds
5.13 Oxidation Numbers
5.14 Calculating Oxidation Numbers
5.15 Oxidation–Reduction Reactions
5.16 Nomenclature

5.1 The Active Metals

The major concept of this chapter is that the three types of bonding are not separate categories but instead are interrelated depending on how electrons are shared by two atoms. Bonding in compounds is a combination of two or even all three types of bonds. Bond-type triangles are used to help students visualize this interrelationship.

ChemActivities 23–25 in Chemistry: A Guided Inquiry *are designed to accompany this chapter.*

As a general rule, the elements toward the bottom-left corner of the periodic table are metals, and the elements toward the upper-right corner are nonmetals. There are no abrupt changes in the physical properties of the elements, however, as we go across a row of the periodic table or down a column. As a result, the change from metal to nonmetal must be gradual. Instead of arbitrarily dividing elements into metals and nonmetals, it is better to describe some elements as being more metallic and others as more nonmetallic. Metals have low AVEE values, which indicates that valence electrons are relatively easy to remove and that valence subshells are close in energy. Nonmetals have high AVEE values, which means that valence electrons are more difficult to remove and that the subshells are widely spaced in energy.

Elements become less metallic and more nonmetallic as we go across a row of the periodic table from left to right.

Metallic character decreases ⟶
Na Mg Al Si P S Cl Ar
Nonmetallic character increases ⟶

Because AVEE values decrease as we go down a column of the periodic table, the metallic character of an element increases.

↓				↑
		C		
	Metallic	Si	**Nonmetallic**	
	character	Ge	**character**	
	increases	Sn	**increases**	
		Pb		

➤ CHECKPOINT

Arrange the following metals in increasing order of metallic character: K, Ca, and Rb.

The primary difference between the various metals in the periodic table is the ease with which they undergo chemical reactions. The elements toward the bottom-left corner of the periodic table, which have the most metallic character, are the metals that are the most **active** in the sense of being the most **reactive** (see Figure 5.1). Lithium, sodium, and potassium all react with water, for example. As we go down the column, the elements react more vigorously because they become more active as they become more metallic.

Metals are often divided into four classes on the basis of their activity, as shown in Table 5.1. The most active metals are so reactive that they readily combine with the O_2 and H_2O in the atmosphere and must therefore be stored under an inert

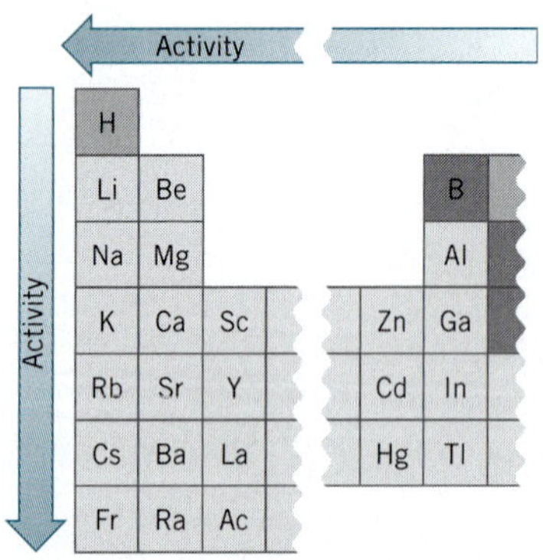

Fig. 5.1 The main-group metals in the bottom-left corner of the periodic table are known as the active metals because they are so reactive.

Table 5.1

Common Metals Divided into Classes on the Basis of Activity

Class I Metals: The Active Metals	
Li, Na, K, Rb, Cs	(Group IA)
Ca, Sr, Ba	(Group IIA)
Class II Metals: The Less Active Metals	
Mg, Al, Zn, Mn	
Class III Metals: The Structural Metals	
Cr, Fe, Sn, Pb, Cu	
Class IV Metals: The Coinage Metals	
Ag, Au, Pt	

liquid, such as mineral oil. Those metals are found in Groups IA and IIA of the periodic table.

Metals in the second class are slightly less active than Class I metals. They don't react with water at room temperature, but they react rapidly with acids. The less active metals can be used for a variety of purposes. Aluminum, for example, is used for aluminum foil and beverage cans. It is important to protect these metals from exposure to acids, however, so aluminum cans are lined with a plastic coating to prevent contact with the acidic soft drinks or fruit juices they contain.

The third class contains metals such as iron, tin, and lead that react only with strong acids. Some cereals that are described as "iron-enriched" contain tiny pieces of iron metal, not Fe^{3+} salts, as the source of the iron they provide because the iron metal dissolves in the strongly acidic stomach fluid to form Fe^{3+} ions.

Metals in the fourth class are so unreactive they are essentially inert at room temperature. These metals are ideal for making jewelry and coins because they don't react with the majority of the substances with which they come into daily contact. As a result, they are often called the coinage metals.

Aluminum metal is more "active" than iron. So much energy is given off when aluminum metal reacts with Fe_2O_3 to give iron metal and Al_2O_3 that this reaction is called the "thermite" reaction.

5.2 Main-Group Metals and Their Ions

GROUP IA: THE ALKALI METALS

We begin our discussion of the formation of ions by examining three separate areas of the periodic table: main-group metals, main-group nonmetals, and transition metals. The **main-group elements** in the periodic table are made up of the elements in Groups IA through VIIIA. The electron configurations introduced in Chapter 3 and the periodic table can be used to explain the reactions in which metals form positive ions.

The metals in Group IA include lithium (Li), sodium (Na), potassium (K), rubidium (Rb), cesium (Cs), and francium (Fr). These elements are called the **alkali metals** because they form compounds, such as NaOH, that were once known as *alkalies*. Sodium and potassium are relatively common elements. In fact, they are among the eight most abundant elements in the earth's crust, as shown in Table 5.2. The other alkali metals are much less abundant. Discussions of the chemistry of alkali metals therefore focus on sodium and potassium.

The electron configuration of a neutral sodium atom can be written as follows:

$$\text{Na} \quad 1s^2\,2s^2\,2p^6\,3s^1 \qquad \text{or} \qquad [\text{Ne}]\,3s^1$$

Sodium atoms lose their $3s$ electron to form positively charged Na^+ ions with the electron configuration: $1s^2\,2s^2\,2p^6$. The electron configuration of the Na^+ ion is the same as that of a neutral Ne atom. An Ne atom and an Na^+ ion are said to be **isoelectronic** because they have the same number of electrons. Only one electron is lost from sodium during the formation of its ion because the electrons in the second shell are closer to the nucleus and therefore more strongly attracted to the nucleus than the $3s$ electron.

The electron configurations of the alkali metals are all characterized by a single valence electron.

Li	[He] $2s^1$	Rb	[Kr] $5s^1$
Na	[Ne] $3s^1$	Cs	[Xe] $6s^1$
K	[Ar] $4s^1$	Fr	[Rn] $7s^1$

Group IA elements all have similar valence-shell electron configuration (xs^1) and are also characterized by unusually small AVEE values. This means that it is easier

Table 5.2

Percent by Weight of the Most Common Elements in the Earth's Crust

Element	Percent by Weight
O	46.60
Si	27.72
Al	8.13
Fe	5.00
Ca	3.63
Na	2.83
K	2.59
Mg	2.09

to remove an electron from these elements than from almost any other elements in the periodic table. Consequently, the chemistry of these elements is dominated by their tendency to lose an electron to form positively charged ions (Li^+, Na^+, K^+).

The periodic table can be used to predict the charge on many metal ions. As a general rule, the maximum charge on a metal ion can be found from the group number for that element. The alkali metals in Group IA form ions with a +1 charge.

GROUP IIA: THE ALKALINE EARTH METALS

The elements in Group IIA (Be, Mg, Ca, Sr, Ba, and Ra) are all metals, and all but Be and Mg are active metals. These elements are called the **alkaline earth metals.** The term *alkaline* reflects the fact that many compounds of these metals can neutralize acids. The term *earth* was historically used to describe the fact that many of these compounds are insoluble in water.

Most of the chemistry of the alkaline earth metals (Group IIA) can be predicted from the behavior of the alkali metals (Group IA). Once again, these elements are characterized by relatively small average valence electron energies. As a result, the chemistry of these elements is dominated by their tendency to form positive ions. In this case, however, each element has two valence electrons, which means that they tend to form M^{2+} ions.

$$\text{Mg} \quad [\text{Ne}]\ 3s^2 \qquad\qquad \text{Mg}^{2+} \quad [\text{Ne}]$$

> **CHECKPOINT**
>
> Why do atoms of Group IIA elements tend to lose two electrons while those in Group IA only lose one?

The alkaline earth metals in Group IIA are significantly less reactive than the alkali metals in Group IA. They are also harder and have higher melting points than the Group IA metals.

GROUP IIIA METALS

Group IIIA contains a semimetal (B) and four metals (Al, Ga, In, and Tl). Discussions of the chemistry of the Group IIIA metals often focus on aluminum because gallium, indium, and thallium are relatively scarce and therefore of limited interest. Aluminum, on the other hand, is the third most abundant element in the earth's crust (see Table 5.2).

The average valence electron energy for aluminum is only slightly larger than that of the metals in Groups IA and IIA. Thus, the chemistry of aluminum is dominated by the tendency of its atoms to lose electrons to form positive ions. Because the electron configuration of aluminum contains three valence electrons, this metal forms the Al^{3+} ion in many if not most of its compounds.

$$\text{Al} \quad [\text{Ne}]\ 3s^2\ 3p^1 \qquad\qquad \text{Al}^{3+} \quad [\text{Ne}]$$

5.3 Main-Group Nonmetals and Their Ions

In Chapter 3, we noted that the average valence electron energy of the elements increases as we go from left to right across a row of the periodic table. As we go across a row, we therefore reach the point at which it takes too much energy to remove enough electrons from a neutral atom to form an ion with a noble gas electron configuration. The atoms of the main-group elements toward the right side of the periodic table therefore tend to gain electrons to form negatively charged ions with one of the noble gas electron configurations. Ionic compounds, or salts, that contain both positive and negative ions are often formed when main-group elements at opposite sides of the periodic table combine.

The alkali metals react with the nonmetals in Group VIIA (F_2, Cl_2, Br_2, I_2, and At_2) to form ionic compounds, or **salts.** Chlorine, for example, reacts with sodium metal to produce sodium chloride (table salt).

$$2\ Na(s) + Cl_2(g) \longrightarrow 2\ NaCl(s)$$

Because they form salts with so many metals, the elements in Group VIIA are known as the **halogens.** This name comes from the Greek word for "salt" (*hals*) and the Greek word meaning "to produce" (*gennan*). The salts formed by the halogens are called **halides.** These salts include **fluorides** (LiF), **chlorides** (NaCl), **bromides** (KBr), and **iodides** (NaI). The halide ions all carry a charge of −1 because the halogens need only one more electron to achieve a filled-shell electron configuration.

$$Cl\ \ [Ne]\ 3s^2\ 3p^5 \qquad\qquad Cl^-\ \ [Ne]\ 3s^2\ 3p^6 = [Ar]$$

The halogens exist in their free elemental forms as diatomic molecules (F_2, Cl_2, Br_2, I_2, and At_2). However, the halogens are never found in nature in their elemental form. They are found as part of ionic compounds in which they have a −1 charge. Fluoride ions are found in minerals such as fluorite (CaF_2) and cryolite (Na_3AlF_6). Chloride ions are found in rock salt (NaCl), in oceans (which are ~2% Cl^- by weight), and in lakes that have a high salt content, such as the Great Salt Lake in Utah. Both bromide and iodide ions are found at low concentrations in the oceans, as well as in brine wells.

Hydrogen is the only element in the periodic table that is listed in more than one group. In most periodic tables it can be found among the metals in Group IA and the nonmetals in Group VIIA. This can be understood by looking at the position of hydrogen in the three-dimensional version of the periodic table shown in Figure 4.9. Hydrogen is almost exactly in the middle of the elements in terms of its AVEE or its electronegativity. When it reacts with an element that is less electronegative, it tends to pick up one electron, like the halogens, to form a negative ion with the electron configuration of the noble gas helium.

$$H\ \ 1s^1 \qquad\qquad H^-\ \ 1s^2 = [He]$$

Compounds that contain the H^- ion are known as **hydrides.** Thus, potassium reacts with hydrogen to form potassium hydride.

$$2\ K(s) + H_2(g) \longrightarrow 2\ KH(s)$$

The elements in Group VIA often react with metals to form compounds in which they contain a negatively charged ion with a −2 charge. Thus, oxygen forms **oxides** that contain the O^{2-} ion.

$$O\ \ [He]\ 2s^2\ 2p^4 \qquad\qquad O^{2-}\ \ [He]\ 2s^2\ 2p^6 = [Ne]$$

Oxygen is the most abundant element on the planet. The earth's crust is 46.6% oxygen by weight, the oceans are 86% oxygen, and the atmosphere is 21% oxygen. In the earth's crust oxygen is present primarily as a −2 ion in ionic compounds. In the oceans oxygen is found in the covalently bonded H_2O molecule. Oxygen exists in its elemental form as a diatomic molecule (O_2) in the atmosphere.

Like oxygen, sulfur tends to form a −2 ion. Compounds that contain S^{2-} are called **sulfides.**

$$S \quad [Ne]\ 3s^2\ 3p^4 \qquad\qquad S^{2-} \quad [Ne]\ 3s^2\ 3p^6 = [Ar]$$

Sulfur also forms compounds with most of the other elements in the periodic table. Elemental sulfur usually consists of cyclic S_8 molecules in which each atom fills its valence shell by forming single bonds to two neighboring atoms. Because sulfur forms unusually strong S—S single bonds, it is capable of forming cyclic molecules that contain 6, 7, 8, 10, and 12 sulfur atoms.

Although N_2 is virtually inert to chemical reactions at room temperature, the most active metals react with nitrogen to form **nitrides,** such as lithium nitride, Li_3N. The nitride ion carries a charge of −3 because it takes three electrons to reach a filled-shell electron configuration.

$$N \quad 1s^2\ 2s^2\ 2p^3 \qquad\qquad N^{3-} \quad 1s^2\ 2s^2\ 2p^6 = [Ne]$$

Nitrogen is an essential component of the proteins, nucleic acids, vitamins, and hormones that make life possible. Nitrogen in the atmosphere, however, exists as diatomic N_2 molecules that are more or less inert to chemical reactions at room temperature. Some plants contain bacteria that carry an enzyme that catalyzes a reaction known as "nitrogen fixation." Most other plants require nitrogen in the form of a fertilizer. Animals pick up the nitrogen they need from plants or other animals in their diet.

Exercise 5.1

Predict the most stable ion and the corresponding electron configuration for the following atoms.

(a) Ca

(b) P

Solution

(a) Calcium will lose two electrons from its valence shell to achieve the electron configuration of argon.

$$Ca^{2+} \quad 1s^2\ 2s^2\ 2p^6\ 3s^2\ 3p^6 = [Ar]$$

(b) Phosphorus will gain three electrons in order to achieve the electron configuration of argon.

$$P^{3-} \quad 1s^2\ 2s^2\ 2p^6\ 3s^2\ 3p^6 = [Ar]$$

5.4 Transition Metals and Their Ions

The **transition metals** that lie between the main-group elements on either side of the periodic table form a variety of positively charged ions, such as Ag^+, Cu^{2+}, and Fe^{3+}. They form these ions by losing electrons from either valence-shell *s* or *d* orbitals. Iron, for example, can form a +2 ion by losing its 4*s* electrons.

$$\begin{array}{ccc} \text{Fe} & \longrightarrow & \text{Fe}^{2+} + 2\,e^- \\ [\text{Ar}]\,4s^2\,3d^6 & & [\text{Ar}]\,3d^6 \end{array}$$

Iron can also form a +3 ion by losing one of the 3*d* electrons to form an electron configuration in which the *d* orbitals are half-filled, that is, with one electron in each of the five orbitals in the subshell.

$$\begin{array}{ccc} \text{Fe} & \longrightarrow & \text{Fe}^{3+} + 3\,e^- \\ [\text{Ar}]\,4s^2\,3d^6 & & [\text{Ar}]\,3d^5 \end{array}$$

Transition metals are like main-group metals in many ways. They look like metals, they are malleable and ductile, they conduct heat and electricity, and they form positive ions.

Exercise 5.2

The most stable ions of titanium are Ti^{2+} and Ti^{4+}. Predict the electron configuration of the two ions.

Solution

Titanium has the following electron configuration:

$$\text{Ti} \quad [\text{Ar}]\,4s^2\,3d^2$$

This element forms Ti^{2+} ions by the loss of the two electrons in the valence 4*s* orbital.

$$\text{Ti}^{2+} \quad [\text{Ar}]\,3d^2$$

Ti^{4+} ions are formed by the loss of the two valence electrons from the 4*s* subshell plus the loss of the two 3*d* electrons to give the noble gas electron configuration of argon.

$$\text{Ti}^{4+} \quad [\text{Ar}]$$

5.5 Predicting the Products of Reactions That Produce Ionic Compounds

The products of many reactions between main-group metals and other elements can be predicted from the electron configurations of the elements. Consider the reaction between sodium and chlorine to form sodium chloride, for example.

Students are asked to write Lewis structures for ionic compounds here to prepare them for discussions of the ionic bond and the calculation of oxidation numbers.

$$2\,\text{Na}(s) + \text{Cl}_2(g) \longrightarrow 2\,\text{NaCl}(s)$$

The net effect of the reaction is to transfer an electron from a neutral sodium atom to a neutral chlorine atom to form Na^+ and Cl^- ions that have filled-shell configurations.

$$\text{Na}\cdot + \cdot\ddot{\underset{\cdot\cdot}{\text{Cl}}}: \longrightarrow [\text{Na}]^+[:\ddot{\underset{\cdot\cdot}{\text{Cl}}}:]^-$$

We can often predict the product of reactions that form ionic compounds by using average valence electron energies (AVEE) and electron configurations of the elements involved in the reaction. Consider the reaction between potassium metal and hydrogen gas, for example.

$$K(s) + H_2(g) \longrightarrow$$

Potassium and hydrogen have similar valence electron configurations.

$$K \quad [Ar]\ 4s^1 \qquad\qquad H \quad 1s^1$$

To form an ionic compound an electron has to be transferred from one element to the other. We can decide which element should lose an electron by comparing the AVEE for potassium (0.42 MJ/mol) with that for hydrogen (1.31 MJ/mol). These data suggest that each potassium atom consumed in the reaction should lose an electron to a hydrogen atom to form K^+ and H^- ions.

$$K\cdot + \cdot H \longrightarrow [K]^+[:H]^-$$

The balanced equation for the reaction is written as follows.

$$2\ K(s) + H_2(g) \longrightarrow 2\ KH(s)$$

Exercise 5.3

Lithium metal is used in high-energy, lightweight batteries. The batteries are sealed to prevent the lithium metal from reacting with the oxygen in the air. Write a balanced equation for the following reaction.

$$Li(s) + O_2(g) \longrightarrow$$

Solution

We start by using the electron configurations of the elements to predict the ions that form in this reaction.

$$Li \quad [He]\ 2s^1 \qquad\qquad O \quad [He]\ 2s^2\ 2p^4$$

Everything we've learned about the structure of the atom in Chapter 3 suggests that it should be easier to remove an electron from lithium than from oxygen. We can confirm this by comparing the average valence electron energies for lithium (0.52 MJ/mol) and oxygen (1.91 MJ/mol). We therefore expect that lithium will lose electrons and oxygen will gain electrons to form the Li^+ and O^{2-} ions.

$$Li^+ \quad [He] \qquad\qquad O^{2-} \quad [He]\ 2s^2\ 2p^6 = [Ne]$$

We then combine these ions to predict the product of the reaction—Li_2O—and write a balanced equation for the reaction, as follows.

$$4\ Li(s) + O_2(g) \longrightarrow 2\ Li_2O(s)$$

5.6 Oxides, Peroxides, and Superoxides

The method used to predict the products of reactions of the main-group metals is remarkably powerful. Exceptions to its predictions arise, however, when very active metals react with oxygen, which is one of the most reactive nonmetals.

Lithium is well behaved. It reacts with O_2 to form an oxide, as predicted in Exercise 5.3.

$$4\,\mathrm{Li}(s) + \mathrm{O_2}(g) \longrightarrow 2\,\mathrm{Li_2O}(s)$$

Sodium, however, reacts with O_2 under normal conditions to form a compound that contains twice as much oxygen.

$$2\,\mathrm{Na}(s) + \mathrm{O_2}(g) \longrightarrow \mathrm{Na_2O_2}(s)$$

Compounds that are unusually rich in oxygen are called **peroxides.** The prefix *per-* means "above normal" or "excessive." Na_2O_2 is a peroxide because it contains more than the expected amount of oxygen.

The difference between oxides and peroxides can be understood by remembering the basic assumption behind the chemistry of the alkali metals: These elements form compounds in which the metal is present as a +1 ion. Oxides, such as Li_2O, contain the O^{2-} ion.

$$[\mathrm{Li}]^+[:\ddot{\underset{\cdot\cdot}{\mathrm{O}}}:]^{2-}[\mathrm{Li}]^+$$

Peroxides, such as Na_2O_2, contain the polyatomic O_2^{2-} ion, which is known as the peroxide ion.

$$[\mathrm{Na}]^+[:\ddot{\underset{\cdot\cdot}{\mathrm{O}}}-\ddot{\underset{\cdot\cdot}{\mathrm{O}}}:]^{2-}[\mathrm{Na}]^+$$

The formation of sodium peroxide can be explained by recognizing that sodium is so reactive that the metal is consumed before each O_2 molecule can combine with enough sodium to form Na_2O. This explanation is supported by the fact that sodium reacts with O_2 in the presence of a large excess of the metal, or in the presence of a limited amount of O_2, to form the oxide expected when the reaction goes to completion.

$$4\,\mathrm{Na}(s) + \mathrm{O_2}(g) \longrightarrow 2\,\mathrm{Na_2O}(s)$$

It is also consistent with the fact that the very active alkali metals—potassium, rubidium, and cesium—react so rapidly with oxygen they form **superoxides,** in which the alkali metal reacts with O_2 in a 1:1 mole ratio. Superoxides contain the polyatomic O_2^- ion.

$$2\,\mathrm{K}(s) + \mathrm{O_2}(g) \longrightarrow 2\,\mathrm{KO_2}(s)$$

Potassium superoxide forms on the surface of potassium metal, even when the metal is stored under an inert solvent. Thus old pieces of potassium metal are potentially dangerous. When someone tries to cut the metal, the pressure of the knife pushing down on the area where the superoxide touches the metal can induce the following reaction.

$$\mathrm{KO_2}(s) + 3\,\mathrm{K}(s) \longrightarrow 2\,\mathrm{K_2O}(s)$$

Because potassium oxide is more stable than potassium superoxide, the reaction gives off enough energy to boil potassium metal off the surface, which reacts explosively with the oxygen and water vapor in the atmosphere.

The reactions between the alkali metals and oxygen raise an important point. We can predict the product of any reaction between a main-group metal and a main-group nonmetal from the AVEE values for the elements and their electron configurations. But we always have to check our predictions with an experiment because other factors, such as the rate at which one of the reactants is consumed, may affect the formula of the substance actually produced in the reaction.

➤ CHECKPOINT

What is the formula of the peroxide of strontium, Sr?

The alkaline earth metals also react with oxygen to form oxides and peroxides. Because the activity of the alkaline earth metals increases as we go down the column, magnesium forms the oxide (MgO), whereas barium forms the peroxide (BaO_2).

5.7 The Ionic Bond

The ionic bond is introduced here so that students will be able to understand the classifications of bond types later in the chapter. In addition, when the calculation of oxidation numbers is introduced, the methodology of assigning electrons to the most electronegative atoms will not be new to them. A more complete discussion of ionic compounds is in Chapter 9.

Ionic compounds are held together by the force of attraction between ions of opposite charge. The magnitude of the **electrostatic** or **coulombic** attraction between the particles depends on the product of the charge on the two ions and the square of the distance between the ions.

$$F = \frac{q_+ \times q_-}{r^2}$$

The bond that forms when the ions are brought together is called, logically enough, an **ionic bond.** In the formation of an ionic bond between a metal and nonmetal, there is a transfer of electron density from the metal to the nonmetal. This can be described using Lewis structures, which show the valence-shell electrons.

$$\mathrm{K}\cdot + \cdot\ddot{\underset{\cdot\cdot}{\mathrm{I}}}: \longrightarrow [\mathrm{K}]^+[:\ddot{\underset{\cdot\cdot}{\mathrm{I}}}:]^-$$

$$2\,\mathrm{Li}\cdot + \cdot\ddot{\underset{\cdot}{\mathrm{S}}}: \longrightarrow [\mathrm{Li}]^+[:\ddot{\underset{\cdot\cdot}{\mathrm{S}}}:]^{2-}[\mathrm{Li}]^+$$

There is a fundamental difference between the covalent compounds discussed in the previous chapter and the ionic compounds in this chapter. Covalent compounds usually exist in the form of molecules that contain a limited number of atoms. Even compounds as complex as glucose ($C_6H_{12}O_6$) and vitamin B_{12} ($C_{63}H_{88}N_{14}O_{14}PCo$) consist of distinct molecules that contain a well-defined number of atoms.

The same can't be said of ionic compounds. They exist as huge, three-dimensional networks of ions of opposite charge that are held together by ionic bonds, as shown in Figure 5.2. Each Na^+ ion in this structure is surrounded by six Cl^- ions, and each Cl^- ion is surrounded by six Na^+ ions. The simplest whole-number ratio of sodium ions to chloride ions is therefore 1:1, so the formula of the compound is written as NaCl.

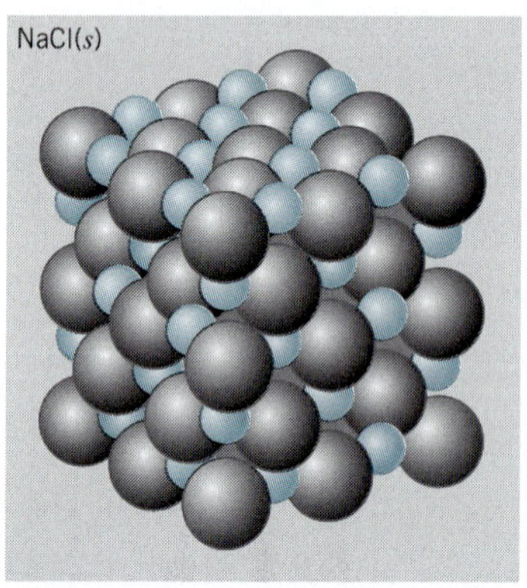

Fig. 5.2 Three-dimensional network structure of NaCl. The smaller spheres represent Na^+ ions and the larger spheres represent Cl^- ions.

5.8 Structures of Ionic Compounds

When we want to describe the structure of a covalent compound, we draw a picture of its molecules. When we want to envision the structure of an ionic compound, we draw a picture of the simplest repeating unit, or **unit cell,** in the three-dimensional network of its ions. By describing the size, shape, and contents of the unit cell—and the way the repeating units stack to form a three-dimensional solid—we can

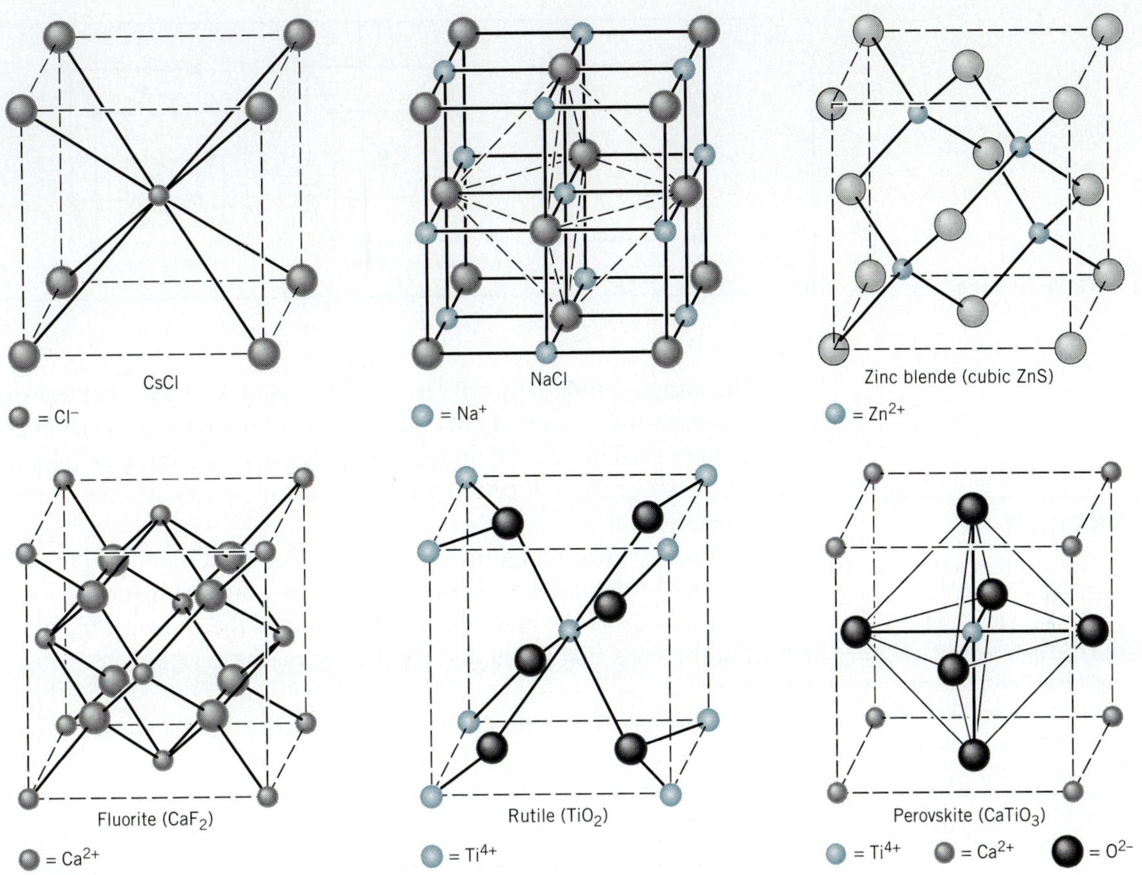

Fig. 5.3 Unit cells of CsCl, NaCl (rock salt), ZnS (zinc blende), CaF_2 (fluorite), TiO_2 (rutile), and $CaTiO_3$ (perovskite).

unambiguously describe the structure of the compound. Figure 5.3 shows the unit cells for several common ionic compounds. Because a unit cell is a part of a larger three-dimensional structure, ions at the corners, edges, and faces are shared by neighboring unit cells. Only when this is taken into consideration does the ionic formula match the ratio of ions in the unit cell.

In 1850 Auguste Bravais showed that every crystal, no matter how complex its structure, could be classified as one of 14 unit cells that meet the following criteria.

- The unit cell is the simplest repeating unit in the crystal.
- Opposite faces of a unit cell are parallel.
- The edge of the unit cell connects equivalent points.

This chapter focuses on only one kind of unit cell, the cubic unit cell. There are three types of cubic unit cells—simple cubic, body-centered cubic, and face-centered cubic—shown in Figures 5.4 and 5.5. Cubic unit cells are important for two reasons. First, many metals, ionic solids, and intermetallic compounds crystallize in cubic unit cells. Second, these unit cells are relatively easy to visualize because the cell-edge lengths are all the same and the cell angles are all 90°. Unit cells are defined in terms of **lattice points,** which are points at which a particle can be found in the lattice structure of the crystal. This particle can be an atom, an ion, or a molecule.

Unit cells are defined so that the cell edges always connect equivalent points. Therefore, an identical particle must be found at each of the eight corners of a cubic unit cell. The unit cell of NaCl shown in Figure 5.3, for example, contains an equivalent Cl^- ion at each of the eight corners of the unit cell.

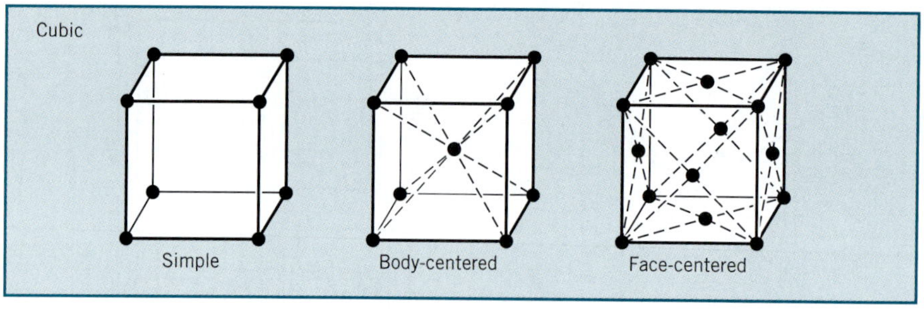

Fig. 5.4 Three common unit cells.

The **simple cubic** unit cell, as its name implies, is the simplest of all unit cells. It consists of a minimum of eight equivalent particles at the eight corners of a cube. Other particles can be present on the edges or the faces of the unit cell, or within the body of the unit cell. But the unit cell must contain equivalent particles at the eight corners.

The **body-centered cubic** unit cell also has eight identical particles on the eight corners of the unit cell. However, there is a ninth identical particle in the center of the body of the unit cell. If the particle at the center of the unit cell differs from the ones that define the eight corners, the crystal is classified as simple cubic.

The **face-centered cubic** unit cell starts with eight identical particles on the eight corners of the cube. But the structure also contains the same particles in the centers of the six faces of the unit cell, for a total of 14 identical lattice points.

> **CHECKPOINT**
>
> Cubic cells exist as part of a network of cells in ionic crystals. The atoms at the corners of a unit cell don't belong exclusively to one unit cell but are shared by eight unit cells. Sketch a diagram to show that an atom at the corner of a simple cubic cell is actually part of eight simple cubic cells.

5.9 Metallic Bonds

The covalent bond that binds one chlorine atom to another to form a Cl_2 molecule has one thing in common with the ionic bond between the Na^+ and Cl^- ions in NaCl. In both cases, the electrons in the bond are **localized.** The electrons either

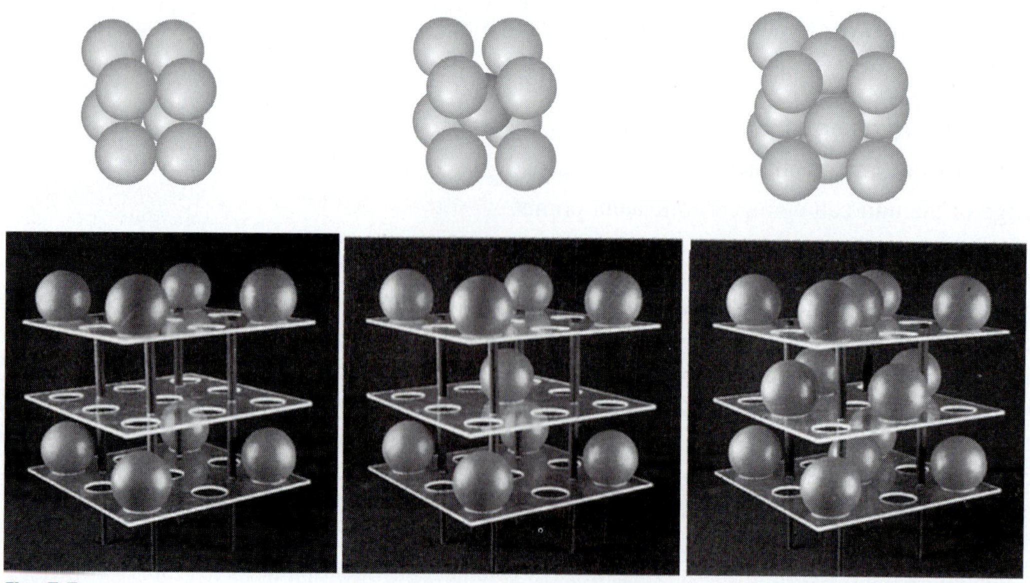

Fig. 5.5 Models of simple cubic (left), body-centered cubic (middle), and face-centered cubic unit cells (right).

reside on one of the atoms or ions that form the bond or they are shared by a pair of atoms.

Metal atoms don't have enough valence electrons to reach a filled-shell configuration by sharing electrons with their neighbors. They are also relatively large and have relatively small AVEE values. This means that the valence electrons can be shared with many neighboring atoms, not just one. In effect, these valence electrons are **delocalized** over a number of metal atoms.

Because the delocalized valence electrons aren't tightly bound to individual atoms, they are free to move through the metal. A useful picture of the structure of metals therefore envisions the metal atoms as positive ions locked in a crystal lattice surrounded by a sea of valence electrons that move among the ions, as shown in Figure 5.6. The force of attraction between the positive metal ions and the sea of mobile negative electrons forms a **metallic bond** that holds the particles together.

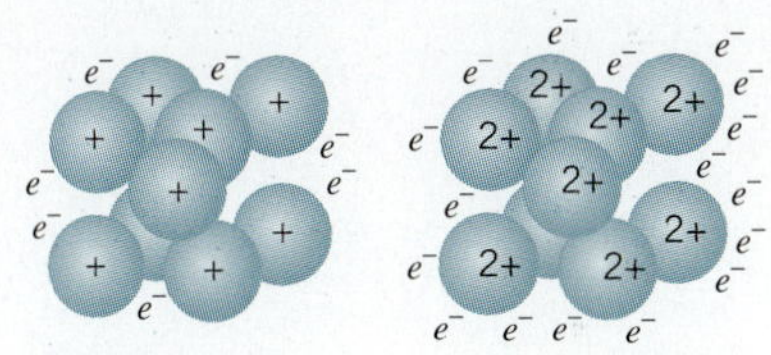

Fig. 5.6 The force of attraction between the positively charged metal ions and the surrounding sea of electrons in a metal is called a metallic bond. The electrons are not bound to individual atoms but may move throughout the metal.

Metals exist as extended three-dimensional arrays of atoms, which pack so that each atom can touch as many neighboring atoms as possible. When a metal is heated, or beaten with a hammer, the planes of atoms that form the structure can slip past one another, which explains why metals are malleable and ductile (see Section 1.13). Each atom in the structure has a limited number of loosely held valence electrons that it can share with its neighbors. This is only possible when the atoms are relatively close together. Due to the mobility of the electrons, metals can easily transfer kinetic energy from one atom to another. As a result, they are good conductors of heat.

As we have seen, atoms become larger as we go toward the bottom left-hand corner of the periodic table. Here we find atoms that have relatively small AVEE values. Therefore, elements in that region of the periodic table have metallic character. This means the electrons are not tightly held and the valence subshells are close in energy.

Because the valence electrons in a metal are delocalized instead of being strictly associated with a single atom, they are free to move from one atom to another. The energy levels in isolated atoms differ from the energy levels of atoms that are bonded to one another. As a result, in a metal there is even a smaller difference in energy between the *s*, *p*, and *d* subshells within the valence shell of bonded metal atoms as compared to isolated atoms. This allows electrons to move between the subshells of an atom or adjacent atoms. The electrons that freely move through the subshells are said to form **conduction bands.** If we draw the metal into a thin wire, and connect the wire to a source of an electric current, electrons that enter the wire displace electrons that were already present on the atoms closest to the source of the current. Electrons flow through the conduction band until they eventually displace electrons from the other side of the wire. Metals are therefore good conductors of electricity.

> **CHECKPOINT**
>
> Covalent and ionic compounds do not conduct electricity in the solid state. These compounds are called *insulators*. Metals do conduct electricity. Explain why substances exhibiting metallic bonding are conductors while covalent and ionic compounds are not.

Metals in their elemental form (Na, Cu, Fe) do not consist of individual atoms but instead are composed of a three-dimensional network of positive metal ions in a sea of electrons. Because a pure metal in its elemental state, such as iron, consists only of the element iron, we write the chemical formula as Fe. There are also metallic compounds made up of more than one element. The chemical formulas of metallic compounds represent the smallest whole-number ratio of the atoms present in the compound. An example is $CuAl_2$, an intermetallic compound with a fixed composition. Pure aluminum metal is too weak to be used as a structural metal in cars or airplanes. However, the addition of microcrystals of $CuAl_2$ to aluminum strengthens the aluminum metal by interfering with the way planes of atoms slip past each other. The result is a metal that is both harder and stronger than pure aluminum.

> **CHECKPOINT**
>
> How does the bond in $CuAl_2$ differ from the bond in NaCl?

5.10 The Relationship Among Ionic, Covalent, and Metallic Bonds

There are enormous differences in the physical properties of sodium chloride, sodium metal, and chlorine, as shown in Table 5.3. These differences result from differences between the structure and bonding in these substances. Chemists often classify compounds as ionic, metallic, or covalent on the basis of macroscopic and physical properties. These classifications are closely related to the type of bonding between the atoms that form each substance. Thus, compounds with primarily metallic bonding are called metallic compounds and those with predominantly ionic bonding are called ionic compounds.

As we have seen, each Na^+ ion in NaCl is surrounded by six Cl^- ions, and vice versa. Removing an ion from the compound therefore involves breaking at least six bonds. Ionic compounds such as NaCl tend to have high melting points and boiling points. Ionic compounds are therefore usually solids at room temperature.

Each Na atom in sodium metal is bound to as many neighboring atoms as possible; this plus strong attractive forces between metal atoms means that metals are usually solids at room temperature. Sodium metal has a very high boiling point because it takes a great deal of energy to disrupt all of the bonds necessary to remove individual Na atoms from the metal.

$Cl_2(g)$ consists of molecules in which one chlorine atom is tightly bound to another chlorine atom to form a covalent bond. The covalent bonds within the molecules are at least as strong as ionic bonds, but the melting and boiling points of chlorine are much lower than those for ionic compounds such as sodium chloride. The strong Cl—Cl bonds within a Cl_2 molecule aren't broken when chlorine melts or boils. It is only necessary to break the relatively weak forces of attraction that hold one chlorine molecule to another for chlorine to melt or boil.

The electronegativity scale used in this text is derived from average valence electron energies. Because this scale is derived from PES data in the text, students have a more concrete basis for understanding electronegativity. Through Chapters 4 and 5 we have used the terms AVEE *and* electronegativity *interchangeably. For the remaining portion of the text we will use primarily* electronegativity *because this is the term traditionally used by chemists.*

The difference between the physical properties of NaCl and Cl_2 is so large that it is easy to believe that the bond between two atoms is either ionic or covalent. G. N. Lewis, who introduced the concept of the covalent bond, recognized that this isn't true. In the paper in which he first described bonds based on the sharing of electrons, Lewis argued that words such as *ionic* and *covalent* referred to the extremes at either end of a continuous spectrum of bonding.

To see how he came to that conclusion, let's compare what happens when covalent and ionic bonds form. When two chlorine atoms come together to form a covalent bond, each atom contributes one electron to form a pair of electrons shared equally by the two chlorine atoms. The electrons are shared equally because both chlorine atoms have the same AVEE or electronegativity.

$$:\ddot{\underset{..}{Cl}}\cdot + \cdot\ddot{\underset{..}{Cl}}: \longrightarrow :\ddot{\underset{..}{Cl}}-\ddot{\underset{..}{Cl}}:$$

The electronegativity scale developed from AVEE in Chapter 4 will be used throughout the rest of the text. Chemists use the term *electronegativity* rather than *AVEE.*

Table 5.3

Some Physical Properties of NaCl, Na, and Cl_2

	NaCl	Na	Cl_2
Phase at room temperature	Solid	Solid	Gas
Density (g/cm^3)	2.2	0.97	0.0032
Melting point	801°C	97.81°C	−100.98°C
Boiling point	1413°C	882.9°C	−34.6°C

If the electronegativities of the atoms in a compound are about the same and the elements are from the right side of the periodic table, the atoms share electrons, and the substance is considered to be **covalent.** Examples of covalent compounds are methane (CH_4) and nitrogen dioxide (NO_2).

CH_4		NO_2	
C	$EN = 2.54$	O	$EN = 3.61$
H	$EN = 2.30$	N	$EN = 3.07$
	$\Delta EN = 0.24$		$\Delta EN = 0.54$

These compounds, which consist of discrete molecules, have relatively low melting points (MP) and boiling points (BP), and they are both gases at room temperature. These are characteristic properties of low-molecular-weight covalent compounds.

	CH_4	NO_2
MP	−182.5°C	−163.6°C
BP	−161.5°C	−151.8°C

If the electronegativities of the atoms in a substance are about the same and if the atoms come from the left side of the periodic table, the substance will be *metallic*. Sodium, although made up of only one element, is a metallic substance because the sodium atoms are held together by metallic bonds. The properties of sodium given in Table 5.3 represent a typical metallic substance. The compound CdLi, with $\Delta EN = 0.6$, is also classified as metallic.

When a sodium atom combines with a chlorine atom to form an ionic bond, each atom contributes one electron to form a pair of electrons, but that pair of electrons spends most of its time on the more electronegative chlorine.

$$\mathrm{Na}\cdot + \cdot\underset{\cdot\cdot}{\ddot{\mathrm{Cl}}}: \longrightarrow [\mathrm{Na}]^+[:\underset{\cdot\cdot}{\ddot{\mathrm{Cl}}}:]^-$$

When the difference between the electronegativities of the elements in a compound is relatively large, the compound is best classified as *ionic*. Large differences in electronegativity are normally associated with bonding between metals from the left side of the periodic table and nonmetals from the right side. NaCl and SrF_2 are good examples of ionic compounds. In each case, the electronegativity of the nonmetal is at least two units larger than that of the metal.

NaCl		SrF_2	
Cl	$EN = 2.87$	F	$EN = 4.19$
Na	$EN = 0.87$	Sr	$EN = 0.96$
	$\Delta EN = 2.00$		$\Delta EN = 3.23$

We can therefore assume a net transfer of electrons from the metal to the nonmetal to form positive and negative ions that form a three-dimensional network. We write the following Lewis structures for these compounds.

$$\mathrm{NaCl} \qquad [\mathrm{Na}]^+[:\underset{\cdot\cdot}{\ddot{\mathrm{Cl}}}:]^-$$

$$\mathrm{SrF_2} \qquad [:\underset{\cdot\cdot}{\ddot{\mathrm{F}}}:]^-[\mathrm{Sr}]^{2+}[:\underset{\cdot\cdot}{\ddot{\mathrm{F}}}:]^-$$

These compounds have high melting points and boiling points, which are characteristic properties of ionic compounds.

	NaCl	*SrF*$_2$
MP	801°C	1473°C
BP	1413°C	2489°C

When ionic compounds dissolve in water, they break apart to form the ions that compose them. The resulting ions are free to move throughout the solution, and therefore the aqueous solution can conduct an electric current.

$$NaCl(s) \xrightarrow{H_2O} Na^{+}(aq) + Cl^{-}(aq)$$

Similarily when ionic compounds melt, ions are formed and, therefore, molten ionic compounds also conduct electricity.

Inevitably, there must be compounds that fall between the extremes of covalent and ionic compounds. For such compounds, the difference between the electronegativities of the elements is large enough to be significant, but not large enough to classify the compound as ionic. Consider water, for example.

$$\begin{array}{ll} & H_2O \\ O & EN = 3.61 \\ H & EN = 2.30 \\ \hline & \Delta EN = 1.31 \end{array}$$

Water is neither purely ionic nor purely covalent. It doesn't contain positive and negative ions, as indicated by the first Lewis structure in Figure 5.7. But the electrons aren't shared equally, as indicated by the second Lewis structure in Figure 5.7. Water is best described as a **polar covalent compound.** One end, or pole, of the molecule has a partial positive charge ($\delta+$), and the other end has a partial negative charge ($\delta-$).

Ionic and covalent bonds differ in the extent to which a pair of electrons is shared by the atoms that form the bond. When one of the atoms is much better at drawing electrons toward itself than the other, the bond is *ionic*. When the atoms are approximately equal in their ability to draw electrons toward themselves, the atoms share the pair of electrons more or less equally, and the bond is *covalent*. Unfortunately, the terms *ionic* and *covalent* describe the extremes of a continuum of bonding. There is some covalent character in even the most ionic compounds. The chemistry of magnesium oxide, for example, can be understood if we assume that MgO contains Mg^{2+} and O^{2-} ions. But no compounds are 100% ionic. There is experimental evidence, for example, that the partial charge on the magnesium and oxygen atoms in MgO is +1.5 and −1.5, respectively.

Electronegativity is a powerful concept that summarizes the tendency of an atom of an element to gain, lose, or share electrons when it combines with another

> **➤ CHECKPOINT**
>
> What is the difference in electronegativity between aluminum and chlorine? Using only the difference in electronegativity, would you expect $AlCl_3$ to exhibit primarily ionic or covalent bonding?

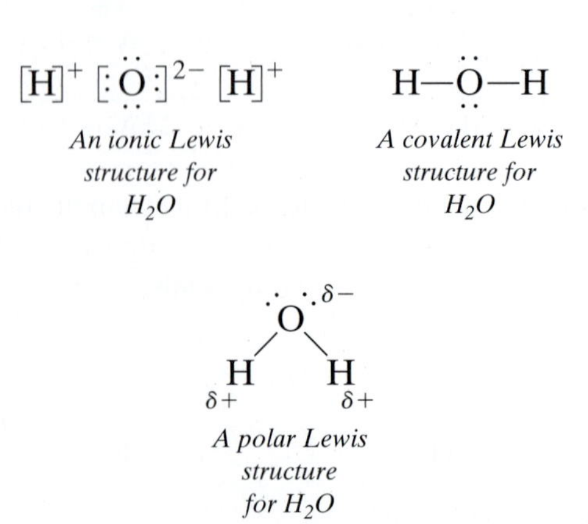

Fig. 5.7 Water is neither purely ionic nor purely covalent; it is a polar covalent compound.

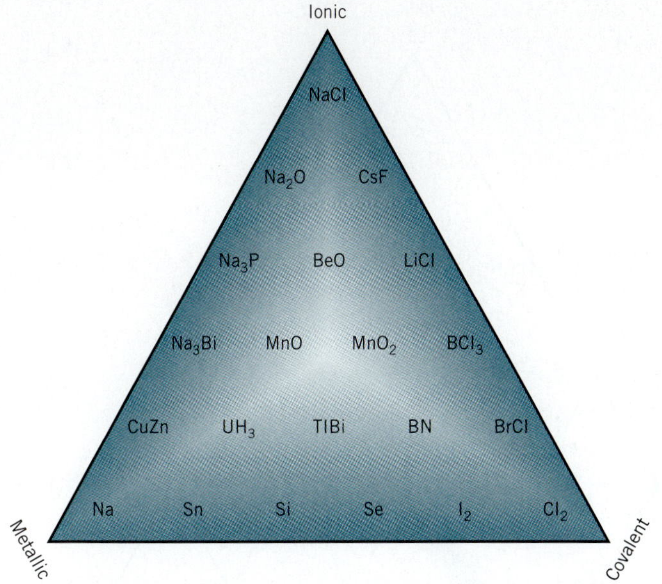

Fig. 5.8 The three different forms of chemical bonds—ionic, covalent, and metallic—form a two-dimensional plane, not a linear continuum.

atom. The difference between the electronegativity of the atoms in a bond (ΔEN), however, isn't enough information to categorize the bond between atoms as primarily ionic, covalent, or metallic. Consider BF_3 ($\Delta EN = 2.14$) and SiF_4 ($\Delta EN = 2.27$), for example. The difference between the electronegativities of the atoms leads us to expect the compounds to behave as if they were ionic, but both compounds are best classified as covalent. They are both gases at room temperature, and their boiling points are $-99.9°C$ and $-86°C$, respectively.

Consider the following compounds: MgH_2 ($\Delta EN = 1.0$) and CO_2 ($\Delta EN = 1.1$). The value of ΔEN is nearly the same in these compounds. But MgH_2 is usually assumed to be an ionic compound that contains Mg^{2+} and H^- ions, whereas CO_2 is assumed to be a covalent molecule held together by carbon-to-oxygen double bonds.

The problem of predicting bond type results from the fact that all three types of bonds are based on the force of attraction between particles of opposite charge and the bonds all result from the transfer or sharing of electrons.

One approach to the problem was taken in the 1970s by William L. Jolly, who constructed the diagram shown in Figure 5.8.[1] This figure assumes that the three different bond types—ionic, covalent, and metallic—form a two-dimensional triangle, not a linear continuum. Some elements combine to form ionic bonds, others form metallic bonds, and still others form covalent bonds.

In the 1990s, Leland Allen, William Jensen, and Gordon Sproul returned to a triangular representation to describe the bonding in systems that don't fit the extremes of ionic, covalent, or metallic bonding.[2,3,4] Whereas Jolly's triangle was descriptive, the more recent bond-type triangles have predictive power. Figure 5.9 shows the basic structure of a bond-type triangle constructed for the elements between lithium and fluorine. In this bond-type triangle, the difference between electronegativity of the atoms in a bond is plotted on the vertical axis, and the average electronegativity of these atoms is plotted along the base of the triangle (the horizontal axis).

In both triangles, the vertex on the left represents pure metallic bonding between identical metal atoms. The vertex on the right corresponds to a pure covalent bond

[1]W. L. Jolly, *The Principles of Inorganic Chemistry,* McGraw-Hill, New York, 1974, p. 187.

[2]L. C. Allen, *Journal of the American Chemical Society,* **114,** 1510 (1992).

[3]W. B. Jensen, *Bulletin of the History of Chemistry,* **13–14,** 47 (1992–1993).

[4]G. Sproul, *Journal of Physical Chemistry,* **98,** 13221 (1994).

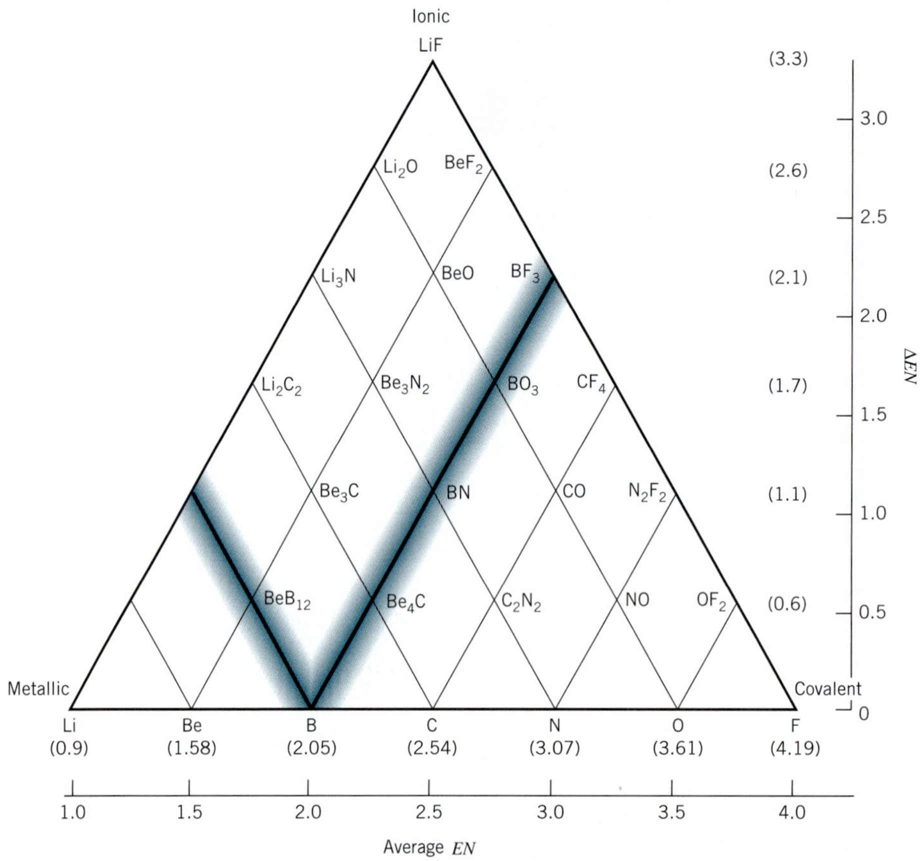

Fig. 5.9 A bond-type triangle for the second-row elements. ΔEN is plotted on the vertical axis, and the average electronegativity is plotted along the base. The bond type varies from ionic at the top of the triangle to metallic at the bottom left to covalent at the bottom right. A compound can be classified according to bond type by locating it within a bond-type triangle. The dark, shadowy lines separate the triangle into three regions: metallic, ionic, and covalent.

between identical nonmetal atoms. Compounds near the top of the triangle represent ideal ionic bonds, such as LiF, in which the transfer of electrons is essentially complete.

Compounds near the three vertices are predominantly metallic, covalent, or ionic, respectively. Those that lie toward the middle of the diagram are the most difficult to describe in terms of any one of the three categories of chemical bonds. Compounds in the middle area have properties intermediate between the three types of bonding.

The elements to the left of boron on the baseline are metals, whereas those on the right of boron are nonmetals. The two dark, shadowy lines that intersect at boron in Figure 5.9 divide the triangle into metallic, ionic, and covalent regions. The principal reason for categorizing substances by type of bonding is to predict more accurately the properties of the compound.

The position of a compound within a bond-type triangle can be used to visualize the ionic, covalent, or metallic character of the compound. As we proceed down the right side of this diagram, from LiF to F_2, we reach a point at which we encounter compounds that have some of the characteristics of both ionic and covalent compounds. As we go down the left side of the triangle, we encounter an interface between ionic and metallic compounds. Some compounds that may at first glance appear to be ionic may in fact be metallic. As we go up the triangle from the base to the vertex, there is a steady increase in ionic character.

5.11 Bond-Type Triangles

The **bond-type triangle** in Figure 5.9 was constructed by calculating both the electronegatiity difference (ΔEN) and the average electronegativity for a series of **binary compounds**—compounds that contain two elements—of the second-row elements. The vertical position of each compound in this diagram reflects the difference between the electronegativities of the elements. The horizontal position is determined by the average of these electronegativities. Because LiF has the largest electronegativity difference of any of the compounds in this figure, it is the most ionic compound in this triangle. It therefore appears at the top of the triangle.

Other compounds are placed in the triangle by calculating their ΔEN and the average EN and plotting them at the appropriate positions. For example, the compound BN is located above the base of the triangle midway between boron and nitrogen. The electronegativity at this point on the base of the triangle (2.56) is the average of the electronegativity of boron (2.05) and the electronegativity of nitrogen (3.07). ΔEN, which gives the vertical distance from the base, is 1.02. The location of compounds in the triangle gives information about the type of bonding. CO, for example, is best described as covalent, but perhaps not as covalent as NO, which is closer to the right corner of the triangle.

Figure 5.9 doesn't contain formulas for compounds that might be formed by combining lithium with either boron or beryllium because no binary compounds of these elements are known. However, the possible properties of such compounds could be predicted from their position on the triangle.

An enlarged and more detailed bond type triangle is shown in Figure 5.10. This triangle has been expanded to include many of the most common elements and their binary compounds. By plotting more compounds on the triangle we can make a better estimate of the position of the borders between the three types of bonding. The dark, shadowy lines separating the regions into ionic, metallic, covalent, and semimetallic are based on a classification of more than 300 binary compounds by Sproul.[4]

A diagonal stairstep line is used to separate the metals and nonmetals on most periodic tables. In Chapter 3 we noted that AVEE provides a rationale for the position of this line. Tellurium and arsenic, which have about the same electronegativities (AVEE values), define the right-hand limit of metallic behavior in the periodic table, whereas aluminum represents the left-hand limits of nonmetallic behavior. In Figure 5.10 the same elements are used to separate the triangle into different types of bonding. Dark, shadowy lines parallel to the sides of the triangle have been drawn in Figure 5.10 to separate elements below Al from those above Te.

This expanded representation of the bond type triangle not only allows classification of a variety of binary compounds but also makes clear that changing atom combinations can dramatically alter the physical properties of a substance. A series of compounds of fluorine can be seen along the right side of the triangle. As we go down this line, we pass from ionic to covalent compounds. Compounds of Cs can be found along the left side of the triangle. As we go down the line, the compounds change from ionic to metallic bonding.

The area between tellurium and aluminum in Figure 5.10 is called the metalloid or semimetal region.[5] In this region of the triangle a change is occurring from metallic to covalent bonding, and these compounds have properties that are intermediate between the two bonding types. Many compounds that fall in this region have electrical conductivities between conductors and insulators and are called semiconductors. Other compounds that fall outside this region but are close to the ionic–covalent border are also semiconductors.

The bond-type triangle shows why the value of ΔEN doesn't always provide enough information to categorize a compound. Both the average electronegativities of the elements in the compounds and the electronegativity differences must be taken into consideration. The bond-type triangle in Figure 5.10, for example, clearly shows both BF_3 and SiF_4 within the covalent region of the triangle.

[5]L. C. Allen, *Journal of the American Chemical Society,* **114,** 1510 (1992).

➤ CHECKPOINT

In the previous checkpoint you were asked to use only ΔEN to describe the type of bonding in $AlCl_3$. Now use the bond-type triangle to describe the bonding in $AlCl_3$.

In order to provide lines of demarcation between the bonding regions, the bond type of some compounds must be known. Sproul has listed some 300 binary compounds classified according to Wells.[1,2] *When plotted, the division between bond types is readily apparent. This list of classified compounds in the* Instructor's Manual *is helpful when selecting example compounds or preparing exams.*

Sproul's work has shown that lines drawn in the triangle parallel to the sides of the triangle at electronegativities corresponding roughly to 1.6 and 2.2 will allow classification of bond type with a 96% degree of confidence.[2]

[1]G. Sproul, *J. Chem. Educ.* **78,** 387 (2001).

[2]G. Sproul, *J. Phy. Chem.* **98,** 6699 (1994).

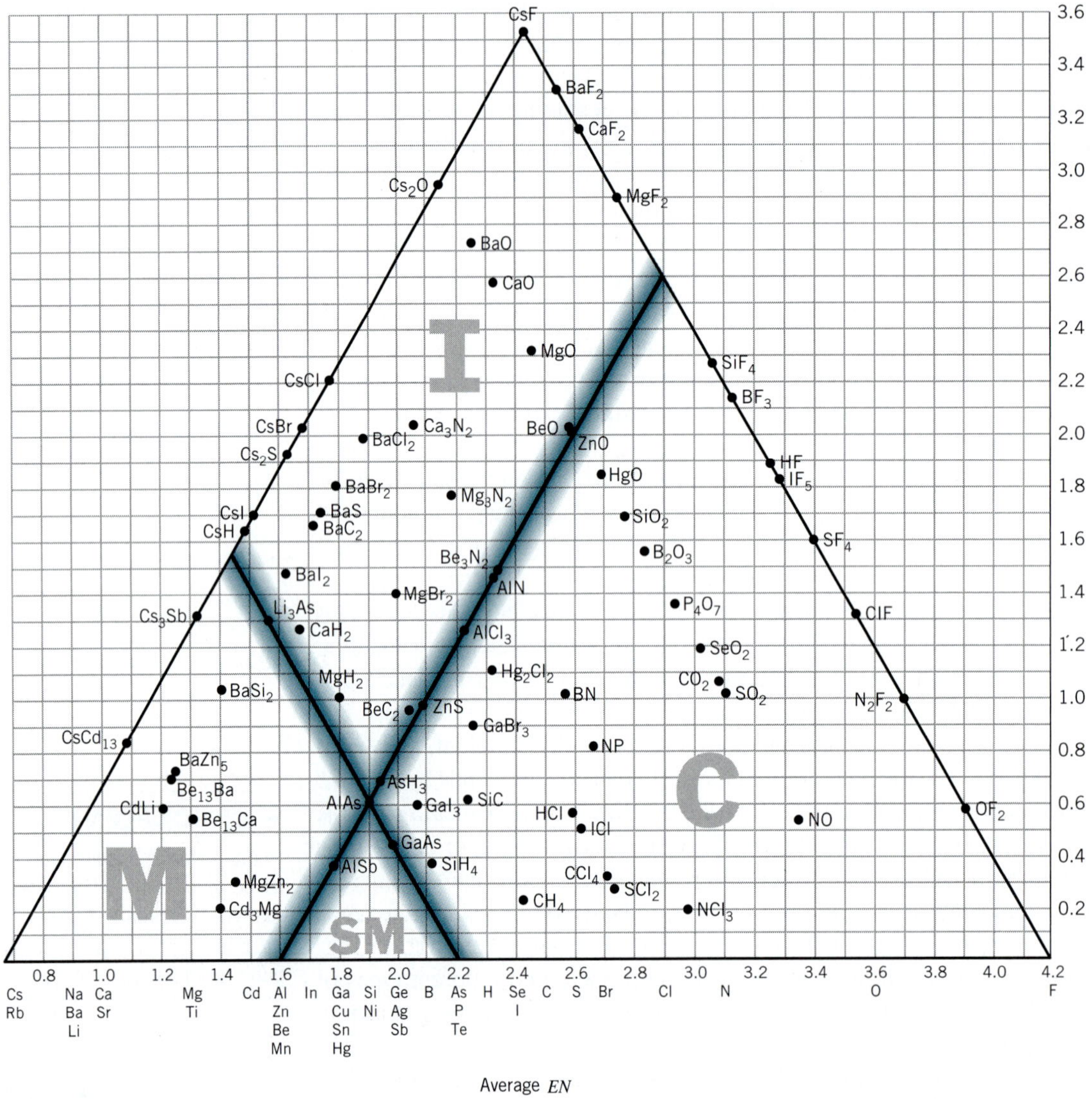

Fig. 5.10 Bond-type triangle for elements with electronegativities ranging from Cs to F. The lines separating the regions into ionic (I), metallic (M), covalent (C), and semimetallic compounds (SM) were determined empirically. These lines are parallel to the sides of the triangle and start at the positions occupied by the elements Al and Te on the base of the triangle.

Average EN is a measure of the covalent character of bonding electrons along the metallic-to-covalent continuum, while ΔEN relates to the ionic character of a bond. Low values of average EN correspond to closely spaced energy levels and low attraction for electrons; thus a small value of average EN indicates more metallic character in a bond. However, a large ΔEN could confer ionic character, and the position in the bond-type triangle will aid in describing the bond type.

Predominantly covalent substances are formed when atoms that have relatively large electronegativities and small ΔEN values combine. Metallic substances are formed when we combine elements that have small electronegativities and a small ΔEN. Ionic compounds occur, as we might expect, when elements with very different electronegativities are combined.

The bond-type triangle enables us to think about bonds in terms of three contributing factors. Each compound on the triangle can be described in terms of the relative contributions of the three bonding types to the overall bond.

As we move from left to right on the triangle, the bonding electrons change from delocalization in metallic compounds to localized shared electrons in covalent compounds. The average electronegativity increases from left to right on the bond-type triangle and thus is a measure of the degree of covalent character of the bond. As we move from the bottom to the top of the triangle, the transfer of bonding electrons becomes more complete. The difference in electronegativity increases from the bottom to the top of the triangle and thus is a measure of the ionic character of a bond.

In the previous section, we encountered a pair of compounds that have more or less the same ΔEN but very different bonding: MgH_2 and CO_2. Because they have similar values of ΔEN, these compounds lie on the same horizontal line in Figure 5.10. But MgH_2 lies in the ionic region, whereas CO_2 lies in the covalent region, as we would expect.

Consider the compounds CdLi, AlAs, SiC, NO, and OF_2, which have a value of ΔEN of about 0.6 and therefore lie along a narrow band parallel to the base of the triangle in Figure 5.10. The bond-type triangle suggests that these compounds have very different bonding characteristics and therefore very different properties. CdLi is a metallic compound, AlAs lies on the metallic–covalent border, SiC is just beyond the metallic–covalent border, and NO and OF_2 are predominantly covalent compounds. As you move to the right along this horizontal line, the covalent character of the bonding increases while the difference in electronegativity remains about the same.

Let's now consider a series of compounds that have more or less the same average electronegativities but different values of ΔEN: $GaBr_3$, Hg_2Cl_2, and CaO. The average electronegativity for these compounds is approximately 2.3. ΔEN increases, however, from about 0.9 for $GaBr_3$ to 1.1 for Hg_2Cl_2 and 2.6 for CaO. $GaBr_3$ lies toward the bottom of Figure 5.10, in the covalent region. Hg_2Cl_2 lies close to the border between ionic and covalent compounds, and is therefore much more ionic. CaO is toward the top of the triangle and is predominantly ionic.

Classification of bonding is most difficult for compounds near the borders between the three types of bonds. Compounds lying close to these borders have properties intermediate between the bonding classifications separated by the dividing line. They therefore exhibit some of the most interesting and exciting properties and are areas of current chemical research. These include semiconductors, high-temperature superconductors, and ceramics for automobile engines.

Students may also use bond type to categorize materials as covalent, ionic, or metallic and hence predict properties of new atom combinations and understand those of known materials. This is treated in more detail in Chapter 9.

Observable properties such as conductivity in the solid, molten, and aqueous states are used to help distinguish the interface between different bond types. Recently, Nelson has suggested a way to classify substances by electrical character.[3]

The properties of ceramics, glasses, and other materials can be put into perspective by bond-type triangles. Students can predict what binary atomic combination might produce compounds with certain desired characteristics.

[3] P. G. Nelson, *J. Chem. Educ.* **71,** 24 (1994).

Exercise 5.4

Use electronegativities from Table B.7 in the appendix and the bond-type triangle in Figure 5.10 to describe the bonding in the following compounds.

(a) $AlBr_3$

(b) $SnCl_4$

(c) CaS

(d) InNa

Solution

(a) $EN_{Al} = 1.61$ $EN_{Br} = 2.69$ Avg. $EN = 2.15$ $\Delta EN = 1.08$
Find the average electronegativity between Al and Br (2.15) on the base of the triangle and then move up the triangle to the difference in electronegativity (1.08). This point falls on the border between ionic and covalent bonding. Therefore, $AlBr_3$ would be expected to have characteristics of both ionic and covalent bonding.

(b) $EN_{Sn} = 1.82$ $EN_{Cl} = 2.87$ Avg. $EN = 2.35$ $\Delta EN = 1.05$
This places $SnCl_4$ in the area for predominately covalent bonding.

(c) $EN_{Ca} = 1.03$ $EN_S = 2.59$ Avg. $EN = 1.81$ $\Delta EN = 1.56$
CaS has primarily ionic bonding.

(d) $EN_{In} = 1.66$ $EN_{Na} = 0.87$ Avg. $EN = 1.27$ $\Delta EN = 0.79$
The bond in InNa is primarily metallic.

5.12 Properties of Metallic, Covalent, and Ionic Compounds

Bond-type triangles can be used to stimulate student interest and promote student inquiry, while at the same time providing a firm grounding in bonding characteristics and how these characteristics affect properties. Jensen has recently suggested other ways students may profit from this method of classifying compounds as to bond type.[4]

➤ CHECKPOINT

Classify the following compounds as ionic, covalent, or metallic. Which would be the best compound for an insulator that would stand up to high heat? $BaSi_2$, $BaBr_2$, $GaBr_3$.

The bond-type triangle allows the classification of compounds according to the way the compounds share electrons in a bond. The classification of bonding type is important because the properties of the compound depend on the degree of metallic, covalent, or ionic character in the bond between the atoms. A compound that is primarily ionic has a high melting point and will not conduct electricity as a solid but will conduct in the molten state or when dissolved in water. A covalent compound has a low melting point and, like a solid ionic compound, is an insulator. Metallic compounds have a range of melting points and conduct heat and electricity in the solid state.

By knowing the properties associated with a particular bond type it is possible to select those atoms whose combination may produce a material having certain desired characteristics. Thus if we are interested in finding a substance that will conduct electricity as a solid, we would seek a material containing metallic bonds. If an insulator is desired, a compound located in the ionic region of the bond-type triangle would be appropriate. A low-melting insulator could be prepared from atoms whose ΔEN and $\overline{EN}$ place it in the lower-right region of the triangle. Semiconductors are found in the lower-middle of the bond-type triangle in the area corresponding to the crossover from metallic to covalent bonding.

5.13 Oxidation Numbers

The bond type triangle helps us understand why it is so difficult to decide whether to talk about ZnS as if it contained Zn^{2+} and S^{2-} ions or polar covalent bonds in which zinc and sulfur atoms share a pair of electrons almost, but not quite, equally.

Both the traditional rules for assigning oxidation states plus a second method that relates oxidation states to structure and shows students how electrons are assigned in the molecule to determine the oxidation state of each atom are given. This makes the teaching of redox reactions of organic compounds much more straightforward.

Because it is sometimes difficult to distinguish between ionic and covalent compounds, chemists have developed the concept of **oxidation number** or **oxidation state.** Oxidation numbers treat all compounds as if they were ionic.

It doesn't matter whether the compound actually contains ions. The oxidation number is the charge an atom would have *if the compound were ionic*. Both the strontium atom in SrF_2 and the carbon atom in CO, for example, are assigned oxidation numbers of +2 or are described as being in the +2 oxidation state, even though one of the compounds is predominantly ionic and the other is predominantly covalent. As we will see, oxidation numbers are nothing more than a bookkeeping method for keeping track of the flow of electrons in a chemical reaction.

For the active metals in Groups IA and IIA, oxidation numbers give a good description of the charge that the metal has in its compounds. The main-group metals in Groups IIIA and IVA, however, form compounds that have a significant amount of covalent character. Although we assign an oxidation number of +3 to

[4]W. B. Jensen, *J. Chem. Educ.* **72,** 395 (1995).

aluminum and −1 to bromine, it is misleading to assume that aluminum bromide contains Al^{3+} and Br^- ions. It actually exists as Al_2Br_6 molecules.

The problem becomes even more severe when we turn to the chemistry of the transition metals. MnO, for example, is ionic enough to be considered a salt that contains Mn^{2+} and O^{2-} ions. Mn_2O_7, on the other hand, is a covalent compound that boils at room temperature. It is therefore more useful to think about Mn_2O_7 as if it contained manganese in a +7 oxidation state, not Mn^{7+} ions.

For many years, chemists have used a set of general rules to determine the oxidation number of an element.

- The oxidation number is 0 in any neutral substance that contains atoms of only one element. Aluminum foil, iron metal, and the H_2, O_2, O_3, P_4, and S_8 molecules all contain atoms that have an oxidation number of 0.
- The oxidation number is equal to the charge on the ion for ions that contain only a single atom. The oxidation number of the Na^+ ion, for example, is +1, whereas the oxidation number for the Cl^- ion is −1.
- The oxidation number of hydrogen is +1 when it is combined with a *more electronegative element*. Hydrogen is therefore in the +1 oxidation state in CH_4, NH_3, H_2O, and HCl.
- The oxidation number of hydrogen is −1 when it is combined with a *less electronegative element*. Hydrogen is therefore in the −1 oxidation state in LiH, NaH, CaH_2, and $LiAlH_4$.
- The elements in Groups IA and IIA form compounds in which the metal atoms have oxidation numbers of +1 and +2, respectively.
- Oxygen usually has an oxidation number of −2. Exceptions include molecules and polyatomic ions that contain O—O bonds, such as O_2, O_3, H_2O_2, and the O_2^{2-} ion.
- Elements in Group VIIA have an oxidation number of −1 when the atom is bonded to a less electronegative element.
- The sum of the oxidation numbers of the atoms in a neutral substance is zero.

$$H_2O \quad (2 \text{ hydrogen} \times +1) + (1 \text{ oxygen} \times -2) = 0$$

- The sum of the oxidation numbers in a polyatomic ion is equal to the charge on the ion.

$$OH^- \quad (1 \text{ oxygen} \times -2) + (1 \text{ hydrogen} \times +1) = -1$$

- The least electronegative element is assigned a positive oxidation state. Sulfur carries a positive oxidation state in SO_2, for example, because it is less electronegative than oxygen.

$$SO_2 \quad (1 \text{ sulfur} \times +4) + (2 \text{ oxygen} \times -2) = 0$$

Figure 5.11 shows the common oxidation numbers for many of the elements in the periodic table. There are several clear patterns in the data.

- Elements in the same group often have the same oxidation numbers.
- The largest, or most positive, oxidation number of an atom is often equal to the group number of the element (Group VIIIB atoms are an exception). The largest oxidation number for phosphorus, in Group VA, for example, is +5.

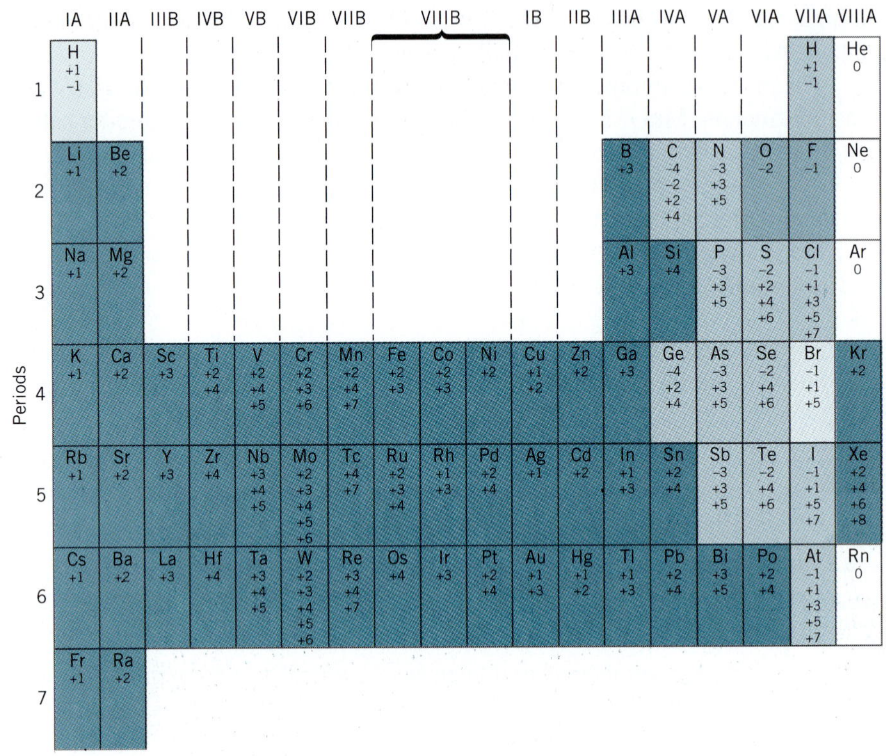

Fig. 5.11 Common oxidation numbers of the elements. Different shades are used to distinguish between elements that form only positive oxidation states, only negative oxidation states, and those that exhibit both positive and negative oxidation states.

- The smallest, or most negative, oxidation number of a nonmetal often can be found by subtracting 8 from the group number. The most negative oxidation state of phosphorus, for example, is $5 - 8 = -3$.

For compounds that contain transition metal ions, such as the Fe^{2+} and Fe^{3+} ions, these rules provide the easiest way of determining oxidation numbers. In the next section an alternate method of determining oxidation numbers is introduced.

Many students get the impression that partial charge, formal charge, and oxidation numbers represent three distinct and unrelated concepts. These concepts are in fact very closely related. The next section shows that the separation of charge between atoms is determined by electronegativities and that formal charge and oxidation numbers are only constructs that chemists have found to be useful tools.

Exercise 5.5

Assign oxidation numbers to the atoms in the following compounds.

(a) Aluminum oxide or alumina (Al_2O_3), which is used in pigments, ceramics, and abrasives.

(b) Xenon tetrafluoride (XeF_4), one of the first rare gas compounds.

(c) Potassium dichromate ($K_2Cr_2O_7$), which is one of the principal components in the Breathalyzer test used to determine whether someone has been drinking alcohol.

Solution

(a) The sum of the oxidation numbers in Al_2O_3 must be zero because the compound is neutral. If we assume that oxygen is in the -2 oxidation state, the oxidation state of the aluminum must be $+3$.

$$Al_2O_3 \quad (2 \text{ aluminum} \times +3) + (3 \text{ oxygen} \times -2) = 0$$

(b) Because the oxidation number of the fluorine is -1, the xenon atom must be present in the $+4$ oxidation state.

$$XeF_4 \quad (1 \text{ xenon} \times +4) + (4 \text{ fluorine} \times -1) = 0$$

(c) Assigning oxidation numbers in $K_2Cr_2O_7$ is simplified if we recognize that it is an ionic compound that contains K^+ and $Cr_2O_7^{2-}$ ions. The oxidation state of the potassium is +1 in the K^+ ion. Because the oxidation number of oxygen is usually −2, the oxidation state of the chromium in the $Cr_2O_7^{2-}$ ion must be +6.

$$Cr_2O_7^{2-} \quad (2 \text{ chromium} \times +6) + (7 \text{ oxygen} \times -2) = -2$$

Exercise 5.6

Arrange the following compounds in order of increasing oxidation state for the carbon atom.

(a) CO, carbon monoxide
(b) CO_2, carbon dioxide
(c) H_2CO, formaldehyde
(d) CH_3OH, methanol
(e) CH_4, methane

Solution

$$\underset{-4}{CH_4} < \underset{-2}{CH_3OH} < \underset{0}{H_2CO} < \underset{+2}{CO} < \underset{+4}{CO_2}$$

5.14 Calculating Oxidation Numbers

Oxidation numbers for atoms in a compound can be calculated using a relation based on the following equation, which was used to calculate partial charge in Chapter 4.

$$\delta_a = V_a - N_a - B_a\left(\frac{EN_a}{EN_a + EN_b}\right)$$

In this equation, δ_a is the partial charge on atom *a*. V_a is the number of valence electrons on a neutral atom of element *a*. N_a is the number of nonbonding electrons on atom *a*. B_a is the number of bonding electrons on the atom. *EN* is the electronegativity of the atom, and the symbol *b* is used to represent the atom to which *a* is bonded.

Formal charge was calculated in Chapter 4 by assuming that the electrons in a covalent bond are shared equally by the atoms that form the bond. In other words, the atoms were treated as if $EN_a = EN_b$. When calculating B_a, the number of bonding electrons, the bonding electrons are divided equally between the atoms that form covalent bonds. Since EN_a and EN_b are assumed to be equal, the value of $EN_a/(EN_a + EN_b)$ is equal to 1/2. This gives us the following equation for formal charge.

$$FC_a = V_a - N_a - \frac{B_a}{2}$$

When we calculate the partial charge on each atom in a bond, we try to estimate the fraction of the electrons in the bond that should be assigned to each atom. When we calculate formal charge we divide the bonding electrons equally between the atoms that form covalent bonds. When we calculate *oxidation numbers* we treat

The equations used to calculate formal charges and oxidation numbers are derived directly from the equation for determining partial charge.[5] *The relationship between formal charge, oxidation numbers, and partial charge is made more concrete by the use of this equation. The partial charge is assumed to show the actual electron distribution between atoms while formal charge assumes both atoms share electrons equally, i.e., both atoms have the same electronegativity.*

The method of calculating the partial charge produces partial charges for diatomic molecules that compare quite well with more sophisticated calculations.[5] *The ratio of electronegativities gives a partitioning of electron density which lies between the extremes of ionic and pure covalent bonding. See the* Instructor's Manual *for an example calculation.*

[5] L. C. Allen, *J. Am. Chem. Soc.* **111,** 9115 (1989).

each bond as if it were an ionic bond in which the electrons in the bond are transferred to the more electronegative element. In other words, we treat the atoms as if $EN_a \gg EN_b$. If the electronegativity of one element, say atom *a*, in a bond is much larger than the other element, the value of $EN_a/(EN_a + EN_b)$ becomes equal to 1. After the electrons in a bond are transferred to the more electronegative atom, the only electrons left to be counted are nonbonding electrons. Thus, the calculation of oxidation number is based on only two pieces of information from the partial charge equation: (1) the number of valence electrons on a neutral atom of the element (V) and (2) the number of electrons assigned to the atom in the Lewis structure of the molecule (N) after electrons of a bond are assigned to the more electronegative atom. The oxidation number can therefore be calculated with the following simplified equation.

$$OX = V - N$$

The relationship among the three ways of describing the charge associated with atoms within a compound is summarized below. It should be remembered that partial charge is the best description of the actual distribution of electron density in a molecule. Formal charge and oxidation numbers are invented bookkeeping methods. Formal charge, which treats all bonds as if they were purely covalent, is particularly useful in assigning Lewis structures. Oxidation numbers, which treat all bonds as purely ionic, are useful in describing a class of reactions called oxidation–reduction reactions.

$$\delta_a = V_a - N_a - B_a\left(\frac{EN_a}{EN_a + EN_b}\right)$$

$$\swarrow \qquad\qquad \searrow$$

$$EN_a = EN_b \qquad\qquad EN_a \gg EN_b$$

$$FC_a = V_a - N_a - (B_a/2) \qquad OX_a = V_a - N_a$$

Oxidation numbers assign the most electronegative atom, an electronegativity that is infinitely large compared to its bond partner, so that all bonding electron density is concentrated on the most electronegative atom.

The three approaches to assigning charge represent a continuum ranging from a complete transfer of electrons (oxidation numbers) to an unequal sharing of electrons (partial charge) to an equal sharing of electrons (formal charge). The best description of the actual charge distribution is partial charge.

As we have seen, oxidation numbers can be assigned to relatively simple compounds on the basis of the rules outlined in the previous section. The equation we've derived for calculating oxidation numbers is particularly useful in molecules in which an element has more than oxidation state, such as ethanol. Ethanol is the alcohol in "alcoholic" beverages. It has the following Lewis structure.

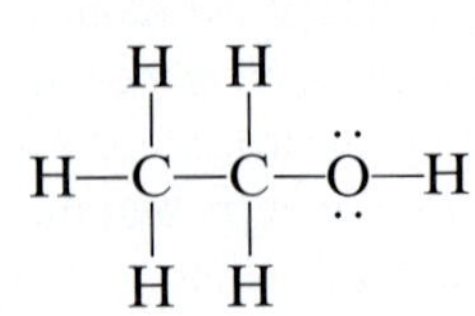

We start by recognizing that O ($EN = 3.61$) is more electronegative than C ($EN = 2.54$), which is more electronegative than H ($EN = 2.30$). We therefore assign the electrons in each covalent bond to the more electronegative atom. When a covalent bond exists between a pair of equivalent atoms, the electrons in the bond are divided equally between the atoms. In the case of the bond between the two carbon atoms, one electron is assigned to each carbon since both atoms have the same electronegativity.

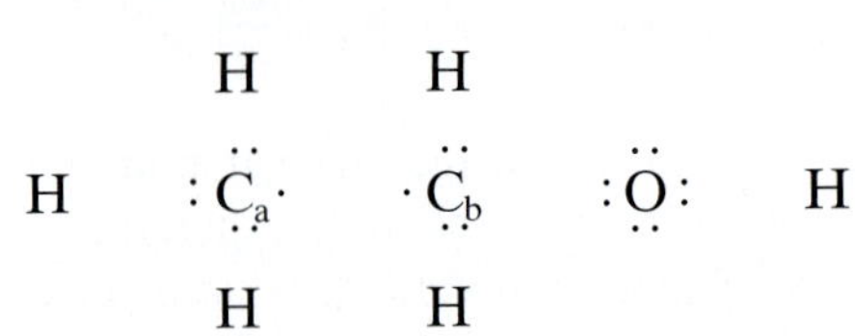

Oxygen is in Group VIA of the periodic table, which means that a neutral atom has six valence electrons. Oxygen is assigned eight electrons in the above structure. The oxidation number of the oxygen atom is therefore −2.

Often minimal coverage of the oxidation states of organic compounds is given because carbon often has more than one oxidation state within a molecule and its oxidation states cannot be easily predicted using the typical rules for assigning oxidation numbers. However, redox reactions of organic compounds are some of the most interesting and relevant redox reactions.

$$OX_{\text{oxygen}} = V - N = 6 - 8 = -2$$

Hydrogen, which is in Group IA, has no valence electrons once the electrons in the bonds are assigned to the more electronegative oxygen or carbon atoms. The oxidation number of the hydrogen atom is therefore +1.

$$OX_{\text{hydrogen}} = V - N = 1 - 0 = +1$$

The two carbon atoms in the structure of ethanol have different oxidation numbers. Carbon is in Group IVA and has four valence electrons. C_a is assigned seven electrons in the above structure, giving it an oxidation number of −3.

$$OX_{\text{carbon } a} = V - N = 4 - 7 = -3$$

C_b has only five electrons assigned to it in the above structure. Thus it will have an oxidation number of −1.

$$OX_{\text{carbon } b} = V - N = 4 - 5 = -1$$

Note that the sum of the oxidation numbers for all of the nine atoms in the CH_3CH_2OH molecule is zero, which is consistent with the fact that there is no charge on this molecule.

Exercise 5.7

Acetic acid, $C_2H_4O_2$, is the compound that gives vinegar its sour taste. The Lewis structure of acetic acid is shown below. Use the method described in this section to determine the oxidation numbers for all atoms in acetic acid.

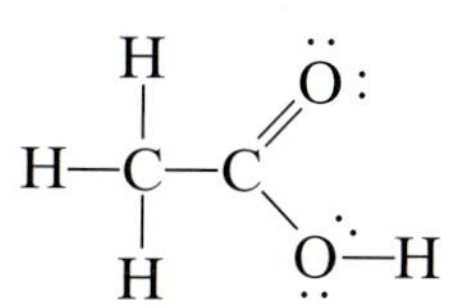

Solution

The first step in the calculation involves assigning the electrons in each bond to the more electronegative element.

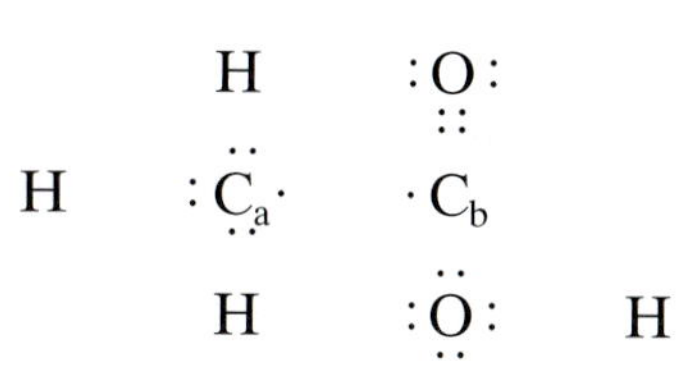

Note that all four electrons in the carbon–oxygen double bond are assigned to oxygen. Both oxygen atoms therefore end up with eight nonbonding electrons. In the carbon–carbon bond one electron is assigned to each carbon since they have equal electronegativities.

Oxygen	$OX = 6 - 8 = -2$
Hydrogen	$OX = 1 - 0 = +1$
Carbon$_a$	$OX = 4 - 7 = -3$
Carbon$_b$	$OX = 4 - 1 = +3$

CHECKPOINT

The partial charge gives the best description of the charge on atoms in a compound. What do formal charge and oxidation numbers represent?

The sum of the oxidation numbers for all seven atoms is zero. If the molecule had a charge, the sum of the oxidation numbers would be equal to that charge.

5.15 Oxidation–Reduction Reactions

A common and important class of chemical reactions involve **oxidation** and **reduction.** The principal use of oxidation numbers is to help us understand oxidation–reduction reactions. Consider the reaction that occurs when magnesium metal burns in the presence of oxygen.

$$2\,Mg(s) + O_2(g) \longrightarrow 2\,MgO(s)$$

The term *oxidation* was originally used to describe reactions, such as this, in which an element combines with oxygen.

$$2\,Mg(s) + O_2(g) \longrightarrow 2\,MgO(s)$$

oxidation

The term *reduction* comes from the Latin stem meaning "to lead back." Anything that leads back to magnesium metal therefore involves reduction. The reaction between magnesium oxide and carbon at 2000°C to form magnesium metal and carbon monoxide is an example of the reduction of magnesium oxide to magnesium metal.

$$MgO(s) + C(s) \longrightarrow Mg(s) + CO(g)$$

reduction

The classification of reactions as oxidation–reduction is based on an arbitrary definition.[6] For this reason the transfer of electrons and the transfer of an oxygen atom are both given as examples of redox reactions. The description of redox reactions is covered more fully in Chapter 12.

The oxidation of magnesium metal to form magnesium oxide can be described in terms of the transfer of electrons from magnesium to oxygen. From this perspective, the reaction between magnesium and oxygen would be written as follows.

$$2\,Mg + O_2 \longrightarrow 2\,[Mg]^{2+}[O]^{2-}$$

In the course of the reaction, each magnesium atom loses two electrons to form an Mg^{2+} ion.

$$Mg \longrightarrow Mg^{2+} + 2\,e^-$$

Each O_2 molecule, on the other hand, gains four electrons to form a pair of O^{2-} ions.

$$O_2 + 4\,e^- \longrightarrow 2\,O^{2-}$$

[6] R. Stewart, *Oxidation Mechanisms,* W. A. Benjamin, Inc., New York, 1964.

Because electrons are neither created nor destroyed in a chemical reaction, oxidation and reduction are linked. The four electrons gained by oxygen require that four electrons be lost by magnesium, causing two Mg atoms to form two Mg^{2+} ions. The number of electrons gained during reduction must equal the number of electrons lost during oxidation. It is impossible to have oxidation without the reduction, as shown in Figure 5.12.

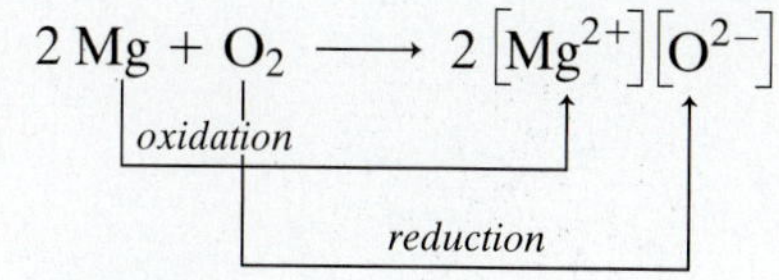

Fig. 5.12 Oxidation cannot occur in the absence of reduction.

As their understanding of chemical reactions improved, chemists recognized that oxidation–reduction reactions don't always involve the transfer of electrons. They can also occur by the transfer of atoms. Consider the following reaction, for example.

$$CO(g) + H_2O(g) \longrightarrow CO_2(g) + H_2(g)$$

The total number of electrons in the valence shell of each atom remains the same in the reaction.

$$:C{\equiv}O: + H{-}\ddot{\underset{..}{O}}{-}H \longrightarrow \ddot{O}{=}C{=}\ddot{O} + H{-}H$$

What changes in the reaction is the oxidation state of the atoms. The oxidation state of carbon increases from +2 to +4, while the oxidation state of the hydrogen decreases from +1 to 0.

$$\underset{+2}{CO} + \underset{+1}{H_2O} \longrightarrow \underset{+4}{CO_2} + \underset{0}{H_2}$$

Oxidation and reduction are therefore best defined as follows.

> ***Oxidation* occurs when the oxidation number of an atom becomes more positive.**
> ***Reduction* occurs when the oxidation number of an atom becomes more negative.**

Thus in the reaction of CO with H_2O the oxidation state of carbon increases and CO is said to be oxidized. The oxidation state of the hydrogen atoms in water decreases and water is said to be reduced.

What can we deduce from the fact that the oxidation state of the nitrogen atom is +5 in nitric acid? Because nitrogen is found in Group VA of the periodic table, an oxidation state of +5 is the highest oxidation state to which a nitrogen atom can be oxidized. We can therefore conclude that the nitrogen atom in nitric acid cannot be oxidized; it can only be reduced. Nitric acid is therefore likely to react with substances that can reduce the nitrogen atom to a lower oxidation state. It reacts with copper metal, for example, to form Cu^{2+} ions and either NO or NO_2, depending on the concentration of the acid.

The use of oxidation numbers allows chemical reactions to be classified as oxidation–reduction reactions or as reactions that are not. The electron densities around atoms change in reactions that are not classified as oxidation–reduction (i.e. acid-base, complex formation). However, this change in electron density is only apparent with partial charge. The use of oxidation numbers obscures the change in partial charge.[7]

[7]J. N. Butler, *Ionic Equilibrium*, John Wiley & Sons, Inc., New York, 1998.

➤ CHECKPOINT

Are the following reactions oxidation–reduction reactions?

$$H_3PO_4(aq) + NH_3(aq) \longrightarrow (NH_4)H_2PO_4(aq)$$

$$4\,NH_3(g) + 5\,O_2(g) \longrightarrow 4\,NO(g) + 6\,H_2O(g)$$

$$CH_3OH(l) + CO(g) \longrightarrow CH_3C({=}O){-}OH(l)$$

5.16 Nomenclature

Long before chemists knew the formulas for chemical compounds, they developed a system of **nomenclature** (from the Latin words *nomen,* "name," and *calare,* "to call") that gave each compound a unique name. Today we often use chemical formulas, such as NaCl, $C_{12}H_{22}O_{11}$, and $Co(NO)_6(ClO_4)_3$, to describe chemical compounds. But we still need unique names that unambiguously identify each compound.

COMMON NAMES

Some compounds have been known for so long that a systematic nomenclature cannot compete with well-established common names. Examples of compounds for which common names are used include water (H_2O), ammonia (NH_3), and methane (CH_4).

NAMING IONIC COMPOUNDS, OR SALTS

The names of ionic compounds are written by listing the name of the positive ion followed by the name of the negative ion.

$NaCl$	sodium chloride
$(NH_4)_2SO_4$	ammonium sulfate
Fe_2O_3	iron(III) oxide
$NaHCO_3$	sodium hydrogen carbonate
$Al(ClO_4)_3$	aluminum perchlorate

We therefore need a series of rules that allow us to unambiguously name the positive and negative ions before we can name ionic compounds.

NAMING POSITIVE IONS

Positive ions that consist of a single atom carry the name of the element from which they are formed.

Na^+	sodium	Zn^{2+}	zinc
Ca^{2+}	calcium	H^+	hydrogen
K^+	potassium	Sr^{2+}	strontium

As we have seen, some metals—particularly the transition metals—form positive ions in more than one oxidation state. One of the earliest methods of distinguishing between these ions used the suffixes *-ic* and *-ous* added to the Latin name of the element to represent the higher and lower oxidation states, respectively.

Fe^{2+}	ferrous	Fe^{3+}	ferric
Sn^{2+}	stannous	Sn^{4+}	stannic
Cu^+	cuprous	Cu^{2+}	cupric

Chemists now use a simpler method, in which the charge on the ion is indicated by a Roman numeral in parentheses immediately after the name of the element.

Fe^{2+}	iron(II)	Fe^{3+}	iron(III)
Sn^{2+}	tin(II)	Sn^{4+}	tin(IV)
Cu^+	copper(I)	Cu^{2+}	copper(II)

There are only a limited number of polyatomic positive ions. These ions often have common names ending with the suffix *-onium*.

H_3O^+	hydronium	NH_4^+	ammonium

NAMING NEGATIVE IONS

Negative ions that consist of a single atom are named by adding the suffix *-ide* to the stem of the name of the element.

F^-	fluoride	O^{2-}	oxide
Cl^-	chloride	S^{2-}	sulfide
H^-	hydride	P^{3-}	phosphide

RULES FOR NAMING POLYATOMIC NEGATIVE IONS

The names of the common polyatomic negative ions are given in Table 1.5. At first glance, the nomenclature of these ions seems hopeless. There are several rules, however, that can bring some order to this apparent chaos.

- The name of the ion usually ends in *-ite* or *-ate*.
- The *-ite* ending indicates a low oxidation state. Thus, the NO_2^- ion is the nitrite ion.
- The *-ate* ending indicates a high oxidation state. Thus, the NO_3^- ion is the nitrate ion.
- When more than two polyatomic ions exist with the same central atom, the prefix *per-* (as in *hyper-*) and the prefix *hypo-* are used to indicate the very largest and very smallest number of oxygens. Consider the ClO^-, ClO_2^-, ClO_3^-, and ClO_4^- ions, for example. The lowest oxidation state for the chlorine atom is formed in ClO^-, which is known as the hypochlorite ion. ClO_2^- and ClO_3^- are the chlorite and chlorate ions, respectively. The ion with the largest oxidation state for the chlorine atom is ClO_4^-, which is the perchlorate ion.

There are a handful of exceptions to these generalizations. The names of the hydroxide (OH^-), cyanide (CN^-), and peroxide (O_2^{2-}) ions, for example, have the *-ide* ending because they were once thought to be monatomic ions.

Exercise 5.8

Name the following ionic compounds.

(a) $FePO_4$, which is used in the automobile industry as an anticorrosion film that improves the adherence of paint to metal.

(b) $SrCO_3$, which is used as an X-ray absorber in the glass faceplate of color television tubes.

(c) $Ca(ClO)_2$, which is used in bleaching and sanitizing applications. It is the solid bleach in Clorox 2.

Solution

(a) Iron ions can exist with two charges, Fe^{2+} and Fe^{3+}. Since the phosphate ion has a -3 charge, the compound $FePO_4$ contains the Fe^{3+} ion. The compound is therefore known as iron(III) phosphate.

(b) Because strontium always has a $+2$ charge in its compounds, $SrCO_3$ is known as strontium carbonate.

(c) $Ca(ClO)_2$ contains the Ca^{2+} and ClO^- ions. Because calcium only forms Ca^{2+} ions, the compound is known as calcium hypochlorite.

NAMING SIMPLE COVALENT COMPOUNDS

One of the most difficult tasks faced in determining the name of a compound is deciding whether to name it as an ionic or covalent compound. A bond-type triangle can be used to determine the predominant type of bonding in binary compounds. As a general rule, nonmetals bonded together are usually classified as covalent compounds.

Chemists write formulas in which the least electronegative element is written first, followed by the more electronegative element(s). The suffix *-ide* is then added to the stem of the name of the more electronegative atom.

HCl	hydrogen chloride
NO	nitrogen oxide

The number of atoms of an element in simple covalent compounds is indicated by adding one of the following Greek prefixes to the name of the element.

1	mono-	6	hexa-
2	di-	7	hepta-
3	tri-	8	octa-
4	tetra-	9	nona-
5	penta-	10	deca-

The prefix *mono-* is seldom used because it is redundant. The principal exception to this rule is carbon monoxide (CO).

Exercise 5.9

Name the following compounds.

(a) NO_2, which is a component of a series of reactions responsible for Los Angeles smog.

(b) SF_4, which is a highly reactive colorless gas.

Solution

(a) Nitrogen dioxide

(b) Sulfur tetrafluoride

NAMING ACIDS

Some covalent compounds that contain hydrogen, such as HCl, HBr, and HCN, dissolve in water to produce acids. The solutions are named by replacing the hydrogen in the name of the compound with the prefix *hydro-* and then replacing the suffix *-ide* with *-ic*. For example, hydrogen chloride (HCl) dissolves in water to form hydrochloric acid, hydrogen bromide (HBr) forms hydrobromic acid, and hydrogen cyanide (HCN) forms hydrocyanic acid.

Many of the oxygen-rich polyatomic negative ions in Table 1.5 form acids that are named by replacing the suffix *-ate* with *-ic* and the suffix *-ite* with *-ous*.

$CH_3CO_2^-$	acetate	CH_3CO_2H	acetic acid
CO_3^{2-}	carbonate	H_2CO_3	carbonic acid
BO_3^{3-}	borate	H_3BO_3	boric acid
NO_3^-	nitrate	HNO_3	nitric acid
NO_2^-	nitrite	HNO_2	nitrous acid
SO_4^{2-}	sulfate	H_2SO_4	sulfuric acid
SO_3^{2-}	sulfite	H_2SO_3	sulfurous acid
ClO_4^-	perchlorate	$HClO_4$	perchloric acid
ClO_3^-	chlorate	$HClO_3$	chloric acid
ClO_2^-	chlorite	$HClO_2$	chlorous acid
ClO^-	hypochlorite	$HClO$	hypochlorous acid
PO_4^{3-}	phosphate	H_3PO_4	phosphoric acid
MnO_4^-	permanganate	$HMnO_4$	permanganic acid
CrO_4^{2-}	chromate	H_2CrO_4	chromic acid

Salts containing anions with acidic hydrogens can be named by indicating the presence of the acidic hydrogen as follows.

$NaHCO_3$ sodium hydrogen carbonate (also known as sodium bicarbonate)
$NaHSO_3$ sodium hydrogen sulfite (also known as sodium bisulfite)
KH_2PO_4 potassium dihydrogen phosphate

Exercise 5.10

Name the following compounds.

(a) $NaClO_3$
(b) $Al_2(SO_4)_3$
(c) P_4S_3
(d) SCl_4

Solution

The key to naming the compounds is recognizing that the first two are ionic compounds and the last two are covalent compounds.

(a) sodium chlorate
(b) aluminum sulfate
(c) tetraphosphorus trisulfide
(d) sulfur tetrachloride

Key Terms

Active metal
Alkali metal
Alkaline earth metal
Binary compound
Body-centered cubic
Bond-type triangle
Bromide
Chloride
Conduction band
Coulombic
Covalent
Delocalized
Electronegativity
Electrostatic
Face-centered cubic
Fluoride
Halide
Halogen
Hydride
Iodide
Ionic bond
Isoelectronic
Lattice point
Localized
Main-group element
Metallic bond
Nitride
Nomenclature
Oxidation
Oxidation number
Oxidation–reduction reaction
Oxidation state
Oxide
Peroxide
Polar covalent compound
Reactive
Reduction
Salt
Simple cubic
Sulfide
Superoxide
Transition metal
Unit cell

Problems

The Active Metals

1. List the elements in the third row of the periodic table in order of decreasing metallic character. List the elements in order of increasing nonmetallic character. Explain the similarities between the trends. Identify each element as either a metal, nonmetal, or semimetal.
2. List the elements in Group VA of the periodic table in terms of increasing metallic character. Identify each element in the group as a metal, nonmetal, or semimetal.
3. Which of the following sets of elements is arranged in order of increasing nonmetallic character?
(a) $Sr < Al < Ga < N$ (b) $K < Mg < Rb < Si$
(c) $Ge < P < As < N$ (d) $Al < B < N < F$

4. Describe the general trends in the activity, or reactivity, of the main-group metals.
5. What do we mean when we say that sodium is more metallic than lithium?
6. For each of the following pairs of metals, which is more active?
   (a) Mg or Ca (b) Na or Mg
   (c) K or Mg (d) Mg or Al
7. Which metal in each of the following pairs would you expect to react more rapidly with water?
   (a) Na or K (b) Na or Mg
   (c) Mg or Ca (d) Ca or Al

Main-Group Metals and Their Ions

Group IA: The Alkali Metals

8. What are the two most abundant alkali metals?
9. Why is the chemistry of the alkali metals dominated by the singly charged positive ion?
10. Give the electron configurations for each of the +1 ions of the alkali metals.
11. The Cs^+ ion is isoelectronic with what electrically neutral atom?
12. Which is more difficult to remove, an electron from K^+ or Ar? Explain.
13. How do the AVEE and the first ionization energy vary in Group IA from Li through Fr? Explain.
14. How does the size of the alkali metals change from Li to Fr? Explain.
15. How would you expect the second ionization energy of Li to compare to its first ionization energy? Explain.

Group IIA: The Alkaline Earth Metals

16. Why is the chemistry of the akaline earth metals dominated by the doubly charged positive ion?
17. Give the electron configurations for the +2 ions of the alkaline earth metals.
18. The Ca^{2+} ion is isoelectronic with what electrically neutral atom?
19. Which is more difficult to remove, an electron from Mg or Na?
20. How do the AVEE and second ionization energy vary in Group IIA from Be through Ra?
21. How does the size of the alkaline earth metals vary from Be to Ra? Explain.
22. How would you expect the third ionization energy of Ba to compare to the second ionization energy? Explain.

Group IIIA Metals

23. Which of the Group IIIA metals is the most abundant?
24. Rationalize why Al forms the +3 ion in most of its compounds.
25. Give the trend in AVEE and first ionization energy for B and Al in Group IIIA.
26. What is the trend in size through Group IIIA for the atoms B through Tl? Explain.
27. Give the electron configuration for Ga, Ga^+, Ga^{2+}, and Ga^{3+}. How would you expect the fourth ionization energy for Ga to compare to the third? Explain.
28. What is the most likely charge on the positive ion formed by B?

Main-Group Nonmetals and Their Ions

29. Calculate the charge on the negative ions formed by the following elements.
    (a) As (b) Te (c) Se
30. Calculate the charge on the negative ions formed by the following elements.
    (a) C (b) P (c) S (d) I
31. Describe the difference between the hydrogen atoms in metal hydrides such as LiH and nonmetal hydrides such as CH_4 and H_2O.
32. Write the formulas for the fluoride, hydride, sulfide, nitride, and oxide of barium.
33. Why do the halides tend to form ions with a −1 charge?
34. Define the terms *hydride, nitride,* and *oxide.*
35. What is the sign of the charge that is usually found on the ions of the main-group elements on the right-hand side of the periodic table? Explain.
36. In what elemental form do the halogens occur in nature?
37. Why is hydrogen listed in two groups on the periodic table?

Transition Metals and Their Ions

38. Where are the transition metals located on the periodic table?
39. Give the electron configuration for Zn. What is the most likely charge on the zinc ion? Explain.
40. Give the electron configuration for Cr. What is the charge on the chromium ion that is most likely to form? Explain.
41. Give the electron configurations for Ni and Ni^{2+}.
42. Cobalt commonly forms two ions. What would you expect the charges on these ions to be? Explain.

Predicting the Products of Reactions That Produce Ionic Compounds

43. Which one of the following elements is most likely to form an oxide with the formula XO and also a hydride with the formula XH_2?
    (a) Na (b) Mg (c) Al (d) Si (e) P
44. A main-group metal reacts with hydrogen and oxygen to form compounds with the formulas XH_4 and XO_2. In which group of the periodic table does the element belong?

45. In which column of the periodic table do we find main-group metals that form sulfides with the formula M_2S_3 and react with acid to form M^{3+} ions and H_2 gas?
46. Explain why sodium reacts with chlorine to form NaCl and not $NaCl_2$ or $NaCl_3$.
47. Explain why magnesium reacts with chlorine to form $MgCl_2$ and not $MgCl_3$.
48. Describe why lithium reacts with nitrogen to give Li_3N and not a compound with another formula.
49. Describe the most important factors in determining the formula of an ionic compound such as NaCl or $MgCl_2$.
50. Predict the product of the reaction between aluminum and nitrogen.
51. Predict the product of the reaction between strontium metal and phosphorus.
52. The light meters in automatic cameras are based on the sensitivity of gallium arsenide to light. Predict the formula of gallium arsenide.
53. Write balanced equations for the reactions of sodium metal with each of the following elements.
    (a) F_2 (b) O_2 (c) H_2 (d) S_8 (e) P_4
54. Write balanced equations for the reactions of calcium metal with each of the following elements.
    (a) H_2 (b) O_2 (c) S_8
    (d) F_2 (e) N_2 (f) P_4
55. Predict the product of the reactions of F_2 with each of the following metals.
    (a) Zn (b) Al (c) Sn (d) Mg (e) Bi

Oxides, Peroxides, and Superoxides

56. What is the difference between an oxide and a peroxide?
57. What is the difference between an oxide and a superoxide?
58. Write the formulas for the oxide and peroxide of Cs.
59. Give the Lewis dot structures for Rb, Rb^+, and O^{2-}. Use the dot structures to predict the formulas of the superoxide and oxide of rubidium.
60. What is the formula of the oxide of beryllium?

The Ionic Bond

61. NaCl is classified as an ionic compound. What forces act to hold the atoms of this compound together?
62. Define and describe an ionic bond.
63. Use Lewis dot structures for Na and Cl to write the Lewis structure for NaCl.
64. Write Lewis structures for $BaCl_2$, Na_2S, and CaS.
65. What is one fundamental difference between a covalent and an ionic compound?
66. The attractive force, F, that exists between two oppositely charged ions can be expressed as:

$$F = \frac{q_+ \times q_-}{r^2}$$

In an ionic compound the q's refer to the charges on the ions and r is the distance between them.
    (a) What factors does the attractive force depend on?
    (b) For a fixed distance, is the force of attraction greater for two ions of charge +1 and −1 or for two ions of charge +2 and −2?
    (c) If q_+ is +1 and q_- is −1, is the force of attraction greater at a smaller distance or at a larger distance between the ions? Explain using the relationship given above.
    (d) Which ionic compound would have the strongest attractive forces, NaCl or KCl? Refer to Tables 3.7 and 3.8. Explain.
    (e) Refer to Appendix B, Table B.4. Which ionic compound, Li_2O or CaO, is most tightly held together? Explain.

Structures of Ionic Compounds

67. Define the term *unit cell.* Describe the common properties of all unit cells.
68. Describe the difference between simple cubic, body-centered cubic, and face-centered cubic unit cells.
69. Ca metal forms a face-centered cubic closest-packed unit cell. Roughly sketch this unit cell.
70. Sodium hydride crystallizes in a face-centered cubic unit cell of H^- ions with Na^+ ions at the center of the unit cell and in the center of each edge of the unit cell. How many Na^+ ions does each H^- ion touch? How many H^- ions does each Na^+ ion touch?

Metallic Bonds

71. What similarities do covalent bonds and ionic bonds share?
72. What is the difference between localized and delocalized electrons?
73. Define *metallic bond.*
74. What are the characteristics of atoms that tend to enter into metallic bonding?
75. There are two characteristics of an atom that are reflected in the AVEE that favor metallic bonding. What are they?
76. Roughly sketch the unit cell for Na metal. Show the electrons and the charge on the atoms. Do the same for Ba. Both have a body-centered cubic unit cell.
77. What are the differences between a metallic bond and an ionic bond? Between a metallic bond and a covalent bond?

The Relationship Among Ionic, Covalent, and Metallic Bonds

78. Explain why chemists believe that ionic and covalent are the two extremes of a continuum of differences in the bonding between two atoms.

79. Which of the following elements forms bonds with fluorine that are the most covalent?
(a) P (b) Ca (c) Al (d) O (e) Se

80. Which one of the following elements forms the most covalent bond with oxygen?
(a) Sr (b) In (c) Sb (d) Te (e) Se

81. Which of the following compounds are best described as ionic?
(a) CO (b) H_2O (c) BeF_2
(d) $MgBr_2$ (e) AlI_3 (f) ZnS
(g) CdLi

82. Which of the following compounds are best described as metallic?
(a) CaH_2 (b) BrF_3 (c) NF_3
(d) $SiCl_4$ (e) AsH_3 (f) $MgZn_2$

Answer problems 83–86 using the difference in electronegativity, ΔEN, only.

83. Which of the following compounds should be ionic?
(a) ZnS (b) $AlCl_3$ (c) SnF_2
(d) BH_3 (e) H_2S

84. Which of the following compounds should be covalent?
(a) CH_4 (b) CO_2 (c) $SrCl_2$
(d) NaH (e) SF_4

85. Which of the following pairs of elements should combine to give ionic compounds?
(a) Mg + O_2 (b) S_8 + F_2
(c) Na + Hg (d) K + I_2

86. Which of the following pairs of elements should combine to give metallic compounds?
(a) N_2 + O_2 (b) Cl_2 + F_2 (c) Cl_2 +Cs
(d) S_8 + Na (e) Cu + Sn

Bond-Type Triangles

87. Predict whether the following compounds are ionic or covalent by using both the general rule that metals combine with nonmetals to form ionic compounds and a bond-type triangle.
(a) OF_2 (b) CS_2 (c) MgO (d) ZnS

88. Predict whether the following compounds are ionic or covalent by using both the general rule that metals combine with nonmetals to form ionic compounds and a bond-type triangle.
(a) IF_3 (b) $SiCl_4$ (c) BF_3 (d) Na_2S

89. Construct a bond-type triangle for the elements of the third period, Na through Cl. Electronegativities are given in Figure 4.8. First draw a horizontal baseline starting with the electronegativity of Na and ending with Cl. Place all other elements of the period according to their electronegativities along the line. Locate the midpoint along the baseline between Na and Cl. At a distance above the midpoint corresponding to the ΔEN for Na and Cl, draw the two equal sides of the triangle. Determine the chemical formula of the binary compounds formed by as many of the pairs of elements as you can. Plot the compounds on the triangle using the average electronegativity and the difference in electronegativity, ΔEN, for the elements in each binary compound. How would you classify each compound?

90. Use Figure 5.10 to determine whether there is an electronegativity difference, ΔEN, above which a compound can always be predicted to be ionic.

91. Use Figure 5.10 to determine whether there is an electronegativity difference, ΔEN, below which a binary compound can always be classified as covalent.

92. For each of the following characteristics, select three binary compounds from Figure 5.10 which exhibit the characteristic.
(a) ΔEN is relatively the same for the three compounds, but the bonding type changes from metallic to semimetallic to covalent.
(b) ΔEN ranges from about 3 to about 1, but all compounds are ionic.
(c) The average EN of the atoms in the compounds stays the same, but the ionic character increases.
(d) The ΔEN of the atoms in the compounds stays the same, but the covalent character increases.
(e) ΔEN is about the same, but the bonding type changes from metallic to ionic to covalent.

93. Classify the following compounds according to bonding type:
(a) HgS (b) GaSb (c) Li_3N
(d) NaBr (e) $SnBr_4$ (f) Na_3P
(g) InP (h) InN (i) TeO_2

94. Why are two variables, ΔEN and average EN, required to classify a compound according to bonding type?

95. Make a plot of ΔEN versus the minimum electronegativity for the compounds of Figure 5.9. Add a few of the compounds given in Figure 5.10 to the plot and interpret the results.

96. Using the compounds found in Figure 5.9, make a plot with electronegativity on both the x and y axes. Plot the electronegativity of the atom with the higher electronegativity on the x axis. Plot the electronegativity of the atom with the lower electronegativity on the y axis. Add a few of the compounds given in Figure 5.10 and interpret the results.

Properties of Metallic, Covalent, and Ionic Compounds

97. For each of the following properties, list a binary solid compound that exhibits the property.
(a) conducts electricity in the solid state
(b) is an insulator
(c) is a semiconductor
(d) has a very high melting point
(e) has a low melting point
(f) electrons are shared unequally between the atoms composing the compound

98. The French have recently reported a new compound composed of Mn and Al. Is this compound a conductor, insulator, or semiconductor? Explain.

99. Select two compounds from Figure 5.10 that have about the same ΔEN, of which one is a conductor of electricity and one is an insulator.

100. (a) Describe the bonding in the following compounds as ionic, metallic, or covalent: BBr_3, Li_3P, SrF_2, $SrZn_5$, Cd_3N_2.
(b) From the above compounds, pick the ones that satisfy the following criteria:
(i) high melting and boiling points; dissolves in water to give solutions that conduct electricity
(ii) low melting and boiling points; localized electrons
(iii) low melting point and high boiling point; both solid and liquid conduct electricity

101. Consider the following three compounds: P_4O_6, Mg_3N_2, Mg_3Sb_2.
(a) One of these compounds melts at 24°C and the liquid phase does not conduct a electric current.
(b) One melts at 961°C and the solid phase does conduct an electric current.
(c) One melts at 800°C and when melted conducts an electric current.

Identify each compound and show how you arrived at your conclusions.

102. Which of the following compounds, Mg_3N_2, $BaSi_2$, B_2O_3:
(a) are insulators?
(b) has a very high melting point but does not conduct electricity in the solid state?
(c) conducts electricity in the solid state?

Explain your answers and show all calculations.

Oxidation Numbers

103. An area of active research interest in recent years has involved ions such as $Re_2Cl_8^{2-}$, $Cr_2Cl_9^{3-}$, and $Mo_2Cl_8^{4-}$ that contain bonds between metal atoms. Calculate the oxidation number of the metal atom in each compound.

104. Determine the oxidation state of phosphorus in the following compounds or ions.
(a) K_3P (b) Na_3PO_4 (c) PO_3^{3-} (d) P_2Cl_4

105. Calculate the oxidation number of the aluminum atom in the following compounds or ions.
(a) $LiAlH_4$ (b) $Al(H_2O)_6^{3+}$ (c) $Al(OH)_4^-$

106. The active ingredient in Rolaids™ antacid tablets has the formula $NaAl(OH)_2CO_3$. Calculate the oxidation state of the aluminum atom in the compound.

107. Which of the following compounds contain hydrogen in a negative oxidation state?
(a) H_2S (b) H_2O (c) NH_3 (d) H_3PO_4
(e) $LiAlH_4$ (f) HF (g) CaH_2 (h) CH_4

108. Calculate the oxidation number of the chlorine atom in the following compounds or ions.
(a) Cl_2 (b) Cl^- (c) ClO^-
(d) ClO_2^- (e) ClO_3^- (f) ClO_4^-

109. Calculate the oxidation number of the iodine atom in the following compounds. Group together any compounds in which iodine has the same oxidation number.
(a) HI (b) KI (c) I_2 (d) HOI
(e) KIO_3 (f) I_2O_5 (g) KIO_4 (h) H_5IO_6

110. The noble gases rarely form compounds. Therefore, their most common oxidation number is zero. However, Xe does form a limited number of compounds. Calculate the oxidation number of the xenon atom in the following compounds or ions and describe any trends in the oxidation states.
(a) XeF_2 (b) XeF_4 (c) $XeOF_2$
(d) XeF_6 (e) $XeOF_4$ (f) XeO_3
(g) XeO_4 (h) XeO_6^{4-}

111. Calculate the oxidation number of barium in BaO. If it is assumed that barium is in its normal oxidation state, what does this suggest about the oxidation state of oxygen in this compound?

112. Carbon can have any oxidation number between −4 and +4. Calculate the oxidation number of carbon in the following compounds. Group compounds with the same oxidation number and describe any trends.
(a) CCl_4 (b) $COCl_2$ (c) CO (d) CO_2
(e) CS_2 (f) CH_3Li (g) CH_4 (h) H_2CO
(i) Na_2CO_3 (j) HCO_2H

113. Sulfur can have any oxidation number between +6 and −2. Calculate the oxidation number of sulfur in the following species. Group species with the same oxidation number and describe any trends.
(a) S_8 (b) H_2S (c) ZnS
(d) SF_4 (e) SF_6 (f) SO_2
(g) SO_3 (h) SO_3^{2-} (i) SO_4^{2-}
(j) H_2SO_3 (k) H_2SO_4

114. Calculate the oxidation number of manganese in the following compounds. Group compounds with the same oxidation number and describe any trends.
(a) MnO (b) Mn_2O_3 (c) MnO_2
(d) MnO_3 (e) Mn_2O_7 (f) $Mn(OH)_2$
(g) $Mn(OH)_3$ (h) H_2MnO_4 (i) $HMnO_4$
(j) $CaMnO_3$ (k) $MnSO_4$

115. Calculate the oxidation number of titanium in the following compounds. Group compounds with the same oxidation number and describe any trends.
(a) TiO (b) TiO_2 (c) Ti_2O_3
(d) Ti_2S_3 (e) $TiCl_3$ (f) $TiCl_4$
(g) K_2TiO_3 (h) H_2TiCl_6 (i) $Ti(SO_4)_2$

116. Prussian blue is a pigment with the formula $Fe_4[Fe(CN)_6]_3$. If the compound contains the $Fe(CN)_6^{4-}$ ion, what is the oxidation state of the other four iron

atoms? Turnbull's blue is a pigment with the formula $Fe_3[Fe(CN)_6]_2$. If the compound contains the $Fe(CN)_6^{3-}$ ion, what is the oxidation state of the other three iron atoms?

Calculating Oxidation Numbers

117. Determine the oxidation state of each atom in the following organic molecules.

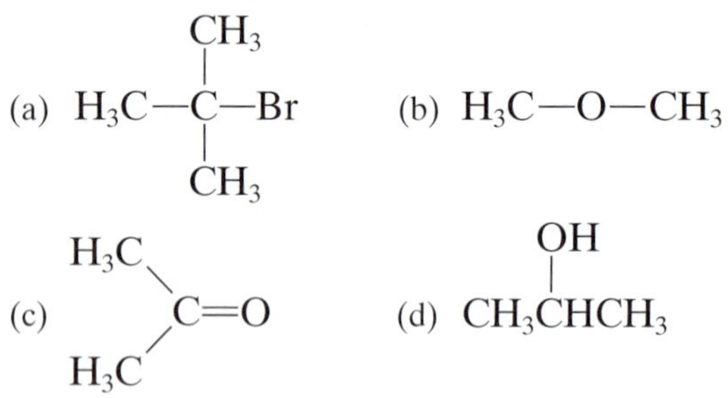

118. Determine the oxidation number of each atom in the following organic molecules.

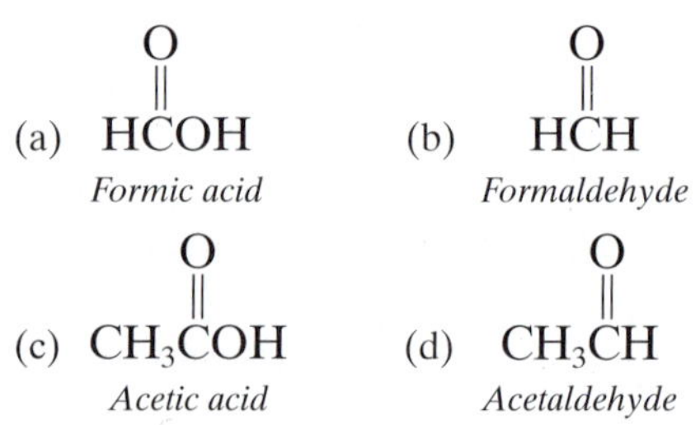

119. Determine the oxidation number of each atom in the following organic molecules.

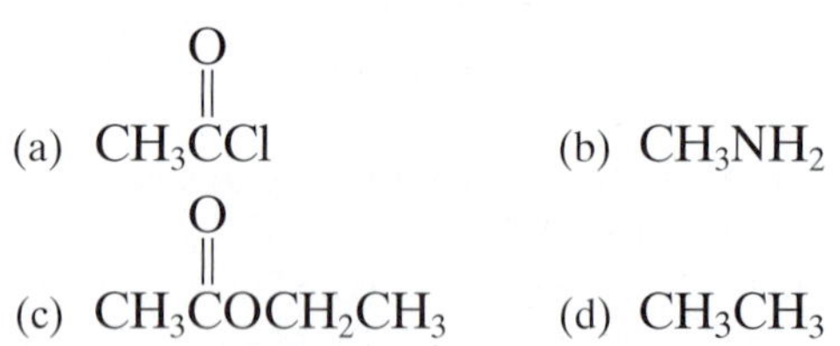

(e) $CH_2{=}CH{-}CH_3$ (f) $HC{\equiv}CH$

120. Describe the difference between partial charge, formal charge, and oxidation number.

121. Calculate the partial charge on the atoms in HCl using the methods of Section 4.11. Find the oxidation numbers of H and Cl in HCl. Give the formal charge on all the atoms in HCl. Compare and contrast these three results.

Oxidation–Reduction Reactions

122. Decide whether each of the following reactions involves oxidation–reduction. If it does, identify what is oxidized and what is reduced.
(a) $CO_2(g) + H_2O(l) \longrightarrow H_2CO_3(aq)$
(b) $Fe_2O_3(s) + 3\,CO(g) \longrightarrow 2\,Fe(s) + 3\,CO_2(g)$
(c) $CO_2(g) + H_2(g) \longrightarrow CO(g) + H_2O(g)$
(d) $CO(g) + 2\,H_2(g) \longrightarrow CH_3OH(l)$

123. Decide whether each of the following reactions involves oxidation–reduction. If it does, identify what is oxidized and what is reduced.
(a) $Mg(s) + 2\,HCl(g) \longrightarrow MgCl_2(s) + H_2(g)$
(b) $I_2(s) + 3\,Cl_2(g) \longrightarrow 2\,ICl_3(s)$
(c) $NaOH(aq) + HCl(aq) \longrightarrow NaCl(aq) + H_2O(aq)$
(d) $2\,Na(s) + 2\,H_2O(l) \longrightarrow 2\,NaOH(aq) + H_2(g)$

124. Decide whether each of the following reactions involves oxidation–reduction. If it does, identify what is oxidized and what is reduced.
(a) $P_4O_{10}(g) + 10\,C(s) \longrightarrow 10\,CO(g) + P_4(g)$
(b) $P_4(s) + 5\,O_2(g) \longrightarrow P_4O_{10}(s)$

125. What is the change in oxidation state of C and S in the following reaction? Which is oxidized and which is reduced?

$$4\,C(s) + S_8(l) \longrightarrow 4\,CS_2(l)$$

126. The Ag_2S that forms when silver tarnishes can be removed by polishing the silver with a source of cyanide ion or by wrapping it in aluminum foil and immersing it in salt water.

$$Ag_2S(s) + 4\,CN^-(aq) \longrightarrow 2\,Ag(CN)_2^-(aq) + S^{2-}(aq)$$

$$3\,Ag_2S(s) + 2\,Al(s) \longrightarrow 6\,Ag(s) + Al_2S_3(s)$$

Which of the reactions involves oxidation–reduction?

Nomenclature

127. Describe what is wrong with the common names for the following compounds and write a better name for each compound.
(a) phosphorus pentoxide (P_2O_5)
(b) iron oxide (Fe_2O_3)
(c) chlorine monoxide (Cl_2O)
(d) copper bromide ($CuBr_2$)

128. Explain why calcium bromide is a satisfactory name for $CaBr_2$, but $FeBr_2$ must be called iron(II) bromide.

129. Write the formulas for the following compounds.
(a) tetraphosphorus trisulfide
(b) silicon dioxide
(c) carbon disulfide
(d) carbon tetrachloride
(e) phosphorus pentafluoride

130. Write the formulas for the following compounds.
(a) silicon tetrafluoride (b) sulfur hexafluoride
(c) oxygen difluoride (d) dichlorine heptoxide
(e) chlorine trifluoride

131. Write the formulas for the following compounds.
(a) tin(II) chloride (b) mercury(II) nitrate
(c) tin(IV) sulfide (d) chromium(III) oxide
(e) iron(II) phosphide

132. Write the formulas for the following compounds.
(a) beryllium fluoride (b) magnesium nitride
(c) calcium carbide (d) barium peroxide
(e) potassium carbonate

133. Write the formulas for the following compounds.
(a) cobalt(III) nitrate
(b) iron(III) sulfate
(c) gold(III) chloride
(d) manganese(IV) oxide
(e) tungsten(VI) chloride

134. Name the following compounds.
(a) KNO_3 (b) Li_2CO_3
(c) $BaSO_4$ (d) PbI_2

135. Name the following compounds.
(a) $AlCl_3$ (b) Na_3N (c) Ca_3P_2
(d) Li_2S (e) MgO

136. Name the following compounds.
(a) NH_4OH (b) H_2O_2 (c) $Mg(OH)_2$
(d) $Ca(ClO)_2$ (e) NaCN

137. Name the following compounds.
(a) Sb_2S_3 (b) $SnCl_2$ (c) SF_4
(d) $SrBr_2$ (e) $SiCl_4$

138. Write the formulas of the following common acids.
(a) acetic acid (b) hydrochloric acid
(c) sulfuric acid (d) phosphoric acid
(e) nitric acid

139. Write the formulas of the following less common acids.
(a) carbonic acid (b) hydrocyanic acid
(c) boric acid (d) phosphorous acid
(e) nitrous acid

140. If sodium carbonate is Na_2CO_3 and sodium hydrogen carbonate is $NaHCO_3$, what are the formulas for sodium sulfite and sodium hydrogen sulfite?

141. The prefix *thio-* describes compounds in which sulfur replaces an oxygen, for example, cyanate (OCN^-) and thiocyanate (SCN^-). If SO_4^{2-} is the sulfate ion, what is the formula for the thiosulfate ion?

142. Name the compound in each of the following minerals.
(a) fluorite (CaF_2) (b) galena (PbS)
(c) quartz (SiO_2) (d) rutile (TiO_2)
(e) hematite (Fe_2O_3)

143. Name the compound in each of the following minerals.
(a) calcite ($CaCO_3$) (b) barite ($BaSO_4$)

Integrated Problems

144. The atoms composing the compounds SeO_2, CaH_2, and Cs_3Sb all have about the same difference in electronegativity, ΔEN. Yet only one of the compounds conducts electricity in the solid state, only one has a high melting point and dissolves in water to give a solution that conducts electricity, and only one has a relatively low melting point. Identify each of the compounds and explain your identification.

145. The common oxidation numbers for iron are +2 and +3. Iron reacts with $O_2(g)$ and $Cl_2(g)$ to form two different oxides and two different chlorides. What would be the formula of the four compounds formed? Classify the compounds according to bond type. Explain your answer.

146. Compare and contrast (what is the same and what is different) the distribution of electrons in covalent, ionic, and metallic bonds.

147. Partial charge, formal charge, and oxidation number are all used to describe the arrangement of electrons or charge associated with atoms within a compound. Determine the formal charge, partial charge, and oxidation number on both atoms in IBr. Give a short description for each of the three calculations describing what information the calculated value gives you. Which of the three calculations most accurately describes the distribution of charge between I and Br? What is the utility of the two calculations that are not the best description of the distribution of charge?

148. Describe what happens to the *distribution of electrons* in a bond between two elements as you move from left to right on a bond-type triangle. Describe what happens to the *distribution of electrons* in a bond between two elements as you move from the bottom to the top of a bond-type triangle.

149. Describe the types of bonding in $Ba(NO_3)_2$. What is different about the bonding in barium nitrate as compared to other compounds described in this chapter? Draw a Lewis structure for the cation and anion in this compound.

150. The following reaction is used in the Breathalyzer to determine the amount of ethyl alcohol on the breath of individuals suspected of driving while under the influence.

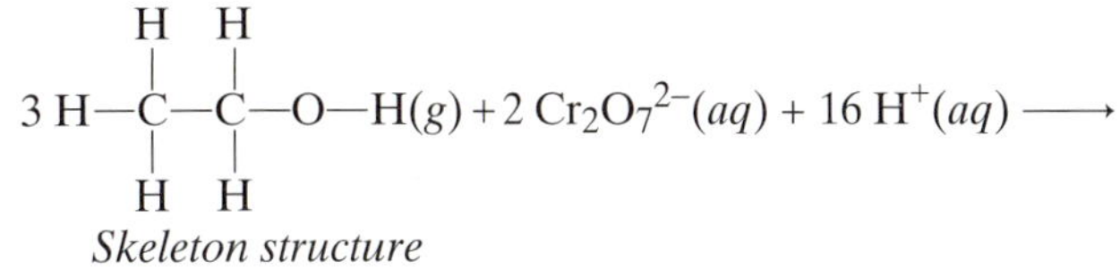

Skeleton structure

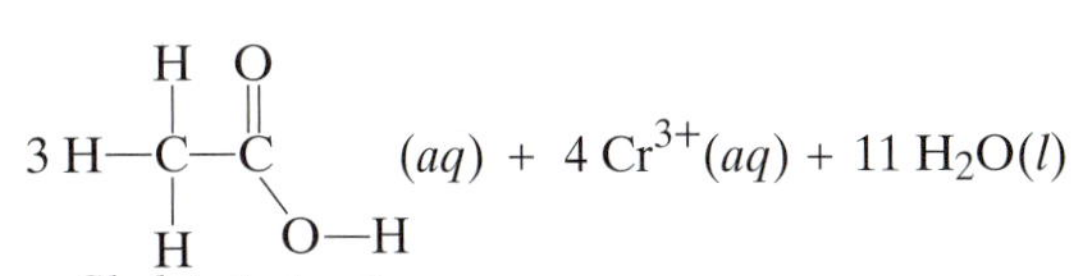

Skeleton structure

(a) If the reaction is an oxidation–reduction reaction, identify what is oxidized and what is reduced.
(b) If the reaction is not an oxidation–reduction reaction, explain how you reached your conclusion.

151. The industrial manufacture of acrylonitrile, C_3H_3N, an important chemical used for the production of plastics, synthetic rubber, and fibers is shown below.

$$4\ C_3H_6(g) + 6\ NO(g) \longrightarrow 4\ C_3H_3N(g) + 6\ H_2O(g) + N_2(g)$$

The skeleton structures for C_3H_3N and C_3H_6 are

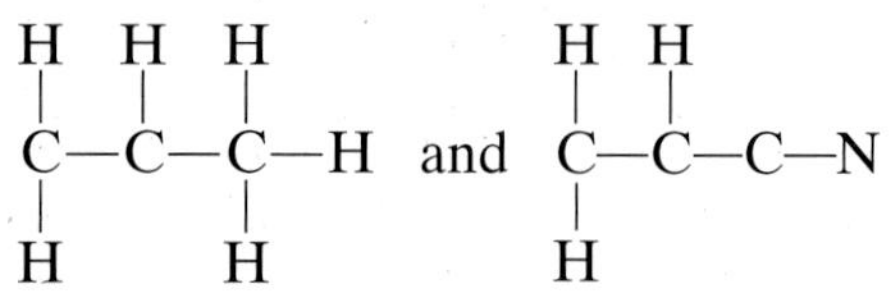

(a) Draw the Lewis structures for C_3H_6 and C_3H_3N and give the bond angle around each carbon atom.

(b) Which of the following statements concerning this reaction are true? Explain each.

(i) NO is oxidized.

(ii) C_3H_6 is reduced.

(iii) Both C_3H_6 and NO are reduced.

(iv) None of the above.

152. Which one of the following solid compounds has the highest melting point? Explain fully and show all calculations. See Problem 66.

HgO, ZrO, SrO, SeO_2

153. (a) Lead(II) sulfide solid reacts with oxygen molecules to form lead(II) oxide solid and sulfur dioxide gas. Write a balanced reaction for this chemical change.

(b) Is the reaction in (a) an oxidation–reduction reaction? If so, what is oxidized and what is reduced?

154. Name all of the species in the following reaction:

$$4\ NaCl(s) + 2\ H_2SO_4\ (aq) + MnO_2(s) \longrightarrow 2\ Na_2SO_4(aq) + MnCl_2(aq) + 2\ H_2O(l) + Cl_2(g)$$

Is the reaction an oxidation–reduction reaction? If so, what is oxidized and what is reduced?

155. Which of the following solid compounds would have a high melting point and not conduct electricity?

HgO, MgO, HgCl

Chapter Six

GASES

6.1 Temperature
6.2 Temperature as a Property of Matter
6.3 The States of Matter
6.4 Elements or Compounds That Are Gases at Room Temperature
6.5 The Properties of Gases
6.6 Pressure Versus Force
6.7 Atmospheric Pressure
6.8 Boyle's Law
6.9 Amontons' Law
6.10 Charles' Law
6.11 Gay-Lussac's Law
6.12 Avogadro's Hypothesis
6.13 The Ideal Gas Equation
6.14 Ideal Gas Calculations: Part I
6.15 Ideal Gas Calculations: Part II
6.16 Dalton's Law of Partial Pressures
6.17 The Kinetic Molecular Theory
6.18 How the Kinetic Molecular Theory Explains the Gas Laws

Special Topics

6A.1 Graham's Laws of Diffusion and Effusion
6A.2 Deviations from Ideal Gas Law Behavior: The van der Waals Equation
6A.3 Analysis of the van der Waals Constants

The gas laws are presented using experimental data to emphasize the process of science. The approach is more historical than usual, but because the development of concepts from experimental data is stressed, the experimental relationships which preceded the ideal gas law and kinetic molecular theory are given.

ChemActivity 34 in Chemistry: A Guided Inquiry *is designed to accompany this chapter.*

The philosophy of developing the gas laws by using experimental evidence has been followed. Depending on the background of students, some instructors may wish to move directly to the calculations using the ideal gas law in Section 6.13.

The Nobel Prize–winning physicist Richard Feynman once asked, "If, in some cataclysm, all of scientific knowledge were to be destroyed, and only one sentence passed on to the next generations of creatures, what statement would contain the most information in the fewest words?"[1] Feynman then answered the question, "I believe it is the atomic hypothesis that all things are made of atoms—little particles that move about in perpetual motion, attracting each other when they are a little distance apart, but repelling upon being squeezed into one another." This is, as Feynman observed, a statement of enormous power. So far we have developed the idea that all things are made of atoms, but we haven't discussed the other parts of Feynman's statement, that atoms move about in perpetual motion and that particles attract and repel each other.

Atoms are never at rest. Even atoms which are bonded together to form a molecule move relative to one another. The sodium and chloride ions of sodium chloride jiggle even though they are locked into an ordered three-dimensional structure. Gaseous molecules, such as the O_2 molecules in the atmosphere, speed rapidly about the room at several hundred miles per hour. This raises the question: What is the origin of this perpetual motion?

6.1 Temperature

We don't need to be told that the temperature is 95°F (35°C) on a summer afternoon to know it is hot. Nor do we need to be told that it is −15°F (−26°C) on a clear winter night to realize it is cold. For most purposes, we can rely on our senses to distinguish between hot and cold.

But there are times when our senses can be misled. Imagine you are getting out of bed in the middle of a cold winter night, stepping onto a "cold" floor and then onto a small throw rug. Your senses tell you that the rug is warmer than the floor. Unfortunately, your senses are wrong. The floor isn't colder than the rug. Both objects are just as warm (or just as cold) as the air in the room. When you reach into the freezer, metal ice-cube trays feel colder than plastic trays. But this can't be true; all the trays in the freezer are equally cold. Then why do they feel different?

Just as some materials conduct electricity better than others, some materials conduct heat better than others. Metals are good conductors of heat. The metal atoms are held in place by a sea of electrons, which are free to move throughout the metal. As a metal is heated, its atoms vibrate more rapidly and the electrons transport heat from one part of the metal to another. Materials whose electrons are more tightly bound do not conduct heat as well as or in the same way as metals. Covalent and ionic compounds conduct heat by passing the heat along through vibrating strings of atoms and ions instead of through electrons. Different substances in contact with one another transfer heat only when the atoms in one substance acquire sufficient heat to cause them to vibrate so wildly that they collide with neighboring atoms in the other substance, causing its atoms to vibrate more. Good thermal conductors feel cooler than poor thermal conductors at room temperature because good conductors transfer more heat away from our hand, making our hand feel cool.

Because our senses can be misled, it is useful to have a reliable measure of the degree to which an object is either hot or cold. This quantity is called the **temperature** of an object.

[1]From *The Feynman Lectures on Physics*, Volume 1, pp. 1–2, by R. Feynman and R. Leighton; Copyright © 1963, by The California Institute of Technology. Reprinted by permission of Addison-Wesley-Longman, Reading, Massachusetts.

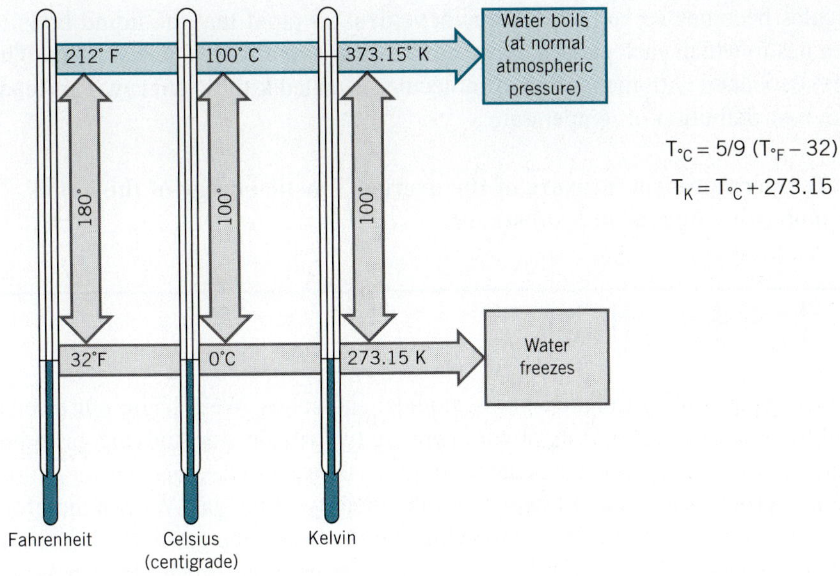

Fig. 6.1 The three common temperature scales. Note that water boils at 212°F, 100°C, or 373.15 K and freezes at 32°F, 0°C, or 273.15 K.

6.2 Temperature as a Property of Matter

Temperature is nothing more than a quantitative measure of the degree to which an object is either "hot" or "cold." Temperature can be measured on either relative or absolute scales (Figure 6.1). The Celsius (°C) and Fahrenheit (°F) scales measure *relative* temperatures. These scales compare the temperature of a system with arbitrary standards such as boiling water and an ice-water bath. The Kelvin (K) scale measures *absolute* temperatures. An object that has a temperature of 0°C has a temperature of 32°F and a temperature of 273 K.

At absolute zero on the Kelvin scale (−273.15°C), the atoms, molecules, or ions in a substance are packed as close together as they can get, and any motion of the units is at a minimum. What happens when a substance at absolute zero is warmed? The atoms, molecules, or ions begin to move—to vibrate or jiggle more vigorously. The warmer we make the substance, the more its components jostle about.

Consider what happens when ice melts, for example. Ice melts when it is heated because the average motion of the water molecules increases with temperature. Molecules can move in three ways: (1) vibration, (2) rotation, and (3) translation (Figure 6.2). Water molecules *vibrate* when H—O bonds are stretched or bent. *Rotation* involves the motion of a molecule around its center of gravity. *Translation* literally means to change from one place to another and describes the motion of molecules through space.

As the system becomes warmer, the motion of the water molecules eventually becomes too large to allow the molecules to be locked into the rigid structure of ice. At this point, the solid melts to form a liquid. Eventually, the kinetic energy of the

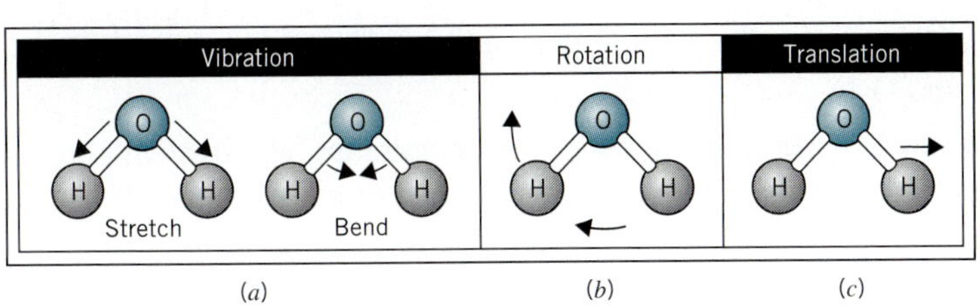

Fig. 6.2 A water molecule can move in three ways. (*a*) Vibration occurs when O—H bonds stretch or bend. (*b*) Rotation involves movement around the center of gravity of the molecule. (*c*) Translation occurs when a water molecule moves through space.

> **➤ CHECKPOINT**
>
> How does the average kinetic energy of the water molecules compare in samples of liquid water and gaseous water (or steam) at 100°C?

molecules becomes so large and their movement so rapid that the liquid boils to form a gas in which each particle moves more or less randomly through space. The energy associated with the motion of molecules is called **kinetic energy.** This leads us to a new definition of temperature:

Temperature is a measure of the average kinetic energy of the atoms, molecules, or ions in a substance.

6.3 The States of Matter

There are three **states** of matter: gases, liquids, and solids. We'll begin our discussion of the states of matter with gases. There are two reasons for studying gases before liquids and solids. First, the behavior of gases is easier to describe, because most of the properties of gases don't depend on the identity of the gas. We can therefore develop a model for a gas without worrying about whether the gas is O_2, N_2, H_2, or a mixture of gases. Second, a relatively simple yet powerful model known as the **kinetic molecular theory** is available that explains most of the behavior of gases.

The term *gas* comes from the Greek word for "chaos" because gases consist of a chaotic collection of particles in constant, random motion. In the course of our discussion of gases, we will provide the basis for answering the following questions.

- Why does popcorn "pop" when we heat it?
- Why does a hot-air balloon rise when the air in the balloon is heated?
- Why does a balloon filled with helium rise?
- Why does a balloon filled with CO_2 sink?
- At 25°C and 1 atmosphere of pressure, which weighs more: a liter of dry air or air that is saturated with water vapor?
- Is the volume of the O_2 in the room in which you are sitting the same as the volume of the N_2? Is the pressure of the O_2 the same as the pressure of the N_2?

6.4 Elements or Compounds That Are Gases at Room Temperature

Before examining the chemical and physical properties of gases, it may be useful to ask: What kinds of elements or compounds are gases at room temperature? To help answer the question, a list of some common compounds that are gases at room temperature is given in Table 6.1.

There are several patterns in the data in Table 6.1.

- Common gases at room temperature include both elements (such as H_2 and O_2) and compounds (such as CO_2 and NH_3).
- Elements that are gases at room temperature are all *nonmetals* (such as He, Ar, N_2, and O_2).
- Compounds that are gases at room temperature are all *covalent compounds* (such as CO_2, SO_2, and NH_3) that contain two or more nonmetals.
- With rare exception, these gases have relatively small atomic or molecular weights.

Table 6.1
Common Gases at Room Temperature

Element or Compound	Molecular Weight (g/mol)
H_2 (hydrogen)	2.02
He (helium)	4.00
CH_4 (methane)	16.04
NH_3 (ammonia)	17.03
Ne (neon)	20.18
HCN (hydrogen cyanide)	27.03
CO (carbon monoxide)	28.01
N_2 (nitrogen)	28.01
NO (nitrogen oxide)	30.01
C_2H_6 (ethane)	30.07
O_2 (oxygen)	32.00
PH_3 (phosphine)	34.00
H_2S (hydrogen sulfide)	34.08
HCl (hydrogen chloride)	36.46
F_2 (fluorine)	38.00
Ar (argon)	39.95
CO_2 (carbon dioxide)	44.01
N_2O (dinitrogen oxide)	44.01
C_3H_8 (propane)	44.10
NO_2 (nitrogen dioxide)	46.01
O_3 (ozone)	48.00
C_4H_{10} (butane)	58.12
SO_2 (sulfur dioxide)	64.07
BF_3 (boron trifluoride)	67.81
Cl_2 (chlorine)	70.91
Kr (krypton)	83.80
CF_2Cl_2 (dichlorodifluoromethane)	120.92
Xe (xenon)	131.29
SF_6 (sulfur hexafluoride)	146.05

As a general rule, elements and compounds that consist of relatively light, covalent molecules are most likely to be gases at room temperature.

6.5 The Properties of Gases

Gases have three characteristic properties: (1) They are easy to compress, (2) they expand to fill their containers, and (3) they occupy much more space than the equivalent mass of a liquid or solid under normal conditions.

COMPRESSIBILITY

The internal combustion engine found in most cars provides a good example of the ease with which gases can be compressed. In a typical four-stroke engine, the piston is first pulled out of the cylinder to create a partial vacuum, which draws a mixture of gasoline vapor and air into the cylinder (Figure 6.3). The piston is then pushed into the cylinder, compressing the gasoline–air mixture to a fraction of the original volume.

The ratio of the volume of the gas in the cylinder after the first stroke to its volume after the second stroke is the *compression ratio* of the engine. Modern cars

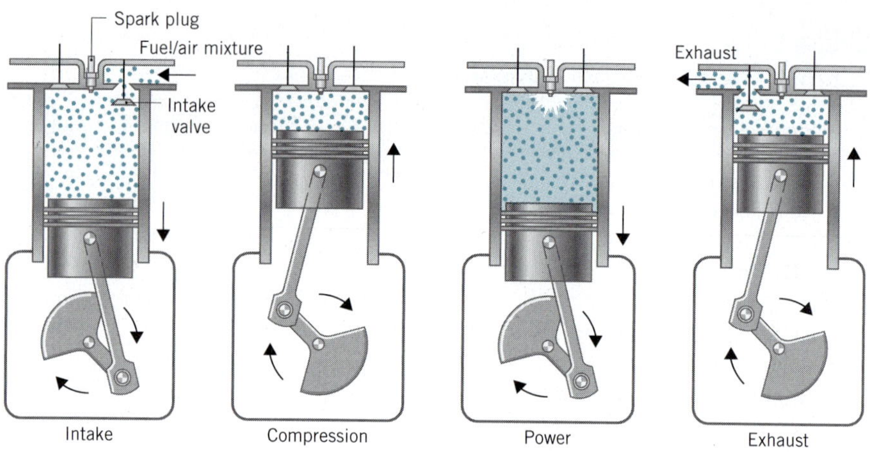

Fig. 6.3 The operation of a four-stroke engine can be divided into four cycles: intake, compression, power, and exhaust stages.

run at compression ratios of about 9:1, which means the gasoline–air mixture in the cylinder is compressed by a factor of 9 in the second stroke. After the gasoline–air mixture is compressed, the spark plug at the top of the cylinder fires, and the resulting explosion pushes the piston out of the cylinder in the third stroke. Finally, the piston is pushed back into the cylinder in the fourth stroke, clearing out the exhaust gases.

Liquids are much harder to compress than gases. They are so difficult to compress that the hydraulic brake systems used in most cars operate on the principle that there is essentially no change in the volume of the brake fluid when pressure is applied to the liquid. Most solids are even harder to compress. The only exceptions belong to a unique class of compounds that includes natural and synthetic rubber. Most rubber balls that are easy to compress, such as a racquetball, are filled with air, which is compressed when the ball is squeezed.

EXPANDABILITY

Anyone who has walked into a kitchen where bread is baking has experienced the fact that gases expand to fill their containers, as the air in the kitchen becomes filled with wonderful aromas. Unfortunately, the same thing happens when someone breaks open a rotten egg and the characteristic odor of hydrogen sulfide (H_2S) rapidly diffuses through the room. Because gases expand to fill their containers, the volume of a gas is equal to the volume of its container.

VOLUMES OF GASES VERSUS VOLUMES OF LIQUIDS OR SOLIDS

The difference between the volume of a gas and the volume of the liquid or solid from which it forms can be illustrated with the following examples. One gram of liquid oxygen at its boiling point (−183°C) has a volume of 0.894 mL. The same amount of O_2 gas at −183°C and atmospheric pressure has a volume of over 200 mL. Similar results are obtained when the volumes of solids and gases are compared. One gram of solid CO_2 at −79°C has a volume of 0.641 mL. At −79°C and atmospheric pressure, the same amount of CO_2 gas has a volume of over 300 mL. As a general rule, the volume of a liquid or solid increases by a factor of several hundred when it forms a gas at room temperature and atmospheric pressure.

The consequences of the enormous change in volume that occurs when a liquid or solid is transformed into a gas are frequently used to do work. The steam engine, which brought about the industrial revolution, is based on the fact that water boils to form a gas (steam) which has a much larger volume. The gas therefore escapes from the container in which it has been generated, and the escaping steam can be made to do work. The same principle is at work when dynamite is used to blast rocks. In 1867, the Swedish chemist Alfred Nobel discovered that the highly dangerous liquid explosive known as nitroglycerin could be absorbed onto clay or sawdust to produce a solid that was much more stable and therefore safer to use. When dynamite is detonated, the nitroglycerin decomposes to produce a mixture of CO_2, H_2O, N_2, and O_2 gases.

$$4\ C_3H_5N_3O_9(l) \longrightarrow 12\ CO_2(g) + 10\ H_2O(g) + 6\ N_2(g) + O_2(g)$$

Because 29 mol of gas are produced for every 4 mol of liquid that decomposes, and each mole of gas occupies a volume hundreds of times larger than a mole of liquid, the reaction produces a shock wave that destroys anything in its vicinity.

The same phenomenon occurs on a much smaller scale when we pop popcorn. When kernels of popcorn are heated in oil, the liquids inside the kernel turn into gases. The pressure that builds up inside the kernel is enormous and eventually causes the kernel to explode.

Popcorn "pops" because of the enormous difference between the volume of the liquids inside the kernel and the volume of the gases these liquids produce when they boil.

6.6 Pressure Versus Force

The volume of a gas is one of its characteristic properties. Another characteristic property is the **pressure** the gas exerts on its surroundings. Many of us got our first exposure to the pressure of a gas when we rode to the neighborhood gas station to check the pressure in our bicycle tires. Depending on the kind of bicycle we had, we added air to the tires until the pressure gauge read between 30 and 70 pounds per square inch (70 lb/in^2), or psi. Two important properties of pressure can be gleaned from this example.

- The pressure of a gas increases as more gas is added to the container (as long as the volume is constant and the temperature does not change).
- Pressure is measured in units (such as lb/in^2) that describe the **force** exerted by the gas divided by the **area** over which the force is distributed.[2]

The first conclusion can be summarized in the following relationship, where P is the pressure of the gas and n is the number of moles of gas in the container.

$$P \propto n \quad (T \text{ and } V \text{ constant})$$

The symbol $\propto$ means "proportional to," and the relationship is read as "pressure is proportional to the number of moles of gas." Because the pressure increases as gas is added to a container with a fixed volume at a given temperature, P is directly proportional to n.

The second conclusion describes the relationship between pressure and force. Pressure is defined as the force exerted on an object divided by the area over which the force is distributed.

$$\text{Pressure} = \frac{\text{force}}{\text{area}}$$

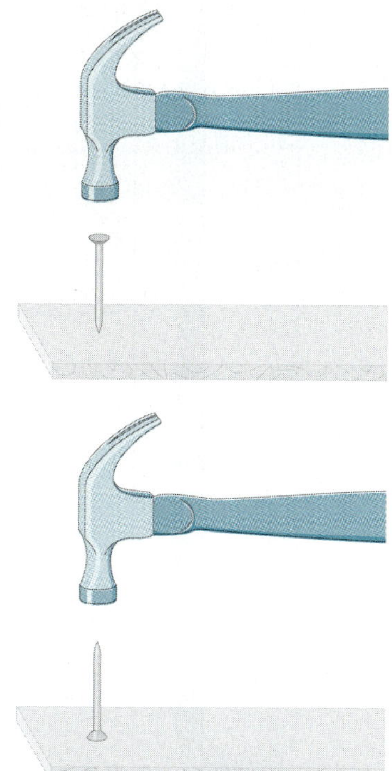

Fig. 6.4 The *force* exerted by a hammer hitting a nail is the same regardless of whether the hammer hits the nail on the head or on the point. But the *pressure* exerted on the wood is very different.

The difference between pressure and force can be illustrated with an analogy based on a 10-penny nail, a hammer, and a piece of wood (Figure 6.4). By resting the nail on its point and hitting the head with the hammer, we can drive the nail into the

[2]The SI unit for pressure is the pascal (Pa), which is defined as a force of 1 newton averaged over an area of 1 m^2.

wood. But what happens if we turn the nail over and rest the head of the nail against the wood? If we hit the nail with the same force, we can't get the nail to penetrate into the wood.

When we hit the nail on the head, the force of the blow is applied to the very small area of the wood in contact with the point of the nail, and the nail slips easily into the wood. But when we turn the nail over, and hit it on the point, the same force is distributed over a much larger area. The force is now distributed over the surface of the wood that touches any part of the nail head. As a result, the pressure applied to the wood is much smaller and the nail just bounces off the wood.

➤ CHECKPOINT

Is the pressure exerted on the floor by a woman wearing a spike heel greater than, less than, or equal to the pressure exerted when the same person wears a flat sandal? Why?

Exercise 6.1

(a) Calculate the pressure exerted by the shoes of a 200-lb man wearing size 10 shoes, if each shoe makes contact with an area of the floor that is 20 in^2.

(b) Calculate the pressure exerted by each heel of a 100-lb woman in high heels, if the area beneath the heel of each shoe is 0.25 in^2.

Solution

(a) The pressure is calculated by dividing the force by the area over which it is distributed. Because one-half of the weight of the man is applied to each shoe, the pressure in this case is 5.0 lb/in^2.

$$\text{Pressure} = \frac{\text{force}}{\text{area}} = \frac{100 \text{ lb}}{20 \text{ in}^2} = 5.0 \text{ lb/in}^2$$

(b) We can assume that about one-fourth of the weight of the woman is applied to each heel, if her weight is evenly divided between the heel and sole of each shoe.

$$\text{Pressure} = \frac{\text{force}}{\text{area}} = \frac{25 \text{ lb}}{0.25 \text{ in}^2} = 1.0 \times 10^2 \text{ lb/in}^2$$

The pressure exerted by the heels of the 100-lb woman is 20 times greater than the pressure exerted by the man, even though he weighs twice as much.

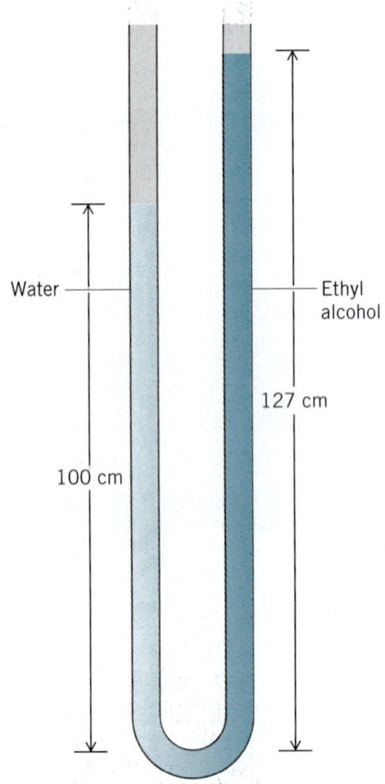

Fig. 6.5 Because of the difference between the densities of water and ethyl alcohol, the weight of a column of water 100 cm long balances the weight of a column of ethyl alcohol 127 cm long in the other arm of a U-tube.

6.7 Atmospheric Pressure

What would happen if we bent a long piece of glass tubing into the shape of the letter U and then carefully filled one arm of the U-tube with water and the other arm with ethyl alcohol? Most people expect the height of the columns of liquid in the two arms of the tube would be the same. Experimentally, we find the results shown in Figure 6.5. A 100-cm column of water balances a 127-cm column of ethyl alcohol, regardless of the diameter of the glass tubing.

We can explain this observation by comparing the densities of water (1.00 g/cm^3) and ethyl alcohol (0.789 g/cm^3). A column of water 100 cm tall exerts a pressure proportional to 100 g/cm^2.[3]

$$100 \text{ cm} \times \frac{1.00 \text{ g}}{\text{cm}^3} = 100 \text{ g/cm}^2$$

[3]Pressure is defined as force/area. The force is the mass times the acceleration of gravity (ma). Thus the actual pressure is 100g/cm^2 $\times$ 980 cm/sec^2.

A column of ethyl alcohol 127 cm tall exerts the same pressure.

$$127 \text{ cm} \times \frac{0.789 \text{ g}}{\text{cm}^3} = 100 \text{ g/cm}^2$$

Because the pressure of the water pushing down on one arm of the U-tube is equal to the pressure of the alcohol pushing down on the other arm of the tube, the system is in balance. This demonstration provides the basis for understanding how a mercury barometer can be used to measure the pressure of the atmosphere.

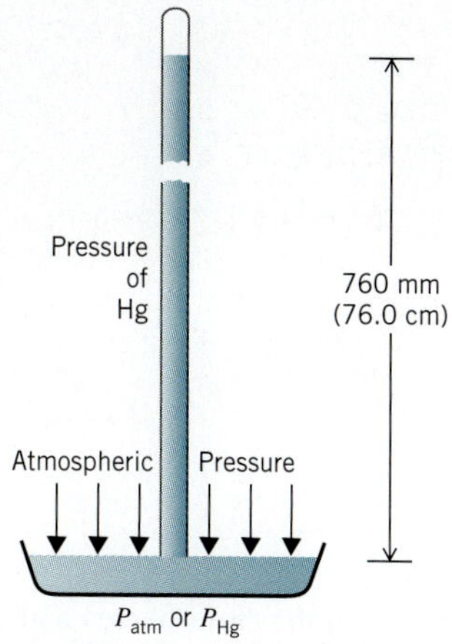

Fig. 6.6 On a sunny day, at sea level, the pressure of a 760-mm-tall column of mercury inside a glass tube balances the pressure of the atmosphere pushing down on the pool of mercury that surrounds the tube. The pressure of the atmosphere is therefore said to be equivalent to 760 mmHg.

THE DISCOVERY OF THE BAROMETER

In the early 1600s, Galileo argued that suction pumps were able to draw water from a well because of the "force of vacuum" inside the pump. After Galileo's death, the Italian mathematician and physicist Evangelista Torricelli (1608–1647) proposed another explanation. He suggested that the air in the atmosphere has weight and that the force of the atmosphere pushing down on the surface of the water drives the water into the suction pump when it is evacuated.

In 1646 Torricelli described an experiment in which a glass tube about 1 m long was sealed at one end, filled with mercury, and then inverted into a dish filled with mercury, as shown in Figure 6.6. Some, but not all, of the mercury drained out of the glass tube into the dish. Torricelli explained this by assuming that mercury drains from the glass tube until the pressure of the column of mercury pushing down on the *inside* of the tube exactly balances the pressure of the atmosphere pushing down on the surface of the liquid *outside* the tube.

Torricelli predicted that the height of the mercury column would change from day to day as the pressure of the atmosphere changed. Today, his apparatus is known as a *barometer,* from the Greek word *baros,* meaning "weight," because it literally measures the weight of the atmosphere. Repeated experiments showed that the average pressure of the atmosphere at sea level is equal to the pressure of a column of mercury 760 mm tall. Thus, a standard unit of pressure known as the *atmosphere* was defined as follows.

$$1 \text{ atm} = 760 \text{ mmHg} \qquad \text{(exactly)}$$

To recognize Torricelli's contributions, some scientists describe pressure in units of torr, defined as follows.

$$1 \text{ torr} = 1 \text{ mmHg}$$

Exercise 6.2

Calculate the pressure in units of atmospheres on a day when a barometer gives a measurement of 745.8 mmHg.

Solution

The conversion between mmHg and atmospheres is based on the following definition.

$$1 \text{ atm} = 760 \text{ mmHg}$$

Using that equality to generate an appropriate unit factor gives the following result.

$$745.8 \text{ mmHg} \times \frac{1 \text{ atm}}{760 \text{ mmHg}} = 0.9813 \text{ atm}$$

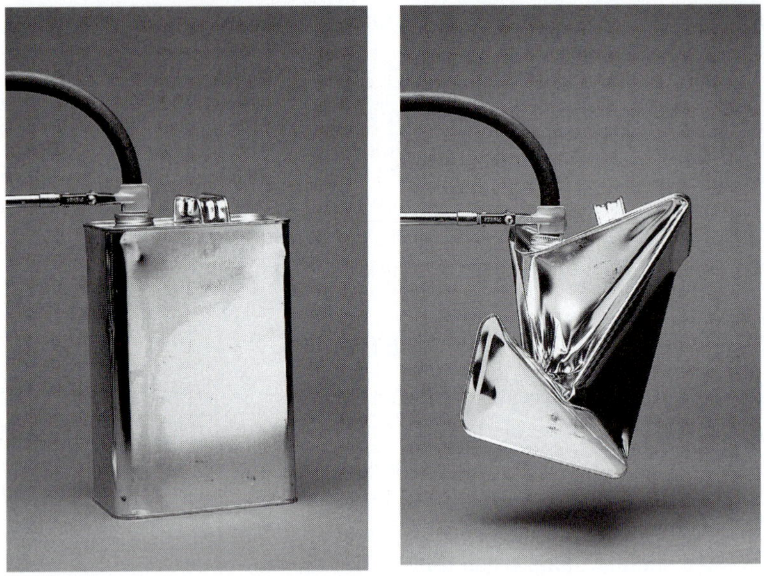

The pressure of the atmosphere can be demonstrated by connecting an empty paint-thinner can to a vacuum pump. Within seconds of turning on the vacuum pump, the can collapses.

Although chemists still work with pressures in units of atm or mmHg, neither unit is accepted in the SI system (see appendix page A-4). The SI unit of pressure is the pascal (Pa). The relationship between one standard atmosphere pressure and the pascal is given by the following.

$$1 \text{ atm} = 101{,}325 \text{ Pa} = 101.325 \text{ kPa} = 0.101325 \text{ MPa}$$

The pressure of the atmosphere can be demonstrated by connecting a 1-gallon can to a vacuum pump. Normally the pressure inside the can balances the pressure of the atmosphere pushing on the outside of the can. When the vacuum pump is turned on, however, the can rapidly collapses as it is evacuated. The surface area of a 1-gallon can is about 250 in^2. At 14.7 lb/in^2 (1 atmosphere), this corresponds to a total force over the surface of the can of about 3700 lb. For the sake of comparison, it might be noted that each of the 18 wheels of a 70,000-lb truck carries only about 3900 lb.

THE DIFFERENCE BETWEEN PRESSURE OF A GAS AND PRESSURE RESULTING FROM WEIGHT

There is an important difference between the pressure of a gas and the other examples of pressure discussed in this section. The pressure exerted by a 70,000-pound truck is directional. The truck exerts all of its pressure on the surface beneath its wheels. In contrast, gas pressure is the same in all directions. To demonstrate this, we can fill a glass cylinder with water and rest a glass plate on top of the cylinder. When we turn the cylinder over, the plate doesn't fall to the floor because the pressure of the air outside the cylinder pushing up on the bottom of the plate is larger than the pressure exerted by the water in the cylinder pushing down on the plate. It would take a column of water 33.9 feet tall to produce as much pressure as the gas in the atmosphere.

Students are asked to use data to discover each gas law, such as Boyle's law. The approach is not new, but considerable detail on early experiments is given. This follows the philosophy of having students move through the development of a concept for themselves using experimental data.

6.8 Boyle's Law

Torricelli's work with the mercury barometer caught the eye of the British scientist Robert Boyle. Boyle's most famous experiments were done in a J-tube apparatus similar to the one shown in Figure 6.7. By adding mercury to the open end of the tube, Boyle was able to trap a small volume of air in the sealed end.

Boyle was interested in a phenomenon he called the "spring of air." His experiments were based on the fact that gases are *elastic*. (They return to their original size and shape after being stretched or squeezed.) When he added more mercury to the open end of the J-tube, Boyle noticed that the air in the sealed end was compressed into a smaller volume.

Table 6.2 contains some of the experimental data he reported in his book, *New Experiments Physico-Mechanicall, Touching the Spring of Air, and its Effects . . .*, published in 1662. The first column in Table 6.2 lists the volume of the gas in the sealed end of the J-tube, in arbitrary units. The second column is the difference in heights of the mercury in the sealed and open arms of the J-tube, to the nearest $\pm 1/16$ inch. The third column is the product of the volume of the gas (V) and the pressure (P) calculated from Boyle's data. The number of moles of gas and temperature were held constant for the experiments.

The product of the pressure times the volume for any measurement in the table is equal to the product of the pressure times the volume for any other measurement.

$$P_1V_1 = P_2V_2$$

This expression, or its equivalent,

$$P \propto 1/V \qquad (T \text{ and } n \text{ constant})$$

is now known as **Boyle's law.**

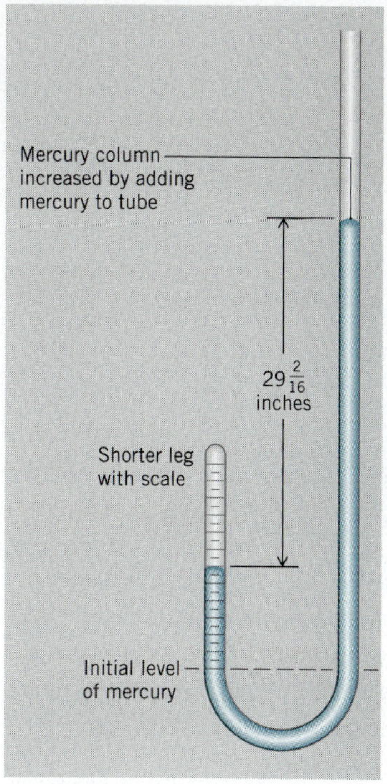

Fig. 6.7 Boyle's law is based on data obtained with a J-tube apparatus. The temperature and number of moles of gas are constant. The volume is in arbitrary units and the height is in inches.

6.9 Amontons' Law

Toward the end of the 1600s, the French physicist Guillaume Amontons built a thermometer based on the fact that the pressure of a gas is directly proportional to its temperature. The relationship between the pressure and the temperature of a gas is therefore known as **Amontons' law.**

$$P \propto T \qquad (V \text{ and } n \text{ constant})$$

Amontons' law explains why car manufacturers recommend adjusting the pressure of your tires before you start on a trip. The flexing of the tire as you drive inevitably raises the temperature of the air in the tire. When this happens, the pressure of the gas inside the tires increases.

Amontons' law can be demonstrated with the apparatus shown in Figure 6.8. Data obtained with this apparatus at various temperatures are given in Table 6.3.

The relationship between the temperature and pressure of a gas provided the first definition for absolute zero on the temperature scale. In 1779 Joseph Lambert defined **absolute zero** as the temperature at which the pressure of a gas becomes zero. When the data in Table 6.3 are plotted, the pressure of a gas approaches zero when the temperature is about $-270°C$, as shown in Figure 6.9. When more accurate measurements are made, the pressure of a gas extrapolates to zero when the temperature is $-273.15°C$. Absolute zero on the Celsius scale is therefore $-273.15°C$.

> **CHECKPOINT**
>
> Does the diameter of the tube used in Boyle's experiments affect the results obtained?

Table 6.2

Boyle's Data on the Dependence of the Volume of a Gas on the Pressure of the Gas

Volume	Pressure	$P \times V$
48	$29\frac{2}{16}$	1.4×10^3
46	$30\frac{9}{16}$	1.4×10^3
44	$31\frac{15}{16}$	1.4×10^3
42	$33\frac{8}{16}$	1.4×10^3
40	$35\frac{5}{16}$	1.4×10^3
38	37	1.4×10^3
36	$39\frac{4}{16}$	1.4×10^3
34	$41\frac{10}{16}$	1.4×10^3
32	$44\frac{3}{16}$	1.4×10^3
30	$47\frac{1}{16}$	1.4×10^3
⋮	⋮	⋮

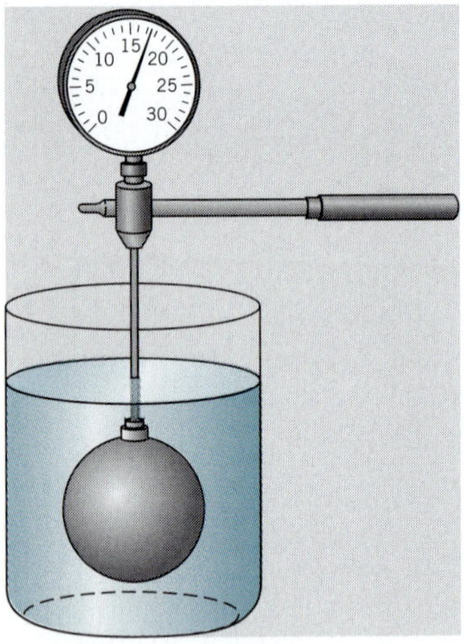

Fig. 6.8 The apparatus for demonstrating Amontons' law consists of a pressure gauge connected to a metal sphere of constant volume, which is immersed in water at different temperatures.

The relationship between the temperature and pressure data in Table 6.3 can be simplified by converting the temperatures from the Celsius to the Kelvin scale.

$$T_K = T_{°C} + 273.15$$

When this is done, a plot of the temperature versus the pressure of a gas gives a straight line that passes through the origin. Any two points along the line therefore fit the following equation.

$$\frac{P_1}{P_2} = \frac{T_1}{T_2}$$

It is important to remember that this equation is valid only if the temperatures are converted from the Celsius to the Kelvin scale before calculations are done, and the volume of gas and number of moles of gas must be constant.

➤ CHECKPOINT

What relationship can be derived by combining Boyle's and Amontons' laws?

6.10 Charles' Law

On June 5, 1783, Joseph and Etienne Montgolfier used a fire to inflate a spherical balloon about 30 feet in diameter that traveled about 1.5 miles before it came back to earth. News of this remarkable achievement spread throughout France, and Jacques-Alexandre-César Charles immediately tried to duplicate the performance. As a result of his work with hot air balloons, Charles noticed that the volume of a gas is directly proportional to its temperature when the pressure and number of moles of gas are fixed.

$$V \propto T \qquad (P \text{ and } n \text{ constant})$$

This relationship between the temperature and volume of a gas, which became known as **Charles' law,** can be used to explain how a hot-air balloon works. Ever since the third century B.C., it has been known that an object floats when it weighs less than the fluid it displaces. If a gas expands when heated, then a given weight of hot air occupies a larger volume than the same weight of cold air. Hot air is therefore less dense than cold air. Once the air in a balloon gets hot enough, the net weight of the balloon plus the hot air is less than the weight of an equivalent volume of cold air, and the balloon starts to rise. When the gas in the balloon is allowed to cool, the balloon returns to the ground.

Table 6.3
The Dependence of the Pressure of a Gas on Its Temperature

Temperature (°C)	Pressure (lb/in^2)
100	18.1
74	16.7
24	14.5
0	13.2
−47	10.8

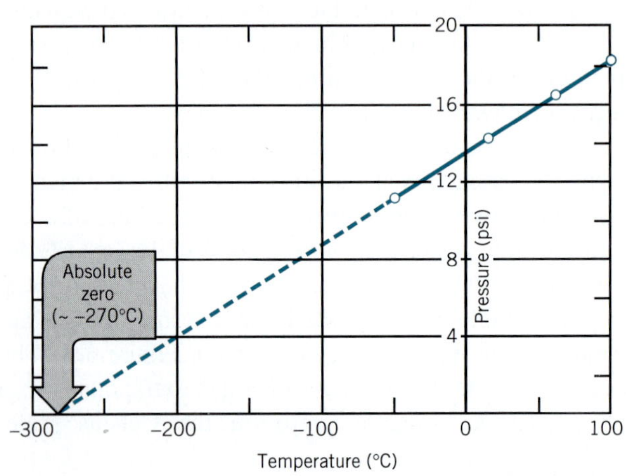

Fig. 6.9 When data obtained with the Amontons' law apparatus in Figure 6.8 are extrapolated, the pressure of a gas approaches zero when the temperature of the gas is approximately −270°C.

Charles' law can be demonstrated with the apparatus shown in Figure 6.10. A 30-mL syringe and a thermometer are inserted through a rubber stopper into a flask that has been cooled to 0°C. The ice bath is then removed and the flask is immersed in a warm-water bath. The gas in the flask expands as it warms, slowly pushing the piston out of the syringe. The total volume of the gas in the system is equal to the volume of the flask plus the volume of the syringe. Table 6.4 contains typical data obtained with this apparatus.

The data in Table 6.4 are plotted in Figure 6.11. This graph provides us with another way of defining absolute zero on the temperature scale: It is the temperature at which the volume of a gas becomes zero when the plot of volume versus temperature for a gas is extrapolated. As expected, the value of absolute zero obtained by extrapolating the data in Table 6.4 is essentially the same as the value obtained from the graph of pressure versus temperature in the preceding section. Absolute zero can therefore be more accurately defined as the temperature at which the pressure and the volume of a gas extrapolate to zero.

When the temperatures in Table 6.4 are converted from the Celsius to the Kelvin scale, a plot of the volume versus the temperature of a gas becomes a straight line that passes through the origin. Any two points along the line can therefore be used to construct the following equation, which is known as Charles' law.

$$\frac{V_1}{V_2} = \frac{T_1}{T_2}$$

Before you use this equation, however, it is important to remember to convert temperatures from °C to K.

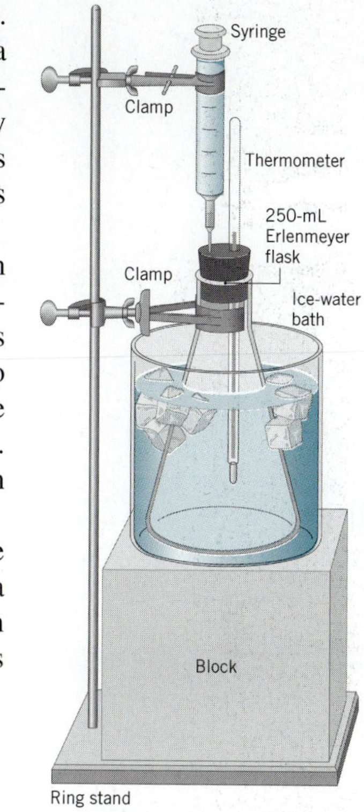

Fig. 6.10 Charles' law can be demonstrated with a simple apparatus. When the flask is removed from the ice bath and placed in a warm-water bath, the gas in the flask expands, slowly pushing up on the piston of the syringe.

6.11 Gay-Lussac's Law

Joseph Louis Gay-Lussac (1778–1850) studied the volume of gases consumed or produced in a chemical reaction because he was interested in the reaction between hydrogen and oxygen to form water. Gay-Lussac found that 199.89 parts by volume of hydrogen were consumed for every 100 parts by volume of oxygen. Thus, hydrogen and oxygen seemed to combine in a simple 2:1 ratio by volume.

$$\underset{2\ volumes}{\text{hydrogen}} + \underset{1\ volume}{\text{oxygen}} \longrightarrow \text{water}$$

Gay-Lussac found similar whole-number ratios for the reactions between other pairs of gases. He also obtained similar results when he analyzed the volumes of gases given off when compounds decomposed. Ammonia, for example, decomposed to give three times as much hydrogen by volume as nitrogen.

$$\text{ammonia} \longrightarrow \underset{1\ volume}{\text{nitrogen}} + \underset{3\ volumes}{\text{hydrogen}}$$

On December 31, 1808, Gay-Lussac announced his results at a meeting of the Societé Philomatique in Paris in terms of a **law of combining volumes.** At the time, he summarized the law as follows: Gases combine among themselves in very simple proportions. Today, **Gay-Lussac's law** is stated as follows: The ratio of the volumes of gases consumed or produced in a chemical reaction is equal to a ratio of simple whole numbers when temperature and pressure are constant.

➤ CHECKPOINT

Could Charles' law have been predicted from Boyle's and Amontons' laws, or is it independent of those laws?

Table 6.4

Dependence of the Volume of a Gas on Its Temperature

Temperature (°C)	Volume (mL)
0	107.9
5	109.7
10	111.7
15	113.6
20	115.5
25	117.5
30	119.4
35	121.3
40	123.2

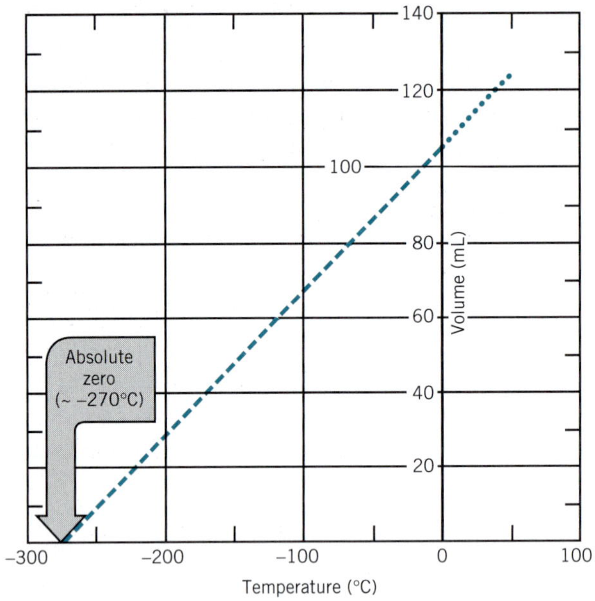

Fig. 6.11 When data obtained with the Charles' law apparatus in Figure 6.10 are extrapolated, the volume of the gas approaches zero when the temperature of the gas is approximately −270°C.

6.12 Avogadro's Hypothesis

Gay-Lussac's law of combining volumes was announced only a few years after John Dalton proposed his atomic theory. The link between the two ideas was first recognized by the Italian physicist Amadeo Avogadro three years later, in 1811. Avogadro argued that Gay-Lussac's law of combining volumes could be explained by assuming that equal volumes of different gases collected under similar conditions contain the same number of particles.

HCl and NH_3 combine in a 1:1 ratio by volume, for example, because one molecule of HCl is consumed for every molecule of NH_3 in the reaction and equal volumes of the gases contain the same number of molecules.

$$NH_3(g) + HCl(g) \longrightarrow NH_4Cl(s)$$

Anyone who has blown up a balloon can accept the notion that the volume of a gas is proportional to the number of particles in the gas.

$$V \propto n \qquad (T \text{ and } P \text{ constant})$$

The more air you add to a balloon, the bigger it gets. Unfortunately this example doesn't test **Avogadro's hypothesis,** that equal volumes of *different gases* contain the same number of particles. The best way to prove the validity of the hypothesis is to measure the number of molecules in a given volume of different gases, which can be done with the apparatus shown in Figure 6.12. The temperature and pressure are held constant.

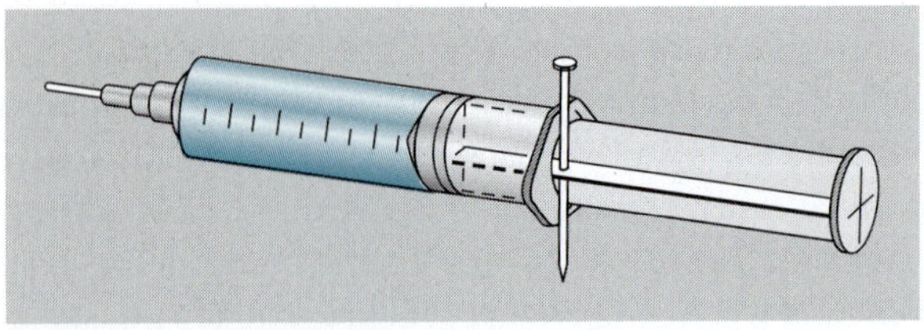

Fig. 6.12 Apparatus used to demonstrate Avogadro's hypothesis.

A small hole is drilled through the plunger of a 50-mL plastic syringe. The plunger is pushed into the syringe, and the syringe is sealed with a syringe cap. The plunger is then pulled out of the syringe until the volume reads 50 mL, and a nail is inserted through the hole in the plunger so that the plunger is not drawn back into the barrel of the syringe. The "empty" syringe is then weighed, the syringe is filled with 50 mL of a gas, and the syringe is reweighed. The difference between the measurements is the mass of 50 mL of the gas.

The results of experiments with six gases are given in Table 6.5. The number of molecules in a 50-mL sample of any one of the gases can be calculated from the mass of the sample, the molecular weight of the gas, and the number of molecules in a mole. Consider the following calculation of the number of H_2 molecules in 50 mL of hydrogen gas, for example.

$$0.005 \text{ g } H_2 \times \frac{1 \text{ mol } H_2}{2.02 \text{ g } H_2} \times \frac{6.02 \times 10^{23} \text{ molecules}}{1 \text{ mol}} = 1 \times 10^{21} \text{ } H_2 \text{ molecules}$$

The last column in Table 6.5 summarizes the results obtained when the calculation is repeated for each gas. The number of significant figures in the answer changes from one calculation to the next. But the number of molecules in each sample is the same, within experimental error. We therefore conclude that equal volumes of different gases collected under the same conditions of temperature and pressure do in fact contain the same number of particles.

Table 6.5

Experimental Data for the Mass of 50-mL Samples of Different Gases (*T* and *P* Are Constant)

Compound	Mass of 50 mL Gas (g)	Molecular Weight of Gas (g/mol)	Number of Molecules in 50 mL Gas
H_2	0.005	2.02	1×10^{21}
N_2	0.055	28.01	1.2×10^{21}
O_2	0.061	32.00	1.1×10^{21}
CO_2	0.088	44.01	1.2×10^{21}
C_4H_{10}	0.111	58.12	1.15×10^{21}
CCl_2F_2	0.228	120.91	1.14×10^{21}

6.13 The Ideal Gas Equation

So far, gases have been described in terms of four variables: pressure (P), volume (V), temperature (T), and the number of moles of gas (n). In the course of this chapter, five relationships between pairs of these variables have been discussed. In each case, two of the variables were allowed to change while the other two were held constant. The discussion of the bicycle tire, for example, showed that the pressure of a gas is directly proportional to the moles of gas when the temperature and volume of the gas are held constant.

$$P \propto n \qquad (T \text{ and } V \text{ constant})$$

Other relationships between pairs of variables include the following.

Boyle's law:	$P \propto 1/V$	(T and n constant)
Amontons' law:	$P \propto T$	(V and n constant)
Charles' law:	$V \propto T$	(P and n constant)
Avogadro's hypothesis:	$V \propto n$	(P and T constant)

Each of these relationships is a special case of a more general relationship known as the **ideal gas equation.**

$$PV = nRT$$

In this equation, R is a proportionality constant known as the **ideal gas constant** and T is the absolute temperature. The value of R depends on the units used to express the four variables P, V, n, and T. By convention, most chemists use the following set of units.

P: atmospheres	T: kelvins
V: liters	n: moles

According to the **ideal gas law,** the product of pressure times volume of an ideal gas divided by the product of amount of gas times absolute temperature is a constant.

$$\frac{PV}{nT} = R$$

We can calculate the value of R from experimental data similar to those presented in Table 6.4. In these experiments the number of moles of gas and the pressure are held constant. For exactly 1 mol of gas at 1.0000 atm and 0°C (standard temperature and pressure, **STP**), the volume is found to be 22.414 L. The value of R can therefore be calculated as follows.

$$\frac{(1.0000 \text{ atm})(22.414 \text{ L})}{(1.0000 \text{ mol})(273.15 \text{ K})} = 0.082057 \text{ L-atm/mol-K}$$

For most ideal gas calculations, four significant figures are sufficient. The standard value of the ideal gas constant is therefore 0.08206 L-atm/mol-K.

6.14 Ideal Gas Calculations: Part I

The ideal gas equation can be used to predict the value of any one of the variables that describes a gas from known values of the other three.

Exercise 6.3

Many gases are available for use in the laboratory in compressed gas cylinders in which they are stored at high pressures. Calculate the mass of O_2 that could be stored at 21°C and 170 atm in a cylinder with a volume of 60.0 L.

Solution

For this problem we know three of the four variables in the ideal gas equation. We know the pressure (170 atm), volume (60.0 L), and temperature (21°C) of the gas, as shown in Figure 6.13. This suggests that the ideal gas equation will play an important role in solving the problem.

$$PV = nRT$$

Before we can use the equation, we have to convert the temperature from °C to K.

$$T_K = T_{°C} + 273 = 294 \text{ K}$$

We can then substitute the known information into the ideal gas equation:

$$(170 \text{ atm})(60.0 \text{ L}) = (n)(0.08206 \text{ L-atm/mol-K})(294 \text{ K})$$

and then solve the equation for the number of moles of gas in the container.

$$n = \frac{(170 \text{ atm})(60.0 \text{ L})}{(0.08206 \text{ L-atm/mol-K})(294 \text{ K})} = 423 \text{ mol}$$

We then use the mass of a mole of O_2 to calculate the number of grams of oxygen that could be stored in the cylinder.

$$423 \text{ mol } O_2 \times \frac{32.00 \text{ g } O_2}{1 \text{ mol } O_2} = 1.35 \times 10^4 \text{ g } O_2$$

According to this calculation, 13.5 kg of O_2 can be stored in a 60.0-L compressed gas cylinder at 21°C and 170 atm.

Fig. 6.13 Compressed gas cylinder.

The key to solving ideal gas problems often involves recognizing what is known and deciding how to use the information.

Exercise 6.4

Calculate the mass of the air in a hot-air balloon that has a volume of 4.00×10^5 L when the temperature of the gas is 30°C and the pressure is 748 mmHg. Assume that the air in the atmosphere is a mixture of roughly 78% N_2 (28.0 g/mol), 21% O_2 (32.0 g/mol), and 1% Ar (39.9 g/mol) and therefore has an average molecular weight of about 29.0 g/mol.

Solution

We know the pressure (748 mmHg), volume (4.00×10^5 L), and temperature (30°C) of the gas, as shown in Figure 6.14. We might therefore consider using the ideal gas equation to calculate the number of moles of gas in the balloon.

$$PV = nRT$$

Before we can do that, however, we have to convert the pressure to units of atmospheres,

$$748 \text{ mmHg} \times \frac{1 \text{ atm}}{760 \text{ mmHg}} = 0.984 \text{ atm}$$

and the temperature to kelvins.

$$T_K = T_{°C} + 273 = 303 \text{ K}$$

Fig. 6.14 Hot-air balloon.

We can then substitute this information into the ideal gas equation and solve for the variable we don't know.

$$n = \frac{PV}{RT} = \frac{(0.984\ \text{atm})(4.00 \times 10^5\ \text{L})}{(0.08206\ \text{L-atm/mol-K})(303\ \text{K})} = 1.58 \times 10^4\ \text{mol}$$

We then go back and reread the problem to see whether we are any closer to its solution.

The problem asks for the mass of the air in the balloon, which can be calculated from the number of moles of gas and the average molecular weight of air.

$$1.58 \times 10^4\ \text{mol} \times \frac{29.0\ \text{g}}{1\ \text{mol}} = 4.58 \times 10^5\ \text{g}$$

The balloon therefore contains just over 1000 pounds of air.

The ideal gas equation can even be applied to problems that don't seem to ask for one of the variables in this equation.

Exercise 6.5

Calculate the molecular weight of butane if 0.5813 g of the gas fills a 250.0-mL flask at a temperature of 24.4°C and a pressure of 742.6 mmHg.

Solution

We know something about the pressure (742.6 mmHg), volume (250.0 mL), and temperature (24.4°C) of the gas, as shown in Figure 6.15. We might therefore start by calculating the number of moles of gas in the sample. To do this, we need to convert the pressure, volume, and temperature into appropriate units.

$$742.6\ \text{mmHg} \times \frac{1\ \text{atm}}{760\ \text{mmHg}} = 0.9771\ \text{atm}$$

$$250.0\ \text{mL} \times \frac{1\ \text{L}}{1000\ \text{mL}} = 0.2500\ \text{L}$$

$$T_{\text{K}} = T_{°\text{C}} + 273.15 = 297.6\ \text{K}$$

Solving the ideal gas equation for the number of moles of gas and substituting the known values of the pressure, volume, and temperature into the equation give the following result.

$$n = \frac{PV}{RT} = \frac{(0.9771\ \text{atm})(0.2500\ \text{L})}{(0.08206\ \text{L-atm/mol-K})(297.6\ \text{K})} = 0.01000\ \text{mol}$$

The problem asks for the molecular weight of the gas. At this point, we know the mass of the gas in the sample and the number of moles of gas in the sample. We can therefore calculate the molecular weight of butane by dividing the mass of the sample by the number of moles of gas in the sample.

$$\frac{0.5813\ \text{g}}{0.01000\ \text{mol}} = 58.13\ \text{g/mol}$$

The molecular weight of butane is therefore 58.13 g/mol.

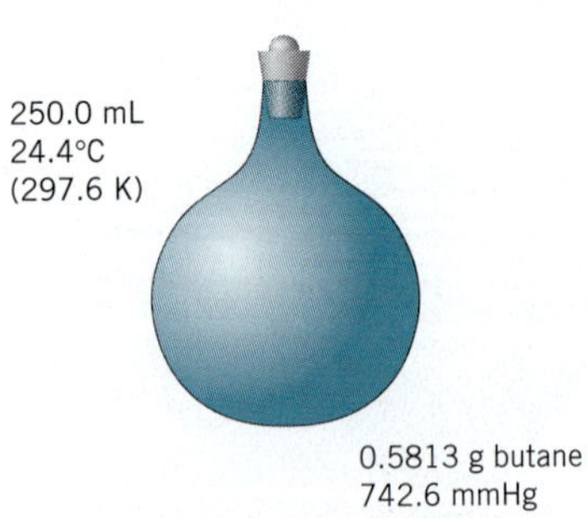

Fig. 6.15 Stoppered round-bottom flask containing butane.

The ideal gas equation can even be used to solve problems that don't seem to contain enough information.

Exercise 6.6

Calculate the density in grams per liter of O_2 gas at 0°C and 1.00 atm.

Solution

This time we have information about only two of the variables in the ideal gas equation.

$$P = 1.00 \text{ atm} \qquad T = 0°\text{C}$$

What can we calculate from this information? One way to answer that question is to rearrange the ideal gas equation, putting the knowns on one side and the unknowns on the other.

$$\frac{n}{V} = \frac{P}{RT}$$

We now have enough information to calculate the number of moles of gas per liter.

$$\frac{n}{V} = \frac{P}{RT} = \frac{(1.00 \text{ atm})}{(0.08206 \text{ L-atm/mol-K})(273 \text{ K})} = 0.0446 \text{ mol } O_2/\text{L}$$

The problem asks for the density of the gas in grams per liter. Because we know the number of moles of O_2 per liter, we can use the molar mass of O_2 to calculate the number of grams per liter.

$$\frac{0.0446 \text{ mol } O_2}{1 \text{ L}} \times \frac{32.00 \text{ g } O_2}{1 \text{ mol } O_2} = 1.43 \text{ g } O_2/\text{L}$$

The density of O_2 gas at 0°C and 1.00 atm is therefore 1.43 g/L.

Exercise 6.7

Explain why balloons filled with helium at room temperature (21°C) and 1.00 atm float, whereas balloons filled with CO_2 sink. The average molecular weight of dry air is 29.0 g/mol.

Solution

Avogadro's hypothesis suggests that equal volumes of different gases at the same temperature and pressure contain the same number of gas particles. But this hypothesis is nothing more than a special case of the ideal gas law, which predicts that the number of moles of gas per liter at room temperature (21°C) and 1 atm does not depend on the identity of the gas.

$$\frac{n}{V} = \frac{P}{RT} = \frac{(1.00 \text{ atm})}{(0.08206 \text{ L-atm/mol-K})(294 \text{ K})} = 0.0414 \text{ mol/L}$$

The number of grams per liter, however, depends on the mass of a mole of each gas. Helium is less dense than air, whereas CO_2 is more dense than air.

$$d_{He} = \frac{0.0414 \text{ mol}}{1 \text{ L}} \times \frac{4.00 \text{ g He}}{1 \text{ mol}} = 0.166 \text{ g He/L}$$

$$d_{air} = \frac{0.0414 \text{ mol}}{1 \text{ L}} \times \frac{29.0 \text{ g air}}{1 \text{ mol}} = 1.20 \text{ g air/L}$$

$$d_{CO_2} = \frac{0.0414 \text{ mol}}{1 \text{ L}} \times \frac{44.0 \text{ g } CO_2}{1 \text{ mol}} = 1.82 \text{ g } CO_2/\text{L}$$

➤ CHECKPOINT

Use the difference between the average molecular weights of air (29.0 g/mol) and water (18.0 g/mol) to explain why 1 L of dry air weighs more than 1 L of air that has been saturated with water vapor.

A balloon filled with helium therefore weighs significantly less than the air it displaces, and it floats. A balloon filled with CO_2, on the other hand, weighs more than the air it displaces, so it sinks.

6.15 Ideal Gas Calculations: Part II

Gas law problems often ask you to predict what happens when one or more changes are made in the variables that describe the gas. The most powerful approach to such problems is based on the fact that the ideal gas constant is, in fact, a constant.

We start by solving the ideal gas equation for the ideal gas constant.

$$R = \frac{PV}{nT}$$

Thus, the ratio PV/nT under one set of conditions must be equal to the ratio under another set of conditions.

$$\frac{P_1V_1}{n_1T_1} = \frac{P_2V_2}{n_2T_2}$$

We then substitute the known values of pressure, temperature, volume, and amount of gas into the equation and solve for the appropriate unknown.

Exercise 6.8

A sample of NH_3 gas fills a 27.0-L container at -15°C and 2.58 atm. Calculate the volume of the gas at 21°C and 751 mmHg.

Solution

We can start by listing what we know and what we don't know about the system.

Initial Conditions	*Final Conditions*
$P_1 = 2.58$ atm	$P_2 = 751$ mmHg
$V_1 = 27.0$ L	$V_2 = ?$
$T_1 = -15$°C	$T_2 = 21$°C
$n_1 = ?$	$n_2 = ?$

We convert the data to a consistent set of units for use in the ideal gas equation.

Initial Conditions	*Final Conditions*
$P_1 = 2.58$ atm	$P_2 = 0.988$ atm
$V_1 = 27.0$ L	$V_2 = ?$
$T_1 = 258$ K	$T_2 = 294$ K
$n_1 = ?$	$n_2 = ?$

We can then substitute the information into the following equation.

$$\frac{P_1V_1}{n_1T_1} = \frac{P_2V_2}{n_2T_2}$$

When this is done we get one equation with three unknowns: n_1, n_2, and V_2.

$$\frac{(2.58 \text{ atm})(27.0 \text{ L})}{(n_1)(258 \text{ K})} = \frac{(0.988 \text{ atm})(V_2)}{(n_2)(294 \text{ K})}$$

It is impossible to solve one equation with three unknowns. But a careful reading of the problem suggests that the number of moles of NH_3 is the same before and after the temperature and pressure change.

$$n_1 = n_2$$

Substituting this equality into the unknown equation gives the following.

$$\frac{(2.58 \text{ atm})(27.0 \text{ L})}{(n_1)(258 \text{ K})} = \frac{(0.988 \text{ atm})(V_2)}{(n_1)(294 \text{ K})}$$

Multiplying both sides of the equation by n_1 gives one equation with one unknown.

$$\frac{(2.58 \text{ atm})(27.0 \text{ L})}{(258 \text{ K})} = \frac{(0.988 \text{ atm})(V_2)}{(294 \text{ K})}$$

We can therefore solve for the only unknown left in the equation.

$$V_2 = \frac{(2.58 \text{ atm})(27.0 \text{ L})(294 \text{ K})}{(258 \text{ K})(0.988 \text{ atm})} = 80.3 \text{ L}$$

According to the calculation, the volume of the gas increases from 27.0 L at $-15°C$ and 2.58 atm to 80.3 L at 21°C and 751 mmHg.

6.16 Dalton's Law of Partial Pressures

The *CRC Handbook of Chemistry and Physics* describes the atmosphere as 78.084% N_2, 20.946% O_2, 0.934% Ar, and 0.033% CO_2 by volume, after the water vapor has been removed. What image does this description evoke in your mind? Do you believe that only 20.946% of the room in which you are sitting contains O_2? Or do you believe that the atmosphere in your room is a more or less homogeneous mixture of the gases?

In Section 6.5 we argued that gases expand to fill their containers. The volume of O_2 in your room is therefore the same as the volume of N_2. Both gases expand to fill the room.

What about the pressure of the different gases in your room? Is the pressure of the O_2 in the atmosphere the same as the pressure of the N_2? We can answer the question by rearranging the ideal gas equation as follows.

$$P = n \times \frac{RT}{V}$$

According to this equation, the pressure of a gas is proportional to the number of moles of gas, if the temperature and volume are held constant. Because the temperature and volume of the O_2 and N_2 in the atmosphere are the same, the pressure of each gas is proportional to the number of the moles of the gas. Because there is more N_2 in the atmosphere than O_2, the contribution to the total pressure of the atmosphere from N_2 is larger than the contribution from O_2.

John Dalton was the first to recognize that the total pressure of a mixture of gases is the sum of the contributions of the individual components of the mixture. The part of the total pressure of a mixture that results from one component is called the **partial pressure** of that component. **Dalton's law of partial pressures** states that the total pressure (P_T) of a mixture of gases is the sum of the partial pressures of the various components.

$$P_T = P_1 + P_2 + P_3 + \cdots$$

Exercise 6.9

Calculate the total pressure of a mixture that contains 1.00 g of H_2 and 1.00 g of He in a 5.00-L container at 21°C.

Solution

We start, as always, by listing what we know about the problem.

$$V = 5.00 \text{ L} \qquad T = 21°\text{C}$$

We also know the masses of H_2 and He in the flask, however, so we can calculate the number of moles of each gas in the container.

$$1.00 \text{ g } H_2 \times \frac{1 \text{ mol } H_2}{2.016 \text{ g } H_2} = 0.496 \text{ mol } H_2$$

$$1.00 \text{ g He} \times \frac{1 \text{ mol He}}{4.003 \text{ g He}} = 0.250 \text{ mol He}$$

Because we know the volume, the temperature, and the number of moles of each component of the mixture, we can calculate the partial pressures of H_2 and He. We start by rearranging the ideal gas equation as follows.

$$P = \frac{nRT}{V}$$

We then use that equation to calculate the partial pressure of each gas.

$$P_{H_2} = \frac{(0.496 \text{ mol } H_2)(0.08206 \text{ L-atm/mol-K})(294 \text{ K})}{5.00 \text{ L}} = 2.39 \text{ atm}$$

$$P_{He} = \frac{(0.250 \text{ mol He})(0.08206 \text{ L-atm/mol-K})(294 \text{ K})}{5.00 \text{ L}} = 1.21 \text{ atm}$$

The total pressure in the mixture is the sum of the partial pressures of the two components.

$$P_T = P_{hydrogen} + P_{helium} = 3.60 \text{ atm}$$

Dalton derived the law of partial pressures from his work on the amount of water vapor that could be absorbed by air at different temperatures. It is therefore fitting that this law is used most often to correct for the amount of water vapor picked up when a gas is collected by displacing water. Suppose, for example, that we want to collect a sample of O_2 prepared by heating potassium chlorate until it decomposes.

$$2\ KClO_3(s) \longrightarrow 2\ KCl(s) + 3\ O_2(g)$$

The gas given off in this reaction can be collected by filling a flask with water, inverting the flask in a trough, and then letting the gas bubble into the flask.

Because some of the water in the flask will evaporate during the experiment, the gas that collects in the flask is going to be a mixture of O_2 and water vapor. The total pressure of the gas is the sum of the partial pressures of the two components.

$$P_T = P_{oxygen} + P_{water}$$

The total pressure of the mixture must be equal to atmospheric pressure. (If it were any larger, the gas would push water out of the container. If it were any smaller, water would be forced into the container.) If we had some way to estimate the partial pressure of the water in the system, we could therefore calculate the partial pressure of the oxygen gas.

The partial pressure of the gas that collects in a closed container above a liquid is known as the **vapor pressure** of the liquid. Table B.3 in the appendix gives the vapor pressure of water at various temperatures. If we know the temperature at which a gas is collected by displacing water, and we assume that the gas is saturated with water vapor at that temperature, we can calculate the partial pressure of the gas by subtracting the vapor pressure of water from the total pressure of the mixture of gases collected in the experiment.

Exercise 6.10

Calculate the number of grams of O_2 that can be collected by displacing water from a 250-mL bottle at 21°C and 746.2 mmHg.

Solution

It is often useful to work a problem backward. In this case, our goal is the mass of O_2 in the flask. To reach this goal we need to calculate the number of moles of O_2 in the sample. To do this, we need to know the pressure, volume, and temperature of the O_2. We already know some of the information, as shown in Figure 6.16.

We know the volume of the O_2 because a gas expands to fill its container. We also know the temperature of the O_2. But we don't know the pressure of the O_2.

$$V_{oxygen} = 250 \text{ mL} \qquad T_{oxygen} = 21°\text{C} \qquad P_{oxygen} = ?$$

All we know is the *total pressure* of the oxygen plus the water vapor that collects in the bottle.

$$P_T = P_{oxygen} + P_{water} = 746.2 \text{ mmHg}$$

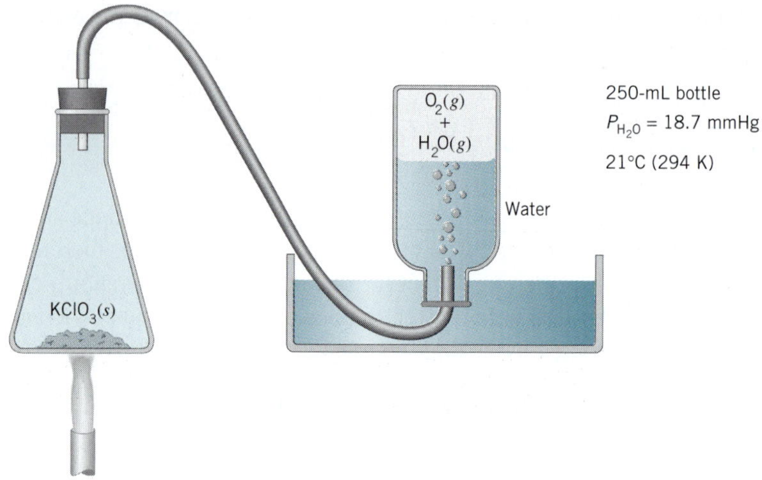

Fig. 6.16 Gases, such as O_2, that aren't very soluble in water can be collected by displacing water from a container.

Table B.3 in the appendix gives the vapor pressure of water at 21°C as 18.7 mmHg. If we assume that the gas collected in the experiment is saturated with water vapor, we can write the following.

$$P_T = P_{oxygen} + P_{water}$$
$$746.2 \text{ mmHg} = P_{oxygen} + 18.7 \text{ mmHg}$$

In other words, the partial pressure of O_2 must be less than the total pressure in the bottle by 18.7 mmHg.

$$P_{oxygen} = 746.2 \text{ mmHg} - 18.7 \text{ mmHg} = 727.5 \text{ mmHg}$$

We now know the pressure (727.5 mmHg), volume (250 mL), and temperature (21°C) of the O_2 collected in the experiment. Once we convert the measurements to appropriate units, we can use the ideal gas equation to calculate the number of moles of O_2 collected.

$$n = \frac{PV}{RT} = \frac{(0.9572 \text{ atm})(0.250 \text{ L})}{(0.08206 \text{ L-atm/mol-K})(294 \text{ K})} = 0.00992 \text{ mol } O_2$$

We now have enough information to calculate the number of grams of O_2 collected.

$$0.00992 \text{ mol } O_2 \times \frac{32.00 \text{ g } O_2}{1 \text{ mol}} = 0.317 \text{ g}$$

6.17 The Kinetic Molecular Theory

The experimental observations about the behavior of gases discussed so far can be explained with a theoretical model known as the *kinetic molecular theory*. This theory is based on the following six postulates, or assumptions.

1. Gases are composed of a large number of particles that behave like hard, spherical objects in a state of constant, random motion.
2. These particles move in a straight line until they collide with another particle or the walls of the container.

3. The particles are much smaller than the distance between particles. Most of the volume of a gas is therefore empty space.
4. There is no force of attraction between gas particles or between the particles and the walls of the container.
5. Collisions between gas particles or collisions with the walls of the container are perfectly elastic. Energy can be transferred from one particle to another during a collision, but the total kinetic energy of the particles after the collision is the same as it was before the collision.
6. The average kinetic energy of a collection of gas particles depends on the temperature of the gas and nothing else.

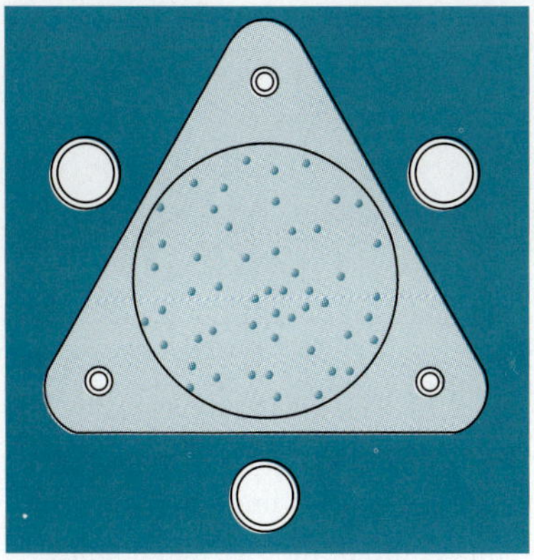

Fig. 6.17 The six postulates of the kinetic molecular theory can be demonstrated with a molecular dynamics simulator.

The assumptions of the kinetic molecular theory can be illustrated with the apparatus shown in Figure 6.17, which consists of a glass plate surrounded by walls mounted on a vibrating motor. A handful of steel ball bearings is placed on top of the glass plate to represent the gas particles.

When the motor is turned on, the glass plate vibrates, which makes the ball bearings move in a constant, random fashion (postulate 1). Each ball moves in a straight line until it collides with another ball or with the walls of the container (postulate 2). Although collisions are frequent, the average distance between the ball bearings is much larger than the diameter of the balls (postulate 3). There is no force of attraction between the individual ball bearings or between the ball bearings and the walls of the container (postulate 4).

The collisions that occur in the apparatus are very different from those that occur when a rubber ball is dropped on the floor. Collisions between the rubber ball and the floor are *inelastic,* as shown in Figure 6.18. A portion of the energy of the ball is lost each time it hits the floor, until it eventually rolls to a stop. In our apparatus, however, the collisions are perfectly *elastic*. The balls have just as much kinetic energy—on the average—after a collision as before (postulate 5).

Any object in motion has a *kinetic energy* that is defined as one-half of the product of its mass times its velocity squared.

$$KE = \frac{1}{2}mv^2$$

At any time, some of the ball bearings on the apparatus are moving faster than others, but the system can be described by an *average kinetic energy*. When we increase the "temperature" of the system by increasing the voltage to the motors, we find that the average kinetic energy of the ball bearings increases (postulate 6).

➤ CHECKPOINT

The kinetic molecular theory assumes that the temperature of a gas on the macroscopic scale is directly proportional to the average kinetic energy of the particles on the atomic scale. Use this assumption to explain what happens to the particles of a gas when the temperature of a gas increases.

We have described gaseous systems in terms of an "average" kinetic energy. The use of the term "average" implies that not all particles are moving with the same kinetic energy. Some particles must be moving with less than the average kinetic energy and some with more kinetic energy than the average. In other words, there is a distribution of kinetic energies among the particles. Collisions of rapidly moving particles with slower ones result in a transfer of energy from the rapid particle to the slow particle. About 10^{27} collisions per second take place in 1 mL of gas at 1 atm pressure and room temperature. This assures that there is continuous exchange of both kinetic energy and velocities between the particles of a gas. At any given instant a few particles are moving very slowly, a few are moving very rapidly, and the bulk of the particles have velocities and hence kinetic energies somewhere between the two extremes.

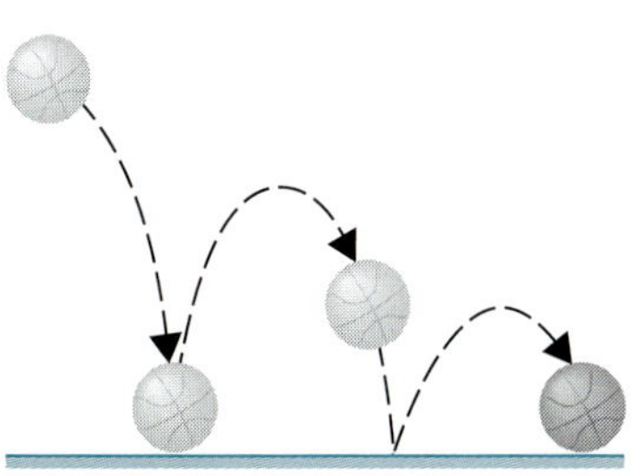

Fig. 6.18 Most collisions are inelastic. Some energy is transferred each time a ball collides with the floor, for example.

Figure 6.19 shows a plot of kinetic energy along the x axis versus the relative number of particles which possess a particular kinetic energy along the y axis. When temperature increases, the number of particles having a greater kinetic energy also increases, as indicated in Figure 6.19 by the shift in the distribution curve to the right.

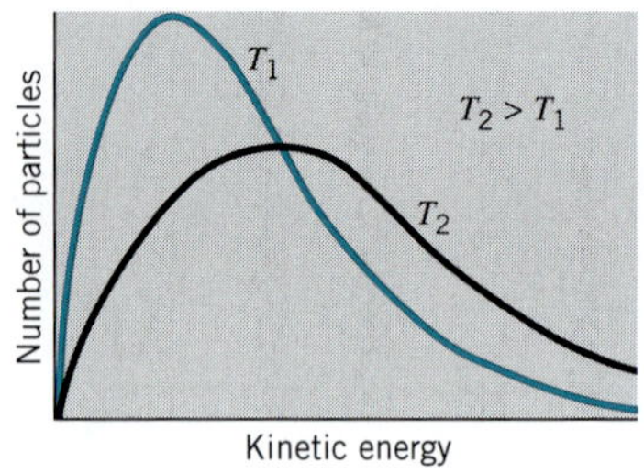

Fig. 6.19 The kinetic molecular theory states that the *average* kinetic energy of a gas is proportional to the temperature of the gas, and nothing else. At any given temperature, however, some of the gas particles are moving faster than others.

6.18 How the Kinetic Molecular Theory Explains the Gas Laws

The kinetic molecular theory can be used to explain each of the experimentally determined gas laws. The pressure of a gas results from collisions between the gas particles and the walls of the container. Each time a gas particle hits the wall, it exerts a force on the wall. The frequency of collisions with the walls and the mass and velocity of the gas particles determine the magnitude of the force. Because pressure is force per unit area, a change in force or surface area results in a change in pressure.

THE LINK BETWEEN P AND n (T AND V CONSTANT)

If the temperature is held constant, the average kinetic energy of the gas particles remains the same. If volume is held constant, the surface area also remains the same. Any increase in the number of gas particles in the container increases the frequency of collisions with the walls and therefore the pressure of the gas.

AMONTONS' LAW ($P \propto T$; n AND V CONSTANT)

The last postulate of the kinetic molecular theory states that the average kinetic energy of a gas particle depends only on the temperature of the gas. Thus, the average kinetic energy of the gas particles increases as the gas becomes warmer. Because the mass of the particles is constant, their kinetic energy can increase only if the average velocity of the particles increases. The faster the particles are moving when they hit the wall and the greater the frequency of collisions with the wall the greater the force they exert on the wall. Because the volume is held constant, the surface area is constant. The impact per collision and the number of collisions become larger as the temperature increases, the pressure of the gas must increase as well.

BOYLE'S LAW ($P \propto 1/V$; T AND n CONSTANT)

Gases can be compressed because most of the volume of a gas is empty space. If we compress a gas without changing its temperature, the average kinetic energy of the gas particles stays the same. There is no change in the speed with which the particles move, but the volume of the container is smaller. Thus, the particles travel from one end of the container to the other in a shorter time. This means that they hit the walls more often. Any increase in the frequency of collisions with the walls must lead to an increase in the pressure of the gas. Thus, the pressure of a gas becomes larger as the volume of the gas becomes smaller.

CHARLE'S LAW ($V \propto T$; P AND n CONSTANT)

The average kinetic energy of the particles in a gas is proportional to the temperature of the gas. Because the mass of the particles is constant, the particles must move faster as the gas becomes warmer. If they move faster, the particles will have a greater impact on the container each time they hit the walls and they will strike the walls more frequently. These two factors lead to an increase in the pressure of the gas. If the walls of the container are flexible, the container will expand until the pressure of the gas once again balances the pressure of the atmosphere. The volume of the gas in a flexible container therefore increases as the temperature of the gas increases.

AVOGADRO'S HYPOTHESIS ($V \propto n$; T AND P CONSTANT)

As the number of gas particles increases, the frequency of collisions with the walls of the container must increase. This, in turn, leads to an increase in the pressure of the gas. Flexible containers, such as a balloon, will expand until the pressure of the gas inside the balloon once again balances the pressure of the gas outside. Thus, the volume of the gas is proportional to the number of gas particles.

DALTON'S LAW OF PARTIAL PRESSURES ($P_T = P_1 + P_2 + P_3 + \cdots$; V AND T CONSTANT)

Imagine what would happen if six ball bearings of a different size were added to the ball bearings all ready in the apparatus in Figure 6.17. The total pressure would increase because there would be more collisions with the walls of the container. But the pressure resulting from the collisions between the original ball bearings and the walls of the container would remain the same. There is so much empty space in the container that each type of ball bearing hits the walls of the container as often in the mixture as it would if there was only one kind of ball bearing on the glass plate. The total number of collisions with the wall in the mixture is therefore equal to the sum of the collisions that would occur when each size of ball bearing is present by itself. In other words, the total pressure of a mixture of gases is equal to the sum of the partial pressures of the individual gases.

RELATIVE VELOCITIES OF MOLECULES

The last postulate of the kinetic theory states that the temperature of a system is proportional to the average kinetic energy of its particles and nothing else. In other words, the temperature of a system increases if and only if there is an increase in the average kinetic energy of its particles.

Two gases at the same temperature, such as H_2 and O_2, must have the same average kinetic energy. This can be represented by the following equation.

Graham's laws are found in the Special Topics section at the end of the chapter. They are not included in the core because these calculations can be applied only in very special cases. The idea of relative speeds is, however, important and so we have included this in the core.

$$\frac{1}{2} m_{H_2} v_{H_2}^2 = \frac{1}{2} m_{O_2} v_{O_2}^2$$

This equation can be simplified by multiplying both sides by two.

$$m_{\mathrm{H}_2} v_{\mathrm{H}_2}^2 = m_{\mathrm{O}_2} v_{\mathrm{O}_2}^2$$

It can then be rearranged to give the following.

$$\frac{v_{\mathrm{H}_2}^2}{v_{\mathrm{O}_2}^2} = \frac{m_{\mathrm{O}_2}}{m_{\mathrm{H}_2}}$$

Taking the square root of both sides of the equation gives a relationship between the ratio of the velocities at which the two gases move and the square root of the ratio of their masses.

$$\frac{v_{\mathrm{H}_2}}{v_{\mathrm{O}_2}} = \sqrt{\frac{m_{\mathrm{O}_2}}{m_{\mathrm{H}_2}}}$$

Because mass is proportional to molecular weight, this relationship may also be written in terms of molecular weight.

$$\frac{v_{\mathrm{H}_2}}{v_{\mathrm{O}_2}} = \sqrt{\frac{MW_{\mathrm{O}_2}}{MW_{\mathrm{H}_2}}}$$

Exercise 6.11

Calculate the average velocity of an H_2 molecule at 0°C if the average velocity of an O_2 molecule at that temperature is 425 m/s.

Solution

The relative velocities of the H_2 and O_2 molecules at a given temperature are described by the following equation.

$$\frac{v_{H_2}}{v_{O_2}} = \sqrt{\frac{MW_{O_2}}{MW_{H_2}}}$$

Substituting the molecular weights of H_2 and O_2 and the average velocity of an O_2 molecule into the equation gives the following result.

$$\frac{v_{H_2}}{425 \text{ m/s}} = \sqrt{\frac{32.0 \text{ g/mol}}{2.02 \text{ g/mol}}}$$

Solving the equation for the average velocity of an H_2 molecule gives a value of about 1690 m/s, or about 3780 mi/h.

Because it is easy to make mistakes when setting up a ratio problem such as this, it is important to check the answer to see whether it makes sense. This relationship suggests that light molecules move faster on the average than heavy molecules. In this case, the answer makes sense because H_2 molecules are much lighter than O_2 molecules and they should therefore travel much faster.

➤ CHECKPOINT

According to kinetic molecular theory, if two different gases are at the same temperature, they have the same kinetic energy. How can this be true if the particles composing the two gases are moving at different velocities?

Key Terms

Absolute zero
Amontons' law
Area
Avogadro's hypothesis
Boyle's law
Charles' law
Compressibility
Dalton's law of partial pressures
Expandability
Force
Gay-Lussac's law
Ideal gas constant
Ideal gas equation
Ideal gas law
Kinetic energy
Kinetic molecular theory
Law of combining volumes
Partial pressure
Pressure
States
STP
Temperature
Vapor pressure

Problems

Temperature

1. Why do some objects known to be at the same temperature feel warmer or colder to the touch?
2. Could you use touch to arrange objects in order of increasing warmth or coldness? Explain why or why not.
3. Define *temperature*.

Temperature as a Property of Matter

4. What are the two most common temperature scales in everyday use? What temperature scale do scientists usually use?
5. What is the freezing point of water on three different temperature scales?
6. Describe what happens to the motion of water molecules as water at 25°C is heated to 373 K.
7. What is meant by kinetic energy?
8. If the average kinetic energies of the atoms composing two different substances are the same, what can be said about the temperatures of the two substances?

The States of Matter

9. What are the three states of matter?
10. Why are gases easier to study than other states of matter?

11. Liquid water and gaseous water coexist at 100°C and 1 atm pressure. How do the average kinetic energies of the molecules of water in these states compare under these conditions?

Elements or Compounds That Are Gases at Room Temperature

12. Which of the following elements and compounds are most likely to be gases at room temperature?
(a) Ar (b) CO (c) CH_4 (d) $C_{10}H_{22}$
(e) Cl_2 (f) Fe_2O_3 (g) Na (h) NaCl
(i) Pt (j) S_8
13. List all the elements from Table 6.1 that are gases at room temperature.
14. Is it true that all elements that have a molecular weight below 40 g/mol are gases at room temperature? Support your answer.
15. In general, common gases have two characteristics. What are they?

The Properties of Gases

16. What are three characteristic properties of gases?
17. Why is the volume of a gas the same as the volume of its container?
18. How does the volume of a mole of liquid water compare to that of a mole of gaseous water both at 25°C and 1 atm?
19. A helium atom is smaller than a xenon atom. Yet the volume of a mole of helium and xenon is the same at 25°C and 1 atm. Explain.
20. Predict what will happen to the weight of an evacuated cylinder when it is filled with helium gas. Will it increase, decrease, or remain the same?

Pressure Versus Force

21. How is pressure related to force?
22. In a hurricane-strength wind, which has the greater pressure on it, a billboard or a stop sign? Which has the greater force on it?
23. If two marbles are pushed together, a great pressure can be produced with little applied force. Explain.

Atmospheric Pressure

24. What is the pressure in units of atmospheres when a barometer reads 745.8 mmHg? What is the pressure in units of pascals?
25. One atmosphere pressure will support a column of mercury 760 mm tall in a barometer with a tube 1.00 cm in diameter. What would be the height of the column of mercury if the diameter of the tube were twice as large?
26. Atmospheric pressure is announced during weather reports in the United States in units of inches of mercury. How many inches of mercury would exert a pressure of 1.00 atm? In Canada, atmospheric pressure is reported in units of pascals. What is 1.00 atm in pascals?
27. If the directions that come with your car tell you to inflate the tires to 200 kPa pressure and you have a tire pressure gauge calibrated in pounds per square inch (psi), what pressure in psi will you use?
28. The vapor pressure of the mercury gas that collects at the top of a barometer is 2×10^{-3} mmHg. Calculate the vapor pressure of the gas in atmospheres.
29. Calculate the force exerted on the earth by the atmosphere if atmospheric pressure is 14.7 lb/in^2 and the surface area of the planet is 5.1×10^8 km^2.

Boyle's Law

30. A 425-mL sample of O_2 gas was collected at 742.3 mmHg. What would be the pressure in mmHg if the gas were allowed to expand to 975 mL at constant temperature?
31. What would happen to the volume of a balloon filled with 0.357 L of H_2 gas collected at 741.3 mmHg if the atmospheric pressure increased to 758.1 mmHg? T is constant.
32. What is the volume of a scuba tank if it takes 2000 L of air collected at 1 atm to fill the tank to a pressure of 150 atm? T is constant.
33. Calculate the volume of a balloon that could be filled at 1.00 atm with the helium in a 2.50-L compressed gas cylinder in which the pressure is 200 atm at 25°C.

Amontons' Law

34. A can is filled with 5.00 atm of a gas at 21°C. Calculate the pressure in the can when it is stored in a warehouse on a hot summer day when the temperature reaches 38°C.
35. An automobile tire was inflated to a pressure of 32 lb/in^2 at 21°C. At what temperature would the pressure reach 60 psi?
36. At 25°C, four-fifths of the pressure of the atmosphere is due to N_2 and one-fifth is due to O_2. At 100°C, what fraction of the pressure is due to N_2?

Charles' Law

37. Calculate the percent change in the volume of a toy balloon when the gas inside is heated from 22°C to 75°C in a hot-water bath.
38. A sample of O_2 gas with a volume of 0.357 liter was collected at 21°C. Calculate the volume of the gas when it is cooled to 0°C if the pressure remains constant.
39. Two balloons at room temperature are filled to the same pressure, one with 1 mol of He gas and one with 1 mol of Xe gas. If they are both cooled to the freezing point of water, which balloon will have the greater change in volume?

Gay-Lussac's Law

40. Calculate the ratio of the volumes of sulfur dioxide and oxygen produced when sulfuric acid decomposes. Temperature and pressure are constant.

$$2\ H_2SO_4(aq) \longrightarrow 2\ SO_2(g) + O_2(g) + 2\ H_2O(l)$$

41. Calculate the volume of H_2 and N_2 gas formed when 1.38 L of NH_3 decomposes at a constant temperature and pressure.

$$2\ NH_3(g) \longrightarrow N_2(g) + 3\ H_2(g)$$

42. Ammonia burns in the presence of oxygen to form nitrogen oxide and water.

$$4\ NH_3(g) + 5\ O_2(g) \longrightarrow 4\ NO(g) + 6\ H_2O(g)$$

What volume of NO can be prepared when 15.0 L of ammonia reacts with excess oxygen, if all measurements are made at the same temperature and pressure?

43. Acetylene burns in oxygen to form CO_2 and H_2O.

$$2\ C_2H_2(g) + 5\ O_2(g) \longrightarrow 4\ CO_2(g) + 2\ H_2O(g)$$

Calculate the total volume of the products formed when 15.0 L of C_2H_2 burns in the presence of 15.0 L of O_2, if all measurements are made at the same temperature and pressure.

44. Methane reacts with steam to form hydrogen and carbon monoxide.

$$CH_4(g) + H_2O(g) \longrightarrow CO(g) + 3\ H_2(g)$$

It can also react with steam to form carbon dioxide.

$$CH_4(g) + 2\ H_2O(g) \longrightarrow CO_2(g) + 4\ H_2(g)$$

What are the products of a reaction if 1.50 L of methane is found by experiment to react with 1.50 L of water vapor? T and P are constant.

Avogadro's Hypothesis

45. Which weighs more, dry air at 25°C and 1 atm, or air at that temperature and pressure that is saturated with water vapor? (Assume that the average molecular weight of air is 29.0 g/mol.)

46. Which of the following samples would have the largest volume at 25°C and 750 mmHg?
(a) 100 g CO_2 (b) 100 g CH_4
(c) 100 g NO (d) 100 g SO_2

47. Nitrous oxide decomposes to form nitrogen and oxygen. Use Avogadro's hypothesis to determine the formula for nitrous oxide if 2.36 L of the compound decomposes to form 2.36 L of N_2 and 1.18 L of O_2. T and P are constant.

48. Two equal-volume containers at the same temperature are filled with different gases to the same pressure. One contains N_2 and one contains H_2. Which has the greater number of molecules?

The Ideal Gas Equation

49. Predict the shape of the following graphs for an ideal gas (assume all other variables are held constant).
(a) pressure versus volume
(b) pressure versus temperature
(c) volume versus temperature
(d) kinetic energy versus temperature
(e) pressure versus the number of moles of gas

50. Which of the following graphs could not give a straight line for an ideal gas?
(a) V versus T (b) T versus P (c) P versus $1/V$
(d) n versus $1/T$ (e) n versus $1/P$

51. Which of the following statements is always true for an ideal gas?
(a) If the temperature and volume of a gas both increase at constant pressure, the moles of gas must also increase.
(b) If the pressure increases and the temperature decreases for a constant number of moles of gas, the volume must decrease.
(c) If the volume and the moles of gas both decrease at constant temperature, the pressure must decrease.

52. Calculate the value of the ideal gas constant in units of mL-psi/mol-K if 1.00 mol of an ideal gas at 0°C occupies a volume of 22,400 mL at 14.7 lb/in^2 pressure.

Ideal Gas Calculations: Part I

53. Nitrogen gas sells for roughly 50 cents per 100 cubic feet at 0°C and 1 atm. What is the price per gram of nitrogen?

54. Calculate the pressure in atmospheres of 80 g of CO_2 in a 30-L container at 23°C.

55. Calculate the temperature at which 1.5 g of O_2 has a pressure of 740 mmHg in a 1.0-L container.

56. Calculate the number of kilograms of O_2 gas that can be stored in a compressed gas cylinder with a volume of 40 L when the cylinder is filled at 150 atm and 21°C.

57. Calculate the pressure in an evacuated 250-mL container at 0°C when the O_2 in 1.00 cm^3 of liquid oxygen evaporates. Liquid oxygen has a density of 1.118 g/cm^3.

58. A 1.00-L flask was evacuated, 5.00 g of liquid NH_3 was added to the flask, and the flask was sealed with a cork. If it takes 7.10 atmosphere of pressure in the flask to blow out the cork, at what temperature will the cork be blown out?

59. Calculate the density of CH_2Cl_2 in the gas phase at 40°C and 1.00 atm. Compare this with the density of liquid CH_2Cl_2 (1.336 g/cm^3).

60. Calculate the density of methane gas, CH_4, in kilograms per cubic meter at 25°C and 956 mmHg.

61. Calculate the density of helium at 0°C and 1.00 atm and compare this with the density of air (1.29 g/L) at 0°C and 1.00 atm. Explain why 1.00 ft^3 of helium can lift a weight of 0.076 lb under these conditions.

62. Calculate the ratio of the densities of H_2 and O_2 at 0°C and 100°C at 1.00 atm.

63. Which of the noble gases in Group VIIIA of the periodic table has a density of 3.7493 g/L at 0°C and 1.00 atm?

64. Calculate the average molecular weight of air assuming a sample of air weighs 1.700 times as much as an equivalent volume of ammonia, NH_3.

Ideal Gas Calculations: Part II

65. What is the volume of the gas in a balloon at −195°C if the balloon has been filled to a volume of 5.0 L at 25°C?

66. CO_2 gas with a volume of 25.0 L was collected at 25°C and 0.982 atm. Calculate the pressure of the gas if it is compressed to a volume of 0.150 L and heated to 350°C.

67. Calculate the pressure of 4.80 g of ozone, O_3, in a 2.45-L flask at 25°C. Assume that the ozone completely decomposes to molecular oxygen.

$$2\ O_3(g) \longrightarrow 3\ O_2(g)$$

Calculate the pressure inside the flask once the reaction is complete.

68. One mole of an ideal gas occupies a volume of 22.4 L at 0°C and 1.00 atm. What would be the volume of a mole of an ideal gas at 22°C and 748.8 mmHg?

69. 10.0 L of O_2 gas was collected at 120°C and 749.3 mmHg. Calculate the volume of the gas when it is cooled to 0°C and stored in a container at 1.00 atm.

70. 5.0 L of CO_2 gas was collected at 25°C and 2.5 atm. At what temperature would the gas have to be stored to fill a 10.0-L flask at 0.978 atm?

71. Two 10-L samples of O_2 collected at 120°C and 749.3 mmHg are combined and stored in a 1.25-L flask at 27°C. Calculate the pressure of the gas.

Dalton's Law of Partial Pressures ($P_T = P_1 + P_2 + P_3 + \cdots$)

72. Calculate the partial pressure of propane in a mixture that contains equal weights of propane (C_3H_8) and butane (C_4H_{10}) at 20°C and 746 mmHg.

73. Calculate the partial pressure of helium in a 1.00-L flask that contains equal numbers of moles of N_2, O_2, and He at a total pressure of 7.5 atm.

74. Calculate the total pressure in a 10.0-L flask at 27°C of a sample of gas that contains 6.0 g of H_2, 15.2 g of N_2, and 16.8 g of He.

75. A 1-L flask is filled with carbon monoxide at 27°C until the pressure is 0.200 atm. Calculate the total pressure after 0.450 g of carbon dioxide have been added to the flask.

76. Water was added to a 1-L flask at 25°C, and the pressure of the water that evaporated was found to be 23.8 mmHg at that temperature. What would be the pressure of the water if the experiment were repeated in a 0.5-L flask?

77. Calculate the volume of the hydrogen obtained when the water vapor is removed from 289 mL of H_2 gas collected by displacing water from a flask at 15°C and 0.988 atm.

The Kinetic Molecular Theory

78. What would happen to a balloon if the collisions between gas molecules were not perfectly elastic?

79. What would happen to a balloon if the gas molecules were in a state of constant motion, but the motion was not random?

80. Explain why the pressure of a gas is evidence for the assumption that gas particles are in a state of constant random motion.

81. Use the kinetic molecular theory to explain why the pressure of a gas is proportional to the number of gas particles and the temperature of the gas but inversely proportional to the volume of the gas.

82. Which of the following are true according to the kinetic molecular theory of gases? Explain your answer in each case.
   (a) All of the molecules of a gas are moving with the same kinetic energy at a given temperature.
   (b) The kinetic energy of a gas can be increased by increasing the pressure while holding the temperature constant.
   (c) The average kinetic energy of the molecules increases as temperature increases.
   (d) Most of the volume of a gas is empty space.

How the Kinetic Molecular Theory Explains the Gas Laws

83. Two identical flasks are labeled A and B. Flask A contains $NH_3(g)$ at 50°C and flask B contains $O_2(g)$ at 50°C. The average kinetic energy of O_2 is 7×10^{-21} J/molecule.
   (a) What is the average kinetic energy of an NH_3 molecule? Explain.
   (b) Which molecule, NH_3 or O_2, is moving the most rapidly? Explain.
   (c) With the information provided, can you tell in which flask the pressure is the greatest? Explain.

(d) How could the average kinetic energy of the O_2 molecules be doubled?

(e) If the only change made were to double the volume of the flask containing O_2 molecules, explain what would happen to the
  (i) average kinetic energy
  (ii) pressure

(f) Give three ways the pressure in the flask containing NH_3 could be doubled.

84.

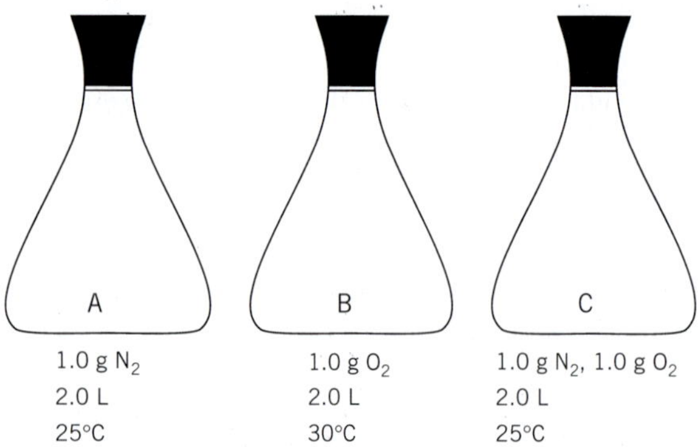

(a) $N_2(g)$ and $O_2(g)$ are placed into flasks A and B as indicated above. In which flask is the pressure the greatest? Show all calculations.

(b) In which flask, A or B, is the average kinetic energy the greatest? Explain.

(c) If 1.0 g of N_2 and 1.0 g of O_2 are placed into the 2.0-L flask C at 25°C, what will be the total pressure? Explain.

(d) Which molecules are moving the fastest in flask C? Explain.

(e) Which molecules, N_2 or O_2, are colliding most frequently with the walls of the container in flask C? Explain.

85. Three flasks of identical volume each contain argon gas.

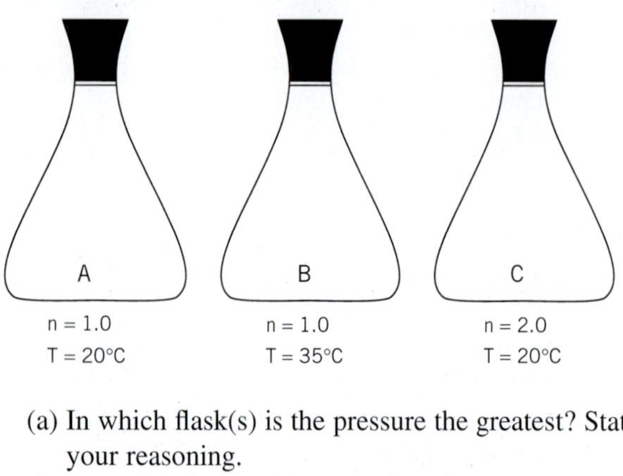

(a) In which flask(s) is the pressure the greatest? State your reasoning.

(b) In which flask(s) is the average kinetic energy of the argon atoms the least? Explain.

(c) In which flask(s) are the argon atoms moving the most rapidly? Provide a clear explanation.

(d) In which flask(s) are the argon atoms colliding the most frequently with the walls of the flask? Explain your answer.

86. Two identical flasks, one containing He and one containing Ne, are at the same temperature but different pressures.

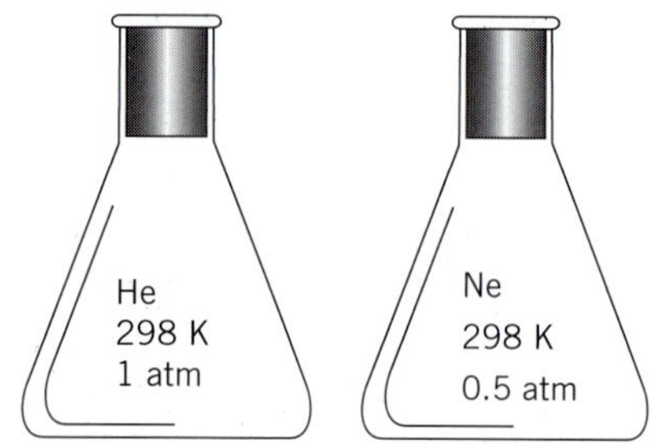

Indicate whether the following statements are true or false and explain your reasoning for each.

(a) Both flasks contain the same number of atoms.

(b) The atoms in the flask containing the gas at the highest pressure are moving more rapidly than the atoms in the low-pressure flask.

(c) The kinetic energy of the atoms is greatest in the flask with the lowest pressure.

87. A 1.0-L flask contains CH_4 gas at 0°C and 1.0 atm.

(a) Will the kinetic energy of the molecules increase, decrease, or remain the same if
  (i) the temperature is increased?
  (ii) the pressure is decreased at 0°C?
  (iii) the volume is decreased at 0°C?
  (iv) the number of moles of CH_4 is doubled and the temperature decreased to -10°C?

(b) Will the CH_4 molecules move more rapidly, less rapidly, or remain the same if the temperature is unchanged but the volume of the flask is doubled?

(c) Will the pressure increase, decrease, or remain the same if
  (i) the temperature is increased while n and V are unchanged?
  (ii) the temperature is unchanged while doubling the volume and number of moles?
  (iii) the volume is doubled and the number of moles is halved at 0°C?

(d) A 1.0-L flask is filled with He gas at 0°C and 1.5 atm. An identical flask at 0°C and 1.0 atm is filled with CH_4. Which flask contains the most molecules?

Integrated Problems

88. What are the molecular formulas for phosphine, PH_x, and diphosphine, P_2H_y, if the densities of the gases are 1.517 and 2.944 g/L, respectively, at 0°C and 1.00 atm?

89. Calculate the weight of magnesium that would be needed to generate 500 mL of hydrogen gas at 0°C and 1.00 atm.

$$Mg(s) + 2\,HCl(aq) \longrightarrow Mg^{2+}(aq) + 2\,Cl^-(aq) + H_2(g)$$

90. Calculate the formula of the oxide formed when 10.0 g of chromium metal reacts with 6.98 L of O_2 at 20°C and 0.994 atm.

91. Calculate the volume of CO_2 gas measured at 756 mmHg and 23°C given off when 150 kg of limestone is heated until it decomposes.

$$CaCO_3(s) \longrightarrow CaO(s) + CO_2(g)$$

92. Calculate the volume of O_2 that would have to be inhaled at 20°C and 1.00 atm to consume 1.00 kg of fat, $C_{57}H_{110}O_6$.

$$2\ C_{57}H_{110}O_6(s) + 163\ O_2(g) \longrightarrow 114\ CO_2(g) + 110\ H_2O(l)$$

93. Calculate the volume of CO_2 gas collected at 23°C and 0.991 atm that can be prepared by reacting 10.0 g of calcium carbonate with excess acid.

$$CaCO_3(aq) + 2\ H^+(aq) \longrightarrow Ca^{2+}(aq) + CO_2(g) + H_2O(l)$$

94. Determine the identity of an unknown metal if 1.00 g of the metal reacts with excess acid according to the following equation to produce 374 mL of H_2 gas at 25°C and 1.00 atm.

$$M(s) + 2\ H^+(aq) \longrightarrow M^{2+}(aq) + H_2(g)$$

95. Imagine two identical flasks at the same temperature. One contains 2 g of H_2 and the other contains 28 g of N_2. Which of the following properties are the same for the two flasks?
(a) pressure (b) average kinetic energy (c) density
(d) number of molecules per container
(e) weight of the container

96. At 25°C, four-fifths of the pressure of the atmosphere is due to N_2 and one-fifth is due to O_2. What volume of a room will N_2 occupy?

97. Which of the noble gases in Group VIIIA of the periodic table has a density of 5.86 g/L at 0°C and 1.00 atm?

98. Two flasks with the same volume and temperature are connected by a valve. One gram of hydrogen is added to one flask, and 1 g of oxygen is added to the other.
(a) In which flask is the average kinetic energy of the gas molecules the greatest?
(b) In which flask are the gas molecules moving the most rapidly?

99. When a meteorologist reports precipitation in the weather forecast, this is usually associated with a low-pressure system. Explain why precipitation would be associated with a low atmospheric pressure. (Hint: See Problem 45.)

100. If equal weights of O_2 and N_2 are placed in identical containers at the same temperature, which of the following statements is true?
(a) Both flasks contain the same number of molecules.
(b) The pressure in the flask that contains the N_2 will be greater than the pressure in the flask that contains the O_2.
(c) There will be more molecules in the flask that contains O_2 than the flask that contains N_2.
(d) This question cannot be answered unless we know the weights of O_2 and N_2 in the flask.
(e) None of the above are correct.

101. N_2H_4 decomposes at a fixed temperature in a closed container to form $N_2(g)$ and $H_2(g)$. If the reaction goes to completion, the final pressure will be:
(a) the same as the initial pressure
(b) twice the initial pressure
(c) three times the initial pressure
(d) one-half of the initial pressure
(e) one-third of the initial pressure

102. For the apparatus diagramed below, what will be the final partial pressures of O_2 and N_2 after the stopcock is opened? The temperature of both flasks is the same and does not change after opening the stopcock. What will be the final pressure in the apparatus?

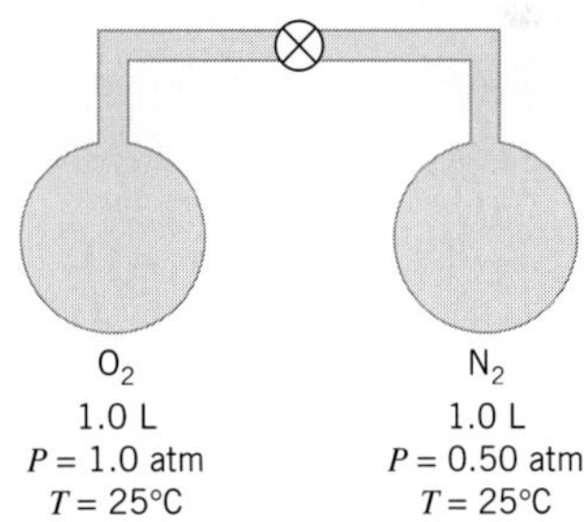

103. N_2 gas is in a 2.0-L container at 298 K and 1 atm. Explain what will happen to both the kinetic energy and the frequency of collisions of the nitrogen molecules for each of the following conditions.
(a) The number of moles of N_2 is halved.
(b) The temperature is changed to 15°C.
(c) The temperature is changed to 1500°C.
(d) The volume is decreased to 1 L.

104. Equal volumes of oxygen and an unknown gas at the same temperature and pressure weigh 3.00 g and 7.50 g, respectively. Which of the following is the unknown gas?
(a) CO (b) CO_2 (c) NO
(d) NO_2 (e) SO_2 (f) SO_3

105. What is the molecular weight of acetone if 0.520 g of acetone occupies a volume of 275.5 mL at 100°C and 756 mmHg?

106. Boron forms a number of compounds with hydrogen, including B_2H_6, B_4H_{10}, B_5H_9, B_5H_{11}, and B_6H_{10}. For which compound would a 1.00-g sample occupy a volume of 390 cm^3 at 25°C and 0.993 atm?

107. Cyclopropane is an anesthetic that is 85.63% carbon and 14.37% hydrogen by mass. What is the molecular formula of the compound if 0.45 L of cyclopropane reacts with excess oxygen at 120°C and 0.72 atm to form 1.35 L of carbon dioxide and 1.35 L of water vapor?

108. Calculate the molecular formula of diazomethane, assuming the compound is 28.6% C, 4.8% H, and 66.6% N by mass and the density of the gas is 1.72 g/L at 25°C and 1.00 atm.

Chapter Six

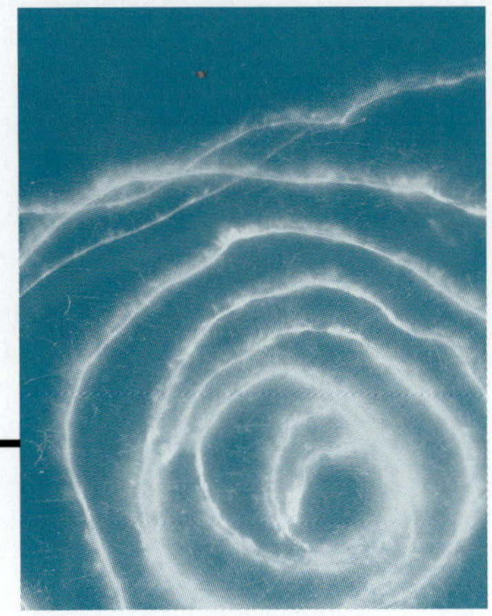

SPECIAL TOPICS

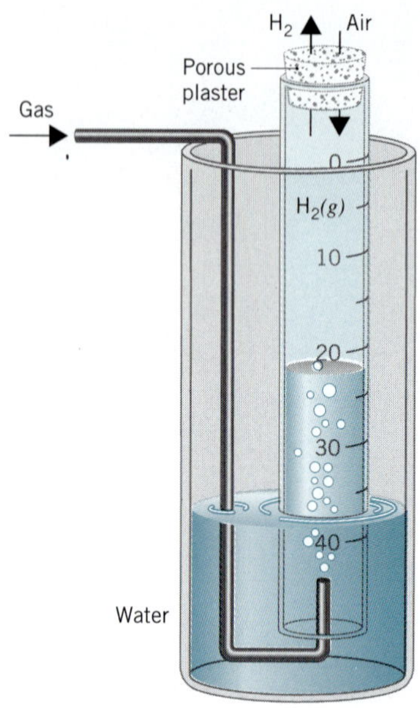

Fig. 6A.1 The rate of diffusion, or mixing of a gas with air, can be studied with the apparatus shown. If the gas escapes from the tube faster than air enters the tube, the amount of water in the tube will increase. If air enters the tube faster than the gas escapes, water will be displaced from the tube.

6A.1 Graham's Laws of Diffusion and Effusion

Most of the physical properties of a gas don't depend on the identity of the gas. But a few do. One of these physical properties can be seen when the movement of gases is studied.

In 1829 Thomas Graham used an apparatus similar to the one shown in Figure 6A.1 to study the **diffusion** of gases—the rate at which two gases mix. The apparatus consists of a glass tube sealed at one end with plaster that has holes large enough to allow a gas to enter or leave the tube. When the tube is filled with H_2 gas, the level of water in the tube slowly rises because the H_2 molecules inside the tube escape through the holes in the plaster more rapidly than the molecules in air can enter the tube. By studying the rate at which the water level in the apparatus changed, Graham was able to obtain data on the rate at which different gases mixed with air.

Graham found that the rates at which gases diffuse are inversely proportional to the square root of their densities.

$$\text{Rate}_{\text{diffusion}} \propto \frac{1}{\sqrt{\text{density}}}$$

This relationship eventually became known as **Graham's law of diffusion.**

To understand the importance of this discovery, we have to remember that equal volumes of different gases at the same temperature and pressure contain the same number of particles. As a result, the number of moles of gas per liter at a given temperature and pressure is constant. This means that the density of a gas is directly proportional to its molecular weight (*MW*). Graham's law of diffusion can therefore also be written as follows.

$$\text{Rate}_{\text{diffusion}} \propto \frac{1}{\sqrt{MW}}$$

Similar results were obtained when Graham studied the rate of **effusion** of a gas, which is the rate at which the gas escapes through a pinhole into a vacuum. The rate of effusion of a gas is also inversely proportional to the square root of either the density or the molecular weight of the gas.

$$\text{Rate}_{\text{effusion}} \propto \frac{1}{\sqrt{\text{density}}} \propto \frac{1}{\sqrt{MW}}$$

Graham's law of effusion can be demonstrated with the apparatus in Figure 6A.2. A thick-walled filter flask is evacuated with a vacuum pump. A syringe is filled with 25 mL of gas, and the time required for the gas to escape through the syringe needle into the evacuated filter flask is measured with a stopwatch.

The experimental data in Table 6A.1 were obtained by using a special needle with a very small (0.015-cm) hole through which the gas could escape.

As we can see when the data are graphed in Figure 6A.3, the *time* required for 25-mL samples of different gases to escape into a vacuum is proportional to the square root of the molecular weight of the gas. The *rate* at which the gases effuse is therefore inversely proportional to the square root of the molecular weight. Graham's observations about the rate at which gases diffuse (mix) or effuse (escape through a pinhole) suggest that relatively light gas particles, such as H_2 molecules and He atoms, move faster than relatively heavy gas particles, such as CO_2 and SO_2 molecules.

Table 6A.1

Time Required for 25-mL Samples of Gases to Escape Through a 0.015-cm Hole into a Vacuum

Compound	Time (s)	Molecular Weight (g/mol)
H_2	5.1	2.02
He	7.2	4.00
NH_3	14.2	17.0
Air	18.2	29.0
O_2	19.2	32.0
CO_2	22.5	44.0
SO_2	27.4	64.1

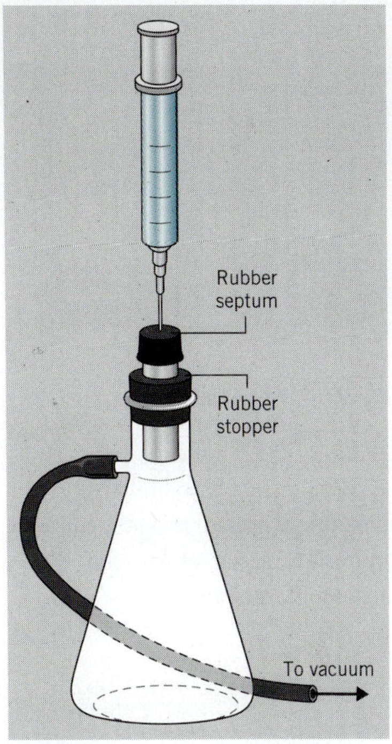

Fig. 6A.2 The rate of effusion of a gas can be demonstrated with the apparatus shown. The time required for the gas in the syringe to escape into an evacuated flask is measured. The faster the gas molecules effuse, the less time it takes for a given volume of the gas to escape into the flask.

Graham's experimental observations support the kinetic molecular theory development of the relation between velocities and molecular weight given in Section 6.18. If the rate of effusion of a gas is directly related to the molecular velocity, then Graham's law can be obtained from

$$\frac{v_1}{v_2} = \sqrt{\frac{MW_2}{MW_1}} = \frac{\text{Rate}_1}{\text{Rate}_2}$$

6A.2 Deviations from Ideal Gas Law Behavior: The van der Waals Equation

The behavior of real gases usually agrees with the predictions of the ideal gas equation to within ±5% at normal temperatures and pressures. At low temperatures or high pressures, real gases deviate significantly from ideal gas behavior. In 1873, while searching for a way to link the behavior of liquids and gases, the Dutch physicist Johannes van der Waals developed an explanation for these deviations and an equation that was able to fit the behavior of real gases over a much wider range of pressures.

van der Waals realized that two of the assumptions of the kinetic molecular theory were questionable. The kinetic theory assumes that gas particles occupy a negligible fraction of the total volume of the gas. It also assumes that the force of attraction between gas molecules is zero.

The first assumption works at pressures close to 1 atm. But it becomes increasingly less valid as the gas is compressed. Imagine for the moment that the atoms or molecules in a gas were all clustered in one corner of a cylinder, as shown in Figure 6A.4. At normal pressures, the volume occupied by the particles is a negligibly small fraction of the total volume of the gas. But at high pressures, this is no longer true. For O_2, for example, the gas molecules occupy 0.13% of the total volume at 1.00 atm but 17% of the volume at 100 atm. As a result, real gases are not as compressible at high pressures as an ideal gas. The volume of a real gas is therefore larger than expected from the ideal gas equation at high pressures.

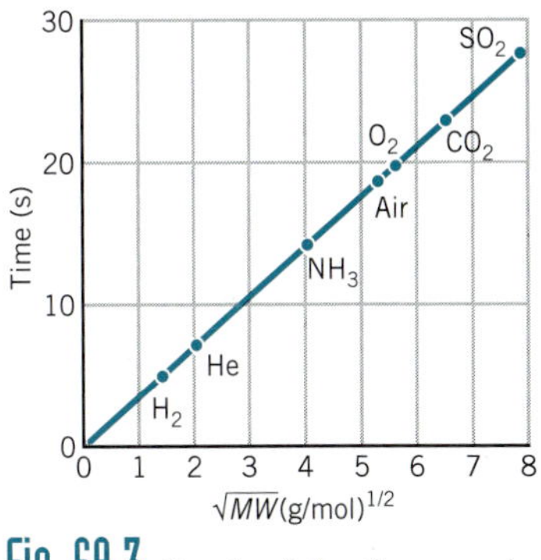

Fig. 6A.3 Graph of the time required for 25-mL samples of different gases to escape into an evacuated flask versus the square root of the molecular weight of the gas. Relatively heavy molecules move more slowly, and it takes more time for the gas to escape.

$$V_{\text{real}} \geq V_{\text{ideal}}$$

van der Waals recognized that we can correct for the fact that the volume of a real gas is too large at high pressures by *subtracting* a term from the volume of

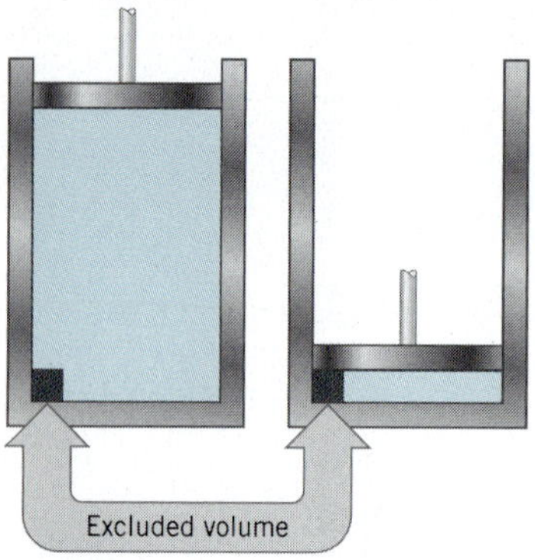

Fig. 6A.4 The volume occupied by the particles in a gas is relatively small at low pressures, but it can be a significant fraction of the total volume at high pressures.

the real gas before we substitute it into the ideal gas equation. He therefore introduced a constant (b) into the ideal gas equation that was related to the volume actually occupied by a mole of gas particles. Because the volume of the gas particles depends on the number of moles of gas in the container, the term that is subtracted from the real volume of the gas is equal to the number of moles of gas times b.

$$P(V - nb) = nRT$$

When the pressure is relatively low and the volume is reasonably large, the nb term is too small to make any difference in the calculation. But at high pressures, when the volume of the gas is small, the nb term corrects for the fact that the volume of a real gas is larger than expected from the ideal gas equation.

The assumption that there is no force of attraction between gas particles can't be true. If it were, gases would never condense to form liquids. In reality, there is a small force of attraction between gas molecules that tends to hold the molecules together. This force of attraction has two consequences: (1) Gases condense to form liquids at low temperatures, and (2) the pressure of a real gas is sometimes smaller than expected for an ideal gas.

$$P_{\text{real}} \leq P_{\text{ideal}}$$

To correct for the fact that the pressure of a real gas is smaller than expected from the ideal gas equation, van der Waals *added* a term to the pressure in the equation. The term contained a second constant (a) and has the form an^2/V^2. The complete **van der Waals equation** is therefore written as follows.

$$\left(P + \frac{an^2}{V^2}\right)(V - nb) = nRT$$

At normal temperatures and pressures, we can use the ideal gas law to describe the behavior of most gases.

$$P_{\text{ideal}}\, V = nRT$$

Under other conditions, we can use the van der Waals equation.

$$\left(P_{\text{real}} + \frac{an^2}{V^2}\right)(V - nb) = nRT$$

The van der Waals equation is something of a mixed blessing. It provides a much better fit with the behavior of a real gas than the ideal gas equation. But it does this at the cost of a loss in generality. The ideal gas equation is equally valid for any gas, whereas the van der Waals equation contains a pair of constants (a and b) that change from gas to gas. Values of the van der Waals constants for gases are given in Table 6A.2.

The ideal gas equation predicts that a plot of PV versus P for a gas at constant n and T would be a horizontal line because PV should be constant. Experimental data for PV versus P for H_2 and N_2 gas at 0°C and CO_2 at 40°C are given in Figure 6A.5. As the pressure increases, the product of pressure times volume for N_2 and CO_2 first falls below the line expected from the ideal gas equation and then rises above the line.

This behavior can be understood by comparing the results of calculations using the ideal gas equation and the van der Waals equation for 1.00 mol of CO_2 at 0°C in containers of different volumes. Let's start with a 22.4-L container. According to the ideal gas equation, the pressure of the gas should be 1.00 atm.

$$P = \frac{nRT}{V} = \frac{(1.00\text{ mol})(0.08206\text{ L-atm/mol-K})(273\text{ K})}{(22.4\text{ L})} = 1.00\text{ atm}$$

Table 6A.2

van der Waals Constants for Various Gases

Compound	$a(L^2 \times atm/mol^2)$	b(L/mol)
He	0.03412	0.02370
Ne	0.2107	0.01709
H_2	0.2444	0.02661
Ar	1.345	0.03219
O_2	1.360	0.03803
N_2	1.390	0.03913
CO	1.485	0.03985
CH_4	2.253	0.04278
CO_2	3.592	0.04267
NH_3	4.170	0.03707

Substituting what we know about CO_2 into the van der Waals equation gives a much more complex equation.

$$\left(P + \frac{an^2}{V^2}\right)(V - nb) = nRT$$

$$\left(P + \frac{(3.592\ \text{L}^2\text{-atm/mol}^2)(1.00\ \text{mol})^2}{(22.4\ \text{L})^2}\right)(22.4\ \text{L} - (1.00\ \text{mol})(0.04267\ \text{L/mol}))$$
$$= (1.00\ \text{mol})(0.08206\ \text{L-atm/mol-K})(273\ \text{K})$$

The equation can be solved, however, for the pressure of the gas.

$$P = 0.995\ \text{atm}$$

At normal temperatures and pressures, the ideal gas and van der Waals equations give essentially the same results.

Let's now repeat the calculation, assuming that the gas is compressed so that it fills a container which has a volume of only 0.200 L. According to the ideal gas equation, the pressure would have to increase to 112 atm to compress 1.00 mol of CO_2 at 0°C to a volume of 0.200 L.

$$P = \frac{nRT}{V} = \frac{(1.00\ \text{mol})(0.08206\ \text{L-atm/mol-K})(273\ \text{K})}{(0.200\ \text{L})} = 112\ \text{atm}$$

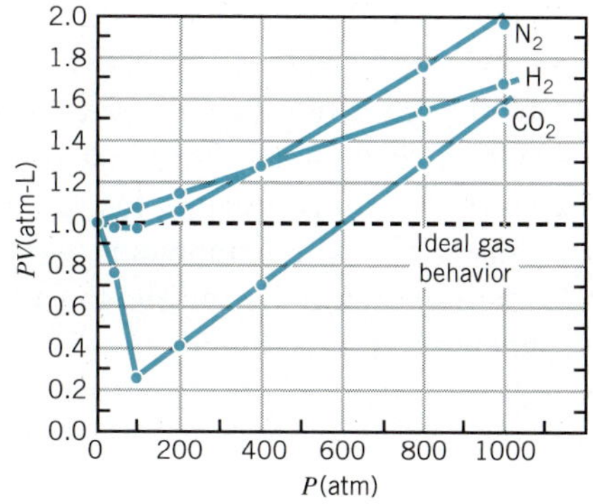

Fig. 6A.5 Plot of the product of pressure times volume for samples of H_2, N_2, and CO_2 gases versus the pressure of the gases, at constant temperature.

The van der Waals equation, however, predicts that the pressure will only have to increase to 52.6 atm to achieve the same results.

$$\left(P + \frac{(3.592 \text{ L}^2\text{-atm/mol}^2)(1.00 \text{ mol})^2}{(0.200 \text{ L})^2}\right)(0.200 \text{ L} - (1.00 \text{ mol})(0.04267 \text{ L/mol}))$$
$$= (1.00 \text{ mol})(0.08206 \text{ L-atm/mol-K})(273 \text{ K})$$
$$P = 52.6 \text{ atm}$$

As the pressure of CO_2 increases, the van der Waals equation initially gives pressures that are *smaller* than the ideal gas equation, as shown in Figure 6A.5, because of the strong force of attraction between CO_2 molecules.

Let's now compress the gas even further, raising the pressure until the volume of the gas is only 0.0500 L. The ideal gas equation predicts that the pressure would have to increase to 448 atm to compress 1.00 mol of CO_2 at 0°C to a volume of 0.0500 L.

$$P = \frac{nRT}{V} = \frac{(1.00 \text{ mol})(0.08206 \text{ L-atm/mol-K})(273 \text{ K})}{(0.0500 \text{ L})} = 448 \text{ atm}$$

The van der Waals equation predicts that the pressure will have to reach 1.62×10^3 atm to achieve the same results.

$$\left(P + \frac{(3.592 \text{ L}^2\text{-atm/mol}^2)(1.00 \text{ mol})^2}{(0.0500 \text{ L})^2}\right)(0.0500 \text{ L} - (1.00 \text{ mol})(0.04267 \text{ L/mol}))$$
$$= (1.00 \text{ mol})(0.08206 \text{ L-atm/mol-K})(273 \text{ K})$$
$$P = 1.62 \times 10^3$$

The van der Waals equation gives results that are *larger* than the ideal gas equation at very high pressures, as shown in Figure 6A.5, because of the volume occupied by the CO_2 molecules.

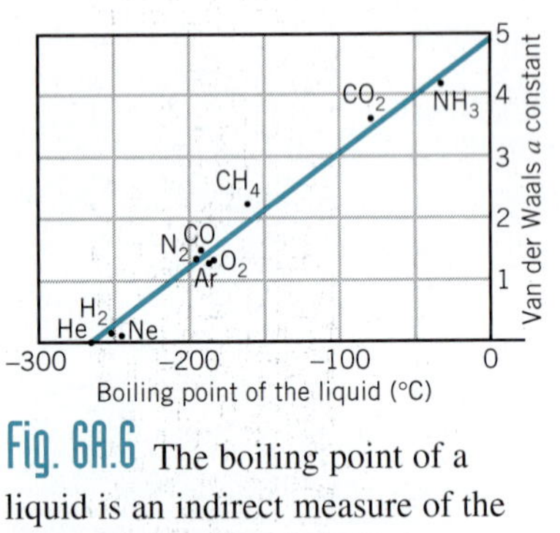

Fig. 6A.6 The boiling point of a liquid is an indirect measure of the force of attraction between its molecules. Thus, it isn't surprising to find a correlation between the value of the van der Waals constant *a*, which measures the force of attraction between gas particles, and the boiling point of simple compounds that are gases at room temperature.

6A.3 Analysis of the van der Waals Constants

The van der Waals equation contains two constants, *a* and *b*, that are characteristic properties of a particular gas. The first of the constants corrects for the force of attraction between gas particles. Compounds for which the force of attraction between particles is strong have large values for *a*. If you think about what happens when a liquid boils, you might expect compounds with large values of *a* to have higher boiling points. (As the force of attraction between gas particles becomes stronger, we have to go to higher temperatures before we can break the forces of attraction between the molecules in the liquid to form a gas.) It isn't surprising to find a correlation between the value of the *a* constant in the van der Waals equation and the boiling points of a number of simple compounds, as shown in Figure 6A.6. Gases with very small values of *a*, such as H_2 and He, must be cooled to almost absolute zero before they condense to form a liquid.

The other van der Waals constant, *b*, is a rough measure of the size of a gas particle. As the volume of the container becomes smaller and smaller, the volume occupied by the molecules becomes more significant.

Problems

Graham's Laws of Diffusion and Effusion

6A-1. Define *diffusion of a gas*.

6A-2. Define *effusion of a gas*.

6A-3. What do Graham's observations suggest about the relative speeds of molecules that have different molecular weights?

6A-4. List the following gases in order of increasing rate of diffusion.
(a) Ar (b) Cl_2 (c) CF_2Cl_2 (d) SO_2 (e) SF_6

6A-5. Bromine vapor is roughly five times as dense as oxygen gas. Calculate the relative rates at which $Br_2(g)$ and $O_2(g)$ diffuse.

6A-6. Two flasks with the same volume are connected by a valve. One gram of hydrogen is added to one flask, and 1 g of oxygen is added to the other. What happens to the weight of the gas in the flask filled with hydrogen when the valve is opened?

6A-7. What happens to the relative amounts of N_2, O_2, Ar, CO_2, and He in air as air diffuses from one flask to another through a pinhole?

6A-8. N_2O and NO are often known by the common names nitrous oxide and nitric oxide. Associate the correct formula with the appropriate common name if nitric oxide diffuses through a pinhole 1.21 times as fast as nitrous oxide.

6A-9. If it takes 6.5 s for 25.0 cm^3 of helium gas to effuse through a pinhole into a vacuum, how long would it take for 25.0 cm^3 of CH_4 to escape under the same conditions?

6A-10. Calculate the molecular weight of an unknown gas if it takes 60.0 s for 250 cm^3 of the gas to escape through a pinhole in a flask into a vacuum and if it takes 84.9 s for the same volume of oxygen to escape under identical conditions.

6A-11. A lecture hall has 50 rows of seats. If laughing gas (N_2O) is released from the front of the room at the same time ammonia (NH_3) is released from the back of the room, in which row (counting from the front) will students first begin to laugh and smell the NH_3? (Assume Graham's law of diffusion is valid.)

6A-12. The atomic weight of radon was first estimated by comparing its rate of diffusion with that of mercury vapor. What is the atomic weight of radon if mercury vapor diffuses 1.082 times as fast?

Deviations from Ideal Gas Law Behavior: The van der Waals Equation

6A-13. Predict whether the force of attraction between particles makes the volume of a real gas larger or smaller than that of an ideal gas.

6A-14. Predict whether the fact that the volume of gas particles is not zero makes the volume of a real gas larger or smaller than that of an ideal gas.

6A-15. Describe the conditions under which significant deviations from ideal gas behavior are observed.

6A-16. Calculate the fraction of empty space in CO_2 gas, assuming 1 L of the gas at 0°C and 1.00 atm can be compressed until it changes to a liquid with a volume of 1.26 cm^3.

6A-17. The following data were obtained in a study of the pressure and volume of a sample of acetylene.

P (atm):	1	45.8	84.2	110.5	176.0	282.2	398.7
V (L):	1	0.01705	0.00474	0.00411	0.00365	0.00333	0.00313

Calculate the product of pressure times volume for each measurement. Plot PV versus P and explain the shape of the curve.

6A-18. Calculate the pressure of 1.00 mol of O_2 at 0°C in 1.0-L, 0.10-L, and 0.010-L containers using both the ideal gas equation and the van der Waals equation.

Analysis of the van der Waals Constants

6A-19. Use the van der Waals constants for helium, neon, and argon to calculate the relative sizes of the atoms of these gases.

6A-20. Identify the term in the van der Waals equation used to explain why gases become cooler when they are allowed to expand rapidly.

6A-21. Refer to Figure 6A.5. At 100 atm pressure, which term in the van der Waals equation, an^2/V^2 or nb, is most important for CO_2? Which term is most important for H_2 at the same pressure? Explain.

Chapter Seven

MAKING AND BREAKING OF BONDS

7.1 Energy

We need energy to run our automobiles, power our batteries, heat our homes, and fuel our bodies. But what do we mean when we say "energy," and what exactly is the origin of this energy?

It has often been stated that the United States runs on petroleum. Petroleum is oil trapped underground. It is from petroleum that gasoline is made, and in fact petroleum is our society's main source of energy. From petroleum it is possible to separate out various chemical compounds. Most of these compounds are called *hydrocarbons* because they are composed of only carbon and hydrogen atoms. A few common hydrocarbons found in petroleum are given in Table 7.1

When these hydrocarbons are mixed with oxygen and ignited in the cylinder of an automobile by a spark, a chemical reaction occurs. When octane is burned, for example, the following reaction takes place:

$$2\ C_8H_{18}(g) + 25\ O_2(g) \longrightarrow 16\ CO_2(g) + 18\ H_2O(g)$$

The hot gases produced in this reaction expand, moving the pistons in the cylinders of the engine, and the automobile's wheels turn. The burning of hydrocarbons in a furnace supplies the heat for our homes.

The carbohydrates, or "sugars," in the food we ingest are metabolized by the body in a series of complex reactions. The overall reaction between the sugar known as glucose and oxygen in the body can be summarized by the following chemical reaction.

$$C_6H_{12}O_6(s) + 6\ O_2(g) \longrightarrow 6\ CO_2(g) + 6\ H_2O(l)$$

As a result of this reaction, the body is provided with energy to do work and to maintain an appropriate temperature.

In Chapters 7 and 13 a different approach to thermodynamics is introduced. Chemical reactions are thought of as rearrangements of atoms, that is, a process in which bonds are broken and bonds are made. Standard enthalpies of atom combination are used in place of enthalpies of formation for calculating enthalpy changes. ChemActivities 31–32 in Chemistry: A Guided Inquiry *are designed to accompany this chapter.*

Instructors accustomed to using heats of formation may find enthalpies of atom combination unsettling at first. Although this approach seems awkward to those who are accustomed to using "products minus reactants," students readily accept it and after a while even oldtimers come to realize the pedagogical power of teaching thermodynamics in this way.

Table 7.1

Common Hydrocarbons Found in Petroleum

Name	Lewis Structure	Line Structure
Butane	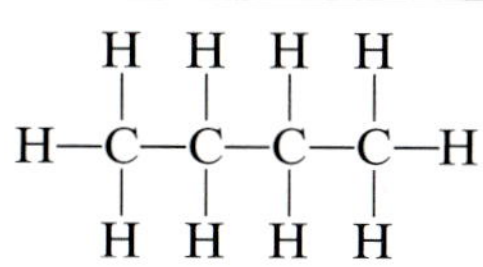	$CH_3—CH_2—CH_2—CH_3$
Isopentane (2-methyl butane)	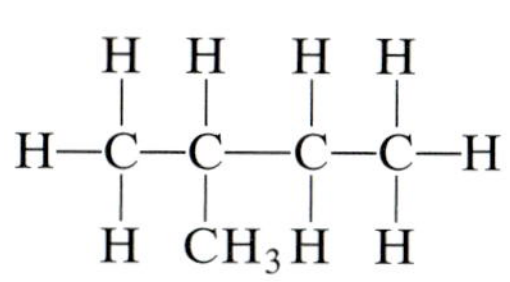	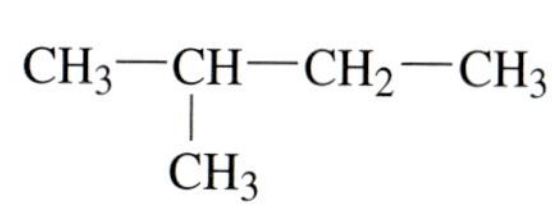
Octane	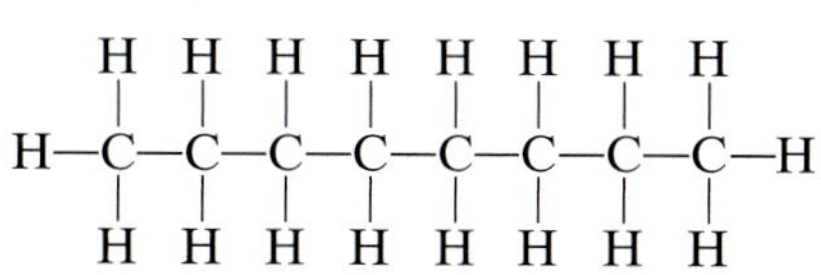	$CH_3—CH_2—CH_2—CH_2—CH_2—CH_2—CH_2—CH_3$
Isooctane (2,2,4-trimethyl pentane)	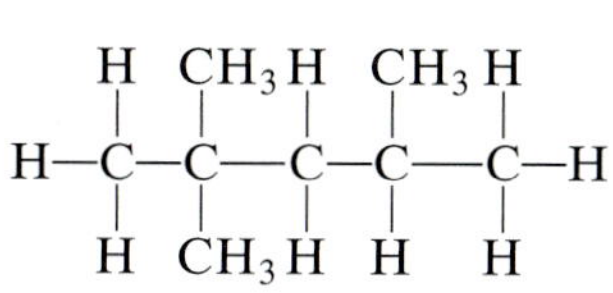	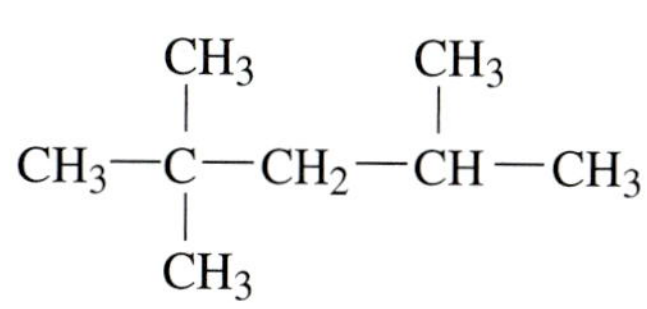

Instructors who prefer the more traditional use of enthalpies of formation to calculate changes in enthalpy may substitute Sections 7.15 and 7.16 for sections 7.11–7.13. A complete table of standard enthalpies of formation is found in Appendix B.16 at the end of the text.

A flashlight battery supplies electrons, which pass through a wire in a bulb and thereby produce light. The electrons are driven through the wire as a result of a chemical reaction that takes place in the battery.

All of the processes just mentioned occur because the energy needed to move pistons, drive electrons through wires, and heat our bodies and homes can be derived from energy-producing chemical reactions.

What is the ultimate source of the energy produced in those reactions? How much energy is available from such reactions? How can the reactions be used to obtain the maximum amount of energy from a given process? The answers to these questions form a major part of the remainder of this book.

Because this is a course in chemistry, the examples given above all relate to chemical processes. There are other sources of energy, however, and some understanding of the different ways energy may be produced is essential to a more complete understanding of chemical energy.

Energy may be classified as either kinetic or potential energy. **Kinetic energy** (mechanical energy) is the energy of motion. Molecules that are moving through space (translational motion) or rotating around their center of gravity (rotational motion) possess kinetic energy. **Potential energy** is the energy of position. A box lifted up a ladder is at a higher potential energy than a box left on the ground because the position of the box on the ladder is higher in the earth's gravitational field than a box on the ground.

A vibrating molecule has both kinetic and potential energy. The kinetic portion of the energy results from the fact that the atoms in the molecule are moving relative to one another. The potential energy results from the changes in the distance between atoms.

Most substances expand when heated. The energy supplied by heat not only causes the atoms of the molecules to move more vigorously but also causes the molecules to move away from each other. Thus, both potential and kinetic energy change during heat expansion because adding energy increases motion and the increased motion leads to increased distance between the atoms and molecules.

Energy can be transferred from one object to another. When a billiard ball is struck by a cue stick, for example, the mechanical energy of the motion of the cue stick is transformed into the kinetic energy of the billiard ball.

In Chapter 3 we described the technique of photoelectron spectroscopy (PES). In PES, photons are used to transfer the electromagnetic energy contained in electromagnetic waves to an electron, causing the electron to be ejected from an atom. The resulting kinetic energy of the electron is measured and used to calculate the energy required to remove the electron from the atom.

Energy can be converted from one form to another. A car's battery, for example, converts chemical energy into electrical energy, which is then converted into mechanical energy. The spectrum of hydrogen studied in Chapter 3 provides another example. In this case, the energy associated with moving an electron from one energy level to another is absorbed or emitted as electromagnetic radiation.

No matter what form energy takes or how it is transferred, the total energy before a process takes place and the total energy after the process is completed are the same. In other words, energy is conserved. **Conservation of energy** means that energy cannot be created or destroyed. In an elastic collision between a moving cue ball and a stationary billiard ball, the billiard ball moves away with increased kinetic energy. The cue ball, however, loses energy and therefore slows down. When a photon strikes an atom in a PES experiment, causing an electron to be ejected, we use the following equation to calculate the energy required to remove the electron from the atom.

$$h\nu = \mathrm{IE} + \mathrm{KE}$$

The equation is based on conservation of energy. The kinetic energy (KE) of the ejected electron plus the energy required to dislodge the electron (IE) from the atom is equal to the total energy of the photon.

This chapter is concerned with the energy transfers and conversions associated with chemical reactions. These processes are part of the area of study known as *thermodynamics.* Chemical energy is the energy associated with the force of attraction between electrons and nuclei in molecules and metals and between oppositely charged ions in ionic compounds. In other words, chemical energy is the energy that is due to chemical bonds.

As a general rule: *Energy changes that occur during a chemical reaction are due to the making and breaking of chemical bonds.*[1] The amount of energy associated with a chemical reaction is directly related to the strengths of the chemical bonds that are broken and formed during a chemical reaction. The relationship between energy and the strengths of chemical bonds can be illustrated by the following hypothetical formation of molecules from individual gaseous atoms.

Imagine a collection of atoms in the gas phase consisting of 2 mol of carbon atoms, 6 mol of hydrogen atoms, and 1 mol of oxygen atoms. The atoms can be used to make two structures that are consistent with the rules developed for Lewis structures, one of which is dimethyl ether.

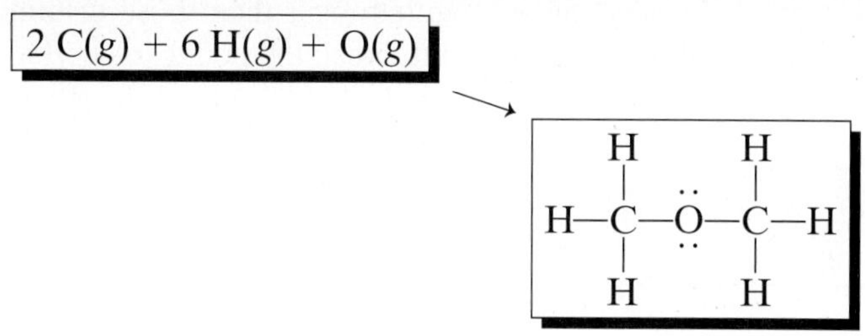

Dimethyl ether

If these atoms are brought together to make 1 mol of gaseous dimethyl ether, 3151 kJ of energy is released. What is the origin of the energy? The energy is produced by making six C—H bonds and two C—O bonds, for a total of eight bonds. This process gives off energy because: *Energy is always released during the formation of chemical bonds.*

The starting atoms can be arranged in a different combination to form a different compound, ethyl alcohol.

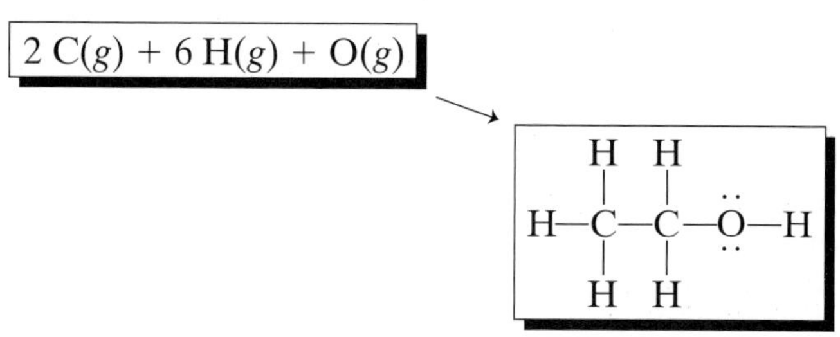

Ethyl alcohol

This compound contains five C—H bonds, one C—C bond, one C—O bond, and one O—H bond for a total, once again, of eight bonds. The formation of 1 mol of gaseous ethyl alcohol from its gaseous atoms releases 3204 kJ of energy. Thus, more energy is released in the formation of ethyl alcohol than in the formation of

[1]There are also other attractive forces that may be formed or overcome that can produce energy changes, as discussed later in this chapter and in Chapter 8.

dimethyl ether. We can therefore conclude that the eight bonds in ethyl alcohol are stronger than the eight bonds in dimethyl ether.

The effect of bond strength on the energy released during a chemical reaction can be further illustrated with a reaction more practical than the formation of compounds from their gaseous atoms. Consider the reactions that occur when ethyl alcohol and dimethyl ether are burned, or combusted, in the presence of oxygen. In each case, 1 mol of the alcohol or ether reacts with 3 mol of oxygen, and produces 2 mol of carbon dioxide and 3 mol of water.

$$\underset{\textit{Ethyl alcohol}}{CH_3CH_2OH(g)} + 3\ O_2(g) \longrightarrow 2\ CO_2(g) + 3\ H_2O(g)$$

$$\underset{\textit{Dimethyl ether}}{CH_3OCH_3(g)} + 3\ O_2(g) \longrightarrow 2\ CO_2(g) + 3\ H_2O(g)$$

The combustion of 1 mol of ethyl alcohol releases 1275 kJ of energy, whereas the combustion of 1 mol of dimethyl ether releases 1327 kJ of energy. The same products are formed in both reactions, but different bonds must be broken in the two reactant molecules.

> **CHECKPOINT**
>
> In the following reaction are bonds broken or are they made?
>
> $$O(g) + O(g) \longrightarrow O_2(g)$$
>
> Will this reaction release energy or require an input of energy?

Breaking chemical bonds always requires an input of energy. As we have seen, the eight bonds in ethyl alcohol are stronger than those in dimethyl ether. More energy must therefore be expended to break the bonds in ethyl alcohol. As a result, less energy is given off when ethyl alcohol is burned. In other words, in a combustion reaction dimethyl ether would be a better source of energy than ethyl alcohol.

7.2 Heat

Heat is energy in transit. Heat is one way in which energy can be transferred from one object to another. The transfer of heat is usually associated with a change in temperature. Although heat and temperature are related to one another, they are not the same thing. **Temperature,** as discussed in Chapter 6, is a measure of the "hotness" or "coldness" of an object and may be measured using the Fahrenheit, Celsius, or Kelvin scales. Because heat is defined as the transfer of energy, it must be measured in the same units as energy (joules). Although heat is associated with energy transfer, it is incorrect to consider a system or object as *containing* heat energy. A system that contains one of the forms of energy discussed in the previous section can transfer some of that energy to another object by means of heat.

If a hot brick is placed in contact with a cold brick, energy will be transferred as heat from the hot brick to the cold one. The hot brick will get cooler and the cool brick will get warmer. Eventually the two bricks will come to the same temperature. The hot brick has lost energy and the cold brick has gained energy.

7.3 Heat and the Kinetic Molecular Theory

Chemists often divide the universe into a system and its surroundings. The **system** is that small portion of the universe in which we are interested. It may consist of the water in a beaker or a gas trapped in a cylinder by a piston, as shown in Figure 7.1. The **surroundings** are everything else—in other words, the rest of the universe.

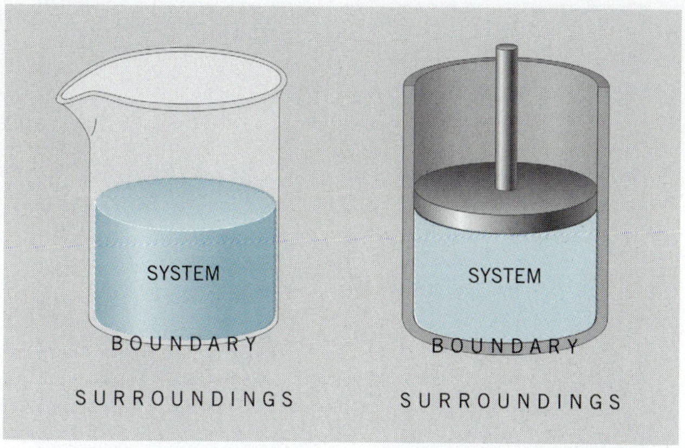

Fig. 7.1 In the kinetic theory, heat is transferred across the boundary between a system and its surroundings.

The system and its surroundings are separated by a **boundary.** The boundary can be as real as the glass in a beaker or the walls of a balloon. It can also be imaginary, such as a line 200 nm from the surface of a metal that arbitrarily divides the air close to the metal surface from the rest of the atmosphere. The boundary can be rigid or it can be elastic. In the kinetic theory, heat is transferred across the boundary between a system and its surroundings.

If a hot brick is placed in contact with a cold brick, kinetic energy will be transferred from the hot brick to the cold one. The particles in the hot brick are jiggling about more rapidly than those of the cold brick. The motion of the particles of the hot brick transfers some energy to the cold brick. Eventually the particles in the two bricks come to the same average energy; in other words, the temperatures of the two bricks become the same.

The amount of energy needed to increase the temperature of a system by a given amount depends on the nature of the system. Materials are composed of atoms, molecules, or ions arranged in different ways. They may consist of a single state, or several states of matter. Energy supplied to a substance can increase the translational, rotational, and vibrational motion of the particles that form the substance. The increased motion can cause the particles in the substance to move a little farther apart from one another. Because there is always at least some force of attraction between these particles, moving them farther apart increases their potential energy—much like stretching a rubber band increases its potential energy. Thus, the transfer of energy by means of heat can increase both the kinetic and potential energy of the system. In some cases, for example, when there is a change in the state of matter, only the potential energy may be changed by heat and the temperature may not change. The **kinetic theory of heat** can be summarized as follows: *Heat, when it enters a system causing a change in temperature, produces an increase in the average motion with which the particles of the system move.*

The idea that atoms, molecules, and ions are in continuous random motion plays a very important role in much of the chemistry that is to follow. In addition to providing a means of understanding temperature and heat, the kinetic theory explains many properties of liquids, solids, and gases and the factors that influence how fast a chemical reaction occurs.

> **CHECKPOINT**
>
> Use the kinetic theory of heat to explain what happens to the particles of a gas in a balloon when heat enters the balloon from its surroundings.

7.4 The First Law of Thermodynamics

The term *law* is used in two fundamentally different ways by scientists. When they talk about Boyle's law, or Charles' law, they are describing a mathematical equation

Energy is neither created nor destroyed. The energy absorbed by the system (the ice cubes) in this example is exactly equal to the energy lost by its surroundings (the tea).

that fits experimental data reasonably well. When they talk about the laws of thermodynamics, they are referring to statements for which there are no exceptions. They are statements of universal validity. The first law of thermodynamics can be summarized in the following statement: *Energy is conserved.* In other words, energy cannot be created or destroyed. Thus, the total energy before and after any process is carried out must be the same. By total energy we mean the energy of everything that may conceivably be altered as a result of the process. The **first law of thermodynamics** cannot be derived from general principles, but no experiment has yet been carried out that contradicts this law. In addition to being conserved, energy is additive. That is, the total energy of a system is the sum of the energy of its parts.

This is an unconventional presentation of the relationship between system, surroundings, and the first law. This presentation incorporates concrete examples which make an understanding of the first law more accessible to students.

Consider the following experiment. A cylinder fitted with a piston is filled with an ideal gas, as shown in Figure 7.2. The piston is secured by stops, which prevent its movement. A hot brick is then brought into contact with the apparatus. The brick is hotter than the gas in the piston–cylinder apparatus. As a result, the average kinetic energy of the particles in the brick is larger than that of the molecules of gas in the container. Some of the kinetic energy of the brick particles is transferred to the wall of the container and then, through collisions with the wall, to the molecules of the gas. As a result, the temperature of the gas increases, the motion of the gas molecules increases, collisions of the gas molecules with the walls of the container are more frequent and more forceful, and the pressure in the container increases. The brick has lost some energy, which has been transferred as heat to the gas and container.

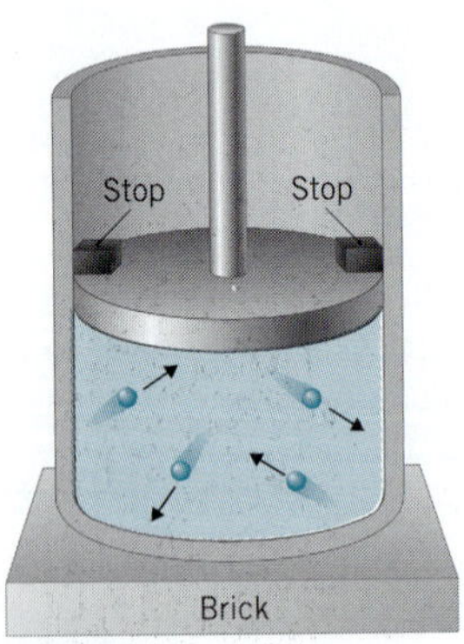

Fig. 7.2 A piston-and-cylinder apparatus containing an ideal gas in which the piston is held in place by stops.

In this experiment, the gas is the system and the brick, container, and piston are the surroundings. The first law of thermodynamics assures us that the total energy change, ΔE_{total}, may be written as follows.

$$\Delta E_{total} = \Delta E_{sys} + \Delta E_{surr} = 0$$

The subscripts "sys" and "surr" stand for the system and its surroundings, respectively. The additive nature of energy allows the energy changes to be expressed as a sum of all of the individual energy changes that are part of a process, and the conservation condition assures that ΔE_{total} must be zero.

In our piston experiment, all the energy that entered the system went directly to the increased energy of the molecules of gas. The piston didn't move despite the increased pressure in the container because the piston was held in place. The energy of the surroundings and that of the gas could and did change. We may summarize the energy changes in the following way. Before the brick was brought into contact with the container the total energy was

$$(E_{\text{total}})_{\text{before}} = (E_{\text{gas}})_{\text{before}} + (E_{\text{surr}})_{\text{before}}$$

and after the brick exchanged energy with the gas and container the total energy was

$$(E_{\text{total}})_{\text{after}} = (E_{\text{gas}})_{\text{after}} + (E_{\text{surr}})_{\text{after}}$$

The total change in energy can be related to the energy before and after the process as follows.

$$\Delta E_{\text{total}} = (E_{\text{total}})_{\text{after}} - (E_{\text{total}})_{\text{before}} = 0$$

Substituting what we know about the relationship between the total energy and the energy of the system and surroundings gives the following.

$$\Delta E_{\text{total}} = (E_{\text{gas}} + E_{\text{surr}})_{\text{after}} - (E_{\text{gas}} + E_{\text{surr}})_{\text{before}} = 0$$

This equation can be rearranged as follows.

$$\Delta E_{\text{total}} = [(E_{\text{gas}})_{\text{after}} - (E_{\text{gas}})_{\text{before}}] + [(E_{\text{surr}})_{\text{after}} - (E_{\text{surr}})_{\text{before}}] = 0$$

The first term on the right side of this equation is equal to the change in the energy of the system (gas), and the second term is equal to the change in the energy of the surroundings.

$$\Delta E_{\text{total}} = \Delta E_{\text{gas}} + \Delta E_{\text{surr}} = 0$$

According to the first law, there is no change in the total energy of the system and its surroundings. Thus, the change in the energy of the surroundings must be equal in magnitude and opposite in sign to the change in the energy of the system. In other words, the energy lost by the surroundings was gained by the gas in the system.

$$\Delta E_{\text{gas}} = -\Delta E_{\text{surr}}$$

Note that ΔE for either the system or the surroundings is positive when energy is gained and negative when energy is lost.

Now imagine a second experiment in which the process is the same but the stops on the piston are removed, as shown in Figure 7.3. Just as before, we must take account of everything that may have its energy changed as a result of the process. With the stops removed, the piston can move upward. In order for the piston to move, some energy will have to be used to lift the weight of the piston. When the piston moves up because of the increased pressure in the container, it will be higher in the earth's gravitational field and thus have increased energy. We'll call this energy change ΔE_{piston}.

Once again, the total change in energy is equal to the change in the energy of the system plus the change in the energy of the surroundings.

$$\Delta E_{\text{total}} = \Delta E_{\text{sys}} + \Delta E_{\text{surr}}$$

➤ CHECKPOINT

A hot brick is thrown into cold water. According to the first law of thermodynamics, could the brick get hotter and the water colder?

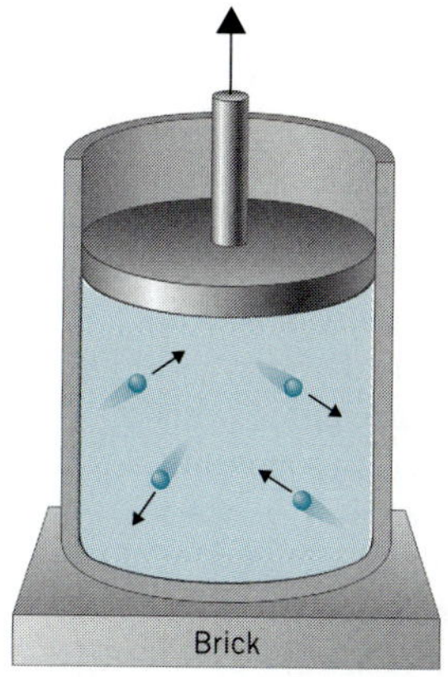

Fig. 7.3 A piston-and-cylinder apparatus with an ideal gas with the stops removed, so that the piston can move.

Here the system consists of the gas. The piston, brick, and container are the surroundings.

$$\Delta E_{\text{total}} = \Delta E_{\text{gas}} + \Delta E_{\text{piston}} + \Delta E_{\text{brick + container}} = 0$$

The energy of the piston is increased by $mg\Delta h$, where m is the mass of the piston, g is the acceleration due to gravity, and Δh is the height to which the piston has been raised. The raising of a weight requires work, w. The law of conservation of energy dictates that the energy gained by the piston must be lost by the system. Chemists therefore define ΔE_{piston} as $-w$ to show that if the system does work, the energy of the system must decrease. Chemists often write the energy change of the piston as follows.

$$\Delta E_{\text{piston}} = mg\Delta h = -\text{work} = -w$$

Thus the first law becomes

$$\Delta E_{\text{total}} = \Delta E_{\text{gas}} - w + \Delta E_{\text{brick + container}} = 0$$

$$\Delta E_{\text{gas}} - w = -\Delta E_{\text{brick + container}}$$

In this case the energy lost by the brick and container appears in the increased temperature of the gas and the increased height of the piston. Not all of the energy lost has gone into increasing the temperature of the gas molecules, as was the case in the first experiment. In the second experiment some of the energy went into increasing the temperature of the gaseous molecules, but some also went into the work of lifting the piston. If the same amount of energy is supplied to the gas in both experiments, the final temperature of the gas in the first experiment will be higher than in the second.

In the first experiment, the container was kept at a *constant volume*. In the second experiment, the volume of the gas changed, but the container was kept at *constant pressure*. The piston moved to maintain a balance between the pressure produced by the gas in the container and the pressure pushing down on the gas from the piston.

Experiments done in the laboratory are usually done in open flasks and not in piston–cylinder types of containers. An expanding system in the laboratory, however, does push against something, if not a piston. What is this something that must be pushed back if the system is to expand? It is the atmosphere. The weight pushing down on the system under investigation is thus the weight of the atmosphere. What happens when a system contracts? The atmosphere slips a little closer to the earth, and the system can actually gain energy as a result of this process. In either case, for processes carried out under conditions of constant pressure, the first law is written as follows.

$$\Delta E_{\text{total}} = \Delta E_{\text{sys}} - w + \Delta E_{\text{surr}} = 0$$

A new term is defined, called the **enthalpy** (H), which is a measure both of the energy change in the system and any work done to or by the system. For example, an expanding gas raising a weight represents work done by the gas.

$$(\Delta E_{\text{sys}} - w) = \Delta H_{\text{sys}}$$

In the two experiments just described, energy was supplied to the system from the surroundings. Now consider a process that is of considerable interest to a chemist.

The same apparatus is used as in the first and second experiments, except the hot brick is removed and the ideal gas is replaced with a mixture of ethane gas and

➤ CHECKPOINT

If the same quantity of heat is transferred from the brick to the gas in a fixed-volume piston cylinder (Figure 7.2) and in a movable piston cylinder (Figure 7.3), in which container will the temperature of the gas increase the most?

oxygen gas (Figure 7.4). A spark is then used to induce the following combustion reaction.

$$2\ C_2H_6(g) + 7\ O_2(g) \longrightarrow 4\ CO_2(g) + 6\ H_2O(g)$$

The temperature in the container increases as a result of the energy produced by the breaking and making of chemical bonds in the reaction. Heat is given off by the chemical reaction because the sum of the bond strengths in the products is greater than the sum of the reactant bond strengths. Part of this heat goes into heating the container and the gases; part goes into raising the piston. When this reaction is done under conditions of constant pressure, the heat given off is equal to the change in the enthalpy of the system that occurs during the reaction.

The first law of thermodynamics summarizes the changes that take place in the experiment.

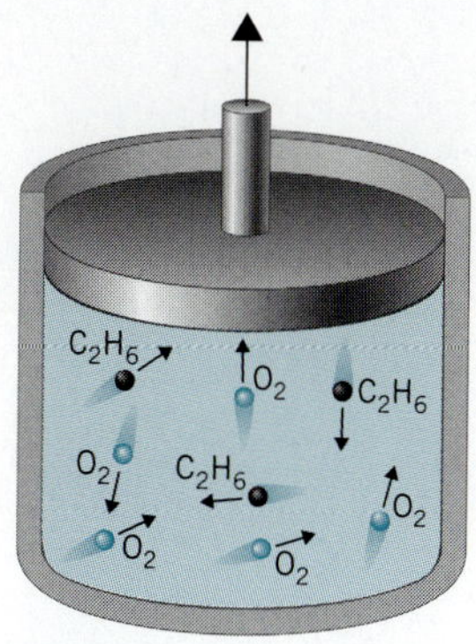

Fig. 7.4 The ideal gas in the piston–cylinder apparatus has been replaced with a mixture of ethane and oxygen. There are no stops on the piston, and the source of external heat, the brick, has been removed.

$$\Delta E_{\text{total}} = \Delta E_{\text{sys}} + \Delta E_{\text{container}} + \Delta E_{\text{piston}} = 0$$

Once again, we can specify that the system is the gaseous mixture in the cylinder and that the container and piston are the surroundings. The system is heated because of the energy released in the chemical reaction; the product gases expand and raise the piston. Thus we identify ΔE_{sys} with ΔE_{rxn}, and ΔE_{piston} as the work.

$$\Delta E_{\text{total}} = \Delta E_{\text{rxn}} + \Delta E_{\text{container}} - w = 0$$

The enthalpy change for the reaction, ΔH_{rxn}, measures the energy change of the reaction, ΔE_{rxn}, and the work of lifting the piston.

$$\Delta H_{\text{rxn}} = \Delta E_{\text{rxn}} - w$$

Thus,

$$\Delta E_{\text{total}} = \Delta H_{\text{rxn}} + \Delta E_{\text{container}} = 0$$

Because ΔE_{total} is always equal to zero, the enthalpy of reaction is equal in magnitude and opposite in sign to the change in the energy of the container.

$$\Delta H_{\text{rxn}} = -\Delta E_{\text{container}}$$

Table 7.2 gives the heat produced at constant pressure when several hydrocarbons burn in the presence of oxygen. Because these reactions are run under conditions of constant pressure, the heat produced is equal to the enthalpy change for the reactions.

Exercise 7.1

We previously saw that breaking bonds requires an input of energy and that heat is released when bonds are formed. What does this tell us about the relative strengths of the bonds in the products and reactants in the combustion of the hydrocarbons in Table 7.2?

Solution

In each case, these hydrocarbons burn to form CO_2 and H_2O.

$$C_5H_{12}(g) + 8\ O_2(g) \longrightarrow 5\ CO_2(g) + 6\ H_2O(l)$$

The combustion reactions all release heat. Therefore, the sum of the bond strengths must be larger in the products than in the reactants. If this were not the case, more heat would be required to break the bonds in the reactants than was returned on formation of the products, and this would result in a net input of heat.

7.5 State Functions

Every system can be described in terms of certain measurable properties. A gas, for example, can be described in terms of the number of moles of particles it contains, its temperature, its pressure, its volume, its mass, or its density. Those properties describe the **state** of the system at a particular moment in time. Some of the properties depend on the size of the sample, such as mass and volume, and are therefore examples of **extensive properties** of the system. Others, such as temperature and density, are **intensive properties** that don't depend on the size of the sample being studied.

Properties such as enthalpy, energy, temperature, or pressure can also be classified on the basis of whether they are **state functions.** A property of a system is a

Table 7.2

Enthalpy of Reaction for the Combustion of Common Hydrocarbons at 298 K and 1 atm

Hydrocarbon	Name	Structure	Heat Released per Mole of Hydrocarbon that Reacts with Oxygen (kJ)
$CH_4(g)$	Methane	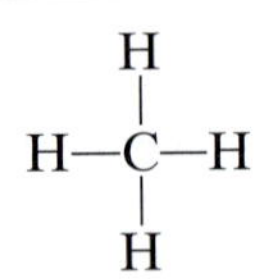	890
$C_2H_6(g)$	Ethane	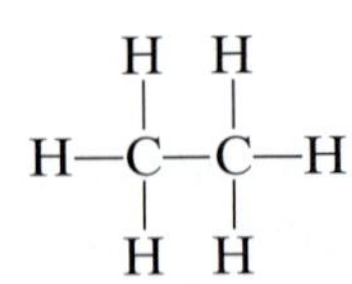	1560
$C_3H_8(g)$	Propane	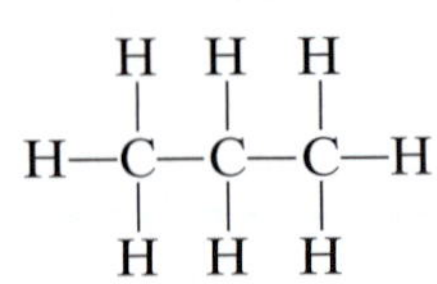	2222
$C_4H_{10}(g)$	Butane	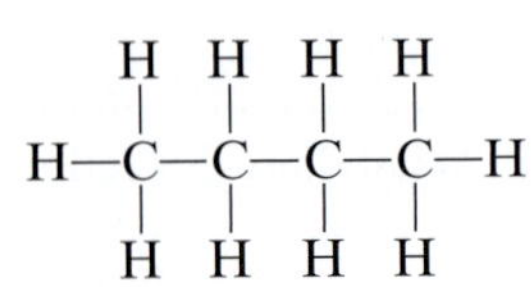	2877
$C_5H_{12}(g)$	Pentane	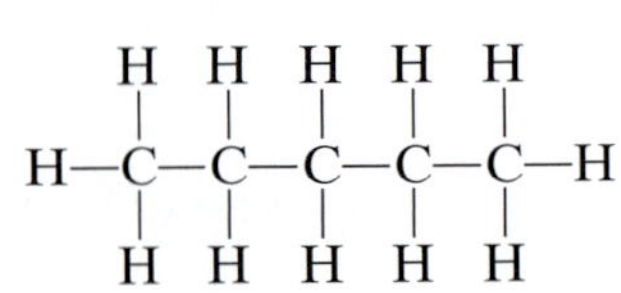	3540

state function if it depends only on the present condition of the system, and not on the path used to get to that condition.

Consider the temperature of a liquid. The fact that the temperature is 75.1°C at some moment doesn't tell us anything about the history of the system. It doesn't tell us how often the liquid has been heated or cooled before we take the measurement. Temperature is therefore a state function because it only reflects the state of the system at the moment at which it is measured. *Energy is also a state function;* it depends only on the state of the system, not on the path used to get to that state.

An analogy can be used to illustrate that energy is a state function. Suppose we have a crate on the first floor of a New York skyscraper and we wish to take the crate to the tenth floor. To do that we will have to expend someone's or something's energy. The energy of the crate will change because the crate's potential energy is greater on the tenth floor than on the first by an amount that depends on the mass of the crate, the acceleration of gravity, and the difference in height between the tenth and first floors. The only variable factor influencing the energy of the crate is height. If the crate were transported to the fifteenth floor, the energy of the crate would be larger than it was on the tenth or first floor. How the crate gets from one floor to the other doesn't change the energy of the crate. The energy change undergone by the crate depends only on the initial floor and the final floor.

This is what we mean by the term *state function.* Only the initial and final conditions matter. Suppose the crate was transported to the tenth floor from the first and then returned to the first floor. What would be the energy change of the crate? Because the initial and final conditions are the same, the energy of the crate hasn't changed. Even though the energy of the crate is unchanged by this process, some work had to be expended in lifting the crate. If the crate was moved from the first to the fifth floor and then returned to the first, the energy is again unchanged but the amount of work done is different. That is, work is not a state function. The amount of work done depends on the pathway used to get from the initial to the final state.

➤ CHECKPOINT

Suppose a crate is dropped from the tenth floor of a building. When at rest on the tenth floor, what kind of energy does the crate have? During the fall, what kind of energy does the crate have? What happens to the energy acquired by the falling crate when it comes to rest on the ground?

7.6 The Enthalpy of a System

All chemical reactions, no matter how simple or complex, have one thing in common. They all involve the breaking and/or reforming of bonds between atoms or ions. At some step, for example, the reaction between sodium metal and chlorine gas must involve breaking bonds between atoms in sodium metal and between chlorine atoms in Cl_2 gas. And the reaction must eventually involve the formation of the ionic bonds that hold the Na^+ and Cl^- ions together in NaCl.

$$2\ \mathrm{Na}(s) + \mathrm{Cl_2}(g) \longrightarrow 2\ \mathrm{NaCl}(s)$$

Because the making and breaking of bonds plays such a central role in chemistry, a way to determine the energy consumed or produced in a chemical reaction is of extreme importance to chemists. The first law of thermodynamics provides us with a way to monitor changes in the energy that accompanies a chemical reaction.

$$\Delta E_{\mathrm{total}} = \Delta E_{\mathrm{sys}} + \Delta E_{\mathrm{surr}} = 0$$

All we have to do is find a way to measure the heat transferred to or absorbed from the surroundings. If we can achieve that goal, the heat given off or consumed can be used to calculate the change in energy due to the chemical reaction.

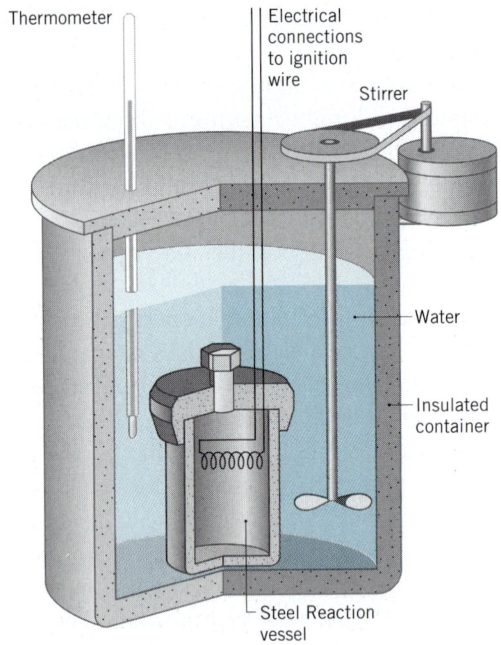

Fig. 7.5 A constant-volume bomb calorimeter. Because the volume of the system is constant, no work of expansion can be done. As a result, $\Delta E_{surr} = -\Delta E_{sys}$.

Experiments involving measurement of heat changes can be done in a **calorimeter,** shown in Figure 7.5. Because the volume of the container in which the reaction is run can't change, the heat given off or absorbed during the reaction goes entirely to produce change in the energy of the surroundings. By monitoring the change in temperature of a water bath that surrounds the reaction container, the energy change of the surroundings that results from the making and breaking of chemical bonds can be calculated. Because $\Delta E_{surr} = -\Delta E_{sys}$, ΔE_{sys} can be found.

For many years, thermodynamicists have used the symbol q to represent the heat transferred from the system to its surroundings or vice versa. In a calorimeter, at constant volume, the heat given off or absorbed in a chemical reaction is equal to the change in the energy of the system. If the energy of the system decreases, q will be negative because heat is given off. If ΔE is positive, q is positive, because heat has been absorbed by the system.

$$\Delta E_{sys} = q \qquad \text{(at constant volume)}$$

Chemists, however, usually carry out reactions in containers such as beakers or flasks that are open to the atmosphere. These reactions occur under conditions of *constant pressure*. When the volume of a system can change, all of the heat supplied to or taken from the system doesn't go into producing a temperature change. When describing a system at constant pressure, chemists therefore look for changes in the *enthalpy* of the system, ΔH_{sys}.

Enthalpy, like energy, is a state function. The change in enthalpy that occurs during a chemical reaction at constant pressure is exactly equal to the heat given off or absorbed by the reaction. A constant-pressure calorimeter is used to collect data to allow the calculation of ΔH.

$$\Delta H = q \qquad \text{(at constant pressure)}$$

Calorimeters are routinely used to determine the caloric content of food. The food in question may be sucrose, which is table sugar. Sucrose is reacted with oxygen in the calorimeter, and the heat released from the reaction is measured by determining the increase in temperature of the surrounding water.

$$C_{12}H_{22}O_{11}(s) + 12\ O_2(g) \longrightarrow 12\ CO_2(g) + 11\ H_2O(l)$$

In this reaction, 5645 kJ of heat is released by the combustion of 1 mol of sucrose. The reaction that occurs in the calorimeter is the same as the overall reaction that takes place in the body when sucrose is eaten. Thus an equivalent amount of energy, 5645 kJ, would be supplied to or stored by the body. This energy is generally referred to in terms of Calories (with a capital C) rather than kilojoules when talking about nutritional value. One Calorie (Cal) is equivalent to 4.184 kJ. Thus the caloric value of 1 mol of sucrose is about 1349 Cal. It is important to note that a Calorie (with a capital C) is equal to 1000 calories (with a small c).

One mole of sucrose weighs 342 g, so that if one teaspoon (about 5 g) of sucrose is ingested it will provide about 20 Cal, or about 4 Cal per gram. If that energy is not expended by work or exercise, it remains in the body. Slow walking expends about 150 Cal per hour, so an 8-minute walk will use up the calories from a teaspoon of sugar.

➤ CHECKPOINT

A tablespoon of sugar can provide several Calories of energy. What does this tell us about how the strengths of the bonds in the reactants compare to those of the products?

7.7 Specific Heat

The results of experiments done by Joseph Black between 1759 and 1762 showed what happens when liquids at different temperatures are mixed. When equal volumes of water at 100°F and 150°F were mixed, Black found that the temperature of the mixture was the average of the two samples (125°F), as shown in Figure 7.6*a*. He concluded that the amount of heat lost by the sample that was at 150°F was equal to the amount of heat absorbed by the sample that was at 100°F.

When water at 100°F was mixed with an equal volume of mercury at 150°F, however, the temperature of the liquids after mixing was only 115°F (Figure 7.6*b*). The temperature of the mercury fell by 35°F, but the temperature of the water increased by only 15°F. Black assumed that the heat lost by the mercury as it cooled down was equal to the heat gained by the water as it became warmer. He therefore concluded that the temperature of the water changed by a smaller amount because water has a larger "capacity for heat." In other words, it takes more heat to produce a given change in the temperature of water than it does to produce the same change in the temperature of an equivalent volume of mercury. Subsequent experiments have

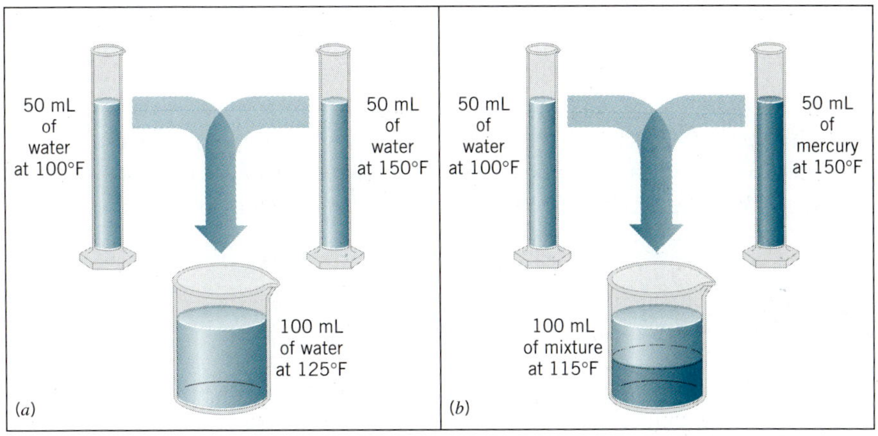

Fig. 7.6 (*a*) The temperature of a mixture of equal volumes of water is the average of the temperatures before the samples are combined. (*b*) When equal volumes of water and mercury are mixed, the temperature of the mixture is much closer to the temperature of the water before mixing.

Table 7.3

Specific Heats and Molar Heat Capacities of Common Substances (25°C)

Substance	Specific Heat (J/g·K)	Molar Heat Capacity (J/mol·K)
Al(*s*)	0.901	24.3
C(*s*)	0.709	8.52
Cu(*s*)	0.3844	24.43
$H_2O(l)$	4.18	75.3
Fe(*s*)	0.449	25.1
Hg(*l*)	0.1395	27.98
$O_2(g)$	0.9172	29.35
$N_2(g)$	1.040	29.12
NaCl(*s*)	0.8641	50.50

shown that it is not only the capacity for heat but also the amount of matter that determines the temperature change of a substance when it is heated.

Because water has a larger capacity for heat than mercury, it takes more heat to raise the temperature of a given mass of liquid water by one degree than it does to raise the temperature of the same mass of mercury by one degree. Any measurement of the capacity for heat must take into account not only the mass of the sample being heated and the change in the temperature observed, but also the identity of the substance being heated.

By convention, the heat needed to raise the temperature of *one gram* of substance by *one degree Celsius* is known as the **specific heat,** *S*. The units of specific heat were originally cal/g-°C. Because one degree on the Celsius scale is equal to one kelvin, specific heats can also be described in units of cal/g-K.

As might be expected, it is also possible to describe the heat required to raise the temperature of *one mole* of a substance by one degree. When this is done, we get the **molar heat capacity,** *C*, in units of either cal/mol-°C or cal/mol-K. Molar heat capacities are generally used by chemists because they compare equal numbers of particles.

When the SI system (Appendix A) was introduced, the approved unit of heat became the *joule,* which is related to the calorie by the following equality.

$$4.184 \text{ J} = 1 \text{ cal}$$

The approved unit of temperature in this system is the kelvin. In the SI system, the units of specific heat are J/g·K and the units of molar heat capacity are J/mol·K.

Absorption of heat by a substance may increase the kinetic energy of the particles that compose the substance. Absorption of equal amounts of heat by equal moles of substances will increase the temperature of the substance with the lower molar heat capacity more than that of a substance with a higher molar heat capacity. Thus, substances with high molar heat capacities require more thermal energy to give their particles increased motion than do those of low molar heat capacity.

Table 7.3 lists specific heats and molar heat capacities of a variety of substances. Note in particular the high specific heat and molar heat capacity of liquid water.

Exercise 7.2

If 10 mol each of aluminum, copper, and iron absorb equivalent amounts of heat, which one of the metals will experience the largest increase in temperature?

Solution

Since samples with an equal number of moles of atoms are being compared, molar heat capacity with units based on the number of moles (J/mol·K) rather than specific heat with units based on mass (J/g·K) should be used. The molar heat capacities of aluminum, copper, and iron are given in Table 7.3. The molar heat capacity of aluminum is smaller than the molar heat capacities of copper and iron. Therefore, aluminum will experience the largest increase in temperature.

> **CHECKPOINT**
>
> Table 7.3 lists iron as having a smaller specific heat than aluminum but a larger molar heat capacity. Explain why this is the case.

We are now ready to derive a formula that can be used to calculate the amount of heat given off or absorbed (q) by a substance. We need to know three quantities: the number of moles (n) or mass in grams (m), the molar heat capacity (C) or the specific heat (s), and the change in temperature (ΔT).

$$q = nC\Delta T$$
$$q = sm\Delta T$$

The heat given off or absorbed by a chemical reaction is measured in the calorimeter shown in Figure 7.5 by measuring the change in the temperature of the water that surrounds the reaction chamber.

Exercise 7.3

(a) How much heat is required to raise the temperature of 500.0 g of an iron bar from 25.0°C to 50.0°C?

(b) The hot iron at 50.0°C is placed into 500.0 g of water at 25.0°C. If the iron loses 5.1×10^3 J of heat, what will be the final temperature of the water?

Solution

(a) The heat required to raise the temperature of iron can be found by application of the relationship among heat, number of moles or grams, and heat capacity.

$$q = sm\Delta T$$
$$q = 500 \text{ g} \times 0.449 \text{ J/g}\cdot\text{K} \times (323.2 - 298.2)\text{ K}$$
$$= 5.61 \times 10^3 \text{ J}$$

Note that ΔT is the same whether °C or the Kelvin scale is used.

The molar heat capacity could also have been used.

$$q = 500 \text{ g} \times \frac{1 \text{ mol}}{55.85 \text{ g}} \times 25.1 \text{ J/mol}\cdot\text{K} \times 25.0 \text{ K}$$
$$= 5.61 \times 10^3 \text{ J}$$

(b) If the hot iron is placed into colder water, the iron will lose heat and the water will gain heat. Thus, the final temperature of the iron will be less than its initial temperature and the final water temperature will be greater than its initial temperature. Both the iron and the water will equilibrate at the same final temperature. All the heat lost by the iron is assumed to be gained by the water.

$$q_{H_2O} = nC\Delta T$$

$$5.1 \times 10^3 \text{ J} = 500 \text{ g} \times \frac{1 \text{ mol}}{18 \text{ g}} \times 75.3 \text{ J/mol}\cdot\text{K} \times \Delta T \text{ K}$$

$$\Delta T = 2.4 \text{ K}$$

The final temperature of the iron and the water will be 300.6 K or 27.4°C.

Reaction diagrams are used extensively in Chapters 7 and 13 as a powerful visual model to emphasize that the change in enthalpy associated with a chemical reaction is due to the making and breaking of chemical bonds (or other forces of attraction).

7.8 Enthalpies of Reaction

Chemical reactions are divided into two classes on the basis of whether they give off or absorb heat from their surroundings. **Exothermic** reactions give off heat to the surroundings. **Endothermic** reactions absorb heat from the surroundings.

The heat given off or absorbed in a chemical reaction at constant pressure is known as the **enthalpy of reaction.** When a reaction gives off heat to its surroundings, the energy of the system decreases. As a result, the enthalpy of the system decreases. Exothermic reactions are therefore characterized by negative values of ΔH.

Exothermic reactions: ΔH is negative ($\Delta H < 0$)

Endothermic reactions, on the other hand, take in heat from their surroundings. As a result, the enthalpy of the system increases. Endothermic reactions are therefore characterized by positive values of ΔH.

Endothermic reactions: ΔH is positive ($\Delta H > 0$)

The reaction between NH_4SCN and $Ba(OH)_2$ is an example of a spontaneous endothermic reaction. This reaction absorbs so much heat from its surroundings that it can freeze a beaker to a wooden board if the outside of the beaker is moistened.

An example of an exothermic reaction occurs when a balloon filled with hydrogen gas is ignited in the presence of oxygen. The reaction is accompanied by a large ball of fire and loud boom. The following chemical equation describes the reaction.

$$2\,H_2(g) + O_2(g) \xrightarrow{\Delta H = -483.64 \text{ kJ/mol}_{rxn}} 2\,H_2O(g)$$

When 2 mol of hydrogen react with 1 mol of oxygen to produce 2 mol of water, 483.64 kJ of heat are released. Changes in enthalpy are written in units of kilojoules per mole of reaction (kJ/mol_{rxn}), where "mol_{rxn}" represents the balanced chemical equation as a whole unit. To have meaning, a change in enthalpy must be associated with a specific chemical equation. Although the term "mol_{rxn}" is used to describe the reaction, it does not necessarily mean that there is only 1 mol of reactant or product. In the above chemical equation there are 3 mol of reactants (2 mol of H_2 and 1 mol of O_2) and 2 mol of product, but the reaction as a whole is referred to as one unit or one mol of chemical reaction.

If the chemical equation is changed, there must be an accompanying change made in the enthalpy. For example, if the coefficients in the above chemical equation are all doubled, the change in enthalpy will also be doubled. The coefficients

➤ CHECKPOINT

In a constant-pressure calorimetry experiment, the temperature of the water bath surroundings went down by 2.0 K. Was heat released or absorbed as a consequence of the chemical reaction occurring in the calorimeter? What is the sign of the enthalpy change for the reaction?

can be thought of as giving the number of moles of a substance per mole of reaction, for example, 4 mol H_2/mol_{rxn}.

Enthalpy changes are given in units of J/mol_{rxn} or kJ/mol_{rxn}, where mol_{rxn} represents the balanced chemical equation as a whole unit. To have meaning, a change in a thermodynamic parameter such as enthalpy must therefore be associated with a specific chemical equation. For a more detailed discussion see the article by Norman Craig in J. Chem. Ed., ***64,*** *669 (1987).*

$4\ H_2(g) + 2\ O_2(g)$

↘ $\Delta H = -967.28$ kJ/mol_{rxn}

$4\ H_2O(g)$

Because twice as many moles of reactants undergo the reaction, twice the amount of energy is released. The term "mol_{rxn}" in the enthalpy change of −967.28 kJ/mol_{rxn} now refers to the second chemical equation.

$$4\ H_2(g) + 2\ O_2(g) \longrightarrow 4\ H_2O(g)$$

The reaction of hydrogen and oxygen to produce water is exothermic and therefore releases heat. If the reaction is considered in the reverse direction, an input of heat would be required.

The vertical positions of the reactants, atoms, and products represent their relative energies. Because energy is always required to break molecules apart into their gaseous atoms and always given off when products are formed, the reaction diagrams are consistent for all chemical reactions. This would not be the case for reaction diagrams using enthalpies of formation.

$2\ H_2(g) + O_2(g)$

↗ $\Delta H = +483.64$ kJ/mol_{rxn}

$2\ H_2O(g)$

The decomposition reaction has a positive ΔH and is therefore endothermic. Whenever a chemical equation is reversed, the magnitude of the change in enthalpy will remain constant but the sign will change. In reaction diagrams, when a process is endothermic the arrow points up and when the process is exothermic the arrow points down.

➤ CHECKPOINT

What is the sign of the enthalpy change for the combustion reactions in Table 7.2?

The enthalpy change at 298 K for the following reaction is −92.2 kJ/mol_{rxn}.

$$N_2(g) + 3\ H_2(g) \longrightarrow 2\ NH_3(g)$$

That enthalpy change applies to the reaction in which 1 mol of nitrogen reacts with 3 mol of hydrogen to produce 2 mol of ammonia.

What would be the enthalpy change for the reaction if 0.200 mol of nitrogen reacts? We note that stoichiometry requires that if 0.200 mol of nitrogen reacts, 0.600 mol of hydrogen must be consumed and 0.400 mol of ammonia must be produced. We see from the equation that when 1 mol of nitrogen reacts, 92.2 kJ of heat is produced. This is equivalent to stating that the enthalpy change of the reaction is −92.2 kJ/mol N_2 reacted. If only 0.200 mol of N_2 reacts, the change in enthalpy is given by

The unit, per mol_{rxn}, has been introduced because using just kJ/mol as a unit requires that a specific reactant or product be named, and therefore more than one value can be associated with a single chemical reaction. When enthalpies are multiplied by coefficients from the balanced chemical equation, the coefficients are unitless.

$$(0.200\ \text{mol}\ N_2) \times \left(\frac{-92.2\ \text{kJ}}{1\ \text{mol}\ N_2}\right) = -18.4\ \text{kJ}$$

We could do the same calculation for hydrogen, for which −92.2 kJ of heat is produced for every 3 mol of hydrogen reacted.

$$(0.600\ \text{mol}\ H_2) \times \left(\frac{-92.2\ \text{kJ}}{3\ \text{mol}\ H_2}\right) = -18.4\ \text{kJ}$$

Similarly, for NH_3 we could write:

$$(0.400 \text{ mol NH}_3) \times \left(\frac{-92.2 \text{ kJ}}{2 \text{ mol NH}_3}\right) = -18.4 \text{ kJ}$$

Exercise 7.4

If just kJ is used to represent the change in enthalpy for an entire reaction, a problem results because changes in enthalpy and free energy have the same units. When changes in free energy are used to calculate an equilibrium constant for a reaction using $\Delta G° = -RT \ln K$, and R is given as 8.314 J/mol K and T is in units of K, $\Delta G°$ must be in units of J/mol in order for the units to correctly cancel.

Pentaborane, B_5H_9, burns in the presence of oxygen to form B_2O_3 and water vapor.

$$2\ B_5H_9(g) + 12\ O_2(g) \longrightarrow 5\ B_2O_3(s) + 9\ H_2O(g)$$

At 298 K the enthalpy change for the reaction is -8686.6 kJ/mol$_{rxn}$. Calculate the change in enthalpy when 0.600 mol of pentaborane is consumed.

Solution

As the reaction is written, 8686.6 kJ of heat is produced from the combustion of 2 mol of B_5H_9. Therefore,

$$(0.600 \text{ mol B}_5\text{H}_9) \times \left(\frac{-8686.6 \text{ kJ}}{2 \text{ mol B}_5\text{H}_9}\right) = -2.61 \times 10^3 \text{ kJ}$$

If 0.600 mol of B_5H_9 is consumed, 3.60 mol of O_2 must be consumed and 1.50 mol of B_2O_3 and 2.70 mol of H_2O are produced. It doesn't matter which chemical species is chosen for the calculation of ΔH because stoichiometric constraints relate the number of moles of all species. We could use the product water to calculate the change in enthalpy.

$$(2.70 \text{ mol H}_2\text{O}) \times \left(\frac{-8686.6 \text{ kJ}}{9 \text{ mol H}_2\text{O}}\right) = -2.61 \times 10^3 \text{ kJ}$$

Exercise 7.5

Water is far more abundant than petroleum. Would it be possible to run an automobile with the following reaction?

$$2\ H_2O(g) \longrightarrow 2\ H_2(g) + O_2(g)$$

Cars are now being tested that use hydrogen as a fuel. These cars produce less pollution, and they don't consume petroleum. Explain why the following reaction can be used to power an automobile.

$$2\ H_2(g) + O_2(g) \longrightarrow 2\ H_2O(g)$$

Explain the origin of the heat released or absorbed in these reactions.

Solution

The reaction $2\ H_2O(g) \longrightarrow 2\ H_2(g) + O_2(g)$ is endothermic and couldn't be used to run an engine. The reverse reaction $2\ H_2(g) + O_2(g) \longrightarrow 2\ H_2O(g)$ is exothermic and could supply energy for running a car. The origin of the heat is in the making and breaking of bonds. The Lewis structures of the products and reactants are written as follows:

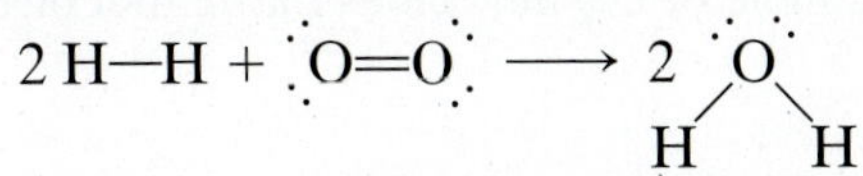

Two H—H bonds and one O═O double bond must be broken to form two $H_2O(g)$ molecules that each contain two H—O bonds. Because the reaction is exothermic, the H—O bonds in water must be stronger than those broken in H_2 and O_2. The reverse reaction is endothermic because more enthalpy is required to break the bonds in water than is gained by the formation of the bonds in H_2 and O_2.

7.9 Enthalpy as a State Function

Both the energy and the enthalpy of a system are state functions. They depend only on the state of the system at any moment, not its history. To examine the consequences of this fact, let's consider the following reaction.

$$H_2(g) + Cl_2(g) \longrightarrow 2\ HCl(g)$$

Because enthalpy is a state function, we can visualize the reaction as occurring by two simple processes. First, we break the bonds in the starting materials to form hydrogen and chlorine atoms in the gas phase.

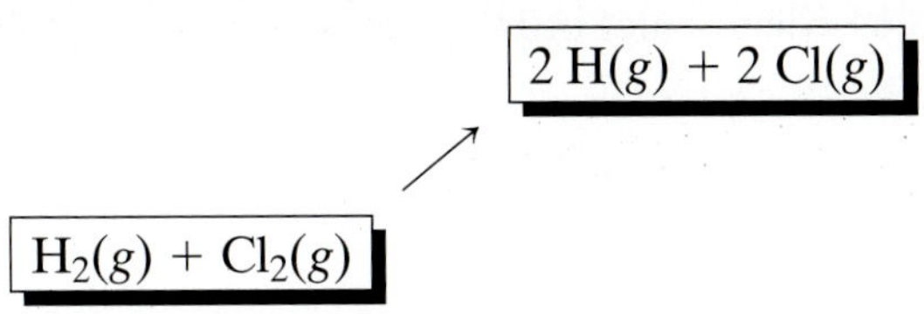

The atoms are then recombined to form the product of the reaction.

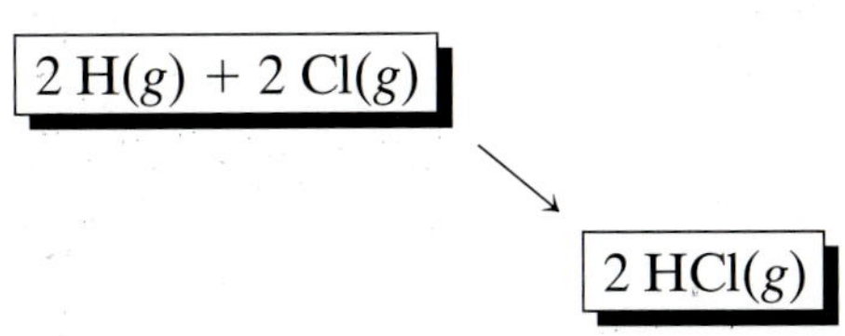

It doesn't matter whether the reaction actually occurs by these steps. Because enthalpy is a state function, ΔH for the hypothetical reaction will be exactly equal to ΔH for the reaction whatever the pathway by which the starting materials are transformed into the products.

To break the covalent bonds in a mole of Cl_2 molecules to form 2 moles of chlorine atoms, 243.4 kJ is required.

$$Cl_2(g) \longrightarrow 2\ Cl(g) \qquad \Delta H = 243.4\ \text{kJ/mol}_{\text{rxn}}$$

It takes 435.3 kJ to break apart a mole of H_2 molecules to form 2 mol of hydrogen atoms.

$$H_2(g) \longrightarrow 2\ H(g) \qquad \Delta H = 435.3\ \text{kJ/mol}_{\text{rxn}}$$

We therefore have to invest a total of 678.7 kJ in the system to transform a mole of H_2 molecules and a mole of Cl_2 molecules into 2 mol of H atoms and 2 mol of Cl atoms.

Bond breaking:

$$\begin{array}{rll} Cl_2(g) \longrightarrow 2\ Cl(g) & & \Delta H = 243.4\ \text{kJ/mol}_{\text{rxn}} \\ H_2(g) \longrightarrow 2\ H(g) & & \Delta H = 435.3\ \text{kJ/mol}_{\text{rxn}} \\ \hline H_2(g) + Cl_2(g) \longrightarrow 2\ H(g) + 2\ Cl(g) & & \Delta H = 678.7\ \text{kJ/mol}_{\text{rxn}} \end{array}$$

We now turn to the process by which the atoms recombine to form HCl molecules. The bond between hydrogen and chlorine atoms is relatively strong. We get back 431.6 kJ for each mole of H—Cl bonds that are formed. Because we have 2 mol of hydrogen atoms and 2 mol of chlorine atoms, we get 2 mol of HCl molecules. Thus, the bond-making process gives off a total of 863.2 kJ.

Bond making:

$$2\ H(g) + 2\ Cl(g) \longrightarrow 2\ HCl(g) \qquad \Delta H = -863.2\ \text{kJ/mol}_{\text{rxn}}$$

The overall change in the enthalpy of the system that occurs during the reaction can be calculated by combining ΔH for the two hypothetical steps in the reaction.

$$\begin{array}{rl} 678.7\ \text{kJ/mol}_{\text{rxn}} & \text{Bond breaking} \\ -863.2\ \text{kJ/mol}_{\text{rxn}} & \text{Bond making} \\ \hline -184.5\ \text{kJ/mol}_{\text{rxn}} & \end{array}$$

According to the calculation, the overall reaction is exothermic (ΔH is negative) by a total of −184.5 kJ when 1 mol of H_2 reacts with a mole of Cl_2 to form 2 mol of HCl.

$$H_2(g) + Cl_2(g) \longrightarrow 2\ HCl(g) \qquad \Delta H = -184.5\ \text{kJ/mol}_{\text{rxn}}$$

Does the reaction proceed through the hypothetical steps? It doesn't matter whether it does or it doesn't. At the heart of reaction thermodynamics is the concept that the enthalpy of a system is a state function. As a result, *the value of ΔH for a reaction doesn't depend on the path used to convert the starting materials into the products of the reaction.* It depends only on the initial and final conditions—the reactants and products of the reaction.

7.10 Standard-State Enthalpies of Reaction

The heat given off or absorbed by a chemical reaction depends on the conditions of the reaction. Three factors are important: (1) the amounts of the starting materials and products involved in the reaction, (2) the temperature at which the reaction is run, and (3) the pressure of any gases involved in the reaction. The reaction in which methane is burned can be used to illustrate why the reaction conditions must be specified.

Assume that we start with a mixture of CH_4 and O_2 at 25°C in a container just large enough so that the pressure of each gas is 1 atm.

$$CH_4(g) + 2\ O_2(g) \longrightarrow CO_2(g) + 2\ H_2O(g)$$

Under these conditions, the reaction gives off 802.4 kJ/mol$_{rxn}$. If we start with the reactants at 1000°C and 1 atm pressure, however, and generate the products at 1000°C and 1 atm, the reaction gives off only 800.5 kJ/mol$_{rxn}$. This illustrates the importance of specifying the conditions under which a reaction occurs when reporting thermodynamic data.

Thermodynamic data are often collected at 25°C (298 K). Measurements taken at other temperatures are identified by adding a subscript specifying the temperature in kelvins. The data collected for the combustion of methane at 1000°C, for example, would be reported as follows: $\Delta H_{1273} = -800.5$ kJ/mol$_{rxn}$.

The effect of pressure and amount of materials on the heat given off or absorbed in a chemical reaction is taken into account by defining a set of standard conditions for thermodynamic experiments. The **standard state** for thermodynamic measurements is defined in terms of the pure substances and the concentrations of any solutions involved in the reaction at 1 bar of pressure.

- The partial pressure of any gas is 1 bar, or 0.9869 atm. This can be rounded to 1 atm for all but the most exact calculations.
- The concentration of any solution is 1 *M*.

Enthalpy measurements done under the standard-state condition are indicated by adding a superscript "°" to the symbol for enthalpy. The *standard-state* enthalpy of reaction for the combustion of methane at 25°C, for example, would be reported as follows.

$$\Delta H^\circ = -802.4 \text{ kJ/mol}_{rxn}$$

➤ CHECKPOINT

What does the symbol ΔH°_{373} mean?

7.11 Calculating Enthalpies of Reaction

The origin of the enthalpy change that accompanies chemical reactions can be seen more clearly by examining the bonds in the products and reactants in the following reaction.

$$CO(g) + H_2O(g) \longrightarrow CO_2(g) + H_2(g)$$

We can use Lewis structures to visualize the reaction.

Once again, we can imagine a process for converting reactants into products. We break all of the bonds in the starting materials to form isolated atoms in the gas phase. Note that a carbon–oxygen triple bond and two oxygen–hydrogen single bonds are broken.

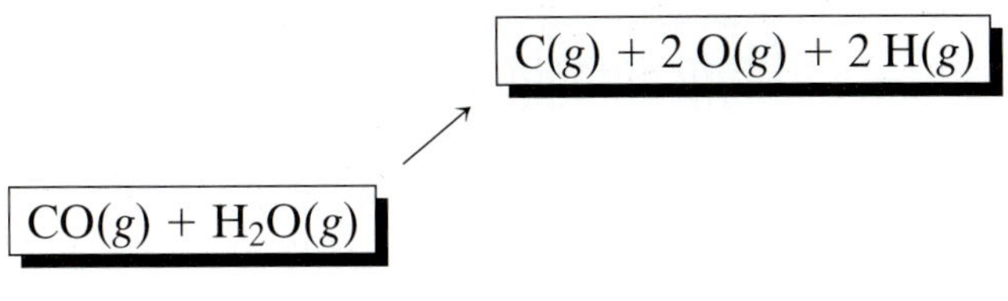

We then allow the atoms to recombine to form the products of the reaction.

$$C(g) + 2\,O(g) + 2\,H(g) \longrightarrow CO_2(g) + H_2(g)$$

Experimentally, we find that it takes 1076.4 kJ/mol$_{rxn}$ to break apart CO molecules to form isolated C and O atoms in the gas phase. It takes 926.3 kJ/mol$_{rxn}$ to break the bonds in H_2O molecules to form H and O atoms. Thus, the bond-breaking process takes a total of 2002.7 kJ.

$$\begin{array}{llll} \text{Bond breaking:} & CO(g) \longrightarrow C(g) + O(g) & & \Delta H^\circ = 1076.4 \text{ kJ/mol}_{rxn} \\ & H_2O(g) \longrightarrow 2\,H(g) + O(g) & & \Delta H^\circ = 926.3 \text{ kJ/mol}_{rxn} \\ \hline & CO(g) + H_2O(g) \longrightarrow C(g) + 2\,H(g) + 2\,O(g) & & \Delta H^\circ = 2002.7 \text{ kJ/mol}_{rxn} \end{array}$$

When the atom combination approach is used, the student sees directly that reaction enthalpies are determined by the difference in the sum of the bond strengths in the reactants and the sum of the bond strengths in the products.

When the isolated atoms in the gas phase are recombined to form new bonds, we see that two C=O double bonds are formed, giving off a total of 1608.5 kJ per mole of CO_2. An H—H single bond is also created. We have already seen that we get 435.3 kJ per mole of H_2 molecules formed when hydrogen atoms combine. We therefore get a total of 2043.8 kJ back when the C, H, and O atoms combine to form one mole of CO_2 and H_2.

$$\begin{array}{llll} \text{Bond making:} & C(g) + 2\,O(g) \longrightarrow CO_2(g) & & \Delta H^\circ = -1608.5 \text{ kJ/mol}_{rxn} \\ & 2\,H(g) \longrightarrow H_2(g) & & \Delta H^\circ = -435.3 \text{ kJ/mol}_{rxn} \\ \hline & C(g) + 2\,H(g) + 2\,O(g) \longrightarrow CO_2(g) + H_2(g) & & \Delta H^\circ = -2043.8 \text{ kJ/mol}_{rxn} \end{array}$$

When we combine the change in the enthalpy of the system for the two hypothetical steps, we conclude that the overall enthalpy of reaction is relatively small, compared with the enthalpy change associated with either bond breaking or bond making.

$$\begin{array}{rl} 2002.7 \text{ kJ/mol}_{rxn} & \text{Bond breaking} \\ -2043.8 \text{ kJ/mol}_{rxn} & \text{Bond making} \\ \hline -41.1 \text{ kJ/mol}_{rxn} & \end{array}$$

The reaction is exothermic, and it gives off heat because the sum of bond strengths in the products is larger than that in the reactants.

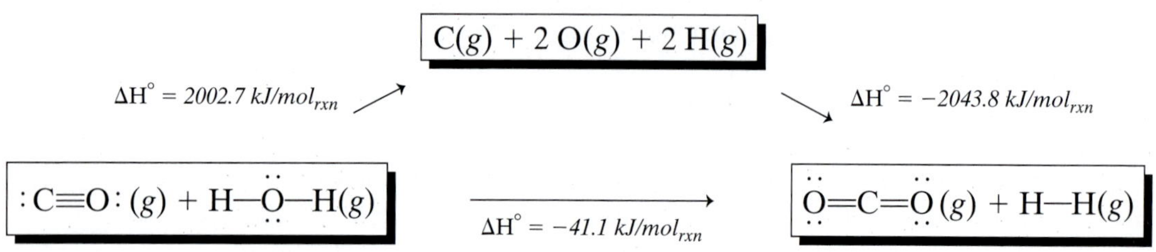

7.12 Enthalpies of Atom Combination

Chemists are often interested in knowing whether a reaction gives off or absorbs heat—and how much heat is given off or absorbed. The heat released or absorbed at a constant pressure is equivalent to the enthalpy change associated with the reaction and can be determined experimentally in the laboratory or calculated as was done in the previous section. It is important to remember that even the calculations are based on experimentally measured values.

There is no clear way for students to relate tabulated values of enthalpies of formation to the bonding situation in a given species. Atom combination parameters, however, have an inherent ease of conceptualization. Once students understand atom combination, the origin of enthalpy changes accompanying chemical reactions will be more readily apparent.

To apply the technique used in the previous section to other reactions, we need a set of data that allows us to predict how much heat will be absorbed when we transform the starting materials into their isolated atoms in the gas phase and how much heat will be given off when the atoms recombine to give the products of the reaction. There are several ways the data could be compiled. Consider, for example, the compound ammonia, NH_3. We could choose to give the enthalpy change for the reaction

$$NH_3(g) \longrightarrow N(g) + 3\ H(g)$$

for which the enthalpy required to break ammonia apart into its atoms is +1171.76 kJ/mol_{rxn}. When the reaction is written in this way, the enthalpy change is called the **enthalpy of atomization** because the reaction refers to the breaking apart of the compound into its atoms.

Another way to arrange the data in a table is based on the reverse of the reaction just shown.

$$N(g) + 3\ H(g) \longrightarrow NH_3(g)$$

In this case ammonia is formed from its atoms, and the enthalpy change for the reaction is −1171.76 kJ/mol_{rxn}. Such enthalpy changes are called **enthalpies of atom combination** because gaseous atoms combine to form a chemical substance. It makes no difference how we tabulate the data as long as we specify to which process, atomization or atom combination, the enthalpies refer. The authors of this text have chosen to use the atom combination reaction, and that is how data are compiled in Appendix B. The enthalpy of atom combination (ac) for $NH_3(g)$ can be found in Table B.13 by looking for the data for compounds of the element nitrogen, and then scanning these data until we find the entry for NH_3 as a gas. By definition, the symbol ΔH°_{ac} represents the enthalpy change under standard-state conditions when 1 mol of the compound listed in the table is formed from its atoms in the gas phase.

$$N(g) + 3\ H(g) \longrightarrow NH_3(g)$$

The value of ΔH°_{ac} for $NH_3(g)$ at one atm pressure and 25°C is −1171.76 kJ/mol_{rxn}.

When we start with a substance that is a gas at 25°C and 1 atm, the enthalpy of atom combination involves nothing more than forming the bonds within that substance. The value of ΔH°_{ac} for methane, for example, reflects the heat that is released when the C—H bonds in a mole of CH_4 molecules are formed from the gaseous atoms of carbon and hydrogen.

$$C(g) + 4\ H(g) \longrightarrow CH_4(g) \qquad \Delta H^\circ_{ac} = -1662.09\ kJ/mol_{rxn}$$

➤ CHECKPOINT

Write the chemical equations for the atom combination reactions of $N(g)$, $H(g)$, and $NH_3(g)$. What are the values of ΔH°_{ac} for $N(g)$ and $H(g)$? Explain your answers. What is the value of ΔH°_{ac} of $NH_3(g)$? Explain your answer.

The data in Appendix B, Table B.13, can be used to calculate ΔH for many processes without having to measure the heat given off or absorbed in the process experimentally. It is possible, for example, to use nothing more than the data in Table B.13 to calculate the economics of one fuel compared to another based on energy production.

Relatively few reactions involve only starting materials and products that exist as gases at room temperature and atmospheric pressure. The enthalpy of atom combination for liquid methanol (CH_3OH), for example, includes two terms in its measurement. The first is the heat released when we convert the carbon, hydrogen, and oxygen atoms into methanol in the gas phase.

$$C(g) + 4\ H(g) + O(g) \longrightarrow CH_3OH(g) \qquad \Delta H^\circ_{ac} = -2037.11\ \text{kJ/mol}_{rxn}$$

The origin of this heat is the formation of the bonds of gaseous methanol. The covalent bonds, formed by the sharing of electrons between atoms, are called *intramolecular bonds* because they are within the molecule itself.

The second part of the enthalpy of atom combination is the heat given off when the gas condenses to form liquid methanol, which can be measured by calorimetry.

$$CH_3OH(g) \longrightarrow CH_3OH(l) \qquad \Delta H^\circ = -38.00\ \text{kJ/mol}_{rxn}$$

The release of heat during a change of state is not due to formation of new intramolecular bonds. The covalent bonds in liquid and gaseous methanol are essentially the same. What then is responsible for the enthalpy change when gaseous methanol is condensed to the liquid state?

Enthalpies of atom combination are used to show that when methanol is formed from its atoms, a different enthalpy change is found depending on whether gaseous or liquid methanol is produced. This particular calculation shows that the molecules of methanol are the same whether they are in the gaseous or liquid state and that the bonding pattern does not change.

Enthalpy changes associated with phase changes of molecules are due to the making and breaking of intermolecular forces rather than the making and breaking of bonds. Chapter 8 gives a more detailed discussion of the origin of intermolecular interactions.

In the gaseous state the methanol molecules are far apart and only touch when they collide. In the liquid state the molecules of methanol are in constant contact with one another. Molecules contain electrons distributed between atoms according to the electronegativities of the atoms. The partial charges developed on the atoms can give a charge separation that produces an electric dipole moment. These dipoles are significant only when molecules are close together, and they serve to attract molecules to one another. Similar types of attractive forces exist between ions or other particles that make up a liquid or solid substance. Such forces are called *intermolecular forces*[2] when they exist between molecules and *interionic forces* when they occur between ions.

When a gas is condensed, its particles come close enough that they are attracted to one another, and as a consequence heat is released. Thus, the enthalpies of atom combination for liquids and solids are the sum of the contributing bond enthalpies and any additional interactions by intermolecular or interionic forces.

The enthalpy of atom combination for gaseous methanol from its gaseous atoms is -2037.11 kJ/mol$_{rxn}$, whereas the enthalpy of atom combination for liquid methanol from its gaseous atoms is -2075.11 kJ/mol$_{rxn}$.

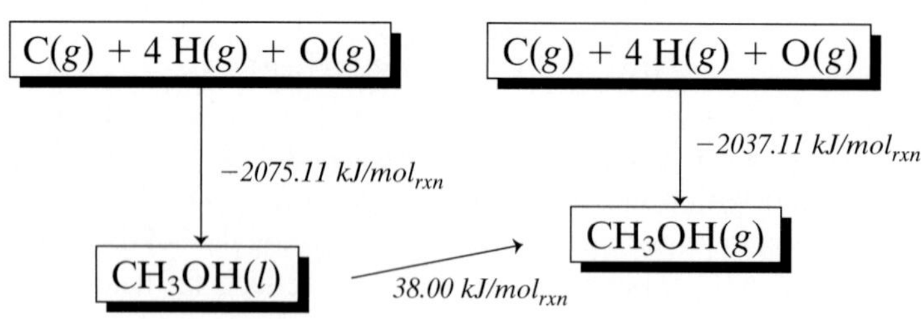

The difference between the two values (38.00 kJ/mol$_{rxn}$) represents the enthalpy which must be absorbed to evaporate one mole of liquid methanol to form gaseous methanol or the energy released when a mole of gaseous methanol condenses to form liquid methanol. The change in enthalpy is associated with forming or overcoming intermolecular forces.

Enthalpy of atom combination data for more than 200 substances are given in the table of standard-state enthalpies, free energies, and entropies of atom combination found in Appendix B. The first column of Table B.13 lists the chemical formulas and

[2]One way of remembering the difference between *intramolecular* and *intermolecular* forces is to remember that many college students participate in *intramural* sports, but the number who participate in *intercollegiate* sports is significantly smaller.

Table 7.4

Enthalpies of Atom Combination for Several Gaseous Organic Compounds[a]

Name	Formula	ΔH°_{ac}, kJ/mol$_{rxn}$ at 298.15 K
Methane	$CH_4(g)$	−1662.09
Ethane	$C_2H_6(g)$	−2823.94
Propane	$C_3H_8(g)$	−3992.9
n-Butane	$C_4H_{10}(g)$	−5169.38
Isobutane	$C_4H_{10}(g)$	−5177.75
n-Pentane	$C_5H_{12}(g)$	−6337.9
n-Hexane	$C_6H_{14}(g)$	−7509.1
Methanol	$CH_3OH(g)$	−2037.11
Ethanol	$CH_3CH_2OH(g)$	−3223.53
n-Propanol	$CH_3CH_2CH_2OH(g)$	−4394.2
n-Butanol	$CH_3CH_2CH_2CH_2OH(g)$	−5564.5
n-Pentanol	$CH_3CH_2CH_2CH_2CH_2OH(g)$	−6735.9
Dimethyl ether	$CH_3OCH_3(g)$	−3171.3
Ethylmethyl ether	$CH_3OCH_2CH_3(g)$	−4354.6
Diethyl ether	$CH_3CH_2OCH_2CH_3(g)$	−5541.4
Dipropyl ether	$CH_3CH_2CH_2OCH_2CH_2CH_3(g)$	−7883.1
Ethylene	$CH_2{=}CH_2(g)$	−2251.70
Propene	$CH_3CH{=}CH_2(g)$	−3432.6
1-Butene	$CH_2{=}CHCH_2CH_3(g)$	−4604.9
1-Pentene	$CH_2{=}CHCH_2CH_2CH_3(g)$	−5777.4
1-Hexene	$CH_2{=}CHCH_2CH_2CH_2CH_3(g)$	−6947.7
1,3-Butadiene	$CH_2{=}CHCH{=}CH_2(g)$	−4058.9
Benzene	$C_6H_6(g)$	−5523.07
Toluene	$C_6H_5CH_3(g)$	−6690.0
Ethylbenzene	$C_6H_5CH_2CH_3(g)$	−7870.3
1,3,5-Trimethylbenzene	$C_6H_3(CH_3)_3(g)$	−9067.2

[a]From J. D. Cox and G. Pilcher, *Thermochemistry of Organic and Organometallic Compounds,* Academic Press, New York, 1970.

physical states of a variety of substances. The second column lists the enthalpies of atom combination associated with forming all of the bonds in 1 mol of the substance from isolated atoms in the gas phase. In addition, Table 7.4 gives the enthalpies of atom combination data of several gas phase organic compounds. These data can be used to determine the change in enthalpy associated with any chemical reaction if data for all of the reactants and products are available in the table.

To illustrate how these data are used, let's consider the synthesis of ammonia from nitrogen and hydrogen.

$$N_2(g) + 3\,H_2(g) \longrightarrow 2\,NH_3(g)$$

Because enthalpy is a state function, we can divide the reaction into a sequence of bond breaking and bond making. Heat would have to be absorbed to break the N≡N triple bonds in a mole of N_2 and the H—H single bonds in three moles of H_2 to form isolated nitrogen and hydrogen atoms in the gas phase.

The usual way of teaching thermodynamics involves standard free energies and enthalpies of formation which rely upon standard states involving the elements in their stable states of aggregation, usually at 298 K. Standard entropies introduce yet another standard state, that of the crystalline material at absolute zero. There is no conceptually easy way to relate any of these parameters to bond making and breaking.

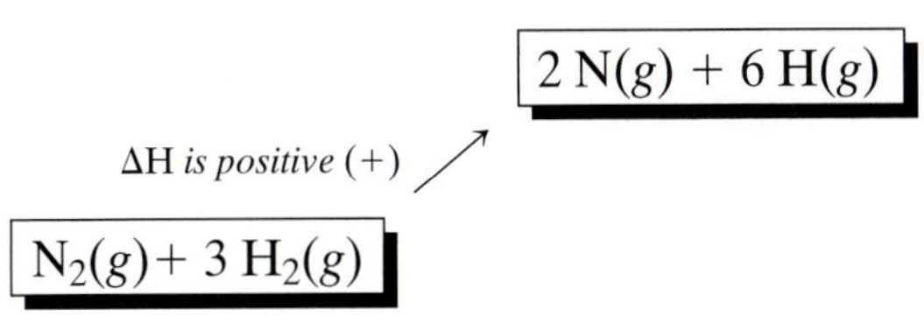

Heat is then given off when the atoms come together to form NH_3 molecules in the gas phase.

$$2\,N(g) + 6\,H(g) \xrightarrow{\Delta H \text{ is negative } (-)} 2\,NH_3(g)$$

The primary reason for introducing the sign convention is to remind us that endothermic reactions, which absorb heat from the surroundings, are represented by *positive* values of ΔH. Exothermic reactions, which give off heat, are described by *negative* values of ΔH. If we wish to know the enthalpy change for breaking the bonds in a compound, we must reverse the sign on the enthalpy given in Appendix B.13.

Enthalpies of atom combination treat all species the same way by describing the formation of each species from its atoms in the monatomic gaseous state. Thus for all thermodynamic parameters the same standard state is used. The conceptual obscurity of a standard state related to molecular species rather than to the atoms which compose a material is alleviated by the use of atom combination parameters.

It always takes energy to break the bonds in a molecule to form isolated atoms in the gas phase. When we calculate the heat that must be absorbed to break the bonds for this hypothetical reaction, we must therefore use positive values for the enthalpy for each of the starting materials consumed in the step. Heat is always given off when atoms recombine to form molecules. Negative enthalpies are therefore used to calculate the heat given off when the products of the reaction are formed.

The energetics of the bond breaking in the synthesis of ammonia can be calculated from the enthalpies tabulated in Table B.13 for H_2 ($\Delta H^\circ_{ac} = -435.30$ kJ/mol$_{rxn}$) and N_2 ($\Delta H^\circ_{ac} = -945.41$ kJ/mol$_{rxn}$). In the course of this hypothetical step in the reaction, 1 mole of N_2 and 3 moles of H_2 are transformed into hydrogen and nitrogen atoms in the gas phase. The enthalpy of reaction for breaking the N≡N triple bonds in a mole of N_2 is therefore equal to the enthalpy required to break all the bonds in the molecule.

At the same time, 3 moles of H_2 is converted into 6 moles of hydrogen atoms in the gas phase. The enthalpy of reaction for breaking the H—H single bonds in 3 moles of H_2 is equal to three times the enthalpy required for one mole of H_2 molecules.

$$\begin{array}{llll} \text{Bond breaking:} & N_2(g) \longrightarrow 2\,N(g) & \Delta H^\circ = & 945.41 \text{ kJ/mol}_{rxn} \\ & 3\,H_2(g) \longrightarrow 6\,H(g) & \Delta H^\circ = & 1305.9 \text{ kJ/mol}_{rxn} \\ \hline & N_2(g) + 3\,H_2(g) \longrightarrow 2\,N(g) + 6\,H(g) & \Delta H^\circ = & 2251.3 \text{ kJ/mol}_{rxn} \end{array}$$

It is important to stress that enthalpies of atom combination are known to the same precision as enthalpies of formation and allow calculation of enthalpy change to the same precision. Enthalpies of atom combination are not *average bond energies. They are experimentally determined and may be easily calculated from standard enthalpies of formation.*

The heat given off during bond formation can be calculated from the enthalpy of atom combination of NH_3($\Delta H^\circ_{ac} = -1171.76$ kJ/mol$_{rxn}$ in Table B.13 in the appendix). When the isolated nitrogen and hydrogen atoms come together to form NH_3 molecules, 2 mol of NH_3 is formed. The enthalpy of reaction for the formation of 2 mol of NH_3 is therefore twice the enthalpy of atom combination of NH_3.

$$\text{Bond making:} \quad 2\,N(g) + 6\,H(g) \longrightarrow 2\,NH_3(g) \qquad \Delta H^\circ_{ac} = -2343.52 \text{ kJ/mol}_{rxn}$$

Bond breaking is endothermic (ΔH is positive) and that heat is released when the atoms recombine to form bonds, which means that bond making is exothermic (ΔH is negative).

The making and breaking of bonds is emphasized rather than the rote "products minus reactants" rule. This gives students a clearer picture of the source of the change in enthalpy during a chemical reaction.

All we have to do to calculate the overall enthalpy of reaction is add the results of our calculations for the two hypothetical steps in the reaction.

$$\begin{array}{lr} \text{Bond breaking:} & 2251.3 \text{ kJ/mol}_{rxn} \\ \text{Bond making:} & -2343.52 \text{ kJ/mol}_{rxn} \\ \hline & -92.2 \text{ kJ/mol}_{rxn} \end{array}$$

The overall reaction is therefore exothermic.

It is important to remember when doing enthalpy calculations this way that very few chemical reactions actually occur by first breaking all the bonds in the starting materials to form isolated atoms in the gas phase, followed by recombination of the atoms to form the products of the reaction. But, for our purposes, that doesn't matter. If all we want to know is whether the reaction gives off or absorbs heat, we can *assume* that it occurs by atomization and recombination steps. Because enthalpy is a state function, ΔH for a reaction doesn't depend on the path used to convert the starting materials into the products.

Exercise 7.6

Charcoal is mainly carbon. Use enthalpy of atom combination data from Appendix B.13 to predict whether the reaction between charcoal and oxygen to form carbon dioxide is endothermic or exothermic and to calculate the amount of heat given off or absorbed in the reaction below.

$$C(s) + O_2(g) \longrightarrow CO_2(g)$$

Solution

We can imagine the process by using the following diagram.

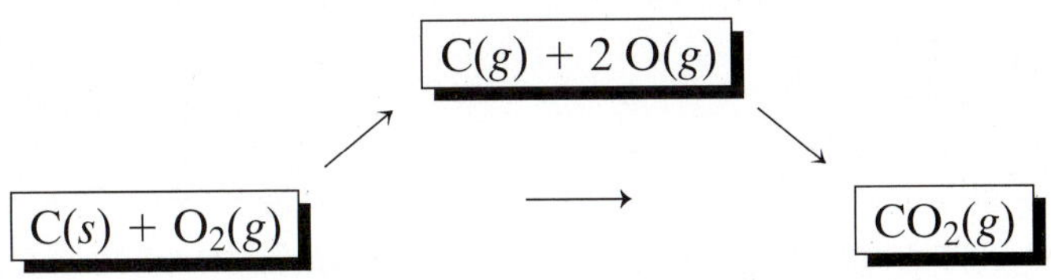

When solid carbon is formed from isolated carbon atoms in the gas phase, a significant release of heat occurs, as shown in the enthalpy of atom combination data.

$$C(g) \longrightarrow C(s) \qquad \Delta H^\circ_{ac} = -716.68 \text{ kJ/mol}_{rxn}$$

Likewise, atom combination data show that when the O=O double bond in an O_2 molecule is formed, 498.34 kJ/mol$_{rxn}$ of heat is released.

$$2\,O(g) \longrightarrow O_2(g) \qquad \Delta H^\circ_{ac} = -498.34 \text{ kJ/mol}_{rxn}$$

Therefore, the hypothetical step in which we break all the bonds in the starting materials requires a large input of heat (enthalpy).

$$\begin{array}{lrcl} \text{Bond breaking:} & C(s) \longrightarrow C(g) & \Delta H^\circ = & 716.68 \text{ kJ/mol}_{rxn} \\ & O_2(g) \longrightarrow 2\,O(g) & \Delta H^\circ = & 498.34 \text{ kJ/mol}_{rxn} \\ \hline & C(s) + O_2(g) \longrightarrow C(g) + 2\,O(g) & \Delta H^\circ = & 1215.02 \text{ kJ/mol}_{rxn} \end{array}$$

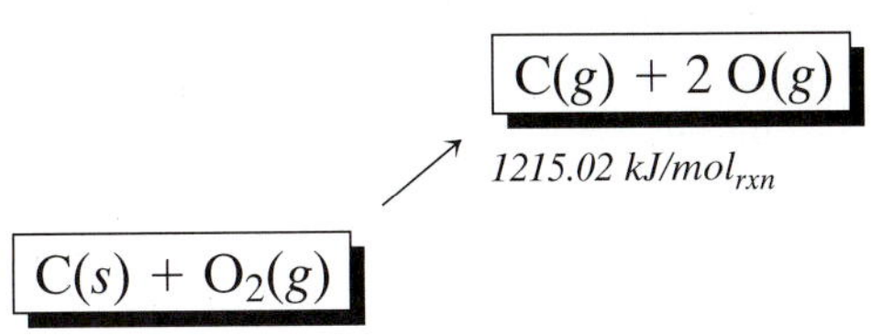

➤ CHECKPOINT

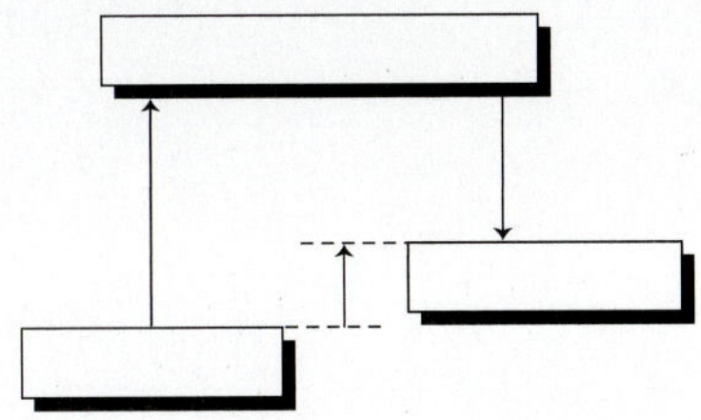

For the purposes of calculating enthalpy changes, a chemical reaction can be diagramned as shown above. If the total bond strengths in the products are less than that in the reactants, label the boxes above as *reactants, products,* and *gaseous atoms*. Draw and label a similar diagram if the total bond strengths in the products are greater than that in the reactants. Also label the arrows that represent the *enthalpy to break bonds, the enthalpy to form bonds*, and the *enthalpy change* for the overall reaction.

Students find the use of enthalpies of atom combination combined with the use of reaction diagrams that show the making and breaking of bonds to be very straightforward. Students have traditionally had difficulty with both understanding thermochemical calculations and determining the correct sign for the change in enthalpy.

The use of enthalpies of atom combination makes the conceptual understanding of the processes that lead to changes in enthalpy much more concrete than the algorithm employed in enthalpy of formation calculations. Having mastered enthalpies of atom combination, students seem to have no difficulty moving to calculations employing enthalpies of formation.

But the C═O double bonds in a CO_2 molecule are unusually strong ($\Delta H^\circ_{ac} = -1608.53$ kJ/mol). Thus, a great deal of energy is given off when the isolated atoms come together to form CO_2 molecules.

Bond making: $C(g) + 2\ O(g) \longrightarrow CO_2(g)$ $\Delta H^\circ_{ac} = -1608.53\ \text{kJ/mol}_{rxn}$

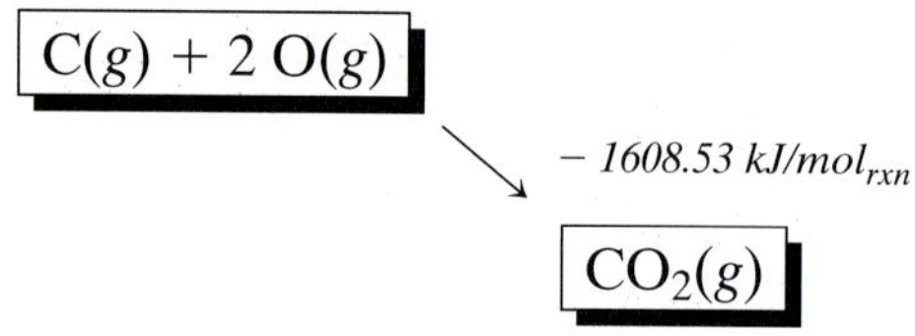

The overall reaction is therefore strongly exothermic.

Bond breaking:	$1215.02\ \text{kJ/mol}_{rxn}$
Bond making:	$-1608.53\ \text{kJ/mol}_{rxn}$
	$-393.51\ \text{kJ/mol}_{rxn}$

The heat produced by this reaction is used in backyard grills to cook hamburgers.

Exercise 7.7

The reason that charcoal can be used in a grill is that the C═O double bonds in CO_2 are so strong that heat is given off in the reaction. Carbon from charcoal can also react with oxygen to produce carbon monoxide (CO). Is the reaction endothermic or exothermic?

Solution

The reaction is represented by the following equation.

$$2\ C(s) + O_2(g) \longrightarrow 2\ CO(g)$$

The enthalpy change can be understood in terms of the following diagram.

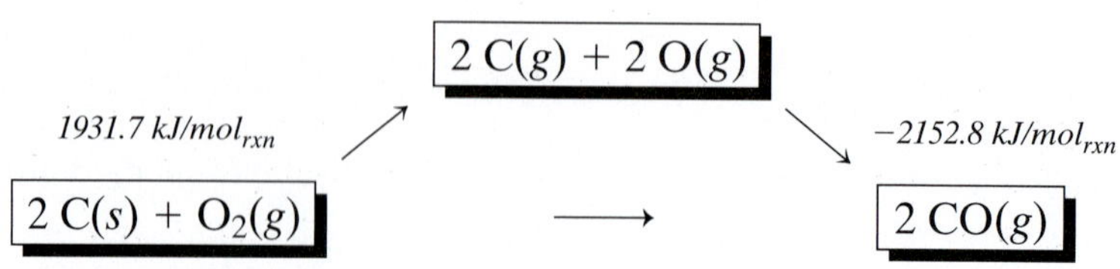

The overall enthalpy of reaction can be calculated by adding the enthalpy of reaction for the two steps.

$$\Delta H^\circ = 1931.7\ \text{kJ/mol}_{rxn} + (-2152.8\ \text{kJ/mol}_{rxn}) = -221.1\ \text{kJ/mol}_{rxn}$$

The reaction is exothermic, but not as exothermic as the reaction in which charcoal burns to form carbon dioxide.

$$C(s) + O_2(g) \longrightarrow CO_2(g) \qquad \Delta H^\circ = -393.51\ \text{kJ/mol}_{rxn}$$

The reason is that the C≡O triple bond in CO isn't as strong as the two C=O double bonds in CO_2. The reaction of carbon with oxygen can and does produce carbon monoxide, particularly if inadequate supplies of oxygen are available for conversion to CO_2. Carbon monoxide can combine with the iron in human blood and is toxic. To cut down on the production of CO from automobiles, catalytic converters are used to convert CO to CO_2 by the reaction

$$2\ CO(g) + O_2(g) \longrightarrow 2\ CO_2(g)$$

➤ CHECKPOINT

Is the following reaction endothermic or exothermic?

$$2\ CO(g) + O_2(g) \longrightarrow 2\ CO_2(g)$$

7.13 Using Enthalpies of Atom Combination to Probe Chemical Reactions

Enthalpies of atom combination can be used in ways other than to calculate the enthalpy change for a reaction. Values of these enthalpies measure the total bond strengths for gaseous compounds. They therefore provide a direct means of comparing the strengths of the bonds holding a compound together.

In 1833, Jöns Jacob Berzelius suggested that compounds with the same formula but different structures should be called **isomers** (literally, "equal parts"). The following compounds are isomers because they both contain the same number of carbon and hydrogen atoms.

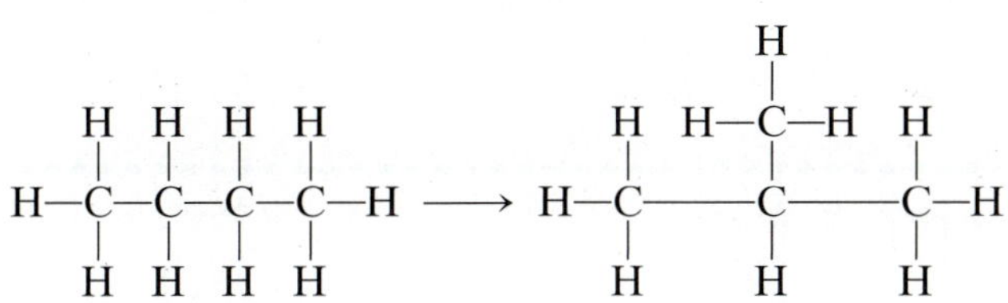

Butane *Isobutane*

Butane is known as a straight-chain hydrocarbon because it contains one continuous chain of C—C bonds. Isobutane, on the other hand, is an example of a branched hydrocarbon. The so-called "butane lighter fluid" actually contains more isobutane than anything else.

Because they are isomers, these compounds must have the same molecular formula: C_4H_{10}. But there is an even greater similarity between the two compounds. Each compound contains 3 C—C bonds and 10 C—H bonds. Are the bonds all of the same strength? Enthalpy of atom combination data from Table 7.4 show us that butane ($\Delta H^\circ_{ac} = -5169.38$ kJ/mol$_{rxn}$) has a slightly less negative enthalpy of atom combination than isobutane ($\Delta H^\circ_{ac} = -5177.75$ kJ/mol$_{rxn}$). If we form the compounds from their atoms, more heat will be released when the atoms combine to form isobutane. Thus, isobutane(g) has stronger bonds than butane(g).

We can also calculate the enthalpy change for the reaction given above.

Bond breaking in butane:	$\Delta H^\circ =$	5169.38 kJ/mol$_{rxn}$
Bond making in isobutane:	$\Delta H^\circ_{ac} =$	−5177.75 kJ/mol$_{rxn}$
	$\Delta H^\circ =$	−8.37 kJ/mol$_{rxn}$

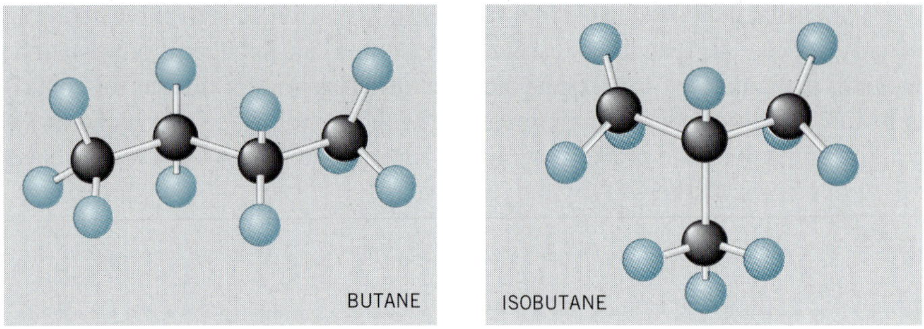

Fig. 7.7 Two isomers with the formula C_4H_{10} are possible: butane and isobutane. Butane is a straight-chain hydrocarbon, and isobutane is a branched hydrocarbon. All carbon atoms in both hydrocarbons have 109° bond angles with hydrogen.

We see that the overall enthalpy change for the conversion of butane(g) into isobutane(g) is negative (-8.37 kJ/mol$_{rxn}$), which means that the bonds formed in the product, isobutane, are stronger than those broken in the reactant, butane. This information tells us that it may be possible to convert straight-chain hydrocarbons to branched chains. Experimental data confirm this. At high temperatures (500–600°C) and high pressures (25–50 atm), straight-chain hydrocarbons isomerize to form branched hydrocarbons. The reaction can be run under more moderate conditions in the presence of a catalyst, such as a mixture of silica (SiO_2) and alumina (Al_2O_3). Reactions such as this play a vital role in the refining of gasoline, because branched alkanes burn more evenly than straight-chain alkanes and are therefore less likely to cause an engine to "knock."

Although we tend to describe butane as a straight-chain hydrocarbon, it is important to remember that the ED model would predict that the geometry around each carbon atom in the molecule would be tetrahedral. The shapes of the butane and isobutane molecules are shown in Figure 7.7.

Exercise 7.8

Enthalpies of atom combination give a direct measurement of the strengths of the bonds in reactants and products. It is therefore possible for students to gain a deeper understanding of the origin of the change in enthalpy during a chemical reaction.

Fumaric acid and maleic acid are isomers that differ only in the orientation of substituents around the C=C double bond in these compounds.

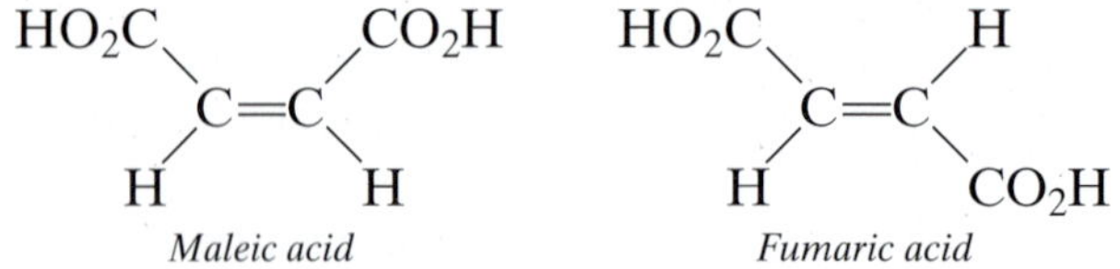

Maleic acid *Fumaric acid*

In maleic acid, the two hydrogen atoms and the two —CO_2H groups are on the same side of a horizontal plane passing through the atoms in the C=C double bond. In fumaric acid, they are on opposite sides of the bond. Use the enthalpies of atom combination of fumaric acid ($\Delta H^\circ_{ac} = -5545.03$ kJ/mol$_{rxn}$) and maleic acid ($\Delta H^\circ_{ac} = -5524.53$ kJ/mol$_{rxn}$) to predict whether we should be able to convert fumaric acid to maleic acid, or maleic acid to fumaric acid.

Solution

More heat is released when fumaric acid is formed from its atoms than when maleic acid forms. Thus fumaric acid has *stronger bonds* than maleic acid. This suggests that it may be possible to convert maleic acid to fumaric acid by some chemical process. Again we can also calculate the enthalpy change for the conversion of maleic acid to fumaric acid:

Bond breaking in maleic acid:	ΔH° =	5524.53 kJ/mol_{rxn}
Bond making in fumaric acid:	ΔH°_{ac} =	−5545.03 kJ/mol_{rxn}
	ΔH° =	−20.50 kJ/mol_{rxn}

Experimentally we find that maleic acid can be converted to fumaric acid by heating maleic acid in the presence of a strong acid for about 30 minutes.

➤ CHECKPOINT

Compare the enthalpies of atom combination of $CH_3CH_2OH(g)$ and $CH_3CH_2OH(l)$. What can you conclude from the difference in the two enthalpies?

Exercise 7.9

The enthalpy change for the reaction of HF with SiO_2

$$4\ HF(g) + SiO_2(s) \longrightarrow SiF_4(g) + 2\ H_2O(g)$$

is −103.4 kJ/mol_{rxn}. That for the reaction with HCl

$$4\ HCl(g) + SiO_2(s) \longrightarrow SiCl_4(g) + 2\ H_2O(g)$$

is +139.6 kJ/mol_{rxn}.

How can we account for the difference in the enthalpy change for these two similar reactions? What is the average bond enthalpy for Si—Cl and Si—F bonds?

Solution

The magnitude and sign of the enthalpy change are determined by the strengths of the bonds in the products and reactants. These two reactions have two species in common, SiO_2 and H_2O, so the difference in the enthalpy change must be due to the difference in bond strengths for HF and SiF_4 in one reaction and HCl and $SiCl_4$ in the other. Appendix B.13 provides the necessary data.

	ΔH°_{ac} (kJ/mol_{rxn})
$HF(g)$	−567.7
$SiO_2(s)$	−1864.9
$SiF_4(g)$	−2386.5
$H_2O(g)$	−926.29
$HCl(g)$	−431.64
$SiCl_4(g)$	−1599.3

We note that although the bond strength of HF (567.7 kJ/mol_{rxn}) is greater than that of HCl (431.64 kJ/mol_{rxn}), the reaction with HF liberates more heat (−103.4 kJ/mol_{rxn}) than the reaction with HCl (139.6 kJ/mol_{rxn}). Therefore the bond strengths in the product SiF_4 must be greater than the bond strengths in $SiCl_4$. The data confirm this. The Si—F bonds in SiF_4 are much stronger than the Si—Cl bonds in $SiCl_4$. Despite the lower bond strength of HCl as compared to HF, the bonds formed by the product SiF_4 more than compensate for the enthalpy required to break the HF bond.

The average bond enthalpy for Si—F and Si—Cl bonds may be estimated from ΔH°_{ac} data. For SiF_4 with four Si—F bonds, the average Si—F bond strength is $2386.5/4 = 597$ kJ/mol, and that for the Si—Cl bond is $1599.3/4 = 400$ kJ/mol. Thus almost 200 kJ more heat is released when a mole of Si—F bonds is formed than when a mole of Si—Cl bonds is formed.

Exercise 7.10

Frequently chemists need approximate data for compounds for which no thermodynamic data are available. One way of estimating the bond strengths in a compound is to use average bond enthalpies. Estimate the sum of the bond strengths in Cl_3SiOH.

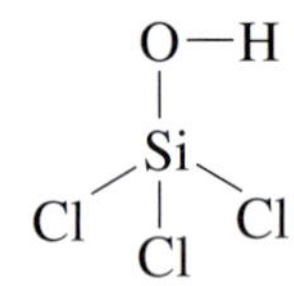

Solution

In the previous exercise we saw that the average Si—Cl bond enthalpy was 400 kJ/mol. The O—H bond enthalpy may be estimated from the data for H_2O.

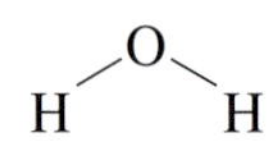

$$926.29/2 = 463 \text{ kJ per mole of O—H bonds}$$

The SiO_2 (silica) Si—O bond is somewhat more difficult. The empirical formula is SiO_2 but silica is actually a network of Si—O bonds. Each silicon atom is bonded to four oxygen atoms.

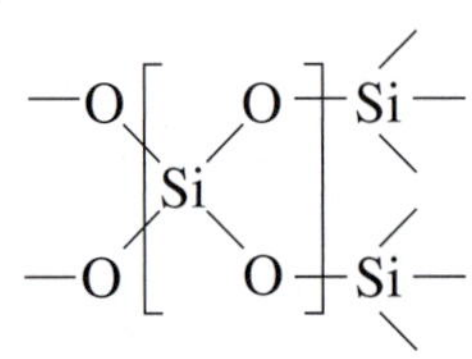

Thus the average Si—O bond enthalpy is $1864.9/4 = 466$ kJ/mol.

Now we may estimate the bond strengths in Cl_3SiOH. There are three Si—Cl bonds, one Si—O bond, and one O—H bond for a total of

$$3 \times (400) + (466) + (463) = 2129 \text{ kJ/mol}$$

➤ CHECKPOINT

Estimate the sum of the bond strengths in F_3SiOH.

7.14 Bond Length and the Enthalpy of Atom Combination

The relationship between bond length and enthalpies of atom combination is more easily recognizable than the relationship with enthalpies of formation.

In Chapters 3 and 4 we noted that the distance between neighboring atoms, the bond length, could often be estimated from the relative size of atoms. We expect a C—O single bond to be longer than a C—H single bond, for example, because the covalent radius of oxygen is larger than that of hydrogen. We can now relate the strengths of certain bonds to the bond length. Figure 3.26 gives covalent radii from which the bond lengths for the series of halogen acids given in Table 7.5 can be

Table 7.5

Bond Lengths and Enthalpies of Atom Combination for the Hydrogen Halides

Molecule	Bond Length (nm)	ΔH°_{ac} (kJ/mol$_{rxn}$)
HF(*g*)	0.101	−567.7
HCl(*g*)	0.136	−431.64
HBr(*g*)	0.151	−365.93
HI(*g*)	0.170	−298.01

estimated. The enthalpies of atom combination taken from Appendix B.13 are also given in Table 7.5.

The strongest hydrogen–halogen bond, H—F, has the shortest bond length. The other halogen compounds with hydrogen follow the trend of increasing bond length with decreasing bond strength. In general, if the atoms composing a bond are similar, with the same types of bonds being formed, the atom pair with the largest internuclear distance will have the weakest bond strength.

> **➤ CHECKPOINT**
>
> Predict which molecule, ClF(*g*) or BrF(*g*), has the weakest bond.

Exercise 7.11

The bond length in NO(*g*) is about the same as the bond length in CO(*g*), but the enthalpy of atom combination for CO(*g*) is almost double that for NO(*g*). How can this be explained?

Solution

On the basis of atomic covalent radii, we would expect the bond length of a single bond in NO to be smaller than that of a single bond in CO. That the bond lengths are about the same suggests that different types of bonding may be present. The very much larger enthalpy of bond formation confirms this. Therefore, a close look at the structures of the two compounds is required. The Lewis structures for these compounds can be written as follows.

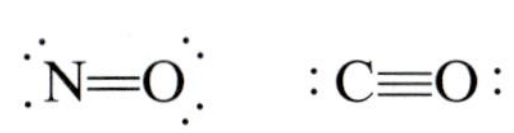

These structures make clear that in CO the bond is a triple bond, whereas in NO the bond is a relatively weaker double bond.

7.15 Hess's Law

As we have seen, enthalpy is a state function. As a result, the value of ΔH for a reaction doesn't depend on the path used to go from one of the states to the other.

In 1840, Germain Henri Hess, professor of chemistry at the University and Artillery School in St. Petersburg, Russia, came to the same conclusion on the basis of experiment and proposed a general rule known as **Hess's law,** which states that ΔH_{rxn} is the same regardless of whether a reaction occurs in one step or in several steps. Thus we can calculate the enthalpy of reaction by adding the enthalpies associated with a series of hypothetical steps into which the reaction can be broken.

Exercise 7.12

The heat given off under standard-state conditions when water is formed from its elements as both a liquid and a gas has been measured.

$$H_2(g) + \tfrac{1}{2}\,O_2(g) \longrightarrow H_2O(l) \qquad \Delta H^\circ = -285.83 \text{ kJ/mol}_{rxn}$$
$$H_2(g) + \tfrac{1}{2}\,O_2(g) \longrightarrow H_2O(g) \qquad \Delta H^\circ = -241.82 \text{ kJ/mol}_{rxn}$$

Use these data and Hess's law to calculate ΔH° for the following reaction.

$$H_2O(l) \longrightarrow H_2O(g)$$

Solution

The key to this problem is finding a way to combine the two reactions for which experimental data are known to give the reaction for which ΔH° is unknown. We can do this by reversing the direction in which the first reaction is written and then adding it to the equation for the second reaction. We assume that 1 mol of water is decomposed into its elements in the first reaction and that a mole of water vapor is formed from its elements in the second reaction.

$$\begin{array}{rcll} H_2O(l) & \longrightarrow & H_2(g) + \tfrac{1}{2}\,O_2\,(g) & \Delta H^\circ = 285.83 \text{ kJ/mol}_{rxn} \\ H_2(g) + \tfrac{1}{2}\,O_2(g) & \longrightarrow & H_2O(g) & \Delta H^\circ = -241.82 \text{ kJ/mol}_{rxn} \\ \hline H_2O(l) & \longrightarrow & H_2O(g) & \Delta H^\circ = 44.01 \text{ kJ/mol}_{rxn} \end{array}$$

When we reversed the direction in which the first reaction was written, we had to change the sign of ΔH° for the reaction. $H_2(g)$ and $\frac{1}{2}\,O_2(g)$ don't appear in the final chemical equation because they are on *both* sides of the chemical equations added together to obtain the final chemical equation. These terms therefore cancel.

There was just enough information in Exercise 7.12 to solve the problem. The next exercise forces us to choose, from a wealth of information, the reactions that will be combined.

Exercise 7.13

Before pipelines were built to deliver natural gas, individual towns and cities contained plants that produced a fuel known as "town gas" by passing steam over red-hot charcoal.

$$C(s) + H_2O(g) \longrightarrow CO(g) + H_2(g)$$

Calculate ΔH° for the reaction from the following information.

$$C(s) + \tfrac{1}{2}\,O_2(g) \longrightarrow CO(g) \qquad \Delta H^\circ = -110.53 \text{ kJ/mol}_{rxn}$$
$$C(s) + O_2(g) \longrightarrow CO_2(g) \qquad \Delta H^\circ = -393.51 \text{ kJ/mol}_{rxn}$$
$$CO(g) + \tfrac{1}{2}\,O_2(g) \longrightarrow CO_2(g) \qquad \Delta H^\circ = -282.98 \text{ kJ/mol}_{rxn}$$
$$H_2(g) + \tfrac{1}{2}\,O_2(g) \longrightarrow H_2O(g) \qquad \Delta H^\circ = -214.82 \text{ kJ/mol}_{rxn}$$

Solution

In this case, the desired equation can be found by using the first reaction to generate $CO(g)$ from $C(s)$ and the reverse of the fourth reaction to generate $H_2(g)$ from $H_2O(g)$.

$$\begin{array}{rll} C(s) + \frac{1}{2}O_2(g) \longrightarrow CO(g) & \Delta H^\circ = & -110.53 \text{ kJ/mol}_{rxn} \\ H_2O(g) \longrightarrow H_2(g) + \frac{1}{2}O_2(g) & \Delta H^\circ = & 214.82 \text{ kJ/mol}_{rxn} \\ \hline C(s) + H_2O(g) \longrightarrow CO(g) + H_2(g) & \Delta H^\circ = & 131.29 \text{ kJ/mol}_{rxn} \end{array}$$

7.16 Enthalpies of Formation

Hess's law suggests that we can save a great deal of work measuring enthalpies of reaction by using a little imagination in choosing the reactions for which measurements are made. The question is: What is the best set of reactions to study so that we get the greatest benefit from the smallest number of experiments?

In Exercise 7.13 we calculated ΔH° for the reaction

$$C(s) + H_2O(g) \longrightarrow CO(g) + H_2(g)$$

by combining enthalpy of reaction measurements for the following reactions.

$$C(s) + \tfrac{1}{2}O_2(g) \longrightarrow CO(g)$$
$$H_2(g) + \tfrac{1}{2}O_2(g) \longrightarrow H_2O(g)$$

These reactions have one thing in common. Each reaction leads to the formation of a compound from the elements in their most thermodynamically stable form at 25°C. The enthalpy of reaction for each of the reactions is therefore the **enthalpy of formation** of the compound, ΔH_f°. By definition, ΔH_f° is the enthalpy associated with the reaction that forms 1 mole of a substance from its elements in their most thermodynamically stable states at 25°C and 1 atm. Enthalpies of formation are similar to enthalpies of atom combination in that both represent the formation of a substance from its elements. However, the values differ because enthalpies of formation represent formation from the elements in their *most thermodynamically stable states* as compared to enthalpies of atom combination in which compounds are formed from *gaseous atoms*. Because the enthalpy change of a reaction is a state function, tabulated enthalpies of formation in conjunction with Hess's law can be used to calculate enthalpy changes of reaction in a manner similar to that used with enthalpies of atom combination.

After students have learned to use enthalpies of atom combination, ΔH°_{ac}, and they understand what changes in enthalpy mean, it is not difficult for them to understand that enthalpies of formation are just another way of tabulating data for the purpose of calculating ΔH for a reaction. Instructors should consider coverage of this section if their students will be going on to courses which utilize enthalpies of formation.

The text emphasizes that most chemical reactions do not proceed by the breaking of all bonds in the reactants and the subsequent formation of the bonds in the product. Equally important is the recognition that reactions do not proceed by a step-by-step process as indicated by the use of Hess's law.

By incorporating both enthalpies of atom combination and heats of formation for thermochemistry calculations, the concept that change in enthalpy is a state function which is independent of the reaction pathway is reinforced.

Exercise 7.14

Which of the following equations describes a reaction for which ΔH° is equal to the enthalpy of formation of a compound, ΔH_f°?

(a) $Mg(s) + \frac{1}{2}O_2(g) \longrightarrow MgO(s)$

(b) $MgO(s) + CO_2(g) \longrightarrow MgCO_3(s)$

(c) $Mg(s) + C(s) + \frac{3}{2}O_2(g) \longrightarrow MgCO_3(s)$

(d) $Mg(g) + O(g) \longrightarrow MgO(s)$

Solution

Equations (a) and (c) describe enthalpy of formation reactions. Those reactions result in the formation of a compound from the most thermodynamically stable form of its elements. Equation (b) can't be an enthalpy of formation reaction because the product of the reaction isn't formed from its elements. Equation (d) is an enthalpy of atom combination reaction, not an enthalpy of formation reaction.

Exercise 7.15

Use Hess's law to calculate $\Delta H°$ for the reaction

$$MgO(s) + CO_2(g) \longrightarrow MgCO_3(s)$$

from the following enthalpy of formation data.

$$\begin{aligned} Mg(s) + \tfrac{1}{2}\,O_2(g) &\longrightarrow MgO(s) & \Delta H_f^\circ &= -601.70\ \text{kJ/mol}_{rxn} \\ C(s) + O_2(g) &\longrightarrow CO_2(g) & \Delta H_f^\circ &= -393.51\ \text{kJ/mol}_{rxn} \\ Mg(s) + C(s) + \tfrac{3}{2}\,O_2(g) &\longrightarrow MgCO_3(s) & \Delta H_f^\circ &= -1095.8\ \text{kJ/mol}_{rxn} \end{aligned}$$

Solution

The reaction in which we are interested converts two reactants (MgO and CO_2) into a single product ($MgCO_3$). We might therefore start by reversing the direction in which we write the enthalpy of formation reactions for MgO and CO_2, thereby decomposing these substances into their elements in their most thermodynamically stable form. We can then add the enthalpy of formation reaction for $MgCO_3$, thereby forming the product from its elements in their most stable form.

$$\begin{aligned} MgO(s) &\longrightarrow Mg(s) + \tfrac{1}{2}\,O_2(g) & \Delta H° &= 601.70\ \text{kJ/mol}_{rxn} \\ CO_2(g) &\longrightarrow C(s) + O_2(g) & \Delta H° &= 393.51\ \text{kJ/mol}_{rxn} \\ Mg(s) + C(s) + \tfrac{3}{2}\,O_2(g) &\longrightarrow MgCO_3(s) & \Delta H_f^\circ &= -1095.8\ \text{kJ/mol}_{rxn} \\ \hline MgO(s) + CO_2(g) &\longrightarrow MgCO_3(s) & \Delta H° &= -100.6\ \text{kJ/mol}_{rxn} \end{aligned}$$

Adding the three equations gives the desired unknown reaction. $\Delta H°$ for the reaction is therefore the sum of the enthalpies of the three hypothetical steps. We recognize that this chemical reaction does not occur as a result of these steps. Because ΔH is independent of path, any combination of reactions can be used to yield the overall reaction.

The procedure used in Exercise 7.15 works, no matter how complex the reaction. All we have to do as the reaction becomes more complex is add more intermediate steps. This approach works because enthalpy is a state function. Thus, $\Delta H°$ is the same regardless of the path used to get from the starting materials to the products of the reaction. Instead of running the reaction in a single step,

$$MgO(s) + CO_2(g) \longrightarrow MgCO_3(s)$$

we can split it into two steps. In the first step, the starting materials are converted to the elements from which they form in their most thermodynamically stable states.

$$MgO(s) + CO_2(g) \longrightarrow Mg(s) + C(s) + \tfrac{3}{2}\,O_2(g)$$

In the second step, the elements combine to form the products of the reaction.

$$Mg(s) + C(s) + \tfrac{3}{2}\,O_2(g) \longrightarrow MgCO_3(s)$$

If we analyze the technique used to solve the problem in Exercise 7.15, we find that we calculated ΔH°_{rxn} for the reaction in which MgO and CO_2 combine to form $MgCO_3$ by *adding* the enthalpy of formation for the products of this reaction and *subtracting* the enthalpy of formation for each of the reactants. In other words, we can calculate ΔH° for any reaction

$$aA + bB \longrightarrow cC + dD$$

(where A, B, C, and D represent chemical species and a, b, c, and d represent coefficients in the balanced chemical equation) from enthalpy of formation (ΔH°_f) data using the following equation.

$$\Delta H^\circ = [c(\Delta H^\circ_f)_C + d(\Delta H^\circ_f)_D] - [a(\Delta H^\circ_f)_A + b(\Delta H^\circ_f)_B]$$

Standard-state enthalpy of formation data for a variety of elements and compounds can be found in Table B.16 in the appendix. One point needs to be understood before this table can be used effectively. By definition, the enthalpy of formation of any element in its most thermodynamically stable form under standard-state conditions is zero.

Under standard-state conditions, the most thermodynamically stable form of oxygen, for example, is the diatomic molecule in the gas phase: $O_2(g)$. By definition, the enthalpy of formation of this substance is equal to the enthalpy associated with the reaction in which it is formed from its elements in their most thermodynamically stable form. For O_2 molecules in the gas phase, ΔH°_f is therefore equal to the heat given off or absorbed in the following reaction.

$$O_2(g) \longrightarrow O_2(g)$$

Because the initial and final states of the reaction are identical, no heat can be given off or absorbed, so ΔH°_f for $O_2(g)$ is zero.

We are now ready to use standard-state enthalpy of formation data to predict enthalpies of reaction.

Exercise 7.16

In Exercise 7.4 we calculated the enthalpy change associated with the combustion of 0.600 mol of pentaborane, B_5H_9. Use the enthalpy of formation data below to calculate the heat given off when a mole of B_5H_9 reacts with excess oxygen according to the following equation.

$$2\,B_5H_9(g) + 12\,O_2(g) \longrightarrow 5\,B_2O_3(s) + 9\,H_2O(g)$$

Solution

We start by looking up the appropriate data in Table B.16 in the appendix.

Compound	ΔH_f° (kJ/mol$_{rxn}$)
$B_5H_9(g)$	73.2
$B_2O_3(s)$	−1272.77
$O_2(g)$	0
$H_2O(g)$	−241.82

We then use these data with the above equation.

$$\Delta H^\circ = [5(\Delta H_f^\circ)_{B_2O_3} + 9(\Delta H_f^\circ)_{H_2O}] - [2\,(\Delta H_f^\circ)_{B_5H_9} + 12\,(\Delta H_f^\circ)_{O_2}]$$

$$\Delta H^\circ = [5\,(-1272.77\text{ kJ/mol}_{rxn}) + 9\,(-241.82\text{ kJ/mol}_{rxn})] - [2\,(73.2\text{ kJ/mol}_{rxn}) + 12\,(0\text{ kJ/mol}_{rxn})]$$

$$\Delta H^\circ = -8686.6\text{ kJ/mol}_{rxn}$$

According to the balanced equation, this is the enthalpy change when 2 mol of B_5H_9 is consumed. ΔH° is therefore −4343.3 kJ for the consumption of 1 mol of B_5H_9. This is five times the molar enthalpy of reaction for the combustion of CH_4. On a per-gram basis, it is about 20% larger than the energy released when methane burns. It isn't surprising that B_5H_9 was once studied as a potential rocket fuel

Key Terms

Boundary
Calorimeter
Conservation of energy
Endothermic
Enthalpy
Enthalpy of atom combination
Enthalpy of atomization
Enthalpy of formation
Enthalpy of reaction
Exothermic
Extensive property
First law of thermodynamics
Heat
Hess's Law
Intensive property
Isomer
Kinetic energy
Kinetic theory of heat
Molar heat capacity
Potential energy
Specific heat
Standard state
State
State function
Surroundings
System
Temperature

Problems

Energy

1. Define *kinetic* and *potential* energy.
2. A ball is attached to a spring and supported from the ceiling. When the ball is at rest, does it have kinetic or potential energy or both? If the ball is pulled down and released, does it have kinetic or potential energy or both?
3. What is one way energy can be transferred from one object to another?
4. Give an example of how energy can be converted from one form into another.
5. What is meant by the term *thermodynamics*?
6. When certain chemical reactions occur, energy may be released. What is the source of this energy?

Heat

7. What is the difference between heat and temperature?
8. Temperature may be measured in units of degrees using any one of several scales. In what units is heat measured?
9. If a lead ball is dropped from the top of a building to the sidewalk, it will be observed that the ball and the sidewalk will be warmer. What is the source of this heat?

Heat and the Kinetic Molecular Theory

10. Neon gas at 25°C is contained in a flask that is connected to an identical flask that contains neon gas at 50°C. If the connection between the flasks is removed, describe in molecular terms what will happen to the speed of the molecules. What will be the final temperature if equal quantities of Ne are in both flasks?
11. An iron bar at 25°C is placed in contact with an identical iron bar at 50°C. Describe on a microscopic level what will happen to the temperature of the bars.
12. Define *heat* according to the kinetic theory.
13. What physical properties on both the atomic and macroscopic scale change when a balloon filled with helium is heated? Use the kinetic theory of heat to explain each of the changes.
14. Define the terms *system, surroundings*, and *boundary*. Give three examples of a system separated from its surroundings by a boundary, either real or imaginary.
15. It is often believed that things that are hot contain a lot of heat. Use the thermodynamic concepts of system, surroundings, and boundaries to explain why this notion is incorrect.

The First Law of Thermodynamics

16. An ideal gas in a fixed, volume container is heated. Describe what happens to the gas molecules and their energy.
17. The first law of thermodynamics is often described as saying that "energy is conserved." Describe why it is incorrect to assume that the first law suggests that the "energy of a *system* is conserved."
18. Describe what happens to the energy of a system when the system does work on its surroundings. What happens to the energy of the system when it loses heat to its surroundings?
19. Give examples of both a system doing work on its surroundings and a system losing heat to its surroundings. Describe what happens to the energy of the system in each case.
20. Give examples of both a system having work done on it by its surroundings and a system gaining heat from its surroundings. What happens to the energy of the system in each case?
21. Explain why the first law of thermodynamics is often described as suggesting that there is no such thing as a free lunch.
22. Give a verbal definition of the first law of thermodynamics.
23. Describe what happens to the energy of a system when an exothermic reaction is run under conditions of constant volume.
24. The enthalpy change of a system is a measure of two quantities. What are they?

State Functions

25. Give examples of at least five physical properties that are state functions.
26. Which of the following descriptions of a trip are state functions?
    (a) work done (b) energy expended
    (c) cost (d) distance traveled
    (e) tire wear (f) gasoline consumed
    (g) change in location of the car
    (h) elevation change (i) latitude change
    (j) longitude change
27. Which of the following are state functions?
    (a) temperature (b) energy (c) pressure
    (d) volume (e) heat (f) work

The Enthalpy of a System

28. What one thing do chemical reactions have in common?
29. The reaction between hydrogen and oxygen to produce water gives off heat. What is the origin of this heat? How could this heat be measured?
30. What is meant by the change in enthalpy that accompanies a chemical reaction?
31. What is one important use for calorimetric data?

Specific Heat

32. What is the difference between specific heat and molar heat capacity?
33. The same quantity of heat is added to equal numbers of moles of CCl_4 and H_2O. The temperature increases more in the CCl_4 than in the H_2O. Which has the highest heat capacity? Explain.
34. If 1 mol of water at 20°C is placed in contact with 1 mol of Hg(l) at 50°C, will the final temperature be 35°C, greater than 35°C, or less than 35°C? Explain.
35. In a calorimeter experiment done at constant pressure in which all the heat from a chemical reaction was absorbed by the surrounding water bath, the temperature of the water went up by 2.31 K. If the size of the water bath was 200 g, what was the amount of heat transferred to the water? If the chemical reaction was

$$CH_4(g) + 2\ O_2(g) \longrightarrow CO_2(g) + 2H_2O(l)$$

and 2.17×10^{-3} mol of CH_4 was reacted, what is ΔH for the reaction?

Enthalpies of Reaction

36. Oxyacetylene torches are fueled by the combustion of acetylene, C_2H_2.

$$2\ C_2H_2(g) + 5\ O_2(g) \longrightarrow 4\ CO_2(g) + 2\ H_2O(g)$$

If the enthalpy change for the reaction is -2511.14 kJ/mol$_{rxn}$, how much heat can be produced by the reaction of
(a) 2 mol of C_2H_2 (b) 1 mol of C_2H_2
(c) 0.500 mol of C_2H_2 (d) 0.2000 mol of C_2H_2
(e) 10 g of C_2H_2

37. If the enthalpy change for the following reaction

$$C(s) + H_2O(g) \longrightarrow CO(g) + H_2(g)$$

is 131.29 kJ/mol$_{rxn}$, how much heat will be absorbed by the reaction of
(a) 1 mol of $H_2O(g)$ (b) 2 mol of $H_2O(g)$
(c) 0.0300 mol of $H_2O(g)$ (d) 0.0500 mol of $C(s)$

38. How much heat is released when 1 mol of nitrogen reacts with 2 mol of O_2 to give 2 mol of $NO_2(g)$, if $\Delta H°$ for the reaction is 33.2 kJ/mol$_{rxn}$?

$$N_2(g) + 2\ O_2(g) \longrightarrow 2\ NO_2(g)$$

39. Calculate the standard-state enthalpy change for the following reaction, if 1.00 g of magnesium gives off 46.22 kJ of heat when it reacts with excess fluorine.

$$Mg(s) + F_2(g) \longrightarrow MgF_2(s)$$

40. Calculate $\Delta H°$ for the following reaction, assuming that 1.00 g of hydrogen gives off 4.65 kJ of heat when it reacts with 1.00 g of calcium.

$$Ca(s) + H_2(g) \longrightarrow CaH_2(s)$$

Enthalpy as a State Function

41. Nitrogen and oxygen can react directly with one another to produce nitrogen dioxide according to

$$N_2(g) + 2\ O_2(g) \longrightarrow 2\ NO_2(g) \qquad (1)$$

The reaction may also be imagined take place by first producing nitrogen oxide

$$N_2(g) + O_2(g) \longrightarrow 2\ NO(g) \qquad (2)$$

which then produces NO_2

$$2\ NO(g) + O_2(g) \longrightarrow 2\ NO_2(g) \qquad (3)$$

The overall reaction is found by summing reactions (2) and (3) to give

$$N_2(g) + 2\ O_2(g) \longrightarrow 2\ NO_2(g) \qquad (4)$$

How does the enthalpy change for reaction (1) compare to that for reaction (4)? Does this illustrate that enthalpy is a state function? Explain.

42. Consider the following process. Argon gas is heated to 35°C and then cooled to 10°C. The gas is then brought back to its original state. Has the gas undergone an enthalpy change? Explain.

43. The enthalpy change for a chemical reaction does not depend on the way the reaction is carried out. What two things must be known in order to determine the enthalpy change for a reaction?

Standard-State Enthalpies of Reaction

44. What factors affect the quantity of heat given off or absorbed by a given chemical reaction?

45. At what temperature and pressure are thermodynamic data usually reported?

46. What does the superscript "°" on the symbol $\Delta H°$ tell us about the conditions under which the enthalpy change is reported?

47. Explain why there is only one value of $\Delta H°$ for a reaction at 25°C, but many values of ΔH.

Calculating Enthalpies of Reaction

48. For the reaction

$$2\ H_2(g) + O_2(g) \longrightarrow 2\ H_2O(g)$$

draw the Lewis structures for each species. What bonds are broken in the course of the reaction? What bonds are formed? The enthalpy change for this reaction is -484 kJ/mol$_{rxn}$. Are the bonds in the reactants or products the strongest? Explain. What would be the enthalpy change for the reaction

$$2\ H_2O(g) \longrightarrow 2\ H_2(g) + O_2(g)$$

Explain.

49. If a chemical reaction is exothermic, what can be said about the sums of the bond strengths in the products and reactants? What does an endothermic reaction tell us about the relative bond strengths in products and reactants?

Enthalpies of Atom Combination

50. Why is separating atoms in molecules an endothermic process?

51. What does the sign of ΔH tell you about the net flow of energy in a given chemical reaction?

52. Why is the energy released when a bond is formed precisely the same as the amount of energy needed to break the bond?

53. Predict whether each of the following reactions would be exothermic or endothermic.
(a) $CO(g) \longrightarrow C(g) + O(g)$
(b) $2\ H(g) + O(g) \longrightarrow H_2O(g)$
(c) $Na^+(g) + Cl^-(g) \longrightarrow NaCl(s)$
What is the sign of ΔH in each of the reactions?

54. What is the value of ΔH for the overall process of separating 1 mol of CH_4 into its constituent atoms and then reforming 1 mol of CH_4?

55. Does the amount of enthalpy released when a molecule is formed from its gaseous atoms depend on the amount of substance formed? For example, how much enthalpy is released when 2 mol of $CH_4(g)$ is formed as opposed to 1 mol?

56. If the sum of the enthalpies of atom combination for all of the reactants is more negative than that for the products, will the value of ΔH be positive or negative?

57. Predict without using tables which of the following reactions would be endothermic.
(a) $H_2(g) \longrightarrow 2\ H(g)$ (b) $H_2O(g) \longrightarrow H_2O(l)$

58. Predict without using tables which of the following reactions would be endothermic.
(a) $2\ C_8H_{18}(g) + 25\ O_2(g) \longrightarrow 16\ CO_2(g) + 18\ H_2O(g)$
(b) $Na(g) + Cl(g) \longrightarrow NaCl(s)$
(c) $Na^+(g) + e^- \longrightarrow Na(g)$

59. Use the enthalpy of atom combination data in Appendix B.13 to determine whether heat is given off or absorbed when limestone is converted to lime and carbon dioxide.

$$CaCO_3(s) \longrightarrow CaO(s) + CO_2(g)$$

60. Calculate ΔH° for the following reaction from the enthalpy of atom combination data in Appendix B.13.

$$CO(g) + NH_3(g) \longrightarrow HCN(g) + H_2O(g)$$

61. Phosphine (PH_3) is a foul-smelling gas, which often burns on contact with air. Use the enthalpy of atom combination data in Appendix B.13 to calculate ΔH° for the reaction, to obtain an estimate of the amount of energy given off when the compound burns.

$$PH_3(g) + 2\ O_2(g) \longrightarrow H_3PO_4(s)$$

62. Carbon disulfide (CS_2) is a useful, but flammable, solvent. Calculate ΔH° for the following reaction from the enthalpy of atom combination data in Appendix B.13.

$$CS_2(l) + 3\ O_2(g) \longrightarrow CO_2(g) + 2\ SO_2(g)$$

63. The disposable lighters that so many smokers carry use butane as a fuel. Calculate ΔH° for the combustion of butane from the enthalpy of atom combination data in Appendix B.13 and Table 7.4.

$$2\ C_4H_{10}(g) + 13\ O_2(g) \longrightarrow 8\ CO_2(g) + 10\ H_2O(g)$$

64. The first step in the synthesis of nitric acid involves burning ammonia. Calculate ΔH° for the following reaction from the enthalpy of atom combination data in Appendix B.13.

$$4\ NH_3(g) + 5\ O_2(g) \longrightarrow 4\ NO(g) + 6\ H_2O(g)$$

65. Lavoisier believed that all acids contained oxygen because so many compounds he studied that contained oxygen form acids when they dissolve in water. Calculate ΔH° for the reaction between tetraphosphorus decaoxide and water to form phosphoric acid from the enthalpy of atom combination data in Appendix B.13.

$$P_4O_{10}(s) + 6\ H_2O(l) \longrightarrow 4\ H_3PO_4(aq)$$

66. Calculate ΔH° for the decomposition of hydrogen peroxide.

$$2\ H_2O_2(aq) \longrightarrow 2\ H_2O(l) + O_2(g)$$

67. Calculate ΔH° for the thermite reaction.

$$Fe_2O_3(s) + 2\ Al(s) \longrightarrow 2\ Fe(s) + Al_2O_3(s)$$

68. Calculate ΔH° for the reaction of Al with Cr_2O_3 and compare to Problem 67 to predict which reaction will liberate the most heat per mole of Al consumed.

$$Cr_2O_3(s) + 2\ Al(s) \longrightarrow 2\ Cr(s) + Al_2O_3(s)$$

69. The first step in extracting iron ore from pyrite, FeS_2, involves roasting the ore in the presence of oxygen to form iron(III) oxide and sulfur dioxide.

$$4\ FeS_2(s) + 11\ O_2(g) \longrightarrow 2\ Fe_2O_3(s) + 8\ SO_2(g)$$

Calculate ΔH° for the reaction.

70. In which of the following reactions is the sum of the bond strengths greater in the products than in the reactants?
(a) $CH_3OH(l) \longrightarrow HCHO(g) + H_2(g)$
(b) $2\ CH_3OH(l) \longrightarrow 2\ CH_4(g) + O_2(g)$
(c) $CH_3OH(l) \longrightarrow CO(g) + 2\ H_2(g)$

71. Which compound, P_4 or P_2, has the strongest average P—P bonds?

72. Use the enthalpy of atom combination data given in Table 7.4 and Appendix B.13 to calculate the enthalpy changes for the combustion reactions of each of the hydrocarbons listed in Table 7.2. In all cases the hydrocarbons are gaseous, and the products of the reaction are $CO_2(g)$ and $H_2O(l)$.

Using Enthalpies of Atom Combination to Probe Chemical Reactions

73. Find the enthalpies of atom combination for the following species in Appendix B.13.

$$H_2(g),\ H_2O(g),\ CO(g),\ CH_4(g),\ CO_2(g),\ O_2(g)$$

From these data calculate the average bond strength for H—H, O—H, C—H, C≡O, C=O. List these bonds in order of increasing bond strength. Use your average bond strengths to calculate $\Delta H°$ for

$$CH_4(g) + 2\ O_2(g) \longrightarrow CO_2(g) + 2\ H_2O(g)$$

74. Would you expect the following reaction to be exothermic or endothermic? Explain. Use your data from Problem 73.

$$CO(g) + 2\ H_2(g) \longrightarrow CH_3OH(g)$$

75. Calculate the enthalpy change for the reaction

$$2\ ZnS(s) + 3\ O_2(g) \longrightarrow 2\ ZnO(s) + 2\ SO_2(g)$$

In terms of bond strengths, explain the sign of the enthalpy reaction.

76. Explain the sign of the enthalpy change for

$$2\ NO_2(g) \longrightarrow 2\ NO(g) + O_2(g)$$

by using bond strengths.

Bond Length and the Enthalpy of Atom Combination

77. (a) Predict the order of increasing bond length for
    (i) $H_2(g)$ (ii) $I_2(g)$ (iii) $F_2(g)$
    (b) Rank the same molecules in order of increasing bond strength.
    (c) Refer to Appendix B.13 to find the bond strengths for each molecule. Is your answer in (b) supported by these data?
78. Which of the carbon-to-carbon bonds in the following is shortest? Longest?
    (a) H_3CCH_3 (b) H_2CCH_2 (c) HCCH
    Which carbon–carbon bond is strongest? Explain.
79. Is the following statement always true? Explain. *The shorter the bond, the stronger the bond.*

Hess's Law

80. Explain how Hess's law is a direct consequence of the fact that the enthalpy of a system is a state function.
81. Use the following data to calculate $\Delta H°$ for the conversion of graphite into diamond.

$$C(s,\ \text{graphite}) + O_2(g) \longrightarrow CO_2(g) \quad \Delta H° = -393.51\ \text{kJ/mol}_{\text{rxn}}$$

$$C(s,\ \text{diamond}) + O_2(g) \longrightarrow CO_2(g) \quad \Delta H° = -395.41\ \text{kJ/mol}_{\text{rxn}}$$

82. Use the following data

$$H_2(g) + \tfrac{1}{2}\ O_2(g) \longrightarrow H_2O(l) \quad \Delta H° = -285.83\ \text{kJ/mol}_{\text{rxn}}$$

$$H_2(g) + O_2(g) \longrightarrow H_2O_2(aq) \quad \Delta H° = -191.17\ \text{kJ/mol}_{\text{rxn}}$$

to calculate $\Delta H°$ for the decomposition of hydrogen peroxide.

$$2\ H_2O_2(aq) \longrightarrow 2\ H_2O(l) + O_2(g)$$

83. In the presence of a spark, nitrogen and oxygen react to form nitrogen oxide.

$$N_2(g) + O_2(g) \longrightarrow 2\ NO(g)$$

Calculate $\Delta H°$ for the reaction from the following data.

$$\tfrac{1}{2}\ N_2(g) + O_2(g) \longrightarrow NO_2(g) \quad \Delta H° = 33.2\ \text{kJ/mol}_{\text{rxn}}$$

$$NO(g) + \tfrac{1}{2}\ O_2(g) \longrightarrow NO_2(g) \quad \Delta H° = -57.1\ \text{kJ/mol}_{\text{rxn}}$$

84. Enthalpy of reaction data can be combined to determine $\Delta H°$ for reactions that are difficult, if not impossible, to study directly. Nitrogen and oxygen, for example, do not react directly to form dinitrogen pentoxide.

$$2\ N_2(g) + 5\ O_2(g) \longrightarrow 2\ N_2O_5(g)$$

Use the following data to determine $\Delta H°$ for the hypothetical reaction in which nitrogen and oxygen combine to form N_2O_5.

$$N_2(g) + 3\ O_2(g) + H_2(g) \longrightarrow 2\ HNO_3(aq) \quad \Delta H° = -414.7\ \text{kJ/mol}_{\text{rxn}}$$

$$N_2O_5(g) + H_2O(l) \longrightarrow 2\ HNO_3(aq) \quad \Delta H° = -140.24\ \text{kJ/mol}_{\text{rxn}}$$

$$2\ H_2(g) + O_2(g) \longrightarrow 2\ H_2O(l) \quad \Delta H° = -571.7\ \text{kJ/mol}_{\text{rxn}}$$

85. Use the following data

$$3\ C(s) + 4\ H_2(g) \longrightarrow C_3H_8(g) \quad \Delta H° = -103.85\ \text{kJ/mol}_{\text{rxn}}$$

$$C(s) + O_2(g) \longrightarrow CO_2(g)$$
$$\Delta H° = -393.51 \text{ kJ/mol}_{rxn}$$
$$H_2(g) + \tfrac{1}{2} O_2(g) \longrightarrow H_2O(g)$$
$$\Delta H° = -241.82 \text{ kJ/mol}_{rxn}$$

to calculate the heat of combustion of propane, C_3H_8.

$$C_3H_8(g) + 5\ O_2(g) \longrightarrow 3\ CO_2(g) + 4\ H_2O(g)$$

86. Use the following data

$$C_4H_9OH(l) + 6\ O_2(g) \longrightarrow 4\ CO_2(g) + 5\ H_2O(g)$$
$$\Delta H° = -2456.1 \text{ kJ/mol}_{rxn}$$
$$(C_2H_5)_2O(l) + 6\ O_2(g) \longrightarrow 4\ CO_2(g) + 5\ H_2O(g)$$
$$\Delta H° = -2510.0 \text{ kJ/mol}_{rxn}$$

to calculate $\Delta H°$ for the following reaction.

$$(C_2H_5)_2O(l) \longrightarrow C_4H_9OH(l)$$

Enthalpies of Formation

87. For which of the following substances is $\Delta H_f°$ equal to zero?
(a) $P_4(s)$ (b) $H_2O(g)$ (c) $H_2O(l)$
(d) $O_3(g)$ (e) $Cl(g)$ (f) $F_2(g)$
(g) $Na(g)$

88. Use the enthalpy of formation data in Appendix B.16 to calculate the enthalpy change for the following reaction. Compare the results of the calculation to that of Problem 59.

$$CaCO_3(s) \longrightarrow CaO(s) + CO_2(g)$$

89. Calculate $\Delta H°$ for the following reaction from the enthalpy of formation data in Appendix B.16. Compare to Problem 60.

$$CO(g) + NH_3(g) \longrightarrow HCN(g) + H_2O(g)$$

90. Use the enthalpy of formation data in Appendix B.16 to calculate $\Delta H°$ for the combustion of PH_3. Compare the answer to Problem 61.

$$PH_3(g) + 2\ O_2(g) \longrightarrow H_3PO_4(s)$$

91. Carbon disulfide (CS_2) is a useful, but flammable, solvent. Calculate $\Delta H°$ for the following reaction from the enthalpy of formation data in Appendix B.16.

$$CS_2(l) + 3\ O_2(g) \longrightarrow CO_2(g) + 2\ SO_2(g)$$

92. Calculate $\Delta H°$ for the combustion of butane from the enthalpy of formation data in Appendix B.16. The enthalpy of formation of butane is -126.2 kJ/mol$_{rxn}$.

$$2\ C_4H_{10}(g) + 13\ O_2(g) \longrightarrow 8\ CO_2(g) + 10\ H_2O(g)$$

93. The first step in the synthesis of nitric acid involves burning ammonia. Calculate $\Delta H°$ for the following reaction from the enthalpy of formation data in Appendix B.16.

$$4\ NH_3(g) + 5\ O_2(g) \longrightarrow 4\ NO(g) + 6\ H_2O(g)$$

94. Calculate $\Delta H°$ for the following from the enthalpy of formation data in Appendix B.16. Compare the answer to Problem 65.

$$P_4O_{10}(s) + 6\ H_2O(l) \longrightarrow 4\ H_3PO_4(aq)$$

95. Small quantities of oxygen can be prepared in the laboratory by heating potassium chlorate ($KClO_3$) until it decomposes. Calculate $\Delta H°$ for the following reaction from the enthalpy of formation data in Appendix B.16. The enthalpy of formation of $KClO_3(s)$ is -391.2 kJ/mol.

$$2\ KClO_3(s) \longrightarrow 2\ KCl(s) + 3\ O_2(g)$$

96. Use enthalpies of formation to predict which of the following reactions gives off the most heat per mole of aluminum consumed.

$$Fe_2O_3(s) + 2\ Al(s) \longrightarrow 2\ Fe(s) + Al_2O_3(s)$$
$$Cr_2O_3(s) + 2\ Al(s) \longrightarrow 2\ Cr(s) + Al_2O_3(s)$$

Integrated Problems

97. Use the enthalpy of combustion for methane, given below, to estimate the energy released when 100 ft^3 of natural gas is burned.

$$CH_4(g) + 2\ O_2(g) \longrightarrow CO_2(g) + 2\ H_2O(g)$$
$$\Delta H° = -802.36 \text{ kJ/mol}_{rxn}$$

98. Which do you predict has the stronger bond, C—H or C—Cl? Calculate the average C—H bond enthalpy in CH_4 from $\Delta H_{ac}°$. Calculate the average C—Cl bond enthalpy in CCl_4 from $\Delta H_{ac}°$. Compare the two bond enthalpies. Is this the result you predicted?

99. If the enthalpy change for breaking all the bonds in the reactants is greater than the enthalpy change for making the bonds of the products, what is the sign of ΔH?

100. If the enthalpy change for breaking all the bonds in the reactants is less than the enthalpy change for making the bonds of the products, what is the sign for ΔH?

101. A measure of the forces which operate between molecules in a liquid can be obtained by comparing the enthalpy required to separate the molecules to the gaseous phase. In which of the following liquids are the intermolecular forces strongest?
(a) CH_3COOH (b) CH_3CH_2OH
(c) C_6H_6 (d) CCl_4

102. Determine the average C—H bond enthalpy for the following compounds (see Table 7.4).
(a) CH_4 (b) C_2H_6
(c) C_3H_8 (d) C_4H_{10}, *n*-butane

103. The enthalpy change per mole of hydrocarbon combusted with oxygen is given in Table 7.2. Calculate the amount of heat released per mole of covalent bonds broken in each hydrocarbon listed. Is there a relation between molecular structure and the heat released? If so, what is it?

104. Isomers are compounds that have the same number and kinds of atoms but have a different arrangement of the atoms. The enthalpies of atom combination for several pairs of gaseous isomers are given below. For each pair, decide which has the strongest bonds.

(a) $CH_3CH_2CH_2OH$ and CH_3CHOH (with CH_3 attached to the central C)
$-4394.2\ kJ/mol_{rxn}$ $\quad$ $-4410.7\ kJ/mol_{rxn}$

(b) $CH_2{=}CHCH_2CH_3$ and $CH_3CH{=}CHCH_3$
$-4604.9\ kJ/mol_{rxn}$ $\quad$ $-4611.9\ kJ/mol_{rxn}$

(c) $H_2C{=}CHCH{=}CHCH_3$ and $H_2C{=}CHCH_2CH{=}CH_2$
$-5243.3\ kJ/mol_{rxn}$ $\quad$ $-5213.5\ kJ/mol_{rxn}$

105. When hydrocarbons are bonded only by single bonds, they are said to be saturated; pentane, $CH_3CH_2CH_2CH_2CH_3$, is an example. If carbon–carbon double or triple bonds are present, the compound is said to be *unsaturated.* Unsaturated hydrocarbons are generally better for human nutrition, hence the claims made by manufacturers to have reduced saturated fats in foods such as margarine. An unsaturated hydrocarbon can be saturated by adding hydrogen across the double bond.

$$CH_3CH{=}CHCH_3(g) + H_2(g) \longrightarrow CH_3CH_2CH_2CH_3(g)$$

Is the sum of the bond strengths greater in the products or the reactants in the above reaction? Draw Lewis structures for the reactants and products and give all bond angles. Which species are planar? The C═C bond is rigid and therefore the —CH_3 groups attached to those carbon atoms can appear either on the same side (in which case the compound is called *cis*-2-butene) or on opposite sides (in which case the compound is named *trans*-2-butene).

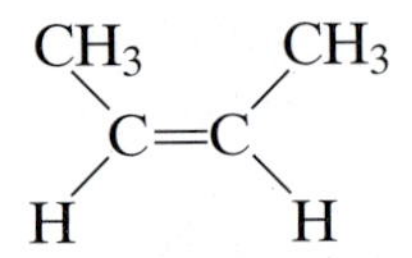

cis-2-*Butene*
$\Delta H^\circ_{ac} = -4611.86\ kJ/mol_{rxn}$

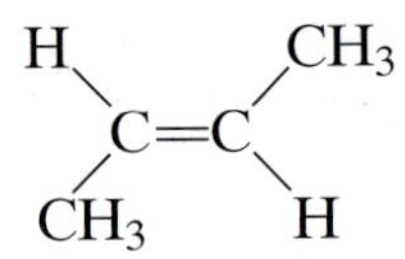

trans-2-*Butene*
$\Delta H^\circ_{ac} = -4616.58\ kJ/mol_{rxn}$

Given the enthalpies of atom combination above and data from Appendix B.13, suggest a way that a chemical reaction could be used to differentiate between the two forms of 2-butene.

106. Both dimethyl ether, CH_3—O—CH_3, and ethyl alcohol, CH_3CH_2OH, have been suggested as possible fuels. When reacted with oxygen, O_2, both compounds yield $CO_2(g)$ and $H_2O(g)$. The reactions are called *combustion reactions.*
(a) Write balanced chemical equations that describe the combustion reaction between dimethyl ether(*g*) and O_2. Write a second reaction for the combustion reaction between ethyl alcohol(*g*) and O_2.
(b) Calculate ΔH for both reactions. (See Table 7.4.) Why is the heat different from that given in Section 7.1?
(c) Which is the better fuel? In other words, which releases the most heat on combustion with O_2?
(d) Which molecule, dimethyl ether or ethyl alcohol, has the stronger bonds? Explain.

107. For the following reaction

$$SiBr_4(g) + 2\ Cl_2(g) \longrightarrow SiCl_4(g) + 2\ Br_2(g)$$

(a) Calculate ΔH° for the reaction. [ΔH°_{ac} for $SiBr_4(g)$ is $-1272\ kJ/mol_{rxn}$.]
(b) Calculate the average Si—Br, Si—Cl, Cl—Cl, and Br—Br bond enthalpies.
(c) Which do you expect to be the stronger bond, Si—Br or Si—Cl? Explain. Which do you expect to be the stronger bond, Br—Br or Cl—Cl? Explain. Do your predictions agree with the calculations in (b)?
(d) Is the reaction endothermic or exothermic? Explain the sign of ΔH°.

108. When carbon is burned in air, the following reaction takes place and releases heat.

$$C(s) + O_2(g) \longrightarrow CO_2(g)$$

Which of the following is responsible for the heat produced?
(a) breaking oxygen–oxygen bonds
(b) making carbon–oxygen bonds
(c) breaking carbon–carbon bonds
(d) both (a) and (c) are correct
(e) (a), (b), and (c) are correct
(f) none of the above are correct

109. In which molecule would you expect the nitrogen–nitrogen bond strengths to be the greatest? Explain.
(a) $H_2N—NH_2$ (b) $F_2N—NF_2$
(c) $HN{=}NH$ (d) $N{\equiv}N$

110. Determine the average bond strengths in N_2, H_2, and NH_3. Would you expect the following reaction to be exothermic or endothermic? Use the average bond strengths to support your answer.

$$N_2(g) + 3\ H_2(g) \longrightarrow 2\ NH_3(g)$$

111. For the reaction

$$2\ H_2(g) + O_2(g) \longrightarrow 2\ H_2O(g)$$

determine the average bond strengths for hydrogen–hydrogen, oxygen–oxygen, and oxygen–hydrogen bonds. Do you expect the reaction to be exothermic or endothermic? Explain.

112. In terms of the bonds made and the bonds broken, explain why the following reaction is endothermic.

$$Si(s) + 2\ H_2(g) \longrightarrow SiH_4(g)$$

Use enthalpies of atom combination (Table B.13) to support your answer.

113. Methane, CH_4, is commonly used in the laboratory as a fuel for Bunsen burners.

$$CH_4(g) + O_2(g) \longrightarrow CO_2(g) + H_2O(g)$$

(a) Balance the equation.
(b) Give the Lewis structures of all products and reactants.
(c) Calculate the enthalpy change for the combustion of CH_4.
(d) Are the bonds stronger in the reactants or products?
(e) On average which is stronger, a C—H or a C=O bond?

114. Iodine reacts with the halogens to form a wide variety of compounds. Two reactions are shown below.

$$I_2(g) + Cl_2(g) \longrightarrow 2\ ICl(g)$$
$$I_2(g) + Br_2(g) \longrightarrow 2\ IBr(g)$$

(a) Based on bond lengths, which do you expect to have the strongest bond, Cl_2 or Br_2? Explain.
(b) Based on bond lengths, which do you expect to have the weakest bonds, ICl or IBr? Explain.
(c) Are your predictions consistent with ΔH°_{ac} data?
(d) Which of the two reactions above is the most exothermic?
(e) Explain why, in terms of bonds made and bonds broken, one of these reactions is more exothermic than the other.

115. Consider the reaction

$$SiCl_4(g) + 2\ H_2O(g) \longrightarrow SiO_2(s) + 4\ HCl(g)$$

(a) Find ΔH° for this reaction. Show all work.
(b) How do the strengths of the bonds in the products compare to those of the reactants? Explain.
(c) How would you expect ΔH° for the above reaction to compare to ΔH° for the following?

$$SiF_4(g) + 2\ H_2O(g) \longrightarrow SiO_2(s) + 4\ HF(g)$$

Explain.
(d) How would the bond strengths in $SiCl_4(g)$ compare to those in $SiF(g)$? Explain.

116. Magnesium reacts with chlorine according to the following equation.

$$Mg(s) + Cl_2(g) \longrightarrow MgCl_2(s)$$

(a) Classify the bonding type of each substance in this reaction according to whether it is ionic, covalent, metallic, or metalloid. Explain how you made your classification.
(b) Calculate ΔH for this reaction.
(c) Explain the sign and magnitude of ΔH for this reaction.

Chapter Eight

LIQUIDS AND SOLUTIONS

Special Topics

8.1 The Structure of Gases, Liquids, and Solids

Liquids, solids, and gases constitute three phases of matter. Most of the pure substances that are liquids at room temperature and pressure are composed of molecules. Ionic and metallic compounds are generally solids at room temperature and pressure. Although ionic and metallic substances can be melted to form the liquid state, we shall confine our discussion in this chapter to molecular liquids.

This chapter builds on the concepts of molecular structure and bonding that were developed early in the text. ChemActivity 26 in Chemistry: A Guided Inquiry *is designed to accompany this chapter.*

The kinetic molecular theory explains the characteristic properties of gases by assuming that gas particles are in a state of constant, random motion and that the diameter of the particles is very small compared with the distance between particles (Figure 8.1*c*). Because most of the volume of a gas is empty space, the simplest analogy might compare the particles of a gas to fruit flies in a jar.

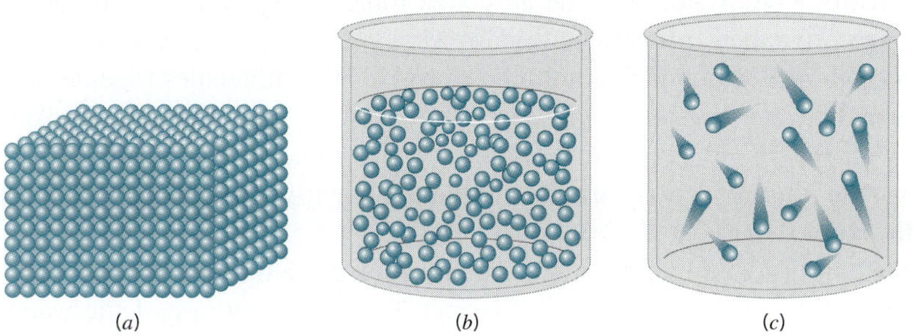

Fig. 8.1 Particles in a solid (*a*) are packed tightly in a regular pattern. The particles in a liquid (*b*) do not pack as tightly as they do in a solid. The structure of liquids also contains small, particle-sized holes that enable the liquid to flow so that it can conform to the shape of its container. Gas particles (*c*) are in random motion.

Many of the properties of solids have been captured in the way the term *solid* is used in English. It describes something that holds its shape, such as a solidly constructed house. It implies continuity; there is a definite position for each particle. It implies the absence of empty space, as in a solid chocolate Easter bunny. Finally, it describes things that occupy three dimensions, as in solid geometry. A solid might be compared to a brick wall, in which the individual bricks form a regular structure and the amount of empty space is kept to a minimum (Figure 8.1*a*).

In this chapter, intermolecular forces and the kinetic theory of liquids are used to interpret properties of liquids. Vapor pressure, melting, freezing, and boiling points are all treated from this point of view. Enthalpies of vaporization and fusion are used to show how intermolecular forces are related to a change in state.

Liquids have properties between the extremes of gases and solids. Like gases, they flow to conform to the shape of their containers (Figure 8.1*b*). Like solids, they can't expand to fill their containers, and they are very difficult to compress. The structure of a liquid might be compared to a bag full of marbles being shaken vigorously, back and forth. A model for liquids therefore assumes that there are small, particle-sized holes randomly distributed through the liquid. Particles that are close to one of these holes behave in much the same way as particles in a gas; those that are far from a hole act more like the particles in a solid.

The difference in the structures of gases, liquids, and solids might best be understood by comparing the densities of substances in the three phases. As shown by the data in Table 8.1, typical solids are about 20% more dense than the corresponding liquid, whereas the liquid is about 800 times as dense as the gas. The models shown in Figure 8.1 are consistent with the data.

Table 8.1
Densities of Solid, Liquid, and Gaseous Forms of Three Elements

	Solid (g/cm^3)	Liquid (g/cm^3)	Gas (g/cm^3)
Ar	1.65	1.40	0.001784
N_2	1.026	0.8081	0.001251
O_2	1.426	1.149	0.001429

Because water is the only substance that we routinely encounter as a solid, a liquid, and a gas, it may be useful to consider what happens to water as we change the temperature. At low temperatures, water is a solid in which the individual molecules are locked into a rigid structure. As we raise the temperature, the average kinetic energy of the molecules increases, due to an increase in the motion with which these molecules move about their lattice positions in the solid.

To understand the effect of molecular motion, we need to differentiate between intramolecular bonds and intermolecular forces, as shown in Figure 8.2.

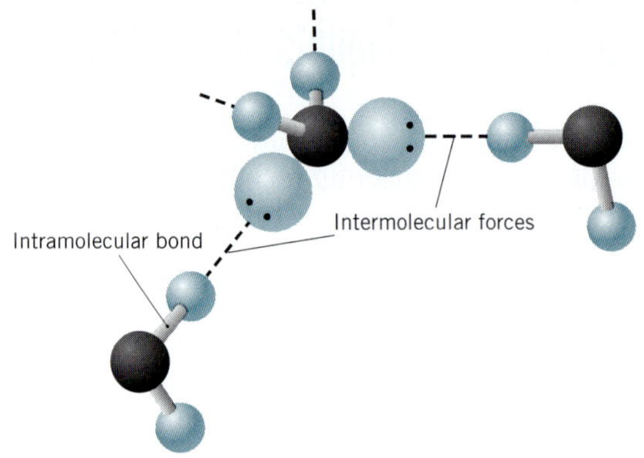

Fig. 8.2 The covalent bonds between the hydrogen and oxygen atoms in a water molecule are called *intramolecular bonds.* The attractive forces between water molecules are called *intermolecular forces.*

Molarity was first introduced in Section 2.18 of Chapter 2 for instructors who need it for the lab. We have chosen to use molarity as the only unit of concentration in the core chapters. Molality and mole fraction are used in a Special Topics appendix at the end of this chapter, which covers freezing point depression, boiling point elevation, and Raoult's Law.

The covalent bonds between the hydrogen and oxygen atoms in a water molecule are called **intramolecular bonds.**[1] The attractive forces between the neighboring water molecules are called **intermolecular forces.**[2]

The *intramolecular* bonds that hold the atoms in H_2O molecules together are much stronger than the *intermolecular* forces between water molecules. It takes 463 kJ to break the H—O bonds in a mole of water molecules, but only about 50 kJ to break the intermolecular forces that hold a mole of water molecules to one another in the solid state.

Increased motion disrupts the attractive intermolecular forces between water molecules. As solid water (ice) becomes warmer, the kinetic energy of the water molecules eventually becomes too large to allow the molecules to be locked into the rigid structure of ice. At this point, the solid melts to form a liquid, in which the intermolecular forces between molecules now are sufficiently weakened that the molecules may move through the entire volume of the liquid. As the temperature continues to increase, the kinetic energy of the water molecules becomes so large, and they move so rapidly, that most of the attractive intermolecular forces are overcome, and the liquid boils to form a gas in which each particle moves more or less randomly through space. At no point is the increased motion sufficient to overcome the strengths of the covalent bonds. Therefore, water exists as a molecular species in the gaseous state up to very high temperatures.

At low temperatures molecules have less kinetic energy to overcome the attractive forces and hence approach each other more closely. At higher temperatures the molecules possess sufficient energy to allow for a larger separation. The distance between molecules is one factor that determines the strength of intermolecular forces. Thus how closely molecules may approach one another is an important factor in determining how strongly molecules attract each other. It follows then that, in addition to temperature, the size and shape of molecules become significant for the strength of intermolecular attractive forces.

We now have a means of rationalizing why a substance forms a solid, liquid, or gas at room temperature. The difference between the three phases of matter is based on a competition between the strength of intermolecular forces and the kinetic energy of the system. When the force of attraction between the particles is relatively weak, the substance is likely to be a gas at room temperature. When the force of attraction is strong, the substance is more likely to be a solid. As might be expected, a substance is a liquid at room temperature when the intermolecular

[1]The prefix *intra-* comes from a Latin stem meaning "within or inside." Intramural sports, for example, match teams from within the same institution.

[2]The prefix *inter-* comes from a Latin stem meaning "between." It is used in words such as *interact, intermediate, international,* and *intercollegiate.*

forces are neither too strong nor too weak. For a given intermolecular force strength, the higher the temperature, the more likely the substance is to be a gas.

8.2 Intermolecular Forces

The kinetic theory of gases assumes that there is no force of attraction between the particles in a gas. If that assumption were correct, gases would never condense to form liquids and solids at low temperatures. In 1873 the Dutch physicist Johannes van der Waals derived an equation that not only included the force of attraction between gas particles but also corrected for the fact that the volume of the particles becomes a significant fraction of the total volume of the gas at high pressures.

The van der Waals equation described in the Special Topics section of Chapter 6 gives a better description of the experimental data for real gases than can be obtained with the ideal gas equation. But that wasn't van der Waals' goal. He was trying to develop a model that would explain the behavior of liquids by including terms that reflected the size of the atoms or molecules in the liquid and the strength of the forces between the atoms or molecules.

The weak intermolecular forces in liquids and solids are therefore often called **van der Waals forces.** Intermolecular forces can be divided into four categories: (1) dipole–dipole, (2) dipole–induced dipole, (3) induced dipole–induced dipole, and (4) hydrogen bonding. A given substance, depending on its structure, may exhibit more than one of these intermolecular forces. Because the forces of attraction between particles vary greatly from one substance to another, it isn't possible to describe the properties of liquids or solids with just one equation ($PV = nRT$), as we did with gases in Chapter 6.

DIPOLE–DIPOLE FORCES

Many molecules are held together by intramolecular bonds that fall between the extremes of pure ionic and pure covalent bonds. The difference between the electronegativities of the atoms in the molecules is large enough that the electrons aren't shared equally, and yet small enough that the electrons aren't drawn exclusively to one of the atoms to form positive and negative ions. The bonds in the molecules are said to be polar because they have positive and negative ends, or poles. When polar bonds and nonbonding pairs of electrons are distributed in a molecule such that there is a separation of charge within the molecule, the molecule will be **polar,** with a positive and a negative pole.

The acetone molecules in nail polish remover have a **dipole moment** (2.88 D) because the carbon atom in the C═O double bond has a slight positive charge and the oxygen atom in this bond has a slight negative charge. The net result is a force of attraction between the positive end of one molecule and the negative end of another, as shown in Figure 8.3.

The dipole–dipole interaction in acetone is relatively weak; it requires only a small amount of energy to pull acetone molecules apart from one another. In contrast, the covalent bonds (C—H, C—C, and C═O) in acetone are much stronger. The strength of the dipole–dipole attraction depends on the magnitude of the dipole

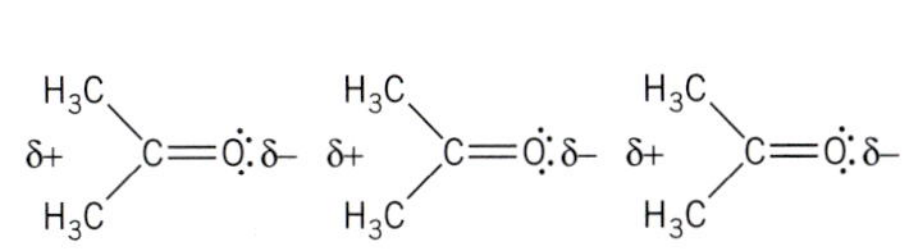

Fig. 8.3 A dipole–dipole force exists between adjacent acetone molecules. The dipole moment is generated by the polar C═O bond.

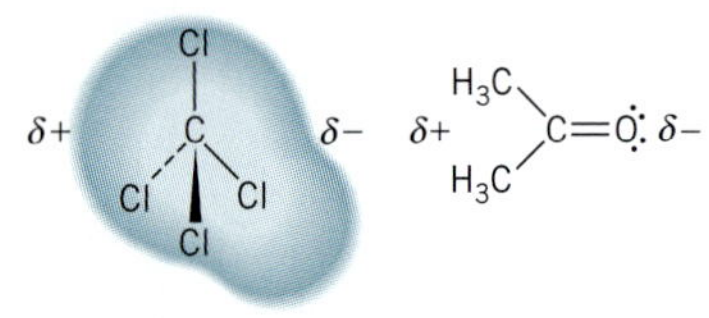

Fig. 8.4 When a CCl_4 molecule comes close to a polar acetone molecule, the distribution of electrons around carbon tetrachloride is distorted. The shaded area around the carbon tetrachloride molecule represents the electron density. The electron density is attracted toward the partial positive charge of the acetone. A small dipole moment is induced in the carbon tetrachloride molecule, which allows a weak dipole–induced dipole force of attraction.

moment and on how closely the molecules approach one another. The closer the approach, the greater are the intermolecular attractive forces. If molecules approach too closely, however, a repulsive force occurs.

DIPOLE–INDUCED DIPOLE FORCES

What would happen if we mixed acetone with carbon tetrachloride, which has no dipole moment? The individual C—Cl bonds in carbon tetrachloride are polar due to the difference in electronegativity between carbon and chlorine atoms. Because the molecule has a symmetric tetrahedral molecular geometry, the polarities of the four bonds cancel one another to yield a uniform charge distribution with a dipole moment of zero. The electrons in a molecule aren't static, however, but are in constant motion. When a carbon tetrachloride molecule comes close to a polar acetone molecule, the electrons in carbon tetrachloride can shift to one side of the molecule to produce a very small dipole moment, as shown in Figure 8.4.

By distorting the distribution of electrons in carbon tetrachloride, the polar acetone molecule induces a small dipole moment in the CCl_4 molecule, which creates a dipole–induced dipole force of attraction between the acetone and carbon tetrachloride molecules. The strength of dipole–induced dipole interactions increases as the molecules approach one another and weaken rapidly as the molecules move apart. In general, a polar molecule can distort the electron cloud of any neighboring molecule, producing a dipole–induced dipole interaction.

INDUCED DIPOLE–INDUCED DIPOLE FORCES

Bromine has no dipole and therefore has neither dipole–dipole nor dipole–induced dipole forces, yet bromine, Br_2, is a liquid at room temperature. A liquid can exist only if there are sufficient forces of attraction between adjacent molecules to hold the molecules together.

The electrons in bromine are in constant motion. Thus, there is some probability that for an instant in time there may be more electron density on one side of the molecule than on the other, causing a temporary dipole. The temporary dipole can induce a temporary dipole in an adjacent bromine molecule, as shown in Figure 8.5. Such fluctuations in electron density occur constantly, creating temporary induced dipole–induced dipole forces of attraction—also known as **dispersion forces** or **London forces**—between pairs of molecules or atoms throughout a liquid. All molecules experience some degree of induced dipole–induced dipole interaction. This is the only type of intermolecular force present in substances, such as Br_2, that are made up of nonpolar molecules.

Because intermolecular forces must be overcome to melt a molecular solid or to boil a liquid, melting points and boiling points can serve as a measure of the relative strengths of the intermolecular forces which hold the molecules together. Table 8.2 shows the melting points and boiling points of chlorine, bromine, and

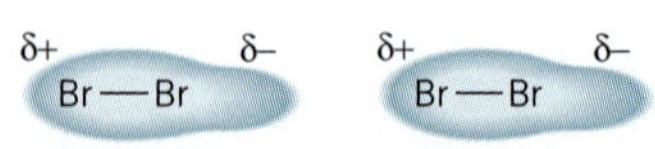

Fig. 8.5 Fluctuations in electron density occur around the nuclei of neighboring bromine molecules, creating induced dipole–induced dipole forces of attraction.

Table 8.2

Melting Points and Boiling Points of Three Substances That Have Only Dispersion Forces

	Molecular Weight (g/mol)	Melting Point (°C)	Boiling Point (°C)
Cl_2	70.91	−101.0	−34.6
Br_2	159.81	−7.2	58.8
I_2	253.81	113.5	184.4

iodine. All three molecules are nonpolar and have only dispersion forces, but they have significantly different melting and boiling points. At room temperature and pressure chlorine is a gas, bromine a liquid, and iodine a solid. The data in Table 8.2 indicate that the dispersion forces in iodine are much stronger than the dispersion forces in bromine, which are stronger than the dispersion forces in chlorine.

The magnitude of an induced dipole moment depends on the ease with which a molecule can be polarized—in other words, the ease with which electrons can be moved in the molecule to form a dipole moment. The polarizability of a molecule depends on the number of electrons, how tightly they are held, and the shape of the molecule. For similar substances, dispersion force interactions increase with increasing molecular weight. There is an increase in molecular weight as we move from Cl_2 to Br_2 to I_2, which indicates an increase in the strength of the dispersion forces between adjacent molecules. This increase is the result of an increase in the distance of the valence electrons from the nucleus as we move down the periodic table from Cl to Br to I. As the distance from the nucleus increases, the electrons are held less tightly and can more easily be pulled to one side of the molecule. In general, polarizability and induced dipole forces increase as the number of electrons on an atom, ion, or molecule increases.

The shape of a molecule is also a factor in determining the magnitude of dispersion forces. This factor can be demonstrated using three molecules with equal molecular weights. As we have seen, molecules that contain the same atoms but have different structures are called *isomers*. The data in Figure 8.6 show how the shape of a molecule influences the boiling point of a compound. One of the compounds, neopentane, has very symmetrical molecules, with four identical CH_3 groups arranged in a tetrahedral pattern around a central carbon atom. Consequently, the symmetrical neopentane doesn't have as much surface area in contact with neighboring molecules as does the straight-chain *n*-pentane molecule. As a result, the dispersion forces are weaker and the boiling point is lower than for *n*-pentane. Isopentane, being a branched molecule, falls in between neopentane and *n*-pentane in how closely it can approach its neighbors, and therefore isopentane has a boiling point between that of neopentane and *n*-pentane.

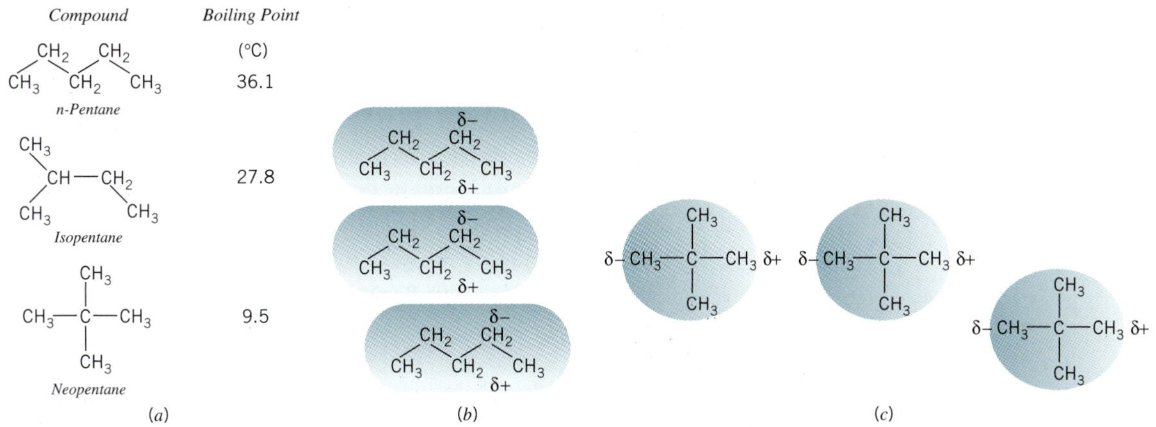

Fig. 8.6 (*a*) The structure and boiling points for three isomers of C_5H_{12}. (*b*) Cylindrically shaped *n*-pentane molecules can approach one another along their entire length, allowing the formation of temporary dispersion force interactions. These interactions are larger than those found in neopentane due to the increased polarizability of their electron clouds. (*c*) Spherically shaped neopentane molecules do not have as much surface area as *n*-pentane and undergo smaller dispersion force interactions.

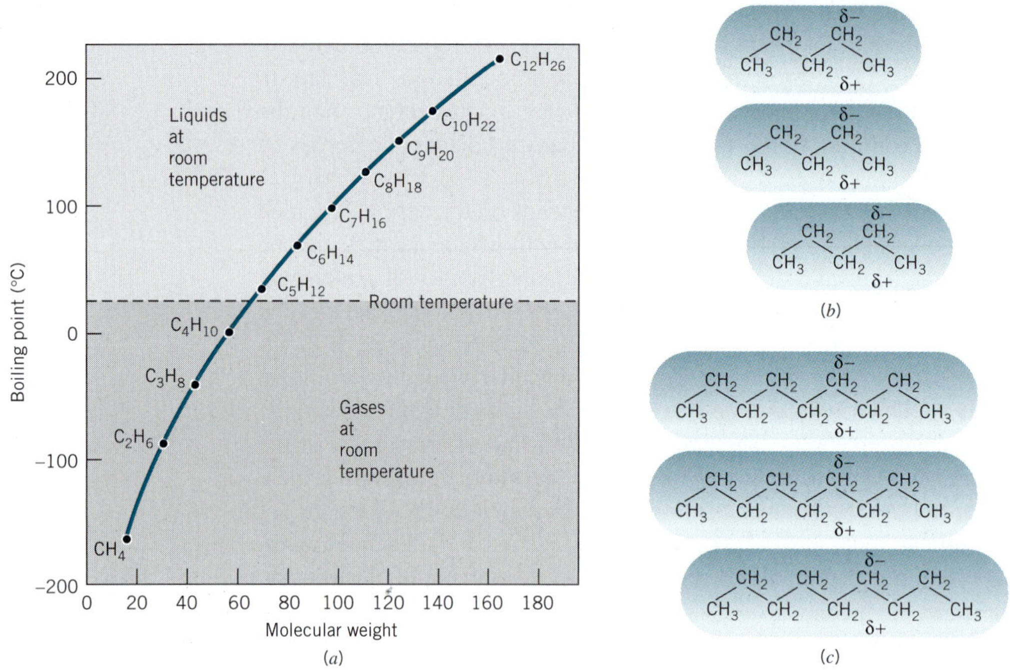

Fig. 8.7 (*a*) There is a gradual increase in the boiling points of compounds with the generic formula C_nH_{2n+2} as the molecules in the series increase in chain length. (*b*) C_5H_{12} molecules approach one another and undergo temporary dispersion force interactions. (*c*) C_9H_{20} molecules approach one another and are subject to even greater dispersion force interactions due to the greater polarizability of the more extended electron cloud.

The relationship between the molecular weight of a compound and its boiling point is shown in Figure 8.7. The compounds in Figure 8.7 all have the same generic formula, C_nH_{2n+2}, and are all straight-chain hydrocarbons. The only differences between the compounds are their sizes and molecular weights. As shown in Figure 8.7, the relationship between the molecular weights of the compounds and their boiling points isn't a straight line, but it is a remarkably smooth curve.

The molecules shown in Figure 8.7 increase in length from 1 to 12 carbons. Because the molecules are nonpolar, the only intermolecular interactions between them are due to dispersion interactions. With each additional increase in the length of the chain of carbon atoms, an additional contribution to the dispersion interactions is made. Increased molecular weight corresponds to additional polarizable electrons and hence to increased intermolecular interactions. For similar compounds, dispersion interactions increase directly with increased molecular weight and length of the molecule. Hence, the attractive forces holding the molecules together as a liquid increase. This leads to an increase in the boiling point. Dispersion forces increase as a species becomes more polarizable.

➤ CHECKPOINT

Arrange the elements in Group VIIIA (He, Ne, Ar, Kr) in order of increasing dispersion interactions.

HYDROGEN BONDING

The intermolecular force called **hydrogen bonding** is actually a type of dipole–dipole interaction. Hydrogen bonds are separated from other examples of van der Waals forces because they are unusually strong: 15–25 kJ/mol. However, it should be noted that even though the name of the interaction is hydrogen *bonding,* a hydrogen bond is not a covalent, ionic, or metallic bond. It is an intermolecular force between adjacent molecules.

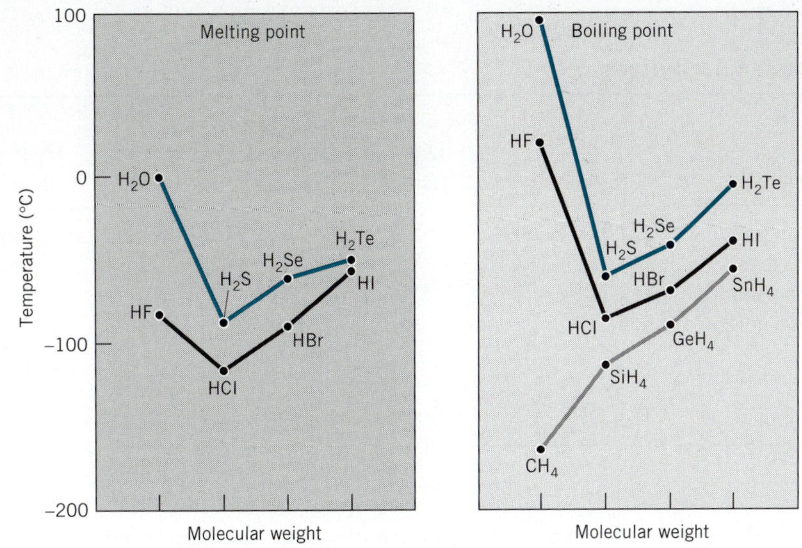

Fig. 8.8 Plot of the melting points and boiling points of hydrides of elements in Groups IVA, VIA, and VIIA. The melting points and boiling points of HF and H_2O are anomalously large because of the strength of the hydrogen bonds between molecules in those compounds.

Molecules that can form hydrogen bonds have relatively polar H—*X* bonds, such as NH_3, H_2O, and HF. The hydrogen bond is created when a hydrogen atom forms a bridge between two very electronegative atoms. The hydrogen is covalently bonded to one of the atoms and hydrogen bonded to the other. The H—*X* bond must be polar to create the partial positive charge on the hydrogen atom that allows the bridging interactions to exist. As the *X* atom in the H—*X* bond becomes more electronegative, hydrogen bonding between molecules becomes more important. Hydrogen bonding is most important when the hydrogen atom is bonded to N, O, or F atoms. An illustration of hydrogen bonding is shown as the dashed lines between hydrogen and oxygen atoms on adjacent water molecules in Figure 8.2.

> **➤ CHECKPOINT**
>
> Explain why hydrogen bonds between NH_3 molecules are weaker than those between H_2O molecules.

As previously discussed, melting and boiling points can be used to compare the strengths of intermolecular forces. Figure 8.8 shows the relationship between the melting points and boiling points of the hydrides of elements in Groups IVA, VIA, and VIIA. The boiling points of the hydrides of Group IVA (CH_4, SiH_4, GeH_4, and SnH_4) increase in a somewhat linear fashion with molecular weight, as would be predicted from our discussion of molecular weights and dispersion forces. However, the melting points of H_2O and HF don't follow the expected trends for the hydrides of elements in Groups VIA and VIIA. The unusually high melting and boiling points of H_2O and HF are due to the strong hydrogen bonds formed between water molecules and between HF molecules. The effect of hydrogen bonds in water is discussed in more detail in Section 8.9.

8.3 Relative Strengths of Intermolecular Forces

We can now classify the forces of attraction between particles in terms of four categories: dipole–dipole, dipole–induced dipole, induced dipole–induced dipole or dispersion forces, and hydrogen bonds. Consider the compounds in Table 8.3, for example.

Note that dispersion forces are present in all the compounds. In fact, dispersion forces must always be present in all molecular substances because these forces depend on induced dipoles resulting from electronic motion. All compounds contain electrons, and therefore all compounds will have dispersion forces. The other forces, dipole–dipole, dipole–induced dipole, and hydrogen

Table 8.3

Intermolecular Forces

Molecule	Structure	Dipole–Dipole	Dipole–Induced Dipole	Dispersion Forces	Hydrogen Bonding
Propyl alcohol	$CH_3CH_2CH_2OH$	Yes	Yes	Yes	Yes
Diethyl ether	$CH_3CH_2OCH_2CH_3$	Yes	Yes	Yes	No
Ethyl fluoride	CH_3CH_2F	Yes	Yes	Yes	No
Tin tetrachloride	$SnCl_4$	No	No	Yes	No
Acetic acid	$CH_3C(=O)OH$	Yes	Yes	Yes	Yes

bonding, depend on the structure of the molecule, and their presence can be deduced only if a suitable structure such as the Lewis structure for the compound is known.

Table 8.4 compares three classes of organic compounds: alkanes, aldehydes, and carboxylic acids. There are two trends in the data in Table 8.4. The first is the relationship between boiling point (BP) and molecular weight. For each class of compounds, as molecular weight increases, there is a corresponding increase in boiling point ($BP_{butanal} < BP_{pentanal} < BP_{hexanal}$). Figure 8.7 shows the same trend graphically for the alkanes. When comparing similar molecular compounds of significantly different molecular weights, the compounds with the highest molecular weight will generally have the strongest intermolecular forces and, therefore, the highest boiling point. This relationship is due to the dispersion force interactions that increase with increasing molecular weight.

Experimental data such as melting points are used to make the discussion of intermolcular forces more concrete and to reinforce the relationship between experimental data and models for explaining what is happening on the molecular level.

The second trend that can be observed in Table 8.4 is shown by comparing the boiling points of molecules from different categories that have similar molecular weights. Heptane, hexanal, and pentanoic acid have similar molecular weights (100.2, 100.2, and 102.1 g/mol, respectively), but their respective boiling points are 98.4, 128, and 186°C. To understand the difference in boiling points, we must examine the structure of the molecules and determine the types of intermolecular forces between molecules.

Table 8.4

Boiling Points of Three Classes of Organic Compounds

Alkane	MW (g/mol)	BP (°C)	Aldehyde	MW (g/mol)	BP (°C)	Carboxylic Acid	MW (g/mol)	BP (°C)
Butane $CH_3(CH_2)_2CH_3$	58.1	−0.5	Butanal $CH_3(CH_2)_2CHO$	72.1	75.7	Butanoic acid $CH_3(CH_2)_2COOH$	88.1	164
Pentane $CH_3(CH_2)_3CH_3$	72.2	36.1	Pentanal $CH_3(CH_2)_3CHO$	86.1	103	Pentanoic acid $CH_3(CH_2)_3COOH$	102.1	186
Hexane $CH_3(CH_2)_4CH_3$	86.2	69.0	Hexanal $CH_3(CH_2)_4CHO$	100.2	128	Hexanoic acid $CH_3(CH_2)_4COOH$	116.2	205
Heptane $CH_3(CH_2)_5CH_3$	100.2	98.4	Heptanal $CH_3(CH_2)_5CHO$	114.2	153	Heptanoic acid $CH_3(CH_2)_5COOH$	130.2	223
Octane $CH_3(CH_2)_6CH_3$	114.2	126	Octanal $CH_3(CH_2)_6CHO$	128.2	171	Octanoic acid $CH_3(CH_2)_6COOH$	144.2	239

The structures of the three compounds are given in Figure 8.9. Alkanes such as heptane are composed of only carbon and hydrogen. Carbon and hydrogen have relatively similar electronegativities, so that the covalent bonds aren't very polar. In addition, the symmetric structure of alkanes leads to a uniform distribution of charge within the molecule and results in a molecule with essentially no dipole moment. The addition of electronegative oxygen to form aldehydes and carboxylic acids causes compounds such as hexanal and pentanoic acid to have a significant dipole moment. Both aldehydes and carboxylic acids are polar and can have dipole–dipole interactions. This causes aldehydes and carboxylic acids to have higher boiling points than alkanes of similar molecular weights.

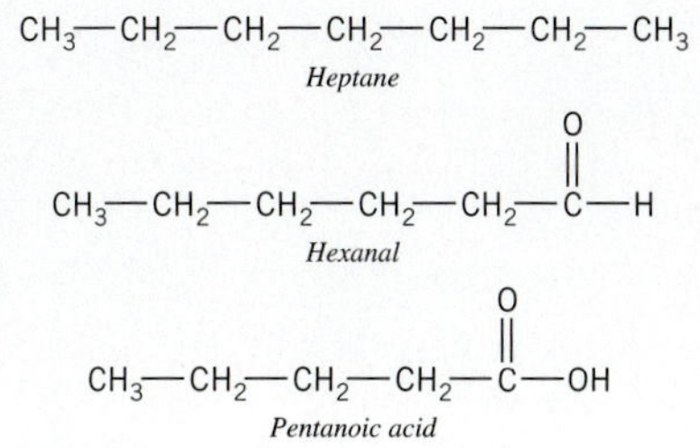

Fig. 8.9 Structures of heptane, hexanal, and pentanoic acid.

The boiling point of pentanoic acid, however, is considerably higher than that of hexanal. This indicates that the intermolecular forces attracting pentanoic acid molecules to one another are stronger than the forces found in hexanal. The structure of pentanoic acid in Figure 8.9 shows that a hydrogen atom is attached to a very electronegative oxygen atom. This allows for the formation of hydrogen bonds between this hydrogen atom on one molecule and an oxygen atom on an adjacent molecule. Therefore, pentanoic acid has dispersion, dipole–induced dipole, dipole–dipole, and hydrogen bonding intermolecular forces, giving it a high boiling point.

Exercise 8.1

The following compounds have similar molecular weights. Arrange them in order of increasing boiling point.

Formaldehyde: $H_2C{=}O$ Methanol: $CH_3{-}OH$ Ethane: CH_3CH_3

Solution

Since the compounds all have similar molecular weights, they would be expected to have similar dispersion forces. The structures indicate that the following intermolecular forces would be present.

Formaldehyde: Formaldehyde is polar and therefore would have dispersion, dipole–dipole, and dipole–induced dipole interactions.

Methanol: Methanol is polar and also has a hydrogen atom covalently bonded to an electronegative oxygen that is capable of forming hydrogen bonds with neighboring molecules. Therefore, it would be expected to have dispersion, dipole–dipole, dipole–induced dipole, and hydrogen bonding interactions.

Ethane: Ethane is nonpolar and would have only dispersion forces.

The order of increasing boiling points therefore would be ethane $<$ formaldehyde $<$ methanol.

➤ CHECKPOINT

Arrange the following compounds in order of increasing boiling point.

(a) carbon tetrachloride (CCl_4), acetone (C_3H_6O), and tetrabromobutane ($C_4H_6Br_4$)

Arrange the following compounds in order of decreasing boiling point.

(b) octanoic acid $CH_3(CH_2)_6$ COOH, decane $CH_3(CH_2)_8CH_3$, and nonanal $CH_3(CH_2)_7CHO$

The relative strengths of intermolecular forces and intramolecular bonds is summarized in Table 8.5. The energies listed for ion–dipole, ionic, and metallic interactions involve breaking multiple intermolecular forces or bonds. All other energies in the table refer to the interaction of two atoms or molecules. Intermolecular forces play a major role in determining the properties of liquids. In the following sections we will discuss these properties.

Table 8.5 gives quantitative values for the strengths of intermolecular forces, ion–dipole forces, and the three primary types of bonds. The purpose of the table is to show the relative strengths of each of these interactions. Note that the values reported for ion–dipole and ionic bonds involve attraction between several particles rather than individual forces of attraction as given for the other interactions.

Table 8.5

Relative Strengths of Intermolecular Forces and Bonds

Intermolecular Force	Example	Energy (kJ/mol)
Dipole–Dipole	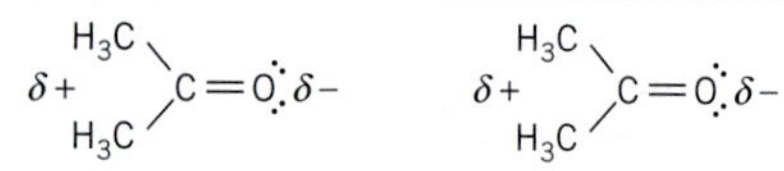	~5
Dipole–induced dipole	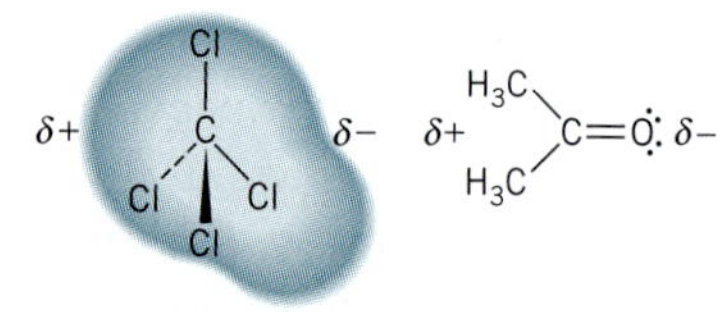	~2
Induced dipole–induced dipole (dispersion)	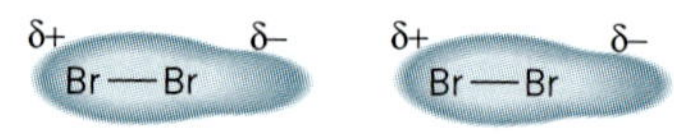	~5
Ion–dipole	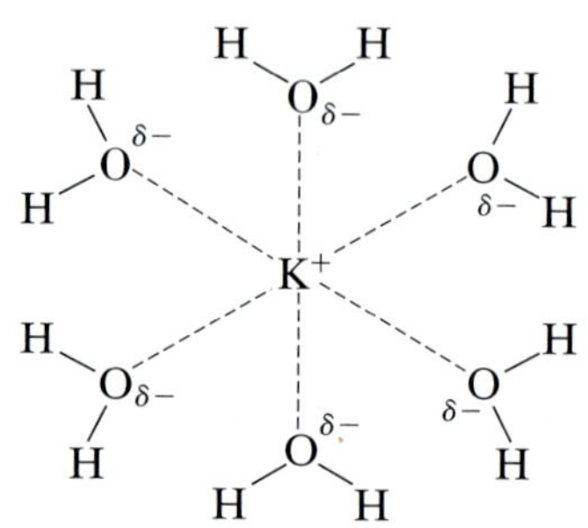	~342
Hydrogen bond	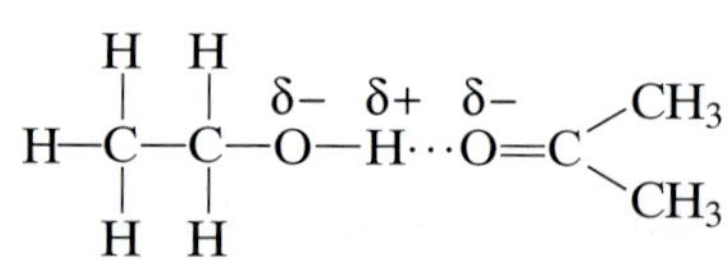	20

Bond	Example	Energy (kJ/mol)
Covalent	:C̤̈l—C̤̈l:	243
Ionic	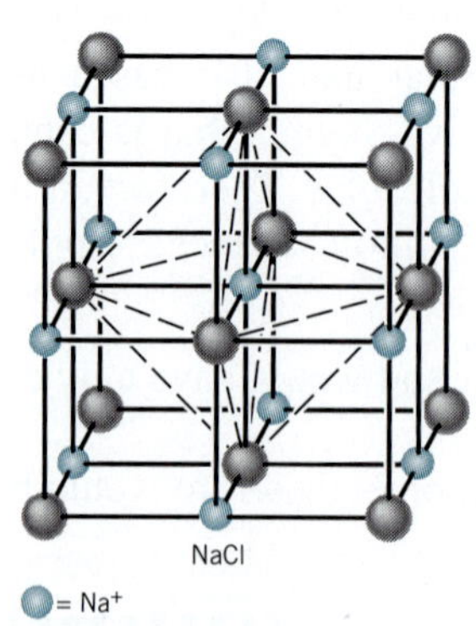	787
Metallic	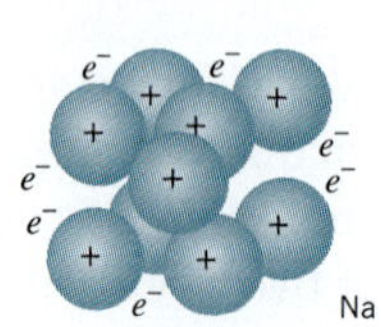	107

8.4 The Kinetic Theory of Liquids

The kinetic theory of gases discussed in Chapter 6 can be extended to liquids. Liquids and solids are more complex, but, like gases, the particles in a liquid are in constant motion. The *average kinetic energy* of the particles in a liquid is also the same as the average kinetic energy of the particles in a gas at the same temperature. We must include the term *average* in the statement because all molecules at the same temperature don't have the same kinetic energy. There is an enormous range of kinetic energies possessed by molecules at a given temperature.

The force of attraction between the particles in a liquid, however, is large enough to hold the particles relatively close together. As a result, collisions between particles in a liquid occur much more frequently than in a gas. In a typical gas, a particle will collide with its neighbors about 10^9 times per second, whereas in a liquid the frequency of collisions is 10^{13} times per second.

In a liquid the particles undergo almost constant collisions with their neighbors to form clusters of particles that stick together for a moment, and then come apart. The force of attraction between the particles in a liquid can be estimated by measuring the amount of heat required to transform a given quantity of a liquid into the corresponding gas (or "vapor") at a given temperature. The result of the measurement is known as the **enthalpy of vaporization** of the liquid. ΔH^0_{vap}. The larger the force of attraction between the particles, the larger the enthalpy of vaporization.

Chapter 7 described enthalpies of atom combination as the energy released due to the forces of attraction that result when gaseous atoms combine to form a compound. These forces of attraction consist of both intramolecular bonds and intermolecular or interionic forces. Therefore, enthalpies of atom combination can be used as a direct means of comparing intermolecular forces between molecules.

The enthalpies of atom combination for liquid and gaseous carbon tetrachloride, for example, are -1338.84 and -1306.3 kJ/mol$_{rxn}$, respectively. The covalent intramolecular bonds in the liquid and the gaseous CCl_4 are the same (covalent C—Cl bonds). Any difference between the enthalpies of atom combination for the liquid and gas must therefore be due to intermolecular forces. The more negative enthalpy of atom combination for liquid CCl_4 shows that the intermolecular forces are stronger in the liquid than in the gaseous CCl_4.

Reaction diagrams like those in Chapter 7 are used to calculate changes in enthalpy associated with phase changes. We emphasize that these diagrams are useful for visualizing enthalpy changes but that no bonds are actually broken during a phase change for a molecular species.

As we have seen in Chapter 7, the enthalpy change at 298 K for the reaction $CCl_4(l) \longrightarrow CCl_4(g)$ may be calculated from the data in the following diagram.

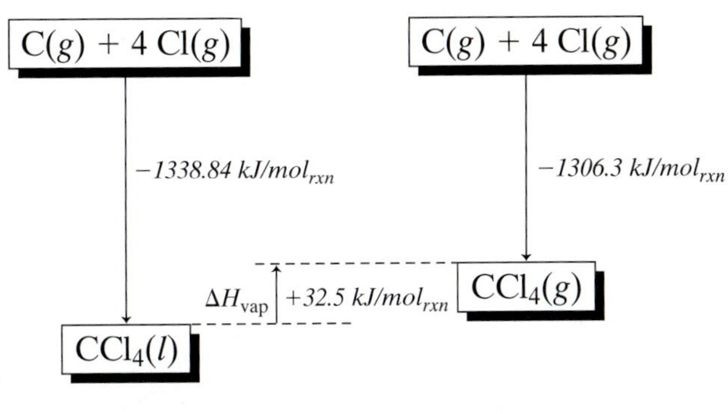

Remember that this diagram is only used to illustrate how the change in enthalpy can be calculated from the enthalpies of atom combination found in Appendix B.13. It is *not* meant to imply that the CCl_4 phase change from liquid to gas involves the breaking of bonds to form atoms. No C—Cl bonds are broken during the phase change.

The enthalpy change, which is also the enthalpy of vaporization, is 32.5 kJ/mol$_{rxn}$. The enthalpy change is positive, indicating that heat must be put into $CCl_4(l)$ to change it to the gaseous state; that is, intermolecular forces that attract the CCl_4 molecules to each other in the liquid state must be overcome.

> **➤ CHECKPOINT**
>
> The enthalpy of fusion for water is 6 kJ/mol$_{rxn}$ at 0°C.
>
> $$H_2O(s) \longrightarrow H_2O(l)$$
>
> Which phase change, liquid to gas or solid to liquid, involves breaking the stronger intermolecular forces? Which state—solid, liquid, or gas—has the strongest intermolecular forces?

A similar approach may be taken to determine the relative magnitude of intermolecular forces in solids. In this case the reaction involves converting a solid to a liquid, and the enthalpy change that accompanies the process is called the **enthalpy of fusion,** ΔH^0_{fus}. The major difference is that for a liquid to gas phase change essentially all of the intermolecular forces in the liquid must be broken when forming the gas because there is very little interaction of particles in the gas phase. During melting, however, not all intermolecular forces in the solid phase must be broken to form the liquid. Intermolecular interactions are still important in the liquid phase.

The kinetic theory of liquids can explain the effect of changes in the temperature of a liquid on many of its characteristic properties. Consider the density of a liquid, for example. The density of a substance is determined by the mass and shape of its particles and how close the particles are together. As the temperature of a liquid increases, the average kinetic energy of its particles increases. The increased thermal motion will cause the particles to move farther apart. As a result, the density of a liquid usually decreases with increasing temperature.

8.5 The Vapor Pressure of a Liquid

When chemists discuss a liquid to gas phase change, the gaseous state is often described as a *vapor.* The phase change from a liquid to a gas is referred to as either *boiling* or *evaporation,* depending on the conditions under which the phase change occurs. The phase change from a gas to a liquid is referred to as **condensation.**

A liquid doesn't have to be heated to its boiling point before it can become a gas. Water, for example, evaporates from an open container at room temperature (20°C), even though the boiling point of water is 100°C. According to the kinetic molecular theory, the *average kinetic energy* of particles in the liquid, solid, and gas states depends on the temperature of the substance. However, not all molecules have the same energy at a given temperature. Instead, they possess a range of kinetic energies.

Figure 8.10 illustrates the fraction of the molecules in the liquid phase that have specific kinetic energies at two different temperatures. As the temperature increases

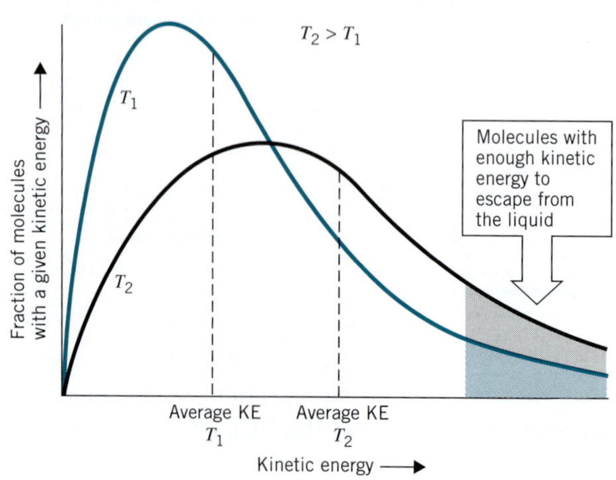

Fig. 8.10 At a given temperature, some of the particles in a liquid have enough energy to form a gas. As the temperature increases, the fraction of the molecules moving fast enough to escape from the liquid increases. As a result, the vapor pressure of the liquid also increases.

from T_1 to T_2, the shape of the curve changes, and there is an overall increase in the average kinetic energy of the molecules. Even at temperatures well below the boiling point of the liquid, some of the particles are moving fast enough to escape from the liquid. The shaded portion of this graph begins at the minimum energy a molecule must possess in order to escape into the vapor state. Any kinetic energy larger than that value will be sufficient to allow the molecule to escape from the liquid state and move into the gaseous state. At temperature T_2 a larger number of molecules have sufficient energy to move into the vapor state than at temperature T_1.

When a liquid evaporates, the average kinetic energy of the molecules in the liquid decreases. As a result, the liquid becomes cooler. It therefore absorbs energy from its surroundings until it returns to thermal equilibrium. But as soon as that happens, some of the molecules in the liquid state acquire enough energy to escape from the liquid. In an open container, the process continues until all of the liquid evaporates.

Figure 8.11 illustrates what happens to a liquid placed in a closed container that is maintained at a constant temperature with respect to time. Water is used here as the example, but other liquids would behave in the same way. There is initially a vacuum in the container, with a pressure of 0 atm (no water has yet evaporated). As time passes (t_0), some of the molecules begin to escape from the surface of the liquid to form a gas. The pressure exerted by the water vapor is called the partial pressure of the water vapor. At time t_1, enough water vapor has accumulated so that some of the molecules in the gas phase begin to condense to form the liquid state. But the rate at which the liquid is evaporating to form a vapor is still larger than the rate at which the vapor condenses to form a liquid.

By time t_2, the pressure exerted by the water vapor has continued to increase due to the evaporation of additional water. The rate at which the liquid evaporates to form a gas, however, has become equal to the rate at which the gas condenses to form the liquid (illustrated by an equal number of molecules entering the liquid state as enter the gas state). At this point, the system is said to be in **equilibrium** (from Latin, "a state of balance"). The space above the liquid is saturated with water vapor. The number of water molecules in the vapor phase remains constant, and hence the pressure in the container will no longer change as long as the temperature is constant. The pressure due to the water vapor in the closed container at equilibrium is called the **vapor pressure.**

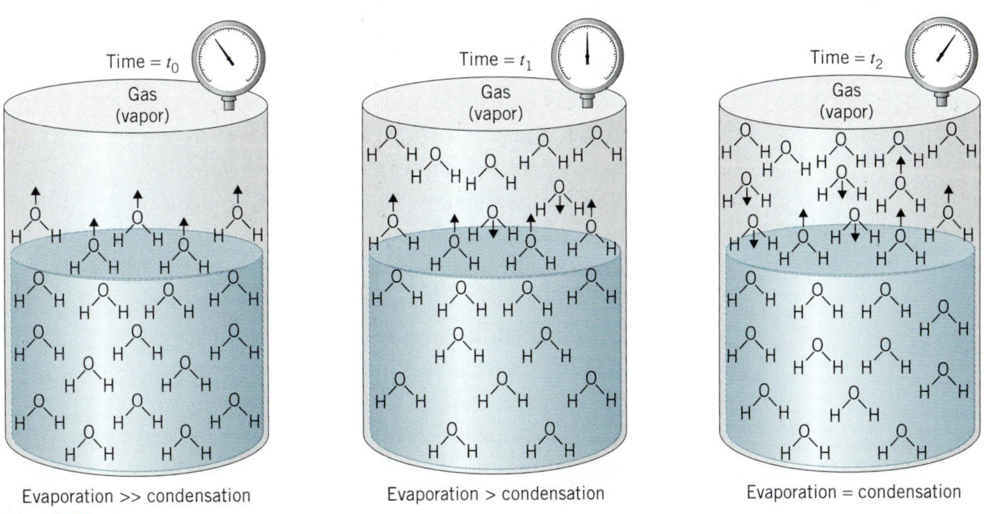

Fig. 8.11 Change in the pressure exerted by a vapor with respect to time. The vapor pressure of the liquid is the equilibrium partial pressure of the gas (or vapor) that collects above the liquid in a closed container at a given temperature.

The vapor pressure is the maximum pressure that can be exerted by a vapor at a given temperature. Individual molecules continue to move between the liquid and gas states, but at equilibrium the total number of molecules in the liquid and gas states remain constant and therefore the pressure remains constant. Note that the number of molecules in the liquid and gas states don't have to be equal. What is equal is the number of molecules in the gas phase entering the liquid state and the number of molecules in the liquid phase entering the gas state.

Exercise 8.2

Suppose instead of a vacuum in the closed container of Figure 8.11, the initial pressure in the container was 1 atm. The 1 atm of pressure can be assumed to be due to dry air composed primarily of nitrogen and oxygen. Describe what happens to the pressure with time as the liquid evaporates.

Solution

The pressure at t_0 would be 1 atm prior to evaporation. As time progresses to t_1, the water evaporates, and the total pressure in the container becomes equal to 1 atm plus the partial pressure of the water vapor. As equilibrium is established at t_2, the total pressure in the container will be 1 atm plus the vapor pressure of water at that temperature.

We have used two terms that are quite similar but are very important to differentiate. In the above discussion we used the term *pressure exerted by the vapor* to describe the partial pressure exerted by the water vapor in the closed container. The *pressure exerted by the vapor* changes with respect to time. We then used the term *vapor pressure* to describe the pressure exerted by a vapor under the very special condition of equilibrium with its liquid. *Vapor pressure* is a constant for a given liquid at a given temperature.

The kinetic theory suggests that the vapor pressure of a liquid depends on its temperature because the fraction of the molecules that have enough energy to escape from a liquid increases with the temperature of the liquid, as shown in Figure 8.10. The vapor pressure of a liquid is determined by the strength of the intermolecular forces that hold the molecules of the liquid together. The stronger the attraction between molecules, the smaller the tendency for the molecules to escape from the liquid into the gas phase, and hence the lower the vapor pressure. Any increase in the temperature of the liquid, however, will increase the kinetic energy of its molecules. The increased motion of the molecules offsets the intermolecular attractive forces. Thus, the vapor pressure of the liquid will increase with increasing temperature.

> **➤ CHECKPOINT**
>
> A liquid is placed in an evacuated container (initial pressure equals 0) and allowed to reach equilibrium with its vapor. What will happen to the pressure exerted by the vapor if the volume of the container is decreased without a change in temperature? Describe what must happen on the molecular level.

The vapor pressures of water at temperatures from 0°C to 50°C are given in Table B.3 in the appendix. Figure 8.12 shows that the relationship between vapor pressure and temperature is not linear. The vapor pressure of water increases more rapidly than the temperature of the system.

Below the surface of a liquid, the force of **cohesion** (literally, "sticking together") between particles is the same in all directions, as shown in Figure 8.13. Particles on the surface of the liquid feel a net force attracting them back toward the body of the liquid. As a result, the liquid takes on the shape that has the smallest possible surface area. The force that controls the shape of the liquid is called the **surface tension.** The stronger the force of attraction between the particles in the

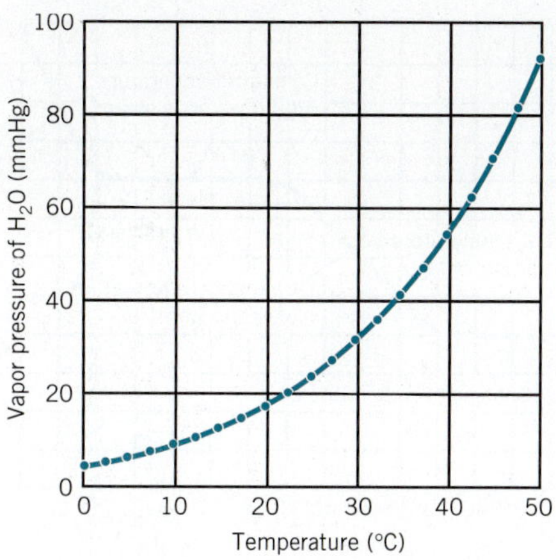

Fig. 8.12 Plot of the vapor pressure of water versus temperature.

liquid, the larger the surface tension. As the temperature increases, the increased motion of the particles in a liquid partially overcomes the attractions between the particles. As a result, the surface tension of the liquid decreases with increasing temperature.

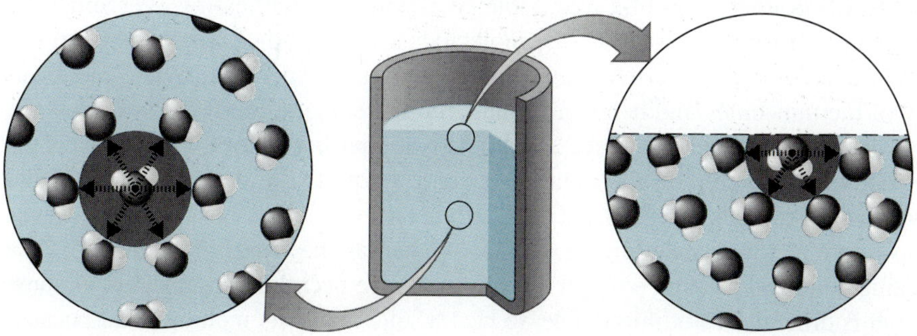

Fig. 8.13 Molecules in a liquid feel a force of cohesion that pulls them into the body of the liquid. For molecules in the interior of the liquid, the forces are exerted from all directions, resulting in a net force of zero. However, molecules at the surface are attracted only by molecules to the side or below, resulting in a net force in the downward direction.

8.6 Melting Point and Freezing Point

Does water always become hotter when it is heated? Think about what happens when you heat a pot of water. At first, the water gets hotter, but eventually it starts to boil. From that moment on, the temperature of the water remains the same (100°C at 1 atm), regardless of how much heat is added to the water, until all of the liquid has boiled away.

Imagine another experiment in which a thermometer is immersed in a snowbank on a day when the temperature finally gets above 0°C. The snow gradually melts as it gains heat from the air above it. But the temperature of the snow remains at 0°C until the last snow melts.

Apparently some heat can enter a system without changing its temperature. This is encountered whenever there is a change in the state of matter. Heat can enter or leave a sample without any detectable change in its temperature when a solid melts, when a liquid freezes or boils, or when a gas condenses to form a liquid.

Figure 8.14 shows the results obtained when ice initially at −100°C is heated in an expandable but closed container at 1 atm pressure. Initially, the heat that enters the system is used to increase the temperature of the ice from −100°C to 0°C. At 0°C the heat entering the system is used to melt the ice, and there is no change in the temperature until all of the ice is gone. The amount of heat required to melt the ice is called the *enthalpy of fusion*.

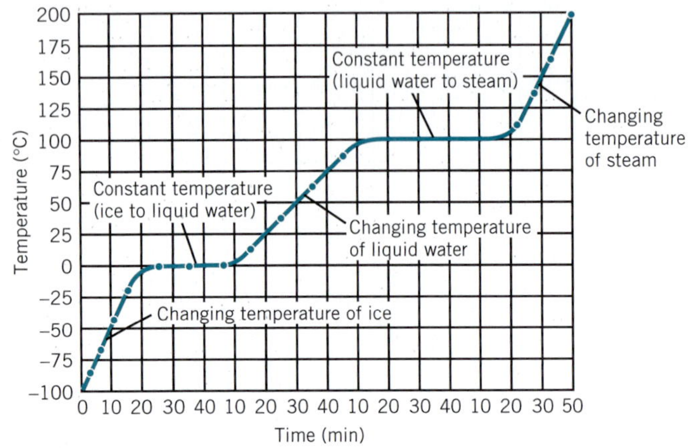

Fig. 8.14 Plot showing the change in temperature and changes in phase as ice is heated to form liquid water and then steam at 1 atm pressure.

Once the ice melts, the temperature of the water slowly increases from 0°C to 100°C. But once the water starts to boil, the heat that enters the sample is used to convert the liquid to a gas, and the temperature of the sample remains constant until the liquid has boiled away. The amount of heat required to vaporize a liquid is called the *enthalpy of vaporization*. As heat continues to be added, the steam in the closed container increases in temperature, causing the container to expand even further.

At the boiling point of a liquid, the energy input from heating overcomes the intermolecular forces that attract the molecules to one another. Since the energy is used to overcome the intermolecular forces, there is no increase in the kinetic energy of the molecules and hence no change in the temperature of the system.

Pure, crystalline solids have a characteristic **melting point** (MP), which is the temperature at which the solid melts to become a liquid. The melting point of solid oxygen, for example, is −218.4°C at 1 atm pressure. Liquids have a characteristic temperature at which they turn into solids, known as the **freezing point** (FP). The freezing point and melting point of a given substance occur at the same temperature.

It is difficult, if not impossible, to heat a solid above its melting point because the heat that enters the solid at its melting point is used to convert the solid to a liquid. It is possible, however, to cool some liquids to temperatures below their freezing points without forming a solid. When this is done, the liquid is said to be *supercooled.*

Because it is difficult to heat solids to temperatures above their melting points, and because pure solids melt over a very small temperature range, melting points are often used to help identify compounds. We can distinguish between the three sugars known an *glucose* (MP = 150°C), *fructose* (MP = 103–105°C), and *sucrose* (MP = 185–186°C), for example, by determining the melting point of a small sample.

Measurements of the melting point of a solid can also provide information about the purity of the substance. Pure, crystalline solids melt over a very narrow range of temperatures, whereas mixtures melt over a broad temperature range. Mixtures also tend to melt at temperatures below the melting points of the pure solids. A common example is provided by the addition of salt to ice, which lowers the melting point of the ice.

8.7 Boiling Point

Boiling occurs when the intermolecular forces that hold molecules in the liquid phase to one another are broken and the molecules move into the gaseous state. Figure 8.15 shows a diagram illustrating the interactions that must be overcome to boil water. Note that no covalent bonds are broken during the boiling process. When a liquid is heated, it eventually reaches a temperature at which the vapor

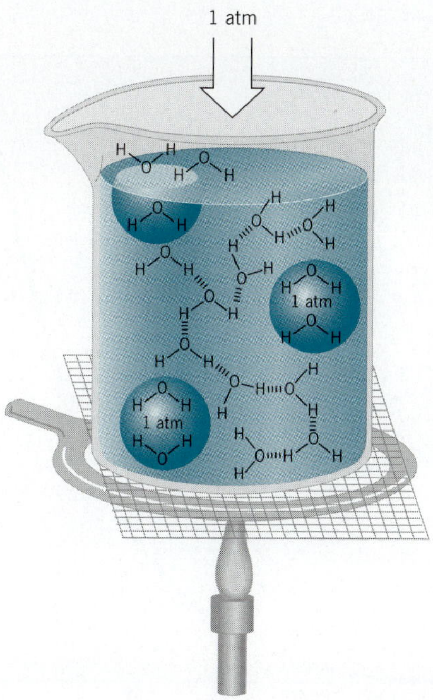

Fig. 8.15 Molecular scale diagram of water boiling. Bubbles of water vapor form inside the liquid.

pressure is large enough that bubbles of vapor form inside the body of the liquid. This temperature is called the **boiling point.** Once the liquid starts to boil, the temperature remains constant until all of the liquid has been converted to a gas.

The *normal boiling point* of a liquid is defined as the temperature at which the liquid boils at 1 atm pressure. The normal boiling point of water is 100°C. However, liquids can boil at many different temperatures depending on the pressure exerted on the liquid. If you try to cook an egg in boiling water while camping in the Rocky Mountains at an elevation of 10,000 feet, you will find that it takes longer for the egg to cook because water boils at only 90°C at that elevation.

Before microwave ovens became popular, pressure cookers were used to decrease the amount of time it took to cook food. In a typical pressure cooker, water can remain a liquid at temperatures as high as 120°C, and food cooks in as little as one-third the normal time.

To explain why water boils at 90°C in the mountains and at 120°C in a pressure cooker, even though the normal boiling point of water is 100°C, we have to understand why a liquid boils.

A liquid boils when the pressure of the vapor escaping from the liquid is equal to the pressure exerted on the liquid by its surroundings.

We recognize that a liquid is boiling by the formation of bubbles in the liquid. The bubbles are balls of vapor formed by molecules that have acquired sufficient energy to enter the vapor phase. The vapor in the bubble is in equilibrium with the liquid. Therefore, the pressure exerted by the vapor within the bubble is equal to the vapor pressure of the liquid at the temperature of the liquid. If the external pressure is larger than the vapor pressure developed in the gaseous pockets, the pockets will be crushed and no visible evidence of boiling will be seen. If, however, the vapor pressure in the bubbles is equal to the external pressure, they won't collapse and will be seen to rise to the surface.

This means that we can cause a liquid to boil in two ways: by increasing the temperature of the liquid or by decreasing the pressure exerted on the liquid. As the temperature of a liquid is increased, there is a corresponding increase in the vapor pressure, as shown for water in Figure 8.16. When the vapor pressure has increased such that it equals the pressure exerted on the surface of the liquid, boiling will occur.

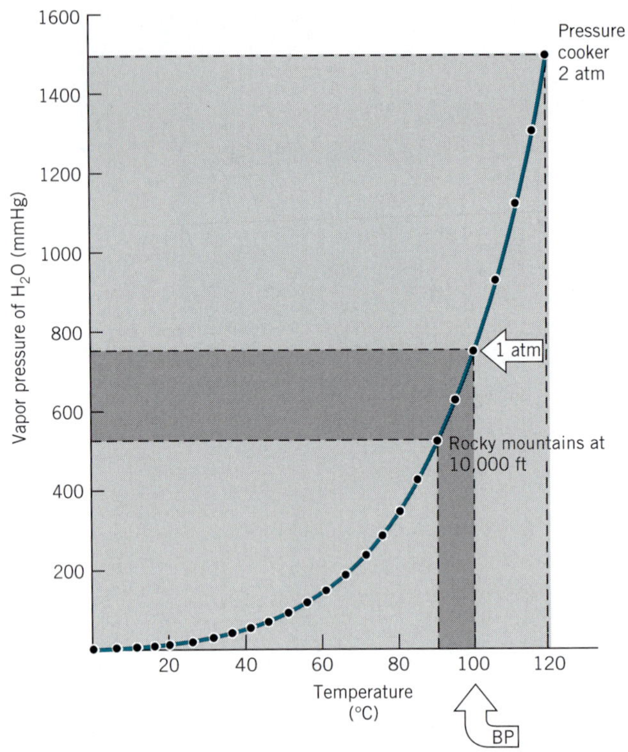

Fig. 8.16 A liquid boils when its vapor pressure is equal to the pressure exerted on the liquid by the surroundings. The normal boiling point of water is 100°C. In the mountains, atmospheric pressure is less than 1 atm, and water boils at a temperature below 100°C. In a pressure cooker at 2 atm, water doesn't boil until the temperature reaches 120°C.

The normal boiling point of water is 100°C because this is the temperature at which the vapor pressure of water is 760 mmHg, or 1 atm. At 10,000 feet above sea level, the pressure of the atmosphere is only 526 mmHg. At that elevation, water boils when its vapor pressure is 526 mmHg. Figure 8.16 shows that water needs to be heated to only 90°C for its vapor pressure to reach 526 mmHg.

Pressure cookers are equipped with a valve that lets gas escape when the pressure inside the pot exceeds some fixed value. This valve is often set at 15 pounds/in^2 (psi), which, combined with the usual prevailing atmospheric pressure of approximately 15 psi, means that the water vapor inside the pot must reach a pressure of 2 atm before it can escape. Because water doesn't reach a vapor pressure of 2 atm until the temperature is 120°C, it boils in the pressurized container at 120°C.

A second way to cause a liquid to boil is to reduce the pressure exerted on the surface of the liquid. The vapor pressure of water is roughly 20 mmHg at room temperature. We can therefore make water boil at room temperature by reducing the pressure in its container to less than 20 mmHg.

Anything that influences the vapor pressure of a liquid also has an effect on the boiling point. Thus, for substances composed of molecules with similar shapes and masses, the liquid with the strongest intermolecular forces of attraction will have the highest boiling point.

8.8 Phase Diagrams

Figure 8.17 shows an example of a **phase diagram,** which describes the state or phase of a substance at different combinations of temperature and pressure. This diagram is divided into three areas, which represent the solid, liquid, and gaseous states of the substance.

The best way to remember which area corresponds to each of these states is to remember the conditions of temperature and pressure that are most likely to be

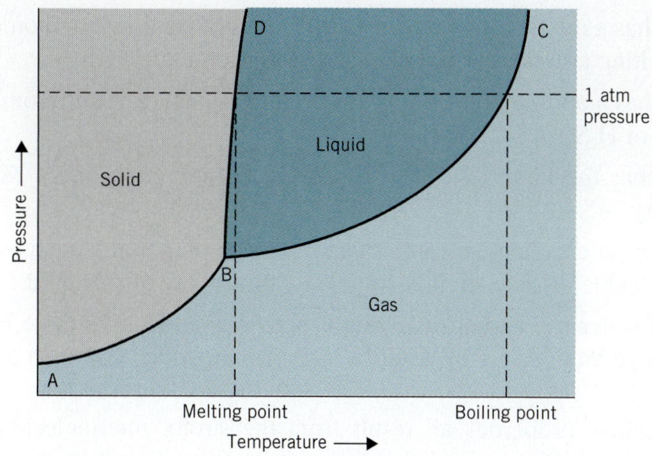

Fig. 8.17 A phase diagram, which describes the state of a substance at any possible combination of temperature and pressure. A horizontal line across a phase diagram at a pressure of 1 atm crosses the curve separating solids and liquids at the melting point of the solid, and it crosses the curve separating liquids and gases at the boiling point of the liquid.

associated with a solid, a liquid, and a gas. Low temperatures and high pressures favor the formation of a solid. Gases, on the other hand, are most likely to be found at high temperatures and low pressures. Liquids lie between these extremes.

Phase diagrams can be used in several ways. We can focus on the regions separated by the lines in these diagrams and get some idea of the conditions of temperature and pressure that are most likely to produce a gas, a liquid, or a solid. Or we can focus on the lines that divide the diagram into states, which represent the combinations of temperature and pressure at which two states are in equilibrium.

The points along the line connecting points *A* and *B* in the phase diagram in Figure 8.17 represent the combinations of temperature and pressure at which the solid is in equilibrium with the gas. The solid line between points *B* and *C* is identical to the plot of the temperature dependence of the vapor pressure of the liquid shown in Figure 8.12. It contains all of the combinations of temperature and pressure at which the liquid boils. The solid line between points *B* and *D* contains the combinations of temperature and pressure at which the solid and liquid are in equilibrium.

The *BD* line is almost vertical because the melting point of a solid isn't particularly sensitive to changes in pressure. For most compounds, this line has a small positive slope, as shown in Figure 8.17. The slope of this line is slightly negative for water, however. As a result, water can melt at temperatures near but below its freezing point when subjected to increased pressure.

Figure 8.17 shows what happens when we draw a horizontal line across a phase diagram at a pressure of exactly 1 atm. This dashed line crosses the line between points *B* and *D* at the normal melting point of the substance because solids normally melt at the temperature at which the solid and liquid are in equilibrium at 1 atm pressure. The dashed line crosses the line between points *B* and *C* at the boiling point of the substance because the normal boiling point of a liquid is the temperature at which the liquid and gas are in equilibrium at 1 atm pressure and the vapor pressure of the liquid is therefore equal to 1 atm.

8.9 Hydrogen Bonding and the Anomalous Properties of Water

We are so familiar with the properties of water that it is difficult to appreciate the extent to which its behavior is unusual.

- Most solids expand when they melt. Water expands when it *freezes.*
- Most solids are more dense than the corresponding liquids. Ice (0.917 g/cm^3) is not as dense as water; therefore, ice floats on liquid water.

- Water has a melting point at least 100°C higher than expected on the basis of the melting points of H_2S, H_2Se, and H_2Te (see Figure 8.8).
- Water has a boiling point almost 200°C higher than expected from the boiling points of H_2S, H_2Se, and H_2Te.
- Water has the largest surface tension of any common liquid except liquid mercury.
- Water is an excellent solvent. It can dissolve compounds, such as NaCl, that are insoluble or only slightly soluble in almost any other liquid.
- Liquid water has an unusually high specific heat. It takes more heat to raise the temperature of 1 g of water by 1°C than any other common liquid.

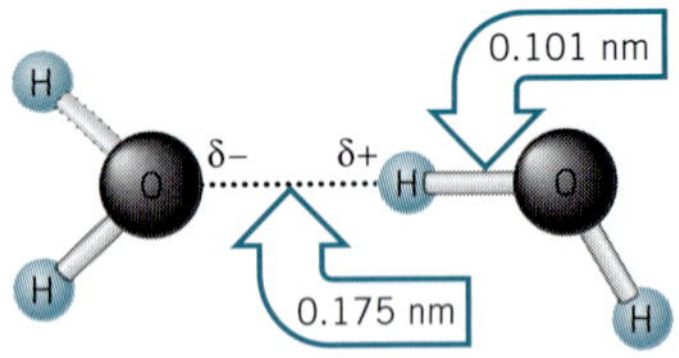

Fig. 8.18 *Hydrogen bonding* is the primary attraction between neighboring water molecules. The polarity of water molecules creates this unusually strong intermolecular force.

These anomalous properties all result from the strong intermolecular forces in water. In Section 5.10 we concluded that water is best described as a polar molecule in which there is a partial separation of charge to give positive and negative poles. The force of attraction between a positive partial charge on the hydrogen atom on one water molecule and the negative partial charge on the oxygen atom on another gives rise to the intermolecular force called hydrogen bonding, as shown in Figure 8.18. The hydrogen bonds in water are particularly important because of the dominant role that water plays in the chemistry of living systems.

The hydrogen bonds between water molecules in ice produce the open structure shown in Figure 8.19. When ice melts, some of the hydrogen bonds are broken, and the structure collapses to form a liquid that is about 10% denser. This unusual property of water has several important consequencs. The expansion of water when it freezes is responsible for the cracking of concrete, which forms potholes in streets and highways. But it also means that ice floats on top of rivers and streams.

As discussed in Section 8.2, water has a much higher boiling point than would be predicted from its molecular weight. Figure 8.8 shows a steady increase in

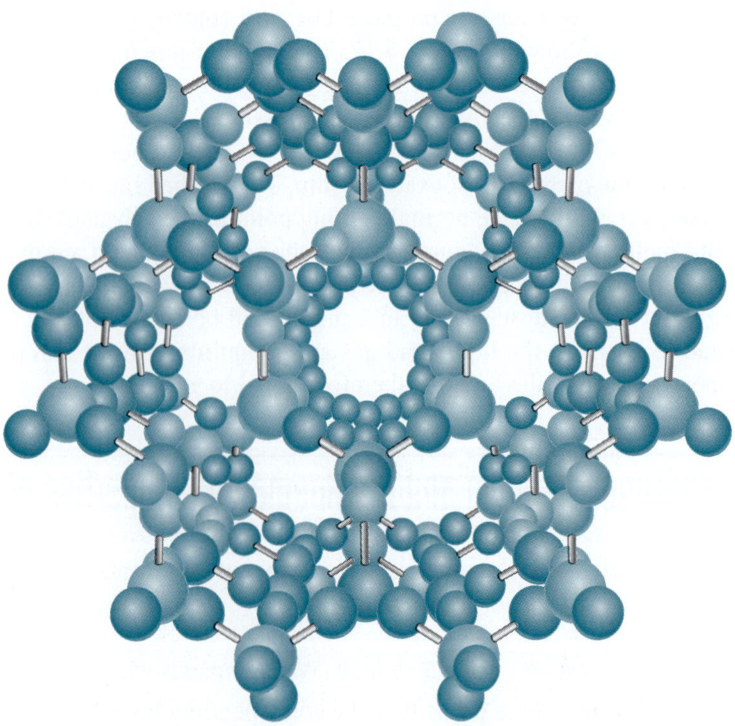

Fig. 8.19 Structure of ice. Note that in each of the hydrogen bonds the hydrogen atoms are closer to one of the oxygen atoms than the other.

boiling point in the series CH_4, SiH_4, GeH_4, and SnH_4. However, in the other two series shown in the figure, the boiling points of H_2O and HF are anomalously large because of the strong hydrogen bonds between molecules in liquid H_2O and HF. If this doesn't seem important, try to imagine what life would be like if water boiled at $-80°C$.

The unusually large specific heat of water discussed in Section 7.7 is also related to the strength of the hydrogen bonds between water molecules. Anything that increases the motion of water molecules, and therefore the temperature of water, must interfere with the hydrogen bonds between the molecules. The fact that it takes so much energy to overcome hydrogen bonds means that water can store enormous amounts of thermal energy. Although the water in lakes and rivers gets warmer in the summer and cooler in the winter, the large specific heat of water limits the range of temperatures; otherwise, extremes of temperature would threaten the life that flourishes in those environments. The specific heat of water is also responsible for the ability of oceans to act as a thermal reservoir that moderates the swings in temperature which occur from winter to summer.

8.10 Solutions: Like Dissolves Like

The initial portion of this chapter has dealt primarily with the structure and properties of pure liquids. The final sections of the chapter deal with solutions. Recall from Chapter 2 that solutions can involve solvents and solutes composed of many states of matter. The discussion of solutions in this chapter, however, is restricted to solvents that are liquids.

Solutes, solvents, and solutions were introduced in Chapter 2. Why solutions form is introduced after the discussion of the liquid state.

Figure 8.20 shows what happens when we add a pair of solutes to a pair of solvents.

Solutes: I_2 and $KMnO_4$
Solvents: H_2O and CCl_4

The solutes have two things in common. They are both solids, and they both have an intense violet or purple color. The solvents are both colorless liquids that don't mix with one another.

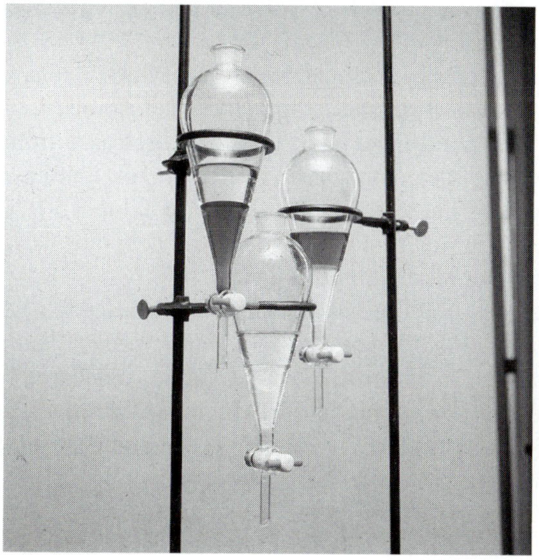

Fig. 8.20 Water and carbon tetrachloride form two separate liquid phases in a separatory funnel (center); $KMnO_4$ dissolves in the water layer on the top of the separatory funnel to form an intensely colored solution (right); I_2 dissolves in the CCl_4 layer on the bottom of the separatory funnel to form an intensely colored solution (left).

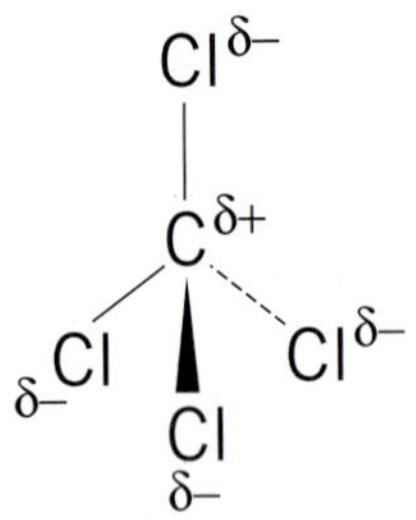

Fig. 8.21 Although there is some separation of charge within the individual bonds in CCl_4, the symmetrical shape of the CCl_4 molecule ensures that there is no net dipole moment. CCl_4 is therefore *nonpolar.*

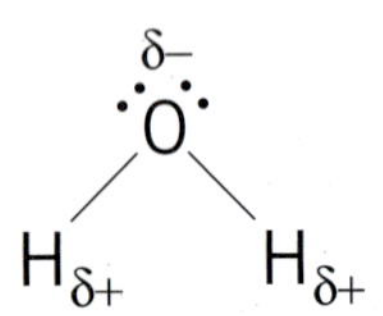

Fig. 8.22 Because water molecules are bent, or angular, they have distinct negative and positive poles. H_2O is therefore an example of a *polar molecule.*

Table 8.6
Solubilities of I_2 and $KMnO_4$ in CCl_4 and Water

	H_2O	CCl_4
I_2	Slightly soluble	Very soluble
$KMnO_4$	Very soluble	Insoluble

The solutes are held together by different forces. Iodine consists of individual I_2 molecules held together by relatively weak intermolecular forces. Potassium permanganate consists of K^+ and MnO_4^- ions held together by the strong force of attraction between ions of opposite charge. The attraction between the ions is much stronger than the dispersion between I_2 molecules. It is therefore much easier to separate the I_2 molecules in solid iodine than it is to separate $KMnO_4$ into its constituent ions.

There is also a significant difference between the solvents, CCl_4 and H_2O. The difference between the electronegativities of the carbon and chlorine ato ms in CCl_4 is so small ($\Delta EN = 0.33$) that there is relatively little ionic character in the C—Cl bonds. Even if there were some separation of charge in the bonds, the CCl_4 molecule wouldn't be polar because it has a symmetrical shape in which the four chlorine atoms point toward the corners of a tetrahedron, as shown in Figure 8.21. CCl_4 is therefore best described as a **nonpolar solvent.**

The difference between the electronegativities of the hydrogen and oxygen atoms in water is much larger ($\Delta EN = 1.31$), and the H—O bonds in the molecule are therefore polar. If the H_2O molecule were linear, the polarity of the two O—H bonds would cancel, and the molecule would have no net dipole moment. Water molecules, however, have a bent, or angular, shape. As a result, they have distinct positive and negative poles, and water is a polar molecule, as shown in Figure 8.22. Water is therefore classified as a **polar solvent.**

Because water and carbon tetrachloride don't mix, two separate liquid phases are clearly visible when these solvents are added to a flask. We can use the densities at 25°C of CCl_4 (1.584 g/cm^3) and H_2O (1.0 g/cm^3) to decide which phase is water and which is carbon tetrachloride. The more dense CCl_4 settles to the bottom of the flask.

When a few crystals of iodine are added and the contents of the flask are shaken, the I_2 dissolves in the CCl_4 layer to form a violet solution. The water layer becomes a very light brown, which suggests that very little I_2 dissolves in water.

When the experiment is repeated with potassium permanganate, the water layer picks up the characteristic dark purple color of the MnO_4^- ion, and the CCl_4 layer remains colorless. This suggests that $KMnO_4$ dissolves in water but not in carbon tetrachloride. The results of the experiment are summarized in Table 8.6. Two important questions are raised. Why does $KMnO_4$ dissolve in water, but not in carbon tetrachloride? Why does I_2 dissolve in carbon tetrachloride but to only a very small extent in water?

It takes a lot of energy to separate K^+ and MnO_4^- ions in potassium permanganate. The ions can form strong attractive interactions with neighboring water molecules, as shown in Figure 8.23. The energy released from the formation of the *ion–dipole* interactions compensates for the energy that has to be invested to separate individual ions in the $KMnO_4$ crystal (See Table 8.5.). No such forces are possible between the K^+ or MnO_4^- ions and the nonpolar CCl_4 molecules. As a result, $KMnO_4$ does not dissolve in CCl_4.

The I_2 molecules in iodine and the CCl_4 molecules in carbon tetrachloride are both held together by weak intermolecular forces. These intermolecular forces must be broken in order to separate solute molecules from one another and solvent molecules from one another. Figure 8.24 shows that similar intermolecular forces are formed, however, between I_2 and CCl_4 molecules when the I_2 is dissolved in CCl_4. I_2 therefore readily dissolves in CCl_4 because the intermolecular forces that are broken in the solute and solvent are very similar to the

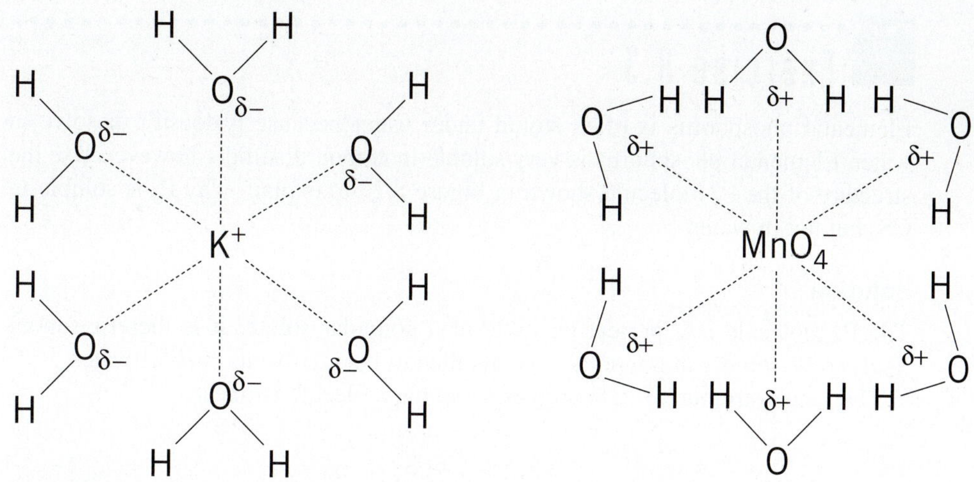

Fig. 8.23 $KMnO_4$ dissolves in water because the energy released when ion–dipole forces form between the K^+ ion and the negative end of neighboring water molecules and between the MnO_4^- ion and the positive end of the solvent molecules compensates for the energy it takes to separate the K^+ and MnO_4^- ions.

intermolecular forces that are formed between the solute and solvent molecules. The molecules in water are held together by hydrogen bonds that are stronger than most intermolecular forces. No interaction between I_2 and H_2O molecules is strong enough to compensate for the hydrogen bonds between water molecules that have to be broken to dissolve iodine in water, so relatively little I_2 dissolves in H_2O.

We can summarize the results of the experiment by noting that *nonpolar solutes* (such as I_2) dissolve in *nonpolar solvents* (such as CCl_4), whereas many ionic solutes (such as $KMnO_4$) dissolve in *polar solvents* (such as H_2O). As a general rule, *like dissolves like*.

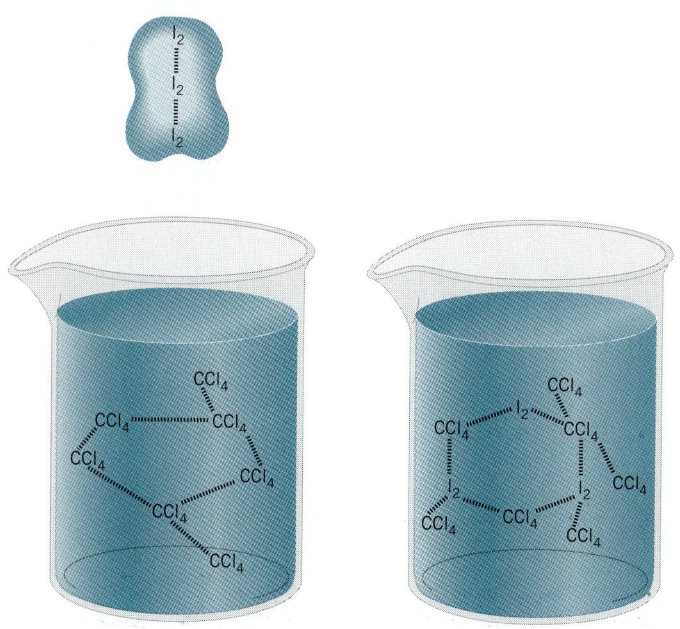

Fig. 8.24 I_2 dissolves in CCl_4 because the intermolecular forces in I_2 and in CCl_4 are similar to the intermolecular forces between CCl_4 and I_2.

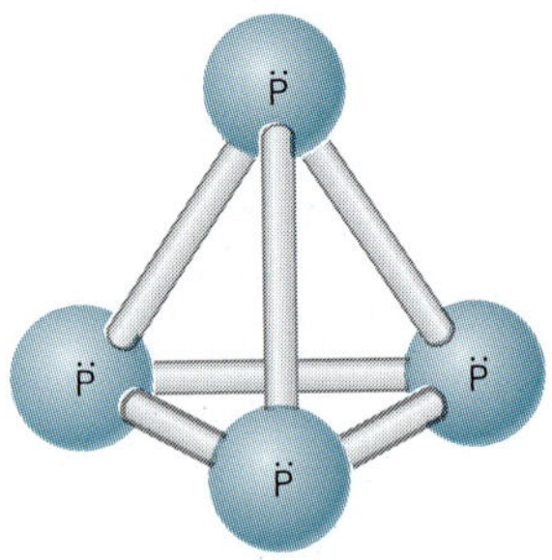

P_4, white phosphorous

Fig. 8.25 Pure elemental phosphorus is a white, waxy solid that consists of tetrahedral P_4 molecules in which the P—P—P bond angle is only 60°.

Exercise 8.3

Elemental phosphorus is often stored under water because it doesn't dissolve in water. Elemental phosphorus is very soluble in carbon disulfide, however. Use the structure of the P_4 molecule shown in Figure 8.25 to explain why P_4 is soluble in CS_2 but not in water.

Solution

The P_4 molecule is a perfect example of a nonpolar solute. It is therefore more likely to be soluble in nonpolar solvents than in polar solvents such as water.

The Lewis structure of CS_2 suggests that the molecule is linear.

$$:\ddot{S}=C=\ddot{S}:$$

Thus, even if there is some separation of charge in the C═S double bond, the molecule would have no net dipole moment, because of its symmetry. The electronegativities of carbon ($EN = 2.54$) and sulfur ($EN = 2.59$), however, suggest that the C═S double bonds are almost perfectly covalent. CS_2 is therefore a nonpolar solvent, which readily dissolves nonpolar P_4.

Exercise 8.4

The iodide ion reacts with iodine in aqueous solution to form I_3^-, or triiodide ion.

$$I^-(aq) + I_2(aq) \longrightarrow I_3^-(aq)$$

What would happen if CCl_4 were added to an aqueous solution that contained a mixture of KI, I_2, and KI_3?

Solution

Two layers would form, with CCl_4 on the bottom and the aqueous layer on top. KI and KI_3 are both ionic compounds. One contains K^+ and I^- ions; the other contains K^+ and I_3^- ions. The ionic compounds are more soluble in polar solvents, such as water, than in nonpolar solvents, such as CCl_4. KI and KI_3 would therefore remain in the aqueous solution. I_2 is a nonpolar molecule, which is more soluble in a nonpolar solvent, such as carbon tetrachloride. Some I_2 would therefore leave the aqueous layer and enter the CCl_4 layer, where it would exhibit the characteristic violet color of solutions of molecular iodine.

8.11 Why Do Some Solids Dissolve in Water?

When asked to describe what happens when a teaspoon of sugar is stirred into a cup of coffee, people often answer, "The sugar initially settles to the bottom of the cup. When the coffee is stirred, it dissolves to produce a sweeter cup of coffee." When asked to extend the description to the molecular level, they hesitate. If they are chemistry students, like yourself, they often ask to be reminded of the formula for sugar. When told that the sugar used in cooking is *sucrose,* $C_{12}H_{22}O_{11}$, they write equations such as the following.

$$C_{12}H_{22}O_{11}(s) \xrightarrow{H_2O} C_{12}H_{22}O_{11}(aq)$$

Although there is nothing wrong with this equation, it doesn't explain what happens at the molecular level when sugar dissolves. Nor does it provide any hints about why sugar dissolves but other solids do not. To understand what happens at the molecular level when a solid dissolves, we have to refer back to the discussion of molecular and ionic solids in Chapter 5.

The sugar we use to sweeten coffee or tea is a *molecular solid,* in which the individual molecules are held together by hydrogen bond intermolecular forces. When sugar dissolves in water, the hydrogen bond intermolecular forces between the individual sucrose molecules are broken, and the individual $C_{12}H_{22}O_{11}$ molecules are released into solution.

It takes energy to break the forces between the $C_{12}H_{22}O_{11}$ molecules in sucrose. It also takes energy to break the hydrogen bonds in water that must be disrupted to insert one of the sucrose molecules into the solution. Sugar dissolves in water because the slightly polar sucrose molecules form hydrogen bonds with the polar water molecules. The intermolecular forces that form between the solute and the solvent help compensate for the energy needed to disrupt the structure of both the pure solute and the solvent. In the case of sugar and water, the process works so well that up to 1800 g of sucrose can dissolve in a liter of water as shown in Figure 8.26.

Ionic solids (or salts) contain positive and negative ions that are held together by the strong force of attraction between particles with opposite charges. When an ionic solid dissolves in water, the ions that form the solid are released into solution, where they are attracted to the polar solvent molecules as shown in Figure 8.27. As a result, we can generally assume that salts dissociate into their ions when they

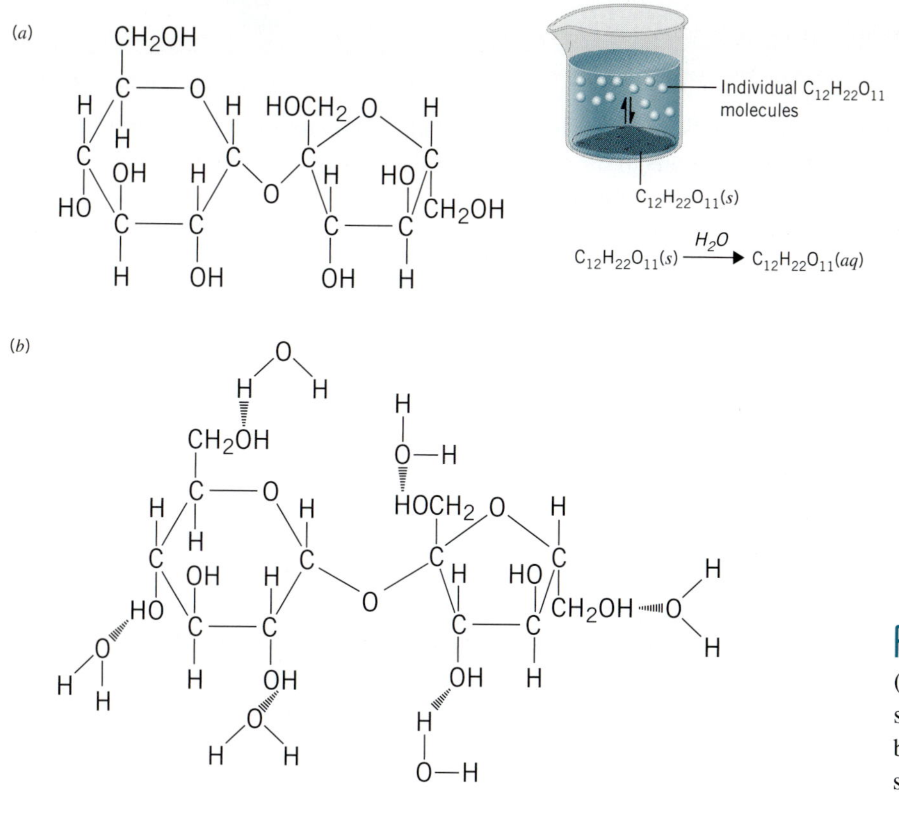

Fig. 8.26 (*a*) Structure of sucrose. (*b*) Hydrogen bonds are broken in both sucrose and water and then formed between sucrose and water in the solution.

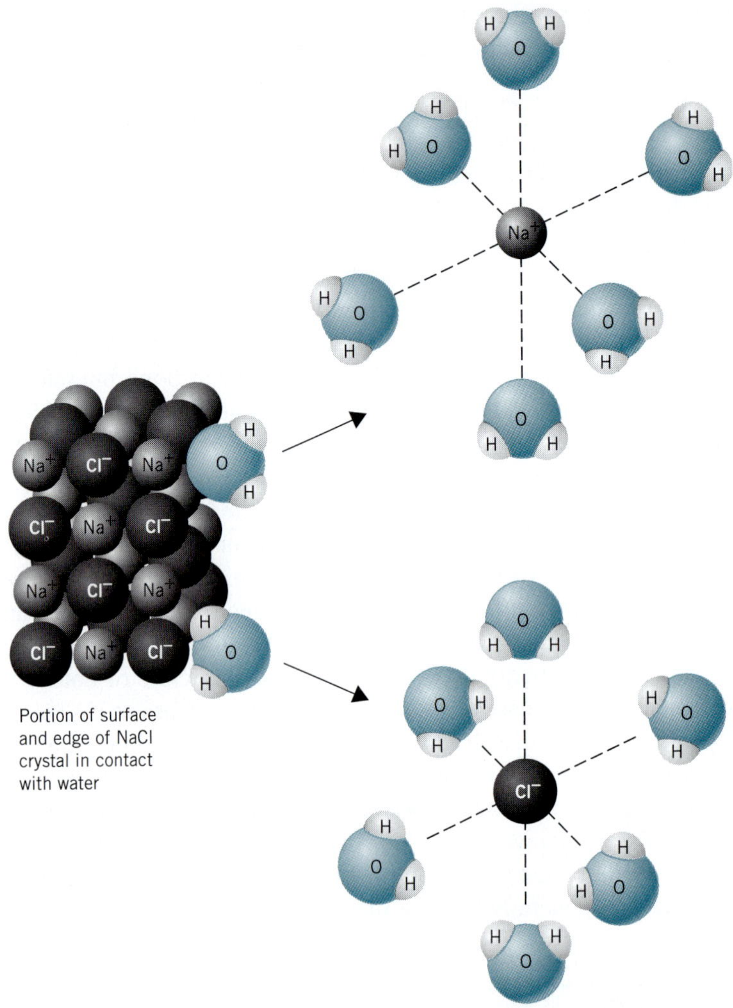

Fig. 8.27 Interaction of water and ionic compounds. The electrostatic interaction between the polar water molecules and the ions is a dipole–ion force. Reprinted with permission from L. J. Malone, *Basic Concepts of Chemistry,* John Wiley & Sons, Inc., New York, 1994, p. 351.

dissolve in water. Ionic compounds dissolve in water if the energy given off when the ions interact with water molecules partially compensates for the energy needed to break the ionic bonds in the solid and for the energy required to separate the water molecules so that the ions can be inserted into the solution.

It takes an enormous amount of heat to break apart an ionic crystal. It takes 2033 kJ/mol, for example, to transform a mole of solid $BaCl_2$ into Ba^{2+} and 2 Cl^- ions in the gas phase.

$$BaCl_2(s) \longrightarrow Ba^{2+}(g) + 2\,Cl^-(g) \quad \Delta H^\circ = 2033 \text{ kJ/mol}_{rxn}$$

It takes so much energy to separate the Ba^{2+} and 2 Cl^- ions in $BaCl_2$ that we might not expect the compound to dissolve in water. The force of attraction between Ba^{2+} and 2 Cl^- ions with water molecules is so large, however, that 2047 kJ/mol$_{rxn}$ of energy is released when the gaseous Ba^{2+} and 2 Cl^- ions interact with water molecules.

$$Ba^{2+}(g) + 2\,Cl^-(g) \xrightarrow{H_2O} Ba^{2+}(aq) + 2\,Cl^-(aq) \quad \Delta H^\circ = -2047 \text{ kJ/mol}_{rxn}$$

> **CHECKPOINT**
>
> Use the atom combination data in Appendix B.13 to calculate ΔH° for the reaction in which $BaCl_2(s)$ dissolves in water to form $Ba^{2+}(aq)$ and 2 $Cl^-(aq)$.

The overall enthalpy of reaction for the process in which solid $BaCl_2$ dissolves in water is therefore exothermic. The interaction of the ions with water more than compensates for the energy needed to break apart the ionic structure.

$$BaCl_2(s) \xrightarrow{H_2O} Ba^{2+}(aq) + 2\,Cl^-(aq) \quad \Delta H^\circ = -14 \text{ kJ/mol}_{rxn}$$

Silver chloride is very insoluble in water. As was previously seen for $BaCl_2$, a lot of heat is required to separate AgCl(*s*) into its gas phase ions.

$$AgCl(s) \longrightarrow Ag^+(g) + Cl^-(g) \qquad \Delta H^\circ = 915.7 \text{ kJ/mol}_{rxn}$$

The heat released when the gaseous Ag^+ and Cl^- ions interact with water is also large.

$$Ag^+(g) + Cl^-(g) \xrightarrow{H_2O} Ag^+(aq) + Cl^-(aq) \qquad \Delta H^\circ = -850.2 \text{ kJ/mol}_{rxn}$$

But the heat released when the ions interact with water isn't large enough to compensate for the heat needed to separate the ions in the crystal. As a result, the overall enthalpy of reaction is very unfavorable.

$$AgCl(s) \xrightarrow{H_2O} Ag^+(aq) + Cl^-(aq) \qquad \Delta H^\circ = 65.5 \text{ kJ/mol}_{rxn}$$

With a positive ΔH° of this magnitude, we would expect relatively little AgCl to dissolve in water. In fact, less than 0.002 g of AgCl dissolves in a liter of water at room temperature. The solubility of silver chloride in water is so small that AgCl is often said to be "insoluble" in water, even though that term is misleading.

> **CHECKPOINT**
>
> Calculate ΔH° for the reaction below using the data from Table B.13 in the appendix.
>
> $$AgCl(s) \xrightarrow{H_2O} Ag^+(aq) + Cl^-(aq)$$

8.12 Solubility Equilibria

Discussions of solubility equilibria are based on the following assumption: *When they dissolve, solids break apart to give solutions of the molecules or ions from which they are formed.* Most molecular solutes dissolve to give individual molecules,

$$C_{12}H_{22}O_{11}(s) \xrightarrow{H_2O} C_{12}H_{22}O_{11}(aq)$$

and ionic solids dissociate to give solutions of the positive and negative ions they contain.

$$NaCl(s) \xrightarrow{H_2O} Na^+(aq) + Cl^-(aq)$$

Solutes such as NaCl that break up almost completely into ions when they dissolve are called **strong electrolytes.** Solutes such as sucrose that don't break up into ions when they dissolve are called **nonelectrolytes.** There are some solutes that partially break up into ions and partially exist as soluble molecules in solution; they are called **weak electrolytes.**

We can detect the presence of Na^+ and Cl^- ions in an aqueous solution with the conductivity apparatus shown in Figure 8.28. This apparatus consists of a light bulb connected to a pair of metal wires that can be immersed in a beaker of water. The circuit in the conductivity apparatus isn't complete. In order for the light bulb to glow when the apparatus is plugged into an electrical outlet there must be a way for electrical charge to flow through the solution from one of the metal wires to the other.

When an ionic solid dissolves in the beaker of water, the resulting positive and negative ions are free to move through the aqueous solution. The net result is a flow of electric charge through the solution that completes the circuit and lets the light bulb glow. As might be expected, the brightness of the bulb is proportional to the concentration of the ions in the solution. Slightly soluble ionic compounds such as calcium sulfate make the light bulb glow dimly. When the wires are immersed in a solution of a very soluble ionic compound, such as NaCl, the light bulb glows brightly.

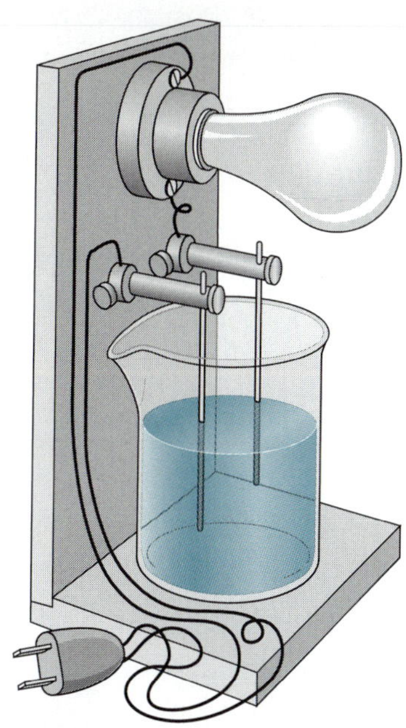

Fig. 8.28 A conductivity apparatus can be used to demonstrate the difference between aqueous solutions of ionic and covalent solids.

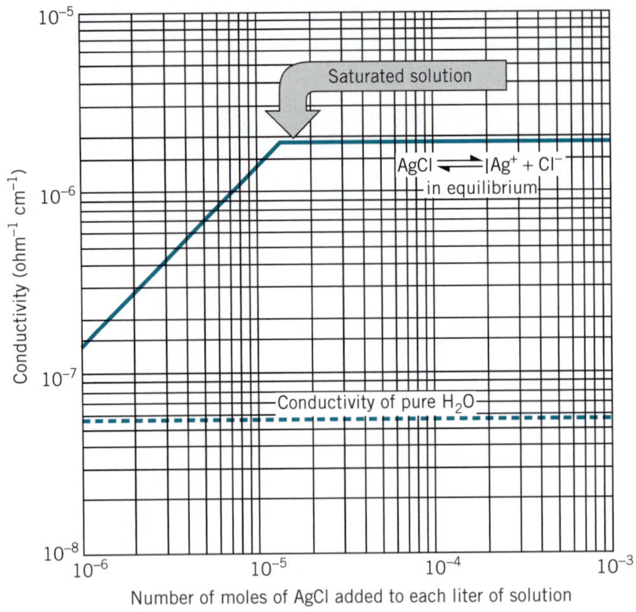

Fig. 8.29 The conductivity of a solution of AgCl in water increases at first as the AgCl dissolves and dissociates into Ag^+ and Cl^- ions. Once the solution has become saturated with AgCl, however, the conductivity remains the same no matter how much solid is added.

The conductivity apparatus in Figure 8.28 gives us only qualitative information about the relative concentrations of the ions in different solutions. It is possible to build a more sophisticated instrument that gives quantitative measurements of the conductivity of a solution, which is directly proportional to the concentration of the ions in the solution. Figure 8.29 shows what happens to the conductivity of water as we gradually add infinitesimally small amounts of AgCl to water and wait for the solid to dissolve before taking measurements.

The system conducts a very small electric current even before any AgCl is added because of small quantities of H_3O^+ and OH^- ions in water. The solution becomes a slightly better conductor when AgCl is added because some of the ionic compound dissolves to give Ag^+ and Cl^- ions, which can carry an electric current through the solution. The conductivity continues to increase as more AgCl is added, until about 0.002 g of the ionic compound has dissolved per liter of solution.

The fact that the conductivity doesn't increase after the solution has reached a concentration of 0.002 g AgCl/liter tells us that there is a limit on the solubility of the ionic compound in water. Once the solution reaches that limit, no more AgCl dissolves, regardless of how much solid we add to the system. This is exactly what we would expect if the solubility of AgCl is controlled by an equilibrium. Once the solution reaches equilibrium, the rate at which AgCl dissolves to form Ag^+ and Cl^- ions is equal to the rate at which the ions recombine to form AgCl.

Figure 8.30 shows what happens to the concentrations of Ag^+ and Cl^- when a large excess of solid silver chloride is added to the water. When the ionic compound is first added, it dissolves and dissociates rapidly. The Ag^+ and Cl^- concentrations increase rapidly at first.

$$AgCl(s) \xrightarrow[\textit{dissociate}]{\textit{dissolve}} Ag^+(aq) + Cl^-(aq)$$

The concentrations of the ions soon become large enough that the reverse reaction starts to compete with the forward reaction, which leads to a decrease in the rate at which Ag^+ and Cl^- ions enter the solution. The reverse reaction is the formation of

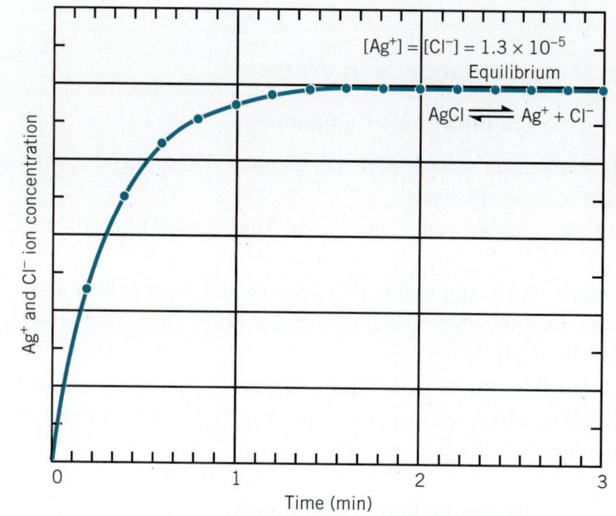

Fig. 8.30 Plot of concentration of Ag^+ and Cl^- ions versus time as solid AgCl dissolves in water.

insoluble silver chloride. Reactions in which soluble species form an insoluble product are called **precipitation reactions.**

$$Ag^+(aq) + Cl^-(aq) \xrightarrow[\textit{precipitate}]{\textit{associate}} AgCl(s)$$

Eventually, the Ag^+ and Cl^- ion concentrations become large enough that the rate at which precipitation occurs exactly balances the rate at which AgCl dissolves. Once that happens, there is no change in the concentration of the ions with time, and the reaction is at equilibrium. When the system reaches equilibrium it is called a **saturated solution** because it contains the maximum concentration of ions that can exist in equilibrium with the solid ionic compound at a given temperature. The amount of substance that must be added to a given volume of solvent to form a saturated solution is called the **solubility** of the substance. Chemists use double arrows to indicate a reaction at equilibrium.

$$AgCl(s) \xrightleftharpoons{H_2O} Ag^+(aq) + Cl^-(aq)$$

8.13 Solubility Rules

There are two primary factors that can be used to predict whether a molecular solute will be soluble in water: similar molecular structure and similar intermolecular forces of attraction. However, predicting whether an ionic compound will be soluble in water is not so straightforward.

There are a number of patterns in the data obtained from measuring the solubilities of different ionic compounds. These patterns form the basis for the rules outlined in Table 8.7, which can guide predictions of whether a given ionic compound will dissolve in water. The rules are based on the following definitions of the terms *soluble, insoluble,* and *slightly soluble* as indicated in Figure 8.31.

- A compound is defined as being soluble if it dissolves in water to give a solution with a concentration of at least 0.1 mol/L at room temperature.
- A compound is defined as being insoluble if the concentration of an aqueous solution is less than 0.001 *M* at room temperature.
- Slightly soluble compounds give solutions that fall between the extremes.

Table 8.7
Solubility Rules for Ionic Compounds in Water

Soluble Ionic Compounds

The Na^+, K^+, and NH_4^+ ions form *soluble ionic compounds.* Thus, NaCl, KNO_3, and $(NH_4)_2CO_3$ are *soluble ionic compounds.*

The nitrate ion (NO_3^-) forms *soluble ionic compounds.* Thus, $Cu(NO_3)_2$ and $Fe(NO_3)_3$ are soluble.

The chloride (Cl^-), bromide (Br^-), and iodide (I^-) ions usually form *soluble ionic compounds.* Exceptions include ionic compounds of the Pb^{2+}, Hg_2^{2+}, Ag^+, and Cu^+ ions. $CuBr_2$ is soluble, but CuBr is not.

The sulfate ion (SO_4^{2-}) usually forms *soluble ionic compounds.* Exceptions include $BaSO_4$, $SrSO_4$, and $PbSO_4$, which are insoluble, and Ag_2SO_4, $CaSO_4$, and Hg_2SO_4, which are slightly soluble.

Insoluble Ionic Compounds

Sulfides (S^{2-}) are usually *insoluble.* Exceptions include Na_2S, K_2S, $(NH_4)_2S$, MgS, CaS, SrS, and BaS.

Oxides (O^{2-}) are usually *insoluble.* Exceptions include Na_2O, K_2O, SrO, and BaO, which are soluble, and CaO, which is slightly soluble.

Hydroxides (OH^-) are usually *insoluble.* Exceptions include NaOH, KOH, $Sr(OH)_2$, and $Ba(OH)_2$, which are soluble, and $Ca(OH)_2$, which is slightly soluble.

Chromates (CrO_4^{2-}), phosphates (PO_4^{3-}), and carbonates (CO_3^{2-}) are usually *insoluble.* Exceptions include ionic compounds of the Na^+, K^+, and NH_4^+ ions, such as Na_2CrO_4, K_3PO_4, and $(NH_4)_2CO_3$.

It is important to remember that there is a limit to the solubility of even the most soluble ionic compounds and that ionic compounds that are labeled "insoluble" will dissolve to a very small extent.

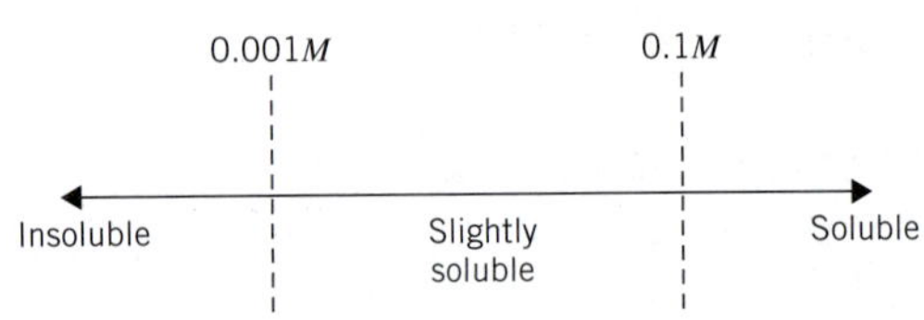

Fig. 8.31 Solubilities of compounds in water cover a wide range that is divided into the categories insoluble, slightly soluble, and soluble.

Writing net ionic equations from molecular equations is introduced now because students are now able to distinguish between molecular and ionic compounds, and by using the solubility rules they can predict which ionic compounds will completely dissociate in dilute solution.

8.14 Net Ionic Equations

Chapter 2 introduced the writing and balancing of chemical equations. Chemical equations can be written in three different forms: molecular, full ionic, and net ionic equations. Most of the equations that you have encountered thus far in this text have been **molecular equations** in which all reactants and products have been written as electrically neutral molecules and compounds. The reaction of aqueous barium chloride with aqueous sodium sulfate is sometimes written as a molecular equation.

$$BaCl_2(aq) + Na_2SO_4(aq) \longrightarrow BaSO_4(s) + 2\,NaCl(aq)$$

Section 8.12 described how strong electrolytes dissociate in aqueous solution to form ions. Strong electrolytes, such as soluble ionic compounds, are considered to dissociate essentially 100% when dissolved in dilute aqueous solution. Therefore, a better description of what is taking place on the molecular level in solution can be obtained from the full ionic equation. **Full ionic equations** are written to show all aqueous strong electrolytes as aqueous ions. Although chemists seldom write full ionic equations when describing chemical reactions, they are very useful as an intermediate step in writing net ionic equations. The above reaction can be written as a full ionic equation.

$$Ba^{2+}(aq) + 2\,Cl^-(aq) + 2\,Na^+(aq) + SO_4^{2-}(aq) \longrightarrow BaSO_4(s) + 2\,Na^+(aq) + 2\,Cl^-(aq)$$

The three aqueous ionic compounds—$BaCl_2$, Na_2SO_4 and NaCl—are all shown as having dissociated into their ions. However, as was shown in Table 8.7, $BaSO_4$ is insoluble in water and is therefore listed in the chemical equation as a solid. Because ions of solid $BaSO_4$ remain attracted to one another as part of a three-dimensional ionic solid, the chemical formula is best described by the electrically neutral compound. Pure liquids and gases are also treated as electrically neutral molecules or compounds.

Note that in the full ionic equation there are some ions that appear on both sides of the chemical equation. Because these ions are present on both the reactant and product sides of the equation, they remain unchanged during the chemical reaction and are referred to as **spectator ions.** Spectator ions may be canceled from both sides of the equation on a one-to-one basis—for every ion eliminated from the product side of the equation, the corresponding ion must be eliminated from the reactant side. This yields the **net ionic equation.**

$$Ba^{2+}(aq) + SO_4^{2-}(aq) \longrightarrow BaSO_4(s)$$

Two sodium ions and two chloride ions have been removed from both sides of the chemical equation. Net ionic equations contain all the information necessary to understand what is happening on the molecular level during a chemical reaction. In all three ways of writing chemical equations, the atoms and the net charge on both sides of the equation are balanced.

The most difficult task in writing net ionic equations is determining which chemical species are strong electrolytes and should be written as ions. Strong electrolytes include soluble ionic compounds (refer to Table 8.7 to determine whether an ionic compound is soluble), strong acids (HCl, HBr, HI, HNO_3, $HClO_4$, and H_2SO_4), and strong bases (LiOH, NaOH, KOH, RbOH, CsOH, $Ca(OH)_2$, $Sr(OH)_2$, and $Ba(OH)_2$). The strong acids and bases will be discussed in Chapter 11.

Exercise 8.5

Write the molecular, full ionic, and net ionic equations that describe the following chemical reactions.

(a) Aqueous solutions of sodium chromate and lead(II) nitrate react to form a yellow precipitate of lead(II) chromate in an aqueous solution of sodium nitrate.

(b) An aqueous solution of the weak acid, acetic acid (CH_3CO_2H), reacts with an aqueous solution of the strong base strontium hydroxide to form the ionic compound strontium acetate and liquid water.

Solution

(a) Molecular equation:

$$Na_2CrO_4(aq) + Pb(NO_3)_2(aq) \longrightarrow PbCrO_4(s) + 2\,NaNO_3(aq)$$

Full ionic:

$$2\ Na^+(aq) + CrO_4^{2-}(aq) + Pb^{2+}(aq) + 2\ NO_3^-(aq) \longrightarrow PbCrO_4(s) + 2\ Na^+(aq) + 2\ NO_3^-(aq)$$

Net ionic:

$$CrO_4^{2-}(aq) + Pb^{2+}(aq) \longrightarrow PbCrO_4(s)$$

(b) Molecular equation:

$$2\ CH_3CO_2H(aq) + Sr(OH)_2(aq) \longrightarrow Sr(CH_3CO_2)_2(aq) + 2\ H_2O(l)$$

Full ionic:

$$2\ CH_3CO_2H(aq) + Sr^{2+}(aq) + 2\ OH^-(aq) \longrightarrow Sr^{2+}(aq) + 2\ CH_3CO_2^-(aq) + 2\ H_2O(l)$$

Acetic acid is not broken into its ions because it is not a strong electrolyte. Water is not broken into ions because it is a liquid.

Only the strontium ion is found on both sides of the equation in exactly the same form, and therefore only the strontium can be eliminated from the full ionic equation. Eliminating the spectator ion yields

$$2\ CH_3CO_2H(aq) + 2\ OH^-(aq) \longrightarrow 2\ CH_3CO_2^-(aq) + 2\ H_2O(l)$$

This can be simplified to the net ionic equation

$$CH_3CO_2H(aq) + OH^-(aq) \longrightarrow CH_3CO_2^-(aq) + H_2O(l)$$

8.15 Hydrophilic and Hydrophobic Molecules

Hydrocarbons, such as the alkanes discussed in Section 8.3, are compounds that contain only carbon and hydrogen. Because of the molecular geometry and small difference between the electronegativities of carbon and hydrogen ($\Delta EN = 0.24$), hydrocarbons are nonpolar. As a result, they don't dissolve in polar solvents such as water. Hydrocarbons are therefore described as insoluble in water.

When one of the hydrogen atoms in a hydrocarbon is replaced with an —OH group, the compound is known as an **alcohol,** as shown in Figure 8.32. Because alcohols contain the same —OH group as water, alcohols have properties between the extremes of hydrocarbons and water. When the hydrocarbon chain is short, the alcohol is soluble in water. Methanol (CH_3OH) and ethanol (CH_3CH_2OH) are infinitely soluble in water, for example. There is no limit on the amount of these alcohols that can dissolve in a given quantity of water. The alcohol in beer, wine, and hard liquors is ethanol, and mixtures of ethanol and water can have any concentration between the extremes of pure alcohol (200 proof) and pure water (0 proof).

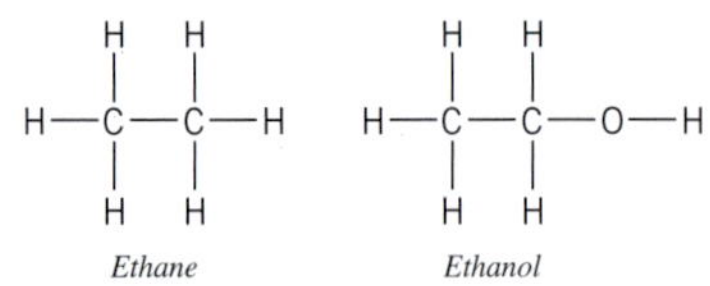

Fig. 8.32 When a hydrogen in a hydrocarbon such as ethane is replaced by an —OH group, the resulting compound is called an alcohol.

As the hydrocarbon chain becomes longer, the alcohol becomes less soluble in water, as shown in Table 8.8. One end of the longer alcohol molecules has so much nonpolar character it is called **hydrophobic** (literally, "water hating"), as

Table 8.8

Solubilities of Alcohols in Water

Formula	Name	Solubility in Water (g/100 g)
CH_3OH	Methanol	Infinitely soluble
CH_3CH_2OH	Ethanol	Infinitely soluble
$CH_3(CH_2)_2OH$	Propanol	Infinitely soluble
$CH_3(CH_2)_3OH$	Butanol	9
$CH_3(CH_2)_4OH$	Pentanol	2.7
$CH_3(CH_2)_5OH$	Hexanol	0.6
$CH_3(CH_2)_6OH$	Heptanol	0.18
$CH_3(CH_2)_7OH$	Octanol	0.054
$CH_3(CH_2)_9OH$	Decanol	Insoluble in water

shown in Figure 8.33. The other end contains an —OH group that can form hydrogen bonds to neighboring water molecules and is therefore said to be **hydrophilic** (literally, "water loving"). As the hydrocarbon chain becomes longer, the hydrophobic character of the molecule increases, and the solubility of the alcohol in water gradually decreases until it becomes essentially insoluble in water.

People encountering the terms *hydrophilic* and *hydrophobic* for the first time sometimes have difficulty remembering which stands for water hating and which stands for water loving. If you remember that Hamlet's girlfriend was named Ophelia (not Ophobia), you can remember that the prefix *philo-* is commonly used to describe love—for example, in *philanthropist, philharmonic, philosopher*—and that *phobia* means "dislike."

➤ CHECKPOINT

Amino acids are classified as either hydrophilic or hydrophobic on the basis of their side chains. Use the structure of the side chains of the following amino acids to justify the classifications shown below.

	Amino Acid	*Side Chain*
Hydrophobic	Alanine	$—CH_3$
	Cysteine	$—CH_2CH_2SCH_3$
Hydrophilic	Lysine	$—CH_2CH_2CH_2CH_2NH_3^+$
	Serine	$—CH_2OH$

Hydrophilic head

$CH_3CH_2CH_2CH_2CH_2CH_2CH_2CH_2CH_2CH_2OH$

Hydrophobic tail

Fig. 8.33 One end of the decanol molecule is nonpolar and therefore *hydrophobic;* the other end is polar and therefore *hydrophilic*.

Table 8.9 shows that the ionic compound NaCl is relatively soluble in water. Water molecules, being polar, are able to cluster around the positively and negatively charged ions formed when NaCl dissolves, as shown in Figure 8.27. As the solvent becomes more like a hydrocarbon, the solubility of NaCl decreases because the longer-chain hydrocarbon solvents do not interact as strongly with Na^+ and Cl^-.

Table 8.9

Solubility of Sodium Chloride in Water and in Alcohols

Formula of Solvent	Solvent Name	Solubility of NaCl (g/100 g solvent)
H_2O	Water	35.92
CH_3OH	Methanol	1.40
CH_3CH_2OH	Ethanol	0.065
$CH_3(CH_2)_2OH$	Propanol	0.012
$CH_3(CH_2)_3OH$	Butanol	0.005
$CH_3(CH_2)_4OH$	Pentanol	0.0018

Key Terms

Alcohol
Boiling point
Cohesion
Condensation
Dipole moment
Dispersion forces (London forces)
Enthalpy of fusion
Enthalpy of vaporization
Equilibrium
Freezing point
Full ionic equation
Hydrocarbons
Hydrogen bonding
Hydrophilic
Hydrophobic
Intermolecular forces
Intramolecular bonds
Melting point
Molecular equation
Net ionic equation
Nonelectrolyte
Nonpolar solvent
Phase diagram
Polar
Polar solvent
Precipitation reaction
Saturated solution
Solubility
Spectator ion
Strong electrolyte
Surface tension
van der Waals forces
Vapor pressure
Weak electrolyte

Problems

The Structure of Gases, Liquids, and Solids

1. Describe the differences in the properties of gases, liquids, and solids on the atomic scale. Explain how the differences give rise to the observed differences in the macroscopic properties of the three states of matter.
2. Describe the difference between *intermolecular* forces and *intramolecular* bonds, giving examples of each. Which are stronger?
3. Solids consist of particles locked into a rigid structure. Describe what happens on a microscopic scale when a solid melts. Describe what happens when a liquid is heated to boiling.
4. Why are most substances solids at very low temperatures? At very high temperatures most substances are gases. Why?
5. What determines whether a substance is a solid, liquid, or gas at room temperature?
6. Why are liquids usually less dense than their corresponding solids?

Intermolecular Forces

7. There are four categories of intermolecular forces. What are they and how do they differ? Give an example of each.
8. Why do induced dipole–induced dipole forces increase as the number of electrons in a molecule increases?
9. What structural features are neccessary for a molecule to exhibit hydrogen bonding?
10. List all of the types of intermolecular forces that each of the following molecules would have.
    (a) SO_2 (b) CH_3OH (c) ICl_3 (d) SF_4
    Note that the Lewis structure of each molecule must be determined.
11. If a molecule is known to have a dipole moment, what types of intermolecular forces must be present?

Relative Strengths of Intermolecular Forces

12. Predict the order in which the boiling points of the following compounds should increase. Explain your reasoning.
    (a) NH_3 (b) PH_3 (c) AsH_3 (d) SbH_3
13. Which compound would you expect to have the highest boiling point? Explain your reasoning.
    (a) methane, CH_4
    (b) chloromethane, CH_3Cl
    (c) dichloromethane, CH_2Cl_2
    (d) chloroform, $CHCl_3$
    (e) carbon tetrachloride, CCl_4
14. Explain why the boiling points of hydrocarbons that have the generic formula C_nH_{2n+2} increase with molecular weight.
15. Explain why propane (C_3H_8) is a gas but pentane (C_5H_{12}) is a liquid at room temperature.
16. Explain why methane (CH_4) is a liquid only over a very narrow range of temperatures.
17. Why is the boiling point of *n*-pentane greater than that of isopentane?
18. The melting points of the hydrogen halides are given below.

	MP, °C	*Dipole Moment (D)*
HCl	−114	1.08
HBr	−87	0.81
HI	−51	0.45

Explain this trend.

The Kinetic Theory of Liquids

19. The average velocity of a molecule in a gas is about 1000 miles per hour. How fast are the same molecules in the liquid phase moving when they are at the same temperature? Explain.

20. If a liquid has a high enthalpy of vaporization, what does this mean about the forces holding the liquid together?
21. What is the difference between an enthalpy of fusion and an enthalpy of vaporization?
22. What effect does decreasing the temperature have on the density of most liquids? Explain.

The Vapor Pressure of a Liquid

23. Explain why it is important to specify the temperature at which the vapor pressure of a liquid is measured.
24. Explain why water eventually evaporates from an open container at room temperature (~20°C), even though it normally boils at 100°C.
25. One postulate of the kinetic theory of gases suggests that the temperature of a gas is directly proportional to the *average kinetic energy* of the particles in the gas. Why is the term *average* used?
26. Use the kinetic molecular theory to explain why the vapor pressure of water becomes greater as the temperature of the water increases.
27. Explain why a cloth soaked in water feels cool when placed on your forehead.
28. What would happen to the vapor pressure of liquid bromine, Br_2, at 20°C if the liquid was transferred from a narrow 10-mL graduated cylinder into a wide petri dish or crystallizing dish? Would it increase, decrease, or remain the same? What would happen to the vapor pressure in a closed container if more liquid were added to the container? Would it increase, decrease, or remain the same?
29. Explain what it means to say that the liquid and vapor in a closed container are in *equilibrium*.
30. Explain why each of the following increases the rate at which water evaporates from an open container.
    (a) increasing the temperature of the water
    (b) increasing the surface area of the water
    (c) blowing air over the surface of the water
    (d) decreasing the atmospheric pressure on the water
31. On a very dry day snow changes directly to a gas through a process known as *sublimation*. Use kinetic molecular theory to explain how this can happen.
32. Why does a pressure cooker cook food more rapidly than an open cooker?
33. The force of cohesion between mercury atoms is much larger than the force of cohesion between water molecules. Conversely, the attractive force between water molecules and glass is much larger than the attractive force between mercury atoms and glass. Use these observations to explain why mercury forms small drops when it spills on glass rather than forming a single large puddle such as water does.
34. Describe how the surface tension of water can be used to explain the fact that a steel sewing needle lies on the surface of water.
35. Explain why a drop of water seems to bead up on the surface of a freshly waxed car.
36. Explain the advantages and disadvantages in a plant of having leaves that have a large surface area. Explain why plants that grow in arid climates seldom have leaves as broad as those found on maple trees.
37. Which of each pair of the following compounds would have the lower boiling point? The lower vapor pressure?
    (a) $CH_3CH_2CH_3$ / CH_3CH_2OH
    (b) CH_3CH_3 / CH_3Br
    (c) $HOCH_2CH_2OH$ / CH_3CH_2OH
    (d) $CH_3CH_2CH_2CH_2CH_3$ / $CH_3CH_2OCH_2CH_3$
38. Predict which of the following liquids should have the lowest boiling point from their vapor pressures (VP) at 0°C.
    (a) acetone, VP = 67 mmHg
    (b) benzene, VP = 24.5 mmHg
    (c) ether, VP = 183 mmHg
    (d) methyl alcohol, VP = 30 mmHg
    (e) water, VP = 4.6 mmHg

Melting Point and Freezing Point

39. Does a solid always get hotter when heat is added? Explain.
40. The enthalpy of fusion of water is 6.0 kJ/mol_{rxn} and that of methanol (CH_3OH) is 3.2 kJ/mol_{rxn}. Which solid, water or methanol, has the greatest intermolecular forces? Predict whether the enthalpy of fusion of CH_3CH_3 will be greater than, less than, or the same as that of methanol. Explain.
41. Why do different substances have different melting points?
42. Explain why a "3-minute egg" cooked while camping in the Rocky Mountains does not taste as good as it does when cooked while camping near the Great Lakes.
43. Explain why it takes more time to boil food in Denver than in Salt Lake City.
44. Explain why water boils when the pressure on the system is reduced.
45. At what temperature does water boil when the pressure is 50 mmHg? Use Figure 8.12.
46. What pressure has to be achieved before water can boil at 20°C? Use Figure 8.12.
47. Increasing the temperature of a liquid will do which of the following?
    (a) increase the boiling point
    (b) increase the melting point
    (c) increase the vapor pressure
    (d) increase the amount of heat required to boil a mole of the liquid
    (e) all of the above
48. Liquid air is composed primarily of liquid oxygen (BP = −183°C) and liquid nitrogen (BP = −196°C).

It can be separated into its component gases by increasing the temperature until one of the gases boils off. Which gas boils off first?

49. According to the data in Table 8.4, butane (C_4H_{10}) should be a gas at room temperature (BP = $-0.5°C$). Use this to explain why you can hear gas escape when a can of butane lighter fluid is opened with the nozzle pointed up. If you shake one of the cans, however, you can hear a liquid bounce against either end of the can. Furthermore, when you open the can with the nozzle pointed down, you can see a liquid escape. Explain how butane is stored as a liquid in cans at room temperature.

50. (a) Which of the following molecules has the most polar bonds? Explain.
 CH_4, CCl_4, CBr_4
 (b) What type of bonding is present in all these compounds?
 (c) Which of the liquids of these molecules has the highest vapor pressure? Which has the lowest? Explain.
 (d) Arrange these compounds in order of increasing boiling point. State which is the highest. Explain.

Phase Diagrams

51. At what conditions of temperature and pressure is the formation of a solid favored?

52. At a low pressure and a high temperature, what state of a substance is most likely to exist?

53. A horizontal dotted line is drawn across the phase diagram of Figure 8.17 at a pressure less than 1 atm. Describe what information can be obtained from the intersection of the dotted line with a solid line on the diagram. How will the freezing point and the boiling point be changed from what they were at 1 atm?

54. According to Figure 8.17, what effect does decreasing the pressure have on the freezing point of most compounds?

Hydrogen Bonding and the Anomalous Properties of Water

55. Why is the boiling point and melting point of water much higher than you would expect from the boiling points and melting points of H_2S, H_2Se, and H_2Te?

56. Why is hydrogen bonding very strong in HF and H_2O. Explain why hydrogen bonding is much weaker in HCl and H_2S?

57. Why is ice less dense than liquid water?

58. Explain why the strength of hydrogen bonds decreases in the following order: HF > H_2O > NH_3.

59. Why does water have an unusually large specific heat?

Solutions: Like Dissolves Like

60. Use a drawing to describe what happens when I_2 molecules dissolve in CCl_4 and when $KMnO_4$ dissolves in water.

61. One way of screening potential anesthetics involves testing whether the compound dissolves in olive oil, because all common anesthetics, including dinitrogen oxide (N_2O), cyclopropane (C_3H_6), and halothane ($CF_3CHBrCl$), are soluble in olive oil. What property do the compounds have in common?

62. Carboxylic acids with the general formula $CH_3(CH_2)_nCO_2H$ have a nonpolar CH_3CH_2 . . . tail and a polar . . . CO_2H head. What effect does increasing the value of n have on the solubility of carboxylic acids in polar solvents, such as water? What is the effect on their solubility in nonpolar solvents, such as CCl_4?

63. Which of the following compounds would be the most soluble in a nonpolar solvent, such as CCl_4?
 (a) H_2O
 (b) CH_3OH
 (c) $CH_3CH_2CH_2OH$
 (d) $CH_3CH_2CH_2CH_2CH_2OH$
 (e) $CH_3CH_2CH_2CH_2CH_2CH_2CH_2OH$

64. Potassium iodide reacts with iodine in aqueous solution to form aqueous potassium triiodide, KI_3.

$$KI(aq) + I_2(aq) \longrightarrow KI_3(aq)$$

What would happen if we added CCl_4 to the reaction mixture?
 (a) The KI and KI_3 would dissolve in the CCl_4 layer.
 (b) The I_2 would dissolve in the CCl_4 layer.
 (c) Both KI and I_2, but not KI_3, would dissolve in the CCl_4 layer.
 (d) Neither KI, KI_3, nor I_2 would dissolve in the CCl_4 layer.
 (e) No distinct CCl_4 layer would form because CCl_4 is soluble in water.

65. Phosphorus pentachloride can react with itself in an equilibrium reaction to form an ionic compound that contains the PCl_4^+ and PCl_6^- ions.

$$2\ PCl_5 \rightleftharpoons PCl_4^+ + PCl_6^-$$

The extent to which the reaction occurs depends on the solvent in which it is run. Predict whether a nonpolar solvent, such as CCl_4, favors the products or the reactants of the reaction. Predict what would happen to the reaction if we used a polar solvent, such as acetonitrile (CH_3CN).

Why Do Some Solids Dissolve in Water?

66. Explain why some molecular solids are soluble in water but others are not.

67. Explain why $BaCl_2$ is soluble in water but AgCl is not.

68. On the microscopic level, explain what happens when an ionic compound dissolves in water.

Solubility Equilibria

69. Explain why the light bulb in the conductivity apparatus in Figure 8.28 glows more brightly when the wires are immersed in a solution of NaCl than when the wires are immersed in tap water.

70. Explain why the addition of a few small crystals of silver chloride makes water a slightly better conductor of electricity. Explain why the conductivity gradually increases as more AgCl is added, until it eventually reaches a maximum. Describe what is happening in the solution when its conductivity reaches the maximum.

Solubility Rules

71. Which of the following ionic compounds are *insoluble* in water?
(a) $Ba(NO_3)_2$ (b) $BaCl_2$ (c) $BaCO_3$
(d) BaS (e) $BaC_2H_3O_2$

72. Which of the following ionic compounds are *insoluble* in water?
(a) $(NH_4)_2SO_4$ (b) K_2CrO_4 (c) Na_2S
(d) $Pb(NO_3)_2$ (e) $Cr(OH)_3$

73. Which of the following ionic compounds are *soluble* in water?
(a) PbS (b) PbO (c) $PbCrO_4$
(d) $PbCO_3$ (e) $Pb(NO_3)_2$

Net Ionic Equations

74. Write net ionic equations for each of the following molecular equations.
(a) $Zn(s) + Cu(NO_3)_2(aq) \longrightarrow Zn(NO_3)_2(aq) + Cu(s)$
(b) $MnCl_2(aq) + (NH_4)_2S(aq) \longrightarrow MnS(s) + 2\ NH_4Cl(aq)$

75. Write molecular and net ionic equations for each of the following chemical reactions.
(a) Magnesium metal reacts with the strong acid HCl to produce an aqueous solution of magnesium chloride and hydrogen gas.
(b) Aqueous solutions of sodium carbonate and calcium nitrate react to form a precipitate of calcium carbonate and an aqueous solution of sodium nitrate.

76. Use the following net ionic equations to write molecular equations. Assume that the nitrate anion and potassium cation are present in solution.
(a) $Ag^+(aq) + Cl^-(aq) \longrightarrow AgCl(s)$
(b) $Ni^{2+}(aq) + CrO_4^{2-}(aq) \longrightarrow NiCrO_4(s)$

Hydrophilic and Hydrophobic Molecules

77. Why are hydrocarbons not soluble in water?

78. Why are some alcohols soluble in water while others are not?

79. Arrange the following compounds in order of increasing solubility in water. Explain your order.
(a) NaCl (b) $CH_3CH_2CH_2CH_3$
(c) CH_3CH_2OH (d) CH_3COOH

Integrated Problems

80. How would you explain the difference between the bonds that are broken when NaCl or diamond (composed entirely of carbon) boils and the forces broken when water boils? How would you explain the difference between the boiling points of NaCl (BP = 1465°C), diamond (BP = 4827°C), and water (BP = 100°C)?

81. Because they are isomers, ethanol (CH_3CH_2OH) and dimethyl ether (CH_3OCH_3) have the same molecular weight. Explain why ethanol (BP = 78.5°C) has a much higher boiling point than dimethyl ether (BP = −23.6°C).

82. The following compounds have the same molecular weight. Explain why one of the compounds has a higher boiling point than the other.

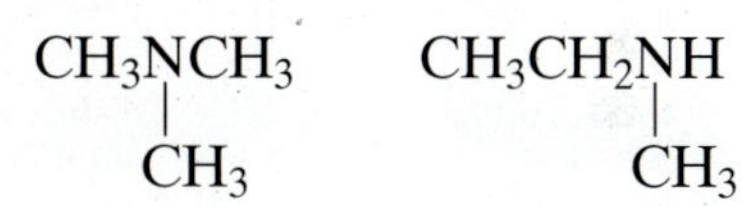

BP = 3°C — *Trimethyl amine*; *BP = 35°C* — *Ethylmethyl amine*

83. You like a really hot cup of tea. Would you prefer to live in Denver or in Salt Lake City if hot tea is your only consideration? Why?

84. A thermometer is taped to the outside of an open flask filled with water. The water is heated to boiling and allowed to continue to boil until it is gone. Sketch a rough plot of how the temperature would change with respect to time for water initially at 25°C. Suppose the same setup is used except that no heat is supplied to the open flask, which is initially at 25°C. If you come back some time later, all of the water will have evaporated. How do you think the temperature recorded by the thermometer would have changed with respect to time? What is the difference between the phase change of water at 100°C and at 25°C?

85. Which illustration most correctly describes the boiling of water at 100°C?

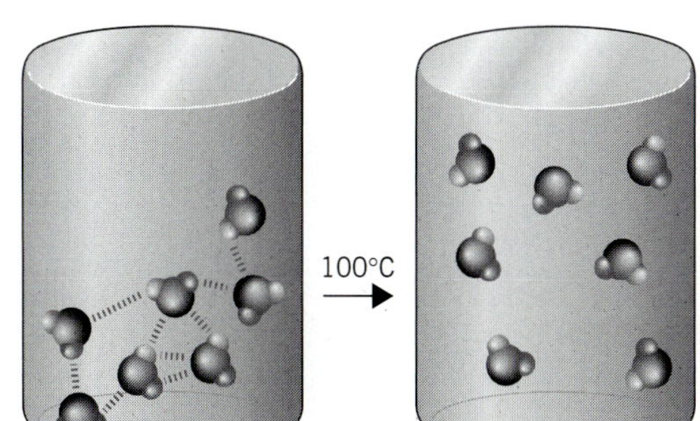

(*a*)

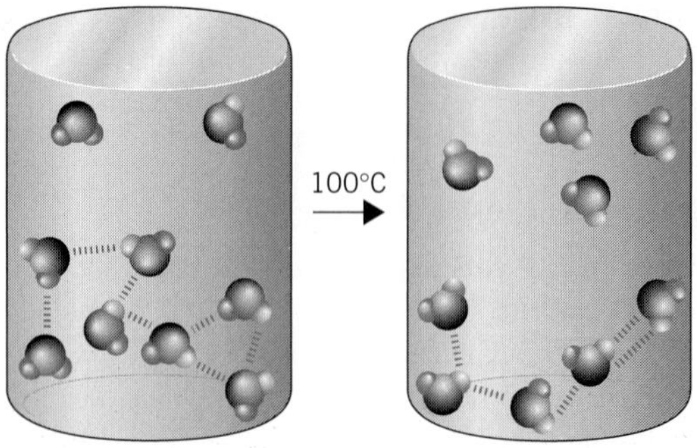

(*b*)

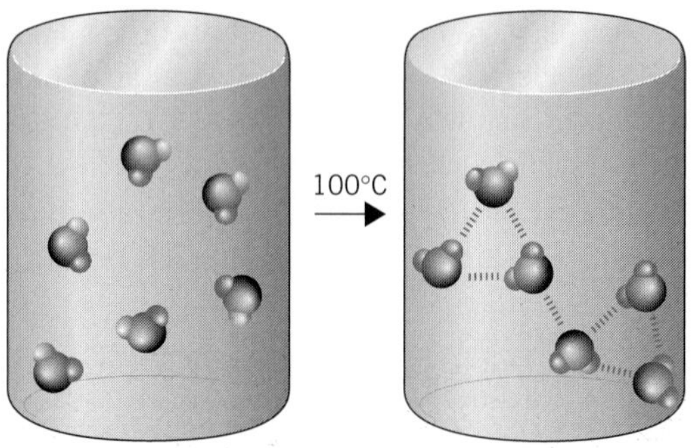

(*c*)

86. Compare the enthalpy change associated with boiling one mole of water (converting 1 mol of water into steam at 100°C) to evaporating 1 mol of water (converting 1 mol of water to water vapor at room temperature).

87. The diagram below shows liquid methanol, CH_3OH, being prepared to be poured into a beaker of water. Draw a figure that illustrates the solution that will result. Clearly show the intermolecular forces. Describe the intermolecular forces that are broken in the solute and solvent and formed in the solution. Do you expect methanol to be very soluble in water?

88. Which of the following statements is true? Explain your reasoning on the molecular level.
    (a) A substance with a high vapor pressure will have a high boiling point.
    (b) A substance with a high vapor pressure will have a low boiling point.
    (c) There can be no relationship between vapor pressure and boiling point.

89. An insoluble solid that forms when two or more soluble species are mixed together is known as a precipitate. Use the solubility rules for ionic compounds in water to predict whether each of the following aqueous mixtures would produce a precipitate. Identify the precipitate in cases where one is formed. (Hint: Begin by determining what compounds might be formed by mixing the solutions together.)
    (a) $FeCl_3$ and KOH
    (b) Na_2CO_3 and $(NH_4)_2SO_4$
    (c) $Pb(NO_3)_2$ and NaCl

90. Liquid benzene, C_6H_6, has a vapor pressure of 325 mmHg at 80°C. If 1.00 g of benzene is placed in an evacuated 1.00-L flask at 80°C, determine the mass of benzene that will evaporate and the final pressure in the container. Determine the same variables if the experiment is repeated using a 500-mL evacuated container. Explain what is different about the two experiments.

91. Use the data from Table 8.4 to answer the following questions.
    (a) 1-Pentanol is an alcohol with the molecular formula $CH_3CH_2CH_2CH_2CH_2OH$ and a molecular mass of 88.2 g/mol. How would you expect the boiling point of pentanol to be related to the boiling points of butanoic acid, pentanal, and hexane? Explain your reasoning.
    (b) Predict the boiling point for nonanal, $CH_3(CH_2)_7CHO$. Explain your reasoning.

92. A sealed container equipped with a movable piston contains a liquid in equilibrium with its vapor. A pressure gauge is attached to the container to measure the pressure of the vapor. Explain what is happening on the molecular level for each of the following experimental observations.
    (a) A valve on the container is opened and some of the gas is allowed to escape. The pressure is observed to immediately drop when the valve is opened. When the valve is closed, however, the pressure is observed to increase and then stabilize. (The volume of the container and the temperature remain constant.)
    (b) The movable piston on the container is depressed, causing the volume to decrease. The pressure increases as the piston moves in but then stabilizes at a constant pressure even as the piston continues to be depressed. (The temperature of the container remains constant.)

93. Octane, dibutyl ether, and 1-octanol have the following structures.

$$CH_3CH_2CH_2CH_2CH_2CH_2CH_2CH_3$$
Octane

$$CH_3CH_2CH_2CH_2—O—CH_2CH_2CH_2CH_3$$
Dibutyl ether

$$CH_3CH_2CH_2CH_2CH_2CH_2CH_2CH_2OH$$
1-Octanol

(a) Arrange the three compounds in order of increasing vapor pressure.
(b) Arrange the three compounds in order of decreasing boiling point.

94. Which of the following properties would you expect to generally increase as temperature increases: vapor pressure, surface tension, and heat of vaporization? Explain.

95. The specific heat of liquid water at room temperature is about 4.2 J/g · K. Would you expect the specific heat of gaseous water to be greater than that of the liquid at room temperature? Explain.

96. The surface tension of liquid water is greater than that of liquid methanol (CH_3OH) at room temperature. How can you explain the difference?

97. Two molecular compounds have the same molecular weight, but one boils at 195°C and the other at 142°C. What factors can account for the difference in boiling point? Which compound has the lowest vapor pressure? Do either or both of the compounds have a dipole moment?

98. Consider the following three compounds:

$$CH_3CH_2CH_2CH_2CH_2CH_3$$
n-Hexane

2-Hexanone

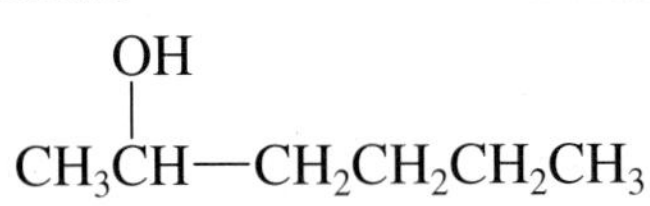

2-Hexanol

(a) What types of intermolecular forces exist between like molecules of each compound?
(b) Which compound will have the highest boiling point? Explain.
(c) Which compound will have the lowest vapor pressure? Explain.
(d) Which compound will be the most soluble in water? Explain.

99. The following data for the hydrides of Group V are given:

	Boiling Point, °C
NH_3	−33
PH_3	−88
AsH_3	−57
SbH_3	−17

(a) List these compounds in order of increasing vapor pressure. Which has the highest vapor pressure? Which has the lowest?
(b) Which of the following would be the best solvent for NH_3? Explain.

hexane, CCl_4, CH_3OH, 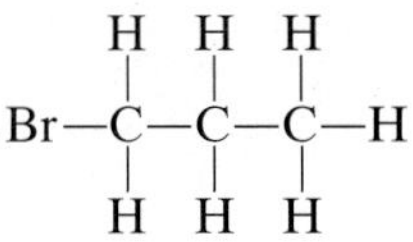

100.

		MW (g/mol)	*μ(D)*
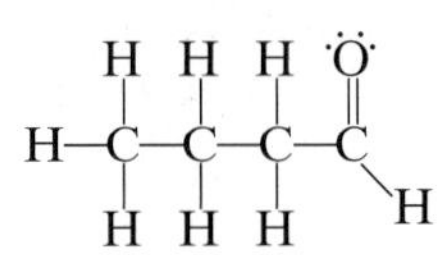	(A)	72	2.5
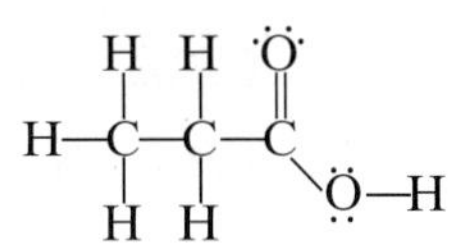	(B)	74	1.7
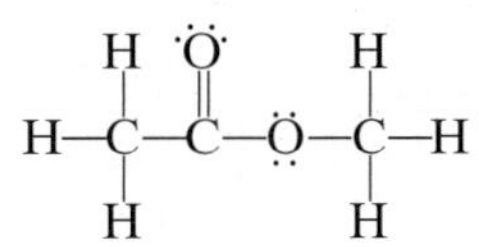	(C)	74	1.6

(a) Arrange these compounds in order of increasing boiling point. Specify which has the highest boiling point and which has the lowest. Explain your reasoning in detail.
(b) Which of the above would have the lowest vapor pressure? The highest vapor pressure? Explain.
(c) Which should be the most soluble in water? Explain.

101. Fig. 8.29 shows a plot of conductivity versus the moles of AgCl (a very slightly soluble ionic compound) added to the solution. Make a rough sketch of this plot. The x axis will be moles of added compound. Add a second plot that describes what would happen to conductivity as NaCl, a soluble salt, is added to a beaker of pure water. Add a third plot that describes what would happen to conductivity as table sugar ($C_{12}H_{22}O_{11}$), a molecular solid, is added to a beaker of pure water.

102. When a molecular solid is heated and melts to form a liquid, must all of the intermolecular forces between the solid molecules be broken? Explain your reasoning.

Chapter Eight

SPECIAL TOPICS

8A.1 Colligative Properties

Dissolving a solute in a solvent results in a solution with physical properties that are different from the pure solute or pure solvent. **Colligative properties** are physical properties of solutions that depend on the number of solute particles in a solution but not on the identity of the solute particles. This means that two solutions that contain different solutes but have the same solvent and the same concentrations of solute particles would exhibit the same colligative properties.

To begin our discussion of colligative properties we must introduce a new unit for measuring concentration, **mole fraction.** By definition, the mole fraction of any component of a solution is the number of moles of that component divided by the total number of moles of solute and solvent. The symbol for mole fraction is the Greek letter chi, χ. The mole fraction of the *solute* is the number of moles of solute divided by the total number of moles of solute and solvent.

$$\chi_{\text{solute}} = \frac{\text{moles of solute}}{\text{moles of solute} + \text{moles of solvent}}$$

Conversely, the mole fraction of the *solvent* is the number of moles of solvent divided by the total number of moles of solute and solvent.

$$\chi_{\text{solvent}} = \frac{\text{moles of solvent}}{\text{moles of solute} + \text{moles of solvent}}$$

In a solution that contains a single solute dissolved in a solvent, the sum of the mole fractions of solute and solvent must be equal to 1.

$$\chi_{\text{solute}} + \chi_{\text{solvent}} = 1$$

Exercise 8A.1

A solution of hydrogen sulfide in water can be prepared by bubbling H_2S gas into water until no more gas dissolves. Calculate the mole fraction of both H_2S and H_2O in this solution if 0.385 g of H_2S gas dissolves in 100 g of water at 20°C and 1 atm.

Solution

The number of moles of solute in the solution can be calculated from the mass of H_2S that dissolves.

$$0.385 \text{ g } H_2S \times \frac{1 \text{ mol } H_2S}{34.08 \text{ g } H_2S} = 0.0113 \text{ mol } H_2S$$

To determine the mole fraction of the solute and solvent, we also need to know the number of moles of water in the solution.

$$100 \text{ g } H_2O \times \frac{1 \text{ mol } H_2O}{18.02 \text{ g } H_2O} = 5.55 \text{ mol } H_2O$$

The mole fraction of the solute is the number of moles of H_2S divided by the total number of moles of both H_2S and H_2O.

$$\chi_{\text{solute}} = \frac{0.0113 \text{ mol } H_2S}{0.0113 \text{ mol } H_2S + 5.55 \text{ mol } H_2O} = 0.00203$$

The mole fraction of the solvent is the number of moles of H_2O divided by the moles of both H_2S and H_2O.

$$\chi_{\text{solvent}} = \frac{5.55 \text{ mol } H_2O}{0.0113 \text{ mol } H_2S + 5.55 \text{ mol } H_2O} = 0.998$$

Note that the sum of the mole fractions of the two components of the solution is 1.

$$\chi_{\text{solute}} + \chi_{\text{solvent}} = 0.00203 + 0.998 = 1.000$$

In Chapter 6, we saw that the ideal gas law was valid only for ideal gases. In much the same way, the equations we'll use to describe the colligative properties of solutions are only valid for **ideal solutions.** An ideal solution is one in which the forces that hold the solute particles together are similar to those that hold the solvent particles together. When this is true, the forces of attraction that must be broken to separate the particles in the solute and in the solvent are similar to the forces formed when the solute and solvent particles interact. This results in a change in enthalpy that is zero for the solution process. This is not the case for most real solutions. As a result, the equations we develop to describe colligative properties will give good approximations for real-world solutions, but not exact answers.

8A.2 Depression of the Partial Pressure of a Solvent

The vapor pressure of a liquid was defined in Chapter 8 as the pressure of a vapor in equilibrium with the corresponding liquid. When a solute is added to a liquid solvent, there is a decrease in the pressure exerted by the vapor of the solvent above the solution. We'll define $P°$ as the vapor pressure of the pure liquid—the solvent—and P as the pressure exerted by the vapor of the solvent over a solution.

$$P < P°$$

Partial pressure of the solvent above a solution	*Vapor pressure above the pure solvent*

Between 1887 and 1888, François-Marie Raoult showed that the pressure of the solvent escaping from a solution is equal to the mole fraction of the solvent times the vapor pressure of the pure solvent. This equation is known as **Raoult's law.**

$$P = \chi_{\text{solvent}} P°$$

Partial pressure of the solvent above a solution	*Vapor pressure above the pure solvent*

When the solvent is pure, and the mole fraction of the solvent is equal to 1, P is equal to $P°$. As the mole fraction of the solvent becomes smaller, the partial pressure of the solvent escaping from the solution also becomes smaller.

Let's assume, for the moment, that the solvent is the only component of the solution volatile enough to have a measurable vapor pressure. This would be true, for example, for a nonvolatile solute, such as NaCl, dissolved in water. The partial pressure of the solution would be equal to the pressure produced by the solvent escaping from the solution. Raoult's law states that the difference between the vapor pressure of the pure solvent and the partial pressure over the solution increases as the mole fraction of the solvent decreases.

When a solute is added to a pure solvent, the change in the partial pressure of the solvent above the solution is the difference between the vapor pressure of the pure solvent and the partial pressure exerted by the solvent above the solution.

$$\Delta P = P^\circ - P$$

Substituting Raoult's law into the above equation gives the following result.

$$\Delta P = P^\circ - \chi_{solvent}P^\circ = (1 - \chi_{solvent})P^\circ$$

This equation can be simplified by remembering the relationship between the mole fraction of the solute and the mole fraction of the solvent.

$$\chi_{solute} + \chi_{solvent} = 1$$

Substituting this relationship into the equation that defines ΔP gives another form of Raoult's law.

$$\Delta P_{solvent} = \chi_{solute}\, P^\circ_{solvent}$$

This equation reminds us that, for an ideal solution, as more solute is dissolved in the solvent, the change in pressure, ΔP, increases.

One of the consequences of vapor pressure depression of solvents by solutes can be seen in Figure 8A.1. This figure shows a beaker containing pure solvent and a second beaker containing a solution with a nonvolatile solute. Both beakers are inside a sealed container. The pure solvent and the solvent in the solution begin to evaporate, attempting to establish equilibrium between the liquid and the vapor phase. However, the vapor pressure associated with the pure solvent will be larger than the pressure exerted by the solvent in the solution. As time passes the solvent will gradually evaporate from the beaker that contains pure solvent and will condense in the beaker that contains the solution.

The vapor that accumulates above a solution that contains a *volatile* solute will have two components: the solvent and the solute. The total pressure of this vapor will be the sum of the partial pressure of the solvent and the partial pressure of the solute.

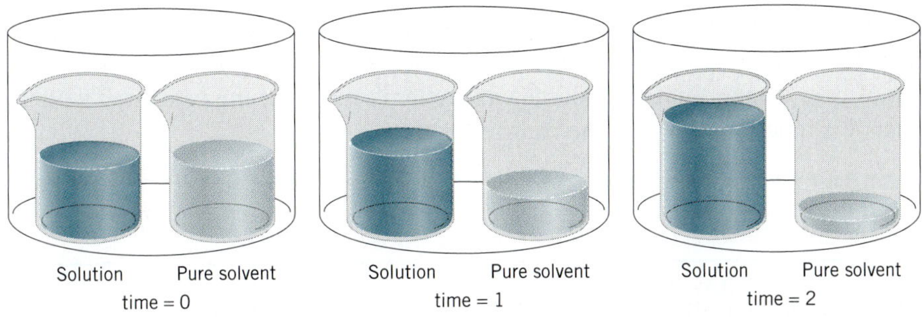

Fig. 8A.1 As evaporation takes place, the sealed container fills with the vapor of the solvent. Due to the difference in the partial pressures of the solvent over the pure solvent and over the solution, containing a nonvolatile solute, the solvent gradually evaporates from the beaker containing the pure solvent while vapor condenses into the beaker containing the solution.

Exercise 8A.2

A solution is prepared by mixing 500 mL of ethanol (C_2H_6O) and 500 mL of water at 25°C. The vapor pressures of pure water and pure ethanol at that temperature are 23.76 and 59.76 mmHg, respectively. The densities of water and ethanol at 25°C are 0.9971 and 0.786 g/mL, respectively. Determine the partial pressure of each component in the solution and the total pressure. Assume the solution is ideal.

Solution

We begin by determining the number of moles of each of the components of the solution.

$$500 \text{ mL } H_2O \times \frac{0.9971 \text{ g } H_2O}{1 \text{ mL}} \times \frac{1 \text{ mol } H_2O}{18.02 \text{ g } H_2O} = 27.7 \text{ mol } H_2O$$

$$500 \text{ mL } C_2H_6O \times \frac{0.786 \text{ g } C_2H_6O}{1 \text{ mL}} \times \frac{1 \text{ mol } C_2H_6O}{46.07 \text{ g } C_2H_6O} = 8.53 \text{ mol } C_2H_6O$$

We then calculate the mole fractions of the two components of the solution.

$$\chi_{\text{water}} = \frac{27.7 \text{ mol water}}{8.53 \text{ mol ethanol} + 27.7 \text{ mol water}} = 0.765$$

$$\chi_{\text{ethanol}} = \frac{8.53 \text{ mol ethanol}}{8.53 \text{ mol ethanol} + 27.7 \text{ mol water}} = 0.235$$

According to Raoult's law, the partial pressure of the water escaping from the solution is equal to the product of the mole fraction of water and the vapor pressure of pure water.

$$P_{\text{water}} = \chi_{\text{water}} P^\circ_{\text{water}} = (0.765)(23.76 \text{ mmHg}) = 18.2 \text{ mmHg}$$

The partial pressure of ethanol is found in a similar fashion.

$$P_{\text{ethanol}} = \chi_{\text{ethanol}} P^\circ_{\text{ethanol}} = (0.235)(59.76 \text{ mmHg}) = 14.0 \text{ mmHg}$$

The total pressure of the gases escaping from the solution is the sum of the partial pressures of the two gases.

$$P_{\text{total}} = P_{\text{water}} + P_{\text{ethanol}} = 32.2 \text{ mmHg}$$

Although the solution is a 50:50 mixture by volume, slightly more than three-quarters of the particles in the solution are water molecules. As a result, the total pressure of the solution more closely resembles the vapor pressure of pure water than it does that of pure ethanol. In addition, the change in partial pressure associated with the ethanol is much greater than the change for water.

8A.3 Boiling Point Elevation

Chapter 8 described how a pure liquid could be boiled by heating the liquid to increase its vapor pressure until the vapor pressure became equal to the pressure pushing down on the surface of the liquid. If a nonvolatile solute is added to the pure liquid, the resulting pressure due to the vapor of the solvent above the solution will be reduced. This means that it will be necessary to increase the temperature even more in order to increase the pressure of the solvent's vapor until it becomes equal to the pressure pushing down on the surface of the solution. Therefore, the boiling point of a solution will be greater than the boiling point of the pure solvent. This is known as boiling point elevation.

Because changes in the boiling point of the solvent (ΔT_{BP}) that occur when a solute is added result from changes in the partial pressure of the solvent, the magnitude of the change in the boiling point is also proportional to the mole fraction of the solute.

In very dilute solutions, the mole fraction of the solute is proportional to the molality of the solution. Molality, *m,* is defined as the number of moles of solute per kilogram of solvent.

$$m = \frac{\text{moles of solute}}{\text{kilograms of solvent}}$$

Molality is similar to molarity except the denominator is *kilograms of solvent* instead of *liters of solution*. Molality has an important advantage over molarity. The molarity of an aqueous solution changes with temperature because the density of water is sensitive to temperature. The molality of a solution doesn't change with temperature because it is defined in terms of the mass of the solvent, not its volume.

The equation that describes the magnitude of the boiling point elevation that occurs when a solute is added to a solvent can be written as follows.

$$\Delta T_{BP} = k_b m$$

Here, ΔT_{BP} is the **boiling point elevation,** that is, the change in boiling point that occurs when a solute dissolves in the solvent, and k_b is a proportionality constant known as the *molal boiling point elevation constant* for the solvent. Molal boiling point elevation constants for selected compounds are given in Table 8A.1.

Because colligative properties depend on the number of particles in solution, but not their identity, colligative properties can be used to determine molecular weights. Consider elemental sulfur, for example. The chemical formula is written as S_8. How do we know that sulfur forms S_8 molecules?

Table 8A.1

Freezing Point Depression and Boiling Point Elevation Constants

Compound	Freezing Point (°C)	k_f (°C/m)
Water	0	1.853
Acetic acid	16.66	3.90
Benzene	5.53	5.12
p-Xylene	13.26	4.3
Naphthalene	80.29	6.94
Cyclohexane	6.54	20.0
Carbon tetrachloride	−22.95	29.8
Camphor	179.8	40
Compound	**Boiling Point (°C)**	**k_b (°C/m)**
Water	100	0.515
Ethyl ether	34.55	2.02
Carbon disulfide	46.23	2.35
Benzene	80.10	2.53
Carbon tetrachloride	76.75	5.03
Camphor	207.42	5.95

Exercise 8A.3

Calculate the molecular weight of sulfur if 35.5 g of sulfur dissolves in 100 g of CS_2 to produce a solution that has a boiling point of 49.48°C.

Solution

The relationship between the boiling point of the solution and the molecular weight of sulfur isn't immediately obvious. We therefore start by asking: What do we know about the problem? We might start by drawing a figure, such as Figure 8A.2, that helps us organize the information in the problem.

We know the boiling point of the solution, so we might start by looking up the boiling point of the pure solvent (Table 8A.1) in order to calculate the change in the boiling point that occurs when the sulfur is dissolved in CS_2.

$$\Delta T_{BP} = 49.48°C - 46.23°C = 3.25°C$$

We also know that the change in the boiling point is proportional to the molality of the solution.

$$\Delta T_{BP} = k_b m$$

Since we know the change in the boiling point (ΔT_{BP}) and we can look up the boiling point elevation constant for the solvent (k_b) in Table 8A.1, we might decide to calculate the molality of the solution at this point.

$$m = \frac{\Delta T_{BP}}{k_b} = \frac{3.25°C}{2.35°C/m} = 1.38\ m$$

In the search for the solution to a problem, it is periodically useful to consider what we have achieved so far. At this point, we know the molality of the solution and the mass of the solvent used to prepare the solution. We can therefore calculate the number of moles of sulfur present in the carbon disulfide solution.

$$\frac{1.38 \text{ mol sulfur}}{1000 \text{ g } CS_2} \times 100 \text{ g } CS_2 = 0.138 \text{ mol sulfur}$$

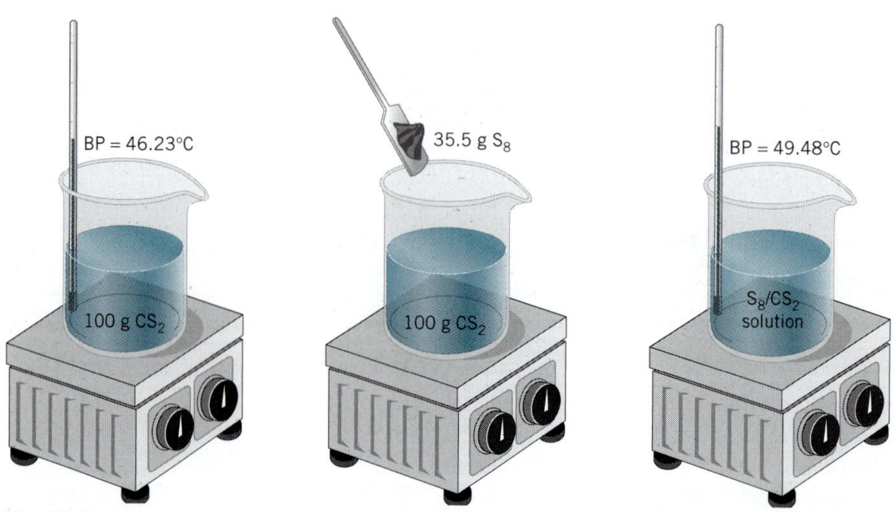

Fig. 8A.2 Diagram for Exercise 8A.3, where 35.5 g of S_8 is dissolved in 100 g of CS_2.

We now know the number of moles of sulfur in the solution and the mass of the sulfur. We can therefore calculate the number of grams per mole of sulfur.

$$\frac{35.5 \text{ g}}{0.138 \text{ mol}} = 257 \text{ g/mol}$$

From the periodic table we know the atomic mass of sulfur is 32.07 g/mol of sulfur atoms. Dividing this into the mass of a mole of sulfur molecules tells us that a molecule of sulfur must contain 8 sulfur atoms.

8A.4 Freezing Point Depression

It is observed that the addition of a solute to a solvent will lower the freezing point of the solution below that of the pure solvent.

Examples of the use of freezing point depression include adding salt to ice to decrease the temperature of melting ice when making homemade ice cream or salting highways to prevent the formation of ice during the winter.

An equation, similar to the boiling point elevation equation, can be written to describe what happens to the freezing point (or melting point) of a solvent when a solute is added to the solvent.

$$\Delta T_{FP} = -k_f m$$

In this equation, ΔT_{FP} is the **freezing point depression,** that is, the change in freezing point that occurs when the solute dissolves in the solvent, and k_f is the *molal freezing point depression constant* for the solvent. A negative sign is used in the equation to indicate that the freezing point of the solvent decreases when a solute is added. Molal freezing point depression constants for selected compounds are given in Table 8A.1.

Exercise 8A.4

Determine the molecular weight of acetic acid if a solution containing 30.0 g of acetic acid in 1000 g of water freezes at −0.93°C. Do the results agree with the assumption that acetic acid has the formula CH_3CO_2H?

Solution

The freezing point depression for the solution is equal to the difference between the freezing point of the solution (−0.93°C) and the freezing point of pure water (0°C).

$$\Delta T_{FP} = -0.93°\text{C} - 0.0°\text{C} = -0.93°\text{C}$$

We now turn to the equation that defines the relationship between freezing point depression and the molality of the solution.

$$\Delta T_{FP} = -k_f m$$

Because we know the change in the freezing point, and because we can find the freezing point depression constant for water in Table 8A.1, we have enough information to calculate the molality of the solution.

$$m = -\frac{\Delta T_{FP}}{k_f} = -\frac{-0.93 \text{ °C}}{1.853 \text{ °C}/m} = 0.50\ m$$

At this point, we might return to the statement of the problem, to see if we are making any progress toward an answer. According to this calculation, there is 0.50 moles of acetic acid per kilogram of solvent. The problem stated that there was 30.0 g of acetic acid per 1000 g of solvent. Because we know the number of grams and the number of moles of acetic acid in the sample, we can calculate the molecular weight of acetic acid.

$$\frac{30.0 \text{ g}}{0.50 \text{ mol}} = 60 \text{ g/mol}$$

The results of the experiment are in good agreement with the molecular weight (60.05 g/mol) expected if the formula for acetic acid is CH_3CO_2H.

Exercise 8A.5

Explain why a 0.100-*m* solution of HCl dissolved in benzene has a freezing point depression of 0.512°C, whereas a 0.100 *m* solution of HCl in water has a freezing point depression of 0.369°C.

Solution

We can predict the change in the freezing point that should occur in the solutions from the freezing point depression constant for the solvent and the molality of the solution. For the 0.100 *m* solution of HCl in benzene, the results of the calculation agree with the experimental value.

$$\Delta T_{FP} = -k_f m = -\ (5.12°\text{C}/m)(0.100\ m) = -0.512°\text{C}$$

For water, however, the calculation gives a predicted value for the freezing point depression that is half of the observed value.

$$\Delta T_{FP} = -k_f m = -\ (1.853°\text{C}/m)(0.100\ m) = -0.185°\text{C}$$

To explain the results, it is important to remember that colligative properties depend on the relative number of solute particles in a solution, not their identity. If the acid dissociates (breaks up into its ions) to an appreciable extent, the solution will contain more solute particles than we might expect from its molality.

If HCl dissociates completely in water, the total concentration of solute particles (H^+ and Cl^- ions) in the solution will be twice as large as the molality of the solution. The freezing point depression for the solution therefore will be twice as large as the change that would be observed if HCl did not dissociate.

$$HCl(g) \xrightarrow{H_2O} H^+(aq) + Cl^-(aq)$$

If we assume that 0.100 *m* HCl dissociates to form H^+ and Cl^- ions in water, the freezing point depression for the solution should be −0.371°C, which is slightly larger than what is observed experimentally.

$$\Delta T_{FP} = -k_f m = -(1.853°\text{C}/m)(2 \times 0.100\ m) = -0.371°\text{C}$$

This exercise suggests that HCl doesn't dissociate into ions when it dissolves in benzene, but dilute solutions of HCl dissociate more or less quantitatively in water.

In 1884 Jacobus Henricus van't Hoff introduced another term into the freezing point depression and boiling point elevation expressions to explain the colligative properties of solutions of compounds that dissociate when they dissolve in water.

$$\Delta T_{\mathrm{FP}} = -k_{\mathrm{f}}(i)m$$

Substituting the experimental value for the freezing point depression of a 0.100-*m* HCl solution into the equation gives a value for the *i* term of 1.99. If HCl did not dissociate in water, *i* would be 1. If it dissociates completely, *i* would be 2. The experimental value of 1.99 suggests that about 99% of the HCl molecules dissociate in the solution.

8A.5 Soaps, Detergents, and Dry-Cleaning Agents

The chemistry behind the manufacture of soap hasn't changed since it was made from animal fat and the ash from wood fires almost 5000 years ago. Solid animal fats (such as the tallow obtained during the butchering of sheep and cattle) and liquid plant oils (such as palm oil and coconut oil) are still heated in the presence of a strong base to form a soft, waxy material that enhances the ability of water to wash away the grease and oil that form on our bodies and our clothes.

Animal fats and plant oils contain compounds known as *fatty acids.* Fatty acids, such as stearic acid (see Figure 8A.3), have small, polar, hydrophilic heads attached to long, nonpolar, hydrophobic tails. Fatty acids are seldom found by themselves in nature. They are usually bound to molecules of glycerol ($HOCH_2CHOHCH_2OH$) to form triglycerides, such as the one shown in Figure 8A.4. The triglycerides break down in the presence of a strong base to form the Na^+ or K^+ salt of the fatty acid, as shown in Figure 8A.5. This reaction is called *saponification*, which literally means "the making of soap."

The cleaning action of soap results from the fact that soaps are *surfactants*—they tend to concentrate on the surface of water. They cling to the surface because they try to orient their polar CO_2^- heads toward water molecules and their nonpolar $CH_3CH_2CH_2\ldots$ tails away from neighboring water molecules.

Water can't wash the soil out of clothes by itself because the soil particles that cling to textile fibers are covered by a layer of nonpolar grease or oil molecules, which repel water. The nonpolar tails of the soap dissolve in the grease or oil that surrounds a soil particle, as shown in Figure 8A.6. The soap therefore disperses or emulsifies the soil particles, which makes it possible to wash the particles out of the clothes.

Most soaps are more dense than water. They can be made to float, however, by incorporating air into the soap during its manufacture. Most soaps are also opaque; they absorb rather than transmit light. Translucent soaps can be made by

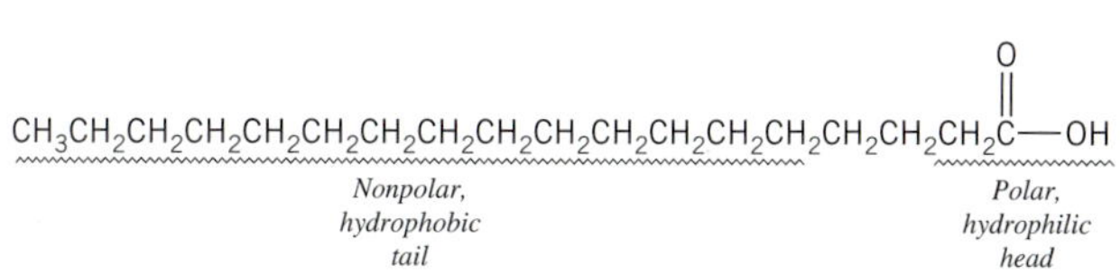

Fig. 8A.3 The hydrocarbon chain on one end of a fatty acid molecule is nonpolar and hydrophobic, whereas the —CO_2H group on the other end of the molecule is polar and hydrophilic.

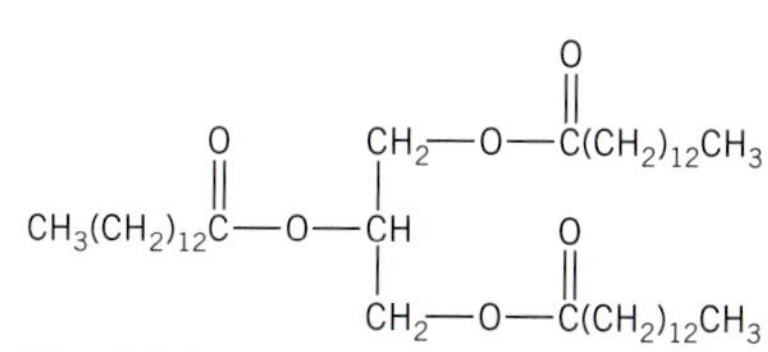

Fig. 8A.4 Structure of the triglyceride known as *trimyristin,* which can be isolated in high yield from nutmeg.

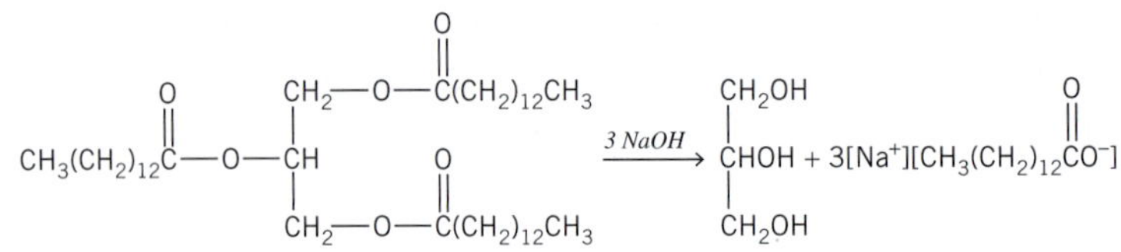

Fig. 8A.5 Saponification of the trimyristin extracted from nutmeg.

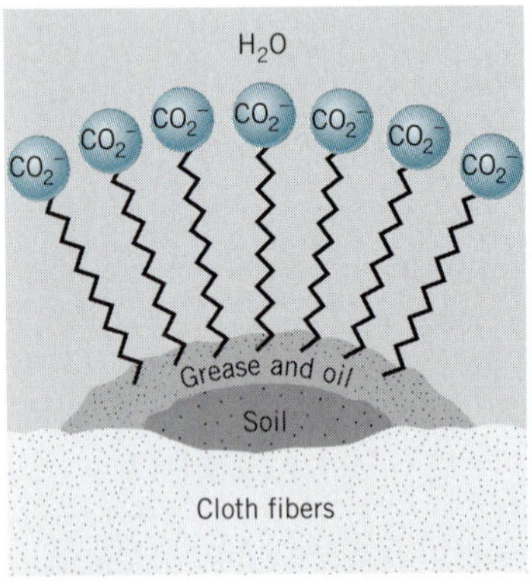

Fig. 8A.6 Soap molecules disperse, or emulsify, soil particles coated with a layer of nonpolar grease or oil molecules.

adding alcohol, sugar, and glycerol, which slow down the growth of soap crystals while the soap solidifies. Liquid soaps are made by replacing the sodium salts of the fatty acids with the more soluble K^+ or NH_4^+ salts.

In the 1950s, more than 90% of the cleaning agents sold in the United States were soaps. Today soap represents less than 20% of the market for cleaning agents. The primary reason for the decline in the popularity of soap is the reaction between soap and "hard" water. The most abundant positive ions in tap water are Na^+, Ca^{2+}, and Mg^{2+}. Water that is particularly rich in Ca^{2+}, Mg^{2+}, or Fe^{3+} ions is said to be hard. Hard water interferes with the action of soap because the Ca^{2+}, Mg^{2+}, and Fe^{3+} ions combine with soap to form insoluble precipitates that have no cleaning power. The precipitates not only decrease the concentration of the soap in solution, they actually bind soil particles to clothing, leaving a dull, gray film.

One way around the problem is to "soften" the water by replacing the Ca^{2+} and Mg^{2+} ions with Na^+ ions. Many water softeners are filled with a resin that contains —SO_3^- ions attached to a polymer, as shown in Figure 8A.7. The resin is treated with NaCl until each —SO_3^- ion picks up an Na^+ ion. When hard water flows over the resin, Ca^{2+} and Mg^{2+} ions bind to the —SO_3^- ions on the polymer chain and Na^+ ions are released into the solution. Periodically, the resin becomes saturated with Ca^{2+} and Mg^{2+} ions. When that happens, the resin has to be regenerated by being washed with a concentrated solution of NaCl.

There is another way around the problem of hard water. Instead of removing Ca^{2+} and Mg^{2+} ions from water, we can find a cleaning agent that doesn't form insoluble salts with those ions. Synthetic detergents are an example of such cleaning agents. Detergents consist of long, hydrophobic hydrocarbon tails attached to polar, hydrophilic —SO_3^- or —OSO_3^- heads, as shown in Figure 8A.8. By themselves, detergents don't have the cleaning power of soap. "Builders" were therefore added to synthetic detergents to increase their strength. The builders were often salts of highly charged ions, such as the triphosphate ion ($P_3O_{10}^{5-}$).

Cloth fibers swell when they are washed in water. This leads to changes in the dimensions of the cloth that can cause wrinkles—which are local distortions in the structure of the fiber—or even more serious damage, such as shrinking. These problems can be avoided by "dry cleaning," which uses a nonpolar solvent that doesn't adhere to, or wet, the cloth fibers. The nonpolar solvents used in dry cleaning

Fig. 8A.7 When a water softener is "charged," it is washed with a concentrated NaCl solution until all of the —SO_3^- ions pick up an Na^+ ion. The softener then picks up Ca^{2+} and Mg^{2-} from hard water, replacing those ions with Na^+ ions.

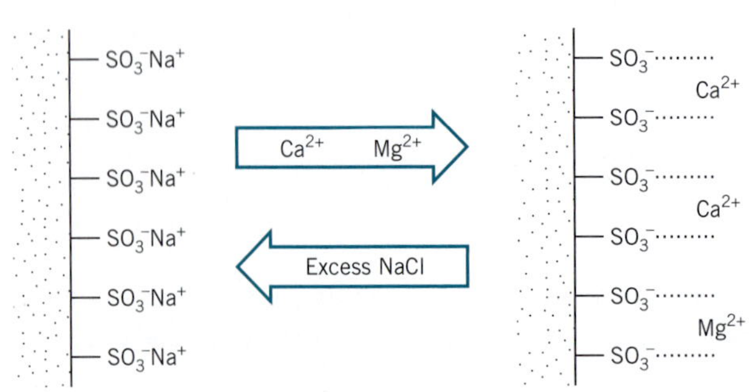

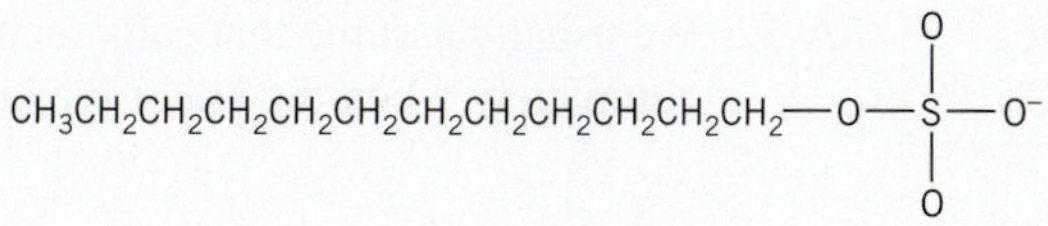

Fig. 8A.8 Structure of one of the components of a synthetic detergent.

dissolve the nonpolar grease or oil layer that coats soil particles, freeing the soil particles to be removed by detergents added to the solvent or by the tumbling action inside the machine. Dry cleaning has the added advantage that it can remove oily soil at lower temperatures than soap or detergent dissolved in water, so it is safer for delicate fabrics.

When dry cleaning was first introduced in the United States between 1910 and 1920, the solvent was a mixture of hydrocarbons isolated from petroleum when gasoline was refined. Over the years, those flammable hydrocarbon solvents have been replaced by halogenated hydrocarbons, such as trichloroethane ($Cl_3C{-}CH_3$), trichloroethylene ($Cl_2C{=}CHCl$), and perchloroethylene ($Cl_2C{=}CCl_2$).

Problems

Colligative Properties

8A-1. Define *colligative properties*.

8A-2. 5.62 g of methanol, CH_3OH, is dissolved in 50.0 g of water. Calculate the mole fraction of water and methanol in the solution. How can you check your answer?

8A-3. Which pairs of the following substances would be expected to mix to form ideal solutions?
(a) hexane and heptane
(b) hexane and hexanoic acid
(c) NaCl and water
(d) methanol and water

Depression of the Partial Pressure of a Solvent

8A-4. Predict what will happen to the rate at which water evaporates from an open flask when salt is dissolved in the water, and explain why the rate of evaporation changes. If you place a beaker of pure water (I) and a beaker of a saturated solution of sugar in water (II) in a sealed bell jar, the level of water in beaker I will slowly decrease, and the level of the sugar solution in beaker II will slowly increase. Explain why.

8A-5. Explain why the vapor pressure of a liquid at a particular temperature is not a colligative property but the change in the partial pressure of the liquid when a solute is added is a colligative property.

8A-6. 2.56 g of the nonvolatile solute sucrose $C_{12}H_{22}O_{11}$ is added to 500 g of water at 25°C. What will be the partial pressure of the water over this solution?

8A-7. If 500 g of pentane is mixed with 500 g of heptane at 20°C, what will be the total pressure above the solution? The vapor pressure of pentane is 420 mm Hg and that of heptane is 36.0 mm Hg at 20°C.

8A-8. The partial pressure of water above a solution of water and a nonvolatile solute at 25°C is 19.5 mm Hg. What is the mole fraction of the solute?

Boiling Point Elevation

8A-9. Explain how the decrease in the vapor pressure of a solvent that occurs when a solute is added to the solvent leads to an increase in the solvent's boiling point.

8A-10. What change in the vapor pressure of a solvent occurs when a solute is added?

8A-11. What is the boiling point of a solution of 10.0 g of P_4 in 25.0 g of carbon disulfide?

8A-12. The boiling point elevation of a solution consisting of 2.50 g of a nonvolatile solute in 100.0 g of benzene is 0.686°C. What is the molecular weight of the unknown solute?

8A-13. What is the boiling point of a solution containing 3.41 g of the nonvolatile solute, I_2, in 50.0 g of carbon tetrachloride?

8A-14. The boiling point elevation of a solution of 0.120 mol of a sugar in 50.0 g of water is 1.23°C. Calculate k_b from these data.

Freezing Point Depression

8A-15. Predict the shape of a plot of the freezing point of a solution versus the molality of the solution.

8A-16. Explain why salt is added to the ice that surrounds the container in which ice cream is made.

8A-17. What is the approximate freezing point of a saturated solution of caffeine ($C_8H_{10}O_2N_4{\cdot}H_2O$) in water if it takes 45.6 g of water to dissolve 1.00 g of caffeine?

8A-18. A 0.100-m solution of sulfuric acid in water freezes at $-0.371°C$. Which of the following statements is consistent with this observation?
(a) H_2SO_4 does not dissociate in water.
(b) H_2SO_4 dissociates into H^+ and HSO_4^- ions in water.
(c) H_2SO_4 dissociates in water to form two H^+ ions and one SO_4^{2-} ion.
(d) H_2SO_4 associates in water to form $(H_2SO_4)_2$ molecules.

8A-19. The "Tip of the Week" in a local newspaper suggested using a fertilizer such as ammonium nitrate or ammonium sulfate instead of salt to melt snow and ice on sidewalks, because salt can damage lawns. Which of the following compounds would give the largest freezing point depression when 100 grams is dissolved in 1 kg of water?
(a) NaCl
(b) NH_4NO_3
(c) $(NH_4)_2SO_4$

8A-20. p-Dichlorobenzene (PDCB) is replacing naphthalene as the active ingredient in mothballs. Calculate the value of k_f for camphor if a 0.260 m solution of PDCB in camphor decreases the freezing point of camphor by 9.8°C.

8A-21. Explain why many cities and states spread salt on icy highways.

8A-22. We usually assume that salts such as KCl dissociate completely when they dissolve in water.

$$KCl(s) \xrightarrow{H_2O} K^+(aq) + Cl^-(aq)$$

Estimate the percentage of the KCl that actually dissociates in water if the freezing point of a 0.100 m solution of the salt in water is $-0.345°C$.

8A-23. Calculate the freezing point of a 0.100 m solution of acetic acid in water if the CH_3CO_2H molecules are 1.33% ionized in the solution.

8A-24. Compare the values of k_f and k_b for water, benzene, carbon tetrachloride, and camphor. Explain why measurements of molecular weight based on freezing point depression might be more accurate than those based on boiling point elevation.

8A-25. If an aqueous solution containing a nonvolatile solute boils at 100.50°C, at what temperature does it freeze?

Soaps, Detergents, and Dry-Cleaning Agents

8A-26. Why are soaps composed of hydrophilic and hydrophobic groups?

8A-27. What is hard water? How can it be softened?

8A-28. What is meant by "dry" cleaning? What advantages does dry cleaning have over soaps or detergents?

Chapter Nine

SOLIDS

9.1 Solids

Chapter 9 is a brief excursion into materials. The properties of solids and the relationship of these properties to structure constitute the bulk of this chapter. ChemActivities 23–25 in Chemistry: A Guided Inquiry *are designed to accompany this chapter.*

Chemists commonly classify solids as metallic, ionic, molecular, or network covalent according to their properties. This classification is based on macroscopic properties that result from differences on the atomic scale. The type of bonding between atoms is determined by electronegativity differences and the average electronegativity of the atoms composing the substance.

The triangular plots called bond-type triangles, introduced in Chapter 5, can be used to predict the type of bonding and, because the type of bonding is closely related to the classification scheme, a bond-type triangle can be used to classify solids.

Metallic solids are expected for those combinations of atoms whose electronegativities and electronegativity differences are found at the lower-left-hand corner of a bond-type triangle. Ionic compounds are found in the center of the triangle, and covalent materials are found in the lower-right-hand corner. Semiconductors, metalloids, and ceramics are located along the boundaries between the regions on the triangle.

Solids can be divided into three categories on the basis of the way the particles pack together. **Crystalline solids** are three-dimensional analogs of a brick wall. They have a regular structure in which particles pack in a repeating pattern, row to row and layer to layer, from one edge of the solid to the other. **Amorphous solids** (literally, "solids without form") such as glass have a random structure with little if any long-range order. Many solids, such as aluminum and steel, have a structure that falls between the two extremes. Such **polycrystalline solids** are aggregates of large numbers of small crystals or grains within which the structure is regular, but the crystals or grains are arranged in a random fashion.

Solids can also be classified on the basis of the forces that hold the particles together. This approach categorizes solids as either molecular, network covalent, ionic, or metallic. As shown in Chapter 5, a bond-type triangle may be used to show how electronegativity and electronegativity differences can be used to determine whether the bonding between atoms is best described as ionic, metallic, or covalent. The triangle shown in Figure 9.1 suggests that the bonding in elemental cesium is primarily metallic, cesium fluoride contains ionic bonds, and elemental fluorine is covalently bonded.

Ionic and covalent bonds are often imagined as if they were opposite ends of a two-dimensional model of bonding in which compounds that contain polar bonds fall somewhere between the two extremes.

Ionic . . . polar . . . covalent

In reality, there are three kinds of bonds between adjacent atoms: ionic, covalent, and metallic. Nonmetals combine to form elements and compounds that contain primarily covalent bonds, such as F_2, HCl, and CH_4. Metals combine with nonmetals to form ionic compounds such as CsF and CaO, which are held together by predominately ionic bonds. The force of attraction between atoms in metals, such

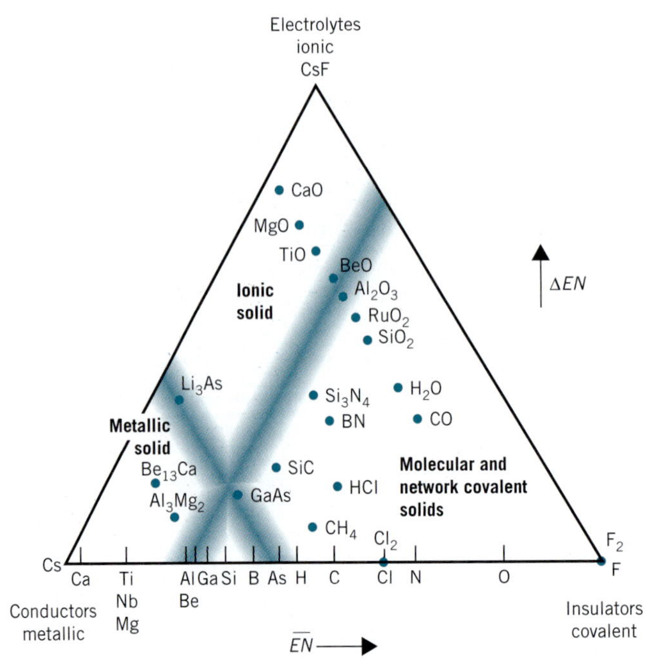

Fig. 9.1 Bond-type triangle for selected compounds. Such triangles may be used to classify solids as molecular, network covalent, ionic, or metallic.

as copper, or between the atoms in alloys, such as brass and bronze, or in compounds, such as Al_3Mg_2, are mainly metallic bonds.

The type of bonding in a compound can be used to classify the solid. Thus, if a compound can be located on a bond-type triangle, the properties of the solid can be predicted.

9.2 Molecular and Network Covalent Solids

MOLECULAR SOLIDS

The iodine (I_2) used to make the antiseptic known as tincture of iodine, the cane sugar ($C_{12}H_{22}O_{11}$) found in a sugar bowl, and the polyethylene used to make garbage bags all have one thing in common. They are all examples of compounds that are **molecular solids** at room temperature. Water and bromine are liquids that form molecular solids when cooled slightly; H_2O freezes at 0°C and Br_2 freezes at −7°C. Many substances that are gases at room temperature will form molecular solids when cooled far enough; F_2, at the extreme right of the bond-type triangle in Figure 9.1, freezes to form a molecular solid at −220°C.

Molecular solids contain both *intramolecular* bonds and *intermolecular* forces. The atoms within the individual molecules are held together by relatively strong intramolecular covalent bonds. Molecular solids are therefore found in the covalent region of a bond-type triangle. The molecules in a molecular solid are held together by much weaker intermolecular forces. Because the forces between these molecules are relatively weak, molecular solids are often soft substances, with low melting points.

Dry ice, or solid carbon dioxide, is a perfect example of a molecular solid. The van der Waals forces holding the CO_2 molecules together are weak enough that at −78°C dry ice **sublimes**—it goes directly from the solid to the gas phase.

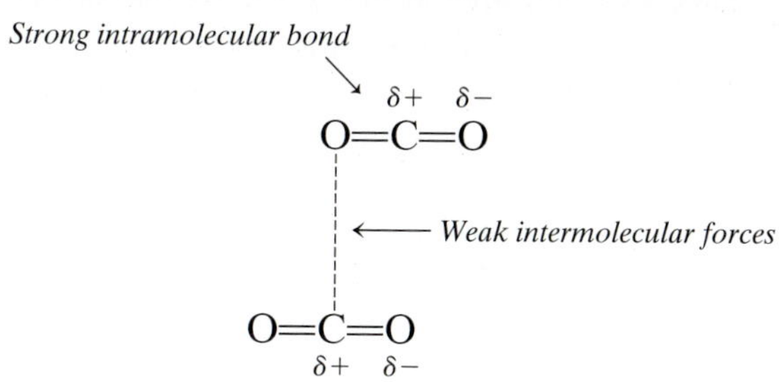

Changes in the strength of the van der Waals forces that hold molecular solids together can have important consequences for the properties of the solid. Polyethylene $(\text{—}CH_2\text{—}CH_2\text{—})_n$ is a soft plastic that melts at relatively low temperatures. Replacing one of the hydrogens on every other carbon atom with a chlorine atom produces a plastic known as poly(vinyl chloride), or PVC, which is hard enough to be used to make the plastic pipes that are slowly replacing metal pipes for plumbing.

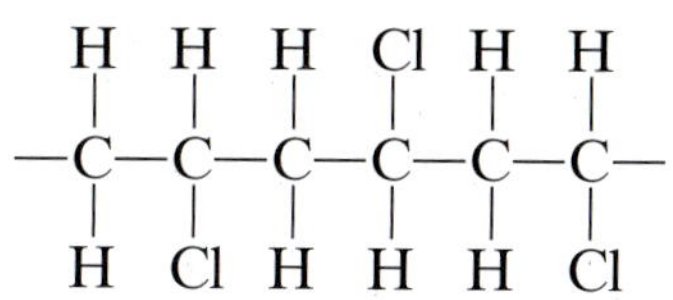

Poly(vinyl chloride) or PVC

Much of the strength of PVC can be attributed to the van der Waals force of attraction between the chains of (—CH_2—CHCl—)$_n$ molecules that form the solid. A copolymer of poly(vinyl chloride) (—CH_2—CHCl—)$_n$ and poly(vinylidene chloride) (—CH_2—CCl_2—)$_n$ is sold under the tradename Saran. The same increase in the force of attraction between chains that makes PVC harder than polyethylene gives a thin film of Saran a tendency to be attracted to itself. Saran wrap therefore clings to itself, whereas the polyethylene in sandwich bags does not.

The family of substances known as the *halogens* (F_2, Cl_2, Br_2, and I_2) can provide a basis for understanding the effect of differences in the strengths of intermolecular forces on the properties of a molecular solid. Consider chlorine, for example, which exists as diatomic Cl_2 molecules in the gas phase at room temperature. When the gas is cooled, the average kinetic energy of the Cl_2 molecules decreases. As the motion of the molecules decreases, the force of attraction between the molecules becomes large enough to hold the molecules together, and the gas condenses to form a liquid. Further cooling transforms the liquid into a molecular solid, as shown by its position on the bond-type triangle (Figure 9.1).

The unit on which the solid is built is still the diatomic Cl_2 molecule. The covalent bond holding one chlorine atom to another in Cl_2 is relatively strong, about 243 kJ/mol. The intermolecular forces that hold one Cl_2 molecule to another are much smaller, only about 18 kJ/mol. Thus, the *intermolecular* forces responsible for the formation of the solid are very different from the *intramolecular* bonds within the molecule.

Because the Cl_2 molecule has no dipole moment, the intermolecular forces that hold Cl_2 molecules together result solely from induced dipole–induced dipole or dispersion forces. The dispersion forces are nondirectional, and the molecules pack in the solid in the geometry that allows them to come as close together as possible.

Covalent molecules with a dipole moment, such as HCl and CO, also form molecular solids when cooled (Figure 9.1) in which the molecules pack as tightly as possible. Polar molecules, however, also have a directional component to the intermolecular forces, namely, dipole–dipole interactions. This force controls the orientation of the HCl and CO molecules as they pack, so that the negative end of one dipole is oriented toward the positive end of the other.

The chief distinction between the solid and liquid phases for covalent molecules is the regular pattern of packing in the solid versus the random structure in the liquid. Consider ice, for example. The individual H_2O molecules in the molecular solid (see Figure 8.19) are held together by a combination of dipole, dispersion, and hydrogen bond forces. Two parameters can be used to estimate the relative strength of these intermolecular forces, the melting point of the compound, and the enthalpy of fusion, ΔH_{fus}. The melting point, as we saw in Section 8.6, is the temperature at which the solid melts at 1 atm pressure. The enthalpy of fusion is the heat required to melt the substance, in units of kilojoules per mole for the reaction as written. The enthalpy of fusion of H_2O is relatively small, only 6.00 kJ/mol$_{rxn}$ ($H_2O(s) \longrightarrow H_2O(l)$). This is only a small fraction of the strength of the hydrogen bonds between water molecules because melting the solid doesn't break all of the hydrogen bonds between the water molecules, only some of them. To break all of the hydrogen bonds we have to boil water; the enthalpy of vaporization of H_2O is 40.88 kJ/mol$_{rxn}$ ($H_2O(l) \rightarrow H_2O(g)$) at the boiling point.

Melting points and enthalpies of fusion are convenient measures of the strengths of the intermolecular interactions that hold molecular solids together. Table 9.1 gives the melting points and the enthalpies of fusion of the halogens. The only forces that hold the crystals together are dispersion forces. Because dispersion forces depend on the number of electrons, as the size of the halogen atoms increases, the dispersion force interactions should also increase. This is reflected in

Table 9.1

Melting Points and Enthalpies of Fusion of the Halogens

Halogen	Molecular Weight (g/mol)	MP (°C)	ΔH_{fus} (kJ/mol_{rxn})
F_2	38	−219.6	0.51
Cl_2	71	−101	6.41
Br_2	160	−7.2	10.8
I_2	254	113.5	15.3

the increase in both the melting point and the enthalpy of fusion with increasing molecular weight.

The effect of adding dipole–dipole and hydrogen bond interactions to the intermolecular forces that hold molecules together can be seen in the data in Table 9.2. Dispersion and dipole forces exist in all three compounds. Two of the compounds, H_2O and CH_3OH, also form hydrogen bonds. As the number of hydrogen atoms that can form hydrogen bonds increases from zero in CH_3OCH_3 to one per molecule in CH_3OH and then two per molecule in H_2O, there is a significant increase in the melting point. The enthalpy of fusion is determined by the intermolecular attractive forces. The number of electrons, and hence the polarizability, of the compounds decreases from CH_3OCH_3 to CH_3OH to H_2O, and it might be expected that water would have the smallest enthalpy of fusion. The fact that water has the highest enthalpy of fusion shows the magnitude of the influence of the hydrogen bonding.

Table 9.2

Melting Points and Enthalpies of Fusion

Compound	Molecular Weight (g/mol)	MP (°C)	ΔH_{fus} (kJ/mol_{rxn})
CH_3OCH_3	46	−141.5	4.94
CH_3OH	32	−97.9	3.18
H_2O	18	0	6.00

➤ CHECKPOINT

Describe the differences on the atomic and macroscopic scales between molecular solids and network covalent solids.

NETWORK COVALENT SOLIDS

Network covalent solids include substances, such as diamond, whose crystals can be viewed as a single giant molecule made up of an almost endless number of covalent bonds. Each carbon atom in diamond is covalently bound to four other carbon atoms oriented toward the corners of a tetrahedron, as shown in Figure 9.2. Diamond is the hardest natural substance, and it melts at 3550°C. Quartz is a network covalent solid composed of SiO_2 and is located in the covalent region of Figure 9.1. Network covalent solids are often very hard, and they are notoriously difficult to melt. Both molecular solids and network covalent solids are located in the covalent region of a bond-type triangle. A bond-type triangle therefore can't be used to distinguish between these two types of solids.

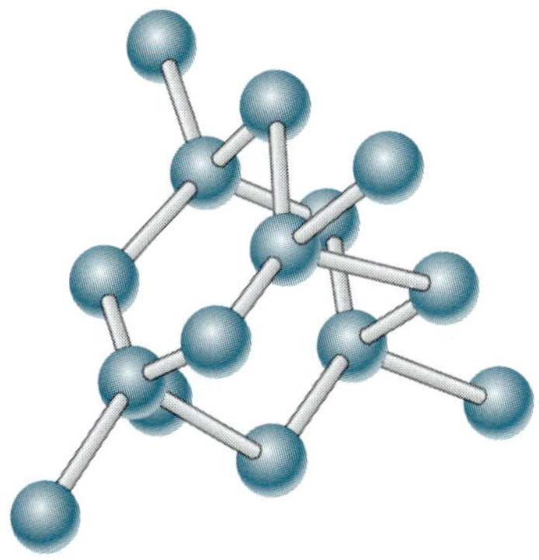

Fig. 9.2 A perfect diamond is a single molecule in which each carbon atom is tightly bound to four neighboring carbon atoms arranged toward the corners of a tetrahedron.

9.3 Ionic Solids

As discussed in Section 5.7, **ionic solids** are salts, such as NaCl (Figure 5.3), that form an extended three-dimensional network of ions that are held together by the strong force of attraction between ions of opposite charge (see Figure 9.3). Because the force of attraction depends inversely on the square of the distance between the positive and negative charges, the strength of an ionic bond depends inversely on the size of the ions that form the solid.

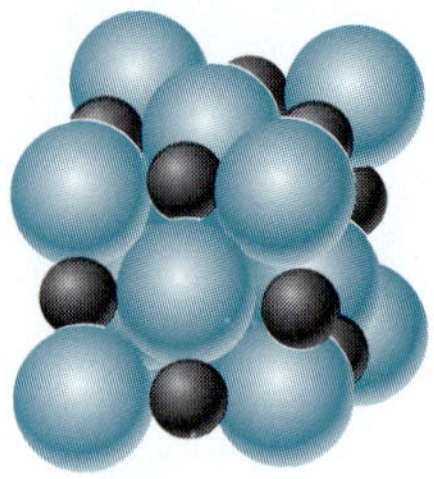

Fig. 9.3 Ionic compounds are made up of a three-dimensional network of positive and negative ions.

$$F = \frac{q_1 \times q_2}{r^2}$$

When the ions are large, the bond is relatively weak. But the ionic bond is still strong enough to ensure that salts have relatively high melting points and boiling points. Sodium chloride, for example, melts at 801°C and boils at 1413°C.

Solids retain their shape, are difficult to compress, and are usually denser than liquids and gases. These characteristic properties suggest that solids contain particles which are packed as tightly as possible. Ionic compounds form solids in which the force of attraction between the ions of opposite charge is maximized by keeping the ions as close together as possible. Ionic solids are located in the ionic region of a bond-type triangle, as shown in Figure 9.1

> **CHECKPOINT**
>
> Describe the differences on the atomic and macroscopic scales between molecular solids and ionic solids.

Some understanding of the strength of the bonding in an ionic compound can be obtained by considering the enthalpy of the process in which the structure of an ionic solid is completely disrupted to form isolated ions in the gas phase.

$$NaCl(s) \longrightarrow Na^+(g) + Cl^-(g)$$

Reaction diagrams like those in Chapter 7 are used to calculate changes in enthalpy associated with phase changes. We emphasize that these diagrams are useful for visualizing enthalpy changes but that no bonds are actually broken during a phase change for a molecular species.

This process can be visualized by first transforming the salt into sodium and chlorine atoms in the gas phase. This requires the input of 640 kJ/mol_{rxn}, which is the enthalpy of atomization of sodium chloride. That value can be determined by reversing the sign of the enthalpy of atom combination given in Appendix B.13.

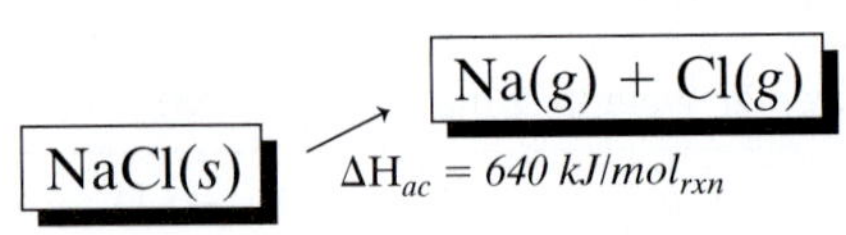

An electron is then removed from sodium, which requires an input of energy corresponding to the first ionization energy of that element.

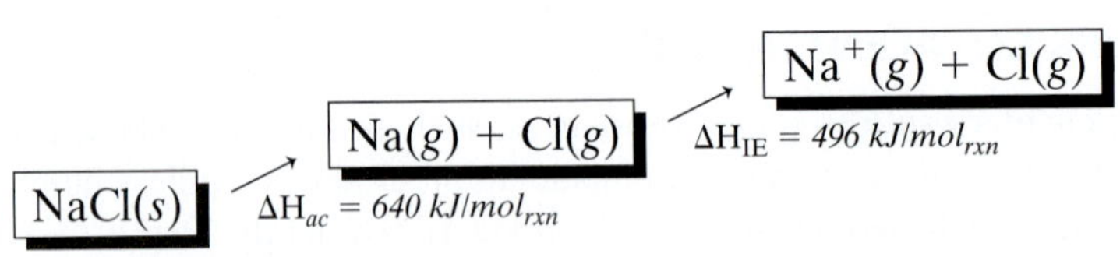

An electron is then added to a neutral chlorine atom to form a Cl^- ion in the gas phase. The energy associated with the step is called the **electron affinity** (Appendix B, Table B.6) of the element. Electron affinity is usually, but not always, exothermic because most neutral atoms will give off heat when they accept an extra electron.

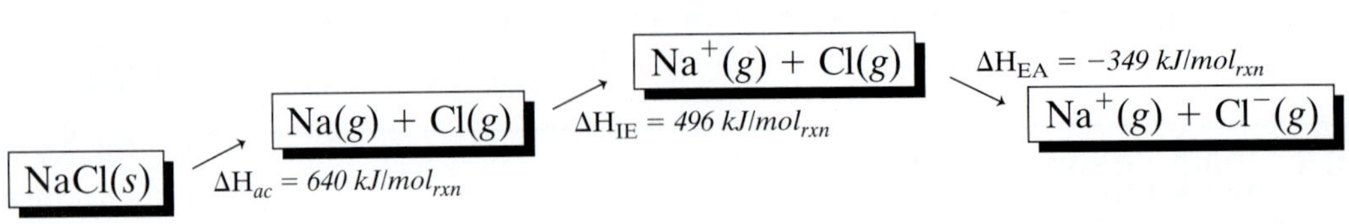

We can now complete the thermodynamic cycle by bringing the Na^+ and Cl^- ions in the gas phase together to form the solid NaCl. Because the force of attraction between the ions is relatively large, this is a strongly exothermic step.

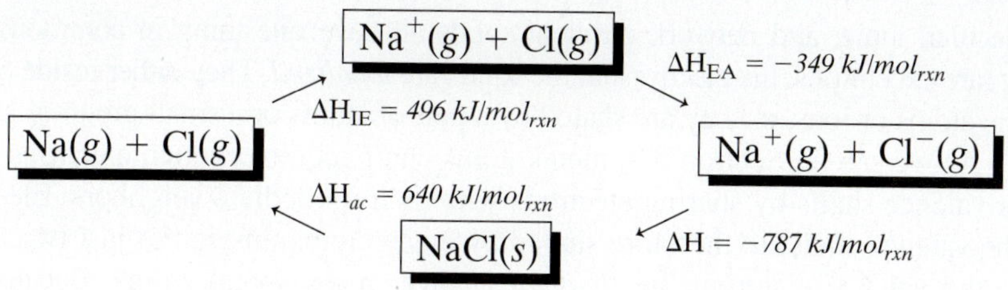

The energy required to break an ionic compound into isolated ions in the gas phase is known as the **lattice energy.** According to the thermodynamic cycle shown above, the lattice energy for NaCl would be +787 kJ/mol_{rxn}.

The lattice energies of compounds of the alkali metals with a halogen are given in Table 9.3. Note that the lattice energies of the compounds decrease as the size of the ions increases because of an increase in the distance between the centers of the positive and negative charges on the ions. It therefore takes less energy to break one of the solids apart as the ions become larger, or less energy is given off when one of the compounds is formed from the corresponding positive and negative ions in the gas phase.

The lattice energies for ionic compounds formed when one of the alkaline earth metals combines with oxygen to form an oxide (MgO, CaO) shows a trend similar to the halides given in Table 9.3. The lattice energy for MgO, however, is about five times as large as the lattice energy for NaCl. Part of the difference can be explained by noting that MgO contains ions with charges of +2 and −2. Thus, the product of the charge on the positive and negative ions is four times larger in MgO than it is in NaCl, which contains +1 and −1 ions. The remainder of the difference results from the fact that the Mg^{2+} ion is smaller than the Na^+ ion, and the O^{2-} ion is smaller than the Cl^- ion.

Table 9.3

Lattice Energies of Alkali Metal Halides (kJ/mol_{rxn})

	F^-	Cl^-	Br^-	I^-
Li^+	1046	861	818	762
Na^+	923	787	747	704
K^+	821	718	682	649
Rb^+	785	689	660	630
Cs^+	740	659	631	604

➤ CHECKPOINT

Which has the larger lattice energy, $MgCl_2$ or MgF_2?

Exercise 9.1

Arrange the following ionic compounds in order of increasing lattice energy: KF, CaF_2, $CaCl_2$, CaO.

Solution

The lattice energy for ionic compounds is directly related to the product of the charges on the ions and inversely related to the distance between the ions.

KF has a larger distance between its K^+ and F^- ions and a smaller product of charges on its ions than does CaF_2 or $CaCl_2$. Therefore KF has a smaller lattice energy than does either CaF_2 or $CaCl_2$.

CaF_2 and $CaCl_2$ have the same charge on their ions, but $CaCl_2$ has a larger interionic distance than CaF_2 and hence a smaller lattice energy than CaF_2.

CaO has ions of charge +2 and −2 and about the same distance between ions as does CaF_2 and thus has a larger lattice energy than CaF_2.

The compounds arranged in order of increasing lattice energy are: KF < $CaCl_2$ < CaF_2 < CaO.

Often the distance between ions can be estimated by using the principles established in Chapter 3, Section 3.21. In some cases reference to Appendix B.4 may be necessary.

9.4 Metallic Solids

Molecular, ionic, and network covalent solids all have one thing in common. With only rare exceptions, the electrons in the solids are *localized.* They either reside on one of the atoms or ions, or they are shared by a pair of atoms or a small group of atoms.

As we saw in Section 5.9, metal atoms can't acquire enough electrons to fill their valence shells by sharing electrons with their immediate neighbors. Electrons in the valence shell are therefore shared by many atoms, instead of just two. In effect, the valence electrons are *delocalized* over many metal atoms. Because the electrons aren't tightly bound to individual atoms, they are free to migrate through the metal. As a result, metals are good conductors of heat and electricity. Electrons that enter the metal at one edge can displace other electrons to give rise to a net flow of electrons through the metal.

The bonds that hold metals together are very different from ionic and covalent bonds and are therefore placed in a category of their own: **metallic bonds.** Metallic bonding occurs when both ΔEN and the average EN of the atoms are relatively small (i.e., in the lower-left corner of Figure 9.1). In a metal, the metal atoms form bonds with many neighboring atoms. Thus metals are usually solids in which each atom is surrounded by as many neighboring atoms as possible. Lithium, for example, crystallizes in a structure in which each atom touches eight nearest neighbors. The distance between the nuclei of the atoms is 0.304 nm.

Lithium has three electrons: $1s^2\ 2s^1$. There is a significant difference between the ease with which an electron can be removed from the $1s$ and $2s$ orbitals on a lithium atom, however. According to the data in Table 3.4, it takes 0.52 MJ/mol to remove an electron from the $2s$ orbital on lithium, but 6.26 MJ/mol to remove one of the electrons from the $1s$ orbital. The core electrons in the $1s$ orbitals on a lithium atom are bound so tightly to the nucleus of the atom that they are unaffected by other atoms. Thus, there is only one valence electron to be shared per lithium atom in the metal, and the electron must be shared with all of the neighboring atoms.

In the gas phase, lithium can form a diatomic Li_2 molecule that is held together by the sharing of a pair of electrons by the two lithium nuclei.[1] The distance between the lithium atoms in the Li_2 molecule is 0.267 nm, which is considerably smaller than the distance between lithium atoms in the metal. This suggests that the covalent bond in the Li_2 molecule is significantly stronger than the metallic bonds in lithium metal. However, there are more bonds per lithium atom in the metal. As a result, the enthalpy of atomization $Li(s) \longrightarrow Li(g)$ ($-\Delta H_{ac}$ from Appendix B.13) for lithium metal is 159 kJ/mol$_{rxn}$, whereas the bond holding the two atoms together in an Li_2 molecule is only 57 kJ/mol$_{rxn}$.

Figure 4.2 and the AVEE values show that as the periodic table is descended, the energy difference between subshells within a given shell becomes smaller. In other words, *s*, *p*, and *d* subshells become closer in energy. When an atom enters into combination with another atom or group of atoms, the difference in energy between the *s*, *p*, and *d* subshells becomes even smaller. Elements that are metals have a common characteristic. The energies of the valence subshells of neighboring metallic atoms are very similar. This allows electrons to move easily between all available subshells from atom to atom. The electrons are said to be *delocalized.* This means that the electrons are not confined to the space between nuclei of atoms. Bonding becomes nondirectional, and the atoms pack together as tightly as possible.

[1] Although dilithium molecules can exist in the gas phase, the famous "dilithium crystals" that fuel the Starship *Enterprise* exist only in the imagination of Gene Roddenberry.

Thus, a substance held together by metallic bonds can be considered to be made up of metallic cations in a sea of electrons.

As one goes down a column of the periodic table, the size of the atoms increase, making it easier to remove outer shell electrons, and the energy gaps between subshells decrease. These two factors explain why metallic behavior increases as we move from top right to bottom left in the periodic table.

> **CHECKPOINT**
> The AVEE value or electronegativity of an atom is made up of two important contributions. What are they, and why are they important for understanding metallic behavior?

9.5 Physical Properties That Result from the Structure of Metals

Metals have certain characteristic physical properties.

- They have a metallic shine, or luster.
- They are usually solids at room temperature.
- They are *malleable* (from the Latin word for "hammer"): They can be hammered, pounded, or pressed into different shapes.
- They are *ductile:* They can be drawn into thin sheets or wires without breaking.
- They conduct heat and electricity.

The structures of metals can be used to explain their characteristic properties.

A chrome-plated surface has a characteristic metallic luster because the metal reflects (literally, "throws back") a significant fraction of the light that hits its surface. Silver is better than any other metal at reflecting light: Roughly 88% of the light that hits the surface of a silver mirror is reflected.

Why are metals solid? Some nonmetals, such as hydrogen and oxygen, are gases at room temperature because the atoms of these elements form molecules that are held together by weak intermolecular forces. Metal cations are held closely together by strong metallic bonds in a three-dimensional network and are therefore solids at room temperature (except mercury).

> **CHECKPOINT**
> Describe the differences on the atomic and macroscopic scales between metallic solids and ionic solids.

Metals are malleable and ductile because they pack in structures that contain planes of atoms. In theory, changing the shape of the metal is simply a matter of applying a force that makes the atoms in one of these planes slide past the atoms in an adjacent plane, as shown in Figure 9.4. In practice, it is easier to do this when the metal is hot. The layers of atomic cores in metals can slip easily over one another because there are no directional forces tending to keep them in locked positions, and thus metals will be ductile and malleable.

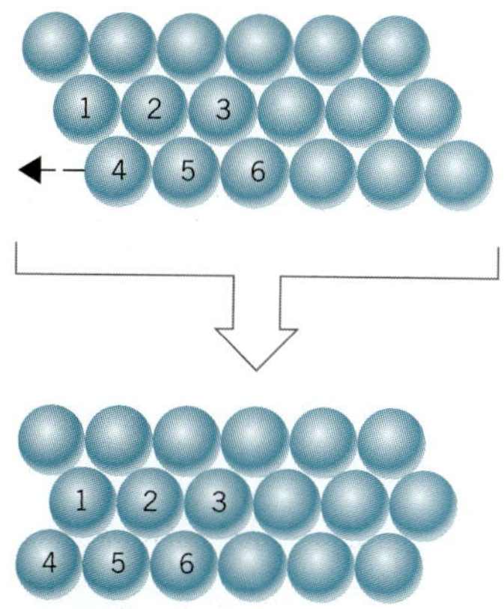

Fig. 9.4 Metals are malleable and ductile because planes of atoms can slip past one another to reach equivalent positions.

Why are metals good conductors of heat and electricity? As we have already seen, the delocalization of valence electrons in a metal allows the solid to conduct an electric current. Metals conduct heat by the movement of electrons. Because the electrons are relatively free to move in metals, the electrons can quickly transport heat throughout the metal.

Table 9.4 summarizes the structure and physical properties of the solids discussed in this chapter.

9.6 Semimetals

The chemist's classification of substances according to the type of bonding—metallic, covalent, or ionic—is not based on a readily measured property. However, there are characteristics that are easily measured and that also serve as a useful way

Students can identify the type of solid from the bond-type triangle and then can describe the structure by recognizing the different interactions between the units of the solid. The bond-type triangle is further used to show how chemists can use electronegativities to predict the properties of a material.

Using the bond-type triangle, students can predict binary atomic combinations that should produce a material having certain characteristics. For example, a combination of Na and Si to produce NaSi gives a compound which falls in the metallic region and is a conducting solid. NaH is ionic and nonconducting. Additional examples of compounds classified by bond type by Sproul can be found in the Instructor's Manual.

Table 9.4

Classifications and Properties of Solids

Classification of the Solid	Primary Type of Bonding	Force Holding Solid Together	Physical Properties	Examples
Molecular	Covalent	Intermolecular forces	Low melting points, electrical insulators	H_2O, Cl_2, HCl, CO_2
Network covalent	Covalent	Covalent bonds	Very high melting points, electrical insulators, very hard	Diamond (C), quartz (SiO_2)
Ionic	Ionic	Ionic bonds	High melting points, electrical conductor in the molten and aqueous state	NaCl, CaO, LiF, $BaCl_2$
Metallic	Metallic	Metallic bonds	Range of melting points, conductors of heat and electricity, lustrous, malleable, ductile	Na, Fe, Al, $CuAl_2$, $BaZn_5$

of categorizing substances. Bond-type triangles take advantage of the correspondence between the bonding classification and observable properties.

The labels *electrolytes, conductors,* and *insulators* that appear in the three corners of the bond-type triangle in Figure 9.1 are observable properties that are closely related to the primary bond type in an element or compound. Metals are conductors because they conduct electricity in both the solid and liquid states. Ionic substances are electrolytes because the ions they contain that are released into solution when they dissolve in water can conduct an electric current. Covalent compounds are poor conductors of electricity and are often insulators because the electrons are localized. They are held tightly between the nuclei that form the covalent bond. Compounds or materials that lie relatively far from one of the vertices of the bond-type triangle can exhibit properties that seem to be a mixture of these categories.

In general, atoms that have small AVEE values (see Chapter 3) have small energy gaps in their valence subshells and tend to form delocalized (metallic) bonds. Atoms that have large AVEE values have a large energy separation of their valence subshells, and these atoms tend to form covalent bonds. Semimetals fall between the extremes of delocalized (metallic) and localized (covalent or ionic) bonding.

Most periodic tables contain a line that separates the elements that are more likely to be metals from those that are more likely to be nonmetals. The elements along the dividing line are called semimetals or metalloids and have properties between those of metals and nonmetals. The semimetals can be found on the bond-type triangle in Figure 9.1 in the region bounded by Al and As. Because these elements lie between those that conduct electricity and those that are insulators, the semimetals are often known as **semiconductors.**

9.7 The Search for New Materials

Materials scientists seek to understand the physical and chemical characteristics of solids and their properties. Often this understanding requires knowledge of chemistry, physics, engineering, and biology. Materials chemistry is concerned with the relationship between structure and the properties and performance of a material. Examples of these materials include piezoelectric crystals that deform when an electric field is applied. Such materials are used in loudspeakers, pressure gauges, and buzzers. Floppy disks and hard drives for computers use other new materials that have come from materials science research. Catalytic converters, sunglasses, and superconductivity all have resulted from an understanding of the atomic structure of solids.

It has been at least 5000 years since the early Egyptians first mixed sand (SiO_2) with the earthy residue left behind when the Nile flooded each year—which contained a mixture of $CaCO_3$, Na_2CO_3, NaCl, and CuO—and then heated the mixture to form **glass.** Glass is a fascinating material. Its structure, in some ways, more closely resembles a liquid than a solid. It has therefore been defined as a material analogous to the liquid state that forms as the result of a reversible change in viscosity as it cools, but which has achieved such a high viscosity that, for all practical purposes, it is rigid. Alternatively, it can be defined as the product of a fusion—or melting process—that has cooled to a rigid condition without forming crystals.

The difference between the structures of sand and glass is one of long-range order. Sand forms a regular crystal. There is an alternating array of Si—O bonds, arranged in a tetrahedral geometry around each silicon atom to form a regular, three-dimensional solid. Each of the oxygen atoms in sand acts as a bridge between two silicon atoms. Because the Si—O bonds are more or less covalent, the result is a network covalent solid—a crystal that can be considered to be a single giant molecule.

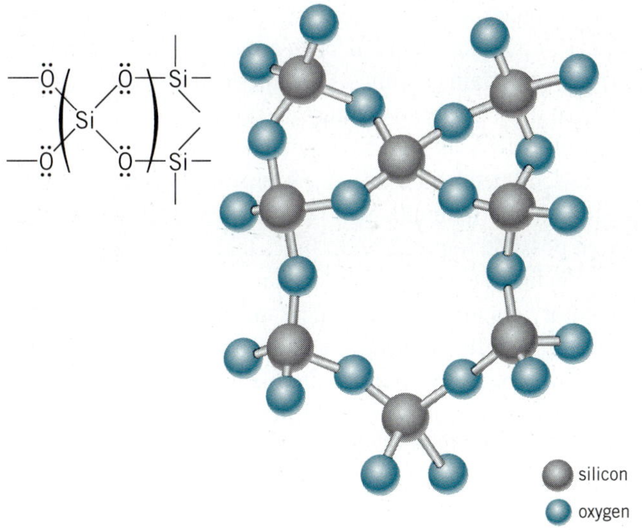

When sand is heated and various impurities are incorporated into the molten liquid that forms, the regular structure of the solid can't re-form. On the average, one in three of the oxygen atoms no longer serves as a bridge between silicon atoms; it is now a terminal oxygen atom. The result is a material that doesn't have the long-range order of a perfect crystal, or even the short-range order of a polycrystalline material. There are enough bonds within the material to make it rigid but not enough to make it crystalline.

Glass is just one example of a family of materials known as **ceramics.** Ceramics are materials that are hard, strong, and light, although they are often brittle. The term *ceramic* comes from a Greek word meaning "burnt stuff," because they are often made in high-temperature reactions. Most ceramics are heat resistant. They are also excellent insulators of both heat and electricity.

Ceramics are usually made from such common minerals as clay and sand under normal conditions and fired at high heat. They are inorganic solids that fall into the nonmetallic classification. As such, the bonding in ceramics can be covalent network or ionic. Ceramics have different properties depending on their atomic structure.

Another basic component for building ceramics that has been used for at least 5000 years is clay, which is essentially a hydrated compound of aluminum and silicon with the empirical formula $H_2Al_2Si_2O_9$. When wet, clay is plastic and can be shaped as desired. The shape is retained when the clay is heated in a kiln. Bricks that are made from clay have a porous structure and fracture easily. If the minerals feldspar ($KAlSi_3O_8$) and quartz (SiO_2) are added to clay before it is fired, a smooth material, known as porcelain, is formed, which has a glasslike surface and is therefore less porous.

In 1891, Edward Acheson synthesized a new class of ceramics by reacting silicon dioxide with an excess of carbon in an electric furnace at 2300 K.

$$SiO_2(s) + 3\ C(s) \longrightarrow SiC(s) + 2\ CO(g)$$

Both the structure and properties of silicon carbide (SiC) are analogous to diamond. Both materials are inert to chemical reactions, except at very high temperatures; both have very high melting points; and both are among the hardest substances known. Shortly after he synthesized silicon carbide, Acheson founded the Carborundum Company to market the material. Then, as now, materials in this class are most commonly used as abrasives.

Most of the ceramics produced each year are used as glass, porcelain, floor tiles, bricks, clay sewage pipes, concrete, cement, or one of the abrasives such as SiC. Since World War II, however, there has been a growing interest in the electrical and magnetic properties of certain ceramics. Ceramics can be made that vary by as much as a factor of 10^{19} in their ability to conduct an electric current. Some ceramics, such as CrO_2, conduct electricity as well as a metal would. Others, such as SiC, located near the covalent–semimetal interface in Figure 9.1, are semiconductors, like the semimetals silicon and germanium. Still others, such as glass and porcelain, are insulators. It is even possible to produce ceramics that have pronounced magnetic field effects yet don't conduct electricity. These ceramics play an important role in the development of memory circuits and permanent magnets.

The properties of solids depend on several factors. One of the first considerations, however, is whether the materials are likely to be ionic, metallic, or covalent solids. There are also intermediate possibilities, such as semimetals, semiconductors, and semielectrolytes. Many of the characteristics of solids can be anticipated on the basis of their position in a bond-type triangle.

The bond-type triangle shown in Figure 9.1 can help us understand the remarkable differences in the properties of solids that might not seem dissimilar from the positions of their elements in the periodic table. Consider BeO and CO, for example. BeO is also known as beryllia; it melts at 2250°C, is very hard, and is a ceramic. Carbon monoxide falls in the covalent region of the triangle and forms a molecular solid at temperatures below −200°C. The electronegativity difference between Be and O is larger than that for C and O, and the average electronegativity is smaller for BeO than for CO. These two conditions place BeO and CO in very different regions of the triangle.

Compounds such as Al_3Mg_2 are located in the metallic region. As we move away from the metallic region, atom combinations produce compounds that become increasingly insulating. Ceramic materials such as SiC, BN, and BeO are found near or along the interface between ionic and covalent areas. Semiconductors are located along the semimetal–covalent boundary, reflecting the changeover from conducting toward insulating materials.

Exercise 9.2

Describe how you would determine what elements might be used to prepare:

(a) a heat resistant, insulating ceramic material

(b) a new semiconductor

(c) a crystalline material that would stand up to high temperatures and not conduct electricity or heat

Solution

(a) The best place to search for such ceramics would be toward the bottom and center of a bond-type triangle such as the one shown in Figure 9.1 or in Figure 5.10. Compounds that lie too close to the right corner would tend to form molecules that lack the long-range order needed to form a ceramic. Compounds that lie close to the left corner would be more likely to form metallic bonds that would make the material a conductor. Compounds that lie close to the top of the triangle would have the long-range order and insulating properties that are desired, but such solids are often too brittle to form useful materials.

Both silicon carbide (SiC) and boron nitride (BN) make ceramic materials. Characteristics of the two ceramics are that they are strong but brittle. Both are poor conductors of electricity, and both are very hard materials. BN is, in fact, comparable in hardness to diamond.

(b) You would look toward the bottom of the triangle. This time you might shift the focus slightly to the left of center. GaAs, for example, lies on the border of the semimetal region and toward the covalent (insulator) region and is a semiconductor (Figure 9.1). The element silicon is a semiconductor used in the manufacture of silicon chips for integrated circuits. GaAs is a new semiconducting material that has certain advantages over silicon. GaP is also promising for use as a semiconductor.

(c) These characteristics are most likely to be met by materials whose atoms have electronegativities that place them in the upper middle regions of a bond-type triangle. Such combinations might be aluminum and oxygen or magnesium and oxygen.

➤ CHECKPOINT

What types of solids are B_4C and MoC? Suggest an application for each of the compounds.

Research for the New Millennium

THE SEARCH FOR HIGH-TEMPERATURE SUPERCONDUCTORS

When an electric current is passed through any material at room temperature, some of the energy of the electrons is dissipated in the form of heat. In metals, resistance to an electric current decreases as the metal is cooled. In 1911, Heike Kamerlingh Onnes found that when mercury is cooled to temperatures below 4.1 K, its resistance falls to zero. Above that temperature, mercury is a conductor of electricity. Below the transition point, it becomes a **superconductor.** By 1913, he found that tin and lead also become superconductors at temperatures below 4 K.

Kamerlingh Onnes recognized the potential of superconductivity for constructing magnets with unusually strong magnetic fields. Standard electromagnets are made by winding a coil of insulated copper wire around an iron alloy core. As the current passes through the copper wire, a magnetic field is created. The field induces an alignment of electrons in the iron alloy core, which in turn produces a magnetic field in the core that is up to 1000 times larger than the field produced by the copper wire. There is an upper limit to the strength of the field that iron alloy magnets can produce, however. The magnets "saturate" at a magnetic field above about 2 Tesla, which is 40,000 times larger than the earth's magnetic field.

Kamerlingh Onnes believed that superconducting magnets could be produced that would achieve much higher fields. Unfortunately, none of the superconducting metals he studied were able to carry enough electric current. It took 50 years before alloys of niobium and tantalum were discovered that could carry the current needed to produce high-field magnets.

Commercial nuclear magnetic resonance spectrometers that used superconducting magnets made from niobium–tantalum alloys became available toward the end of the 1960s. The primary disadvantage of the instruments was the fact that the alloy has to be cooled to the temperature of liquid helium (4.2 K) before it becomes a superconductor. The cost of maintaining one of the instruments could be decreased by as much as a factor of 1000 if it could operate at liquid nitrogen temperatures (77 K).

The search for "high-temperature" superconductors is an important object lesson in the relationship of theory and experiment. At first glance, we might expect ReO_3 and RuO_2 to be insulators, like other metal oxides. In practice, those oxides conduct electricity the way a metal would. In 1964, it was found that other metal oxides, such as NbO and TiO, conduct electricity so well that they become superconductors when cooled to extremely low temperatures (1 K). This is also unexpected, based on their position in a bond-type triangle.

A major step in the evolution of high-temperature superconductors occurred in 1986, when Alex Müller and Georg Bednorz at the IBM Research Laboratory in Zurich discovered that certain ceramic materials that contained lanthanum, barium, copper, and oxygen became superconductors when cooled to temperatures below 35 K. Their results contained two surprises. First, ceramics—such as the plates on which we eat dinner—were considered to be electrical insulators, not conductors. Second, the transition temperature for superconductivity in the new material was higher than that for any known metal or metal alloy.

Within a few years, a family of superconducting ceramics had been discovered that were all based on compounds of copper and oxygen. Müller and Bednorz worked with ceramics that were derivatives of a compound with the formula La_2CuO_4. If forced to assign oxidation states to the compound, most chemists would write it as $[La^{3+}]_2[Cu^{2+}][O^{2-}]_4$. The parent compound is an insulator. When some of the lanthanum atoms are replaced with barium atoms, however, a nonstoichiometric superconductor with the formula $La_{2-x}Ba_xCuO_4$ is obtained.

Applying the concept of oxidation states to the compound, we are formally replacing an La^{3+} ion by a Ba^{2+} ion each time a barium atom is incorporated into the structure. If the net charge on the compound is going to stay the same, the oxidation state of the copper atom must increase. Each time an La^{3+} ion is replaced with a Ba^{2+} ion, a Cu^{2+} ion becomes a Cu^{3+} ion. Müller and Bednorz found that when enough barium had been incorporated to raise the average oxidation state of the copper to +2.2, the compound became a superconductor at low temperatures.

The electron that is formally removed from the copper atom is apparently delocalized and therefore capable of moving through the solid when it is cooled to low temperatures. Similar results can be obtained by incorporating either strontium or calcium into La_2CuO_4. $La_{1.8}Sr_{0.2}CuO_4$ has the highest transition temperature of any member of the family: 40 K.

Superconductivity has also been observed with a family of compounds known as 1-2-3 superconductors. The first member of the family was discovered in 1987, when $YBa_2Cu_3O_7$ was found to be a superconductor when cooled to

95 K—above the temperature of liquid nitrogen. (The common name of the superconductors is based on the fact that there are three metals in a 1:2:3 ratio.) The compound also contains copper in a fractional oxidation state. The yttrium atom can be assumed to exist in the +3 oxidation state. Thus one Y^{3+} and two Ba^{2+} ions contribute a charge of +7 toward balancing the charge of −14 on the seven oxygens. The remaining charge of +7 has to be distributed over the three copper atoms, for an average oxidation state of +2.33.

The most common demonstration of high-temperature superconductors is based on the Meissner effect. In 1933, the German physicists W. Meissner and R. Ochsenfeld found that superconductors repel an external magnetic field. On the macroscopic scale, the superconductor seems to repel the magnet that produced the magnetic field. As a result, when one of the high-temperature superconductors is cooled with liquid nitrogen, it levitates off the surface of a magnet.

9.8 The Structure of Metals and Other Monatomic Solids

The structure of a solid is determined by the way the units that compose the solid are arranged in three-dimensional space. Therefore, several sections of this chapter are devoted to a description of lattices and efficient ways of packing atoms, ions, or molecules.

We can describe the structure of pure metals by assuming that the atoms of the metals are identical perfect spheres. The same model can be used to describe the structure of the solid noble gases (He, Ne, Ar, Kr, Xe) at low temperatures. These substances all crystallize in one of four basic structures, known as simple cubic (SC), body-centered cubic (BCC), hexagonal closest packed (HCP), and cubic closest packed (CCP).

Solids are very difficult to compress because the amount of space between particles in a solid is at a minimum. As a rule, therefore, we can conclude that the most probable structure for a solid is the structure that makes the most efficient use of space. When a solid crystallizes, the particles that form the solid pack as tightly as possible. To illustrate the principle, let's try to imagine the best way to pack spheres, such as Ping-Pong balls, into an empty box.

One approach involves carefully packing the Ping-Pong balls to form a square-packed plane of spheres, as shown in Figure 9.5.

A second plane of spheres can be stacked directly on top of the first. The result is a regular structure in which the simplest repeating unit is a cube of eight spheres, as shown in Figure 9.6. The structure is called **simple cubic packing.** Each sphere in the structure touches four identical spheres in the same plane. It also touches one sphere in the plane above and one in the plane below. Each sphere is therefore said to have a **coordination number** of 6. If the spheres represent atoms, each atom in the structure can form bonds to its six nearest neighbors.

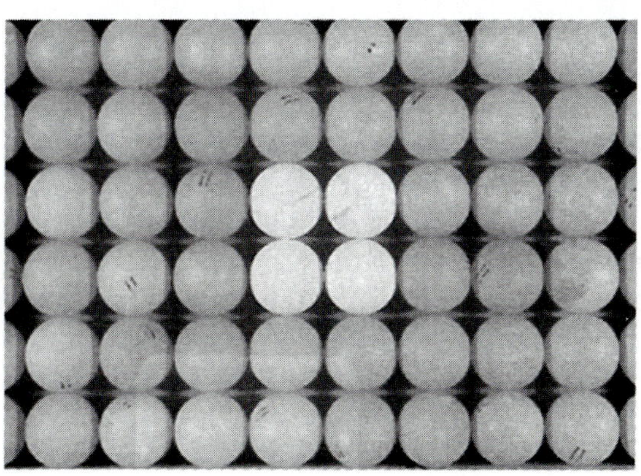

Fig. 9.5 A square-packed plane of spheres.

Fig. 9.6 A simple cubic packing of spheres.

One way to decide whether the simple cubic structure is an efficient way of packing spheres is to ask: What happens when we shake the box? Do the Ping-Pong balls stay in the same positions, or do they settle into a different structure? It is fairly easy to show that a simple cubic structure isn't an efficient way of using space. Only 52% of the available space is actually occupied by the spheres in a simple cubic structure. The rest is empty space. Because the structure is inefficient, only one element—polonium—crystallizes in a simple cubic structure.

This raises an interesting question: How can we use space more efficiently? Another approach starts by separating the spheres to form a square-packed plane in which the spheres do not quite touch each other, as shown in Figure 9.7.

Fig. 9.7 A square-packed plane in which the spheres do not quite touch.

The spheres in the second plane pack above the holes in the first plane, as shown in Figure 9.8. Spheres in the third plane pack above holes in the second plane. Spheres in the fourth plane pack above holes in the third plane, and so on. The result is a structure in which the odd-numbered planes of atoms are identical and the even-numbered planes are identical. The *ABABABAB* . . . repeating structure of square-packed planes is known as **body-centered cubic packing.**

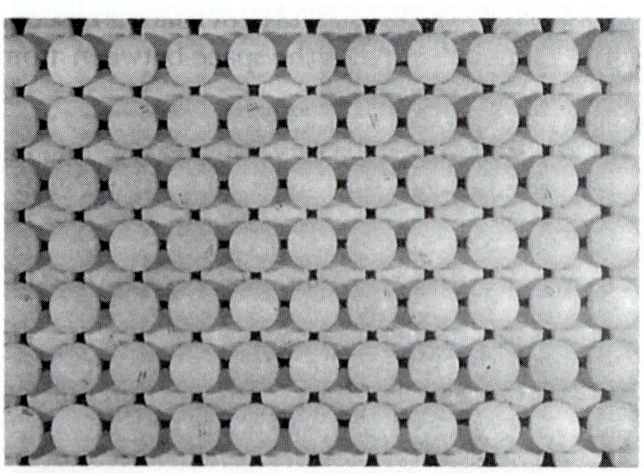

Fig. 9.8 The spheres in the second plane of a body-centered cubic structure pack above the holes in the plane shown in Figure 9.7.

This structure is called *body-centered cubic* because each sphere touches four spheres in the plane above and four more in the plane below, arranged toward the corners of a cube. Thus, the repeating unit in the structure is a cube of eight spheres with a ninth identical sphere in the center of the body—in other words, a body-centered cube, as shown in Figure 9.9. The coordination number in this structure is 8.

Body-centered cubic packing is a more efficient way of using space than simple cubic packing, as 68% of the space in the structure is filled. Body-centered cubic packing is an important structure for metals. All of the metals in Group IA (Li, Na, K, Rb, Cs, and Fr), barium in Group IIA, and a number of the early transition metals (such as V, Cr, Mo, W, and Fe) pack in a body-centered cubic structure.

Two structures pack spheres so efficiently they are called **closest-packed structures.** Both start by packing the spheres in planes in which each sphere touches six others oriented toward the corners of a hexagon, as shown in Figure 9.10.

A second plane is then formed by packing spheres above the triangular holes in the first plane, as shown in Figure 9.11.

What about the next plane of spheres? The spheres in the third plane could pack *directly above the spheres in the first plane* to form an *ABABABAB* . . . repeating structure. Because such a structure is composed of alternating planes of hexagonal closest packed spheres, it is called a **hexagonal closest packed** structure. Each sphere touches three spheres in the plane above, three spheres in the plane below, and six spheres in the same plane, as shown in Figure 9.12. Thus, the coordination number in a hexagonal closest packed structure is 12, and 74% of the space in a hexagonal closest packed structure is filled. No more efficient way of packing spheres is known, and the hexagonal closest packed structure is important for such metals as Be, Co, Mg, and Zn, as well as the rare gas He at low temperatures.

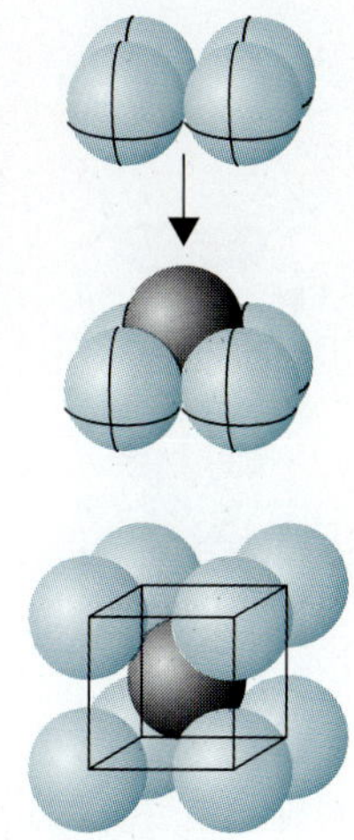

Fig. 9.9 Body-centered cubic structure. All spheres represent identical atoms.

Fig. 9.10 A closest packed plane in which each sphere touches six others oriented toward the corners of a hexagon.

There is another way of stacking hexagonal closest packed planes of spheres. The atoms in the *third plane can be packed above the holes in the first plane* that weren't used to form the second plane. The fourth hexagonal closest packed plane of atoms then packs directly above the first. The net result is an *ABCABCABC* . . .

> **➤ CHECKPOINT**
>
> How does the packing in Figure 9.10 differ from the packing in Figure 9.5?

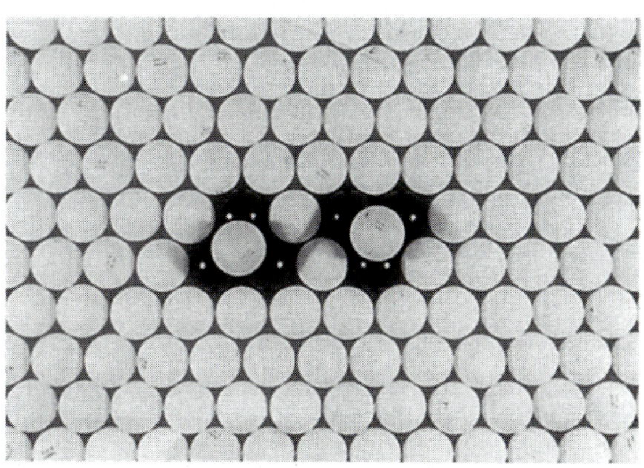

Fig. 9.11 Atoms in the second plane of closest packed structures pack above the triangular holes in the first plane shown in Figure 9.10.

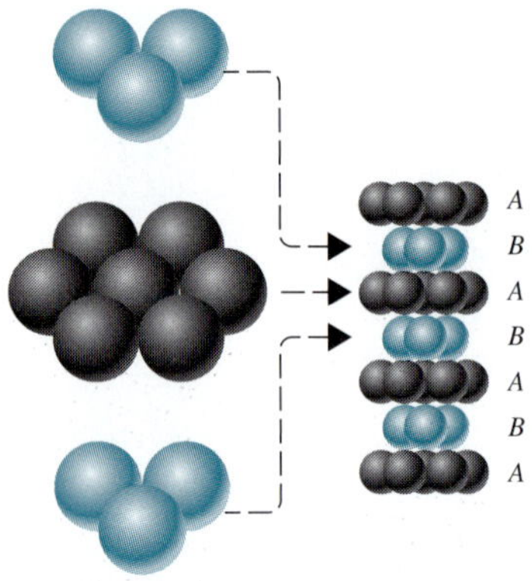

Hexagonal closest packed structure

Fig. 9.12 Each atom in a hexagonal closest packed structure touches six atoms in the same plane, three in the plane above, and three in the plane below. The result is an *ABABAB* . . . repeating pattern of closest packed planes. All spheres represent identical atoms.

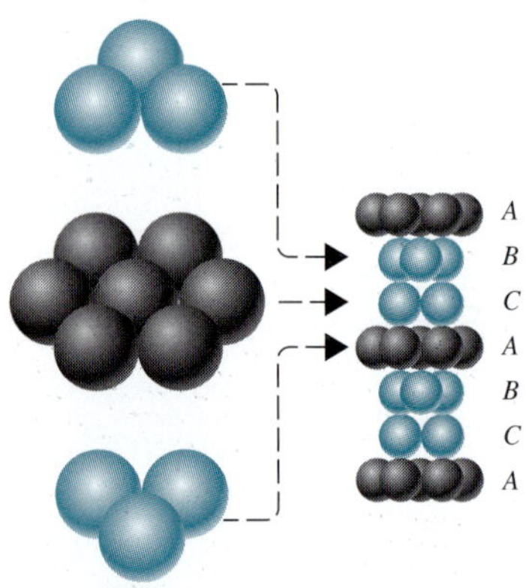

Cubic closest packed structure

Fig. 9.13 Each atom in a cubic closest packed structure also touches six atoms in the same plane, three in the plane above, and three in the plane below. But the atoms in the top plane are rotated by 180° relative to the bottom plane. The planes of atoms therefore form an *ABCABCABC* . . . repeating pattern. All spheres represent identical atoms.

structure, which is called **cubic closest packed.** Each sphere in the structure touches six others in the same plane, three in the plane above, and three in the plane below, as shown in Figure 9.13. Thus, the coordination number is still 12.

The difference between hexagonal and cubic closest packed structures can be understood by comparing Figures 9.12 and 9.13. In the hexagonal closest packed structure, the atoms in the first and third planes lie directly above each other. In the cubic closest packed structure, the atoms in those planes are oriented in different directions.

The cubic closest packed structure is just as efficient as the hexagonal closest packed structure. (Both use 74% of the available space.) Many metals, including Ag, Al, Au, Ca, Cu, Ni, Pb, and Pt, crystallize in a cubic closest packed structure. All the rare gases except helium behave in the same manner when cooled to temperatures low enough to allow solidification.

9.9 Coordination Numbers and the Structures of Metals

The coordination numbers of the four structures described in the preceding section are summarized in Table 9.5.

Table 9.5
Coordination Numbers for Common Crystal Structures

Structure	Coordination Number	Stacking Pattern
Simple cubic	6	*AAAAAAAA* . . .
Body-centered cubic	8	*ABABABAB* . . .
Hexagonal closest packed	12	*ABABABAB* . . .
Cubic closest packed	12	*ABCABCABC* . . .

Some metals pack in hexagonal or cubic closest packed structures. Not only do those structures use space as efficiently as possible, they also have the largest possible coordination numbers, which allows each metal atom to form bonds to the largest number of neighboring metal atoms.

It is less obvious why one-third of the metals pack in a body-centered cubic structure, in which the coordination number is only 8. The popularity of the body-centered cubic structure can be understood by referring to Figure 9.14. The coordination number for body-centered cubic structures given in Table 9.5 counts only the atoms that actually touch a given atom in the structure. Figure 9.14 shows that each atom also *almost touches* four neighbors in the same plane, a fifth neighbor two planes above, and a sixth two planes below. The distance from each atom to the nuclei of the nearby atoms is only 15% larger than the distance to the nuclei of the atoms that it actually touches. Each atom in a body-centered cubic structure therefore interacts with 14 other atoms—eight strong interactions to the atoms that it touches and six weaker interactions to the atoms it almost touches.

This makes it easier to understand why a metal might prefer the body-centered cubic structure to the hexagonal or cubic closest packed structure. Each metal atom in the closest packed structures interacts with 12 neighboring atoms. In the body-centered cubic structure, each atom interacts with 14 neighboring atoms.

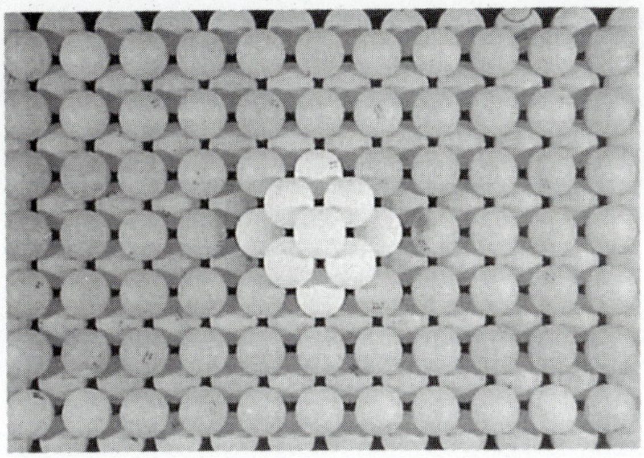

Fig. 9.14 Each atom in a body-centered cubic structure touches four atoms in the plane above and four in the plane below. In addition, each atom *almost* touches six more atoms.

9.10 Unit Cells: The Simplest Repeating Unit in a Crystal

So far, our description of solids has focused on the way the particles pack to fill space. Another way of describing the structures of solids was introduced in Section 5.8. This approach assumes that crystals are three-dimensional analogs of a piece of wallpaper. Wallpaper has a regular repeating design that extends from one edge to the other. Crystals have a similar repeating design, but in this case the design extends in three dimensions from one edge of the solid to the other.

We can unambiguously describe a piece of wallpaper by specifying the size, shape, and contents of the simplest repeating unit in the design. We can describe a three-dimensional crystal by specifying the size, shape, and contents of the simplest repeating unit and the way the repeating units stack to form the crystal. The simplest repeating unit in a crystal is called a **unit cell,** which is defined in terms of **lattice points**—the points in space about which the particles are free to vibrate in a crystal.

This section focuses on the three unit cells shown in Figure 9.15: simple cubic, body-centered cubic, and face-centered cubic. These unit cells are important for two reasons. First, many metals, ionic solids, and intermetallic compounds crystallize in cubic unit cells. Second, because these unit cells have identical edge lengths for a given cubic cell and the cell angles are all 90°, calculations based on these structures are somewhat easier to do than calculations based on more complex unit cells.

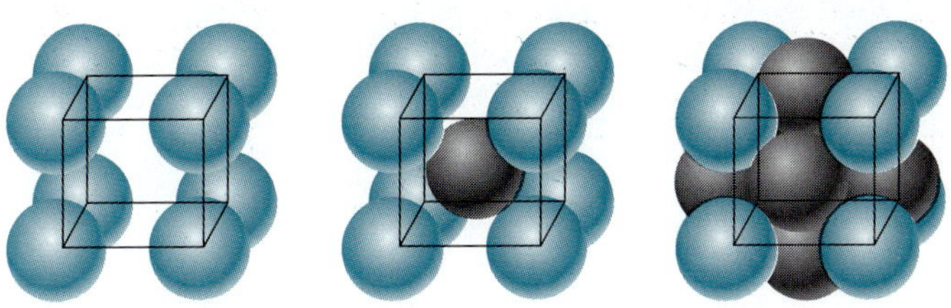

Fig. 9.15 Models of simple cubic (left), body-centered cubic (center), and face-centered cubic unit cells (right). All spheres represent identical atoms.

The **simple cubic unit cell** is the simplest repeating unit in a simple cubic structure. Each corner of the unit cell is defined by a lattice point at which an identical particle can be found. By convention, the edge of a unit cell always connects equivalent points. Each of the eight corners of the unit cell therefore must contain an identical particle.

The **body-centered cubic unit cell** is the simplest repeating unit in a body-centered cubic structure. Once again, there are eight identical particles on the eight corners of the unit cell. In this case, however, there is a ninth identical particle in the center of the body of the unit cell.

The **face-centered cubic unit cell** also starts with identical particles on the eight corners of the cube. But the structure also contains the same particles in the centers of the six faces of the unit cell, for a total of 14 identical lattice points. The face-centered cubic unit cell is the simplest repeating unit in a cubic closest packed structure. In fact, the presence of face-centered cubic unit cells in the structure explains why the structure is known as *cubic* closest packed.

> **CHECKPOINT**
>
> Iron metal and cesium chloride have similar structures. The simplest repeating unit in iron is a cube of eight iron atoms with a ninth iron atom in the center of the body of the cube. The simplest repeating unit in CsCl is a cube of Cl^- ions with a Cs^+ ion in the center of the body. Explain why one of the structures is classified as a body-centered cubic unit cell and the other as a simple cubic unit cell.

9.11 Measuring the Distance Between Particles in a Unit Cell

Nickel was identified in Section 9.8 as one of the metals that crystallizes in a cubic closest packed structure. When we consider that a nickel atom has a mass of only 9.75×10^{-23} g and a radius of only 1.24×10^{-10} m, it is a remarkable achievement for us to be able to describe the structure of the metal. The obvious question is: How do we know that nickel packs in a cubic closest packed structure?

The only way to determine the structure of matter on an atomic scale is to use a probe that is even smaller. As we have seen in Chapter 3, one of the most useful probes for studying matter on the atomic scale is electromagnetic radiation. In 1912, Max von Laue found that X rays which struck the surface of a crystal were diffracted into patterns that resembled the patterns produced when light passes through a very narrow slit. Shortly thereafter, William Lawrence Bragg, who was just completing an undergraduate degree in physics at Cambridge University, explained von Laue's results. Bragg argued that X rays were reflected from planes of atoms near the surface of the crystal, as shown in Figure 9.16. He then concluded that the only way the X rays could stay in phase was if some integer (n) times the wavelength of the radiation (λ) was twice the distance (d) between adjacent planes of atoms times the sine of the angle θ.

$$n\lambda = 2d \sin \theta$$

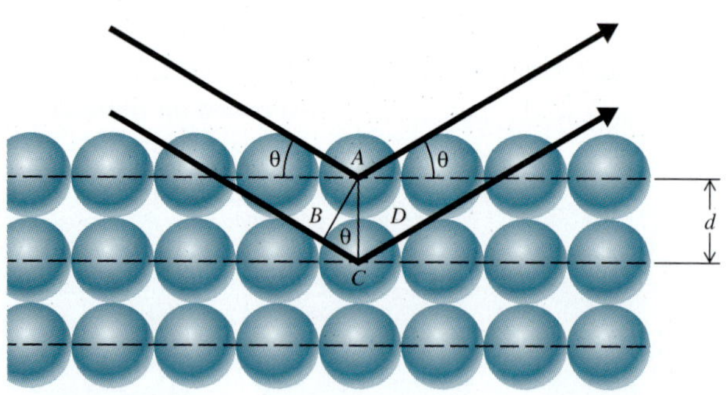

Fig. 9.16 Diffraction of X rays by the first and second planes in a crystal.

This relationship, which became known as the **Bragg equation,** allows us to calculate the distance between planes of atoms in a crystal from the pattern of diffraction of X rays of known wavelength.

The pattern by which X rays are diffracted by nickel metal suggests that the metal packs in a cubic unit cell. The length of the edge of a unit cell in the crystal is 0.3524 nm. Knowing that nickel crystallizes in a cubic unit cell isn't enough to determine the structure of Ni metal. We still have to decide whether it is a simple cubic, body-centered cubic, or face-centered cubic unit cell. As we'll see in the next section, we can do this by measuring the density of the metal.

> **CHECKPOINT**
>
> Why does the diffraction of X rays by solids show that solids consist of an ordered arrangement of particles?

9.12 Determining the Unit Cell of a Crystal

Atoms on the corners, edges, and faces of a unit cell are shared by more than one unit cell, as shown in Figure 9.17. An atom on a face is shared by two unit cells, so only half of the atom belongs to each of the cells. An atom on an edge is shared by four unit cells, and an atom on a corner is shared by eight unit cells. Thus, only one-quarter of an atom on an edge and one-eighth of an atom on a corner can be assigned to each of the unit cells that share the atoms.

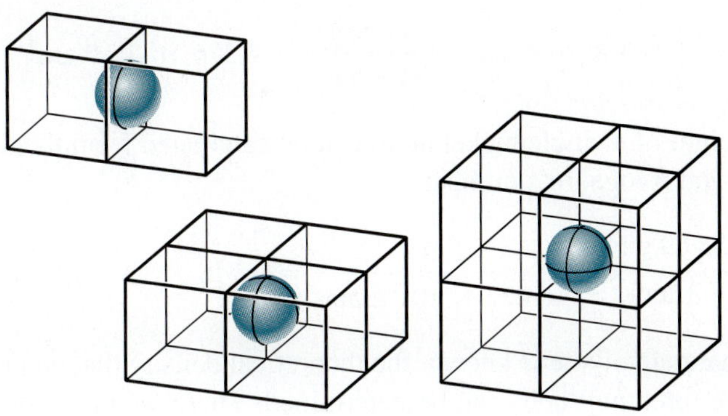

Fig. 9.17 Because an atom on the face of a unit cell is shared by two unit cells, only one-half of the atom belongs to each of the cells. For similar reasons, one-quarter of an atom on the edge of a unit cell, and one-eighth of an atom on the corner of a unit cell, belong to the unit cell.

If nickel crystallized in a simple cubic unit cell, there would be a nickel atom on each of the eight corners of the cell. Because only one-eighth of each atom could be attributed to a given unit cell, each unit cell in a simple cubic structure would have one net nickel atom.

Simple cubic structure:

$$8 \text{ corners} \times \frac{1}{8} = 1 \text{ net atom/unit cell}$$

If nickel formed a body-centered cubic structure, there would be two atoms per unit cell, because the nickel atom in the center of the body wouldn't be shared with any other unit cells.

Body-centered cubic structure:

$$\left(8 \text{ corners} \times \frac{1}{8}\right) + 1 \text{ body} = 2 \text{ net atoms/unit cell}$$

If nickel crystallized in a face-centered cubic structure, the six atoms on the faces of the unit cell would contribute three net nickel atoms, for a total of four atoms per unit cell.

Face-centered cubic structure:

$$\left(8 \text{ corners} \times \frac{1}{8}\right) + \left(6 \text{ faces} \times \frac{1}{2}\right) = 4 \text{ net atoms/unit cell}$$

Because they have different numbers of atoms in a unit cell, each of the structures would have a significantly different density. Let's therefore calculate the theoretical density for nickel on the basis of each structure and the unit cell edge length for nickel given in the previous section: 0.3524 nm. To do this, we need to know the volume of a unit cell in cubic centimeters and the weight of a single nickel atom. The type of cubic cell associated with nickel can now be found by calculating the theoretical density of nickel for each cell. The calculated density that matches the experimental density will give us the correct unit cell for nickel.

The volume (V) of the unit cell is equal to the cell edge length (a) cubed.

$$V = a^3 = (0.3524\text{ nm})^3 = 0.04376\text{ nm}^3$$

Because there are 10^9 nm in a meter and 100 cm in a meter, there must be 10^7 nm in a centimeter.

$$\frac{10^9\text{ nm}}{1\text{ m}} \times \frac{1\text{ m}}{100\text{ cm}} = 10^7\text{ nm/cm}$$

Converting the volume of the unit cell to cubic centimeters gives the following result.

$$4.376 \times 10^{-2}\text{ nm}^3 \times \frac{(1\text{ cm})^3}{(10^7\text{ nm})^3} = 4.376 \times 10^{-23}\text{ cm}^3$$

The weight of a single nickel atom can be calculated from the atomic weight of the metal and Avogadro's number

$$\frac{58.69\text{ g Ni}}{1\text{ mol}} \times \frac{1\text{ mol}}{6.022 \times 10^{23}\text{ atoms}} = 9.746 \times 10^{-23}\text{ g/atom}$$

Now that the volume is known, the theoretical density that nickel would have in each of the three unit cells can be determined. There would be only one nickel atom per unit cell if nickel crystallized in a simple cubic unit cell. The density of nickel, if it crystallized in a simple cubic structure, would therefore be 2.227 g/cm^3.

Simple cubic structure:

$$\frac{9.746 \times 10^{-23}\text{ g/unit cell}}{4.376 \times 10^{-23}\text{ cm}^3\text{/unit cell}} = 2.227\text{ g/cm}^3$$

Because there would be twice as many nickel atoms per unit cell if nickel crystallized in a body-centered cubic structure, the density of nickel in this structure would be twice as large as for the simple cubic.

Body-centered cubic structure:

$$\frac{2(9.746 \times 10^{-23}\text{ g/unit cell})}{4.376 \times 10^{-23}\text{ cm}^3\text{/unit cell}} = 4.454\text{ g/cm}^3$$

There would be four nickel atoms per unit cell in a face-centered cubic structure, and the density of nickel in this structure would be four times as large as for the simple cubic.

Face-centered cubic structure:

$$\frac{4(9.746 \times 10^{-23}\text{ g/unit cell})}{4.376 \times 10^{-23}\text{ cm}^3\text{/unit cell}} = 8.909\text{ g/cm}^3$$

➤ CHECKPOINT

What three factors account for the density of a metal?

The experimental value for the density of nickel is 8.90 g/cm^3. The obvious conclusion is that nickel crystallizes in a face-centered cubic unit cell and therefore has a cubic closest packed structure.

9.13 Calculating the Size of an Atom or Ion

Estimates of the radii of most metal atoms can be found in Appendix B.4. Where do these data come from? How do we know, for example, that the metallic radius of a nickel atom is 0.1246 nm? The starting point for calculating the metallic radius of an atom uses the results of the previous two sections. We now know that nickel crystallizes in a cubic unit cell with a cell edge length of 0.3524 nm, and we know that the unit cell for the crystal is face-centered cubic.

One of the faces of a face-centered cubic unit cell is shown in Figure 9.18. According to Figure 9.18, the diagonal across the face of the unit cell is equal to four times the metallic radius of a nickel atom.

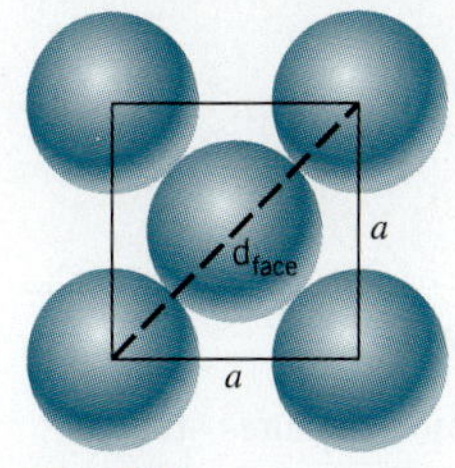

Fig. 9.18 The diagonal across the face of a face-centered cubic unit cell is equal to four times the radius of the atoms that form the cell.

$$d_{face} = 4\ r_{Ni}$$

The Pythagorean theorem states that the long edge of a right triangle is equal to the sum of the squares of the other sides. The diagonal across the face of the unit cell is therefore related to the unit cell edge length, a, by the following equation.

$$d_{face}^2 = a^2 + a^2 = 2a^2$$

Taking the square root of both sides gives the following result.

$$d_{face} = a\sqrt{2}$$

Because the diagonal across the face is four times the metallic radius of a nickel atom, the following substitution can be made.

$$4r_{Ni} = a\sqrt{2}$$

Thus, the metallic radius of a nickel atom is 0.1246 nm.

$$r_{Ni} = \frac{a\sqrt{2}}{4} = \frac{0.3524\text{ nm} \times \sqrt{2}}{4} = 0.1246\text{ nm}$$

Ionic solids have structures similar to those discussed in Section 9.8. However, ionic solids are composed of two or more different ions. We can describe the structure of ionic solids by assuming that the ions are perfect spheres but of different sizes (see Figure 9.3). A similar approach therefore can be taken for estimating the size of an ion. Consider cesium chloride, for example, which crystallizes in a simple cubic unit cell of Cs^+ ions with a Cl^- ion in the center of the body of the cell, as shown in Figure 9.19.

The Cl^- ion in the center of the unit cell must touch the Cs^+ ions at the corners. The diagonal across the body of the CsCl unit cell is therefore equal to the sum of the radii of two Cs^+ ions and two times the radius of a Cl^- ion.

$$d_{body} = 2\ r_{Cs^+} + 2\ r_{Cl^-}$$

The three-dimensional equivalent of the Pythagorean theorem suggests that the square of the diagonal across the body of a cube is the sum of the squares of the three sides.

$$d_{body}^2 = a^2 + a^2 + a^2 = 3a^2$$

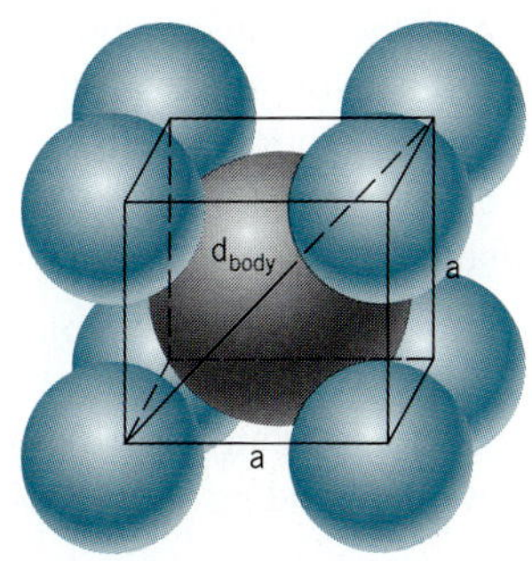

Fig. 9.19 The diagonal across the body of the CsCl unit cell is equal to twice the sum of the radii of the Cs^+ and Cl^- ions.

Taking the square root of both sides of the equation gives the following result.

$$d_{body} = a\sqrt{3}$$

The cell edge length in CsCl has been experimentally determined to be 0.4123 nm. Therefore, the diagonal across the body in the unit cell is 0.7141 nm.

$$d_{body} = a\sqrt{3} = 0.4123\text{ nm} \times \sqrt{3} = 0.7141\text{ nm}$$

The sum of the ionic radii of Cs^+ and Cl^- ions is half that distance, or 0.3571 nm.

$$r_{Cs^+} + r_{Cl^-} = \frac{d_{body}}{2} = \frac{0.7141 \text{ nm}}{2} = 0.3571 \text{ nm}$$

If we had an estimate of the size of either the Cs^+ or Cl^- ion, we could use the results of our calculation to estimate the size of the other ion. By combining the analysis of many ionic compounds, it is possible to create a set of consistent data for the size of the ions that form the crystals. Some of the data were reported in Section 3.21; a more complete set can be found in Appendix B.4. The small discrepancy between the sum of the ionic radii of the Cs^+ ion (0.169 nm) and the Cl^- ion (0.181 nm) reported in tables of ionic radii and the results of the calculation for CsCl reflect the fact that ionic radii seem to vary slightly from one crystal to another.

➤ CHECKPOINT

For the three cubic structures studied, which has the simplest relationship between the radii of the ions and the length of the edge of the unit cell?

Key Terms

Amorphous solid
Body-centered cubic unit cell
Body-centered cubic packing
Bragg equation
Ceramic
Closest-packed structure
Coordination number
Crystalline solid
Cubic closest packing
Electron affinity
Face-centered cubic unit cell
Glass
Hexagonal closest packing
Ionic solid
Lattice energy
Lattice points
Metallic bond
Metallic solid
Molecular solid
Network covalent solid
Polycrystalline solid
Semiconductor
Simple cubic packing
Simple cubic unit cell
Sublimes
Superconductor
Unit cell

Problems

Solids

1. What are the three categories of solids that are based on the forces that hold the particles together?
2. What three types of bonds can hold particles together?
3. What category of compounds is formed by nonmetals? By metals? By metals combining with nonmetals?

Molecular and Network Covalent Solids

4. Which of the following solids are held together by an extended network of covalent bonds?
   (a) sodium chloride (b) $CuAl_2$
   (c) gold (d) calcium carbonate
   (e) diamond (f) dry ice (solid CO_2)
5. Which force must be overcome to sublime dry ice, solid CO_2?
   (a) metallic bonding (b) ionic bonding
   (c) covalent bonding (d) dispersion forces
6. Which force must be overcome to melt solid pentane, C_5H_{12}?
   (a) metallic bonding (b) ionic bonding
   (c) covalent bonding (d) dispersion forces

Ionic Solids

7. Define the term *lattice energy.*
8. The lattice energy of NaCl refers to which of the following reactions?
   (a) $2\ Na(s) + Cl_2(s) \longrightarrow 2\ NaCl(s)$
   (b) $NaCl(s) \longrightarrow Na(g) + Cl(g)$
   (c) $Na(g) + Cl(g) \longrightarrow NaCl(g)$
   (d) $NaCl(s) \longrightarrow Na^+(g) + Cl^-(g)$
   (e) $Na^+(g) + Cl^-(g) \longrightarrow NaCl(s)$
9. Which of the following salts has the largest lattice energy?
   (a) LiF (b) LiCl (c) LiBr (d) LiI
10. Which of the following salts has the largest lattice energy?
    (a) NaCl (b) NaI (c) KI
    (d) MgO (e) MgS
11. Explain the following trends in lattice energies.
    (a) $MgF_2 > MgCl_2 > MgBr_2 > MgI_2$
    (b) $BeF_2 > MgF_2 > CaF_2 > SrF_2 > BaF_2$
12. Use the *CRC Handbook of Chemistry and Physics*[2] to determine the solubility in water of NaF, NaCl, NaBr, and NaI. Describe the relationship between the solubilities of the salts and their lattice energies.

[2]CRC Press, Boca Raton, Florida.

13. Use lattice energies to explain why MgO is much less soluble in water than is CaO.
14. Using the enthalpy data in Appendices B.13 and B.5 and knowing that the enthalpy change for the reaction $O(g) + 2\,e^- \longrightarrow O^{2-}(g)$ is +448 kJ/mol$_{rxn}$, calculate the lattice enthalpies for MgO, CaO, and BaO. Explain the trend you observe.
15. One of the simplest ways of distinguishing between two covalent compounds is to measure their melting points or boiling points. Naphthalene melts at 80.5°C and camphor melts at 179.8°C, for example. Would you expect the melting points and boiling points of ionic compounds to be higher, lower, or about the same as covalent compounds? Explain.

Metallic Solids

16. What are delocalized electrons?
17. What is a metallic bond? What types of atoms are most likely to form metallic bonds?
18. Describe the bonding in a substance that is held together by metallic bonds.
19. Molecular, ionic, and network covalent solids all have one characteristic in common that makes them different from metallic solids. What is this characteristic?

Physical Properties That Result from the Structure of Metals

20. Explain why metals are usually solids at room temperature.
21. Explain why metals are malleable and ductile.
22. Explain why metals conduct heat and electricity.
23. Which of the following categories is most likely to contain a compound that is a poor conductor of electricity when solid but a very good conductor when molten?
    (a) molecular solids (b) covalent solids
    (c) ionic solids (d) metallic solids

Semimetals

24. Why do ionic compounds conduct electricity better in the liquid state than the solid state?
25. Which one(s) of the following compounds should conduct an electric current when dissolved in water?
    (a) $MgCl_2$ (b) CO_2 (c) CH_3OH
    (d) KNO_3 (e) Ca_3P_2
26. Which of the following would you expect to conduct an electric current?
    (a) solid Na metal (b) liquid Na metal
    (c) solid NaCl (d) liquid NaCl
    (e) NaCl dissolved in water
27. Use Figures 5.10 and 9.1 to describe the characteristics of a semimetal.

The Search for New Materials

28. Where on the periodic table are the atoms that are most likely to be involved in the formation of ceramics located?
29. Where on the periodic table are the atoms that are most likely to form semiconductors located?
30. Use a bond-type triangle to classify the following compounds and describe the characteristics of each.
    (a) CrO_2 (b) SiC (c) GaP (d) BeO
31. Suggest two elements that might be combined to produce the following.
    (a) a material with a high melting point that is not an electrical conductor
    (b) an insulating material that has a low melting point
    (c) a conductor of electricity in the solid state

The Structure of Metals and Other Monatomic Solids

32. Describe the difference in the way planes of atoms stack to form *hexagonal closest packed, cubic closest packed, body-centered cubic,* and *simple cubic* structures.
33. Explain why the structure of polonium is called *simple cubic;* why the structure of iron is called *body-centered cubic;* and why the structure of cobalt is called *hexagonal closest packed.*
34. Determine the coordination numbers of the metal atoms in each of the following structures.
    (a) cubic closest packed aluminum
    (b) hexagonal closest packed magnesium
    (c) body-centered cubic chromium
    (d) simple cubic polonium
35. In which of the following structures would a xenon atom form the largest number of induced dipole–induced dipole interactions?
    (a) simple cubic
    (b) body-centered cubic
    (c) cubic closest packed
    (d) hexagonal closest packed
36. Sodium crystallizes in a structure in which the coordination number is 8. Which structure best describes the crystal?
    (a) simple cubic
    (b) body-centered cubic
    (c) cubic closest packed
    (d) hexagonal closest packed

Coordination Numbers and the Structure of Metals

37. List three common structures for metals.
38. Define *coordination number.*
39. Roughly sketch a simple cubic packing of spheres. Show how to find the coordination number of a representative sphere.

40. Roughly sketch a body-centered cubic structure. Show how to find the coordination number of a representative sphere.
41. Why might a given metal prefer one type of structure to another?

Unit Cells: The Simplest Repeating Unit in a Crystal

42. What is a lattice point?
43. Define *unit cell.*
44. There are three common types of unit cells. What are they? What are the coordination numbers in each?
45. Draw the unit cell for CsCl and for Cs. Classify each according to bond type. What are the differences between the two unit cells? Specify the identity of the particles that occupy the lattice points for each.

Measuring the Distance Between Particles in a Unit Cell

46. In addition to knowing the length of an edge of a unit cell, what else must be known to calculate the size of an atom in the cell?
47. Why is the density of a metal related to how close the particles are to one another?
48. If the particles composing a solid are not ordered, would there be a pattern in X-ray diffraction data?

Determining the Unit Cell of a Crystal

49. Silver crystallizes in a face-centered cubic unit cell with an edge length of 0.40862 nm. Calculate the density of Ag metal in grams per cubic centimeter.
50. Potassium crystallizes in a cubic unit cell with an edge length of 0.5247 nm. The density of potassium is 0.856 g/cm^3. Determine whether the element crystallizes in a simple cubic, a body-centered cubic, or a face-centered cubic unit cell.
51. Determine whether calcium crystallizes in a simple cubic, a body-centered cubic, or a face-centered cubic unit cell, given that the cell edge length is 0.5582 nm and the density of the metal is 1.55 g/cm^3.
52. Determine whether molybdenum crystallizes in a simple cubic, a body-centered cubic, or a face-centered cubic unit cell, given that the cell edge length is 0.3147 nm and the density of the metal is 10.2 g/cm^3.
53. Which of the following metals crystallizes in a face-centered cubic unit cell with an edge length of 0.3608 nm if the density of the metal is 8.95 g/cm^3?
(a) Na (b) Ca (c) Tl (d) Cu (e) Au
54. CdO crystallizes in a cubic unit cell with a cell edge length of 0.4695 nm. Calculate the number of Cd^{2+} and O^{2-} ions per unit cell. The density of the crystal is 8.15 g/cm^3.
55. LiF crystallizes in a cubic unit cell with a cell edge length of 0.4017 nm. Calculate the number of Li^+ and F^- ions per unit cell, if the density of the salt is 2.640 g/cm^3.
56. The metallic radius of a vanadium atom is 0.1321 nm. What is the density of vanadium if the metal crystallizes in a body-centered cubic unit cell?

Calculating the Size of an Atom or Ion

57. Chromium metal (d = 7.20 g/cm^3) crystallizes in a body-centered cubic unit cell. Calculate the volume of the unit cell and the radius of a chromium ion.
58. Calculate the atomic radius of an Ar atom, if argon crystallizes at low temperature in a face-centered cubic unit cell with a density of 1.623 g/cm^3.
59. Barium crystallizes in a body-centered cubic structure in which the cell edge length is 0.5025 nm. Calculate the shortest distance between neighboring barium ions in the crystal.
60. NaH crystallizes in a structure similar to that of NaCl. If the cell edge length in the crystal is 0.4880 nm, what is the average length of the Na—H bond?
61. TlI crystallizes in a structure similar to that of CsCl with a cell edge length of 0.4198 nm. Calculate the average Tl—I bond length in the crystal. If the ionic radius of an I^- ion is 0.216 nm, what is the ionic radius of the Tl^+ ion?
62. Calculate the ionic radius of the Cs^+ ion, if the cell edge length for CsCl is 0.4123 nm and the ionic radius of a Cl^- ion is 0.181 nm.

Integrated Problems

63. From the enthalpy data in Appendix B.13, calculate the enthalpy change required to break the bond in the following: $F_2(g)$, $Cl_2(g)$, $Br_2(g)$, and $I_2(g)$. Compare the enthalpies to the enthalpy required to *melt* each of the halogens given in Table 9.1. What conclusions can you reach concerning the forces that hold the atoms together and those that hold the molecules to one another?
64. From the data in Appendix B, calculate the lattice energy of $BaCl_2$. Compare the lattice energy of $BaCl_2$ to that of NaCl and account for any differences.
65. From the data in Appendix B, calculate the lattice energy of $BeCl_2$ and compare it to that of $BaCl_2$. Explain any differences.
66. The enthalpies of fusion of the alkali metals are given below.

Metal	ΔH_{fus} (kJ/mol_{rxn})
Li	2.9
Na	2.6
K	2.4
Rb	2.2

Identify the type of solid formed and explain the trend seen in the enthalpy required to melt the solids. Arrange the solids in order of increasing melting point.

67. For each of the properties (A through J) listed below, choose the appropriate electronegativity characteristic [(a) through (f) below] for a binary compound.
   A. A good conductor of electricity
   B. A hard material that conducts electricity in the melted state
   C. An insulator
   D. A material that conducts electricity when dissolved in water
   E. A semiconductor
   F. A ceramic
   G. A molecular crystal
   H. A metallic compound
   I. An ionic compound
   J. A hard material that is an insulator
      (a) large ΔEN, low EN for both atoms
      (b) small ΔEN, high EN for both atoms
      (c) moderate ΔEN, moderate EN for both atoms
      (d) moderate ΔEN, high EN for both atoms
      (e) small ΔEN, low EN for both atoms
      (f) large ΔEN

68. Explain why binary materials may be separated into the ionic, covalent, semimetal, or metallic regions by a bond-type triangle.

69. Classify the following binary compounds as primarily metallic, molecular, or ionic. Figures 9.1 and 5.10 may be helpful.

(a) B_2H_6	(b) B_4C	(c) InAs	(d) HgI_2
(e) Hg_2Na_3	(f) K_2S	(g) Cd_3Mg	(h) KBr
(i) MgH_2	(j) GaS	(k) LiH	(l) Be_3P_2

70. (a) Two of the following compounds have very high melting points. Which two? Explain your reasoning.

BaO, MgO, HgO

(b) Of the two compounds that have very high melting points, which one would have the higher melting point? Explain.

Chapter Ten

AN INTRODUCTION TO KINETICS AND EQUILIBRIUM

Special Topics

10.1 Reactions That Don't Go to Completion

Suppose that you were asked to describe the steps involved in calculating the mass of the finely divided white solid produced when a 2.00-g strip of magnesium metal is burned. You might organize your work as follows.

While the topics of kinetics and equilibrium are found in all other texts, this text combines an introduction to kinetics with equilibrium into a single chapter which shows their interrelationship. ChemActivities 36–41 in Chemistry: A Guided Inquiry *are designed to accompany this chapter.*

- Assume that the magnesium reacts with oxygen in the atmosphere when it burns.
- Predict that the formula of the product is MgO.
- Use the chemical formula to generate the following balanced equation.

$$2\ Mg(s) + O_2(g) \longrightarrow 2\ MgO(s)$$

- Use the mass of a mole of magnesium to convert grams of magnesium into moles of magnesium.

$$2.00 \text{ g Mg} \times \frac{1 \text{ mol Mg}}{24.31 \text{ g Mg}} = 0.0823 \text{ mol Mg}$$

- Use the balanced equation to convert moles of magnesium into moles of magnesium oxide.

$$0.0823 \text{ mol Mg} \times \frac{2 \text{ mol MgO}}{2 \text{ mol Mg}} = 0.0823 \text{ mol MgO}$$

- Use the mass of a mole of magnesium oxide to convert moles of MgO into grams of MgO.

$$0.0823 \text{ mol MgO} \times \frac{40.30 \text{ g MgO}}{1 \text{ mol MgO}} = 3.32 \text{ g MgO}$$

Before you read any further, ask yourself, "How confident am I in the answer?"

Before we can trust the answer, we have to consider whether there are any hidden assumptions behind the calculation and then check the validity of the assumptions. Three assumptions have been made in the calculation.

- We assumed that the metal strip was pure magnesium.
- We assumed that the magnesium reacts with only the oxygen in the atmosphere to form MgO, ignoring the possibility that some of the magnesium might react with the nitrogen in the atmosphere to form Mg_3N_2.
- We assumed that the reaction didn't stop until all of the magnesium metal had been consumed.

It is relatively easy to correct for the fact that the starting material may not be pure magnesium by weight. We can also correct for the fact that as much as 5% of the product of the reaction is Mg_3N_2 instead of MgO. But it is the third assumption that is of particular importance in this chapter.

The assumption that chemical reactions go to completion is fostered by calculations, such as predicting the amount of MgO produced by burning a known amount of magnesium. It is reinforced by demonstrations such as the reaction in which a copper penny dissolves in concentrated nitric acid, which seems to continue until the copper penny disappears.

However, chemical reactions don't always go to completion. The following equation provides an example of a chemical reaction that seems to stop prematurely.

$$2\ NO_2(g) \rightleftharpoons N_2O_4(g)$$

At 25°C, when 1 mol of NO_2 is added to a 1.00-L flask, the reaction seems to stop when 95% of the NO_2 has been converted to N_2O_4. Once it has reached that point, the reaction doesn't go any further. As long as the reaction is left at 25°C, about 5% of the NO_2 that was present initially will remain in the flask. Reactions that seem to stop before the limiting reagent is consumed are said to reach **equilibrium.**

It is useful to recognize the difference between reactions that come to equilibrium and those that stop when they run out of the limiting reagent. The reaction between a copper penny and nitric acid is an example of a reaction that continues until it has essentially run out of the limiting reagent. We indicate this by writing the equation for the reaction with a single arrow pointing from the reactants to the products.

$$Cu(s) + 4\ HNO_3(aq) \longrightarrow Cu(NO_3)_2(aq) + 2\ NO_2(g) + 2\ H_2O(l)$$

We indicate that a reaction comes to equilibrium by writing a pair of arrows pointing in opposite directions between the two sides of the equation.

$$2\ NO_2(g) \rightleftharpoons N_2O_4(g)$$

To work with reactions that come to equilibrium, we need a way to specify the amount of each reactant or product that is present in the system at any moment in time. By convention, this information is specified in terms of the concentration of each component of the system in units of moles per liter. This quantity is indicated by a symbol that consists of the formula for the reactant or product written in parentheses. For example,

(NO_2) = concentration of NO_2 in moles per liter at some moment in time

We then need a way to describe the system when it is at equilibrium. This is done by writing the symbols for each component of the system in square brackets.

$[NO_2]$ = concentration of NO_2 in moles per liter if, and only if, the reaction is at equilibrium

The fact that some reactions come to equilibrium raises a number of interesting questions.

- Why do these reactions seem to stop before all of the starting materials are converted to the products of the reaction?
- What is the difference between reactions that seem to go to completion and reactions that reach equilibrium?
- Is there any way to predict whether a reaction will go to completion or reach equilibrium?
- How does a change in the conditions of the reaction influence the amount of product formed?

Before we can understand how and why a chemical reaction comes to equilibrium, we have to build a model of the factors that influence the rate of a chemical reaction. We'll then apply this model to the simple chemical reactions that occur in the gas phase. The next chapter will then include interactions between the components of the reaction and the solvent in which the reaction is run.

➤ CHECKPOINT

If $N_2O_4(g)$ is added to an evacuated container at 25°C, what compounds would be present in the container at equilibrium?

10.2 Gas Phase Reactions

The simplest chemical reactions are those that occur in the gas phase in a single step, such as the following reaction in which a compound known as *cis*-2-butene is converted into its isomer, *trans*-2-butene.

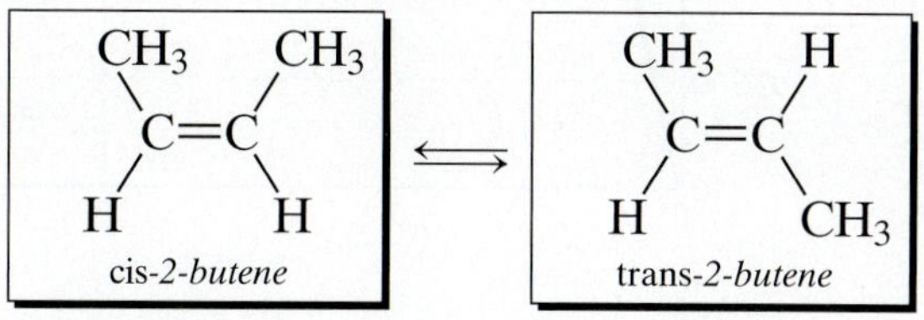

The two isomers of 2-butene have different physical properties. *cis*-2-butene melts at −139°C, whereas the *trans* form melts at −106°C. The *cis* isomer boils above 4°C, whereas the *trans* form boils at 1°C. One of these isomers can be converted into the other by rotating one end of the C═C double bond relative to the other. Rotation around C═C double bonds, however, doesn't occur at room temperature. We can therefore obtain a pure sample of the starting material—*cis*-2-butene—and wait as it is transformed to the product—*trans*-2-butene—when we heat the sample.

Gas phase reactions were chosen to introduce kinetics and equilibria to avoid the complication of a solvent. The collision theory model of chemical reactions is used to explain why a reaction comes to equilibrium.

Table 10.1 shows the amount of both *cis*- and *trans*-2-butene present in the system when a 1.000-mol sample of the *cis* isomer is heated to 400°C in a 10.0-L flask. Initially, there is no *trans*-2-butene in the system. With time, however, the amount of *cis*-2-butene gradually decreases as the compound is converted into the *trans* isomer.

The amount of the *trans* isomer present at any moment in time can be calculated from the amount of the *cis* isomer that remains in the system and vice versa. The total number of moles of the two isomers must always be the same as the number of moles of the *cis* isomer that were present at the start of the reaction. In this case, we started with 1.000 mol of the *cis* isomer. At any moment in time, the number of moles of the *cis* isomer that remain in the system is 1.000 − *x*,

$$n_{cis\text{-2-butene}} = 1.000 - x$$

and the number of moles of the *trans* isomer is equal to *x*.

$$n_{trans\text{-2-butene}} = x$$

Table 10.1

Experimental Data for the Isomerization of *cis*-2-Butene to *trans*-2-Butene at 400°C in a 10.0-L Flask

Time	Moles of *cis*-2-Butene	Moles of *trans*-2-Butene
0	1.000	0
5.00 days	0.919	0.081
10.00 days	0.848	0.152
15.00 days	0.791	0.209
20.00 days	0.741	0.259
40.00 days	0.560	0.440
60.00 days	0.528	0.472
120.00 days	0.454	0.546
1 year	0.441	0.559
2 years	0.441	0.559
3 years	0.441	0.559

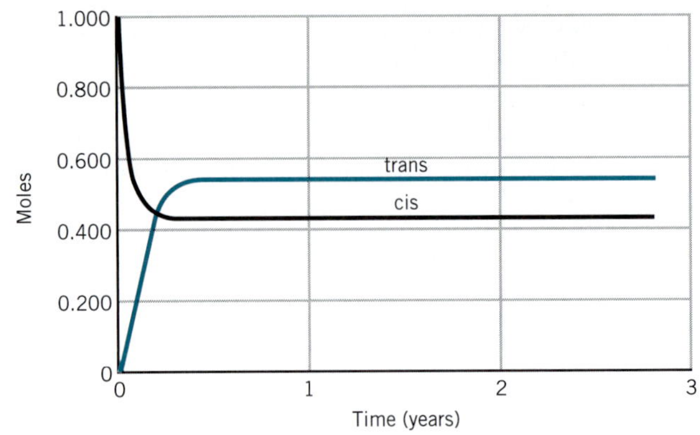

Fig. 10.1 A plot of the number of moles of the *cis* and *trans* isomers of 2-butene versus time at 400°C. Initially there is no *trans*-2-butene, but as time passes the concentration of the *cis* isomer decreases and the concentration of the *trans* isomer increases.

Figure 10.1 shows a plot of the number of moles of the *cis* and *trans* isomers of 2-butene versus time. Figure 10.1 clearly shows that there is no change in the number of moles once the reaction reaches the point at which the system contains 0.441 mol of *cis*-2-butene and 0.559 mol of *trans*-2-butene. No matter how long we wait, no more *cis*-2-butene is converted into *trans*-2-butene. As we have seen, this is an indication that the reaction has come to equilibrium.

Chemical reactions at equilibrium are described in terms of the number of moles per liter of each component of the system, not the moles of each component. For example,

$$[\textit{cis}\text{-2-butene}] = \text{concentration of } \textit{cis}\text{-2-butene at equilibrium in moles per liter}$$
$$[\textit{trans}\text{-2-butene}] = \text{concentration of } \textit{trans}\text{-2-butene at equilibrium in moles per liter}$$

The concentrations of *cis*- and *trans*-2-butene at equilibrium depend on the initial conditions of the experiment. But, at a given temperature, the ratio of the equilibrium concentrations of the two components of the reaction is always the same. It doesn't matter whether we start with a great deal of *cis*-2-butene or a relatively small amount, or whether we start with a pure sample of *cis*-2-butene or one that already contains some of the *trans* isomer. When the reaction reaches equilibrium at 400°C, the concentration of the *trans* isomer divided by that of the *cis* isomer is always 1.27.

The equation that describes the relationship between the concentrations of the two components of the reaction at equilibrium is known as the **equilibrium constant expression,** where K_c is the **equilibrium constant** for the reaction.

> **➤ CHECKPOINT**
>
> If 1.0 mol of *trans*-2-butene is placed into an empty flask at 400°C, what will be the equilibrium ratio of *trans*-2-butene to *cis*-2-butene?

$$K_c = \frac{[\textit{trans}\text{-2-butene}]}{[\textit{cis}\text{-2-butene}]}$$

The subscript "c" in the equilibrium constant indicates that the constant has been calculated from the concentrations of the reactants and products in units of moles per liter.

Exercise 10.1

Calculate the equilibrium constant for the conversion of *cis*-2-butene to *trans*-2-butene for the following sets of experimental data.

(a) A 5.00-mol sample of the *cis* isomer was added to a 10.0-L flask and heated to 400°C until the reaction came to equilibrium. At equilibrium, the system contained 2.80 mol of the *trans* isomer.

(b) A 0.100-mol sample of the *cis* isomer was added to a 25.0-L flask and heated to 400°C until the reaction came to equilibrium. At equilibrium, the system contained 0.0559 mol of the *trans* isomer.

Solution

(a) If 5.00 mol of *cis*-2-butene forms 2.80 mol of *trans*-2-butene, then 2.20 mol of the *cis* isomer must remain after the reaction reaches equilibrium. Because the experiment was done in a 10.0-L flask, the equilibrium concentrations of the two components of the reaction have the following values.

$$[trans\text{-2-butene}] = \frac{2.80 \text{ mol}}{10.0 \text{ L}} = 0.280\ M$$

$$[cis\text{-2-butene}] = \frac{2.20 \text{ mol}}{10.0 \text{ L}} = 0.220\ M$$

The equilibrium constant, K_c, for the reaction is therefore 1.27.

$$K_c = \frac{[trans\text{-2-butene}]}{[cis\text{-2-butene}]} = \frac{0.280\ M}{0.220\ M} = 1.27$$

(b) This system comes to equilibrium at the following concentrations of the *cis* and *trans* isomers.

$$[trans\text{-2-butene}] = \frac{0.0559 \text{ mol}}{25.0 \text{ L}} = 0.00224\ M$$

$$[cis\text{-2-butene}] = \frac{0.0441 \text{ mol}}{25.0 \text{ L}} = 0.00176\ M$$

Even though the equilibrium concentrations of the two isomers are very different from the values obtained in the previous experiment, the ratio of the concentrations is exactly the same.

$$K_c = \frac{[trans\text{-2-butene}]}{[cis\text{-2-butene}]} = \frac{0.00224\ M}{0.00176\ M} = 1.27$$

10.3 The Rate of a Chemical Reaction

Experiments such as the one that gave us the data in Table 10.1 are classified as measurements of **chemical kinetics** (from a Greek stem meaning "to move"). The goal of these experiments is to describe the **rate of reaction,** that is, the rate at which the reactants are transformed into the products of the reaction.

The term *rate* is often used to describe the change in a quantity that occurs per unit of time. The rate of inflation, for example, is the change in the average cost of a collection of standard items per year. The rate at which an object travels through space is the distance traveled per unit of time, such as miles per hour or kilometers per second.

In chemical kinetics, the distance traveled is the change in the concentration of one of the components of the reaction. The rate of a reaction is therefore the change in the concentration of one of the components, $\Delta(X)$, that occurs during a given time, Δt. The concentration of X is written in parentheses here because the system isn't at equilibrium.

$$\text{Rate} = \frac{\Delta(X)}{\Delta t} \qquad \text{(where } X \text{ is one of the products)}$$

By convention, the symbol Δ represents a change calculated by subtracting the initial conditions from the final conditions. Thus, $\Delta(X)$ represents the final concentration minus the initial concentration.

$$\Delta(X) = (X)_{\text{fiina}} - (X)_{\text{initial}}$$

If X is one of the products of the reaction, then $\Delta(X)$ is positive. If X is a reactant, $\Delta(X)$ is negative and a minus sign is added to the rate equation to turn the rate into a positive number.

$$\text{Rate} = \frac{-\Delta(X)}{\Delta t} \qquad \text{(when } X \text{ is one of the reactants)}$$

Let's use the data in Table 10.1 to calculate the rate at which *cis*-2-butene is transformed into *trans*-2-butene during each of the following periods.

- During the first time interval, when the number of moles of *cis*-2-butene in the 10.0-L flask falls from 1.000 to 0.919.
- During the second interval, when the amount of *cis*-2-butene falls from 0.919 to 0.848 mol.
- During the third interval, when the amount of *cis*-2-butene falls from 0.848 to 0.791 mol.

Before we can calculate the rate of the reaction during each of the time intervals, we have to remember that the rate of reaction is defined in terms of changes in the number of moles per liter (M) of one of the components of the reaction, not the number of moles of that reactant. Thus, we have to recognize that 1.000 mol of *cis*-2-butene in a 10.0-L flask corresponds to a concentration of 0.1000 M.

During the first time period, the rate of the reaction is 1.6×10^{-3} M/day

$$\text{Rate} = \frac{-\Delta(X)}{\Delta t} = \frac{-(0.0919\ M - 0.1000\ M)}{(5.00 \text{ days} - 0 \text{ days})} = 1.6 \times 10^{-3}\ M/\text{day}$$

During the second time period, the rate of reaction is slightly smaller.

$$\text{Rate} = \frac{-\Delta(X)}{\Delta t} = \frac{-(0.0848\ M - 0.0919\ M)}{(10.00 \text{ days} - 5.00 \text{ days})} = 1.4 \times 10^{-3}\ M/\text{day}$$

During the third time period, the rate of reaction is even smaller.

$$\text{Rate} = \frac{-\Delta(X)}{\Delta t} = \frac{-(0.0791\ M - 0.0848\ M)}{(15.00 \text{ days} - 10.00 \text{ days})} = 1.1 \times 10^{-3}\ M/\text{day}$$

These calculations illustrate an important point: The rate of the reaction isn't constant; it changes with time. The rate of the reaction gradually decreases as the starting materials are consumed, which means that the rate of reaction changes while it is being measured.

We can minimize the error this introduces into our measurements by measuring the rate of reaction over periods of time that are short compared with the time it takes for the reaction to occur. We might try, for example, to measure the infinitesimally

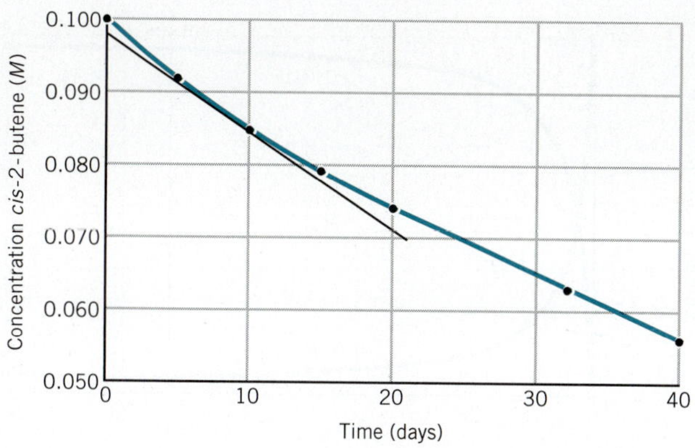

Fig. 10.2 The rate of reaction at a given time for the isomerization of *cis*-2-butene is the negative of the slope of a tangent drawn to the concentration curve at that particular point in time.

small change in reactant concentration, $d(X)$, that occurs over an infinitesimally short period of time, dt. This ratio is known as the *instantaneous rate of reaction.*

$$\text{Rate} = \frac{-d(X)}{dt}$$

The instantaneous rate of reaction at any moment in time can be calculated from a graph of the concentration of the reactant (or product) versus time. Figure 10.2 shows how the rate of reaction for the isomerization of *cis*-2-butene can be calculated from such a graph. The rate of reaction at any moment is equal to the negative of the slope of a tangent drawn to the curve at that moment.

An interesting result is obtained when the instantaneous rate of reaction is calculated at various points along the curve in Figure 10.2. The rate of reaction at every point on the curve is directly proportional to the concentration of *cis*-2-butene at that moment in time.

$$\text{Rate} = k(cis\text{-2-butene})$$

This equation, which is determined from experimental data, describes the rate of the reaction. It is therefore called the **rate law** for the reaction. The proportionality constant k is known as the **rate constant.**

> **CHECKPOINT**
>
> The following data were obtained for the rate constant for the decomposition of one of the metabolites that supplies energy in a biochemical system.
>
> | *Temperature (°C)* | *Rate Constant* (s^{-1}) |
> |---|---|
> | 15 | 2.5×10^{-2} |
> | 20 | 4.5×10^{-2} |
> | 25 | 8.1×10^{-2} |
> | 30 | 1.6×10^{-1} |
>
> What do these data suggest happens to the rate of the reaction as the temperature increases?

10.4 The Collision Theory of Gas Phase Reactions

One way to understand why some reactions come to equilibrium is to consider a simple gas phase reaction that occurs in a single step, such as the transfer of a chlorine atom from $ClNO_2$ to NO to form NO_2 and ClNO.

$$ClNO_2(g) + NO(g) \rightleftharpoons NO_2(g) + ClNO(g)$$

This reaction can be understood by writing the Lewis structures for the four components of the reaction. Because they contain an odd number of electrons, both NO and NO_2 can combine with a neutral chlorine atom to form a molecule in which all of the electrons are paired. The reaction therefore involves the transfer of a chlorine atom from one molecule to another, as shown in Figure 10.3.

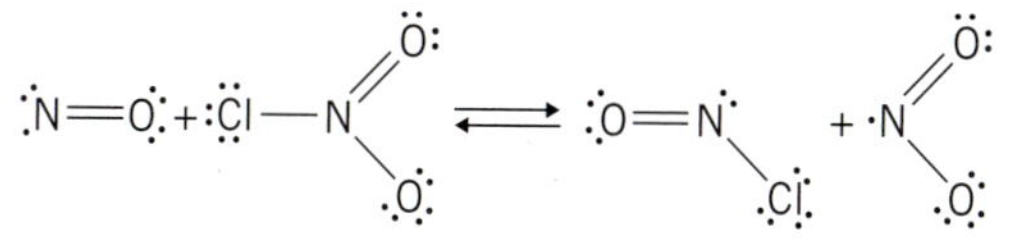

Fig. 10.3 The reaction between $ClNO_2$ and NO to form NO_2 and ClNO is a simple, one-step reaction that involves the transfer of a chlorine atom.

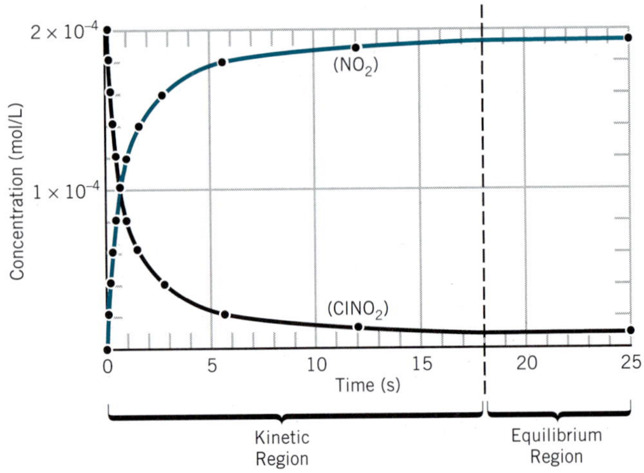

Fig. 10.4 Plot of the change in the concentration of $ClNO_2$ superimposed on a plot of the change in concentration of NO_2 as $ClNO_2$ reacts with NO to produce NO_2 and ClNO. The graph can be divided into a kinetic and an equilibrium region.

Figure 10.4 combines a plot of the concentration of $ClNO_2$ versus time as this reactant is consumed with a plot of the concentration of NO_2 versus time as this product is formed in the reaction. The data in Figure 10.4 are consistent with the following rate law for the reaction.

$$\text{Rate} = k(ClNO_2)(NO)$$

According to this rate law, the rate at which $ClNO_2$ and NO are converted into NO_2 and ClNO is proportional to the product of the concentrations of the two reactants. Initially, the rate of reaction is fast. As the reactants are converted to products, however, the $ClNO_2$ and NO concentrations become smaller, and the reaction slows down.

We might expect the reaction to stop when it runs out of either $ClNO_2$ or NO. In practice, the reaction seems to stop before this happens. This is a very fast reaction—the concentration of $ClNO_2$ drops by a factor of 2 in a second. And yet, no matter how long we wait, some residual $ClNO_2$ and NO remain in the reaction flask.

Figure 10.4 divides the plot of the change in the concentrations of NO_2 and $ClNO_2$ into a **kinetic region** and an **equilibrium region.** The kinetic region marks the period during which the concentrations of the components of the reaction are constantly changing. The equilibrium region is the period after which the reaction seems to stop, when there is no further significant change in the concentrations of the components of the reaction.

The fact that the following reaction seems to stop before all of the reactants are consumed can be explained with the **collision theory** of chemical reactions. The collision theory assumes that $ClNO_2$ and NO molecules must collide before a chlorine atom can be transferred from one molecule to the other.

$$ClNO_2(g) + NO(g) \rightleftharpoons NO_2(g) + ClNO(g)$$

That assumption explains why the rate of the reaction is proportional to the concentration of both $ClNO_2$ and NO.

$$\text{Rate} = k(ClNO_2)(NO)$$

The number of collisions per second between $ClNO_2$ and NO molecules depends on their concentrations. As $ClNO_2$ and NO are consumed in the reaction, the number of collisions per second between the molecules becomes smaller, and the reaction slows down.

Suppose that we start with a mixture of $ClNO_2$ and NO, but no NO_2 or ClNO. The only reaction that can occur at first is the transfer of a chlorine atom from $ClNO_2$ to NO.

$$ClNO_2(g) + NO(g) \longrightarrow NO_2(g) + ClNO(g)$$

Eventually, NO_2 and ClNO build up in the reaction flask, and those molecules begin to collide as well. Collisions between the molecules can result in the transfer of a chlorine atom in the opposite direction.

$$ClNO_2(g) + NO(g) \longleftarrow NO_2(g) + ClNO(g)$$

The collision theory of chemical reactions assumes that the rate of a simple, one-step reaction is proportional to the product of the concentrations of the substances consumed in that reaction. The rate of the forward reaction is therefore proportional to the product of the concentrations of the two starting materials.

$$\text{Rate}_{\text{forward}} = k_f(ClNO_2)(NO)$$

The rate of the reverse reaction, on the other hand, is proportional to the concentrations of the products of the reaction.

$$\text{Rate}_{\text{reverse}} = k_r(NO_2)(ClNO)$$

Initially, the rate of the forward reaction is much larger than the rate of the reverse reaction, because the system contains $ClNO_2$ and NO but virtually no NO_2 and ClNO.

$$\text{Initially:} \quad \text{Rate}_{\text{forward}} \gg \text{Rate}_{\text{reverse}}$$

As $ClNO_2$ and NO are consumed, the rate of the forward reaction slows down. At the same time, NO_2 and ClNO accumulate, and the reverse reaction speeds up.

The system eventually reaches a point at which the rates of the forward and reverse reactions are the same.

$$\text{Eventually:} \quad \text{Rate}_{\text{forward}} = \text{Rate}_{\text{reverse}}$$

At this point, the concentrations of reactants and products remain the same. $ClNO_2$ and NO are consumed in the forward reaction at the same rate at which they are produced in the reverse reaction. The same thing happens to NO_2 and ClNO. When the rates of the forward and reverse reactions are the same, there is no longer any change in the concentrations of the reactants or products of the reaction. In other words, the reaction is at equilibrium.

We can now see that there are two ways to define equilibrium.

- A system in which there is no change in the concentrations of the reactants and products of a reaction
- A system in which the rates of the forward and reverse reactions are the same

The first definition is based on the results of experiments that tell us that some reactions seem to stop prematurely—they reach a point at which no more reactants are converted to products before the limiting reagent is consumed. The other definition is based on a theoretical model of chemical reactions that explains why reactions reach equilibrium.

➤ CHECKPOINT

If the rate of the reverse reaction is greater than the rate of the forward reaction at a given moment in time, does this mean that the reverse rate constant is larger than the forward rate constant?

10.5 Equilibrium Constant Expressions

Reactions don't stop when they come to equilibrium. But the rate of the forward and reverse reactions are the same, so there is no net change in the concentrations of the reactants or products, and on the macroscopic scale the reactions appear to stop. Chemical equilibrium is an example of a *dynamic* balance between the forward and reverse reactions, not a static balance.

Let's look at the logical consequences of the assumption that the reaction between $ClNO_2$ and NO eventually reaches equilibrium at a given temperature.

$$ClNO_2(g) + NO(g) \rightleftharpoons NO_2(g) + ClNO(g)$$

The rates of the forward and reverse reactions are the same when the system is at equilibrium.

$$\text{At equilibrium:} \qquad \text{Rate}_{\text{forward}} = \text{Rate}_{\text{reverse}}$$

Substituting the rate laws for the forward and reverse reactions into the equality gives the following result.

$$\text{At equilibrium:} \qquad k_f(ClNO_2)(NO) = k_r(NO_2)(ClNO)$$

But this equation is valid only when the system is at equilibrium, so we should replace the ($ClNO_2$), (NO), (NO_2), and (ClNO) terms with symbols which indicate that the reaction is at equilibrium. By convention, we use square brackets for this purpose. The equation describing the balance between the forward and reverse reactions when the system is at equilibrium should therefore be written as follows.

$$\text{At equilibrium:} \qquad k_f[ClNO_2][NO] = k_r[NO_2][ClNO]$$

Rearranging the equation gives the following result.

$$\frac{k_f}{k_r} = \frac{[NO_2][ClNO]}{[ClNO_2][NO]}$$

Because k_f and k_r are constants, the ratio of k_f divided by k_r must also be a constant. This ratio is the **equilibrium constant** for the reaction, K_c. As we have seen, the ratio of the concentrations of the reactants and products is known as the **equilibrium constant expression.**

Equilibrium constant expression

$$K_c = \frac{k_f}{k_r} = \frac{[NO_2][ClNO]}{[ClNO_2][NO]}$$

Equilibrium constant

No matter what combination of concentrations of reactants and products we start with, the reaction reaches equilibrium when the ratio of the concentrations defined by the equilibrium constant expression is equal to the equilibrium constant for the reaction at the chosen temperature. We can start with a lot of $ClNO_2$ and very little NO, or a lot of NO and very little $ClNO_2$. It doesn't matter. When the reaction reaches equilibrium, the relationship between the concentrations of the reactants and products described by the equilibrium constant expression will always be the same. At 25°C, this reaction always reaches equilibrium when the ratio of the concentrations is 1.3×10^4. K_c is always reported without units. However, any calculations using K_c require the concentration of the products and reactants to be in units of molarity (moles/liter).

$$K_c = \frac{[NO_2][ClNO]}{[ClNO_2][NO]} = 1.3 \times 10^4 \qquad \text{(at 25°C)}$$

What happens if we approach equilibrium from the other direction? We start with a system that contains the products of the reaction—NO_2 and ClNO—and then let the reaction come to equilibrium. The rate laws for the forward and reverse reactions will still be the same.

$$\text{Rate}_{\text{forward}} = k_f(\text{ClNO}_2)(\text{NO})$$
$$\text{Rate}_{\text{reverse}} = k_r(\text{NO}_2)(\text{ClNO})$$

Now, however, the rate of the forward reaction initially will be much smaller than the rate of the reverse reaction.

$$\text{Initially:} \qquad \text{Rate}_{\text{forward}} \ll \text{Rate}_{\text{reverse}}$$

But, as time passes, the rate of the reverse reaction will slow down and the rate of the forward reaction will speed up until they become equal. At that point, the reaction will have reached equilibrium.

$$\text{At equilibrium:} \qquad k_f[\text{ClNO}_2][\text{NO}] = k_r[\text{NO}_2][\text{ClNO}]$$

Rearranging the equation gives us the same equilibrium constant expression.

$$K_c = \frac{k_f}{k_r} = \frac{[\text{NO}_2][\text{ClNO}]}{[\text{ClNO}_2][\text{NO}]} = \frac{[\text{products}]}{[\text{reactants}]}$$

We get the same equilibrium constant expression and the same equilibrium constant no matter whether we start with only reactants, only products, or a mixture of reactants and products.

Exercise 10.2

The rate constants for the forward and reverse reactions in the following equilibrium have been measured. At 25°C, k_f is 7.3 × 10³ liters per mole-second and k_r is 0.55 liter per mole-second. Calculate the equilibrium constant for the reaction.

$$\text{ClNO}_2(g) + \text{NO}(g) \rightleftharpoons \text{NO}_2(g) + \text{ClNO}(g)$$

Solution

We start by recognizing that the rates of the forward and reverse reactions at equilibrium are the same.

$$\text{At equilibrium:} \qquad \text{Rate}_{\text{forward}} = \text{Rate}_{\text{reverse}}$$

We then substitute the rate laws for the reaction into the equality.

$$\text{At equilibrium:} \qquad k_f[\text{ClNO}_2][\text{NO}] = k_r[\text{NO}_2][\text{ClNO}]$$

We then rearrange the equation to get the equilibrium constant expression for the reaction.

$$K_c = \frac{k_f}{k_r} = \frac{[\text{NO}_2][\text{ClNO}]}{[\text{ClNO}_2][\text{NO}]}$$

The equilibrium constant for the reaction is therefore equal to the rate constant for the forward reaction divided by the rate constant for the reverse reaction.

$$K_c = \frac{k_f}{k_r} = \frac{7.3 \times 10^3 \text{ L/mol-s}}{0.55 \text{ L/mol-s}} = 1.3 \times 10^4$$

➤ CHECKPOINT

If the forward rate constant for a given reaction is twice the reverse rate constant, what will be the value of the equilibrium constant for the reaction?

Any reaction that reaches equilibrium, no matter how simple or complex, has an equilibrium constant expression that satisfies the following rules.

Rules for Writing Equilibrium Constant Expressions

- Even though chemical reactions that reach equilibrium occur in both directions, the substances on the right side of the equation are assumed to be the "products" of the reaction and the substances on the left side of the equation are assumed to be the "reactants."
- The products of the reaction are always written above the line, in the numerator.
- The reactants are always written below the line, in the denominator.
- For systems in which all species are either gaseous or aqueous, the equilibrium constant expression contains a term for every reactant and every product of the reaction.
- The numerator of the equilibrium constant expression is found by multiplying the concentrations of each product of the reaction raised to a power equal to the coefficient for the component in the balanced equation for the reaction.
- The denominator of the equilibrium constant expression is the product of the concentrations of each reactant raised to a power equal to the coefficient for the component in the balanced equation for the reaction.

Exercise 10.3

Write equilibrium constant expressions for the following reactions.

(a) $2\ NO_2(g) \rightleftharpoons N_2O_4(g)$

(b) $2\ SO_3(g) \rightleftharpoons 2\ SO_2(g) + O_2(g)$

(c) $N_2(g) + 3\ H_2(g) \rightleftharpoons 2\ NH_3(g)$

Solution

In each case, the equilibrium constant expression is the product of the concentrations of the species on the right side of the equation divided by the product of the concentrations of those on the left side of the equation. All concentrations are raised to the power equal to the coefficient for the species in the balanced equation.

(a) $K_c = \dfrac{[N_2O_4]}{[NO_2]^2}$

(b) $K_c = \dfrac{[SO_2]^2[O_2]}{[SO_3]^2}$

(c) $K_c = \dfrac{[NH_3]^2}{[N_2][H_2]^3}$

What happens to the magnitude of the equilibrium constant for a reaction when we turn the equation around? Consider the following reaction, for example.

$$ClNO_2(g) + NO(g) \rightleftharpoons NO_2(g) + ClNO(g)$$

The equilibrium constant expression for the equation is written as follows.

$$K_c = \frac{[NO_2][ClNO]}{[ClNO_2][NO]} = 1.3 \times 10^4 \qquad \text{(at 25°C)}$$

Because this is an equilibrium reaction, it can also be represented by an equation written in the opposite direction.

$$NO_2(g) + ClNO(g) \rightleftharpoons ClNO_2(g) + NO(g)$$

The equilibrium constant expression is now written as follows.

$$K_c' = \frac{[ClNO_2][NO]}{[NO_2][ClNO]}$$

Each of the equilibrium constant expressions is the inverse of the other. We can therefore calculate K_c' by dividing K_c into 1.

$$K_c' = \frac{1}{K_c} = \frac{1}{1.3 \times 10^4} = 7.7 \times 10^{-5}$$

We can also calculate equilibrium constants by combining two or more reactions for which the values of K_c are known. For example, we know the equilibrium constants for the following gas phase reactions at 200°C.

$$N_2(g) + O_2(g) \rightleftharpoons 2\,NO(g) \qquad K_{c1} = 2.3 \times 10^{-19} = \frac{[NO]^2}{[N_2][O_2]}$$

$$2\,NO(g) + O_2(g) \rightleftharpoons 2\,NO_2(g) \qquad K_{c2} = 3 \times 10^6 = \frac{[NO_2]^2}{[NO]^2[O_2]}$$

We can combine the reactions to obtain an overall equation for the reaction between N_2 and O_2 to form NO_2.

$$\begin{array}{rl} N_2(g) + O_2(g) \rightleftharpoons 2\,NO(g) & \\ +\ 2\,NO(g) + O_2(g) \rightleftharpoons 2\,NO_2(g) & \\ \hline N_2(g) + 2\,O_2(g) \rightleftharpoons 2\,NO_2(g) & \quad K_c = ? \end{array}$$

The equilibrium constant expression for the overall reaction is equal to the product of the equilibrium constant expressions for the two steps in the reaction.

$$K_c = K_{c1} \times K_{c2} = \frac{[NO]^2}{[N_2][O_2]} \times \frac{[NO_2]^2}{[NO]^2[O_2]} = \frac{[NO_2]^2}{[N_2][O_2]^2}$$

The equilibrium constant for the overall reaction is therefore equal to the product of the equilibrium constants for the individual reactions.

$$K_c = K_{c1} \times K_{c2} = (2.3 \times 10^{-19})(3 \times 10^6) = 7 \times 10^{-13}$$

Exercise 10.4

The sugar glucose is metabolized in the body to produce a compound called glucose 6-phosphate. One possible way the reaction could proceed is by the direct reaction of inorganic phosphate, labeled P_i by biochemists, with glucose.

$$\text{glucose}(aq) + P_i(aq) \rightleftharpoons \text{glucose 6-phosphate}(aq)$$

Write the equilibrium constant for the reaction in terms of concentrations.

Once the concept of the equilibrium constant is developed, the reaction quotient is introduced. If students can learn to think about the reaction quotient as a snapshot of the reaction at a given moment in time, many of the misunderstandings about what constitutes a "spontaneous" reaction can be avoided.

Solution

No matter whether a reaction occurs in the gas phase or, as in this reaction, in a cell in the body, the same principles apply. All the species in the reaction exist in a cellular solution, and hence all can be expressed as concentrations in moles per liter.

$$K_c = \frac{[\text{glucose-6-phosphate}]}{[\text{glucose}][P_i]}$$

10.6 Reaction Quotients: A Way to Decide Whether a Reaction Is at Equilibrium

The difference between the reaction quotient and the equilibrium constant can be used to decide in which direction the reaction must shift to reach equilibrium. The comparison of the reaction quotient to the equilibrium constant is used in subsequent chapters as well, to determine which way a reaction proceeds under a given set of concentrations.

In Chapter 13, when free energy is introduced, students are prepared to think about determining the direction of a reaction. They also know that ΔG° applies to a very particular set of concentrations and are less likely to be confused by what a positive or negative value of ΔG° really means.

We now have a model that describes what happens when a reaction reaches equilibrium. At the molecular level, the rate of the forward reaction is equal to the rate of the reverse reaction. Because the reaction proceeds in both directions at the same rate, there is no apparent change in the concentrations of the reactants or the products on the macroscopic scale (i.e., the level of objects visible to the naked eye). The model can also be used to predict the direction in which a reaction has to shift to reach equilibrium.

If the concentrations of the reactants are too large for the reaction to be at equilibrium, the rate of the forward reaction will be faster than the reverse reaction, and some of the reactants will be converted to products until equilibrium is achieved. Conversely, if the concentrations of the reactants are too small, the rate of the reverse reaction will exceed that of the forward reaction, and the reaction will convert some of the excess products back into reactants until the system reaches equilibrium.

We can determine the direction in which a reaction has to shift to reach equilibrium by calculating the **reaction quotient (Q_c)** for the reaction. The reaction quotient is written in the same way as the equilibrium constant, but in Q_c the concentrations refer to any moment in time. The equilibrium constant expression refers only to the moment in time when the system is at equilibrium. To illustrate how the reaction quotient is used, consider the following gas phase reaction.

$$H_2(g) + I_2(g) \rightleftharpoons 2\ HI(g)$$

The equilibrium constant expression for the reaction is written as follows.

$$K_c = \frac{[HI]^2}{[H_2][I_2]} = 60 \qquad (\text{at } 350°C)$$

By analogy, we can write the expression for the reaction quotient as follows.

$$Q_c = \frac{(HI)^2}{(H_2)(I_2)}$$

Q_c can take on any value between zero and infinity. If the system contains a great deal of HI and very little H_2 and I_2, the reaction quotient is very large. If the system contains relatively little HI and a great deal of H_2 and/or I_2, the reaction quotient is very small.

At any moment in time, there are three possibilities.

- **Q_c is smaller than K_c.** The system contains too much reactant and not enough product to be at equilibrium. The value of Q_c must increase in order for the reaction to reach equilibrium. Thus, the reaction has to convert some of the reactants into products to come to equilibrium.

- **Q_c is equal to K_c.** If this is true, then the reaction is at equilibrium.
- **Q_c is larger than K_c.** The system contains too much product and not enough reactant to be at equilibrium. The value of Q_c must become smaller before the reaction can come to equilibrium. Thus, the reaction must convert some of the products into reactants to reach equilibrium.

Exercise 10.5

Assume that the concentrations of H_2, I_2, and HI can be measured for the following reaction at any moment in time.

$$H_2(g) + I_2(g) \rightleftharpoons 2\ HI(g) \qquad K_c = 60 \quad \text{(at 350°C)}$$

For each of the following sets of concentrations, determine whether the reaction is at equilibrium. If it isn't, decide in which direction it must go to reach equilibrium.

(a) $(H_2) = (I_2) = (HI) = 0.010\ M$

(b) $(HI) = 0.30\ M$, $(H_2) = 0.010\ M$, $(I_2) = 0.15\ M$

(c) $(H_2) = (HI) = 0.10\ M$, $(I_2) = 0.0010\ M$

Solution

(a) The best way to decide whether the reaction is at equilibrium is to compare the reaction quotient with the equilibrium constant for the reaction.

$$Q_c = \frac{(HI)^2}{(H_2)(I_2)} = \frac{(0.010)^2}{(0.010)(0.010)} = 1.0 < K_c$$

The reaction quotient in this case is smaller than the equilibrium constant. The only way to get the system to equilibrium is to increase the magnitude of the reaction quotient. This can be done by converting some of the H_2 and I_2 into HI. The reaction therefore has to shift to the right to reach equilibrium.

(b) The reaction quotient for this set of concentrations is equal to the equilibrium constant for the reaction.

$$Q_c = \frac{(HI)^2}{(H_2)(I_2)} = \frac{(0.30)^2}{(0.010)(0.15)} = 60 = K_c$$

The reaction is therefore at equilibrium.

(c) The reaction quotient for this set of concentrations is larger than the equilibrium constant for the reaction.

$$Q_c = \frac{(HI)^2}{(H_2)(I_2)} = \frac{(0.10)^2}{(0.10)(0.0010)} = 1.0 \times 10^2 > K_c$$

To reach equilibrium, the concentrations of the reactants and products must be adjusted until the reaction quotient is equal to the equilibrium constant. This involves converting some of the HI back into H_2 and I_2. Thus, the reaction has to shift to the left to reach equilibrium.

Exercise 10.6

The equilibrium constant for the metabolism of glucose to glucose-6-phosphate is 6×10^{-3} at 25°C. The P_i (phosphate) concentration in a typical cell is approximately 1×10^{-2} *M* and that of glucose-6-phosphate is about 1×10^{-4} *M*. What is the minimum concentration of glucose needed to push the reaction forward toward the formation of glucose-6-phosphate?

Solution

We can start by using the value of K_c for the equilibrium constant expression developed in Exercise 10.4.

$$K_c = \frac{[\text{glucose-6-phosphate}]}{[\text{glucose}][\text{P}_i]} = 6 \times 10^{-3}$$

We can then substitute the known concentrations of glucose-6-phosphate and the phosphate ion into this equation,

$$\frac{[1 \times 10^{-4}]}{[\text{glucose}][1 \times 10^{-2}]} = 6 \times 10^{-3}$$

and calculate the concentration of glucose if this reaction is at equilibrium.

$$[\text{glucose}] \approx 2\,M$$

The reaction will go forward to generate more product only if Q_c for the reaction is smaller than K_c. Thus, if the concentration of glucose is larger than 2 *M*, Q_c will be less than K_c, and glucose will react with P_i to give more product. Is 2 *M* a reasonable concentration of glucose for a cell? Certainly not. A sugar concentration of 2 *M* could never be tolerated by the body. Indeed, the blood sugar concentration is closer to 5×10^{-3} *M*.

This means that the reaction we have just hypothesized for the metabolism of glucose by the body doesn't occur. How, then, is glucose metabolized? Another species, adenosine triphosphate (ATP), reacts with the glucose to provide an alternate way to convert glucose to glucose-6-phosphate. An application of equilibrium principles has allowed us to reject a possible direct reaction of glucose and seek another way to carry out the reaction.

10.7 Changes in Concentration That Occur as a Reaction Comes to Equilibrium

The relative sizes of Q_c and K_c for a reaction tell us whether the reaction is at equilibrium at any moment in time. If it isn't, the relative sizes of Q_c and K_c tell us the direction in which the reaction must shift to reach equilibrium. Now we need a way to predict how far the reaction has to go to reach equilibrium. Suppose that you are faced with the following problem.

Phosphorus pentachloride decomposes to phosphorus trichloride and chlorine when heated.

$$PCl_5(g) \rightleftharpoons PCl_3(g) + Cl_2(g)$$

The equilibrium constant for the reaction is 0.030 at 250°C. If the initial concentration of PCl_5 is 0.100 mol/L and there is no PCl_3 or Cl_2 in the system, we can calculate the concentrations of PCl_5, PCl_3, and Cl_2 at equilibrium.

The first step toward solving the problem involves organizing the information so that it provides clues as to how to proceed. The problem contains four pieces of information: (1) a balanced equation, (2) an equilibrium constant for the reaction, (3) a description of the initial conditions, and (4) an indication of the goal of the calculation, namely, to figure out the equilibrium concentrations of the three components of the reaction.

The following format offers a useful way to summarize this information.

	$PCl_5(g)$	$\rightleftharpoons$ $PCl_3(g)$	+ $Cl_2(g)$	$K_c = \dfrac{[PCl_3][Cl_2]}{[PCl_5]} = 0.030$ (at 250°C)
Initial	0.100 *M*	0	0	
Equilibrium	?	?	?	

We start with the balanced equation and the equilibrium constant for the reaction and then add what we know about the initial and equilibrium concentrations of the various components of the reaction. Initially, the flask contains 0.100 mol/L of PCl_5 and no PCl_3 or Cl_2. Our goal is to calculate the equilibrium concentrations of the three substances.

Before we do anything else, we have to decide whether the reaction is at equilibrium. We can do this by comparing the reaction quotient for the initial conditions with the equilibrium constant for the reaction.

$$Q_c = \frac{(PCl_3)(Cl_2)}{(PCl_5)} = \frac{(0)(0)}{(0.100)} = 0 < K_c$$

Although the equilibrium constant is small ($K_c = 3.0 \times 10^{-2}$), the reaction quotient is even smaller ($Q_c = 0$) The only way for the reaction to get to equilibrium is for some of the PCl_5 to decompose into PCl_3 and Cl_2.

Because the reaction isn't at equilibrium, one thing is certain: The concentrations of PCl_5, PCl_3, and Cl_2 will all change as the reaction comes to equilibrium. Because the reaction has to shift to the right to reach equilibrium, the PCl_5 concentration will become smaller, while the PCl_3 and Cl_2 concentrations will become larger.

At first glance, the problem appears to have three unknowns: the equilibrium concentrations of PCl_5, PCl_3, and Cl_2. And we have only one equation, the equilibrium constant expression. Because it is impossible to solve one equation for three unknowns, we need to look for relationships between the unknowns that can simplify the problem. One way of achieving this goal is to look at the relationship between the changes that occur in the concentrations of PCl_5, PCl_3, and Cl_2 as the reaction approaches equilibrium.

Exercise 10.7

Calculate the increase in the PCl_3 and Cl_2 concentrations that occur as the following reaction comes to equilibrium if the concentration of PCl_5 decreases by 0.042 mol/L.

$$PCl_5(g) \rightleftharpoons PCl_3(g) + Cl_2(g)$$

Solution

The decomposition of PCl_5 has a 1:1:1 stoichiometry, as shown in Figure 10.5. For every mole of PCl_5 that decomposes, we get 1 mol of PCl_3 and 1 mol of Cl_2. Thus, the change in the concentration of PCl_5 that occurs as the reaction comes to

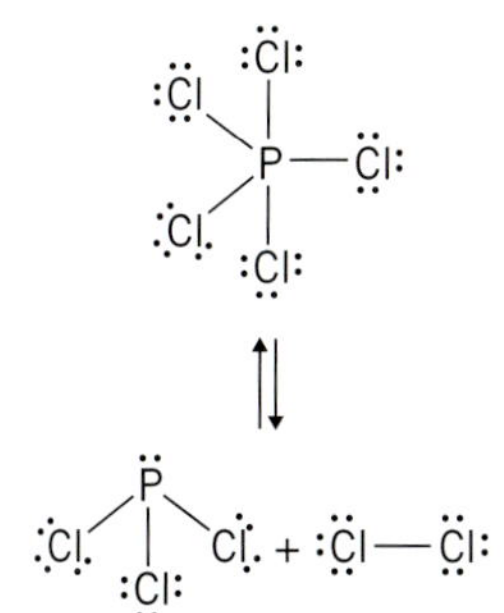

Fig. 10.5 The decomposition of PCl_5 to form PCl_3 and Cl_2 is a reversible reaction with a 1:1:1 stoichiometry.

> **➤ CHECKPOINT**
>
> Assume that 1.00 mol of $PCl_3(g)$ and 1.00 mol of $Cl_2(g)$ are added to an empty 1.00-L container. Furthermore, assume that the concentration of PCl_3 decreases by 0.96 mol/L as the reaction comes to equilibrium. What are the concentrations of PCl_5 and Cl_2 at equilibrium?

equilibrium is equal to the changes in the PCl_3 and Cl_2 concentrations. If 0.042 mol/L of PCl_5 is consumed as the reaction comes to equilibrium, 0.042 mol/L each of PCl_3 and Cl_2 must be formed at the same time.

Exercise 10.7 raises an important point. There is a relationship between the *change in the concentrations* of the three components of the reaction as it comes to equilibrium because of the stoichiometry of the reaction. We now continue to examine the problem posed at the beginning of Section 10.7.

It would be useful to have a symbol to represent the change that occurs in the concentration of one of the components of a reaction as it goes from the initial conditions to equilibrium. For chemical reactions we use $\Delta(X)$ to represent the magnitude of the change that occurs in the molar concentration of X as the reaction comes to equilibrium. For example, $\Delta(PCl_5)$ is the magnitude of the change in the concentration of PCl_5 that occurs as the compound decomposes to form PCl_3 and Cl_2. We find that the concentration of PCl_5 at equilibrium is equal to the initial concentration of PCl_5 minus the amount of PCl_5 consumed as the reaction comes to equilibrium.

$[PCl_5]$	$=$	$(PCl_5)_i$	$-$	$\Delta(PCl_5)$
Concentration at equilibrium		*Initial concentration*		*PCl_5 consumed as reaction comes to equilibrium*

We can then define $\Delta(PCl_3)$ and $\Delta(Cl_2)$ as the changes that occur in the concentrations of PCl_3 and Cl_2 as the reaction comes to equilibrium. The concentrations of both substances at equilibrium will be larger than their initial concentrations.

$$[PCl_3] = (PCl_3)_i + \Delta(PCl_3)$$
$$[Cl_2] = (Cl_2)_i + \Delta(Cl_2)$$

Because of the 1:1:1 stoichiometry of the reaction, the magnitude of the change in the concentration of PCl_5 as the reaction comes to equilibrium is equal to the changes in the concentrations of PCl_3 and Cl_2, as we saw in Exercise 10.7.

$$\Delta(PCl_5) = \Delta(PCl_3) = \Delta(Cl_2)$$

We can therefore rewrite the equations that define the equilibrium concentrations of PCl_5, PCl_3, and Cl_2 in terms of a single unknown: ΔC.

$$[PCl_5] = (PCl_5)_i - \Delta C$$
$$[PCl_3] = (PCl_3)_i + \Delta C$$
$$[Cl_2] = (Cl_2)_i + \Delta C$$

Substituting what we know about the initial concentrations of PCl_5, PCl_3, and Cl_2 into the equations gives the following result.

$$[PCl_5] = 0.100 - \Delta C$$
$$[PCl_3] = [Cl_2] = 0 + \Delta C$$

We can now summarize what we know about the reaction as follows.

	$PCl_5(g)$ $\rightleftharpoons$	$PCl_3(g)$ +	$Cl_2(g)$
Initial	$0.100\ M$	0	0
Change	$-\Delta C$	ΔC	ΔC
Equilibrium	$0.100 - \Delta C$	ΔC	ΔC

We now have only one unknown, ΔC, and we need only one equation to solve for one unknown. The obvious equation to turn to is the equilibrium constant expression for the reaction.

$$K_c = \frac{[PCl_3][Cl_2]}{[PCl_5]} = 0.030$$

Substituting what we know about the equilibrium concentrations of PCl_5, PCl_3, and Cl_2 into the equation gives the following result.

$$\frac{[\Delta C][\Delta C]}{[0.100 - \Delta C]} = 0.030$$

This equation can be expanded and then rearranged to give a quadratic equation:

$$[\Delta C]^2 + 0.030[\Delta C] - 0.0030 = 0$$

which can be solved with the quadratic formula.

$$\Delta C = \frac{-b \pm \sqrt{b^2 - 4ac}}{2a} = \frac{-(0.030) \pm \sqrt{(0.030)^2 - 4(1)(-0.0030)}}{2(1)}$$

$$\Delta C = 0.042 \text{ or } -0.072$$

Although two answers come out of the calculation, only the positive root makes any physical sense because we can't have a negative concentration. Thus, the change in the concentrations of PCl_5, PCl_3, and Cl_2 as the reaction comes to equilibrium is 0.042 mol/L.

$$\Delta C = 0.042\ M$$

Substituting the value of ΔC back into the equations that define the equilibrium concentrations of PCl_5, PCl_3, and Cl_2 gives the following results for the question posed at the beginning of Section 10.7.

$$[PCl_5] = 0.100 - 0.042 = 0.058\ M$$
$$[PCl_3] = [Cl_2] = 0 + 0.042 = 0.042\ M$$

In other words, slightly less than half of the PCl_5 present initially decomposes into PCl_3 and Cl_2 when the reaction comes to equilibrium.

To check whether the results of the calculation represent legitimate values for the equilibrium concentrations of the three components of this reaction, we can substitute the values into the equilibrium constant expression.

$$\frac{[PCl_3][Cl_2]}{[PCl_5]} = \frac{[0.042][0.042]}{[0.058]} = 0.030$$

Because the equilibrium constant calculated from the concentrations is equal to the value of K_c given in the problem, we have confidence in our results.

Exercise 10.8

Suppose that 1.00 mol of butane was initially placed in a 1.00-L flask containing no isobutane at 25°C. What would be the equilibrium concentrations of butane and isobutane if $K_c = 2.5$ for the reaction in which butane is converted into isobutane?

Solution

We start, as always, by representing the information in the problem in the following format.

	butane(g) $\rightleftharpoons$	isobutane(g)
Initial	1.00 M	0 M
Equilibrium	$1.00 - \Delta C$	ΔC

We then write the equilibrium constant expression for the reaction:

$$K_c = \frac{[\text{isobutane}]}{[\text{butane}]} = 2.5$$

Substituting the expressions for the equilibrium concentrations of butane and isobutane into this equation gives:

$$\frac{[\text{isobutane}]}{[\text{butane}]} = \frac{[\Delta C]}{[1.00 - \Delta C]} = 2.5$$

We then solve for ΔC and use the results of this calculation to determine the concentrations of butane and isobutane at equilibrium.

$$[\text{isobutane}] = \Delta C = 0.71$$
$$[\text{butane}] = 1.00 - \Delta C = 0.29$$

CHECKPOINT

Do the equilibrium concentrations found in Exercise 10.8 correctly reproduce the K_c?

10.8 Hidden Assumptions That Make Equilibrium Calculations Easier

Suppose that you were asked to solve a slightly more difficult problem.

Sulfur trioxide decomposes to give sulfur dioxide and oxygen with an equilibrium constant of 1.6×10^{-10} at 300°C.

$$2\ SO_3(g) \rightleftharpoons 2\ SO_2(g) + O_2(g)$$

Calculate the equilibrium concentrations of the three components of the system if the initial concentration of SO_3 is 0.100 M.

Once again, the first step in the problem involves building a representation of the information in the problem.

	$2\ SO_3(g) \rightleftharpoons$	$2\ SO_2(g)$ +	$O_2(g)$	$K_c = 1.6 \times 10^{-10}$
Initial	0.100 M	0	0	
Equilibrium	?	?	?	

We then compare the reaction quotient for the initial conditions with the equilibrium constant for the reaction.

$$Q_c = \frac{(SO_2)^2\,(O_2)}{(SO_3)^2} = \frac{(0)^2\,(0)}{(0.100)^2} = 0 < K_c$$

Because the initial concentrations of SO_2 and O_2 are zero, the reaction has to shift to the right to reach equilibrium. As might be expected, some of the SO_3 has to decompose to SO_2 and O_2.

The stoichiometry of the reaction is more complex than the reaction in the previous section, but the changes in the concentrations of the three components of the reaction are still related, as shown in Figure 10.6. For every 2 mol of SO_3 that decomposes we get 2 mol of SO_2 and 1 mol of O_2.

The signs of the ΔC terms in the problem are determined by the fact that the reaction has to shift from left to right to reach equilibrium. The coefficients in the ΔC terms mirror the coefficients in the balanced equation for the reaction. Because twice as many moles of SO_2 are produced as moles of O_2, the change in the concentration of SO_2 as the reaction comes to equilibrium must be twice as large as the change in the concentration of O_2. Because 2 mol of SO_3 is consumed for every mole of O_2 produced, the change in the SO_3 concentration must be twice as large as the change in the concentration of O_2.

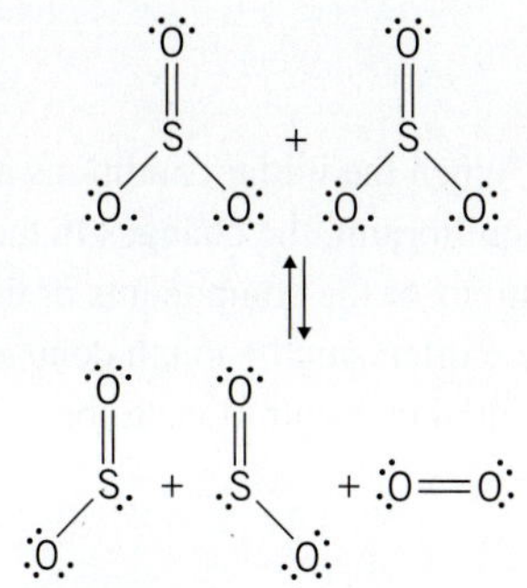

Fig. 10.6 The stoichiometry of this reaction requires that the change in concentrations of both SO_3 and SO_2 must be twice as large as the change in the concentration of O_2 that occurs as the reaction comes to equilibrium.

	$2\ SO_3(g) \rightleftharpoons$	$2\ SO_2(g)\ +$	$O_2(g)$	$K_c = 1.6 \times 10^{-10}$
Initial	0.100 M	0	0	
Change	$-2\Delta C$	$+2\Delta C$	$+\Delta C$	
Equilibrium	$0.100 - 2\Delta C$	$2\Delta C$	ΔC	

Substituting what we know about the problem into the equilibrium constant expression for the reaction gives the following equation.

$$K_c = \frac{[SO_2]^2\,[O_2]}{[SO_3]^2} = \frac{[2\Delta C]^2\,[\Delta C]}{[0.100 - 2\Delta C]^2} = 1.6 \times 10^{-10}$$

CHECKPOINT

The three lines of information given under the balanced chemical equation help to organize the concentrations of reactants and products when solving an equilibrium problem. Which relationship between concentrations (initial, change, equilibrium) must follow the stoichiometry of the balanced chemical equation?

This equation is a bit more of a challenge to expand, but it can be rearranged to give the following cubic equation.

$$4[\Delta C]^3 - (6.4 \times 10^{-10})[\Delta C]^2 + (6.4 \times 10^{-11})[\Delta C] - (1.6 \times 10^{-12}) = 0$$

Solving cubic equations is difficult, however. This problem is therefore an example of a family of problems that are difficult, if not impossible, to solve exactly. Such problems are solved with a general strategy that consists of making an assumption or approximation that turns them into simpler problems.

What assumption can be made to simplify the problem? Let's go back to the first thing we did after building a representation for the problem. We started our calculation by comparing the reaction quotient for the initial concentrations with the equilibrium constant for the reaction.

$$Q_c = \frac{(SO_2)^2\,(O_2)}{(SO_3)^2} = \frac{(0)^2\,(0)}{(0.100)^2} = 0 < K_c$$

We then concluded that the reaction quotient ($Q_c = 0$) was smaller than the equilibrium constant ($K_c = 1.6 \times 10^{-10}$) and decided that some of the SO_3 would have to decompose in order for the reaction to come to equilibrium.

But what about the relative sizes of the reaction quotient and the equilibrium constant for the reaction? The initial values of Q_c and K_c are both relatively small, which means that the initial conditions are reasonably close to equilibrium, as shown in Figure 10.7. As a result, the reaction doesn't have far to go to reach equilibrium. It is therefore reasonable to assume that ΔC is relatively small in this problem.

It is essential to understand the nature of the assumption being made. We aren't assuming that ΔC is zero. If we did that, some of the unknowns would disappear from the equation! We are only assuming that ΔC is so small compared with the initial concentration of SO_3 that it doesn't make a significant difference when $2\Delta C$ is subtracted from that number. We can write the assumption as follows.

$$0.100 - 2\Delta C \approx 0.100$$

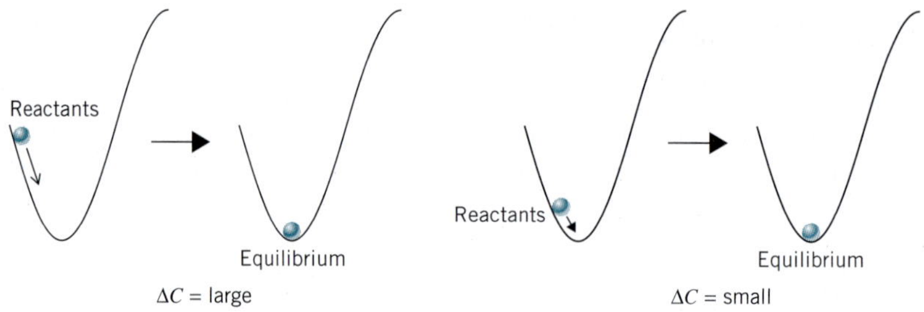

Fig. 10.7 When the initial conditions are close to equilibrium, the changes in the concentrations of the components of the reaction are often small enough compared with the initial concentrations to be ignored.

Let's now go back to the equation we are trying to solve.

$$\frac{[2\Delta C]^2\,[\Delta C]}{[0.100 - 2\Delta C]^2} = 1.6 \times 10^{-10}$$

By assuming that $2\Delta C$ is very much smaller than 0.100, we can replace this equation with the following approximate equation.

$$\frac{[2\Delta C]^2\,[\Delta C]}{[0.100]^2} \approx 1.6 \times 10^{-10}$$

The ΔC values in the numerator cannot be neglected in the equation because they are multiplied. Only when adding or subtracting ΔC to a larger number can ΔC be neglected.

Expanding this gives an equation that is much easier to solve for ΔC.

$$4[\Delta C]^3 \approx 1.6 \times 10^{-12}$$
$$\Delta C \approx 7.4 \times 10^{-5}\,M$$

Before we can go any further, we have to check our assumption that $2\Delta C$ is so small compared with 0.100 that it doesn't make a significant difference when it is subtracted from that number. Is the assumption valid? Is $2\Delta C$ small enough compared with 0.100 to be ignored?

$$0.100 - 2(0.000074) \approx 0.100$$

Using the rules of significant figures, when 0.00015 is subtracted from the initial concentration of 0.100, we are left with a result equal to the initial concentration. Therefore, the assumption is valid. In general, the change in concentration is small enough to be ignored if the change in concentration is less than 5% of the initial concentration. In this example,

$$\frac{2\Delta C}{0.100} \times 100\% = \frac{2(0.000074)}{0.100} \times 100\% = 0.15\% < 5\%$$

We can now use the approximate value of ΔC to calculate the equilibrium concentrations of SO_3, SO_2, and O_2.

$$[SO_3] = 0.100 - 2\Delta C \approx 0.100\,M$$
$$[SO_2] = 2\Delta C \approx 1.5 \times 10^{-4}\,M$$
$$[O_2] = \Delta C \approx 7.4 \times 10^{-5}\,M$$

The equilibrium between SO_3 and mixtures of SO_2 and O_2 therefore strongly favors SO_3, not SO_2.

We can check the results of our calculation by substituting these results into the equilibrium constant expression for the reaction.

$$K_c = \frac{[SO_2]^2\,[O_2]}{[SO_3]^2} = \frac{[1.5 \times 10^{-4}]^2\,[7.4 \times 10^{-5}]}{[0.100]^2} = 1.7 \times 10^{-10}$$

The value of the equilibrium constant that comes out of the calculation agrees closely with the value given in the problem. Our assumption that $2\Delta C$ is negligibly small compared with the initial concentration of SO_3 is therefore valid, and we can feel confident in the answers it provides.

We can also use the equilibrium expression to solve for the concentration of products and reactants at equilibrium when both products and reactants are present initially. Consider the same reaction, in which SO_3 decomposes to form SO_2 and O_2. But this time, let's assume that the initial concentrations of both SO_3 and O_2 are 0.100 *M*. We start, as always, by arranging the relevant information in the problem in the following format.

	$2\,SO_3(g)$	$\rightleftharpoons$	$2\,SO_2(g)$	+	$O_2(g)$	$K_c = 1.6 \times 10^{-10}$
Initial	0.100 *M*		0		0.100 *M*	
Change	$-2\Delta C$		$+2\Delta C$		$+\Delta C$	
Equilibrium	$0.100 - 2\Delta C$		$2\Delta C$		$0.100 + \Delta C$	

$$Q_c = \frac{(SO_2)^2\,(O_2)}{(SO_3)^2} = \frac{(0)^2\,(0.100)}{(0.100)^2} = 0 < K_c$$

The reaction must proceed to the right to reach equilibrium because there is no SO_2 present initially. Substituting what we know about the concentrations of the three components of the reaction at equilibrium into the equilibrium constant expression gives:

$$K_c = \frac{[SO_2]^2[O_2]}{[SO_3]^2} = \frac{[2\Delta C]^2[0.100 + \Delta C]}{[0.100 - 2\Delta C]^2} = 1.6 \times 10^{-10}$$

Because K_c is relatively small, we can assume that ΔC and $2\Delta C$ will be small compared with the initial concentrations of SO_3 and O_2.

$$\frac{[2\Delta C]^2[0.100]}{[0.100]^2} \approx 1.6 \times 10^{-10}$$

Rearranging this equation gives the following result,

$$4\,[\Delta C]^2 = 1.6 \times 10^{-11}$$

which can be solved for the approximate value of ΔC.

$$\Delta C = 2.0 \times 10^{-6}\,M$$

Note that $\Delta C \ll (O_2)_{initial}$ and $\Delta C \ll (SO_3)_{initial}$, confirming the validity of our approximation. We can now use this approximate value of ΔC to calculate the equilibrium concentrations of the three components of the reaction.

$$[SO_3] = 0.100 - 2\Delta C \approx 0.100\,M$$
$$[SO_2] = 2\Delta C \approx 4.0 \times 10^{-6}\,M$$
$$[O_2] = 0.100 + \Delta C \approx 0.100\,M$$

> **CHECKPOINT**
>
> For which of the following equilibrium constants could it be safely assumed that ΔC is small compared with the initial concentration of A when A decomposes to form B and C?
>
> $$A(g) \rightleftharpoons B(g) + C(g)$$
>
> (a) $K = 1.0 \times 10^5$
> (b) $K = 1.0 \times 10^{-5}$
> (c) $K = 1.0 \times 10^{-1}$
> (d) $K = 1.0 \times 10^{-10}$

The concentration of SO_2 (4.0×10^{-6} M) produced in this calculation when O_2 was initially present is considerably smaller than the concentration of SO_2 (1.5×10^{-4} M) produced in the previous calculations when there was no O_2 or SO_2 initially present.

Exercise 10.9

The equilibrium constant for the reaction

$$H_2(g) + Br_2(g) \rightleftharpoons 2\ HBr(g)$$

is 2.18×10^6 at 730°C.

(a) One mol of $H_2(g)$ and 1 mol of $Br_2(g)$ are placed into an empty container at 730°C and allowed to react. Make an estimate of how many moles of HBr would be produced without doing detailed equilibrium calculations. Explain why your estimate is reasonable.

(b) If 3.20 mol of HBr is placed into an empty 12.0-L container at 730°C, what will be the concentrations of HBr, Br_2, and H_2 at equilibrium? State all assumptions. A calculation is required.

Solution

(a) The reaction will proceed to the right, and because the equilibrium constant is very large, 2.18×10^6, the reaction proceeds almost completely to the formation of products. Thus, it is not unreasonable to assume that very little reactant remains. If 1 mol each of H_2 and Br_2 completely react, 2 mol of HBr will be formed. At equilibrium we expect that almost all the product that can be formed will be formed.

(b) The first step in solving this problem might involve calculating the initial concentration of HBr.

$$\frac{3.20 \text{ mol HBr}}{12.0 \text{ L}} = 0.267\ M \text{ HBr}$$

Because we start with only the product of the reaction, and none of the reactants, the reaction quotient is very much larger than the equilibrium constant for the reaction.

$$Q_c = \frac{(\text{HBr})^2}{(\text{H}_2)(\text{Br}_2)} = \frac{(0.267)^2}{(0)(0)} = \infty > K_c$$

The reaction therefore has to shift to the left to reach equilibrium. It might be more convenient to write the equilibrium in the other direction.

$$2\ HBr(g) \rightleftharpoons H_2(g) + Br_2(g)$$

We can then set up the problem as follows.

	$2\ HBr(g)$ $\rightleftharpoons$	$H_2(g)$ +	$Br_2(g)$	$K'_c = ?$
Initial	0.267 M	0	0	
Change	$-2\Delta C$	ΔC	ΔC	
Equilibrium	$0.267 - 2\Delta C$	ΔC	ΔC	

We now have to determine the value of the equilibrium constant for the reaction when it is written in this direction. Transforming the reactants into the products, and vice versa, gives a new equilibrium constant expression that is the inverse of the original expression. Thus, the equilibrium constant for the decomposition of HBr must be the inverse of the equilibrium constant for the formation of HBr from its elements.

$$K_c' = \frac{[H_2][Br_2]}{[HBr]^2} = \frac{[\Delta C][\Delta C]}{[0.267 - 2\Delta C]^2} = \frac{1}{2.18 \times 10^6} = 4.59 \times 10^{-7}$$

Because K_c is relatively small, we can assume that $2\Delta C$ is small compared to 0.267 *M*.

$$\frac{[\Delta C][\Delta C]}{[0.267]^2} \approx 4.59 \times 10^{-7}$$

Solving this approximate equation gives the following result.

$$\Delta C \approx 1.81 \times 10^{-4}\,M$$

Was it valid to assume that $2\Delta C$ was small compared with the initial concentration of HBr?

$$\frac{2(1.81 \times 10^{-4})}{0.267} \times 100\% = 0.136\% < 5\%$$

Yes, it was. The equilibrium concentrations of the three components of the reaction are therefore:

$$[HBr] = 0.267 - 2\Delta C = 0.267\,M$$
$$[Br_2] = [H_2] = \Delta C = 1.81 \times 10^{-4}\,M$$

10.9 The Effect of Temperature on an Equilibrium Constant

The temperature at which the reaction was run has been reported each time an equilibrium constant has been given in this chapter. If the equilibrium constant is really constant, why do we have to worry about the temperature of the reaction?

The value of K_c for a reaction is constant at a given temperature, but it can change with temperature. Consider the reaction between nitrogen and oxygen to produce nitrogen monoxide.

$$N_2(g) + O_2(g) \rightleftharpoons 2\,NO(g)$$

The equilibrium constant for this reaction increases with temperature, as shown in Table 10.2.

Because the equilibrium constant is a ratio of product concentrations divided by reactant concentrations, an increase in the equilibrium constant dictates a shift in the reaction equilibrium toward the product, NO. Because equilibrium constants are temperature dependent, it is important to make sure that the correct equilibrium constant is used when calculating the extent of a chemical reaction.

The equilibrium constant for the reaction at 25°C is so small that virtually no nitrogen monoxide is formed by the reaction between the atmospheric gases, N_2 and O_2. However, at high temperatures, such as those that result from a lightning flash or from combustion inside the cylinder of an automobile engine, the equilibrium constant is orders of magnitude larger. This results in the formation of NO in the atmosphere during electrical storms and from human activities such as driving a car. NO further reacts with O_2 in the atmosphere to produce NO_2, which can combine with water to produce nitric acid, HNO_3. The nitric acid then falls with rain to produce acid rain.

Table 10.2

Temperature Dependence of the Equilibrium Constant for the Formation of NO

Temperature (°C)	K_c
25	4.3×10^{-31}
227	2.7×10^{-18}
727	7.5×10^{-9}
1727	4.0×10^{-4}
2727	1.5×10^{-2}

It is also important to recognize that the equilibrium constant can decrease with increasing temperature. For example, for the reaction

$$2\ NO_2(g) \rightleftharpoons N_2O_4(g)$$

the equilibrium constant is 170 at 25°C but falls to 2.1 at 100°C.

10.10 Le Châtelier's Principle

In 1884, the French chemist and engineer Henry-Louis Le Châtelier proposed one of the central concepts of chemical equilibria. **Le Châtelier's principle** can be stated as follows:

> **A change in one of the variables that describes a system at equilibrium produces a shift in the position of the equilibrium that counteracts the effect of the change.**

Our attention so far has been devoted to describing what happens as a system comes to equilibrium. Le Châtelier's principle describes what happens to a system when something momentarily takes it away from equilibrium. This section focuses on three ways in which we can change the conditions of a chemical reaction at equilibrium: (1) changing the concentration of one of the components of the reaction, (2) changing the pressure on the system, and (3) changing the temperature.

CHANGES IN CONCENTRATION

To illustrate what happens when we change the concentration of one of the reactants or products of a reaction at equilibrium, let's consider a system that consists of 0.50 mol of butane in a 1.0-L flask at 25°C.

	$CH_3CH_2CH_2CH_3(g)$ ⇌	$CH_3CH(CH_3)CH_3(g)$	$K_c = 2.5$
	Butane	*Isobutane*	
Initial	0.50 *M*	0	
Equilibrium	$0.50 - \Delta C$	ΔC	

Substituting the expressions for the equilibrium concentrations of butane and isobutane into the equilibrium constant expression gives:

$$K_c = \frac{[\text{isobutane}]}{[\text{butane}]} = \frac{[\Delta C]}{[0.50 - \Delta C]} = 2.5$$

This equation can then be solved for the equilibrium concentrations of butane and isobutane.

$$[\text{isobutane}] = \Delta C = 0.36\ M$$
$$[\text{butane}] = 0.50 - \Delta C = 0.14\ M$$

Now suppose 0.20 mol of isobutane is added to the reaction while it is at equilibrium. The system is no longer at equilibrium because the concentration of isobutane is now 0.56 *M*. The reaction quotient at the instant the isobutane is added is larger than the equilibrium constant for the reaction.

$$Q_c = \frac{(\text{isobutane})}{(\text{butane})} = \frac{0.56}{0.14} = 4.0 > K_c$$

The system must shift to reestablish the equilibrium ratio of 2.5, and so some of the isobutane will have to be converted into butane. When equilibrium is reestablished,

	butane(g) $\rightleftharpoons$	isobutane(g)
Initial	0.14M	0.56 M
Equilibrium	$0.14 + \Delta C$	$0.56 - \Delta C$

Substituting the new conditions for equilibrium into the equilibrium constant expression gives the following equation:

$$K_c = \frac{[\text{isobutane}]}{[\text{butane}]} = \frac{[0.56 - \Delta C]}{[0.14 + \Delta C]} = 2.5$$

from which ΔC can be found to be 0.060.

This can be solved for the new equilibrium concentrations of butane and isobutane.

$$[\text{isobutane}] = 0.56 - 0.060 = 0.50\ M$$
$$[\text{butane}] = 0.14 + 0.060 = 0.20\ M$$

The calculation can be checked to determine if the ratio [isobutane]/[butane] correctly gives K_c.

$$K_c = \frac{[0.50]}{[0.20]} = 2.5$$

By comparing the new equilibrium concentrations with those obtained before adding isobutane, we can see the effect of increasing the concentration of the isobutane on the equilibrium mixture.

Before	*After*
[isobutane] = 0.36 M	[isobutane] = 0.50 M
[butane] = 0.14 M	[butane] = 0.20 M

The addition of isobutane shifted the equilibrium in such a way as to produce more butane and remove some of the added isobutane. Addition of more butane would have the opposite effect, causing the equilibrium to shift toward the products.

The addition of a reactant or product will shift the equilibrium to reduce the amount of added reactant or product. The removal of a reactant or product will cause the equilibrium to shift to produce that reactant or product.

CHANGES IN PRESSURE

Sometimes it is convenient to discuss gas phase chemical reactions in terms of the partial pressures of individual species rather than their concentrations. However, this makes no difference to the general conclusions about equilibrium that have been discussed. The reason is that partial pressures are related directly to concentrations through the ideal gas law equation.

The effect of changing the pressure on a gas phase reaction depends on the stoichiometry of the reaction. We can demonstrate this by looking at the result of increasing the total pressure on the following reaction at equilibrium.

$$N_2(g) + 3\ H_2(g) \rightleftharpoons 2\ NH_3(g)$$

Let's start with a system that initially contains 2.5 atm of N_2 and 7.5 atm of H_2 at 500°C, and allow the reaction to come to equilibrium. Let's then compress the

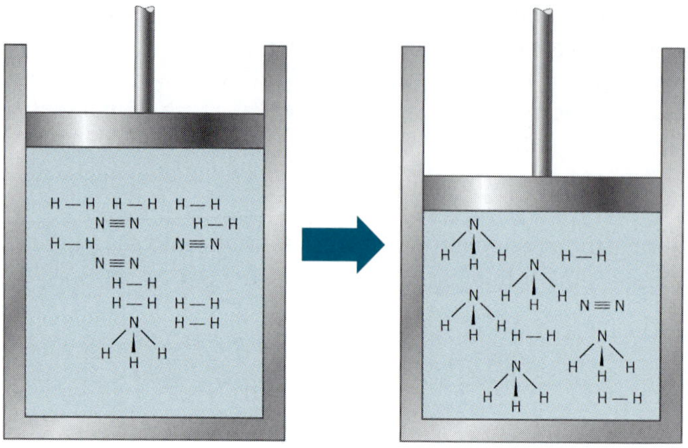

Fig. 10.8 The total number of molecules in the system decreases when N_2 reacts with H_2 to form NH_3. Shifting the equilibrium toward NH_3 decreases the total pressure of the gaseous mixture.

system by increasing the pressure by a factor of 10 and allow the system to return to equilibrium. The partial pressures at equilibrium of all three components of the reaction change when the system is compressed.

Before Compression	*After Compression*
$P_{NH_3} = 0.12$ atm	$P_{NH_3} = 8.4$ atm
$P_{N_2} = 2.4$ atm	$P_{N_2} = 21$ atm
$P_{H_2} = 7.3$ atm	$P_{H_2} = 62$ atm

Before the system was compressed, the partial pressure of NH_3 was only about 1% of the total pressure. After the system is compressed, the partial pressure of NH_3 is almost 10% of the total. These data provide another example of Le Châtelier's principle. A reaction at equilibrium was subjected to a stress—an increase in the total pressure on the system. The reaction then shifted in the direction that minimized the effect of the stress. The reaction shifted toward the products because this reduced the total number of molecules in the gaseous mixture. This in turn decreased the total pressure exerted by the gases, as shown in Figure 10.8.

Whenever the pressure exerted on a system containing gaseous reactants or products is changed, the equilibrium will shift to minimize the pressure change. If the pressure is increased, the equilibrium will shift in the direction of fewer moles of gas. If the pressure is decreased, the equilibrium will shift to produce more moles of gas.

CHECKPOINT

Which way will the equilibrium for the following reaction shift if more P_2 is added to the system at equilibrium? Which way will the equilibrium shift if the pressure is increased? The temperature remains constant.

$$2\ P_2(g) \rightleftharpoons P_4(g)$$

CHANGES IN TEMPERATURE

Changes in the concentrations of the reactants or products of a reaction shift the position of the equilibrium, but they don't change the equilibrium constant for the reaction. Similarly, a change in the pressure on a reaction shifts the position of the equilibrium without changing the magnitude of the equilibrium constant. Changes in the temperature of the system, however, affect the position of the equilibrium by changing the magnitude of the equilibrium constant for the reaction.

Chemical reactions usually give off heat to their surroundings or absorb heat from their surroundings. If we consider heat to be one of the reactants or products of a reaction, we can understand the effect of changes in temperature on the equilibrium. Increasing the temperature of a reaction that gives off heat is the same as adding more of one of the products of the reaction. This places a stress on the reaction, which must be alleviated by converting some of the products back to reactants.

The reaction in which NO_2 dimerizes to form N_2O_4 provides an example of the effect of changes in temperature on the equilibrium constant for a reaction. The reaction is exothermic.

$$2\ NO_2(g) \rightleftharpoons N_2O_4(g) \qquad \Delta H^0 = -57.2\ \text{kJ/mol}_{\text{rxn}}$$

Thus, raising the temperature of the system is equivalent to adding excess product to the system. The equilibrium constant therefore decreases with increasing temperature. For an endothermic reaction, such as the formation of NO from N_2 and O_2 discussed in Section 10.9, an increase in temperature will cause an increase in the equilibrium constant.

Exercise 10.10

Predict the effect of the following changes on the equilibrium of SO_3 with SO_2 and O_2.

$$2\ SO_3(g) \rightleftharpoons 2\ SO_2(g) + O_2(g) \qquad \Delta H^0 = 197.84\ \text{kJ/mol}_{\text{rxn}}$$

(a) Increasing the temperature of the reaction at equilibrium

(b) Increasing the pressure on the reaction at equilibrium

(c) Adding more O_2 when the reaction is at equilibrium

(d) Removing some O_2 from the system when the reaction is at equilibrium

Solution

(a) Because this is an endothermic reaction, which absorbs heat from its surroundings, an increase in the temperature of the reaction leads to an increase in the equilibrium constant and therefore a shift in the position of the equilibrium toward the products.

(b) There is a net increase in the number of molecules in the system as the reactants are converted to products, which leads to an increase in the pressure of the system. The system can minimize the effect of an increase in pressure by shifting the position of the equilibrium toward the reactants, thereby converting some of the SO_2 and O_2 to SO_3.

(c) Adding more O_2 to the system will shift the position of the equilibrium toward the reactants.

(d) Removing O_2 from the system has the opposite effect to part (c); it shifts the equilibrium toward the products of the reaction.

Exercise 10.11

Exercise 10.6 showed that the reaction for the direct metabolism of glucose to glucose-6-phosphate wasn't likely to happen in a typical cell.

$$\text{glucose}(aq) + P_i(aq) \rightleftharpoons \text{glucose-6-phosphate}(aq)$$

The enthalpy change for the reaction is $+35$ kJ/mol$_{rxn}$. What would happen to the equilibrium constant for the reaction if the temperature were increased?

Solution

For an endothermic reaction, an increase in temperature will increase the equilibrium constant. Thus, an increase in temperature will increase the equilibrium constant for the reaction and make the reaction more favorable.

10.11 Le Châtelier's Principle and the Haber Process

Ammonia has been commercially produced from N_2 and H_2 since 1913, when Badische Anilin und Soda Fabrik (BASF) built a plant that used the Haber process to make 30 metric tons of synthetic ammonia per day.

$$N_2(g) + 3\ H_2(g) \rightleftharpoons 2\ NH_3(g) \qquad \Delta H^0 = -92.2\ \text{kJ/mol}_{rxn}$$

Until that time, the principal source of nitrogen for use in farming had been animal and vegetable waste. Today, almost 20 million tons of ammonia worth \$2.5 billion is produced in the United States each year, about 80% of which is used for fertilizers.

The **Haber process** is an example of the use of Le Châtelier's principle to optimize the yield of an industrial chemical. An increase in the pressure at which the reaction is run favors the products of the reaction because there is a net reduction in the number of molecules in the system as N_2 and H_2 combine to form NH_3. Because the reaction is exothermic, the equilibrium constant increases as the temperature of the reaction decreases.

Table 10.3 shows the mole percent of NH_3 at equilibrium when the reaction is run at different combinations of temperature and pressure. The mole percent of

A photograph of the first high-pressure reactor for the synthesis of ammonia by the Haber process.

Table 10.3

Mole Percentage of NH_3 at Equilibrium

	Pressure (atm)			
Temperature (°C)	200	300	400	500
400	38.74	47.85	58.87	60.61
450	27.44	35.93	42.91	48.84
500	18.86	26.00	32.25	37.79
550	12.82	18.40	23.55	28.31
600	8.77	12.97	16.94	20.76

NH_3 under a particular set of conditions is equal to the number of moles of NH_3 at equilibrium divided by the total number of moles of all three components of the reaction times 100. As the data in Table 10.3 demonstrate, the best yields of ammonia are obtained at low temperatures and high pressures.

Unfortunately, low temperatures slow down the rate of the reaction, and the cost of building plants rapidly escalates as the pressure at which the reaction is run is increased. When commercial plants are designed, a temperature is chosen that allows the reaction to proceed at a reasonable rate without decreasing the equilibrium concentration of the product by too much. The pressure is also adjusted so that it favors the production of ammonia without excessively increasing the cost of building and operating the plant. The optimum conditions for running the reaction at present are a pressure between 140 and 340 atm and a temperature between 400°C and 600°C.

Despite all efforts to optimize reaction conditions, the percentages of hydrogen and nitrogen converted to ammonia are still relatively small. Another form of Le Châtelier's principle is therefore used to drive the reaction to completion. Periodically, the reaction mixture is cycled through a cooling chamber. The boiling point of ammonia (BP = −33°C) is much higher than that of either hydrogen (BP = −252.8°C) or nitrogen (BP = −195.8°C). Ammonia can be removed from the reaction mixture, forcing the equilibrium to the right. The remaining hydrogen and nitrogen gases are then recycled through the reaction chamber, where they react to produce more ammonia.

10.12 Equilibrium Reactions That Involve Pure Solids

If we add solid LiF to enough water to make a liter of solution, about 2 g of the compound will dissolve. Continued addition of LiF will result in a buildup of solid LiF on the bottom of the flask. The chemical equation that describes this process can be written as follows.

$$\mathrm{LiF}(s) \rightleftharpoons \mathrm{Li}^+(aq) + \mathrm{F}^-(aq)$$

How do we write the equilibrium constant expression for such a reaction? The Li^+ and F^- ions are dissolved in water, so we can give their equilibrium concentrations in moles per liter, as $[Li^+]$ and $[F^-]$. LiF(s) is more difficult to deal with. We could calculate the number of moles of LiF per liter of the solid LiF. But the result (102 mol per liter) is a constant; the number of moles of LiF in a given volume of LiF is always the same. We therefore incorporate this constant into the equilibrium constant as follows:

$$K_c = \frac{[Li^+][F^-]}{[LiF]} = \frac{[Li^+][F^-]}{102}$$

or

$$K_{sp} = [Li^+][F^-] = 102 \times K_c$$

The new equilibrium constant (**K_{sp}**) is given a subscript "sp" because it is known as a **solubility product equilibrium constant.**

The concentration of any pure solid does not change as the solid dissolves,

$$M_a\,X_b(s) \rightleftharpoons a\,M^{b+}(aq) + b\,X^{a-}(aq)$$

and the equilibrium expression may be written as

$$K_{sp} = [M^{b+}]^a\,[X^{a-}]^b$$

for the dissolution of any sparingly soluble solid.

K_{sp} can be used to predict whether a solid will **precipitate** when solutions containing ions are mixed. If a solution of $LiNO_3$ (that contains the Li^+ and NO_3^- ions) is added to a solution of NaF (that contains the Na^+ and F^- ions), the initial concentrations of Li^+ and F^- in the mixture can be used to determine a reaction quotent, Q.

$$Q_{sp} = (Li^+)(F^-)$$

If $Q_{sp} > K_{sp}$ then a precipitate of solid LiF will form.

$$Li^+(aq) + F^-(aq) \rightleftharpoons LiF(s)$$

Exercise 10.12

What is the equilibrium constant, K_{sp}, for the following reaction if 2 g of LiF will dissolve in enough water to give 1 L of the saturated solution?

$$LiF(s) \rightleftharpoons Li^+(aq) + F^-(aq)$$

Solution

We start by writing the equilibrium constant expression for LiF.

$$K_{sp} = [Li^+][F^-]$$

We then calculate the molar concentrations of the Li^+ and F^- ions in the solution. We know that 2 g of LiF dissolved, which corresponds to 0.08 mol of LiF.

$$2\text{ g LiF} \times \frac{1\text{ mol LiF}}{25.9\text{ g LiF}} = 0.08\text{ mol LiF}$$

For each mole of LiF that dissolves, 1 mol of Li^+ and 1 mol of F^- are formed. Thus if 0.08 mol of LiF dissolves, then 0.08 mol of the Li^+ ion and 0.08 mol of the F^- ion are formed. The concentrations of both Li^+ and F^- are therefore

0.08 mol/L, or 0.08 *M*. We now have enough information to calculate the value of K_{sp} for LiF.

$$K_{sp} = [0.08][0.08] = 6 \times 10^{-3}$$

Exercise 10.13

Enough solid magnesium hydroxide, $Mg(OH)_2$, is added to pure water to form a liter of a saturated solution. Use the K_{sp} for $Mg(OH)_2$ to estimate the molar concentrations of the Mg^{2+} and OH^- ions in this solution. Calculate the solubility of $Mg(OH)_2$ in units of g/100 mL.

$$Mg(OH)_2(s) \rightleftharpoons Mg^{2+}(aq) + 2\ OH^-(aq) \qquad K_{sp} = 1.8 \times 10^{-11}$$

Solution

The solubility product equilibrium expression for $Mg(OH)_2$ would be written as follows.

$$K_{sp} = [Mg^{2+}][OH^-]^2 = 1.8 \times 10^{-11}$$

$Mg(OH)_2$ has a 1:2 stoichiometry. For every mole of $Mg(OH)_2$ that dissolves, 1 mol of the Mg^{2+} ion and 2 mol of the OH^- ion are produced. We can therefore summarize the information in this problem as follows:

	$Mg(OH)_2(s) \rightleftharpoons$	$Mg^{2+}(aq)$	$+\ 2\ OH^-(aq)$
Initial		0	0
Equilibrium		ΔC	$2\Delta C$

Substituting what we know about the equilibrium concentrations of the Mg^{2+} and OH^- ions into the solubility product expression gives:

$$K_{sp} = [\Delta C][2\Delta C]^2 = 1.8 \times 10^{-11}$$

Rearranging this equation gives

$$4[\Delta C]^3 = 1.8 \times 10^{-11}$$

which can be solved for ΔC.

$$\Delta C = 1.7 \times 10^{-4}$$

The concentrations of the Mg^{2+} and OH^- ions at equilibrium are therefore:

$$[Mg^{2+}] = \Delta C = 1.7 \times 10^{-4}\ M$$
$$[OH^-] = 2\Delta C = 2(1.7 \times 10^{-4}) = 3.4 \times 10^{-4}\ M$$

The molar concentration of the dissolved $Mg(OH)_2$ is ΔC, $1.7 \times 10^{-4}\ M$. Each liter of this solution therefore contains 1.7×10^{-4} mol or 9.9×10^{-3} g, of $Mg(OH)_2$. The number of moles of $Mg(OH)_2$ that dissolves is the same as the number of moles of Mg^{2+} formed but is one-half of the number of moles of OH^- formed.

$$1.7 \times 10^{-4} \text{ mol } Mg(OH)_2 \times \frac{58.32 \text{ g } Mg(OH)_2}{1 \text{ mol}} = 9.9 \times 10^{-3} \text{ g } Mg(OH)_2$$

This is the amount of $Mg(OH)_2$ that would dissolve to form a liter of the saturated solution. The amount in 100 mL of solution would be one-tenth as large: 9.9×10^{-4} g. The solubility of $Mg(OH)_2$ is therefore 9.9×10^{-4} g/100 mL.

> **➤ CHECKPOINT**
>
> Will a precipitate form in a solution in which the Mg^{2+} ion concentration is 1.7×10^{-4} *M* and the concentration of the OH^- ion is 2.4×10^{-4} *M*?

Exercise 10.14

NaOH is added to water until the OH^- concentration is 0.010 *M*. How many moles of $Mg(OH)_2$ will dissolve in 1.0 L of this solution?

> **➤ CHECKPOINT**
>
> If K_{sp} for the dissolution of LiF(*s*) is 6×10^{-3}, calculate the concentration of the ions formed.

Solution

The equilibrium reaction is

	$Mg(OH)_2(s) \longrightarrow$	$Mg^{2+}(aq)$	$+ \; 2\,OH^-(aq)$
Initial		0	0.010
Change		ΔC	$2\Delta C$
Equilibrium		ΔC	$0.010 + 2\Delta C$

and

$$\begin{aligned} K_{sp} &= [Mg^{2+}][OH^-]^2 = 1.8 \times 10^{-11} \\ &= [\Delta C][0.010 + 2\Delta C]^2 \end{aligned}$$

K_{sp} is small and therefore it seems reasonable to neglect $2\Delta C$ in comparision to 0.010 *M*. Thus

$$K_{sp} = [\Delta C][0.010]^2$$

The Mg^{2+} concentration comes only from the dissolution of $Mg(OH)_2$, but the OH^- concentration is already 0.010 *M*. Only enough $Mg(OH)_2$ can dissolve so that the change in the Mg^{2+} concentration satisfies the equilibrium condition

$$[\Delta C][0.010]^2 = 1.8 \times 10^{-11}$$

Solving this equilibrium expression shows that the Mg^{2+} concentration can only be $[Mg^{2+}] = [\Delta C] = 1.8 \times 10^{-7}$. Thus the maximum amount of $Mg(OH)_2$ that can dissolve is 1.8×10^{-7} mol/L. This is much less than the amount that can dissolve in pure water, 1.7×10^{-4} mol/L, as calculated in Exercise 10.13. It could have been anticipated that $Mg(OH)_2$ would not dissolve appreciably in a solution containing an ion in common with the ions produced by its dissolution. $2\Delta C$ is also seen to be very small in comparison with 0.010 *M*, and the assumption is valid. This is known as the **common ion effect** when treating aqueous ionic equilibria.

Key Terms

- Chemical kinetics
- Collision theory
- Common ion effect
- Equilibrium
- Equilibrium constant (K_c)
- Equilibrium constant expression
- Equilibrium region
- Haber process
- Kinetic region
- Le Châtelier's principle
- Precipitate
- Rate constant
- Rate law
- Rate of reaction
- Reaction quotient (Q_c)
- Solubility product equilibrium constant (K_{sp})

Problems

Reactions That Don't Go to Completion

1. Describe the difference between reactions that go to completion and reactions that come to equilibrium.
2. What is meant by the term *equilibrium?*
3. Describe the meaning of the symbols [NO] and (NO).

Gas Phase Reactions

4. Define the terms *equilibrium constant* and *equilibrium constant expression.*
5. If 10.0 mol of *trans*-2-butene is placed into an empty flask at 400°C, what will be the equilibrium ratio of *trans*-2-butene to *cis*-2-butene? What if 15.0 mol is placed into an empty flask at the same temperature?
6. If K_c is greater than 1 for the reaction $A \longrightarrow B$, will the equilibrium concentrations of the products be less than or greater than the equilibrium concentrations of the reactants? What if K_c is less than 1?
7. Is the following statement true or false? The equilibrium concentrations depend on the initial concentrations, but the ratio of the equilibrium concentrations is independent of the initial concentrations.

The Rate of a Chemical Reaction

8. Describe how the rate of a chemical reaction is analogous to other rate processes, such as the rate at which a car travels or the rate of inflation.
9. Translate the following equation into an English sentence that carries the same meaning.

$$\text{Rate of reaction} = \frac{-\Delta(X)}{\Delta t}$$

10. What does the rate law tell us about a chemical reaction? Does the rate law tell us anything about the ratio of reactants to products at equilibrium?
11. What is a rate constant? How does it differ from the rate law?

The Collision Theory of Gas Phase Reactions

12. For the reaction of $ClNO_2$ with NO to form ClNO and NO_2, use the collision theory to explain why the rate of reaction is dependent on the concentrations of the reactants $ClNO_2$ and NO.
13. Explain how the rates of the forward and reverse reactions change as a reaction proceeds.
14. Sketch a graph of what happens to the concentrations of N_2, H_2, and NH_3 versus time as the following reaction comes to equilibrium.

$$N_2(g) + 3\,H_2(g) \rightleftharpoons 2\,NH_3(g)$$

Assume that the initial concentrations of N_2 and H_2 are both 1.00 mol/L and that no NH_3 is present initially. Label the kinetic and the equilibrium regions of the graph.

15. Give two ways to define *equilibrium.*
16. On the molecular level, do chemical reactions really stop at equilibrium? Explain.

Equilibrium Constant Expressions

17. Which of the following is the correct equilibrium constant expression for the reaction?

$$Cl_2(g) + 3\,F_2(g) \rightleftharpoons 2\,ClF_3(g)$$

(a) $K_c = \dfrac{2[ClF_3]}{[Cl_2] + 3[F_2]}$ (b) $K_c = \dfrac{[Cl_2] + 3[F_2]}{2[ClF_3]}$

(c) $K_c = \dfrac{[ClF_3]}{[Cl_2][F_2]}$ (d) $K_c = \dfrac{[ClF_3]^2}{[Cl_2][F_2]^3}$

(e) $K_c = \dfrac{[Cl_2][F_2]^3}{[ClF_3]^2}$

18. Which of the following is the correct equilibrium constant expression for the reaction?

$$2\,NO_2(g) \rightleftharpoons 2\,NO(g) + O_2(g)$$

(a) $K_c = \dfrac{[NO_2]}{[NO][O_2]}$ (b) $K_c = \dfrac{[NO][O_2]}{[NO_2]}$

(c) $K_c = \dfrac{[NO_2]^2}{[NO]^2[O_2]}$ (d) $K_c = \dfrac{[NO]^2[O_2]}{[NO_2]^2}$

(e) $K_c = \dfrac{[2\,NO]^2[O_2]}{[2\,NO_2]^2}$

19. Write equilibrium constant expressions for the following reactions.
(a) $O_2(g) + 2\,F_2(g) \rightleftharpoons 2\,OF_2(g)$
(b) $2\,SO_2(g) + O_2(g) \rightleftharpoons 2\,SO_3(g)$
(c) $2\,SO_3(g) + 2\,Cl_2(g) \rightleftharpoons 2\,SO_2Cl_2(g) + O_2(g)$
20. Write equilibrium constant expressions for the following reactions.
(a) $2\,NO(g) + 2\,H_2(g) \rightleftharpoons N_2(g) + 2\,H_2O(g)$
(b) $2\,NOCl(g) \rightleftharpoons 2\,NO(g) + Cl_2(g)$
(c) $2\,NO(g) + O_2(g) \rightleftharpoons 2\,NO_2(g)$
21. Write equilibrium constant expressions for the following reactions.
(a) $2\,CO(g) + O_2(g) \rightleftharpoons 2\,CO_2(g)$
(b) $CO_2(g) + H_2(g) \rightleftharpoons CO(g) + H_2O(g)$
(c) $CO(g) + 2\,H_2(g) \rightleftharpoons CH_3OH(g)$

Rules for Writing Equilibrium Constant Expressions

22. Write equilibrium constant expressions for the following reactions.
(a) $2\,NO_2(g) \rightleftharpoons 2\,NO(g) + O_2(g)$
(b) $2\,NO(g) + O_2(g) \rightleftharpoons 2\,NO_2(g)$

Calculate the value of K_c at 500 K for reaction (a) if the value of K_c for reaction (b) is 6.2×10^5 at 500 K.

23. Use the equilibrium constants for reactions (a) and (b) at 200°C to calculate the equilibrium constant for reactions (c) at that temperature.
(a) $2\ NO(g) \rightleftharpoons N_2(g) + O_2(g)$ $K_c = 4.3 \times 10^{18}$
(b) $2\ NO_2(g) \rightleftharpoons 2\ NO(g) + O_2(g)$ $K_c = 3.4 \times 10^{-7}$
(c) $2\ NO_2(g) \rightleftharpoons N_2(g) + 2\ O_2(g)$ $K_c = ?$

24. Use the equilibrium constants for reactions (a) and (b) at 1000 K to calculate the equilibrium constant for reaction (c), the water–gas shift reaction, at that temperature.
(a) $CO(g) + \frac{1}{2}O_2(g) \rightleftharpoons CO_2(g)$ $K_c = 1.1 \times 10^{11}$
(b) $H_2O(g) \rightleftharpoons H_2(g) + \frac{1}{2}O_2(g)$ $K_c = 7.1 \times 10^{-12}$
(c) $CO(g) + H_2O(g) \rightleftharpoons CO_2(g) + H_2(g)$ $K_c = ?$

25. Calculate K_c for the following reaction at 400 K if 1.000 mol/L of NOCl decomposes at that temperature to give equilibrium concentrations of 0.0222 *M* NO, 0.0111 *M* Cl_2, and 0.978 *M* NOCl.

$$2\ NOCl(g) \rightleftharpoons 2\ NO(g) + Cl_2(g)$$

26. Taylor and Crist [*Journal of the American Chemical Society*, **63,** 1381 (1941)] studied the reaction between hydrogen and iodine to form hydrogen iodide.

$$H_2(g) + I_2(g) \rightleftharpoons 2\ HI(g)$$

They obtained the following data for the concentrations of H_2, I_2, and HI at equilibrium in units of moles per liter.

Trial	*[H_2]*	*[I_2]*	*[HI]*
I	0.0032583	0.0012949	0.015869
II	0.0046981	0.0007014	0.013997
III	0.0007106	0.0007106	0.005468

Calculate the value of K_c for each of the trials. Realizing that there will be deviation due to experimental error, is K_c constant for the reaction?

Reaction Quotients: A Way to Decide Whether a Reaction Is at Equilibrium

27. Suppose that the reaction quotient (Q_c) for the following reaction at some moment in time is 1.0×10^{-8} and the equilibrium constant for the reaction (K_c) at the same temperature is 3×10^{-7}.

$$2\ NO_2(g) \rightleftharpoons 2\ NO(g) + O_2(g)$$

Which of the following is a valid conclusion?
(a) The reaction is at equilibrium.
(b) The reaction must shift toward the products to reach equilibrium.
(c) The reaction must shift toward the reactants to reach equilibrium.

28. Which of the following statements correctly describes a system for which Q_c is larger than K_c?
(a) The reaction is at equilibrium.
(b) The reaction must shift to the right to reach equilibrium.
(c) The reaction must shift to the left to reach equilibrium.
(d) The reaction can never reach equilibrium.

29. Under which set of conditions will the following reaction shift to the right to reach equilibrium?

$$2\ SO_2(g) + O_2(g) \rightleftharpoons 2\ SO_3(g)$$

(a) $K_c < 1$ (b) $K_c > 1$ (c) $Q_c < K_c$
(d) $Q_c = K_c$ (e) $Q_c > K_c$

30. Carbon monoxide reacts with chlorine to form phosgene.

$$CO(g) + Cl_2(g) \rightleftharpoons COCl_2(g)$$

The equilibrium constant, K_c, for the reaction is 1.5×10^4 at 300°C. Is the system at equilibrium at the following concentrations: 0.0040 *M* $COCl_2$, 0.00021 *M* CO, and 0.00040 *M* Cl_2? If not, in which direction does the reaction have to shift to reach equilibrium?

Changes in Concentration That Occur as a Reaction Comes to Equilibrium

31. Describe the relationship between the initial concentration of a reactant (*X*), the concentration of the reactant at equilibrium [*X*], and the change in the concentration of *X* that occurs as the reaction comes to equilibrium, $\Delta(X)$.

32. Explain why the change in the N_2 concentration that occurs when the following reaction comes to equilibrium is related to the change in the H_2 concentration.

$$N_2(g) + 3\ H_2(g) \rightleftharpoons 2\ NH_3(g)$$

Derive an equation that describes the relationship between the changes in the concentrations of the two reagents.

33. When confronted with the task in the previous problem, the following incorrect answer is often given.

$$\Delta(N_2) = 3\Delta(H_2)$$

Explain why the equation is wrong. Write the correct form of the relationship.

34. Calculate the changes in the CO and Cl_2 concentrations that occur if the concentration of $COCl_2$ decreases by 0.250 mol/L as the following reaction comes to equilibrium.

$$COCl_2(g) \rightleftharpoons CO(g) + Cl_2(g)$$

35. Calculate the changes in the N_2 and H_2 concentrations that occur if the concentration of NH_3 decreases by 0.234 mol/L as the following reaction comes to equilibrium.

$$2\ NH_3(g) \rightleftharpoons N_2(g) + 3\ H_2(g)$$

36. Which of the following equations describes the relationship between the magnitude of the changes in the NO_2 and O_2 concentrations as the following reaction comes to equilibrium?

$$2\,NO(g) + O_2(g) \rightleftharpoons 2\,NO_2(g)$$

(a) $\Delta(NO_2) = \Delta(O_2)$ (b) $\Delta(NO_2) = 2\Delta(O_2)$
(c) $\Delta(O_2) = 2\Delta(NO_2)$

37. Which of the following equations correctly describes the relationship between the changes in the Cl_2 and F_2 concentrations as the following reaction comes to equilibrium?

$$Cl_2(g) + 3\,F_2(g) \rightleftharpoons 2\,ClF_3(g)$$

(a) $\Delta(Cl_2) = \Delta(F_2)$ (b) $\Delta(Cl_2) = 2\Delta(F_2)$
(c) $\Delta(Cl_2) = 3\Delta(F_2)$ (d) $\Delta(F_2) = 2\Delta(Cl_2)$
(e) $\Delta(F_2) = 3\Delta(Cl_2)$

38. Which of the following describes the change that occurs in the concentration of H_2O when ammonia reacts with oxygen to form nitrogen oxide and water according to the following equation if the change in the NH_3 concentration is ΔC?

$$4\,NH_3(g) + 5\,O_2(g) \rightleftharpoons 4\,NO(g) + 6\,H_2O(g)$$

(a) ΔC (b) $1.5\Delta C$ (c) $2\Delta C$ (d) $4\Delta C$
(e) $6\Delta C$

39. Calculate the concentrations of H_2 and NH_3 at equilibrium if a reaction that initially contained 1.000 *M* concentrations of both N_2 and H_2 is found to have an N_2 concentration of 0.922 *M* at equilibrium.

	$N_2(g)$ +	$3\,H_2(g)$ $\rightleftharpoons$	$2\,NH_3(g)$
Initial	1.000 *M*	1.000 *M*	0 *M*
Equilibrium	0.922 *M*	?	?

40. Calculate the equilibrium constant for the reaction in the previous problem.

Hidden Assumptions That Make Equilibrium Calculations Easier

41. Calculate the equilibrium concentrations of N_2O_4 and NO_2 when 0.100 *M* N_2O_4 decomposes to form NO_2 at 25°C.

$$N_2O_4(g) \rightleftharpoons 2\,NO_2(g) \qquad K_c = 5.8 \times 10^{-5}$$

42. Without detailed equilibrium calculations, estimate the equilibrium concentration of N_2O_4 present when 1.00 *M* NO_2 reacts to form N_2O_4 at 25°C.

$$N_2O_4(g) \rightleftharpoons 2\,NO_2(g) \qquad K_c = 5.8 \times 10^{-5}$$

43. Calculate the equilibrium concentrations of N_2, H_2, and NH_3 present when a mixture that was initially 0.10 *M* N_2, 0.10 *M* H_2, and 0.10 *M* NH_3 comes to equilibrium at 500°C.

$$N_2(g) + 3\,H_2(g) \rightleftharpoons 2\,NH_3(g)$$
$$K_c = 0.040 \text{ (at 500°C)}$$

44. Calculate the equilibrium concentrations of CO, H_2O, CO_2, and H_2 present in the water–gas shift reaction at 800°C if the initial concentrations of CO and H_2O are 1.00 *M*.

$$CO(g) + H_2O(g) \rightleftharpoons CO_2(g) + H_2(g)$$
$$K_c = 0.72 \text{ (at 800°C)}$$

45. What initial equal concentrations of CO and H_2O would be needed to reach an equilibrium concentration of 1.00 *M* CO_2 in the water–gas shift reaction described in the previous problem?

46. Calculate the equilibrium concentrations of N_2, O_2, and NO present when a mixture that was initially 0.100 *M* in N_2 and 0.090 *M* in O_2 comes to equilibrium at 600°C.

$$N_2(g) + O_2(g) \rightleftharpoons 2\,NO(g) \qquad K_c = 3.3 \times 10^{-10}$$

47. Sulfuryl chloride decomposes to sulfur dioxide and chlorine. Calculate the concentrations of the three components of the system at equilibrium if 6.75 g of SO_2Cl_2 in a 1.00-L flask decomposes at 25°C.

$$SO_2Cl_2(g) \rightleftharpoons SO_2(g) + Cl_2(g)$$
$$K_c = 1.4 \times 10^{-5}$$

48. Without detailed equilibrium calculations, estimate the concentrations of NO and NOCl at equilibrium if a mixture that was initially 0.50 *M* in NO and 0.10 *M* in Cl_2 combined to form nitrosyl chloride, NOCl.

$$2\,NO(g) + Cl_2(g) \rightleftharpoons 2\,NOCl(g)$$
$$K_c = 2.1 \times 10^3 \text{ (at 500 K)}$$

49. Calculate the concentrations of PCl_5, PCl_3, and Cl_2 that are present when the following gas phase reaction comes to equilibrium. Calculate the percent of the PCl_5 that decomposes when the reaction comes to equilibrium. Explain the difference between the results of this calculation and those obtained in Section 10.7.

	$PCl_5(g)$ $\rightleftharpoons$	PCl_3 +	$Cl_2(g)$
Initial	1.00 *M*	0	0

$$K_c = 0.0013 \text{ (at 450 K)}$$

50. Calculate the concentrations of PCl_5, PCl_3, and Cl_2 present when the following gas phase reaction comes to equilibrium. Calculate the percent decomposition in the reaction and explain any difference between the results of this calculation and the results obtained in Section 10.7.

	$PCl_5(g)$ $\rightleftharpoons$	$PCl_3(g)$ +	$Cl_2(g)$
Initial	1.00 *M*	0	0.20 *M*

$$K_c = 0.0013 \text{ (at 450 K)}$$

51. Calculate the concentrations of NO, NO_2, and O_2 present when the following gas phase reaction reaches equilibrium.

$$2\ NO_2(g) \rightleftharpoons 2\ NO(g) + O_2(g)$$

	$2\ NO_2(g)$	$2\ NO(g)$	$O_2(g)$
Initial	0.100 M	0	0

$$K_c = 3.4 \times 10^{-7} \text{ (at 200°C)}$$

52. Calculate the concentrations of NO, NO_2, and O_2 present when the following gas phase reaction reaches equilibrium.

$$2\ NO_2(g) \rightleftharpoons 2\ NO(g) + O_2(g)$$

	$2\ NO_2(g)$	$2\ NO(g)$	$O_2(g)$
Initial	0.10 M	0	0.050 M

$$K_c = 3.4 \times 10^{-7} \text{ (at 200°C)}$$

53. Calculate the equilibrium concentrations of SO_3, SO_2, and O_2 present when 0.100 mol of SO_3 in a 250-mL flask at 300°C decomposes to form SO_2 and O_2.

$$2\ SO_3(g) \rightleftharpoons 2\ SO_2(g) + O_2(g)$$
$$K_c = 1.6 \times 10^{-10} \text{ (at 300°C)}$$

54. Without detailed equilibrium calculations, estimate the equilibrium concentration of SO_3 when a mixture of 0.100 mol of SO_2 and 0.050 mol of O_2 in a 250-mL flask at 300°C combine to form SO_3.

$$2\ SO_2(g) + O_2(g) \rightleftharpoons 2\ SO_3(g)$$
$$K_c = 6.3 \times 10^9 \text{ (at 300°C)}$$

55. Sometimes the technique used in this chapter to simplify equilibrium problems is stated as follows: "Assume that ΔC is zero." Explain why this is wrong. What is the correct way of describing the assumption?

56. What is the advantage of setting up equilibrium problems so that ΔC is small compared with the initial concentrations?

57. Describe how to test whether ΔC is small enough compared with the initial concentrations to be legitimately ignored.

58. At 600°C the equilibrium constant for

$$N_2(g) + O_2(g) \rightleftharpoons 2\ NO(g)$$

is 3.3×10^{-10}.
(a) Is ΔC likely to be small or large for this reaction? Explain.
(b) Find K_c for

$$2\ NO(g) \rightleftharpoons N_2(g) + O_2(g)$$

Is ΔC likely to be small or large for the decomposition of NO?

The Effect of Temperature on an Equilibrium Constant

59. Why is an equilibrium constant always reported at a specific temperature?

60. If an equilibrium constant gets smaller as temperature increases, will increasing the temperature favor the products or the reactants?

61. If K_c decreases with decreasing temperature, will increasing the temperature favor the reactants or products?

Le Châtelier's Principle

62. Le Châtelier's principle has been applied to many fields, ranging from economics to psychology to political science. Give an example of Le Châtelier's principle in a field outside the physical sciences.

63. Predict the effect of increasing the pressure at constant T on the following reactions at equilibrium.
(a) $2\ SO_3(g) + 2\ Cl_2(g) \rightleftharpoons 2\ SO_2Cl_2(g) + O_2(g)$
(b) $O_2(g) + 2\ F_2(g) \rightleftharpoons 2\ OF_2(g)$
(c) $2\ NO(g) + O_2(g) \rightleftharpoons 2\ NO_2(g)$

64. Predict the effect of decreasing the pressure at constant T on the following reactions at equilibrium.
(a) $N_2O_4(g) \rightleftharpoons 2\ NO_2(g)$
(b) $N_2(g) + O_2(g) \rightleftharpoons 2\ NO(g)$
(c) $NO(g) + NO_2(g) \rightleftharpoons N_2O_3(g)$

65. Predict the effect of increasing the concentration of the reagent indicated in boldface on each of the following reactions at equilibrium. T and P are constant.
(a) $\mathbf{2\ NO_2(g)} \rightleftharpoons N_2O_4(g)$
(b) $2\ SO_3(g) \rightleftharpoons 2\ SO_2(g) + \mathbf{O_2(g)}$
(c) $\mathbf{PF_5(g)} \rightleftharpoons PF_3(g) + F_2(g)$

66. Predict the effect of decreasing the concentration of the boldface reagent on each of the following reactions at equilibrium. T and P are constant.
(a) $N_2(g) + O_2(g) \rightleftharpoons \mathbf{2\ NO(g)}$
(b) $\mathbf{3\ O_2(g)} \rightleftharpoons 2\ O_3(g)$
(c) $Cl_2(g) + \mathbf{3\ F_2(g)} \rightleftharpoons 2\ ClF_3(g)$

67. Use Le Châtelier's principle to predict the effect of an increase in pressure on the solubility of a gas in water. T is constant.

Le Châtelier's Principle and the Haber Process

68. List as many ways as possible of increasing the yield of ammonia in the Haber process.

$$N_2(g) + 3\ H_2(g) \rightleftharpoons 2\ NH_3(g)$$

69. Explain why an increase in pressure favors the formation of ammonia in the Haber process.

70. Predict how an increase in the volume of the container by a factor of 2 would affect the concentrations of ammonia and oxygen in the following reaction. T is constant.

$$4\ NH_3(g) + 5\ O_2(g) \rightleftharpoons 4\ NO(g) + 6\ H_2O(g)$$

71. According to the data in Table 10.3, what are the worst conditions of temperature and pressure under which to carry out the production of ammonia?

72. How are the data in Table 10.3 consistent with Le Châtelier's principle? Consider a fixed temperature with a changing pressure and a fixed pressure with a changing temperature.

Equilibrium Reactions That Involve Pure Solids

73. Write an equation that describes the relationship between the concentrations of the Ag^+ and CrO_4^{2-} ions in a saturated solution of Ag_2CrO_4.

74. Write an equation that describes the relationship between the concentrations of the Bi^{3+} and S^{2-} ions in a saturated solution of Bi_2S_3.

75. Calculate the K_{sp} constant for the dissolution of strontium fluoride if the solubility of SrF_2 in water is 0.107 g/L.

76. Silver acetate, $Ag(CH_3CO_2)$, is marginally soluble in water. What is the K_{sp} for silver acetate if 1.190 g of $Ag(CH_3CO_2)$ dissolves in 99.40 mL of water?

77. People who have the misfortune of going through a series of X rays of the gastrointestinal tract are often given a suspension of solid barium sulfate in water to drink. $BaSO_4$ is used instead of other Ba^{2+} salts, which also reflect X rays, because it is relatively insoluble in water. (Thus the patient is exposed to the minimum amount of toxic Ba^{2+} ion.) What is the equilibrium constant for the dissolution of barium sulfate if 1.0 g of $BaSO_4$ dissolves in 400,000 g of water?

78. What is the solubility of silver sulfide in water in grams per 100 mL if the K_{sp} for Ag_2S is 6.3×10^{-50}?

79. What is the solubility in water for each of the following salts in grams per 100 mL?
(a) Hg_2S ($K_{sp} = 1.0 \times 10^{-47}$)
(b) HgS ($K_{sp} = 4 \times 10^{-53}$)

80. What is the solubility of Hg_2S in mol/L in a solution that contains an S^{2-} concentration of 0.10 *M*? What is the solubility of HgS in a solution that contains an S^{2-} concentration of 0.10 *M*? See Problem 79 for the K_{sp}.

81. How many grams of AgBr will dissolve in 1.0 L of water containing a Br^- concentration of 0.050 *M*? Table B.10 contains solubility product constant values.

82. What is the solubility of Ag_2CO_3 in water in mol/L? What is the solubility in mol/L of Ag_2CO_3 in a solution containing an Ag^+ concentration of 0.15 *M*? See Table B.10 in the appendix.

Integrated Problems

83. Which of the following diagrams best represents the concentrations of the reactants and products for the following reaction at equilibrium? Explain what is wrong with each incorrect diagram. (● represents isobutane, and ● represents *n*-butane.)

$$\underset{\textit{Isobutane}}{CH_3CH(CH_3)CH_3(g)} \rightleftharpoons \underset{\textit{n-Butane}}{CH_3CH_2CH_2CH_3(g)} \quad K_c = 0.4$$

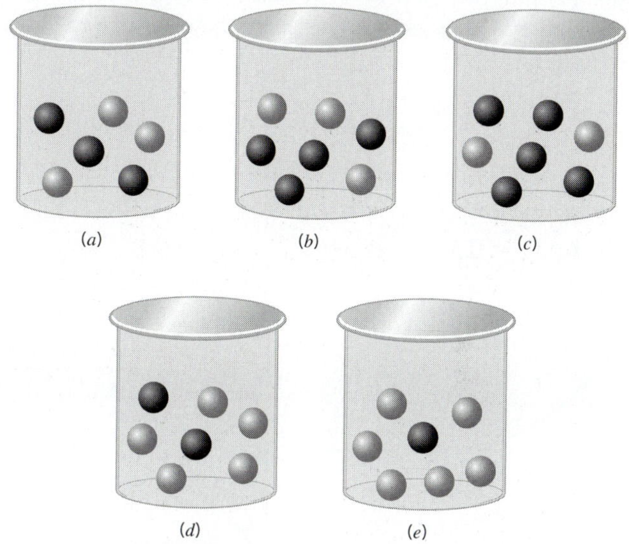

(*a*) (*b*) (*c*) (*d*) (*e*)

84. A sparingly soluble hypothetical ionic compound, MX_2, is placed into a beaker of distilled water. Which of the following diagrams *best* describes what happens in solution? Explain what is wrong with each incorrect diagram.

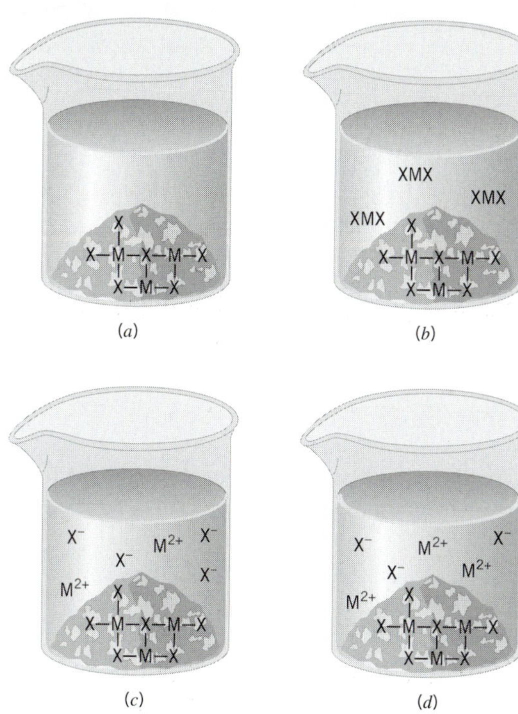

(*a*) (*b*) (*c*) (*d*)

85. Describe the relationship between k_f and k_r for the following one-step reaction at equilibrium.

$$Z(g) + X(g) \underset{k_r}{\overset{k_f}{\rightleftharpoons}} Y(g) \quad K_c = 1 \times 10^{-3}$$

Which is true: $k_f = k_r$, $k_f < k_r$, or $k_f > k_r$? Explain your reasoning.

86. For the reaction $A \rightleftharpoons B$, match the graphs of concentration versus time to the appropriate set of rate constants.

$$\text{Rate}_{\text{forward}} = k_A(A) \quad \text{Rate}_{\text{reverse}} = k_B(B)$$

(a) $k_A = k_B$
(b) $k_A = 1.0/\text{s}$, $k_B = 0.5/\text{s}$
(c) $k_A = 0.5/\text{s}$, $k_B = 1.0/\text{s}$

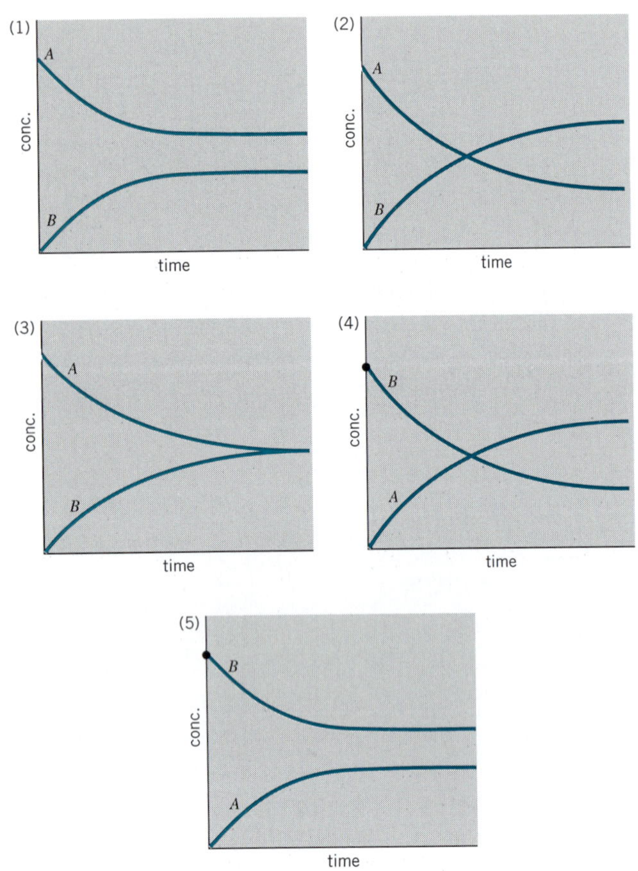

87. Write the equilibrium constant expression for the dissolving of strontium fluoride in water.

$$SrF_2(s) \rightleftharpoons Sr^{2+}(aq) + 2\,F^-(aq)$$

(a) When $SrF_2(s)$ is placed in water the compound dissolves to produce an equilibrium Sr^{2+} concentration of 5.8×10^{-4} mol/L. What is K_{sp} for the reaction?

(b) If 50.00 mL of 0.100 M $Sr(NO_3)_2$ is mixed with 50.00 mL of 0.100 M NaF, will a precipitate form? Explain your answer.

88. Several plots of concentration versus time for the reaction $A \rightleftharpoons B$ are given below. $K_c = 2$. Only one of the plots can be correct. Which one is it? Explain what is wrong with each of the incorrect plots.

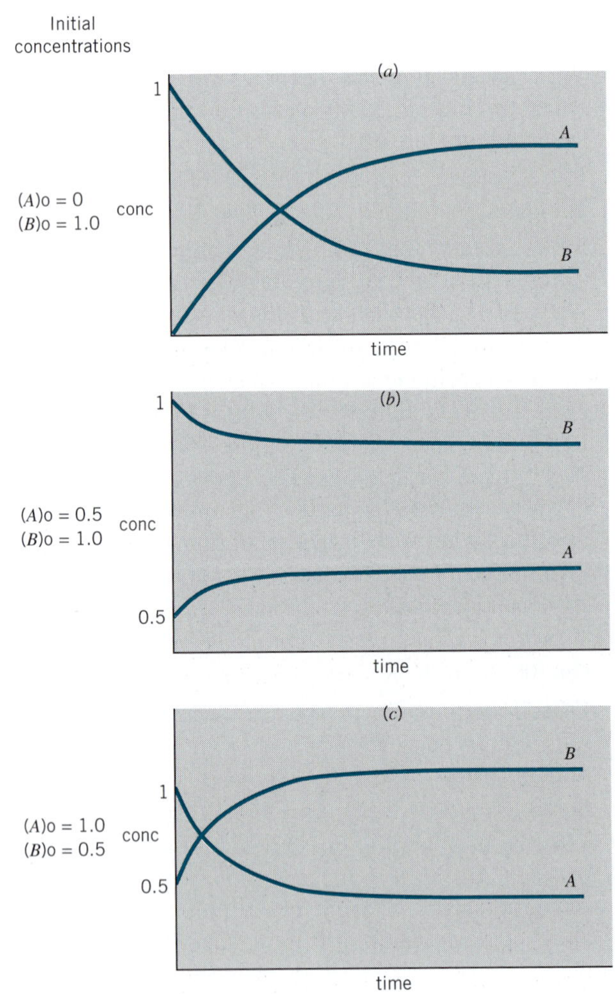

89. Molecular iodine dissociates into iodine atoms at 1000 K

$$I_2(g) \longrightarrow 2I(g)$$

K_c is 3.8×10^{-5}.

(a) If 0.456 mol of I_2 is placed into a 2.30-L flask at 1000 K, what will be the equilibrium concentrations of I_2 and I?

(b) If 0.912 mol of I is placed into a 2.30-L flask containing no I_2 at 1000 K, estimate the concentration of I_2 at equilibrium. Do no detailed equilibrium calculations, but clearly explain your answer.

90. When equilibrium is reached for the dissolving of solid calcium sulfate in water at 25°C

$$CaSO_4(s) \rightleftharpoons Ca^{2+}(aq) + SO_4^{2-}(aq)$$

it is found that $[Ca^{2+}] = [SO_4^{2-}] = 4.9 \times 10^{-3}\,M$.

(a) Calculate the equilibrium constant (K_{sp}) for the above reaction.

(b) A solution contains Ca^{2+} (aq) at a concentration of $3.6 \times 10^{-3}\,M$ and SO_4^{2-} (aq) at a concentration of $8.0 \times 10^{-3}\,M$. Will solid $CaSO_4$ be formed? Show your calculations.

(c) $CaSO_4$ is allowed to dissolve in 1.0 L of water at 25°C until equilibrium is reached. Then the water is allowed to evaporate to half its original volume. What are the equilibrium concentrations of Ca^{2+} and SO_4^{2-} in this solution? Explain.

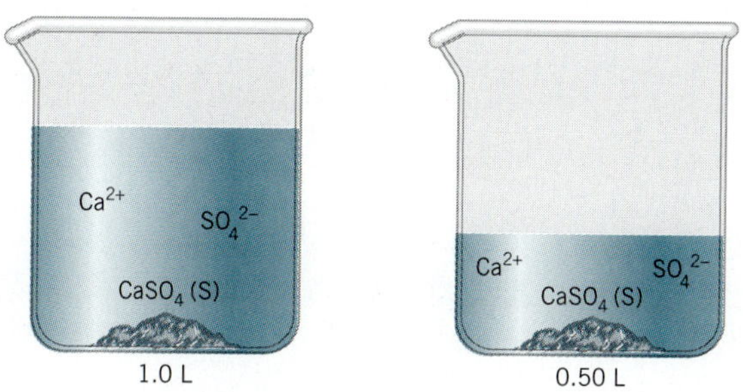

91. At 35°C, for the reaction

$$2\,NOCl(g) \rightleftharpoons 2\,NO(g) + Cl_2(g)$$

$K_c = 1.6 \times 10^{-5}$.
If 1.0 mol of NOCl is placed into an empty 1.0-L flask, what will be the equilibrium concentrations of all species? State all assumptions and show all work. Provide a justification for any assumptions.

92. For the reaction

$$cis\text{-2-butene}(g) \rightleftharpoons trans\text{-2-butene}(g)$$

the rate constant in the forward direction, k_f, is 2.10×10^{-7}/s and that in the reverse direction, K_r, is 1.65×10^{-7}/s.

Which of the following graphs could represent the change in concentrations with time? More than one graph could be correct. Explain your reasoning.

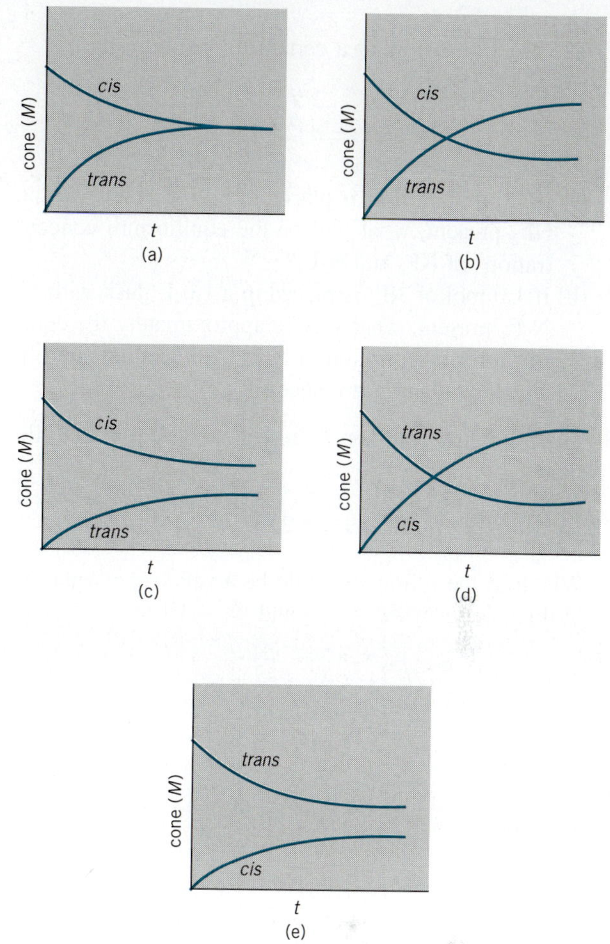

93. At 25°C, 1.5×10^{-2} mol of Ag_2SO_4 dissolves in 1.0 L of water.

$$Ag_2SO_4(s) \rightleftharpoons 2\,Ag^+(aq) + SO_4^{2-}(aq)$$

(a) How many moles of Ag^+ are present?

(b) How many moles of SO_4^{2-} are present?

(c) Calculate the equilibrium constant for the reaction.

94. For the following reaction at 230°C, the concentrations of the species at equilibrium were found to be as follows: $[NO] = 0.0542\,M$, $[O_2] = 0.127\,M$, $[NO_2] = 15.5\,M$.

$$2\,NO(g) + O_2(g) \rightleftharpoons 2\,NO_2(g)$$

(a) What does it mean to say that a reaction has come to equilibrium? Must all reactions eventually come to equilibrium?

(b) What is the equilibrium constant, K_c, for the reaction?

(c) If sufficient O_2 and NO_2 are added to increase $[O_2]$ and $[NO_2]$ to 1.127 and 16.5 M, respectively, while keeping [NO] at 0.0542 M, in which direction will the reaction proceed?

95. The reaction below at a certain temperature has $K_c = 5.0 \times 10^{-9}$.

$$N_2F_4(g) \rightleftharpoons 2\ NF_2(g)$$

(a) If 1.0 mol of N_2F_4 is placed in a 1.0-L flask with no NF_2 present, what will be the equilibrium concentrations of NF_2 and N_2F_4?

(b) If 1.0 mol of NF_2 is placed in a 1.0-L flask with no N_2F_4 present, what will be approximately the equilibrium concentration of N_2F_4? No detailed equilibrium calculations are necessary.

96. The following equation represents a system at equilibrium.

$$Cl_2(g) \rightleftharpoons 2\ Cl(g)$$

Which of the following could be a valid representation of this system? (= Cl_2 and = Cl.)

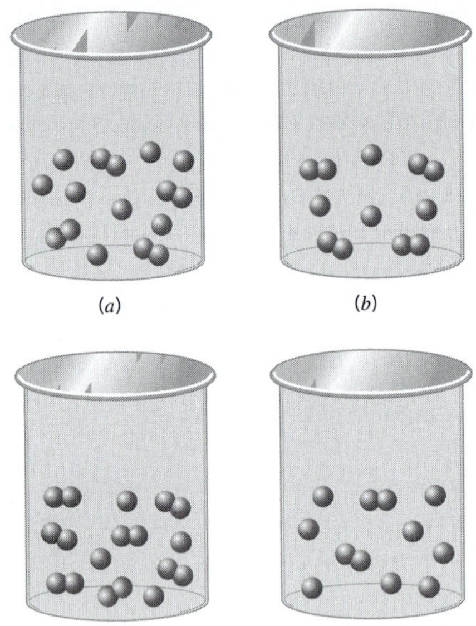

(a) (b) (c) (d) (e) all of the above

Chapter Ten

SPECIAL TOPICS

10A.1 A Rule of Thumb for Testing the Validity of Assumptions

For those instructors who wish a rigorous solution to the extent of gas phase equilibrium reactions, the Special Topics appendix can be included. This appendix shows how to test the validity of assumptions made to simplify the mathematics of equilibrium calculations.

When the simplifying assumptions fail, a second method of calculation is given. In many cases reactions can be pushed in one direction or another and allowed to come to equilibrium from a direction which makes certain mathematical simplifications possible.

There was no doubt about the validity of the assumption that ΔC was small compared with the initial concentration of SO_3 in Section 10.8. The value of ΔC was so small that $2\Delta C$ was an order of magnitude smaller than the experimental error involved in measuring the initial concentration of SO_3.

In general, we can get some idea of whether ΔC might be small enough to be ignored by comparing the initial reaction quotient with the equilibrium constant for the reaction. If Q_c and K_c are both much smaller than 1, or both much larger than 1, the reaction doesn't have very far to go to reach equilibrium, and the assumption that ΔC is small enough to be ignored is probably legitimate.

This raises an interesting question: How do we decide whether it is valid to assume that ΔC is small enough to be ignored? The answer to this question depends on how much error we are willing to let into our calculation before we no longer trust the results. As a rule of thumb, chemists typically assume that ΔC is negligibly small so long as what is added to or subtracted from the initial concentrations of the reactants or products is less than 5% of the initial concentration. The best way to decide whether the assumption meets this rule of thumb in a particular calculation is to try it and see if it works.

Exercise 10A.1

Ammonia is made from nitrogen and hydrogen by the following reaction.

$$N_2(g) + 3\,H_2(g) \rightleftharpoons 2\,NH_3(g)$$

If the initial concentration of N_2 is 0.100 mol/L and the initial concentration of H_2 is 0.100 mol/L. Calculate the equilibrium concentrations of the three components of the reaction at 500°C if the equilibrium constant for the reaction at that temperature is 0.040.

Solution

We start, as always, by building a representation for the problem based on the following general format.

	$N_2(g)$	$+\ 3\,H_2(g) \rightleftharpoons$	$2\,NH_3(g)$	$K_c = 0.040$
Initial	0.100 *M*	0.100 *M*	0	
Equilibrium	?	?	?	

We then calculate the initial reaction quotient and compare it with the equilibrium constant for the reaction.

$$Q_c = \frac{(NH_3)^2}{(N_2)(H_2)^3} = \frac{(0)^2}{(0.100)(0.100)^3} = 0$$

The reaction quotient ($Q_c = 0$) is smaller than the equilibrium constant ($K_c = 0.040$), so the reaction has to shift to the right to reach equilibrium. This will result in a decrease in the concentrations of N_2 and H_2 and an increase in the NH_3 concentration.

The relationship between the changes in the concentrations of N_2, H_2, and NH_3 as the reaction comes to equilibrium is determined by the stoichiometry of the reaction, as shown in Figure 10A.1. We can therefore set up the problem as follows.

	$N_2(g)$	+	$3\ H_2(g)$	$\rightleftharpoons$	$2\ NH_3(g)$	$K_c = 0.040$
Initial	$0.100\ M$		$0.100\ M$		0	
Change	$-\Delta C$		$-3\Delta C$		$+2\Delta C$	
Equilibrium	$0.100 - \Delta C$		$0.100 - 3\Delta C$		$2\Delta C$	

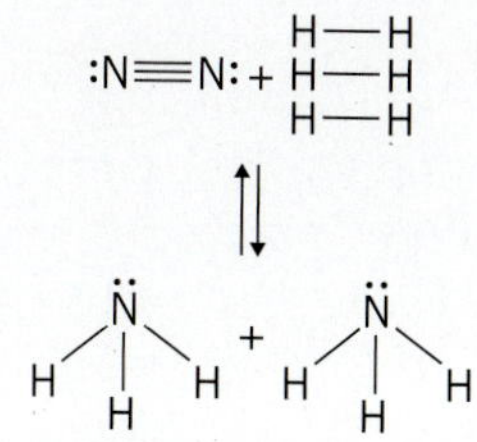

Fig. 10A.1 The stoichiometry of this reaction determines the relationship between the magnitude of the changes in the concentrations of N_2, H_2, and NH_3 as this reaction comes to equilibrium.

Substituting this information into the equilibrium constant expression for the reaction gives the following equation.

$$K_c = \frac{[NH_3]^2}{[N_2][H_2]^3} = \frac{[2\Delta C]^2}{[0.100 - \Delta C][0.100 - 3\Delta C]^3} = 0.040$$

Because Q_c for the initial concentrations and K_c for the reaction are both smaller than 1, let's try the assumption that ΔC is small enough that subtracting it from 0.100—or even subtracting $3\Delta C$ from 0.100—doesn't make a significant change. This assumption gives the following approximate equation.

$$\frac{[2\Delta C]^2}{[0.100][0.100]^3} \approx 0.040$$

Solving the equation for ΔC gives the following result.

$$\Delta C \approx 1.0 \times 10^{-3}\ M$$

Now we have to check our assumptions. Is ΔC significantly smaller than 0.100? Yes, ΔC is about 1% of the initial concentration of N_2.

$$\frac{0.0010}{0.100} \times 100\% = 1.0\%$$

Is $3\Delta C$ significantly smaller than 0.100? Once again, the answer is yes, $3\Delta C$ is only about 3% of the initial concentration of H_2.

$$\frac{3(0.0010)}{0.100} \times 100\% = 3.0\%$$

We can therefore use the approximate value of ΔC to determine the equilibrium concentrations of N_2, H_2, and NH_3.

$$[NH_3] = 2\Delta C \approx 0.0020\ M$$
$$[N_2] = 0.100 - \Delta C \approx 0.099\ M$$
$$[H_2] = 0.100 - 3\Delta C \approx 0.097\ M$$

Only 1% of the nitrogen is converted into ammonia under these conditions.

We can check the validity of our calculation by substituting the information back into the equilibrium constant expression.

$$K_c = \frac{[NH_3]^2}{[N_2][H_2]^3} = \frac{[0.0020]^2}{[0.099][0.097]^3} = 0.044$$

Once again, we have reason to accept the assumption that ΔC is small compared with the initial concentrations because the equilibrium constant calculated from the data agrees well with the value of K_c given in the problem.

10A.2 What Do We Do When the Approximation Fails?

It is easy to envision a problem in which the assumption that ΔC is small compared with the initial concentrations can't possibly be valid. All we have to do is construct a problem in which there is a large difference between the values of Q_c for the initial concentrations and K_c for the reaction at equilibrium. Consider the following problem, for example.

Nitrogen oxide reacts with oxygen to form nitrogen dioxide.

$$2\ NO(g) + O_2(g) \rightleftharpoons 2\ NO_2(g)$$

The equilibrium constant for the reaction is 3.0×10^6 at 200°C. Assume initial concentrations of 0.100 M *for NO and 0.050* M *for O_2. Calculate the concentrations of the three components of the reaction at equilibrium.*

We start, once again, by representing the information in the problem as follows.

	$2\ NO(g)$	+	$O_2(g)$	$\rightleftharpoons$	$2\ NO_2(g)$	$K_c = 3.0 \times 10^6$
Initial	0.100 *M*		0.050 *M*		0	
Equilibrium	?		?		?	

The first step is always the same: Compare the initial value of the reaction quotient with the equilibrium constant.

$$Q_c = \frac{(NO_2)^2}{(NO)^2(O_2)} = \frac{(0)^2}{(0.100)^2(0.050)} = 0 \ll K_c$$

The relationship between the initial reaction quotient ($Q_c = 0$) and the equilibrium constant ($K_c = 3.0 \times 10^6$) tells us something we may already have suspected: the reaction must shift to the right to reach equilibrium.

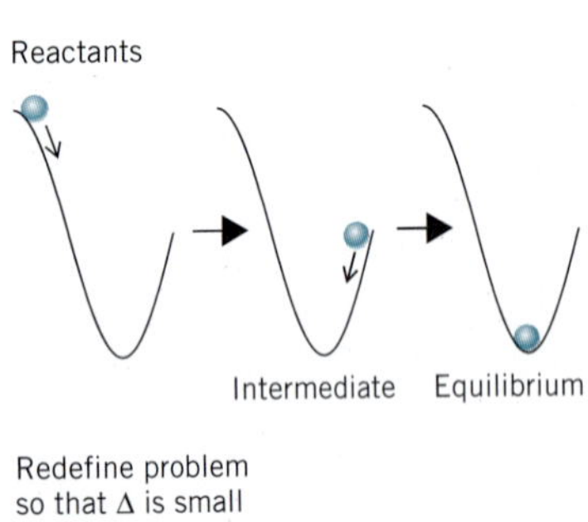

Fig. 10A.2 When the initial conditions are very far from equilibrium, it is often useful to redefine the problem. This involves driving the reaction as far as possible in the direction favored by the equilibrium constant. When the reaction returns to equilibrium from the intermediate conditions, changes in the concentrations of the components of the reaction are often small enough compared with the initial concentration to be ignored.

Some might ask, "Why calculate the initial value of the reaction quotient for the reaction? Isn't it obvious that the reaction has to shift to the right to produce at least some NO_2?" Yes, it is. But calculating the value of Q_c for the reaction does more than tell us in which direction it has to shift to reach equilibrium. It also gives us an indication of how far the reaction has to go to reach equilibrium.

In this case, Q_c is so very much smaller than K_c for the reaction that we have to conclude that the initial conditions are very far from equilibrium. It would therefore be a mistake to assume that ΔC is small.

We can't assume that ΔC is negligibly small in this problem, but we can redefine the problem so that the assumption becomes valid. The key to achieving this goal is to remember the conditions under which we can assume that ΔC is small enough to be ignored. This assumption is valid only when Q_c is of the same order of magnitude as K_c (i.e., when Q_c and K_c are both much larger than 1 or much smaller than 1). We can solve problems for which Q_c isn't close to K_c by redefining the initial conditions so that Q_c becomes close to K_c (see Figure 10A.2). To show how this can be done, let's return to the problem given in this section.

The equilibrium constant for the reaction between NO and O_2 to form NO_2 is much larger ($K_c = 3.0 \times 10^6$) than Q_c. This means that the equilibrium favors the products of the reaction. The best way to handle the problem is to drive the reaction as far as possible to the right, and then let it come back to equilibrium. Let's therefore define an intermediate set of conditions that correspond to what would happen if we push the reaction as far as possible to the right.

	$2\ NO(g)$ +	$O_2(g)$	$\rightleftharpoons 2\ NO_2(g)$	$K_c = 3.0 \times 10^6$
Initial	0.100 *M*	0.050 *M*	0	
Change	−0.100 *M*	−0.050 *M*	+0.100 *M*	
Intermediate	0	0	0.100 *M*	

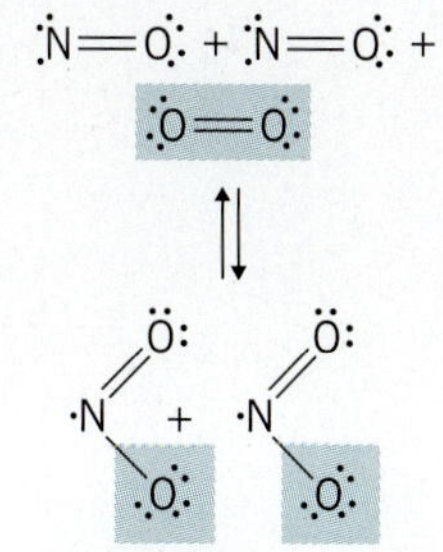

Fig. 10A.3 Once again, the stoichiometry of the reaction determines the relationship among the magnitude of the changes in the concentrations of the three components of the reaction as it comes to equilibrium.

We can see where this gets us by calculating the reaction quotient for the intermediate conditions.

$$Q_c = \frac{(NO_2)^2}{(NO)^2(O_2)} = \frac{(0.100)^2}{(0)^2(0)} = \infty$$

The reaction quotient is now larger than the equilibrium constant, and the reaction has to shift back to the left to reach equilibrium. Some of the NO_2 must now decompose to form NO and O_2. The relationship between the changes in the concentrations of the three components of this reaction is determined by the stoichiometry of the reaction, as shown in Figure 10A.3. We therefore set up the problem as follows.

	$2\ NO(g)$ +	$O_2(g)$	$\rightleftharpoons 2\ NO_2(g)$	$K_c = 3.0 \times 10^6$
Intermediate	0	0	0.100 *M*	
Change	$+2\Delta C$	$+\Delta C$	$-2\ \Delta C$	
Equilibrium	$2\Delta C$	ΔC	$0.100 - 2\Delta C$	

We now substitute what we know about the reaction into the equilibrium constant expression.

$$K_c = \frac{[NO_2]^2}{[NO]^2[O_2]} = \frac{[0.100 - 2\Delta C]^2}{[2\Delta C]^2[\Delta C]} = 3.0 \times 10^6$$

Because the reaction quotient for the intermediate conditions and the equilibrium constant are both relatively large, we can assume that the reaction doesn't have very far to go to reach equilibrium. In other words, we assume that $2\Delta C$ is small compared with the intermediate concentration of NO_2 and derive the following approximate equation.

$$\frac{[0.100]^2}{[2\Delta C]^2[\Delta C]} \approx 3.0 \times 10^6$$

We then solve the equation for an approximate value of ΔC.

$$\Delta C \approx 9.4 \times 10^{-4}\ M$$

We now check our assumption that $2\Delta C$ is small enough compared with the intermediate concentration of NO_2 to be ignored.

$$\frac{2(0.00094)}{0.100} \times 100\% = 1.9\%$$

The value of $2\Delta C$ is less than 2% of the intermediate concentration of NO_2, which means that it can be legitimately ignored in the calculation.

Since the approximation is valid, we can use the new value of ΔC to calculate the equilibrium concentrations of NO_2, NO, and O_2.

$$[NO_2] = 0.100 - 2\Delta C \approx 0.098\ M$$
$$[NO] = 2\Delta C \approx 0.0019\ M$$
$$[O_2] = \Delta C \approx 0.00094\ M$$

In general, the assumption that ΔC is small compared with the initial concentrations of the reactants or products works best under the following conditions.

- When $K_c \ll 1$ and we approach equilibrium from left to right. (We start with excess reactants and form some products.)
- When $K_c \gg 1$ and we approach equilibrium from right to left. (We start with excess products and form some reactants.)

Problems

A Rule of Thumb for Testing the Validity of Assumptions

10A-1. Describe what happens if you make the assumption that ΔC is zero in the following equation.

$$\frac{[0.125 - \Delta C][2.40 - 2\Delta C]^2}{[0.200 + 2\Delta C]^2} = 1.3 \times 10^{-8}$$

Explain how to get around this problem.

10A-2. Explain why ΔC is relatively small when the reaction quotient (Q_c) is reasonably close to the equilibrium constant for the reaction (K_c).

10A-3. Explain why the assumption that ΔC is small compared with the initial concentrations of the reactants and products is doomed to fail when the reaction quotient (Q_c) is very different from the equilibrium constant for the reaction (K_c).

What Do We Do When the Approximation Fails?

10A-4. Describe the technique used to solve problems for which the reaction quotient is very different from the equilibrium constant.

10A-5. Before we can solve the following problem, we have to define a set of intermediate conditions under which the concentration of one of the reactants or products is zero.

	$2\ NO_2(g)$	$\rightleftharpoons$	$2\ NO(g)$	$+\ O_2(g)$
Initial	$0.10\ M$		$0.10\ M$	$0.005\ M$

$$K_c = 5.3 \times 10^{-6} \text{ (at 250°C)}$$

Which of the following goals determines whether we push the reaction as far as possible to the right or as far as possible to the left?

(a) To make both Q_c and K_c large

(b) To make both Q_c and K_c small

(c) To bring Q_c as close as possible to K_c

(d) To make the difference between Q_c and K_c as large as possible

Chapter Eleven

ACIDS AND BASES

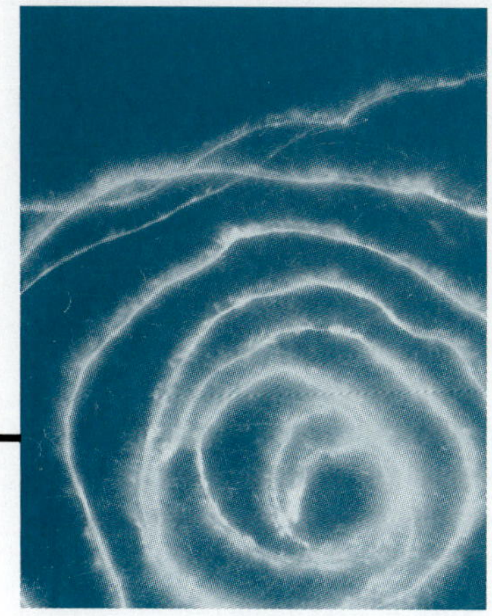

Special Topics

11.1 Properties of Acids and Bases

This chapter is a traditional presentation of acid–base chemistry. The main portion of the chapter focuses on monoprotic acids and bases. Polyprotic acids and bases are discussed in the chapter appendix Special Topics. ChemActivities 42–47 in Chemistry: A Guided Inquiry *are designed to accompany this chapter.*

For more than 300 years, chemists have classified substances that behave like vinegar as **acids** and substances that have properties like wood ash as **bases** (or **alkalies**). The name "acid" comes from the Latin word *acidus,* which means "sour," and refers to the sharp odor and sour taste of many acids. Vinegar, for example, tastes sour because it is a dilute solution of acetic acid in water. Lemon juice tastes sour because it contains citric acid. Milk turns sour when it spoils because lactic acid is formed, and the unpleasant, sour odor of rotten meat or butter can be attributed to compounds such as butyric acid that form when fat spoils.

One of the characteristic properties of acids is its ability to dissolve most metals. Zinc metal, for example, rapidly reacts in hydrochloric acid to form an aqueous solution of $ZnCl_2$ and hydrogen gas.

$$Zn(s) + 2\ HCl(aq) \longrightarrow ZnCl_2(aq) + H_2(g)$$

Another characteristic property of acids is the ability to change the color of vegetable dyes, such as litmus. Litmus is a mixture of blue dyes that turns red in the presence of acid. Litmus has been used to test for acids for at least 300 years.

Bases also have characteristic properties. They taste bitter and often feel slippery. They change the color of litmus from red to blue, thereby reversing the change in color that occurs when litmus comes in contact with an acid. Bases become less alkaline when they are combined with acids, and acids lose their characteristic sour taste and ability to dissolve metals when they are mixed with alkalies.

11.2 The Arrhenius Definition of Acids and Bases

In 1887, Svante Arrhenius took a major step toward answering the important question, "What factors determine whether a compound is an acid or a base?" Arrhenius suggested that acids are compounds that *ionize* when they dissolve in water to give H^+ ions and a corresponding negative ion. According to this model, hydrogen chloride is an acid because it dissociates, or ionizes, when it dissolves in water to give hydrogen (H^+) and chloride (Cl^-) ions (Figure 11.1). This aqueous solution is known as hydrochloric acid and is often written as HCl(*aq*). It is important to recognize, however, that HCl dissociates almost completely to form the H^+ and Cl^- ions when it dissolves in water.

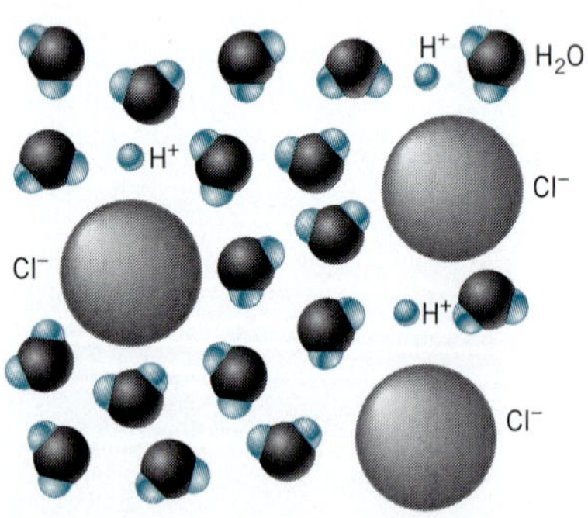

Fig. 11.1 The Arrhenius model assumes that HCl dissociates into H^+ and Cl^- ions when it dissolves in water.

$$HCl(g) \xrightarrow{H_2O} H^+(aq) + Cl^-(aq)$$

Arrhenius argued that bases are compounds that dissociate in water to give OH^- and positive ions. NaOH is an Arrhenius base because it dissociates in water to give the hydroxide (OH^-) and sodium (Na^+) ions.

$$NaOH(s) \xrightarrow{H_2O} Na^+(aq) + OH^-(aq)$$

An **Arrhenius acid** is therefore any substance that ionizes when it dissolves in water to give the hydrogen ion, H^+. An **Arrhenius base** is any substance that gives the hydroxide ion, OH^-, when it dissolves in water. Arrhenius acids include compounds such as HCl, HCN, and H_2SO_4 that ionize in water to give the H^+ ion. Arrhenius bases include ionic compounds that contain the OH^- ion, such as NaOH, KOH, and $Ca(OH)_2$.

> **➤ CHECKPOINT**
>
> Classify the following compounds as Arrhenius acids or bases: HNO_3, $Mg(OH)_2$, CH_3CO_2H.
>
> $$HNO_3(aq) \xrightarrow{H_2O} H^+(aq) + NO_3^-(aq)$$
>
> $$Mg(OH)_2(s) \overset{H_2O}{\rightleftharpoons} Mg^{2+}(aq) + 2\,OH^-(aq)$$
>
> $$CH_3CO_2H(aq) \overset{H_2O}{\rightleftharpoons} H^+(aq) + CH_3CO_2^-(aq)$$

11.3 The Brønsted–Lowry Definition of Acids and Bases

In 1923, Johannes Brønsted and Thomas Lowry independently proposed a more powerful set of definitions of acids and bases. The Brønsted, or Brønsted–Lowry, model is based on the assumption that acids donate H^+ ions to another ion or molecule, which acts as a base. According to this model, HCl doesn't dissociate in water to form H^+ and Cl^- ions. Instead, an H^+ ion is transferred from HCl to a water molecule to form an H_3O^+ ion and a Cl^- ion.

$$HCl(aq) + H_2O(l) \longrightarrow H_3O^+(aq) + Cl^-(aq)$$

The H_3O^+ ion is known as the **hydronium ion.** The Brønsted model of the reaction between HCl and water is shown in Figure 11.2.

Because it is a proton, an H^+ ion is several orders of magnitude smaller than the smallest atom. As a result, the charge on an isolated H^+ ion is distributed over such a small surface area that the H^+ ion is attracted toward any source of negative charge that exists in the solution. Thus, the instant that an H^+ ion is created in an aqueous solution, it bonds to the electronegative oxygen atom of a water molecule. The Brønsted model, in which H^+ ions are transferred from one ion or molecule to another, therefore seems more reasonable than the Arrhenius model, which assumes that H^+ ions exist in aqueous solution.

Even the Brønsted model is naive, however. Each H^+ ion that an acid donates to water is actually bound in a complex of four neighboring water molecules, as shown in Figure 11.3. A more realistic formula for the substance produced when an acid loses an H^+ ion in water is therefore $H(H_2O)_4^+$, or $H_9O_4^+$. For practical purposes, however, this substance can be represented as the H_3O^+ ion.

The reaction between HCl and water provides the basis for understanding the definitions of a Brønsted acid and a Brønsted base. According to the Brønsted model, when HCl dissociates in water, HCl acts as an H^+ ion donor and H_2O acts as an H^+ ion acceptor.

$$\underset{\textit{H}^+\textit{ion donor}}{HCl(aq)} + \underset{\textit{H}^+\textit{ion acceptor}}{H_2O(l)} \longrightarrow H_3O^+(aq) + Cl^-(aq)$$

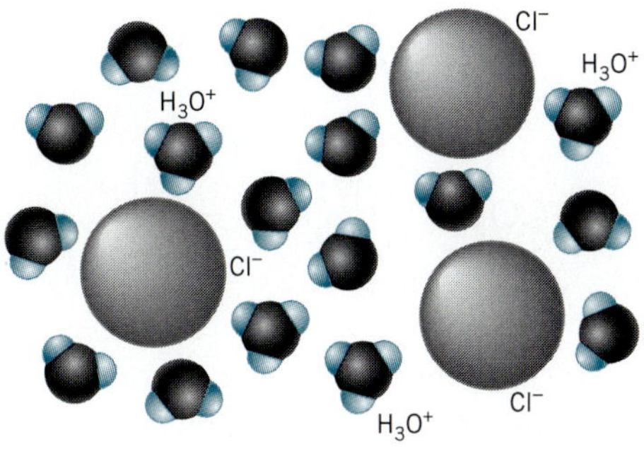

Fig. 11.2 The Brønsted model assumes that HCl molecules donate an H^+ ion to water molecules to form H_3O^+ and Cl^- ions when HCl dissolves in water.

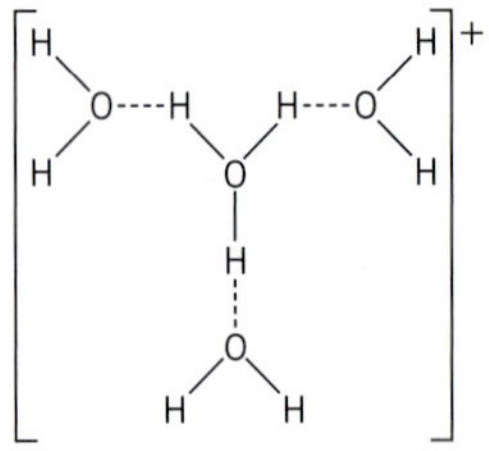

Fig. 11.3 Structure of the $H(H_2O)_4^+$ ion formed when an acid reacts with water. For practical purposes, the ion can be thought of as an H_3O^+ ion.

A **Brønsted acid** is therefore any substance (such as HCl) that can donate an H^+ ion to a base. A **Brønsted base** is any substance (such as H_2O) that can accept an H^+ ion from an acid.

There are two ways of naming the H^+ ion. Some chemists call it a hydrogen ion; others call it a proton. As a result, Brønsted acids are known as either **hydrogen-ion donors** or **proton donors.** Brønsted bases are **hydrogen-ion acceptors** or **proton acceptors.** In the main body of this chapter we will deal primarily with **monoprotic acids.** A monoprotic acid has a single H^+ ion that it can donate. The Special Topics section at the end of the chapter will discuss more complex acids.

From the perspective of the Brønsted model, reactions between acids and bases always involve the transfer of an H^+ ion from a proton donor to a proton acceptor. Acids can be uncharged molecules.

$$\underset{Acid}{HCl(aq)} + \underset{Base}{NH_3(aq)} \longrightarrow Cl^-(aq) + NH_4^+(aq)$$

They can also be positive ions,

$$\underset{Acid}{NH_4^+(aq)} + \underset{Base}{OH^-(aq)} \rightleftharpoons NH_3(aq) + H_2O(l)$$

or negative ions.

$$\underset{Acid}{H_2PO_4^-(aq)} + \underset{Base}{H_2O(l)} \rightleftharpoons HPO_4^{2-}(aq) + H_3O^+(aq)$$

Brønsted bases can be identified from their Lewis structures. According to the Brønsted model, a base is any ion or molecule that can accept a proton. To understand the implications of this definition, look at how the prototypical base, the OH^- ion, accepts a proton.

$$H^+ + :\ddot{\underset{..}{O}}-H^- \longrightarrow H-\ddot{\underset{..}{O}}-H$$

The only way to accept an H^+ ion is to form a covalent bond to it. In order to form a covalent bond to an H^+ ion that has no valence electrons, the base must provide both of the electrons needed to form the bond. Thus, only compounds that have pairs of nonbonding valence electrons can act as H^+ ion acceptors, or Brønsted bases. The following compounds, for example, can all act as Brønsted bases because they all contain nonbonding pairs of electrons.

NH_3

H_2O

CO_3^{2-}

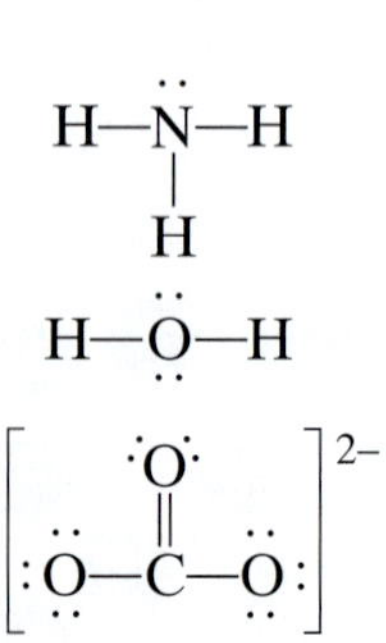

The Brønsted model includes within the category of bases any ion or molecule that contains one or more pairs of nonbonding valence electrons that can accept a proton. There are many molecules and ions that satisfy the definition of a

Brønsted base and relatively few, such as the following, that do not. Substances that do not behave as a Brønsted base have no nonbonding electron pairs.

CH_4 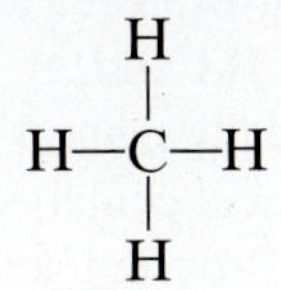

H_2 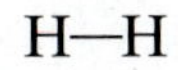

NH_4^+ 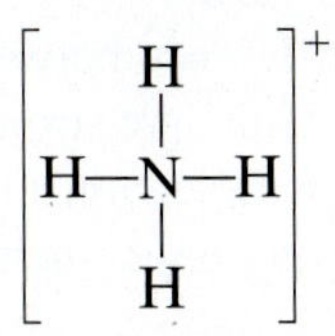

In order to determine whether a substance behaves as a Brønsted acid or base, the substance must be examined in the context of the chemical reaction in which it participates.

Exercise 11.1

Identify the reactant that behaves as a Brønsted acid and the reactant that behaves as a Brønsted base in each of the following reactions.

(a) $HF(aq) + OH^-(aq) \rightleftharpoons H_2O(l) + F^-(aq)$

(b) $CH_3CO_2H(aq) + H_2O(l) \rightleftharpoons CH_3CO_2^-(aq) + H_3O^+(aq)$

(c) $C_6H_5NH_2(aq) + HNO_3(aq) \rightleftharpoons C_6H_5NH_3^+(aq) + NO_3^-(aq)$

Solution

(a) acid: HF base: OH^-

(b) acid: CH_3CO_2H base: H_2O

(c) acid: HNO_3 base: $C_6H_5NH_2$

11.4 Conjugate Acid–Base Pairs

The Brønsted model of acids and bases creates a link between acids and bases. Every time a Brønsted acid acts as an H^+ ion donor, it forms a conjugate base. Imagine a generic acid, H*A*, where *A* represents any anion. When the acid donates an H^+ ion to water, one product of the reaction is the A^- ion, which in the reverse process is an H^+ ion acceptor or a Brønsted base.

$$\underset{\text{Acid}}{HA(aq)} + H_2O(l) \rightleftharpoons H_3O^+(aq) + \underset{\text{Conjugate base}}{A^-(aq)}$$

Conversely, every time a base gains an H^+ ion, the product is a Brønsted acid, H*A*, which can donate a proton in the reverse direction.

$$\underset{\text{Base}}{A^-(aq)} + H_2O(l) \rightleftharpoons \underset{\text{Conjugate acid}}{HA(aq)} + OH^-(aq)$$

Acids and bases in the Brønsted model therefore exist as **conjugate acid–base pairs** whose formulas are related by the gain or loss of a hydrogen ion. The term

Table 11.1
Typical Brønsted Acids and Their Conjugate Bases

Acid	Base
H_3O^+	H_2O
H_2O	OH^-
HCl	Cl^-
H_2SO_4	HSO_4^-
HSO_4^-	SO_4^{2-}
NH_4^+	NH_3

➤ CHECKPOINT
Phosphoric acid, H_3PO_4, is a common additive in soft drinks. Write the chemical equation for the dissociation of phosphoric acid in water and predict the chemical formula of its conjugate base. Aniline, $C_6H_5NH_2$, is a base used in the manufacture of dyes. Write the chemical equation for the reaction of aniline with water and predict the chemical formula of its conjugate acid.

conjugate comes from the Latin stem meaning "joined together" and refers to things that are joined, particularly in pairs. It is therefore the perfect term to describe the relationship between Brønsted acids and bases.

Our use of the symbols H*A* and *A*$^-$ for a conjugate acid–base pair doesn't mean that all acids are electrically neutral molecules or that all bases are negative ions. It signifies only that the acid contains an H^+ ion that isn't present in the conjugate base. As noted earlier, Brønsted acids and bases can be electrically neutral molecules, positive ions, or negative ions. Various Brønsted acids and their conjugate bases are given in Table 11.1.

There is a very large difference between the strengths of the various acids in Table 11.1. HCl, H_2SO_4, and H_3O^+ are strong acids, whereas H_2O and the NH_4^+ ion are weak acids, as we will see in Section 11.8. There is also a very large difference between the strength of a strong base, such as the OH^- ion, and very weak bases, such as the Cl^- ion or water.

It is also important to recognize that some compounds can be both a Brønsted acid and a Brønsted base. H_2O and HSO_4^-, for example, can be found in both columns in Table 11.1. Water is the perfect example of this behavior because it simultaneously acts as an acid and a base when it reacts with itself to form the H_3O^+ and OH^- ions.

$$H_2O(l) + H_2O(l) \rightleftharpoons H_3O^+(aq) + OH^-(aq)$$

The concept of conjugate acid–base pairs plays a vital role in explaining reactions between acids and bases. According to the Brønsted model, an acid always reacts with a base to form the conjugate base and conjugate acid. Consider the following reaction, for example.

$$\underset{\textit{Acid}_1}{HNO_3(aq)} + \underset{\textit{Base}_2}{NH_3(aq)} \longrightarrow \underset{\textit{Conjugate acid}_2}{NH_4^+(aq)} + \underset{\textit{Conjugate base}_1}{NO_3^-(aq)}$$

In the course of this reaction, nitric acid donates an H^+ ion to form its conjugate base, the nitrate ion (NO_3^-). At the same time, ammonia acts as a base, accepting an H^+ ion to form its conjugate acid, the ammonium ion (NH_4^+).

The products of the reaction are often combined and written as an aqueous solution of an ionic compound, or salt.

$$HNO_3(aq) + NH_3(aq) \longrightarrow NH_4NO_3(aq)$$

Because the products of the reaction are neither as acidic as nitric acid nor as basic as ammonia, the reaction is often called a **neutralization reaction.** This doesn't imply that the products have no acid or base properties. It only suggests that the products are less acidic and less basic than the starting materials.

Water is often one of the products of a neutralization reaction. Consider the reaction between formic acid and sodium hydroxide, for example.

$$\underset{\textit{Acid}_1}{HCO_2H(aq)} + \underset{\textit{Base}_2}{NaOH(aq)} \longrightarrow \underset{\textit{Conjugate acid}_2}{H_2O(l)} + Na^+(aq) + \underset{\textit{Conjugate base}_1}{HCO_2^-(aq)}$$

The salt produced in this reaction is sodium formate, $NaHCO_2$. The formate ion, HCO_2^-, is the conjugate base of formic acid, and water is the conjugate acid of the hydroxide ion in sodium hydroxide.

Exercise 11.2

Write the chemical equation for the reaction of chlorous acid, $HClO_2$, with potassium hydroxide. Identify the acid, base, conjugate acid, and conjugate base in the reaction.

Solution

$$HClO_2(aq) + KOH(aq) \longrightarrow H_2O(l) + K^+(aq) + ClO_2^-(aq)$$

acid:	$HClO_2$	conjugate base:	ClO_2^-
base:	KOH	conjugate acid:	H_2O

11.5 The Role of Water in the Brønsted Model

The Lewis structure of water can help us understand why H_3O^+ and OH^- ions play such an important role in the chemistry of aqueous solutions. The Lewis structure suggests that the hydrogen and oxygen atoms in a water molecule are bound together by sharing a pair of electrons. But oxygen ($EN = 3.61$) is much more electronegative than hydrogen ($EN = 2.30$), so the electrons in the covalent bonds are not shared equally by the hydrogen and oxygen atoms. The electrons are drawn toward the oxygen atom in the center of the molecule and away from the hydrogen atoms on either end. As a result, the water molecule is *polar.* The oxygen atom carries a partial negative charge (−0.44), and each hydrogen atom carries a partial positive charge (+0.22).

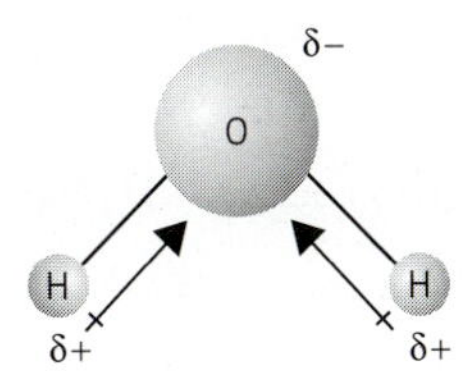

When some water molecules dissociate to form ions in water, positively charged H_3O^+ ions and negatively charged OH^- ions are formed.

$$2\ H_2O(l) \longrightarrow H_3O^+(aq) + OH^-(aq)$$

The opposite reaction can also occur: H_3O^+ ions can combine with OH^- ions to form neutral water molecules.

$$H_3O^+(aq) + OH^-(aq) \longrightarrow 2\ H_2O(l)$$

The fact that water molecules dissociate to form H_3O^+ and OH^- ions, which can then recombine to form water molecules, is indicated by writing a pair of arrows between the "reactants" and "products" that indicate that the reaction occurs in both directions and will establish an equilibrium.

$$2\ H_2O(l) \rightleftharpoons H_3O^+(aq) + OH^-(aq)$$

The Brønsted model explains water's role in acid–base reactions.

- Water dissociates to form ions by transferring an H^+ ion from one molecule acting as an acid to another molecule acting as a base.

$$\underset{Acid}{H_2O(l)} + \underset{Base}{H_2O(l)} \rightleftharpoons H_3O^+(aq) + OH^-(aq)$$

- Acids react with water by donating an H^+ ion to a neutral water molecule acting as a base to form the H_3O^+ ion.

$$\underset{Acid}{HF(aq)} + \underset{Base}{H_2O(l)} \rightleftharpoons H_3O^+(aq) + F^-(aq)$$

- Bases react with water by accepting an H^+ ion from a water molecule acting as an acid to form the OH^- ion.

$$\underset{Base}{NH_3(aq)} + \underset{Acid}{H_2O(l)} \rightleftharpoons NH_4^+(aq) + OH^-(aq)$$

- Water molecules can act as intermediates in acid–base reactions by gaining H^+ ions from an acid

$$\underset{Acid}{CH_3CO_2H(aq)} + \underset{Base}{H_2O(l)} \rightleftharpoons H_3O^+(aq) + CH_3CO_2^-(aq)$$

and then losing the H^+ ions to a base.

$$\underset{Base}{NH_3(aq)} + \underset{Acid}{H_3O^+(aq)} \rightleftharpoons NH_4^+(aq) + H_2O(l)$$

Adding these reactions so that species that are on both sides of the equation cancel gives the overall equation for the acid–base reaction.

$$\begin{array}{c} CH_3CO_2H(aq) + H_2O(l) \rightleftharpoons H_3O^+(aq) + CH_3CO_2^-(aq) \\ NH_3(aq) + H_3O^+(aq) \rightleftharpoons NH_4^+(aq) + H_2O(l) \\ \hline \underset{Acid}{CH_3CO_2H(aq)} + \underset{Base}{NH_3(aq)} \rightleftharpoons \underset{Conjugate\ acid}{NH_4^+(aq)} + \underset{Conjugate\ base}{CH_3CO_2^-(aq)} \end{array}$$

11.6 To What Extent Does Water Dissociate to Form Ions?

At 25°C, the density of water is 0.9971 g/cm^3, or 0.9971 g/mL. The concentration of pure water is therefore 55.35 *M*.

$$\frac{0.9971 \text{ g } H_2O}{1 \text{ mL}} \times \frac{1000 \text{ mL}}{1 \text{ L}} \times \frac{1 \text{ mol } H_2O}{18.015 \text{ g } H_2O} = 55.35 \text{ mol } H_2O/L$$

The concentration of the H_3O^+ and OH^- ions formed by the dissociation of pure neutral H_2O at that temperature is only 1.0×10^{-7} mol/L. The ratio of the concentration of the H_3O^+ (or OH^-) ions to the concentration of the neutral H_2O molecules is therefore 1.8×10^{-9}.

$$\frac{1.0 \times 10^{-7} \, M \, H_3O^+}{55.35 \, M \, H_2O} = 1.8 \times 10^{-9}$$

In other words, only about 2 parts per billion (ppb) of the water molecules dissociate into ions at room temperature.

It is difficult to imagine what 2 ppb means. One way to visualize this number is to assume that 2500 letters appear on a typical page of this book. If typographical errors occurred with a frequency of 2 ppb, the book would have to be 400,000 pages long in order to contain two errors. Figure 11.4 shows a model of 20 water molecules, one of which has dissociated to form H_3O^+ and OH^- ions. If this illustration were a high-resolution photograph of the structure of water, we would encounter a pair of H_3O^+ and OH^- ions on the average of only once for every 25 million such photographs.

The rate at which water molecules react to form the H_3O^+ and OH^- ions in pure water is equal to the rate at which these ions combine to form a pair of neutral water molecules.

$$2\ H_2O(l) \rightleftharpoons H_3O^+(aq) + OH^-(aq)$$

This reaction is therefore said to be at equilibrium. The extent to which water dissociates to form ions can be described in terms of the following equilibrium constant expression.

$$K_c = \frac{[H_3O^+][OH^-]}{[H_2O]^2}$$

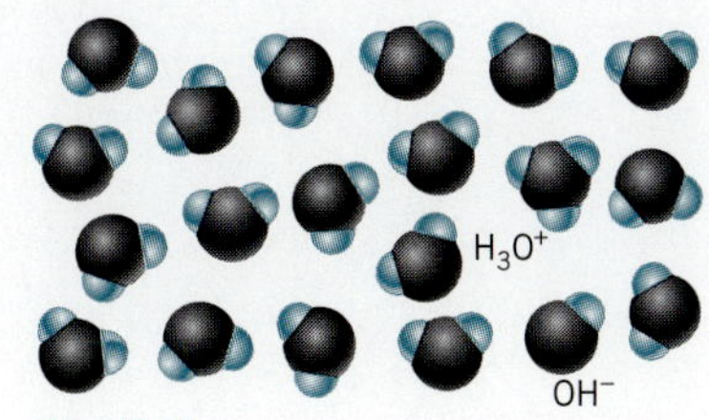

Fig. 11.4 If this figure is thought of as a high-resolution snapshot of 20 H_2O molecules, one of which has dissociated to form H_3O^+ and OH^- ions, we would have to look at an average of 25 million such snapshots before we encountered another pair of these ions.

As we have seen, the concentration of water at 25°C is 55.35 mol/L. The concentration of the H_3O^+ and OH^- ions, on the other hand, is only about 1.0×10^{-7} *M*. The concentration of H_2O molecules is therefore so much larger than the concentration of the H_3O^+ and OH^- ions that it remains effectively constant when water dissociates to form the ions. Chemists therefore rearrange the equilibrium constant expression for the dissociation of water to give the following equation.

$$K_c \times [H_2O]^2 = [H_3O^+][OH^-]$$

This effectively places two variables—the H_3O^+ and OH^- ion concentrations—on one side of the equation and two constants—K_c and $[H_2O]^2$—on the other side of the equation. We then replace the terms on the left-hand side of the equation with a constant known as the **water dissociation equilibrium constant, K_w.**

$$K_w = [H_3O^+][OH^-]$$

Because the H_3O^+ and OH^- concentrations in pure water at 25°C are each 1.0×10^{-7} *M*, the value of K_w at that temperature is 1.0×10^{-14}.

$$K_w = [1.0 \times 10^{-7}][1.0 \times 10^{-7}] = 1.0 \times 10^{-14} \quad \text{(at 25°C)}$$

Equilibrium constants are determined experimentally and are subject to some uncertainty. Therefore, equilibrium constants in this text are generally reported to only two significant figures. The equilibrium constant expression for the dissociation of water is still useful, however, for obtaining approximate concentrations of the H_3O^+ and OH^- ions in water. Regardless of the source of the H_3O^+ and OH^- ions in water, the product of the concentrations of the ions at equilibrium at 25°C is approximately 1.0×10^{-14}.

What happens to the H_3O^+ and OH^- ion concentrations that come from the dissociation of water when we add a strong acid to water? Suppose, for example, that we add enough acid to a beaker of water to raise the H_3O^+ concentration to 0.010 *M*. Note that there are now two sources of H_3O^+. One source is the acid and the other is the dissociation of water.

$$HA(aq) + H_2O(l) \longrightarrow H_3O^+(aq) + A^-(aq)$$
$$2\ H_2O(l) \rightleftharpoons H_3O^+(aq) + OH^-(aq)$$

The development of the equilibrium expressions (K_a, K_b, K_w) in this text differs from the usual presentation. Section 11.6 develops the K_w expression for the dissociation of water from K_c

$$K_c = \frac{[H_3O^+][OH^-]}{[H_2O]^2}$$

Rather than dropping the concentration of water from the equation, we have attempted to demonstrate to students that because the concentration of water is 55 M and the concentration of H_3O^+ and $OH^- \approx 1 \times 10^{-7}$ M, there is no significant change in the concentration of water as equilibrium is established. Therefore the $[H_2O]^2$ term can be viewed as a constant and K_w is defined as $K_c[H_2O]^2$.

According to Le Châtelier's principle, the additional H_3O^+ from the acid should drive the equilibrium between water and its ions to the left, reducing the number of H_3O^+ and OH^- ions from the dissociation of water. Adding an acid to water therefore decreases the extent to which the water dissociates into ions.

Because the dissociation of water shifts to the left in the presence of an acid, the concentration of H_3O^+ and OH^- ions from the dissociation of water will be smaller than the 1.0×10^{-7} *M* concentration in pure water. The total concentration of the H_3O^+ ion will be equal to the H_3O^+ ion from dissociation of the acid (0.010 *M*) plus the H_3O^+ ion from the dissociation of water ($<1.0 \times 10^{-7}$ *M*). The H_3O^+ ion concentration from the dissociation of water, however, is so small that it is negligible when compared to the amount from the acid. Therefore, when the system returns to equilibrium, the H_3O^+ ion concentration is still about 0.010 *M*. Furthermore, when the reaction returns to equilibrium, the product of the H_3O^+ and OH^- ion concentrations is once again approximated by K_w.

$$[H_3O^+][OH^-] = 1.0 \times 10^{-14}$$

If the concentration of the H_3O^+ ion is 0.010 *M*, the concentration of the OH^- ion when the system returns to equilibrium is only 1.0×10^{-12} *M*.

$$[OH^-] = \frac{K_w}{[H_3O^+]} = \frac{1.0 \times 10^{-14}}{0.01} = 1.0 \times 10^{-12}\ M$$

Some of the H_3O^+ ions in the solution come from the dissociation of water; others come from the acid that has been added to the water. All of the OH^- ions, on the other hand, come from the dissociation of water. In pure water, dissociation of the H_2O molecules gives us an OH^- ion concentration of 1.0×10^{-7} *M*. In the acid solution, the OH^- ion concentration is five orders of magnitude smaller. This means that adding enough acid to water to increase the concentration of the H_3O^+ ion to 0.010 *M* decreases the dissociation of water by a factor of about 100,000.

CHECKPOINT

When an acid is added to water, the equilibrium between water molecules and their ions shifts to the left, decreasing the $[H_3O^+]$ from the dissociation of water.

$$2\ H_2O(l) \rightleftharpoons H_3O^+(aq) + OH^-(aq)$$

Does K_w still equal 10^{-14}?

Adding an acid to water therefore has an effect on the concentration of both the H_3O^+ and OH^- ions. Some H_3O^+ is already present as a result of the dissociation of water. Adding an acid to water increases the concentration of the ion. However, adding an acid to water decreases the extent to which water dissociates and therefore leads to a significant decrease in the concentration of the OH^- ion.

As might be expected, the opposite effect is observed when a base is added to water. Because we are adding a base, the OH^- ion concentration increases. Once the system returns to equilibrium, the product of the H_3O^+ and OH^- ion concentrations is once again equal to K_w. The only way this can be achieved, of course, is by decreasing the concentration of the H_3O^+ ion.

11.7 pH as a Measure of the Concentration of H_3O^+ Ion

Many of the values determined from equilibrium calculations or pH calculations based on log $[H_3O^+]$ are at best good approximations and are at times incorrect. See the discussions by Stephen Hawkes in J. Chem. Educ. ***71,*** *747 (1994) and* J. Chem. Educ. ***69,*** *542 (1992). In general, calculated pH values in the text are reported to only one digit past the decimal.*

The range of concentrations of the H_3O^+ and OH^- ions in aqueous solutions is enormous. Typical solutions in the laboratory have concentrations of H_3O^+ or OH^- ion as large as 1.0 *M* or as small as 1×10^{-14} *M*. Perhaps the best way to bring a range of 14 orders of magnitude into perspective is to note that it is comparable to the difference between 4¢ and a national debt of $4 trillion, or the difference between the radius of a gold atom and a distance of 8 miles.

In 1909, the Danish biochemist S. P. L. Sørenson proposed a way of conveniently describing a large range of concentrations. Sørenson worked at a laboratory set up by the Carlsberg Brewery to apply scientific methods to the study of the fermentation reactions in brewing beer. Faced with the task of constructing graphs of the activity of the malt versus the H_3O^+ ion concentration, Sørenson suggested using logarithmic mathematics to condense the range of H_3O^+ and OH^- concentrations to a more convenient scale. By definition, the logarithm of a number is the power to which a base must be raised to obtain that number. The logarithm to the base 10 of 10^{-7}, for example, is -7.

$$\log(10^{-7}) = -7$$

Because the concentrations of the H_3O^+ and OH^- ions in aqueous solutions are usually smaller than 1 *M*, the logarithms of the concentrations are negative numbers. Because he considered positive numbers more convenient, Sørenson suggested that the sign of the logarithm should be changed after it had been calculated. He therefore introduced the *symbol "p" to indicate the negative of the logarithm of a number*. Thus, **pH** is the negative of the logarithm of the H_3O^+ ion concentration.

$$\text{pH} = -\log[H_3O^+]$$

Similarly, **pOH** is the negative of the logarithm of the OH^- ion concentration.

$$\text{pOH} = -\log[OH^-]$$

However, the equation is valid only for very dilute solutions of acid in pure water. Most solutions used in the laboratory do not meet this criterion, so the above equation can only be used to approximate the pH of real solutions. Since pH values from this calculation are approximations, they normally are reported to no more than two digits after the decimal. In this text, *calculated* pH values will be normally reported to only one digit past the decimal. However, experimentally measured values of pH using a pH meter can be accurately determined to two digits after the decimal.

Exercise 11.3

What is the pH of Pepsi Cola if the concentration of the H_3O^+ ion in the solution is 0.0035 *M*?

Solution

The pH of a solution is the negative of the logarithm of the H_3O^+ ion concentration.

$$\begin{aligned} \text{pH} &= -\log[H_3O^+] \\ &= -\log(3.5 \times 10^{-3}) \\ &= -(-2.5) = 2.5 \end{aligned}$$

The advantage of the pH scale is illustrated in Figure 11.5*a*, which shows the possible combinations of H_3O^+ ion and OH^- concentrations in an aqueous solution. In this graph the concentration of OH^- is plotted on the vertical axis and the concentration of H_3O^+ is plotted on the horizontal axis. For the plot to be a reasonable size, it is restricted to 1 order of magnitude of concentration.

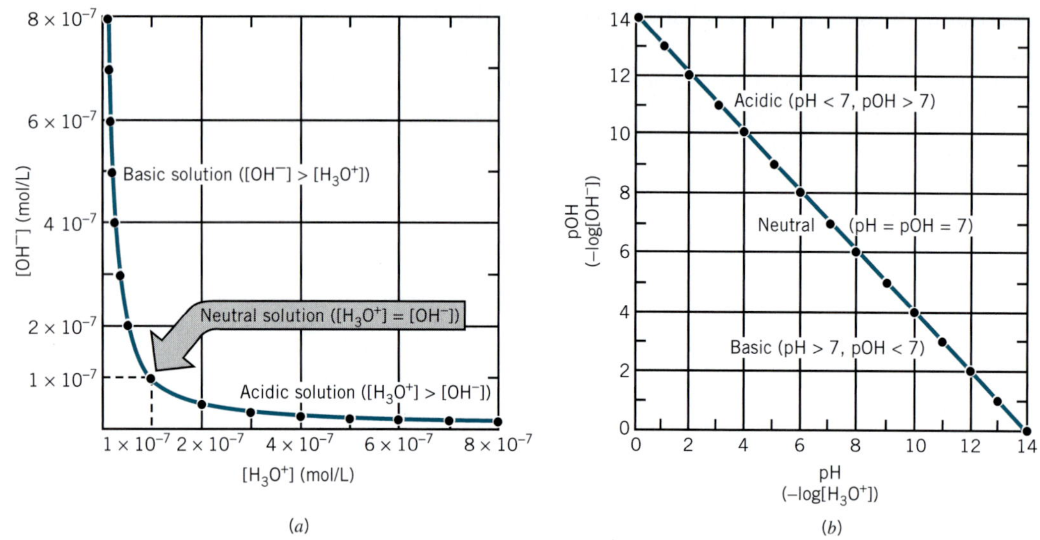

Fig. 11.5 (*a*) A small fraction of data showing the relationship between the H_3O^+ and OH^- ion concentrations in aqueous solutions. Every point on the solid line represents a pair of H_3O^+ and OH^- concentrations when the solution is at equilibrium. (*b*) We can fit the entire range of H_3O^+ and OH^- concentrations from Table 11.2 on a single graph by plotting pH versus pOH. Any point on the solid line corresponds to a solution at equilibrium.

The concept of pH compresses the range of H_3O^+ concentrations into a scale that is much easier to handle. As the H_3O^+ concentration decreases from roughly 1 *M* to 10^{-14} *M*, the pH of the solution increases from 0 to 14. The relationship between the H_3O^+ concentration, OH^- concentration, and the pH of a solution is shown in Table 11.2. The entire range of data in Table 11.2 is plotted in the graph of pH versus pOH in Figure 11.5*b*. Note that not only is the scale compressed in Figure 11.5*b*, but the shape of the graph has changed from a curve in Figure 11.5*a* to a straight line in Figure 11.5*b*.

Table 11.2

Pairs of Equilibrium Concentrations of the H_3O^+ and OH^- Ions with Their Respective pH Values in Water at 25°C

$[H_3O^+]$ (mol/L)	$[OH^-]$ (mol/L)	pH	
1	1×10^{-14}	0	Acidic solution
1×10^{-1}	1×10^{-13}	1	
1×10^{-2}	1×10^{-12}	2	
1×10^{-3}	1×10^{-11}	3	
1×10^{-4}	1×10^{-10}	4	
1×10^{-5}	1×10^{-9}	5	
1×10^{-6}	1×10^{-8}	6	
1×10^{-7}	1×10^{-7}	7	Neutral solution
1×10^{-8}	1×10^{-6}	8	Basic solution
1×10^{-9}	1×10^{-5}	9	
1×10^{-10}	1×10^{-4}	10	
1×10^{-11}	1×10^{-3}	11	
1×10^{-12}	1×10^{-2}	12	
1×10^{-13}	1×10^{-1}	13	
1×10^{-14}	1	14	

The concentration of the H_3O^+ ion in pure water at 25°C is 1.0×10^{-7} *M*. Thus, the pH of pure water is 7.

$$\text{pH} = -\log[\text{H}_3\text{O}^+] = -\log(1.0 \times 10^{-7}\ M) = 7.0$$

If the pH is given, the H_3O^+ ion concentration can be calculated with the following equation.

$$[\text{H}_3\text{O}^+] = 10^{-\text{pH}}$$

Solutions for which the concentrations of the H_3O^+ and OH^- ions are equal are said to be *neutral*.[1] Solutions in which the concentration of the H_3O^+ ion is larger than 1×10^{-7} *M* at 25°C are described as *acidic*. Those in which the concentration of the H_3O^+ ion is smaller than 1×10^{-7} *M* are *basic*. Thus, at 25°C when the pH of a solution is less than 7, the solution is acidic. When the pH is more than 7, the solution is basic.

$$\begin{array}{lll} \text{Acidic} & [\text{H}_3\text{O}^+] > 1 \times 10^{-7}\ M & \left.\begin{array}{l}\text{pH} < 7 \\ \text{pH} > 7\end{array}\right\} \text{at } 25^\circ\text{C} \\ \text{Basic} & [\text{H}_3\text{O}^+] < 1 \times 10^{-7}\ M & \end{array}$$

Measurements of pH in the laboratory were historically done with **acid–base indicators,** which are weak acids or weak bases that change color when they gain or lose an H^+ ion. Acid–base indicators are still used in the laboratory for rough pH measurements. An example of an acid–base indicator is litmus, which turns pink in solutions whose pH is below 5 and turns blue when the pH is above 8. To a large extent, these indicators have been replaced by pH meters, which are more accurate. The actual measuring device in a pH meter is an electrode that consists of a resin-filled tube with a thin glass bulb at one end. When the electrode is immersed in a solution to be measured, it produces an electric potential that is related to the H_3O^+ ion concentration in the solution.

Adding an acid to water increases the H_3O^+ concentration and decreases the OH^- concentration. Adding a base does the opposite. Regardless of what is added to water, however, the product of the concentrations of the ions at equilibrium is 1.0×10^{-14} at 25°C.

$$[\text{H}_3\text{O}^+][\text{OH}^-] = 1.0 \times 10^{-14}$$

The relationship between the pH and pOH of an aqueous solution can be derived by taking the logarithm of both sides of the K_w expression.

$$\log([\text{H}_3\text{O}^+][\text{OH}^-]) = \log(1.0 \times 10^{-14})$$

The log of the product of two numbers is equal to the sum of their logs. Thus, the sum of the logs of the H_3O^+ and OH^- ion concentrations is equal to the log of 10^{-14}.

$$\log[\text{H}_3\text{O}^+] + \log[\text{OH}^-] = -14.0$$

Both sides of the equation can now be multiplied by -1.

$$-\log[\text{H}_3\text{O}^+] - \log[\text{OH}^-] = 14.0$$

[1]The term *neutral* has been used previously in the text to describe particles with no net electric charge. (In other words, *electrically neutral*). The term *neutral* is used in this chapter to describe the acid–base properties of a substance or solution.

Substituting the definitions of pH and pOH into the equation gives the following result.

$$pH + pOH = 14.0$$

This equation can be used to convert from pH to pOH, and vice versa, for any aqueous solution at 25°C, regardless of how much acid or base has been added to the solution.

> **➤ CHECKPOINT**
>
> Describe what happens to the pH of a solution as the concentration of the H_3O^+ ion in the solution increases.

Exercise 11.4

The pH values of samples of lemon juice and vinegar are quite similar: 2.2 and 2.5, respectively. Calculate the H_3O^+ and OH^- concentrations for both solutions and compare H_3O^+ concentrations.

Solution

Lemon juice: $pH = -\log[H_3O^+] = 2.2$ $\quad pOH = -\log[OH^-] = 11.8$

$[H_3O^+] = 10^{-2.2} = 6 \times 10^{-3}\ M$ $\quad [OH^-] = 10^{-11.8} = 2 \times 10^{-12}\ M$

Vinegar: $pH = -\log\ [H_3O^+] = 2.5$ $\quad pOH = -\log[OH^-] = 11.5$

$[H_3O^+] = 10^{-2.5} = 3 \times 10^{-3}\ M$ $\quad [OH^-] = 10^{-11.5} = 3 \times 10^{-12}\ M$

The sample of lemon juice has twice as much $[H_3O^+]$ as the vinegar sample, even though they have very similar pH values.

11.8 Relative Strengths of Acids and Bases

Many hardware stores sell muriatic acid—a 6 *M* solution of hydrochloric acid, HCl(*aq*)—to clean bricks and concrete and control the pH of swimming pools. Grocery stores sell vinegar, which is a 1 *M* solution of acetic acid, CH_3CO_2H. Although both substances are acids, you wouldn't use muriatic acid in salad dressing, and vinegar is ineffective in cleaning bricks or concrete.

The difference between the two acids can be explored with the apparatus shown in Figure 11.6. Two metal electrodes connected to a source of electricity are dipped into a beaker that contains a solution of one of these acids. If the solution can conduct an electric current, it completes the electric circuit and the light bulb starts to glow. The intensity with which the bulb glows depends on the ability of the solution to conduct a current. This, in turn, depends on the concentration of the positive and negative ions in the solution that carry the electric current. As a result, the brightness of the light bulb is directly related to the number of ions in solution.

When the electrodes are immersed in pure water, the light bulb doesn't glow. This isn't surprising because the concentrations of H_3O^+ and OH^- ions in pure water are very small, only about 10^{-7} *M* at room temperature. When the electrodes are immersed in a 1 *M* solution of acetic acid, the bulb glows, but only dimly. When the electrodes are immersed in a 1 *M* solution of hydrochloric acid, the light bulb glows very brightly. Although the concentration of the acid in both solutions is the same, 1 *M*, hydrochloric acid contains far more ions than the equivalent acetic acid solution.

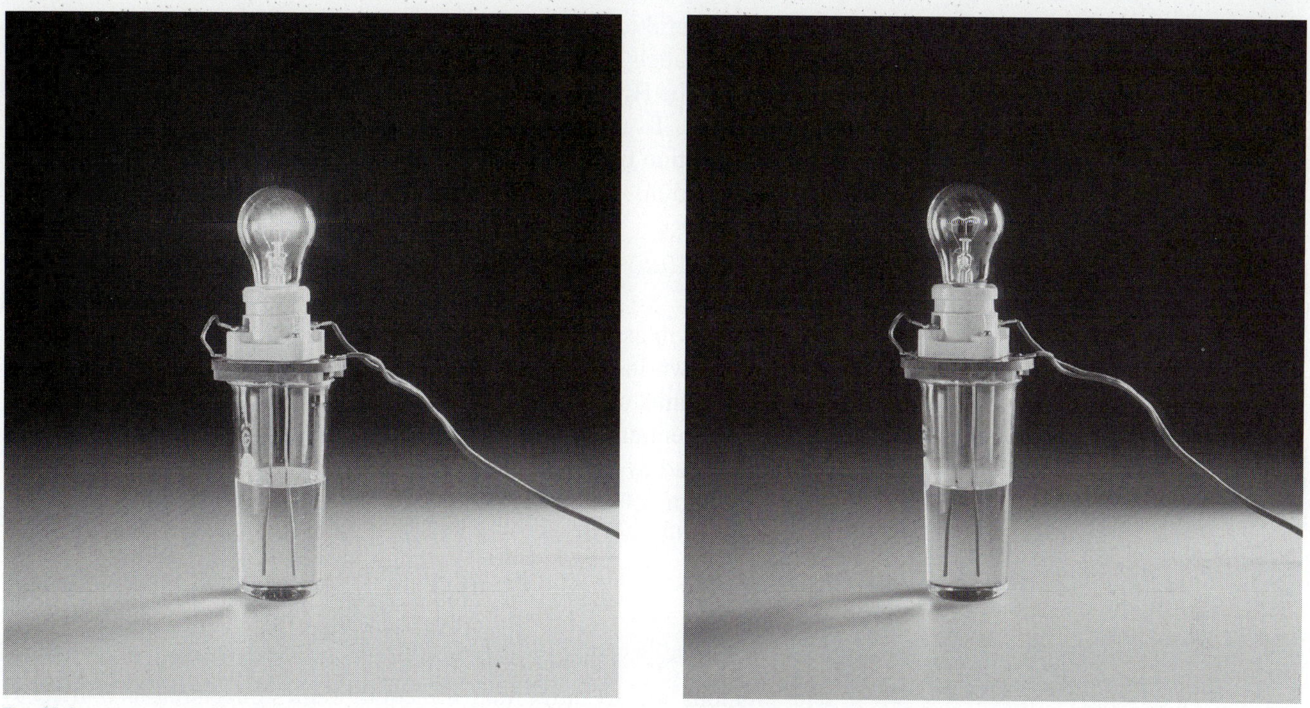

Fig. 11.6 The conductivity apparatus shows that even though the solution of HCl on the left and the solution of acetic acid on the right have the same concentrations, the acid in the solution on the left with the brightly glowing bulb has dissociated to produce more ions.

The difference between these acids is a result of differences in the abilities of the two acids to donate a proton to water. The hydrochloric acid in muriatic acid is a **strong acid,** and the acetic acid in vinegar is a **weak acid.** Hydrochloric acid is strong because it is very good at transferring an H^+ ion to a water molecule. Molecules of HCl react with water to form H_3O^+ and Cl^- ions.

$$HCl(aq) + H_2O(l) \longrightarrow H_3O^+(aq) + Cl^-(aq)$$

In a 0.1 *M* solution of hydrochloric acid very few HCl molecules remain in solution. Hydrochloric acid is therefore a strong acid indeed.

Note that a strong acid is not the same thing as a concentrated acid. Concentration is measured as molarity. Even a weak acid can have a high concentration. A strong acid is one that dissociates essentially completely when dissolved in water. A very dilute solution of HCl will dissociate essentially 100% in water and is therefore a strong acid even though it has a low concentration.

Acetic acid (often abbreviated as HOAc) is a weak acid because it is not very good at transferring H^+ ions to water. In a 1 *M* solution, less than 0.4% of the CH_3CO_2H molecules react with water to form H_3O^+ and $CH_3CO_2^-$ ions.

$$CH_3CO_2H(aq) + H_2O(l) \rightleftharpoons H_3O^+(aq) + CH_3CO_2^-(aq)$$

More than 99.6% of the acetic acid molecules remain undissociated.

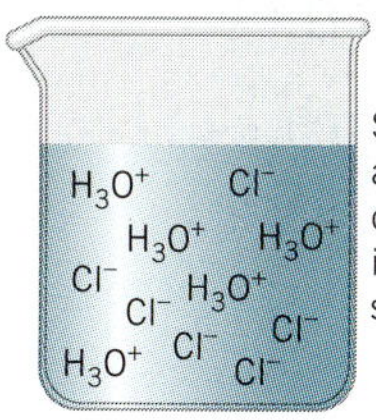

Strong acids are almost totally dissociated into ions in dilute solutions.

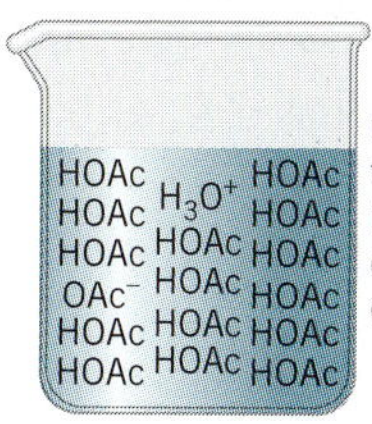

Only about 0.4% of the HOAc molecules in a 1.0 *M* solution of acetic acid dissociate into ions.

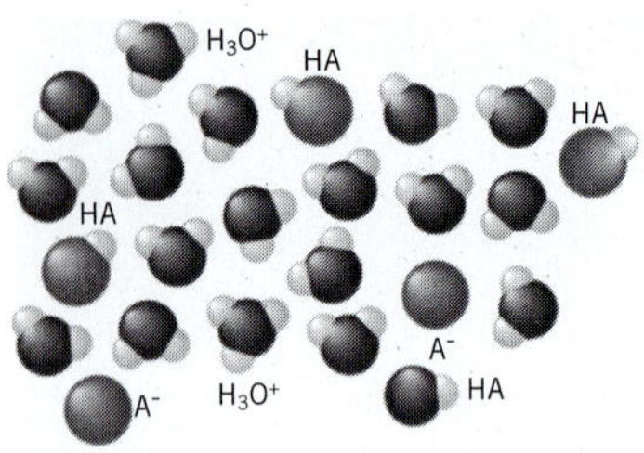

Fig. 11.7 Some, but not all, of the HA molecules in a typical acid react with water to form H_3O^+ and A^- ions when the acid dissolves in water. In a strong acid, there are very few undissociated HA molecules. In a weak acid, most of the HA molecules remain.

It would be useful to have a quantitative measure of the relative strengths of acids to replace the labels *strong* and *weak*. The extent to which an acid dissociates in water to produce the H_3O^+ ion is described in terms of an acid dissociation equilibrium constant, K_a. To understand the nature of this equilibrium constant, let's assume that the reaction between an acid and water can be represented by the following generic equation.

$$HA(aq) + H_2O(l) \rightleftharpoons H_3O^+(aq) + A^-(aq)$$

In other words, we will assume that some of the HA molecules react to form H_3O^+ and A^- ions, as shown in Figure 11.7. By convention, the equilibrium concentrations of the ions in units of moles per liter are represented by the symbols $[H_3O^+]$ and $[A^-]$. The concentration of the undissociated HA molecules that remain in solution is represented by the symbol [HA].

The equilibrium constant expression for the reaction between HA and water would be written as follows.

$$K_c = \frac{[H_3O^+][A^-]}{[HA][H_2O]}$$

Most acid solutions are so dilute that the equilibrium concentration of H_2O is effectively the same as before the acid was added. The equilibrium constant for the reaction therefore is written as follows.

The K_a expression is developed in a manner similar to the K_w in Section 11.6. Because H_2O is in such a large excess, its concentration remains essentially constant as equilibrium is established. Therefore $K_c[H_2O] = K_a$.

$$\frac{[H_3O^+][A^-]}{[HA]} = K_c \times [H_2O]$$

The result is an equilibrium constant for the equation known as the **acid dissociation equilibrium constant, K_a.**

The same treatment of equilibrium used for development of the K_a and K_w allows us to describe the equilibrium for the dissociation of hydronium ion in water. K_c for the dissociation of hydronium ion must have a value of 1. Treating this equilibrium in the same manner as for K_a gives the following relationship: $[H_2O] \times K_c = K_a = 55$.

$$\frac{[H_3O^+][A^-]}{[HA]} = K_c \times [H_2O] = K_a$$

When a strong acid dissolves in water, it reacts extensively with water to form H_3O^+ and A^- ions. (Essentially 100% of the strong acid dissociates, leaving only a very small residual concentration of HA molecules in solution.) The product of the concentrations of the H_3O^+ and A^- ions is therefore much larger than the concentration of the HA molecules, so K_a for a strong acid is greater than 1. Hydrochloric acid, for example, has a K_a of roughly 1×10^6.

$$\frac{[H_3O^+][Cl^-]}{[HCl]} = 1 \times 10^6$$

Weak acids, on the other hand, react only slightly with water. The product of the concentrations of the H_3O^+ and A^- ions is therefore smaller than the concentration of the residual HA molecules. As a result, K_a for a weak acid is less than 1. Acetic acid, for example, has a K_a of only 1.8×10^{-5}.

$$\frac{[H_3O^+][CH_3CO_2^-]}{[CH_3CO_2H]} = 1.8 \times 10^{-5}$$

K_a can therefore be used to distinguish between strong acids and weak acids.

Strong acids	$K_a > 1$
Weak acids	$K_a \ll 1$

The K_a for the dissociation of water is treated in the same manner as the K_a for a weak acid. This yields $K_a = [H_2O] \times K_c = \frac{[H_3O^+][OH^-]}{[H_2O]} = 1.8 \times 10^{-16}$. Note that this is the K_a, not the K_w, of water.

The hydronium ion, H_3O^+, is the strongest acid that is stable in water because the strong acids all dissociate essentially 100% to give H_3O^+. The equilibrium expression that describes the dissociation of H_3O^+ in water is defined in the same way as the K_a for strong and weak acids shown above.

$$H_3O^+(aq) + H_2O(l) \rightleftharpoons H_3O^+(aq) + H_2O(l)$$

$$K_c = \frac{[H_3O^+][H_2O]}{[H_3O^+][H_2O]} = 1$$

$$\frac{[H_3O^+][H_2O]}{[H_3O^+]} = [H_2O] \times K_c = K_a = 55$$

Because K_c is equal to 1 and $[H_2O]$ is 55, the K_a for the dissociation of H_3O^+ in water is equal to 55.

A list of common acids and their acid dissociation constants, K_a, is given in Table 11.3. A more complete list can be found in Appendix B.8. Also included in Appendix B.8 are pK_a values (p$K_a = -\log K_a$). Commonly accepted K_a values are reported for the strong acids. Experimental determination of K_a values for strong acids is difficult, resulting in considerable variation in the values reported in the literature.

The ionization of bases in water can be described in the same manner as the dissociation of acids. A generic base, B, reacting with water can be described by the following chemical equation.

$$B(aq) + H_2O(l) \rightleftharpoons BH^+(aq) + OH^-(aq)$$

Strict adherence to the rules for writing equilibrium constant expressions gives the following equation.

$$K_c = \frac{[BH^+][OH^-]}{[B][H_2O]}$$

As we saw with the acids, the concentration of water at equilibrium is essentially constant, allowing the equilibrium expression to be written as follows.

$$K_b = \frac{[BH^+][OH^-]}{[B]} = K_c \times [H_2O]$$

The new equilibrium constant is known as the **base ionization equilibrium constant, K_b.** Just as K_a values describe the relative strengths of acids, K_b values describe the relative strengths of bases. The values of K_b for a limited number of bases are given in Table B.9 in the appendix. This table also lists the pK_b values of these bases (p$K_b = -\log K_b$).

Strong bases are defined as ions or molecules that ionize more or less completely in water to produce the OH^- ion. Examples of strong bases include the following.

- Group IA metal hydroxides: LiOH, NaOH, KOH, RbOH, and CsOH

$$NaOH(s) \longrightarrow Na^+(aq) + OH^-(aq)$$

➤ CHECKPOINT

Write the equilibrium expressions that describe the K_a and the K_w of water. Show how the two expressions are different.

Table 11.3

Common Acids and Their Acid Dissociation Equilibrium Constants for the Loss of One Proton

	K_a
Strong Acids	
HI	3×10^9
HBr	1×10^9
$HClO_4$	1×10^8
HCl	1×10^6
H_2SO_4	1×10^3
H_3O^+	**55**
HNO_3	28
H_2CrO_4	9.6
Weak Acids	
H_3PO_4	7.1×10^{-3}
HF	7.2×10^{-4}
Citric acid	7.5×10^{-4}
CH_3CO_2H	1.8×10^{-5}
H_2S	1.0×10^{-7}
H_2CO_3	4.5×10^{-7}
H_3BO_3	7.3×10^{-10}
H_2O[a]	1.8×10^{-16}

[a]Note that this is the K_a of water as determined by the acid dissociation equilibrium expression, not the K_w described in Section 11.6.

CHECKPOINT

Which solution has the smallest pH, a solution with a low concentration of an acid with a small K_a or a solution with a low concentration of an acid with a high K_a?

The identical treatment of the dissociation of weak acids, hydronium ion, and water to describe K_a allows students to compare K_a values in Table 11.3. Because these K_a values are directly comparable, students can see the relative strengths of the acids, hydronium ion, and water. It would be inappropriate to compare an acid's K_a with the K_w of water because these values represent equilibria expressions determined in different ways.

- Group IIA metal hydroxides that are soluble or slightly soluble in water: $Ca(OH)_2$ (slightly soluble), $Sr(OH)_2$, and $Ba(OH)_2$

$$Ca(OH)_2(aq) \longrightarrow Ca^{2+}(aq) + 2\ OH^-(aq)$$

- Soluble metal oxides: Li_2O, Na_2O, K_2O, and CaO

$$Li_2O(s) + H_2O(l) \longrightarrow 2\ Li^+(aq) + 2\ OH^-(aq)$$

11.9 Relative Strengths of Conjugate Acid–Base Pairs

The relationship between the strength of an acid and its conjugate base can be understood by considering the implications of the fact that HCl is a strong acid. If it is a strong acid, HCl must be a good proton donor. HCl can only be a good proton donor, however, if the Cl^- ion is a poor proton acceptor. Thus, the Cl^- ion must be a **weak base.**

$$\underset{\text{Strong acid}}{HCl(aq)} + H_2O(l) \longrightarrow H_3O^+(aq) + \underset{\text{Weak base}}{Cl^-(aq)}$$

Because HCl is a strong acid, it dissociates essentially completely, and there is little tendency for the reaction to go in the reverse direction. Thus, the chloride anion is such a very weak conjugate base that it has essentially no base properties. None of the anions that result from the dissociation of strong monoprotic acids (HI, HBr, HCl, HNO_3, $HClO_4$) have appreciable basic properties in water.

Let's now consider the relationship between the strength of the ammonium ion (NH_4^+) and its conjugate base, ammonia (NH_3). The NH_4^+ ion is a weak acid, and its conjugate base ammonia is a reasonably good base.

$$\underset{\text{Weak acid}}{NH_4^+(aq)} + H_2O(l) \rightleftharpoons H_3O^+(aq) + \underset{\text{Good base}}{NH_3(aq)}$$

The results of the two examples can be summarized in the form of the following general rules:

The stronger the acid, the weaker is the conjugate base.
The stronger the base, the weaker is the conjugate acid.

11.10 Relative Strengths of Pairs of Acids and Bases

We have already seen how the value of K_a can be used to decide whether an acid is a strong acid or a weak acid. At times it is also useful to compare the relative strengths of a pair of acids to decide which is stronger. Consider HCl and the H_3O^+ ion, for example.

$$\begin{array}{ll} HCl & K_a = 1 \times 10^6 \\ H_3O^+ & K_a = 55 \end{array}$$

The K_a values suggest both are strong acids, but HCl is a stronger acid than the H_3O^+ ion.

We can now understand why such a large proportion of the HCl molecules in an aqueous solution react with water to form H_3O^+ and Cl^- ions. The Brønsted model suggests that every acid–base reaction converts an acid into its conjugate base and a base into its conjugate acid.

$$\underset{\textit{Acid}_1}{HCl(aq)} + \underset{\textit{Base}_2}{H_2O(l)} \longrightarrow \underset{\textit{Acid}_2}{H_3O^+(aq)} + \underset{\textit{Base}_1}{Cl^-(aq)}$$

Note that there are two acids and two bases in the reaction. The stronger acid, however, is on the left side of the equation.

$$\underset{\textit{Stronger acid}}{HCl(aq)} + H_2O(l) \longrightarrow \underset{\textit{Weaker acid}}{H_3O^+(aq)} + Cl^-(aq)$$

Strong acids are expected to dissociate essentially 100%. We have defined strong acids as having K_a values greater than 1. Only acids with a K_a greater than the K_a of hydronium ion are expected to completely dissociate. Nitric and chromic acids have K_a values greater than 1 but less that of hydronium ion. We will refer to nitric and chromic acid as strong acids while realizing that they do not dissociate to the extent of acids with K_a values greater than hydronium ion.

What about the two bases: H_2O and the Cl^- ion? The general rules given in the previous section suggest that the stronger of a pair of acids must form the weaker of a pair of conjugate bases. The fact that HCl is a stronger acid than the H_3O^+ ion implies that the Cl^- ion is a weaker base than water.

Acid strength	$HCl > H_3O^+$
Base strength	$Cl^- < H_2O$

Thus, the equation for the reaction between HCl and water can be written as follows.

$$\underset{\substack{\textit{Stronger}\\\textit{acid}}}{HCl(aq)} + \underset{\substack{\textit{Stronger}\\\textit{base}}}{H_2O(l)} \longrightarrow \underset{\substack{\textit{Weaker}\\\textit{acid}}}{H_3O^+(aq)} + \underset{\substack{\textit{Weaker}\\\textit{base}}}{Cl^-(aq)}$$

It isn't surprising that almost all of the HCl molecules in a 6 *M* solution react with water to give H_3O^+ ions and Cl^- ions. The stronger of a pair of acids should react with the stronger of a pair of bases to form a weaker acid and a weaker base. *Strong acids with K_a values greater than the K_a of H_3O^+ (55) can be assumed to dissociate completely (100%) in water.*

Let's now look at the relative strengths of acetic acid and the H_3O^+ ion.

CH_3CO_2H	$K_a = 1.8 \times 10^{-5}$
H_3O^+	$K_a = 55$

The K_a values suggest that acetic acid is a much weaker acid than the H_3O^+ ion, which explains why acetic acid is a weak acid in water. Once again, the reaction between the acid and water must convert the acid into its conjugate base and the base into its conjugate acid.

$$\underset{\textit{Acid}_1}{CH_3CO_2H(aq)} + \underset{\textit{Base}_2}{H_2O(l)} \rightleftharpoons \underset{\textit{Acid}_2}{H_3O^+(aq)} + \underset{\textit{Base}_1}{CH_3CO_2^-(aq)}$$

But in this case, the stronger acid and the stronger base are on the right side of the equation.

$$\underset{\text{Weaker acid}}{CH_3CO_2H(aq)} + \underset{\text{Weaker base}}{H_2O(l)} \rightleftharpoons \underset{\text{Stronger acid}}{H_3O^+(aq)} + \underset{\text{Stronger base}}{CH_3CO_2^-(aq)}$$

As a result, only a few of the CH_3CO_2H molecules actually donate an H^+ ion to a water molecule to form the H_3O^+ and $CH_3CO_2^-$ ions. At equilibrium an acid–base reaction should lie predominantly on the side of the chemical equation with the weaker acid and weaker base.

The magnitude of K_a can also be used to explain why some compounds that qualify as Brønsted acids or bases don't act like acids or bases when they dissolve in water. As long as K_a for the acid is significantly larger than the value of K_a for water (1.8×10^{-16}), the acid will ionize to some extent. As the K_a value for the acid approaches the K_a for water, the compound becomes more like water in its acidity. Although it is still a Brønsted acid, it is so weak that we may be unable to detect the acidity in aqueous solution.

Measurements of the pH of dilute solutions are also good indicators of the relative strengths of acids and bases. Experimental values of the pH of 0.10 *M* solutions of a number of common acids and bases are given in Table 11.4. The equation $pH = -\log[H_3O^+]$ is a good model for the behavior of acids but is not exact. Table 11.4 shows that the actual pH of some 0.10 *M* acids vary from what would be calculated with the pH equation. For example, the pH equation would predict that the pH of a 0.10 *M* HCl solution would be 1.0, but the measured value is observed to be 1.1.

➤ CHECKPOINT

HOBr is a weaker acid than HOCl. Which base, OBr^- or OCl^-, is strongest?

Table 11.4

pH of 0.10 *M* Solutions of Common Acids and Bases

Compound	pH
HCl (hydrochloric acid)	1.1
H_2SO_4 (sulfuric acid)	1.2
$NaHSO_4$ (sodium hydrogen sulfate)	1.4
H_2SO_3 (sulfurous acid)	1.5
H_3PO_4 (phosphoric acid)	1.5
HF (hydrofluoric acid)	2.1
CH_3CO_2H (acetic acid)	2.9
H_2CO_3 (carbonic acid)	3.8 (saturated solution)
H_2S (hydrogen sulfide)	4.1
NaH_2PO_4 (sodium dihydrogen phosphate)	4.4
NH_4Cl (ammonium chloride)	4.6
HCN (hydrocyanic acid)	5.1
NaCl (sodium chloride)	6.4
H_2O (freshly boiled distilled water)	7.0
$NaCH_3CO_2$ (sodium acetate)	8.4
$NaHCO_3$ (sodium hydrogen carbonate)	8.4
Na_2HPO_4 (sodium hydrogen phosphate)	9.3
Na_2SO_3 (sodium sulfite)	9.8
NaCN (sodium cyanide)	11.0
NH_3 (aqueous ammonia)	11.1
Na_2CO_3 (sodium carbonate)	11.6
Na_3PO_4 (sodium phosphate)	12.0
NaOH (sodium hydroxide, lye)	13.0

Students are surprised and think that they have made an error in lab when they prepare a 0.10 M *solution of an acid and find that the measured pH is significantly different from the predicted value. Without going into a discussion of activities, we emphasize that equilibrium expression calculations are a mathematical model for what is happening in solution and are often not exact.*

THE LEVELING EFFECT OF WATER

Because all strong acids dissociate in water to produce the same H_3O^+ ion, they all seem to have the same strength when dissolved in water, regardless of the value of K_a. This phenomenon is known as the **leveling effect** of water—the tendency of water to limit the strengths of strong acids and bases. Close to 100% of the HCl molecules in hydrochloric acid react with water to form H_3O^+ and Cl^- ions, for example. Thus a 0.10 *M* solution of HCl produces approximately 0.10 *M* H_3O^+. In the same manner, a 0.10 *M* solution of the strong acid $HClO_4$ will also produce 0.10 *M* H_3O^+.

$$HCl(aq) + H_2O(l) \longrightarrow H_3O^+(aq) + Cl^-(aq)$$
$$HClO_4(aq) + H_2O(l) \longrightarrow H_3O^+(aq) + ClO_4^-(aq)$$

The reason that both acids give a 0.10 *M* H_3O^+ concentration is that the strengths of strong acids are limited by the strength of the acid (H_3O^+) formed when water molecules accept an H^+ ion.

A similar phenomenon occurs in solutions of strong bases. Once the base reacts with water to form the OH^- ion, the solution cannot become any more basic. The strength of a strong base is limited by the strength of the base (OH^-) formed when water molecules lose an H^+ ion.

11.11 Relationship of Structure to Relative Strengths of Acids and Bases

Several factors influence the probability of having a heart attack. Heart attacks occur more often among those who smoke than among those who don't, among the elderly more often than the young, among those who are overweight more than among those who aren't, and among those who never exercise more than among those who exercise regularly. It is possible to sort out the relative importance of these factors, however, by trying to keep as many as possible of the other factors constant.

The same approach can be used to sort out the factors that control the relative strengths of acids and bases. Although structure alone doesn't determine the strength of an acid, many trends can be rationalized from the Lewis structure of an acid. Three factors related to structure affect the acidity of the *X*—H bond in an acid: (1) the polarity of the bond, (2) the size of the atom *X*, and (3) the charge on the ion or molecule. Structure can also be used to help understand the relative acidity of the oxyacids, acids with an *X*—OH group.

THE POLARITY OF THE *X*—H BOND

When all other factors are kept constant, acids become stronger as the *X*—H bond becomes *more polar*. Consider the following electrically neutral compounds, for example, which become more acidic as the difference between the electronegativities of the *X* and H atoms (ΔEN) increases. δ_H is the partial charge on the hydrogen atom in these compounds. HF is the strongest of the four acids, and CH_4 is the weakest.

	K_a	ΔEN	δ_H	*X*—H Bond Length (nm)
HF	7.2×10^{-4}	1.9	+0.29	0.101
H_2O	1.8×10^{-16}	1.3	+0.22	0.103
NH_3	1×10^{-33}	0.8	+0.14	0.107
CH_4	1×10^{-49}	0.2	+0.05	0.114

In all four of the compounds, the size of the central atom, *X*, to which the H is bonded is about the same. This can be seen from the similar bond lengths. As a general rule, similar bond lengths will give similar bond strengths. However, it can also be seen in the above data that there are large differences in ΔEN among the four compounds. The molecule containing the central atom with the largest electronegativity, F, also has the H with the most positive partial charge, 0.29. As the central atom becomes more electronegative, electron density is pulled away from the hydrogen, resulting in a more positive partial charge on the hydrogen and a more polar H—*X* bond. In the case of HF, the difference between the electronegativities of the two atoms is so great that the partial charge on the hydrogen atom is significant even before the acid dissociates into ions.

When these compounds act as acids the H—*X* bond is broken to form H^+ and X^- ions. From the data above it is apparent that the more polar this bond, the larger is the K_a, and hence the easier it is to form the ions. Thus, the more polar the bond, the stronger is the acid.

The data in Table 11.4 illustrate the magnitude of this effect. A 0.10 *M* HF solution is moderately acidic. Water is much less acidic, and the acidity of ammonia is so low that the chemistry of aqueous solutions of NH_3 is dominated by the ability of NH_3 to act as a base.

0.10 *M* HF	pH = 2.1
H_2O	pH = 7
0.10 *M* NH_3	pH = 11.1

THE SIZE OF THE *X* ATOM

We might expect HF, HCl, HBr, and HI to become weaker acids as we go down the column of the periodic table because the *X*—H bond becomes less polar. Experimentally, we find the opposite trend. The acids actually become stronger as we go down the column.

This can be explained by recognizing that the size of the *X* atom also influences the acidity of the *X*—H bond. Acids become stronger as the *X*—H bond becomes weaker, and bonds generally become weaker as the atoms get larger (see Figure 11.8). The K_a data for HF, HCl, HBr, and HI reflect the fact that the enthalpy of atom combination (ΔH^0_{ac}) of the compounds becomes less negative as the *X* atom becomes larger. A less negative enthalpy of atom combination indicates a lower bond strength and thus a stronger acid because the hydrogen is more easily lost. The data below show that bond strength is more important than bond polarity for these acids.

> **CHECKPOINT**
>
> Which acid is stronger, H_2O or H_2S? Why?

	K_a	ΔH^0_{ac} (kJ/mol$_{rxn}$)	δ_H	*X*—H Bond Length (nm)
HF	7.2×10^{-4}	−567.7	+0.29	0.101
HCl	1×10^{6}	−431.6	+0.11	0.136
HBr	1×10^{9}	−365.9	+0.08	0.151
HI	3×10^{9}	−298.0	+0.01	0.170

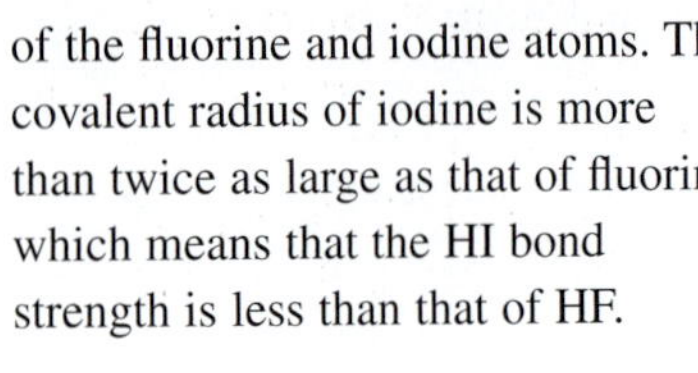

Fig. 11.8 HI is a much stronger acid than HF because of the relative sizes of the fluorine and iodine atoms. The covalent radius of iodine is more than twice as large as that of fluorine, which means that the HI bond strength is less than that of HF.

THE CHARGE ON THE ACID OR BASE

The charge on a molecule or ion can influence its ability to act as an acid or a base. This is clearly shown when the pH of 0.1 *M* solutions of H_3PO_4 and the $H_2PO_4^-$, HPO_4^{2-}, and PO_4^{3-} ions are compared.

H_3PO_4	pH = 1.5
$H_2PO_4^-$	pH = 4.4
HPO_4^{2-}	pH = 9.3
PO_4^{3-}	pH = 12.0

Compounds become less acidic and more basic as the negative charge increases.

Acidity $H_3PO_4 > H_2PO_4^- > HPO_4^{2-}$

Basicity $H_2PO_4^- < HPO_4^{2-} < PO_4^{3-}$

The decrease in acidity with increasing negative charge occurs because it is much easier to remove a positive H^+ ion from a neutral H_3PO_4 molecule than it is to remove an H^+ ion from a negatively charged $H_2PO_4^-$ ion. It is even harder to remove an H^+ ion from a negatively charged HPO_4^{2-} ion.

The increase in basicity with increasing negative charge can be similarly explained. There is a strong force of attraction between the negative charge on a PO_4^{3-} ion and the positive charge on an H^+ ion. As a result, PO_4^{3-} is a reasonably good base. A comparison of HPO_4^{2-} and PO_4^{3-} indicates that the force of attraction between the HPO_4^{2-} ion and an H^+ ion is smaller, because HPO_4^{2-} carries a smaller charge. The HPO_4^{2-} ion is therefore a weaker base than PO_4^{3-}. The charge on the $H_2PO_4^-$ ion is even smaller than that of HPO_4^{2-}, so this ion is an even weaker base than HPO_4^{2-}.

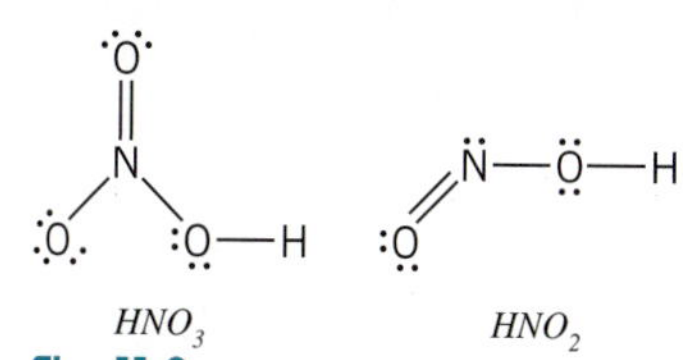

Fig. 11.9 Lewis structures of HNO_3 and HNO_2.

RELATIVE STRENGTHS OF OXYACIDS

There is no difference in either the size of the atom bonded to hydrogen or the charge on the acid when we compare the O—H bonds in oxyacids of the same element, such as H_2SO_4 and H_2SO_3 or HNO_3 and HNO_2 (see Figure 11.9), yet there is a significant difference in the strengths of the acids. Consider the following K_a data, for example.

H_2SO_4	$K_a = 1 \times 10^3$	HNO_3	$K_a = 28$
H_2SO_3	$K_a = 1.7 \times 10^{-2}$	HNO_2	$K_a = 5.1 \times 10^{-4}$

The acidity of the oxyacids increases significantly as the number of oxygen atoms attached to the central atom increases. H_2SO_4 is a much stronger acid than H_2SO_3, and HNO_3 is a much stronger acid than HNO_2.

This trend is easiest to see in the four oxyacids of chlorine. At first glance, we might expect these acids to have more or less the same acidity. The same O—H bond is broken in each case, and the O atom in this bond is bound to a Cl atom in all four compounds. Oxygen, however, is second only to fluorine in electronegativity, and it tends to draw electron density toward itself. As more oxygen atoms are added to the central atom, more electron density is drawn toward the oxygens. This draws electrons away from the O—H bond in the acids, as shown in Figure 11.10. Pulling electron density away from the hydrogen results in the partial charge on the hydrogen, δ_H, becoming more positive, as can be seen below. The O—H bond in HOCl is less polar than that in $HOClO_3$, so perchloric acid is the stronger acid.

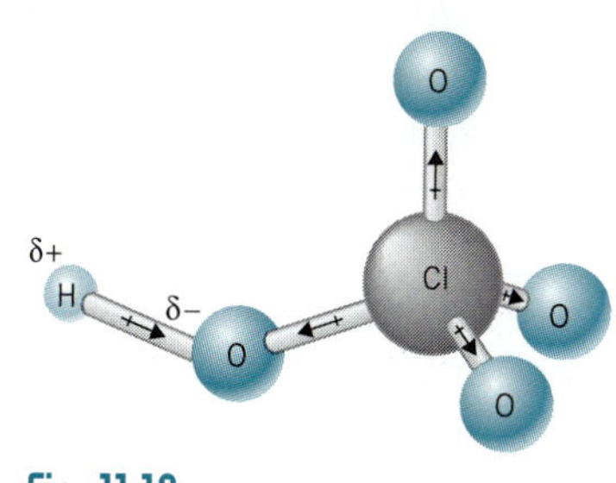

Fig. 11.10 The large difference between the K_a values for HOCl and $HOClO_3$ is the result of the addition of oxygen atoms to the chlorine. The electronegative oxygen atoms draw electron density toward themselves, resulting in electron density being pulled away from the hydrogen and toward the oxygen in the O—H bond. The net result is an increase in the polarity of the O—H bond, which leads to an increase in the acidity of the compound.

	K_a	Number of Oxygen atoms	δ_H	Name of Oxyacid
HOCl	2.9×10^{-8}	1	+0.24	Hypochlorous acid
HOClO	1.1×10^{-2}	2	+0.29	Chlorous acid
$HOClO_2$	5.0×10^{2}	3	+0.31	Chloric acid
$HOClO_3$	1×10^{8}	4	+0.33	Perchloric acid

Exercise 11.5

For each of the following pairs, predict which compound is the stronger acid and explain why.

(a) H_2O or NH_3
(b) NH_4^+ or NH_3
(c) NH_3 or PH_3
(d) H_2SeO_4 or H_2SeO_3

Solution

(a) Oxygen and nitrogen are about the same size, and H_2O and NH_3 are both neutral molecules. The difference between the compounds lies in the polarity of the *X*—H bond. Because oxygen is more electronegative, the O—H bond is more polar, and H_2O ($K_a = 1.8 \times 10^{-16}$) is a much stronger acid than NH_3 ($K_a = 1 \times 10^{-33}$).

(b) The difference between these compounds is the charge. The NH_4^+ ion is a stronger acid than NH_3 because it is easier to remove an H^+ ion from an NH_4^+ ion than from a neutral NH_3 molecule.

(c) PH_3 is a stronger acid than NH_3. Molecules become more acidic as the size of the atom holding the hydrogen atom increases. Longer bonds are weaker bonds and therefore more easily broken.

(d) The compounds are both oxyacids, which differ in the number of oxygen atoms attached to the selenium atom. More electron density is drawn away from the O—H bond in H_2SeO_4 because of the extra oxygen atom. Thus the O—H bond in H_2SeO_4 is more polar than that in H_2SeO_3, and therefore H_2SeO_4 is a stronger acid.

➤ CHECKPOINT

Which acid is stronger, HOCl or HOI? Why?

The relative strengths of Brønsted bases can be predicted from the relative strengths of their conjugate acids combined with the general rule that the stronger of a pair of acids always has the weaker conjugate base.

Exercise 11.6

For each of the following pairs of compounds, predict which compound is the weaker base.

(a) OH^- or NH_2^-
(b) NH_3 or NH_2^-

(c) NH_2^- or PH_2^-

(d) NO_3^- or NO_2^-

Solution

(a) The OH^- ion is the conjugate base of water, and the NH_2^- ion is the conjugate base of ammonia. Because H_2O is a stronger acid than NH_3, the OH^- ion must be a weaker base than the NH_2^- ion.

(b) Because it carries a negative charge, the NH_2^- ion is a stronger base than NH_3.

(c) PH_3 is a stronger acid than NH_3, which means the PH_2^- ion must be a weaker base than the NH_2^- ion.

(d) HNO_3 is a stronger acid than HNO_2, which means that the NO_3^- ion is the weaker base.

11.12 Strong Acid pH Calculations

The simplest acid–base equilibria are those in which a strong acid (or base) is dissolved in water. Consider the calculation of the pH of a solution formed by adding a single drop of 2 *M* hydrochloric acid to 100 mL of water, for example. The key to the calculation is remembering that HCl is a strong acid ($K_a = 10^6$) and that acids as strong as HCl can be assumed to dissociate almost completely in water.

$$HCl(aq) + H_2O(l) \longrightarrow H_3O^+(aq) + Cl^-(aq)$$

The H_3O^+ ion concentration at equilibrium is therefore essentially equal to the initial concentration of the acid.

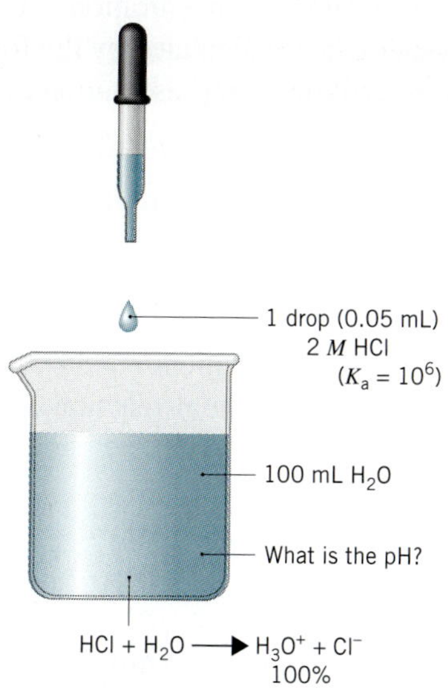

A useful rule of thumb says that there are about 20 drops in each milliliter. One drop of 2 *M* HCl therefore has a volume of 0.05 mL, or 5×10^{-5} L. Now that we know the volume of the acid added to the beaker of water and the concentration of the acid, we can calculate the number of moles of HCl that were added to the water.

$$\frac{2 \text{ mol HCl}}{1 \text{ L}} \times 5 \times 10^{-5} \text{ L} = 1 \times 10^{-4} \text{ mol HCl}$$

Once we know the number of moles of HCl that have been added to the beaker and the volume of water in the beaker, we can calculate the concentration of solution prepared by adding one drop of 2 *M* HCl to 100 mL of water.

$$\frac{1 \times 10^{-4} \text{ mol HCl}}{0.100 \text{ L}} = 1 \times 10^{-3}\, M \text{ HCl}$$

According to the calculation, the initial concentration of HCl is 1×10^{-3} *M*.

If we assume that the acid dissociates completely, the H_3O^+ concentration at equilibrium is 1×10^{-3} *M*. We therefore expect the pH of the solution prepared by adding one drop of 2 *M* HCl to 100 mL of water to be 3.0.

> **➤ CHECKPOINT**
>
> When a strong acid is added to water, why can the pH usually be determined directly from the quantity of strong acid added?

$$\text{pH} = -\log[H_3O^+]$$
$$\text{pH} = -\log(1 \times 10^{-3}) = -(-3.0) = 3.0$$

Calculations such as this give results that are close to the experimental values for solutions that are reasonably dilute.

11.13 Weak Acid pH Calculations

Unlike the strong acids, which dissociate almost completely, weak acids only dissociate slightly in water. An aqueous solution of a weak acid therefore contains significant amounts of both the dissociated ($H_3O^+ + A^-$) and undissociated (*HA*) forms of the acid. Equilibrium problems involving weak acids can be solved by applying the techniques developed in Chapter 10. Consider, for example, the dissociation of acetic acid, CH_3CO_2H, in water. The formula of acetic acid is often abbreviated as HOAc and that of its conjugate base, the acetate ion ($CH_3CO_2^-$), as OAc^-. The H_3O^+, OAc^-, and HOAc concentrations at equilibrium in a 0.10 *M* solution of acetic acid in water can be calculated by the following process. We start the calculation, as always, by building a representation of what we know about the reaction.

	$HOAc(aq) + H_2O(l) \rightleftharpoons$	$H_3O^+(aq)$ +	$OAc^-(aq)$	$K_a = 1.8 \times 10^{-5}$
Initial	0.10 *M*	≈0	0	
Equilibrium	?	?	?	

We then compare the initial reaction quotient (Q_a) with the equilibrium constant (K_a) for the reaction. In this case, the initial reaction quotient is smaller than the equilibrium constant.

$$Q_a = \frac{(H_3O^+)(OAc^-)}{(HOAc)} = \frac{(0)\,(0)}{0.10} = 0 < K_a$$

We therefore conclude that the reaction must shift to the right to reach equilibrium.

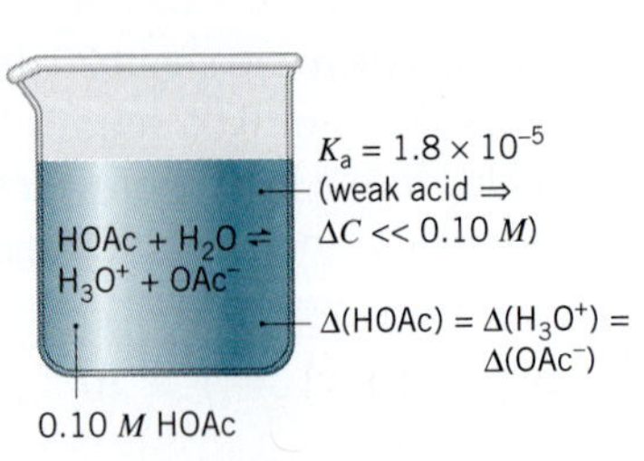

The balanced equation for the reaction states that we get one H_3O^+ ion and one OAc^- ion each time an HOAc molecule dissociates. This allows us to write equations for the equilibrium concentrations of the three components of the reaction. The change in the concentration of acetic acid as the reaction comes to equilibrium, ΔC, will be equal to the difference between the initial concentration of the acid, 0.10 *M*, and the concentration at equilibrium, [HOAc].

$$\Delta C = 0.10\,M - [\text{HOAc}]$$

Rearranging the equation gives the following result.

$$[\text{HOAc}] = 0.10\,M - \Delta C$$

Because there is very little H_3O^+ ion in pure water, we can assume that the initial concentration of this acid is effectively zero (≈ 0). Because we get one H_3O^+ ion and one OAc^- ion every time an acetic acid molecule dissociates, the concentrations of the ions at equilibrium are equal to the change in the concentration of acetic acid as the acid dissociates.

$$[H_3O^+] = \Delta C$$
$$[OAc^-] = \Delta C$$

We can therefore represent the problem that needs to be solved as follows.

	$HOAc(aq) + H_2O(l) \rightleftharpoons$	$H_3O^+(aq)$	$+ OAc^-(aq)$	$K_a = 1.8 \times 10^{-5}$
Initial	0.10 *M*	0	0	
Equilibrium	$0.10 - \Delta C$	ΔC	ΔC	

Substituting what we know about the system at equilibrium into the K_a expression gives the following equation.

$$K_a = \frac{[H_3O^+][OAc^-]}{[\text{HOAc}]} = \frac{[\Delta C][\Delta C]}{[0.10 - \Delta C]} = 1.8 \times 10^{-5}$$

Although we could rearrange the equation and solve it with the quadratic formula, let's test the assumption that acetic acid is a relatively weak acid. In other words, we can assume that ΔC is very small compared with the initial concentration of acetic acid. If this assumption is true, then subtracting ΔC from the initial concentration of the acid (0.10 *M*) will not significantly reduce that value.

$$\frac{[\Delta C][\Delta C]}{[0.10]} \approx 1.8 \times 10^{-5}$$

Or,

$$\frac{[\Delta C]^2}{[0.10]} \approx 1.8 \times 10^{-5}$$

And therefore,

$$[\Delta C]^2 \approx 1.8 \times 10^{-6}$$

We then solve the approximate equation for the value of ΔC.

$$\Delta C \approx 0.0013\,M$$

We now test our assumption. Is ΔC small enough to be ignored in this problem? Yes, because it is less than 5% of the initial concentration of acetic acid.

$$\frac{0.0013}{0.10} \times 100\% = 1.3\%$$

Our assumption is valid because subtracting 0.0013 from 0.10 would give us 0.10, to two significant figures. We can therefore use the value of ΔC obtained from this calculation to estimate the equilibrium concentrations of H_3O^+, OAc^-, and HOAc.

$$[HOAc] = 0.10 - \Delta C \approx 0.10\ M$$
$$[H_3O^+] = [OAc^-] = \Delta C \approx 0.0013\ M$$

Two assumptions were made in the above calculation.

- We assumed that the amount of acid that dissociates is small compared with the initial concentration of the acid.
- We assumed that enough acid dissociates to allow us to ignore the contribution to the H_3O^+ ion concentration from the dissociation of water.

In this case, both assumptions are valid. Only about 1.3% of the acetic acid molecules actually dissociate at equilibrium. But this gives us an H_3O^+ concentration that is 4 orders of magnitude larger than the concentration of the ion in pure water. Thus, we can legitimately ignore the dissociation of water.

The assumptions made in the calculation work for acids that are neither too strong nor too weak. Fortunately, many acids fall into that category. If the acid is too strong to assume that ΔC is small compared with the initial concentration, the problem can always be solved with the quadratic equation. If the acid is too weak or too dilute to ignore the dissociation of water, variations in the pH of the water from one sample to another because of impurities dissolved in the water are probably as large as the effect on the pH of adding the very weak acid.

To illustrate the importance of paying attention to the assumptions that are made in pH calculations, consider the task of calculating the pH of a solution prepared by dissolving 1.0×10^{-8} mol of a strong acid, such as hydrochloric acid, in a liter of water. It is tempting to assume that the acid dissociates completely in water to give a solution with the following H_3O^+ ion concentration.

$$[H_3O^+] = 1.0 \times 10^{-8}\ M$$

If this concentration is used with the definition of pH, we get the following result.

$$pH = -\log[H_3O^+] = -\log(1.0 \times 10^{-8}) = 8.0$$

The result of the calculation, however, does not make sense. It is impossible to create a basic solution (pH $>$ 7) by adding an acid to water!

The mistake, of course, was ignoring the concentration of the H_3O^+ ion from the dissociation of water. In pure water, we start with $1.0 \times 10^{-7}\ M$ H_3O^+ ion. If we add 10^{-8} mol/L from the strong acid, the total concentration of H_3O^+ from both sources will be about $1.1 \times 10^{-7}\ M$. Thus, the pH of the solution is actually slightly less than 7.

$$pH = -\log[H_3O^+] = -\log(1.1 \times 10^{-7}) = 6.96$$

Two factors control the concentration of the H_3O^+ ion in a solution of a weak acid: (1) the acid dissociation equilibrium constant, K_a, of the acid and (2) the concentration of the solution. When we compare solutions of equivalent concentrations, the amount of the H_3O^+ ion at equilibrium increases as the value of K_a of the acid increases, as shown in Exercise 11.7.

> **➤ CHECKPOINT**
>
> Assume that you have two solutions. Solution 1 has a low concentration of an acid with a large K_a. Solution 2 has a high concentration of an acid with a small K_a. Do you have enough information to determine which solution has the largest pH? Explain your answer.

Exercise 11.7

Determine the approximate pH of 0.10 *M* solutions of the following acids.

(a) hypochlorous acid, HOCl, $K_a = 2.9 \times 10^{-8}$

(b) hypobromous acid, HOBr, $K_a = 2.4 \times 10^{-9}$

(c) hypoiodous acid, HOI, $K_a = 2.3 \times 10^{-11}$

Solution

The first calculation can be set up as follows.

	$HOCl(aq) + H_2O(l) \rightleftharpoons$	$H_3O^+(aq)$	$+\ OCl^-(aq)$
Initial	0.10 *M*	≈ 0	0
Equilibrium	$0.10 - \Delta C$	ΔC	ΔC

Substituting the information into the K_a expression gives the following equation.

$$K_a = \frac{[H_3O^+][OCl^-]}{[HOCl]} = \frac{[\Delta C][\Delta C]}{[0.10 - \Delta C]} = 2.9 \times 10^{-8}$$

Because the equilibrium constant for the reaction is relatively small, we can try the assumption that ΔC is small compared with the initial concentration of the acid.

$$\frac{[\Delta C][\Delta C]}{[0.10]} \approx 2.9 \times 10^{-8}$$

Solving the equation for ΔC gives the following result.

$$\Delta C \approx 5.4 \times 10^{-5}\ M$$

Both assumptions made in the calculation are legitimate. ΔC is small compared with the initial concentration of HOCl, but it is several orders of magnitude larger than the H_3O^+ ion concentration from the dissociation of water. We can therefore use the value of ΔC to calculate the pH of the solution.

$$\text{pH} = -\log[H_3O^+] = -\log(5.4 \times 10^{-5}) \approx 4.3$$

Repeating the calculation with the same assumptions for hypobromous and hypoiodous acid gives the following results.

HOCl	$[H_3O^+] \approx 5.4 \times 10^{-5}\ M$	$\text{pH} \approx 4.3$
HOBr	$[H_3O^+] \approx 1.5 \times 10^{-5}\ M$	$\text{pH} \approx 4.8$
HOI	$[H_3O^+] \approx 1.5 \times 10^{-6}\ M$	$\text{pH} \approx 5.8$

As expected, the H_3O^+ ion concentration at equilibrium—and therefore the pH of the solution—depends on the value of K_a for the acid. The H_3O^+ ion concentration decreases and the pH of the solution increases as the value of K_a becomes smaller. The next exercise shows how the H_3O^+ ion concentration at equilibrium depends on the initial concentration of the acid.

Exercise 11.8

Determine the H_3O^+ ion concentration and pH of acetic acid solutions with the following concentrations: 1.0 *M*, 0.10 *M*, and 0.010 *M*.

Solution

The first calculation can be set up as follows.

	$HOAc(aq) + H_2O(l) \rightleftharpoons$	$H_3O^+(aq)$ +	$OAc^-(aq)$	$K_a = 1.8 \times 10^{-5}$
Initial	1.0 *M*	≈ 0	0	
Equilibrium	$1.0 - \Delta C$	ΔC	ΔC	

Substituting this information into the K_a expression gives the following equation.

$$K_a = \frac{[H_3O^+][OAc^-]}{[HOAc]} = \frac{[\Delta C][\Delta C]}{[1.0 - \Delta C]} = 1.8 \times 10^{-5}$$

We now assume that ΔC is small compared with the initial concentration of acetic acid.

$$\frac{[\Delta C][\Delta C]}{[1.0]} \approx 1.8 \times 10^{-5}$$

We then solve the equation for an approximate value of ΔC.

$$\Delta C \approx 0.0042$$

Once again, both of the characteristic assumptions of weak acid equilibrium calculations are legitimate in this case. ΔC is small compared with the initial concentration of acid, but large compared with the concentration of H_3O^+ ion from the dissociation of water. We can therefore use the value of ΔC to calculate the pH of the solution.

$$\text{pH} = -\log[H_3O^+] = -\log(4.2 \times 10^{-3}) = 2.4$$

Repeating the calculation for the different initial concentrations gives the following results.

1.0 *M* HOAc	$[H_3O^+] \approx 4.2 \times 10^{-3}$ *M*	pH $\approx$ 2.4
0.10 *M* HOAc	$[H_3O^+] \approx 1.3 \times 10^{-3}$ *M*	pH $\approx$ 2.9
0.010 *M* HOAc	$[H_3O^+] \approx 4.2 \times 10^{-4}$ *M*	pH $\approx$ 3.4

The concentration of the H_3O^+ ion in an aqueous solution of a weak acid gradually decreases and the pH of the solution increases as the solution becomes more dilute.

➤ CHECKPOINT

Which of the acetic acid solutions in Exercise 11.8 is the most acidic?

11.14 Base pH Calculations

With minor modifications, the techniques applied to equilibrium calculations for acids are valid for solutions of bases in water. Consider, for example, the calculation of the pH of a 0.10 *M* NH_3 solution. We can start by writing an equation for the reaction between ammonia and water.

$$NH_3(aq) + H_2O(l) \rightleftharpoons NH_4^+(aq) + OH^-(aq)$$

The ionization of NH_3 in water can be described by the following equilibrium expression.

$$K_b = \frac{[NH_4^+][OH^-]}{[NH_3]}$$

For NH_3, pK_b is 4.74.

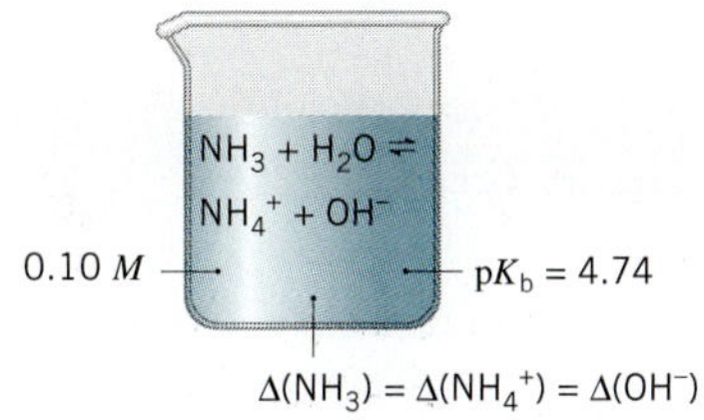

We can organize what we know about the equilibrium with the format we used for equilibria involving acids.

	$NH_3(aq)$ + $H_2O(l)$ $\rightleftharpoons$	$NH_4^+(aq)$ +	$OH^-(aq)$	$pK_b = 4.74$
Initial	0.10 *M*	0	≈ 0	
Equilibrium	$0.10 - \Delta C$	ΔC	ΔC	

We convert pK_b into K_b and substitute the information into the equilibrium constant expression to give the following equation.

$$pK_b = -\log K_b = 4.74 \qquad K_b = 1.8 \times 10^{-5}$$

$$K_b = \frac{[NH_4^+][OH^-]}{[NH_3]} = \frac{[\Delta C][\Delta C]}{[0.10 - \Delta C]} = 1.8 \times 10^{-5}$$

K_b for ammonia is small enough to consider the assumption that ΔC is small compared with the initial concentration of the base.

$$\frac{[\Delta C][\Delta C]}{[0.10]} \approx 1.8 \times 10^{-5}$$

Solving the approximate equation gives the following result.

$$\Delta C \approx 1.3 \times 10^{-3}$$

The value of ΔC is small enough compared with the initial concentration of NH_3 to be ignored and yet large enough compared with the OH^- ion concentration in

> **CHECKPOINT**
>
> Which is more basic, a 1 *M* solution of a base that has a small K_b or a 1 *M* solution of a base that has a large K_b?

water to ignore the dissociation of water. ΔC can therefore be used to calculate the pOH of the solution.

$$pOH = -\log(1.3 \times 10^{-3}) \approx 2.9$$

This, in turn, can be used to calculate the pH of the solution.

$$pH = 14.0 - pOH \approx 11.1$$

RELATIONSHIP BETWEEN K_a AND K_b

Equilibrium problems involving bases can be solved if the value of K_b for the base is known. Values of K_b are listed in Table B.9 for only a limited number of compounds, however. The first step in many base equilibrium calculations therefore involves determining the value of K_b for the reaction from the value of K_a for the conjugate acid.

The relationship between K_a and K_b for a conjugate acid–base pair can be understood by using a specific example of the relationship. Consider a solution containing the acetate ion (OAc^-), for example. We already know the value of K_a for acetic acid.

$$HOAc(aq) + H_2O(l) \rightleftharpoons H_3O^+(aq) + OAc^-(aq) \qquad K_a = 1.8 \times 10^{-5}$$

What is the value of K_b for the conjugate base, OAc^-?

$$OAc^-(aq) + H_2O(l) \rightleftharpoons HOAc(aq) + OH^-(aq) \qquad K_b = ?$$

The K_a and K_b expressions for acetic acid and its conjugate base both contain the ratio of the equilibrium concentrations of the acid and its conjugate base. K_a is proportional to $[OAc^-]$ divided by [HOAc], and K_b is proportional to [HOAc] divided by $[OAc^-]$.

$$K_a = \frac{[H_3O^+][OAc^-]}{[HOAc]} \qquad K_b = \frac{[HOAc][OH^-]}{[OAc^-]}$$

Two changes need to be made to derive the K_b expression from the K_a expression: We need to remove the $[H_3O^+]$ term and introduce an $[OH^-]$ term. We can do this by multiplying the top and bottom of the K_a expression by the OH^- ion concentration.

$$K_a = \frac{[H_3O^+][OAc^-]}{[HOAc]} \times \frac{[OH^-]}{[OH^-]}$$

Rearranging the equation gives the following result.

$$K_a = \frac{[OAc^-]}{[HOAc][OH^-]} \times [H_3O^+][OH^-]$$

The two terms on the right side of the equation should look familiar. The first is the inverse of the K_b expression, and the second is the expression for K_w.

$$K_a = \frac{1}{K_b} \times K_w$$

Rearranging the equation gives the following result.

$$K_aK_b = K_w$$

According to this equation, the value of K_b for the reaction between the acetate ion and water can be calculated from the K_a for acetic acid.

$$K_b = \frac{K_w}{K_a} = \frac{1.0 \times 10^{-14}}{1.8 \times 10^{-5}} = 5.6 \times 10^{-10}$$

The relationship between K_a and K_b for a conjugate acid–base pair can be used to explain the following general rules given in Section 11.9.

The stronger the acid, the weaker is the conjugate base.
The stronger the base, the weaker is the conjugate acid.

The product of K_a for any Brønsted acid times K_b for its conjugate base is equal to K_w for water.

$$K_aK_b = K_w$$

But K_w is a relatively small number (1.0×10^{-14}). As a result, both K_a and K_b can't be large. When K_a is relatively large—a strong acid—K_b for the conjugate base must be relatively small—a weak base. Conversely, if K_b is large—a strong base—K_a for the conjugate acid must be small—a weak acid.

ACID–BASE PROPERTIES OF A SALT

A *salt* is an ionic compound with a cation other than H^+ and an anion other than OH^- or O^{2-}. Dilute solutions of soluble salts are assumed to completely ionize in water. The acid–base properties of a salt are therefore determined by the acid–base properties of the aqueous ions that result from the dissolution of the salt in water. Group I and II cations (Li^+, Na^+, K^+, Rb^+, Cs^+, Ca^{2+}, Sr^{2+}, Ba^{2+}) have no acid or base properties. The conjugate bases of strong monoprotic acids (Cl^-, Br^-, I^-, ClO_4^-, NO_3^-) are so extremely weak that they have essentially no basic properties in water.

To examine the acid–base properties of the salt potassium fluoride, KF, we begin by assuming that it dissociates completely when it dissolves in water.

$$KF(aq) \longrightarrow K^+(aq) + F^-(aq)$$

The potassium ion is neither an acid nor a base. However, the F^- ion is the conjugate base of a weak acid, HF. The fluoride ion will therefore behave as a base.

$$F^-(aq) + H_2O(l) \rightleftharpoons HF(aq) + OH^-(aq)$$

A solution of KF in water is therefore basic, and potassium fluoride is referred to as a basic salt.

Ammonium nitrate, NH_4NO_3, is an example of an acidic salt. A dilute solution will ionize completely when it dissolves in water.

$$NH_4NO_3(aq) \longrightarrow NH_4^+(aq) + NO_3^-(aq)$$

The nitrate anion, NO_3^-, is the conjugate base of the strong acid HNO_3 and therefore has essentially no base properties. The ammonium cation, NH_4^+, however, is the conjugate acid of a weak base ammonia (NH_3) and should therefore behave as an acid in water. An aqueous solution of NH_4NO_3 is therefore acidic.

$$NH_4^+(aq) + H_2O(l) \rightleftharpoons H_3O^+(aq) + NH_3(aq)$$

Sodium chloride, NaCl, is an example of a salt that produces essentially a neutral solution. In dilute aqueous solutions sodium chloride ionizes completely.

$$NaCl(aq) \longrightarrow Na^+(aq) + Cl^-(aq)$$

The chloride ion is the conjugate base of a strong acid and is therefore much too weak a base to be noticed. The sodium ion is neither an acid nor a base in the Brønsted model of acids and bases. Aqueous solutions of NaCl are therefore essentially neutral.

Exercise 11.9

Calculate the HOAc, OAc^-, and OH^- concentrations at equilibrium in a 0.10 M NaOAc solution. (For HOAc, $K_a = 1.8 \times 10^{-5}$.)

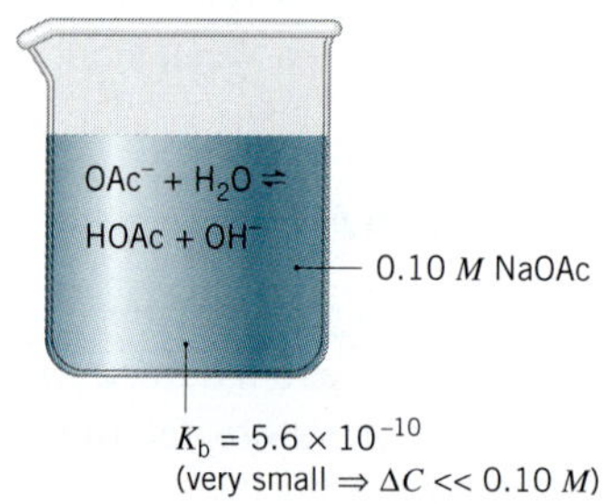

Solution

We start with a solution of an ionic salt—NaOAc—dissolved in water. We assume that the salt ionizes completely.

$$NaOAc(aq) \longrightarrow Na^+(aq) + OAc^-(aq)$$

Because the Na^+ ion is neither an acid nor a base, only the acetate ion will significantly affect the pH. We therefore start, as always, by building a representation of the information in the statement of the problem.

	$OAc^-(aq)$ + $H_2O(l)$ $\rightleftharpoons$	$HOAc(aq)$ +	$OH^-(aq)$	$K_b = ?$
Initial	0.10 M	0	≈ 0	
Equilibrium	$0.10 - \Delta C$	ΔC	ΔC	

The only other information we need to solve the problem is the value of K_b for the reaction between the acetate ion and water, which can be calculated from the value of K_a for acetic acid.

$$K_b = \frac{K_w}{K_a} = \frac{1.0 \times 10^{-14}}{1.8 \times 10^{-5}} = 5.6 \times 10^{-10}$$

Substituting what we know into the equilibrium constant expression gives the following equation.

$$K_b = \frac{[\text{HOAc}][\text{OH}^-]}{[\text{OAc}^-]} = \frac{[\Delta C][\Delta C]}{[0.10 - \Delta C]} = 5.6 \times 10^{-10}$$

Because the value of K_b for OAc^- ion is relatively small, we can feel reasonably confident that ΔC is small compared with the initial concentration of base.

$$\frac{[\Delta C][\Delta C]}{[0.10]} = 5.6 \times 10^{-10}$$

We can then solve the approximate equation the assumption generates.

$$\Delta C \approx 7.5 \times 10^{-6}\ M$$

ΔC is very much smaller than the initial concentration of the base, which means that one of the key assumptions of acid–base equilibrium calculations is legitimate. But $[OH^-]$ is still at least 75 times larger than the concentration of the OH^- ion in pure water, which means that we can also legitimately ignore the dissociation of water in the calculation. We can therefore use the value of ΔC to calculate the equilibrium concentrations of HOAc, OAc^-, and OH^-.

$$[\text{OAc}^-] = 0.10 - \Delta C \approx 0.10\ M$$
$$[\text{HOAc}] = [\text{OH}^-] = \Delta C \approx 7.5 \times 10^{-6}\ M$$

➤ CHECKPOINT

If K_a for HNO_2 is 5.1×10^{-4}, what is K_b for the NO_2^- ion?

11.15 Mixtures of Acids and Bases

So far, we have examined 0.10 *M* solutions of both acetic acid and sodium acetate.

0.10 *M* HOAc	$[H_3O^+] \approx 1.3 \times 10^{-3}\ M$	pH ≈ 2.9
0.10 *M* NaOAc	$[OH^-] \approx 7.5 \times 10^{-6}\ M$	pH ≈ 8.9

What would happen if we added enough sodium acetate to an acetic acid solution so that the solution is simultaneously 0.10 *M* in both HOAc and NaOAc?

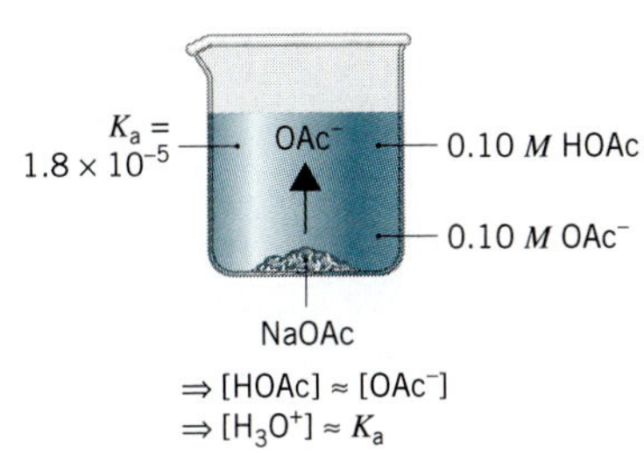

The first step toward answering the question is recognizing that there are two sources of the OAc^- ion in the solution. Acetic acid, of course, dissociates to give the H_3O^+ and OAc^- ions.

$$\text{HOAc}(aq) + \text{H}_2\text{O}(l) \rightleftharpoons \text{H}_3\text{O}^+(aq) + \text{OAc}^-(aq)$$

The salt, sodium acetate, however, also dissociates in water to give the OAc^- ion.

$$NaOAc(s) \xrightarrow{H_2O} Na^+(aq) + OAc^-(aq)$$

These reactions share a common ion: the OAc^- ion. Le Châtelier's principle predicts that adding sodium acetate to acetic acid should shift the equilibrium between HOAc and the H_3O^+ and OAc^- ions to the left. Thus, adding NaOAc reduces the extent to which HOAc dissociates. The mixture is therefore less acidic than 0.10 *M* HOAc, by itself.

In Exercise 11.9 we saw that OAc^- behaves as a base. The K_b for OAc^- is several orders of magnitude smaller than K_a for HOAc. Therefore, the reaction of OAc^- with water is small compared to the reaction of HOAc with water.

We can quantify the discussion by setting up the problem as follows:

	$HOAc(aq) + H_2O(l) \rightleftharpoons$	$H_3O^+(aq)$ +	$OAc^-(aq)$	$K_a = 1.8 \times 10^{-5}$
Initial	0.10 *M*	≈0	0.10 *M*	
Equilibrium	?	?	?	

We don't have any basis for predicting whether the dissociation of water can be ignored in this calculation, so let's make that assumption and check its validity later.

If most of the H_3O^+ ion concentration at equilibrium comes from dissociation of the acetic acid, the reaction has to shift to the right to reach equilibrium.

	$HOAc(aq) + H_2O(l) \rightleftharpoons$	$H_3O^+(aq)$ +	$OAc^-(aq)$	$K_a = 1.8 \times 10^{-5}$
Initial	0.10 *M*	≈0	0.10 *M*	
Equilibrium	$0.10 - \Delta C$	ΔC	$0.10 + \Delta C$	

Substituting the information into the K_a expression gives the following result.

$$K_a = \frac{[H_3O^+][OAc^-]}{[HOAc]} = \frac{[\Delta C][0.10 + \Delta C]}{[0.10 - \Delta C]} = 1.8 \times 10^{-5}$$

Let's assume that ΔC is small compared with the initial concentrations of HOAc and OAc^-.

$$\frac{[\Delta C][0.10]}{[0.10]} \approx 1.8 \times 10^{-5}$$

We then solve the approximate equation the assumption generates.

$$\Delta C \approx 1.8 \times 10^{-5}\ M$$

Are our two assumptions legitimate? Is ΔC small enough compared with 0.10 to be ignored? Is ΔC large enough so that the dissociation of water can be ignored? The answer to these questions is yes. We can therefore use that value of ΔC to calculate the H_3O^+ ion concentration at equilibrium and the pH of the solution.

$$[H_3O^+] \approx \Delta C \approx 1.8 \times 10^{-5}$$
$$pH = -\log(1.8 \times 10^{-5}) \approx 4.7$$

The solution is therefore acidic, but less acidic than HOAc by itself. Le Châtelier's principle predicts that the equilibrium for the dissociation of acetic acid in water would be shifted toward the reactants by the addition of acetate ion. This would result in a decrease in the amount of H_3O^+. In other words, the HOAc solution is less acidic due to the addition of a base, OAc^-.

11.16 Buffers and Buffer Capacity

Mixtures of a weak acid and its conjugate base are called **buffers.** The term *buffer* means "to lessen or absorb shock." These solutions are buffers because they lessen or absorb the drastic change in pH that occurs when small amounts of acids or bases are added to water.

In Section 11.12 we concluded that adding a single drop of 2 *M* HCl to 100 mL of pure water would change the pH by four units, from pH 7 to pH 3. If we add the same amount of HCl to a buffer solution that was 0.10 *M* in both acetic acid (HOAc) and a salt of the acetate ion (OAc^-), there would be no observable change in the pH of the solution. In fact, we can add 10 times as much 2 *M* HCl to the buffer solution without noticing a significant change in the pH of the solution. Thus, the pH of the buffer solution is truly "buffered" against the effect of small amounts of acid or base.

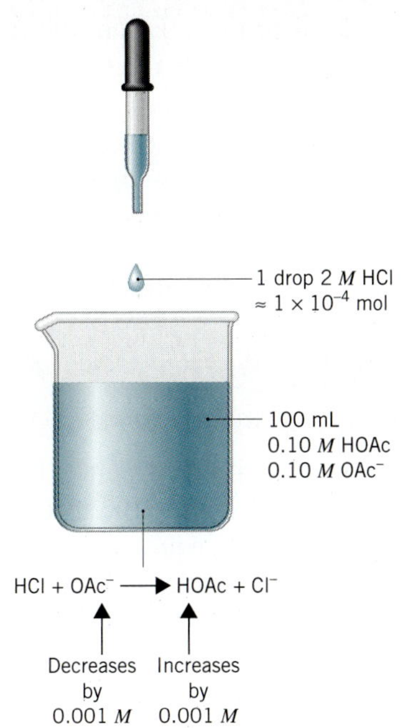

To understand how the buffer works, it is important to recognize that a buffer is made by mixing a weak acid and its conjugate base so that significant amounts of both a weak acid and weak base are present in solution.

$$\underset{\text{Weak acid}}{HOAc(aq)} + H_2O(l) \rightleftharpoons H_3O^+(aq) + \underset{\text{Conjugate base}}{OAc^-(aq)}$$

At equilibrium, the concentrations of acetic acid and the H_3O^+ and OAc^- ions in the solution are related by the following equation.

$$K_a = \frac{[H_3O^+][OAc^-]}{[HOAc]}$$

If we rearrange the equation, by solving for the H_3O^+ ion concentration at equilibrium, we get the following result.

$$[H_3O^+] = K_a \times \frac{[HOAc]}{[OAc^-]}$$

This equation suggests that the H_3O^+ ion concentration in the solution will stay more or less constant as long as there is no significant change in the ratio of the concentrations of the undissociated HOAc molecules and the OAc^- ions. The buffer solution was prepared by adding enough sodium acetate (NaOAc) to acetic acid (HOAc) to make the concentration of both the acid (HOAc) and its conjugate base (OAc^-) the same. (The solution is simultaneously 0.10 *M* HOAc and 0.10 *M* OAc^-.)

If a small amount of a strong acid, H_3O^+, is added to the solution, it will react with the OAc^- ion, which behaves as a Brønsted base by accepting a proton to form a bit more acetic acid.

$$OAc^-(aq) + H_3O^+(aq) \longrightarrow HOAc(aq) + H_2O(l)$$

The net result is a slight increase in the concentration of the undissociated HOAc molecules and a slight decrease in the concentration of the OAc^- ion. But, if only a small amount of acid is added to the solution, the ratio of the concentrations would be roughly the same. As a result, the H_3O^+ ion concentration in the solution remains essentially the same, which means there is no change in the pH of the solution. In other words, the addition of a strong acid, H_3O^+, to the buffer results in the strong acid being converted into an equivalent amount of weaker acid, HOAc.

If a small amount of a strong base, OH^-, is added to the solution, it will react with some of the undissociated HOAc molecules to form a bit more of the OAc^- ion.

$$HOAc(aq) + OH^-(aq) \longrightarrow OAc^-(aq) + H_2O(l)$$

But, once again, if we add only a small amount of the base, the ratio of the concentrations of HOAc and OAc^- in the solution will be more or less the same. This means that the pH of the solution is more or less the same. The addition of a strong base, OH^-, to the buffer converts the strong base into an equivalent amount of weak base, OAc^-.

Because it contains both HOAc molecules and the OAc^- ion, the solution can resist a change in pH when reasonable amounts of either a strong acid or a strong base are added to the solution. The concentrations of acetic acid and the acetate ion will change when the acid or base is added to the buffer solution, but the ratio of the concentrations of these two components of the equilibrium will not change by a large enough amount to produce a significant change in the pH of the solution.

As long as the concentrations of HOAc and the OAc^- ion in the buffer are larger than the amount of acid or base added to the solution, the pH remains constant. Table 11.5 compares the effects of adding different amounts of hydrochloric acid to water and two buffer solutions of different concentrations. The first column gives the number of moles of HCl added per liter. The second column shows the effect of this much acid on the pH of water. The third column shows the effect on a buffer solution that contains 0.10 *M* HOAc and 0.10 *M* NaOAc. The fourth column shows the result of adding the acid to a more concentrated buffer that contains twice as much acetic acid and twice as much sodium acetate. Even when as much as 0.01 mol/L of acid is added to the first buffer, there is very little change in the pH of the buffer solution, as shown in Figure 11.11.

Table 11.5

Effect of Adding Hydrochloric Acid to Water and Two Buffer Solutions

Mol HCl/L Added to Each Solution	pH of Water	pH of Buffer 1 (0.10 *M* HOAc/ 0.10 *M* OAc^-)	pH of Buffer 2 (0.20 *M* HOAc/ 0.20 *M* OAc^-)
0	7	4.74	4.74
0.000001	6	4.74	4.74
0.00001	5	4.74	4.74
0.0001	4	4.74	4.74
0.001	3	4.74	4.74
0.01	2	4.66	4.70
0.1	1	2.72	4.27

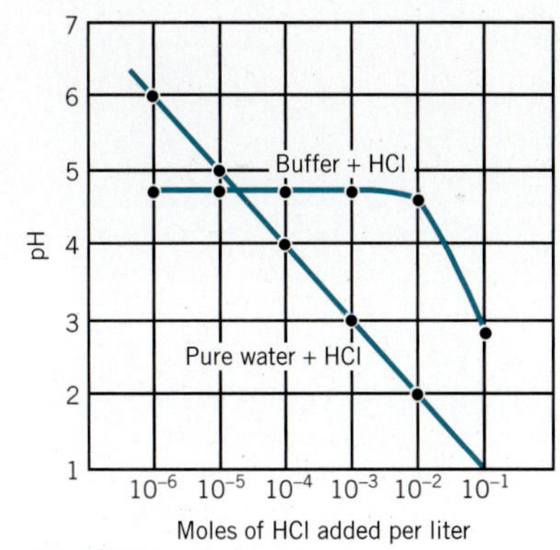

Fig. 11.11 The pH of water is very sensitive to the addition of hydrochloric acid. Adding as little as 10^{-3} mol of HCl to a liter of water can drop the pH by 4 units. The pH of a buffer, however, remains constant until relatively large amounts of hydrochloric acid have been added. The buffer contains 0.10 *M* OAc^- and HOAc.

By definition, the **buffer capacity** of a buffer is the amount of strong acid or base that can be added before the pH of the solution changes significantly. The capacity of the second buffer solution in Table 11.5 is even larger than that of the first. Buffer 2 protects against quantities of acid or base almost as large as 0.1 mol/L.

Buffers can also be made from a weak base and its conjugate acid, such as ammonia and a salt of the ammonium ion.

$$\underset{\text{Conjugate acid}}{NH_4^+(aq)} + H_2O(l) \rightleftharpoons H_3O^+(aq) + \underset{\text{Weak base}}{NH_3(aq)}$$

There is an important difference, however, between this buffer and one made by mixing sodium acetate and acetic acid. Mixtures of HOAc and OAc^- ion form an **acidic buffer,** with a pH below 7. Mixtures of NH_3 and the NH_4^+ ion form a **basic buffer,** with a pH above 7.

We can predict whether a buffer will be acidic or basic by comparing the values of K_a and K_b for the conjugate acid–base pair. K_a for acetic acid is significantly larger than K_b for the acetate ion. We therefore expect mixtures of the conjugate acid–base pair to be acidic.

$$\begin{array}{ll} \text{HOAc} & K_a = 1.8 \times 10^{-5} \\ \text{OAc}^- & K_b = 5.6 \times 10^{-10} \end{array}$$

K_b for ammonia, on the other hand, is much larger than K_a for the ammonium ion. Mixtures of this acid–base pair are therefore basic.

$$\begin{array}{ll} NH_3 & K_b = 1.8 \times 10^{-5} \\ NH_4^+ & K_a = 5.6 \times 10^{-10} \end{array}$$

In general, if K_a for the acid is larger than 1×10^{-7}, the buffer will be acidic. If K_b is larger than 1×10^{-7}, the buffer is basic.

Buffers are used extensively in the chemistry laboratory. Buffers can be purchased from chemical suppliers as solutions with known concentrations and pH values. They may also be purchased as packets of a mixture of a solid conjugate acid–base pair. Dissolving the packet in water yields a buffer with a pH equal to the value stated on the package. In addition, buffers may be prepared by mixing measured amounts of an appropriate conjugate acid–base pair and then adding strong acid or base to adjust the pH to the desired value. The pH of the buffer is normally monitored with a pH meter as the strong acid or strong base is added.

➤ CHECKPOINT

The pH of a buffered solution is 7.4. What happens to the pH if an acid is added to the buffer? What happens to the pH if a base is added to the buffer? Does the resulting pH depend on how much acid or base is added?

BUFFERS IN THE BODY

Buffers are very important in living organisms for maintaining the pH of biological fluids within the very narrow ranges necessary for the biochemical reactions of life processes. Three primary buffer systems maintain the pH of blood in the human body within the limits 7.36 to 7.42. One of these buffers is composed of carbonic acid, H_2CO_3, and its conjugate base HCO_3^-, the hydrogen carbonate ion. Some of the CO_2 from respiring tissue that enters a red blood cell reacts with water and is converted to carbonic acid.

$$CO_2(g) + H_2O(l) \rightleftharpoons H_2CO_3(aq)$$

The carbonic acid then partially dissociates to form an equilibrium mixture with the hydrogen carbonate (or bicarbonate) ion.

$$H_2CO_3(aq) + H_2O(l) \rightleftharpoons HCO_3^-(aq) + H_3O^+(aq) \quad \text{buffer reaction}$$

When the concentration of the H_3O^+ ion in the blood is too high (too low a pH), a condition known as *acidosis* results. This can be caused by a disease such as diabetes mellitus, in which large amounts of acids are produced. As the concentration of H_3O^+ in blood increases, the H_2CO_3/HCO_3^- equilibrium shifts to the left to maintain the proper pH.

A low concentration of H_3O^+ in the blood (too high a pH) is known as *alkalosis*. This condition can result during hyperventilation (rapid breathing) when excess amounts of CO_2 are lost from the body through the lungs. The loss of CO_2 shifts the CO_2/H_2CO_3 equilibrium to the left, resulting in a decrease in the concentration of carbonic acid in the blood. This disrupts the H_2CO_3/HCO_3^- buffer by shifting the equilibrium to the left and therefore reducing the amount of H_3O^+ in the blood. The resulting increase in pH is alkalosis.

> **➤ CHECKPOINT**
>
> Lactic acid, $HC_3H_5O_3$, is produced in the body during strenuous exercise. Write a chemical reaction that can occur in blood to maintain the proper pH when lactic acid is produced.

11.17 Acid–Base Reactions

STRONG ACID–STRONG BASE

The reaction between an acid and a base can produce a salt and water. Consider, for example, the reaction of the strong acid hydrochloric acid with the strong base sodium hydroxide.

$$HCl(aq) + NaOH(aq) \longrightarrow H_2O(l) + NaCl(aq)$$

The full ionic equation for this reaction is

$$H^+(aq) + Cl^-(aq) + Na^+(aq) + OH^-(aq) \longrightarrow H_2O(l) + Na^+(aq) + Cl^-(aq)$$

For acid and base reactions, $H^+(aq)$ is best represented by $H_3O^+(aq)$, and the net ionic equation for this reaction is usually written as follows.

$$\underset{\textit{Stronger acid}}{H_3O^+(aq)} + \underset{\textit{Stronger base}}{OH^-(aq)} \longrightarrow \underset{\textit{Weaker acid}}{H_2O(l)} + \underset{\textit{Weaker base}}{H_2O(l)}$$

H_3O^+ is a stronger acid than H_2O, and OH^- is a stronger base than H_2O; as shown in Section 11.10, the equilibrium in the above reaction lies almost entirely to the right. The equilibrium constant for this reaction is the reciprocal of K_w, or

1.0×10^{14}. No matter which strong acid or strong base is reacted, the net ionic reaction is always that given by the preceeding equation with $K = 1.0 \times 10^{14}$.

When just enough strong acid has been added to completely react with a strong base, the resultant solution will have equal concentrations of H_3O^+ and OH^- and thus give a pH of 7.0.

WEAK ACID–STRONG BASE

If a weak acid is reacted with a strong base, the reaction will go almost to completion.

$$\underset{\textit{Weak acid}}{HA(aq)} + \underset{\textit{Strong base}}{OH^-(aq)} \rightleftharpoons H_2O(l) + A^-(aq)$$

In contrast to the reaction of a strong acid with a strong base, the pH of a solution produced from the complete reaction of a weak acid and strong base will not be 7.0, but will depend on the interaction of A^- with water.

The weak acid, HNO_2, reacts with a sodium hydroxide solution according to

$$\underset{\textit{Weak acid}}{HNO_2(aq)} + \underset{\textit{Strong base}}{OH^-(aq)} \longrightarrow H_2O(l) + NO_2^-(aq)$$

When this reaction is essentially complete, the pH of the resulting solution is due to the interaction of the weak base NO_2^- with H_2O.

The equilibrium constant for this reaction can be calculated from tabulated values of the equilibrium constants for HNO_2 (Table B.8) and K_w.

$$HNO_2(aq) + H_2O(l) \rightleftharpoons H_3O^+(aq) + NO_2^-(aq) \qquad K_a = 5.1 \times 10^{-4}$$
$$H_3O^+(aq) + OH^-(aq) \rightleftharpoons 2H_2O(l) \qquad K = 1.0 \times 10^{14}$$

where K for the second reaction is the inverse of K_w. Summing the two reactions gives

$$HNO_2(aq) + OH^-(aq) \rightleftharpoons H_2O(l) + NO_2^-(aq)$$
$$K_{rxn} = K_a \times K = 5.1 \times 10^{-4} \times 1.0 \times 10^{14} = 5.1 \times 10^{10}$$

The equilibrium constant is so large that the reaction is essentially complete. In general the reaction of a weak acid with a strong base proceeds, as in this example, very far to the right.

STRONG ACID–WEAK BASE

The weak base methyl amine, CH_3NH_2, reacts with a strong acid such as HI according to

$$CH_3NH_2(aq) + HI(aq) \longrightarrow CH_3NH_3^+(aq) + I^-(aq)$$

or

$$CH_3NH_2(aq) + H_3O^+(aq) \longrightarrow CH_3NH_3^+(aq) + H_2O(l)$$

The equilibrium constant for this reaction can be found by combining appropriate reactions to give the desired reaction for which the equilibrium constants are known. Tables B.8 and B.9 contain a tabulation of many of these useful equilibrium constants.

$$CH_3NH_2(aq) + H_2O(l) \rightleftharpoons CH_3NH_3^+(aq) + OH^-(aq) \qquad K_b = 4.8 \times 10^{-4}$$
$$H_3O^+(aq) + OH^-(aq) \rightleftharpoons 2H_2O(l) \qquad K = 1.0 \times 10^{14}$$
$$\overline{CH_3NH_2(aq) + H_3O^+(aq) \rightleftharpoons CH_3NH_3^+(aq) + H_2O(l) \qquad K_{rxn} = K_b \times K = 4.8 \times 10^{10}}$$

The equilibrium constant is so large that the above reaction proceeds nearly to completion. In general, a strong acid will react completely with a weak base. The pH of the resulting mixture will depend on the nature of the acid, in this case $CH_3NH_3^+$, produced in the reaction.

WEAK BASE–WEAK ACID

The reaction between the weak base methyl amine and the weak acid nitrous acid

$$\underset{\textit{Weak base}}{CH_3NH_2(aq)} + \underset{\textit{Weak acid}}{HNO_2(aq)} \longrightarrow CH_3NH_2^+(aq) + NO_2^-(aq)$$

will proceed to an extent that depends on the magnitude of the equilibrium constant. For such reactions in general, K_{rxn} can be found by the following procedure.

$$\begin{array}{rll} HA(aq) + H_2O(l) \rightleftharpoons H_3O^+(aq) + A^-(aq) & K_a \\ B(aq) + H_2O(l) \rightleftharpoons BH^+(aq) + OH^-(aq) & K_b \\ H_3O^+(aq) + OH^-(aq) \rightleftharpoons 2H_2O(l) & K = 1.0 \times 10^{14} \\ \hline HA(aq) + B(aq) \rightleftharpoons BH^+(aq) + A^-(aq) & K_{rxn} = K_a \times K_b = 1.0 \times 10^{14} \end{array}$$

For the CH_3NH_2 and HNO_2 reaction, Tables B.8 and B.9 give the necessary K_a and K_b values and

$$K_{rxn} = 5.1 \times 10^{-4} \times 4.8 \times 10^{-4} \times 1.0 \times 10^{14} = 2.5 \times 10^7$$

In this case K_{rxn} is quite large and the reaction proceeds nearly to completion. However, this is not always the case. For the reaction of HCN with hydrazine, H_2NNH_2, the equilibrium constant is only 0.072; thus this reaction hardly proceeds at all. Weak acid–weak base reactions are difficult to generalize, and in each case an examination of the equilibrium constants is necessary.

11.18 pH Titration Curves

We can measure the concentration of a solution by a technique known as **titration.** A solution of known concentration slowly is added to a known quantity of a reagent with which it reacts until we observe something that tells us that exactly equivalent numbers of moles of the reagents are present. Titrations therefore depend on the existence of a class of compounds known as *indicators*. A common type of titration is an acid–base titration in which an acid of unknown concentration is titrated with a base that has an accurately known concentration. As a result of the titration, it is possible to determine the concentration of the acid solution.

Phenolphthalein is an example of an indicator for acid–base reactions. Phenolphthalein is colorless in acidic solutions. In the presence of a base, it has a light pink color. Phenolphthalein can therefore indicate when enough base has been added to consume the acid that was initially present in the solution because it turns from colorless to pink *just after* the point when equivalent numbers of moles of acid and base have been added to the solution.

The **end point** of an acid–base titration is the point at which the indicator turns color. The **equivalence point** is the point at which exactly enough base has been added to neutralize the acid. Chemists try to use indicators for which the end point is as close as possible to the point at which equivalent amounts of acid and base are present.

Phenolphthalein provides an example of how chemists compensate for the fact that the indicator doesn't always turn color exactly at the equivalence point of the titration. Acid–base titrations that use phenolphthalein as the indicator are stopped at the point at which a pink color appears and remains for 10 seconds as the solution is stirred. Because this is one drop before the indicator turns a permanent color, it is as close as we can get to the equivalence point of the titration.

Exercise 11.10

The reaction between sulfuric acid (H_2SO_4) and potassium hydroxide (KOH) can be described by the following equation. Because this is a reaction between a strong acid and a strong base, the reaction can be assumed to go to completion.

$$H_2SO_4(aq) + 2\ KOH(aq) \longrightarrow 2\ K^+(aq) + SO_4^{2-}(aq) + 2\ H_2O(l)$$

A 25.00-mL sample of sulfuric acid solution requires 42.61 mL of a 0.1500 *M* KOH solution to titrate to the phenolphthalein end point. Calculate the concentration of the sulfuric acid solution.

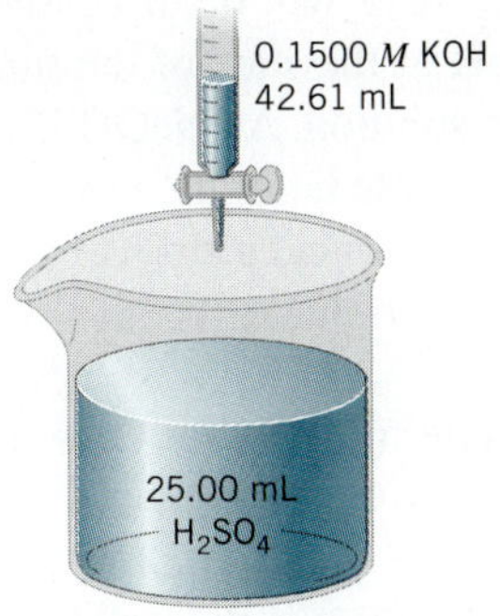

Solution

We only know the volume of the sulfuric acid solution, but we know both the volume and the concentration of the KOH solution. It therefore seems reasonable to start by calculating the number of moles of KOH in this solution.

$$\frac{0.1500 \text{ mol KOH}}{1 \text{ L}} \times 0.04261 \text{ L} = 6.392 \times 10^{-3} \text{ mol KOH}$$

We now know the number of moles of KOH consumed in the reaction, and we have a balanced chemical equation for the reaction that occurs when the two solutions are mixed. We can therefore calculate the number of moles of H_2SO_4 needed to consume that much KOH.

$$6.392 \times 10^{-3} \text{ mol KOH} \times \frac{1 \text{ mol } H_2SO_4}{2 \text{ mol KOH}} = 3.196 \times 10^{-3} \text{ mol } H_2SO_4$$

We now know the number of moles of H_2SO_4 in the sulfuric acid solution and the volume of the solution. We can therefore calculate the number of moles of sulfuric acid per liter, or the molarity of the solution.

$$\frac{3.196 \times 10^{-3} \text{ mol } H_2SO_4}{0.02500 \text{ L}} = 0.1278\ M\ H_2SO_4$$

The sulfuric acid solution therefore has a concentration of 0.1278 mole per liter.

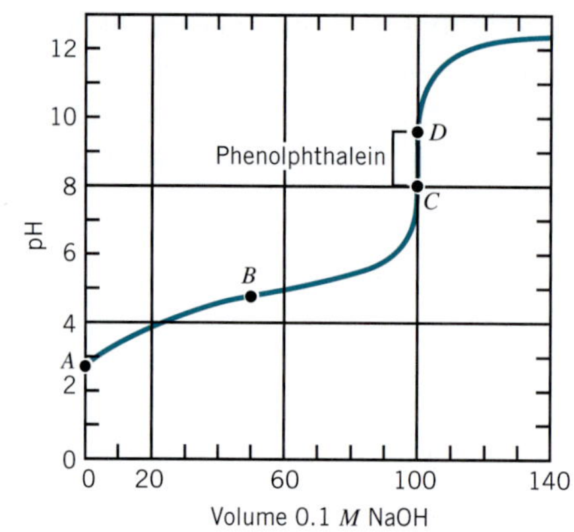

Fig. 11.12 The pH of a 0.10 *M* solution of acetic acid as it is titrated with 0.10 *M* sodium hydroxide. The pH increases at first, as some of the HOAc is converted into OAc^- ions. This creates a buffer, which resists changes in pH until enough NaOH has been added to completely react with the acetic acid. At this point the pH increases rapidly, then slowly levels off as the solution begins to behave more like a 0.10 *M* NaOH solution. Points *A*–*D* are explained in the text.

Acid–base titrations can be described by using a titration curve, a plot of solution pH versus amount of titrant added. Buffers and buffering capacity have a lot to do with the shape of the titration curve in Figure 11.12, which shows the pH of a solution of acetic acid as it is titrated with a strong base, sodium hydroxide. Four points (*A*, *B*, *C*, and *D*) on the curve will be discussed in some detail.

Point *A* represents the pH at the start of the titration. The pH at this point is 2.9, the pH of a 0.10 *M* HOAc solution. As NaOH is added to the solution, some of the acid is converted to its conjugate base, the OAc^- ion. The equilibrium constant for the reaction between a weak acid and a strong base is large. Therefore, the equilibrium lies almost entirely to the right.

$$\mathrm{HOAc}(aq) + \mathrm{OH^-}(aq) \longrightarrow \mathrm{OAc^-}(aq) + \mathrm{H_2O}(l)$$

Point *B* is the point at which exactly half of the HOAc molecules have been converted to OAc^- ions. The $[H_3O^+]$ in the solution is dependent on the equilibrium between HOAc and OAc^-. This equilibrium is described by the dissociation of HOAc in water.

$$\mathrm{HOAc}(aq) + \mathrm{H_2O} \rightleftharpoons \mathrm{OAc^-}(aq) + \mathrm{H_3O^+}(aq)$$

$$K_a = \frac{[\mathrm{H_3O^+}][\mathrm{OAc^-}]}{[\mathrm{HOAc}]}$$

At point *B,* the concentration of HOAc molecules is equal to the concentration of OAc^- ions. Because the ratio of the HOAc and OAc^- concentrations is 1:1, the concentration of the H_3O^+ ion at point *B* is equal to the value of K_a for the acid.

$$K_a = [\mathrm{H_3O^+}] \qquad \text{at point } B$$

This provides us with a way of measuring K_a for an acid. We can titrate a sample of the acid with a strong base, plot the titration curve, and then find the point along the curve at which exactly half of the acid has been consumed. The H_3O^+ ion concentration at that point will be equal to the value of K_a for the acid.

Point *C* is the *equivalence point* for the titration, the point at which enough base has been added to the solution to consume the acid present at the start of the titration. The goal of any titration is to find the equivalence point. What is actually observed is the *end point* of the titration—point *D*—the point at which the indicator, phenolphthalein, changes color.

Every effort is made to bring the end point as close as possible to the equivalence point of a titration. Suppose that phenolphthalein is used as the indicator for this titration, for example. Phenolphthalein turns from colorless to pink as the pH of the solution changes from 8 to 10. Because of the large change in pH at the equivalence point we look for the one drop of titrant that will convert the solution from colorless to pink.

Note the shape of the pH titration curve in Figure 11.12. The pH rises rapidly at first because we are adding a strong base to a weak acid and the base neutralizes some of the acid. The curve then levels off, and the pH remains more or less constant as we add base because some of the HOAc present initially is converted into OAc^- ions to form a buffer solution. The pH of the buffer solution stays relatively constant until most of the acid has been converted to its conjugate base. At that point the pH rises rapidly because essentially all of the HOAc in the system has been converted into OAc^- ions and the buffer is exhausted. The pH then gradually levels off as the solution begins to look like a 0.10 *M* NaOH solution.

The titration curve for a weak base, such as ammonia, titrated with a strong acid, such as hydrochloric acid, would be analogous to the curve in Figure 11.12. The principal difference is that the pH value is high at first and decreases as acid is added.

➤ CHECKPOINT

Describe what happens to the pH of an HCl solution as NaOH is slowly added.

Key Terms

Acid–base indicators
Acid dissociation equilibrium constant (K_a)
Acidic buffer
Acids
Alkalies
Arrhenius acid/base
Base ionization equilibrium constant (K_b)
Bases
Basic buffer
Brønsted acid/base
Buffer
Buffer capacity
Conjugate acid–base pair
End point
Equivalence point
Hydrogen ion acceptor/donor
Hydronium ion
Leveling effect
Monoprotic acid
Neutralization reaction
pH
pOH
Proton acceptor/donor
Strong acid
Strong base
Titration
Water dissociation equilibrium constant (K_w)
Weak acid
Weak base

Problems

Properties of Acids and Bases

1. Describe how acids and bases differ. Give examples of substances from daily life that fit each category.
2. Many metals dissolve in acids. Write an equation for the dissolution of magnesium metal in hydrochloric acid.
3. How could litmus be used to distinguish between an acid and a base?
4. What happens when acids and bases are combined?

The Arrhenius Definition of Acids and Bases

5. Which of the following compounds are Arrhenius acids?
   (a) HBr (b) H_2SO_4
   (c) $FeCl_3$ (d) $Fe(OH)_3$

$$HBr(aq) \xrightarrow{H_2O} H^+(aq) + Br^-(aq)$$
$$H_2SO_4(aq) \xrightarrow{H_2O} HSO_4^-(aq) + H^+(aq)$$
$$FeCl_3(s) \xrightarrow{H_2O} Fe^{3+}(aq) + 3\ Cl^-(aq)$$
$$Fe(OH)_3(s) \overset{H_2O}{\rightleftharpoons} Fe^{3+}(aq) + 3\ OH^-(aq)$$

6. Which of the following compounds are Arrhenius bases?
   (a) LiOH (b) H_2S
   (c) Na_2SO_4 (d) NH_4Cl

$$LiOH(s) \xrightarrow{H_2O} Li^+(aq) + OH^-(aq)$$
$$H_2S(aq) \xrightarrow{H_2O} H^+(aq) + HS^-(aq)$$
$$Na_2SO_4(s) \xrightarrow{H_2O} 2Na^+(aq) + SO_4^{2-}(aq)$$
$$NH_4Cl(s) \xrightarrow{H_2O} NH_4^+(aq) + Cl^-(aq)$$

The Brønsted–Lowry Definition of Acids and Bases

7. Write balanced equations to show what happens when hydrogen bromide dissolves in water to form an acidic solution and when ammonia dissolves in water to form a basic solution.
8. Give an example of an acid whose formula carries a positive charge, an acid that is electrically neutral, and an acid that carries a negative charge.

9. The following compounds are all oxyacids. In each case, the acidic hydrogen atoms are bound to oxygen atoms. Write the skeleton structures for the acids.
(a) H_3PO_4 (b) HNO_3 (c) $HClO_4$
(d) H_3BO_3 (e) H_2CrO_4

10. Describe how the Brønsted definition of an acid can be used to explain why compounds that contain hydrogen with an oxidation number of +1 are often acids.

11. Which of the following compounds cannot be Brønsted bases?
(a) H_3O^+ (b) MnO_4^- (c) BH_4^-
(d) CN^- (e) S^{2-}

12. Which of the following compounds cannot be Brønsted bases?
(a) O_2^{2-} (b) CH_4 (c) PH_3
(d) SF_4 (e) CH_3^+

13. Label the Brønsted acids and bases in the following reactions.
(a) $HSO_4^-(aq) + H_2O(l) \rightleftharpoons H_3O^+(aq) + SO_4^{2-}(aq)$
(b) $CH_3CO_2H(aq) + OH^-(aq) \rightleftharpoons CH_3CO_2^-(aq) + H_2O(l)$
(c) $CaF_2(s) + H_2SO_4(aq) \rightleftharpoons CaSO_4(aq) + 2\ HF(aq)$
(d) $HNO_3(aq) + NH_3(aq) \rightleftharpoons NH_4NO_3(aq)$
(e) $LiCH_3(l) + NH_3(l) \rightleftharpoons CH_4(g) + LiNH_2(s)$

14. Write the chemical equation for the Brønsted bases, reacting with water.
(a) NH_3 (b) CH_3NH_2 (c) IO^-

Conjugate Acid–Base Pairs

15. Identify the conjugate acid–base pairs in the following reactions.
(a) $HCl(aq) + H_2O(l) \rightleftharpoons H_3O^+(aq) + Cl^-(aq)$
(b) $HCO_3^-(aq) + H_2O(l) \rightleftharpoons H_2CO_3(aq) + OH^-(aq)$
(c) $NH_3(aq) + H_2O(l) \rightleftharpoons NH_4^+(aq) + OH^-(aq)$
(d) $CaCO_3(s) + 2\ HCl(aq) \rightleftharpoons Ca^{2+}(aq) + 2\ Cl^-(aq) + H_2CO_3(aq)$

16. Which of the following is not a conjugate acid–base pair?
(a) NH_4^+/NH_3 (b) H_2O/OH^-
(c) H_3O^+/OH^- (d) CH_4/CH_3^-

17. Write Lewis structures for the following Brønsted acids and their conjugate bases.
(a) formic acid, HCO_2H (b) methanol, CH_3OH

18. Write Lewis structures for the following Brønsted bases and their conjugate acids.
(a) methylamine, CH_3NH_2
(b) acetate ion, $CH_3CO_2^-$

19. Identify the conjugate base of each of the following Brønsted acids.
(a) H_3O^+ (b) H_2O (c) OH^- (d) NH_4^+

20. Identify the conjugate base of each of the following Brønsted acids.
(a) HPO_4^{2-} (b) $H_2PO_4^-$
(c) HCO_3^- (d) HS^-

21. Identify the conjugate acid of each of the following Brønsted bases.
(a) O^{2-} (b) OH^- (c) H_2O (d) NH_2^-

22. Acids (such as hydrochloric acid and sulfuric acid) react with bases (such as sodium hydroxide and ammonia) to form salts (such as sodium chloride, sodium sulfate, ammonium chloride, and ammonium sulfate). Write balanced equations for each acid reacting with each base.

The Role of Water in the Brønsted Model

23. True or false? Water is both an acid and a base. Provide support for your answer.

24. Write an equation that shows how a base reacts with water. Do the same for an acid.

25. Give the Lewis structures for water and acetic acid, CH_3CO_2H. Write an equation that describes the interaction between water and acetic acid using Lewis structures.

To What Extent Does Water Dissociate to Form Ions?

26. Explain why the concentrations of the H_3O^+ ion and the OH^- ion in pure water are the same.

27. Explain why it is impossible for water to be at equilibrium when it contains large quantities of both the H_3O^+ and OH^- ions.

28. The dissociation of water is an endothermic reaction.

$$2\ H_2O(l) \rightleftharpoons H_3O^+(aq) + OH^-(aq)$$
$$\Delta H^0 = 55.84\ \text{kJ/mol}_{rxn}$$

Use Le Châtelier's principle to predict what should happen to the fraction of water molecules that dissociate into ions as the temperature of water increases.

29. Use Le Châtelier's principle to explain why adding either an acid or a base to water suppresses the dissociation of water.

30. What is the difference between K_c and K_w for water? Calculate a value of K_c for the dissociation of H_2O to H_3O^+ and OH^-.

31. When an acid is added to pure water, there are two sources for H_3O^+. What are they? Similarly, when a base is added to pure water, there are two sources for OH^-. What are they?

32. Does the product of $[H_3O^+]$ and $[OH^-]$ always have to be 10^{-14} at 25°C? Explain.

33. If the $[H_3O^+]$ concentration in an aqueous solution is 1.0×10^{-12}, what is the $[OH^-]$?

pH as a Measure of the Concentration of H_3O^+ Ion

34. Calculate the number of H_3O^+ and OH^- ions in 1.00 mL of pure water.
35. Calculate the H_3O^+ and OH^- ion concentrations in a solution that has a pH of 3.72.
36. Explain why the pH of a solution decreases when the H_3O^+ ion concentration increases.
37. What happens to the concentration of the H_3O^+ ion when a strong acid is added to water? What happens to the concentration of the OH^- ion? What happens to the pH of the solution?
38. What happens to the OH^- ion concentration when a strong base is added to water? What happens to the concentration of the H_3O^+ ion? What happens to the pH of the solution?
39. Which of the following solutions is the most acidic?
   (a) 0.10 *M* acetic acid, pH = 2.9
   (b) 0.10 *M* hydrogen sulfide, pH = 4.1
   (c) 0.10 *M* sodium acetate, pH = 8.4
   (d) 0.10 *M* ammonia, pH = 11.1
40. Which of the following solutions is the most acidic?
   (a) 0.10 *M* H_3PO_4, pH = 1.5
   (b) 0.10 *M* $H_2PO_4^-$, pH = 4.4
   (c) 0.10 *M* HPO_4^{2-}, pH = 9.3
   (d) 0.10 *M* PO_4^{3-}, pH = 12.0
41. Calculate the pH and pOH of a solution for which $[H_3O^+]$ is 1.5×10^{-6}.
42. The pH of a 0.10 *M* solution of Na_2CO_3 is 11.6. What are $[H_3O^+]$ and $[OH^-]$?
43. The pH of a saturated solution of H_2CO_3 is 3.8. What are $[H_3O^+]$ and $[OH^-]$?
44. Calculate the pH of the following solutions.
   (a) 1.0 *M* HOAc $[H_3O^+] = 4.2 \times 10^{-3}$ *M*
   (b) 0.10 *M* HOAc $[H_3O^+] = 1.3 \times 10^{-3}$ *M*
   (c) 0.010 *M* HOAc $[H_3O^+] = 4.2 \times 10^{-4}$ *M*
45. The pH of a 0.10 *M* solution of ammonia is 11.1. Find the hydroxide concentration.

Relative Strengths of Acids and Bases

46. Define the following terms: *weak acid, strong acid, weak base*, and *strong base*.
47. Describe the difference between strong acids (such as hydrochloric acid) and weak acids (such as acetic acid) and the difference between strong bases (such as sodium hydroxide) and weak bases (such as ammonia).
48. Describe the relationship between the acid dissociation equilibrium constant for an acid, K_a, and the strength of the acid.
49. Use the table of acid dissociation equilibrium constants in Appendix B.8 to classify the following acids as either strong or weak.
   (a) acetic acid, CH_3CO_2H (b) boric acid, H_3BO_3
   (c) chromic acid, H_2CrO_4 (d) formic acid, HCO_2H
   (e) hydrobromic acid, HBr
50. Which of the following is the weakest Brønsted acid?
   (a) $H_2S_2O_3$, $K_a = 0.3$
   (b) H_2CrO_4, $K_a = 9.6$
   (c) H_3BO_3, $K_a = 7.3 \times 10^{-10}$
   (d) C_6H_5OH, $K_a = 1.0 \times 10^{-10}$
51. Which of the following solutions is the most acidic?
   (a) 0.10 *M* CH_3CO_2H, $K_a = 1.8 \times 10^{-5}$
   (b) 0.10 *M* HCO_2H, $K_a = 1.8 \times 10^{-4}$
   (c) 0.10 *M* $ClCH_2CO_2H$, $K_a = 1.4 \times 10^{-3}$
   (d) 0.10 *M* Cl_2CHCO_2H, $K_a = 5.1 \times 10^{-2}$
52. Which of the following compounds is the strongest base?
   (a) $CH_3CO_2^-$ (for CH_3CO_2H, $K_a = 1.8 \times 10^{-5}$)
   (b) HCO_2^- (for HCO_2H, $K_a = 1.8 \times 10^{-4}$)
   (c) $ClCH_2CO_2^-$ (for $ClCH_2CO_2H$, $K_a = 1.4 \times 10^{-3}$)
   (d) $Cl_2CHCO_2^-$ (for Cl_2CHCO_2H, $K_a = 5.1 \times 10^{-2}$)
53. List acetic acid, chlorous acid, hydrofluoric acid, and nitrous acid in order of increasing strength if 0.10 *M* solutions of the acids contain the following equilibrium concentrations.

Acid	*[HA]*	*[A⁻]*	*[H₃O⁺]*
HOAc	0.099 *M*	0.0013 *M*	0.0013 *M*
HOClO	0.072 *M*	0.028 *M*	0.028 *M*
HF	0.092 *M*	0.0081 *M*	0.0081 *M*
HNO_2	0.093 *M*	0.0069 *M*	0.0069 *M*

Relative Strengths of Conjugate Acid–Base Pairs

54. What is the relationship between the strength of a base and its conjugate acid?
55. Which of the following conjugate bases have essentially no base properties in water solvent? Explain. I^-, ClO_4^-, F^-, NO_2^-, NO_3^-
56. Is the conjugate base of a weak acid a relatively strong or weak base? Explain.
57. In the Brønsted model of acid–base reactions, what does the statement that HCl is a stronger acid than H_2O mean in terms of the following reaction?

$$HCl(aq) + H_2O(l) \longrightarrow H_3O^+(aq) + Cl^-(aq)$$

58. In the Brønsted model of acid–base reactions, what does it mean to say that NH_3 is a weaker acid than H_2O?
59. What can you conclude from experiments that suggest that the following reaction proceeds far to the right, as written?

$$HBr(aq) + H_2O(l) \longrightarrow H_3O^+(aq) + Br^-(aq)$$

60. What can you conclude from experiments that suggest that the following reaction proceeds far to the right, as written?

$$HCl(aq) + NH_3(aq) \longrightarrow NH_4^+(aq) + Cl^-(aq)$$

61. What can you conclude from experiments that suggest that the following reaction does not occur to an appreciable extent?

$$HCO_2^-(aq) + H_2O(l) \not\rightarrow HCO_2H(aq) + OH^-(aq)$$

Relative Strengths of Pairs of Acids and Bases

62. Refer to Table 11.3 and Appendix Tables B.8 and B.9. For each pair of reactants, state whether the reaction proceeds to nearly completion, makes only a small amount of product, or doesn't proceed to a detectable amount.
    (a) $H_2SO_4(aq) + NaOH(aq)$
    (b) $H_2O(aq) + HNO_3(aq)$
    (c) $HF(aq) + NaOH(aq)$
    (d) $HCl(aq) + NH_3(aq)$
    (e) $C_5H_5NH(aq) + HNO_2(aq)$
63. Describe what is meant by the leveling effect of water.
64. Why does a 0.10 *M* solution of HNO_3 contain the same $[H_3O^+]$ as a 0.10 *M* solution of HBr? Would a 0.10 *M* solution of HNO_3 have the same $[H_3O^+]$ as a 0.10 *M* solution of HF? Explain.
65. Explain why the H_3O^+ ion concentration in a strong acid solution depends on the concentration of the solution, but not on the value of K_a for the acid.

Relationship of Structure to Relative Strengths of Acids and Bases

66. Use the structures of the acids to explain the following observations.
    (a) H_2SO_4 is a stronger acid than HSO_4^-.
    (b) HNO_3 is a stronger acid than HNO_2.
    (c) H_2S is a stronger acid than H_2O.
    (d) H_2S is a stronger acid than PH_3.
67. Arrange the following in order of increasing acid strength. Explain.
    (a) H_3AsO_4 (b) $H_2AsO_4^-$
    (c) $HAsO_4^{2-}$ (d) AsO_4^{3-}
68. Arrange the following in order of increasing acid strength. Explain.
    (a) SiH_4 (b) HCl (c) H_2S (d) PH_3
69. Which of the following is the strongest Brønsted acid? Explain.
    (a) H_2Se (b) H_2O (c) H_2S (d) H_2Te
70. Which of the following is the weakest Brønsted acid? Explain.
    (a) $HBrO_2$ (b) HBrO
    (c) $HBrO_4$ (d) $HBrO_3$
71. Arrange the following acids in order of increasing acidity. Explain. $HOCH_3$, HOCl, HOI, HOBr
72. Which acid, NH_4^+ or PH_4^+, has the largest K_a? Explain.
73. Arrange the following acids in order of increasing K_a. Explain your order.

$$CH_3COOH, CH_2ClCOOH, CCl_3COOH$$

Strong Acid pH Calculations

74. Calculate the pH and pOH of a 0.035 *M* HCl solution.
75. Calculate the pH and pOH of a solution that contains 0.568 g of HCl per 250 mL of solution.
76. What would be the pH of a 0.010 *M* solution of any strong acid? Explain.
77. What would be the pH of the following solutions?
    (a) 1.0 *M* $HClO_4$
    (b) 0.568 g of $HClO_4$ per 250 mL of solution
    (c) 0.568 g of HNO_3 per 250 mL of solution
    (d) 1.14 g of HBr per 500 mL of solution
78. Calculate the pH of a 0.056 *M* solution of hydroiodic acid, HI.
79. Nitric acid is often grouped with sulfuric acid and hydrochloric acid as one of the strong acids. Calculate the pH of 0.10 *M* nitric acid, assuming that it is a strong acid that dissociates completely. Calculate the pH of the solution using the value of K_a for the acid (for HNO_3, $K_a = 28$).

Weak Acid pH Calculations

80. Describe the two assumptions that are commonly made in weak acid equilibrium problems. Describe how you can test whether the assumptions are valid for a particular calculation.
81. Explain why the techniques used to calculate the equilibrium concentrations of the components of a weak acid solution can't be used for either strong acids or very weak acids.
82. Explain why the H_3O^+ ion concentration in a weak acid solution depends on both the value of K_a for the acid and the concentration of the acid.
83. Which of the following solutions has the largest H_3O^+ ion concentration?
    (a) 0.10 *M* HOAc (b) 0.010 *M* HOAc
    (c) 0.0010 *M* HOAc
84. Calculate the percentage of HOAc molecules that dissociate in 0.10 *M*, 0.010 *M*, and 0.0010 *M* solutions of acetic acid. What happens to the percent ionization as the solution becomes more dilute? (For HOAc, $K_a = 1.8 \times 10^{-5}$.)
85. Formic acid (HCO_2H) was first isolated by the destructive distillation of ants. In fact, the name comes from the Latin word for ants, *formi*. Calculate the HCO_2H, HCO_2^-, and H_3O^+ concentrations in a 0.100 *M* solution of formic acid in water. (For HCO_2H, $K_a = 1.8 \times 10^{-4}$.)
86. Hydrogen cyanide (HCN) is a gas that dissolves in water to form hydrocyanic acid. Calculate the H_3O^+, HCN, and CN^- concentrations in a 0.174 *M* solution of hydrocyanic acid. (For HCN, $K_a = 6 \times 10^{-10}$.)
87. The first disinfectant used by Joseph Lister was called *carbolic acid*. The substance is now known as *phenol*. Calculate the H_3O^+ ion concentration in a 0.0167 *M* solution of phenol (C_6H_5OH). ($K_a = 1.0 \times 10^{-10}$.)

88. Calculate the concentration of acetic acid that would give an H_3O^+ ion concentration of 2.0×10^{-3} *M*. (For HOAc, $K_a = 1.8 \times 10^{-5}$.)
89. Calculate the value of K_a for ascorbic acid (vitamin C) if 2.8% of the ascorbic acid molecules in a 0.100 *M* solution dissociate.
90. Calculate the value of K_a for nitrous acid (HNO_2) if a 0.100 *M* solution is 7.1% dissociated at equilibrium.
91. Why is it wrong to assume that the pH of a 10^{-8} *M* HCl solution is 8?
92. The pH of a 0.10 *M* solution of HOAc is 2.9. Calculate K_a for acetic acid.
93. The $[H_3O^+]$ of a 0.10 *M* solution of HOCl is 5.4×10^{-5} *M*. Calculate the pH of this solution and the K_a for HOCl. How would the pH of this solution compare to that of a 0.10 *M* solution of HNO_3?
94. The pOH of a 0.10 *M* solution of HF is 11.9. Calculate K_a for HF.

Base pH Calculations

95. Which of the following equations correctly describes the relationship between K_b for the formate ion (HCO_2^-) and K_a for formic acid (HCO_2H)?
(a) $K_a = K_w \times K_b$ (b) $K_b = K_a/K_w$
(c) $K_b = K_w/K_a$ (d) $K_b = K_w + K_a$
(e) $K_b = K_w - K_a$
96. Use the relationship between K_a for an acid and K_b for its conjugate base to explain why strong acids have weak conjugate bases and weak acids have strong conjugate bases.
97. Calculate the HCO_2H, OH^-, and HCO_2^- ion concentrations in a solution that contains 0.020 mol of sodium formate ($NaHCO_2$) in 250 mL of solution. (For HCO_2H, $K_a = 1.8 \times 10^{-4}$.)
98. Calculate the OH^-, HOBr, and OBr^- ion concentrations in a solution that contains 0.050 mol of sodium hypobromite (NaOBr) in 500 mL of solution. (For HOBr, $K_a = 2.4 \times 10^{-9}$.)
99. Calculate the pH of a 0.756 *M* solution of NaOAc. (For HOAc, $K_a = 1.8 \times 10^{-5}$.)
100. A solution of NH_3 dissolved in water is known as both aqueous ammonia and ammonium hydroxide. Use the value of K_b for the following reaction to explain why aqueous ammonia is the better name.

$$NH_3(aq) + H_2O(l) \rightleftharpoons NH_4^+(aq) + OH^-(aq)$$
$$K_b = 1.8 \times 10^{-5}$$

101. At 25°C, a 0.10 *M* aqueous solution of methylamine (CH_3NH_2) is 6.8% ionized.

$$CH_3NH_2(aq) + H_2O(l) \rightleftharpoons CH_3NH_3^+(aq) + OH^-(aq)$$

Calculate K_b for methylamine. Is methylamine a stronger base or a weaker base than ammonia?

102. What is the molarity of an aqueous ammonia solution that has an OH^- ion concentration of 1.0×10^{-3} *M*?
103. What is the pH of a 0.10 *M* solution of methylamine? See Problem 101.
104. Calculate K_b for hydrazine (H_2NNH_2) if the pH of a 0.10 *M* aqueous solution of the rocket fuel is 10.54.
105. Calculate the pH of a 0.016 *M* aqueous solution of calcium acetate, $Ca(OAc)_2$. (For HOAc, $K_a = 1.8 \times 10^{-5}$.)
106. Arrange the following 0.10 *M* solutions in order of increasing acidity.
(a) KCl (b) KF (c) KOAc
107. Which of the following 0.10 *M* solutions will be acidic? Explain.
(a) NH_4Cl (b) NaF (c) $RbNO_3$
108. Which of the following 0.10 *M* solutions will be basic? Explain.
(a) Li_2SO_4 (b) $KClO_4$ (c) $NaNO_2$
109. Identify the conjugate acid produced by the dissociation of the anion of each of the following salts.
(a) Na_2HPO_4 (b) $NaHCO_3$
(c) Na_2SO_4 (d) $NaNO_2$

Mixtures of Acids and Bases

110. What is meant by the term *common ion*? If NaOAc is added to a solution of HOAc, which way does the acid equilibrium shift? What happens to the pH?
111. What is the pH of a 0.50 *M* solution of HF? What is the pH of a solution containing 0.50 *M* HF and 0.50 *M* NaF?
112. Determine the pH of a mixture containing 0.10 *M* formic acid and 0.10 *M* sodium formate.
113. Determine the pH of a mixture containing 0.10 *M* ammonia and 0.10 *M* ammonium chloride.
114. Determine the pH of a solution formed by adding 15 g of benzoic acid and 10 g of sodium benzoate to water to form 1.0 l of solution.

Buffers and Buffer Capacity

115. Explain how buffers resist changes in pH.
116. Explain why a mixture of HOAc and NaOAc is an acidic buffer, but a mixture of NH_3 and NH_4Cl is a basic buffer. (For HOAc, $K_a = 1.8 \times 10^{-5}$; for NH_4^+, $K_a = 5.6 \times 10^{-10}$.)
117. Which of the following mixtures would make the best buffer?
(a) HCl and NaCl (b) NaOAc and NH_3
(c) HOAc and NH_4Cl (d) NaOAc and NH_4Cl
(e) NH_3 and NH_4Cl
118. Which of the following solutions is an acidic buffer?
(a) 0.10 *M* HCl and 0.10 *M* NaOH
(b) 0.10 *M* HCl and 0.10 *M* NaCl

(c) 0.10 M HCO_2H and 0.10 M $NaHCO_2$
(d) 0.10 M NH_3 and 0.10 M NH_4Cl

119. Which of the following solutions is a basic buffer?
(a) 0.10 M HCl and 0.10 M NaOH
(b) 0.10 M HCl and 0.10 M NaCl
(c) 0.10 M HCO_2H and 0.10 M $NaHCO_2$
(d) 0.10 M NH_3 and 0.10 M NH_4Cl

120. What is the best way to increase the capacity of a buffer made by dissolving $NaHCO_2$ in an aqueous solution of HCO_2H? Explain.
(a) Increase the concentration of HCO_2H.
(b) Increase the concentration of $NaHCO_2$.
(c) Increase the concentrations of both HCO_2H and $NaHCO_2$.
(d) Increase the ratio of the concentration of HCO_2H to the concentration of $NaHCO_2$.
(e) Increase the ratio of the concentration of $NaHCO_2$ to the concentration of HCO_2H.

Acid–Base Reactions

121. Which of the following reactions would be expected to go nearly to completion?
(a) strong acid–strong base
(b) weak acid–weak base
(c) weak base–strong acid
(d) weak acid–strong base

122. For the reaction

$$HClO_4(aq) + KOH(aq) \longrightarrow H_2O(l) + KCl(aq)$$

it is possible to predict the resulting pH by knowing only the concentrations and volumes of the reactants. Why can this be done?

123. Suppose 100 mL each of 0.10 M $HClO_4$ and KOH are mixed. What will be the pH of the solution?

124. If 50 mL of 0.10 M $HClO_4$ is mixed with 100 mL of 0.10 M KOH, what will be the resulting pH?

125. Write a net ionic equation for the reaction of
(a) HOBr and NaOH
(b) HOAc and NH_3
(c) HBr and KOH
(d) HCO_2H and $Ba(OH)_2$
Which of these reactions will go to completion?

126. What will be the equilibrium constant for the reaction between HOI(aq) and $C_5H_5N(aq)$ (pyridine)? What does the magnitude of the equilibrium constant tell us about this reaction?

pH Titration Curves

127. Define the following: *titration, indicator, end point, equivalence point.*

128. Describe in detail the experiment you would use to measure the value of K_a for formic acid, HCO_2H.

129. Sketch a titration curve of 0.10 M NH_3($K_b = 1.8 \times 10^{-5}$) being titrated with a 0.10 M HCl solution. Label the equivalence point and the point at which the OH^- ion concentration is equal to K_b for the base.

Integrated Problems

130. Explain why pure (nonpolluted) rainwater has a pH of 5.6. Explain why boiling water to drive off the CO_2 raises the pH.

131. On April 10, 1974, at Pitlochny, Scotland, the rain was found to have a pH of 2.4. Calculate the H_3O^+ ion concentration in the rain and compare it with the H_3O^+ concentration in 0.10 M acetic acid, which has a pH of 2.9.

132. Which of the following statements are true for a 0.10 M solution of the weak acid CH_3CO_2H in water? Explain what is wrong with any incorrect answers.
(a) pH = 1.00
(b) $[H_3O^+] \gg [CH_3COO^-]$
(c) $[H_3O^+] = [CH_3COO^-]$
(d) pH < 1.00

133. Which of the following statements are true for a 1.0 M solution of the strong acid HCl in water? Explain what is wrong with any incorrect answers.
(a) $[Cl^-] > [H_3O^+]$ (b) pH = 0.00
(c) $[H_3O^+] = 1.0$ M (d) [HCl] = 1.0 M

134. Which of the following statements are true for a 1.0 M solution of the weak base NH_3 in water? Explain what is wrong with any incorrect answers.
(a) $[OH^-] = [H_3O^+]$
(b) $[NH_4{}^+] = [OH^-]$
(c) pH < 7.00
(d) $[NH_3] = 1.0$ M

135. The pH of a 0.10 M solution of formic acid is 2.37.

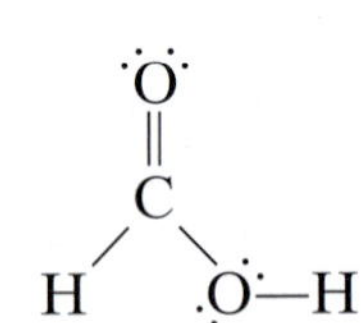

Formic acid

(a) Which of the two hydrogens in the formic acid structure is the "acidic" hydrogen? Explain.
(b) Write a chemical equation that describes what happens when formic acid is placed in water.
(c) Calculate the K_a of formic acid.
(d) Estimate the pH of a 0.10 M solution of the following acid in water. Explain your estimate.

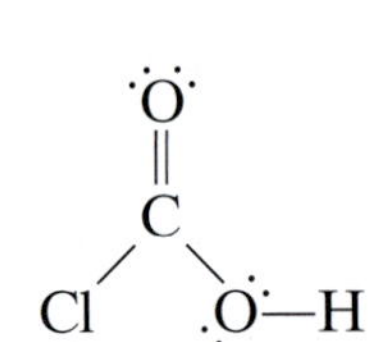

136. Methylamine, CH_3NH_2, is a weak base with $K_b = 3.6 \times 10^{-4}$.
(a) Write a chemical equation that describes what happens when methylamine is placed into water.
(b) What is the pOH of a 0.10 M solution of methylamine?
(c) What is the pH of a 0.10 M solution of methylamine?

137. Rank the following solutions in order of increasing pH. Explain your reasoning.
(a) 0.10 M $HOBrO_2$ (b) 0.10 M HOBrO
(c) 0.10 M HBr (d) 0.10 M NaBr
(e) 0.10 M $NaBrO_2$ (f) 0.10 M $NaBrO_3$

138. Describe a neutral (in terms of acid–base characteristics) solution. The K_w for the self-hydrolysis of H_2O at room temperature (25°C) is 1×10^{-14} and that at body temperature (37°C) is 2.5×10^{-14}.
(a) Describe the quantitative difference in the self-hydrolysis of pure water at 25°C and 37°C.
(b) Determine the approximate pH of pure water at 25°C and 37°C.
(c) Again describe a neutral solution. Is this the same description that you gave at the beginning of the problem?

139. Match each of the following to its corresponding particulate representation. Water molecules are not shown. Explain your reasoning.
(a) a dilute solution of a strong acid
(b) a concentrated solution of a weak acid
(c) a good buffer

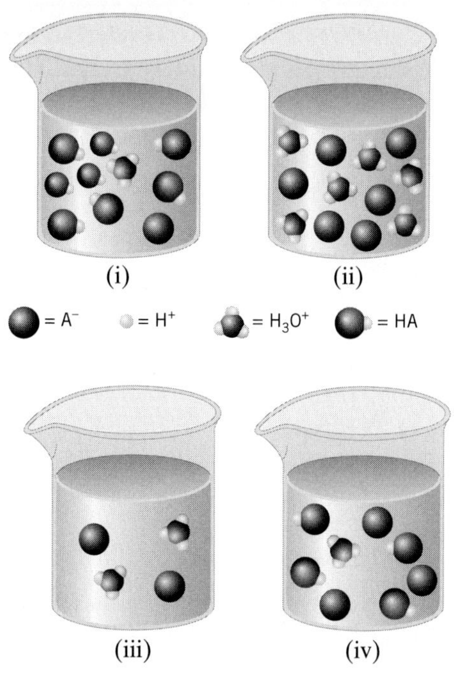

140. When NH_4Cl is dissolved in water, the following species will be present: NH_4^+, Cl^-, H_2O, OH^-, H_3O^+, and NH_3. Using the symbols $\ll$, $<$, and $\approx$, arrange the species in order of increasing concentration in the solution.

141. The weak acid HB can be represented as ∘●, where ● represents B^- and ∘ represents H^+. Match the conjugate base of HB and the conjugate acid of HB with its appropriate particulate representation.
(a) ∘●∘
(b) ●∘●
(c) ●
(d) ●∘●∘
(e) ∘

142. Which of the following compounds could dissolve in water to give a 0.10 M solution with a pH of about 5?
(a) NH_3 (b) NaCl (c) HCl
(d) KOH (e) NH_4Cl

143. Write balanced chemical equations for the following acid–base reactions.
(a) $Al_2O_3(s) + HCl(aq) \longrightarrow$
(b) $CaO(s) + H_2SO_4(aq) \longrightarrow$
(c) $Na_2O(s) + H_2O(l) \longrightarrow$
(d) $MgCO_3(s) + HCl(aq) \longrightarrow$
(e) $NaOH(s) + H_3PO_4(aq) \longrightarrow$

144. Which of the following factors influences the value of K_a for the dissociation of formic acid?

$$HCO_2H(aq) + H_2O(l) \rightleftharpoons HCO_2^-(aq) + H_3O^+(aq)$$

(a) temperature (b) pressure (c) pH
(d) the initial concentration of HCO_2H
(e) the initial concentration of the HCO_2^- ion

145. A. 1.0 mol of the acid HBr is placed in sufficient water to make 1.0 L of solution. $K_a = 1 \times 10^9$.
(a) Write an equation that describes the reaction that takes place.
(b) What will be $[H_3O^+]$ in the solution?
(c) What is the pH?
(d) What will be $[OH^-]$ in the solution?
(e) What is the conjugate base of HBr?

B. 1.0 mol of the base ammonia, NH_3, is placed in sufficient water to make 1.0 L of solution. $K_a = 1.8 \times 10^{-5}$.
(a) Write an equation that describes the reaction that takes place.
(b) What is $[OH^-]$?
(c) What is $[NH_3]$?
(d) What is the concentration of the conjugate acid of ammonia?
(e) What is the pH?

C. 1.0 mol of NaBr is placed in sufficient water to make 1.0 L of solution.

$$NaBr(s) \rightleftharpoons Na^+(aq) + Br^-(aq) \qquad K_c \gg 1$$

(a) What is $[Br^-]$ in this solution?
(b) Write an equation describing the possible reaction of Br^- with water. What do you think

the equilibrium constant for this reaction might be? Explain your reasoning. Part A of this question might be useful.

(c) What is the pH of this solution?

146. The following K_a values have been experimentally determined for formic and acetic acids.

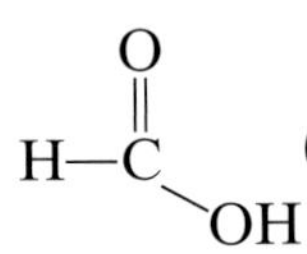

(HCO_2H)

Formic acid
$K_a = 1.8 \times 10^{-4}$

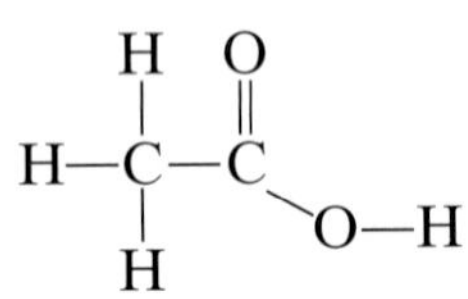

(CH_3CO_2H)

Acetic acid
$K_a = 1.8 \times 10^{-5}$

(a) Which is the stronger acid? State your reasoning.

(b) Write a chemical equation that describes what happens when formic acid is placed into water. Use Lewis structures for products and reactants similar to those above.

(c) What is the pH of a 0.10 *M* solution of formic acid? Show all calculations. State and justify any assumptions made.

(d) Which is the stronger base, HCO_2^- or $CH_3CO_2^-$? Explain your answer.

(e) Will the equilibrium constant for the reaction

$$CH_3CO_2H(aq) + HCO_2^-(aq) \rightleftharpoons CH_3CO_2^-(aq) + HCO_2H$$

be greater than 1 or less than 1? Explain.

(f) Which of the following acids is the strongest? Explain.

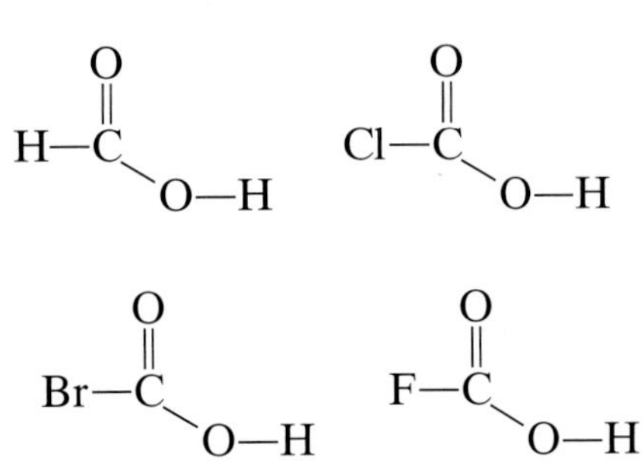

147. K_a for HNO_2 is 5.1×10^{-4}.

(a) Write an equation that describes the reaction of HNO_2 with H_2O.

(b) What is the pH of a 1.0 *M* solution of HNO_2?

(c) What is the pH of a 1.0 *M* solution of $NaNO_2$?

(d) What is the pH of a 1.0 *M* solution of HNO_3?

(e) Which of the following diagrams best describes a 1.0 *M* solution of HNO_3? Explain.

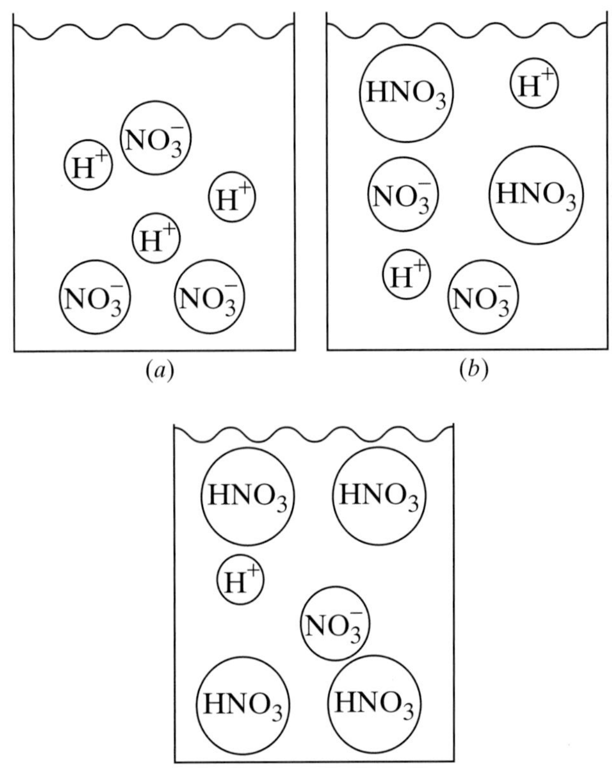

(*a*) (*b*) (*c*)

148. Arrange the following 0.10 *M* solutions in order of increasing pH. Show clearly which has the highest and which has the lowest pH. Explain.
CsCl
HOCl
HOClO
NaOCl
NaOClO

149. Estimate the pH of each of the following solutions. Explain in each case.
(a) 0.10 *M* HCl (b) 0.10 *M* NaCl
(c) 0.10 *M* HOCl (d) 0.10 *M* NaF
(e) 0.10 *M* NaOH

150. Calculate the pH of the following solutions.
(a) A 2.0 *M* solution of HOCl; $K_a = 2.9 \times 10^{-8}$
(b) A 1.0 *M* solution of HF; $K_a = 7.2 \times 10^{-4}$
(c) A 0.02 *M* solution of HCl
(d) A 1.0 *M* solution of NaF

151. Arrange the following 0.10 *M* solutions in order of increasing acidity. Explain.
KI
$KClO_3$
$HOClO_2$ ($HClO_3$)

KOH
NH_4ClO_4
$HOClO_3$ ($HClO_4$)

152. 50.0 mL of 0.100 *M* HCl is titrated with 0.100 *M* NaOH. What is the pH at the start of the titration? What will be the pH after the addition of 20.0, 50.0, and 60.0 mL of NaOH? What is the pH at the equivalence point?

153. 50.0 mL of 0.100 *M* HOAc is titrated with 0.100 *M* NaOH. Calculate the pH at the beginning of the titration and after the addition of 20.0, 50.0, and 60.0 mL of NaOH.

154. Roughly sketch the titration curve for the titration of 50.0 mL of 0.250 *M* NH_3 with 0.500 *M* HCl. What will be the initial pH? What volume of HCl is required to reach the equivalence point? What will be the pH at the equivalence point?

155. Which of the following ions is the conjugate base of a strong acid?
(a) OH^- (b) HSO_4^- (c) NH_2^-
(d) S^{2-} (e) H_3O^+

156. Many insects leave small quantities of formic acid, HCO_2H, behind when they bite. Explain why the itching sensation can be relieved by treating the bite with an aqueous solution of ammonia (NH_3) or baking soda ($NaHCO_3$).

Chapter Eleven

SPECIAL TOPICS

11A.1 Diprotic Acids

The acid equilibrium problems discussed so far have focused on **monoprotic acids** that each have a single H^+ ion, or proton, that can be donated when they act as Brønsted acids. Hydrochloric acid (HCl), acetic acid (CH_3CO_2H or HOAc), and nitric acid (HNO_3) are all monoprotic acids.

Several important acids can be classified as **polyprotic acids,** which can lose more than one H^+ ion when they act as Brønsted acids. **Diprotic acids,** such as sulfuric acid (H_2SO_4), carbonic acid (H_2CO_3), hydrogen sulfide (H_2S), chromic acid (H_2CrO_4), and oxalic acid ($H_2C_2O_4$), have two acidic hydrogen atoms. **Triprotic acids,** such as phosphoric acid (H_3PO_4) and citric acid ($C_6H_8O_7$), have three.

There is usually a large difference in the ease with which polyprotic acids lose the first and second (or second and third) protons. When sulfuric acid is classified as a strong acid, it is often assumed that it loses both of its protons when it reacts with water. That isn't a legitimate assumption. Sulfuric acid is a strong acid because K_a for the loss of the first proton is much larger than 1. We therefore assume that essentially all the H_2SO_4 molecules in an aqueous solution lose the first proton to form HSO_4^-, or hydrogen sulfate ion.

$$H_2SO_4(aq) + H_2O(l) \longrightarrow H_3O^+(aq) + HSO_4^-(aq) \qquad K_{a1} = 1 \times 10^3$$

But K_a for the loss of the second proton is only 1.2×10^{-2}, and only 10% of the H_2SO_4 molecules in a 1 *M* solution lose a second proton.

$$HSO_4^-(aq) + H_2O(l) \rightleftharpoons H_3O^+(aq) + SO_4^{2-}(aq) \qquad K_{a2} = 1.2 \times 10^{-2}$$

H_2SO_4 only loses both H^+ ions when it reacts with a base stronger than water, such as ammonia.

Table 11A.1 gives values of K_a for some common polyprotic acids. The large difference between the values of K_a for the sequential loss of protons by a polyprotic acid is important because it means we can assume that the acids dissociate one step at a time, an assumption known as **stepwise dissociation.**

Let's look at the consequence of the assumption that polyprotic acids lose protons one step at a time by examining the chemistry of a saturated solution ($\approx$0.10 *M*) of H_2S in water. Hydrogen sulfide is the foul-smelling gas that gives

Table 11A.1

Acid Dissociation Equilibrium Constants for Common Polyprotic Acids

Acid	K_{a1}	K_{a2}	K_{a3}
Sulfuric acid (H_2SO_4)	1.0×10^3	1.2×10^{-2}	
Chromic acid (H_2CrO_4)	9.6	3.2×10^{-7}	
Oxalic acid ($H_2C_2O_4$)	5.4×10^{-2}	5.4×10^{-5}	
Sulfurous acid (H_2SO_3)	1.7×10^{-2}	6.4×10^{-8}	
Phosphoric acid (H_3PO_4)	7.1×10^{-3}	6.3×10^{-8}	4.2×10^{-13}
Citric acid ($C_6H_8O_7$)	7.5×10^{-4}	1.7×10^{-5}	4.0×10^{-7}
Carbonic acid (H_2CO_3)	4.5×10^{-7}	4.7×10^{-11}	
Hydrogen sulfide (H_2S)	1.0×10^{-7}	1.3×10^{-13}	

rotten eggs their unpleasant odor. It is an excellent source of the S^{2-} ion, however, and is therefore commonly used in introductory chemistry laboratories. H_2S is a weak acid that dissociates in steps. Some of the H_2S molecules lose a proton in the first step to form HS^-, or hydrogen sulfide ion.

$$\text{First step} \qquad H_2S(aq) + H_2O(l) \rightleftharpoons H_3O^+(aq) + HS^-(aq)$$

A small fraction of the HS^- ions formed in the reaction then go on to lose additional H^+ ions in a second step.

$$\text{Second step} \qquad HS^-(aq) + H_2O(l) \rightleftharpoons H_3O^+(aq) + S^{2-}(aq)$$

Since there are two steps in the reaction, we can write two equilibrium constant expressions.

$$K_{a1} = \frac{[H_3O^+][HS^-]}{[H_2S]} = 1.0 \times 10^{-7}$$

$$K_{a2} = \frac{[H_3O^+][S^{2-}]}{[HS^-]} = 1.3 \times 10^{-13}$$

Although each of the equations contains three terms, there are only four unknowns—$[H_3O^+]$, $[H_2S]$, $[HS^-]$, and $[S^{2-}]$—because the $[H_3O^+]$ and $[HS^-]$ terms appear in both equations. The $[H_3O^+]$ term represents the total H_3O^+ ion concentration from both steps and therefore must have the same value in both equations. Similarly, the $[HS^-]$ term, which represents the balance between the HS^- ions formed in the first step and the HS^- ions consumed in the second step, must have the same value for both equations.

It takes four equations to solve for four unknowns. We already have two equations: the K_{a1} and K_{a2} expressions. We are going to have to find either two more equations or a pair of assumptions that can generate two equations. We can base one assumption on the fact that the value of K_{a1} for H_2S is almost a million times larger than the value of K_{a2}.

$$K_{a1} \gg K_{a2}$$

This means that only a small fraction of the HS^- ions formed in the first step go on to dissociate in the second step. If this is true, most of the H_3O^+ ions in the solution come from the dissociation of H_2S, and most of the HS^- ions formed in this reaction remain in solution. As a result, we can assume that the H_3O^+ and HS^- ion concentrations are more or less equal.

$$\text{First assumption} \qquad [H_3O^+] \approx [HS^-]$$

We need one more equation, and therefore one more assumption. Note that H_2S is a weak acid ($K_{a1} = 1.0 \times 10^{-7}$; $K_{a2} = 1.3 \times 10^{-13}$). Thus, we can assume that most of the H_2S that dissolves in water will still be present when the solution reaches equilibrium. In other words, we can assume that the equilibrium concentration of H_2S is approximately equal to the initial concentration. Because a saturated solution of H_2S in water has an initial concentration of about 0.10 *M*, we can summarize the second assumption as follows.

$$\text{Second assumption} \qquad [H_2S] \approx 0.10\ M$$

We now have four equations in four unknowns.

$$K_{a1} = \frac{[H_3O^+][HS^-]}{[H_2S]} = 1.0 \times 10^{-7}$$

$$K_{a2} = \frac{[H_3O^+][S^{2-}]}{[HS^-]} = 1.3 \times 10^{-13}$$

$$[H_3O^+] \approx [HS^-]$$

$$[H_2S] \approx 0.10\ M$$

Because there is always a unique solution to four equations in four unknowns, we are now ready to calculate the H_3O^+, H_2S, HS^-, and S^{2-} concentrations at equilibrium in a saturated solution of H_2S in water.

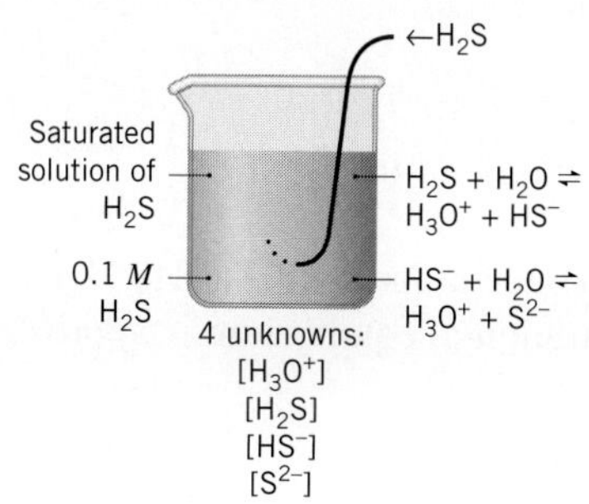

K_{a1} is so much larger than K_{a2} for the acid that we can invoke the stepwise dissociation assumption. We assume that we can work with the equilibrium expression for the first step without worrying about the second step for the moment. We therefore start with the expression for K_{a1}.

$$K_{a1} = \frac{[H_3O^+][HS^-]}{[H_2S]} = 1.0 \times 10^{-7}$$

We then invoke one of our assumptions.

$$[H_2S] \approx 0.10\ M$$

Substituting the approximation into the K_{a1} expression gives the following equation.

$$\frac{[H_3O^+][HS^-]}{[0.10]} \approx 1.0 \times 10^{-7}$$

We then invoke the other assumption.

$$[H_3O^+] \approx [HS^-] \approx \Delta C$$

Substituting that approximation into the K_{a1} expression gives the following result.

$$\frac{[\Delta C][\Delta C]}{[0.10]} \approx 1.0 \times 10^{-7}$$

We now solve the approximate equation for ΔC.

$$\Delta C \approx 1.0 \times 10^{-4}$$

If our assumptions are valid, we are three-fourths of the way to our goal. We know the H_2S, H_3O^+, and HS^- concentrations.

$$[H_2S] \approx 0.10\ M$$
$$[H_3O^+] \approx [HS^-] \approx 1.0 \times 10^{-4}\ M$$

Having extracted the values of three unknowns from the first equilibrium expression, we turn to the second equilibrium expression.

$$K_{a2} = \frac{[H_3O^+][S^{2-}]}{[HS^-]} = 1.3 \times 10^{-13}$$

Substituting the known values of the H_3O^+ and HS^- ion concentrations into the expression gives the following equation.

$$\frac{[1.0 \times 10^{-4}][S^{2-}]}{[1.0 \times 10^{-4}]} = 1.3 \times 10^{-13}$$

Because the equilibrium concentrations of the H_3O^+ and HS^- ions are about the same, the S^{2-} ion concentration at equilibrium is approximately equal to the value of K_{a2} for this acid.

$$[S^{2-}] \approx 1.3 \times 10^{-13}\ M$$

It is now time to check our assumptions. Is the dissociation of H_2S small compared with the initial concentration? Yes. The HS^- and H_3O^+ ion concentrations obtained from the calculation are $1.0 \times 10^{-4}\ M$, which is 0.1% of the initial concentration of H_2S. The following assumption is therefore valid.

$$[H_2S] \approx 0.10\ M$$

Is the difference between the S^{2-} and HS^- ion concentrations large enough to allow us to assume that essentially all of the H_3O^+ ions at equilibrium are formed in the first step and that essentially all of the HS^- ions formed in this step remain in solution? Yes. The S^{2-} ion concentration obtained from the calculation is 10^9 times smaller than the HS^- ion concentration. Thus, our other assumption is also valid.

$$[H_3O^+] \approx [HS^-]$$

We can therefore summarize the concentrations of the various components of the equilibrium as follows.

$$[H_2S] \approx 0.10\ M$$
$$[H_3O^+] \approx [HS^-] \approx 1.0 \times 10^{-4}\ M$$
$$[S^{2-}] \approx 1.3 \times 10^{-13}\ M$$

11A.2 Diprotic Bases

The techniques we have used with diprotic acids can be extended to diprotic bases. The only challenge is calculating the values of K_b for the base. Suppose we are given the following problem:

Calculate the H_2CO_3, HCO_3^-, CO_3^{2-}, and OH^- concentrations at equilibrium in a solution that is initially 0.10 M in Na_2CO_3. (For H_2CO_3, $K_{a1} = 4.5 \times 10^{-7}$ and $K_{a2} = 4.7 \times 10^{-11}$.)

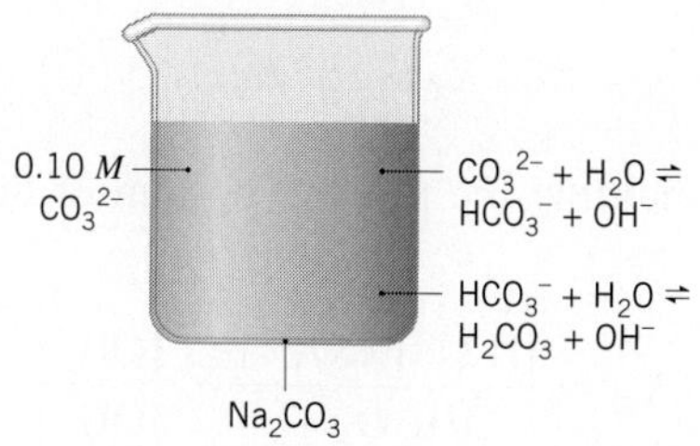

Sodium carbonate dissociates into its ions when it dissolves in water.

$$Na_2CO_3(aq) \xrightarrow{H_2O} 2\,Na^+(aq) + CO_3^{2-}(aq)$$

The carbonate ion then acts as a base toward water, picking up a pair of protons (one at a time) to form the bicarbonate ion, HCO_3^-, and then eventually carbonic acid, H_2CO_3.

$$CO_3^{2-}(aq) + H_2O(l) \rightleftharpoons HCO_3^-(aq) + OH^-(aq) \qquad K_{b1} = ?$$
$$HCO_3^-(aq) + H_2O(l) \rightleftharpoons H_2CO_3(aq) + OH^-(aq) \qquad K_{b2} = ?$$

The first step in solving the problem involves determining the values of K_{b1} and K_{b2} for the carbonate ion. We start by comparing the K_b expressions for the carbonate ion with the K_a expressions for carbonic acid.

$$K_{b1} = \frac{[HCO_3^-][OH^-]}{[CO_3^{2-}]} \qquad K_{a2} = \frac{[H_3O^+][CO_3^{2-}]}{[HCO_3^-]}$$
$$K_{b2} = \frac{[H_2CO_3][OH^-]}{[HCO_3^-]} \qquad K_{a1} = \frac{[H_3O^+][HCO_3^-]}{[H_2CO_3]}$$

The expressions for K_{b1} and K_{a2} have something in common: They both depend on the concentrations of the HCO_3^- and CO_3^{2-} ions. The expressions for K_{b2} and K_{a1} also have something in common: They both depend on the HCO_3^- and H_2CO_3 concentrations. We can therefore calculate K_{b1} from K_{a2} and K_{b2} from K_{a1}.

We start by multiplying the top and bottom of the K_{a1} expression by the OH^- ion concentration to introduce the $[OH^-]$ term.

$$K_{a1} = \frac{[H_3O^+][HCO_3^-]}{[H_2CO_3]} \times \frac{[OH^-]}{[OH^-]}$$

We then group terms in the equation as follows.

$$K_{a1} = \frac{[HCO_3^-]}{[H_2CO_3][OH^-]} \times [H_3O^+][OH^-]$$

The first term in the equation is the inverse of the K_{b2} expression, and the second term is the K_w expression.

$$K_{a1} = \frac{1}{K_{b2}} \times K_w$$

Rearranging the equation gives the following result.

$$K_{a1}K_{b2} = K_w$$

Similarly, we can multiply the top and bottom of the K_{a2} expression by the OH^- ion concentration.

$$K_{a2} = \frac{[H_3O^+][CO_3^{2-}]}{[HCO_3^-]} \times \frac{[OH^-]}{[OH^-]}$$

Collecting terms gives the following equation.

$$K_{a2} = \frac{[CO_3^{2-}]}{[HCO_3^-][OH^-]} \times [H_3O^+][OH^-]$$

The first term in the equation is the inverse of K_{b1}, and the second term is K_w.

$$K_{a2} = \frac{1}{K_{b1}} \times K_w$$

The equation can therefore be rearranged as follows.

$$K_{a2}K_{b1} = K_w$$

We can now calculate the values of K_{b1}, and K_{b2} for the carbonate ion.

$$K_{b1} = \frac{K_w}{K_{a2}} = \frac{1.0 \times 10^{-14}}{4.7 \times 10^{-11}} = 2.1 \times 10^{-4}$$

$$K_{b2} = \frac{K_w}{K_{a1}} = \frac{1.0 \times 10^{-14}}{4.5 \times 10^{-7}} = 2.2 \times 10^{-8}$$

We are finally ready to do the calculations. We start with the K_{b1} expression because the CO_3^{2-} ion is the strongest base in the solution and is therefore the best source of the OH^- ion.

$$K_{b1} = \frac{[HCO_3^-][OH^-]}{[CO_3^{2-}]}$$

The difference between K_{b1} and K_{b2} for the carbonate ion is large enough to suggest that most of the OH^- ions come from the first step and most of the HCO_3^- ions formed in the reaction remain in solution.

$$[OH^-] \approx [HCO_3^-] \approx \Delta C$$

The value of K_{b1} is small enough to assume that ΔC is small compared with the initial concentration of the carbonate ion. If this is true, the concentration of the CO_3^{2-} ion at equilibrium will be roughly equal to the initial concentration of Na_2CO_3, which was described in the statement of the problem as 0.10 *M*.

$$[CO_3^{2-}] \approx 0.10\ M$$

Substituting this information into the K_{b1} expression gives the following result.

$$\frac{[\Delta C][\Delta C]}{[0.10]} \approx 2.1 \times 10^{-4}$$

This approximate equation can now be solved for ΔC.

$$\Delta C \approx 0.0046\ M$$

We then use the value of ΔC to calculate the equilibrium concentrations of the OH^-, HCO_3^-, and CO_3^{2-} ions.

$$[CO_3^{2-}] \approx 0.10\ M$$
$$[OH^-] \approx [HCO_3^-] \approx \Delta C \approx 0.0046\ M$$

We now turn to the K_{b2} expression.

$$K_{b2} = \frac{[H_2CO_3][OH^-]}{[HCO_3^-]}$$

Substituting what we know about the OH^- and HCO_3^- ion concentrations into this equation gives the following result.

$$K_{b2} = \frac{[H_2CO_3][0.0046]}{[0.0046]} \approx 2.2 \times 10^{-8}$$

According to this equation, the H_2CO_3 concentration at equilibrium is approximately equal to K_{b2} for the carbonate ion.

$$[H_2CO_3] \approx 2.2 \times 10^{-8}\ M$$

Summarizing the results of our calculations allows us to test the assumptions made generating the results.

$$[CO_3^{2-}] \approx 0.10\ M$$
$$[OH^-] \approx [HCO_3^-] \approx 4.6 \times 10^{-3}\ M$$
$$[H_2CO_3] \approx 2.2 \times 10^{-8}\ M$$

All of our assumptions are valid. The extent of the reaction between the CO_3^{2-} ion and water to give the HCO_3^- ion is less than 5% of the initial concentration of Na_2CO_3. Furthermore, most of the OH^- ion comes from the first step, and most of the HCO_3^- ion formed in the first step remains in solution.

11A.3 Triprotic Acids

Our techniques for working diprotic acid and diprotic base equilibrium problems can be applied to triprotic acids and bases as well. To illustrate this, let's calculate the H_3O^+, H_3PO_4, $H_2PO_4^-$, HPO_4^{2-}, and PO_4^{3-} concentrations at equilibrium in a

0.100 M H_3PO_4 solution, for which $K_{a1} = 7.1 \times 10^{-3}$, $K_{a2} = 6.3 \times 10^{-8}$, and $K_{a3} = 4.2 \times 10^{-13}$.

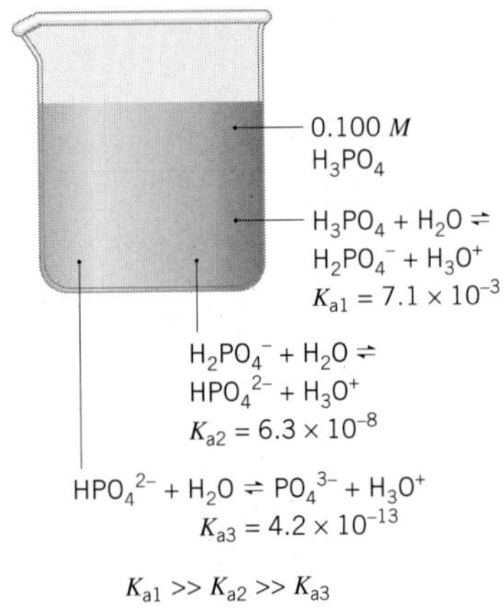

Let's assume that the acid dissociates by steps and analyze the first step—the most extensive reaction.

$$H_3PO_4(aq) + H_2O(l) \rightleftharpoons H_2PO_4^-(aq) + H_3O^+(aq)$$

$$K_{a1} = \frac{[H_3O^+][H_2PO_4^-]}{[H_3PO_4]} = 7.1 \times 10^{-3}$$

We now assume that the difference between K_{a1} and K_{a2} is large enough that most of the H_3O^+ ions come from the first step and most of the $H_2PO_4^-$ ions formed in the step remain in solution.

$$[H_3O^+] \approx [H_2PO_4^-] \approx \Delta C$$

Substituting the assumption into the K_{a1} expression gives the following equation.

$$\frac{[\Delta C][\Delta C]}{[0.100 - \Delta C]} = 7.1 \times 10^{-3}$$

The assumption that ΔC is small compared with the initial concentration of the acid fails in this problem. But we don't really need the assumption because we can use the quadratic formula to solve this equation.

$$\Delta C = 0.023\ M$$

We can then use the value of ΔC to obtain the following information.

$$[H_3PO_4] \approx 0.100 - \Delta C \approx 0.077\ M$$
$$[H_3O^+] \approx [H_2PO_4^-] \approx 0.023\ M$$

We now turn to the second strongest acid in the solution.

$$K_{a2} = \frac{[H_3O^+][HPO_4^{2-}]}{[H_2PO_4^-]} = 6.3 \times 10^{-8}$$

Substituting what we know about the H_3O^+ and $H_2PO_4^-$ ion concentrations into the expression gives the following equation.

$$\frac{[0.023][HPO_4^{2-}]}{[0.023]} \approx 6.3 \times 10^{-8}$$

If our assumptions so far are correct, the HPO_4^{2-} ion concentration in the solution is equal to K_{a2}.

$$[HPO_4^{2-}] \approx 6.3 \times 10^{-8}$$

We have only one more equation, the equilibrium expression for the weakest acid in the solution.

$$K_{a3} = \frac{[H_3O^+][PO_4^{3-}]}{[HPO_4^{2-}]} = 4.2 \times 10^{-13}$$

Substituting what we know about the concentrations of the H_3O^+ and HPO_4^{2-} ions into the expression gives the following equation.

$$\frac{[0.023][PO_4^{3-}]}{[6.3 \times 10^{-8}]} \approx 4.2 \times 10^{-13}$$

This equation can be solved for the phosphate ion concentration at equilibrium.

$$[PO_4^{3-}] \approx 1.2 \times 10^{-18}\ M$$

Summarizing the results of the calculations helps us check the assumptions made along the way.

$$[H_3PO_4] \approx 0.077\ M$$
$$[H_3O^+] \approx [H_2PO_4^-] \approx 0.023\ M$$
$$[HPO_4^{2-}] \approx 6.3 \times 10^{-8}\ M$$
$$[PO_4^{3-}] \approx 1.2 \times 10^{-18}\ M$$

The only approximation used in working the problem was the assumption that the acid dissociates one step at a time. Is the difference between the concentrations of the $H_2PO_4^-$ and HPO_4^{2-} ions large enough to justify the assumption that essentially all of the H_3O^+ ions come from the first step? Yes. Is it large enough to justify the assumption that essentially all of the $H_2PO_4^-$ formed in the first step remains in solution? Yes.

You may never encounter an example of a polyprotic acid for which the differences between successive values of K_a are too small to allow us to assume stepwise dissociation. The assumption works even when we might expect it to fail.

Exercise 11A.1

Calculate the H_3O^+, H_3Cit, H_2Cit^-, $HCit^{2-}$, and Cit^{3-} concentrations in a 1.00 M solution of citric acid. (For H_3Cit, $K_{a1} = 7.5 \times 10^{-4}$, $K_{a2} = 1.7 \times 10^{-5}$, and $K_{a3} = 4.0 \times 10^{-7}$.)

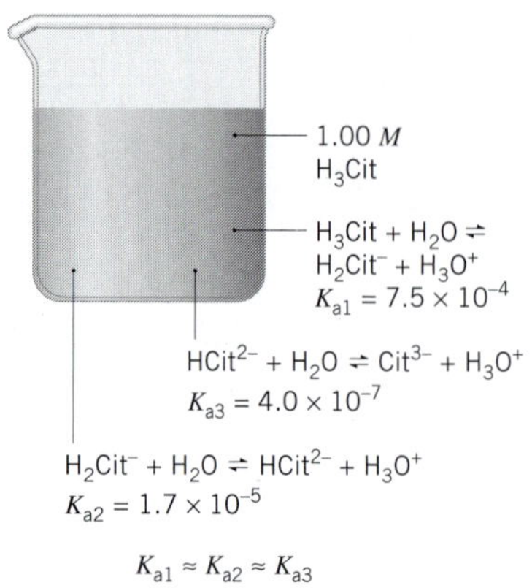

Solution

At first glance, K_{a1}, K_{a2}, and K_{a3} for the acid might seem to be too close to each other to assume stepwise dissociation. The only way to tell for sure is to try the assumption and see whether it works. We start with the expression for the first step.

$$K_{a1} = \frac{[H_3O^+][H_2Cit^-]}{[H_3Cit]} = 7.5 \times 10^{-4}$$

Even though we might be suspicious of its validity, let's assume that most of the H_3O^+ ions at equilibrium come from the citric acid. Let's also assume that most of the H_2Cit^- ions formed in the first step remain in solution.

If the assumptions are valid, the concentrations of the H_3O^+ and H_2Cit^- ions at equilibrium are equal.

$$[H_3O^+] \approx [H_2Cit^-] \approx \Delta C$$

Substituting the assumption into the K_{a1} expression gives the following equation.

$$\frac{[\Delta C][\Delta C]}{[H_3Cit]} \approx 7.5 \times 10^{-4}$$

Because citric acid is a weak acid, we can assume that the equilibrium concentration of the undissociated H_3Cit molecules is roughly equal to the initial concentration of the acid.

$$[H_3Cit] \approx 1.00\ M$$

Substituting this assumption into the K_{a1} expression gives the following result.

$$\frac{[\Delta C][\Delta C]}{[1.00]} \approx 7.5 \times 10^{-4}$$

We can now solve the approximate equation for ΔC.

$$\Delta C \approx 0.027\ M$$

The value of ΔC can now be used to calculate the following concentrations.

$$[H_3Cit] \approx 1.00\ M$$
$$[H_3O^+] \approx [H_2Cit^-] \approx 0.027\ M$$

We can now turn to the K_{a2} expression.

$$K_{a2} = \frac{[H_3O^+][HCit^{2-}]}{[H_2Cit^-]} = 1.7 \times 10^{-5}$$

Substituting what we know about the H_3O^+ and H_2Cit^- concentrations into the expression gives the following equation.

$$\frac{[0.027][HCit^{2-}]}{[0.027]} \approx 1.7 \times 10^{-5}$$

If our assumptions are valid, the $HCit^{2-}$ ion concentration is equal to K_{a2} for the acid.

$$[HCit^{2-}] \approx 1.7 \times 10^{-5}\ M$$

We now turn to the K_{a3} expression.

$$K_{a3} = \frac{[H_3O^+][Cit^{3-}]}{[HCit^{2-}]} = 4.0 \times 10^{-7}$$

Substituting the approximate values for the H_3O^+ and $HCit^{2-}$ ion concentrations into the expression gives the following equation.

$$\frac{[0.027][Cit^{3-}]}{[1.7 \times 10^{-5}]} \approx 4.0 \times 10^{-7}$$

We now solve for the Cit^{3-} ion concentration.

$$[Cit^{3-}] \approx 2.5 \times 10^{-10}\ M$$

Summarizing the results of the calculation allows us to test the assumptions made in deriving them.

$$[H_3Cit] \approx 1.00\ M$$
$$[H_3O^+] \approx [H_2Cit^-] \approx 0.027\ M$$
$$[HCit^{2-}] \approx 1.7 \times 10^{-5}\ M$$
$$[Cit^{3-}] \approx 2.5 \times 10^{-10}\ M$$

Is it legitimate to assume that the amount of H_3Cit that dissociates in the first step is small compared with the initial concentration of the acid? Yes. Is it legitimate to assume that most of the H_3O^+ ion comes from the first step? Yes. Is it legitimate to assume that most of the H_2Cit^- ion formed in the first step remains in solution? Yes. Since all of the assumptions made in the calculation are valid, the results are also valid.

11A.4 Compounds That Could Be Either Acids or Bases

Sometimes the hardest part of a calculation is deciding whether the compound is an acid or a base. Consider sodium bicarbonate, for example, which dissolves in water to give the bicarbonate ion.

$$NaHCO_3(s) \xrightarrow{H_2O} Na^+(aq) + HCO_3^-(aq)$$

In theory, the bicarbonate ion can act as either a Brønsted acid or a Brønsted base toward water.

$$HCO_3^-(aq) + H_2O(l) \rightleftharpoons H_3O^+(aq) + CO_3^{2-}(aq)$$
$$HCO_3^-(aq) + H_2O(l) \rightleftharpoons H_2CO_3(aq) + OH^-(aq)$$

Which reaction predominates? Is the HCO_3^- ion more likely to act as an acid or as a base?

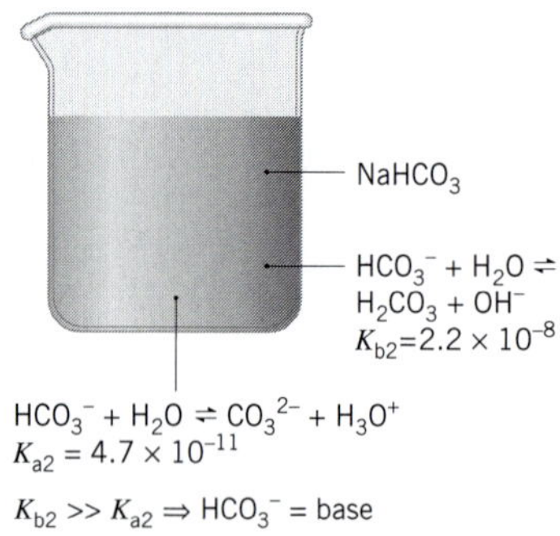

We can answer those questions by comparing the equilibrium constants for the reactions. The equilibrium in which the HCO_3^- ion acts as a Brønsted acid is described by K_{a2} for carbonic acid.

$$K_{a2} = \frac{[H_3O^+][CO_3^{2-}]}{[HCO_3^-]} = 4.7 \times 10^{-11}$$

The equilibrium in which the HCO_3^- acts as a Brønsted base is described by K_{b2} for the carbonate ion.

$$K_{b2} = \frac{[H_2CO_3][OH^-]}{[HCO_3^-]} = 2.2 \times 10^{-8}$$

Since K_{b2} is significantly larger than K_{a2}, the HCO_3^- ion is a stronger base than it is an acid.

Exercise 11A.2

Phosphoric acid, H_3PO_4, is obviously an acid. The phosphate ion, PO_4^{3-}, is a base. Predict whether solutions of the $H_2PO_4^-$ and HPO_4^{2-} ions are more likely to be acidic or basic.

Solution

Let's start by looking at the stepwise dissociation of phosphoric acid.

$$H_3PO_4(aq) + H_2O(l) \rightleftharpoons H_3O^+(aq) + H_2PO_4^-(aq) \qquad K_{a1} = 7.1 \times 10^{-3}$$
$$H_2PO_4^-(aq) + H_2O(l) \rightleftharpoons H_3O^+(aq) + HPO_4^{2-}(aq) \qquad K_{a2} = 6.3 \times 10^{-8}$$
$$HPO_4^{2-}(aq) + H_2O(l) \rightleftharpoons H_3O^+(aq) + PO_4^{3-}(aq) \qquad K_{a3} = 4.2 \times 10^{-13}$$

We can then look at the steps by which the PO_4^{3-} ion picks up protons from water to form phosphoric acid.

$$PO_4^{3-}(aq) + H_2O(l) \rightleftharpoons HPO_4^{2-}(aq) + OH^-(aq) \qquad K_{b1} = ?$$
$$HPO_4^{2-}(aq) + H_2O(l) \rightleftharpoons H_2PO_4^-(aq) + OH^-(aq) \qquad K_{b2} = ?$$
$$H_2PO_4^-(aq) + H_2O(l) \rightleftharpoons H_3PO_4(aq) + OH^-(aq) \qquad K_{b3} = ?$$

Applying the procedure used for sodium carbonate in Section 11A.2 gives the following relationships among the values of the six equilibrium constants.

$$K_{a1}K_{b3} = K_w$$
$$K_{a2}K_{b2} = K_w$$
$$K_{a3}K_{b1} = K_w$$

We can predict whether the $H_2PO_4^-$ ion should be an acid or a base by looking at the values of the K_a and K_b constants that characterize its reactions with water.

$$H_2PO_4^-(aq) + H_2O(l) \rightleftharpoons H_3O^+(aq) + HPO_4^{2-}(aq) \qquad K_{a2} = 6.3 \times 10^{-8}$$
$$H_2PO_4^-(aq) + H_2O(l) \rightleftharpoons H_3PO_4(aq) + OH^-(aq) \qquad K_{b3} = 1.4 \times 10^{-12}$$

Because K_{a2} is significantly larger than K_{b3}, solutions of the $H_2PO_4^-$ ion in water should be acidic.

We can use the same method to decide whether HPO_4^{2-} is an acid or a base when dissolved in water. We start by looking at the values of K_a and K_b that characterize its reactions with water.

$$HPO_4^{2-}(aq) + H_2O(l) \rightleftharpoons H_3O^+(aq) + PO_4^{3-}(aq) \qquad K_{a3} = 4.2 \times 10^{-13}$$
$$HPO_4^{2-}(aq) + H_2O(l) \rightleftharpoons H_2PO_4^-(aq) + OH^-(aq) \qquad K_{b2} = 1.6 \times 10^{-7}$$

Because K_{b2} is significantly larger than K_{a3}, solutions of the HPO_4^{2-} ion in water should be basic.

Measurements of the pH of 0.10 *M* solutions of the species yield the following results, which are consistent with the predictions of this exercise.

H_3PO_4	$H_2PO_4^-$	HPO_4^{2-}	PO_4^{3-}
pH = 1.5	pH = 4.4	pH = 9.3	pH = 12.0

Problems

Diprotic Acids

11A–1. Calculate the H_3O^+, CO_3^{2-}, HCO_3^-, and H_2CO_3 concentrations at equilibrium in a solution that initially contained 0.100 mol of carbonic acid per liter. (For H_2CO_3, $K_{a1} = 4.5 \times 10^{-7}$ and $K_{a2} = 4.7 \times 10^{-11}$.)

11A–2. Calculate the equilibrium concentrations of the important components of a 0.250 *M* malonic acid ($HO_2CCH_2CO_2H$) solution. Use the symbol H_2M as an abbreviation for malonic acid and assume stepwise dissociation of the acid. (For H_2M, $K_{a1} = 1.4 \times 10^{-5}$ and $K_{a2} = 2.1 \times 10^{-8}$.)

11A–3. Check the validity of the assumption in the previous problem that malonic acid dissociates in a stepwise fashion by comparing the concentrations of the HM^- and M^{2-} ions obtained in the problem. Is this assumption valid?

11A–4. Glycine, the simplest of the amino acids found in proteins, is a diprotic acid with the formula $HO_2CCH_2NH_3^+$. If we symbolize glycine as H_2G^+, we can write the following equations for its stepwise dissociation.

$$H_2G^+(aq) + H_2O(l) \rightleftharpoons HG(aq) + H_3O^+(aq) \qquad K_{a1} = 4.5 \times 10^{-3}$$
$$HG(aq) + H_2O(l) \rightleftharpoons G^-(aq) + H_3O^+(aq) \qquad K_{a2} = 2.5 \times 10^{-10}$$

Calculate the concentrations of H_3O^+, G^-, HG, and H_2G^+ in a 2.00 *M* solution of glycine in water.

11A–5. Which of the following equations accurately describes a 2.0 *M* solution of glycine?
(a) $[H_3O^+] \approx [H_2G^+]$ (b) $[H_3O^+] < [H_2G^+]$
(c) $[H_3O^+] > [HG]$ (d) $[H_3O^+] \approx [HG]$
(e) $[H_3O^+] < [HG]$

11A–6. Which of the following equations results from the fact that glycine is a weak diprotic acid?
(a) $[G^-] \approx [H_2G^+]$ (b) $[G^-] \approx K_{a2}$
(c) $[G^-] \approx [HG]$ (d) $[G^-] > [HG]$
(e) $[G^-] \approx [H_3O^+]$

11A–7. Oxalic acid ($H_2C_2O_4$) has been implicated in diseases such as gout and kidney stones. Calculate the H_3O^+, $H_2C_2O_4$, $HC_2O_4^-$, and $C_2O_4^{2-}$ concentrations in a 1.25 *M* solution of oxalic acid.

$$H_2C_2O_4(aq) + H_2O(l) \rightleftharpoons H_3O^+(aq) + HC_2O_4^-(aq) \quad K_{a1} = 5.4 \times 10^{-2}$$

$$HC_2O_4^-(aq) + H_2O(l) \rightleftharpoons H_3O^+(aq) + C_2O_4^{2-}(aq) \quad K_{a2} = 5.4 \times 10^{-5}$$

Diprotic Bases

11A–8. Which of the following sets of equations can be used to calculate K_{b1} and K_{b2} for sodium oxalate ($Na_2C_2O_4$) from K_{a1} and K_{a2} for oxalic acid ($H_2C_2O_4$)?
(a) $K_{b1} = K_w \times K_{a1}$ and $K_{b2} = K_w \times K_{a2}$
(b) $K_{b1} = K_w \times K_{a2}$ and $K_{b2} = K_w \times K_{a1}$
(c) $K_{b1} = K_w/K_{a1}$ and $K_{b2} = K_w/K_{a2}$
(d) $K_{b1} = K_w/K_{a2}$ and $K_{b2} = K_w/K_{a1}$
(e) $K_{b1} = K_{a1}/K_w$ and $K_{b2} = K_{a2}/K_w$

11A–9. Calculate the pH of a 0.028 *M* solution of sodium oxalate ($Na_2C_2O_4$) in water.

$$H_2C_2O_4(aq) + H_2O(l) \rightleftharpoons H_3O^+(aq) + HC_2O_4^-(aq) \quad K_{a1} = 5.4 \times 10^{-2}$$

$$HC_2O_4^-(aq) + H_2O(l) \rightleftharpoons H_3O^+(aq) + C_2O_4^{2-}(aq) \quad K_{a2} = 5.4 \times 10^{-5}$$

11A–10. Calculate the H_3O^+, OH^-, H_2CO_3, HCO_3^-, and CO_3^{2-} concentrations in a 0.150 *M* solution of sodium carbonate (Na_2CO_3) in water. (For H_2CO_3, $K_{a1} = 4.5 \times 10^{-7}$ and $K_{a2} = 4.7 \times 10^{-11}$.)

Triprotic Acids

11A–11. Calculate the pH of a 0.200 *M* solution of H_3PO_4.

11A–12. Arsenic acid, H_3AsO_4, has

$$K_{a1} = 6 \times 10^{-3}, \quad K_{a2} = 1.1 \times 10^{-7}, \quad K_{a3} = 3 \times 10^{-12}$$

Calculate the H_3O^+, H_3AsO_4, $H_2AsO_4^-$, $HAsO_4^{2-}$, and AsO_4^{3-} concentrations is a 1.0 *M* solution.

Compounds That Could Be Either Acids or Bases

11A–13. According to the data in Table 11.4, the pH of a 0.10 *M* $NaHCO_3$ solution is 8.4. Explain why the HCO_3^- ion forms solutions that are basic, not acidic.

11A–14. Which of the following statements concerning sodium hydrogen sulfide (NaSH) is correct? (For H_2S, $K_{a1} = 1.0 \times 10^{-7}$ and $K_{a2} = 1.3 \times 10^{-13}$.)
(a) NaSH is an acid because K_{a1} for H_2S is much larger than K_{a2}.
(b) NaSH is an acid because K_{a1} for H_2S is smaller than K_{b1} for Na_2S.
(c) NaSH is a base because K_{b1} for Na_2S is much larger than K_{b2}.
(d) NaSH is a base because K_{b2} for Na_2S is larger than K_{a2} for H_2S.

11A–15. Calculate the pH of 0.100 *M* solutions of H_3PO_4, NaH_2PO_4, Na_2HPO_4, and Na_3PO_4. (For H_3PO_4, $K_{a1} = 7.1 \times 10^{-3}$, $K_{a2} = 6.3 \times 10^{-8}$, and $K_{a3} = 4.2 \times 10^{-13}$.)

11A–16. Predict whether an aqueous solution of sodium hydrogen sulfate ($NaHSO_4$) should be acidic, basic, or neutral. (For H_2SO_4, $K_{a1} = 1 \times 10^3$ and $K_{a2} = 1.2 \times 10^{-2}$.)

11A–17. Predict whether an aqueous solution of sodium hydrogen sulfite ($NaHSO_3$) should be acidic, basic, or neutral. (For H_2SO_3, $K_{a1} = 1.7 \times 10^{-2}$ and $K_{a2} = 6.4 \times 10^{-8}$.)

Chapter Twelve

OXIDATION–REDUCTION REACTIONS

Special Topics

12.1 Common Oxidation–Reduction Reactions

This chapter introduces redox reactions and cell potentials. Voltaic cells are introduced as a means of studying redox reactions. Examples of oxidation–reduction reactions of both inorganic and organic substances are included in the chapter. ChemActivities 48–51 in Chemistry: A Guided Inquiry *are designed to accompany this chapter.*

Oxidation–reduction reactions can be described in several ways: transfer of an electron, transfer of a hydrogen atom, or transfer of an oxygen atom. We have introduced all three descriptions to prepare students for when they encounter more than just the electron transfer description in other courses such as organic chemistry. The change in oxidation number is the connection between the three descriptions that allows students to identify an oxidation–reduction reaction.

Oxidation–reduction or **redox reactions** are the most common type of chemical reactions encountered on a daily basis. When natural gas (methane, CH_4) burns, for example, an oxidation–reduction reaction occurs that releases more than 800 kJ/mol_{rxn} of energy.

$$CH_4(g) + 2\ O_2(g) \longrightarrow CO_2(g) + 2\ H_2O(g)$$

Within our bodies, a sequence of oxidation–reduction reactions are used to burn sugars such as glucose ($C_6H_{12}O_6$) and the fatty acids in the fats we eat.

$$C_6H_{12}O_6(aq) + 6\ O_2(g) \longrightarrow 6\ CO_2(g) + 6\ H_2O(l)$$
$$CH_3(CH_2)_{16}CO_2H(aq) + 26\ O_2(g) \longrightarrow 18\ CO_2(g) + 18\ H_2O(l)$$

Silver metal is oxidized when it comes in contact with trace quantities of H_2S or SO_2 in the atmosphere or with foods, such as eggs, that are rich in sulfur compounds.

$$4\ Ag(s) + 2\ H_2S(g) + O_2(g) \longrightarrow 2\ Ag_2S(s) + 2\ H_2O(l)$$

The tarnishing of silver is just one example of a broad class of oxidation–reduction reactions that fall under the general heading of **corrosion.** Another example is the series of reactions that occur when iron or steel rusts.

Oxidation–reduction reactions play an important role in biological systems. Green plants convert carbon dioxide and water into carbohydrates and oxygen by a series of oxidation–reduction reactions called **photosynthesis.**

$$6\ CO_2(g) + 6\ H_2O(l) \longrightarrow C_6H_{12}O_6(aq) + 6\ O_2(g)$$

Additional common oxidation–reduction reactions include the production of electrical energy with batteries, the removal of stains from clothes with bleach, and photography.

Oxidation–reduction reactions (see Chapter 5) of ionic compounds involve the transfer of electrons. The element or compound that gains electrons is said to undergo **reduction.** The element or compound that loses electrons undergoes **oxidation.** The reaction below takes place when sodium metal is reacted with chlorine gas. Sodium loses electrons to form Na^+ ions. These electrons are gained by the chlorine to form Cl^- ions.

$$2\ Na + Cl_2 \longrightarrow 2\ [Na^+][Cl^-]$$

Oxidation (Na → Na^+)

Reduction (Cl_2 → Cl^-)

Redox reactions are often studied by considering the processes of oxidation and reduction separately. The oxidation or reduction reactions written by themselves are referred to as **half-reactions.** In actual chemical reactions, reduction and oxidation must always accompany one another. The electrons that are gained by the substance being reduced are lost by the substance being oxidized. The number of electrons gained and lost in a chemical reaction must be equal because no electrons can be created or destroyed during a reaction. In the above reaction, the two electrons lost by the two sodium atoms are gained by the chlorine molecule to form two chloride ions. The reduction and oxidation half-reactions for the redox reactions of Na and Cl_2 are shown below. The addition of these two half-reactions gives the overall reaction.

Oxidation half-reaction	$2\ Na \longrightarrow 2\ Na^+ + 2\ e^-$
Reduction half-reaction	$Cl_2 + 2\ e^- \longrightarrow 2\ Cl^-$

The above definitions of oxidation and reduction do not accurately characterize all redox reactions. Chemists often think about oxidation–reduction in terms of the transfer of oxygen atoms

$$CH_3\overset{\overset{\displaystyle O}{\|}}{C}H(l) + H_2O_2(l) \longrightarrow CH_3\overset{\overset{\displaystyle O}{\|}}{C}OH(l) + H_2O(l)$$

or the transfer of hydrogen atoms.

$$C_2H_4(g) + H_2(g) \longrightarrow C_2H_6(g)$$

The concept of **oxidation numbers** or **oxidation states** gives us a tool to identify redox reactions. Any reaction that results in a change in the oxidation number, whether it involves transfer of electrons, transfer of oxygen atoms, or transfer of hydrogen atoms, is classified as a redox reaction.

> **CHECKPOINT**
> Is it possible to have oxidation without reduction?

12.2 Determining Oxidation Numbers

When assigning oxidation numbers to atoms in a compound, the compound is treated as if it were ionic, and the shared electrons in each bond are assigned to the more electronegative element. Frequently chemists refer to an atom that has been assigned an oxidation number as having a particular *oxidation state*. Two methods for assigning oxidation numbers of atoms were introduced in Chapter 5. In Section 5.14 Lewis structures were used in which all bonding electrons between nonequivalent atoms were assigned to the more electronegative atom. Oxidation numbers were then assigned to each atom in the structure using the equation

Two methods which were introduced in Chapter 5 are used to determine oxidation numbers. The first method uses a calculation that requires students to draw the Lewis structure and then assign bonding electrons to the more electronegative atom in each bond. The advantage of this method is that it works well with organic molecules. The second method uses the traditional set of rules for assigning oxidation numbers that is found in most introductory texts.

$$OX_a = V_a - N_a$$

where OX_a is the oxidation number of atom *a*, V_a is the number of valence electrons of the atom, and N_a is the number of electrons assigned to the atom in the Lewis structure after assigning bonding electrons to the more electronegative atom. This method of assigning oxidation numbers is particularly useful for organic compounds (compounds containing primarily C and H) in which carbon atoms have several different oxidation numbers. For example, 1-propanol ($CH_3CH_2CH_2OH$) has three carbon atoms, each assigned a different oxidation number.

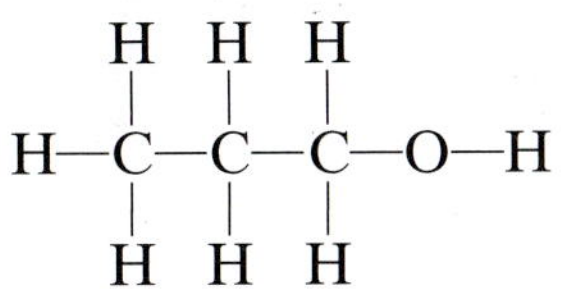

Assigning bonding electrons to the most electronegative atom in each bond yields

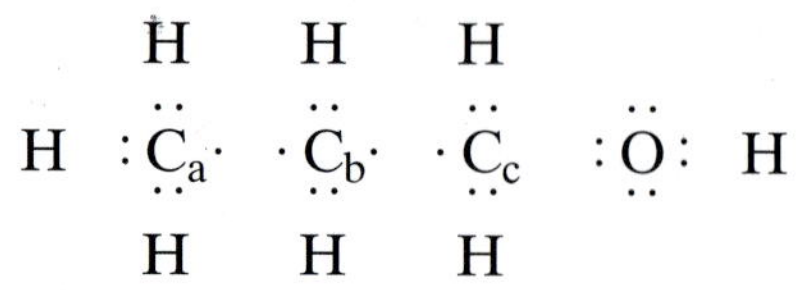

The oxidation state of each of the carbons can now be determined:

$$OX_a = V_a - N_a$$
$$OX_{Ca} = 4 - 7 = -3$$
$$OX_{Cb} = 4 - 6 = -2$$
$$OX_{Cc} = 4 - 5 = -1$$

Carbon has low (more negative) oxidation numbers when bonded to hydrogen. As the number of bonds between carbon and hydrogen decreases, as in the case of C_a (3 bonds to H) and C_b (2 bonds to H), the oxidation number increases ($C_a = -3$ and $C_b = -2$). In addition, an increase in the number of bonds to an electronegative element such as oxygen increases the oxidation number of carbon ($C_c = -1$).

A second method for assigning oxidation numbers employs a series of rules introduced in Section 5.13. The free atoms are used as a reference for measuring the extent of oxidation or reduction and are therefore assigned an oxidation number of zero. If an atom loses electrons to become a positively charged cation, it is assigned an oxidation number equal to the charge on the cation (the number of electrons lost). Similarly, when electrons are gained by an atom to form an anion, the atom is assigned an oxidation number equal to the charge (the number of electrons gained).

When assigning oxidation numbers to inorganic species, it is often easier to apply the rules than to draw the Lewis structure and do the calculation. For example, we can use the rules to find the oxidation number of nitrogen in the nitrate anion. In electrically neutral molecules or polyatomic ions, the sum of the oxidation numbers of the individual atoms must be equal to the overall charge on the molecule or ion. In the polyatomic nitrate ion, NO_3^-, oxygen is assigned its typical oxidation number of -2. The oxidation state of nitrogen can be determined by adding the oxidation numbers of the individual atoms to equal the charge on the nitrate ion.

➤ CHECKPOINT

What are the oxidation numbers for all the atoms in the following compounds?

$FeCl_3$, CH_4, O_2, MnO_4^-, $HCNH_2$ (with O double-bonded to C)

$$OX_N + 3(OX_O) = -1$$
$$OX_N + 3(-2) = -1$$
$$OX_N = +5$$

12.3 Recognizing Oxidation–Reduction Reactions

The most powerful model of oxidation–reduction reactions is based on the following definitions:

Oxidation occurs when the oxidation number of an atom increases.
Reduction occurs when the oxidation number of an atom decreases.

Using this model of oxidation–reduction, we can describe a reaction such as the following in which electrons are transferred in the formation of an ionic compound. The oxidation number for each atom is given directly beneath the atom.

$$\underset{0}{2\,\mathrm{Na}} + \underset{0}{\mathrm{Cl_2}} \longrightarrow 2[\underset{+1}{\mathrm{Na^+}}][\underset{-1}{\mathrm{Cl^-}}]$$

Oxidation

Reduction

Sodium's oxidation number increases from zero to +1, indicating that sodium is oxidized. Chlorine's oxidation number decreases from zero to −1, indicating that it is reduced.

This model for redox reactions is particularly useful for describing reactions such as the following. The oxidation numbers are given beneath each atom.

$$\underset{-4\,+1}{\mathrm{CH_4}(g)} + \underset{0}{2\,\mathrm{O_2}(g)} \longrightarrow \underset{+4\,-2}{\mathrm{CO_2}(g)} + \underset{+1\,-2}{2\,\mathrm{H_2O}\,(g)}$$

In this reaction the oxidation number of carbon increases from −4 to +4, indicating that carbon is oxidized. The oxygen is reduced because its oxidation number decreases from zero to −2. Oxidation–reduction reactions may be identified by recognizing chemical reactions that lead to a change in the oxidation number of one or more atoms.

Exercise 12.1

Indicate which of the following reactions are oxidation–reduction reactions. Indicate what species are being oxidized and which species are being reduced.

(a) $Cu(s) + 2\,Ag^+(aq) \rightleftharpoons Cu^{2+}(aq) + 2\,Ag(s)$

(b) $CH_3CO_2H(aq) + OH^-(aq) \rightleftharpoons CH_3CO_2^-(aq) + H_2O(l)$

(c) $SF_4(g) + F_2(g) \rightleftharpoons SF_6(g)$

(d) $CuSO_4(aq) + BaCl_2(aq) \rightleftharpoons BaSO_4(s) + CuCl_2(aq)$

Solution

(a) This is an oxidation–reduction reaction. The copper is oxidized from copper 0 to copper +2. The silver is reduced from silver +1 to silver 0.

(b) This is a Brønsted acid–base reaction. There is no change in the oxidation numbers of any of the atoms.

(c) This is an oxidation–reduction reaction. The SF_4 is oxidized due to the change in the oxidation number of sulfur. Sulfur changes from +4 in SF_4 to +6 in SF_6. The fluorine molecule, F_2, is reduced as the oxidation number decreases from zero in F_2 to −1 in SF_6.

(d) There is no change in oxidation number of any of the elements in this reaction.

Oxidation–reduction reactions of organic compounds can often be quickly recognized by a change in the number of bonds between carbon and hydrogen or oxygen (or another electronegative element). When carbon is oxidized in an organic molecule, the number of carbon–hydrogen bonds decreases and/or the number of carbon–oxygen bonds increases. For example, when a bottle of wine is opened and allowed to sit in contact with air, the wine will eventually turn to vinegar. The reaction involves the oxidation of the alcohol in wine, ethanol, to form acetic acid.

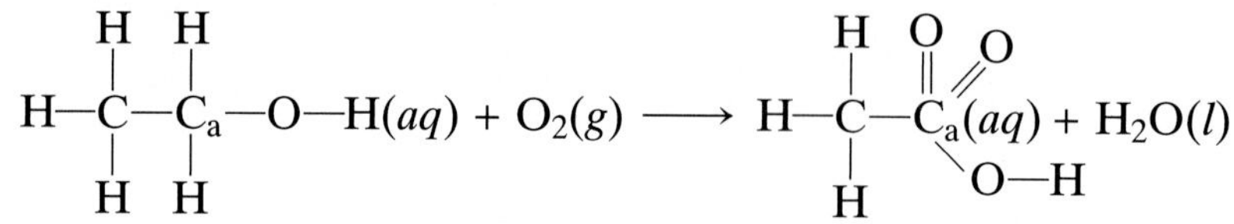

Ethanol *Acetic acid*

In this reaction C_a has two bonds to hydrogen and one bond to oxygen in ethanol. In the product, acetic acid, C_a has no bonds to hydrogen and a single and double bond to oxygen. The carbon atom, C_a, has been oxidized. Even though only the one carbon atom within the molecule was oxidized, we still refer to the entire molecule, ethanol, as having been oxidized. Molecular oxygen is reduced.

We can examine the reaction more quantitatively by assigning oxidation numbers to the atoms that are oxidized or reduced.

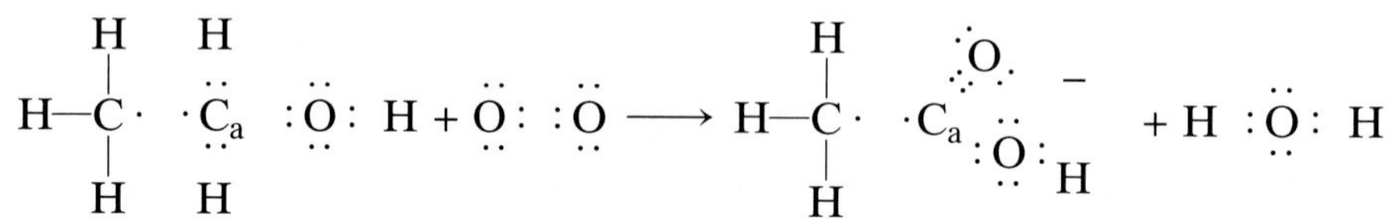

After assigning electrons in each bond of the Lewis structure to the more electronegative atom, the carbon atom, C_a, has five electrons in ethanol and only one electron in acetic acid.

$$OX_{\text{Ca in ethanol}} = 4 - 5 = -1$$
$$OX_{\text{Ca in acetic acid}} = 4 - 1 = +3$$

The oxidation number of the carbon has increased from -1 to $+3$. At the same time the oxidation number of oxygen has decreased from zero in elemental oxygen to -2 for the oxygen atom in acetic acid and the oxygen in water.

$$OX_{\text{elemental oxygen}} = 6 - 6 = 0$$
$$OX_{\text{oxygen in acetic acid}} = 6 - 8 = -2$$
$$OX_{\text{oxygen in water}} = 6 - 8 = -2$$

The method of writing oxidation–reduction equations by showing only the organic reactant and product with the oxidizing or reducing agent over the arrow is introduced here because this method of writing equations is common in organic chemistry and biochemistry.

Many times in the chemical literature, oxidation–reduction reactions of organic compounds are not written as balanced chemical equations as shown above but instead show only the initial organic reactant and the final organic product. In the laboratory, acetic acid can be prepared from ethanol by oxidation with potassium dichromate, $K_2Cr_2O_7$. The reaction is often written as

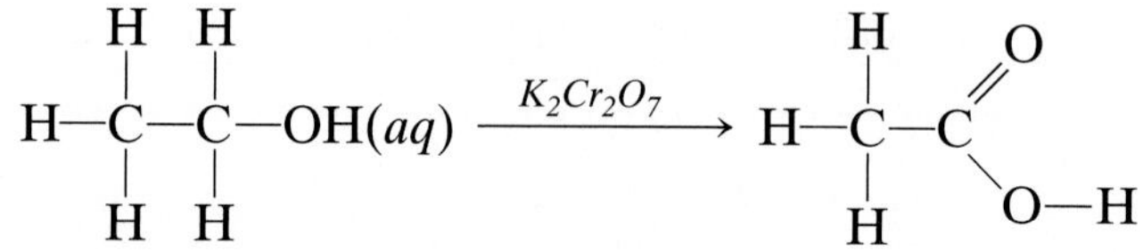

The reactant ethanol and the product acetic acid are shown. The substance that undergoes the reduction half-reaction (or oxidation half-reaction if the organic compound is reduced) plus any additional reaction conditions are written above the arrow. The number of hydrogen and oxygen atoms on both sides of the equation are not balanced.

Exercise 12.2

Identify which of the following reactions (unbalanced) represent oxidation or reduction of an organic compound. Indicate whether the organic compound is oxidized or reduced. Describe the change in oxidation number of the carbon atom which is oxidized or reduced.

(a) $H_2C{=}CH_2(g) \xrightarrow{H_2} H_3C{-}CH_3(g)$

Ethene → *Ethane*

(b) $HC(=O){-}OH(aq) + NaOH(aq) \longrightarrow HC(=O){-}O^-Na^+(aq) + H_2O(l)$

Formic acid → *Sodium formate*

(c) $CH_3CH_2C(=O){-}H(aq) \xrightarrow{KMnO_4} CH_3CH_2C(=O){-}OH(aq)$

Propanal → *Propanoic acid*

Solution

(a) The ethene is reduced to ethane. The oxidation number of both carbon atoms changes from -2 in ethene to -3 in ethane. The H_2 is oxidized.

(b) This is an acid–base reaction. There is no change in oxidation number for any of the atoms in the reaction.

(c) Propanal is oxidized to propanoic acid. The oxidation number of the carbon bonded to oxygen changes from $+1$ in propanal to $+3$ in propanoic acid. The $KMnO_4$ is reduced.

The method of listing redox reactions by showing only one reactant and its product is commonly used to describe biochemical processes. An example, shown in Figure 12.1, describes the citric acid (or Krebs) cycle. The citric acid cycle, a vital metabolic pathway, is made up of a series of reactions that serve as the principal source of metabolic energy for an organism. Nutrients from carbohydrates, proteins, and lipids are oxidized by the citric acid cycle to yield carbon dioxide and water. The carbon in carbon dioxide is in a $+4$ oxidation state, carbon's highest oxidation state. Carbon in this state has been completely oxidized.

We will examine just one of the reactions from the cycle. Reaction 3 represents the first oxidation–reduction reaction in the cycle in which isocitrate is oxidized to α-ketoglutarate. The circled numbers are the oxidation numbers of the carbon atoms.

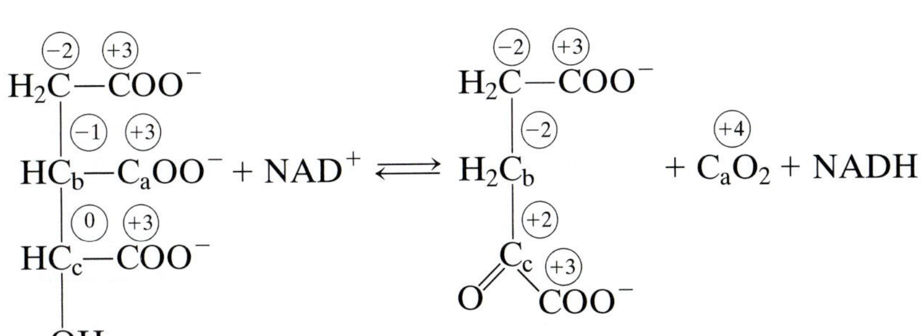

Isocitrate *α-Ketoglutarate*

Three carbon atoms (labeled C_a, C_b, and C_c) undergo changes in oxidation number during the reaction. Carbon atoms C_a and C_c are oxidized from $+3$ to $+4$ and 0 to $+2$,

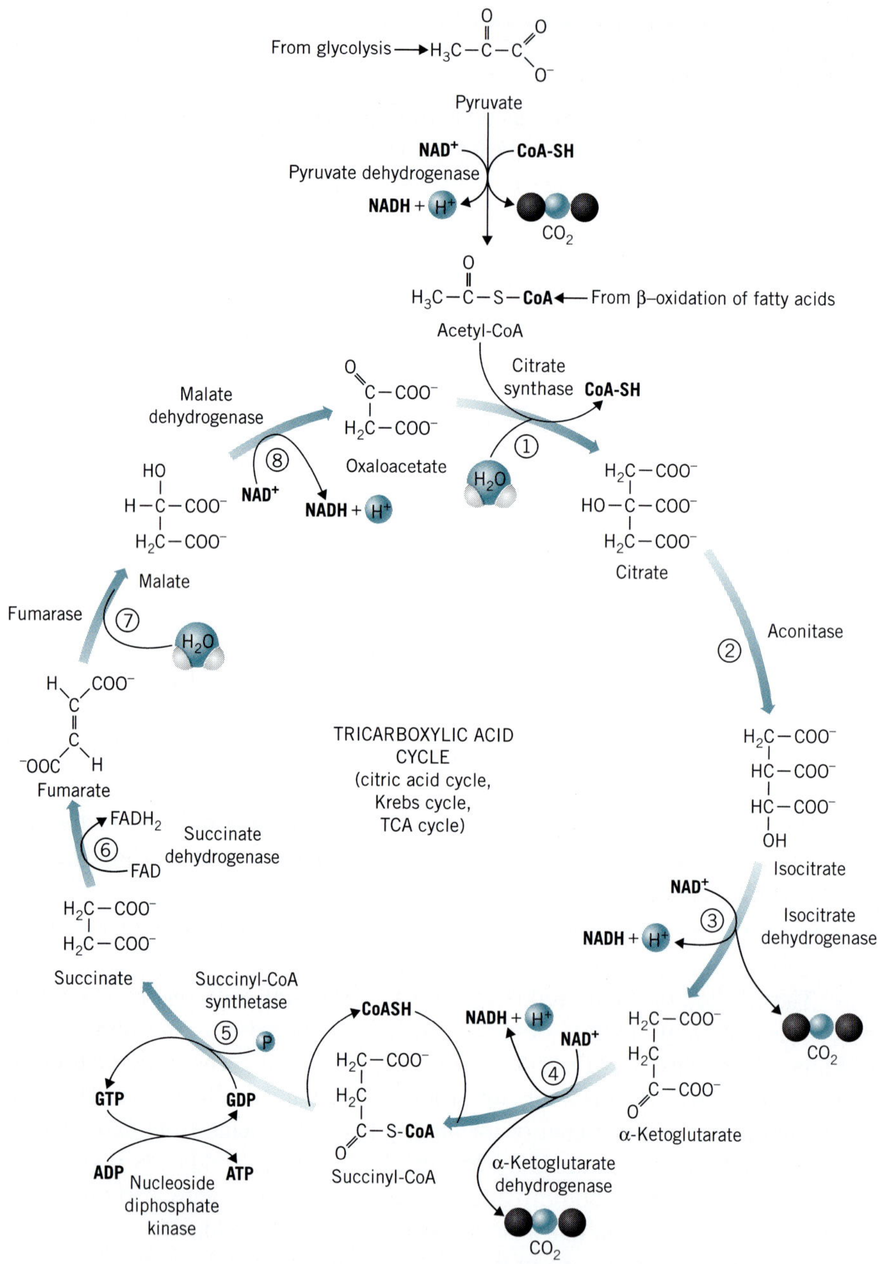

Fig. 12.1 The citric acid cycle. Reprinted from R. H. Garrett and C. M. Grisham, *Biochemistry,* Harcourt Brace, New York, 1994, p. 602.

respectively. Carbon atom C_b is reduced from -1 to -2. There is a $+3$ increase in oxidation numbers and a -1 decrease for a net change within the molecule of $+2$, indicating that the isocitrate has been oxidized. Accompanying this oxidation is the reduction of NAD^+ (nicotinamide adenine dinucleotide) to produce its reduced form, NADH. NAD^+ undergoes a -2 change in oxidation number to balance the $+2$ change of isocitrate. The α-ketoglutarate and NADH produced in the reaction now participate in additional metabolic reactions.

12.4 Voltaic Cells

A closer examination of oxidation–reduction reactions can be simplified by physically separating the process of oxidation from the process of reduction. This can be accomplished using a **voltaic cell** (also known as a **galvanic cell**). A voltaic cell is an electrochemical cell in which a redox reaction produces an electric current. In other words, the cell serves as a simple battery.

Figure 12.2 shows a voltaic cell formed by immersing a strip of zinc metal into a 1 *M* $Zn(NO_3)_2$ solution and immersing a piece of silver wire into a solution of 1 *M* $AgNO_3$. The zinc metal and silver wire are connected with an electrical conductor. The circuit is completed with a **salt bridge,** a U-tube filled with a saturated solution of a soluble salt such as KNO_3. Oxidation (loss of electrons) takes place in the left beaker at the zinc electrode.

$$Zn(s) \rightleftharpoons Zn^{2+}(aq) + 2\,e^- \qquad (12.1)$$

As the reaction proceeds, the zinc metal electrode dissolves away as the zinc metal is converted into aqueous zinc ions. Electrons are not stable in solution, and therefore the electrons produced cause a buildup of negative charge on the surface of the zinc metal, thus causing the electrode to be negative. Chemical equations in which electrons appear as a product or reactant are called *half-reactions*. Equation 12.1 represents an oxidation half-reaction. An oxidation half-reaction cannot occur without an accompanying reduction half-reaction.

Reduction takes place at the silver electrode, and the reduction half-reaction is

$$Ag^+(aq) + e^- \rightleftharpoons Ag(s) \qquad (12.2)$$

Silver ions in solution migrate to the surface of the silver electrode and accept electrons. As a result, the silver ions are converted into silver metal that plates onto the surface of the silver electrode. The silver electrode supplies electrons for the reduction reaction and is positive relative to the zinc electrode.

Because the two electrodes are connected by an electrical conductor, the excess electrons at the zinc electrode move through the wire connecting the two electrodes toward the positively charged silver electrode. This produces an electric current that can be used to light a light bulb or run an electric motor. If a voltmeter is connected between the two electrodes, a **cell potential** for the voltaic cell can be measured. The cell potential is the potential of the cell to do work on its surroundings

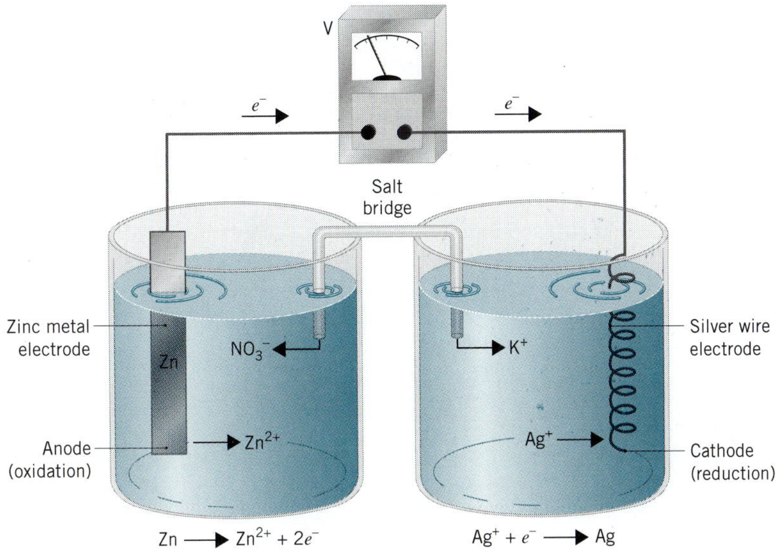

Fig. 12.2 The reaction between zinc metal and silver ions produces an electric potential resulting in the flow of electrons from the zinc electrode to the silver electrode. K^+ ions move from the salt bridge into the solution to replace silver ions as the silver ions are converted to silver metal at the cathode. NO_3^- ions move from the salt bridge into solution to balance the charge of the Zn^{2+} produced from the zinc metal at the anode.

by driving an electric current through a wire. By definition, a potential of 1 volt is produced when 1 joule of energy is used to transport 1 coulomb, C, of electrical charge across the potential.

$$V = \frac{J}{C}$$

The cell potential for the voltaic cell in Figure 12.2 is 1.5624 V. The magnitude of the cell potential is a measure of the driving force behind an electrochemical reaction.

The function of the salt bridge is to maintain electric neutrality in the cell. As the zinc metal is oxidized, it forms zinc ions, which dissolve in solution, producing a positively charged solution. At the other electrode silver ions are reduced to silver metal, which causes the solution to become negatively charged. Negatively charged nitrate ions therefore diffuse from the salt bridge into the zinc solution to balance the excess positive charge. Positively charged potassium ions diffuse into the silver solution to balance the excess positive charge in this solution.

The electrode at which oxidation takes place in an electrochemical cell is called the **anode.** The electrode at which reduction occurs is called the **cathode.** The identity of the cathode and anode can be remembered by recognizing that positive ions, or **cations,** flow toward the cathode, and negative ions, or **anions,** flow toward the anode. In the voltaic cells shown in Figure 12.2, the zinc electrode is the anode and the silver electrode is the cathode.

➤ CHECKPOINT

What is the sign of the anode in Figure 12.2? What is the sign of the cathode?

Illustrations of the voltaic pile prepared from strips of zinc and silver metal separated by moistened cardboard published by Volta in 1800.

BALANCING OXIDATION–REDUCTION EQUATIONS

Although the concepts of oxidation, reduction, and the equal gain and loss of electrons are vital to an understanding of chemistry, the same cannot be said of the process of balancing redox equations using a memorized algorithm. Therefore, balancing redox reactions in acidic and basic solutions has been omitted from the core but may be found in the Special Topics section at the end of Chapter 12 for those who wish to introduce this.

The overall oxidation–reduction reaction in the voltaic cell shown in Figure 12.2 consists of the oxidation of zinc (reaction 12.1) and the reduction of silver (reaction 12.2). Using the law of conservation of mass, it would be tempting to write the following complete oxidation–reduction reaction:

$$Ag^+(aq) + Zn(s) \rightleftharpoons Zn^{2+}(aq) + Ag(s) \qquad (12.3)$$

There are an equal number of silver and zinc atoms on both sides of the equation, so atoms are conserved. However, there is not a balance of electric charge on the two sides of the equation. The implication of reaction 12.3 as written is that for every atom of Zn that is oxidized, two electrons will be released. At the same time, the reduction of a silver ion requires only one electron. If the reaction proceeded as written in reaction 12.3, twice as many electrons would be produced at the zinc electrode as are used at the silver electrode. Because neither matter nor charge is created or destroyed in a chemical reaction, more electrons cannot be produced than are used. In a balanced chemical reaction not only must atoms be conserved, but also electrons. Writing a balanced oxidation–reduction reaction requires that the number of electrons gained during reduction must equal the number of electrons lost during oxidation. Therefore, for every zinc atom that produces two electrons, *two* silver ions must be reduced.

$$\begin{array}{rcl} Zn(s) & \rightleftharpoons & Zn^{2+}(aq) + 2\,e^- \\ 2 \times [Ag^+(aq) + e^- & \rightleftharpoons & Ag(s)] \\ \hline Zn(s) + 2\,Ag^+(aq) & \rightleftharpoons & Zn^{2+}(aq) + 2\,Ag(s) \end{array} \qquad (12.4)$$

This process of writing a balanced oxidation–reduction reaction is known as the *half-reaction method*. The individual half-reactions are written and then multiplied by coefficients such that the electrons lost are equal to the electrons gained. The two half-reactions may then be added together to give the balanced oxidation–reduction reaction.

> **➤ CHECKPOINT**
>
> Is the following oxidation–reduction reaction balanced? Explain your answer.
>
> $$Zn^{2+}(aq) + Cr(s) \rightleftharpoons Cr^{3+}(aq) + Zn(s)$$

12.5 Oxidizing and Reducing Agents

So far we have focused on what happens when a particular substance undergoes a change in oxidation state in an oxidation–reduction reaction. Let's now consider the role that each substance plays in the reaction.

When zinc metal reacts with Ag^+ ions, the zinc metal donates electrons to the Ag^+ ions and therefore reduces the Ag^+ ions. Zinc therefore acts as a **reducing agent** in the reaction. The same electrons that are lost by the zinc are gained by the silver.

$$\underset{\text{Reducing agent}}{Zn(s)} + 2\,Ag^+(aq) \rightleftharpoons Zn^{2+}(aq) + 2\,Ag(s)$$

The Ag^+ ions, on the other hand, gain electrons from the zinc metal and thereby oxidize the zinc. The Ag^+ ion is therefore an **oxidizing agent.**

$$Zn(s) + \underset{\text{Oxidizing agent}}{2\,Ag^+(aq)} \rightleftharpoons Zn^{2+}(aq) + 2\,Ag(s)$$

In general, oxidizing and reducing agents can be defined as follows.

> ***Oxidizing agents* undergo a decrease in oxidation number and cause the oxidation of another species.**
> ***Reducing agents* undergo an increase in oxidation number and cause the reduction of another species.**

Thus, an oxidizing agent undergoes a reduction half-reaction and a reducing agent undergoes an oxidation half-reaction.

Zinc metal is the reducing agent in the voltaic cell shown in Figure 12.2 because it reduces Ag^+ ions by providing electrons to the silver electrode. The Ag^+ ion is the oxidizing agent in the reaction because it oxidizes the zinc metal by removing electrons from the metal.

When zinc metal loses electrons, it forms a conjugate oxidizing agent that could gain electrons if the reaction were reversed.

$$\underset{\substack{\textit{Reducing}\\ \textit{agent}}}{Zn} \rightleftharpoons \underset{\substack{\textit{Oxidizing}\\ \textit{agent}}}{Zn^{2+}} + 2\,e^-$$

Conversely, when the silver ion gains an electron, it forms a conjugate reducing agent. This reducing agent could lose electrons if the reaction went in the opposite direction.

$$\underset{\substack{\textit{Oxidizing}\\ \textit{agent}}}{Ag^+} + e^- \rightleftharpoons \underset{\substack{\textit{Reducing}\\ \textit{agent}}}{Ag}$$

Oxidizing agents and reducing agents are therefore linked or coupled in much the same way that Brønsted acids and Brønsted bases are linked or coupled (see Section 11.4). Every oxidizing agent has a conjugate reducing agent, and vice versa.

The relationship between the strength of a reducing agent and its conjugate oxidizing agent can be understood by extending the argument used in Section 11.9 to link the strengths of Brønsted acids and bases. If zinc metal is a relatively good reducing agent, we must conclude that the Zn^{2+} ion is a relatively weak oxidizing agent.

$$\underset{\substack{\textit{Stronger}\\ \textit{reducing}\\ \textit{agent}}}{Zn(s)} \rightleftharpoons \underset{\substack{\textit{Weaker}\\ \textit{oxidizing}\\ \textit{agent}}}{Zn^{2+}(aq)} + 2\,e^-$$

If, on the other hand, the Ag^+ ion is a relatively good oxidizing agent, then silver metal must be a relatively weak reducing agent.

$$\underset{\substack{\textit{Stronger}\\ \textit{oxidizing}\\ \textit{agent}}}{Ag^+(aq)} + e^- \rightleftharpoons \underset{\substack{\textit{Weaker}\\ \textit{reducing}\\ \textit{agent}}}{Ag(s)}$$

As was seen in Figure 5.11, many atoms have more than one oxidation number. The ability of an element to behave as an oxidizing or reducing agent is related to its oxidation state.

> When an element is in its highest oxidation state, it can act only as an oxidizing agent.
>
> When an element is in its lowest oxidation state, it can act only as a reducing agent.

When an element is in an intermediate oxidation state, it can act as an oxidizing or reducing agent.

Carbon can have oxidation numbers of -4, -3, -2, -1, 0, $+1$, $+2$, $+3$, and $+4$. In carbon dioxide, CO_2, carbon has an oxidation number of $+4$. The carbon in carbon dioxide cannot be further oxidized. Carbon dioxide can only be reduced. In other words, the carbon in carbon dioxide can act only as a oxidizing agent. Conversely, in the molecule methane, CH_4, carbon has an oxidation number of -4 and can only undergo an oxidation reaction. Methane can act only as a reducing agent in a redox reaction. Elemental carbon in the form of graphite (zero oxidation number) can be either oxidized or reduced depending on the substance with which it is reacting.

➤ CHECKPOINT

Identify the oxidizing and reducing agents in the following oxidation–reduction reaction.

$$Sn(s) + 4\ HNO_3(aq) \rightleftharpoons SnO_2(s) + 4\ NO_2(g) + 2\ H_2O(l)$$

12.6 Relative Strengths of Oxidizing and Reducing Agents

Several questions remain concerning the voltaic cell shown in Figure 12.2. How do we know which substance will be oxidized and which will be reduced? How do we decide what reaction occurs? The following rule can be used to predict whether a redox reaction should occur and which substance should be oxidized and which reduced if a reaction does occur.

> **Oxidation–reduction reactions should occur when they convert the stronger of a pair of oxidizing agents and the stronger of a pair of reducing agents into a weaker oxidizing agent and a weaker reducing agent.**

Given that reaction 12.4 occurs in the voltaic cell, zinc must be a stronger reducing agent than silver metal and silver ion must be a stronger oxidizing agent than Zn^{2+} ion.

$$Zn(s) + 2\ Ag^+(aq) \rightleftharpoons Zn^{2+} + 2\ Ag(s)$$

$Zn(s)$	$2\ Ag^+(aq)$	Zn^{2+}	$2\ Ag(s)$
Stronger reducing agent	*Stronger oxidizing agent*	*Weaker oxidizing agent*	*Weaker reducing agent*

On the basis of many such experiments, the common oxidation–reduction half-reactions have been organized into a table in which the strongest reducing agents are at one end and the strongest oxidizing agents are at the other, as shown in Table 12.1. By convention, all of the half-reactions are written as reduction half-reactions. The relationship between the half-reactions is quantified by listing the *reduction potential* for each half-reaction. *Potential* is defined as a measure of the driving force behind an electrochemical reaction and is measured in units of volts, V. Standard reduction potentials for additional species are listed in Appendix B.12.

The cell potential is dependent on the concentrations of any species present in solution, the partial pressures of any gases involved in the reaction, and the temperature at which the reaction is run. To provide a basis for comparing one half-reaction with another, the following set of **standard conditions** for electrochemical measurements has been defined:

1. For solutions, the standard potential is based on a standard state of 1 *M* concentration.
2. All gases have a partial pressure of 1 bar, or 0.9869 atm. This can be rounded to 1 atm for all but the most exact calculations.

Although measurements of standard potentials can be made at any temperature, they are often determined at 25°C. Potentials for reduction half-reactions

Table 12.1

Standard Reduction Potentials, E°_{Red}

	Half-Reaction[a]	E°_{Red}	
	$K^+ + e^- \rightleftharpoons K$	−2.924	Best
	$Ca^{2+} + 2\,e^- \rightleftharpoons Ca$	−2.76	reducing
	$Na^+ + e^- \rightleftharpoons Na$	−2.7109	agents
	$Mg^{2+} + 2\,e^- \rightleftharpoons Mg$	−2.375	
	$Al^{3+} + 3\,e^- \rightleftharpoons Al$	−1.706	
	$Mn^{2+} + 2\,e^- \rightleftharpoons Mn$	−1.18	
	$Zn^{2+} + 2\,e^- \rightleftharpoons Zn$	−0.7628	
	$Cr^{3+} + 3\,e^- \rightleftharpoons Cr$	−0.74	
	$S + 2\,e^- \rightleftharpoons S^{2-}$	−0.508	
	$Cr^{3+} + e^- \rightleftharpoons Cr^{2+}$	−0.41	
	$Fe^{2+} + 2\,e^- \rightleftharpoons Fe$	−0.409	
	$Co^{2+} + 2\,e^- \rightleftharpoons Co$	−0.28	
	$Ni^{2+} + 2\,e^- \rightleftharpoons Ni$	−0.23	
	$Sn^{2+} + 2\,e^- \rightleftharpoons Sn$	−0.1364	
	$Pb^{2+} + 2\,e^- \rightleftharpoons Pb$	−0.1263	
	$Fe^{3+} + 3\,e^- \rightleftharpoons Fe$	−0.036	
	$2\,H^+ + 2\,e^- \rightleftharpoons H_2$	0.0000 . . .	
Oxidizing	$Sn^{4+} + 2\,e^- \rightleftharpoons Sn^{2+}$	0.15	↑
power	$Cu^{2+} + e^- \rightleftharpoons Cu^+$	0.158	Reducing
increases	$Cu^{2+} + 2\,e^- \rightleftharpoons Cu$	0.3402	power
↓	$O_2 + 2\,H_2O + 4\,e^- \rightleftharpoons 4\,OH^-$	0.401	increases
	$Cu^+ + e^- \rightleftharpoons Cu$	0.522	
	$MnO_4^- + 2\,H_2O + 3\,e^- \rightleftharpoons MnO_2 + 4\,OH^-$	0.588	
	$O_2 + 2\,H^+ + 2\,e^- \rightleftharpoons H_2O_2$	0.682	
	$Fe^{3+} + e^- \rightleftharpoons Fe^{2+}$	0.770	
	$Hg_2^{2+} + 2\,e^- \rightleftharpoons 2\,Hg$	0.7961	
	$Ag^+ + e^- \rightleftharpoons Ag$	0.7996	
	$Hg^{2+} + 2\,e^- \rightleftharpoons Hg$	0.851	
	$HNO_3 + 3\,H^+ + 3\,e^- \rightleftharpoons NO + 2\,H_2O$	0.96	
	$Br_2(aq) + 2\,e^- \rightleftharpoons 2\,Br^-$	1.087	
	$CrO_4^{2-} + 8\,H^+ + 3\,e^- \rightleftharpoons Cr^{3+} + 4\,H_2O$	1.195	
	$O_2 + 4\,H^+ + 4\,e^- \rightleftharpoons 2\,H_2O$	1.229	
	$Cr_2O_7^{2-} + 14\,H^+ + 6\,e^- \rightleftharpoons 2\,Cr^{3+} + 7\,H_2O$	1.33	
	$Cl_2(g) + 2\,e^- \rightleftharpoons 2\,Cl^-$	1.3583	
	$PbO_2 + 4\,H^+ + 2\,e^- \rightleftharpoons Pb^{2+} + 2\,H_2O$	1.467	
	$MnO_4^- + 8\,H^+ + 5\,e^- \rightleftharpoons Mn^{2+} + 4\,H_2O$	1.491	
	$Au^+ + e^- \rightleftharpoons Au$	1.68	
Best	$Co^{3+} + e^- \rightleftharpoons Co^{2+}$	1.842	
oxidizing	$O_3(g) + 2\,H^+ + 2\,e^- \rightleftharpoons O_2(g) + H_2O$	2.07	
agents	$F_2(g) + 2\,H^+ + 2\,e^- \rightleftharpoons 2\,HF(aq)$	3.03	

[a] In the tabulations of standard reduction potential, the symbol "H^+" is used instead of "H_3O^+."

measured under these conditions are referred to as **standard reduction potentials, E°_{Red}.**

The magnitude of E°_{Red} can be interpreted as the tendency for the reduction half-reaction to occur. The more positive the potential, the greater the tendency for the reaction to occur. The standard reduction potentials found in Table 12.1 and Appendix B.12 are measured relative to a standard hydrogen electrode. The standard hydrogen half-reaction has been assigned a reduction potential of zero.

$$2\,H^+ + 2\,e^- \rightleftharpoons H_2 \qquad E^\circ_{\text{Red}} = 0.000\ldots\text{ V}$$

The E°_{Red} values for the voltaic cell reaction can be found in Table 12.1.

$$Zn^{2+} + 2\,e^- \rightleftharpoons Zn \qquad E^\circ = -0.7628 \text{ V}$$
$$Ag^+ + e^- \rightleftharpoons Ag \qquad E^\circ = +0.7996 \text{ V}$$

Because the silver reduction half-reaction has a much more positive potential (+0.7996 V) than the zinc half-reaction (−0.7628 V), the silver ion is much more likely to undergo the reduction half-reaction. If the silver ion is to be reduced and act as an oxidizing agent, then something must be oxidized. The zinc half-reaction in Table 12.1 is written as a reduction half-reaction and therefore must be reversed to be an oxidation half-reaction. *When a reaction is reversed, the magnitude of its potential remains the same but its sign changes.*

$$Zn \rightleftharpoons Zn^{2+} + 2\,e^- \qquad E^\circ = +0.7628 \text{ V}$$

The oxidation half-reaction now has a positive potential, indicating that it is much more likely to occur than the reduction half-reaction of Zn^{2+} ion. Zinc metal will undergo the oxidation half-reaction. Therefore zinc metal is a stronger reducing agent than silver metal and Ag^+ is a better oxidizing agent than Zn^{2+}.

The same reasoning can be extended to Table 12.1. The strongest reducing agents (substances that tend to undergo an oxidation half-reaction) are at the top of the table. The strongest reducing agent listed in Table 12.1 is potassium metal. The strongest reducing agent is not K^+ ion. The half-reaction is written as a reduction. Since a reducing agent must itself be oxidized, the half-reaction must be reversed.

$$K \rightleftharpoons K^+ + e^- \qquad E^\circ = +2.924 \text{ V}$$

Writing the half-reaction as an oxidation changes the sign of the potential listed in the table and gives a large positive potential. This indicates the strong tendency for this reaction to occur. Potassium is one of the most reactive metals—it reacts violently when added to water, for example. Experimentally we find that neutral metal atoms act as reducing agents and are oxidized in all of their chemical reactions. Thus potassium readily gives up its electron to form K^+ ion.

The strongest oxidizing agent is found at the bottom of Table 12.1.

$$F_2(g) + 2\,H^+ + 2\,e^- \rightleftharpoons 2\,HF(aq) \qquad E^\circ = +3.03 \text{ V}$$

The large positive potential associated with this reduction half-reaction indicates its strong tendency to occur. Fluorine molecules therefore act as oxidizing agents by readily accepting electrons and undergoing this half-reaction. Fluorine is the most electronegative element in the periodic table and therefore has a strong tendency to gain electrons and be reduced.

Exercise 12.3

Arrange the following oxidizing and reducing agents in order of increasing strength.

Reducing agents Cl^-, Cu, H_2, HF, Pb, and Zn
Oxidizing agents Cr^{3+}, $Cr_2O_7^{2-}$, Cu^{2+}, H^+, O_2, O_3, and Na^+

Solution

According to Table 12.1, these reducing agents become stronger in the following order.

$$HF < Cl^- < Cu < H_2 < Pb < Zn$$

The oxidizing agents become stronger in the following order.

$$Na^+ < Cr^{3+} < H^+ < Cu^{2+} < O_2 < Cr_2O_7^{2-} < O_3$$

Exercise 12.4

Use Table 12.1 to predict whether the following oxidation–reduction reactions should occur as written.

(a) $2\ Ag(s) + S(s) \longrightarrow Ag_2S(s)$

(b) $2\ Ag(s) + Cu^{2+}(aq) \longrightarrow 2\ Ag^+(aq) + Cu(s)$

(c) $MnO_4^-(aq) + 3\ Fe^{2+}(aq) + 2\ H_2O(l) \longrightarrow MnO_2(s) + 3\ Fe^{3+}(aq) + 4\ OH^-(aq)$

(d) $MnO_4^-(aq) + 5\ Fe^{2+}(aq) + 8\ H^+(aq) \longrightarrow Mn^{2+}(aq) + 5\ Fe^{3+}(aq) + 4\ H_2O(l)$

Solution

(a) No. The S^{2-} ion of Ag_2S is a better reducing agent than Ag, and the Ag^+ ion of Ag_2S is a better oxidizing agent than S.

(b) No. Cu is a better reducing agent than Ag, and the Ag^+ ion is a better oxidizing agent than Cu^{2+} ion.

(c) No. MnO_4^- in basic solution is not a strong enough oxidizing agent to oxidize Fe^{2+} to Fe^{3+}.

(d) Yes. MnO_4^- in acidic solution is a strong enough oxidizing agent to oxidize Fe^{2+} to Fe^{3+}.

12.7 Standard Cell Potentials

The voltaic cell in Figure 12.2 has a potential of 1.5624 V. The concentrations of the Ag^+ ions and Zn^{2+} ions are both 1 *M* in the cell. Since these concentrations are at standard conditions, this potential is referred to as a **standard cell potential, $E°$.** The larger the difference between the oxidizing and reducing strengths of the reactants and products, the larger is the cell potential. To obtain a relatively large cell potential, we have to react a strong reducing agent with a strong oxidizing agent. The overall cell potential for a reaction is the sum of the potentials for the oxidation and reduction half-reactions.

$$E^\circ_{\text{reaction}} = E^\circ_{\text{Ox}} + E^\circ_{\text{Red}}$$

The following values of E°_{Red} for the reduction of the Zn^{2+} ion to zinc metal and the reduction of the Ag^+ ion to silver metal can be found in Table 12.1.

$$Zn^{2+} + 2\,e^- \rightleftharpoons Zn \qquad E^\circ_{\text{Red}} = -0.7628\text{ V}$$
$$Ag^+ + e^- \rightleftharpoons Ag \qquad E^\circ_{\text{Red}} = +0.7996\text{ V}$$

In order to calculate the standard cell potential, we must first determine which species is oxidized and which species is reduced. The reduction potential for strong oxidizing agents (such as the Ag^+ ion) is more positive than the reduction potential for weak oxidizing agents (such as the Zn^{2+} ion). Because the silver reduction half-reaction has a much more positive potential (+0.7996 V) than the zinc half-reaction (−0.7628 V), the Ag^+ ion is much more likely to undergo reduction than the Zn^{2+} ion.

The standard cell potential for the voltaic cell in Figure 12.2 can be calculated from standard half-cell potentials of the two half-reactions.

$$\begin{array}{lr} Zn(s) \rightleftharpoons Zn^{2+}(aq) + 2\,e^- & E^\circ_{\text{Ox}} = +0.7628\text{ V} \\ 2\,Ag^+(aq) + 2\,e^- \rightleftharpoons 2\,Ag(s) & E^\circ_{\text{Red}} = +0.7996\text{ V} \\ \hline Zn(s) + 2\,Ag^+(aq) \rightleftharpoons Zn^{2+}(aq) + 2\,Ag(s) & E^\circ_{\text{reaction}} = +1.5624\text{ V} \end{array}$$

The units of half-cell potentials are volts, not volts per mole or volts per electron. When combining half-reactions, it is only necessary to add the two half-cell potentials. We do not multiply the potentials by the integers used to balance the number of electrons transferred in the reaction. In the case of the silver half-reaction, it was necessary to multiply the half-reaction by 2 in order for the electrons gained by the silver to equal the electrons lost by the zinc. However, the standard half-cell potential, E°_{Red}, was *not* multiplied by 2. The value +0.7996 V was taken directly from Table 12.1.

The *magnitude* of the cell potential is a measure of the driving force behind a reaction. The larger the value of the cell potential, the further the reaction is from equilibrium. The *sign* of the cell potential tells us the direction in which the reaction must shift to reach equilibrium. The fact that E° is positive for the zinc–silver cell tells us that when the system is present at standard conditions, it has to shift to the right to reach equilibrium. Reactions for which E° is positive therefore have equilibrium constants that favor the formation of the products of the reaction. A reaction with a positive E° will occur naturally and is referred to as **spontaneous.**

Exercise 12.5

Use the overall cell potentials to predict which of the following reactions are spontaneous.

(a) $Cu(s) + 2\,Ag^+(aq) \longrightarrow Cu^{2+}(aq) + 2\,Ag(s)$ $\quad E^\circ = 0.46\text{ V}$

(b) $2\,Fe^{3+}(aq) + 2\,Cl^-(aq) \longrightarrow 2\,Fe^{2+}(aq) + Cl_2(g)$ $\quad E^\circ = -0.59\text{ V}$

(c) $2\,Fe^{3+}(aq) + 2\,I^-(aq) \longrightarrow 2\,Fe^{2+}(aq) + I_2(aq)$ $\quad E^\circ = 0.24\text{ V}$

(d) $2\,H_2O_2(aq) \longrightarrow 2\,H_2O(l) + O_2(aq)$ $\quad E^\circ = 1.09\text{ V}$

(e) $Cu(s) + 2\,H^+(aq) \longrightarrow Cu^{2+}(aq) + H_2(g)$ $\quad E^\circ = -0.34\text{ V}$

Solution

Any reaction for which the overall cell potential is positive is spontaneous. Reactions (a), (c), and (d) are therefore spontaneous.

Exercise 12.6

Use the standard cell potential for the following reaction

$$Cu(s) + 2\,H^+(aq) \rightleftharpoons Cu^{2+}(aq) + H_2(g) \qquad E° = -0.34\text{ V}$$

to predict the standard cell potential for the opposite reaction

$$Cu^{2+}(aq) + H_2(g) \rightleftharpoons Cu(s) + 2\,H^+(aq) \qquad E° = ?$$

Solution

Turning the reaction around doesn't change the relative strengths of Cu^{2+} and H^+ ions as oxidizing agents, or copper metal and H_2 as reducing agents. The *magnitude* of the potential therefore must remain the same. But turning the equation around changes the *sign* of the cell potential and can therefore turn an unfavorable reaction into one that is spontaneous, or vice versa. The standard cell potential for the reduction of Cu^{2+} ions by H_2 gas is therefore +0.34 V.

$$Cu^{2+}(aq) + H_2(g) \rightleftharpoons Cu(s) + 2\,H^+(aq) \qquad E° = -(-0.34\text{ V}) = +0.34\text{ V}$$

➤ CHECKPOINT

What happens to the cell potential when the direction of the reaction is reversed?

Exercise 12.7

Use cell potential data to explain why copper metal does not dissolve in a 1 *M* solution of a typical strong acid, such as hydrochloric acid,

$$Cu(s) + 2\,H^+(aq) \not\rightarrow$$

but will dissolve in 1 *M* nitric acid.

$$3\,Cu(s) + 2\,HNO_3(aq) + 6\,H^+(aq) \longrightarrow 3\,Cu^{2+}(aq) + 2\,NO(g) + 4\,H_2O(l)$$

Solution

Copper does not dissolve in a typical strong acid because the overall cell potential for the oxidation of copper metal to Cu^{2+} ions coupled with the reduction of H^+ ions to H_2 is negative.

$$\begin{array}{rcl} Cu \rightleftharpoons Cu^{2+} + 2\,e^- & & E°_{Ox} = -(0.34\text{ V}) \\ 2\,H^+ + 2\,e^- \rightleftharpoons H_2 & & E°_{Red} = 0.000\ldots\text{ V} \\ \hline Cu(s) + 2\,H^+(aq) \rightleftharpoons Cu^{2+}(aq) + H_2(g) & & E° = E°_{Ox} + E°_{Red} = -0.34\text{ V} \end{array}$$

Copper dissolves in nitric acid because the reaction at the cathode now involves the reduction of nitric acid to NO gas, and the potential for that half-reaction is strong enough to overcome the half-cell potential for oxidation of copper metal to Cu^{2+} ions.

$$\begin{array}{rcl} 3\,(Cu \rightleftharpoons Cu^{2+} + 2\,e^-) & & E°_{Ox} = -(0.34\text{ V}) \\ 2\,(HNO_3 + 3\,H^+ + 3\,e^- \rightleftharpoons NO + 2\,H_2O) & & E°_{Red} = 0.96\text{ V} \\ \hline 3\,Cu(s) + 2\,HNO_3(aq) + 6\,H^+(aq) \rightleftharpoons 3\,Cu^{2+}(aq) + 2\,NO(g) + 4\,H_2O(l) & & \\ & & E° = E°_{Ox} + E°_{Red} = 0.62\text{ V} \end{array}$$

12.8 Electrochemical Cells at Nonstandard Conditions: The Nernst Equation

In 1836, a chemistry professor from London named John Frederic Daniell developed the first voltaic cell stable enough to be used as a battery. The following chemical equation describes the chemical reaction of this Daniell cell.

$$Zn(s) + Cu^{2+}(aq) \rightleftharpoons Zn^{2+}(aq) + Cu(s)$$

What happens when the above oxidation–reduction reaction is used to do work?

- The zinc electrode becomes lighter as zinc atoms are oxidized to Zn^{2+} ions, which go into solution.
- The copper electrode becomes heavier as Cu^{2+} ions in the solution are reduced to copper metal that plates out on the copper electrode.
- The concentration of Zn^{2+} ions at the anode increases, and the concentration of the Cu^{2+} ions at the cathode decreases.
- Negative ions flow from the salt bridge toward the anode to balance the charge on the Zn^{2+} ions produced at that electrode.
- Positive ions flow from the salt bridge toward the cathode to compensate for the Cu^{2+} ions consumed in the reaction.

An important property of the cell is missing from the list. Over a period of time, the cell runs down and eventually has to be replaced. Let's assume that our cell is initially a standard cell in which the concentrations of the Zn^{2+} and Cu^{2+} ions are both 1 *M*. As the reaction goes forward—as copper ions are consumed and zinc ions are produced—the driving force behind the reaction must become weaker. Therefore, the cell potential must become smaller.

In Section 12.7, we showed that the magnitude of the cell potential measures the driving force behind a reaction. The cell potential is therefore zero if and only if the reaction is at equilibrium. When the reaction is at equilibrium, there is no net change in the amount of zinc ions or copper ions in the system, so no electrons flow from the anode to the cathode. If there is no longer a net flow of electrons, the cell can no longer do electrical work. Its potential for doing work must therefore be zero.

In 1889 Hermann Walther Nernst showed that the potential for an electrochemical reaction is described by the following equation.

$$E = E^\circ - \frac{RT}{nF} \ln Q_c$$

In the **Nernst equation,** E is the cell potential at some moment in time, E° is the cell potential when the reaction is at standard conditions, R is the ideal gas constant in units of joules per mol-K, T is the temperature in Kelvin, n is the number of moles of electrons transferred in the balanced equation for the reaction, F is the charge on a mole of electrons, and Q_c is the reaction quotient at that moment in time. The symbol "ln" indicates a natural logarithm to the base e, where e is an irrational number equal to 2.71828. . . .

Two factors in the Nernst equation are constants: R and F. The ideal gas constant expressed in joules is 8.314 J/mol-K, and the charge on a mole of electrons can be calculated from Avogadro's number and the charge on a single electron.

$$\frac{6.022137 \times 10^{23}\ e^-}{1\ \text{mol}} \times \frac{1.60217733 \times 10^{-19}\ \text{C}}{1\ e^-} = 96{,}485.31\ \text{C/mol} = F$$

The temperature is usually 25°C. Substituting this information into the Nernst equation gives the following.

$$E = E° - \frac{0.02569}{n} \ln Q_c$$

Three of the remaining factors in this equation are characteristics of a particular reaction: n, $E°$, and Q_c. The standard potential for the Daniell cell is 1.10 V. Two moles of electrons are transferred from zinc metal to Cu^{2+} ions in the balanced equation for the reaction, so n is 2 for the cell. Because we never include the concentrations of solids in either reaction quotient or equilibrium constant expressions, Q_c for the reaction is equal to the concentration of the Zn^{2+} ion divided by the concentration of the Cu^{2+} ion.

$$Q_c = \frac{(Zn^{2+})}{(Cu^{2+})}$$

Substituting what we know about the Daniell cell into the Nernst equation gives the following result, which represents the cell potential for the Daniell cell at 25°C at any moment in time.

$$E = 1.10\ \text{V} - \frac{0.02569}{2} \ln \frac{(Zn^{2+})}{(Cu^{2+})}$$

Figure 12.3 shows a plot of the potential for the Daniell cell as a function of the natural logarithm of the reaction quotient. When the reaction quotient is very small, the cell potential is positive and relatively large. This isn't surprising, because the reaction is far from equilibrium and the driving force behind the reaction should be relatively large. When the reaction quotient is very large, the cell potential is negative. This means that the reaction would have to shift back toward the reactants to reach equilibrium.

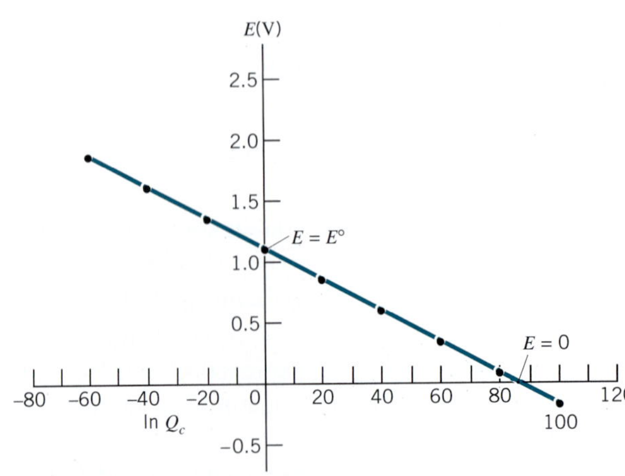

Fig. 12.3 Plot of ln Q_c versus cell potential for the Daniell cell. The cell potential is equal to $E°$ when the cell is at standard conditions. The cell potential is equal to zero if and only if the reaction is at equilibrium.

The most important property of the Nernst equation is the fact that we can use it to calculate the potential of a cell that operates at nonstandard conditions. The Nernst equation can also be used to measure the equilibrium constant for a reaction. To understand how this is done, we have to recognize what happens to the cell potential as an oxidation–reduction reaction comes to equilibrium. Because the reaction is at equilibrium, the reaction quotient is equal to the equilibrium constant ($Q_c = K_c$) and the overall cell potential for the reaction is zero ($E = 0$). At equilibrium, the Nernst equation therefore takes the following form.

$$0 = E^\circ - \frac{RT}{nF} \ln K_c$$

Rearranging the equation gives the following result.

$$\text{At equilibrium} \qquad nFE^\circ = RT \ln K_c$$

According to this equation, we can calculate the equilibrium constant for any oxidation–reduction reaction from its standard cell potential.

$$K_c = e^{nFE^\circ/RT}$$

Exercise 12.8

Calculate the potential at 25°C for the following reaction in which the concentration of Ag^+ is 4.8×10^{-3} *M* and the concentration of Cu^{2+} is 2.4×10^{-2} *M*.

$$Cu(s) + 2\,Ag^+(aq) \rightleftharpoons Cu^{2+}(aq) + 2\,Ag(s)$$

Solution

We start by looking up the standard potentials for the two half-reactions and calculating the standard cell potential.

Oxidation	$Cu \rightleftharpoons Cu^{2+} + 2\,e^-$	$E^\circ_{Ox} = -(0.3402\text{ V})$
Reduction	$Ag^+ + e^- \rightleftharpoons Ag$	$E^\circ_{Red} = 0.7996\text{ V}$
		$E^\circ = E^\circ_{Red} + E^\circ_{Ox} = 0.4594\text{ V}$

We now set up the Nernst equation for the cell, noting that *n* is 2 because two electrons are transferred in the balanced equation for the reaction.

$$E = E^\circ - \frac{0.02569}{2} \ln \left[\frac{(Cu^{2+})}{(Ag^+)^2}\right]$$

Substituting the concentrations of the Cu^{2+} and Ag^+ ions into the equation gives the value for the cell potential.

$$E = 0.4594\text{ V} - \frac{0.02569}{2} \ln \frac{(0.024)}{(0.0048)^2} = 0.37\text{ V}$$

Exercise 12.9

Calculate the equilibrium constant at 25°C for the reaction between zinc metal and acid.

$$Zn(s) + 2\ H^+(aq) \rightleftharpoons Zn^{2+}(aq) + H_2(g)$$

Solution

We start by calculating the overall standard potential for the reaction.

$$\begin{array}{rll} Zn \rightleftharpoons Zn^{2+} + 2\ e^- & E^\circ_{Ox} = -(-0.7628\ V) \\ 2\ H^+ + 2\ e^- \rightleftharpoons H_2 & E^\circ_{Red} = 0.0000\ V \\ \hline Zn(s) + 2\ H^+(aq) \rightleftharpoons Zn^{2+}(aq) + H_2(g) & E^\circ_{cell} = 0.7628\ V \end{array}$$

At equilibrium, $Q_c = K_c$ and $E = 0$.

$$0 = E^\circ - \frac{RT}{nF} \ln K_c$$

We then rearrange the equation

$$nFE^\circ = RT \ln K_c$$

and solve for the natural logarithm of the equilibrium constant.

$$\ln K_c = \frac{nFE^\circ}{RT}$$

Substituting what we know about the reaction into the equation gives the following result. The product of Coulombs time volts is equal to Joules (C × V = J).

$$\ln K_c = \frac{(2)(96{,}485\ C/mol)(0.7628\ V)}{(8.314\ J/mol\text{-}K)(298\ K)} = 59.4$$

The equilibrium constant for the reaction can therefore be calculated by raising e to the power of 59.4.

$$K_c = e^{59.4} = 6 \times 10^{25}$$

This is a very large equilibrium constant, which means that equilibrium lies heavily on the side of the products. The equilibrium constant is so large that the equation for the reaction is written as if it proceeds to completion.

$$Zn(s) + 2\ H^+(aq) \longrightarrow Zn^{2+}(aq) + H_2(g)$$

➤ CHECKPOINT

Would you expect the following reaction run under nonstandard conditions to have a potential greater than, equal to, or less than its standard potential, E°? Explain your reasoning.

$$Ni(s) + Cu^{2+}(aq) \rightleftharpoons Ni^{2+}(aq) + Cu(s)$$

Initial concentrations of Cu^{2+} and Ni^{2+} are 1.5 and 0.010 M, respectively.

12.9 Electrolysis and Faraday's Law

The cells discussed so far have one thing in common: They all use a spontaneous chemical reaction to produce an electric current in an external circuit. These voltaic cells are important because they are the basis for the batteries in modern society. But they are not the only kind of electrochemical cell. It is also possible

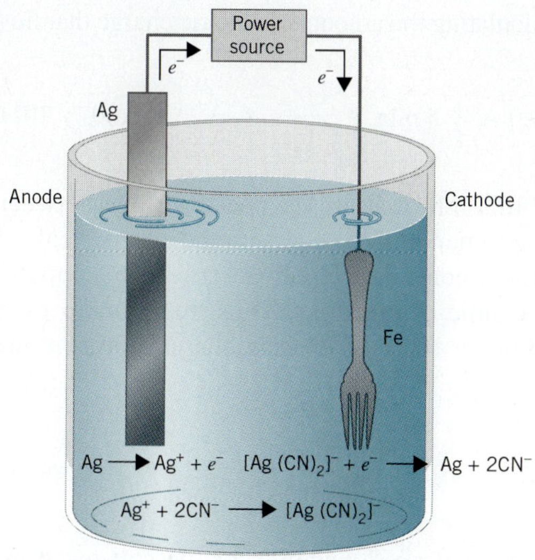

Fig. 12.4 The plating of silver onto a fork.

to construct a cell that does work on a chemical system by driving an electric current through the system. These cells are called **electrolytic cells.** Electrolysis is used to drive an oxidation–reduction reaction in the direction in which it doesn't occur spontaneously.

An application of an electrolytic process is the electroplating of one metal onto another. This is done both to protect against corrosion and to improve appearance. For example, a thin layer of silver can be plated onto jewelry or tableware made from a less expensive metal. Figure 12.4 is a schematic diagram of the plating of silver onto a fork. Oxidation occurs at the silver anode to produce Ag^+ in solution. The Ag^+ reacts with CN^- ion in solution to form the $[Ag(CN)_2]^-$ complex ion. It has been found that the formation of a complex ion for reduction at the cathode produces a more uniform plating than the direct reduction of a metal cation. The $[Ag(CN)_2]^-$ is reduced at the cathode to form a thin layer of silver on the iron fork. The half-reaction at the cathode is given below.

$$[Ag(CN)_2]^- + e^- \rightleftharpoons Ag + 2\,CN^-$$

The relationship between the amount of current passed through a solution and the mass of the substance decomposed or produced by the current was developed by Michael Faraday, a famous nineteenth-century English chemist. **Faraday's law** of electrolysis can be stated as follows.

> **The amount of a substance consumed or produced at one of the electrodes in an electrolytic cell is directly proportional to the amount of electricity that passes through the cell.**

In order to use Faraday's law, we need to recognize the relationship between current, time, and the amount of electric charge that flows through a circuit. By definition, one coulomb of charge is transferred when a 1-ampere (A) current is passed for 1 second.

$$C = A \times s$$

To illustrate how Faraday's law can be used, let's calculate the mass of silver metal that will form at the cathode when a 10.0-A current is passed through the electrolysis cell shown in Figure 12.4 for a period of 5.00 minutes.

We start by calculating the amount of electric charge that flows through the cell.

$$10.0 \text{ A} \times 5 \text{ min} \times \frac{60 \text{ s}}{1 \text{ min}} \times \frac{\text{C}}{\text{A} \cdot \text{s}} = 3.00 \times 10^3 \text{ C}$$

Before we can use this information, we need a bridge between the macroscopic quantity and the phenomenon that occurs on the atomic scale. This bridge is represented by Faraday's constant, which describes the number of coulombs of charge carried by a mole of electrons. Thus, the number of moles of electrons transferred when 3.00×10^3 C of electric charge flows through the cell can be calculated as follows.

$$3.00 \times 10^3 \text{ C} \times \frac{1 \text{ mol } e^-}{96{,}485 \text{ C}} = 3.11 \times 10^{-2} \text{ mol } e^-$$

According to the balanced equation for the half-reaction that occurs at the cathode of the cell, we get 1 mol of silver for every mole of electrons. Thus, we get 3.11×10^{-2} mol, or 3.35 g, of silver in 5.00 minutes.

$$3.11 \times 10^{-2} \text{ mol Ag} \times \frac{107.9 \text{ g Ag}}{1 \text{ mol Ag}} = 3.35 \text{ g Ag}$$

Exercise 12.10

Calculate the volume of H_2 gas at 25°C and 1.00 atm that will collect at the cathode when water is electrolyzed for 2.00 h with a 10.0-A current.

$$2\,H_2O(l) \xrightarrow{\text{Electrolysis}} 2\,H_2(g) + O_2(g)$$

Solution

We start by calculating the amount of electrical charge that passes through the solution.

$$10.0 \text{ A} \times 2.00 \text{ h} \times \frac{60 \text{ min}}{1 \text{ hr}} \times \frac{60 \text{ s}}{1 \text{ min}} \times \frac{\text{C}}{\text{A} \cdot \text{s}} = 7.20 \times 10^4 \text{ C}$$

We then calculate the number of moles of electrons that carry this charge.

$$7.20 \times 10^4 \text{ C} \times \frac{1 \text{ mol } e^-}{96{,}485 \text{ C}} = 0.746 \text{ mol } e^-$$

The balanced equation for the half-reaction that produces H_2 gas at the cathode indicates that we get a mole of H_2 gas for every 2 mol of electrons.

$$\text{Cathode}(-) \quad 2\,H_2O + 2\,e^- \rightleftharpoons H_2 + 2\,OH^-$$

We therefore get 1 mol of H_2 gas at the cathode for every 2 mol of electrons that flows through the cell.

$$0.746 \text{ mol } e^- \times \frac{1 \text{ mol } H_2}{2 \text{ mol } e^-} = 0.373 \text{ mol } H_2$$

We now have the information we need to calculate the volume of the gas produced in this reaction.

$$V = \frac{nRT}{P} = \frac{(0.373 \text{ mol})(0.08206 \text{ L} \cdot \text{atm/mol-K})(298 \text{ K})}{1.00 \text{ atm}} = 9.12 \text{ L}$$

We can extend the general pattern outlined in this section to answer questions that might seem impossible at first glance.

Exercise 12.11

Determine the oxidation number of the chromium in an unknown salt if electrolysis of a molten sample of the salt for 1.50 h with a 10.0-A current deposits 9.71 grams of chromium metal at the cathode.

Solution

We start, as before, by calculating the number of moles of electrons that passed through the cell during electrolysis.

$$10.0 \text{ A} \times 1.50 \text{ h} \times \frac{60 \text{ min}}{1 \text{ h}} \times \frac{60 \text{ s}}{1 \text{ min}} \times \frac{\text{C}}{\text{A} \cdot \text{s}} = 5.40 \times 10^4 \text{ C}$$

$$5.40 \times 10^4 \text{ C} \times \frac{1 \text{ mol } e^-}{96{,}485 \text{ C}} = 0.560 \text{ mol } e^-$$

Since we don't know the balanced equation for the reaction at the cathode in this cell, it doesn't seem obvious how we are going to use this information. We therefore write down this result in a conspicuous location, then go back to the original statement of the question to see what else can be done.

The problem tells us the mass of chromium deposited at the cathode. We need to transform the mass into the number of moles of chromium metal generated.

$$9.71 \text{ g Cr} \times \frac{1 \text{ mol Cr}}{51.996 \text{ g Cr}} = 0.187 \text{ mol Cr}$$

We now know the number of moles of chromium metal produced and the number of moles of electrons it took to produce the metal. We might therefore look at the relationship between the moles of electrons consumed in the reaction and the moles of chromium produced.

$$\frac{0.560 \text{ mol } e^-}{0.187 \text{ mol Cr}} = \frac{3}{1}$$

Three moles of electrons are consumed for every mole of chromium metal produced. The only way to explain this is to assume that the net reaction at the cathode involves reduction of Cr^{3+} ions to chromium metal.

$$\text{Cathode}(-) \quad Cr^{3+} + 3\,e^- \rightleftharpoons Cr$$

Thus, the oxidation number of chromium in the unknown salt must be +3.

Key Terms

Anion
Anode
Cathode
Cation
Cell potential
Corrosion
Electrolytic cells
Faraday's law
Galvanic cell
Half-reaction
Nernst equation
Oxidation
Oxidation number
Oxidation state
Oxidizing agent
Photosynthesis
Redox reactions
Reducing agent
Reduction
Salt bridge
Spontaneous
Standard cell potentials
Standard conditions
Standard reduction potential
Voltaic cell

Problems

Common Oxidation–Reduction Reactions

1. Define *oxidation* and *reduction*.
2. What is a half-reaction?
3. Which of the following are oxidation half-reactions and which are reduction half-reactions?
   (a) $Zn^{2+}(aq) + 2\,e^- \rightleftharpoons Zn(s)$
   (b) $Ag(s) \rightleftharpoons Ag^+(aq) + e^-$
   (c) $Cu^{2+}(aq) + 2\,e^- \rightleftharpoons Cu(s)$
   (d) $Al(s) \rightleftharpoons Al^{3+}(aq) + 3\,e^-$
4. Which of the following reactions contain both an oxidation and a reduction?
   (a) $Zn^{2+}(aq) + Cu(s) \rightleftharpoons Cu^{2+}(aq) + Zn(s)$
   (b) $Zn^{2+}(aq) + Cu^{2+}(aq) \rightleftharpoons Cu(s) + Zn(s)$
   (c) $2\,Ag^+(aq) + Zn(s) \rightleftharpoons Zn^{2+}(aq) + 2\,Ag(s)$
   (d) $Al^{3+}(aq) + Ag^+(aq) \rightleftharpoons Al(s) + Ag(s)$

Determining Oxidation Numbers

5. Find the oxidation number of chromium in the following compounds and ions.
   (a) CrO_4^{2-} (b) $Cr_2O_7^{2-}$ (c) CrO_2
6. Calculate the oxidation number of the aluminum atom in the following compounds and ions.
   (a) $Al(OH)_3$ (b) $AlCl_3$
   (c) $AlCl_4^-$ (d) $Al_2(SO_4)_3$
7. Which of the following compounds or ions contain hydrogen with a negative oxidation number?
   (a) HCl (b) NH_4^+ (c) LiH
   (d) C_2H_6 (e) H_2O
8. Arrange the following compounds in order of increasing oxidation number of the carbon atom.
   (a) C (b) CO
   (c) CO_2 (d) H_2CO
   (e) CH_3OH (f) CH_4
9. Carbon can have any oxidation number between −4 and +4. Calculate the oxidation number of carbon in the following compounds.

   (a) $CH_3{-}CH_2{-}\overset{\overset{\displaystyle O}{\|}}{C}{-}H$

   (b) $CH_3{-}\underset{\underset{\displaystyle CH_3}{|}}{\overset{\overset{\displaystyle CH_3}{|}}{C}}{-}OH$

   (c) 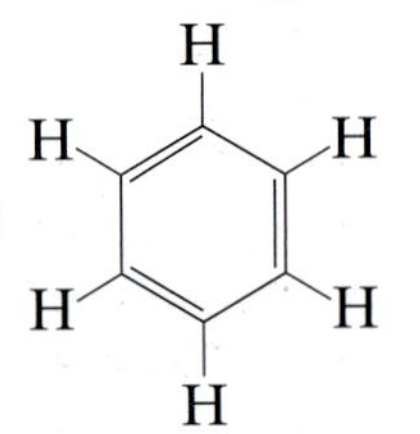 (C_6H_6)

10. Sulfur can have any oxidation number between +6 and −2. Calculate the oxidation number of sulfur in the following compounds.
    (a) Na_2S (b) MgS (c) CS_2
    (d) H_2SO_4 (e) SF_6 (f) $BaSO_3$
    (g) $NaHSO_4$ (h) SCl_2
11. Calculate the oxidation number of each atom in the following compounds.

    (a) $CH_3{-}\overset{\overset{\displaystyle O}{\|}}{C}{-}NH_2$

    (b) $CH_2{=}CH{-}CH_2{-}OH$

    (c) $CH_3{-}CH_2{-}\overset{\overset{\displaystyle O}{\|}}{C}{-}O{-}CH_3$

12. What is the relationship between the oxidation number of chlorine in each of the following acids and the relative strength of the acids? $HClO_4$, $HClO_3$, $HClO_2$, HClO.

Recognizing Oxidation–Reduction Reactions

13. Which statement(s) correctly describes the following reaction?

$$3\ Sn^{2+}(aq) + Cr_2O_7^{2-}(aq) + 14\ H^+(aq) \rightleftharpoons 3\ Sn^{4+}(aq) + 2\ Cr^{3+}(aq) + 7\ H_2O(l)$$

(a) Both the Sn^{2+} and H^+ ions are reduced.
(b) The $Cr_2O_7^{2-}$ ion is reduced.
(c) The Sn^{2+} ion is reduced.
(d) None of the above are true.

14. Hydronium ion participates in reactions other than acid–base reactions. Which of the following involves oxidation or reduction of H_3O^+?

(a) $Mg(s) + 2\ H_3O^+(aq) \rightleftharpoons Mg^{2+}(aq) + H_2(g) + 2\ H_2O(l)$

(b) $H_3O^+(aq) + NH_3(aq) \rightleftharpoons NH_4^+(aq) + H_2O(l)$

15. Oxidation–reduction reactions are often used to prepare organic compounds. Determine if the organic reactants are oxidized or reduced. The reactions are *not* written as complete balanced reactions. They show only the major organic reactant and product.

(a) $CH_3OH(aq) \xrightarrow{K_2Cr_2O_7,\ H^+} H\overset{\overset{O}{\|}}{C}H(aq)$

(b) $CH_3CH_2CH_2OH(aq) \xrightarrow{K_2Cr_2O_7,\ H^+} CH_3CH_2\overset{\overset{O}{\|}}{C}OH(aq)$

16. Decide whether each of the following reactions involves oxidation–reduction. If it does, identify what is oxidized and what is reduced.

(a) $4\ CH_3\overset{\overset{O}{\|}}{C}CH_3 + LiAlH_4 + 4\ H_2O \rightleftharpoons 4\ CH_3\overset{\overset{OH}{|}}{C}HCH_3 + LiOH + Al(OH)_3$

(b) $CH_3CH_2OH \xrightarrow{H_2SO_4} CH_2{=}CH_2 + H_2O$

(c) $CH_3\overset{\overset{O}{\|}}{C}{-}OH + CH_3NH_2 \rightleftharpoons CH_3\overset{\overset{O}{\|}}{C}{-}O^- + CH_3NH_3^+$

17. Use Lewis structures to explain what happens when CO_2 reacts with water to form carbonic acid, H_2CO_3. Is this an oxidation–reduction reaction?

18. Use a Lewis structure to assign oxidation numbers to all the atoms in SF_6.

19. Use Lewis structures to assign oxidation numbers to all the atoms in the following reaction and show that this reaction is a reduction half-reaction.

$$NO_3^-(aq) + 4\ H^+(aq) + 3\ e^- \rightleftharpoons NO(g) + 2\ H_2O(l)$$

20. Assign oxidation numbers to all the atoms in the following reaction using the methods introduced in Section 5.14. Use Lewis structures to assign oxidation numbers to all atoms.

$$2\ SO_3^{2-}(aq) + O_2(g) \rightleftharpoons 2\ SO_4^{2-}(aq)$$

21. Organic chemists often think about oxidation in terms of the loss of a pair of hydrogen atoms. Ethyl alcohol, CH_3CH_2OH, is oxidized to acetaldehyde, CH_3CHO, for example, by removing a pair of hydrogen atoms. Show that this reaction obeys the general rule that oxidation is any process in which the oxidation number of an atom increases.

22. Organic chemists often think about reduction in terms of the gain of a pair of hydrogen atoms. Ethylene, C_2H_4, is reduced when it reacts with H_2 to form ethane, C_2H_6, for example. Show that this reaction obeys the general rule that reduction is any process in which the oxidation number of an atom decreases.

23. Step 6 of the citric acid cycle in Figure 12.1 is the conversion of succinate to fumarate. Show that this reaction is an oxidation–reduction reaction.

24. Step 7 in the citric acid cycle in Figure 12.1 is the conversion of fumarate to malate. Is this an oxidation–reduction reaction? Why or why not?

25. One of the first steps in the corrosion of iron is

$$2\ Fe(s) + O_2(aq) + 2\ H_2O(l) \rightleftharpoons 2\ Fe^{2+}(aq) + 4\ OH^-(aq)$$

Is this a redox reaction?

Voltaic Cells

26. Describe the function of a salt bridge in an electrochemical cell. Explain what happens when the salt bridge is removed from the system and why.

27. Describe the relationships between pairs of the following terms: *cathode, anode, cation,* and *anion.*

28. Explain why cations move from the salt bridge into the solution containing the cathode of a voltaic cell.

29. Explain what happens when oxidation occurs at the anode and reduction occurs at the cathode in a voltaic cell.

30. Explain how an oxidation–reduction reaction can occur even though the reactants and products are physically separated from one another.

31. Balance the following reactions:
    (a) $Al(s) + Cr^{3+}(aq) \rightleftharpoons Al^{3+}(aq) + Cr(s)$
    (b) $Fe^{2+}(aq) + I_2(aq) \rightleftharpoons Fe^{3+}(aq) + I^-(aq)$
    (c) $Cr(s) + Fe^{3+}(aq) \rightleftharpoons Cr^{3+}(aq) + Fe^{2+}(aq)$
    (d) $Zn(s) + H^+(aq) \rightleftharpoons Zn^{2+}(aq) + H_2(g)$
32. For each of the reactions in Problem 31, write the corresponding half-reactions.

Oxidizing and Reducing Agents

33. In each of the following reactions, state which species is oxidized and which species is reduced.
    (a) $Al(s) + Cr^{3+}(aq) \longrightarrow Al^{3+}(aq) + Cr(s)$
    (b) $3\,Cr^{2+}(aq) \longrightarrow Cr(s) + 2\,Cr^{3+}(aq)$
    (c) $Fe(s) + Cr^{3+}(aq) \longrightarrow Cr(s) + Fe^{3+}(aq)$
    (d) $3\,H_2(g) + 2\,Cr^{3+}(aq) \longrightarrow 2\,Cr(s) + 6\,H^+(aq)$
    Identify the oxidizing and reducing agents for each reaction.
34. Identify the oxidizing and reducing agents in each of the following reactions.
    (a) $Co(s) + 2\,Cr^{3+}(aq) \longrightarrow Co^{2+}(aq) + 2\,Cr^{2+}(aq)$
    (b) $H_2(g) + Zn^{2+}(aq) \longrightarrow Zn(s) + 2\,H^+(aq)$
    (c) $Sn(s) + 2\,H^+(aq) \longrightarrow Sn^{2+}(aq) + H_2(g)$
    (d) $3\,Na(l) + AlCl_3(l) \longrightarrow 3\,NaCl(l) + Al(l)$
35. If carbon is in the oxidation state +4, can it behave as an oxidizing agent? Explain.
36. Refer to Figure 5.11. Some of the following can serve as oxidizing agents, some as reducing agents, and some as both. Identify each.

Element	*Oxidation State*
C	−4
N	+5
S	+2
Ge	+4
As	+3
Sb	+5
N	−3
I	+7

37. Which of the following can't be an oxidizing agent? (Refer to Figure 5.11.)
    (a) Cl^- (b) Br_2 (c) Fe^{3+}
    (d) Zn (e) CaH_2
38. Which of the following can't be a reducing agent? (Refer to Figure 5.11.)
    (a) H_2 (b) Cl_2 (c) Fe^{3+} (d) Al (e) LiH
39. Which of the following can be both an oxidizing agent and a reducing agent? (Refer to Figure 5.11.)
    (a) H_2 (b) I_2 (c) H_2O_2 (d) P_4 (e) S_8
40. Identify the oxidizing agents on both sides of the following unbalanced equation.

$$S_2O_3^{2-}(aq) + MnO_4^-(aq) \longrightarrow SO_4^{2-}(aq) + Mn^{2+}(aq)$$

41. Why can an atom in its lowest oxidation state act only as a reducing agent?

Relative Strengths of Oxidizing and Reducing Agents

42. Which of the following transition metals is the strongest reducing agent?
    (a) Cr (b) Mn (c) Fe (d) Co (e) Ni
43. Which of the following oxidation–reduction reactions should occur as written?
    (a) $Co(s) + 2\,Cr^{3+}(aq) \longrightarrow Co^{2+}(aq) + 2\,Cr^{2+}(aq)$
    (b) $H_2(g) + Zn^{2+}(aq) \longrightarrow Zn(s) + 2\,H^+(aq)$
    (c) $Sn(s) + 2\,H^+(aq) \longrightarrow Sn^{2+}(aq) + H_2(g)$
44. Which of the following oxidation–reduction reactions should occur as written? See Appendix B.12.
    (a) $Pb(s) + Mn^{2+}(aq) \longrightarrow Pb^{2+}(aq) + Mn(s)$
    (b) $6\,Fe^{2+}(aq) + 14\,H^+(aq) + Cr_2O_7^{2-}(aq) \longrightarrow 6\,Fe^{3+}(aq) + 2\,Cr^{3+}(aq) + 7H_2O(l)$
    (c) $2\,H_2O_2(aq) \longrightarrow 2\,H_2O(l) + O_2(g)$
45. Use Table 12.1 to predict the products of the following oxidation–reduction reactions.
    (a) $CrO_4^-(aq) + H^+(aq) + Fe(s) \longrightarrow$
    (b) $Mg(s) + Cr^{3+}(aq) \longrightarrow$
    (c) $MnO_4^-(aq) + H^+(aq) + Au^+(aq) \longrightarrow$
46. Which of the following pairs of ions will react when in aqueous solution?
    (a) Cr^{2+} and MnO_4^- (b) Fe^{3+} and $Cr_2O_7^{2-}$
    (c) Ag^+ and Cu^{2+} (d) Fe^{2+} and Cr^{3+}
47. Which of the following pairs of ions will react in aqueous solution? See Appendix B.12.
    (a) Na^+ and $S_2O_8^{2-}$ (b) Hg^{2+} and Cl^-
    (c) Cr^{2+} and I_3^-
48. Solutions of the Sn^{2+} ion are unstable because they are slowly oxidized by air to the Sn^{4+} ion. Explain why adding small pieces of tin metal to a solution of the Sn^{2+} can produce a solution with a reasonable shelf life.
49. Which of the following solutions contains the strongest oxidizing agent?
    (a) $CrO_4^{2-}(aq) + H^+(aq)$
    (b) $Cr_2O_7^{2-}(aq) + H^+(aq)$
    (c) $MnO_4^-(aq) + H^+(aq)$
    (d) $F_2(g) + H^+(aq)$
    (e) $O_2(g) + H^+(aq)$
50. What do the following half-cell reduction potentials tell us about the relative strengths of zinc and copper metal as reducing agents? What do they tell us

about the relative strengths of Zn^{2+} and Cu^{2+} ions as oxidizing agents?

$$Zn^{2+} + 2\,e^- \rightleftharpoons Zn \qquad E° = -0.7628\ V$$
$$Cu^{2+} + 2\,e^- \rightleftharpoons Cu \qquad E° = 0.3402\ V$$

51. Describe an experiment that would allow you to determine the relative strengths of copper and iron metal as reducing agents.
52. Which of the following is the strongest reducing agent?
(a) Zn (b) Fe (c) H_2 (d) Cu (e) Ag
53. Which of the following is the strongest oxidizing agent? See Appendix B.12.
(a) Zn^{2+} (b) Zn (c) Sn^{4+} (d) Sn^{2+} (e) Cl_2
54. Which of the following is the strongest reducing agent? See Appendix B.12.
(a) H^+ (b) H_2 (c) H^- (d) H_2O (e) O_2
55. Which is the better reducing agent? See Appendix B.12.
(a) H^- or K (b) Sn or Fe^{2+} (c) Ag^+ or Au^+

Standard Cell Potentials

56. Describe the difference between E and $E°$ for a cell.
57. Describe the conditions under which the cell potential must be determined in order for the measurement to be equal to $E°$.
58. Describe what the magnitude of $E°$ for an oxidation–reduction reaction tells us about the reaction.
59. Explain why cell potentials measure the relative strengths of a pair of oxidizing agents and the relative strengths of a pair of reducing agents but not their absolute strengths.
60. Describe what the sign of $E°$ for an oxidation–reduction reaction tells us about the reaction.
61. Describe what happens to the sign and magnitude of $E°$ for an oxidation–reduction reaction when the direction in which the reaction is written is reversed.
62. Which of the following oxidation–reduction reactions should occur as written when run under standard conditions?
(a) $Al(s) + Cr^{3+}(aq) \longrightarrow Al^{3+}(aq) + Cr(s)$ $E° = 0.966$ V
(b) $3\,Cr^{2+}(aq) \longrightarrow Cr(s) + 2\,Cr^{3+}(aq)$ $E° = 0.33$ V
(c) $Fe(s) + Cr^{3+}(aq) \longrightarrow Cr(s) + Fe^{3+}(aq)$ $E° = -0.70$ V
(d) $3\,H_2(g) + 2\,Cr^{3+}(aq) \longrightarrow 2\,Cr(s) + 6\,H^+(aq)$ $E° = -0.74$
63. Which of the following oxidation–reduction reactions should occur as written when run under standard conditions?
(a) $2\,Fe^{2+}(aq) + H_2O_2(aq) \longrightarrow 2\,Fe^{3+}(aq) + 2\,OH^-(aq)$ $E° = 0.11$ V
(b) $2\,Fe^{2+}(aq) + Cl_2(aq) \longrightarrow 2\,Fe^{3+}(aq) + 2\,Cl^-(aq)$ $E° = 0.588$ V
(c) $2\,Fe^{2+}(aq) + Br_2(aq) \longrightarrow 2\,Fe^{3+}(aq) + 2\,Br^-(aq)$ $E° = 0.317$ V
(d) $2\,Fe^{2+}(aq) + I_2(aq) \longrightarrow 2\,Fe^{3+}(aq) + 2\,I^-(aq)$ $E° = -0.235$ V
64. Use the results of the previous problem to determine the relative strengths of H_2O_2, Cl_2, Br_2, I_2, and the Fe^{3+} ion as oxidizing agents.
65. Predict which of the following reactions should occur spontaneously as written when run under standard conditions.
(a) $Zn(s) + 2\,H^+(aq) \longrightarrow Zn^{2+}(aq) + H_2(g)$
(b) $Cr(s) + 3\,Fe^{3+}(aq) \longrightarrow Cr^{3+}(aq) + 3\,Fe^{2+}(aq)$
(c) $Mn(s) + Mg^{2+}(aq) \longrightarrow Mn^{2+}(aq) + Mg(s)$
66. Predict which of the following reactions should occur spontaneously as written when run under standard conditions. See Appendix B.12.
(a) $5\,Fe^{3+}(aq) + Mn^{2+}(aq) + 4\,H_2O(l) \longrightarrow 5\,Fe^{2+}(aq) + MnO_4^-(aq) + 8\,H^+(aq)$
(b) $3\,Cu(s) + 2\,HNO_3(aq) + 6\,H^+(aq) \longrightarrow 3\,Cu^{2+}(aq) + 2\,NO(g) + 4\,H_2O(l)$
67. Describe what happens inside the voltaic cell that employs the following chemical reaction to produce a potential.

$$Zn(s) + Cu^{2+}(aq) \rightleftharpoons Zn^{2+}(aq) + Cu(s)$$

68. Calculate $E°$ for the following reaction and predict whether the reaction should occur spontaneously as written when run under standard conditions.

$$6\,Fe^{2+}(aq) + Cr_2O_7^{2-}(aq) + 14\,H^+(aq) \longrightarrow 6\,Fe^{3+}(aq) + 2\,Cr^{3+}(aq) + 7\,H_2O(l)$$

69. Calculate $E°$ for the following reaction and predict whether the reaction should occur spontaneously as written when run under standard conditions. See Appendix B.12.

$$Al^{3+}(aq) + Fe(s) \longrightarrow Fe^{3+}(aq) + Al(s)$$

70. Some metals react with water at room temperature, and others do not. Calculate $E°$ for the following reaction and predict whether the reaction should occur spontaneously as written when run under standard conditions. See Appendix B.12.

$$Ca(s) + 2\,H_2O(l) \longrightarrow Ca^{2+}(aq) + 2\,OH^-(aq) + H_2(g)$$

71. Calculate $E°$ for the following reaction to predict whether the Cu^+ ion should spontaneously undergo disproportionation under standard conditions.

$$2\ Cu^+(aq) \longrightarrow Cu(s) + Cu^{2+}(aq)$$

72. Use standard half-cell reduction potentials to predict whether an $Fe^{2+}(aq)$ solution should react with $H^+(aq)$ solution to produce $Fe^{3+}(aq)$ and $H_2(g)$ under standard conditions.
73. Use half-cell potentials to predict which of the following metals will be most reactive with strong acids (H^+).
(a) Al (b) Mg (c) Zn
74. Exercise 12.7 described why Cu metal dissolves in nitric acid but not in hydrochloric acid. Use half-cell potentials to explain why Zn metal will dissolve in both acids.
75. Explain why copper, gold, mercury, platinum, and silver can be found in the metallic state in nature. Explain why Li is not found in the metallic state in nature.

Electrochemical Cells at Nonstandard Conditions: The Nernst Equation

76. Explain why there is only one value of $E°$ for a cell at a given temperature but many different values of E.
77. Explain why the cell potential becomes smaller as the cell comes closer to equilibrium.
78. What does it mean when E for a cell is equal to zero? What does it mean when $E°$ is equal to zero?
79. Which of the following describes an oxidation–reduction reaction at equilibrium?
(a) $E = 0$ (b) $E° = 0$ (c) $E = E°$
(d) $Q_c = 0$ (e) $\ln K_c = 0$
80. Write the Nernst equation for the following reaction and calculate the cell potential, assuming that the Al^{3+} ion concentration is 1.2 M and the Fe^{3+} ion concentration is 2.5 M at 298 K.

$$Al^{3+}(aq) + Fe(s) \rightleftharpoons Al(s) + Fe^{3+}(aq)$$

Calculate the equilibrium constant for this reaction. Is $Fe(s)$ oxidized by Al^{3+}?

81. Calculate $E°$ for the following reaction. See Appendix B.12.

$$2\ Fe^{2+}(aq) + H_2O_2(aq) \rightleftharpoons 2\ Fe^{3+}(aq) + 2\ OH^-(aq)$$

Write the Nernst equation for the reaction and calculate the cell potential for a system in a pH 10 buffer for which all other components are present at standard conditions.

82. Assume that we start with a Daniell cell at standard conditions.

$$Zn(s) + Cu^{2+}(aq) \rightleftharpoons Zn^{2+}(aq) + Cu(s)$$

Calculate the cell potential under the following set of conditions.
(a) Ninety-nine percent of the zinc metal and Cu^{2+} ions have been consumed.
(b) The reaction has reached 99.99% completion.
(c) The reaction has reached 99.9999% completion.
(d) The Cu^{2+} ion concentration is only $1 \times 10^{-8}\ M$.
(e) The reaction has reached equilibrium.

83. Calculate the standard cell potential for the voltaic cell built around the following oxidation–reduction reaction. See Appendix B.12.

$$2\ MnO_4^-(aq) + 5\ H_2O_2(aq) + 6\ H^+(aq) \rightleftharpoons 2\ Mn^{2+}(aq) + 8\ H_2O(l) + 5\ O_2(g)$$

Use the Nernst equation to predict the effect on the cell potential of an increase in the pH of the solution. Predict the effect of an increase in the H_2O_2 concentration.

84. What would be the effect of an increase in temperature on the potential of the Daniell cell? Would it increase, decrease, or remain the same?
85. Predict the standard cell potential for the following reaction at 25°C.

$$Zn(s) + Fe^{2+}(aq) \rightleftharpoons Zn^{2+}(aq) + Fe(s)$$

Calculate the equilibrium constant for this reaction.

86. Calculate the equilibrium constant at 25°C for the Daniell cell. Use the results of the calculation to explain why a single arrow rather than a double-headed arrow could be used in the following equation.

$$Zn(s) + Cu^{2+}(aq) \longrightarrow Zn^{2+}(aq) + Cu(s)$$

Electrolysis and Faraday's Law

87. What is the difference between an electrolytic cell and a voltaic cell?
88. What reactions would take place at the cathode and the anode if molten KF were electrolyzed?
89. Could an electrolytic cell be used to run an electric motor? Explain.
90. Describe an experiment that uses Faraday's law to determine the number of moles of electrons used in an electrolysis reaction.
91. Calculate the mass of silver metal that can be prepared from Ag^+ in solution with 1.0000 C of electric charge.
92. At a current of 1 amp, how many seconds will it take to prepare 1 mol of Na metal from molten NaCl?

93. If a 5.0 A current were used, how long would it take to prepare 1.0 mol of sodium metal? Of magnesium metal? Of aluminum metal?

94. Predict the products of the electrolysis of molten LiBr and calculate the number of grams of each product formed by electrolysis for 1.0 h with a 2.5 A current.

95. Calculate the number of coulombs necessary to produce a metric tonne (1000 kilograms) of Cl_2 gas from Cl^-.

96. Calculate the mass of copper metal deposited after a 2.0 *M* solution of $CuSO_4$ has been electrolyzed for 2.5 h with a 4.5 A current.

97. Calculate the ratio of the mass of O_2 to the mass of H_2 produced when water is electrolyzed for 2.56 h with a 1.34 A current.

98. Under a particular set of conditions, electrolysis of an aqueous $AgNO_3$ solution generated 1.00 g of silver metal. How many grams of I_2 would be produced by the electrolysis of an aqueous solution of NaI under the same conditions?

99. Calculate the ratio of the mass of Br_2 that collects at the anode to the mass of aluminum metal plated out at the cathode when molten $AlBr_3$ is electrolyzed for 10.5 h with a 20.0 A current.

100. Assume that aqueous solutions of the following salts are electrolyzed for 20.0 min with a 10.0 A current. Which solution will deposit the most grams of metal at the cathode?
(a) $ZnCl_2$ (b) $ZnBr_2$ (c) WCl_6
(d) $ScBr_3$ (e) $HfCl_4$

101. Which of the following ionic compounds will give the most grams of metal when a 10.0 A current is passed through molten samples of the salts for 2.0 h?
(a) KCl (b) $CaCl_2$ (c) $ScCl_3$

102. An electric current is passed through a series of three cells filled with aqueous solutions of $AgNO_3$, $Cu(NO_3)_2$, and $Fe(NO_3)_3$. How many grams of each metal will be deposited by a 1.58 A current for 3.54 h?

103. What are the values of *x* and *y* in the formula Ce_xCl_y if electrolysis of the molten salt for 16.5 h with a 1.00 A current deposits 21.6 g of cerium metal?

104. What is the oxidation number of the manganese in an unknown salt if electrolysis of an aqueous solution of the salt for 30 min with a 10.0 A current generates 2.56 g of manganese metal?

105. Gold forms compounds in the +1 and +3 oxidation states. What is the oxidation number of gold in a compound that deposits 1.53 g of gold metal when electrolyzed for 15 min with a 2.50 A current?

Integrated Problems

106. An electrochemical cell is constructed from a Zn/Zn^{2+} electrode and a Cr/Cr^{3+} electrode as shown below. The voltage measured for the cell is 0.02 V. When the switch is closed the chromium electrode increases in mass.

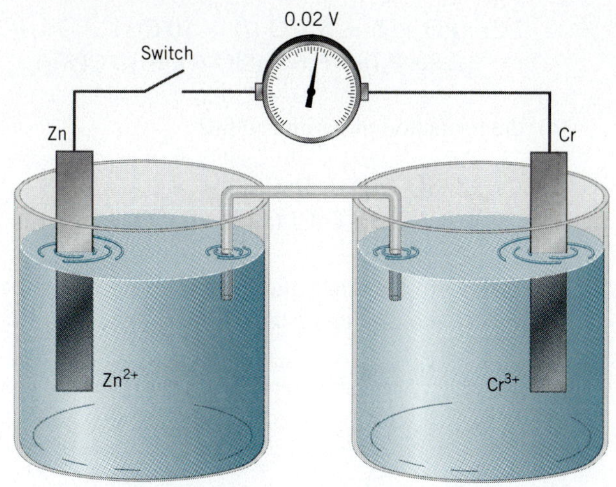

(a) Which electrode is the anode and which is the cathode?
(b) What is the direction of electron flow?
(c) Which electrode is positive? Which electrode is negative?
(d) What are the chemical reactions occurring at each electrode?
(e) What is the overall chemical reaction of the cell?
(f) If the Zn/Zn^{2+} couple is replaced by a standard hydrogen electrode ($2\ H^+(aq) + 2\ e^- \rightleftharpoons H_2(g)$) which is then connected to the Cr/Cr^{3+} couple, what will be the direction of the electron flow and the spontaneous overall chemical reaction?

107. An iron/tin galvanic cell is constructed. An iron strip is placed into a 1.0 *M* solution of $Fe(NO_3)_3$, and this electrode is connected by a salt bridge to an electrode consisting of a strip of tin in a 1.0 *M* $Sn(NO_3)_2$ solution. The tin strip is observed to dissolve.
(a) Roughly sketch the galvanic cell.
(b) Identify the anode and cathode.
(c) Which electrode is positive and which is negative?
(d) What is the overall spontaneous chemical reaction?
(e) What is the standard potential of the cell?

108. The heroine of a movie has been captured by the villains and thrown into a storage room of an old mansion. There are steel bars on the windows, a silver door knob on the door, brass hinges on the skylight, and exposed copper plumbing in the room. In the corner of the room she finds a container of muriatic acid (6 *M* HCl), used for cleaning bricks. Suggest some options for her escape. Explain your reasoning.

109. Which of the following reactions are oxidation–reduction reactions? In each case identify what is oxidized and what is reduced.
(a) the production of iron from its ore

$$Fe_2O_3(s) + 3\ CO(g) \rightleftharpoons 2\ Fe(l) + 3\ CO_2(g)$$

(b) the production of silicon by heating $Ca_3(PO_4)_2$ with carbon and sand

$$2\ Ca_3(PO_4)_2(l) + 6\ SiO_2(l) + 10\ C(s) \rightleftharpoons P_4(g) + 6\ CaSiO_3(l) + 10\ CO(g)$$

(c) the formation of barium sulfate

$$ZnSO_4(aq) + BaCl_2(aq) \rightleftharpoons ZnCl_2(aq) + BaSO_4(s)$$

110. The following electrochemical cell is set up. When the circuit is closed, the tin electrode gains weight.

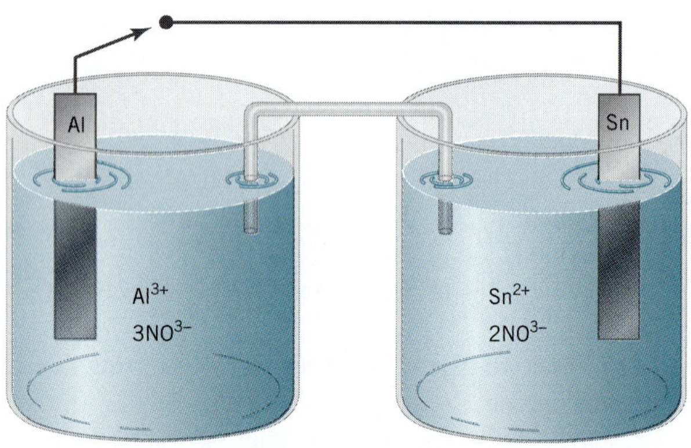

(a) Write an equation that describes the reaction that occurs at the cathode.
(b) Write an equation that describes the reaction that occurs at the anode.
(c) What is the overall cell reaction?
(d) What is the direction of electron flow in the external circuit?
(e) If the aluminum reduction potential is −1.66 V and the cell voltage is 1.52 V, what is the reduction potential for the tin electrode?

111. Which of the following reactions are oxidation–reduction reactions? Explain your reasoning. State what is oxidized and what is reduced.
(a) $CH_4(g) + 2\ O_2(g) \longrightarrow CO_2(g) + 2\ H_2O(g)$
(b) $PbO(s) + CO(g) \longrightarrow Pb(s) + CO_2(g)$
(c) $SiCl_4(l) + 2\ H_2O(l) \longrightarrow 4\ HCl(aq) + SiO_2(s)$
(d) $H_2C{=}CH_2(g) + H_2(g) \longrightarrow CH_3CH_3(g)$

112. Given the following half-cell reduction potentials:

	$E°$(v)
$Ni^{2+}(aq) + 2\ e^- \rightleftharpoons Ni(s)$	−0.23
$Pt^{2+}(aq) + 2\ e^- \rightleftharpoons Pt(s)$	+1.2
$Pd^{2+}(aq) + 2\ e^- \rightleftharpoons Pd(s)$	0.99

(a) Roughly sketch the cell for which the overall cell potential is the greatest.
(b) Identify the cathode, anode, and show the direction of electron flow for the cell in (a).
(c) Will Pt(s) reduce $Pd^{2+}(aq)$? Explain.
(d) What will happen if a strip of Ni(s) is placed into a solution containing $Pt^{2+}(aq)$? Explain.

113. Which of the following statements are true for the reaction

$$5\ CrO_4^{\ 2-}(aq) + 3\ Mn^{2+}(aq) + 16\ H^+(aq) \longrightarrow 5\ Cr^{3+}(aq) + 3\ MnO_4^{\ -}(aq) + 8\ H_2O(l)$$

(a) Both Mn^{2+} and H^+ are oxidizing agents.
(b) $CrO_4^{\ 2-}$ is the oxidizing agent.
(c) Mn^{2+} is reduced.
(d) Cr^{3+} is a reducing agent.
(e) $MnO_4^{\ -}$ is an oxidizing agent.

114. Identify which of the following reactions are acid–base reactions and which are oxidation–reduction reactions. For those redox reactions state, what is oxidized and what is reduced?
(a) $3\ HNO_2(aq) \longrightarrow HNO_3(aq) + 2\ NO(g) + H_2O(l)$
(b) $2\ HNO_3(aq) + Ba(OH)_2(aq) \longrightarrow Ba(NO_3)_2(aq) + 2\ H_2O(l)$
(c) $2\ HBr(aq) + Cl_2(aq) \longrightarrow 2\ HCl(aq) + Br_2(aq)$
(d)

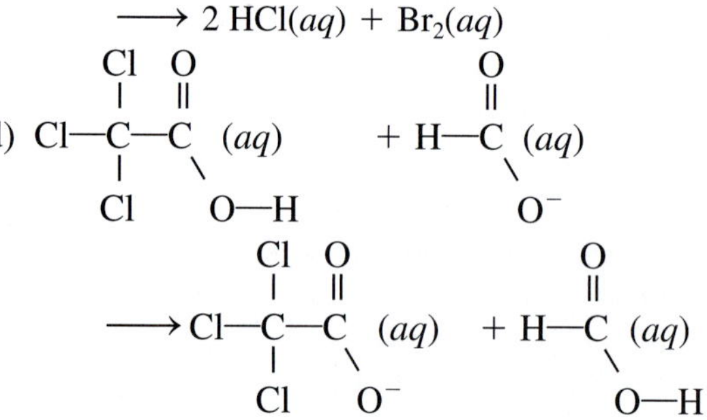

115. The following electrochemical cell is set up.

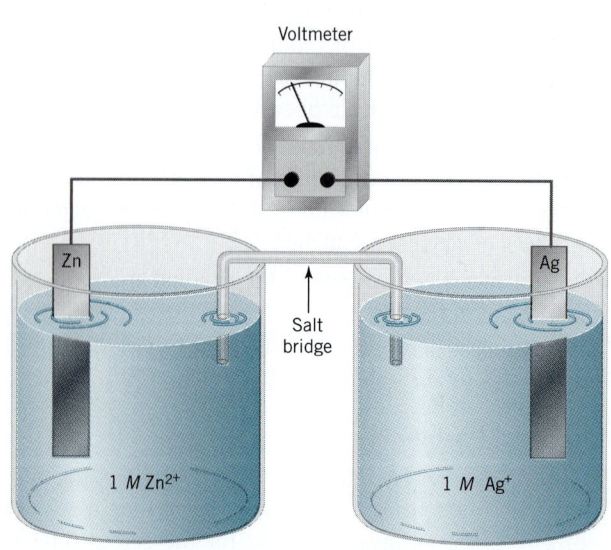

	$E°$(V)
$Zn^{2+}(aq) + 2\,e^- \rightleftharpoons Zn(s)$	−0.76
$Ag^+(aq) + e^- \rightleftharpoons Ag(s)$	+0.80

(a) Which bar, Ag or Zn, will gain weight when the circuit is closed?
(b) What is the voltage of the cell?
(c) Identify the anode and the cathode.
(d) What is the direction of electron flow in the circuit?
(e) What is the overall chemical reaction?
(f) Calculate the equilibrium constant for the overall reaction.

116. The stoichiometry of the reaction between hydrogen peroxide and the MnO_4^- ion depends on the pH of the solution in which the reaction is run.

$$2\,MnO_4^-(aq) + 5\,H_2O_2(aq) + 6\,H^+(aq) \rightleftharpoons 2\,Mn^{2+}(aq) + 5\,O_2(g) + 8\,H_2O(l)$$

$$2\,MnO_4^-(aq) + 3\,H_2O_2(aq) \rightleftharpoons 2\,MnO_2(s) + 3\,O_2(g) + 2\,OH^-(aq) + 2\,H_2O(l)$$

Assume that it took 32.45 mL of 0.145 *M* $KMnO_4$ to titrate 28.46 mL of 0.248 *M* H_2O_2. Was the reaction run in acidic or basic solution?

117. Oxalic acid reacts with the chromate ion in acidic solution to give CO_2 and Cr^{3+} ions.

$$10\,H^+(aq) + 3\,H_2C_2O_4(aq) + 2\,CrO_4^{2-}(aq) \overset{H^+}{\rightleftharpoons} 6\,CO_2(g) + 2\,Cr^{3+}(aq) + 8\,H_2O(l)$$

If 10.0 mL of oxalic acid consumes 40.0 mL of 0.0250 *M* CrO_4^{2-} solution, state the molarity of the oxalic acid solution.

118. A 20.00-mL sample of a $K_2C_2O_4$ solution was acidified and then titrated with an acidic 0.256 *M* $KMnO_4$ solution. What is the molarity of the oxalate solution if it took 14.6 mL of the MnO_4^- solution to reach the end point of the titration?

$$5\,C_2H_2O_4(aq) + 2\,MnO_4^-(aq) + 6\,H^+(aq) \rightleftharpoons 10\,CO_2(g) + 2\,Mn^{2+}(aq) + 8\,H_2O(l)$$

119. Calculate the number of grams of iron (II) chloride ($FeCl_2$) that can be oxidized by 3.2 mL of 3.0 *M* $KMnO_4$.

$$5\,Fe^{2+}(aq) + MnO_4^-(aq) + 8\,H^+(aq) \rightleftharpoons Mn^{2+}(aq) + 4\,H_2O(l) + 5\,Fe^{3+}(aq)$$

Chapter Twelve

SPECIAL TOPICS

12A.1 Balancing Oxidation–Reduction Equations

The special topics include the balancing of redox reactions in acidic and basic solution plus practical applications of oxidation–reduction reactions. The practical applications include batteries, cathodic protection, and electrolysis.

A trial-and-error approach to balancing chemical equations was introduced in Chapter 2. This approach involves playing with the equation—adjusting the ratio of the reactants and products—until the following goals have been achieved.

Goals for Balancing Chemical Equations

- The same number of atoms of each element is found on both sides of the equation, and therefore mass is conserved. (Mass is conserved because atoms are neither created nor destroyed in a chemical reaction.)
- The sum of the positive and negative charges is the same on both sides of the equation, and therefore charge is conserved. (Charge is conserved because electrons are neither created nor destroyed in the reaction.)

At times, an equation is too complex to be solved by trial and error. Consider the following reaction, for example.

$$3\ Cu(s) + 8\ HNO_3(aq) \rightleftharpoons 3\ Cu^{2+}(aq) + 2\ NO(g) + 6\ NO_3^-(aq) + 4\ H_2O\ (l)$$

At other times, more than one balanced equation can be written. The following are just a few of the balanced equations that can be written for the reaction between the permanganate ion and hydrogen peroxide, for example.

$$2\ MnO_4^-(aq) + H_2O_2(aq) + 6\ H^+(aq) \rightleftharpoons 2\ Mn^{2+}(aq) + 3\ O_2(g) + 4\ H_2O(l)$$

$$2\ MnO_4^-(aq) + 3\ H_2O_2(aq) + 6\ H^+(aq) \rightleftharpoons 2\ Mn^{2+}(aq) + 4\ O_2(g) + 6\ H_2O(l)$$

$$2\ MnO_4^-(aq) + 5\ H_2O_2(aq) + 6\ H^+(aq) \rightleftharpoons 2\ Mn^{2+}(aq) + 5\ O_2(g) + 8\ H_2O(l)$$

$$2\ MnO_4^-(aq) + 7\ H_2O_2(aq) + 6\ H^+(aq) \rightleftharpoons 2\ Mn^{2+}(aq) + 6\ O_2(g) + 10\ H_2O(l)$$

Equations such as these have to be balanced by a more systematic approach than trial and error.

12A.2 Redox Reactions in Acidic Solutions

A powerful technique for balancing oxidation–reduction equations involves dividing the reactions into separate oxidation and reduction half-reactions. We then balance the half-reactions, one at a time, and combine them so that electrons are neither created nor destroyed in the reaction.

Consider the oxidation of sulfur dioxide by the dichromate ion in acidic solution.

$$SO_2(aq) + Cr_2O_7^{2-}(aq) \xrightarrow{H^+} SO_4^{2-}(aq) + Cr^{3+}(aq)$$

The reason why this equation is inherently more difficult to balance has nothing to do with the ratio of moles of SO_2 to moles of $Cr_2O_7^{2-}$. It results from the fact that the solvent takes an active role in both half-reactions. Let's therefore use this reaction to illustrate the steps in the half-reaction method for balancing equations.

STEP 1: Write a skeleton equation and assign oxidation numbers to atoms on both sides of the equation. When the rules for assigning oxidation numbers described in Section 5.13 are applied to the skeleton equation for the reaction, we get the following results.

$$\underset{+4\ -2}{SO_2} + \underset{+6\ -2}{Cr_2O_7^{2-}} \longrightarrow \underset{+6\ -2}{SO_4^{2-}} + \underset{+3}{Cr^{3+}}$$

STEP 2: Determine which atoms are oxidized and which are reduced. In the course of the reaction, there is an increase in the oxidation number of the sulfur atom and a decrease in the oxidation number of the chromium. Thus, sulfur is oxidized and chromium is reduced.

$$\underset{+4}{SO_2} + \underset{+6}{Cr_2O_7^{2-}} \longrightarrow \underset{+6}{SO_4^{2-}} + \underset{+3}{Cr^{3+}}$$

Oxidation (+4 → +6)

Reduction (+6 → +3)

STEP 3: Divide the reaction into oxidation and reduction half-reactions and balance the half-reactions. The reaction can be divided into the following half-reactions.

Oxidation	$\underset{+4}{SO_2}$	$\longrightarrow$	$\underset{+6}{SO_4^{2-}}$
Reduction	$\underset{+6}{Cr_2O_7^{2-}}$	$\longrightarrow$	$\underset{+3}{Cr^{3+}}$

It doesn't matter which half-reaction we balance first, so let's start with reduction. Because the $Cr_2O_7^{2-}$ ion contains two chromium atoms that must be reduced from the +6 to the +3 oxidation state, six electrons are consumed in the half-reaction.

$$\text{Reduction} \qquad Cr_2O_7^{2-} + 6\,e^- \longrightarrow 2\,Cr^{3+}$$

What happens to the oxygen atoms when the chromium atoms are reduced? The seven oxygen atoms in the $Cr_2O_7^{2-}$ ion are in the −2 oxidation state. If these atoms were released into solution in the −2 oxidation state when the chromium was reduced, they would be present as O^{2-} ions. But the O^{2-} ion is a very strong base, and the reaction is being run in a strong acid. Any O^{2-} ions liberated when the dichromate ion is reduced would instantly react with the H^+ ions in the acid solution to form water. Because there are seven oxygen atoms in a −2 oxidation state in the dichromate ion, 14 H^+ ions are consumed and seven H_2O molecules are produced in this half-reaction.

$$\text{Reduction} \qquad Cr_2O_7^{2-} + 14\,H^+ + 6\,e^- \rightleftharpoons 2\,Cr^{3+} + 7\,H_2O$$

We can now turn to the oxidation half-reaction, and start by noting that two electrons are given off when sulfur is oxidized from the +4 to the +6 oxidation state.

$$\text{Oxidation} \qquad SO_2 \longrightarrow SO_4^{2-} + 2\,e^-$$

The charge on both sides of this equation is balanced by remembering that the reaction is run in acid, which contains H^+ ions and H_2O molecules. We can therefore add H^+ ions or H_2O molecules to either side of the equation, as needed. The only way to balance the charge on both sides of the equation is to add four H^+ ions to the products of the reaction.

$$\text{Oxidation} \qquad SO_2 \longrightarrow SO_4^{2-} + 2\,e^- + 4\,H^+$$

We then balance the number of hydrogen and oxygen atoms on both sides of the equation by adding a pair of H_2O molecules to the reactants.

$$\text{Oxidation} \qquad SO_2 + 2\,H_2O \rightleftharpoons SO_4^{2-} + 2\,e^- + 4\,H^+$$

STEP 4: Combine the two half-reactions so that electrons are neither created nor destroyed. Six electrons are consumed in the reduction half-reaction and two electrons are given off in the oxidation half-reaction. We can combine the half-reactions so that electrons are conserved by multiplying the oxidation half-reaction by 3.

$$Cr_2O_7^{2-} + 14\,H^+ + 6\,e^- \rightleftharpoons 2\,Cr^{3+} + 7\,H_2O$$
$$3\,(SO_2 + 2\,H_2O \rightleftharpoons SO_4^{2-} + 2\,e^- + 4\,H^+)$$
$$\overline{Cr_2O_7^{2-} + 3\,SO_2 + 14\,H^+ + 6\,H_2O \rightleftharpoons 2\,Cr^{3+} + 3\,SO_4^{2-} + 12\,H^+ + 7\,H_2O}$$

Although the equation appears balanced, we aren't quite finished with it. We can simplify the equation by subtracting 12 H^+ ions and 6 H_2O molecules from each side to generate the following balanced equation. We indicate that the reaction is reversible by writing a double-headed arrow between the left and right sides of the equation.

$$Cr_2O_7^{2-}(aq) + 3\,SO_2(aq) + 2\,H^+(aq) \rightleftharpoons 2\,Cr^{3+}(aq) + 3\,SO_4^{2-}(aq) + H_2O(l)$$

➤ CHECKPOINT

On an exam, students were asked to use half-reactions to write a balanced equation for the reaction between SO_2 and the $Cr_2O_7^{2-}$ ion in the presence of a strong acid.

$$SO_2(g) + Cr_2O_7^{2-}(aq) \xrightarrow{H^+} SO_4^{2-}(aq) + Cr^{3+}(aq)$$

One student wrote the following half-reactions.

$$\text{Oxidation} \quad SO_2 + 4\,OH^- \longrightarrow SO_4^{2-} + 2\,H_2O + 2\,e^-$$
$$\text{Reduction} \quad Cr_2O_7^{2-} + 7\,H_2O + 6\,e^- \longrightarrow 2\,Cr^{3+} + 14\,OH^-$$

The student then wrote the following overall equation.

$$3\,SO_2(g) + Cr_2O_7^{2-}(aq) + H_2O(l) \rightleftharpoons 3\,SO_4^{2-}(aq) + 2\,Cr^{3+}(aq) + 2\,OH^-(aq)$$

Explain why the instructor gave no credit for this answer, even though the overall equation is balanced.

Exercise 12A.1

As we have seen, an endless number of balanced equations can be written for the reaction between the permanganate ion and hydrogen peroxide in acidic solution to form the Mn^{2+} ion and oxygen gas.

$$MnO_4^-(aq) + H_2O_2(aq) \xrightarrow{H^+} Mn^{2+}(aq) + O_2(g)$$

Use the half-reaction method to determine the correct stoichiometry for the reaction.

Solution

STEP 1: Write a skeleton equation and assign oxidation numbers to atoms on both sides of the equation.

$$\underset{+7\;-2}{MnO_4^-} + \underset{+1\;-1}{H_2O_2} \longrightarrow \underset{+2}{Mn^{2+}} + \underset{0}{O_2}$$

STEP 2: Determine which atoms are oxidized and which are reduced.

$$\underset{+7}{\mathrm{MnO_4^-}} + \underset{-1}{\mathrm{H_2O_2}} \longrightarrow \underset{+2}{\mathrm{Mn^{2+}}} + \underset{0}{\mathrm{O_2}}$$

Reduction (+7 → +2); *Oxidation* (−1 → 0)

STEP 3: Divide the reaction into oxidation and reduction half-reactions and balance the half-reactions. This reaction can be divided into the following half-reactions.

$$\text{Oxidation} \qquad \underset{-1}{\mathrm{H_2O_2}} \longrightarrow \underset{0}{\mathrm{O_2}}$$

$$\text{Reduction} \qquad \underset{+7}{\mathrm{MnO_4^-}} \longrightarrow \underset{+2}{\mathrm{Mn^{2+}}}$$

Let's balance the reduction half-reaction first. It takes five electrons to reduce manganese from +7 to +2.

$$\text{Reduction} \qquad \mathrm{MnO_4^-} + 5\,e^- \longrightarrow \mathrm{Mn^{2+}}$$

Because the reaction is run in acid, we can add H^+ ions or H_2O molecules to either side of the equation, as needed. There are two hints that tell us which side of the equation gets H^+ ions and which side gets H_2O molecules. The only way to balance the charge on both sides of the equation is to add eight H^+ ions to the left side of the equation.

$$\text{Reduction} \qquad \mathrm{MnO_4^-} + 8\,\mathrm{H^+} + 5\,e^- \longrightarrow \mathrm{Mn^{2+}}$$

The only way to balance the number of oxygen atoms is to add four H_2O molecules to the right side of the equation.

$$\text{Reduction} \qquad \mathrm{MnO_4^-} + 8\,\mathrm{H^+} + 5\,e^- \rightleftharpoons \mathrm{Mn^{2+}} + 4\,\mathrm{H_2O}$$

To balance the oxidation half-reaction, we have to remove two electrons from a pair of oxygen atoms in the −1 oxidation state to form a neutral O_2 molecule.

$$\text{Oxidation} \qquad \mathrm{H_2O_2} \longrightarrow \mathrm{O_2} + 2\,e^-$$

We can then add a pair of H^+ ions to the products to balance both charge and mass in the half-reaction.

$$\text{Oxidation} \qquad \mathrm{H_2O_2} \rightleftharpoons \mathrm{O_2} + 2\,\mathrm{H^+} + 2\,e^-$$

STEP 4: Combine the two half-reactions so that electrons are neither created nor destroyed. Two electrons are given off during oxidation and five electrons are consumed during reduction. We can combine the half-reactions so that electrons are conserved by using the lowest common multiple of 5 and 2.

$$\begin{array}{r} 2(MnO_4^- + 8\,H^+ + 5\,e^- \rightleftharpoons Mn^{2+} + 4\,H_2O) \\ 5\,(H_2O_2 \rightleftharpoons O_2 + 2\,H^+ + 2\,e^-) \\ \hline 2\,MnO_4^- + 5\,H_2O_2 + 16\,H^+ \rightleftharpoons 2\,Mn^{2+} + 5\,O_2 + 10\,H^+ + 8\,H_2O \end{array}$$

The simplest balanced equation for the reaction is obtained when 10 H^+ ions are subtracted from each side of the equation derived in the previous step.

$$2\,MnO_4^-(aq) + 5\,H_2O_2(aq) + 6\,H^+(aq) \rightleftharpoons 2\,Mn^{2+}(aq) + 5\,O_2(g) + 8\,H_2O(l)$$

12A.3 Redox Reactions in Basic Solutions

The key to balancing half-reactions in basic solutions is to recognize that basic solutions contain H_2O molecules and OH^- ions. We can therefore add water molecules or hydroxide ions to either side of the equation, as needed.

In the previous section, we obtained the following equation for the reaction between the permanganate ion and hydrogen peroxide in an acidic solution.

$$2\,MnO_4^-(aq) + 5\,H_2O_2(aq) + 6\,H^+(aq) \rightleftharpoons 2\,Mn^{2+}(aq) + 5\,O_2(g) + 8\,H_2O(l)$$

It might be interesting to see what happens when the reaction occurs in a basic solution.

Exercise 12A.2

Write a balanced equation for the reaction between the permanganate ion and hydrogen peroxide in a basic solution to form manganese dioxide and oxygen.

$$MnO_4^-(aq) + H_2O_2(aq) \xrightarrow{OH^-} MnO_2(s) + O_2(g)$$

Solution

STEP 1: Write a skeleton equation and assign oxidation numbers to atoms on both sides of the equation.

$$\underset{+7\ -2}{MnO_4^-} + \underset{+1\ -1}{H_2O_2} \longrightarrow \underset{+4\ -2}{MnO_2} + \underset{0}{O_2}$$

STEP 2: Determine which atoms are oxidized and which are reduced.

$$\underset{+7}{MnO_4^-} + \underset{-1}{H_2O_2} \longrightarrow \underset{+4}{MnO_2} + \underset{0}{O_2}$$

Reduction (+7 → +4)

Oxidation (−1 → 0)

STEP 3: Divide the reaction into oxidation and reduction half-reactions and balance the half-reactions. The reaction can be divided into the following half-reactions.

$$\text{Reduction} \quad \underset{+7}{MnO_4^-} \longrightarrow \underset{+4}{MnO_2}$$

$$\text{Oxidation} \quad \underset{-1}{H_2O_2} \longrightarrow \underset{0}{O_2}$$

Let's start by balancing the reduction half-reaction. It takes three electrons to reduce manganese from +7 to the +4 oxidation state.

$$\text{Reduction} \quad MnO_4^- + 3\,e^- \longrightarrow MnO_2$$

We now try to balance the number of atoms and the charge on both sides of the equation. Because the reaction is run in basic solution, we can add either OH^- ions or H_2O molecules to either side of the equation, as needed. The only way to balance the net charge of −4 on the left side of the equation is to add four OH^- ions to the products.

$$\text{Reduction} \quad MnO_4^- + 3\,e^- \longrightarrow MnO_2 + 4\,OH^-$$

We can then balance the number of hydrogen and oxygen atoms by adding two H_2O molecules to the reactants.

$$\text{Reduction} \quad MnO_4^- + 3\,e^- + 2\,H_2O \rightleftharpoons MnO_2 + 4\,OH^-$$

We now turn to the oxidation half-reaction. Two electrons are lost when hydrogen peroxide is oxidized to form O_2 molecules.

$$\text{Oxidation} \quad H_2O_2 \longrightarrow O_2 + 2\,e^-$$

We can balance the charge on the half-reaction by adding a pair of OH^- ions to the reactants.

$$\text{Oxidation} \quad H_2O_2 + 2\,OH^- \longrightarrow O_2 + 2\,e^-$$

The only way to balance the number of hydrogen and oxygen atoms is to add two H_2O molecules to the products.

$$\text{Oxidation} \quad H_2O_2 + 2\,OH^- \rightleftharpoons O_2 + 2\,H_2O + 2\,e^-$$

STEP 4: Combine the two half-reactions so that electrons are neither created nor destroyed. Two electrons are given off during the oxidation half-reaction and three electrons are consumed in the reduction half-reaction. We can therefore combine the half-reactions by using the lowest common multiple of 2 and 3.

$$2\,(MnO_4^- + 3\,e^- + 2\,H_2O \rightleftharpoons MnO_2 + 4\,OH^-)$$

$$3\,(H_2O_2 + 2\,OH^- \rightleftharpoons O_2 + 2\,H_2O + 2\,e^-)$$

$$2\,MnO_4^- + 3\,H_2O_2 + 6\,OH^- + 4\,H_2O \rightleftharpoons 2\,MnO_2 + 3\,O_2 + 8\,OH^- + 6\,H_2O$$

The simplest balanced equation is obtained by subtracting four H_2O molecules and six OH^- ions from each side of the equation derived in the previous step.

$$2\,MnO_4^-(aq) + 3\,H_2O_2(aq) \rightleftharpoons 2\,MnO_2(s) + 3\,O_2(g) + 2\,OH^-(aq) + 2\,H_2O(l)$$

Note that the ratio of moles of MnO_4^- to moles of H_2O_2 consumed is different in acidic and basic solutions. This difference results from the fact that MnO_4^- is reduced all the way to Mn^{2+} in acid, but the reaction stops at MnO_2 in base.

12A.4 Molecular Redox Reactions

Lewis structures can play a vital role in understanding oxidation–reduction reactions of covalent molecules. Consider the following reaction, for example, which is used in the Breathalyzer to determine the amount of ethyl alcohol or ethanol on the breath of individuals who are suspected of driving while under the influence.

$$3\,CH_3CH_2OH(g) + 2\,Cr_2O_7^{2-}(aq) + 16\,H^+(aq) \rightleftharpoons 3\,CH_3CO_2H(aq) + 4\,Cr^{3+}(aq) + 11\,H_2O(l)$$

We start by assigning oxidation numbers to each of the atoms in the Lewis structures shown in Figure 12A.1.

The carbon in the CH_3—group has the same oxidation state in both ethanol and acetic acid. There is a change in the oxidation number of the other carbon atom, however, from −1 to +3. The oxidation half-reaction therefore corresponds to the loss of four electrons by one of the carbon atoms.

$$\text{Oxidation} \qquad CH_3CH_2OH \longrightarrow CH_3CO_2H + 4\,e^-$$

Because the reaction is run in acidic solution, we can add H^+ and H_2O as needed to balance the equation.

$$\text{Oxidation} \qquad CH_3CH_2OH + H_2O \rightleftharpoons CH_3CO_2H + 4\,e^- + 4\,H^+$$

The other half of the reaction involves a six-electron reduction of the $Cr_2O_7^{2-}$ ion in acidic solution to form a pair of Cr^{3+} ions.

$$\text{Reduction} \qquad Cr_2O_7^{2-} + 6\,e^- \longrightarrow 2\,Cr^{3+}$$

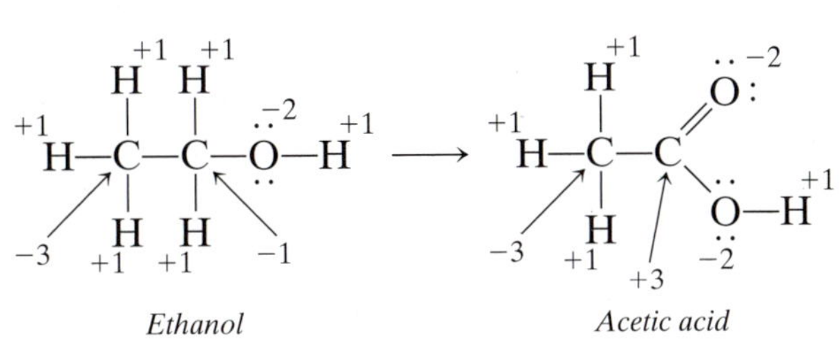

Fig. 12A.1 Oxidation of ethanol to acetic acid.

Adding H^+ ions and H_2O molecules as needed gives the following balanced equation for the reduction half-reaction.

$$\text{Reduction} \qquad Cr_2O_7^{2-} + 14\,H^+ + 6\,e^- \rightleftharpoons 2\,Cr^{3+} + 7\,H_2O$$

We are now ready to combine the two half-reactions by recognizing that electrons are neither created nor destroyed in the reaction.

$$3\,(CH_3CH_2OH + H_2O \rightleftharpoons CH_3CO_2H + 4\,e^- + 4\,H^+)$$
$$2\,(Cr_2O_7^{2-} + 14\,H^+ + 6\,e^- \rightleftharpoons 2\,Cr^{3+} + 7\,H_2O)$$
$$3\,CH_3CH_2OH + 2\,Cr_2O_7^{2-} + 28\,H^+ + 3\,H_2O \rightleftharpoons 3\,CH_3CO_2H + 4\,Cr^{3+} + 12\,H^+ + 14\,H_2O$$

Simplifying the equation by removing 3 H_2O molecules and 12 H^+ ions from both sides of the equation gives the balanced equation for the reaction.

$$3\,CH_3CH_2OH(g) + 2\,Cr_2O_7^{2-}(aq) + 16\,H^+(aq) \rightleftharpoons 3\,CH_3CO_2H(aq) + 4\,Cr^{3+}(aq) + 11\,H_2O(l)$$

Exercise 12A.3

Methyllithium (CH_3Li) can be used to form bonds between carbon and either main-group metals or transition metals.

$$HgCl_2(s) + 2\,CH_3Li(l) \longrightarrow Hg(CH_3)_2(l) + 2\,LiCl(s)$$
$$WCl_6(s) + 6\,CH_3Li(l) \longrightarrow W(CH_3)_6(l) + 6\,LiCl(s)$$

It can also be used to form bonds between carbon and other nonmetals

$$PCl_3(l) + 3\,CH_3Li(l) \longrightarrow P(CH_3)_3(l) + 3\,LiCl(s)$$

or between carbon atoms.

$$CH_3Li(l) + H_2CO(g) \rightleftharpoons [CH_3CH_2OLi] \xrightarrow{H^+} CH_3CH_2OH(l)$$

Use Lewis structures to explain the stoichiometry of the oxidation–reduction reaction used to synthesize methyllithium.

$$CH_3Br(l) + 2\,Li(s) \rightleftharpoons CH_3Li(l) + LiBr(s)$$

Solution

The active metals are reducing agents in all of their chemical reactions. In this reaction, lithium metal reduces a carbon atom from an oxidation state of −2 to −4, as shown in Figure 12A.2. The lithium is oxidized from a zero oxidation state in its

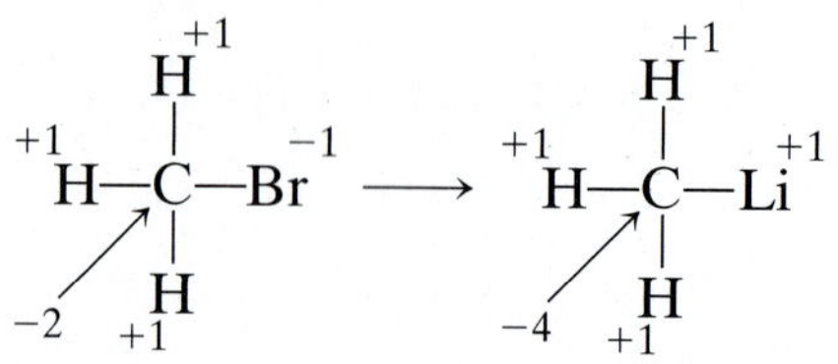

Fig. 12A.2 Reduction of CH_3Br with lithium metal.

elemental form to +1 in methyllithium. Therefore, the carbon atom has gained two electrons and each lithium atom has lost one electron. Thus, 2 mol of lithium is consumed for each mole of CH_3Br in the reaction. One of the Li^+ ions formed in the reaction combines with the negatively charged carbon atom to form methyllithium, and the other precipitates from solution with the Br^- ion as LiBr.

12A.5 Batteries

The term *battery* is used to describe a set of similar or connected items, such as an artillery battery or a battery of telephones. How, then, did it also come to mean a device that converts chemical energy into electrical energy? The first voltaic cell stable enough to be used in a battery was based on the following reaction between zinc metal and Cu^{2+} ions, and it produced a standard-state cell potential of 1.10 V.

$$\mathrm{Zn}(s) + \mathrm{Cu}^{2+}(aq) \rightleftharpoons \mathrm{Zn}^{2+}(aq) + \mathrm{Cu}(s)$$

When a number of these cells are connected in series, we get a "battery" of cells with a voltage equal to 1.10 V times the number of cells.

Until recently, the market for disposable batteries was dominated by batteries based on the *Leclanché cell,* which was introduced in 1860. The cathode (electrode where reduction occurs) in the cell is a mixture of solid MnO_2 and carbon. The anode (electrode where oxidation occurs) is a sheet of zinc metal that is amalgamated with a trace of mercury. The electrolytic solution is a mixture of NH_4Cl and $ZnCl_2$. The cell reactions that occur during the discharge of a manganese dioxide–zinc primary battery can be written as follows.

Cathode (+)	$3\,\mathrm{MnO_2} + 2\,\mathrm{H_2O} + 4\,e^- \rightleftharpoons \mathrm{Mn_3O_4} + 4\,\mathrm{OH^-}$
Anode (−)	$\mathrm{Zn} + 2\,\mathrm{OH^-} \rightleftharpoons \mathrm{ZnO} + \mathrm{H_2O} + 2\,e^-$
Overall reaction	$2\,\mathrm{Zn}(s) + 3\,\mathrm{MnO_2}(s) \rightleftharpoons \mathrm{Mn_3O_4}(s) + 2\,\mathrm{ZnO}(s)$

The initial cell potential of a single dry cell is 1.54 V. The popularity of the Leclanché cell is based on the fact that it is relatively inexpensive, available in many voltages and sizes, and suitable for intermittent use.

In 1949, the first "alkaline" dry cell battery was produced. This cell also uses a mixture of MnO_2 and carbon as the cathode and amalgamated zinc as the anode. However, KOH is used as the electrolyte. Alkaline dry cells (see Figure 12A.3) have several advantages. They can be used over a wider range of temperatures because the electrolyte is more stable. They also require very little electrolyte and can therefore be very compact. More importantly, the batteries maintain a constant voltage for a longer period of time and therefore last longer.

LEAD–ACID BATTERY

The lead storage battery used in cars was first demonstrated to the French Academy of Sciences by Gaston Planté in 1860. It contained nine cells in parallel and was able to deliver extraordinarily large currents. The cells were constructed from lead plates separated by layers of flannel that were immersed in a 10% H_2SO_4 solution. As might be expected, the system soon became known as the *lead–acid battery*.

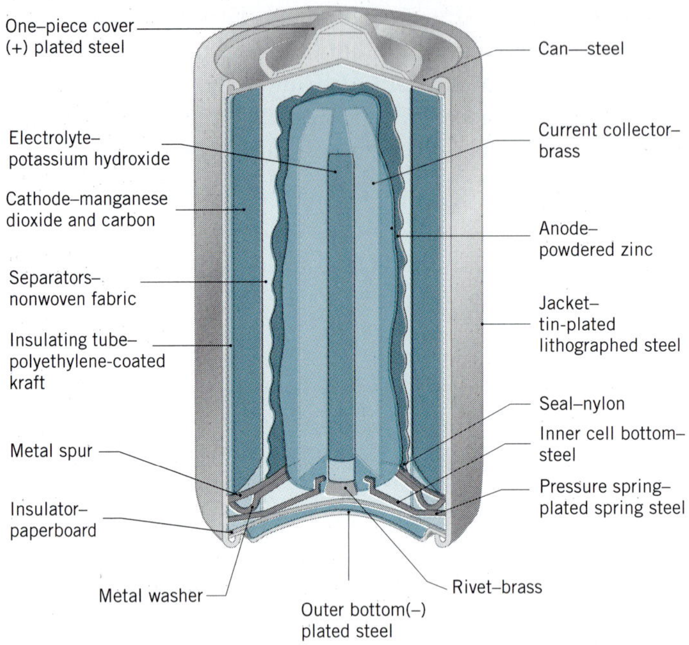

Fig. 12A.3 Cutaway drawing of a typical alkaline battery.

In spite of many studies of the charging and discharging of lead electrodes in acid solution in the last 140 years, there is still some doubt about the exact mechanism of the reactions in the lead–acid storage battery, shown in Figure 12A.4. The most common model suggests that the following half-reaction occurs at the PbO_2 plate.

$$PbO_2(s) + 4\,H^+(aq) + SO_4^{2-}(aq) + 2\,e^- \underset{charge}{\overset{discharge}{\rightleftharpoons}} 2\,H_2O(l) + PbSO_4(aq)$$

When the battery discharges, PbO_2 is reduced to $PbSO_4$. When it is charged, the reaction is reversed. The standard electrode potential for the half-reaction is 1.682 V.

The half-reaction at the lead metal plate can be described by the following equation, for which the standard potential is 0.3858 V.

$$Pb(s) + SO_4^{2-}(aq) \underset{charge}{\overset{discharge}{\rightleftharpoons}} PbSO_4(aq) + 2\,e^-$$

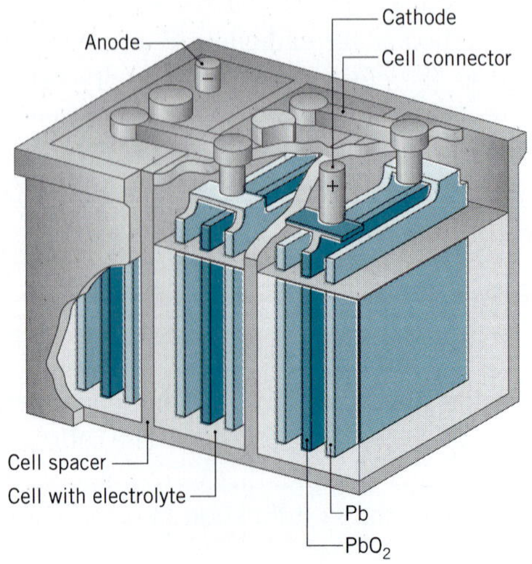

Fig. 12A.4 Cutaway drawing of a typical lead–acid battery.

The total cell reaction can therefore be written as follows.

$$PbO_2(s) + Pb(s) + 4\,H^+(aq) + 2\,SO_4^{2-}(aq) \underset{charge}{\overset{discharge}{\rightleftharpoons}} 2\,PbSO_4(aq) + 2\,H_2O(l)$$

The magnitude of the standard potential for a single cell in lead–acid batteries is slightly larger than 2 V, although the sign of the potential depends on whether the battery is being charged or discharged. The typical 12-V lead storage battery therefore contains six 2-V cells.

During the charging of a lead–acid storage battery, water can be decomposed into its elements.

$$2\,H_2O(l) \rightleftharpoons 2\,H_2(g) + O_2(g)$$

Lead storage batteries therefore represent a potential threat of hydrogen explosions. Because such explosions can spray the 10% sulfuric acid electrolyte onto the individual who is working on the battery, safety goggles should be worn when working with these batteries.

NICAD BATTERIES

Another battery that has become increasingly popular is the nickel–cadmium or *nicad battery*. This type of battery is commonly used in watches and calculators. The history of the nicad battery traces back to the beginning of the twentieth century. After approximately 10,000 experiments, Thomas Edison found that he could make a rechargeable battery from a nickel–iron cell. The active materials in the cell were $Ni(OH)_2$ and iron metal that was partially oxidized to Fe^{2+}. In the mid 1920s, a German manufacturer found that the risk of hydrogen explosions in the batteries was significantly reduced when the iron metal was replaced with cadmium.

The chemistry of nickel–cadmium batteries still isn't fully understood. A charge–discharge mechanism for the system might be written as follows.

$$2\,NiO(OH)\cdot xH_2O + Cd + 2\,H_2O \underset{charge}{\overset{discharge}{\rightleftharpoons}} 2\,Ni(OH)_2\cdot yH_2O + Cd(OH)_2$$

When the batteries discharge, the nickel is reduced from +3 to +2 and the cadmium is oxidized from 0 to +2, with a net potential of 1.29 V. Both the oxidized and reduced forms of nickel contain water trapped in the crystals, but the amount of water of hydration in the two forms is different.

FUEL CELLS

Ever since the start of the space program, enthusiasm has run high for batteries that are *fuel cells*. By definition, a fuel cell is an electrochemical cell in which a reaction with oxygen is used to generate electrical energy. Fuel cells therefore require the continuous feed of a fuel and oxygen, which produces a low-voltage direct current as long as fuel is provided.

So far, it hasn't been practical to burn hydrocarbons, such as natural gas, in a fuel cell. The fuel cells used in the space program burn hydrogen gas to give water, with a cell potential of 1.10 V, as shown in Figure 12A.5.

$$2\,H_2(g) + O_2(g) \rightleftharpoons 2\,H_2O(g) \qquad E^\circ = 1.10\text{ V}$$

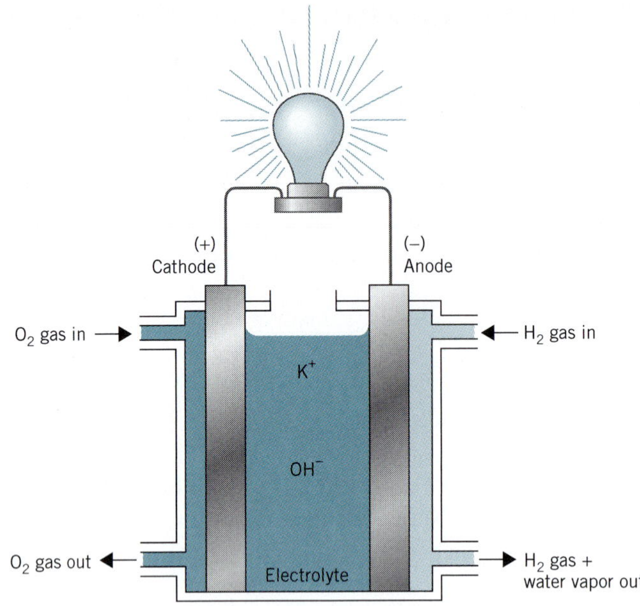

Fig. 12A.5 A hydrogen–oxygen fuel cell.

Fuel cells can also be constructed to burn either ammonia or hydrazine.

$$4\ NH_3(g) + 3\ O_2(g) \rightleftharpoons 2\ N_2(g) + 6\ H_2O(g) \qquad E° = 1.17\ V$$

12A.6 Galvanic Corrosion and Cathodic Protection

The model for electrochemical reactions developed in this chapter can be used to explain the behavior observed when the corrosion of iron metal is studied. Corrosion is the deterioration of a metal by oxidation. In the particular case of iron, the product of corrosion is called rust, a hydrated iron(III) oxide of variable composition.

- **Iron corrodes very slowly in contact with dry air.** Iron metal corrodes in contact with dry air because iron atoms on the surface of the metal slowly react with oxygen in the atmosphere to form a mixture of oxides.

$$2\ Fe(s) + O_2(g) \rightleftharpoons 2\ FeO(s)$$
$$4\ Fe(s) + 3\ O_2(g) \rightleftharpoons 2\ Fe_2O_3(s)$$
$$3\ Fe(s) + 2\ O_2(g) \rightleftharpoons Fe_3O_4(s)$$

  The reaction is uneven, and there are a number of holes in the oxide layer that forms on the surface of the metal. Oxygen atoms migrate through the holes to the metal surface below the oxide layer. Corrosion therefore continues slowly, but surely. At sites where the metal was deformed when it was worked or shaped, the iron oxide often breaks off, exposing a fresh metal surface to further corrosion and pitting.

- **Iron corrodes on contact with strong acids.** Iron metal dissolves in strong acid, as would be expected from its position in Table 12.1. If the acid is boiled to remove any dissolved O_2 gas, the acid reacts with iron to form solutions of the Fe^{2+} ion. This can be understood by comparing the overall potential for the reaction between iron metal and acid to form Fe^{2+} ions with the potential for the reaction to form Fe^{3+} ions.

$$Fe(s) + 2\,H^+(aq) \rightleftharpoons Fe^{2+}(aq) + H_2(g) \qquad E° = 0.409\text{ V}$$
$$2\,Fe(s) + 6\,H^+(aq) \rightleftharpoons 2\,Fe^{3+}(aq) + 3\,H_2(g) \qquad E° = 0.036\text{ V}$$

If O_2 is present, the Fe^{2+} formed when the iron dissolves is oxidized to Fe^{3+}.

$$4\,Fe^{2+}(aq) + O_2(g) + 4\,H^+(aq) \rightleftharpoons 4\,Fe^{3+}(aq) + 2\,H_2O(l) \qquad E° = 0.46\text{ V}$$

- **Iron doesn't rust in contact with water that doesn't contain dissolved O_2 gas.** The standard potential for the reaction between iron metal and the H^+ ion is favorable.

$$Fe(s) + 2\,H^+(aq) \rightleftharpoons Fe^{2+}(aq) + H_2(g) \qquad E° = 0.409\text{ V}$$

However, the concentration of the H^+ ion in water is far from the standard state of 1 *M*. The cell potential for the reaction is lowered due to the low concentration of H^+. This is a large enough change to make the reaction slightly unfavorable.

- **Iron rusts more rapidly in contact with water that contains dissolved O_2 than it does in dry air.** When water and O_2 are both present, the oxidation and reduction reactions no longer have to occur at the same point on the metal surface. The iron metal now simultaneously acts as the anode, the cathode, and the circuit through which electrons flow, as shown in Figure 12A.6.

At one point on the metal surface, iron metal is oxidized to Fe^{2+} ions.

$$\text{Anode} \qquad Fe \rightleftharpoons Fe^{2+} + 2\,e^- \qquad E°_{Ox} = 0.409\text{ V}$$

The electrons released in the reaction flow through the metal to another point, at which O_2 dissolved in the water is reduced.

$$\text{Cathode} \qquad O_2 + 2\,H_2O + 4\,e^- \rightleftharpoons 4\,OH^- \qquad E°_{Red} = 0.401\text{ V}$$

The net result is the formation of an aqueous solution of Fe^{2+} and OH^- ions.

$$2\,Fe(s) + O_2(g) + 2\,H_2O(l) \rightleftharpoons 2\,Fe^{2+}(aq) + 4\,OH^-(aq) E° = 0.810\text{ V}$$

The Fe^{2+} and OH^- ions produced in the half-reactions diffuse toward each other and eventually combine to form a hydrated iron(II) oxide that undergoes further oxidation to form rust, $Fe_2O_3 \cdot 3H_2O$.

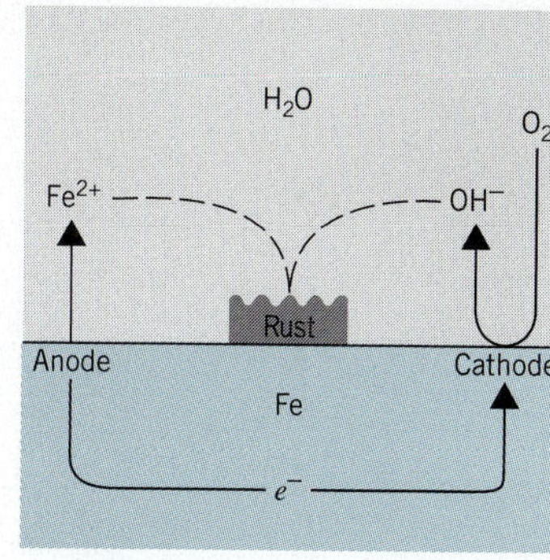

Fig. 12A.6 When iron metal is immersed in water that contains dissolved oxygen, oxidation of the iron and reduction of O_2 can occur at different points on the metal surface. The net effect is the formation of Fe^{2+} and OH^- ions, which diffuse toward each other to form a complex that is further oxidized to form rust, $Fe_2O_3 \cdot 3H_2O$.

- **Iron rusts more rapidly when it comes in contact with copper metal.** A voltaic cell is created at the point at which the two metals touch, as shown in Figure 12A.7.

The iron metal acts as the anode, giving up electrons to form the Fe^{2+} ions.

$$\text{Anode} \qquad Fe \rightleftharpoons Fe^{2+} + 2\,e^- \qquad E°_{Ox} = 0.409\text{ V}$$

The electrons flow to the copper metal, which becomes the cathode at which O_2 dissolved in water is reduced.

$$\text{Cathode} \qquad O_2 + 2\,H_2O + 4\,e^- \rightleftharpoons 4\,OH^- \qquad E°_{Red} = 0.401\text{ V}$$

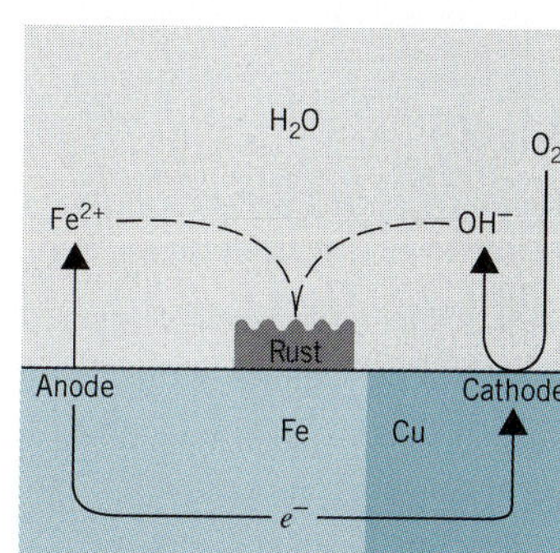

Fig. 12A.7 When iron comes in contact with copper metal, a galvanic cell is created in which the iron is oxidized to Fe^{2+} ions. The electrons given off are transferred to the copper metal, where they are used to reduce O_2 to OH^- ions.

Once again, the Fe^{2+} ions produced at the anode diffuse toward the OH^- ions given off at the cathode and eventually combine with the OH^- ions to form rust. This process is called *galvanic corrosion* because the overall reaction is similar to what would happen if the reaction were run in a galvanic cell.

The copper metal doesn't corrode when it comes in contact with iron because copper metal is a weaker reducing agent than iron. We can see this by imagining the voltaic cell that would be created if copper acted as the anode and the electrons flowed to iron, which would act as the cathode at which O_2 was reduced.

Anode	$Cu \rightleftharpoons Cu^{2+} + 2\,e^-$	$E^\circ_{Ox} = -0.340$ V
Cathode	$O_2 + 2\,H_2O + 4\,e^- \rightleftharpoons 4\,OH^-$	$E^\circ_{Red} = 0.401$ V

The overall potential for the reaction is much smaller than the potential for the reaction that occurs when iron acts as the anode.

$$2\,Cu(s) + O_2(g) + 2\,H_2O(l) \rightleftharpoons 2\,Cu^{2+}(aq) + 4\,OH^-(aq) \qquad E^\circ = 0.061 \text{ V}$$

$$2\,Fe(s) + O_2(g) + 2\,H_2O(l) \rightleftharpoons 2\,Fe^{2+}(aq) + 4\,OH^-(aq) \qquad E^\circ = 0.810$$

The iron therefore corrodes before the copper.

An interesting question is raised when Figures 12A.6 and 12A.7 are compared. The overall reaction for the corrosion of iron is the same, regardless of whether copper metal is in contact with the iron metal. Thus, the overall cell potential for the reactions must be the same. Why, then, does iron corrode so much faster when it comes in contact with copper metal?

The answer seems to rest with the concept of **overvoltage.** The actual potential for the reduction of O_2 on a metal surface is usually larger than the value reported in tables of standard cell potentials. The overvoltage for the reduction of O_2 on copper metal, however, is significantly smaller than the overvoltage on iron. Since overvoltage is defined as the extra voltage that must be applied to a reaction to get it to occur at the same rate, oxidation of iron metal in contact with copper should occur much more rapidly than the corresponding reaction on an isolated iron metal surface.

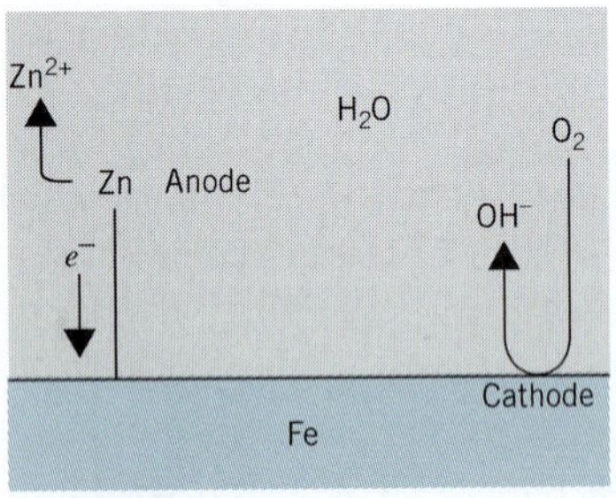

Fig. 12A.8 The corrosion of iron can be prevented by cathodic protection. The iron is either coated with zinc to form galvanized iron or connected to a piece of zinc or magnesium metal. Zinc and magnesium are both stronger reducing agents than iron, so they are oxidized instead of the iron. The iron therefore becomes the cathode at which O_2 is reduced, which protects the iron from corrosion.

- **Iron doesn't rust when it is in contact with zinc or magnesium metal.** Iron can be protected from corrosion if it is allowed to stay in contact with a metal that is a better reducing agent, as shown in Figure 12A.8. This process is called **cathodic protection** because it involves making iron the cathode of a voltaic cell. The zinc or magnesium metal acts as the anode of a voltaic cell.

Anode $\quad Zn \rightleftharpoons Zn^{2+} + 2\,e^- \quad E^\circ_{Ox} = 0.763$ V

The electrons given off in the reaction flow to the iron metal, which acts as the cathode at which O_2 is reduced.

Cathode $\quad O_2 + 2\,H_2O + 4\,e^- \rightleftharpoons 4\,OH^- \quad E^\circ_{Red} = 0.401$ V

The overall potential for the reaction is much larger than the potential for the reaction that would occur if iron were the anode.

$$2\ Zn(s) + O_2(g) + 2\ H_2O(l) \rightleftharpoons 2\ Zn^{2+}(aq) + 4\ OH^-(aq) \qquad E^\circ = 1.164\ V$$

Thus, oxidation of the iron won't occur until the anode is either completely consumed or the electrical contact between the iron and zinc metal is broken. Cathodic protection is the theory behind "galvanized" iron, which is iron metal coated with a thin layer of zinc. Until virtually all of the zinc has corroded, the iron metal that provides the structure for the material remains intact.

12A.7 Electrolysis of Molten NaCl

An idealized cell for the electrolysis of sodium chloride is shown in Figure 12A.9. A source of direct current is connected to a pair of inert electrodes immersed in molten sodium chloride. The Na^+ ions flow toward the negative electrode, and the Cl^- ions flow toward the positive electrode.

When Na^+ ions collide with the negative electrode, the battery provides a large enough potential to force the ions to pick up electrons to form sodium metal. This electrode is called the cathode because reduction always occurs at the cathode, regardless of whether the cell is a voltaic cell or an electrolytic cell.

$$\text{Negative electrode (cathode)} \qquad Na^+ + e^- \longrightarrow Na$$

Cl^- ions that collide with the positive electrode are oxidized to Cl_2 gas, which bubbles off at this electrode. This electrode is called the anode because oxidation always occurs at the anode of an electrochemical cell.

$$\text{Positive electrode (anode)} \qquad 2\ Cl^- \longrightarrow Cl_2 + 2\ e^-$$

The net effect of passing an electric current through the molten salt in the electrolytic cell is to decompose sodium chloride into its elements, sodium metal and chlorine gas.

Electrolysis of NaCl

Cathode (−)	$Na^+ + e^- \longrightarrow Na$
Anode (+)	$2\ Cl^- \longrightarrow Cl_2 + 2\ e^-$

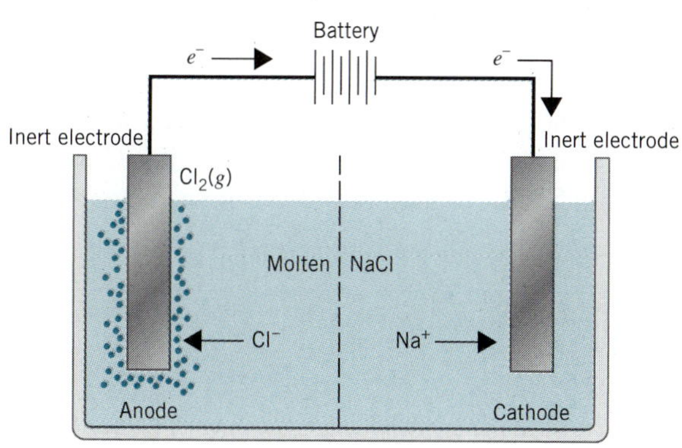

Fig. 12A.9 Electrolysis of molten sodium chloride involves using an electric current to reduce Na^+ ions to sodium metal at the cathode and to oxidize Cl^- ions to Cl_2 gas at the anode.

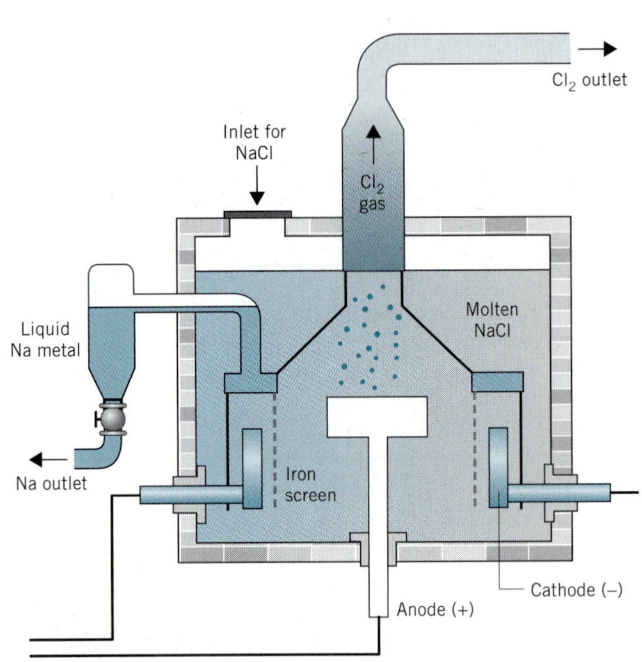

Fig. 12A.10 Cross section of the Downs cell used for the electrolysis of a molten mixture of calcium chloride and sodium chloride.

This example explains why the process is called **electrolysis.** The suffix *-lysis* comes from a Greek stem meaning "to loosen or split up." Electrolysis literally uses an electric current to split a compound into its elements.

$$2\,\text{NaCl}(l) \xrightarrow{\textit{electrolysis}} 2\,\text{Na}(l) + \text{Cl}_2(g)$$

This example also illustrates the difference between voltaic cells and electrolytic cells. Voltaic cells use the energy given off in a spontaneous reaction to do electrical work. Electrolytic cells use electrical work as a source of energy to drive the reaction in the opposite direction.

The dotted vertical line in the center of Figure 12A.9 represents a diaphragm that keeps the Cl_2 gas produced at the anode from coming into contact with the sodium metal generated at the cathode. The function of the diaphragm can be understood by turning to a more realistic drawing of the commercial Downs cell used to electrolyze sodium chloride, shown in Figure 12A.10.

Chlorine gas that forms on the graphite anode inserted into the bottom of the Downs cell bubbles through the molten sodium chloride into a funnel at the top of the cell. Sodium metal that forms at the cathode floats up through the molten sodium chloride into a sodium-collecting ring, from which it is periodically drained. The diaphragm that separates the two electrodes is a screen of iron gauze, which prevents the explosive reaction between sodium metal and chlorine to form sodium chloride that would occur if the products of the electrolysis reaction came in contact.

The feedstock for the Downs cell is a 3:2 mixture by mass of $CaCl_2$ and NaCl. This mixture is used because it has a melting point of 580°C, whereas pure sodium chloride has to be heated to more than 800°C before it melts.

12A.8 Electrolysis of Aqueous NaCl

Figure 12A.11 shows an idealized drawing of a cell in which an aqueous solution of sodium chloride is electrolyzed. Once again, the Na^+ ions migrate toward the

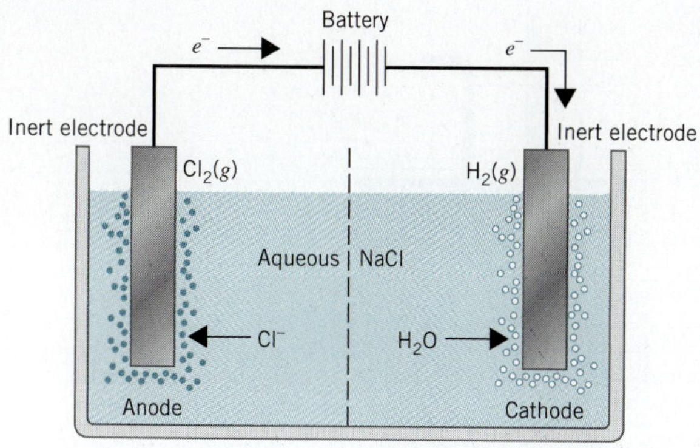

Fig. 12A.11 Electrolysis of aqueous sodium chloride results in reduction of water to form H_2 gas at the cathode and oxidation of Cl^- ions at the anode.

negative electrode and the Cl^- ions migrate toward the positive electrode. But now there are two substances that can be reduced at the cathode: Na^+ ions and water molecules.

Cathode (−)

$$Na^+ + e^- \rightleftharpoons Na \qquad E^\circ_{Red} = -2.71 \text{ V}$$

$$2\,H_2O + 2e^- \rightleftharpoons H_2 + 2\,OH^- \qquad E^\circ_{Red} = -0.83 \text{ V}$$

Because it is much easier to reduce water than Na^+ ions, the products formed at the cathode are hydrogen gas and OH^-.

$$\text{Cathode } (-) \qquad 2\,H_2O(l) + 2\,e^- \rightleftharpoons H_2(g) + 2\,OH^-(aq)$$

There are also two substances that can be oxidized at the anode: Cl^- ions and water molecules.

Anode (+)

$$2\,Cl^- \rightleftharpoons Cl_2 + 2\,e^- \qquad E^\circ_{Ox} = -\ 1.36 \text{ V}$$

$$2\,H_2O \rightleftharpoons O_2 + 4\,H^+ + 4\,e^- \qquad E^\circ_{Ox} = -\ 1.23 \text{ V}$$

The standard potentials for the two half-reactions are so close to each other that we might expect to see a mixture of Cl_2 and O_2 gas collect at the anode. In practice, the only product is Cl_2.

$$\text{Anode } (+) \qquad 2\,Cl^- \rightleftharpoons Cl_2 + 2\,e^-$$

At first glance, it would seem easier to oxidize water ($E^\circ_{Ox} = -1.23$ volts) than Cl^- ions ($E^\circ_{Ox} = -1.36$ volts). It is worth noting, however, that these are standard half-cell potentials, and the cell is never allowed to reach standard conditions. The solution is typically 25% NaCl by mass, which significantly decreases the potential required to oxidize the Cl^- ion. The deciding factor is *overvoltage,* which is the extra voltage that must be applied to a reaction to get it to occur at the rate at which it would occur in an ideal system. Under ideal conditions, a potential of 1.23 volts is large enough to oxidize water to O_2 gas. Under real conditions, however, it can take a much larger voltage to initiate the reaction. (The overvoltage for the oxidation of water can be as large as 1 volt.) By carefully choosing the electrode to maximize the overvoltage for the oxidation of water and then carefully controlling the potential at which the cell operates, we can ensure that only chlorine is produced in the reaction.

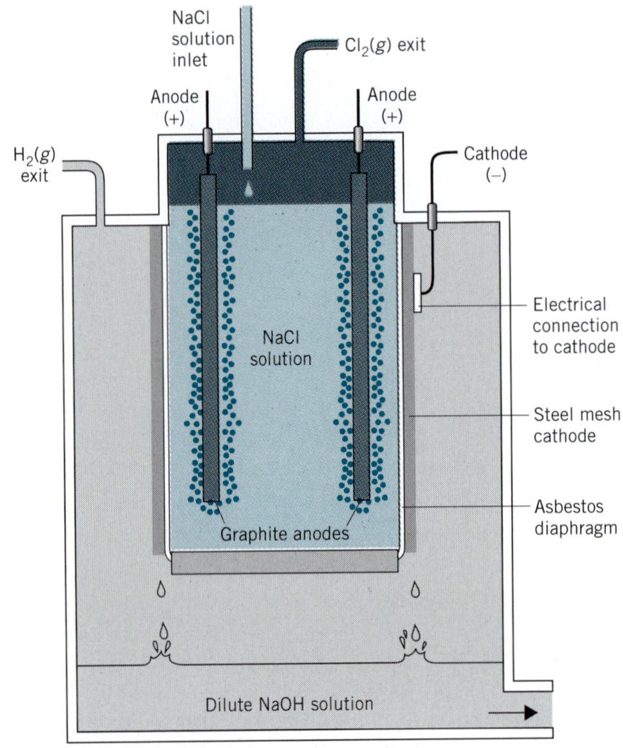

Fig. 12A.12 About half of the commercial cells for the electrolysis of aqueous sodium chloride use the diaphragm cell shown here.

In summary, electrolysis of aqueous solutions of sodium chloride doesn't give the same products as electrolysis of molten sodium chloride. Electrolysis of molten NaCl decomposes the compound into its elements.

$$2\,NaCl(l) \xrightarrow{electrolysis} 2\,Na(l) + Cl_2(g)$$

Electrolysis of aqueous NaCl solutions gives a mixture of hydrogen and chlorine gas and an aqueous sodium hydroxide solution.

$$2\,NaCl(aq) + 2\,H_2O(l) \xrightarrow{electrolysis} 2\,Na^+(aq) + 2\,OH^-(aq) + H_2(g) + Cl_2(g)$$

Because the demand for chlorine is much larger than the demand for sodium, electrolysis of aqueous sodium chloride is a more important process commercially. Electrolysis of an aqueous NaCl solution has two other advantages. It produces H_2 gas at the cathode, which can be collected and sold. It also produces NaOH, which can be drained from the bottom of the electrolytic cell and sold, as shown in Figure 12A.12.

The dotted vertical line in the idealized cell shown in Figure 12A.11 represents a diaphragm that prevents the Cl_2 produced at the anode in the cell from coming into contact with the NaOH that accumulates at the cathode. When the diaphragm is removed from the cell, the products of the electrolysis of aqueous sodium chloride react to form sodium hypochlorite, which is the first step in the preparation of hypochlorite bleaches, such as Clorox.

$$Cl_2(g) + 2\,OH^-(aq) \xrightarrow{electrolysis} Cl^-(aq) + OCl^-(aq) + H_2O(l)$$

12A.9 Electrolysis of Water

Electrolysis can be used to decompose water into its elements.

$$2\,H_2O(l) \xrightarrow{electrolysis} 2\,H_2(g) + O_2(g)$$

A pair of inert electrodes are sealed in opposite ends of a container designed to collect the H_2 and O_2 gas given off in the reaction. The electrodes are then connected to a battery or another source of electric current.

By itself, water is a very poor conductor of electricity. We therefore add an electrolyte to water to provide ions that can flow through the solution, thereby completing the electric circuit. The electrolyte must be soluble in water. It should also be relatively inexpensive. Most importantly, it must contain ions that are harder to oxidize or reduce than water.

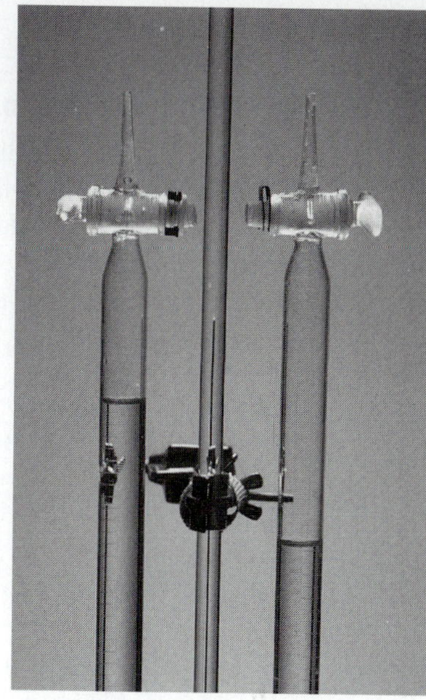

Electrolysis of an aqueous Na_2SO_4 solution results in the reduction of water to form H_2 gas at the cathode and the oxidation of water to form O_2 gas at the anode.

$$2\,H_2O + 2\,e^- \rightleftharpoons H_2 + 2\,OH^- \qquad E^\circ_{Red} = -0.83\ V$$
$$2\,H_2O \rightleftharpoons O_2 + 4\,H^+ + 4\,e^- \qquad E^\circ_{Ox} = -1.23\ V$$

According to Appendix B.12, the following cations are harder to reduce than water: Li^+, Rb^+, K^+, Cs^+, Ba^{2+}, Sr^{2+}, Ca^{2+}, Na^+, and Mg^{2+}. Two of these cations are more likely candidates than the others because they form inexpensive, soluble salts: Na^+ and K^+.

Appendix B.12 suggests that the SO_4^{2-} ion might be the best anion to use because it is one of the most difficult anions to oxidize. The potential for oxidation of the SO_4^{2-} ion to $S_2O_8^{2-}$ ion is -2.05 volts.

$$2\,SO_4^{2-} \rightleftharpoons S_2O_8^{2-} + 2\,e^- \qquad E^\circ_{Ox} = -2.05\ V$$

When an aqueous solution of either Na_2SO_4 or K_2SO_4 is electrolyzed in the apparatus shown in Figure 12A.13, H_2 gas collects at one electrode and O_2 gas collects at the other.

What would happen if we added an indicator such as bromothymol blue to the apparatus? Bromothymol blue turns yellow in acidic solutions (pH $<$ 6) and turns blue in basic solutions (pH $>$ 7.6). According to the equations for the two half-reactions, the indicator should turn yellow at the anode and blue at the cathode.

$$\text{Cathode } (-) \qquad 2\,H_2O + 2\,e^- \rightleftharpoons H_2 + 2\,OH^-$$
$$\text{Anode } (+) \qquad 2\,H_2O \rightleftharpoons O_2 + 4\,H^+ + 4\,e^-$$

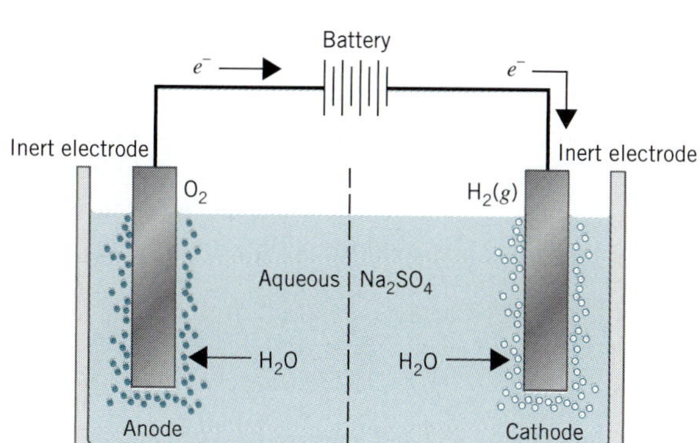

Fig. 12A.13 Electrolysis of an aqueous Na_2SO_4 solution results in the reduction of water to form H_2 gas at the cathode and oxidation of water to form O_2 gas at the anode.

Problems

Balancing Oxidation–Reduction Equations

12A-1. Some of the earliest matches consisted of white phosphorus and potassium chlorate glued to wooden sticks. Write a balanced equation for the following reaction, which occurred when the match was rubbed against a hard surface.

$$P_4(s) + KClO_3(s) \longrightarrow P_4O_{10}(s) + KCl(s)$$

Is this an oxidation–reduction reaction? Explain your answer.

The above equation could be balanced by trial and error. See if you can balance the following equation by trial and error.

$$H^+(aq) + Cr_2O_7^{2-}(aq) + C_2H_5OH(aq) \longrightarrow Cr^{3+}(aq) + H_2O(l) + CO_2(g)$$

Redox Reactions in Acidic Solutions

12A-2. Write balanced equations for the following reactions, in which an acidic solution of the CrO_4^{2-} or $Cr_2O_7^{2-}$ ion is used as an oxidizing agent.

(a) $I^-(aq) + CrO_4^{2-}(aq) \longrightarrow I_2(aq) + Cr^{3+}(aq)$

(b) $Fe^{2+}(aq) + Cr_2O_7^{2-}(aq) \longrightarrow Fe^{3+}(aq) + Cr^{3+}(aq)$

(c) $H_2S(g) + CrO_4^{2-}(aq) \longrightarrow S_8(s) + Cr^{3+}(aq)$

12A-3. Write balanced equations for the following oxidation–reduction reactions, which involve an acidic solution of the MnO_4^- ion.

(a) $Fe^{2+}(aq) + MnO_4^-(aq) \longrightarrow Fe^{3+}(aq) + Mn^{2+}(aq)$

(b) $S_2O_3^{2-}(aq) + MnO_4^-(aq) \longrightarrow SO_4^{2-}(aq) + Mn^{2+}(aq)$

(c) $Mn^{2+}(aq) + PbO_2(s) \longrightarrow MnO_4^-(aq) + Pb^{2+}(aq)$

(d) $SO_2(g) + MnO_4^-(aq) \longrightarrow H_2SO_4(aq) + Mn^{2+}(aq)$

12A-4. Write a balanced equation for the following reaction, which occurs in acidic solution.

$$Br_2(aq) + SO_2(g) \longrightarrow H_2SO_4(aq) + HBr(aq)$$

12A-5. The two most common oxidation states of chromium are Cr(III) and Cr(VI). Balance the following oxidation–reduction equations that involve one of these oxidation states of chromium. Assume that the reactions occur in acidic solution.

(a) $Cr(s) + O_2(g) + H^+(aq) \longrightarrow Cr^{3+}(aq)$

(b) $Fe^{2+}(aq) + Cr_2O_7^{2-}(aq) \longrightarrow Fe^{3+}(aq) + Cr^{3+}(aq)$

(c) $BrO_3^-(aq) + Cr^{3+}(aq) \longrightarrow Br_2(aq) + HCrO_4^-(aq)$

12A-6. Reactions in which a reagent undergoes both oxidation and reduction are known as *disproportionation* reactions. Write balanced equations for the following disproportionation reactions in acidic solution.

(a) $H_2O_2(aq) \longrightarrow H_2O(l) + O_2(g)$

(b) $NO_2(g) + H_2O(l) \longrightarrow HNO_3(aq) + NO(g)$

12A-7. The term *conproportionation* has been proposed to describe reactions that are the opposite of disproportionation reactions. Write balanced equations for the following conproportionation reactions in acidic solution.

(a) $HIO_3(aq) + HI(aq) \longrightarrow I_2(aq)$

(b) $SO_2(g) + H_2S(g) \longrightarrow S_8(s) + H_2O(l)$

12A-8. Acids that are also oxidizing agents play an important role in the preparation of small quantities of the halogens in the laboratory. Write balanced equations for the following oxidation–reduction equations in acidic solution.

(a) $HCl(aq) + HNO_3(aq) \longrightarrow NO(g) + Cl_2(g)$

(b) $HBr(aq) + H_2SO_4(aq) \longrightarrow SO_2(g) + Br_2(aq)$

(c) $HCl(aq) + MnO_4^-(aq) \longrightarrow Cl_2(g) + Mn^{2+}(aq)$

12A-9. The product of the reaction between copper metal and nitric acid depends on the concentration of the acid. In dilute nitric acid, NO is formed. Concentrated nitric acid produces NO_2. Write balanced equations for the following oxidation–reduction reactions between copper metal and nitric acid.

(a) $Cu(s) + HNO_3(aq) \longrightarrow Cu^{2+}(aq) + NO(g)$

(b) $Cu(s) + HNO_3(aq) \longrightarrow Cu^{2+}(aq) + NO_2(g)$

12A-10. Use half-reactions to balance the following oxidation–reduction equation that seems to give only a single product in acid solution.

$$H_2S(g) + H_2O_2(aq) \longrightarrow S_8(s)$$

Redox Reactions in Basic Solutions

12A-11. Balance the following oxidation–reduction equations that occur in basic solution.

(a) $NO(g) + MnO_4^-(aq) \longrightarrow NO_3^-(aq) + MnO_2(s)$

(b) $NH_3(aq) + MnO_4^-(aq) \longrightarrow N_2(g) + MnO_2(s)$

12A-12. Most organic compounds react with MnO_4^- ion in the presence of acid to form CO_2 and water. MnO_4^- in the presence of base, however, is a weaker and therefore more gentle oxidizing agent. Balance the following redox reactions in which the MnO_4^- ion is used to oxidize an organic compound.

(a) $CH_3OH(aq) + MnO_4^-(aq) \xrightarrow{H^+} CO_2(g) + Mn^{2+}(aq)$

(b) $CH_3OH(aq) + MnO_4^-(aq) \xrightarrow{OH^-} HCO_2H(aq) + MnO_2(s)$

12A-13. The Cr^{3+} ion is not soluble in basic solutions because it precipitates as $Cr(OH)_3$. Chromium metal will dissolve in excess base, however, because it can form a soluble $Cr(OH)_4^-$ complex ion. $Cr(OH)_3$ can also be dissolved by oxidizing the chromium to the +6 oxidation state. Write balanced equations for the following reactions that occur in basic solution.

(a) $Cr(s) + OH^-(aq) \longrightarrow Cr(OH)_4^-(aq) + H_2(g)$

(b) $Cr(OH)_3(s) + H_2O_2(aq) \longrightarrow CrO_4^{2-}(aq)$

12A-14. Write balanced equations for the following reactions to see whether there is any difference between the stoichiometry of the reaction in which the sulfite ion is oxidized to the sulfate ion when the reaction is run in acidic versus basic solution.

$$SO_3^{2-}(aq) + O_2(g) \xrightarrow{H^+} SO_4^{2-}(aq)$$

$$SO_3^{2-}(aq) + O_2(g) \xrightarrow{OH^-} SO_4^{2-}(aq)$$

12A-15. Write a balanced equation for the oxidation–reduction reaction used to extract gold from its ores.

$$Au(s) + CN^-(aq) + O_2(g) \longrightarrow Au(CN)_2^-(aq) + OH^-(aq)$$

12A-16. Write a balanced equation for the Raschig process, in which ammonia reacts with the hypochlorite ion in a basic solution to form hydrazine.

$$NH_3(aq) + OCl^-(aq) \longrightarrow N_2H_4(aq) + Cl^-(aq)$$

12A-17. Use half-reactions to balance the following disproportionation reactions that occur in basic solution.

(a) $Cl_2(g) + OH^-(aq) \longrightarrow Cl^-(aq) + OCl^-(aq)$

(b) $Cl_2(g) + OH^-(aq) \longrightarrow Cl^-(aq) + ClO_3^-(aq)$

(c) $P_4(s) + OH^-(aq) \longrightarrow PH_3(g) + H_2PO_2^-(aq)$

(d) $H_2O_2(aq) \longrightarrow O_2(g) + H_2O(l)$

12A-18. The chemistry of nitrogen is unusual because so many of its compounds undergo reactions that give nitrogen in its elemental form. Balance the following oxidation–reduction reaction in which ammonia is converted into molecular nitrogen in basic solution.

$$CuO(s) + NH_3(g) \longrightarrow Cu(s) + N_2(g) + H_2O(l)$$

Molecular Redox Reactions

12A-19. What is oxidized and what is reduced in the following reactions?

(a) $CH_4(g) + 2\,O_2(g) \longrightarrow CO_2(g) + 2\,H_2O(g)$

(b) $CH_3C(H){=}CH_2(g) + H_2(g) \longrightarrow CH_3CH_2CH_3(g)$

12A-20. Which, if any, of the following in each pair of species represents the oxidized species?

(a) CH_3CH_2OH, $CH_3C(=O)H$

(b) $CH_3CH(OH)-CH_3$, $CH_3C(=O)CH_3$

(c) $CH_3CH_2CH_3$, $CH_3CH(CH_3)CH_3$

(d) CH_3OH, CO

Batteries

12A-21. What happens when a lead–acid battery is discharged? How can this battery be charged?

12A-22. Why were Leclanché cell batteries popular?

12A-23. What is the major difference between Leclanché batteries and dry cell batteries?

12A-24. What are the advantages of a dry cell battery compared to a Leclanché battery?

12A-25. What is a fuel cell?

12A-26. What is the danger associated with charging a lead–acid battery?

Galvanic Corrosion and Cathodic Protection

12A-27. Use cell potentials to explain why iron oxidizes to form Fe_2O_3.

12A-28. Use cell potentials to explain why iron dissolves in strong acids.

12A-29. Explain why the H^+ ion concentration in water is too small for iron to dissolve in water from which all oxygen has been removed.

12A-30. Use cell potentials to explain why iron dissolves in water that contains oxygen.

12A-31. Explain why it is a mistake to describe rust as iron (III) oxide, Fe_2O_3.

12A-32. When iron rusts, the metal surface acts as the anode of a voltaic cell. Explain why the iron metal becomes the cathode of a voltaic cell when it is connected to a piece of magnesium or zinc.

12A-33. Explain why galvanized iron, which has a thin coating of zinc metal, corrodes much more slowly than iron by itself.

Electrolysis of Molten NaCl

12A-34. Molten NaCl contains Na^+ and Cl^- ions. Why does Na^+ migrate to the cathode and Cl^- to the anode when a battery supplies a current to the molten salt?

12A-35. What would be produced at the anode and cathode if molten CaF_2 were electrolyzed? Show the half-reactions.

12A-36. How does the electrolysis of molten NaCl illustrate the difference between voltaic and electrolytic cells?

Electrolysis of Aqueous NaCl

12A-37. Explain why electrolysis of aqueous NaCl produces H_2 gas, not sodium metal, at the cathode, and Cl_2 gas, not O_2 gas, at the anode.

12A-38. Calculate the standard cell potential necessary to electrolyze an aqueous solution of NaCl.

12A-39. Which of the following statements best describes what happens when aqueous $MgCl_2$ is electrolyzed?

(a) Mg metal forms at the anode.
(b) Mg^{2+} ions are oxidized at the cathode.
(c) Cl_2 gas is formed at the anode by the oxidation of Cl^-.
(d) Cl^- ions flow toward the cathode.

Electrolysis of Water

12A-40. Explain why an electrolyte, such as Na_2SO_4, has to be added to water before the water can be electrolyzed to H_2 and O_2.

12A-41. Describe what would happen if $NiSO_4$ were used instead of Na_2SO_4 when water was electrolyzed.

12A-42. Why does the solution around the cathode become basic when an aqueous solution of Na_2SO_4 is electrolyzed?

12A-43. What are the most likely products of the electrolysis of an aqueous solution of KBr?

(a) $K^+(aq) + Br^-(aq)$
(b) $K(s) + Br_2(l)$
(c) $OH^-(aq) + H_2(g) + Br_2(l)$
(d) $K^+(aq) + OH^-(aq) + O_2(g) + Br_2(l)$

12A-44. Describe what happens during the electrolysis of the following solutions.

(a) $FeCl_3(aq)$ (b) $NaI(aq)$
(c) $K_2SO_4(aq)$ (d) $H_2SO_4(aq)$

12A-45. Why can't aluminum metal be prepared by the electrolysis of an aqueous solution of the Al^{3+} ion?

12A-46. Which of the following reactions occurs at the anode during electrolysis of an aqueous Na_2SO_4 solution?'

(a) $SO_4^{2-}(aq) \rightleftharpoons SO_2(g) + O_2(g) + 2e^-$
(b) $2\,H_2O(l) \rightleftharpoons O_2(g) + 4\,H^+(aq) + 4\,e^-$
(c) $2\,H_2O(l) + O_2(g) + 4\,e^- \rightleftharpoons 4\,OH^-(aq)$
(d) $SO_4^{2-}(aq) + 4H^+(aq) + 2e^- \rightleftharpoons SO_2(g) + 2H_2O(l)$

Chapter Thirteen

CHEMICAL THERMODYNAMICS

13.1 Spontaneous Chemical and Physical Processes

The concepts of entropy and free energy are added to the concept of enthalpy developed in Chapter 7. Again atom combination thermodynamic parameters are used rather than formation parameters. Atom combination parameters provide an understanding of the entropy of a species in the same way employed to introduce enthalpy. ChemActivities 52–56 in Chemistry: A Guided Inquiry *are designed to accompany this chapter.*

Section 13.15 introduces Gibbs Free Energies of Formation and absolute Entropies for instructors that prefer this approach over the use of atom combination data.

No one is surprised when a cup of coffee gradually loses heat to its surroundings as it cools, or when the ice in a glass of lemonade melts as it absorbs heat from its surroundings. But we would be surprised if a cup of coffee suddenly grew hotter until it boiled, or if the water in a glass of lemonade suddenly froze on a hot summer's day.

A piece of zinc metal dissolves in a strong acid to give bubbles of hydrogen gas.

$$Zn(s) + 2\ H^+(aq) \longrightarrow Zn^{2+}(aq) + H_2(g)$$

But if we saw a video in which H_2 bubbles formed on the surface of a solution and then sank through the solution until they disappeared, while a strip of zinc metal formed in the middle of the solution, we would conclude that the video was being run backward.

Many chemical and physical processes proceed in a direction in which they are said to be **spontaneous.** This raises the question: What factor or factors determine the direction in which a process is spontaneous? What drives a chemical reaction, for example, in one direction and not the other?

So many spontaneous processes are exothermic that it seems reasonable to assume that one of the driving forces that determines whether a reaction is spontaneous is a tendency to give off heat. The oxidation of ethyl alcohol by the body is an example of a spontaneous chemical reaction that is exothermic.

$$CH_3CH_2OH(l) + 3\ O_2(g) \longrightarrow 2\ CO_2(g) + 3\ H_2O(l) \qquad \Delta H^\circ = -1367\ \text{kJ/mol}_{rxn}$$

So is the reaction in which iron is oxidized to iron(III) oxide.

$$4\ Fe(s) + 3\ O_2(g) \longrightarrow 2\ Fe_2O_3(s) \qquad \Delta H^\circ = -1648.4\ \text{kJ/mol}_{rxn}$$

There are spontaneous processes, however, that absorb energy from their surroundings. At 100°C, water boils spontaneously, even though the process is endothermic.

$$H_2O(l) \rightleftharpoons H_2O(g) \qquad \Delta H^\circ_{373} = 40.88\ \text{kJ/mol}_{rxn}$$

Ammonium nitrate dissolves spontaneously in water, even though heat is absorbed when the reaction takes place.

$$NH_4NO_3(s) \overset{H_2O}{\rightleftharpoons} NH_4^+(aq) + NO_3^-(aq) \qquad \Delta H^\circ = 28.1\ \text{kJ/mol}_{rxn}$$

The temperature of the system can also play a role in determining the direction in which a process is spontaneous. At temperatures below 0°C, water spontaneously freezes.

$$T < 0°C \qquad H_2O(l) \longrightarrow H_2O(s)$$

At temperatures above 0°C, the process is spontaneous in the opposite direction.

$$T > 0°C \qquad H_2O(s) \longrightarrow H_2O(l)$$

Thus, the tendency of a process to give off heat can't be the only driving force that determines the direction in which the process is spontaneous. There must be another factor that helps determine whether a process is spontaneous. This factor, known as **entropy,** is a measure of the disorder of the system.

> **CHECKPOINT**
>
> When a teaspoon of table salt is poured into a glass of water the salt dissolves, even though heat is absorbed during the process. Is this a spontaneous process?

13.2 Entropy and Disorder

Water molecules in a rigid structure in ice are more ordered than the water molecules moving about randomly in liquid water. Entropy is commonly considered to be a measure of disorder or randomness. The melting of ice produces an increase in entropy. The increased freedom of motion of water in the liquid state means that liquid water has a higher entropy than solid water.

The equation developed by Boltzmann is not used for any mathematical calculations but is introduced to show the relationship between entropy and disorder.

Disorder can be produced in ways other than the removal of constraints on molecular motion that allows particles to move about more freely. The particles of an ideal gas as discussed in Chapter 6 have a distribution of kinetic energies as shown in Figure 13.1. At the temperature T_1, some of the particles have more than the average kinetic energy and some have less. If the gas is heated without allowing the volume to change, the kinetic energy of the gas particles will increase. At the higher temperature, T_2 the average energy is higher and the energy is shared differently among the particles than at the lower temperature. In Figure 13.1 the energy distribution curve at the higher temperature is broadened, showing that the distribution of energies has changed so that more particles are moving with a higher kinetic energy. The hotter gas has the higher entropy because there is now more energy and there is more randomness in the way the energy is distributed.

In 1877 Ludwig Boltzmann proposed the following relation between entropy and microscopic disorder.

$$S = k \ln W$$

In this equation, S is the entropy, k is a proportionality constant, ln represents a logarithm to the base e, and W is a measure of microscopic disorder. Thus if W increases, the entropy increases, and if W decreases, the entropy decreases. Microscopic disorder can be realized in different ways that are not always obvious. In some cases, such as the melting of a solid or the breaking apart of a larger molecule into smaller molecules, an increase in disorder can be readily inferred. The disorder produced by the heating of a gas is less apparent. Here disorder comes about because the additional energy is shared in more ways by the particles of the heated gas. Because uncertainty and disorder are synonymous, the use of the term **disorder** to represent situations in which disorder is not always obvious has become widespread. We will use the terms *order* and *disorder* to refer to processes for which order or disorder is introduced by a phase change or for which there is a change in the way energy is shared by particles.

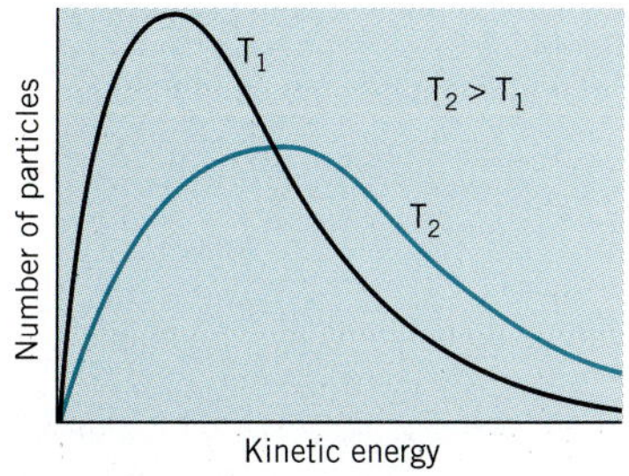

Fig. 13.1 For an ideal gas at a given temperature, not all particles have the same kinetic energy. As temperature increases, the average kinetic energy of the particles increases and the distribution of kinetic energy among the particles is different.

13.3 Entropy and the Second Law of Thermodynamics

Thermodynamics is the study of energy and its transformations. The **first law of thermodynamics** (Chapter 7) provides us with ways of understanding heat changes in chemical reactions, and the **second law of thermodynamics** allows us to determine the direction of processes. Experience has shown us that only those processes that produce a net increase in the disorder of the universe are allowed in nature. The role that disorder plays in determining the direction in which a natural process occurs is summarized by the second law, which states that in order for a process to occur, the entropy of the universe must increase. The term *universe* means everything that might conceivably have its entropy altered as a result of the process.

To see how entropy can affect the direction in which a process is spontaneous, consider the following experiments. Imagine that an ice cube is dropped into one beaker of water at room temperature at the same time that a red-hot penny is dropped into another beaker of water.

The first law of thermodynamics tells us that changes in the energy of the ice cube and penny will be balanced by changes in the energy of the water. However, the first law doesn't tell us which substance loses and which substance gains energy. Experience, on the other hand, tells us that the water in the beaker with the ice cube will become cooler and the water in the beaker with the penny will become hotter.

Although they seem to go in opposite directions, the directions in which these processes occur are determined by the second law of thermodynamics. Both processes occur because they lead to a net increase in the disorder of the universe.

At the moment the ice cube is dropped into the water, the temperature of the ice cube is lower than that of the water that surrounds it. Thus, the average kinetic energy of the molecules in the ice cube is significantly smaller than the average kinetic energy of the molecules in the water. During the melting of the ice cube, the rigid structure of the ice breaks down. After the ice cube melts, the average kinetic energy of the H_2O molecules is the same throughout the system, which is another way of stating that the temperature becomes uniform. The exchange of energy has occurred in such a way that, after the ice cube melts, the spreading of energy has produced more disorder in the universe. The colder water molecules of ice have been warmed and released from the solid structure, thus becoming more disordered. The warmer molecules of liquid water have cooled and become more ordered. Overall, though, there is more disorder due to the breakup of the solid than there is order from the cooling of the warm water.

> **➤ CHECKPOINT**
>
> Assume that a hot brick is thrown into cold water. In the checkpoint in Section 7.4, we noted that the first law of thermodynamics wouldn't prevent the brick from getting hotter and the water from getting cooler. According to the second law of thermodynamics, could the brick get hotter and the water get cooler?

A similar process is at work when a red-hot penny is dropped into a beaker of water. Initially, the average kinetic energy of the atoms in the metal is significantly larger than the average kinetic energy of the molecules in the water. Eventually, the penny and the water that surrounds it will reach the same temperature. The penny cools, the water warms, there are more ways for the energy to be shared, and the universe has gone to a more disordered state. The reverse process in which the penny becomes warmer and the water colder does not occur because that process would result in a more ordered universe. The atoms in the hot penny have acquired more order in cooling, while the cold water molecules have become more random (disordered) due to warming.

Because neither of these processes could occur unless a net increase in the disorder of the universe occurred, we can conclude that these colder objects produce more disorder when brought to a common temperature than do the warmer objects. In other words, there is a total gain in disorder produced by the melting of the ice or the cooling of the penny.

Like the first law of thermodynamics introduced in Section 7.1, the second law describes the results of our experiences while observing the world in which we live. Neither law can be derived mathematically, but no experience that contradicts either law has yet been found. At a time when all other areas of physics seemed open to question, Einstein wrote about thermodynamics as follows, "Thermodynamics is the only science about which I am firmly convinced that, within the framework of the applicability of its basic principles, it will never be overthrown."

As we have seen, the thermodynamic term used to quantify discussions of changes in the amount of disorder in the universe is *entropy,* which has been given the symbol S. When a process produces an increase in the disorder of the universe, the entropy change, ΔS, is *positive.*

It is important to recognize that the second law of thermodynamics describes what happens to the entropy of the *universe,* not the system in which we are interested. By convention, the universe is divided into two components, the system and its surroundings. The change in the entropy of the universe is the sum of the change in the entropy of the system and the surroundings.

$$\Delta S_{univ} = \Delta S_{sys} + \Delta S_{surr}$$

The second law suggests that a process is likely to occur if ΔS_{univ} is positive. Any process, however, for which ΔS_{univ} is negative is highly improbable.

This doesn't mean that ΔS for either the system or its surroundings can't be negative. Consider what happens, once again, when a red-hot penny is dropped into a beaker of water. As the penny cools, the random thermal motion of the metal atoms in the penny decreases. As a result, the metal atoms become more ordered and therefore undergo a decrease in entropy. At the same time, however, the random thermal motion of the water molecules that surround the penny increases, which means that the immediate surroundings become more disordered. Although ΔS for the system (penny) is negative, ΔS for the surroundings (water) is sufficiently positive enough that the overall change in the entropy of the universe is positive. Thus, the overall disorder of the universe increases.

> **➤ CHECKPOINT**
>
> The system becomes more ordered when water freezes. Does this violate the second law of thermodynamics? Why or why not?

Chemists generally choose the chemical reaction as the system in which they are interested. The term *surroundings* is then used to describe the environment that is altered as a consequence of the reaction. Unless otherwise specified, the symbol ΔS will always refer to the system. Entropy, like **enthalpy,** is a state function (Section 7.5). Thus, the value of ΔS for a process depends only on the initial and final conditions of the system.

$\Delta S > 0$ implies that the system becomes *more disordered* during the reaction.
$\Delta S < 0$ implies that the system becomes *less disordered* during the reaction.

The following generalizations can help us decide when a chemical reaction leads to an increase in the disorder of the system.

- Solids have a much more regular structure than liquids. Liquids are therefore more disordered than solids.
- The particles in a gas are in a state of constant, random motion. Gases are therefore more disordered than the corresponding liquids.
- Any process that increases the number of particles increases the amount of disorder in this system.

Exercise 13.1

Which of the following would lead to an increase in the entropy of the system?

(a) The production of a fertilizer, NH_3, from its elements:

$$N_2(g) + 3\ H_2(g) \rightleftharpoons 2\ NH_3(g)$$

(b) The evaporation of water:

$$H_2O(l) \rightleftharpoons H_2O(g)$$

(c) The decomposition of limestone ($CaCO_3$) to produce lime (CaO):

$$CaCO_3(s) \rightleftharpoons CaO(s) + CO_2(g)$$

Solution

(a) The system becomes more ordered in this gas phase reaction because the total number of particles in the system decreases. The entropy of the system therefore decreases.

(b) Gases are more disordered than the corresponding liquids, so the entropy of the system increases.

(c) Reactions in which a compound decomposes into two products generally lead to an increase in entropy because the system becomes more disordered. The increase in entropy in this reaction is even larger because the starting material is a solid and one of the products is a gas.

Section 13.1 suggested that the sign of ΔH for a chemical reaction affects the direction in which the reaction occurs.

Exothermic reactions ($\Delta H < 0$) are often, but not always, spontaneous.

This section introduced the notion that ΔS for a reaction can also play a role in the direction of the reaction.

Reactions that lead to an increase in the disorder of the system ($\Delta S > 0$) are often, but not always, spontaneous.

To decide in which direction a chemical reaction will occur, it is therefore important to consider the effect of changes in both the enthalpy and entropy of the system that occur during the reaction.

Exercise 13.2 uses the reaction of NO_2 to produce N_2O_4 for a qualitative analysis of ΔH and ΔS. This same example is then used in Section 13.6 for a quantitative determination of ΔS°. The reaction is used for a third time for the determination of ΔG° in Exercise 13.5. These three discussions of this reaction can serve as a focal point to connect the concepts of this chapter.

Exercise 13.2

Use the Lewis structures of NO_2 and N_2O_4 and the stoichiometry of the following reaction to decide whether ΔH and ΔS favor the reactants or the products of the reaction.

$$2\ NO_2(g) \rightleftharpoons N_2O_4(g)$$

Solution

NO_2 has an unpaired electron in its Lewis structure.

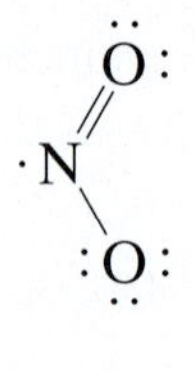

When a pair of NO_2 molecules collide in the proper orientation, a covalent bond can form between the nitrogen atoms to produce the dimer, N_2O_4.

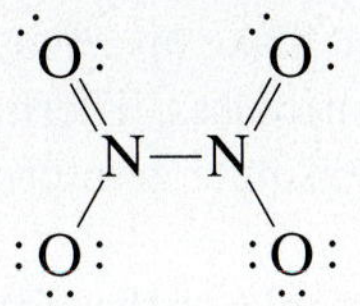

Because a new bond is formed, the reaction should be exothermic. ΔH for the reaction therefore favors the products.

The reaction leads to a decrease in the entropy of the system because the system becomes more ordered when two molecules come together to form a larger molecule. ΔS for the reaction therefore favors the reactants.

> **➤ CHECKPOINT**
>
> Is it possible for a process to proceed if the entropy of the system decreases?

13.4 Standard-State Entropies of Reaction

When the change in the entropy of a system that accompanies a chemical reaction is measured under **standard-state conditions,** the result is the **standard-state entropy of reaction, $\Delta S°$.** By convention, the standard state for thermodynamic measurements is characterized by the following conditions.

Standard-state conditions:

All solutions have concentrations of 1 *M*.

All gases have partial pressures of 1 bar, or 0.9869 atm. This can be rounded to 1 atm for all but the most exact calculations.

Although standard-state entropies can be measured at any temperature, they are often measured at 25°C. Standard-state **entropies of atom combination** at 25°C for a variety of compounds are given in Appendix B.13.

13.5 The Third Law of Thermodynamics

The **third law of thermodynamics** defines absolute zero on the entropy scale.

The entropy of a perfect crystal is zero when the temperature of the crystal is equal to absolute zero (0 K).

The crystal must be perfect, or else there will be some inherent disorder. It also must be at 0 K.

As the crystal warms to temperatures above 0 K, however, the particles in the crystal begin to move more and more vigorously, generating some disorder. The entropy of the crystal gradually increases with temperature as the average kinetic energy of the particles increases. At the melting point, the entropy of the system increases abruptly as the compound is transformed into a liquid, which isn't as well ordered as the solid. The entropy of the liquid gradually increases as the liquid becomes warmer because of the increase in the vibrational, rotational, and translational motion of the particles. There is another abrupt increase in the entropy of the substance at the boiling point, as the liquid is transformed into a random, chaotic gas.

Matter is never at rest. Even at the coldest temperature, absolute zero, there is a residual motion of the atoms that make up all matter. On warming, the motion and hence the disorder of the atoms increases. Thermal entropies are a measure of the increase in disorder over that at absolute zero produced by warming the substance to a higher temperature.

Table 13.1 provides an example of the difference between the entropy of a substance in the solid, liquid, and gaseous phases. Note that the units of entropy are joules per mole kelvin. A typical plot of the entropy of a system versus temperature is shown in Figure 13.2. Figure 13.2 describes the entropy, $S°$, of a single substance at different temperatures. Chemists are usually interested in the difference in entropy, ΔS, between the products and the reactants at a fixed temperature.

Table 13.1
Entropies of Solid, Liquid, and Gaseous Forms of Sulfur Trioxide

Compound	$S°$(J/mol · K)
$SO_3(s)$	70.7
$SO_3(l)$	113.8
$SO_3(g)$	256.76

➤ CHECKPOINT

Why is the entropy of a substance the lowest at absolute zero?

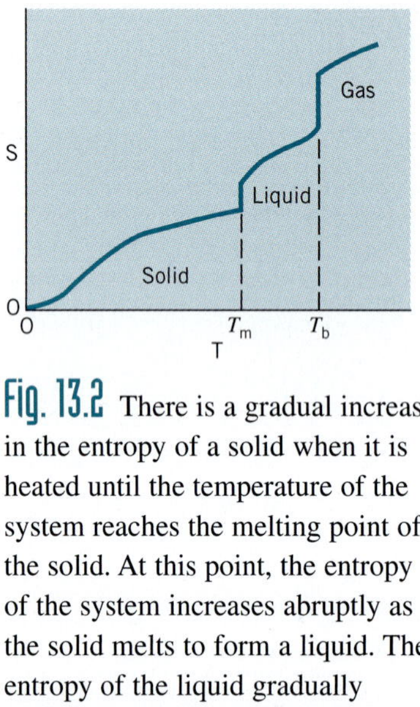

Fig. 13.2 There is a gradual increase in the entropy of a solid when it is heated until the temperature of the system reaches the melting point of the solid. At this point, the entropy of the system increases abruptly as the solid melts to form a liquid. The entropy of the liquid gradually increases until the system reaches the boiling point of the liquid, at which point it increases abruptly once more. The entropy of the gas then increases gradually with temperature.

13.6 Calculating Entropy Changes for Chemical Reactions

In Exercise 13.2 we determined that for the reaction of NO_2 to produce N_2O_4, ΔH favored formation of the products while ΔS favored formation of the reactants. In order to gain additional insight into the meaning of the entropy change that accompanies this chemical reaction, we will calculate a quantitative value for ΔS. The units for ΔS are J/mol$_{rxn}$ · K for the reaction as written.

$$2\ NO_2(g) \rightleftharpoons N_2O_4(g)$$

We will approach the calculation of ΔS using the same method that was used in Chapter 7 for calculating ΔH. Because entropy is a state function, the change in entropy that occurs during a chemical reaction does not depend on the path used to transform the starting reactants into the products of the reaction. This means that we can envision a reaction occurring by the same pathway that we have used for calculating ΔH even though we realize that this is probably not the pathway involved when the chemical reaction occurs. We begin by breaking all of the bonds in the starting materials to form isolated atoms in the gas phase.

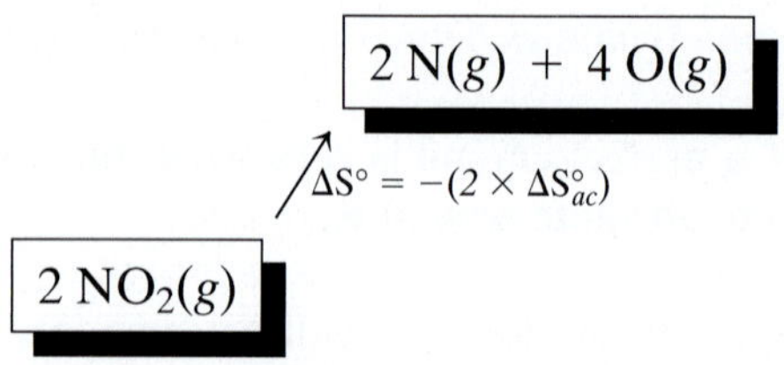

The entropy change produced by this process can be calculated from the entropy of atom combination (ac) data in Appendix B, Table B.13. The tabulated values of ΔS°_{ac} refer to the process whereby 1 mol of the substance listed in the table is formed from its gaseous atoms. Thus, if we want the entropy change for breaking a compound apart into its atoms, we must reverse the sign of the entropy data in the appendix. Because the overall reaction consumes 2 mol of NO_2 for every mole of

N_2O_4 produced, the change in the entropy of the system that accompanies this step in the reaction must be multiplied by 2.

The same reaction diagrams used in Chapter 7 to calculate change in enthalpy are used to develop entropy and free energy. Here students can rely on previous practice with enthalpies of atom combination to learn how to use entropies and free energies of atom combination.

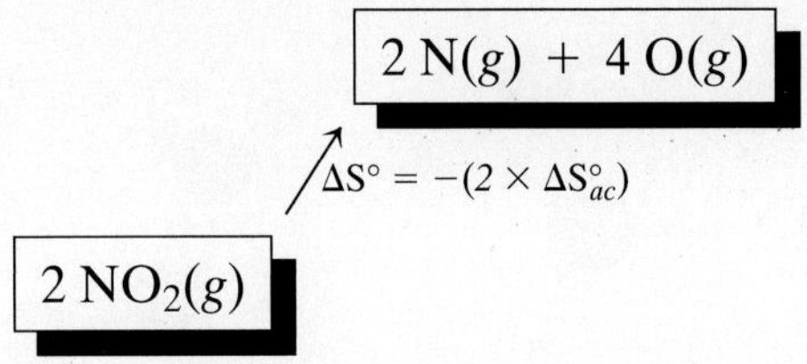

According to Table B. 13, the value of ΔS°_{ac} for 1 mol of NO_2 is -235 J/mol$_{rxn}$ · K. The dissociation of NO_2 into nitrogen and oxygen atoms in the gas phase, on the other hand, must lead to an increase in the disorder of the system. The value of ΔS° for 2 mol of NO_2 in this hypothetical bond-breaking process is therefore $+470$ J/mol$_{rxn}$ · K.

Let's now consider what happens when the isolated atoms in the gas phase recombine to form N_2O_4.

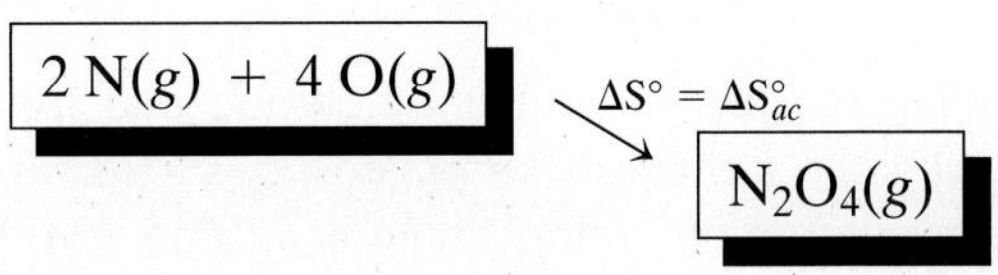

Six isolated atoms in the gas phase have to come together to form each N_2O_4 molecule. As a result, there is a significant decrease in the disorder in the system. The entropy of atom combination for N_2O_4 is relatively large, -647 J/mol$_{rxn}$ · K, and ΔS° for this bond-making process is therefore both large and negative: $\Delta S^\circ = -647$ J/ mol$_{rxn}$ · K.

The overall entropy change for the dimerization of NO_2 to form N_2O_4 can be calculated by adding the values of ΔS° for the breaking and making of bonds in this hypothetical mechanism for the reaction.

Bond breaking	$\Delta S^\circ = 470$ J/mol$_{rxn}$ · K
Bond making	$\Delta S^\circ = -647$ J/mol$_{rxn}$ · K
Overall	$\Delta S^\circ = -177$ J/mol$_{rxn}$ · K

The overall entropy change in the reaction is negative, which isn't surprising because the reaction brings two isolated molecules together to form a more organized molecule.

The entropy of atom combination of a compound can tell us something about the way the atoms are held together to form the compound. Consider ΔS°_{ac} for cyclopentane and 1-pentene in the gas phase, for example.

$5\ C(g) + 10\ H(g) \longrightarrow$ 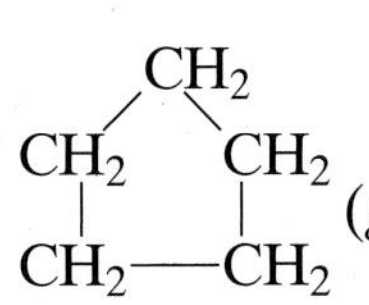 (g)

Cyclopentane

$$\Delta S^\circ_{ac} = -1645 \text{ J/mol}_{rxn} \cdot \text{K}$$

$$5\ C(g) + 10\ H(g) \longrightarrow CH_2{=}CHCH_2CH_2CH_3(g)$$

1-Pentene

$$\Delta S^\circ_{ac} = -1590 \text{ J/mol}_{rxn} \cdot \text{K}$$

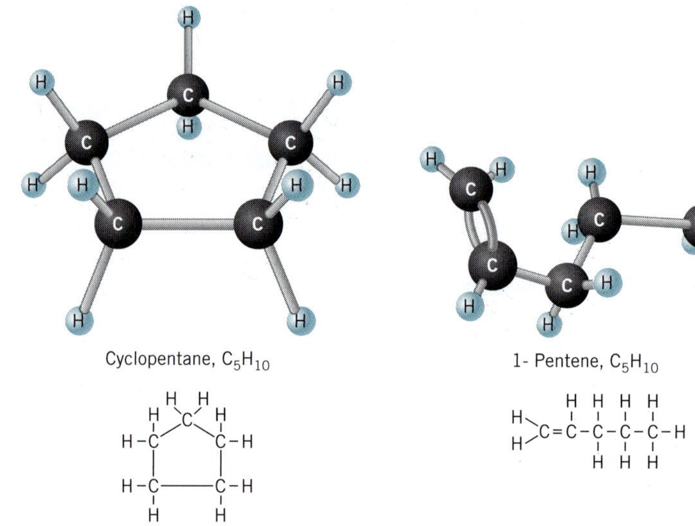

Fig. 13.3 Structures of cyclopentane and 1-pentene.

Both compounds have the same number of carbon and hydrogen atoms, but they differ in the way the atoms are arranged, as shown in Figure 13.3. The entropy of atom combination of cyclopentane is -1645 J/mol$_{rxn}$ · K and that of 1-pentene is -1590 J/mol$_{rxn}$ · K. What do these entropies tell us about the two compounds?

The entropy change associated with a chemical reaction is essentially a measure of the removal or addition of constraints to the motion of atoms, ions, or molecules that occurs in the course of the reaction. Entropies of atom combination are therefore a measure of the freedom of motion lost when ions or molecules are formed from their gaseous atoms.

There are several types of motion associated with a molecule, as shown in Figure 13.4. First, there is the movement of the molecule as a whole from one position in space to another, which is known as *translational* motion. The larger the mass of the molecule, the larger the contribution to entropy from translational motion.

Molecules also tumble and twist as they move from place to place. This motion is called *rotation* and depends on both the molecular weight and the shape of the molecule. A turning or twisting motion occurs when atoms rotate around a

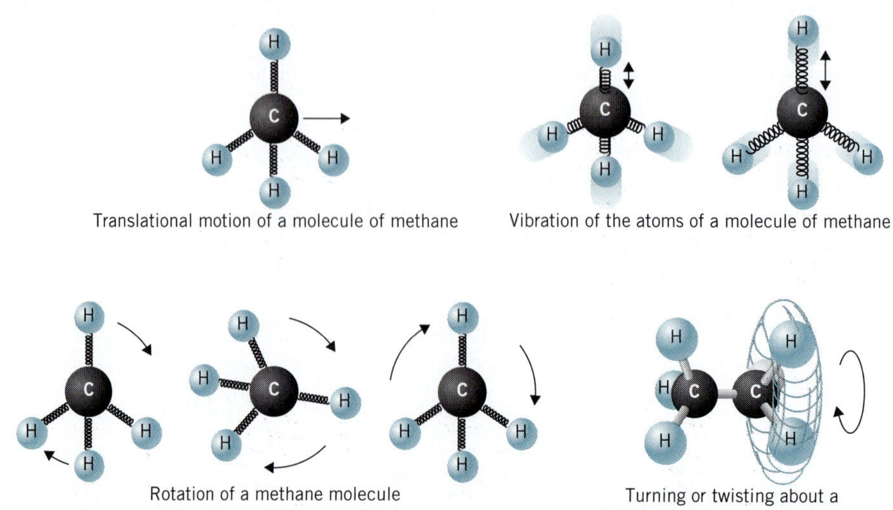

Fig. 13.4 Molecules have three types of motion: translational, rotational, and vibrational.

bond. Molecules that are symmetrical have less rotational freedom than molecules whose atoms are arranged in a less symmetric fashion.

Vibrational motion occurs within the molecule because the bonds holding atoms together behave like a spring. Vibration can occur by the stretching and compressing of a bond, as shown in Figure 13.4, or by changes in the bond angles.

When atoms form molecules, the decrease in the entropy of the system will depend on the number of atoms in the molecule, the symmetry of the molecule, and the nature of the bonds between atoms. Let's consider how these general principles apply to the particular case of cyclopentane and 1-pentene.

Let's first note that the number and kind of atoms in the two compounds are identical. Thus any difference in the entropy of atom combination must result from the way the atoms are held together in the molecule. The more negative value of ΔS°_{ac} for cyclopentane tells us that the atoms in cyclopentane have lost more freedom of motion than those of 1-pentene during the process that leads to formation of these corresponding molecules. This, in turn, means that the atoms of cyclopentane have a more constrained motion than those of 1-pentene.

The carbon atoms in 1-pentene are free to move relative to one another, and the various segments of the molecule twist around the single bonds between carbon atoms. Because cyclopentane is a cyclic structure, the twisting of segments of the molecule would require covalent bonds to break. Thus the carbon atoms in cyclopentane are restricted from undergoing that type of motion.

The entropy change for the reaction

$$\text{Cyclopentane}\,(g) \rightleftharpoons \text{1-Pentene}(g)$$

can be calculated by combining the entropy associated with decomposing cyclopentane into its atoms in the gas phase and the entropy associated with bringing these atoms together to form 1-pentene.

Bond breaking	$\Delta S^\circ = 1645\ \text{J/mol}_{rxn}\cdot\text{K}$
Bond making	$\Delta S^\circ = -1590\ \text{J/mol}_{rxn}\cdot\text{K}$
	$\Delta S^\circ = +55\ \text{J/mol}_{rxn}\cdot\text{K}$

The conversion of cyclopentane to 1-pentene is accompanied by an increase in entropy, which in turn means that there is greater freedom of motion (greater disorder) for the atoms in the molecule 1-pentene than for those of cyclopentane.

Entropies of atom combination give information on how atoms are bonded together. It is important to note, however, that the number of the atoms must be the same for direct comparisons to be made. If more atoms are required to form the compound, more freedom of motion will be lost. The entropies of atom combination of $C_2H_6(g)$ and $CH_4(g)$ are -775 and $-431\ \text{J/mol}_{rxn}\cdot\text{K}$, respectively. C_2H_6 is formed from a total of eight atoms, whereas CH_4 is formed from only five.

ΔS°_{ac} data can provide a useful tool for studying a chemical reaction. There is always a positive entropy change when the bonds in the starting material are broken to produce isolated atoms in the gas phase. The gaseous atoms can then combine to reform the starting materials, or they can rearrange to form the products of the reaction. In either case, the system becomes less disordered. As a result, the entropy change is negative.

This raises an interesting question: Which process is favored by entropy—combination of the atoms in the gas phase to reform the starting materials, or rearrangement of the atoms to form the products of the reaction? Let's consider the two nitrogen atoms and four oxygen atoms in the example discussed at the

➤ CHECKPOINT

What would be the sign of ΔS for the conversion of glucose from its cyclic into its open-chain form?

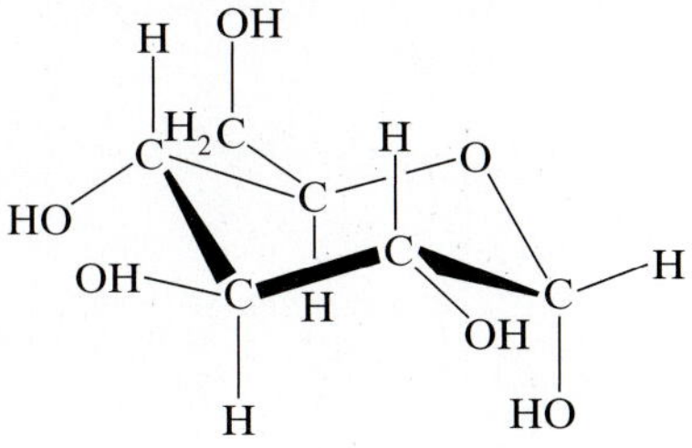

Glucose

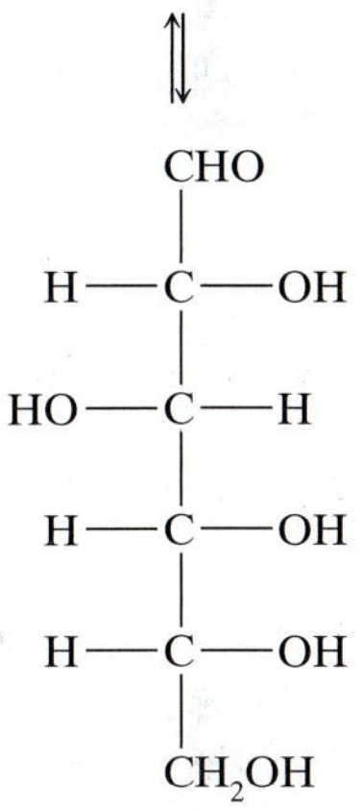

➤ CHECKPOINT

The entropies of atom combination of butane and isobutane are -1469 and $-1485\ \text{J/mol}_{rxn}\cdot\text{K}$, respectively. How can you account for this difference?

$CH_3CH_2CH_2CH_3$ (*Butane*)

$CH_3\underset{\displaystyle |}{\overset{CH_3}{}}\!\!CHCH_3$ — $CH_3CH(CH_3)CH_3$ (*Isobutane*)

beginning of Section 13.6. These atoms may combine to form either two molecules of NO_2 or one molecule of N_2O_4.

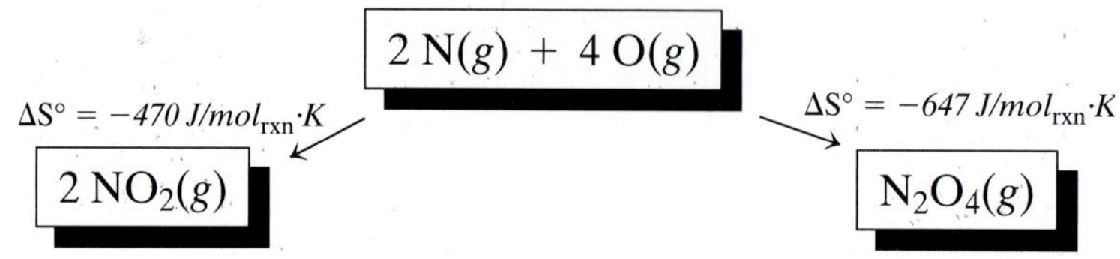

The second law of thermodynamics suggests that natural processes tend to maximum disorder. Entropy therefore favors the process in which the gaseous atoms combine to form the materials that involve the smallest loss of entropy. We can therefore conclude that for this reaction entropy favors the starting materials (NO_2), not the products (N_2O_4).

Exercise 13.3

Use the entropy of atom combination data given below to calculate the value of $\Delta S°$ at 298 K for the following reactions. Explain the sign of $\Delta S°$ for each reaction.

(a) $CH_3OH(l) \rightleftharpoons CH_3OH(g)$

Compound	$\Delta S°_{ac}$ *(J/mol$_{rxn}$ · K)*
$CH_3OH(l)$	−651.2
$CH_3OH(g)$	−538.19

(b) $N_2(g) + O_2(g) \rightleftharpoons 2\ NO(g)$

Compound	$\Delta S°_{ac}$ *(J/mol$_{rxn}$ · K)*
$N_2(g)$	−114.99
$O_2(g)$	−116.97
$NO(g)$	−103.59

Solution

(a) Entropy of atom combination data are used by thinking about chemical reactions as if all bonds in the reactants are broken to form isolated atoms in the gas phase, which then combine to form the products of the reaction. Relatively few reactions occur by this mechanism, but entropy is a state function. This means that the value of ΔS obtained this way is the same as the value of ΔS that would be measured experimentally.

For liquid CH_3OH we imagine a process that involves overcoming the intermolecular forces that hold the molecules together in the liquid and then breaking the intramolecular bonds within the CH_3OH molecules. This leads to an increase in the entropy of the system by 651.2 J/mol$_{rxn}$ · K. If the atoms reassemble to form a mole of CH_3OH gas, the entropy of the system would decrease by 538.19 J/mol$_{rxn}$ · K.

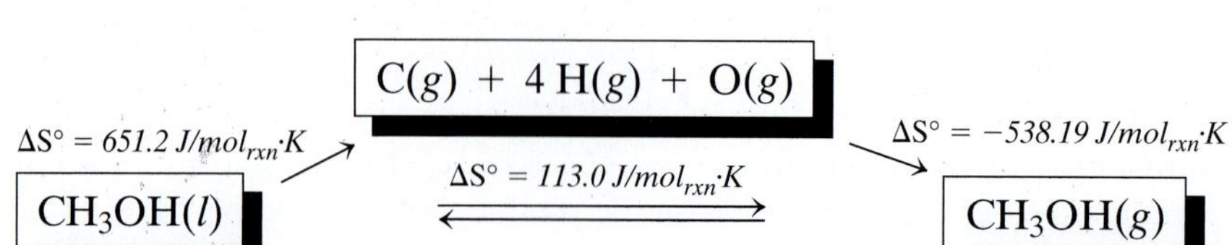

When we combine the change in entropy for the steps in which bonds are broken and made, we find that the overall entropy of reaction is positive.

Bond breaking	$651.2\ \text{J/mol}_{\text{rxn}} \cdot \text{K}$
Bond making	$-538.19\ \text{J/mol}_{\text{rxn}} \cdot \text{K}$
	$113.0\ \text{J/mol}_{\text{rxn}} \cdot \text{K}$

The entropy of the system increases because we are transforming a relatively ordered liquid into a more disordered gas.

(b) The change in entropy of the system that accompanies the process by which a mole of N_2 and a mole of O_2 are transformed into isolated nitrogen and oxygen atoms in the gas phase is equal to the sum of the entropy changes required to break the bonds in the two starting materials.

Bond breaking:

$$N_2(g) \longrightarrow 2\,N(g) \qquad \Delta S^\circ = 114.99\ \text{J/mol}_{\text{rxn}} \cdot \text{K}$$

$$O_2(g) \longrightarrow 2\,O(g) \qquad \Delta S^\circ = 116.97\ \text{J/mol}_{\text{rxn}} \cdot \text{K}$$

$$N_2(g) + O_2(g) \longrightarrow 2\,N(g) + 2\,O(g) \qquad \Delta S^\circ = 231.96\ \text{J/mol}_{\text{rxn}} \cdot \text{K}$$

The magnitude of ΔS° for the step in which the atoms in the gas phase combine to form the two moles of NO generated in the reaction is equal to twice the entropy of atom combination for the product of the reaction. The sign of this step is negative because the system becomes more ordered.

Bond making:

$$2\,N(g) + 2\,O(g) \longrightarrow 2\,NO(g) \qquad \Delta S^\circ_{ac} = -(2 \times 103.59\ \text{J/mol}_{\text{rxn}} \cdot \text{K})$$

The overall change in entropy that accompanies the reaction can be calculated by combining the two steps in the hypothetical reaction.

Bond breaking	$231.96\ \text{J/mol}_{\text{rxn}} \cdot \text{K}$
Bond making	$-207.18\ \text{J/mol}_{\text{rxn}} \cdot \text{K}$
	$24.78\ \text{J/mol}_{\text{rxn}} \cdot \text{K}$

ΔS° for the reaction is small but positive because the rotational freedom is larger in NO because it is less symmetric than N_2 or O_2.

13.7 Gibbs Free Energy

Section 13.3 noted that some reactions are spontaneous because they give off heat ($\Delta H < 0$) and others are spontaneous because they lead to an increase in the disorder of the system ($\Delta S > 0$). The following exercise shows how calculations of ΔH and ΔS can be used to determine the driving force behind a particular reaction.

Exercise 13.4

Use the enthalpy and entropy of atom combination data given below to calculate $\Delta H°$ and $\Delta S°$ for the following reaction. In which direction would each of these factors drive the reaction?

$$N_2(g) + 3\ H_2(g) \rightleftharpoons 2\ NH_3(g)$$

Compound	$\Delta H°_{ac}$ *(kJ/mol$_{rxn}$)*	$\Delta S°_{ac}$ *(J/mol$_{rxn}$ · K)*
$N_2(g)$	−945.41	−114.99
$H_2(g)$	−435.30	−98.74
$NH_3(g)$	−1171.76	−304.99

Solution

Both calculations can be based on the following diagram.

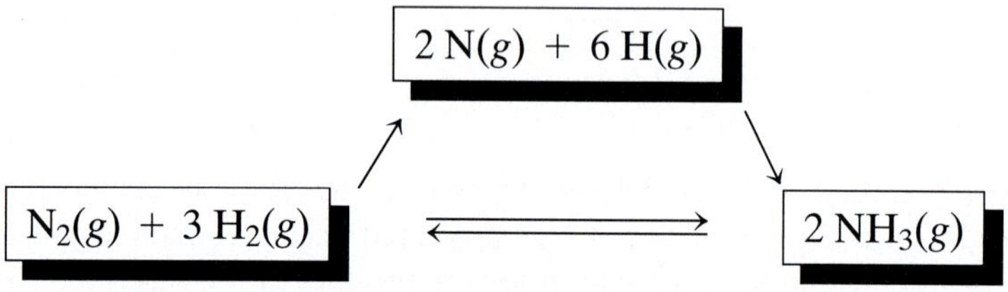

The enthalpy associated with breaking apart a mole of N_2 and 3 mol of H_2 molecules to form isolated nitrogen and hydrogen atoms in the gas phase is equal to the negative of the enthalpy of atom combination of N_2 ($\Delta H°_{ac} = -945.41$ kJ/mol$_{rxn}$) plus three times the negative of the enthalpy of atom combination of H_2 ($\Delta H°_{ac} = -435.30$ kJ/mol$_{rxn}$).

Bond breaking:

$$\begin{array}{rl} N_2(g) \longrightarrow 2\ N(g) & \Delta H° = 945.41\ \text{kJ/mol}_{rxn} \\ 3\ H_2(g) \longrightarrow 6\ H(g) & \Delta H° = (3 \times 435.30\ \text{kJ/mol}_{rxn}) \\ \hline N_2(g) + 3\ H_2(g) \longrightarrow 2\ N(g) + 6\ H(g) & \Delta H° = 2251.3\ \text{kJ/mol}_{rxn} \end{array}$$

The magnitude of the enthalpy given off when the atoms come together to form 2 mol of NH_3 is equal to twice the enthalpy of atom combination of ammonia ($\Delta H°_{ac} = -1171.76$ kJ/mol$_{rxn}$).

Bond making:

$$2\ N(g) + 6\ H(g) \longrightarrow 2\ NH_3 \qquad \Delta H°_{ac} = -(2 \times 1171.76\ \text{kJ/mol}_{rxn})$$

When the bond-breaking and bond-making steps are combined, we find that the overall reaction is exothermic ($\Delta H° < 0$). The sum of the bond strengths in the products must be greater than the sum of the bond strengths in the reactants.

$$\begin{array}{r} 2251.3\ \text{kJ/mol}_{rxn} \\ -2343.52\ \text{kJ/mol}_{rxn} \\ \hline -92.2\ \text{kJ/mol}_{rxn} \end{array}$$

The entropy change associated with transforming a mole of N_2 and 3 mol of H_2 to nitrogen and hydrogen atoms can be found by a similar procedure.

Bond breaking:

$$N_2(g) \longrightarrow 2\,N(g) \qquad \Delta S^\circ = 114.99\ \text{J/mol}_{\text{rxn}} \cdot \text{K}$$
$$3\,H_2(g) \longrightarrow 6\,H(g) \qquad \Delta S^\circ = (3 \times 98.74\ \text{J/mol}_{\text{rxn}} \cdot \text{K})$$
$$N_2(g) + 3\,H_2(g) \longrightarrow 2\,N(g) + 6\,H(g) \qquad \Delta S^\circ = 411.2\ \text{J/mol}_{\text{rxn}} \cdot \text{K}$$

The entropy change associated with the formation of 2 mol of NH_3 can be calculated from the entropy of atom combination of ammonia.

Bond making:

$$2\,N(g) + 6\,H(g) \longrightarrow 2\,NH_3 \qquad \Delta S^\circ_{ac} = -(2 \times 304.99\ \text{J/mol}_{\text{rxn}} \cdot \text{K})$$

The overall entropy of reaction is negative because the reaction transforms 4 mol of reactants into 2 mol of products.

$$\begin{array}{r} 411.2\ \ \text{J/mol}_{\text{rxn}} \cdot \text{K} \\ -609.98\ \text{J/mol}_{\text{rxn}} \cdot \text{K} \\ \hline -198.8\ \ \text{J/mol}_{\text{rxn}} \cdot \text{K} \end{array}$$

According to these calculations, ΔH° for the reaction is -92.2 kJ/mol$_{\text{rxn}}$ and ΔS° for the reaction is -198.8 J/mol$_{\text{rxn}} \cdot$ K. Enthalpy ($\Delta H^\circ < 0$) therefore drives the reaction toward the products, but entropy ($\Delta S^\circ < 0$) drives the reaction toward the reactants.

This exercise raises an important question: What happens when one of these two thermodynamic quantities for a chemical reaction is favorable and the other is not? We can answer this question by defining a new quantity known as the **Gibbs free energy (*G*)** of the system, which reflects the balance between the forces. The name of this quantity recognizes the contributions of J. Willard Gibbs, a professor of mathematical physics at Yale from 1871 until the early 1900s, who is considered by some to be the greatest scientist produced by the United States.

The Gibbs free energy of a system at any moment in time is defined as the enthalpy of the system minus the product of the temperature times the entropy of the system.

$$G = H - TS$$

The Gibbs free energy is a state function because it is defined in terms of thermodynamic properties (enthalpy and entropy) that are state functions. At a given temperature, the change in the Gibbs free energy of the system that occurs during a reaction is therefore equal to the change in the enthalpy of the system minus the product of the temperature times the change in entropy of the system.

$$\Delta G = \Delta H - T\Delta S$$

The beauty of the equation defining the free energy of a system is its ability to determine the relative importance of the enthalpy and entropy terms for a particular reaction. The change in the free energy of the system that occurs during a reaction measures the balance between the two driving forces that determine whether a reaction is spontaneous.

As we have seen, the enthalpy and entropy terms have different sign conventions.

Favorable	*Unfavorable*
$\Delta H < 0$	$\Delta H > 0$
$\Delta S > 0$	$\Delta S < 0$

The entropy term is therefore subtracted from the enthalpy term when calculating ΔG for a reaction.

Because of the way the free energy of the system is defined, ΔG is negative for any reaction for which ΔH is negative and ΔS is positive. ΔG is therefore negative for any reaction that is favored by both the enthalpy and entropy terms. We can therefore conclude that any reaction for which ΔG is negative should be favorable, or spontaneous.

For favorable, or spontaneous, reactions, $\Delta G < 0$.

Conversely, ΔG is positive for any reaction for which ΔH is positive and ΔS is negative. Any reaction for which ΔG is positive is therefore unfavorable.

> **➤ CHECKPOINT**
>
> The checkpoint at the end of Section 13.1 noted that table salt dissolves in water even though heat is absorbed during the process ($\Delta H > 0$). What can we conclude about ΔS for the reaction?

For unfavorable, or nonspontaneous, reactions, $\Delta G > 0$.

Reactions are classified as either **exothermic** ($\Delta H < 0$) or **endothermic** ($\Delta H > 0$) on the basis of whether they give off or absorb heat. Reactions can also be classified as **exergonic** ($\Delta G < 0$) or **endergonic** ($\Delta G > 0$) on the basis of whether the free energy of the system decreases or increases during the reaction.

When a reaction is favored by both enthalpy ($\Delta H < 0$) and entropy ($\Delta S > 0$), there is no need to calculate the value of ΔG to decide whether the reaction should proceed. The same can be said for reactions favored by neither enthalpy ($\Delta H > 0$) nor entropy ($\Delta S < 0$). Free energy calculations are important, however, for reactions favored by only one of these factors.

The change in the free energy of a system that occurs during a reaction can be measured under any set of conditions. If the data are collected under standard-state conditions, the result is the **standard-state free energy of reaction ($\Delta G°$).**

$$\Delta G° = \Delta H° - T\Delta S°$$

As always, the standard state for thermodynamic measurements assumes that all gases are present at 1 atm pressure and that the concentrations of all solutions are 1 *M*.

Exercise 13.5

Use the atom combination data given below to calculate $\Delta H°$ and $\Delta S°$ for the following reaction at 25°C.

$$2\ NO_2(g) \rightleftharpoons N_2O_4(g)$$

Compound	$\Delta H°_{ac}$ *(kJ/mol$_{rxn}$)*	$\Delta S°_{ac}$ *(J/mol$_{rxn}$ · K)*
$NO_2(g)$	−937.86	−235.35
$N_2O_4(g)$	−1932.93	−646.53

Use the results of these calculations to determine the value of $\Delta G°$ for the reaction and to predict whether the reaction will occur as written.

Solution

The following diagram can be used to calculate ΔH° and ΔS° for this reaction from enthalpy and entropy of atom combination data.

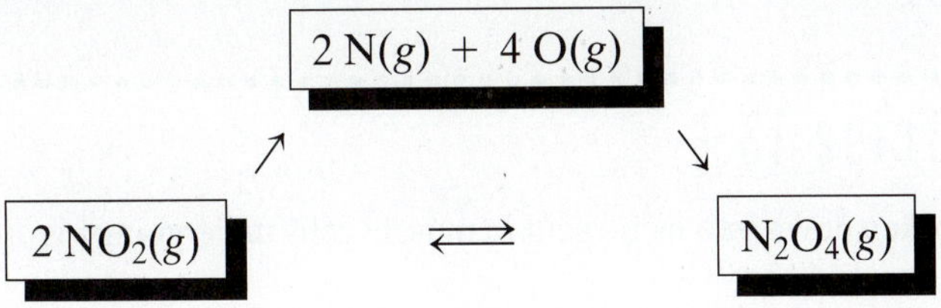

The enthalpy of reaction calculation can be set up as follows:

Bond breaking	$2\,NO_2(g) \longrightarrow 2\,N(g) + 4\,O(g)$	$\Delta H^\circ = (2 \times 937.86\ \text{kJ/mol}_{rxn})$
Bond making	$2\,N(g) + 4\,O(g) \longrightarrow N_2O_4(g)$	$\Delta H^\circ = -1932.93\ \text{kJ/mol}_{rxn}$
		$\Delta H^\circ = -57.20\ \text{kJ/mol}_{rxn}$

The overall reaction is exothermic, and the enthalpy change of reaction is therefore favorable.

The entropy of reaction calculation is done in much the same way.

Bond breaking	$2\,NO_2(g) \longrightarrow 2\,N(g) + 4\,O(g)$	$\Delta S^\circ = (2 \times 235.35\ \text{J/mol}_{rxn}\cdot\text{K})$
Bond making	$2\,N(g) + 4\,O(g) \longrightarrow N_2O_4(g)$	$\Delta S^\circ = -646.53\ \text{J/mol}_{rxn}\cdot\text{K}$
		$\Delta S^\circ = -175.83\ \text{J/mol}_{rxn}\cdot\text{K}$

The reaction leads to a significant decrease in the disorder of the system and therefore isn't favored by the entropy of reaction.

To decide whether the reaction should proceed at 25°C, we have to compare the ΔH° and $T\Delta S^\circ$ terms to see which is larger. Before we can do this, we have to convert the temperature to kelvins.

$$T_K = 25^\circ\text{C} + 273 = 298\ \text{K}$$

We also have to recognize that the units of ΔH° for the reaction are *kilojoules* and the units of ΔS° are *joules per kelvin.* At some point in the calculation, we have to convert these quantities to a consistent set of units. Perhaps the easiest way of doing this is to convert ΔS° to kilojoules. We then multiply the entropy term by the absolute temperature and subtract that quantity from the enthalpy term.

$$\begin{aligned}\Delta G^\circ &= \Delta H^\circ - T\Delta S^\circ \\ &= (-57.2\ \text{kJ/mol}_{rxn}) - (298\ \text{K})(-0.17583\ \text{kJ/mol}_{rxn}\cdot\text{K}) \\ &= (-57.2\ \text{kJ/mol}_{rxn}) + (52.4\ \text{kJ/mol}_{rxn}) \\ &= -4.8\ \text{kJ/mol}_{rxn}\end{aligned}$$

At 25°C, the standard-state free energy for the reaction is negative because the enthalpy term at that temperature is larger in magnitude and opposite in sign to the entropy term, $T\Delta S^\circ$.

It is important to note that standard-state conditions were used in this calculation. As a result, our conclusion that the reaction is spontaneous at 25°C only applies when all gases are present at 1 atm pressure.

Exercise 13.6

Glucose is broken down into lactic acid in muscle cells undergoing vigorous exercise.

$$C_6H_{12}O_6(aq) \rightleftharpoons 2\ CH_3CHOHCO_2H(aq)$$

Use the following thermodynamic data collected at 298.15 K to determine the driving forces (if any) pushing the reaction forward and whether the reaction is favorable at this temperature.

Compound	ΔH°_{ac} *(kJ/mol$_{rxn}$)*	ΔS°_{ac} *(J/mol$_{rxn}$ · K)*
$C_6H_{12}O_6\ (aq)$	−9670.0	−3013.9
$CH_3CHOHCO_2H(aq)$	−4889.7	−1417

Solution

This exercise provides an example of how tables of thermochemical data can be used to determine whether a reaction should occur.

	ΔH° (kJ/mol$_{rxn}$)	ΔS° (J/mol$_{rxn}$ · K)
Bond breaking	9670.0	3013.9
Bond making	−(2 × 4889.7)	−(2 × 1417)
	$\Delta H^\circ = -109.4$ kJ/mol$_{rxn}$	$\Delta S^\circ = 180$ J/mol$_{rxn}$ · K

ΔH° is negative, which means that the bonds in lactic acid are stronger than those in glucose. Thus lactic acid is favored by the enthalpy change. ΔS° is positive, which means the molecular structure of lactic acid has more freedom of motion than that of glucose and hence lactic acid is favored by the entropy change. Both the enthalpy and entropy changes indicate that this reaction can proceed under standard-state conditions.

> **CHECKPOINT**
>
> Think about what happens when cement hardens. (Does it become more or less ordered?) Use the results of predictions of the sign of ΔS for this process to predict whether it is possible to make a cement that will harden without releasing heat. Explain your reasoning.

We could obtain the same conclusion by calculating ΔG° for the reaction.

$$\begin{aligned}\Delta G^\circ &= \Delta H^\circ - T\Delta S^\circ \\ &= (-109.4\ \text{kJ/mol}_{\text{rxn}}) - (298.15\ \text{K} \times 0.180\ \text{kJ/mol}_{\text{rxn}} \cdot \text{K}) \\ &= -163\ \text{kJ/mol}_{\text{rxn}}\end{aligned}$$

The sign of ΔG° is negative, which suggests that the reaction is favorable, and the magnitude of ΔG° is relatively large.

13.8 The Effect of Temperature on the Free Energy of a Reaction

The balance between the contributions of the enthalpy and entropy terms to the free energy of a reaction depends on the temperature at which the reaction is run.

Exercise 13.7

Use the values of ΔH° and ΔS° calculated in Exercise 13.4 to predict whether the following reaction will occur at 25°C and 1 atm partial pressure for each gas.

$$N_2(g) + 3\ H_2(g) \rightleftharpoons 2\ NH_3(g)$$

Solution

According to the values obtained in Exercise 13.4, the reaction is favored by enthalpy but not by entropy.

$$\Delta H^\circ = -92.2\ \text{kJ/mol}_{\text{rxn}} \qquad \text{(favorable)}$$
$$\Delta S^\circ = -198.8\ \text{J/mol}_{\text{rxn}} \cdot \text{K} \qquad \text{(unfavorable)}$$

Before we can compare these terms to see which is larger, we have to incorporate into our calculation the temperature at which the reaction is run.

$$T_K = 25°\text{C} + 273 = 298\ \text{K}$$

We then multiply the entropy of reaction by the absolute temperature and subtract the $T\Delta S^\circ$ term from the ΔH° term.

$$\begin{aligned}\Delta G^\circ &= \Delta H^\circ - T\Delta S^\circ \\ &= (-92.2\ \text{kJ/mol}_{\text{rxn}}) - (298\ \text{K})(-0.1988\ \text{kJ/mol}_{\text{rxn}} \cdot \text{K}) \\ &= (-92.2\ \text{kJ/mol}_{\text{rxn}}) + (59.2\ \text{kJ/mol}_{\text{rxn}}) \\ &= -33.0\ \text{kJ/mol}_{\text{rxn}}\end{aligned}$$

The sign of ΔG° for this reaction is negative, which means the reaction should be spontaneous under standard-state conditions at 25°.

What happens as we raise the temperature of the reaction? The equation used to define free energy suggests that the entropy term will become more important as the temperature increases and may dominate the enthalpy term at high temperatures.

$$\Delta G^\circ = \Delta H^\circ - T\Delta S^\circ$$

Exercise 13.8

Predict whether the following reaction will occur at 500°C and 1 atm partial pressure of each gas.

$$N_2(g) + 3\ H_2(g) \rightleftharpoons 2\ NH_3(g)$$

Assume that the values of ΔH° and ΔS° used in the previous exercise are still valid at that temperature.

Solution

We need to calculate the temperature on the Kelvin scale.

$$T_K = 500°\text{C} + 273 = 773\ \text{K}$$

We then multiply the entropy term by the absolute temperature and subtract the result from the value of $\Delta H°$ for the reaction.

$$\begin{aligned}\Delta G°_{773} &= \Delta H°_{298} - T\Delta S°_{298} \\ &= (-92.2 \text{ kJ/mol}_{rxn}) - (773 \text{ K})(-0.1988 \text{ kJ/mol}_{rxn} \cdot \text{K}) \\ &= (-92.2 \text{ kJ/mol}_{rxn}) - (-154 \text{ kJ/mol}_{rxn}) \\ &= 62 \text{ kJ/mol}_{rxn}\end{aligned}$$

Because the entropy term becomes more important as the temperature increases, the reaction changes from one that is spontaneous at low temperatures to one that isn't spontaneous at high temperatures.

13.9 Beware of Oversimplifications

The calculation of $\Delta G°$ for the following reaction at 25°C in Exercise 13.7 is perfectly legitimate.

$$N_2(g) + 3\ H_2(g) \rightleftharpoons 2\ NH_3(g)$$

The enthalpy and entropy data used in the calculation were standard-state measurements made at 25°C, which can legitimately be used to predict the standard-state free energy of reaction at that temperature. The calculation of $\Delta G°_{773}$ for the reaction at 500°C in Exercise 13.8, however, was based on the assumption that the enthalpy and entropy data measured at 25°C are still valid at 500°C.

This is a useful assumption, but it isn't always valid. Changes in $\Delta H°$ and $\Delta S°$ are often small over moderate temperature ranges. When the temperature range over which the data are extrapolated is large, as it is in the case of Exercise 13.8, the results of the calculation should be taken with a grain of salt. They represent an estimate, not a prediction, of the magnitude of the effect of the change in temperature. In this case, the assumption that $\Delta H°$ and $\Delta S°$ are constant underestimates the change in the magnitude of $\Delta G°$ by about 20%. $\Delta G°_{773}$ for the reaction is 73 kJ/mol$_{rxn}$, not 62 kJ/mol$_{rxn}$.

13.10 Standard-State Free Energies of Reaction

$\Delta G°$ for a reaction can also be calculated from tabulated standard-state Gibbs free energy data, such as the **standard-state free energy of atom combination data, $\Delta G°_{ac}$,** in Table B.13. These values correspond to the formation of 1 mol of the substance listed in the table from its gaseous atoms. **$\Delta G°_{ac}$** is a measure of the strength of the intermolecular forces and intramolecular bonds (ΔH) as well as a measure of the freedom lost on atom combination (ΔS). Consider the following reaction at 298 K and 1 atm partial pressure of each gas.

$$CH_4(g) + 2\ O_2(g) \longrightarrow CO_2(g) + 2\ H_2O(g)$$

Compound	$\Delta G°_{ac}$ *(kJ/mol$_{rxn}$)*
$CH_4(g)$	−1535.00
$O_2(g)$	−463.46
$CO_2(g)$	−1529.08
$H_2O(g)$	−866.80

The following diagram can be used as the basis of calculating the overall change in the free energy of the system that occurs in this reaction from the ΔG°_{ac} data.

$$\begin{array}{ccccc} & \nearrow & C(g) + 4\,H(g) + 4\,O(g) & \searrow & \\ CH_4(g) + 2\,O_2(g) & & \longrightarrow & & CO_2(g) + 2\,H_2O(g) \end{array}$$

The bond-breaking step in the reaction involves atomization of 2 mol of O_2 for each mole of CH_4 consumed in the reaction.

Bond breaking:

$$\begin{array}{ll} CH_4(g) \longrightarrow C(g) + 4\,H(g) & \Delta G^\circ = 1535.00 \text{ kJ/mol}_{rxn} \\ 2\,O_2(g) \longrightarrow 4\,O(g) & \underline{\Delta G^\circ = (2 \times 463.46 \text{ kJ/mol}_{rxn})} \\ & \Delta G^\circ = 2461.92 \text{ kJ/mol}_{rxn} \end{array}$$

The magnitude of the change in the free energy of the system as the isolated atoms in the gas phase combine to form the products of the reaction is equal to the sum of the free energy of atom combination of CO_2 and twice that of H_2O.

Bond making:

$$\begin{array}{ll} C(g) + 2\,O(g) \longrightarrow CO_2(g) & \Delta G^\circ = -1529.08 \text{ kJ/mol}_{rxn} \\ 4\,H(g) + 2\,O(g) \longrightarrow 2\,H_2O(g) & \underline{\Delta G^\circ = -(2 \times 866.80 \text{ kJ/mol}_{rxn})} \\ & \Delta G^\circ = -3262.7 \text{ kJ/mol}_{rxn} \end{array}$$

When the results of the calculations are combined, we find that the overall free energy of reaction is negative. The reaction is therefore spontaneous at 298 K and 1 atm.

$$\begin{array}{lll} \text{Bond breaking} & \Delta G^\circ = & 2461.92 \text{ kJ/mol}_{rxn} \\ \text{Bond making} & \Delta G^\circ = & \underline{-3262.7 \text{ kJ/mol}_{rxn}} \\ & \Delta G^\circ = & -800.8 \text{ kJ/mol}_{rxn} \end{array}$$

If all reactant molecules are atomized to gaseous atoms and if the gaseous atoms may recombine, what factors determine whether the recombination is to form the original reactants or the products? Those molecules having the largest total bond strengths are enthalpically preferred, whereas those molecules having the most freedom associated with their molecular structure are entropically preferred. There is a balance between the stability (enthalpy) and freedom (entropy) that determines the position of the equilibrium.

13.11 Equilibria Expressed in Partial Pressures

Equilibria are often described by equilibrium constant expressions that are written in terms of the concentrations of the reactants and products in units of moles per liter. When this is done, a subscript "c" is added to the symbol for the equilibrium constant, to show that it was calculated from concentrations of the components of the reaction. Consider the following reaction, for example.

$$N_2(g) + 3\ H_2(g) \rightleftharpoons 2\ NH_3(g)$$

The expression for K_c would be written as follows:

$$K_c = \frac{[NH_3]^2}{[N_2][H_2]^3}$$

Equilibrium constant expressions for gas phase reactions can also be written in terms of the partial pressures of the components of the reaction, instead of their concentrations. We can understand why this is possible by rearranging the ideal gas equation

$$PV = nRT$$

to give the following relationship between the pressure of a gas and its concentration in moles per liter, where n/V is the concentration in moles per liter.

$$P = \frac{n}{V} \times RT$$

We can therefore characterize the reaction between N_2 and H_2 to form NH_3 as follows:

$$K_p = \frac{P_{NH_3}{}^2}{P_{N_2} P_{H_2}{}^3}$$

K_p, like K_c, is always reported without units. However, any calculation involving K_p requires the partial pressures of the products and reactants to be in units of atmospheres.

What is the relationship between K_p and K_c for a gas phase reaction? According to the rearranged version of the ideal gas equation, the pressure of a gas is equal to the concentration of the gas times the product of the ideal gas constant and the temperature in units of kelvins.

$$P = \left[\frac{n}{V}\right] \times RT$$

We can therefore calculate the value of K_p for a reaction by multiplying each of the terms in the K_c expression by RT.

$$K_p = \frac{P_{NH_3}{}^2}{P_{N_2} P_{H_2}{}^3} = \frac{([NH_3] \times RT)^2}{([N_2] \times RT)([H_2] \times RT)^3}$$

Collecting terms in this example gives the following result.

$$K_p = K_c \times (RT)^{-2}$$

In general, the value of K_p for a reaction can be calculated from K_c with the following equation.

$$K_p = K_c \times (RT)^{\Delta n}$$

In this equation, Δn is the difference between the number of moles of gaseous products and the number of moles of gaseous reactants in the balanced equation ($\Delta n = n_{prod} - n_{react}$).

Exercise 13.9

Calculate the value of K_p for the following reaction at 650°C if the value of K_c for the reaction at that temperature is 0.040.

$$N_2(g) + 3\,H_2(g) \rightleftharpoons 2\,NH_3(g)$$

Solution

The value of K_p for a reaction can be calculated from K_c with the following equation.

$$K_p = K_c \times (RT)^{\Delta n}$$

To use this equation, we need to know the value of Δn for the reaction. The balanced equation for this gas-phase reaction generates 2 mol of products for every 4 mol of reactants consumed. The value of Δn for the reaction is therefore $2 - 4$, or -2.

$$K_p = K_c \times (RT)^{-2}$$

We now use the known value of the ideal gas constant and the temperature in kelvins to calculate the value of K_p for the reaction at the temperature given.

$$K_p = (0.040) \times [(0.08206 \text{ L} \cdot \text{atm/mol} \cdot \text{K})\,(923 \text{ K})]^{-2} = 7.0 \times 10^{-6}$$

The techniques for working problems using K_p expressions are the same as those described for K_c problems. The equilibrium constant calculated from tabulated standard Gibbs free energies, enthalpies, and entropies will give K_p for reactions which involve only gases or which have only gases and pure liquids or solids as products and reactants. Calculations of an equilibrium constant from the tabulated data will give K_c for reactions that contain only aqueous species and pure liquids and solids. For reactions involving only gaseous species and pure liquids and solids, K_c may be converted into K_p by the relationship described above.

Exercise 13.10

K_p for the following reaction is 60 at 350°C.

$$H_2(g) + I_2(g) \rightleftharpoons 2\,HI(g)$$

If the initial partial pressures of H_2 and I_2 are 1.00 atm each, what will be the equilibrium pressures of HI, H_2, and I_2?

Solution

There is no HI present initially. The reaction will therefore proceed to the right to form HI. The partial pressure of H_2 and I_2 will decrease by an amount that can be represented as ΔP, and the partial pressure of HI will increase by an amount that is twice as large.

	$H_2(g)$	+	$I_2(g)$	$\rightleftharpoons$	$2\,HI(g)$
Initial	1.00 atm		1.00 atm		0 atm
Change	$-\Delta P$		$-\Delta P$		$2\Delta P$
Equilibrium	$1.00 - \Delta P$		$1.00 - \Delta P$		$2\Delta P$

Substituting this information into the equilibrium constant expression gives the following result.

$$K_p = \frac{P_{HI}^{\ 2}}{P_{H_2}P_{I_2}} = \frac{(2\Delta P)^2}{(1.00 - \Delta P)(1.00 - \Delta P)} = 60$$

The equilibrium constant is reasonably large. This means that ΔP isn't likely to be small compared with the initial partial pressures of H_2 and I_2. We are therefore going to have to expand this equation as follows.

$$\frac{4\Delta P^2}{\Delta P^2 - 2\Delta P + 1.00} = 60$$

$$56\Delta P^2 - 120\Delta P + 60 = 0$$

Using the quadratic equation gives the following result.

$$\Delta P = \frac{120 \pm \sqrt{(120)^2 - 4(56)(60)}}{2(56)} = \frac{120 \pm 31}{112}$$

If we choose the positive root, then ΔP is 1.35, which means that more H_2 and I_2 are consumed as the reaction comes to equilibrium than were present initially. Thus, the negative root is used, which gives a ΔP of 0.79 atm. The equilibrium pressure of HI is $2\Delta P = 1.6$ atm, and those of H_2 and I_2 are $1.00 - \Delta P = 0.21$ atm.

13.12 Interpreting Standard-State Free Energy of Reaction Data

We are now ready to ask the question: What does the value of $\Delta G°$ tell us about a reaction? Consider the following reaction, for example.

$$N_2(g) + 3\ H_2(g) \rightleftharpoons 2\ NH_3(g) \qquad \Delta G° = -33.0\ \text{kJ/mol}_{\text{rxn}}$$

By definition, the value of $\Delta G°$ for a reaction measures the difference between the free energies of the reactants and products *when all components of the reaction are present at standard-state conditions.*

$\Delta G°$ therefore describes the reaction only when all three components are present at 1 atm pressure. Note in the following equation that the **reaction quotient, Q_p,** is written using pressures, not concentrations as previously used in Section 10.6. This is because the tabulated values of the thermodynamic parameters for gases are given in units of pressure.

$$\text{Standard state} \qquad Q_p = \frac{P_{NH_3}^{\ 2}}{P_{N_2}P_{H_2}^{\ 3}} = \frac{(1)^2}{(1)\,(1)^3} = 1$$

The *sign* of $\Delta G°$ tells us the direction in which the reaction has to shift from standard-state conditions to come to equilibrium. The fact that $\Delta G°$ is negative for this reaction at 25°C means that the system under standard-state conditions at that temperature would have to shift to the right, converting some of the reactants into products, before it can reach equilibrium. The *magnitude* of $\Delta G°$ for a reaction tells us how far the standard state is from equilibrium. The larger the value of $\Delta G°$, the further the reaction has to go to get from the standard-state conditions to equilibrium.

Assume, for example, that we start with the following reaction under standard-state conditions, as shown in Figure 13.5.

$$N_2(g) + 3\ H_2(g) \rightleftharpoons 2\ NH_3(g)$$

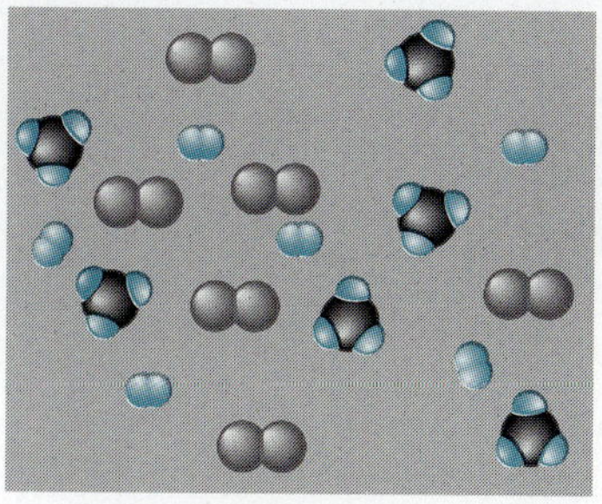

Fig. 13.5 At standard-state conditions, the partial pressures of N_2, H_2, and NH_3 in this system are 1 atm. Thus Q_p for the system is equal to 1.

The value of ΔG at that moment in time is equal to the standard-state free energy for the reaction, ΔG.

When $Q_p = 1$, $\Delta G = \Delta G°$.

As the reaction gradually shifts to the right, converting N_2 and H_2 into NH_3, the value of ΔG for the reaction becomes less negative. If we could find some way to harness the tendency of the reaction to come to equilibrium, we could get the reaction to do work. The free energy of a reaction at any moment in time is therefore said to be a measure of the energy available to do work.

13.13 The Relationship Between Free Energy and Equilibrium Constants

When a reaction leaves the standard state because of a change in the concentrations or partial pressures of the reactants and products, we have to describe the system in terms of a non-standard-state free energy of reaction. The difference between $\Delta G°$ and ΔG for a reaction is important. There is only one value of $\Delta G°$ for a reaction at a given temperature, but there are an infinite number of possible values of ΔG.

Figure 13.6 shows the relationship between ΔG and $\ln Q_p$ for the following reaction.

$$N_2(g) + 3\ H_2(g) \rightleftharpoons 2\ NH_3(g)$$

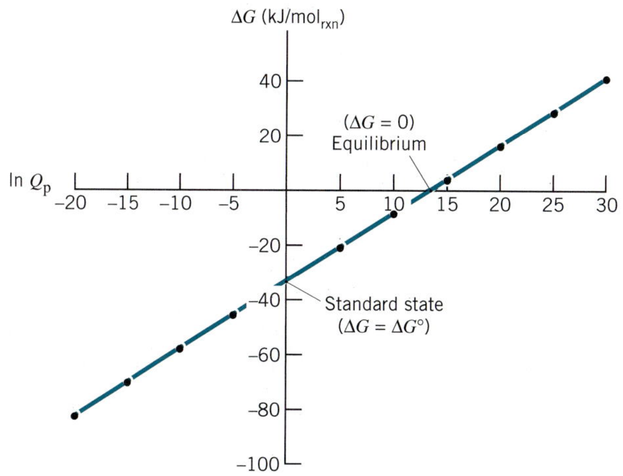

Fig. 13.6 Plot of $\ln Q_p$ versus ΔG for the reaction in which N_2 and H_2 combine to form NH_3. $\Delta G = \Delta G°$ when the reaction quotient is 1 ($\ln Q_p = 0$). $\Delta G = 0$ when the reaction is at equilibrium.

Data on the far left side of Figure 13.6 correspond to relatively small values of Q_p. They therefore describe systems in which there is far more reactant than product. The sign of ΔG for these systems is negative and the magnitude of ΔG is large. The system is therefore relatively far from equilibrium, and the reaction must shift to the right to reach equilibrium.

Data on the far right side above the x axis of Figure 13.6 describe systems in which there is more product than reactant. The sign of ΔG is now positive and the magnitude of ΔG is moderately large. The sign of ΔG tells us that the reaction would have to shift to the left to reach equilibrium. The magnitude of ΔG tells us how far we have to go to reach equilibrium.

The points at which the straight line in Figure 13.6 crosses the horizontal and vertical axes of the diagram are particularly important. The straight line crosses the vertical axis when the reaction quotient for the system is equal to 1 (ln $Q_p = 0$). That point therefore describes the standard-state conditions, and the value of ΔG at that point is equal to the standard-state free energy of reaction, $\Delta G°$.

When $Q_p = 1$, $\Delta G = \Delta G°$.

The point at which the straight line crosses the horizontal axis describes a system for which ΔG is equal to zero. Because there is no driving force behind the reaction, the system must be at equilibrium.

➤ CHECKPOINT

Which way does a reaction proceed when $Q_p > K_p$?

When $Q_p = K_p$, $\Delta G = 0$.

In general, the relationship between the free energy of reaction at any moment in time (ΔG) and the standard-state free energy of reaction ($\Delta G°$) is described by the following equation.

$$\Delta G = \Delta G° + RT \ln Q$$

In this equation, R is the ideal gas constant measured in joules (8.314 J/mol$_{rxn} \cdot$ K), T is the temperature in kelvins, ln represents a logarithm to the base e, and Q is the reaction quotient measured in either pressure or concentration units.

As we have seen, the driving force behind a chemical reaction is zero ($\Delta G = 0$) when the reaction is at equilibrium ($Q = K$).

$$0 = \Delta G° + RT \ln K$$

This equation can be solved for the relationship between $\Delta G°$ and K.

$$\Delta G° = -RT \ln K$$

Because ΔG° is often (incorrectly) used to predict the "spontaneity" of a reaction, the relationship of the reaction quotient to the equilibrium constant should assist in helping students to understand the difference between ΔG and ΔG°. ΔG° is a measure of the magnitude of the equilibrium constant, and therefore, when compared to RTlnQ shows how far the reaction is from equilibrium.

This equation allows us to calculate the equilibrium constant for any reaction from the standard-state free energy of reaction, or vice versa.

To understand the relationship between $\Delta G°$ and K, we must recognize that the magnitude of $\Delta G°$ tells us how far the standard state is from equilibrium. The smaller the absolute value of $\Delta G°$, the closer the standard state is to equilibrium. The larger the absolute value of $\Delta G°$, the further the reaction has to go to reach equilibrium. The relationship between $\Delta G°$ and the equilibrium constant for a chemical reaction is illustrated by the data in Table 13.2.

Table 13.2

Values of $\Delta G°$ and K for Common Reactions at 25°C

Reaction	$\Delta G°$(kJ/mol$_{rxn}$)	K
$2\ SO_3(g) \rightleftharpoons 2\ SO_2(g) + O_2(g)$	141.7	1.5×10^{-25}
$H_2O(l) \rightleftharpoons H^+(aq) + OH^-(aq)$	79.9	1.0×10^{-14}
$AgCl(s) \overset{H_2O}{\rightleftharpoons} Ag^+(aq) + Cl^-(aq)$	55.7	1.7×10^{-10}
$HOAc(aq) \overset{H_2O}{\rightleftharpoons} H^+(aq) + OAc^-(aq)$	27.1	1.8×10^{-5}
$N_2(g) + 3\ H_2(g) \rightleftharpoons 2\ NH_3(g)$	−33.0	6×10^5
$HCl(g) \xrightarrow{H_2O} H^+(aq) + Cl^-(aq)$	−35.9	2×10^6
$Cu^{2+}(aq) + 4\ NH_3(aq) \longrightarrow Cu(NH_3)_4{}^{2+}(aq)$	−70.6	2.3×10^{12}
$Zn(s) + Cu^{2+}(aq) \longrightarrow Zn^{2+}(aq) + Cu(s)$	−212.6	1.8×10^{37}

Exercise 13.11

Calculate the equilibrium constant for the following reaction at 25°C from the value of $\Delta G°$ for this reaction calculated in Exercise 13.7.

$$N_2(g) + 3\ H_2(g) \rightleftharpoons 2\ NH_3(g)$$

Solution

Exercise 13.7 gave the following value for $\Delta G°$ for this reaction at 25°C.

$$\Delta G° = -33.0 \text{ kJ/mol}_{rxn}$$

We now turn to the relationship between $\Delta G°$ for a reaction and the equilibrium constant for the reaction.

$$\Delta G° = -RT \ln K$$

Solving for the natural log of the equilibrium constant gives the following equation.

$$\ln K = -\frac{\Delta G°}{RT}$$

Substituting the known values of $\Delta G°$, R, and T into this equation gives the following result.

$$\ln K_p = -\frac{(-33.0 \times 10^3 \text{ J/mol}_{rxn})}{(8.314 \text{ J/mol}_{rxn} \cdot \text{K})(298 \text{ K})} = 13.3$$

The equilibrium constant for the reaction at 25°C is therefore 6×10^5.

$$K_p = e^{13.3} = 6 \times 10^5$$

Because this is a gas phase reaction, the equilibrium constant that results from this calculation is K_p.

It is easy to make mistakes when handling the sign of the relationship between ΔG° and K. It is therefore a good idea to check the final answer to see whether it makes sense. ΔG° for the reaction in this exercise is negative, and the equilibrium should lie on the side of the products. The equilibrium constant should therefore be much larger than 1, which it is.

Exercise 13.12

Calculate the acid dissociation equilibrium constant (K_a) for formic acid at 25°C from the value of ΔG° given below.

$$HCO_2H(aq) \rightleftharpoons H^+(aq) + HCO_2^-(aq) \qquad \Delta G^\circ = 21.3 \text{ kJ/mol}_{rxn}$$

Solution

We start with the relationship between ΔG° and the equilibrium constant for the reaction

$$\Delta G^\circ = -RT \ln K$$

and solve for the natural logarithm of the equilibrium constant.

$$\ln K = -\frac{\Delta G^\circ}{RT}$$

Substituting the known values of ΔG°, R, and T into the equation gives the following result.

$$\ln K_c = -\frac{(21.3 \times 10^3 \text{ J/mol}_{rxn})}{(8.314 \text{ J/mol}_{rxn} \cdot \text{K})(298 \text{ K})} = -8.60$$

We can now calculate the value of the equilibrium constant.

$$K_c = e^{-8.60} = 1.8 \times 10^{-4}$$

In this case ΔG° is positive and the equilibrium will favor the reactants; in other words, the equilibrium constant will be less than 1.

Exercise 13.13

The amino acid glutamine can be formed by reacting glutamic acid with NH_3 in aqueous solution.

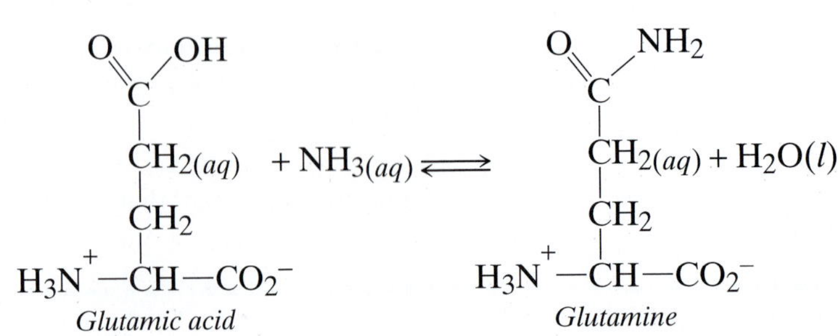

The equilibrium constant for this reaction is 3.5×10^{-3} at 25°C.

(a) Calculate $\Delta G°$ for the reaction.

(b) What does this value for $\Delta G°$ signify?

(c) NH_3 is present in the body at a concentration of about 1×10^{-2} *M*. If a concentration of glutamine of 1×10^{-2} *M* was desired from the reaction, what would have to be the concentration of glutamic acid?

Solution

(a)
$$\begin{aligned}\Delta G° &= -RT \ln K \\ &= -8.314 \text{ J/mol}_{rxn} \cdot \text{K}(298 \text{ K}) \ln 3.5 \times 10^{-3} \\ &= 14 \times 10^3 \text{ J/mol}_{rxn}\end{aligned}$$

(b) The large positive $\Delta G°$ means that the reaction does not proceed very far to the right. This does not mean, however, that the reaction cannot ever proceed spontaneously to make glutamine. It is ΔG that determines the spontaneous direction of the reaction, not $\Delta G°$.

(c) $$\Delta G = \Delta G° + RT \ln Q = \Delta G° + RT \ln \frac{(\text{glutamine})}{(\text{glutamic acid})(NH_3)}$$

$$\Delta G = (14 \times 10^3 \text{ J/mol}_{rxn}) + 8.314 \text{ J/mol}_{rxn} \cdot \text{K}(298 \text{ K}) \ln \frac{(1 \times 10^{-2})}{(\text{glutamic acid})(1 \times 10^{-2})}$$

For the reaction to proceed, ΔG must be negative. In order to find what glutamic acid concentration will make ΔG negative, we first set $\Delta G = 0$. Then

$$\Delta G = 0 = (14 \times 10^3 \text{ J/mol}_{rxn}) + 8.314 \text{ J/mol}_{rxn} \cdot \text{K}(298 \text{ K}) \ln \frac{(1 \times 10^{-2})}{(\text{glutamic acid})(1 \times 10^{-2})}$$

and (glutamic acid) $= 3 \times 10^2$ *M* at equilibrium. That is, in order for the reaction to proceed, the glutamic acid concentration would have to exceed 300 *M*! Thus this reaction could not be used for the biosynthesis of glutamine. Compare the results of this calculation to that of Exercise 10.6. Can you use the procedure of Exercise 10.6 to arrive at the same conclusion for the glutamic acid reaction?

13.14 The Temperature Dependence of Equilibrium Constants

When equilibrium constants were introduced in Chapter 10, we noted that they aren't strictly constant because they change with temperature. We are now ready to understand why.

The standard-state free energy of reaction is a measure of how far the standard state is from equilibrium.

$$\Delta G° = -RT \ln K$$

But the magnitude of $\Delta G°$ depends on the temperature of the reaction.

$$\Delta G° = \Delta H° - T\Delta S°$$

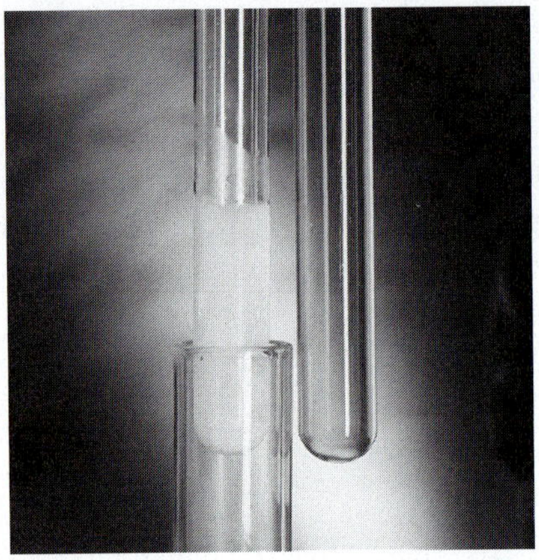

Fig. 13.7 The equilibrium between NO_2 and N_2O_4 changes with temperature. The characteristic brown color of NO_2 gas disappears as a tube containing NO_2 is lowered into a liquid nitrogen bath (−196°C).

As a result, the equilibrium constant must depend on the temperature of the reaction.

An example of this phenomenon is the reaction in which NO_2 dimerizes to form N_2O_4.

$$\underset{\textit{Brown}}{2\ NO_2(g)} \rightleftharpoons \underset{\textit{Colorless}}{N_2O_4(g)}$$

NO_2 is a brown gas and N_2O_4 is colorless. We can therefore monitor the extent to which NO_2 dimerizes to form N_2O_4 by examining the intensity of the brown color in a sealed tube of the gas. What should happen to the equilibrium between NO_2 and N_2O_4 as the temperature is lowered?

Figure 13.7 shows what happens to the intensity of the brown color when a sealed tube containing NO_2 gas in equilibrium with N_2O_4 at 25°C is immersed in liquid nitrogen. There is a drastic decrease in the amount of NO_2 in the tube as it is cooled to −196°C.

The dimerization of NO_2 is exothermic ($\Delta H° = -57.2$ kJ/mol$_{rxn}$). In Chapter 10 we described how decreasing the temperature of an exothermic reaction increases the equilibrium constant. The equilibrium constant for this reaction increases as we decrease the temperature of the system, as shown by the data in Table 13.3. The reaction therefore shifts toward N_2O_4 as we lower the temperature, as predicted by Le Châtelier's principle.

We can now develop the relationship between temperature and the equilibrium constant. We have seen that

$$\Delta G° = -RT \ln K = \Delta H° - T\Delta S°$$

Solving for ln K we find

$$\ln K = \frac{-\Delta H°}{RT} + \frac{\Delta S°}{R}$$

Because both $\Delta H°$ and $\Delta S°$ are essentially constant for moderate changes in temperature, this equation describes the change in K resulting from a change in temperature. Thus ln K is seen to depend on the magnitudes and signs of $\Delta H°$ and $\Delta S°$. However, the significance of the term $-\Delta H°/RT$ is dependent on temperature. For example, as temperature increases this term becomes smaller. If $\Delta H° < 0$, $-\Delta H°/RT$ is positive and contributes to making ln K large no matter what the sign of $\Delta S°$. But as temperature increases this term becomes a smaller contributor and ln K decreases. If $\Delta H° > 0$, $-\Delta H°/RT$ is negative and contributes to making ln K small. At higher temperatures this negative influence becomes less significant and ln K is larger than at low temperatures.

Table 13.3

Temperature Dependence of the Equilibrium Constant for the Dimerization of NO_2

Temperature (°C)	K_c	K_p
100	2.1	0.069
25	170	6.95
0	1.4×10^3	62.5
−78	4.0×10^8	2.5×10^7

Table 13.4

Temperature Dependence of the Equilibrium Constant for the Reaction of Methane to Produce Ethane

Temperature (K)	$\Delta G°(\text{kJ/mol}_{rxn})$	$\Delta G°/T$ (J/mol$_{rxn}$ · K)	ln K	K
298	68.58	230	−27.7	9×10^{-13}
400	69.84	175	−21.0	8×10^{-10}
600	72.28	120	−14.4	6×10^{-7}

Temperature dependence of the equilibrium constant is determined by the relation:

$$\ln K = \frac{-\Delta H°}{RT} + \frac{\Delta S°}{R}$$

For most purposes, ΔH° and ΔS° can be taken to be relatively constant over small temperature intervals. Thus the temperature dependence of lnK is determined by −ΔH°/RT.

It is tempting to predict the effect of temperature on the equilibrium constant by considering only the expression

$$\Delta G° = \Delta H° - T\Delta S°$$

Suppose that in a given situation $\Delta H°$ is positive and $\Delta S°$ is negative. Then $\Delta G°$ would become more positive with increasing temperature, and it might be expected that the equilibrium constant would decrease. However, according to Le Châtelier's principle, a positive $\Delta H°$ means that the equilibrium constant should increase with increasing temperature. This dilemma is resolved by noting that it is not $\Delta G°$ but $\Delta G°/T$ which determines the change of the equilibrium constant with temperature:

$$\Delta G° = -RT \ln K \quad \text{therefore} \quad \ln K = \frac{-\Delta G°}{RT}$$

Note that this equation *cannot* be used to describe the change in K as temperature is changed because $\Delta G°$ is *not* constant with respect to temperature (see Table 13.4).

The reaction of methane to produce ethane

$$2\ CH_4(g) \rightleftharpoons C_2H_6(g) + H_2(g)$$

has $\Delta H° = 64.94\ \text{kJ/mol}_{rxn}$ and $\Delta S° = -12.24\ \text{J/mol}_{rxn} \cdot \text{K}$. If $\Delta H°$ and $\Delta S°$ are assumed not to vary appreciably with temperature, $\Delta G°$ will become more positive with increasing temperature but K will nevertheless increase.

$$\Delta G° = 64.94\ \text{kJ/mol}_{rxn} - T(-0.01224\ \text{J/mol}_{rxn} \cdot \text{K})$$

The data in Table 13.4 illustrate the temperature dependence of the equilibrium constant for the above reaction, whose free energy change becomes more positive with increasing temperature.

Contrary to what is often stated, ΔG° does not determine how K will vary with temperature. For example, if ΔH° > 0 and ΔS° < 0, ΔG° (= ΔH° − TΔS°) will be positive and become more positive with increasing temperature; yet for ΔH° > 0, Le Chatlier's principle predicts that K will increase with increasing temperature. How can a more positive ΔG° lead to a larger K? The answer is that it is not ΔG° which determines how K varies with temperature but −ΔG°/T.

The entropy of the universe equation, $\Delta S_{univ} = \Delta S_{sys} + \Delta S_{surr}$ may be rewritten as

$$\Delta S_{\text{univ}} = \Delta S° - \frac{\Delta H°}{T} = \frac{-\Delta G°}{T}$$

where ΔS_{surr} is identical to −ΔH°/T. It is what happens to the entropy of the surroundings that determines how K varies with temperature. This is a common textbook error and often confuses students when predictions made based on the sign of ΔG° are compared with predictions made from Le Chatlier's principle.

Exercise 13.14

Proteins consist of long polymeric chains of amino acids. The hemoglobin that binds O_2 in the lungs and carries it through the bloodstream to the muscles, for example, consists of four polypeptide chains: two α-chains that contain 141 amino acids each and two β-chains that contain 146 amino acids, for a total molecular weight of roughly 65,000 g/mol. The active form of most proteins is one in which the amino acid chain folds, as shown in Figure 13.8.

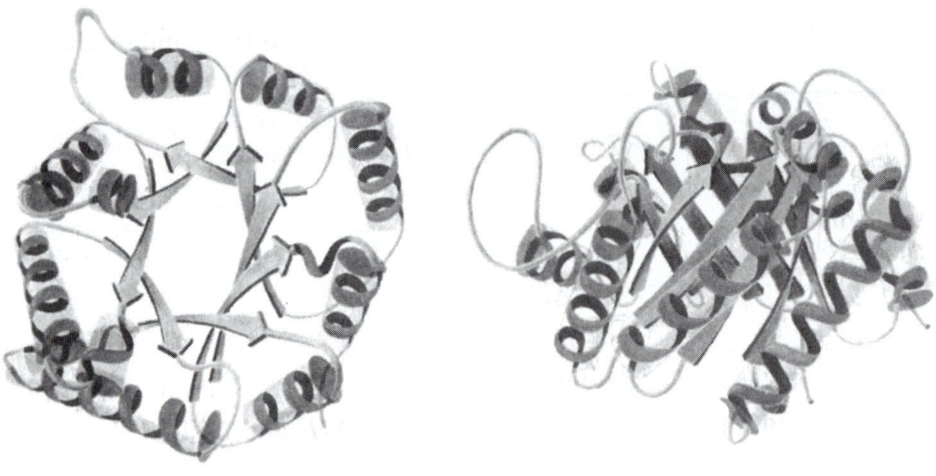

(a) *(b)*

Fig. 13.8 Top view (*a*) and side view (*b*) of the enzyme known as triose phosphate isomerase illustrate the folding and twisting of the protein chain within the molecule. Reprinted with permission from J. R. Holum, *Fundamentals of General, Organic, and Biological Chemistry,* John Wiley & Sons, Inc., New York, 1994, p. 598.

When heated or exposed to a change in pH, many proteins lose their folded structure. When this happens, the protein can't carry out its normal biochemical function and is said to have been denatured.

Trypsin is a type of protein called an enzyme that breaks the peptide bonds between amino acids in other proteins. The equilibrium constant for the denaturation of trypsin is 7.20 at 50°C.

$$\text{Trypsin} \rightleftharpoons \text{denatured trypsin}$$

(a) Calculate ΔG° for the process.

(b) ΔH° for the denaturation of trypsin is $+278\ \text{kJ/mol}_{\text{rxn}}$. What is ΔS° at 50°C?

(c) What do the sign and magnitude of ΔS° imply concerning the difference between denatured trypsin and trypsin?

(d) Under standard conditions, will trypsin denature at 50°C?

(e) What is the maximum temperature under standard conditions at which trypsin is stable?

(f) Use ΔH° and ΔS° to describe what happens when a protein such as trypsin denatures.

(g) As temperature increases, does the equilibrium constant increase, decrease, or remain the same? Will the equilibrium shift and, if so, in which direction?

Solution

(a) ΔG° can be calculated from the equilibrium constant for the denaturation of trypsin.

$$\begin{aligned}\Delta G^\circ &= -RT \ln K_c \\ &= -(8.314\ \text{J/mol}_{\text{rxn}} \cdot \text{K})(323\ \text{K})(\ln 7.20) \\ &= -5.30\ \text{kJ/mol}_{\text{rxn}}\end{aligned}$$

(b) We can start with the equation that defines the relationship among free energy, enthalpy, and entropy.

$$\Delta G^\circ = \Delta H^\circ - T\Delta S^\circ$$

Substituting the known values of $\Delta G°$, $\Delta H°$, and T gives the following result,

$$-5.30 \text{ kJ/mol}_{rxn} = 278 \text{ kJ/mol}_{rxn} - (323 \text{ K})(\Delta S°)$$

which can be solved for the value of $\Delta S°$ for the reaction.

$$\Delta S° = 877 \text{ J/mol}_{rxn} \cdot \text{K}$$

(c) $\Delta S°$ is relatively large and positive. This means that a significant amount of freedom of motion has been gained by the denaturing of trypsin.

(d) $\Delta G°$ is negative for the denaturation of trypsin at 50°C, so heating trypsin to that temperature will cause it to unfold.

(e) Trypsin will denature at any temperature at which $\Delta G°$ is negative. It would therefore be useful to find the temperature at which $\Delta G°$ for this reaction is zero. We start with the equation that defines the relationship among $\Delta G°$, $\Delta H°$, and $\Delta S°$.

$$\Delta G° = \Delta H° - T\Delta S°$$

We then set $\Delta G° = 0$ and solve for T, assuming that $\Delta H°$ and $\Delta S°$ don't change much with temperature.

$$0 = \Delta H° - T\Delta S°$$

$$T = \frac{\Delta H°}{\Delta S°} = \frac{(278 \text{ kJ/mol}_{rxn})}{(0.877 \text{ kJ/mol}_{rxn} \cdot \text{K})} = 317 \text{ K} = 44°\text{C}$$

According to this calculation, the maximum temperature at which trypsin is stable is 44°C. At any temperature *above* 44°C, $\Delta G°$ for the denaturation reaction is negative and the protein should denature.

(f) The positive sign of $\Delta H°$ shows that the protein is more tightly bonded in its folded structure than in its unfolded state. The large positive sign of $\Delta S°$ means that the unfolding process has produced rotational and vibrational freedom of motion not possible in the folded position. Thus denaturation requires the disruption of strong forces of attraction to produce weaker ones. For denaturation to occur, there must be a gain in entropy that offsets the disruption of the forces of attraction.

(g) This is an endothermic reaction ($\Delta H° = 278$ kJ/mol$_{rxn}$). As temperature increases, the equilibrium constant will increase. As the equilibrium constant becomes larger, the reaction will shift to the right to produce more product.

➤ CHECKPOINT

If $\Delta H° > 0$ and $\Delta S° > 0$ for a reaction, which way will the equilibrium shift if the temperature is decreased? If $\Delta H° > 0$ and $\Delta S° < 0$ for a reaction, which way will the equilibrium shift if the temperature is decreased?

13.15 Gibbs Free Energies of Formation and Absolute Entropies

We saw in Chapter 7 that enthalpy changes can be calculated by use of enthalpies of atom combination or enthalpies of formation. Enthalpies of atom combination use the free gaseous atoms as a point of reference, while enthalpies of formation use the element in its most stable state of aggregation at 1 bar and usually at 298 K. Because the bar and the atmosphere differ by, so little, we will as in Chapter 7, assume the standard-state pressure to be 1 atm.

After students have learned to use free energies and entropies of atom combination, it is not difficult for them to understand that enthalpies of formation are just another way of tabulating data for the purpose of thermodynamic calculations. Instructors should consider coverage of this section if their students will be going on to courses which utilize free energies of formation and absolute entropies.

Similarly it is possible to use formation data to calculate ΔG° for a given process. Free energies of formation, just as enthalpies of formation, are based on differences in free energy between the compound of interest and the elements that compose the compound, each element being in its stable state. These elements are assigned a free energy of formation of zero. For oxygen this stable state is $O_2(g)$, for hydrogen $H_2(g)$, bromine is a liquid $Br_2(l)$, and for iron the stable state is the solid, Fe(s). ΔG° for a chemical reaction may be found from free energies of formation (ΔG_f°) by taking the difference between the free energies of reactants and products. For the general reaction

$$\text{aA} + \text{bB} \rightleftharpoons \text{cC} + \text{dD}$$
$$\Delta G^\circ = [\text{c}(\Delta G_f^\circ)_C + \text{d}(\Delta G_f^\circ)_D] - [\text{a}(\Delta G_f^\circ)_A + \text{b}(\Delta G_f^\circ)_B]$$

The superscript zero refers to a standard state of 1 atm for gases, solids, and liquids. A substance in solution is at 1 M concentration. A table of free energies of formation is given in Appendix B.16.

Exercise 13.15

Calculate the free energy change for

$$CH_4(g) + 2O_2(g) \longrightarrow CO_2\,(g) + 2H_2O\,(g)$$

at 298 K and 1 atm from the data tabulated in Appendix B.16.

Solution

The Gibbs free energies of formation from Appendix B.16 are:

	ΔG_f° (kJ/mol$_{rxn}$)
$CH_4(g)$	−50.752
$O_2(g)$	0
$CO_2(g)$	−394.359
$H_2O(g)$	−228.572

ΔG° is found from

$$\Delta G^\circ = [2(\Delta G_f^\circ)_{H_2O(g)} + (\Delta G_f^\circ)_{CO_2(g)}] - [2(\Delta G_f^\circ)_{O_2(g)} + (\Delta G_f^\circ)_{CH_4(g)}]$$
$$\Delta G^\circ = [2(-228.572) - (394.359)] - [0 + (-50.752)]$$
$$= -800.751\ \text{kJ/mol}_{\text{rxn}}$$

Tabulations of entropies calculated from experimental data are also available. These tables are called **absolute entropies** or third law entropies. The standard state is 1 atm and usually 298 K. Entropies, however, are reported as absolute values, S°, rather than as comparisions with stable states of the elements. The entropy change for a process can be calculated directly from S° values. For the general chemical reaction

$$\text{aA} + \text{bB} \rightleftharpoons \text{cC} + \text{dD}$$
$$\Delta S = [\text{c}S_C^\circ + \text{d}S_D^\circ] - [\text{a}S_A^\circ + \text{b}S_B^\circ]$$

entropies are tabulated along with free energies and enthalpies of formation in Appendix B.16.

ΔG°, ΔH°, and ΔS° for processes and chemical reactions can be calculated from either atom combination data or from free energy of formation data, enthalpy of formation data, or absolute entropy data.

Exercise 13.16

Calculate the entropy and the enthalpy change for the reaction

$$CH_4(g) + 2\ O_2(g) \longrightarrow CO_2(g) + 2\ H_2O(g)$$

at 298 K and 1 atm from the data tabulated in Appendix B.16.

Use the enthalpy and entropy change to calculate ΔG° for the above reaction and compare ΔG° to that found in Exercise 13.15.

Solution

Enthalpy of formation data and absolute entropy data from Appendix B.16 are tabulated below.

	ΔH_f° (kJ/mol_{rxn})	$S^\circ (J/mol_{rxn} \cdot K)$
$CH_4(g)$	−74.81	186.264
$O_2(g)$	0	205.138
$CO_2(g)$	−393.509	213.74
$H_2O(g)$	−241.818	188.25

$$\Delta H^\circ = [2\,(\Delta H_f^\circ)_{H_2O(g)} + (\Delta H_f^\circ)_{CO_2(g)}] - [2\,(\Delta H_f^\circ)_{O_2(g)} + (\Delta H_f^\circ)_{CH_4(g)}]$$
$$\Delta H^\circ = [2\,(-241.818) + (-393.509)] - [0 + (-74.81)]$$
$$= -802.34 \text{ kJ/mol}_{rxn}$$

$$\Delta S^\circ = [2\,S^\circ_{H_2O(g)} + S^\circ_{CO_2(g)}] - [2\,S^\circ_{O_2(g)} + S^\circ_{CH_4(g)}]$$
$$= [2\,(188.25) + (213.74)] - [2\,(205.138) + (186.264)]$$
$$= -6.30 \text{ J/mol}_{rxn} \cdot \text{K}$$

$$\Delta G^\circ = \Delta H^\circ - T\Delta S^\circ$$
$$= -802.34 - 298\,(-0.00630)$$
$$= -800.46 \text{ kJ/mol}_{rxn}$$

There is a very small difference (0.04%) in the value of ΔG° calculated by the two methods used in Exercises 13.15 and 13.16. This difference is typical for thermodyanmic calculations of this type and is considered insignificant.

Key Terms

Absolute entropy
Disorder
Endergonic ($\Delta G > 0$)
Endothermic
Enthalpy
Entropy (S)
Entropy of atom combination
Exergonic ($\Delta G < 0$)
Exothermic
First law of thermodynamics
Gibbs free energy (G)
Reaction quotient (Q_p)
Second law of thermodynamics
Spontaneous
Standard-state conditions
Standard-state entropy of reaction (ΔS°)
Standard-state free energy of atom combination (ΔG°_{ac})
Standard-state free energy of formation (ΔG_f°)
Standard-state free energy of reaction (ΔG°)
Third law of thermodynamics

Problems

Spontaneous Chemical and Physical Processes

1. What is meant by the term *spontaneous?*
2. Are exothermic processes spontaneous? Give an example of a spontaneous endothermic process.
3. Are there processes that are spontaneous at some temperatures but not at other temperatures? Give an example to support your answer.

Entropy and Disorder

4. What is the relationship between the entropy of a system and microscopic disorder?
5. Which has a greater entropy, liquid water or gaseous water?
6. A drop of red food-coloring solution is added to a beaker of water. As time passes, the entire beaker of water becomes a very pale red. Has there been an increase or decrease in disorder? Has there been an increase or decrease in entropy?

Entropy and the Second Law of Thermodynamics

7. Describe what happens on a microscopic scale when an ice-cold penny is dropped into water at room temperature.
8. Which of the following reactions leads to an increase in the entropy of the system?
   (a) $H_2(g) + Cl_2(g) \rightleftharpoons 2\ HCl(g)$
   (b) $2\ NO_2(g) \rightleftharpoons N_2O_4(g)$
   (c) $CaO(s) + CO_2(g) \rightleftharpoons CaCO_3(s)$
   (d) $2\ H_2(g) + O_2(g) \rightleftharpoons 2\ H_2O(g)$
   (e) $2\ NH_3(g) \rightleftharpoons N_2(g) + 3\ H_2(g)$
9. In which of the following processes is ΔS negative?
   (a) $2\ H_2O_2(aq) \rightleftharpoons 2\ H_2O(l) + O_2(g)$
   (b) $CO_2(s) \rightleftharpoons CO_2(g)$
   (c) $H_2O(l) \rightleftharpoons H_2O(g)$
   (d) $4\ Al(s) + 3\ O_2(g) \rightleftharpoons 2\ Al_2O_3(s)$
10. In which of the following processes is ΔS positive?
   (a) $2\ NO(g) + Cl_2(g) \rightleftharpoons 2\ NOCl(g)$
   (b) $NaCl(s) \rightleftharpoons NaCl(l)$
   (c) $O_2(g) \rightleftharpoons 2\ O_3(g)$
   (d) $C_2H_4(g) + H_2(g) \rightleftharpoons C_2H_6(g)$
11. Which of the following processes should have the most positive value of ΔS?
   (a) $N_2(g) + O_2(g) \rightleftharpoons 2\ NO(g)$
   (b) $3\ C_2H_2(g) \rightleftharpoons C_6H_6(l)$
   (c) $H_2O(l) \rightleftharpoons H_2O(g)$
   (d) $4\ Al(s) + 3\ O_2(g) \rightleftharpoons 2\ Al_2O_3(s)$
12. Which of the following processes should have the most positive value of ΔS?
   (a) $H_2O(l) \rightleftharpoons H_2O(s)$
   (b) $NaNO_3(s) \overset{H_2O}{\rightleftharpoons} Na^+(aq) + NO_3^-(aq)$
   (c) $2\ HCl(g) \rightleftharpoons H_2(g) + Cl_2(g)$
   (d) $2\ H_2(g) + O_2(g) \rightleftharpoons 2\ H_2O(g)$
13. For some processes, even though the entropy change is favorable, the process doesn't occur. Explain why.
14. For some processes, even though the enthalpy change is favorable, the process doesn't occur. Explain why.

Standard-State Entropies of Reaction

15. What are the standard-state conditions for entropy for a chemical reaction?
16. Explain the difference between the symbols ΔS and $\Delta S°$.

The Third Law of Thermodynamics

17. What is the entropy of a perfect crystal at absolute zero? What happens to the entropy of the crystal as the temperature increases?
18. What happens to the motion of the particles composing a crystal as the temperature is raised from absolute zero to room temperature?
19. Which state of matter for a given substance, solid, liquid, or gas has the highest entropy? Explain.

Calculating Entropy Changes for Chemical Reactions

20. Which favors the formation of products in a reaction, an increase or a decrease in entropy? Explain.
21. If $\Delta S°$ for a reaction is positive, does this mean that the reactants or products have a greater entropy? What if $\Delta S°$ is negative?
22. One of the key steps toward transforming coal into a liquid fuel involves reducing carbon monoxide with H_2 gas to form methanol. Calculate $\Delta S°$ at 298 K for the following reaction to determine whether there is a favorable change in the entropy of the system.

$$CO(g) + 2\ H_2(g) \rightleftharpoons CH_3OH(l)$$

23. Tetraphosphorus decaoxide is often used as a dehydrating agent because of its tendency to pick up water. Calculate $\Delta S°$ at 25°C for the following reaction to determine whether entropy might be a measure of the driving force behind this reaction.

$$P_4O_{10}(s) + 6\ H_2O(l) \rightleftharpoons 4\ H_3PO_4(aq)$$

24. Calculate $\Delta S°$ at 25°C for the following reaction. Comment on both the sign and the magnitude of $\Delta S°$. The entropy of atom combination of $NH_4NO_2(s)$ is $-949.5\ J/mol_{rxn} \cdot K$.

$$NH_4NO_2(s) \rightleftharpoons N_2(g) + 2\ H_2O(g)$$

25. Calculate $\Delta S°$ at 298 K for the following reaction. Is there more disorder in the products or reactants?

$$3\ O_2(g) \rightleftharpoons 2\ O_3(g)$$

26. Compare the entropies of atom combination for the various forms of elemental phosphorus. Explain why $\Delta S°_{ac}$ is more negative for P(*s*) than $P_2(g)$. Why is $\Delta S°_{ac}$ for $P_4(g)$ more negative than for $P_2(g)$?

Compound	$\Delta S°_{ac}$ *(J/mol$_{rxn}$ · K)*
P(*s*)	−122.10
P(*g*)	0
$P_2(g)$	−108.257
$P_4(g)$	−372.79

Gibbs Free Energy

27. Explain the difference between ΔG and $\Delta G°$ for a chemical reaction.
28. What does it mean when ΔG for a reaction is zero?
29. Which of the following combinations of $\Delta H°$ and $\Delta S°$ always indicates a spontaneous reaction?
    (a) $\Delta H° > 0, \Delta S° < 0$ (b) $\Delta H° < 0, \Delta S° > 0$
    (c) $\Delta H° > 0, \Delta S° > 0$ (d) $\Delta H° < 0, \Delta S° < 0$
    (e) $\Delta H° = 0, \Delta S° = 0$
30. Predict the signs of $\Delta H°$ and $\Delta S°$ for the following reactions without referring to a table of thermodynamic data, and explain your predictions.
    Hints: The reaction is explosive under suitable conditions.
    (a) $2\ H_2(g) + O_2(g) \rightleftharpoons 2\ H_2O(g)$
    Hints: Is table salt stable?
    (b) $2\ Na(s) + Cl_2(g) \rightleftharpoons 2\ NaCl(s)$
    Hints: The reaction occurs under standard conditions.
    (c) $N_2(g) + 3\ H_2(g) \rightleftharpoons 2\ NH_3(g)$
    Hints: The reaction occurs spontaneously to produce stable CuO.
    (d) $2\ Cu(NO_3)_2(s) \rightleftharpoons 2\ CuO(s) + 4\ NO_2(g) + O_2(g)$
31. Predict the signs of $\Delta H°$ and $\Delta S°$ for the following reaction without referring to a table of thermodynamic data, and explain your predictions.

$$NH_3(g) \overset{H_2O}{\rightleftharpoons} NH_3(aq)$$

    Explain why the odor of NH_3 gas that collects above an NH_3 solution becomes more intense as the temperature increases.
32. Limestone decomposes to form lime and carbon dioxide when it is heated. Calculate $\Delta H°$ and $\Delta S°$ at 25°C for the decomposition of limestone and decide whether enthalpy or entropy is favorable for the reaction as written.

$$CaCO_3(s) \rightleftharpoons CaO(s) + CO_2(g)$$

33. Calculate $\Delta H°$ and $\Delta S°$ at 298 K for the thermite reaction and determine which is the most important measure of the driving force behind the reaction.

$$Fe_2O_3(s) + 2\ Al(s) \rightleftharpoons Al_2O_3(s) + 2\ Fe(s)$$

34. The first step in extracting iron ore from pyrite, FeS_2, involves roasting the ore in the presence of oxygen to form iron (III) oxide and sulfur dioxide.

$$4\ FeS_2(s) + 11\ O_2(g) \rightleftharpoons 2\ Fe_2O_3(s) + 8\ SO_2(g)$$

    Calculate $\Delta H°$ and $\Delta S°$ at 25°C for the reaction and identify the most important measure of the driving force for the reaction.
35. Calculate $\Delta H°$ and $\Delta S°$ at 298 K for the following reaction. Are products or reactants favored?

$$2\ KMnO_4(s) + 5\ H_2O_2(aq) + 6\ H^+(aq) \longrightarrow 2\ K^+(aq) + 2\ Mn^{2+}(aq) + 5\ O_2(g) + 8\ H_2O(l)$$

36. What determines whether a reaction will be spontaneous?
37. It is possible to make hydrogen chloride by reacting phosphorus pentachloride with water and boiling the HCl out of the solution. Calculate $\Delta H°$ and $\Delta S°$ for this reaction at 25°C. What is the driving force behind the reaction: the enthalpy of reaction, the entropy of reaction, or Le Châtelier's principle?

$$PCl_5(g) + 4\ H_2O(l) \rightleftharpoons H_3PO_4(aq) + 5\ HCl(g)$$

38. Which of the following processes is spontaneous when all species are in their standard states at 298 K?
    (a) $CH_3OH(l) \rightleftharpoons CH_3OH(g)$
    (b) $CH_3OH(l) \rightleftharpoons HCHO(g) + H_2(g)$
    (c) $2\ CH_3OH(l) \rightleftharpoons 2\ CH_4(g) + O_2(g)$
    (d) $CH_3OH(l) \rightleftharpoons CO(g) + 2\ H_2(g)$
39. Calculate $\Delta H°$ and $\Delta S°$ for the following reaction at 298 K. Explain why it is a mistake to use water to put out a fire that contains white-hot iron metal.

$$3\ Fe(s) + 4\ H_2O(l) \longrightarrow Fe_3O_4(s) + 4\ H_2(g)$$

40. Use thermodynamic data to predict whether sodium or silver will react with water at 298 K.

$$2\ Na(s) + 2\ H_2O(l) \longrightarrow 2\ Na^+(aq) + 2\ OH^-(aq) + H_2(g)$$

$$2\ Ag(s) + 2\ H_2O(l) \longrightarrow 2\ Ag^+(aq) + 2\ OH^-(aq) + H_2(g)$$

41. Use thermodynamic data at 25°C to explain why zinc reacts with 1 *M* acid, but not with water.

$$Zn(s) + 2\,H^+(aq) \longrightarrow Zn^{2+}(aq) + H_2(g)$$
$$Zn(s) + 2\,H_2O(l) \longrightarrow Zn^{2+}(aq) + 2\,OH^-(aq) + H_2(g)$$

42. Calculate $\Delta H°$ and $\Delta S°$ at 25°C for the following reaction. Why is ammonium nitrate a potential explosive?

$$2\,NH_4NO_3(s) \longrightarrow 2\,N_2(g) + O_2(g) + 4\,H_2O(g)$$

43. Silane, SiH_4, decomposes to form elemental silicon and hydrogen. Calculate $\Delta H°$ and $\Delta S°$ for this reaction at 25°C and predict the effect on the reaction of an increase in the temperature at which it is run.

$$SiH_4(g) \rightleftharpoons Si(s) + 2\,H_2(g)$$

44. For which of the following reactions would you expect $\Delta H°$ and $\Delta G°$ to be about the same?
   (a) $4\,Fe(s) + 3\,O_2(g) \longrightarrow 2\,Fe_2O_3(s)$
   (b) $2\,Na(s) + 2\,H_2O(l) \longrightarrow 2\,Na^+(aq) + 2\,OH^-(aq) + H_2(g)$
   (c) $Fe_2O_3(s) + 2\,Al(s) \longrightarrow Al_2O_3(s) + 2\,Fe(s)$
   (d) $N_2O_4(g) \rightleftharpoons 2\,NO_2(g)$
   (e) $CaC_2(s) + 2\,H_2O(l) \rightleftharpoons Ca^{2+}(aq) + 2OH^-(aq) + C_2H_2(g)$

The Effect of Temperature on the Free Energy of a Reaction

45. For each of the following reactions, how will the $\Delta G°$change as the temperature increases?
   (a) $2\,CO(g) + O_2(g) \rightleftharpoons 2\,CO_2(g)$
   $\Delta H° = -565.97$ kJ/mol$_{rxn}$,
   $\Delta S° = -173.00$ J/mol$_{rxn} \cdot$ K
   (b) $2\,H_2O(g) \rightleftharpoons 2\,H_2(g) + O_2(g)$
   $\Delta H° = 483.64$ kJ/mol$_{rxn}$,
   $\Delta S° = 90.01$ J/mol$_{rxn} \cdot$ K
   (c) $2\,N_2O(g) \rightleftharpoons 2\,N_2(g) + O_2(g)$
   $\Delta H° = -164.1$ kJ/mol$_{rxn}$,
   $\Delta S° = 148.66$ J/mol$_{rxn} \cdot$ K
   (d) $PbCl_2(s) \overset{H_2O}{\rightleftharpoons} Pb^{2+}(aq) + 2\,Cl^-(aq)$
   $\Delta H° = 23.39$ kJ/mol$_{rxn}$,
   $\Delta S° = -12.5$ J/mol$_{rxn} \cdot$ K

46. Explain why the $\Delta G°$ for the following reaction becomes more positive as the temperature increases.

$$N_2(g) + 3\,H_2(g) \rightleftharpoons 2\,NH_3(g)$$

47. Which of the following diagrams best describes the relationship between $\Delta G°$ and temperature for the following reaction?

$$2\,H_2(g) + O_2(g) \rightleftharpoons 2\,H_2O(g)$$

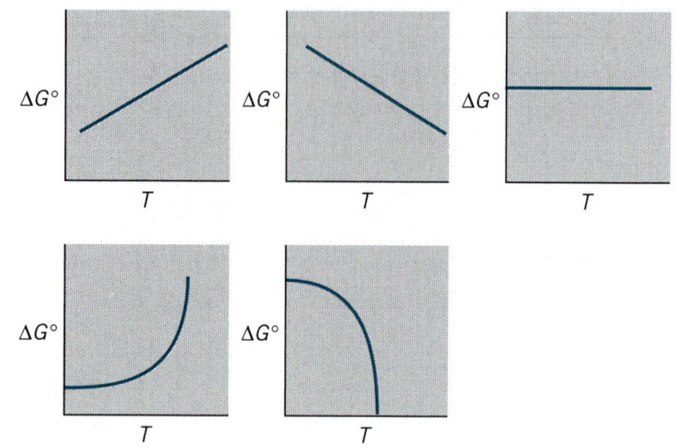

48. Calculate $\Delta G°$ for the following reaction from $\Delta H°$ and $\Delta S°$ data for the reaction at 298 K.

$$CS_2(l) + 3\,O_2(g) \rightleftharpoons CO_2(g) + 2\,SO_2(g)$$

49. What is the sign of $\Delta G°$ for the following process at −10°C, 0°C, and 10°C at 1 atm pressure?

$$H_2O(s) \longrightarrow H_2O(l)$$

50. What is the sign of $\Delta G°$ for the following process at 90°C, 100°C, and 110°C at 1 atm pressure of $H_2O(g)$?

$$H_2O(l) \longrightarrow H_2O(g)$$

51. For the process $H_2O(l) \rightleftharpoons H_2O(g)$, what are the signs of ΔH and ΔS?

Beware of Oversimplifications

52. Is it usually correct to assume that $\Delta H°$ and $\Delta S°$ do not change very much with temperature? When might this assumption be in serious error?

53. For the reaction

$$2\,H_2(g) + O_2(g) \longrightarrow 2\,H_2O(g)$$

$\Delta G°$ is −457 kJ/mol$_{rxn}$, $\Delta H°$ is −484 kJ/mol$_{rxn}$, $\Delta S°$ is −89 kJ/mol$_{rxn}$, all at 298 K. Assume that both $\Delta H°$ and $\Delta S°$ are independent of temperature and calculate $\Delta G°$at 500 K and 1000 K. The experimentally determined values are at 500 K -438 kJ/mol$_{rxn}$ and at 1000 K -385 kJ/mol$_{rxn}$. Compare your calculations with the correct numbers. Why are these values different?

Standard-State Free Energies of Reaction

54. What are standard-state conditions for free energies of atom combination?
55. Using free energies of atom combination, calculate ΔG° for the following reactions.
    (a) $CO(g) + 2\,H_2(g) \rightleftharpoons CH_3OH(l)$
    (b) $3\,O_2(g) \rightleftharpoons 2\,O_3(g)$
    (c) $CaCO_3 \rightleftharpoons CaO(s) + CO_2(g)$
56. Calculate the free energy change for the following reaction using free energies of atom combination.

$$H_2O(l) \rightleftharpoons H_2O(g) \quad 298\text{ K, 1 atm}$$

In which direction will this reaction proceed? What would be the sign of the free energy change for the reverse of this reaction? Explain what this means.

Equilibria Expressed in Partial Pressures

57. Explain why pressure can be used instead of concentration to describe equilibrium constant expressions for gas phase reactions.
58. Which equation correctly describes the relationship between K_p and K_c for the following reaction?

$$Cl_2(g) + 3\,F_2(g) \rightleftharpoons 2\,ClF_3(g)$$

(a) $K_p = K_c$ (b) $K_p = K_c \times (RT)^{-1}$
(c) $K_p = K_c \times (RT)^{-2}$ (d) $K_p = K_c \times (RT)$
(e) $K_p = K_c \times (RT)^2$

59. Which equation correctly describes the relationship between K_p and K_c for the following reaction?

$$2\,NO_2(g) \rightleftharpoons 2\,NO(g) + O_2(g)$$

(a) $K_p = K_c$ (b) $K_p = 1/K_c$
(c) $K_p = K_c(RT)$ (d) $K_p = K_c(RT)^{-1}$

60. Calculate K_p for the decomposition of NOCl at 500 K if 27.3% of a 1.00-atm sample of NOCl decomposes to NO and Cl_2 at equilibrium.

$$2\,NOCl(g) \rightleftharpoons 2\,NO(g) + Cl_2(g)$$

61. Calculate the partial pressures of phosgene, carbon monoxide, and chlorine at equilibrium when a system that was initially 0.124 atm in $COCl_2$ decomposes at 300°C according to the following equation.

$$COCl_2(g) \rightleftharpoons CO(g) + Cl_2(g) \quad K_p = 3.2 \times 10^{-3}$$

62. Without detailed equilibrium calculations, approximate the partial pressure of SO_3 that would be present at equilibrium when a mixture that was initially 0.490 atm in SO_2 and 0.245 atm in O_2 comes to equilibrium at 700 K.

$$2\,SO_2(g) + O_2(g) \rightleftharpoons 2\,SO_3(g) \quad K_p = 6.7 \times 10^4$$

63. Calculate the partial pressures of SO_3, SO_2, and O_2 that would be present at equilibrium when a mixture that was initially 0.490 atm in SO_3 reacts at 700 K.

$$2\,SO_3(g) \rightleftharpoons 2\,SO_2(g) + O_2(g) \quad K_p = 1.5 \times 10^{-5}$$

64. Approximate the partial pressures of N_2 and H_2 present at equilibrium when a mixture that was initially 0.50 atm in N_2, 0.60 atm in H_2, and 0.20 atm in NH_3 comes to equilibrium at a temperature at which K_p for the following reaction is 1.0×10^{-5}.

$$N_2(g) + 3\,H_2(g) \rightleftharpoons 2\,NH_3(g)$$

65. A mixture of NH_3, O_2 NO_2, and H_2O is initially 0.50 atm in each of the four gases. In which direction will the reaction proceed?

$$4\,NO_2(g) + 6\,H_2O(g) \rightleftharpoons 4\,NH_3(g) + 7\,O_2(g)$$
$$K_p = 1.8 \times 10^{-28}$$

66. The partial pressures of N_2, H_2, and NH_3 present at equilibrium at 1000 K are 2.0, 6.0, and 0.018 atm, respectively. Calculate ΔG° for the following reaction.

$$2\,NH_3(g) \rightleftharpoons N_2(g) + 3\,H_2(g)$$

67. Calculate the concentrations of N_2, O_2, and NO present when a mixture that was initially 0.40 *M* N_2 and 0.60 *M* O_2 reacts to form NO at 700°C. (Hint: Look carefully at the symbol for the equilibrium constant for the reaction.)

$$N_2(g) + O_2(g) \rightleftharpoons 2\,NO(g)$$
$$K_p = 4.3 \times 10^{-9} \text{ (at 700°C)}$$

68. Industrial chemicals can be made from coal by a process that starts with the reaction between red-hot coal and steam to form a mixture of CO and H_2. This mixture, which is known as synthesis gas, can be converted to methanol (CH_3OH) in the presence of a ruthenium/cobalt catalyst. The methanol produced in the reaction can then be converted into a host of other products, ranging from acetic acid to gasoline. The

partial pressure of methanol when a mixture of 1.00 atm CO and 2.00 atm H_2 comes to equilibrium at 65°C is 0.98 atm. Calculate the standard free energy of the reaction at 65°C.

$$CO(g) + 2\ H_2(g) \rightleftharpoons CH_3OH(g)$$

Interpreting Standard-State Free Energy of Reaction Data

69. What does ΔG° measure for a chemical reaction?
70. If ΔG° is large, what does this tell us about how far from equilibrium the standard-state conditions are?
71. Under what conditions does $\Delta G = \Delta G^\circ$?

The Relationship Between Free Energy and Equilibrium Constants

72. Which of the following correctly describes a reaction at equilibrium?
(a) $\Delta G = 0$ (b) $\Delta G^\circ = 1$ (c) $\Delta G = \Delta G^\circ$
(d) $Q = 0$ (e) $\ln K = 0$
73. Calculate the equilibrium constant at 25°C for the following reaction from the standard-state free energies of atom combination of the reactants and products.

$$CO(g) + 2\ H_2(g) \rightleftharpoons CH_3OH(g)$$

74. Calculate K_p at 1000 K for the following reaction. Describe the assumptions you had to make to do the calculation.

$$2\ CH_4(g) \rightleftharpoons C_2H_6(g) + H_2(g)$$

75. Calculate the equilibrium constant at 25°C for the following reaction.

$$2\ HI(g) + Cl_2(g) \rightleftharpoons 2\ HCl(g) + I_2(s)$$

76. Calculate ΔH°, ΔS°, and ΔG° for the following reactions at 298 K. Use the data to calculate the values of K_a for hydrochloric and acetic acid. Compare the results of the calculations with the K_a data in Table 11.3 in Chapter 11.

$$HCl(g) \rightleftharpoons H^+(aq) + Cl^-(aq)$$
$$CH_3CO_2H(aq) \rightleftharpoons H^+(aq) + CH_3CO_2^-(aq)$$

77. Calculate ΔH°, ΔS°, and ΔG° for the following reaction at 298 K. Use the data to calculate the value of K_b for ammonia. Compare the result of the calculation with the value of K_b in Appendix B.9.

$$NH_3(aq) + H_2O(l) \rightleftharpoons NH_4^+(aq) + OH^-(aq)$$

The Temperature Dependence of Equilibrium Constants

78. Synthesis gas can be made by reacting red-hot coal with steam. Estimate the temperature at which the equilibrium constant for the reaction is equal to 1.

$$C(s) + H_2O(g) \rightleftharpoons CO(g) + H_2(g)$$

79. Calculate ΔH°, ΔS°, ΔG°, and the equilibrium constant at 25°C for the following reaction. Predict the effect of an increase in the temperature of the system on the equilibrium constant for the reaction.

$$CO_2(g) + H_2(g) \rightleftharpoons CO(g) + H_2O(g)$$

80. What happens to the equilibrium constant for the following reaction as the temperature increases?

$$PCl_5(g) \rightleftharpoons PCl_3(g) + Cl_2(g)$$

81. At approximately what temperature does the equilibrium constant for the following reaction become larger than 1?

$$PCl_5(g) \rightleftharpoons PCl_3(g) + Cl_2(g)$$

82. For which of the following reactions would the equilibrium constant increase with increasing temperature?
(a) $2\ H_2(g) + O_2(g) \rightleftharpoons 2\ H_2O(g)$
(b) $2\ HCl(g) \rightleftharpoons H_2(g) + Cl_2(g)$
(c) $2\ NH_3(g) \rightleftharpoons N_2(g) + 3\ H_2(g)$
83. Calculate the equilibrium constant for the following reaction at 25°C, 200°C, 400°C, and 600°C. Explain any trend you observe. Assume ΔH° and ΔS° do not change with temperature.

$$2\ SO_3(g) \rightleftharpoons 2\ SO_2(g) + O_2(g)$$

84. Calculate ΔH° and ΔS° for the following reaction. Predict what will happen to the equilibrium constant of the reaction as the temperature increases.

$$Cu(s) + 2\ H^+ \rightleftharpoons Cu^{2+}(aq) + H_2(g)$$

Gibbs Free Energies of Formation and Absolute Entropies

85. Calculate ΔS° for the following reactions using absolute entropies.
(a) $2\ H_2\ (g) + O_2(g) \longrightarrow 2\ H_2O\ (g)$
(b) $2\ Na\ (s) + Cl_2(g) \longrightarrow 2\ NaCl\ (s)$
(c) $N_2\ (g) + 3\ H_2(g) \longrightarrow 2\ NH_3\ (g)$

Calculate $\Delta H°$ for the above reactions from enthalpies of formation. Calculate $\Delta G°$ from $\Delta H°$ and $\Delta S°$ obtained from enthalpy of formation data and absolute entropy data. Compare this $\Delta G°$ with $\Delta G°$ calculated from free energy of formation data Compare this to the predictions made in Problem 30.

86. For Problems 48, 55, 56, and 73, calculate $\Delta G°$ from free energies of formation.

87. Calculate $\Delta H°$ and $\Delta S°$ from enthalpies of formation and absolute entropies for the process

$$NH_3\ (g) \longrightarrow NH_3\ (aq)$$

Compare your answer to Problem 31.

Integrated Problems

88. $\Delta S°$ for the following reaction is favorable even though $\Delta H°$ is not.

$$CH_3OH(l) \rightleftharpoons CH_3OH(g)$$

Assume that methanol boils at the temperature at which ΔG for this reaction is equal to zero. Use the values of $\Delta H°$ and $\Delta S°$ at 25°C for the reaction to estimate the boiling point of methanol.

89. Use the equation that describes the relationship between ΔG and $\Delta G°$ for a reaction to construct a graph of ΔG versus ln Q_c at 25°C for the following reactions over a range of values of Q_c between 10^{-10} and 10^{10}.

$$HOAc(aq) \xrightleftharpoons{H_2O} H^+(aq) + OAc^-(aq)$$
$$\Delta G° = 27.1\ \text{kJ/mol}_{\text{rxn}}$$

$$HCl(g) \xrightarrow{H_2O} H^+(aq) + Cl^-(aq)$$
$$\Delta G° = -35.9\ \text{kJ/mol}_{\text{rxn}}$$

Explain the significance of the difference between the points at which the straight lines for the data cross the horizontal and vertical axes.

90. From the following $\Delta H°$ and $\Delta S°$ values, predict whether each of the reactions would lead to $K > 1$ or $K < 1$ at 25°C. If a reaction has $K < 1$, at what temperature might K become greater than 1? Explain your reasoning.
(a) $\Delta H° = 10\ \text{kJ/mol}_{\text{rxn}}$; $\Delta S° = 30\ \text{J/mol}_{\text{rxn}} \cdot \text{K}$
(b) $\Delta H° = 2\ \text{kJ/mol}_{\text{rxn}}$; $\Delta S° = 100\ \text{J/mol}_{\text{rxn}} \cdot \text{K}$
(c) $\Delta H° = -125\ \text{kJ/mol}_{\text{rxn}}$; $\Delta S° = 80\ \text{J/mol}_{\text{rxn}} \cdot \text{K}$
(d) $\Delta H° = -10\ \text{kJ/mol}_{\text{rxn}}$; $\Delta S° = -30\ \text{J/mol}_{\text{rxn}} \cdot \text{K}$

91. For the reaction

$$2\ NO(g) + O_2(g) \rightleftharpoons 2\ NO_2(g)$$

(a) Find $\Delta H°$.
(b) Calculate $\Delta S°$.
(c) What is $\Delta G°$?
(d) At 298 K, are the products or reactants favored? Explain.
(e) Calculate the equilibrium constant for the reaction at 25°C.
(f) If the partial pressures of NO, O_2, and NO_2 are 1.0 atm, in which direction will the reaction proceed at 298 K?
(g) If the temperature is decreased, will the equilibrium constant increase, decrease, or stay the same? Explain.

92. For the following two reactions, the thermodynamic parameters are given at 25°C.

(I) $2\ CO(g) + O_2(g) \rightleftharpoons 2\ CO_2(g)$
$\Delta H° = -565\ \text{kJ/mol}_{\text{rxn}}$
$\Delta S° = -173\ \text{J/mol}_{\text{rxn}} \cdot \text{K}$

(II) $2\ H_2O(g) \rightleftharpoons 2\ H_2(g) + O_2(g)$
$\Delta H° = +484\ \text{kJ/mol}_{\text{rxn}}$
$\Delta S° = +90\ \text{J/mol}_{\text{rxn}} \cdot \text{K}$

(a) For which of these reactions is $\Delta H°$ favorable? Explain.
(b) For which of these reactions is $\Delta S°$ favorable? Explain.
(c) For which reaction will the equilibrium constant be greater than 1 at 25°C?
(d) Explain the sign of $\Delta S°$ for each of these reactions.
(e) If the temperature is increased at constant pressure, which way will the equilibrium shift for each reaction?

93. A reaction responsible for the production of acid rain is

$$2\ NO(g) + O_2(g) \rightleftharpoons 2\ NO_2(g)$$

This reaction occurs in lightning flashes in the upper atmosphere. If the temperature in the lightning flash is much higher than 25°C, will $\Delta G°$ be greater, lesser, or the same as $\Delta G°$ at 25°C? Explain.

94. The following enthalpy and entropy changes are known for the reactions shown at 298 K.

$$Cu(s) + 2H^+(aq) \rightleftharpoons Cu^{2+}(aq) + H_2(g)$$
$$\Delta H° = 65\ \text{kJ/mol}_{\text{rxn}}$$
$$\Delta S° = -2.1\ \text{J/mol}_{\text{rxn}} \cdot \text{K}$$

$$Zn(s) + 2H^+(aq) \rightleftharpoons Zn^{2+}(aq) + H_2(g)$$
$$\Delta H° = -153\ \text{kJ/mol}_{\text{rxn}}$$
$$\Delta S° = -23.1\ \text{J/mol}_{\text{rxn}} \cdot \text{K}$$

(a) Which of these two metals, Zn or Cu, will dissolve appreciably in an acid solution? Explain.
(b) Which of these two reactions has an equilibrium constant that increases with temperature?

(c) Would you have expected that both reactions would have had a negative entropy change? How can you explain $\Delta S < 0$ for these reactions?

95. Use the following thermodynamic data to predict which should be the strongest acid, H_2O or CH_3COOH at 25°C.

$$H_2O(aq) \rightleftharpoons H^+(aq) + OH^-(aq)$$
$$CH_3COOH(aq) \rightleftharpoons H^+(aq) + CH_3COO^-(aq)$$

	ΔH°_{ac} (kJ/mol$_{rxn}$)	ΔS°_{ac} (J/mol$_{rxn}\cdot$K)
$H_2O(aq)$	−970	−321
$H^+(aq)$	−218	−115
$OH^-(aq)$	−696	−286
$CH_3COOH(aq)$	−3288	−919
$CH_3COO^-(aq)$	−3071	−896

Show all calculations and explain your reasoning.

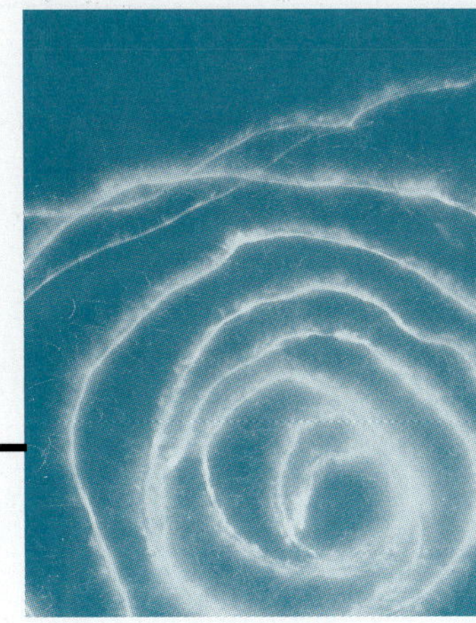

Chapter Fourteen

KINETICS

Special Topics

14.1 The Forces That Control a Chemical Reaction

This chapter describes rate laws and their applications, the collision theory of kinetics, catalysis, and activation energy. The topics are presented in a fairly traditional fashion. ChemActivities 57–62 in Chemistry: A Guided Inquiry *are designed to accompany this chapter.*

We noted in Chapter 5 that sodium reacts with water,

$$2\ Na(s) + 2\ H_2O(l) \longrightarrow 2\ Na^+(aq) + 2\ OH^-(aq) + H_2(g)$$

whereas magnesium metal does not.

$$Mg(s) + H_2O(l) \not\longrightarrow$$

The later part of this text provides a more sophisticated view of chemical reactions. By introducing the idea that chemical reactions can come to equilibrium in Chapter 10, we showed that we can construct a continuum of chemical reactions. Toward one end of the continuum are those reactions that only proceed to a limited extent. Reactions such as the following lie very heavily on the side of the reactants.

$$HOAc(aq) + H_2O(l) \longleftarrow H_3O^+(aq) + OAc^-(aq) \qquad K_a = 1.8 \times 10^{-5}$$

Toward the other end are reactions that proceed almost to completion.

$$Zn(s) + Cu^{2+}(aq) \longrightarrow Zn^{2+}(aq) + Cu(s) \qquad K = 1.8 \times 10^{37}$$

The previous chapter showed how thermodynamics can explain why some reactions occur and not others. In that chapter, we saw that thermodynamics is a powerful tool for predicting what should (or should not) happen in a chemical reaction. It is ideally suited for answering questions that begin, "What if . . . ?" Thermodynamics predicts, for example, that hydrogen should react with oxygen to form water.

$$2\ H_2(g) + O_2(g) \longrightarrow 2\ H_2O(g)$$

It also predicts that iron(III) oxide should react with aluminum metal to form aluminum oxide and iron metal.

$$Fe_2O_3(s) + 2\ Al(s) \longrightarrow Al_2O_3(s) + 2\ Fe(s)$$

Both of these reactions can, in fact, occur. We can experience the first by touching a balloon filled with H_2 gas with a candle tied to the end of a meter stick. We can demonstrate the other by igniting a mixture of Fe_2O_3 and powdered aluminum.

This doesn't mean that H_2 and O_2 burst spontaneously into flame the instant the gases are mixed. In the absence of a spark, mixtures of H_2 and O_2 can be stored for years with no detectable change. Nor does it mean that scrupulous attention must be paid to prevent Fe_2O_3 from coming in contact with aluminum metal, lest some violent reaction take place. In the absence of a source of heat, mixtures of Fe_2O_3 and powdered aluminum are so stable they are sold in 50-pound lots under the trade name Thermite.

It is a mistake to lose sight of the fact that thermodynamics can never tell us what will happen. It can only tell us what should or might happen. Reactions that behave as thermodynamics predicts they should are said to be under **thermodynamic control.** For example, potassium metal reacts violently when it comes in contact with water.

$$2\ K(s) + 2\ H_2O(l) \longrightarrow 2\ K^+(aq) + 2\ OH^-(aq) + H_2(g)$$
$$\Delta G^\circ = -406.77\ \text{kJ/mol}_{\text{rxn}}$$

Other reactions are said to be under **kinetic control.** Thermodynamics predicts that these reactions should occur, but the rate of these reactions is negligibly slow. Examples of reactions under kinetic control include the reactions in which elemental sulfur and its compounds burn.

$$S_8(s) + 8\,O_2(g) \longrightarrow 8\,SO_2(g)$$
$$CS_2(g) + 3\,O_2(g) \longrightarrow CO_2(g) + 2\,SO_2(g)$$
$$2\,H_2S(g) + 3\,O_2(g) \longrightarrow 2\,H_2O(g) + 2\,SO_2(g)$$

Thermodynamics predicts that these reactions should form SO_3, not SO_2, because SO_3 is more stable than SO_2. Under normal conditions, however, these reactions invariably stop at SO_2 because the reaction between SO_2 and O_2 to form SO_3 is extremely slow.

To fully understand chemical reactions, we need to combine the predictions of thermodynamics with studies of the factors that influence the rates of chemical reactions. These factors fall under the general heading of *chemical kinetics* (from a Greek stem meaning "to move"), which is the subject of this chapter.

➤ CHECKPOINT

Thermodynamics predicts that under normal circumstances carbon in the form of graphite is more stable than carbon in the form of diamond. Why don't you have to worry that a diamond ring will turn to graphite?

14.2 Chemical Kinetics

Chemical kinetics can be defined as the study of the following questions.

- What is the rate at which the reactants are converted into the products of a reaction?
- What factors influence the rate of a reaction?
- What is the sequence of steps, or the mechanism, by which the reactants are converted into products?

As we saw in Section 10.3, the term *rate* is used to describe the change in a quantity that occurs per unit of time. The rate at which cars are sold, for example, is equal to the number of cars that change hands divided by the period of time during which this number is counted. The rate at which an object travels through space is the distance traveled per unit of time, such as miles per hour or kilometers per second. In chemical kinetics, the distance traveled is analogous to the change in the concentration of one of the components of the reaction. The rate of a reaction is therefore the change in the concentration of one of the reactants or products, $\Delta(X)$, that occurs during a given period of time, Δt.

$$\text{Rate} = \frac{(X)_{\text{final}} - (X)_{\text{initial}}}{t_{\text{final}} - t_{\text{initial}}} = \frac{\Delta(X)}{\Delta t}$$

One of the first kinetic experiments was done in 1850 by Ludwig Wilhemy, a lecturer in physics at the University of Heidelberg. Wilhemy studied the rate at which sucrose, or cane sugar, reacts with acid to form a mixture of fructose and glucose known as invert sugar.

$$\text{Sucrose}(aq) + H_3O^+(aq) \longrightarrow \underbrace{\text{glucose}(aq) + \text{fructose}(aq)}_{\textit{invert sugar}}$$

Fig. 14.1 Acid–base titration using phenolphthalein as the indicator.

Experimental data obtained from spectroscopic measurements on the fading of phenolphthalein in the presence of excess base are used to reinforce the kinetics calculations introduced in Chapter 10.

Wilhemy found that the rate at which sucrose is consumed in this reaction is directly proportional to its concentration when the pH and temperature of the solution are held constant. The rate of the reaction at any moment in time can therefore be described by the following equation, in which k is a proportionality constant and (sucrose) is the concentration of sucrose in units of moles per liter at that moment.

$$\text{Rate} = k(\text{sucrose})$$

Because the sucrose concentration decreases as sucrose is consumed in the reaction, the rate of the reaction gradually decreases.

14.3 Is the Rate of Reaction Constant?

Acid–base titrations often use phenolphthalein as the endpoint indicator, as shown in Figure 14.1. A solution that contains phenolphthalein in the presence of a large excess of base initially has a pink color. However, the solution gradually turns colorless as the phenolphthalein reacts with the excess base.

Table 14.1 shows what happens to the concentration of phenolphthalein in a solution at 25°C that was initially 0.00500 M in phenolphthalein and 0.61 M in OH^- ion.

Table 14.1
Experimental Data for the Reaction Between Phenolphthalein and Excess Base

Concentration of Phenolphthalein (M)	Time (s)
0.00500	0.0
0.00450	10.5
0.00400	22.3
0.00350	35.7
0.00300	51.1
0.00250	69.3
0.00200	91.6
0.00150	120.4
0.00100	160.9
0.000500	230.2
0.000250	299.5
0.000150	350.7
0.000100	391.2

As you can see when the data are plotted in Figure 14.2, the phenolphthalein concentration decreases by a factor of 10 over a period of about 230 seconds. Because the concentration of OH^- is so large relative to the concentration of the phenolphthalein, the rate of reaction depends only on the concentration of phenolphthalein.

To find the rate of reaction over any period of time, we divide the change in the concentration of phenolphthalein by the time interval over which the change occurred. Because phenolphthalein is consumed in the reaction, the change in the phenolphthalein concentration is a negative number. (The concentration of phenolphthalein is smaller at the end of the time interval than at the beginning of the time interval.) By convention, rates are reported as positive numbers. When the

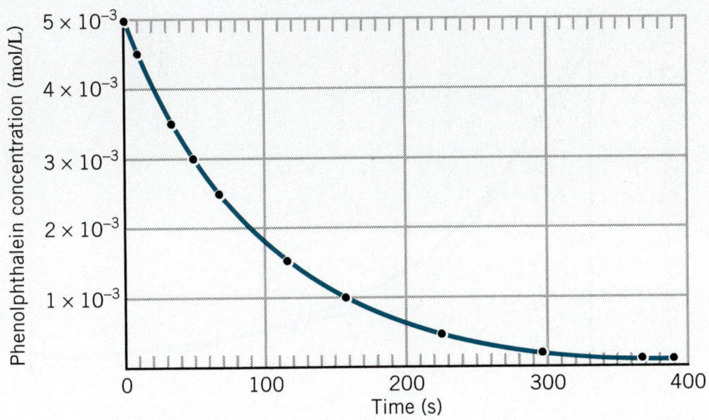

Fig. 14.2 Plot of the concentration of phenolphthalein versus time for the reaction between this acid–base indicator and excess OH^- ion.

rate of a reaction is defined in terms of the rate at which one of the reactants disappears, we therefore add a negative sign to the equation that defines the rate of the reaction.

$$\text{Rate} = -\frac{\Delta(\text{phenolphthalein})}{\Delta t}$$

> **➤ CHECKPOINT**
>
> Use Table 14.1 to determine how much time it takes for the concentration of phenolphthalein to decrease by 0.00050 *M* from 0.00500 to 0.00450 *M*. Determine the time it takes for the concentration to decrease by another 0.00050 *M* from 0.00450 to 0.00400 *M*. As the reaction proceeds, what happens to the time required for the concentration to change by 0.00050 *M*? As the reaction proceeds, what happens to the rate of the reaction?

Exercise 14.1

Use the data in Table 14.1 to calculate the rate at which phenolphthalein reacts with the OH^- ion during each of the following periods.

(a) During the first time interval, when the phenolphthalein concentration falls from 0.00500 *M* to 0.00450 *M*.

(b) During the second interval, when the concentration falls from 0.00450 *M* to 0.00400 *M*.

(c) During the third interval, when the concentration falls from 0.00400 *M* to 0.00350 *M*.

Solution

The rate of reaction is equal to the change in the phenolphthalein concentration divided by the length of time over which the change occurs.

(a) During the first time period, the rate of the reaction is 4.8×10^{-5} mol per liter per second.

$$\text{Rate} = -\frac{\Delta(X)}{\Delta t} = -\frac{(0.00450\ M) - (0.00500\ M)}{(10.5 - 0\ \text{s})} = 4.8 \times 10^{-5}\ M/\text{s}$$

(b) During the second time period, the rate of reaction is slightly slower.

$$\text{Rate} = -\frac{\Delta(X)}{\Delta t} = -\frac{(0.00400\ M) - (0.00450\ M)}{(22.3 - 10.5\ \text{s})} = 4.2 \times 10^{-5}\ M/\text{s}$$

(c) During the third time period, the rate of reaction is even slower.

$$\text{Rate} = -\frac{\Delta(X)}{\Delta t} = -\frac{(0.00350\ M) - (0.00400\ M)}{(35.7 - 22.3\ \text{s})} = 3.7 \times 10^{-5}\ M/\text{s}$$

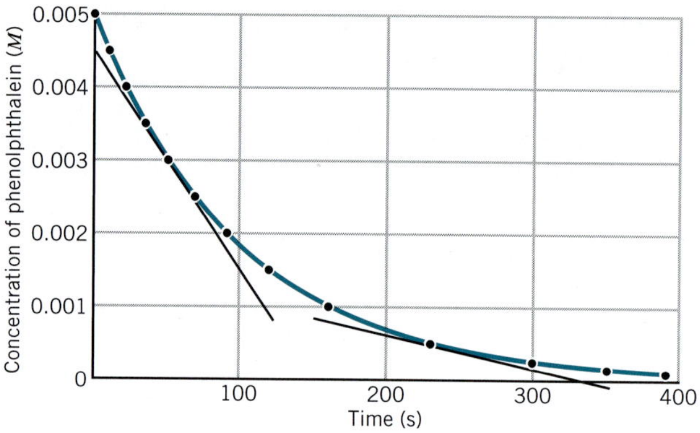

Fig. 14.3 The instantaneous rate of reaction at any moment in time can be calculated from the slope of a tangent to the graph of concentration versus time. At a point 50 s after the beginning of the reaction, for example, the instantaneous rate of this reaction is 2.7×10^{-5} mole per liter per second. At a point 250 s after the beginning of the reaction, the instantaneous rate of reaction is 4.1×10^{-6} mole per liter per second.

14.4 Instantaneous Rates of Reaction

Exercise 14.1 illustrates an important point: The rate of the reaction between phenolphthalein and the OH^- ion isn't constant; it changes with time. Like most reactions, the rate of the reaction gradually decreases as the reactants are consumed.

The rate of reaction not only changes from one time period to another, but it is also constantly changing. This means that the rate of reaction changes while it is being measured. To minimize the error this introduces into our measurements, it seems advisable to measure the rate of reaction over periods of time that are short compared with the time it takes for the reaction to occur. We might try, for example, to measure the infinitesimally small change in concentration, $d(X)$, that occurs over an infinitesimally short period of time, dt. The ratio of these quantities is known as the **instantaneous rate of reaction.**

$$\text{Rate} = -\frac{d(X)}{dt}$$

The instantaneous rate of reaction can be calculated from a graph of the concentration of one of the components of the reaction versus time. Figure 14.3 shows how the instantaneous rate can be calculated for the reaction between phenolphthalein and excess base. The rate of reaction at any moment in time is equal to the slope of a tangent drawn to the curve at that moment.

The instantaneous rate of reaction can be measured at any time between the moment at which the reactants are mixed and the time when the reaction reaches equilibrium. The **initial instantaneous rate of reaction** is the rate measured at the moment the reactants are mixed.

➤ CHECKPOINT

What is the problem with using a large value for dt when determining the rate of a reaction using the following equation?

$$\text{Rate} = -\frac{d(X)}{dt}$$

14.5 Rate Laws and Rate Constants

An interesting result is obtained when the instantaneous rate of reaction is calculated at various points along the curve in Figure 14.3. The rate of reaction at every point on the curve is directly proportional to the concentration of phenolphthalein at that moment in time.

$$\text{Rate} = k(\text{phenolphthalein})$$

Because this equation is an experimental law that describes the rate of the reaction, it is called the **rate law** for the reaction. The proportionality constant, k, is known as the **rate constant.**

➤ CHECKPOINT

Use the rate law for the reaction between phenolphthalein and excess base to explain why the rate of the reaction gradually slows down.

Exercise 14.2

Calculate the rate constant for the reaction between phenolphthalein and the OH^- ion if the instantaneous rate of reaction is 2.5×10^{-5} mol/L · s when the concentration of phenolphthalein is 0.00250 *M*.

Solution

We start with the rate law for the reaction.

$$\text{Rate} = k(\text{phenolphthalein})$$

We then substitute the known rate of reaction and the known concentration of phenolphthalein into the equation.

$$2.5 \times 10^{-5} \frac{\text{mol/L}}{\text{s}} = k(0.00250 \text{ mol/L})$$

Solving for the rate constant gives the following result.

$$k = 0.010 \text{ s}^{-1}$$

Because the rate of reaction is the change in the concentration of phenolphthalein divided by the time over which the change occurs, it is reported in units of moles per liter per second. Because the number of moles of phenolphthalein per liter is the molarity of the solution, the rate can also be reported in terms of the change in molarity per second: *M*/s. The units of the rate constant will depend on the form of the rate law. When used in the rate law, the rate constant must have units that yield a rate in units of *M*/s.

➤ CHECKPOINT

Describe the difference between the *rate* of a chemical reaction and the *rate constant*. How do the rate and rate constant vary with respect to concentration and time?

Exercise 14.3

Use the rate constant for the reaction between phenolphthalein and the OH^- ion calculated in the previous exercise to calculate the initial instantaneous rate of reaction for the data in Table 14.1.

Solution

We start, once again, with the rate law for the reaction.

$$\text{Rate} = k(\text{phenolphthalein})$$

We then substitute into this equation the rate constant for the reaction and the initial concentration of phenolphthalein.

$$\text{Rate} = (0.010 \text{ s}^{-1})(0.00500 \text{ mol/L}) = 5.0 \times 10^{-5} \frac{\text{mol/L}}{\text{s}}$$

➤ CHECKPOINT

If the rate law has the form rate = $k(X)^2$, what units will the rate constant have?

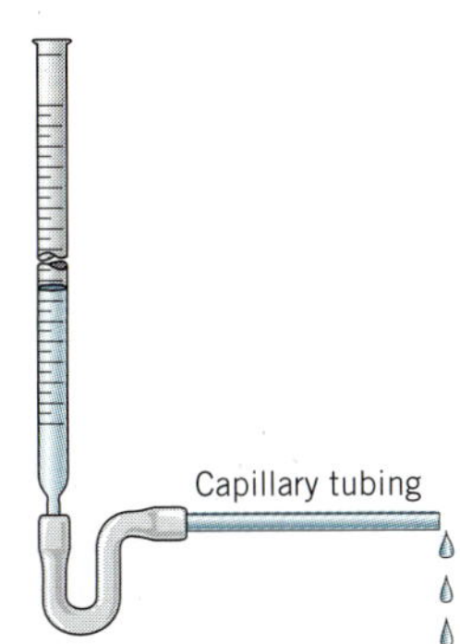

Fig. 14.4 The rate at which water in the buret flows through the capillary tube depends on the height of the water in the buret and on the diameter of the bore of the capillary tube.

14.6 A Physical Analog of Kinetic Systems

A physical model of the kinetics of a chemical reaction can be constructed with a 50-mL buret and a 1-ft length of capillary tubing, as shown in Figure 14.4. The

Table 14.2

Volume of Water in Buret versus Time

Volume of Water in Buret (mL)	Time (s)
50	0
40	19
30	42
20	72
10	116
0	203

A physical analog of first-order kinetics is provided by a Mohr buret connected to a length of 1-mm i.d. capillary tubing.[1] The rate at which water flows from the buret follows first-order kinetics. This analogy provides students with a method of comprehending the meaning of first-order kinetics. The rate at which water flows through the capillary tubing is directly proportional to the pressure of the water in the buret, which depends on the height of the water in the buret.

[1] D. A. Davenport, *J. Chem. Educ.* **52**, 379 (1975).

➤ CHECKPOINT

Explain why the rate at which the buret drains is constant when a source of compressed air is connected to the top of the buret.

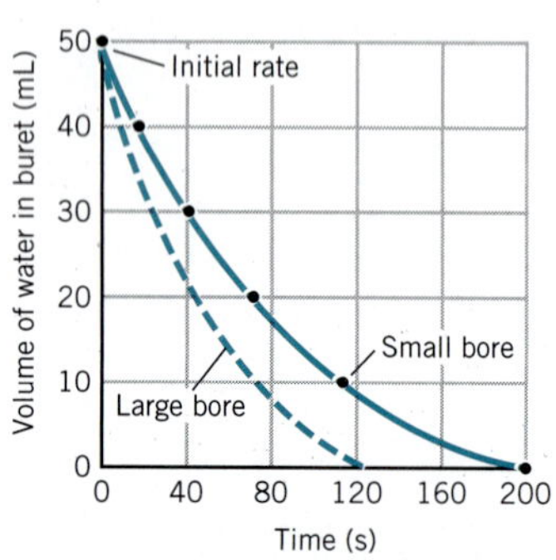

Fig. 14.5 Plot of the volume of water in the buret of the apparatus shown in Figure 14.4 versus time. The solid line represents the rate of water flowing through a small-bore tube, and the dashed line represents the rate of water flowing through a larger-bore tube.

buret is filled with water, and the volume of water is monitored versus time as water gradually flows through the capillary tubing. A typical set of data obtained with the apparatus is given in Table 14.2.

The volume of water in the buret versus time is plotted in Figure 14.5. This graph is similar to the plots of the concentration of phenolphthalein versus time in Figures 14.2 and 14.3. The average rate at which the first 10 mL of water drains from the buret is 0.53 mL per second.

$$\text{First 10 mL} \qquad \text{Rate} = -\frac{(40\ \text{mL} - 50\ \text{mL})}{(19\ \text{s} - 0\ \text{s})} = 0.53\ \text{mL/s}$$

But the average rate steadily decreases for the second, third, fourth, and fifth 10-mL portions.

$$\text{Second 10 mL} \qquad \text{Rate} = -\frac{(30\ \text{mL} - 40\ \text{mL})}{(42\ \text{s} - 19\ \text{s})} = 0.43\ \text{mL/s}$$

$$\text{Third 10 mL} \qquad \text{Rate} = -\frac{(20\ \text{mL} - 30\ \text{mL})}{(72\ \text{s} - 42\ \text{s})} = 0.33\ \text{mL/s}$$

$$\text{Fourth 10 mL} \qquad \text{Rate} = -\frac{(10\ \text{mL} - 20\ \text{mL})}{(116\ \text{s} - 72\ \text{s})} = 0.23\ \text{mL/s}$$

$$\text{Fifth 10 mL} \qquad \text{Rate} = -\frac{(0\ \text{mL} - 10\ \text{mL})}{(203\ \text{s} - 116\ \text{s})} = 0.11\ \text{mL/s}$$

If the rate at which water escaped from the buret were constant, we could calculate the rate at any moment in time and get the same answer. But the rate at which water escapes from the buret depends on the pressure pushing the water through the tube. As the height of the column of water in the buret decreases, less pressure is exerted on the water in the capillary tube, and the water flows through the tube at a slower rate.

Instead of calculating the rate averaged over a 20-s or 80-s period, we would do better to plot the data, draw a smooth curve through the points, and then calculate the instantaneous rate by determining the slope of a tangent to the curve at that moment in time.

Suppose the capillary tubing in Figure 14.4 is replaced by similar tubing of the same length but a larger bore. Now a smaller hydrostatic pressure is needed to force the water through the capillary. Consequently, if the initial volumes of water in the buret are the same, the rate of flow will be increased through the larger-bore capillary. The larger-bore capillary corresponds to a chemical reaction that has a fast rate, and the smaller bore is analogous to a chemical reaction that has a slower rate. Because the rates for both processes change, the only way to compare the rates is to compare them at the same instant in time (see Figure 14.5). A convenient way of doing this is to compare the initial rates.

14.7 The Rate Law Versus the Stoichiometry of a Reaction

In the 1930s, Sir Christopher Ingold and co-workers at the University of London studied the kinetics of substitution reactions such as the following.

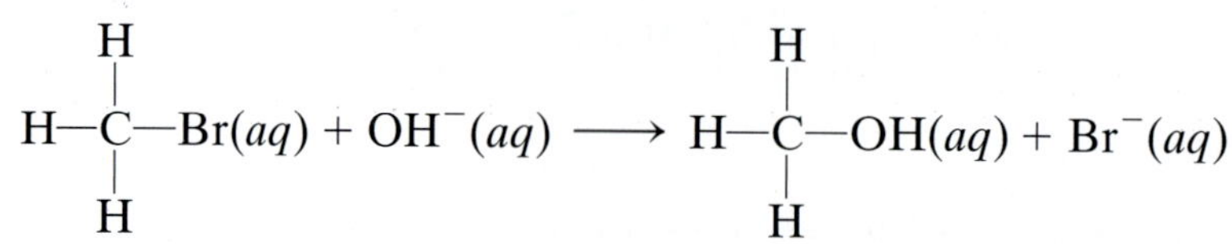

They found that the rate of the reaction is proportional to the concentrations of both reactants.

$$\text{Rate} = k(CH_3Br)(OH^-)$$

When they ran a similar reaction on a slightly different starting material, they got similar products.

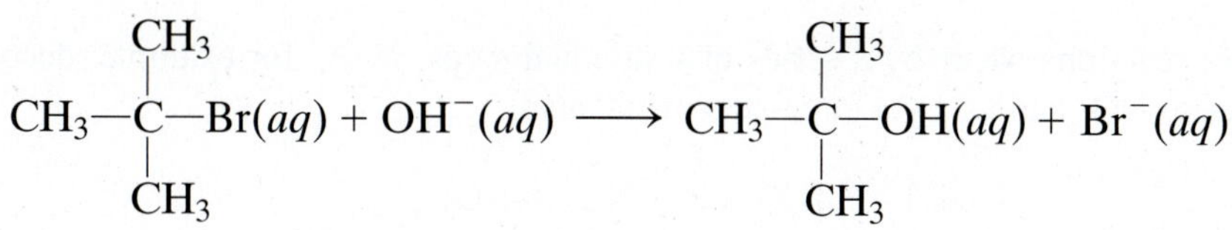

But now the rate of reaction was proportional to the concentration of only one of the reactants.

$$\text{Rate} = k((CH_3)_3CBr)$$

These results illustrate an important point: *The rate law for a reaction can't be predicted from the stoichiometry of the reaction; it must be determined experimentally.* Sometimes, the rate law is consistent with what we expect from the stoichiometry of the reaction. The balanced equation for the following reaction, for example, involves two molecules of HI, and the rate law for the reaction depends on the square of the concentration of HI.

$$2\ HI(g) \longrightarrow H_2(g) + I_2(g) \qquad \text{Rate} = k(HI)^2 = -\frac{d(HI)}{dt}$$

Often, however, the rate law doesn't follow the stoichiometry. The observed rate law for the decomposition of N_2O_5, for example, is proportional to the concentration of N_2O_5, not to the square of the concentration of N_2O_5, as we might expect from the stoichiometry of the reaction.

$$2\ N_2O_5(g) \longrightarrow 4\ NO_2(g) + O_2(g) \qquad \text{Rate} = k(N_2O_5) = -\frac{d(N_2O_5)}{dt}$$

The decomposition of N_2O_5 raises another interesting point. We can study the kinetics of this reaction by monitoring the rate at which N_2O_5 is consumed. Or we could study the rate at which NO_2 or O_2 is formed. We'll get different answers, however, depending on which reagent we choose. According to the balanced equation for this reaction, we get 4 mol of NO_2 for every two moles of N_2O_5 consumed. NO_2 is therefore formed at a rate that is twice as fast as the rate at which N_2O_5 is consumed.

$$\frac{d(NO_2)}{dt} = -2\frac{d(N_2O_5)}{dt}$$

We get only 1 mol of O_2, however, for every 2 mol of N_2O_5 consumed in this reaction. Thus, O_2 is formed at a rate that is only one-half the rate at which N_2O_5 is consumed.

$$\frac{d(O_2)}{dt} = -\frac{1}{2}\frac{d(N_2O_5)}{dt}$$

➤ CHECKPOINT

Predict the relationship between the rate of disappearance of HI and the rate of formation of H_2 in the following reaction.

$$2\ HI(g) \longrightarrow H_2(g) + I_2(g)$$

Is the rate of disappearance of HI faster, slower, or equal to the rate of formation of H_2?

14.8 Order and Molecularity

Some reactions occur in a single step. The reaction in which a chlorine atom is transferred from $ClNO_2$ to NO to form NO_2 and ClNO is a good example of a one-step reaction.

$$ClNO_2(g) + NO(g) \longrightarrow NO_2(g) + ClNO(g)$$

Other reactions occur by a series of individual steps. N_2O_5, for example, decomposes to NO_2 and O_2 by a three-step mechanism.

Step 1 $N_2O_5 \rightleftharpoons NO_2 + NO_3$

Step 2 $NO_2 + NO_3 \longrightarrow NO_2 + NO + O_2$

Step 3 $NO + NO_3 \longrightarrow 2\ NO_2$

The steps in a reaction are classified in terms of their **molecularity,** which describes the number of molecules consumed in that step. When a single molecule is consumed, the step is called **unimolecular.** When two molecules are consumed, it is **bimolecular.**

Exercise 14.4

Determine the molecularity of each step in the reaction by which N_2O_5 decomposes to NO_2 and O_2.

Solution

All we have to do is count the number of molecules consumed in each step in the reaction to decide that the first step is unimolecular and the other two steps are bimolecular.

Step 1 $N_2O_5 \rightleftharpoons NO_2 + NO_3$ (unimolecular)

Step 2 $NO_2 + NO_3 \longrightarrow NO_2 + NO + O_2$ (bimolecular)

Step 3 $NO + NO_3 \longrightarrow 2\ NO_2$ (bimolecular)

Reactions can also be classified in terms of their **order.** The decomposition of N_2O_5 is a **first-order reaction** because the rate of reaction depends on the concentration of N_2O_5 raised to the first power.

$$\text{Rate} = k(N_2O_5)$$

The decomposition of HI is a **second-order reaction** because the rate of reaction depends on the concentration of HI raised to the second power.

$$\text{Rate} = k(\text{HI})^2$$

The decomposition of dinitrogen oxide, N_2O, in the presence of platinum metal is a **zero-order reaction** because the rate of reaction is constant and is independent of the concentration of N_2O.

$$2\ N_2O(g) \xrightarrow{Pt} 2\ N_2(g) + O_2(g)$$
$$\text{Rate} = k\ (N_2O)^0 = k$$

When the rate of a reaction depends on more than one reagent, we classify the reaction in terms of the order of each reagent. The overall order is the sum of the orders of the individual reactants.

In general, for a reaction

$$A + B \longrightarrow C + D$$

the rate law is written in terms of the concentrations of the reactants, each raised to an exponent. The exponent gives the order of the reaction with respect to a particular reactant. Thus, for the general reaction above

$$\text{Rate} = k(A)^m(B)^n$$

where k is the rate constant for the reaction, m is the order with respect to A, and n is the order with respect to B. The overall order of the reaction is given by $m + n$.

Exercise 14.5

When there are only trace quantities of NO in the atmosphere, it doesn't react with O_2 to form NO_2. As the concentration of NO in the atmosphere builds up, however, it reaches a point at which this reaction can and does occur.

$$2\ NO(g) + O_2(g) \longrightarrow 2\ NO_2(g)$$

The experimentally determined rate law for the reaction is

$$\text{Rate} = k(NO)^2(O_2)$$

Classify the order of this reaction.

Solution

The reaction is first-order in O_2, second-order in NO, and third-order overall.

When the rate of a reaction doesn't depend on the concentration of one or more of the substances consumed in that reaction, it is said to be *zero-order* in that substance. In a zero-order reaction, the rate of reaction literally depends on the concentration of that reagent to the zeroth power. Since any quantity raised to the zeroth power is equal to 1, the concentration of the reagent has no effect on the rate of reaction.

We have already seen an example of a reaction that is zero-order in one of the reactants.

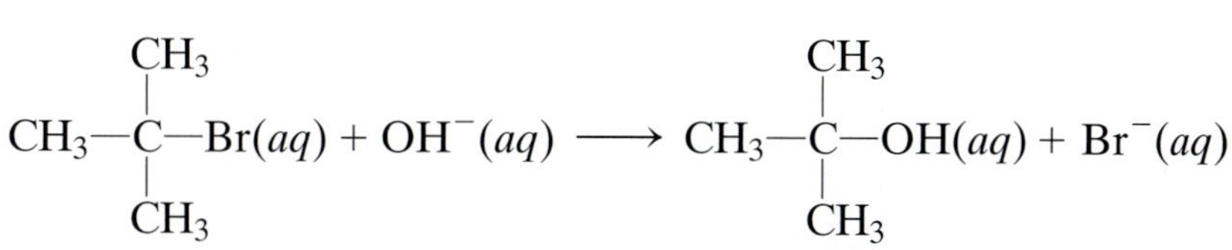

The rate law for this reaction is first-order in $(CH_3)_3CBr$, but it is zero-order in the OH^- ion.

$$\text{Rate} = k((CH_3)_3CBr)^1(OH^-)^0 = k((CH_3)_3CBr)$$

It is important to recognize the difference between the molecularity and the order of a reaction. The molecularity of a reaction, or a step within a reaction, describes what happens on the molecular level. The order of a reaction describes what happens on the macroscopic scale. We determine the order of a reaction experimentally by watching the products of a reaction appear or the starting materials disappear. The molecularity of the reaction is something we deduce to explain the experimental results.

14.9 A Collision Theory of Chemical Reactions

The **collision theory** of chemical reactions introduced in Section 10.4 can be used to explain the observed rate laws for both one-step and multistep reactions. The theory assumes that the rate of any step in a reaction depends on the frequency of collisions between the particles involved in that step. Consider the following simple, one-step reaction, for example.

$$ClNO_2(g) + NO(g) \longrightarrow NO_2(g) + ClNO(g)$$

Figure 14.6 provides a basis for understanding the implications of the collision theory for this reaction. The kinetic molecular theory assumes that the number of collisions per second in a gas depends on the number of particles per unit volume. Thus, the rate at which NO_2 and ClNO are formed in the reaction should be directly proportional to the concentrations of both $ClNO_2$ and NO.

$$\text{Rate} = k(ClNO_2)(NO)$$

The collision theory suggests that the rate of any step in a reaction is proportional to the concentrations of the reagents consumed in that step. The rate law for a one-step reaction should therefore agree with the stoichiometry of the reaction. The following reaction, for example, occurs in a single step.

$$CH_3Br(aq) + OH^-(aq) \longrightarrow CH_3OH(aq) + Br^-(aq)$$

When the reactant molecules collide in the proper orientation, a pair of nonbonding electrons on the OH^- ion can be donated to the carbon atom at the center of the

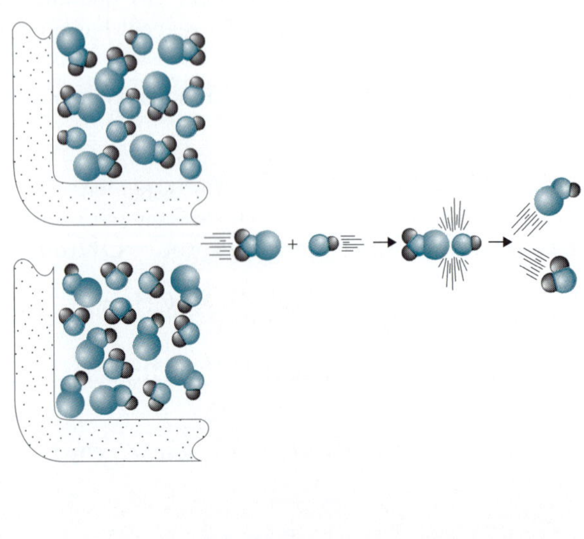

Fig. 14.6 "Snapshot" of a small portion of a container in which $ClNO_2$ reacts with NO to form NO_2 and ClNO. The collision theory of reactions assumes that molecules must collide in order to react. Anything that increases the frequency of the collisions increases the rate of reaction, so the rate of reaction must be proportional to the concentrations of both of the reactants consumed in this reaction.

CH_3Br molecule, as shown in Figure 14.7. When this happens, a carbon–oxygen bond forms at the same time that the carbon–bromine bond is broken. The net result of the reaction is the substitution of an OH^- ion for a Br^- ion.

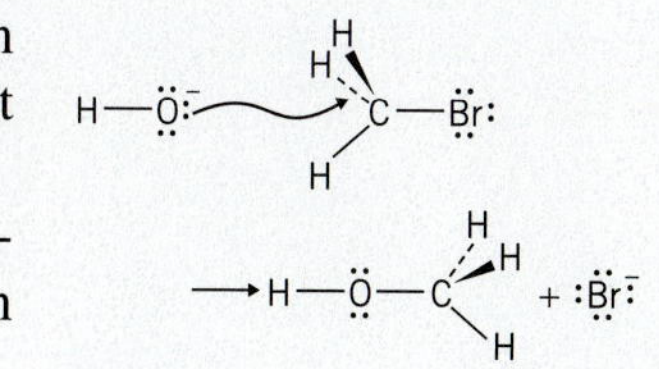

Fig. 14.7 The reaction between CH_3Br and the OH^- ion occurs in a single step, which involves attack by the OH^- ion on the carbon atom. In the course of the reaction, a carbon–oxygen bond forms at the same time that the carbon–bromine bond is broken.

Because the reaction occurs in a single step, which involves collisions between the two reactants, the rate of the reaction is proportional to the concentration of both reactants.

$$\text{Rate} = k(CH_3Br)(OH^-)$$

Not all reactions occur in a single step. The following reaction occurs in three steps.

$$(CH_3)_3CBr(aq) + OH^-(aq) \longrightarrow (CH_3)_3COH(aq) + Br^-(aq)$$

In the first step, the $(CH_3)_3CBr$ molecule dissociates into a pair of ions.

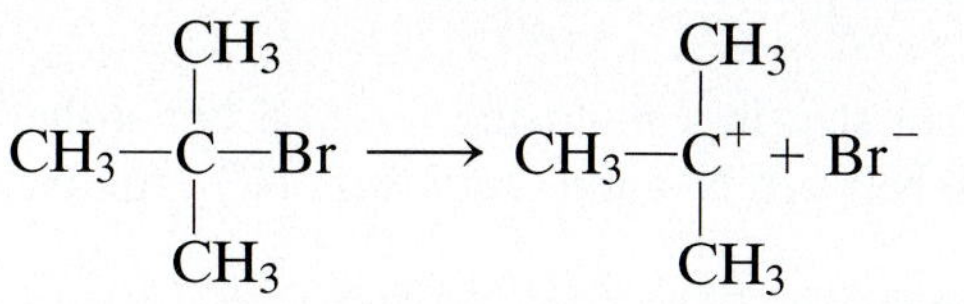

First step

The positively charged $(CH_3)_3C^+$ ion then reacts with water in a second step.

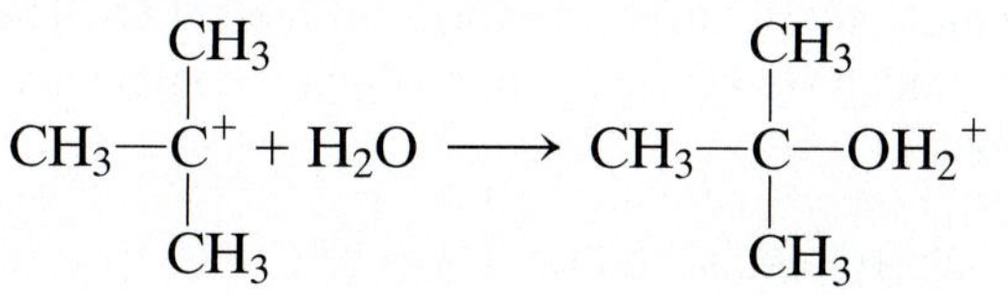

Second step

The product of the reaction then loses a proton to either the OH^- ion or water in the final step.

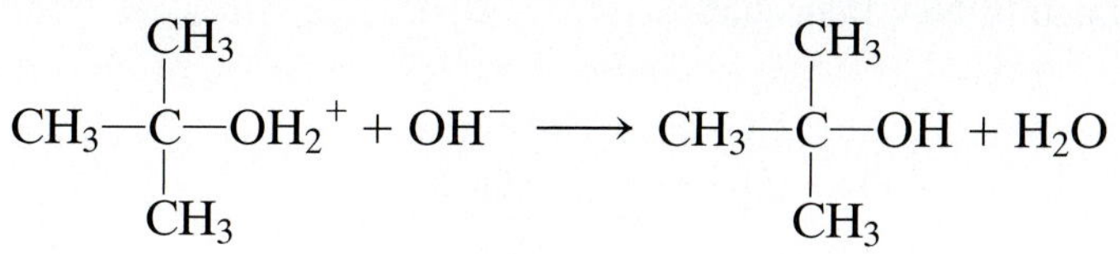

Third step

The second and third steps in the reaction are very much faster than the first.

$(CH_3)_3CBr \longrightarrow (CH_3)_3C^+ + Br^-$	Slow step
$(CH_3)_3C^+ + H_2O \longrightarrow (CH_3)_3COH_2^+$	Fast step
$(CH_3)_3COH_2^+ + OH^- \longrightarrow (CH_3)_3COH + H_2O$	Fast step

The overall rate of this reaction is therefore very close to the rate of the first step. The first step in this reaction is therefore called the **rate-limiting step** because it literally limits the rate at which the products of the reaction can be formed. Because only one reagent is involved in the rate-limiting step, the overall rate of reaction is proportional to the concentration of only that reagent.

$$\text{Rate} = k((CH_3)_3CBr)$$

The rate law for this reaction therefore differs from what we would predict from the stoichiometry of the reaction. Although the reaction consumes both $(CH_3)_3CBr$ and OH^-, the rate of the reaction is proportional only to the concentration of $(CH_3)_3CBr$.

The rate laws for chemical reactions can be explained by the following general rules.

- The rate of any step in a reaction is directly proportional to the concentrations of the reagents consumed in that step.
- The overall rate law for a reaction is determined by the sequence of steps, or the **mechanism,** by which the reactants are converted into the products of the reaction.
- The overall rate law for a reaction is dominated by the rate law for the slowest step in the reaction.

14.10 The Mechanisms of Chemical Reactions

What happens when the first step in a multistep reaction is not the rate-limiting step? Consider the reaction between NO and O_2 to form NO_2, for example.

$$2\ NO(g) + O_2(g) \longrightarrow 2\ NO_2(g)$$

The reaction involves a two-step mechanism. The first step is a relatively fast reaction in which a pair of NO molecules combine to form a dimer, N_2O_2. The product of this step then undergoes a much slower reaction in which it combines with O_2 to form a pair of NO_2 molecules.

$$\text{Step 1} \quad 2\ NO \rightleftharpoons N_2O_2 \quad \text{(fast)}$$
$$\text{Step 2} \quad N_2O_2 + O_2 \longrightarrow 2\ NO_2 \quad \text{(slow)}$$

The net effect of these reactions is the transformation of two NO molecules and one O_2 molecule into a pair of NO_2 molecules.

$$\begin{array}{r}
2\ NO \rightleftharpoons N_2O_2 \\
N_2O_2 + O_2 \longrightarrow 2\ NO_2 \\
\hline
2\ NO + O_2 \longrightarrow 2\ NO_2
\end{array}$$

The mechanism for this reaction explains the observation made in Exercise 14.5. At very low NO concentrations, the first reaction doesn't form enough N_2O_2 to allow the second reaction to occur at a significant rate.

In this reaction, the second step is the rate-limiting step. No matter how fast the first step takes place, the overall reaction cannot proceed any faster than the second step in the reaction. As we have seen, the rate of any step in a reaction is directly proportional to the concentrations of the reactants consumed in that step. The rate law for the second step in this reaction is therefore proportional to the concentrations of both N_2O_2 and O_2.

$$\text{Step 2} \quad \text{Rate}_{2nd} = k(N_2O_2)(O_2)$$

Because the first step in the reaction is much faster, the overall rate of reaction is more or less equal to the rate of the rate-limiting second step.

$$\text{Rate}_{overall} \approx k(N_2O_2)(O_2)$$

This rate law isn't very useful, because it is difficult to measure the concentrations of **intermediates,** such as N_2O_2, that are simultaneously formed and consumed in the reaction. It would be better to have an equation that related the overall rate of reaction to the concentrations of the original reactants.

Let's take advantage of the fact that the first step in the reaction is an equilibrium step.

$$\text{Step 1} \qquad 2\ \text{NO} \rightleftharpoons N_2O_2$$

The collision theory predicts that the rate of the forward reaction in this step will depend on the concentration of NO raised to the second power.

$$\text{Step 1} \qquad \text{Rate}_{\text{forward}} = k_f(\text{NO})^2$$

The rate of the reverse reaction, however, will depend only on the concentration of N_2O_2.

$$\text{Step 1} \qquad \text{Rate}_{\text{reverse}} = k_r(N_2O_2)$$

Because the first step in the reaction is very much faster than the second, the first step should come to equilibrium. When that happens, the rate of the forward and reverse reactions for the first step are the same.

$$k_f(\text{NO})^2 = k_r(N_2O_2)$$

Let's rearrange this equation to solve for one of the terms that appears in the rate law for the second step in the reaction

$$(N_2O_2) = \frac{k_f}{k_r}(\text{NO})^2$$

Substituting this equation for (N_2O_2) in the rate law for the second step gives the following result.

$$\text{Rate}_{\text{2nd}} = \frac{k(k_f)}{k_r}(\text{NO})^2(O_2) = kK_c(\text{NO})^2\,(O_2)$$

Note that k_f/k_r is equal to the equilibrium constant for the reaction as shown in Section 10.5. Since k, k_f, and k_r are all constants, they can be replaced by a single constant, k', to give the experimental rate law for this reaction described in Exercise 14.5.

$$\text{Rate}_{\text{overall}} \approx \text{rate}_{\text{2nd}} = k'(\text{NO})^2(O_2)$$

14.11 Zero-Order Reactions

The collision theory can explain why the rate law for a reaction can be zero-order in one of the substances consumed in that reaction. In order for the reaction to be zero-order in a reactant, the reaction must occur in more than one step and the rate-limiting step for the reaction must not involve that substance. The rate law for the following reaction, for example, is zero-order in the OH^- ion because the OH^- ion isn't involved in the rate-limiting step.

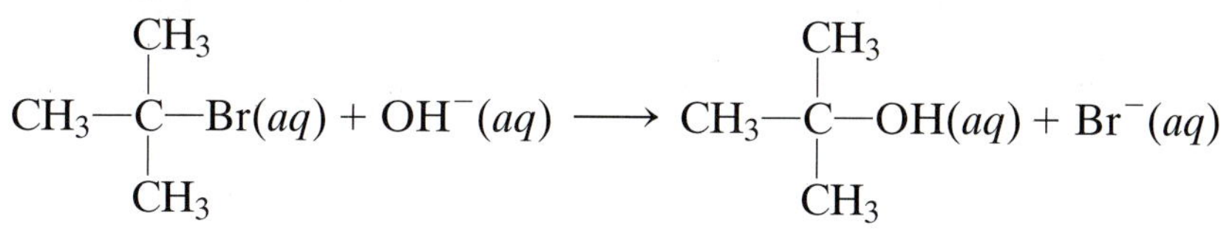

One question remains unanswered: How do we explain the fact that some reactions are zero-order in all of the starting materials? Consider the following reaction, for example, in which nitrogen oxide decomposes to its elements on the surface of a piece of platinum metal.

$$2\,NO(g) \xrightarrow{Pt} N_2(g) + O_2(g)$$

In the presence of a large excess of NO, the rate of the reaction is zero-order in NO.

$$\text{Rate} = k(NO)^0 = k$$

The decomposition of NO (or NO_2) into its elements is an important reaction because of the significant environmental problem that NO and NO_2 represent when the gases accumulate in the atmosphere. Most of the catalytic processes that have been studied to remove NO and NO_2 from the atmosphere are accompanied by the release of other pollutants, such as CO_2. Thus, the search for a solid surface on which NO can be induced to decompose into its elements is of vital importance.[1]

A simple analogy can help us understand why the decomposition of NO to its elements on a solid surface is zero-order in NO, as long as a large excess of NO is present. Assume that you are waiting in line to eat at a popular restaurant. Until there is an empty table, no one else can be seated. The rate at which people are seated doesn't depend on the number of people standing in line. It only depends on the rate at which tables are vacated.

The same thing is true for the decomposition of NO on the platinum metal surface. For NO to react, it must find a site on the metal at which it can bind to the metal surface. No matter how much NO is present, there is only a limited number of sites available at which reaction can occur, and only those NO molecules that occupy one of these sites can undergo a reaction. The reaction is therefore zero-order until virtually all of the NO has been consumed.

Any chemical reaction that occurs by a mechanism by which the reactant has to bind to a catalyst that has a limited number of active sites will exhibit similar behavior. Consider the following reaction, for example.

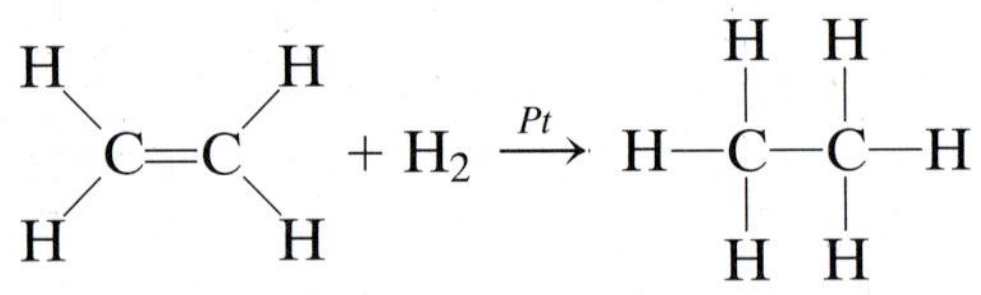

> **CHECKPOINT**
>
> How does the dependence of rate on concentration differ between a zero-order and first-order reaction? Describe what happens to the rate of a reaction with respect to time for both a zero- and a first-order reaction.

Both starting materials must bind to an active site on the platinum metal before the reaction can occur. When both gases are present in excess, the rate of the reaction is zero-order in *both* of the starting materials.

$$\text{Rate} = k(C_2H_4)^0(H_2)^0 = k$$

14.12 Determining the Order of a Reaction from Rates of Reaction

The rate law for a reaction can be determined by studying what happens to the instantaneous rate of reaction when we start with different initial concentrations of

[1] For a review of the role that catalysts such as platinum metal can play in improving the environment, see the paper by John N. Armor in *Chemistry of Materials*, **6,** 730–738 (1994).

Table 14.3

Rate of Reaction Data for the Decomposition of HI

Trial	Initial Concentration of HI (M)	Initial Instantaneous Rate of Reaction (M/s)
Trial 1	1.0×10^{-2}	4.0×10^{-6}
Trial 2	2.0×10^{-2}	1.6×10^{-5}
Trial 3	3.0×10^{-2}	3.6×10^{-5}

the reactants. To show how this is done, let's determine the rate law for the reaction in which hydrogen iodide decomposes to give a mixture of hydrogen and iodine in the gas phase.

$$2\ HI(g) \longrightarrow H_2(g) + I_2(g)$$

Data on initial rates of reaction for three experiments run at different initial concentrations of HI are given in Table 14.3.

The only difference in the three experiments is the initial concentration of HI. Let's compare the experiments two at a time. The difference between trial 1 and trial 2 is a twofold increase in the initial HI concentration, which leads to a fourfold increase in the initial rate of reaction.

$$\frac{\text{Rate for trial 2}}{\text{Rate for trial 1}} = \frac{1.6 \times 10^{-5}\ M/\text{s}}{4.0 \times 10^{-6}\ M/\text{s}} = 4.0$$

The only difference between trials 1 and 3 is a threefold increase in the initial concentration of HI. The initial rate of reaction, however, increases by a factor of 9.

$$\frac{\text{Rate for trial 3}}{\text{Rate for trial 1}} = \frac{3.6 \times 10^{-5}\ M/\text{s}}{4.0 \times 10^{-6}\ M/\text{s}} = 9.0$$

The data suggest that the rate of the reaction is proportional to the *square of the HI concentration.* The reaction is therefore second order in HI, as noted in Section 14.7.

$$\text{Rate} = k(\text{HI})^2$$

Exercise 14.6

Use the following experimental data to determine the rate law for the reaction.

$$NH_4^+(aq) + NO_2^-(aq) \longrightarrow N_2(g) + 2\ H_2O(l)$$

	Initial Concentration of NH_4^+ (M)	*Initial Concentration of* NO_2^- (M)	*Initial Instantaneous Rate of Reaction* (M/s)
Trial 1	5.00×10^{-2}	2.00×10^{-2}	2.70×10^{-7}
Trial 2	5.00×10^{-2}	4.00×10^{-2}	5.40×10^{-7}
Trial 3	1.00×10^{-1}	2.00×10^{-2}	5.40×10^{-7}

Solution

As shown in Section 14.8, the rate law for trial 1 can be written as

$$\text{Rate}_1 = k(NH_4^+)_1^m (NO_2^-)_1^n$$

and the rate law for trial 2 is

$$\text{Rate}_2 = k(\text{NH}_4^+)_2^m(\text{NO}_2^-)_2^n$$

The order with respect to NH_4^+ and NO_2^- can be found from the experimental data if we know how the initial rates of reaction change when the concentrations of only one of the reactants is varied. In trials 1 and 2 the (NH_4^+) remains constant while the (NO_2^-) is doubled. Thus, a ratio of trial 1 to trial 2 can be used to determine how the changing (NO_2^-) affects the initial rate. This allows the determination of the order, n, with respect to (NO_2^-).

$$\frac{\text{Rate}_1}{\text{Rate}_2} = \frac{k(\text{NH}_4^+)_1^m(\text{NO}_2^-)_1^n}{k(\text{NH}_4^+)_2^m(\text{NO}_2^-)_2^n}$$

$$\frac{2.70 \times 10^{-7}}{5.40 \times 10^{-7}} = \frac{k(5.00 \times 10^{-2})^m(2.00 \times 10^{-2})^n}{k(5.00 \times 10^{-2})^m(4.00 \times 10^{-2})^n}$$

The rate constants and the (NH_4^+) terms are the same in the numerator and denominator and therefore cancel out of the equation.

$$\frac{2.70 \times 10^{-7}}{5.40 \times 10^{-7}} = \frac{(2.00 \times 10^{-2})^n}{(4.00 \times 10^{-2})^n} = \left(\frac{2.00 \times 10^{-2}}{4.00 \times 10^{-2}}\right)^n = \left(\frac{1}{2}\right)^n$$

$$\frac{1}{2} = \left(\frac{1}{2}\right)^n$$

$$n = 1$$

> **➤ CHECKPOINT**
>
> Why are trials 2 and 3 not used to determine the order of the reaction in terms of either the NH_4^+ or NO_2^- ion? Can trials 1 and 2 be used to determine the order with respect to NH_4^+?

Similarly, we can use the ratio of trials 1 and 3 to find the order with respect to NH_4^+.

$$\frac{\text{Rate}_1}{\text{Rate}_3} = \frac{k(5.00 \times 10^{-2})^m(2.00 \times 10^{-2})^n}{k(1.00 \times 10^{-1})^m(2.00 \times 10^{-2})^n}$$

$$\frac{2.70 \times 10^{-7}}{5.40 \times 10^{-7}} = \left(\frac{5.00 \times 10^{-2}}{1.00 \times 10^{-1}}\right)^m$$

$$m = 1$$

> **➤ CHECKPOINT**
>
> Determine the value of the rate constant for the reaction in Exercise 14.6.

The rate law is then rate $= k(\text{NH}_4^+)^1(\text{NO}_2^-)^1$, and the overall order is $1 + 1 = 2$.

14.13 The Integrated Form of Zero-, First-, and Second-Order Rate Laws

As we have seen, the rate law for a reaction provides a basis for studying the mechanism of the reaction. It is also useful for predicting how much reactant will remain or how much product will be formed in a given amount of time. For these calculations, we use the **integrated form of the rate laws.** Both the rate laws and their integrated form can be used to determine the order of the reaction and the rate constant for reaction. A derivation of the integrated forms of the rate laws can be found in the Special Topics section at the end of this chapter.

Let's start with the simplest of all rate laws, a reaction that is zero order.

$$-\frac{d(X)}{dt} = k(X)^0 = k$$

When this equation is rearranged and both sides are integrated, we get the following result.

Integrated Form of the Zero-Order Rate Law

$$(X) - (X)_0 = -kt$$

In this equation, (X) is the concentration of X at any moment in time, $(X)_0$ is the initial concentration of the reagent, k is the rate constant for the reaction, and t is the time since the reaction started.

Exercise 14.7

How long will it take for a reactant with an initial concentration of 2.0 M to fall to a concentration of 1.0 M if the reaction is zero-order? The rate constant, k, for the reaction is 1.5×10^{-2} mol/L · s.

Solution

For a zero-order reaction the concentration at any time (X) after the reaction has begun is related to the initial concentration $(X)_0$ by

$$(X) - (X)_0 = -kt$$

or substituting in the values given

$$(1.0 \text{ mol/L}) - (2.0 \text{ mol/L}) = -(1.5 \times 10^{-2} \text{ mol/L} \cdot \text{s}) \times t$$

and solving for t gives

$$t = 1.0/1.5 \times 10^{-2} = 67 \text{ s}$$

Let's now continue with the rate law for a reaction that is first-order in the disappearance of a single reactant, X.

$$-\frac{d(X)}{dt} = k(X)$$

When the equation is rearranged and both sides are integrated, we get the following result.

Integrated Form of the First-Order Rate Law

$$\ln\left[\frac{(X)}{(X)_0}\right] = -kt$$

Once again, (X) is the concentration of X at any moment in time, $(X)_0$ is the initial concentration of the reagent, k is the rate constant for the reaction, and t is the time since the reaction started.

Exercise 14.8

The decomposition of N_2O_5 is a first-order reaction. If the initial concentration of N_2O_5 is 1.0 M, what will be the concentration after 5.00 minutes? The rate constant has been experimentally determined to be 0.420 min^{-1}.

Solution

First-order kinetics means that the integrated form of the rate law is

$$\ln(X)/(X)_0 = -kt$$

or using the properties of logarithms

$$\ln(X) - \ln(X)_0 = -kt$$

Substituting what is known, we find

$$\begin{aligned} \ln(X) - \ln(1.0) &= -0.420 \text{ min}^{-1} \times 5.00 \text{ min} \\ \ln(X) &= -2.10 \\ (X) &= 0.12\ M \end{aligned}$$

To further illustrate the power of the integrated form of the rate law for a reaction, we will use the equation to calculate how long it would take for the ^{14}C in a piece of charcoal to decay to half of its original concentration. All radioactive decay processes are first-order. Therefore ^{14}C decays by first-order kinetics.

$$\text{Rate} = k(^{14}\text{C})$$

The rate constant that is determined from experimental data for this reaction is $1.21 \times 10^{-4}\ y^{-1}$:

The integrated form of the rate law would be written as follows.

$$\ln\left[\frac{(^{14}\text{C})}{(^{14}\text{C})_0}\right] = -kt$$

We are interested in the moment when the concentration of ^{14}C in the charcoal is half of its initial value.

$$(^{14}\text{C}) = \frac{1}{2}(^{14}\text{C})_0$$

Substituting this relationship into the integrated form of the rate law gives the following equation.

$$\ln\left[\frac{\frac{1}{2}(^{14}\text{C})_0}{(^{14}\text{C})_0}\right] = -kt$$

We now simplify the equation

$$\ln\left[\frac{1}{2}\right] = -kt$$

and then solve for t.

$$t = -\frac{\ln(1/2)}{k} = \frac{0.6931}{1.21 \times 10^{-4} \text{ yr}^{-1}} = 5.73 \times 10^3 \text{ yr}$$

It therefore takes 5.73×10^3 years for half of the ^{14}C in the sample to decay. This is called the **half-life** ($t_{1/2}$) of ^{14}C. In general, the half-life for a first-order kinetic process can be calculated from the rate constant as follows.

$$t_{1/2} = -\frac{\ln(1/2)}{k} = \frac{0.6931}{k}$$

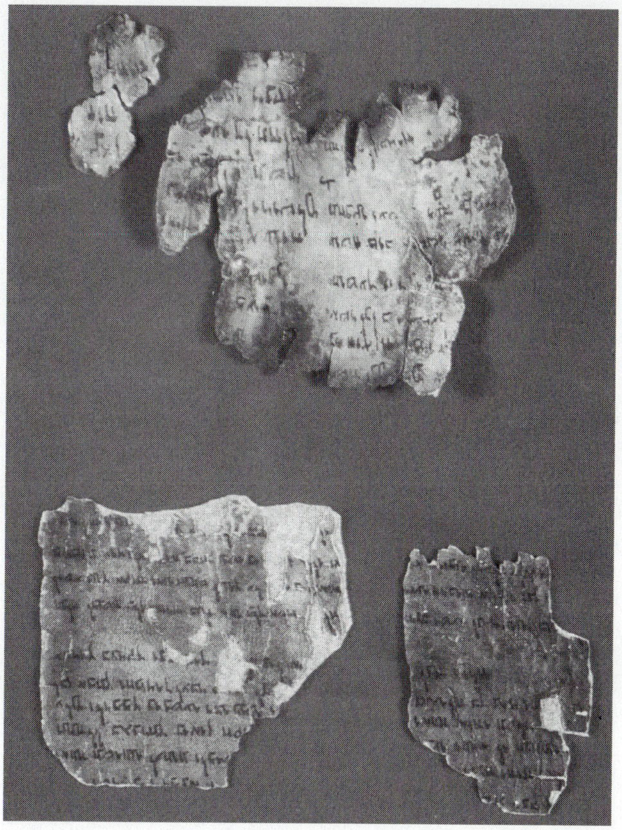

Fragments of the Dead Sea Scrolls, which were authenticated by ^{14}C dating.

Let's now turn to the rate law for a reaction that is second-order in a single reactant, X.

$$-\frac{d(X)}{dt} = k(X)^2$$

The integrated form of the rate law for this reaction is written as follows.

Integrated Form of the Second-Order Rate Law

$$\frac{1}{(X)} - \frac{1}{(X)_0} = kt$$

Once again, (X) is the concentration of X at any moment in time, $(X)_0$ is the initial concentration of X, k is the rate constant for the reaction, and t is the time since the reaction started.

➤ CHECKPOINT

Which radioactive isotope would have the longer half-life, ^{15}O or ^{19}O?

^{15}O	$k = 5.63 \times 10^{-3}\ s^{-1}$
^{19}O	$k = 2.38 \times 10^{-2}\ s^{-1}$

Exercise 14.9

NO_2 reacts to form NO and O_2 by second-order kinetics with a rate constant of 32.6 L/mol min. If the initial NO_2 concentration is 0.15 M, what will be the concentration after 1.0 min?

Solution

The rate law is rate $= k(NO_2)^2$, which integrates to

$$\frac{1}{(X)} - \frac{1}{(X)_0} = kt$$

$(X)_0$, k, and t are known, so solving for (X) we find

$$1/(X) = (32.6 \text{ L/mol min}) \times (1.0 \text{ min}) + 1/(0.15 \text{ L/mol}) = 39$$
$$(X) = 0.026\ M$$

The half-life of a second-order reaction can be calculated from the integrated form of the second-order rate law.

$$\frac{1}{(X)} - \frac{1}{(X)_0} = kt$$

We start by asking, "How long would it take for the concentration of X to decay from its initial value, $(X)_0$, to a value half as large?"

$$\frac{1}{\frac{1}{2}(X)_0} - \frac{1}{(X)_0} = kt_{1/2}$$

The first step in simplifying the equation involves multiplying the top and bottom halves of the first term by 2.

$$\frac{2}{(X)_0} - \frac{1}{(X)_0} = kt_{1/2}$$

Subtracting the second term on the left side of the equation from the first gives the following result.

$$\frac{1}{(X)_0} = kt_{1/2}$$

We can now solve the equation for the half-life of the reaction.

$$t_{1/2} = \frac{1}{k(X)_0}$$

There is an important difference between the equations for calculating the half-lives of zero-order, first-order, and second-order reactions. The half-life of a first-order reaction is a constant, which is inversely proportional to the rate constant for the reaction and independent of the initial concentration. This relationship is particularly useful for first-order reactions such as radioactive decay when the half-life is known and the rate constant is needed.

First-Order Reaction $\quad t_{1/2} = \frac{0.6931}{k}$

The half-life for a zero-order reaction is proportional to the initial concentration of the reactant consumed in the reaction and inversely proportional to the rate constant for the reaction.

Zero-Order Reaction $\quad t_{1/2} = \frac{(X)_0}{2k}$

The half-life for a second-order reaction is inversely proportional to both the rate constant for the reaction and the initial concentration of the reactant that is consumed in the reaction.

Second-Order Reaction $\quad t_{1/2} = \frac{1}{k(X)_0}$

Discussions of reaction half-lives are usually confined to first-order processes because this is the only reaction order for which the half-life is independent of reactant concentration. The equations that describe zero-, first- and second-order reactions are summarized in Table 14.4.

Table 14.4

Equations that Describe Zero-, First-, and Second-Order Reactions

Order	Rate Law	Integrated Form of the Rate Law	Half-Life
Zero	rate = k	$(X) - (X)_0 = -kt$	$t_{1/2} = \frac{(X)_0}{2k}$
First	rate = $k(X)$	$\ln\left[\frac{(X)}{(X)_0}\right] = -kt$	$t_{1/2} = \frac{0.6931}{k}$
Second	rate = $k(X)^2$	$\frac{1}{(X)} - \frac{1}{(X)_0} = kt$	$t_{1/2} = \frac{1}{k(X)_0}$

➤ CHECKPOINT

Suppose a reaction was carried out twice, once with an initial concentration of 1 M and then with an initial concentration of 2 M. How will the half-life for the second trial compare to the half-life of the first trial for a zero-, first-, and second-order reaction?

14.14 Determining the Order of a Reaction with the Integrated Form of Rate Laws

The integrated forms of the rate laws for zero-, first-, and second-order reactions provide another way of determining the order of a reaction. We can use experimental data to examine each rate law to determine the order of the reaction. Let's start by considering a zero-order reaction.

$$\text{Rate} = k$$

If the reaction is zero-order, concentration versus time data for the reaction fit the integrated form of the zero-order rate law.

$$(X) - (X)_0 = -kt$$

This equation contains two variables, (X) and t, and two constants, $(X)_0$ and k. It can therefore be set up in terms of the equation for a straight line.

$$y = mx + b$$
$$(X) = -kt + (X)_0$$

If the reaction is zero-order in X, a plot of the concentration of X versus time will be a straight line with a slope equal to $-k$, as shown in Figure 14.8.

If the plot of (X) versus time isn't a straight line, the reaction can't be zero order in X. Next we can examine the data to see if the reaction is first order. The rate law is

$$\text{Rate} = k(X)$$

If the reaction is first-order, concentration versus time data for the reaction should fit the integrated form of the first-order rate law.

$$\ln\left[\frac{(X)}{(X)_0}\right] = -kt$$

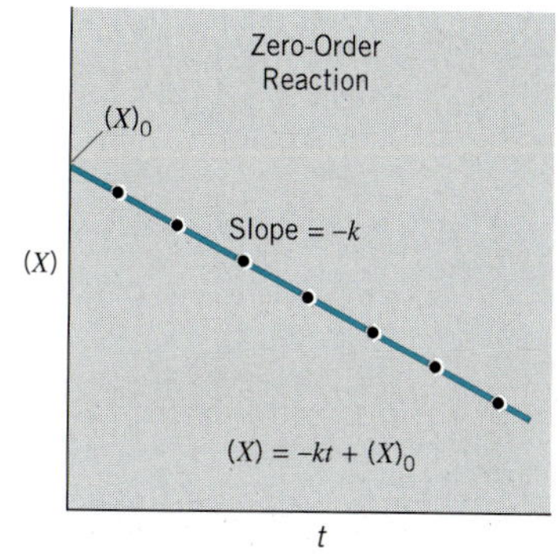

Fig. 14.8 A plot of the concentration of X versus the elapsed time since the reaction started for a zero-order reaction will be a straight line with a slope equal to the negative of the rate constant for the reaction.

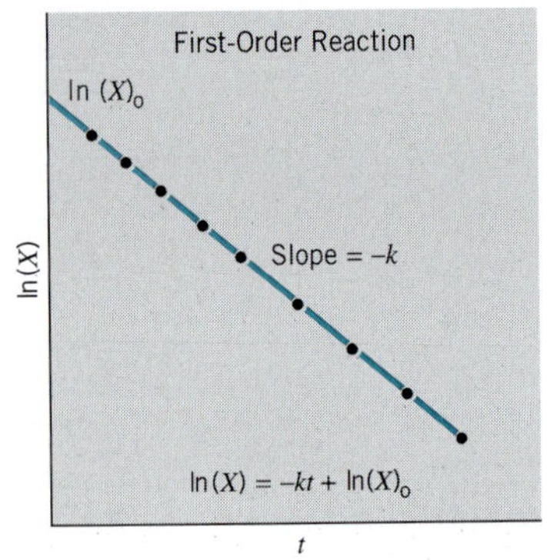

Fig. 14.9 For a reaction that is first order in a single reactant, a plot of the natural log of the concentration of X versus the elapsed time since the reaction started will be a straight line with a slope equal to the negative of the rate constant.

To see how this is done, let's rearrange the integrated form of the first-order rate law as follows.

$$\ln(X) - \ln(X)_0 = -kt$$

We then rearrange the equation for the natural logarithm of the concentration of X at any moment in time.

$$\ln(X) = -kt + \ln(X)_0$$

This equation contains two variables, $\ln(X)$ and t, and two constants, $\ln(X)_0$ and k. It is in the form of an equation for a straight line.

$$y = mx + b$$
$$\ln(X) = -kt + \ln(X)_0$$

If the reaction is first-order in X, a plot of the natural logarithm of the concentration of X versus time will be a straight line with a slope equal to $-k$, as shown in Figure 14.9.

If the plot of $\ln(X)$ versus time isn't a straight line, the reaction can't be first-order in X. Next we can examine the data to see if the reaction is second-order. The rate law is

$$\text{Rate} = k(X)^2$$

If the reaction is second-order, the experimental data will fit the integrated form of the second-order rate law.

$$\frac{1}{(X)} - \frac{1}{(X)_0} = kt$$

This equation also contains two variables, (X) and t, and two constants, $(X)_0$ and k. Thus, it also can be set up in terms of the equation for a straight line.

$$y = mx + b$$
$$\frac{1}{(X)} = kt + \frac{1}{(X)_0}$$

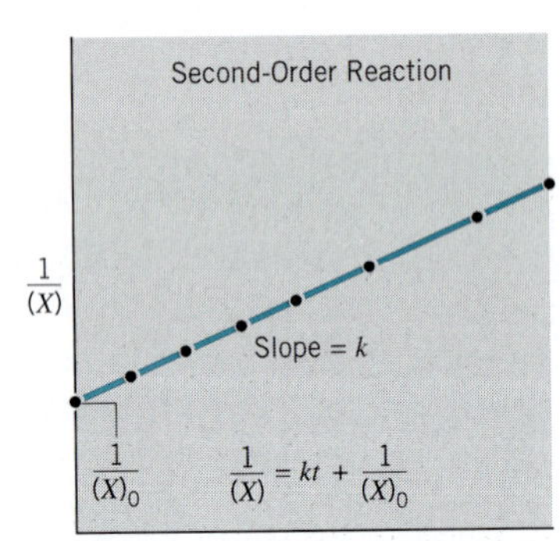

Fig. 14.10 For a reaction that is second order in a single reactant, a plot of the inverse of the concentration of X versus the elapsed time since the reaction started will be a straight line with a slope equal to the rate constant.

If the reaction is second-order in X, a plot of the reciprocal of the concentration of X versus time will be a straight line with a slope equal to k, as shown in Figure 14.10. If the plot of $1/(X)$ versus time is not linear, the reaction cannot be second-order.

Exercise 14.10

Use the data in Table 14.1 to determine whether the reaction between phenolphthalein and the OH^- ion is zero-order, first-order, or second-order with respect to the phenolphthalein. OH^- is in excess.

Solution

The first step in solving the problem involves calculating the natural log of the phenolphthalein concentration, ln(PHTH), and the reciprocal of the concentration, 1/(PHTH), for each point at which a measurement was taken.

(PHTH) (mol/L)	*ln(PHTH)*	*1/(PHTH)*	*Time (s)*
0.00500	−5.298	200	0.0
0.00450	−5.404	222	10.5
0.00400	−5.521	250	22.3
0.00350	−5.655	286	35.7
0.00300	−5.809	333	51.1
0.00250	−5.991	400	69.3
0.00200	−6.215	500	91.6
0.00150	−6.502	667	120.4
0.00100	−6.908	1.00×10^3	160.9
0.000500	−7.601	2.00×10^3	230.2
0.000250	−8.294	4.00×10^3	299.5
0.000150	−8.805	6.67×10^3	350.7
0.000100	−9.210	1.00×10^4	391.2

We then construct graphs of (PHTH) versus *t* (Figure 14.11), of ln(PHTH) versus *t* (Figure 14.12), and of 1/(PHTH) versus *t* (Figure 14.13).

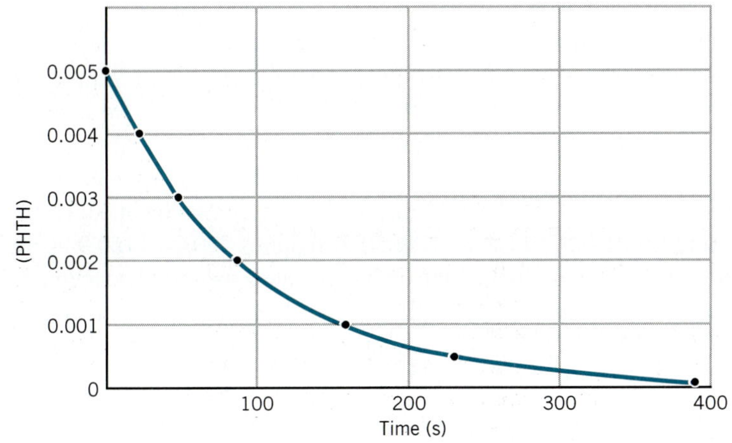

Fig. 14.11 A plot of the concentration of phenolph-thalein versus time for the reaction between phenolphthalein and OH^- ion is *not* a straight line, which shows that the reaction is not a zero-order reaction.

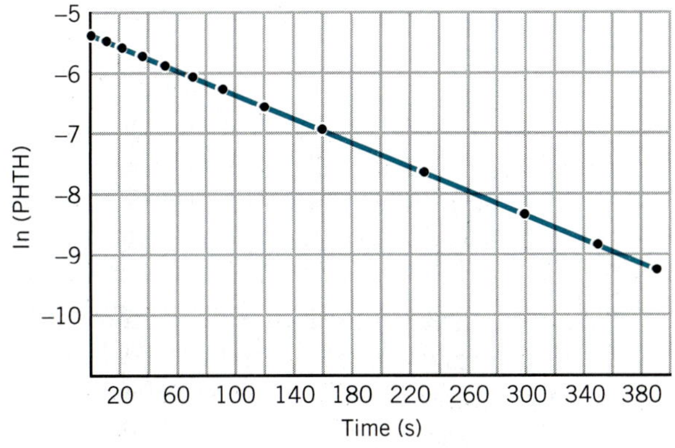

Fig. 14.12 A plot of the natural log of the concentration of phenolphthalein versus time for the reaction between phenolphthalein and OH^- ion is a straight line, which shows that the reaction is first order in phenolphthalein.

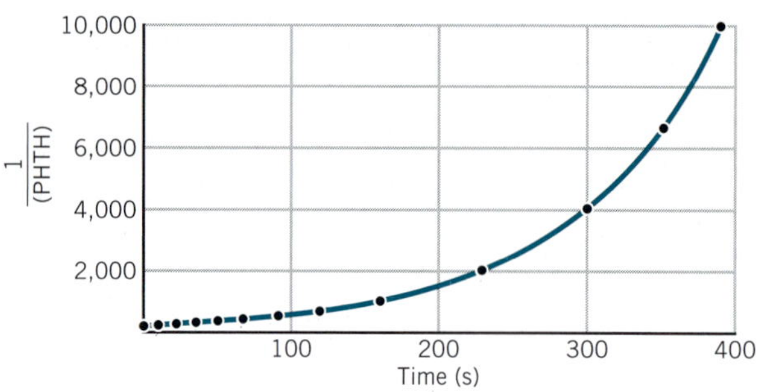

Fig. 14.13 A plot of the reciprocal of the concentration of phenolphthalein versus time for the reaction between phenolphthalein and OH^- ion *isn't* a straight line, which shows that the reaction isn't second-order in phenolphthalein.

Only one of the graphs, Figure 14.12, gives a straight line. We therefore conclude that the data fit a first-order kinetic equation, as noted in Section 14.5.

$$\text{Rate} = k(\text{phenolphthalein})$$

Exercise 14.11

Use the data for the phenophthalein–OH^- reaction in Exercise 14.10 to calculate the half-life and show that this reaction is first-order.

Solution

The data given in Exercise 14.10 can be used to find the half-life for several different initial concentrations. The time taken for the concentration to drop from 0.00400 *M* to 0.00200 *M* is $91.6 - 22.3 = 69.3$ s. We can now imagine that we are starting again with a new concentration of 0.00200 *M* and we can see how long it takes for one-half of that amount to react. The concentration drop from 0.00200 *M* to 0.00100 *M* takes $160.9 - 91.6 = 69.3$ s. Starting again from an initial concentration of 0.00100 *M*, the concentration fall from 0.00100 *M* to 0.00050 *M* requires $230.2 - 160.9 = 69.3$ s.

Thus the half-life of the reaction is independent of the initial concentration. It is only for first-order reactions that the half-life is independent of initial concentration. If the reaction had been zero-or second-order, the half-life would have been different over each concentration interval.

14.15 Reactions That Are First-Order in Two Reactants

What about reactions that are first-order in two reactants, *X* and *Y*, and therefore second-order overall?

$$\text{Rate} = k(X)(Y)$$

A plot of $1/(X)$ versus time won't give a straight line because the reaction isn't second-order in *X*. Unfortunately, neither will a plot of $\ln(X)$ versus time because the reaction isn't strictly first-order in *X*. It is simultaneously first-order in both *X* and *Y*.

One way around the problem is to turn the reaction into a **pseudo-first-order reaction** by making the concentration of one of the reactants so large that it is effectively constant. The rate law for the reaction is still first-order in both reactants, but the initial concentration of one reactant is so much larger than the other that the rate of reaction seems to be sensitive only to changes in the concentration of the reagent present in limited quantities. The reaction between phenolphthalein and excess OH^- ion described in Section 14.3 is pseudo–first-order with respect to phenolphthalein because the OH^- is present in excess.

Assume, for the moment, that the reaction is studied under conditions for which there is a large excess of *Y*. If this is true, the concentration of *Y* will remain essentially constant during the reaction. As a result, the rate of the reaction won't depend on the concentration of the excess reagent. Instead, it will appear to be first-order in the other reactant, *X*. A plot of ln(*X*) versus time will therefore give a straight line with a slope of $-k'$.

$$\text{Rate} = k'(X)$$

If we now run the reaction in the presence of a large excess of *X*, the reaction will appear to be first-order in *Y*. Under these conditions, a plot of ln(*Y*) versus time will be linear with a slope of $-k''$.

$$\text{Rate} = k''(Y)$$

The value of the rate constant obtained from either of these equations won't be the actual rate constant for the reaction. It will be the product of the rate constant for the reaction times the concentration of the reagent that is present in excess.

14.16 The Activation Energy of Chemical Reactions

Only a small fraction of the collisions between reactant molecules convert the reactants into the products of the reaction. This can be understood by turning, once again, to the reaction between $ClNO_2$ and NO.

$$ClNO_2(g) + NO(g) \longrightarrow NO_2(g) + ClNO(g)$$

In the course of the reaction, a chlorine atom is transferred from one nitrogen atom to another. For the reaction to occur, the nitrogen atom in NO must collide with the chlorine atom in $ClNO_2$.

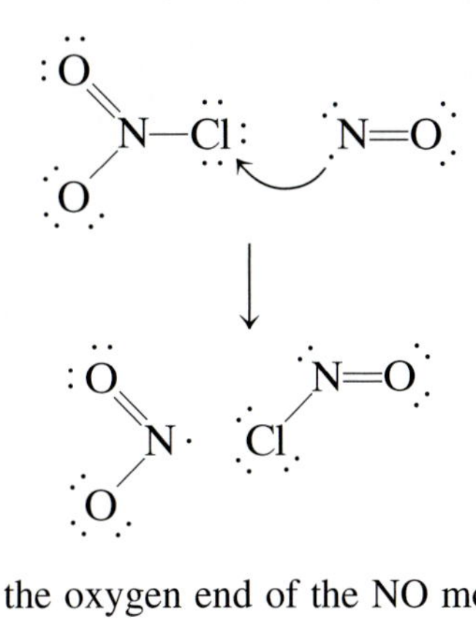

The reaction won't occur if the oxygen end of the NO molecule collides with the chlorine atom on $ClNO_2$.

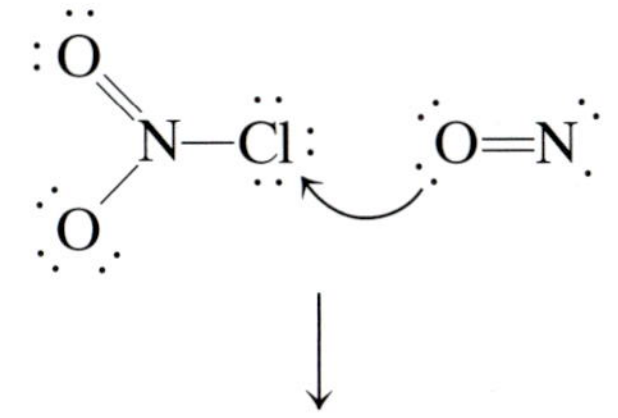

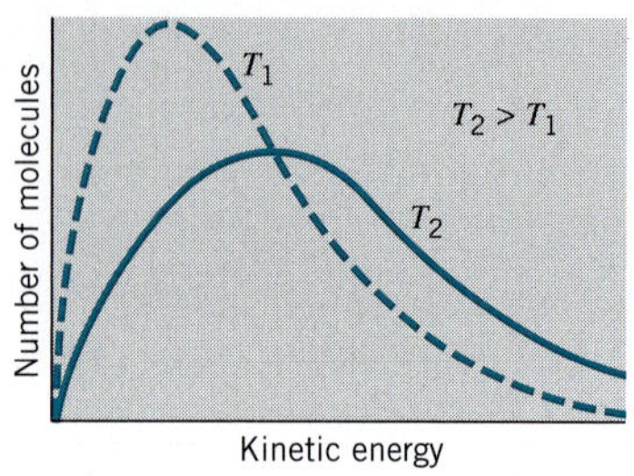

Fig. 14.14 The kinetic molecular theory states that the average kinetic energy of a gas is proportional to the temperature of the gas, and nothing else. At any given temperature, however, some of the gas particles are moving faster than others.

Nor will it occur if one of the oxygen atoms on $ClNO_2$ collides with the nitrogen atom on NO.

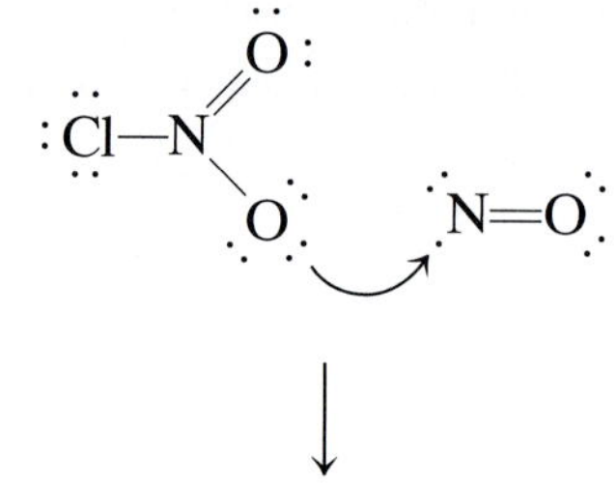

Another factor that influences whether a reaction will occur is the energy the molecules have when they collide. Not all of the molecules have the same kinetic energy, as shown in Figure 14.14. This is important because the kinetic energy molecules have when they collide is the principal source of the energy that must be invested in a reaction to get it started.

The overall standard free energy change for the reaction between $ClNO_2$ and NO is favorable.

$$ClNO_2(g) + NO(g) \longrightarrow NO_2(g) + ClNO(g) \qquad \Delta G^\circ = -23.6 \text{ kJ/mol}_{rxn}$$

But, before the reactants can be converted into products, the collision energy of the system must overcome the **activation energy (E_a)** for the forward reaction, as shown in Figure 14.15. The vertical axis in Figure 14.15 represents the energy as a chlorine atom is transferred from $ClNO_2$ to NO. The horizontal axis represents the sequence of infinitesimally small changes that must occur to convert the reactants into the products of the reaction and is called the *reaction coordinate.*

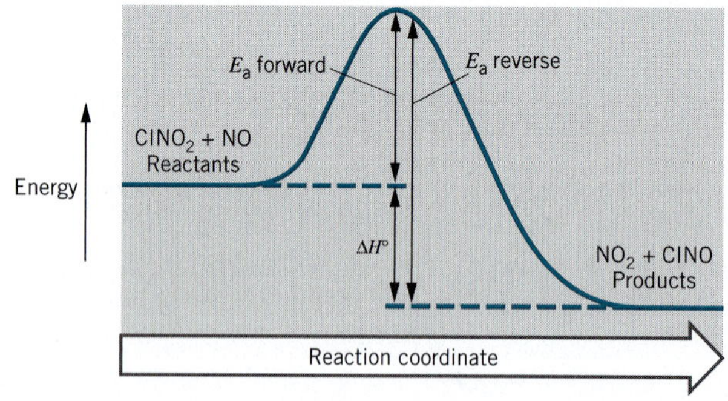

Fig. 14.15 The activation energy for a reaction, $E_{a\,forward}$ is the change in the potential energy of the reactant molecules that must be overcome before the forward reaction can occur. $E_{a\,reverse}$ is the activation energy for the conversion of products back to reactants.

The energy of activation for the reverse reaction, $E_{a\,reverse}$, is the difference between the energy of the products and the top of the energy of activation curve as shown in Figure 14.15. The difference between the energy of the reactants and the products is the enthalpy of reaction, ΔH°, which is also shown in this figure.[2]

[2]For the reaction of $ClNO_2$ and NO, enthalpy and energy are equivalent. This is not true for all reactions, however; the difference between enthalpy and energy is generally very small.

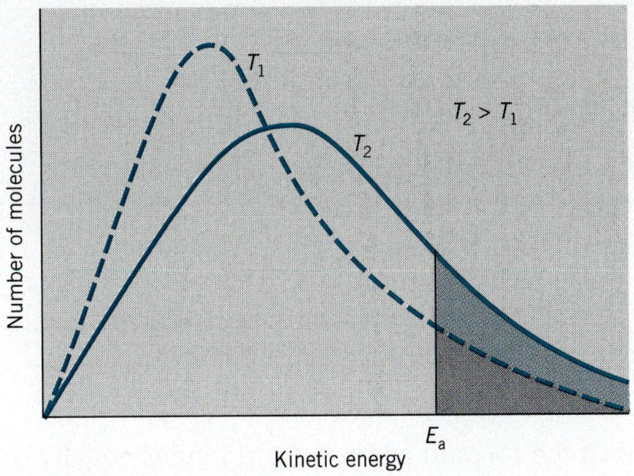

Fig. 14.16 As temperature changes, the activation energy remains essentially constant. As the temperature of the system increases, the number of molecules that have enough energy to react when they collide also increases as shown by the shaded area.

To understand why reactions have an activation energy, consider what has to happen for $ClNO_2$ to react with NO. In order for the reaction to occur, the two molecules have to collide with sufficient energy and they have to collide in exactly the right orientation relative to one another. Some energy also must be invested to begin breaking the Cl—NO_2 bond so that the Cl—NO bond can form.

NO and $ClNO_2$ molecules that collide in the correct orientation, with enough kinetic energy to overcome the activation energy barrier, can react to form NO_2 and ClNO. As the temperature of the system increases, the number of molecules that have enough energy to react ($\geq E_a$) when they collide also increases. See Figure 14.16. The rate of reaction generally increases with temperature. As a rule, the rate of a reaction doubles for every 10°C increase in the temperature of the system.

> **CHECKPOINT**
>
> For an exothermic reaction, $E_{a\ forward} < E_{a\ reverse}$, as shown in Figure 14.15. Is this also true for an endothermic reaction? Draw a reaction coordinate diagram to explain your reasoning.

14.17 Catalysts and the Rates of Chemical Reactions

Aqueous solutions of hydrogen peroxide are stable until we add a small quantity of the I^- ion, a piece of platinum metal, a few drops of blood, or a freshly cut slice of turnip, at which point the hydrogen peroxide rapidly decomposes.

$$2\ H_2O_2(aq) \longrightarrow 2\ H_2O(l) + O_2(g)$$

The reaction therefore provides the basis for understanding the effect of a catalyst on the rate of a chemical reaction. Four criteria must be satisfied in order for a substance to be classified as a **catalyst.**

- Catalysts increase the rate of reaction.
- Catalysts aren't consumed by the reaction.
- A small quantity of catalyst should be able to affect the rate of reaction for a large amount of reactant.
- Catalysts don't change the equilibrium constant, ΔH, or ΔS for the reaction. A catalyst alters the path of the reaction but does not affect the reactants or products. Because ΔH and ΔS are state functions and independent of the path of the reaction, they are not changed by a catalyst.

The first criterion provides the basis for defining a catalyst as something that increases the rate of a reaction. The second reflects the fact that anything consumed in the reaction is a reactant, not a catalyst. The third criterion is a consequence of the second; because catalysts aren't consumed in the reaction, they can catalyze the reaction over and over again. The fourth criterion results from the fact that catalysts

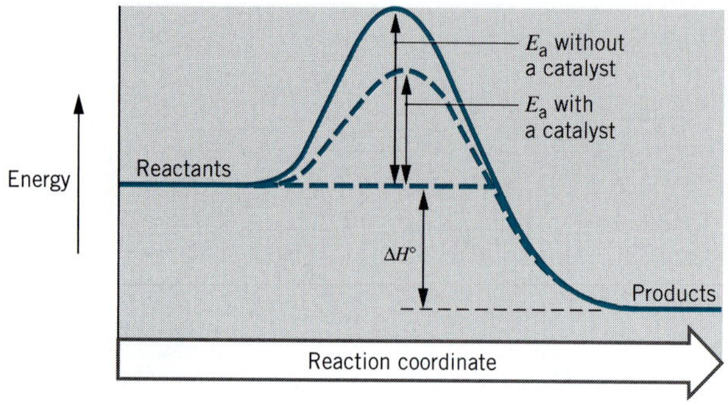

Fig. 14.17 A catalyst increases the rate of a reaction by providing an alternative mechanism that has a smaller activation energy.

speed up the rates of the forward and reverse reactions equally, so the equilibrium constant for the reaction remains the same.

Catalysts increase the rate of reaction by providing a new mechanism that has a smaller activation energy, E_a, as shown in Figure 14.17. A larger proportion of the collisions that occur between reactants now have enough energy to overcome the activation energy for the reaction. As a result, the rate of reaction increases. See Figure 14.18.

The effect of several catalysts on the activation energy for the decomposition of hydrogen peroxide and the relative rate of the reaction is summarized in Table 14.5. Adding a source of I^- ion to the solution decreases the activation energy by 25%, which increases the rate of the reaction by a factor of 2000. A piece of platinum metal decreases the activation energy even further, thereby increasing the rate of reaction by a factor of 40,000. The *catalase* enzyme in blood or turnips decreases the activation energy by almost a factor of 10, which leads to a 600-billion-fold increase in the rate of reaction.

The bombardier beetle uses the enzyme-catalyzed decomposition of hydrogen peroxide as a defense mechanism. When attacked, it mixes the contents of a sac that is about 25% H_2O_2 with a suspension of a crystalline peroxidase enzyme in a turretlike mixing tube. The reaction is so exothermic that the mixture is vaporized and the beetle can direct a hot spray at its attacker. The enzyme causes the reaction to occur so rapidly that the beetle is instantly ready to defend itself when attacked.

To illustrate how a catalyst can decrease the activation energy for a reaction by providing another pathway for the reaction, let's look at the mechanism for the decomposition of hydrogen peroxide catalyzed by the I^- ion. In the presence of the I^- ion, the decomposition of H_2O_2 doesn't have to occur in a single step. It can occur in two steps, both of which are easier and therefore faster than the uncatalyzed reaction. In the first step, the I^- ion is oxidized by H_2O_2 to form the hypoiodite ion, OI^-.

$$H_2O_2(aq) + I^-(aq) \longrightarrow H_2O(l) + OI^-(aq)$$

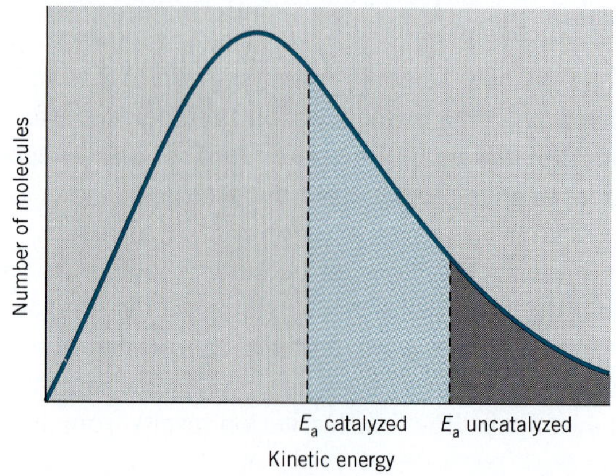

Fig. 14.18 Adding a catalyst decreases the activation energy for a reaction. When the activation energy is lowered, a larger proportion of the molecules will have the necessary energy to react.

Table 14.5

Effect of Catalysts on the Activation Energy for the Decomposition of Hydrogen Peroxide

Catalyst	E_a(kJ/mol$_{rxn}$)	Relative Rate of Reaction
None	75.3	1
I^-	56.5	2.0×10^3
Pt	49.0	4.1×10^4
Catalase	8	6.3×10^{11}

In the second step, the OI^- ion is reduced to I^- by H_2O_2.

$$OI^-(aq) + H_2O_2(aq) \longrightarrow H_2O(l) + O_2(g) + I^-(aq)$$

Because there is no net change in the concentration of the I^- ion as a result of these reactions, the I^- ion satisfies the criteria for a catalyst. Because H_2O_2 and I^- are both involved in the first step in the reaction, and because the first step in the reaction is the rate-limiting step, the overall rate of reaction is first order in both reagents. Because the OI^- ion produced in the first step is consumed in the second step, it doesn't appear in the rate law for the reaction or as one of the products of the reaction.

➤ CHECKPOINT

The reaction coordinate diagrams in Figures 14.15 and 14.17 refer to the same reaction. Locate ΔH on both diagrams. Is ΔH the same for both? Explain.

14.18 Determining the Activation Energy of a Reaction

The rate of a reaction depends on the temperature at which it is run. As the temperature increases, the molecules move faster and therefore collide more frequently. The molecules also have more kinetic energy. Thus, the proportion of collisions that can overcome the activation energy for the reaction increases with temperature.

The relationship between temperature and the rate of a reaction can be explained by assuming that the rate constant depends on the temperature at which the reaction is run. In 1889, Svante Arrhenius showed that the relationship between temperature and the rate constant for a reaction obeyed the following equation.

$$k = Ze^{-E_a/RT}$$

In this equation, k is the rate constant for the reaction, Z is a proportionality constant that varies from one reaction to another, e is the base of natural logarithms, E_a is the activation energy for the reaction, R is the ideal gas constant in joules per mole kelvin, and T is the temperature in kelvins. For a given Z and T, the rate constant, k, is determined by E_a.

The **Arrhenius equation** can be used to determine the activation energy for a reaction. We start by taking the natural logarithm of both sides of the equation.

$$\ln k = \ln Z - \frac{E_a}{RT}$$

We then rearrange the equation to fit the equation for a straight line.

$$y = mx + b$$

$$\ln k = -\frac{E_a}{R}\left(\frac{1}{T}\right) + \ln Z$$

➤ CHECKPOINT

According to the Arrhenius relationship between E_a and k, is k large or small when E_a is large? What does a small k mean concerning how fast the reaction proceeds? Use this relationship to suggest how catalysts speed up reactions.

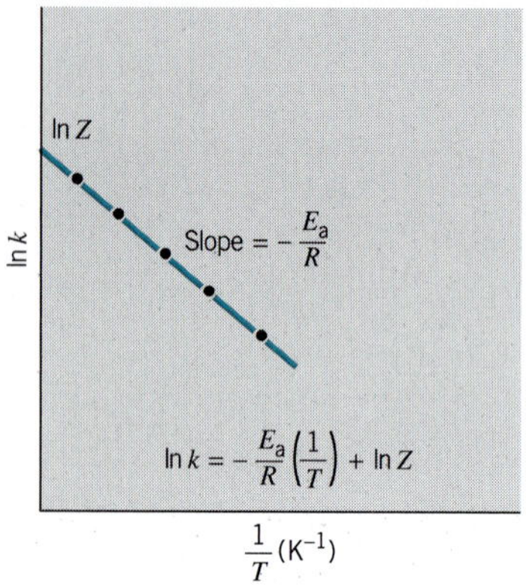

Fig. 14.19 A plot of the natural log of the rate constant for a reaction at different temperatures versus the inverse of the temperature in kelvins is a straight line with a slope equal to $-E_a/R$.

According to this equation, a plot of ln k versus $1/T$ should give a straight line with a slope of $-E_a/R$, as shown in Figure 14.19.

By manipulation of logarithms, it is possible to derive another form of the Arrhenius equation that can be used to predict the effect of a change in temperature on the rate constant for a reaction.

$$\ln\left(\frac{k_1}{k_2}\right) = \frac{E_a}{R}\left(\frac{1}{T_2} - \frac{1}{T_1}\right)$$

Exercise 14.12

Use the following data to determine the forward activation energy for the decomposition of HI.

Temperature (K)	*Rate Constant* (M/s)
573	2.91×10^{-6}
673	8.38×10^{-4}
773	7.65×10^{-2}

Solution

We can determine the activation energy for a reaction from a plot of the natural log of the rate constants versus the reciprocal of the absolute temperature. We therefore start by calculating $1/T$ and the natural logarithm of the rate constants.

ln k	*1/*T *(K^{-1})*
−12.75	0.00175
−7.08	0.00149
−2.57	0.00129

When we construct a graph of the data, we get a straight line with a slope of -2.2×10^4 K. According to the Arrhenius equation, the slope of the line is equal to $-E_a/R$.

$$-2.2 \times 10^4 \text{ K} = -\frac{E_a}{8.314 \text{ J/mol}\cdot\text{K}}$$

When the equation is solved, we get the following value for the activation energy for the reaction.

$$E_a = 1.8 \times 10^2 \text{ kJ/mol}_{rxn}$$

Exercise 14.13

Calculate the rate constant for the decomposition of HI at 600°C.

Solution

We start with the following form of the Arrhenius equation.

$$\ln\left(\frac{k_1}{k_2}\right) = \frac{E_a}{R}\left(\frac{1}{T_2} - \frac{1}{T_1}\right)$$

We then pick any one of the three data points used in the preceding exercise as T_1 and allow the value of T_2 to be 873 K.

$$T_1 = 573\text{ K} \qquad k_1 = 2.91 \times 10^{-6}\ M/\text{s}$$
$$T_2 = 873\text{ K} \qquad k_2 = ?$$

Substituting what we know about the system into the equation given above gives the following result.

$$\ln\left(\frac{2.91 \times 10^{-6}}{k_2}\right) = \frac{1.8 \times 10^5\text{ J/mol}}{8.314\text{ J/mol}\cdot\text{K}}\left(\frac{1}{873\text{ K}} - \frac{1}{573\text{ K}}\right)$$

We can simplify the right-hand side of the equation as follows.

$$\ln\left(\frac{2.91 \times 10^{-6}}{k_2}\right) = -13$$

We then take the antilog of both sides of the equation.

$$\frac{2.91 \times 10^{-6}}{k_2} = 2 \times 10^{-6}$$

Solving for k_2 gives the rate constant for the reaction at 600°C.

$$k_2 = 1\ M/\text{s}$$

Increasing the temperature of the reaction from 573 K to 873 K therefore increases the rate constant for the reaction by a factor of almost a million.

> **➤ CHECKPOINT**
>
> Chapter 8 noted that the boiling point of water depends on atmospheric pressure. As a result, water boils at a lower temperature at 10,000 feet in the Rocky Mountains than at sea level. Use the Arrhenius equation to describe why it takes longer to cook a hard-boiled egg at the top of a mountain than it does at sea level.

14.19 The Kinetics of Enzyme-Catalyzed Reactions

Enzymes are proteins that catalyze reactions in living systems. In 1913, Lenor Michaelis and his student M. L. Menten studied the rate at which an enzyme isolated from yeast catalyzed the hydrolysis of sucrose into fructose and glucose.

$$\text{Sucrose}(aq) + H_2O(l) \xrightarrow{\textit{Enzyme}} \text{fructose}(aq) + \text{glucose}(aq)$$

At first glance, this reaction seems similar to the reaction between sucrose and acid, which had been studied by Ludwig Wilhemy 60 years earlier.

$$\text{Sucrose}(aq) + H_3O^+(aq) \longrightarrow \text{fructose}(aq) + \text{glucose}(aq)$$

Wilhemy found that the reaction between sucrose and acid was first-order in sucrose at constant pH.

$$\text{Rate} = k(\text{sucrose})$$

Michaelis and Menten, however, found that the initial rate of the enzyme-catalyzed reaction was first-order in sucrose only at low concentrations of sucrose. At high concentrations, the initial rate of reaction didn't increase when more sucrose was added to the system.

As we have seen, when the rate of reaction doesn't depend on the concentration of one of the reactants, the reaction is said to be *zero-order* in that reactant.

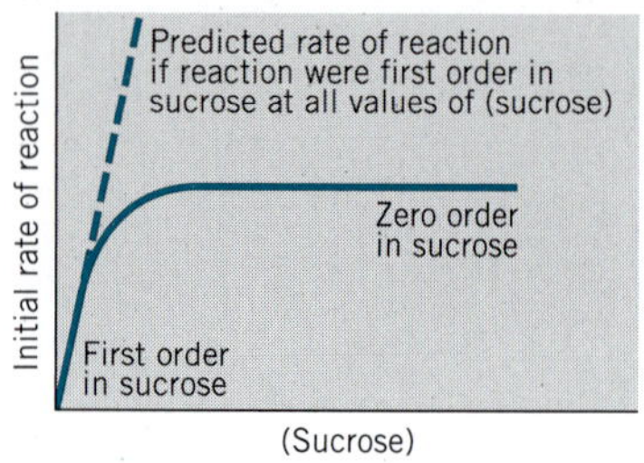

Fig. 14.20 If the reaction in which an enzyme hydrolyzes sucrose to fructose and glucose were first order in the concentration of sucrose, the initial rate of reaction should be directly proportional to the initial concentration of sucrose. Michaelis and Menten, however, showed that the reaction is first order in sucrose at low concentrations of sucrose but zero order in sucrose at high concentrations.

The enzyme-catalyzed hydrolysis of sucrose apparently changes from a first-order reaction to a zero-order reaction as the amount of sucrose in the system increases, as shown in Figure 14.20. There is a maximum initial rate of reaction, rate_{max}, for the enzyme-catalyzed reaction that can't be exceeded no matter how much sucrose we add to the solution. Once the reaction reaches rate_{max}, the only way to increase the rate of reaction is to add more enzyme to the solution.

Michaelis and Menton explained this behavior by assuming that the reaction proceeds through a two-step mechanism. In the first step, the enzyme combines with sucrose to form a complex, *ES*. This reaction is reversible. The rate constant for the forward reaction is often labeled k_1, whereas the rate constant for the reverse reaction is k_{-1}

$$E + S \underset{k_{-1}}{\overset{k_1}{\rightleftharpoons}} ES$$

The enzyme–sucrose complex may decompose to form the products of the reaction, *P*, and regenerate the enzyme.

$$ES \underset{k_{-2}}{\overset{k_2}{\rightleftharpoons}} E + P$$

The slow step in the enzyme-catalyzed reaction is the decomposition of *ES* into enzyme and product. The rate of the overall reaction is therefore determined by the concentration of the enzyme-substrate complex:

$$\text{Rate}_{\text{overall}} = k_2(ES)$$

This means that the maximum rate of reaction will be seen when essentially all of the enzyme is tied up as the *ES* complex.

There is a limit to the rate at which the enzyme can consume sucrose. When there is a large excess of sucrose in the solution, every time the enzyme transforms a molecule of sucrose into the products of the reaction, the enzyme will immediately pick up another sucrose molecule. No matter how much sucrose is added to the solution, the reaction can't occur any faster. The reaction no longer depends on the sucrose concentration and is therefore zero-order in sucrose.

$$\text{Rate} = k(\text{sucrose})^0 = k$$

Enzyme-catalyzed reactions are similar to the metal-catalyzed reactions described in Section 14.11. When there is enough reactant in the system so that essentially all of the catalytic sites are occupied at any moment in time, the reaction will exhibit zero-order kinetics. No matter how much reactant we add to the system, the reaction can't go any faster.

Key Terms

Activation energy (E_a)
Arrhenius equation
Bimolecular
Catalyst
Chemical kinetics
Collision theory
Enzyme
First-order reaction
Half-life
Initial instantaneous rate of reaction
Instantaneous rate of reaction
Integrated form of the rate laws
Intermediates
Kinetic control
Mechanism
Molecularity
Order
Pseudo-first-order reaction
Rate constant
Rate law
Rate-limiting step
Second-order reaction
Thermodynamic control
Unimolecular
Zero-order reaction

Problems

The Forces That Control a Chemical Reaction

1. What is meant by *thermodynamic control of a reaction?* What is meant by *kinetic control of a reaction?* Give an example of each.
2. To understand chemical reactions, two factors are important. What are they?
3. The reaction of H_2 and O_2 to produce water is thermodynamically favorable. What does this mean? Why don't we observe this reaction taking place at room temperature and 1 atm?

Chemical Kinetics

4. Define the study known as *chemical kinetics.*
5. How is the *rate* of a chemical reaction usually expressed?
6. What does the rate equation found for the reaction of sucrose with acid suggest happens to the rate of the reaction as the reaction proceeds?

Is the Rate of Reaction Constant?

7. Why does the expression for the rate of the phenolphthalein reaction contain a negative sign?
8. Describe what happens to the rate at which a reactant is consumed during the course of a reaction. (Does it increase, decrease, or remain the same?)

Instantaneous Rates of Reaction

9. How can the instantaneous rate of reaction be calculated?
10. Define the *instantaneous rate of reaction.*
11. Describe the difference between the rate of reaction measured over a finite period of time and the instantaneous rate of reaction. Explain the advantage of measuring the instantaneous rate of reaction.

Rate Laws and Rate Constants

12. Describe the difference between the terms *rate of reaction, rate law,* and *rate constant.*
13. What are the units of the rate constant for the following reaction?

$$N_2O_4(g) \rightleftharpoons 2\ NO_2(g)$$
$$\text{Rate} = k(N_2O_4)$$

14. What are the units of the rate constant for the following reaction?

$$2\ NO_2(g) \rightleftharpoons N_2O_4(g)$$
$$\text{Rate} = k(NO_2)^2$$

15. What are the units of the rate constant for the following reaction?

$$2\ Br^-(aq) + H_2O_2(aq) + 2\ H^+(aq) \rightleftharpoons Br_2(aq) + 2\ H_2O(aq)$$
$$\text{Rate} = k(Br^-)(H_2O_2)(H^+)$$

16. A reaction is said to be *diffusion controlled* when it occurs as fast as the reactants diffuse through the solution. The following is a good example of a diffusion-controlled reaction.

$$H_3O^+(aq) + OH^-(aq) \rightleftharpoons 2\ H_2O(aq)$$

The rate constant for the reaction is $1.4 \times 10^{11}\ M^{-1}\ s^{-1}$ at 25°C, and the reaction obeys the following rate law. Calculate the rate of reaction in a neutral solution (pH = 7.00).

$$\text{Rate} = k(H_3O^+)(OH^-)$$

A Physical Analog of Kinetic Systems

17. The rate law for a reaction may be written as rate = $k(X)$. In the water analogy, what does the larger bore correspond to in the rate law equation? What is (X)? What are the units on k?

The Rate Law Versus the Stoichiometry of a Reaction

18. Which equation describes the relationship between the rates at which Cl_2 and F_2 are consumed in the following reaction?

$$Cl_2(g) + 3\ F_2(g) \rightleftharpoons 2\ ClF_3(g)$$

(a) $-d(Cl_2)/dt = -d(F_2)/dt$
(b) $-d(Cl_2)/dt = 2[-d(F_2)/dt]$
(c) $2[-d(Cl_2)/dt] = -d(F_2)/dt$
(d) $-d(Cl_2)/dt = 3[-d(F_2)/dt]$
(e) $3[-d(Cl_2)/dt] = -d(F_2)/dt$

19. Which equation describes the relationship between the rates at which Cl_2 is consumed and ClF_3 is produced in the following reaction?

$$Cl_2(g) + 3\ F_2(g) \rightleftharpoons 2\ ClF_3(g)$$

(a) $-d(Cl_2)/dt = d(ClF_3)/dt$
(b) $-d(Cl_2)/dt = 2[d(ClF_3)/dt]$
(c) $2[-d(Cl_2)/dt] = d(ClF_3)/dt$
(d) $-d(Cl_2)/dt = 3[d(ClF_3)/dt]$
(e) $3[-d(Cl_2)/dt] = d(ClF_3)/dt$

20. Calculate the rate at which N_2O_4 is formed in the following reaction at the moment in time when NO_2 is being consumed at a rate of 0.0592 *M*/s.

$$2\ NO_2(g) \rightleftharpoons N_2O_4(g)$$

21. Calculate the rate for the formation of I_2 in the following reaction if the rate for the disappearance of HI is 0.039 $M\ s^{-1}$.

$$2\ HI(g) \rightleftharpoons H_2(g) + I_2(g)$$
$$\text{Rate} = k(\text{HI})^2$$

22. Ammonia burns in the gas phase to form nitrogen oxide and water.

$$4\ NH_3(g) + 5\ O_2(g) \longrightarrow 4\ NO(g) + 6\ H_2O(g)$$

Derive the relationship between the rates at which NH_3 and O_2 are consumed in the reaction. Derive the relationship between the rates at which O_2 is consumed and H_2O is produced.

23. Calculate the rate of formation of NO and the rate of disappearance of NH_3 for the following reaction at the moment in time when the rate of formation of water is 0.040 *M*/s.

$$4\ NH_3(g) + 5\ O_2(g) \longrightarrow 4\ NO(g) + 6\ H_2O(g)$$

24. The instantaneous rate of disappearance of the MnO_4^- ion in the following reaction is 4.56×10^{-3} *M*/s at some moment in time.

$$10\ I^-(aq) + 2\ MnO_4^-(aq) + 16\ H^+(aq) \longrightarrow 2\ Mn^{2+}(aq) + 5\ I_2(aq) + 8\ H_2O(aq)$$

What is the rate of appearance of I_2 at the same moment?

Order and Molecularity

25. Describe the difference between the molecularity and the order of a reaction.
26. Describe the conditions under which the rate law for a reaction is most likely to reflect the stoichiometry of the reaction.
27. List one or more factors that can make the rate law for a reaction differ from what the stoichiometry of the reaction leads us to expect.
28. Describe the difference between unimolecular and bimolecular reactions. Give an example of each.

A Collision Theory of Chemical Reactions

29. In a simple one-step reaction such as the following

$$A + B \longrightarrow C + D$$

what will be the rate law?

30. What is a *rate-limiting step*?
31. Define *mechanism*.
32. If a reaction does not occur in a single step, what must be known in order to write a rate law for the reaction?

The Mechanisms of Chemical Reactions

33. Each of the following reactions was found experimentally to be second order overall. Which of them is most likely to be an elementary reaction that occurs in a single step?
(a) $2\ NO_2(g) + Cl_2(g) \longrightarrow 2\ NO_2Cl(g)$
(b) $N_2O_3(g) \longrightarrow NO(g) + NO_2(g)$
(c) $3\ O_2(g) \longrightarrow 2\ O_3(g)$
(d) $2\ NO(g) \longrightarrow N_2(g) + O_2(g)$

34. NO reacts with H_2 according to the following equation.

$$2\ NO(g) + 2\ H_2(g) \rightleftharpoons N_2(g) + 2\ H_2O(g)$$

The mechanism for the reaction involves two steps.

$$2\ NO + H_2 \longrightarrow N_2 + H_2O_2 \quad \text{(slow step)}$$
$$H_2O_2 + H_2 \longrightarrow 2\ H_2O \quad \text{(fast step)}$$

What is the predicted rate law for the reaction?

35. The following reaction is first order in both NO_2 and F_2.

$$2\ NO_2(g) + F_2(g) \rightleftharpoons 2\ NO_2F(g)$$
$$\text{Rate} = k(NO_2)(F_2)$$

The rate law is consistent with which of the following mechanisms?
(a) $2\ NO_2 + F_2 \rightleftharpoons 2\ NO_2F$
(b) $NO_2 + F_2 \rightleftharpoons NO_2F + F$ (fast step)
$NO_2 + F \longrightarrow NO_2F$ (slow step)
(c) $NO_2 + F_2 \longrightarrow NO_2F + F$ (slow step)
$NO_2 + F \longrightarrow NO_2F$ (fast step)
(d) $F_2 \longrightarrow 2\ F$ (slow step)
$2\ NO_2 + 2F \longrightarrow 2\ NO_2F$ (fast step)

36. The disproportionation of NO to N_2O and NO_2 is third order in NO.

$$3\ NO(g) \rightleftharpoons N_2O(g) + NO_2(g)$$
$$\text{Rate} = k(NO)^3$$

The rate law is consistent with which of the following mechanisms?
(a) $NO + NO + NO \longrightarrow N_2O + NO_2$ (one-step reaction)
(b) $2\ NO \longrightarrow N_2O_2$ (slow step)
$N_2O_2 + NO \longrightarrow N_2O + NO_2$ (fast step)
(c) $2\ NO \rightleftharpoons N_2O_2$ (fast step)
$N_2O_2 + NO \longrightarrow N_2O + NO_2$ (slow step)

37. Is a rate law for the reaction

$$I^-(aq) + OCl^-(aq) \longrightarrow Cl^-(aq) + OI^-(aq)$$

that is first-order in I^- and second-order in OCl^- consistent with the following mechanism?

$$OCl^- + H_2O \rightleftharpoons HOCl + OH^- \quad \text{(fast step)}$$
$$I^- + HOCl \longrightarrow HOI + Cl^- \quad \text{(slow step)}$$
$$HOI + OH^- \longrightarrow OI^- + H_2O \quad \text{(fast step)}$$

38. N_2O_5 decomposes to form NO_2 and O_2.

$$2\ N_2O_5(g) \longrightarrow 4\ NO_2(g) + O_2(g)$$

The rate law for the reaction is first-order in N_2O_5.

$$\text{Rate} = k(N_2O_5)$$

Show how the rate law is consistent with the following three-step mechanism for the reaction.

Step 1
$N_2O_5 \rightleftharpoons NO_2 + NO_3$ (fast step)
Step 2
$NO_2 + NO_3 \longrightarrow NO_2 + NO + O_2$ (slow step)
Step 3
$NO + NO_3 \longrightarrow 2\ NO_2$ (fast step)

39. The following is the experimental rate law for the reaction between hydrogen and bromine to form hydrogen bromide, HBr.

$$\text{Rate} = k(H_2)(Br_2)^{1/2}$$

Explain how the following mechanism is consistent with the rate law.

$$Br_2 \rightleftharpoons 2\ Br \quad \text{(fast step)}$$
$$Br + H_2 \longrightarrow HBr + H \quad \text{(slow step)}$$
$$H + Br_2 \longrightarrow HBr + Br \quad \text{(fast step)}$$

40. Describe the relationship between the forward and reverse rate constants and the equilibrium constant for a one-step reaction.

41. The following is a one-step reaction.

$$CH_3Cl(aq) + I^-(aq) \rightleftharpoons CH_3I(aq) + Cl^-(aq)$$

What is the equilibrium constant for the reaction if the rate constant for the forward reaction is $5.2 \times 10^{-7}\ M^{-1}\ s^{-1}$ and the rate constant for the reverse reaction is $1.5 \times 10^{-11}\ M^{-1}s^{-1}$?

Zero-Order Reactions

42. What effect does decreasing the temperature have on the rate of a zero-order reaction?

43. How can collision theory account for a zero-order reaction?

44. What effect on the rate does increasing the concentrations have on a zero-order reaction that has the following rate law? Rate = $k\ (C_2\ H_4)^0\ (H_2)^0$.

Determining the Order of a Reaction from Rates of Reaction

45. Use the following data to determine the rate law for the reaction between nitrogen oxide and chlorine to form nitrosyl chloride.

$$2\ NO(g) + Cl_2(g) \longrightarrow 2\ NOCl(g)$$

Initial (NO) (M)	*Initial (Cl_2)* (M)	*Initial Rate of Reaction* (M/*s*)
0.10	0.10	0.117
0.20	0.10	0.468
0.30	0.10	1.054
0.30	0.20	2.107
0.30	0.30	3.161

46. Use the results of the preceding problem to determine the rate constant for the reaction. Predict the initial instantaneous rate of reaction when the initial NO and Cl_2 concentrations are both 0.50 *M*.

47. Use the following data to determine the rate law for the reaction between nitrogen oxide and oxygen to form nitrogen dioxide. (Hint: How are pressures related to concentrations?)

$$2\ NO(g) + O_2(g) \rightleftharpoons 2\ NO_2(g)$$

Initial Pressure NO (mmHg)	*Initial Pressure O_2 (mmHg)*	*Initial Rate of Reaction (mmHg/s)*
100	100	0.355
150	100	0.800
250	100	2.22
150	130	1.04
150	180	1.44

48. Use the results of the preceding problem to determine the rate constant for the reaction. Predict the initial rate of reaction when the initial pressures for both NO and O_2 are 250 mmHg.

49. Use the following data to determine the rate law for the reaction between methyl iodide and the OH^- ion in aqueous solution to form methanol and the iodide ion.

$$CH_3I(aq) + OH^-(aq) \rightleftharpoons CH_3OH(aq) + I^-(aq)$$

Initial (CH_3I) (M)	*Initial* (OH^-) (M)	*Initial Rate of Reaction* (M/s)
1.35	0.10	8.78×10^{-6}
1.00	0.10	6.50×10^{-6}
0.85	0.10	5.53×10^{-6}
0.85	0.15	8.29×10^{-6}
0.85	0.25	1.38×10^{-5}

50. Use the results of the preceding question to determine the rate constant for the reaction. Predict the initial instantaneous rate of reaction when the initial CH_3I concentration is 0.10 *M* and the initial OH^- concentration is 0.050 *M*.

The Integrated Form of Zero-, First-, and Second-Order Rate Laws

51. Describe the kinds of problems that are best solved by using the rate law for a reaction, such as the following.

$$\text{Rate} = k(N_2O_5)$$

52. Describe the kinds of problems that are best solved with the integrated form of the rate law.

$$\ln\left[\frac{(N_2O_5)}{(N_2O_5)_0}\right] = -kt$$

53. The decomposition of NH_3 gas on tungsten metal follows zero-order kinetics with a rate constant of 3.4×10^{-6} mol/L · s. If the initial concentration of $NH_3(g)$ is 0.0068 *M*, what will be the concentration after 1000 s?

54. The reaction in which NO_2 forms a dimer is second-order in NO_2.

$$2\ NO_2(g) \rightleftharpoons N_2O_4(g)$$
$$\text{Rate} = k(NO_2)^2$$

(a) Calculate the rate constant for the reaction if it takes 0.0050 s for the initial concentration of NO_2 to decrease from 0.50 *M* to 0.25 *M*.
(b) How long will it take for the concentration to decrease from 0.50 *M* to 0.15 *M*?

55. The decomposition of hydrogen peroxide is first-order in H_2O_2.

$$2\ H_2O_2(aq) \longrightarrow 2\ H_2O(aq) + O_2(g)$$
$$\text{Rate} = k(H_2O_2)$$

(a) How long will it take for half of the H_2O_2 in a 10-gallon sample to be consumed if the rate constant for the reaction is $5.6 \times 10^{-2}\ s^{-1}$?
(b) What will be the concentration of a 1.5 *M* H_2O_2 solution after 20 s?

56. Which graph best describes the rate of the following reaction if the reaction is first-order in N_2O_4?

$$N_2O_4(g) \rightleftharpoons 2\ NO_2(g)$$

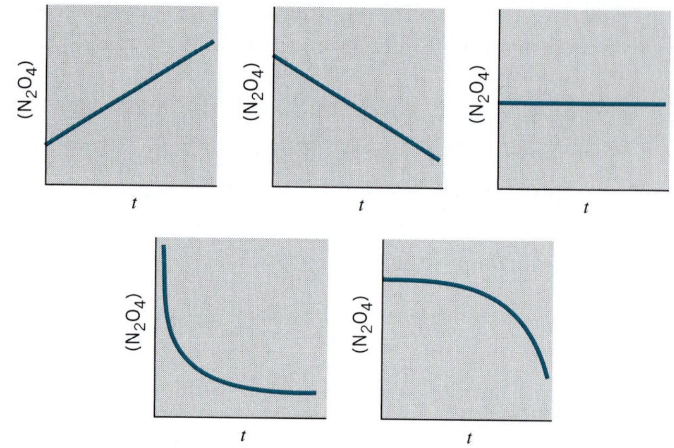

57. Calculate the rate constant for the following acid–base reaction if the half-life for the reaction is 0.0282 s at 25°C and the reaction is first-order in the NH_4^+ ion.

$$NH_4^+(aq) + H_2O(l) \rightleftharpoons NH_3(aq) + H_3O^+(aq)$$

58. The following reaction is second-order in NO_2.

$$2\ NO_2(g) \rightleftharpoons N_2O_4(g)$$

What effect would doubling the initial concentration of NO_2 have on the half-life for the reaction?

59. The reaction of HI(*g*) to produce $H_2(g)$ and $I_2(g)$ is second-order with a rate constant of 9.6×10^{-2} L/mol · min. If the initial concentration of HI is 0.55 *M*, what will be the concentration of HI after 10 minutes?

60. The age of a rock can be estimated by measuring the amount of ^{40}Ar trapped inside. The calculation is based on the fact that ^{40}K decays to ^{40}Ar by a first-order process. It also assumes that none of the ^{40}Ar produced by the reaction has escaped from the rock since the rock was formed.

$$^{40}K + e^- \longrightarrow {}^{40}Ar \qquad k = 5.81 \times 10^{-11}\ \text{year}^{-1}$$

Calculate the half-life of the radioactive decay.

61. Another way of determining the age of a rock involves measuring the extent to which the ^{87}Rb in the rock has decayed to ^{87}Sr.

$$^{87}Rb \longrightarrow {}^{87}Sr + e^- \qquad k = 1.42 \times 10^{-11}\ \text{year}^{-1}$$

What fraction of the ^{87}Rb would still remain in a rock after 1.19×10^{10} years?

62. ^{14}C measurements on the linen wrappings from the book of Isaiah in the Dead Sea Scrolls suggest that the scrolls contain about 79.5% of the ^{14}C expected in living tissue. How old are the scrolls, if the half-life for the decay of ^{14}C is 5.73×10^3 years?

63. The Lascaux cave near Montignac in France contains a series of cave paintings. Radiocarbon dating of charcoal taken from the site suggests an age of 15,520 years. What fraction of the ^{14}C present in living tissue is still present in the sample? (For ^{14}C, $t_{1/2} = 5.73 \times 10^3$ years.)

64. A skull fragment found in 1936 at Baldwin Hills, California, was dated by ^{14}C analysis. Approximately 100 g of bone was cleaned and treated with 1 *M* HCl(*aq*) to destroy the mineral content of the bone. The bone protein was then collected, dried, and pyrolyzed. The CO_2 produced was collected and purified, and the ratio of ^{14}C to ^{12}C was measured. If the sample contained roughly 5.7% of the ^{14}C present in living tissue, how old was the skeleton? (For ^{14}C, $t_{1/2} = 5.73 \times 10^3$ years.)

65. Charcoal samples from Stonehenge in England emit 62.3% of the disintegrations per gram of carbon per minute expected for living tissue. What is the age of the samples? (For ^{14}C, $t_{1/2} = 5.73 \times 10^3$ years.)

66. A lump of beeswax was excavated in England near a collection of Bronze Age objects roughly 2500 to 3000 years of age. Radiocarbon analysis of the beeswax suggests an activity equal to roughly 90.3% of the activity observed for living tissue. Determine whether the beeswax was part of the hoard of Bronze Age objects. (For ^{14}C, $t_{1/2} = 5.73 \times 10^3$ years.)

67. The activity of the ^{14}C in living tissue is 15.3 disintegrations per minute per gram of carbon. The limit for reliable determination of ^{14}C ages is 0.10 disintegration per minute per gram of carbon. Calculate the maximum age of a sample that can be dated accurately by radiocarbon dating. Assume the half-life of ^{14}C is 5.73×10^3 years.

Determining the Order of a Reaction with the Integrated Form of Rate Laws

68. Plot the following data to determine the rate law for the decomposition of N_2O.

$$2\,N_2O(g) \longrightarrow 2\,N_2(g) + O_2(g)$$

(N_2O) (*M*)	0.100	0.086	0.079	0.075	0.066	0.059	0.050	0.025
Time (s)	0	80	120	160	240	320	480	960

Use the relation between half-life and order shown in Table 14.4 to support your answer.

69. Use the results of the preceding problem to calculate the rate constant for the reaction. Predict the concentration of N_2O after 900 s.

70. Plot the following data to determine the rate law for the hydrolysis of the BH_4^- ion.

$$BH_4^-(aq) + 4\,H_2O(l) \longrightarrow B(OH)_4^-(aq) + 4\,H_2(g)$$

(BH_4^-) (*M*)	0.100	0.088	0.077	0.068	0.060	0.052	0.046
Time (h)	0	24	48	72	96	120	144

71. Use the results of the preceding question to calculate the half-life for the reaction.

72. Triphenylphosphine, PPh_3, reacts with nickel tetracarbonyl, $Ni(CO)_4$, to displace a molecule of carbon monoxide.

$$Ni(CO)_4 + PPh_3 \rightleftharpoons Ph_3PNi(CO)_3 + CO$$

The following data were obtained when the reaction was run at 25°C in the presence of a large excess of triphenylphosphine.

$(Ni(CO)_4)$ (*M*)	10.0	7.6	5.8	4.4	3.3	2.5
Time (s)	0	40	80	120	160	200

Use the data to determine the reaction order.

73. The rate of the reaction in the preceding problem does not depend on the concentration of PPh_3. Combine this fact with the results of the preceding problem to determine whether the rate law for the reaction is consistent with the following mechanism.

$$Ni(CO)_4 \longrightarrow Ni(CO)_3 + CO \quad \text{(slow step)}$$
$$Ni(CO)_3 + PPh_3 \rightleftharpoons Ph_3PNi(CO)_3 \quad \text{(fast step)}$$

74. Butadiene, C_4H_6, in the gas phase undergoes a reaction to produce C_8H_{12}. The following data were obtained for the reaction.

t(*min*)	(C_4H_6)
0	1.0
20	0.98
100	0.91
500	0.66
1000	0.49
2000	0.32
3000	0.25

Determine the order of the reaction. Calculate the rate constant for the reaction. How long does it take for the initial concentration to decrease by one-half? How long does it take for this value to decrease to half its value? What does this tell you about the order of the reaction?

75. For the gas phase reaction

$$2\,NO_3 \longrightarrow 2\,NO_2 + O_2$$

the following data were determined.

t(*s*)	(NO_3)
0	0.50
500	0.32
1000	0.24
2000	0.16
3000	0.12

What is the order of the reaction with respect to NO_3? Can you use the relations between half-life and order in Table 14.4 to learn anything about the order of this reaction?

76. For the gas phase reaction in the presence of platinum metal

$$A(g) \longrightarrow B(g) + C(g)$$

the following data were obtained.

$t(s)$	(A)
0	0.50
2	0.47
5	0.43
10	0.35
15	0.28
20	0.20
25	0.13
30	0.05

Use these data to determine the order and the rate constant for the reaction.

77. Dimethyl ether, CH_3OCH_3, decomposes at high temperatures as shown in the following equation.

$$CH_3OCH_3(g) \longrightarrow CH_4(g) + H_2(g) + CO(g)$$

The following data were obtained when the partial pressure of CH_3OCH_3 was studied as the compound decomposed at 500°C. Use the data to determine the order of the reaction. (Hint: How is pressure related to concentration?)

$P_{CH_3OCH_3}$ (mmHg)	312	278	251	227	156	78
Time (s)	0	390	777	1195	3155	6310

Use a plot of the data to determine the order and support your answer by application of the the half-life relationships given in Table 14.4.

Reactions That Are First Order in Two Reactants

78. The following reaction is first-order in both CH_3I and OH^-.

$$CH_3I(aq) + OH^-(aq) \rightleftharpoons CH_3OH(aq) + I^-(aq)$$

Describe how to turn the reaction into one that is pseudo–first-order in CH_3I.

79. $Cr(NH_3)_5Cl^{2+}$ reacts with the OH^- ion in aqueous solution to displace Cl^- from the complex ion.

$$Cr(NH_3)_5Cl^{2+}(aq) + OH^-(aq) \rightleftharpoons Cr(NH_3)_5(OH)^{2+}(aq) + Cl^-(aq)$$

The following data were obtained when the reaction was run at 25°C in a buffer solution at constant pH.

$(Cr(NH_3)_5Cl^{2+})$ (M)	1.00	0.81	0.66	0.53	0.50	0.43	0.35	0.25
Time (min)	0	3	6	9	10	12	15	20

Use the data to determine whether the reaction is first-order or second-order in $Cr(NH_3)_5Cl^{2+}$.

80. The rate of the reaction in the preceding problem is proportional to the pH of the buffer solution in which the reaction is run. Each time the buffer is changed so that the OH^- ion concentration is doubled, the rate of reaction increases by a factor of 2. Combine this observation with the results of the preceding problem to determine the rate law for the reaction.

81. Show that the rate law derived in the preceding problem is consistent with the following mechanism.

$$Cr(NH_3)_5Cl^{2+} + OH^- \longrightarrow Cr(NH_3)_4(NH_2)(Cl)^+ + H_2O \quad \text{(slow step)}$$

$$Cr(NH_3)_4(NH_2)(Cl)^+ \rightleftharpoons Cr(NH_3)_4(NH_2)^{2+} + Cl^- \quad \text{(fast step)}$$

$$Cr(NH_3)_4(NH_2)^{2+} + H_2O \rightleftharpoons Cr(NH_3)_5(OH)^{2+} \quad \text{(fast step)}$$

82. The following reaction is first-order in both reactants and therefore second-order overall.

$$CH_3I(aq) + OH^-(aq) \rightleftharpoons CH_3OH(aq) + I^-(aq)$$

When the reaction is run in a buffer solution, however, it is pseudo–first-order in CH_3I.

$$\text{Rate} = k(CH_3I)$$

What is the half-life of the reaction in a pH 10.00 buffer if the rate constant for this pseudo-first-order reaction is $6.5 \times 10^{-9}\ s^{-1}$?

83. In an experiment such as that described in Problem 78, if the concentration of CH_3I is 0.0010 M, which of the following OH^- concentrations would be best for turning the reaction into a pseudo-first-order reaction in CH_3I? Explain.
(a) 0.0010 M (b) 0.010 M
(c) 0.0001 M (d) 1.0 M

84. Why is it useful to arrange a kinetics experiment so that a reaction exhibits pseudo-order kinetics?

The Activation Energy of Chemical Reactions

85. Use Figure 14.14 to explain why the rate of a reaction generally increases with increasing temperature.

86. The larger the activation energy for a reaction, the slower the rate of the reaction will be. Explain.

87. Describe the factors that determine whether a collision between two molecules will lead to a reaction.

Catalysts and the Rates of Chemical Reactions

88. Describe the four properties of a catalyst. Give an example of a catalyzed reaction and show how the catalyst meets the criteria.
89. If the activation energy for a reaction is lowered, will the rate of the reaction speed up, slow down, or remain the same? Explain.
90. Why is ΔH for a reaction unaffected by a catalyst?
91. Draw a reaction coordinate diagram for which ΔH is endothermic. Add a dotted line showing how a catalyst changes the diagram.
92. Assume that the activation energy was measured for both the forward (E_a = 120 kJ/mol$_{rxn}$) and reverse (E_a = 185 kJ/mol$_{rxn}$) directions of a reversible reaction. What would be the activation energy for the reverse reaction in the presence of a catalyst that decreased the activation energy for the forward reaction to 90 kJ/mol$_{rxn}$?

Determining the Activation Energy of a Reaction

93. The rate constant for the decomposition of N_2O_5 increases from 1.52×10^{-5} s^{-1} at 25°C to 3.83×10^{-3} s^{-1} at 45°C. Calculate the activation energy for the reaction.
94. Calculate the activation energy for the following reaction, if the rate constant for the reaction increases from 87.1 M^{-1} s^{-1} at 500 K to 1.53×10^3 M^{-1} s^{-1} at 650 K.

$$2\ NO_2(g) \rightleftharpoons 2\ NO(g) + O_2(g)$$

95. Calculate the activation energy for the decomposition of NO_2 from the temperature dependence of the rate constant for the reaction.

$$2\ NO_2(g) \longrightarrow N_2(g) + 2\ O_2(g)$$

Temperature (K)	319	329	352	381	389
k (M^{-1} s^{-1})	0.522	0.755	1.70	4.02	5.03

96. Calculate the rate constant at 780 K for the following reaction, if the rate constant for the reaction is 3.5×10^{-7} M^{-1} s^{-1} at 550 K and the activation energy is 183 kJ/mol$_{rxn}$.

$$2\ HI(g) \rightleftharpoons H_2(g) + I_2(g)$$

97. Calculate the rate constant at 75°C for the following reaction, if the rate constant for the reaction is 6.5×10^{-5} M^{-1} s^{-1} at 25°C and the activation energy is 92.9 kJ/mol$_{rxn}$.

$$CH_3I(aq) + OH^-(aq) \rightleftharpoons CH_3OH(aq) + I^-(aq)$$

98. According to the Arrhenius relation, what happens to the rate constant if the activation energy for a particular reaction is lowered, as in the case of the introduction of a catalyst? Explain why the rate of a reaction changes when a catalyst is added.

The Kinetics of Enzyme-Catalyzed Reactions

99. Explain why the rate of the enzyme-catalyzed hydrolysis of sucrose is first order in sucrose at low concentrations of the substance.
100. Explain why the rate of enzyme-catalyzed reactions becomes zero order at very high concentrations of the substrate.

Integrated Problems

101. A 50-mL Mohr buret is connected to a 1-ft length of capillary tubing. The buret is then filled with water, and the volume of water versus time is monitored as the water gradually flows through the capillary tubing. Use the following data to determine whether the rate at which water flows through the capillary tubing fits first-order or second-order kinetics. Determine the rate constant for the process. (See buret in Problem 103.)

V (mL)	50	40	30	20	10	0
t (s)	0	19	42	72	116	203

102. Predict the effect of doubling the length of the capillary tubing on the rate constant for the process described in the previous problem. Compare your prediction with the following experimental results. (See buret in Problem 103.)

V (mL)	50	40	30	20	10	0
t (s)	0	43	97	171	282	523

103. A piece of glass tubing threaded through a one-hole rubber stopper was inserted in the mouth of the Mohr buret. A source of compressed air was then connected to the glass tubing. When a constant pressure was applied to the top of the buret, the following data were obtained.

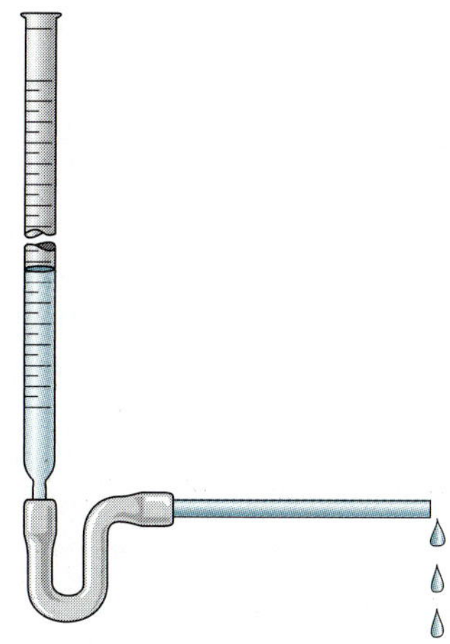

V (mL)	45	40	35	30	25	20	15	10	5	0
t (s)	0	75	150	224	300	376	446	520	595	674

Determine whether the data fit zero-order, first-order, or second-order kinetics.

104. For the reaction

$$2\,N_2O_5(g) \rightleftharpoons 4\,NO_2(g) + O_2(g)$$

the activation energy for the forward reaction is 200 kJ/mol$_{rxn}$. Using enthalpy of atom combination data from Appendix B, Table B.14, answer the following.

(a) What is the activation energy for the reverse reaction?

(b) If a catalyst is added and the activation energy for the forward reaction is reduced to 150 kJ/mol$_{rxn}$, what will be the activation energy for the reverse reaction?

105. I. From the reaction coordinate diagrams below, and assuming constant temperature and Z for all the diagrams, select the diagram for the conversion of reactants to products that has the following:

(a) the smallest rate constant for an endothermic reaction

(b) the largest rate constant for an exothermic reaction

(c) the largest rate constant for a reverse reaction

(d) the most rapid establishment of equilibrium

II. Of diagrams II and IV, which is most likely to have an equilibrium constant greater than 1? Can you decide which of the two reactions proceeds the most rapidly?

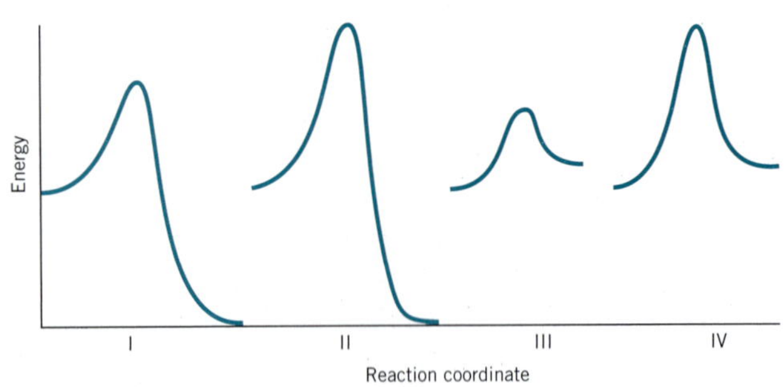

106. Enzymes act as catalysts in biochemical reactions. They are not consumed in the reaction and therefore do not appear as a reactant in the chemical equation. Their function is to provide a site where reactants can be brought together in the proper orientation to lead to a reaction. Predict the order of a biochemical reaction to which a very small amount of the appropriate enzyme has been added. Explain your reasoning.

107. Explain why the rate of each step in a reaction is proportional to the concentrations of the reagents consumed in that step.

108. Which of the following graphs best describes the relationship between the rate of a reaction and the temperature of the reaction?

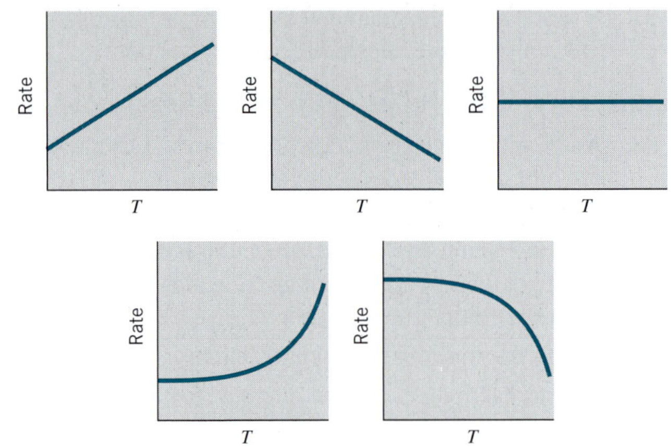

109. Explain why mixtures of H_2 and O_2 gas do not react when stored at room temperature for several years, whereas the reaction is complete within a few days at 300°C, within a few hours at 500°C, and almost instantaneously at 700°C.

110. Which of the following will be affected by the addition of a catalyst to a chemical reaction?

(a) The magnitude of the equilibrium constant.

(b) The value of k_f, the forward rate constant.

(c) The value of k_r, the reverse rate constant.

(d) The value of the forward rate.

(e) The value of the reverse rate.

(f) The value of the ratio k_f/k_r.

(g) $\Delta H°$

(h) $\Delta S°$

(i) $\Delta G°$

111. For the decomposition of hydrogen peroxide, H_2O_2

$$2\,H_2O_2(aq) \longrightarrow 2\,H_2O(l) + O_2(g)$$

the following mechanism is suggested.

$$H_2O_2 \longrightarrow 2\,OH \quad \text{(slow)}$$
$$H_2O_2 + OH \longrightarrow H_2O + HO_2 \quad \text{(fast)}$$
$$HO_2 + OH \longrightarrow H_2O + O_2 \quad \text{(fast)}$$

(a) According to the mechanism, what is the rate law for the reaction? Explain.

(b) The concentration of H_2O_2 was determined at various times as shown below:

conc (M)	*time (sec)*
1.0	0
0.78	300
0.50	835
0.37	1200
0.25	1670
0.13	2460
0.082	3000
0.041	3835

Do these data agree with the rate law suggested by the mechanism? Explain.

(c) What is the half-life of the reaction? Show how you obtain your answer.

(d) If ΔH° for the reaction is < 0, which is larger, the activation energy in the forward direction or the activation energy in the reverse direction? Explain.

(e) If a piece of a turnip is added to $H_2O_2(aq)$, the rate of the decomposition is increased by about 600 billion times. Roughly sketch the reaction coordinate diagrams for the decomposition of H_2O_2 in the presence and absence of turnips. Label the diagram clearly.

112. For the decomposition of benzoyl peroxide [$(C_6H_5COO)_2$]

$$(C_6H_5COO)_2 \longrightarrow 2\ C_6H_5COO$$

the following data are experimentally determined at 100°C.

Concentration of $(C_6H_5COO)_2$	t*(min)*
0.100	0
0.0760	7.83
0.0500	19.8
0.0380	27.6
0.0250	39.6

(a) What is the half-life of the reaction in minutes?

(b) Various plots of the concentration of the reactant versus time are given below.

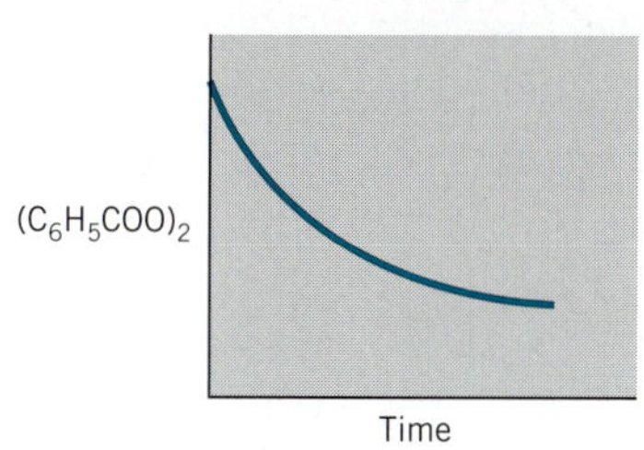

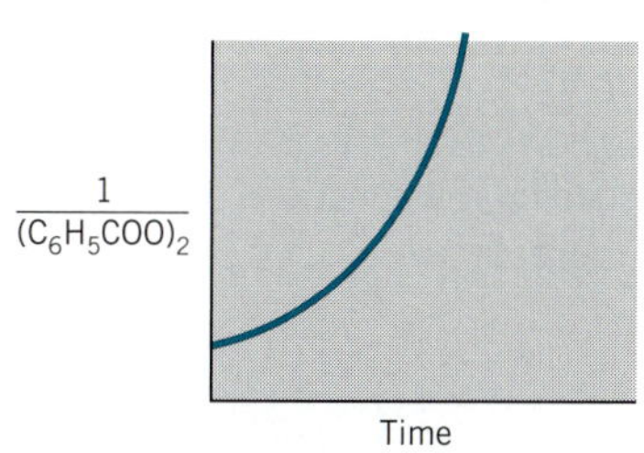

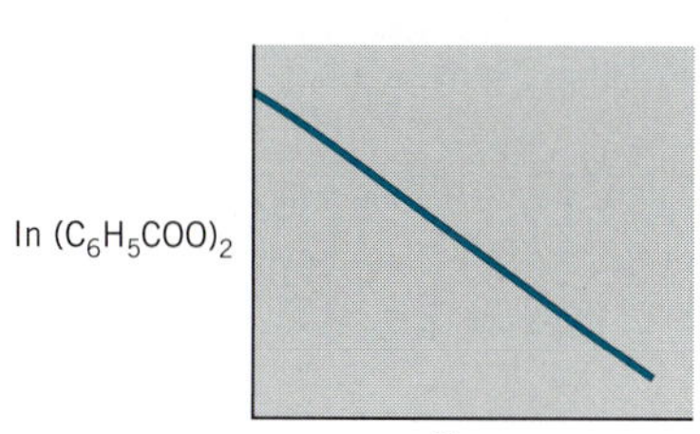

What is the order of the reaction? Explain.

(c) What is the value of the rate constant at 100°C?

(d) Would you expect the half-life of the reaction to be longer, shorter, or the same at 70°C? Explain.

(e) The activation energy in the forward direction is 100 kJ/mol and enthalpy change for the reaction is 25 kJ/mol. What is the activation energy in the reverse direction? Is the equilibrium constant for the reaction greater than 1 or less than 1? Explain.

113. The rate law for the reaction

$$2\ H_2(g) + 2\ NO(g) \longrightarrow N_2(g) + 2\ H_2O(g)$$

is rate $= k(H_2)(NO)^2$. Are any of the following mechanisms consistent with the observed rate law? Explain.

(a) $H_2 + NO \longrightarrow H_2O + N$ (slow)
$N + NO \longrightarrow N_2 + O$ (fast)
$O + H_2 \longrightarrow H_2O$ (fast)

(b) $H_2 + 2\ NO \longrightarrow N_2O + H_2O$ (slow)
$N_2O + H_2 \longrightarrow N_2 + H_2O$ (fast)

(c) $2\ NO \rightleftharpoons N_2O_2$ (fast equilibrium)
$N_2O_2 + H_2 \longrightarrow N_2O + H_2O$ (slow)
$N_2O + H_2 \longrightarrow N_2 + H_2O$ (fast)

114. For the reaction

$$2A + B \longrightarrow \text{products}$$

the experimentally determined rate law is

$$\text{Rate} = k(A)(B)$$

Which of the following mechanisms are possible? Explain fully.

(a) $2A + B \longrightarrow$ products

(b) $A + B \longrightarrow C$ (slow)
$C + A \longrightarrow$ products (fast)

(c) $A \rightleftharpoons C$ (fast)
$C + A + B \longrightarrow$ products (slow)

(d) $A \longrightarrow C$ (slow)
$C + A + B \longrightarrow$ products (fast)

115. The reaction

$$2\ NO(g) + Cl_2(g) \longrightarrow 2\ NOCl(g)$$

was studied at -10°C with the following results.

Initial Concentration (M)		*Initial Rate* (M/s)
$(NO)_0$	$(Cl_2)_0$	
0.10	0.10	0.18
0.10	0.20	0.36
0.20	0.20	1.44

(a) What is the rate law for this reaction? Show all work.
(b) What is the value (with units) of the rate constant?
(c) Is the mechanism

$$NO + Cl_2 \rightleftharpoons NOCl_2 \quad \text{(fast equilibrium)}$$
$$NOCl_2 + NO \longrightarrow 2\ NOCl \quad \text{(slow)}$$

consistent with the rate law? Explain.

116. The activation energy in the forward direction for the reaction

$$CO(g) + NO_2(g) \longrightarrow CO_2(g) + NO(g)$$

is 134 kJ/mol_{rxn}. $\Delta H°$ for the reaction is $-226\ kJ/mol_{rxn}$. The reaction occurs in a single step.
(a) What is the activation energy in the reverse direction?
(b) Give two reasons why not all collisions between molecules are effective in causing a reaction.
(c) Draw the best activated complex for the above reaction and explain why you think this activated complex is best.
(d) What effect does increasing the temperature usually have on the rate of the reaction? Explain.

117. For the reaction

$$2\,A \longrightarrow 2\,B + C$$

the following data were collected. The order of the reaction is known to be either first or second.

Time(s)	*Concentration of* A, M
0	1.000
10	0.800
20	0.667
30	0.571
40	0.500
50	0.444
120	0.250

(a) What is the order of the reaction with respect to *A*?
(b) What is the rate law for the reaction?
(c) Find the rate constant with units.
(d) What effect, if any, would a catalyst have on the rate constant? Explain.

118. $CH_4(g) + Cl_2(g) \longrightarrow$ 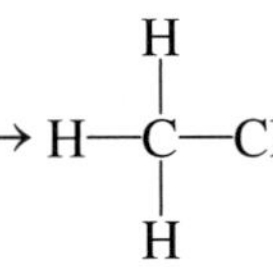 $(g) + HCl(g)$

(a) The above reaction may proceed by the following mechanism:

$$Cl_2 \rightleftharpoons 2\ Cl \quad \text{(fast equilibrium)}$$
$$CH_4 + Cl \longrightarrow CH_3 + HCl \quad \text{(slow step)}$$
$$CH_3 + Cl_2 \longrightarrow CH_3Cl + Cl \quad \text{(fast step)}$$
plus other fast steps

What is the rate law according to this mechanism?
(b) The following data were obtained for this reaction:

Expt. #	$(CH_4)_0$	$(Cl_2)_0$	*Rate* (M/s)
1	0.100	0.100	1.00×10^{-5}
2	0.200	0.100	2.00×10^{-5}
3	0.200	0.200	2.83×10^{-5}

Is the mechanism given in part (a) consistent with these data? Explain.

119. The fluorocarbon C_2F_4 reacts with itself to form a cyclic species in the gas phase. The following data were found for this process.

t(min)	(C_2F_4)
0	1.0
10	0.56
15	0.45
20	0.38
40	0.23
60	0.17

What is the order of the reaction? Calculate the rate constant from these data.

Chapter Fourteen

SPECIAL TOPICS

14A.1 Deriving the Integrated Rate Laws

To derive the integrated form of the zero-order rate law, we start with the equation that describes the rate law for the reaction.

$$-\frac{d(X)}{dt} = k(X)^0 = k$$

We then rearrange the equation, as follows.

$$d(X) = -k\,dt$$

Our goal is to integrate both sides of the equation. Mathematically, this is equivalent to finding the area under the curve that would be produced if the function were graphed. This process is indicated with integral signs, as follows.

$$\int d(X) = \int -k\,dt = -k\int dt$$

We are interested in the area under the curve between the time when the reaction starts ($t = 0$) and some later time (t).

$$\int_{(X)_0}^{(X)} d(X) = -k\int_0^t dt$$

When these integrals are evaluated, we get the following equation.

Integrated Form of the Zero-Order Rate Law

$$(X) - (X)_0 = -kt$$

When using this equation, remember that (X) is the concentration of the reactant at any moment in time, $(X)_0$ is the initial concentration of the reactant, k is the rate constant for the reaction, and t is the time since the reaction started.

The derivation of the integrated form of the first-order rate law also starts with the equation that defines the rate law of the reaction.

$$-\frac{d(X)}{dt} = k(X)$$

We then rearrange the equation as follows.

$$\frac{1}{(X)}\,d\,(X) = -k\,dt$$

We then integrate both sides of the equation.

$$\int_{(X)_0}^{(X)} \frac{1}{(X)}\,d\,(X) = -k\int_0^t dt$$

The integral of $1/(X)\ d(X)$ is equal to the natural logarithm of (X). The integrated form of the first-order rate law can therefore be written as follows.

Integrated Form of the First-Order Rate Law

$$\ln\left[\frac{(X)}{(X)_0}\right] = -kt$$

Once again, (X) is the concentration of the reactant at any moment in time, $(X)_0$ is the initial concentration of the reactant, k is the rate constant for the reaction, and t is the time since the reaction started.

The derivation of the integrated form of the second-order rate law starts with the equation that defines the rate law of the reaction.

$$-\frac{d(X)}{dt} = k(X)^2$$

We start by rearranging the equation as follows.

$$-\frac{1}{(X)^2}\, d\,(X) = k\, dt$$

We then integrate both sides of the equation.

$$-\int_{(X)_0}^{(X)} \frac{1}{(X)^2}\, d\,(X) = k \int_0^t dt$$

The integral of $1/(X)\, d(X)$ is $-1/(X)$. The integrated form of the second-order rate law is therefore written as follows.

Integrated Form of the Second-Order Rate Law

$$\frac{1}{(X)} - \frac{1}{(X)_0} = kt$$

Once again, the (X) term is the concentration of X at any moment in time, $(X)_0$ is the initial concentration of X, k is the rate constant for the reaction, and t is the time since the reaction started.

Chapter Fifteen

CHEMICAL ANALYSIS

How do chemists know what is in the food you eat or the water you drink? How do chemists determine the structure of molecules? How can chemistry be used to solve crimes? This chapter introduces the types of questions that chemists can answer, the strategies that chemists use to solve problems, and the tools that they use to do it. The focus of this chapter is on modern instrumental techniques. Case studies of real-world problems are used as examples of the wide variety of problems that chemists can solve. There are many types of instruments and methods of analysis that chemists use in the laboratory. We begin by looking at several ways in which similar instruments and methods can be categorized.

Although most students in introductory chemistry are not chemistry majors, many of them will have jobs that require a basic understanding of chemistry. Most of the interactions of nonchemists with chemists involve analysis of a sample. It is therefore important for them to understand the types of problems a chemist can help solve. This chapter is designed to provide an introduction to how samples are analyzed and some of the tools which chemists use.

15.1 Methods of Analysis

One way in which methods of chemical analysis can be categorized is by the type of information that they provide. **Qualitative methods** of analysis tell what chemical substances are present in a sample. **Quantitative methods** of analysis tell how much of the *analyte* (the specific chemical substance to be analyzed for) is present in a sample. *Structural analysis* determines the chemical structure of the analyte.

Methods of chemical analysis are also categorized as *wet* or *instrumental* methods of analysis. Wet chemical methods are based on chemical reactions. These techniques are divided into **volumetric, gravimetric,** and *qualitative* analysis methods. Volumetric methods are quantitative methods that measure the volume of a solution, such as in a titration. Gravimetric methods of analysis are quantitative methods based on a measurement of mass; they involve the precipitation of the analyte followed by drying and weighing of the precipitate. Qualitative analysis is based on specific observable reactions (color change, precipitant formation, evolution of gas) that occur when a chemical reagent is added to a sample that contains the analyte. Wet methods of analysis were the primary laboratory techniques available to chemists until the mid-twentieth century. Most traditional introductory chemistry laboratory experiments employ these techniques.

Instrumental methods of analysis involve the use of instruments for a measurement and are often based on physical rather than chemical properties of the analyte. Instrumental methods have largely replaced wet methods of analysis in the modern laboratory because they are faster, less labor intensive, and often more sensitive than wet chemistry methods. Wet chemistry methods of analysis, however, are still used to prepare samples for analysis.

Case studies are used in this chapter to present real-world problems. An effort has been made to present problems from a variety of fields (e.g., consumer products, environmental, forensics, and anthropology). An attempt has also been made to incorporate chemical concepts from all the subdisciplines of chemistry and relate them to the concepts previously covered in the core. Students should find these quite readable.

Several problems that can be used as case studies to generate discussion in class have been included in the instructor's manual. Although a technique has been suggested for each problem, there is no one correct answer and students may suggest several alternate methods of analysis. In addition, discussions of these problems can diverge into sampling, sample preparation, use of standards, detection limits, and interpretation of data.

15.2 Separation of Mixtures

Most of the substances that we encounter on a daily basis are not pure substances but are mixtures consisting of several or many chemical compounds, as noted in Chapter 1. It is often necessary for chemists to separate a mixture into its pure components before the components can be identified or measured.

The most widely used method for separating components of a mixture is **chromatography.** Chromatography includes a wide variety of separation techniques. Some of these require no instrumentation, whereas others involve sophisticated computer-controlled instruments. In common to all types of chromatography is the use of a **stationary phase** and a **mobile phase.** The stationary phase is a solid or liquid coated on a solid that remains stationary, as its name implies. The mobile phase is a gas or liquid that moves either over or through the stationary phase.

A sample mixture to be separated is carried through the stationary phase by the flowing mobile phase. Components of the mixture will interact with both the stationary phase and the mobile phase. Components that interact strongly with the mobile phase move rapidly with the flow of the mobile phase. Components which interact strongly with the stationary phase are retarded in their flow. This provides the mechanism for the separation of the mixture into its components. The principal difference between one type of chromatography and another is the makeup of the stationary and mobile phases.

15.3 The Olive Oil Caper[1]

Most cooks consider olive oil to be the finest vegetable oil in terms of its taste and aroma. Nutritionally it is considered superior to other vegetable oils because it is believed to decrease cholesterol levels in the blood, therefore reducing the risk of heart disease. As a result, olive oil is generally more expensive than other vegetable cooking oils. To increase profits, a few companies have been suspected of mixing less expensive corn, peanut, or soybean oils with their olive oil. In the late 1980s Richard Flor at the U.S. Customs Services laboratory in Washington, D.C., received a sample taken from a recently imported shipment of olive oil. The sample looked like olive oil but had an off-taste. However, taste is subjective and not admissible as evidence in court. With the assistance of his colleague Le Tiet Hecking, Flor developed a technique to determine whether the olive oil sample had been adulterated. Any difference in taste between oils must be based on differences in the chemical composition of the oil.

THE PROBLEM: HOW TO DETERMINE WHETHER THE OLIVE OIL IS PURE?

Oils are a complex mixture of triglycerides, compounds formed from glycerol and three fatty acids. Glycerol is an organic molecule consisting of a three-carbon chain with a hydroxy functional group (OH) attached to each carbon. **Fatty acids** consist of long carbon chains with a carboxylic acid functional group (COOH), as shown below. Fatty acids differ from one another based on the number of carbons and the number and location of double bonds within the carbon chain. Table 15.1 lists common fatty acids found in triglycerides.

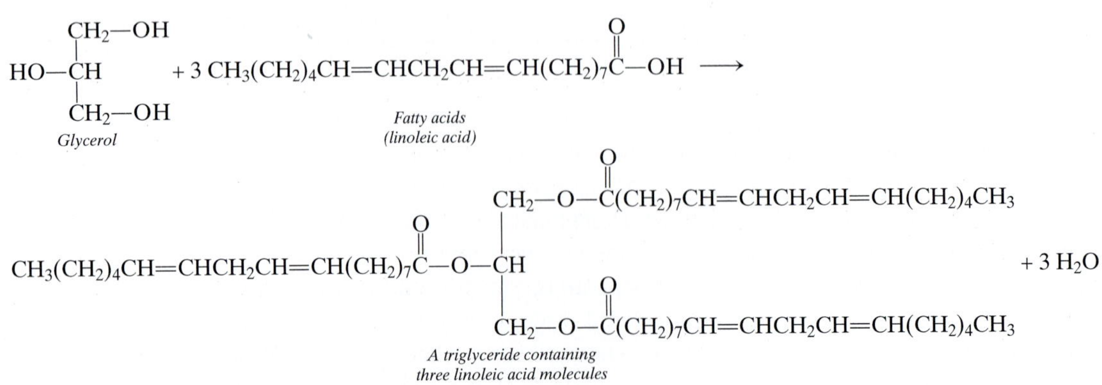

[1]Robin Meadows, Making the grade, *Chem Matters,* **7,** 10–11 (December 1989). Excerpted with permission. Copyright © 1989 American Chemical Society.

Table 15.1

Common Fatty Acids

Acid	Number of Carbons	Number of C—C Double Bonds	Formula
Myristic	14	0	$CH_3(CH_2)_{12}COOH$
Palmitic	16	0	$CH_3(CH_2)_{14}COOH$
Stearic	18	0	$CH_3(CH_2)_{16}COOH$
Oleic	18	1	$CH_3(CH_2)_7CH{=}CH(CH_2)_7COOH$
Linoleic	18	2	$CH_3(CH_2)_4CH{=}CHCH_2CH{=}CH(CH_2)_7COOH$

Triglycerides differ from one another based on the types of fatty acids in the molecule and the location of the fatty acids on the glycerol. They can be represented by a shorthand notation indicating the three fatty acids used to form them. The first triglyceride shown below was formed from two linoleic acids (L) and one oleic acid (O) and is designated LLO. The second triglyceride below was formed from the fatty acids linoleic (L), oleic (O), and palmitic (P) and is designated LOP.

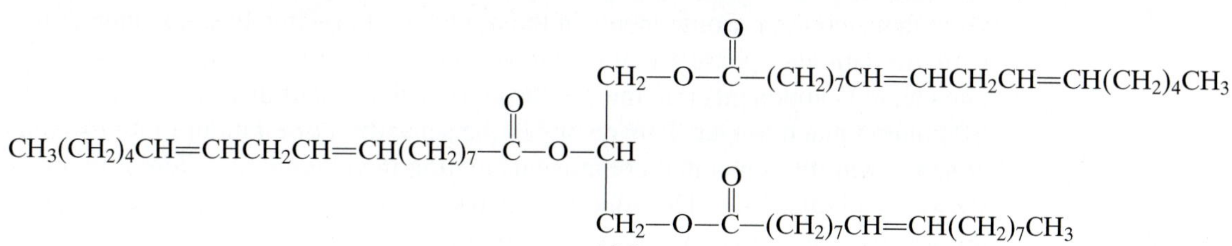

LLO

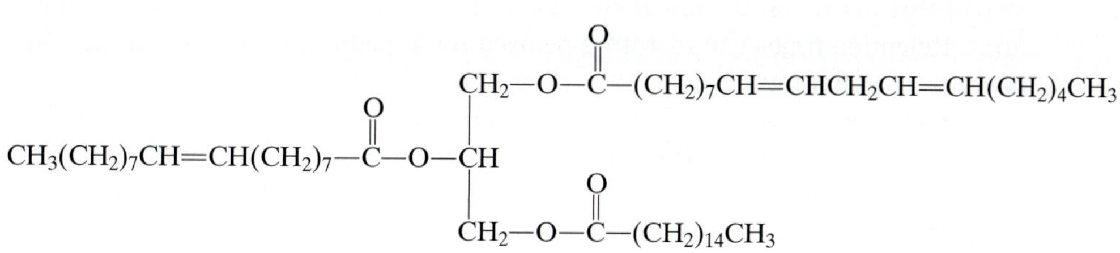

LOP

Cooking oils vary in the types and amounts of triglycerides. Table 15.2 lists characteristic ranges of the fatty acid composition of several common oils. The

Table 15.2

Percent Fatty Acid Composition of Some Common Oils[a]

Oil	Myristic Acid	Palmitic Acid	Stearic Acid	Oleic Acid	Linoleic Acid
Olive	0–1	5–15	1–4	67–84	8–12
Peanut	—	7–12	2–6	30–60	20–38
Corn	1–2	1–11	3–4	25–35	50–60
Cottonseed	0–2	6–10	2–4	20–30	50–58
Soybean	1–2	6–10	2–4	20–30	50–58

[a]Adapted from John R. Holum, *Fundamentals of General, Organic, and Biological Chemistry,* 5th ed. John Wiley & Sons, New York, 1978, p. 570.

Various instrumental techniques are introduced in this chapter. Enough detail is presented to give students some background of the basic theory of operation and applications of a technique. However, this chapter is not meant to be a mini-instrumental analysis chapter.

identification of an oil by its fatty acid composition is complicated by the range of composition based on the specific species of plant and growing conditions. However, if the triglycerides in a sample of oil can be separated from one another and identified, they should be characteristic of a particular type of oil.

THE METHOD OF ANALYSIS: HIGH-PERFORMANCE LIQUID CHROMATOGRAPHY

High-performance liquid chromatography (HPLC) can be used for the separation of a wide array of complex mixtures. The separation takes place in a **chromatography column** packed with an unreactive, finely ground solid coated with a liquid stationary phase. The liquid mobile phase is forced through the column under high pressure. As the mobile phase carries a mixture through the column, components in the mixture interact with the stationary phase. The interaction can be based on solubility, electrostatic attraction (attraction between positive and negative electric charges), adsorption (attractive forces at the interface between two immiscible phases), or other mechanisms. The interactions typically involve attractions resulting from intermolecular forces. Different components in the mixture will interact with the stationary phase to varying degrees, resulting in some components spending more time associated with the stationary phase than other components. The interaction retards the progress of the components through the column. Components of the mixture with relatively weak interactions with the stationary phase are moved through the column quite rapidly by the mobile phase. Components that interact strongly with the stationary phase remain on the column much longer. This results in the separation of a mixture into its components, with the separated components exiting the column at different times as shown in Figure 15.1. The time that it takes a particular component to move through the column is known as the **retention time** and is a characteristic property of that particular chemical substance for a given set of experimental conditions. Retention times can therefore be used for a qualitative analysis of the components of a mixture.

A schematic diagram of an HPLC setup is shown in Figure 15.2. The mobile phase is pumped from a solvent reservoir past an injector port, where the sample mixture is introduced. High pressures are required to push the mobile phase and sample mixture through the column. As the mobile phase exits the column, it passes through a detector capable of measuring the presence of the components of the mixture. The detector produces an electrical response which is observed on a readout device. The mobile phase solvent and mixture components are collected for recycling or disposal. The retention time of a component must be compared to the retention time of pure known solutions for a qualitative identification of the component. The response of the detector (area under a peak in a chromatogram) is proportional to the amount of the component present. Samples of known composition and concentration, called standards, must be run for each component for a quantitative analysis.

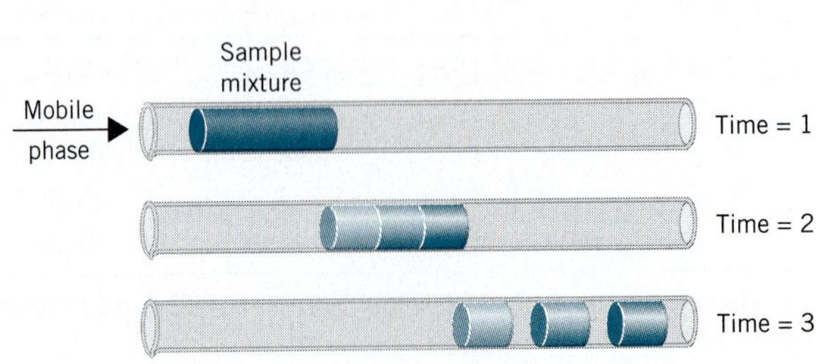

Fig. 15.1 Separation of a mixture on a chromatography column.

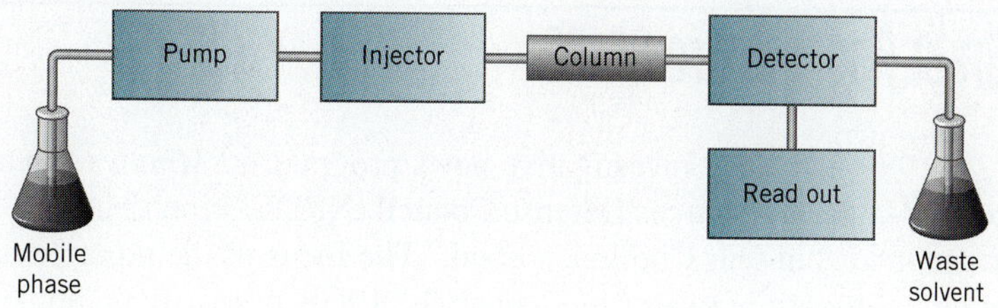

Fig. 15.2 Diagram of a high-performance liquid chromatograph.

THE SOLUTION: ANALYSIS OF OLIVE OIL

Flor and Hecking analyzed samples from a shipment of the suspect olive oil using HPLC. Chromatograms of a sample of good olive oil and olive oil from the suspect shipment are shown in Figure 15.3. The horizontal axis is a plot of time, and the vertical axis represents the response of the detector (in this case absorbance of electromagnetic radiation). Each separated peak represents a component of the mixture. Peaks corresponding to triglycerides in the mixture have been identified using known standards and are labeled according to the fatty acids used to form the triglycerides. The area under a peak is related to the concentration of that triglyceride in the mixture. As can be seen by comparing the two chromatograms, the two mixtures appear to contain the same triglycerides but in different concentrations. From the chromatograms, Flor and Hecking estimated that the suspect olive oil had been adulterated with 25% to 30% corn oil, which has higher concentrations of the triglycerides LLL, LLO, and LLP than olive oil. They verified this by preparing mixtures of corn oil and olive oil in the lab. They found that a mixture of 28% corn oil and 72% olive oil produced a chromatogram essentially identical to the chromatogram of the suspect olive oil. These were results that could be presented as evidence in court.

➤ CHECKPOINT

Using Figure 15.3, describe the differences in the relative amounts of LLL and OOO in the good olive oil and in the suspect sample.

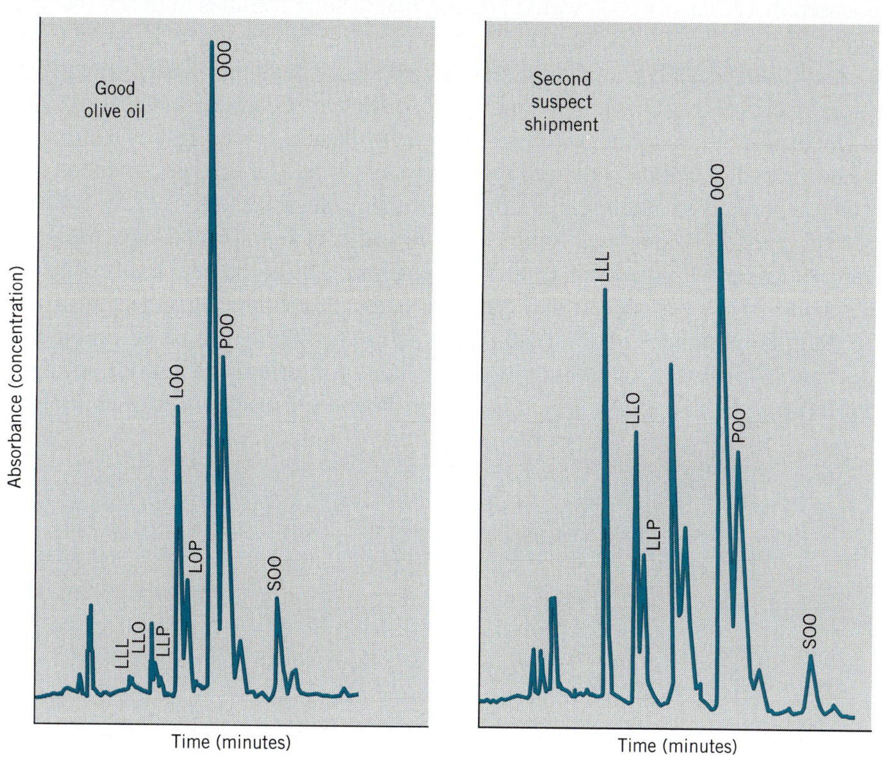

Fig. 15.3 Chromatogram of olive oil sample on the left and sample suspected of adulteration on the right. Each peak is labeled to identify the triglyceride. Reprinted with permission from Robin Meadows, *Chem Matters,* **7,** No. 4, 11, Copyright © 1989 American Chemical Society.

15.4 The Great Apple Scare Of '89

On February 26, 1989, the CBS investigative news program *60 Minutes* reported accusations from a Natural Resources Defense Council (NRDC) report entitled "Intolerable Risk: Pesticides in Our Children's Food." The focus of the report was on Alar (daminozide), a growth regulator used since the 1960s primarily on apples to keep the fruit on the tree longer, resulting in apples that were better shaped, redder, and firmer and had a longer shelf life. The NRDC, an activist group, charged that Alar posed an intolerable cancer risk, particularly to children who ate apples or apple products. The NRDC further attacked the Environmental Protection Agency (EPA) for not having banned the use of Alar. Public reaction to the NRDC report and the media coverage resulted in a 31% decrease in apple juice sales, a 25% decrease in applesauce sales, and the removal of apples and apple products from school lunch cafeterias in the spring of 1989.

Alar was first identified as a risk factor by the EPA in 1985. Apple processors began testing for Alar in their products using a method that could detect Alar at the 1–2 ppm level (1 ppm, or 1 part per million, represents 1 part analyte in a million parts of sample and is equivalent to 1 mg of analyte in a 1-kg sample). The relative size of 1 ppm can be placed in perspective by noting that 1 ppm of the distance between New York City and Los Angeles is about 15 feet. The 1–2 ppm level was well below the 30 ppm maximum level allowed by the federal government at that time. After the 1985 EPA report, the use of Alar by apple growers began a substantial decline.

THE PROBLEM: HOW MUCH ALAR IS IN APPLE PRODUCTS?

On May 14, 1989, *60 Minutes* expanded its report on Alar, describing analytical tests that had detected Alar in apple products down to the 0.02 ppm level. The next day Dr. C. Robert Binkley, vice president of technical services at Knouse Foods in Biglerville, Pennsylvania, met with his staff to view a tape of the program. Binkley announced that Knouse Foods would upgrade their testing facilities to match the level of detection described by *60 Minutes*. Work was immediately begun to design two new laboratory facilities and an analytical services laboratory building. Orders were placed for new analytical instrumentation, and Maurine Rickard, head of analytical services, was sent to observe a laboratory in California where the new procedure was being used. The new procedure not only allowed the determination of Alar at lower detection limits but also minimized interferences from other chemical substances. By July 1989 the analytical services building was complete, new gas chromatograph/mass spectrometers were installed, and Rickard was testing all apple products at the new levels of detection. Knouse Foods was able to state that their products contained no detectable Alar.

Alar, daminozide, is an organic acid and is not known to be carcinogenic. However, hydrolysis of Alar produces unsymmetrical dimethylhydrazine (UDMH) (see Figure 15.4), which has been shown to cause cancer in animals in

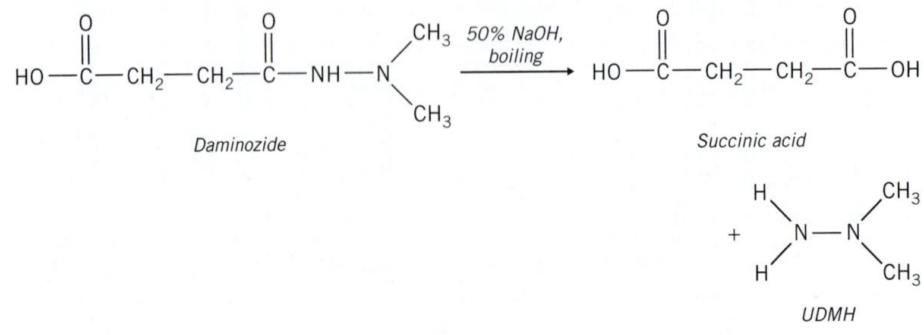

Fig. 15.4 Hydrolysis of Alar (daminozide) to succinic acid and UDMH.

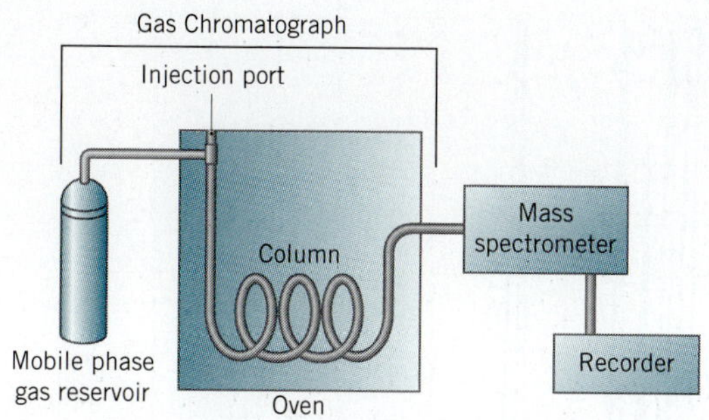

Fig. 15.5 Diagram of a gas chromatograph.

some laboratory tests. Alar is known to penetrate apple skins and therefore cannot be washed off. Heat treatment during apple processing could cause as much as 1% of any Alar present to decompose into unsymmetrical dimethylhydrazine. In addition, it has been estimated that 1% to 5% of any Alar present might be converted to the carcinogenic by-product under physiological conditions in the body.

THE METHOD OF ANALYSIS: GAS CHROMATOGRAPHY/ MASS SPECTROMETRY

Tandem techniques are the combination of two or more types of analytical methods. The capabilities of the resulting instrument are often greater than the simple sum of the abilities of the two individual techniques. One of the most powerful and most widely used tandem techniques is **gas chromatography/mass spectrometry (GC/MS).** A gas chromatograph (GC), like the HPLC previously described, is a chromatographic instrument used to separate a mixture into its components (Figure 15.5). A sample mixture is injected into the GC and separated into its component compounds in a column. As each compound exits the GC column, it moves into the mass spectrometer, which serves as a detector. Not only does the mass spectrometer produce a signal response to each component leaving the column, but it can also produce a spectrum that will allow for the qualitative identification of the component.

The major limitation of GC is that the sample to be analyzed must be volatile and thermally stable so that it can be carried along by the gaseous mobile phase. Because Alar is not thermally stable, samples suspected of containing Alar must be derivatized (converted to another compound) before analysis. Alar is hydrolyzed to UDMH by the reaction shown in Figure 15.4. The UDMH is then reacted with salicylaldehyde to produce the stable product shown in Figure 15.6.

Gas chromatography uses a gaseous mobile phase (such as He or N_2) and a liquid stationary phase. The liquid stationary phase either is coated on unreactive particles in a metal or glass coiled tube called a *packed column* or is coated directly on the inside surface of a small-bore tube called a *capillary column*. The column is placed in an oven to maintain an adequate temperature for

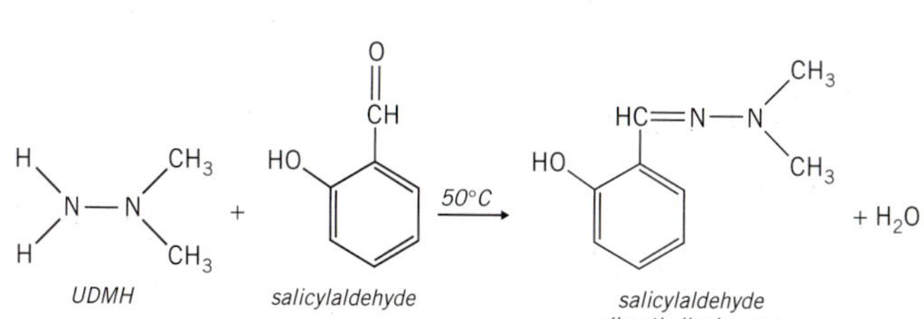

Fig. 15.6 Derivatization of UDMH.

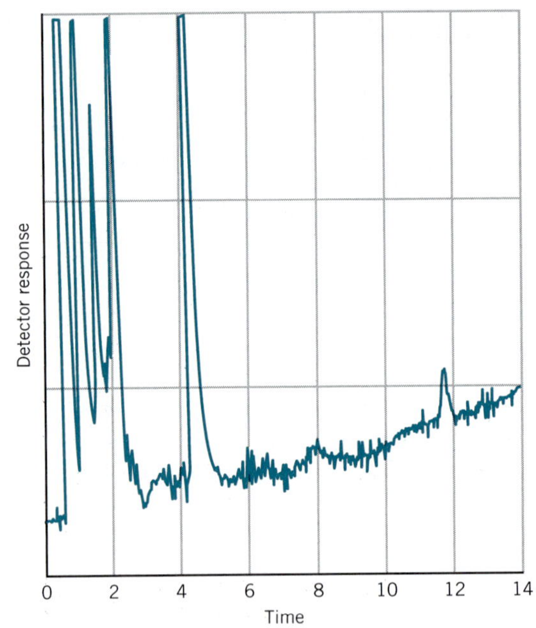

Fig. 15.7 Gas chromatogram of apple juice.

all components to remain in the vapor state. As the mobile phase exits the column, it passes through a detector capable of measuring the presence of the components of the mixture. In this case the mass spectrometer serves as the detector. The resulting mass spectrum and the retention time of the components exiting the column can be used for qualitative analysis, and the response of the detector to each component can be used for quantitative analysis just as with HPLC.

Figure 15.7 shows a chromatogram of an apple juice sample. As with most biological samples, the sample mixture contains many components, and thus the resulting chromatogram is quite complex, with many peaks. When there are many peaks with similar retention times, an exact qualitative identification is difficult. Coupling the GC with a mass spectrometer simplifies the analysis.

The use of a mass spectrometer to determine the atomic masses of elements was described in Section 2.2. When coupled with a GC, the mass spectrometer can be used to identify the chemical component responsible for each peak in the gas chromatogram. Bombardment of a component molecule by high-energy electrons produces a **parent ion** by dislodging an electron from the molecule. The parent ion is often unstable and will tend to fragment into smaller ions. The fragmentation pattern is unique and characteristic of a given compound and therefore useful for qualitative analysis. The weakest bonds in a compound will tend to rupture most easily to form the fragments. Since a compound has a constant structure, the same bonds will always rupture, yielding the same fragmentation pattern. The formation of the parent ion and selected fragment ions of the Alar derivative are shown in Figure 15.8. The resulting mixture of parent ion and fragment ions can be separated according to their *mass-to-charge* ratio. Because most of the ions have a +1 charge, the ions are in effect separated by mass.

The parent and fragment ions are accelerated through an evacuated tube. The method for separating the ions varies among different types of mass spectrometers. The mass spectrometer shown in Figure 15.9 uses a magnetic field to separate the ions. The interaction between the magnetic field and the charge on the moving ions bends the path along which the ions travel. The larger the mass of the ion, the

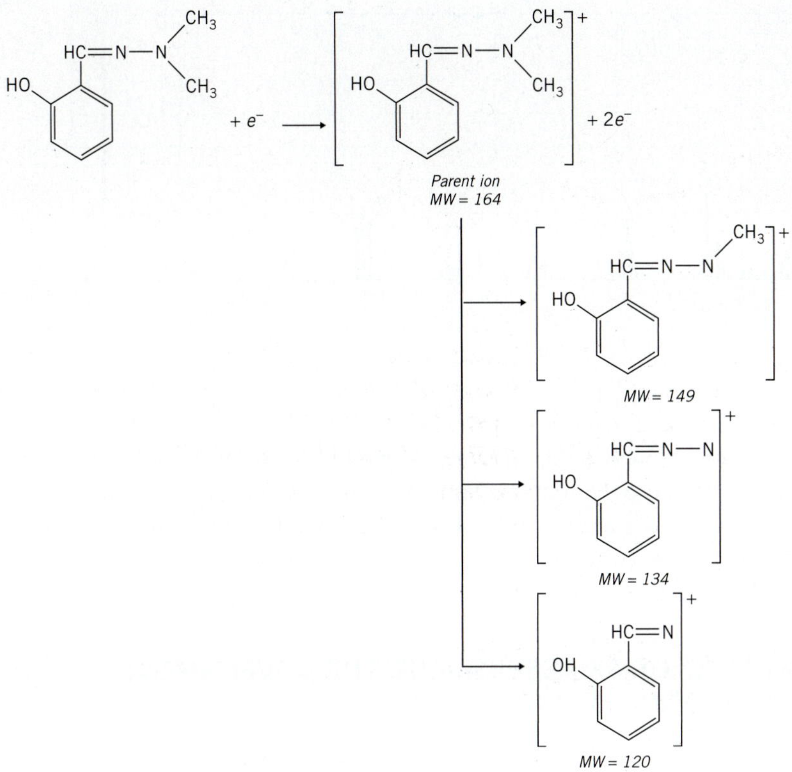

Fig. 15.8 Fragmentation of the Alar derivative.

smaller the angle through which its path is bent before it enters the detector. Because the parent ion and fragments have different masses, they will be separated from one another.

The resulting mass spectrum is a plot of relative intensity (relative number of ions of a given type striking the detector) on the vertical axis versus mass-to-charge ratio on the horizontal axis. Figure 15.10 shows the mass spectrum for the Alar derivative. Several useful pieces of information can be obtained from the mass spectrum to assist in identifying the compound. Mass spectra can be used to determine the mass of a parent ion from the peak or group of peaks with the highest assigned mass in the spectrum (to the right on the x axis). In Figure 15.10 the mass represented by this peak is approximately 164 g/mol, which corresponds to the molecular weight of the Alar derivative. The selected fragmentation ions shown in Figure 15.8 have molecular

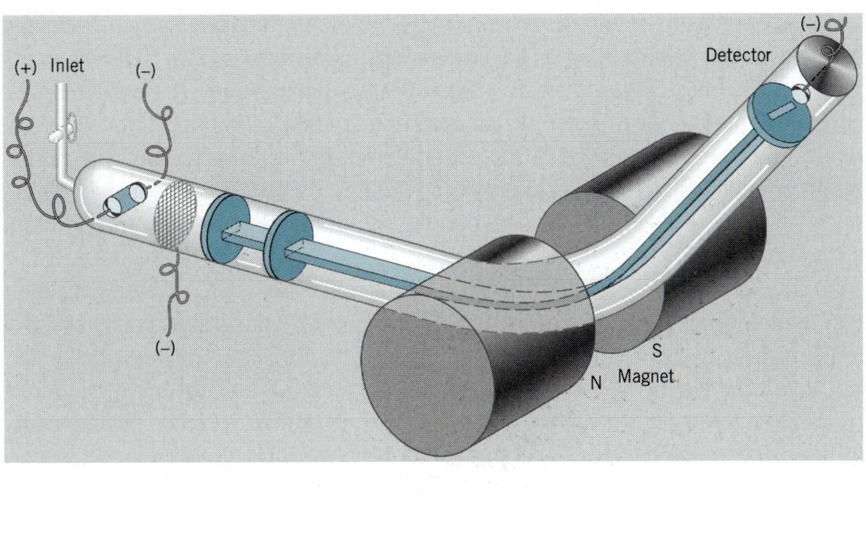

Fig. 15.9 Diagram of a mass spectrometer.

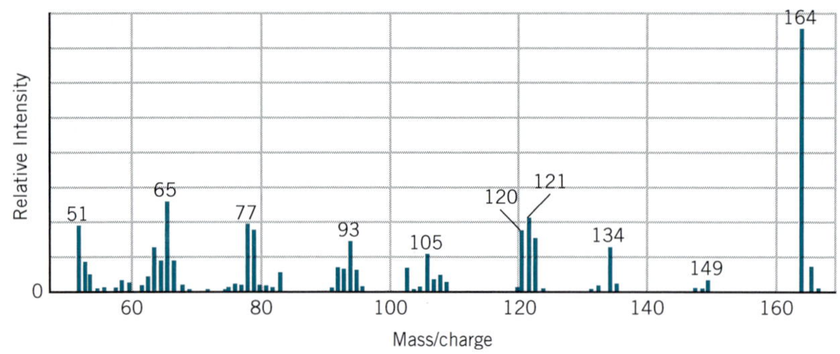

Fig. 15.10 Mass spectrum of the Alar derivative.

> **➤ CHECKPOINT**
>
> Suggest a structure for the fragment ion that would produce the peak shown at a mass value of 93 in Figure 15.10.

weights that correspond to the masses assigned to peaks in the mass spectrum of Figure 15.10. Because the fragmentation pattern does not vary for constant experimental conditions, the mass spectrum of a suspected GC component can be compared to the mass spectrum of a standard, a known compound, in order to verify the identification of the component.

THE SOLUTION: RISK ASSESSMENT, THE GOVERNMENT, THE MEDIA, AND THE PUBLIC

In response to the *60 Minutes* reports, the International Apple Institute took out full-page ads in newspapers across the country describing the health benefits of apples. The NRDC launched a media campaign with actress Meryl Streep as their spokesperson calling for tighter pesticide regulations. The Environmental Protection Agency (EPA), Food and Drug Administration (FDA), and U.S. Department of Agriculture (USDA) issued a joint statement that apples were safe and posed no imminent health hazard to children. The EPA also charged that the study on which the NRDC based its report was misleading and based on poor data.

The Great Apple Scare of '89 became less concerned with Alar in apples than with the interrelationship among science, public opinion, and politics in the regulation of chemicals in foods and the environment. In June 1989 Uniroyal Chemical Company, the producer of Alar, announced that it would voluntarily halt the sale of Alar in the United States for food crop use. In addition it would recall and issue refunds on all Alar meant for food use. A significant factor in the controversy was the public's skepticism that industry and government agencies could be trusted to regulate the level of suspected carcinogens in the environment to protect the public. This distrust was fueled by the inexact process of risk assessment.

Having completed a year of introductory chemistry, students are now prepared for a discussion of the relationship of chemistry to societal issues. Risk–benefit analysis is a starting point for such a discussion.

Risk assessment and the setting of acceptable limits of potentially hazardous chemicals in the environment is based on many assumptions. The NRDC risk estimate was 25 times greater than the EPA estimate. Each group accused the other of using inaccurate, out-of-date, or misinterpreted data. Two important questions arise: In their assessment of the risk of public exposure to Alar, why did the EPA and other agencies discount the study on which the NRDC report was based? Second, why is it so difficult to measure the risk stemming from exposure to pesticides and pesticide residues?

The key to the first question is the difference between advances that have been made in measurements of exposure to potentially hazardous substances (detection) and measurements of their biological effect (risk). Advances in technology have enabled us to detect substances at levels as low as a few picograms (10^{-12} g). Unfortunately, the bioassays used to determine whether the substances are hazardous are insensitive and subject to large experimental errors. As a result, studies of the effect of low dosages consistent with normal exposure to the substances can't be done. High doses are therefore used, and the data are then extrapolated to low-dosage situations.

As a rule, toxicologists try to limit the dose in bioassays to the maximum tolerated dose (MTD) that causes no more than a 10% weight loss and few early deaths within the test animal population. In the study used by the NRDC, the top dose of UDMH was 29 mg/kg of body weight per day. Because the dose used in the study was so high that many animals died early, the EPA concluded that it was not valid to estimate the risk of exposure to UDMH in humans. They noted that the average citizen's exposure to UDMH is only 0.000047 mg/kg and that other studies showed no significant increase in tumors at exposure levels up to 3 mg/kg.

Government regulation of additives and contaminants in food is further complicated by the Delaney clause enacted in 1958 which states that no additive found to induce cancer in humans or animals shall be deemed as safe. Prior to 1985 apple producers used less sensitive methods to assure that their products did not contain Alar above the accepted tolerance level. In 1989 more sophisticated techniques replaced those methods to allow the detection of smaller concentrations of Alar. As chemical analysis techniques become more sensitive and as limits of detection drop below the part per billion (ppb) level, the establishment of regulations for acceptable limits of suspected carcinogens will become more important.

A major theme for research in the twenty-first century will be to improve our ability to assess the risks of exposure to potentially hazardous substances. In their book *In Search of Safety: Chemicals and Cancer Risk* (Harvard University Press, 1988), John Graham, Laura Green, and Marc Roberts wrote, "as scientists discover more of the causal mechanisms underlying the health effects of chemicals . . . they will be able to make quantitative risk assessments with the reliability that the legal/political system pretends is possible today."

Exercise 15.1

Select the chromatographic method (GC or HPLC) that would be best for separating each of the following substances from a mixture.

(a) pentane, C_5H_{12}

(b) vitamin E

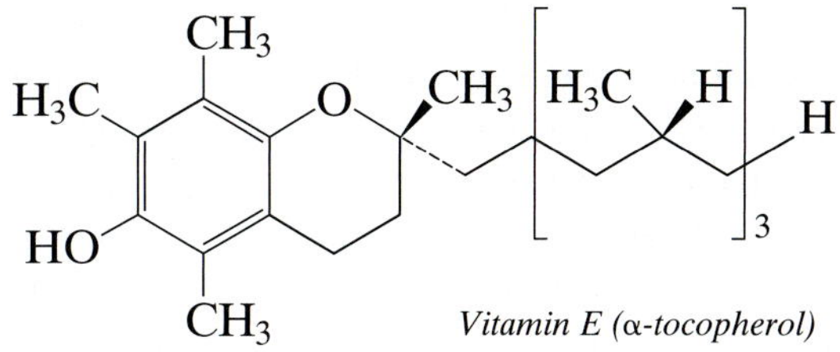

Vitamin E (α-tocopherol)

Solution

(a) Pentane is a relatively low-molecular-weight, nonpolar molecule that is easily volatilized and thus easily separated from a mixture by gas chromatography.

(b) Vitamin E has a high molecular weight and consequently would not be easily volatilized. However, it would be expected to be soluble in a nonpolar solvent and could therefore be separated from a mixture using high-performance liquid chromatography.

Exercise 15.2

A mixture of octane (C_8H_{18}), acetic acid ($CH_3\overset{\overset{O}{\|}}{C}OH$), and methyl chloride ($CH_3Cl$) is separated by high-performance liquid chromatography using a nonpolar stationary phase and a polar aqueous buffer as the mobile phase. Which component will come off the column first and which will come off last?

Solution

Since the mobile phase is a polar aqueous solution and the stationary phase is nonpolar, the primary intermolecular forces are hydrogen bonding in the mobile phase and dispersion forces in the stationary phase. Acetic acid will move through the column most rapidly with the mobile phase because it will be the most soluble in water as a result of its polarity and hydrogen bonding. Octane will be retained on the column longest since it is the only nonpolar component and will therefore interact with the stationary phase the most.

15.5 Fighting Crime with Chemistry

In 1987 results of forensic DNA tests of semen taken from a rape victim were used as part of the evidence presented in court to convict Tommi Lee Andrews of rape, sexual battery, aggravated battery, and armed burglary. This represented one of the first cases in the United States in which DNA tests were used to help obtain a conviction. In 1993 Kirk Bloodsworth was released from the Maryland House of Corrections after his 1985 conviction of rape and murder was overturned based on DNA evidence. Maryland state attorneys stated that Kirkwood would never even have been charged had DNA testing been available in 1985. DNA testing has become an important tool for forensic chemists in solving crimes.

THE PROBLEM: HOW CAN DNA BE USED TO LINK A SUSPECT TO A CRIME SCENE?

Deoxyribonucleic acid (DNA) is the most fundamental molecule of all living organisms. DNA molecules are found in the cells of all living organisms and are responsible for the storage of genetic information. This genetic information is what causes an organism's offspring to have characteristics common to its parents.

DNA is a polymer made up of repeating units known as nucleotides that consist of a phosphate, a sugar, and a nitrogen base. The sugar found in DNA is deoxyribose (Figure 15.11). There are four nitrogen bases found in DNA: adenine (A), guanine (G), cytosine (C), and thymine (T) (Figure 15.12). The phosphate and sugar form the backbone of the DNA polymer strand, with a nitrogen base branching from each sugar (Figure 15.13). It is the sequence of nucleotides along the DNA strand that determines the genetic code of the DNA. A human DNA strand contains several billion nucleotides. The segment of a DNA molecule that contains a sequence of nucleotides which codes for a specific trait is called a *gene.*

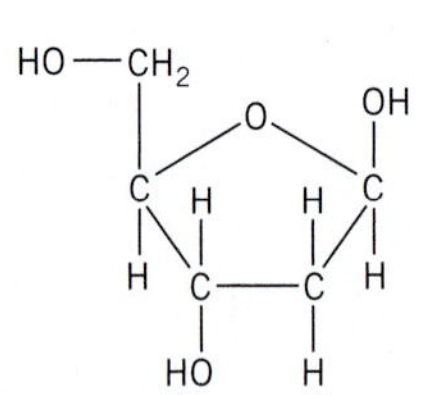

Fig. 15.11 β-D-Deoxyribofuranose, the sugar on which DNA is built.

DNA molecules consist of two polymer strands which are intertwined with one another and held together by hydrogen bonds between the nitrogen bases. This hydrogen bonding always occurs between specific bases. Thymine always links with

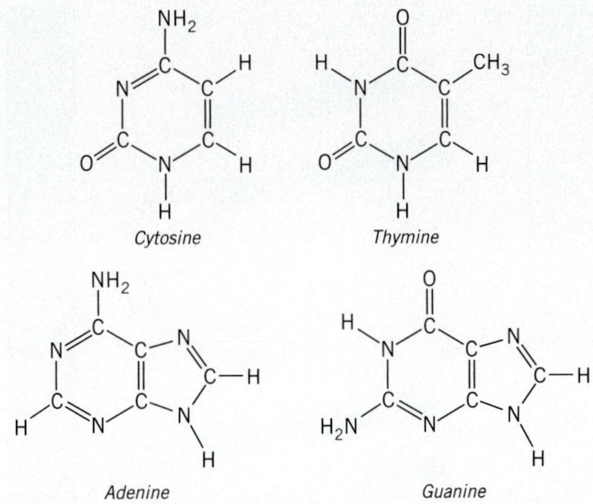

Fig. 15.12 The nitrogen bases found in DNA.

adenine, and cytosine always links to guanine (Figure 15.14). The resulting structure of the double-stranded DNA is referred to as a helix (Figure 15.15). Most of the sequences of nucleotides found within the members of a single species of plant or animal are identical. But there are always areas for each individual where the sequence differs, which is what makes each member of a species unique. With the exception of identical siblings, every human has a unique sequence of nucleotides and therefore unique DNA.

Forensic scientists compare the DNA of a suspect to the DNA found in semen, blood, hair roots, or other body cells found at the scene of a crime. An individual's DNA can be used much like a fingerprint. Since trying to match the sequence of all

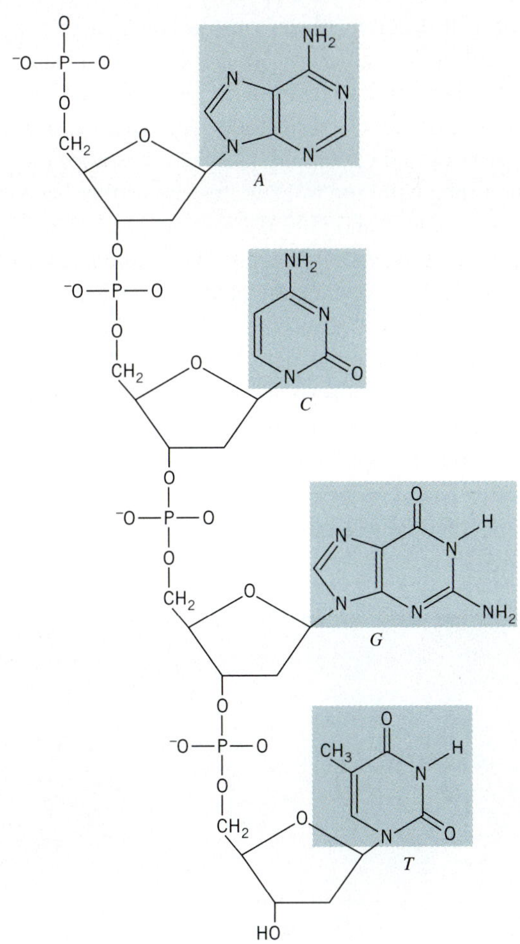

Fig. 15.13 This tetranucleotide provides an example of a short segment of a single strand of DNA.

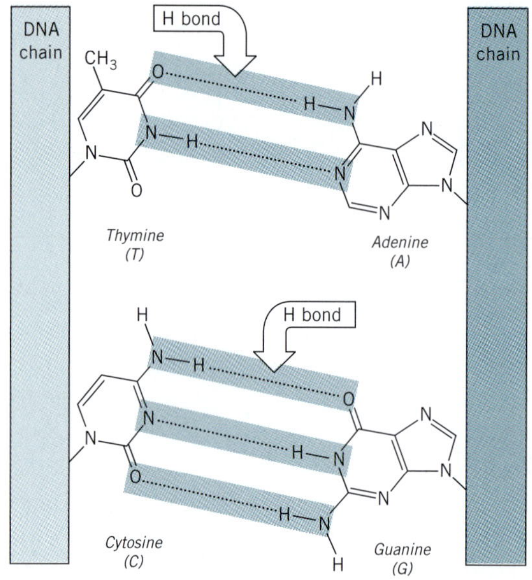

Fig. 15.14 Two strands of DNA are held together by hydrogen bonds between specific pairs of nitrogen bases. This figure shows the hydrogen bonds between A and T and between G and C.

of the billions of nucleotides in DNA samples would be an overwhelming task, forensic scientists analyze only those portions of the DNA strand that vary significantly from one individual to another.

The most powerful DNA identification method is called **restriction fragment length polymorphism (RFLP).** In this technique, restriction enzymes are used to fragment the DNA at the location of a specific sequence of nucleotides. Since the sequence would occur at different places on the DNA of different individuals, the resulting fragments differ in size from one individual to another. The fragments (Figure 15.16*a*) are then separated according to molecular weight and electric charge by **electrophoresis.** A short strand of synthetic DNA called a probe, with a known sequence of nitrogen bases and containing radioactive isotopes, is added to the separated fragments. The probe will bind to DNA fragments that have the complementary series of nitrogen bases (A to T and G to C) as shown in Figure 15.16*b*. The purpose of the probe is to provide a means of detecting the complementary sections of DNA. The radioactive isotopes in the probe undergo radioactive decay, which leaves a pattern on photographic film that corresponds to the pattern of the separated DNA fragments that are attached to probes. The steps of RFLP analysis are shown in Figure 15.17.

> **➤ CHECKPOINT**
>
> Describe the sequence of bases on a probe molecule designed to bind to a strand of DNA containing the following sequence of bases: TTCG.

The pattern of an electrophoresis gel used in a Minnesota murder investigation is shown in Figure 15.18. The pattern on the left labeled D corresponds to DNA from the suspect's blood. The patterns labeled "jeans" and "shirt" correspond to DNA from bloodstains found on the suspect's jeans and shirt. Pattern V shows results from the DNA of the victim's blood. The remaining two patterns are standards

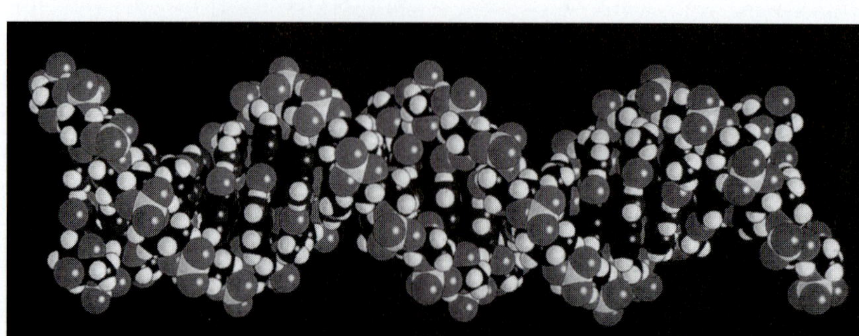

Fig. 15.15 A computer model of the helical structure of double-stranded DNA.

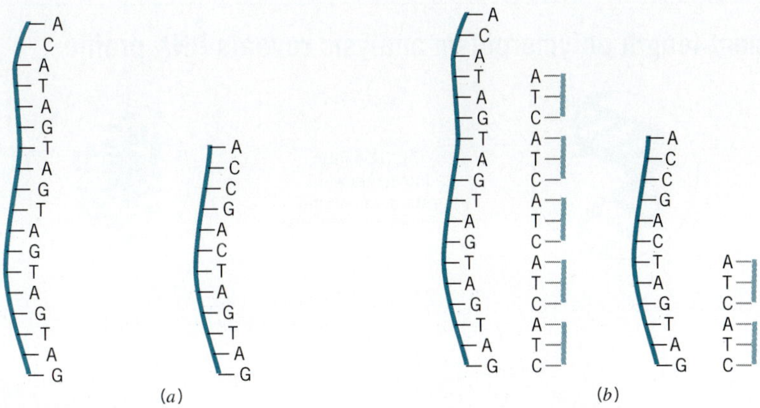

Fig. 15.16 (*a*) The two DNA fragments shown can be separated from one another by electrophoresis. If they have the same electric charge the smaller fragment on the right will move faster through an electrophoresis gel than the larger fragment on the left, allowing them to be separated from one another. (*b*) Fragments of separated DNA which contain a specific sequence of bases (such as TAG shown in the diagram) can be detected using a radioactive probe (shown in a lighter shade). The probe molecules contain a specific sequence of bases (in this example, ATC) which allows them to bind to complementary base sequences (TAG) on the DNA fragments. This illustration has been simplified by showing repeating sequences of only three bases. Actual DNA fragments have sequences of 200 to 300 bases; probe molecules contain 20 to 100 bases. Reprinted with permission from Richard Saferstein *Chem Matters,* **9,** No. 3. Copyright © 1991 American Chemical Society.

of known composition. The gel shows a good match between the victim's blood and the bloodstains on the suspect's clothes. The suspect later confessed to the murder.

THE METHOD OF ANALYSIS: ELECTROPHORESIS

Electrophoresis is a widely used technique in biochemistry and molecular biology for the separation of components in biological samples. The components to be separated must have an electric charge in aqueous solution. Amino acids, proteins, and DNA fragments are examples of components that can be separated by the technique. The sample is placed on a solid substrate such as paper, cellulose acetate, or polyacrylamide gel. An electric field is then applied across the solid substrate by means of a positive and negative electrodes located at opposite ends of the substrate. Components with positive charges are attracted by and move toward the negative electrode. In the same manner, negatively charged components move toward the positive electrode. The rate of migration of the components along or through the substrate is dependent on the magnitude of the charge and the size of the sample components. Thus, the components in the mixture can be separated. By using standards (samples of known composition) simultaneously with the mixture, the components in the mixture can be identified by comparing the distances of migration of the components and standards.

THE SOLUTION: IDENTIFICATION BEYOND A REASONABLE DOUBT

If all of the over 3 billion nucleotides in DNA were compared in a DNA test, there could be no doubt of the match between two samples and therefore the identification of the individual from which a sample came. However, only a small portion of the DNA strand is actually used in the test. Even though the sections used are those that

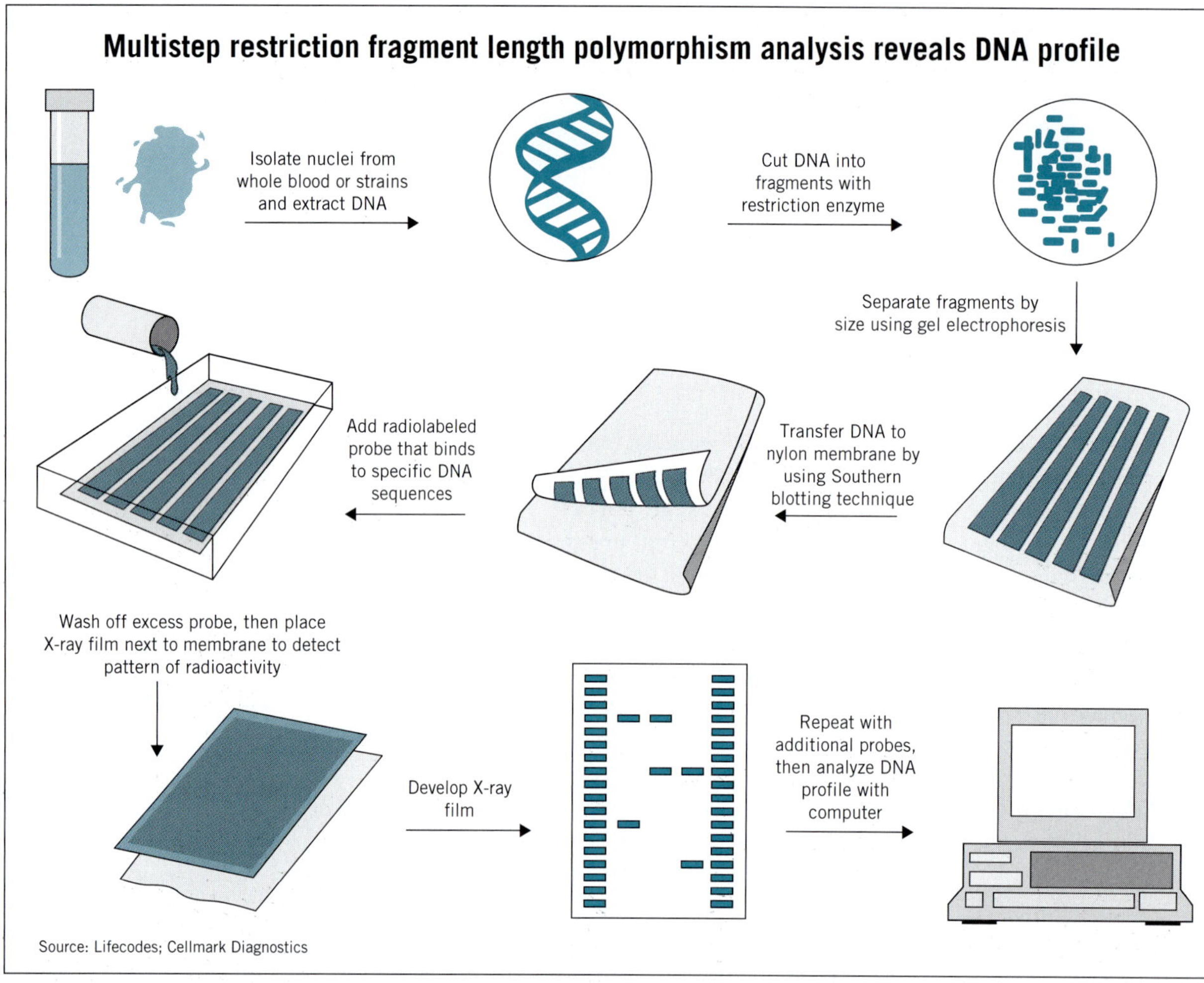

Fig. 15.17 Steps in restriction fragment length polymorphism analysis. Reprinted with permission from P. Zurer, DNA profiling fast becoming an accepted tool for identification?, *C & E News,* **72,** 11 (1994). Copyright 1994 American Chemical Society.

show the greatest variation among individuals, there is some probability of an incorrect match. The probability of a mismatch is the subject of considerable controversy, particularly in court cases. Some experts say a mismatch could occur only once in 10 million, whereas others place the probability at once in several thousand. Currently a great deal of research is being done to establish reliable databases for the calculation of good probability data and for the standardization of DNA testing techniques in forensic laboratories.

15.6 Interaction of Electromagnetic Radiation with Matter: Spectroscopy

Spectroscopy is a method of analysis based on the interaction of electromagnetic radiation with matter. Two common interactions between electromagnetic radiation and matter are *reflection* and *refraction.* When visible light strikes a mirror, it is reflected. When visible light passes through a prism, it is refracted (bent) and broken up into its component wavelengths (colors). The different interactions of electromagnetic radiation with matter allow chemists to measure many properties

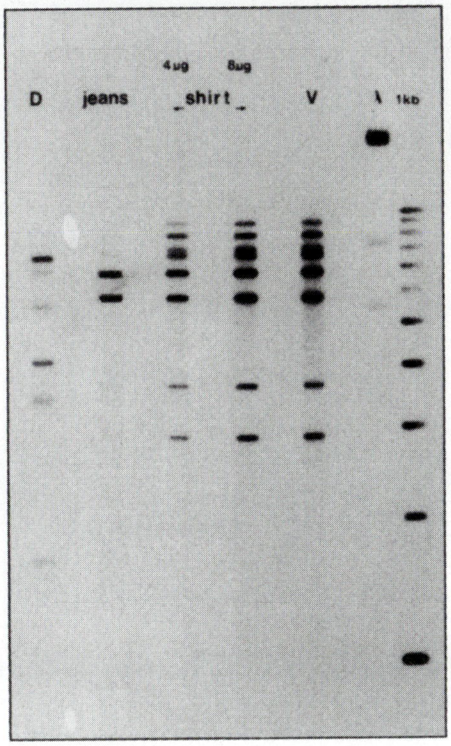

Fig. 15.18 DNA fingerprints from a Minnesota murder case. This single gel was used to compare the DNA of blood taken from the suspect's jeans and shirt with blood samples taken from the suspect, D, and the victim, V. The sample from the jeans and the two shirt samples match the sample from the victim. From R. Saferstein, DNA fingerprinting, *Chem Matters,* **9,** 12 (1991).

of matter. This chapter is primarily concerned with a third type of interaction, namely, **absorption** of electromagnetic radiation. Chapter 3 described how electromagnetic radiation is absorbed by an atom in order for an electron to move from one energy level to a higher energy level. The energy of the radiation absorbed corresponds to the change in energy of the electron. Section 3.3 noted that the energy of the photons (E) of the electromagnetic radiation that produce this transition is related to the wavelength (λ) and frequency (ν) by the following equation.

$$E = h\nu = hc/\lambda$$

Electromagnetic radiation passing through a sample can be described on a macroscopic scale by Figure 15.19, in which the intensity of electromagnetic radiation exiting the sample (I_T) is less than the initial intensity of electromagnetic radiation (I_0) striking it. Only the electromagnetic radiation with an energy corresponding to the difference in energy between two energy levels in the sample molecules is absorbed (I_A).

The relationship between these three variables is given by

$$I_0 = I_T + I_A$$

The **transmittance,** T, and **percent transmittance,** $\%T$, are defined as follows:

$$T = I_T/I_0 \qquad \%T = (I_T/I_0) \times 100$$

The absorbed radiation is known as **absorbance,** A, and is defined as

$$A = \log(I_0/I_T)$$

Since (I_0/I_T) is equal to $1/T$, absorbance and transmittance are related by

$$A = \log(1/T) \qquad \text{and} \qquad A = \log(100/\%T)$$

➤ CHECKPOINT

Which of the following electron configurations could correspond to an atom of zinc that has absorbed electromagnetic radiation?

Zn [Ar] $4s^2\ 3d^{10}$
Zn [Ar] $4s^2\ 3d^9\ 4p^1$
Zn [Ar] $4s^2\ 3d^9\ 5s^1$

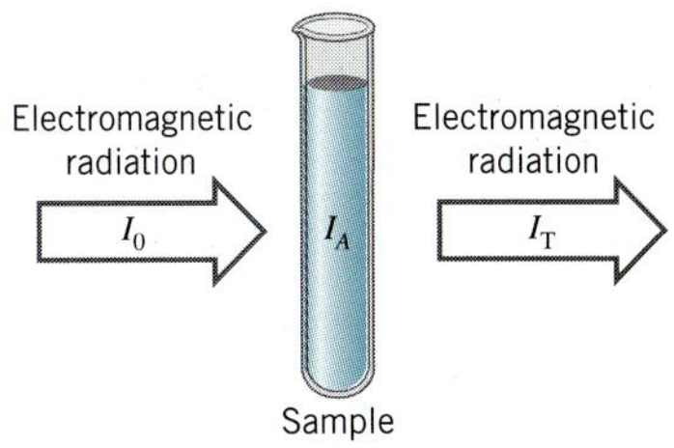

Fig. 15.19 Absorbance of light by a sample.

The second equation is often algebraically rearranged to give a convenient method for converting between absorbance and percent transmittance.

$$A = \log 100 - \log \%T$$
$$A = 2 - \log \%T$$

Figure 15.20 shows a simplified diagram of an absorption spectrometer. The source emits electromagnetic radiation, which is dispersed into its component **wavelengths** by a dispersing element. A selected wavelength of electromagnetic radiation is aimed at the sample. The amount of radiation passing through the sample is measured by the detector. An electrical response from the detector that is proportional to the amount of light striking it is recorded on an output device.

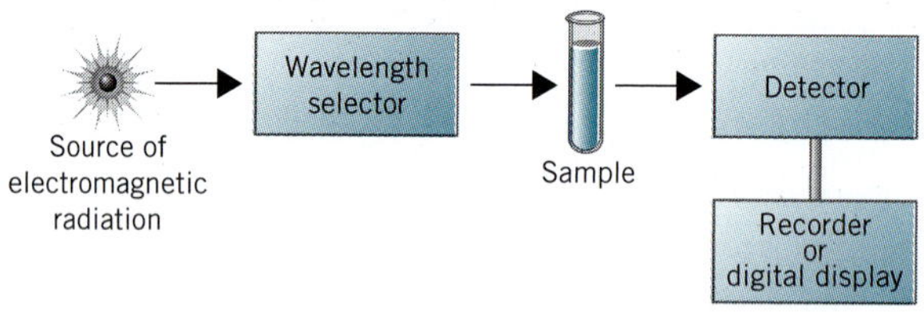

Fig. 15.20 Diagram of a spectrometer.

Spectrophotometric methods of analysis are typically classified according to the region of the electromagnetic spectrum employed (see Figure 3.3). Like other methods of analysis, spectrophotometric methods can be utilized for quantitative, qualitative, and structural analyses. These techniques may also be classified as *elemental* or *molecular* analyses. In elemental analysis the sample is analyzed without regard for the molecule or compound in which the element is found. This is contrasted by molecular analysis, which involves the analysis of a sample for the molecules present. It is important when choosing a spectroscopic technique to select the technique that gives the type of information that is needed.

15.7 The Fox River Mystery[2]

Fish kills, in which large numbers of fish die suddenly, occur occasionally in many bodies of water. They are often due to natural phenomena such as the stress of spawning or a lack of dissolved oxygen. However, in 1986 the number and size of fish kills on a section of the Fox River in Oshkosh, Wisconsin, became so serious that the Wisconsin Department of Natural Resources (DNR) began to monitor the dissolved oxygen in the water. In the spring of 1988 the largest kill occurred, over 30,000 fish. Measurements of the level of dissolved oxygen were found to be normal. Joe Ball, a water resource specialist with the DNR, began to search for the cause of the kills.

THE PROBLEM: WHAT WAS CAUSING THE FISH KILLS?

Typical causes of large fish kills are pesticides, herbicides, and toxic materials discharged by industries. Ball began a systematic investigation, looking at each facility located along the section of the river where the kills occurred. The water near a golf course was checked for herbicides and pesticides, discharge water from a sewage treatment plant was checked for chlorine levels, and water near local industries was checked for toxic organic compounds and heavy metals. But each analysis gave a

[2]Harvey Black, Fox River fish kill, *Chem Matters,* **8,** 6–9 (October 1990). Excerpted with permission. Copyright © 1990 American Chemical Society.

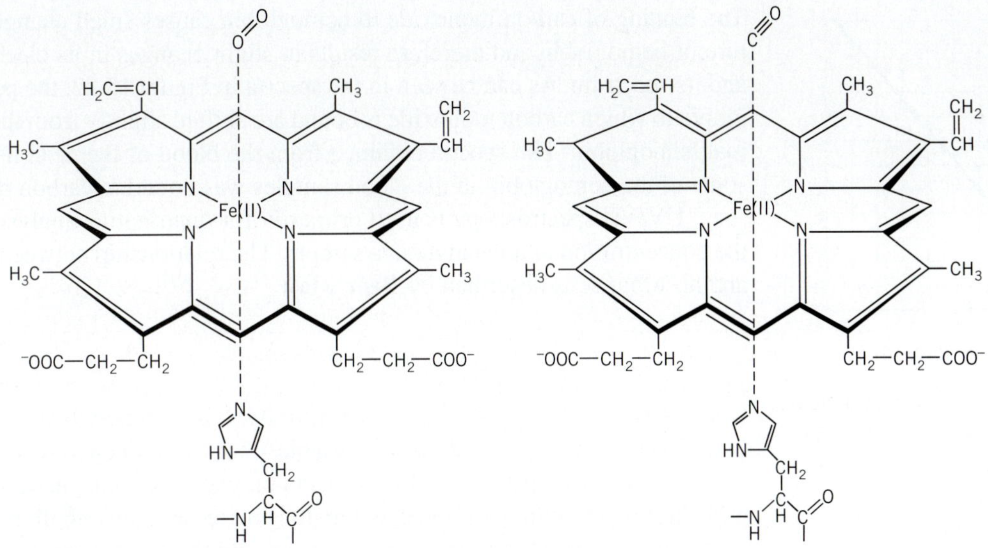

Fig. 15.21 Heme group of a hemoglobin molecule. The iron atom in the heme group on the left is bound to an oxygen molecule. The iron atom in the right-hand heme is bound to carbon monoxide.

negative result. Ball could not find the source of the pollution causing the fish kills. A testing facility for an outboard motor company was investigated next. The facility tested several powerful prototype motors on boats secured in position. No fuel leaks were found. The scientific literature was examined to determine the typical compounds found in the exhaust from motors. These included unburned volatile organics, carbon dioxide, water, and carbon monoxide. The level of volatile organics in the water at the testing facility was found to be within acceptable limits. The carbon dioxide and water from the exhaust do not harm fish. Carbon monoxide, although a lethal gas, is not very soluble in water and disperses rapidly into the air. However, Ball decided to investigate the carbon monoxide more thoroughly.

Carbon monoxide can bind to the hemoglobin in blood. Hemoglobin carries oxygen from the lungs to tissue in the body and then carries carbon dioxide back to the lungs to be exhaled. Carbon monoxide binds so tightly to hemoglobin that the hemoglobin is no longer capable of binding to oxygen and therefore results in suffocation. Figure 15.21 shows the heme group of a hemoglobin molecule. The heme group on the left is bound to oxygen, and the heme group on the right is bound to carbon monoxide.

Since carbon monoxide is difficult to detect in water, Ball decided to examine the blood of some of the dead fish and compare that to the blood of healthy fish. Tom Ecker, a chemist at the Wisconsin State Laboratory of Hygiene, was asked to do the analysis. He analyzed the blood of the fish using visible spectroscopy.

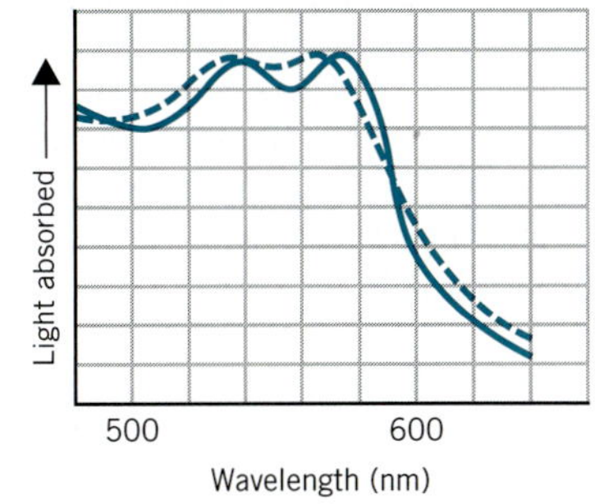

Fig. 15.22 UV/visible spectra of hemoglobin in two blood samples. The solid line is the spectrum for hemoglobin bound to oxygen. The dashed line is the spectrum for hemoglobin bound to carbon monoxide. Reprinted with permission from Harvey Black, *Chem Matters,* **8,** No. 3, 9, Copyright © 1990 American Chemical Society.

THE METHOD OF ANALYSIS: ULTRAVIOLET/VISIBLE SPECTROSCOPY

Ultraviolet/visible (UV/Vis) spectroscopy is an absorption method used for the analysis of molecules and ions. An absorption spectrum is obtained by measuring the absorbance of a sample at different wavelengths. Figure 15.22 shows absorbance spectra for blood samples from both dead and healthy fish. At wavelengths between 500 and 580 nm, a large portion of the incident radiation is absorbed, and at wavelengths longer than 620 nm very little is absorbed. The peaks in the spectrum represent wavelengths of electromagnetic radiation that correspond in energy to possible electron transitions in the sample molecule and are therefore absorbed. Other wavelengths are transmitted by the sample. A normal hemoglobin molecule has a unique set of possible electron transitions, resulting in absorbance of specific wavelengths.

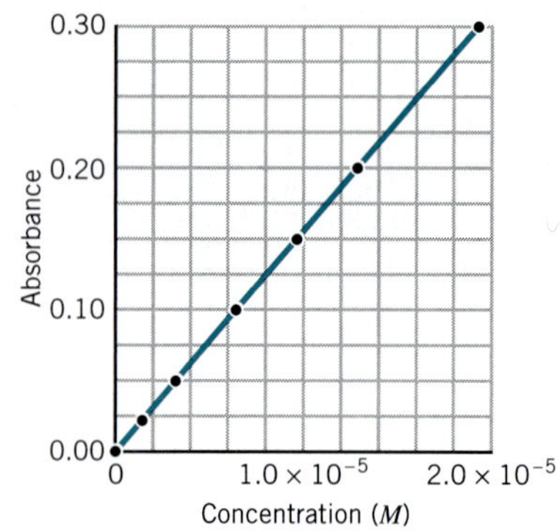

Fig. 15.23 Beer's law standard curve of absorbance versus concentration of a series of hemoglobin standard solutions measured at 576 nm.

The binding of carbon monoxide to hemoglobin causes small changes in the structure of hemoglobin and therefore results in slight changes in its electron transitions and its spectrum. As can be seen in the spectra in Figure 15.22, the peaks for hemoglobin to which carbon monoxide is bound are shifted slightly from the peaks of normal hemoglobin. The spectra obtained from the blood of the dead fish showed that some of the hemoglobin in the blood samples was bound to carbon monoxide.

UV/Vis spectroscopy is used primarily for quantitative analysis to determine the concentration of an analyte in a sample. The relationship between concentration and absorbance is described by **Beer's law,**

$$A = \epsilon bc$$

where A is absorbance, ϵ is molar absorptivity, b is path length, and c is concentration. The equation shows the three variables that can influence absorbance.

The **path length** refers to the distance that electromagnetic radiation must pass through the sample. If the path length is increased, the radiation will encounter more sample molecules as it passes through, and therefore a greater quantity of radiation will be absorbed. Path length is usually measured in units of centimeters.

Absorbance measurements can be used for the determination of the concentration of a sample because of the direct relationship between absorbance and concentration described by Beer's law. As the concentration of a sample increases, radiation passing through the sample will encounter more molecules of the absorbing substance, causing more radiation to be absorbed. Concentration is measured in units of molarity. Absorbance is therefore directly proportional to both path length (b) and concentration (c). Beer's law is written as an equality using a proportionality constant, molar absorptivity (ϵ). **Molar absorptivity** is dependent on the substance being measured and the wavelength of electromagnetic energy being measured. The dependence of hemoglobin's molar absorptivity on wavelength can be seen in the spectrum of hemoglobin shown in Figure 15.22 in which the absorbance is seen to change with wavelength for a sample with a constant hemoglobin concentration and a constant path length. Molar absorptivity is measured in units of molarity^{-1} cm^{-1} (M^{-1}cm^{-1}).

Absorbance data are typically used for quantitative measurements by constructing a standard curve. The absorbances of several standard solutions containing known analyte concentrations are measured at a fixed wavelength. Figure 15.23 shows a plot of absorbance versus the concentration **(standard Beer's law plot)** for the solutions of hemoglobin. The straight-line plot indicates that Beer's law is obeyed. A straight line has a constant slope equal to $\Delta y/\Delta x$. For a Beer's law plot the slope is equal to $\Delta A/\Delta c$. Because the measurements are done at a fixed wavelength and a constant path length, ϵb is a constant. Therefore, slope $= \Delta A/\Delta c = \epsilon b$. The absorbance of an unknown sample can now be measured and the concentration of the unknown determined by using the standard plot.

➤ CHECKPOINT

The UV/Vis spectrum of hemoglobin bound to oxygen shown as the solid line in Figure 15.22 was obtained using a single sample of hemoglobin with a constant concentration and was measured in a sample container with a constant path length. The change in absorbance shown in Figure 15.22 must therefore be due to changes in hemoglobin's molar absorptivity from one wavelength to another. At which of the following wavelengths does hemoglobin bound to oxygen have the largest molar absorptivity?

500 nm 540 nm 575 nm

THE SOLUTION: THE CULPRIT IS IDENTIFIED

The results of Ecker's analysis showed that 60% to 70% of the hemoglobin molecules in the dead fish were bound to carbon monoxide. Additional tests verified the results. As a final test Ball placed a cage of fish in the river near the outboard motor testing site. A second cage of fish was placed at a control site. After just 8 hours the fish at the test site were dead or dying whereas the fish at the control site were healthy. The blood of the dead fish was analyzed and found to have 50% to 80% of its hemoglobin bound to carbon monoxide. This confirmed carbon monoxide from the test facility as the cause of the fish kills. The testing facility installed a piping system to exhaust the motors out through a stack rather than into the water. In addition the company paid the state of Wisconsin \$40,000 for the fish kill and an additional \$20,000 to fund further studies of the river.

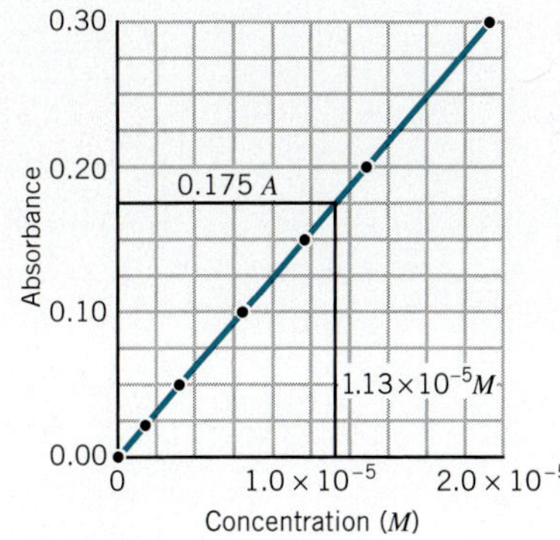

Fig. 15.24 Determining an unknown concentration from a Beer's law standard curve.

Exercise 15.3

The absorbance of a hemoglobin solution is measured in a UV/Vis spectrometer using a cuvette with a 1.00 cm path length. The absorbance at 576 nm is found to be 0.175. Use the standard curve in Figure 15.23 to determine both the concentration of the hemoglobin solution and the molar absorptivity of hemoglobin.

Solution

The absorbance of 0.175 is located on the *y* axis of the spectrum in Figure 15.23. A horizontal line is drawn from this absorbance until it intersects with the Beer's law plot (see Figure 15.24). At this point a vertical line is dropped to the *x* axis to determine the concentration of hemoglobin that corresponds to the absorbance of 0.175. The concentration is found to be $1.13 \times 10^{-5}\ M$. The slope of the Beer's law plot ($\Delta y/\Delta x$) is equal to $\Delta A/\Delta c = \epsilon b$. Since the path length, b, is 1.00 cm in this measurement, the molar absorptivity is equal to the slope of the line. The molar absorptivity is therefore determined to be $15.5 \times 10^{3}\ M^{-1}\ \text{cm}^{-1}$.

15.8 An Off-Color Fatty Alcohol[3]

THE PROBLEM: WHAT IS THE CAUSE OF THE OFF-COLOR IN THE FATTY ALCOHOL?

A company that sells fatty alcohols to the rubber industry received a complaint from one of its customers that the fatty alcohol, stearyl alcohol ($C_{18}H_{37}OH$), had an off-color. The customer returned a sample of the off-color alcohol. The sample was then used to identify the contaminants in the fatty alcohol that were causing the off-color. The contaminants were first separated from the fatty alcohol using chromatography. They were then examined using infrared spectroscopy.

THE METHOD OF ANALYSIS: INFRARED SPECTROSCOPY

Infrared (IR) spectroscopy is a molecular absorbance technique used primarily for qualitative analysis and structural determination. In Chapter 6 it was shown that all matter is in constant motion on the atomic level. The covalent bonds in molecules can be thought of as springs that hold the atoms together rather than the rigid sticks often used in molecular model kits. Two atoms connected by a covalent bond can vibrate just like two balls connected by a spring. The energy associated with the vibration will depend on the mass of the two atoms and the strength of the bond that connects them, as shown in Figure 15.25. Two atoms bonded together will therefore have their own characteristic vibrational frequency. The energy associated with this vibration falls in the infrared portion of the electromagnetic spectrum. Infrared radiation with a frequency that matches the frequency of vibration in a molecule will be absorbed by that molecule. Absorption of infrared radiation can be reported in units of wavelength,

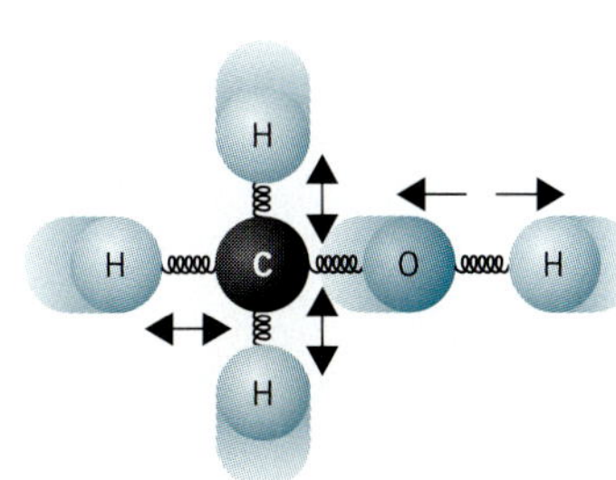

Fig. 15.25 Vibration of atoms in a methanol molecule.

[3] "Professional Analytical Chemists in Industry: A Short Course for Undergraduate Students in Problem Solving," Procter & Gamble.

Table 15.3

Characteristic Infrared Wavelengths and Wavenumbers of Absorption

Bond Type	Range of Wavelengths (μm)	Range of Wavenumbers (cm^{-1})
—C—H	3.38–3.51	2960–2850
=C—H	3.23–3.33	3100–3000
≡C—H	3.03	3300
C=C	5.95–6.17	1680–1620
C≡C	4.49–4.76	2230–2100
O—H	2.74–4.00 (broad)	3650–2500 (broad)
N—H	2.94–3.13	3400–3200
C=O	5.56–6.13	1800–1630
C—O	7.69–10.00	1300–1000
C—N	7.35–9.80	1360–1020

which is related to frequency by the relationship $\lambda = c/\nu$, where c is the speed of electromagnetic radiation in a vacuum. Table 15.3 lists the expected wavelength range for absorption by a number of common covalently bonded atoms.

IR spectroscopy finds its greatest utility in structure determination of organic molecules. Organic chemicals are often classified by **functional groups,** an atom or group of atoms that give a compound its characteristic properties. Organic alcohols have a hydroxyl, OH, as the functional group, whereas ketones have a carbonyl group, C=O, as shown in Figure 15.26. The use of infrared radiation as a probe of the structure of a molecule can be demonstrated with the infrared spectrum of alcohols. Because all organic alcohols contain the hydroxyl functional group, they all will have an O—H bond and, according to Table 15.3, will therefore have an absorption peak in the range between 2.7 and 4.0 μm.

UV/Vis spectra are shown plotted as absorbance versus wavelength as shown in Figure 15.22. Infrared spectra, however, are plotted as percentage transmittance on the y axis rather than absorbance. Because absorbance and percentage transmittance have an inverse relationship (as more radiation is absorbed, less radiation is transmitted through the sample), infrared spectra have a baseline along the top edge of the spectra with absorption peaks pointing down. The x axis may be plotted as wavelength measured in micrometers (microns, 10^{-6} m) or as **wavenumbers,** $\overline{\nu}$, measured in cm^{-1}.

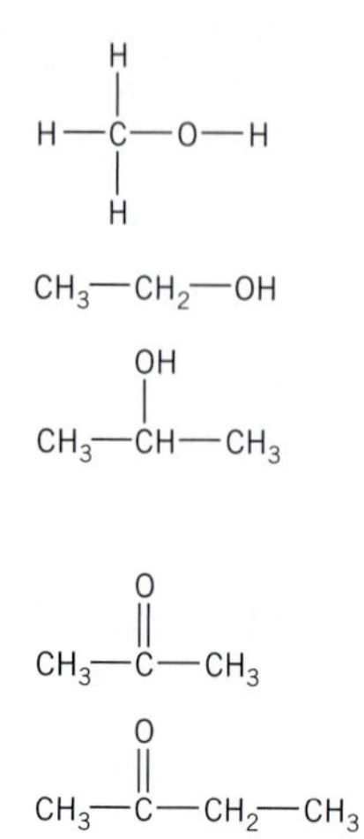

Alcohols

Methyl alcohol
(wood alcohol, a common industrial solvent)

Ethyl alcohol
(the alcohol found in "alcoholic" beverages)

Isopropyl alcohol
(rubbing alcohol)

Ketones

Acetone
(nail polish remover)

Methyl ethyl ketone
(a common industrial solvent)

Fig. 15.26 Common alcohols and ketones.

Wavenumbers are the reciprocal of the wavelength measured in cm and are directly proportional to the frequency of infrared radiation as shown in the following equation.

$$\bar{\nu} = 1/\lambda = \nu/c$$

Wavenumber ranges for absorption by several common covalently bonded atoms accompany the wavelength ranges given in Table 15.3. Often infrared spectra are plotted showing both wavelength and wavenumbers on the *x* axis. In the spectra shown in Figure 15.27, wavelength is plotted along the top of the spectra and wave numbers along the bottom. Figure 15.27 shows the infrared spectra for two alcohols,

> **CHECKPOINT**
>
> NO and NO_2 are examined by infrared spectroscopy. Is the N—O bond strength greater in NO or NO_2? Which molecule would you expect to absorb infrared radiation at the greater frequency? Wavelength? Wavenumber?

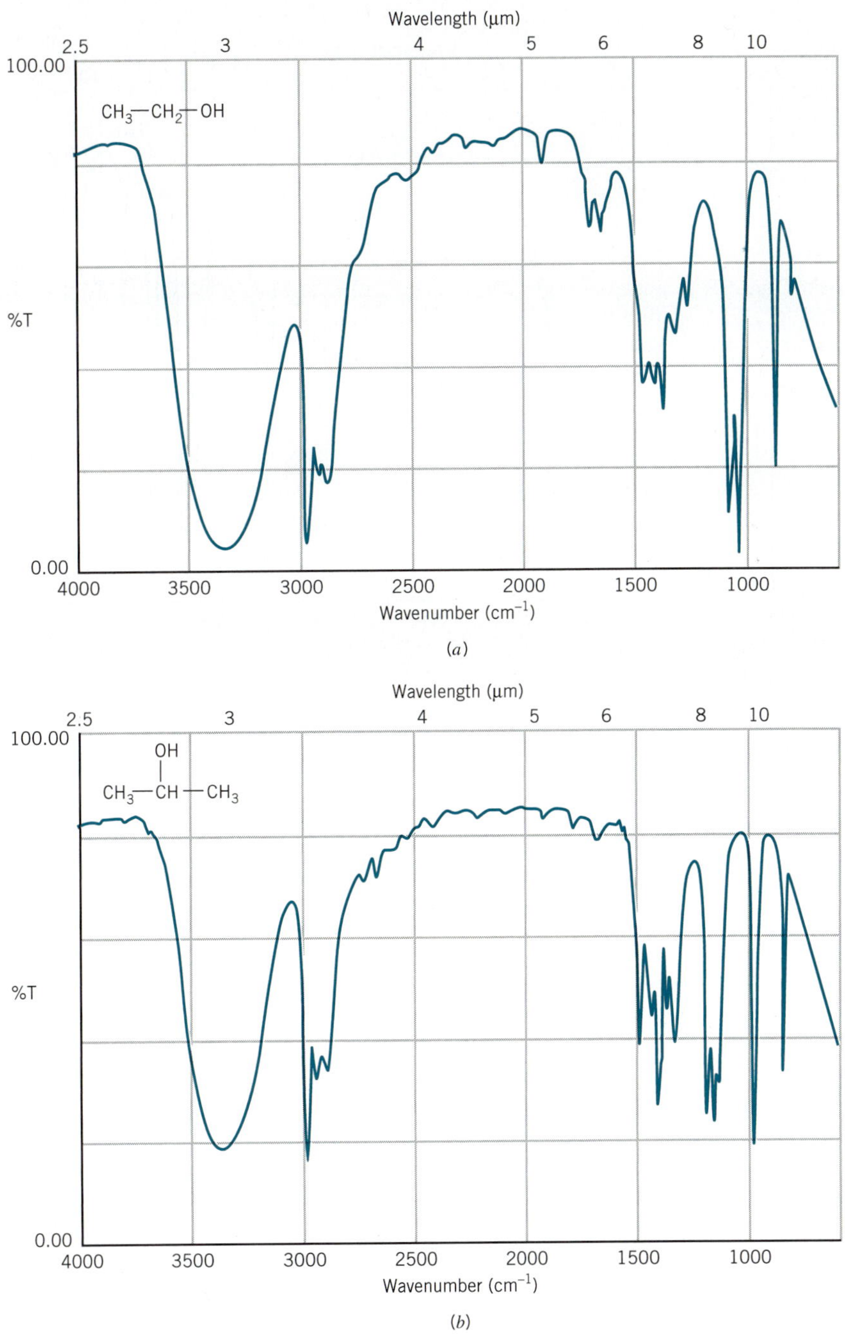

Fig. 15.27 IR spectra of (*a*) ethyl alcohol and (*b*) isopropyl alcohol.

ethyl alcohol and isopropyl alcohol. Note the absorption band for the hydroxyl group between 2.7 and 3.0 μm, (3700–3330 cm^{-1}). Since both molecules have the same functional group, they both absorb infrared radiation of approximately the same frequency or wavelength. By using tables of characteristic wavelengths of absorbance, it is possible to identify the functional groups found in an organic molecule.

Exercise 15.4

The spectrum below is an infrared spectrum of acetone. Use Table 15.3 to identify the peak caused by the carbonyl group, C=O.

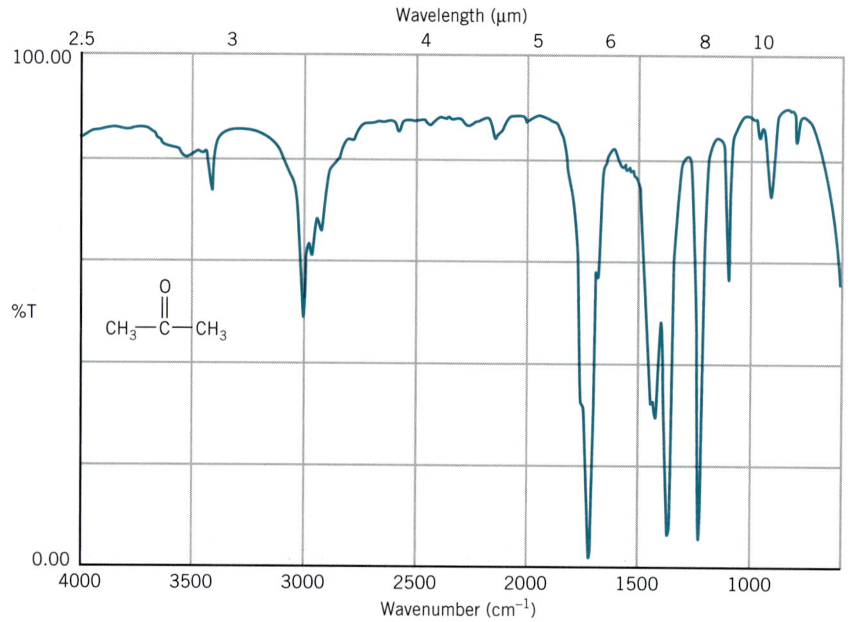

Exercise 15.4

Solution

Table 15.3 lists the wavelength range for a carbonyl absorbance as 5.56–6.13 μm (1800–1630 cm^{-1}). The peak at 5.85 μm (1710 cm^{-1}) in the acetone spectrum corresponds to the carbonyl group absorbance.

Exercise 15.5

A carboxylic acid functional group contains both a carbonyl group and a hydroxyl group. Which of the following spectra would you expect to be due to the carboxylic acid, acetic acid?

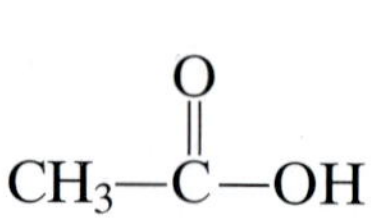

Solution

Spectrum *a* is of butanol, an alcohol. It shows the characteristic broad absorption of a hydroxyl group at 3.00 μm (3330 cm^{-1}) but no peak corresponding to a carbonyl

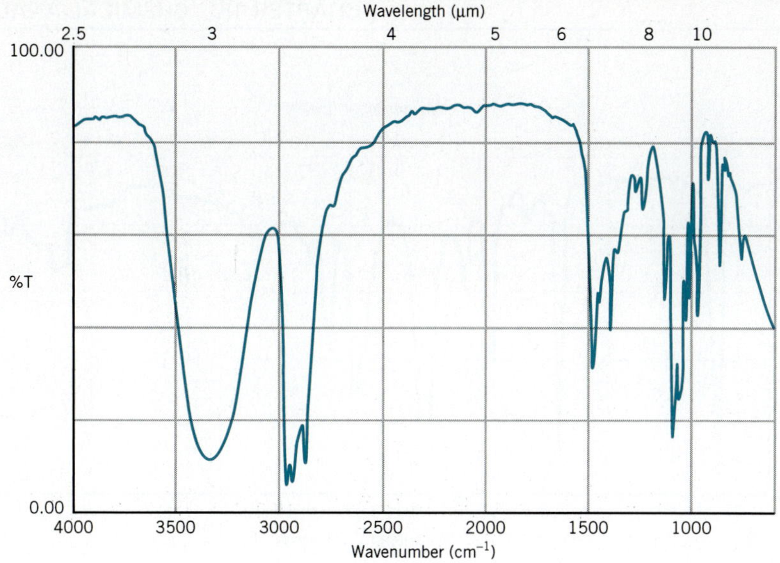

(*a*)

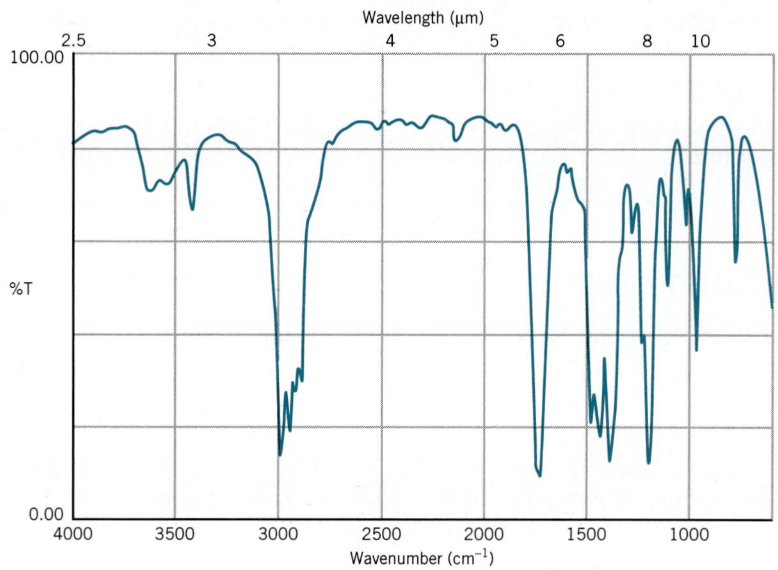

(*b*)

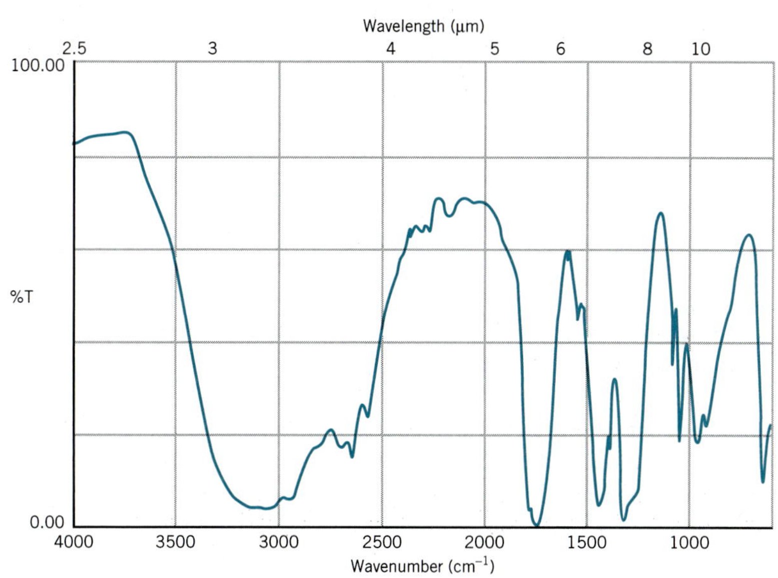

(*c*)

Exercise 15.5

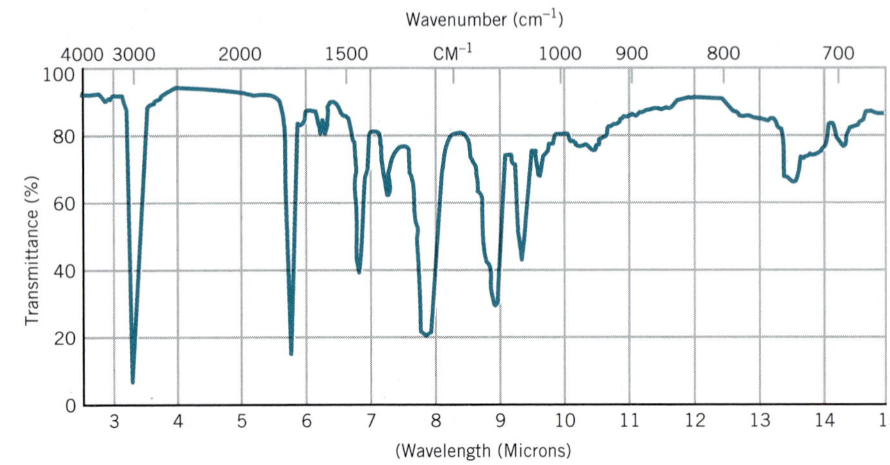

Fig. 15.28 IR spectrum of contaminant from the fatty alcohol sample. The spectrum is consistent with a common component in paint. From "Professional Analytical Chemists in Industry: A Short Course for Undergraduate Students in Problem Solving," Proctor & Gamble.

group. Spectrum *b* is of methyl ethyl ketone. It shows a typical carbonyl peak at 5.82 μm (1720 cm^{-1}) but no hydroxyl peak. Spectrum *c* contains both a broad peak at 3.25 μm (3080 cm^{-1}) corresponding to an OH group and a peak at 5.85 μm (1710 cm^{-1}) corresponding to a carbonyl group. Spectrum *c* is therefore the acetic acid spectrum.

THE SOLUTION: IDENTIFICATION OF THE CONTAMINANT

Because each compound has a characteristic spectrum, IR spectroscopy can be used for qualitative analysis. This technique allows the identification of the *molecules* that are present in a sample. Therefore, spectra of the unknown compounds can be matched to previously obtained spectra of known compounds. This is most easily accomplished in cases where the unknown is suspected to be one of a limited number of possibilities or when a database of known spectra can be searched by computer. Figure 15.28 shows the spectrum obtained for one of the contaminants in the fatty alcohol sample. This spectrum was compared to standard spectra and identified as long oil soya alkyd, a common component in paint. The customer with the off-color fatty alcohol was contacted and confirmed that painters had been working in the plant after the fatty alcohol was received and before the off-color was noted. The off-color was due to volatile organic compounds that had evaporated from the paint and been absorbed by the fatty alcohol.

15.9 The Search for New Compounds

Each year thousands of previously unknown naturally occurring chemical compounds are discovered. These new compounds often have interesting properties with many potential applications. An area of recent research has been the search for new compounds that are potential new medicinal drugs. Researchers have focused much of their search on sources that have been used for medicinal purposes by traditional native cultures. These diverse and sometimes exotic sources of new compounds include algae, plants, marine sponges, and frog or toad skin secretions. The native medicines often contain compounds which are bioactive and thus provide medicinal benefits. By isolating and identifying the bioactive compounds, researchers hope to develop new drugs.

THE PROBLEM: WHAT ARE THE STRUCTURES OF THE BIOACTIVE COMPOUNDS IN *CRYPTOLEPIS SANGUINOLENTA?*

A team of scientists from Ghana, the University of Pittsburgh, and Burroughs Wellcome Company (a pharmaceutical company) investigated *Cryptolepis sanguinolenta,* a shrub found in West Africa.[4] The shrub has long been used by Ghanaian traditional healers for the treatment of fevers. In addition, an extract from the roots has been used at the Center for Scientific Research into Plant Medicine in Ghana for the treatment of malaria and urinary and upper respiratory tract infections. Investigation of the plant involved the isolation and purification of the bioactive components of the plant, followed by identification and structure determination. A major tool for structure determination of chemical compounds is nuclear magnetic resonance spectrometry.

THE METHOD OF ANALYSIS: NUCLEAR MAGNETIC RESONANCE SPECTROSCOPY

Nuclear magnetic resonance (NMR) spectroscopy is one of the most powerful tools available for the structure determination of organic molecules. NMR spectroscopy measures changes in the energy in the nuclei of atoms. The two nuclei that are most important for the determination of the structure of organic molecules are ^{1}H and ^{13}C.

Nuclei have an electric charge due to their protons. Because a moving electric charge creates a magnetic field, a spinning nucleus has a magnetic field associated with it. A spinning nucleus is analogous to a tiny bar magnet (it has a north and south pole). When such a magnet is placed close to another magnet, it will align itself so that its south pole is pointing toward the north pole of the other magnet. This would be the most stable (lowest-energy) state of the magnet. The magnet could be turned until the two north poles of the magnets are aligned with one another. This would be an unstable (high-energy) state, and the magnet when released will return to its original (low-energy) north–south alignment, as shown in Figure 15.29*a*. In a similar way, a spinning

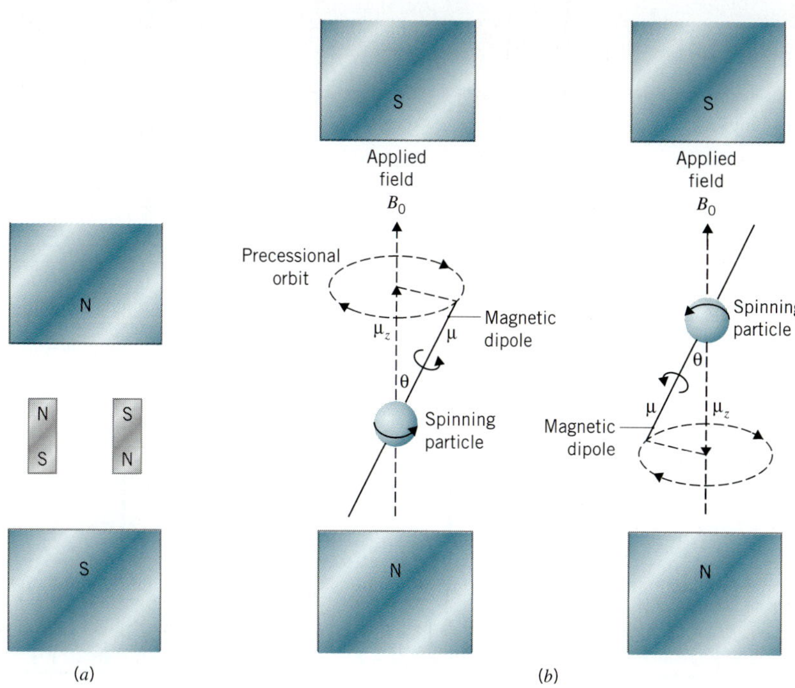

Fig. 15.29 (*a*) Two small bar magnets in a magnetic field. One bar magnet is aligned with the field and the other against the field. (*b*) Two spinning nuclei in an external magnetic field. One nucleus is aligned with the field and the other against it.

[4]A.N. Tackie et al., *Journal of Natural Products,* **56,** 653–670 (1993).

nucleus placed in an external magnetic field will either align with the magnetic field (low-energy state) or against the magnetic field (high-energy state), as shown in Figure 15.29*b*. The difference in energy between the two energy states is quite small and corresponds to energy in the radio wave region of the electromagnetic spectrum.

In NMR spectroscopy the sample is placed in a strong external magnetic field and irradiated with radio waves. The waves have sufficient energy to cause the excitation of the nuclei to higher energy levels. The resulting NMR spectra are plotted as absorbance versus ppm, where ppm represents the parts per million shift in frequency from a reference frequency.

It may appear that NMR spectra would give relatively little useful information (i.e., one peak for the ^1H nuclei present and another peak for ^{13}C nuclei). However, each type of nucleus is capable of giving different signals, depending on the magnetic environment in which the nucleus is found within the molecule. The magnitude of the external magnetic field that the nucleus experiences depends not only on the external magnetic field of the spectrometer but also on magnetic fields generated within the structure of the molecule. Because moving electric charges generate magnetic fields, both moving electrons and spinning nuclei will generate local magnetic fields within the molecule. Thus, the resulting NMR spectrum gives information about the environments in which various nuclei are found. Figure 15.30 shows the ^1H (proton) NMR and ^{13}C NMR spectra of

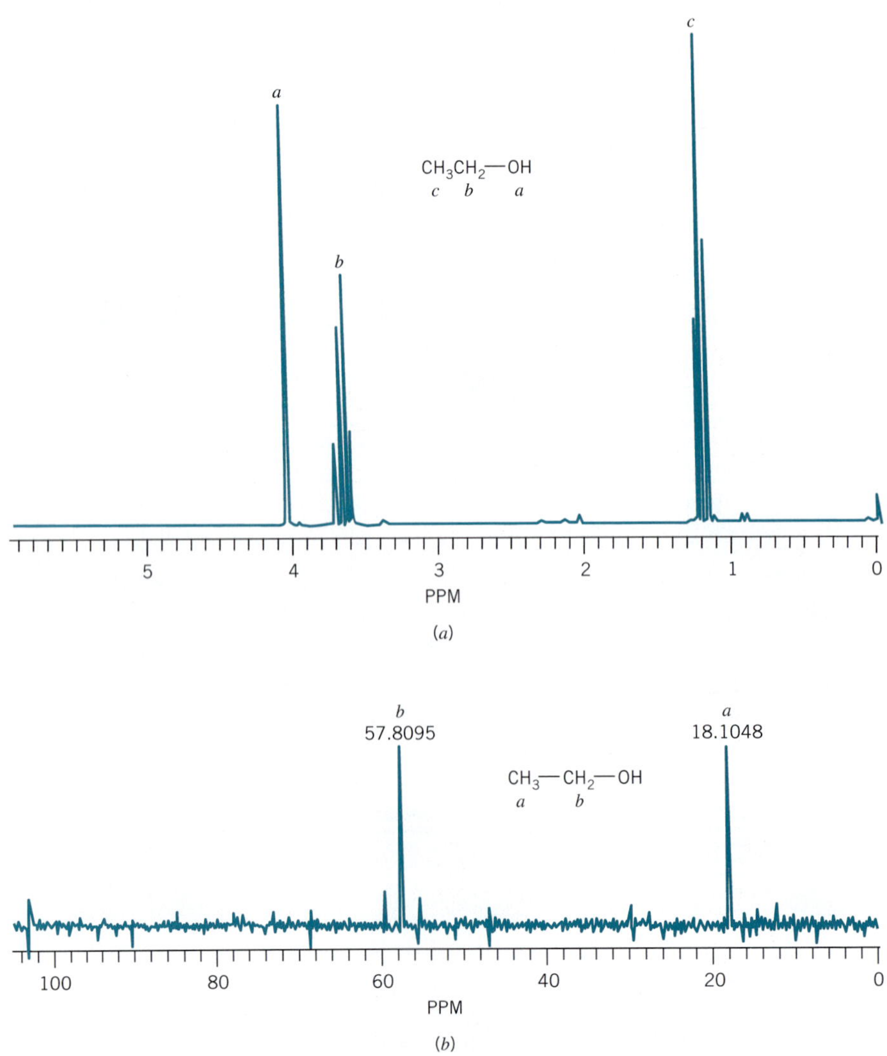

Fig. 15.30 (*a*) ^1H NMR spectrum of ethanol and (*b*) ^{13}C NMR spectrum of ethanol.

ethanol. Each group of peaks in the ^{1}H spectrum corresponds to hydrogen nuclei (protons) in a slightly different magnetic environment. Both ^{1}H and ^{13}C NMR spectra are often obtained for a molecule in order to provide as much data as possible for a structure determination. NMR spectra give three principal pieces of information:

1. The location of a set of peaks along the frequency axis (x axis) is known as the chemical shift and gives information about the electron environment of the nuclei. In Figure 15.30*a* there are three groups of peaks labeled a, b, and c. These three groups of peaks correspond to the hydrogens labeled a, b, and c below. In the ^{13}C NMR spectrum shown in Figure 15.30*b,* there are two peaks corresponding to the two carbon atoms.

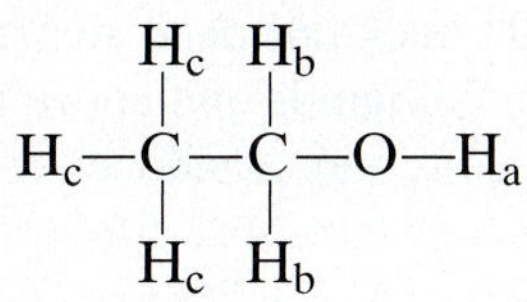

2. The area under a group of peaks corresponds to the relative number of nuclei in each environment. Therefore, the area of the set of peaks labeled b in Figure 15.30*a* is twice as large as the area of peak a, because there is only one hydrogen in the position labeled a on the structure but there are two hydrogens in position b on the structure.
3. The pattern of peaks for a single magnetic environment relates to the number of neighboring nuclei. In Figure 15.30*a* there is a single peak at position a, and there are four peaks at position b and three at position c.

➤ CHECKPOINT

Predict the number of sets (or groups) of peaks in the proton NMR spectra of acetone and acetic acid.

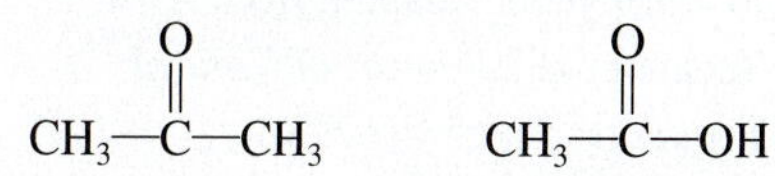

Acetone *Acetic acid*

THE SOLUTION: A NEW COMPOUND IS IDENTIFIED

The investigation of *Cryptolepis sanguinolenta* resulted in the isolation of a new bioactive compound which was named cryptospirolepine. Figure 15.31 shows one of the NMR spectra used in the structure determination, and Figure 15.32 shows the deduced structure of cryptospirolepine.

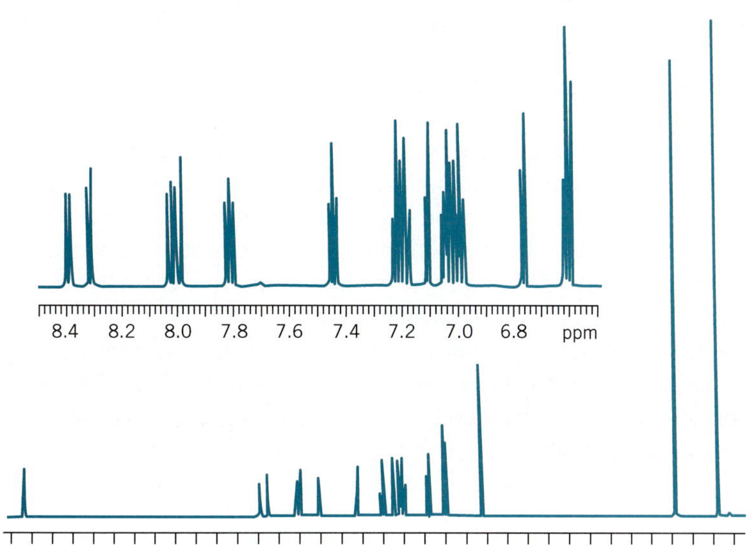

Fig. 15.31 ^{1}H NMR spectrum of cryptospirolepine. The portion of the spectrum from 6.6 to 8.4 ppm has been expanded and is shown in the insert. From A. N. Tackie et al., *Journal of Natural Products,* **56,** 657 (1993).

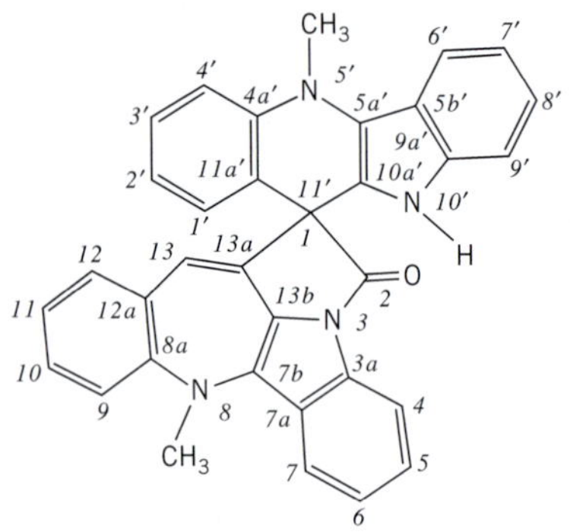

Fig. 15.32 Structure of cryptospirolepine. Adapted from A. N. Tackie et al., *Journal of Natural Products,* **56,** 654 (1993).

15.10 The Search for the Northwest Passage—The Franklin Expedition[5]

On May 19, 1845, Sir John Franklin set sail from England in command of two ships, the H.M.S. *Terror* and the H.M.S. *Erebus*. The ships had a crew of 134 officers and men and were well provisioned with 61,987 kg of flour, 16,749 liters of liquor, 8000 tins (1- to 8-pound capacity) of meats, soups, and vegetables, and 4200 kg of lemon juice to prevent scurvy. The expedition's 3-year mission was to find the Northwest Passage, a navigable sea route above North America to the Pacific Ocean. The ships never returned. Over the next 10 years rescue expeditions were sent in search of the Franklin expedition, but no survivors were ever found. The rescue expeditions brought back horror stories of starvation and cannibalism from the Inuit Indians of King William Island above the Arctic Circle. Later expeditions discovered skeletal remains and abandoned campsites. One expedition discovered skeletal remains by one of the ship's lifeboats mounted on a wooden sled for dragging the boat across the island. The lifeboat contained, among other supplies, silk handkerchiefs, scented soap, sponges, slippers, and combs. These were strange provisions for desperate men pulling a heavy boat.

THE PROBLEM: WHY DID THE FRANKLIN EXPEDITION PERISH?

Although the fate of the Franklin expedition was not in doubt, many questions remained. Why had well-trained and well-equipped explorers been unable to survive where the Inuit Indians had lived for centuries? Why had the crew taken such useless provisions in the lifeboat that they had tried to drag across the island? In 1981 Dr. Owen Beattie, a forensic anthropologist at the University of Alberta, set out to determine what had happened to the Franklin expedition. On King William Island he found bones that he suspected were skeletal remains of the Franklin expedition. He also collected bones of Inuit Indians and caribou from the same geographical area and time period as the Franklin expedition. These samples were subjected to a range of different tests, including tests for heavy metals using an atomic absorption spectrometer.

THE METHOD OF ANALYSIS: ATOMIC ABSORPTION SPECTROMETRY

Atomic absorption (AA) spectrometry is used primarily for quantitative elemental analysis of metals. Unlike previously described spectroscopic techniques, AA spectrometry does not differentiate between a metal as a free atom, an aqueous metal ion, or a metal ion or atom in a compound. It instead measures how much of the element is present regardless of the form or forms in which it is found in the sample.

In AA spectrometry a flame or furnace is used to break compounds apart into their individual atoms. The flame or furnace evaporates the solvent, breaks all bonds to give free atoms, and momentarily suspends the atoms in the path of the electromagnetic radiation. The beam of electromagnetic radiation passes through the flame and is absorbed by the atoms as shown in Figure 15.33. Only those wavelengths of radiation that correspond in energy to changes in electronic energy levels within the atom are absorbed.

[5]Owen Beattie and John Geiger, *Frozen in Time: Unlocking the Secrets of the Franklin Expedition,* E. P. Dutton, New York, 1987.

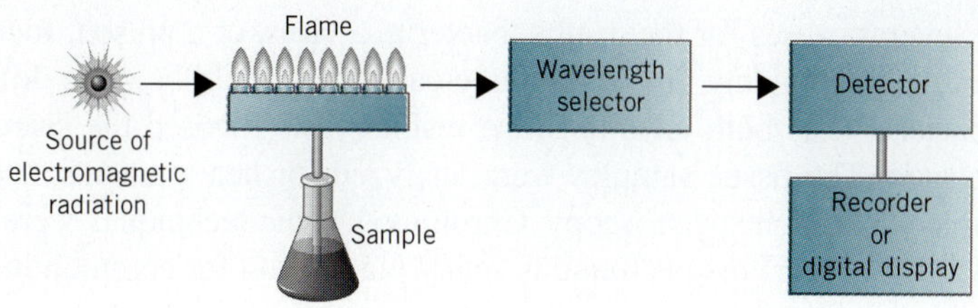

Fig. 15.33 Diagram of atomic absorption spectrometer.

The absorbance of radiation is related to the concentration by Beer's law, $A = \epsilon bc$. The absorbances of **standards** are measured and compared to the absorbance of the same element in the sample in order to find the concentration of an element in the sample.

THE SOLUTION: SURPRISING RESULTS

Results of the analysis for the concentration of lead were quite unexpected. Table 15.4 shows these results. The lead levels in the bone samples from the Franklin expedition were much higher than those of contemporary Inuit Indians, caribou, or modern humans. Lead poisoning is known to cause anorexia, weakness, fatigue, irritability, stupor, paranoia, anemia, and abdominal pain. Lead poisoning could therefore have led to declining health in the crew and could have been a significant factor in the ability of the officers to make appropriate decisions, as in the case of the provisions for the lifeboat. However, lead levels in bone only indicate exposure to lead at some time during the person's life. To establish whether the lead poisoning had occurred during the expedition, it would be necessary to have tissue samples to analyze.

In the summers of 1984 and 1986 Beattie led expeditions to Beechey Island, where the Franklin expedition had spent the winter of 1845. It was known that three crewmen, John Hartnell, William Braine, and John Torrington, had died during that first winter and been buried on Beechey Island. Because the location was north of the Arctic Circle, the graves would be beneath the permafrost and the bodies might still be preserved. Beattie received permission to exhume the bodies and take tissue samples. Results of the analysis of hair samples for lead content are given in Table 15.5. These results indicate that all three crewmen had been exposed to very high levels of lead during the time of the expedition. Lead concentrations of that level would certainly have had detrimental health effects.

A final question remained. What was the source of the lead poisoning? Near the campsite of the Franklin expedition on Beechey Island was a garbage dump where, among other items, the crew members had discarded the empty tin cans used for storing meat and vegetables. At the time of the Franklin expedition in 1845 the canning of foods in tins was relatively new. Beattie found that the tins had been sealed using lead solder. It seems highly probable that the solder was the source of the lead that caused lead poisoning of the crew and contributed to the loss of the Franklin expedition.

Table 15.4
Lead Level in Bone Samples[a]

Sample	Mean Pb Concentration (ppm)
Franklin expedition	138.1
Inuit Indians	5.1
Caribou	2.0
Modern human	29.8

[a]W. A. Kowal, P. M. Krahn, and O. B. Beattie, Lead levels in human tissues from the Franklin forensic project, *Int. J. Environ. Anal. Chem.*, **35,** 119–126 (1989).

Table 15.5
Lead Analysis of Hair Samples[a]

Sample	Mean Lead Concentration (ppm)
Torrington	565
Hartnell	326
Braine	225
Modern human	4

[a]W. A. Kowal, P. M. Krahn, and O. B. Beattie, Lead levels in human tissues from the Franklin forensic project, *Int. J. Environ. Anal. Chem.*, **35,** 119–126 (1989).

15.11 Dead Cats[6]

Cats were dying in the Glen Oaks neighborhood. At first mostly stray cats were found dead, but then family pets began dying. Tissue samples of the dead animals were sent to Michael McClure, a chemist at the state veterinary diagnostic laboratory.

[6]M. W. McClure, The crow's warning, *Chem Matters,* **8,** 7–9 (April 1990). Excerpted with permission. Copyright © 1990 American Chemical Society.

There could be several causes for the deaths: bacteria, a virus, or a poison. Identifying the specific cause within any one of the three categories could be quite difficult. Virology and bacteria tests both gave negative results. This focused the search on some type of poison. The tissue samples were analyzed for heavy metals, such as lead, using atomic absorption spectroscopy. Chromatographic techniques were used to test for alkaloids (organic poisons found in many plants) and for common insecticides. However, none of the tests gave any clue to the cause of death.

THE PROBLEM: WHAT WAS KILLING THE CATS?

A local veterinarian called McClure and asked him to come to Glen Oaks to see another dead cat that had been found. In a vacant lot was the body of another dead cat and nearby was a dead crow. From the condition of the cat's body, it was obvious that the crow had been eating the cat. This suggested that whatever had killed the cat had been passed to the crow. The poison had evidently acted quickly, since the crow was found beside the cat's body. Very few poisons can be passed on in this manner and act this quickly. The field of potential poisons was narrowing. McClure suspected "1080," a rodenticide known as sodium monofluoroacetate.

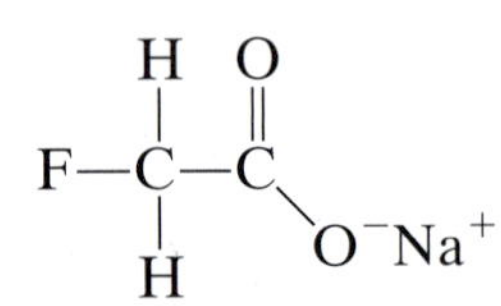

METHOD OF ANALYSIS: POTENTIOMETRY

Potentiometry is based on the relationship between the electric potential developed by an electrode in solution and the concentration of the analyte in solution. Chapter 12 introduced the use of electrodes to measure the potential of an oxidation–reduction reaction and described the potential as a measure of the driving force of a redox reaction. Section 12.8 described the dependence of electric potential on the concentration of the products and reactants. The relationship between potential and concentration allows the use of electrodes for the quantitative analysis of analytes capable of participating in a redox reaction. The most common potentiometric measurement is the measurement of pH using a pH meter. In this measurement the potential of a glass pH electrode is related to the negative log of the concentration of H_3O^+ in solution.

Potentiometric measurements are not limited to pH but can also be used for the determination of the concentrations of a variety of ions. An **ion-selective electrode (ISE)** produces a potential that is proportional to the concentration of a given ion. Ion-selective electrodes are available for the measurement of a number of different ions such as Na^+, K^+, NH_4^+, Ca^{2+}, S^{2-}, F^-, and NO_3^-. A potentiometric measurement requires the use of two electrodes (although they are both often combined into a single physical electrode probe) attached to a pH/pIon meter. The ion-selective electrode develops a potential that depends on the concentration of the analyte. The fluoride ion-selective electrode contains a crystal of LaF_3. A potential develops at the surface of the crystal that is proportional to the negative log of the concentration of F^- in contact with the crystal. The second electrode is a reference electrode, which maintains a constant potential that acts as a reference for the ion-selective electrode. Using standards of known F^- concentrations, the pH/pIon meter is calibrated to give a direct reading of pF:

$$pF = -\log[F^-]$$

THE SOLUTION: THE FLUORIDE LEVEL IS HIGH

A tissue sample was treated so that any 1080 present would be broken down to yield fluoride ion as one of the products. A fluoride ion-selective electrode was then used to measure the F^- concentration in the sample. Tissue samples normally have only a trace amount of fluoride ion, less than 5 ppm. The sample from the dead cat showed a concentration of 35 ppm fluoride ion, suggesting that indeed 1080 was the cause of death.

Key Terms

Absorbance
Absorption
Atomic absorption (AA) spectrometry
Beer's law
Chromatography
Chromatography column
Deoxyribonucleic acid (DNA)
Electrophoresis
Fatty acids
Functional group
Gas chromatography (GC)
Gas chromatography/mass spectrometry (GC/MS)
Gravimetric
High-performance liquid chromatography (HPLC)
Infrared spectroscopy
Ion-selective electrode (ISE)
Mass spectrometry (MS)
Mobile phase
Molar absorptivity
Nuclear magnetic resonance (NMR) spectroscopy
Parent ion
Path length
Percent transmittance
Potentiometry
Qualitative methods
Quantitative methods
Restriction fragment length polymorphism (RFLP)
Retention time
Risk assessment
Spectroscopy
Standard Beer's law plot
Standards
Stationary phase
Transmittance
Ultraviolet/visible spectroscopy (UV/Vis)
Volumetric
Wavelength
Wavenumber

Problems

1. Distinguish between *quantitative, qualitative,* and *structural* methods of analysis.
2. Distinguish between *volumetric* and *gravimetric* methods of analysis.

Chromatography

3. Describe how chromatography is used to separate a mixture. What functions do the stationary and mobile phases have in the separation?
4. Compare and contrast the techniques of gas chromatography and high-performance liquid chromatography. (What is the same and what is different about the two techniques?)
5. How are the intermolecular forces, described in Chapter 8, important in the separation of components on a chromatography column?
6. Hexane, $CH_3(CH_2)_4CH_3$, is found to have a long retention time when separated from a mixture. A column with a C_{18} (18-carbon chain) stationary phase and a methanol–water mobile phase is used for the separation. Explain why hexane has a long retention time.
7. You wish to chromatographically separate a mixture of the following compounds. Which of the two compounds will be most difficult to separate from one another? Why?

(a) $CH_3CH_2CH_2\overset{\overset{O}{\|}}{C}—OH$

(b) $CH_3\overset{\overset{CH_3}{|}}{C}HCH_2CH_2CH_2CH_2CH_2CH_2—OH$

(c) $CH_3CH_2CH_2\overset{\overset{CH_3}{|}}{C}HCH_2CH_2CH_2CH_3$

(d) $CH_3CH_2—O—CH_2CH_2CH_2CH_3$

(e) $CH_3\overset{\overset{CH_3}{|}}{C}HCH_2CH_2CH_2\overset{\overset{OH}{|}}{C}HCH_2CH_3$

8. The mixture below is separated using HPLC with a octadecylsiloxane (nonpolar) stationary phase and a methanol (CH_3OH)/water mobile phase. In what order will the components come off the column?

$CH_3CH_2CH_2CH_2CH_2CH_2CH_3$
Heptane

$CH_3CH_2CH_2CH_2CH_2CH_2CH_2—OH$
Heptanol

$CH_3CH_2CH_2—O—CH_2CH_2CH_2CH_3$
Butyl propyl ether

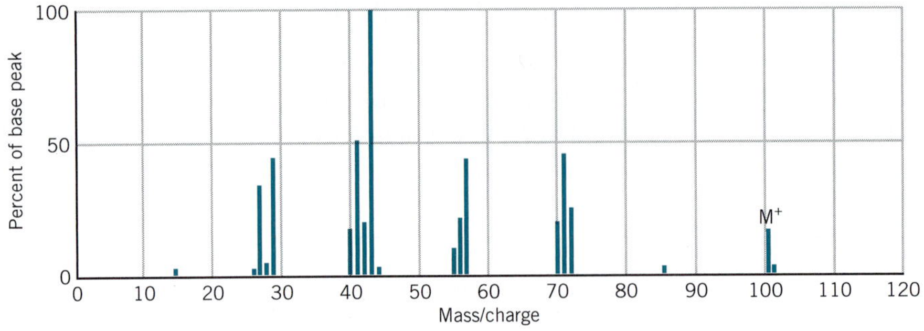

Fig. 15.34 Mass spectrum to accompany Problem 11.

What would happen to the order in which the compounds come off a polar column if pentane ($CH_3CH_2CH_2CH_2CH_3$) is used as the mobile phase?

Mass Spectrometry

9. What is a mass spectrum? What information can be obtained from a mass spectrum?
10. What are the principal functions of the gas chromatograph and the mass spectrometer in a GC/MS system?
11. Which of the following compounds is responsible for the mass spectrum in Figure 15.34? Explain your reasoning.
    (a) octanol, $CH_3CH_2CH_2CH_2CH_2CH_2CH_2CH_2$—OH
    (b) heptane, $CH_3CH_2CH_2CH_2CH_2CH_2CH_3$
    (c) acetic acid, CH_3COOH

Spectroscopy

12. Compare and contrast UV/Vis spectroscopy and atomic absorption spectroscopy. (What is the same and what is different about the two techniques?)
13. Give the Beer's law equation and define all of the terms.
14. Which of the following spectrophotometric techniques are used primarily for quantitative analysis and which are used for qualitative (or structure determination) analysis?
    (a) infrared (b) NMR
    (c) UV/Vis (d) atomic absorption
15. UV/Vis, infrared, and NMR spectroscopy all involve passing electromagnetic radiation through a sample and measuring the resulting absorbance or transmittance. What is different about the three techniques that allows them to provide different information about a molecule?
16. Explain why samples to be analyzed by IR and NMR spectroscopic techniques are normally separated into pure components before the analysis rather than being analyzed as a mixture.
17. The absorbance of a solution of copper nitrate is measured using UV/Vis spectrophotometry. What will happen to the absorbance for each of the conditions given below?
    (a) A cuvette with a longer path length is used.
    (b) The concentration of the copper nitrate is decreased.
    (c) The wavelength of the spectrometer is changed to a value where the molar absorptivity is less.
18. Match the following compounds with their IR spectra shown in Figure 15.35. Explain your reasoning.

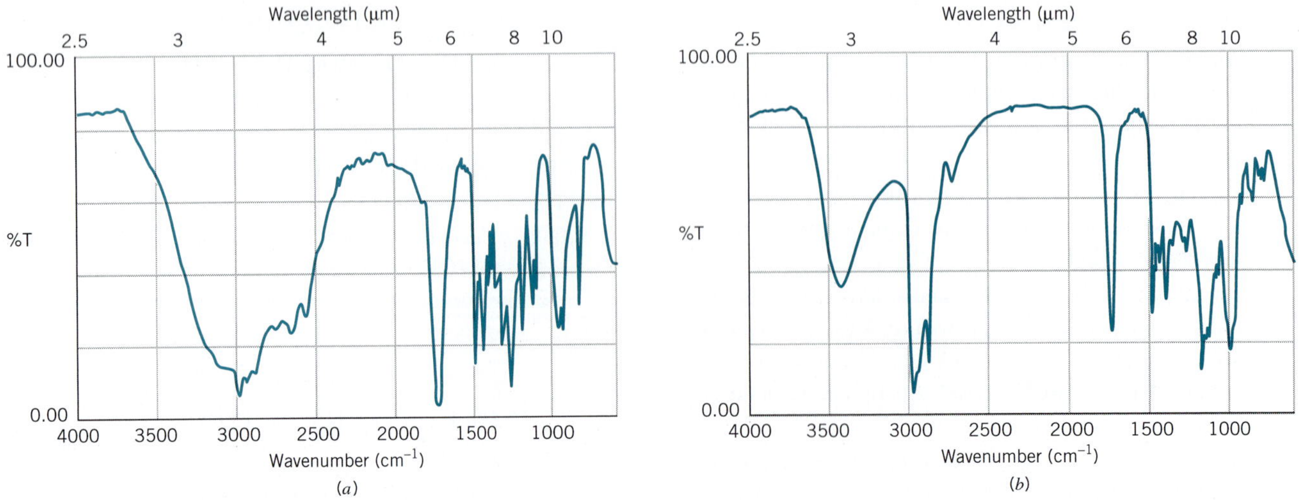

Fig. 15.35 Infrared spectra to accompany Problem 18.

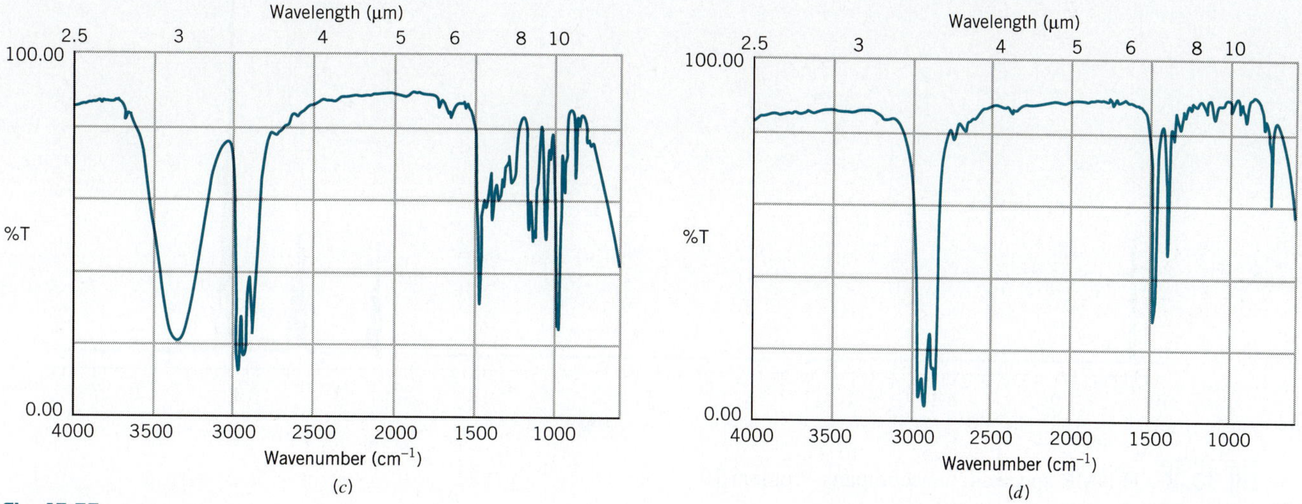

Fig. 15.35 (continued)

(a) octane, $CH_3CH_2CH_2CH_2CH_2CH_2CH_2CH_3$

(b) 3-pentanol, $CH_3CH_2\overset{\overset{\displaystyle OH}{|}}{C}HCH_2CH_2$

(c) isobutyric acid, $CH_3\underset{\underset{\displaystyle CH_3}{|}}{C}H\overset{\overset{\displaystyle O}{\|}}{C}{-}OH$

(d) butyraldehyde, $CH_3CH_2CH_2\overset{\overset{\displaystyle O}{\|}}{C}H$

19. Which of the following compounds is responsible for the NMR spectrum shown in Figure 15.36? Explain your reasoning.

(a) butyraldehyde, $CH_3CH_2CH_2\overset{\overset{\displaystyle O}{\|}}{C}H$

(b) octane, $CH_3CH_2CH_2CH_2CH_2CH_2CH_2CH_3$

(c) formic acid, $H\overset{\overset{\displaystyle O}{\|}}{C}{-}OH$

(d) methanol, CH_3OH

20. Suggest appropriate structures for each of the following:
   (a) A compound with the molecular formula $C_4H_8O_2$ has three sets of peaks in its ^1H NMR spectrum with relative intensities of 2:3:3 and two strong absorbance peaks in the IR at 5.65 μm (1770 cm^{-1}) and 8.05 μm (1240 cm^{-1}).
   (b) A compound with the formula $C_3H_6O_2$ has three sets of peaks in its ^1H NMR spectrum with relative intensities of 3:2:1 and two strong IR absorption peaks: one a broad peak at 3.4 μm (2940 cm^{-1}) and the other a sharper peak at 5.8 μm (1720 cm^{-1}).

21. Titrations are a quantitative volumetric method of analysis in which the volume of titrant necessary to completely react with a sample, the equivalence point, is measured. Absorbance measurements are sometimes used to monitor the progress of a titration in order to determine the equivalence point. A plot of absorbance versus volume of titrant added to the sample will have an inflection point that corresponds to the equivalence point. A generalized chemical equation for the titration can be written as:

$$\text{Titrant} + \text{sample} \longrightarrow \text{products}$$

Which of the following plots of absorbance versus volume of titrant describe the titration when the spectrometer is set to a wavelength at which the sample absorbs but the titrant and products do not? Explain your

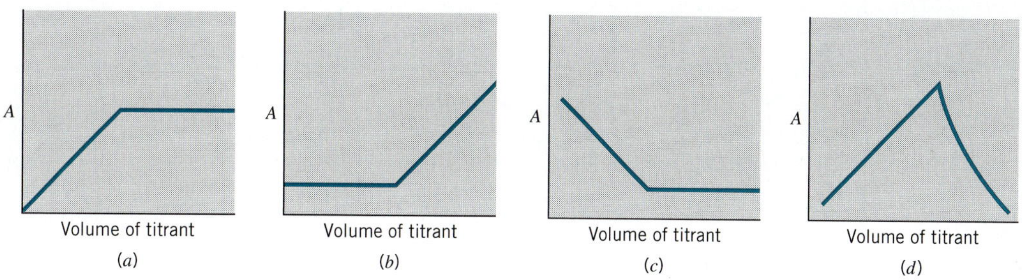

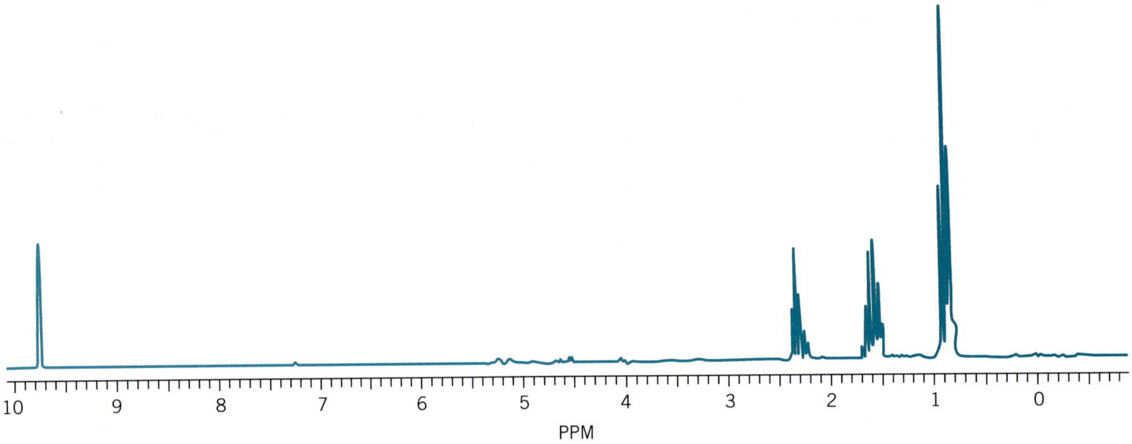

Fig. 15.36 ^{1}H NMR spectrum to accompany Problem 19.

reasoning. Which of the plots describe a titration in which the spectrometer is set to a wavelength at which the products absorb but not the sample or titrant? Explain your reasoning.

22. When making quantitative measurements of hemoglobin solutions using UV/Vis spectrophotometry, a wavelength of 576 nm will often be used. Explain why this wavelength is used as opposed to some other wavelength, such as 620 nm (see Figure 15.22).

23. Calculate the absorbance of a 5.2×10^{-4} *M* solution of the drug tolbutamine measured at 262 nm in a 1.00-cm cell. The molar absorptivity of tolbutamine at 262 nm is 703 M^{-1} cm^{-1}.

24. The concentration of iron in a solution can be determined using UV/Vis spectrophotometry by reacting the iron with 1,10-phenanthroline to produce a colored complex. The following data were obtained for a series of standard solutions of iron/1,10-phenanthroline complex measured in a 1.00-cm cell.

Iron Concentration (M)	*Absorbance*
0.50×10^{-4}	0.109
1.0×10^{-4}	0.218
2.0×10^{-4}	0.436
3.0×10^{-4}	0.656
4.0×10^{-4}	0.872

(a) Prepare a Beer's law plot using the above data.
(b) Calculate the concentration of an iron/1,10-phenanthroline complex solution that has an absorbance of 0.317.
(c) Calculate the molar absorptivity of the iron/1,10-phenanthroline complex.

25. Spectroscopic measurements can be used for kinetic studies by monitoring the absorbance of a reactant with respect to time. A reaction is found to have the following rate law with respect to a reactant, *X*.

$$\text{Rate} = k[X]$$

Which of the following plots of absorbance of *X* versus time corresponds to the above rate law? Explain your reasoning. Which of the plots corresponds to a rate law of rate $= k[X]^0$? Explain.

Potentiometry

26. Enzyme electrodes are used for the quantitative analysis of samples of biological fluids. The electrodes are constructed by placing a membrane with immobilized enzyme directly onto an ion-selective electrode. An enzyme electrode that is sensitive to the concentration of urea is constructed by immobilizing the enzyme urease in a membrane placed onto an ammonium ion-selective

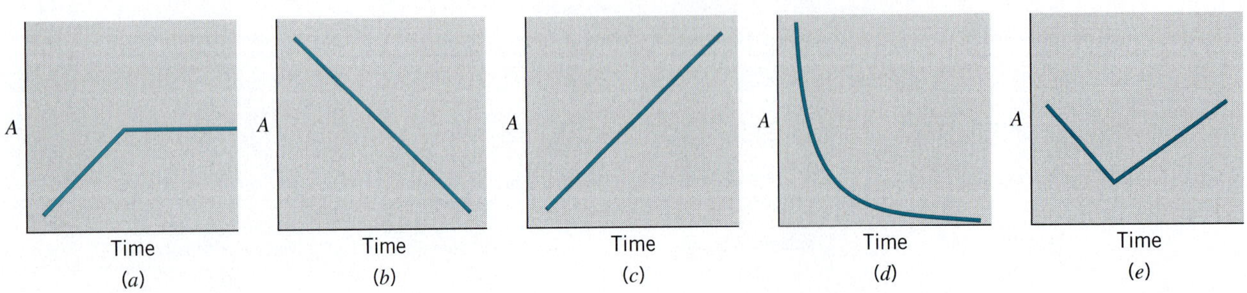

electrode. Urease catalyzes the following reaction:

$$\underset{\textit{Urea}}{NH_2CONH_2(aq)} + 2\,H_2O(l) + H^+(aq) \xrightarrow{\textit{urease}} \underset{\textit{Ammonium}}{2\,NH_4^+(aq)} + HCO_3^-(aq)$$

Explain how this electrode can respond to the presence of urea. On the basis of the discussion of enzymes in Chapter 14, why would the electrode not give a response for urea without the presence of the enzyme?

Integrated Problems

27. Discuss why legislation requiring zero concentration of pesticides in food would be unrealistic.
28. Indicate which of the following are elemental methods of analysis and which are molecular methods.
    (a) mass spectrometry
    (b) gas chromatography
    (c) UV/Vis spectroscopy
    (d) atomic absorption spectrometry
    (e) NMR spectrometry
29. Suggest the steps that a chemist might take to find the cause of an unusual aftertaste in a shipment of ginger ale.
30. Suggest an appropriate method of analysis for each of the following problems and describe why the method is appropriate to solve the problem.
    (a) Separate the components in a sample of perfume.
    (b) Determine the concentration of calcium in a sample of drinking water.
    (c) Determine whether mercury is leaching into the soil at a landfill.
    (d) Identify an organic compound isolated from a sample of milk.
    (e) Determine the structure of an enzyme isolated from beef heart.
    (f) Analyze the fumes from the exhaust of a car.
31. Which of the techniques described in this chapter can be used to distinguish between the following isomers (compounds with the same molecular formula but different structures)? Explain your reasoning.

$$CH_3CH_2CH_2CH_2CH_2{-}OH \qquad CH_3\underset{\displaystyle OH}{\underset{|}{C}}HCH_2CH_2CH_3$$

32. A lab assistant, by mistake, has washed all of the labels off a set of bottles that still contain chemicals. You know what was in the bottles, but you don't know which bottle is which. Describe how NMR, IR, or a combination of the two techniques could be used to distinguish between the chemicals listed in each of the sets below. Explain your reasoning.
    (a) methanol, CH_3OH; ethanol, CH_3CH_2OH; and propanol, $CH_3CH_2CH_2OH$
    (b) propanol, $CH_3CH_2CH_2OH$; propanoic acid, CH_3CH_2COOH; and ethyl methyl ether, $CH_3CH_2{-}O{-}CH_3$
33. A mixture of the following compounds was separated by HPLC with a stationary nonpolar phase and a CH_3OH/water mobile phase.

$$CH_3CH_2CH_2\overset{\displaystyle O}{\overset{\|}{C}}OH$$

A

$$CH_3CH_2CH_2CH_3$$

B

$$CH_3CH_2{-}O{-}CH_2CH_3$$

C

(a) Which of the following chromatograms would you expect to obtain? Explain your reasoning.

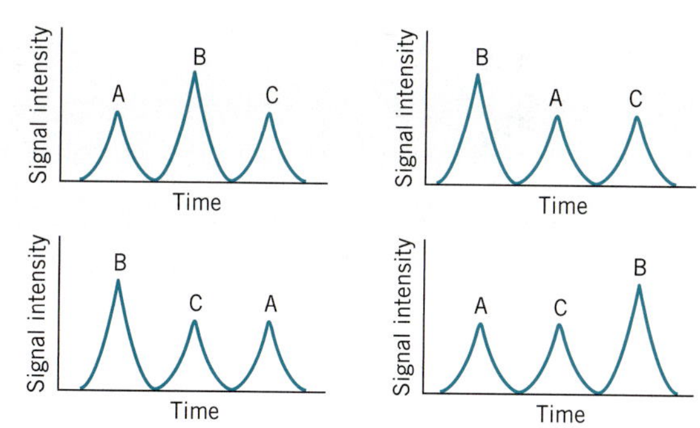

(b) Use the correct chromatogram above to estimate the *relative* amounts of *A*, *B*, and *C* in the original mixture. Explain your reasoning.
(c) Could IR spectroscopy be used to specifically identify *A*, *B*, and *C*? If so, what characteristics of the compounds would allow for this identification? Explain.

Appendix A

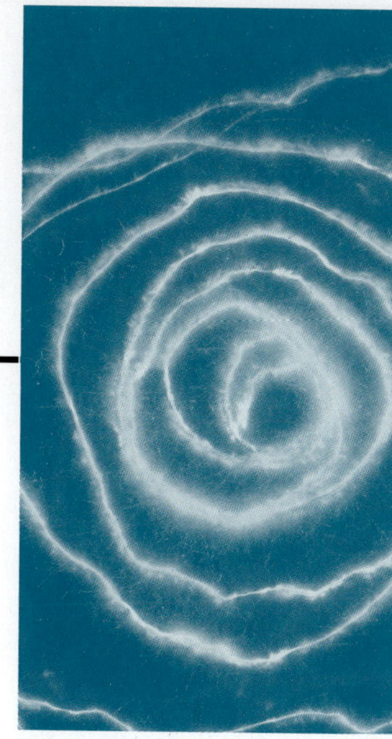

A.1 Systems of Units

All measurements contain a number that indicates the magnitude of the quantity being measured and a set of units that provide a basis for comparing the quantity with a standard reference. There are several systems of units, each containing units for properties such as length, volume, weight, and time.

ENGLISH UNITS OF MEASUREMENT

In the English system of units in use in the United States, the individual units are defined in an arbitrary way. There are 12 inches in a foot, 3 feet in a yard, and 1760 yards in a mile. There are 2 cups in a pint and 2 pints in a quart but 4 quarts in a gallon. There are 16 ounces in a pound but 32 ounces in a quart. The relationships between some of the common units in the English system are given in Table A.1.

Table A.1
The English System of Units

Length: inch (in.), foot (ft), yard (yd), mile (mi)	
12 in. = 1 ft	5280 ft = 1 mi
3 ft = 1 yd	1760 yd = 1 mi
Volume: fluid ounce (oz), cup (c), pint (pt), quart (qt), gallon (gal)	
2 c = 1 pt	32 oz = 1 qt
2 pt = 1 qt	4 qt = 1 gal
Weight: ounce (oz), pound (lb), ton	
16 oz = 1 lb	2000 lb = 1 ton
Time: second (s), minute (min), hour (h), day (d), year (y)	
60 s = 1 min	24 h = 1 d
60 min = 1 h	$365^1/_4$ d = 1 y

THE METRIC SYSTEM

More than 300 years ago, the Royal Society of London discussed replacing the irregular English system of units with one based on decimals. It was not until the French Revolution, however, that a decimal-based system of units was adopted. This **metric system** is based on the fundamental units of measurement for length, volume, and mass shown in Table A.2.

The principal advantage of the metric system is the ease with which the base units can be converted into a unit that is more appropriate for the quantity being measured. This is done by adding a prefix to the name of the base unit. The prefix *kilo-* (k), for example, implies multiplication by a factor of 1000. Thus, a kilometer is equal to 1000 meters.

$$1 \text{ km} = 1000 \text{ m}$$

The prefix *milli-* (m), on the other hand, means division by a factor of 1000. A milliliter (mL) is equal to 0.001 liters.

$$1 \text{ mL} = 0.001 \text{ L}$$

The common metric prefixes are given in Table A.3.

A second advantage of the metric system is the link between the base units of length and volume. By definition, a liter is equal to the volume of a cube exactly

Table A.2
The Fundamental Units of the Metric System

Length: meter (m)	
1 m = 1.094 yd	1 yd = 0.9144 m
Volume: liter (L)	
1 L = 1.057 qt	1 qt = 0.9464 L
Mass: gram (g)	
1 g = 0.002205 lb	1 lb = 453.6 g

Table A.3
Metric System Prefixes

Prefix	Symbol	Meaning
femto-	f	$\times$ 1/1,000,000,000,000,000 (10^{-15})
pico-	p	$\times$ 1/1,000,000,000,000 (10^{-12})
nano-	n	$\times$ 1/1,000,000,000 (10^{-9})
micro-	μ	$\times$ 1/1,000,000 (10^{-6})
milli-	m	$\times$ 1/1,000 (10^{-3})
centi-	c	$\times$ 1/100 (10^{-2})
deci-	d	$\times$ 1/10 (10^{-1})
kilo-	k	$\times$ 1,000 (10^{3})
mega-	M	$\times$ 1,000,000 (10^{6})
giga-	G	$\times$ 1,000,000,000 (10^{9})
tera-	T	$\times$ 1,000,000,000,000 (10^{12})

10 cm tall, 10 cm long, and 10 cm wide. Because the volume of that cube is 1000 cubic centimeters and a liter contains 1000 milliliters, 1 milliliter is equivalent to 1 cubic centimeter.

$$1 \text{ mL} = 1 \text{ cm}^3$$

The third advantage of the metric system is the link between the base units of volume and weight. The gram was originally defined as the mass of 1 mL of water at 4°C. (It is important to specify the temperature because water expands or contracts as the temperature changes.)

SI UNITS OF MEASUREMENT

A series of international conferences on weights and measures has been held periodically since 1875 to refine the metric system. At the 11th conference, in 1960, a new system of units known as the **International System of Units** (abbreviated **SI** in all languages) was proposed as a replacement for the metric system. The seven base units for the SI system are given in Table A.4.

Table A.4
SI Base Units

Physical Quantity	Name of Unit	Symbol
Length	meter	m
Mass	kilogram	kg
Time	second	s
Temperature	kelvin	K
Electric current	ampere	A
Amount of substance	mole	mol
Luminous intensity	candela	cd

DERIVED SI UNITS

The units of every measurement in the SI system, no matter how simple or complex, should be derived from one or more of the seven base units. The preferred unit for volume is the cubic meter, for example, because volume has units of length cubed and the

SI unit for length is the meter. The preferred unit for speed is meters per second because speed is the distance traveled divided by the time it takes to cover this distance.

SI unit of volume: m^3 SI unit of speed: m/s

Some of the common derived SI units are given in Table A.5.

Table A.5

Common Derived SI Units in Chemistry

Physical Quantity	Name of Unit	Symbol	
Density		kg/m^3	
Electric charge	coulomb	C	(1 C = 1 A · s)
Electric potential	volt	V	(1 V = 1 J/C)
Energy	joule	J	(1 J = 1 kg · m^2/s^2)
Force	newton	N	(1 N = 1 kg · m/s^2)
Frequency	hertz	Hz	(1 Hz = 1 s^{-1})
Pressure	pascal	Pa	(1 Pa = 1 N/m^2)
Velocity (speed)	meters per second	m/s	
Volume	cubic meter	m^3	

NON-SI UNITS

Strict adherence to SI units would require changing directions such as "add 250 mL of water to a 1-L beaker" to "add 0.00025 cubic meters of water to an 0.001-m^3 container." Because of this, a number of units that are not strictly acceptable under the SI convention are still in use. Some of the non-SI units are given in Table A.6.

Table A.6

Non-SI Units in Common Use

Physical Quantity	Name of Unit	Symbol	
Volume	liter	L	(1 L = 1 × 10^{-3} m^3)
Length	angstrom	Å	(1 Å = 0.1 nm)
Pressure	atmosphere	atm	(1 atm = 101.325 kPa)
	torr	mmHg	(1 mmHg = 133.32 Pa)
Energy	electron volt	eV	(1 eV = 1.602 × 10^{-19} J)
Temperature	degree Celsius	°C	($T_C = T_K - 273.15$)
Concentration	molarity	*M*	(1 *M* = 1 mol/L)

A.2 Uncertainty in Measurement

There is a fundamental difference between stating that there are 12 inches in a foot and stating that the circumference of the earth at the equator is 24,903.01 miles. The first relationship is based on a *definition.* By convention, there are exactly 12 inches in 1 foot. The second relationship is based on a *measurement.* It reports the circumference of the earth to within the limits of experimental error in an actual measurement.

Many unit factors are based on definitions. There are exactly 5280 feet in a mile and 2.54 centimeters in an inch, for example. Unit factors based on definitions are known with complete certainty. (There is no error or uncertainty associated with the numbers.) Measurements, however, are always accompanied by a finite amount of error or uncertainty, which reflects limitations in the techniques used to make them.

The first measurement of the circumference of the earth, in the third century B.C., for example, gave a value of 250,000 stadia, or 29,000 miles. As the quality of the instruments used to make the measurement improved, the amount of error gradually decreased. But it never disappeared. Regardless of how carefully measurements are made, they always contain an element of uncertainty.

SYSTEMATIC AND RANDOM ERRORS

There are two sources of error in a measurement: (1) limitations in the sensitivity of the instruments used and (2) imperfections in the techniques used to make the measurement. These errors can be divided into two classes: systematic and random.

The idea of **systematic error** can be understood in terms of the bull's-eye analogy shown in Figure A.1. Imagine what would happen if you aimed at a target with a rifle whose sights were not properly adjusted. Instead of hitting the bull's-eye, you would systematically hit the target at another point. Your results would be influenced by a systematic error caused by an imperfection in the equipment being used.

Systematic error can also result from mistakes the individual makes while taking the measurement. In the bull's-eye analogy, a systematic error of this kind might occur if you flinched and pulled the rifle toward you each time it was fired.

To understand **random error,** imagine what would happen if you closed your eyes for an instant just before you fired the rifle. The bullets would hit the target more or less randomly. Some would hit too high, others would hit too low. Some would hit too far to the right, others too far to the left. Instead of an error that systematically gives a result too far in one direction, you now have a random error with random fluctuations.

Random errors most often result from limitations in the equipment or techniques used to make a measurement. Suppose, for example, that you wanted to collect 25 mL of a solution. You could use a beaker, a graduated cylinder, or a buret. Volume measurements made with a 50-mL beaker are accurate to within ±5 mL. In other words, you would be as likely to obtain 20 mL of solution (5 mL too little) as 30 mL (5 mL too much). You could decrease the amount of error by using a graduated cylinder that is capable of measurements to within ±1 mL. The error could be decreased even further by using a buret, which is capable of delivering a volume to within 1 drop, or ±0.05 mL.

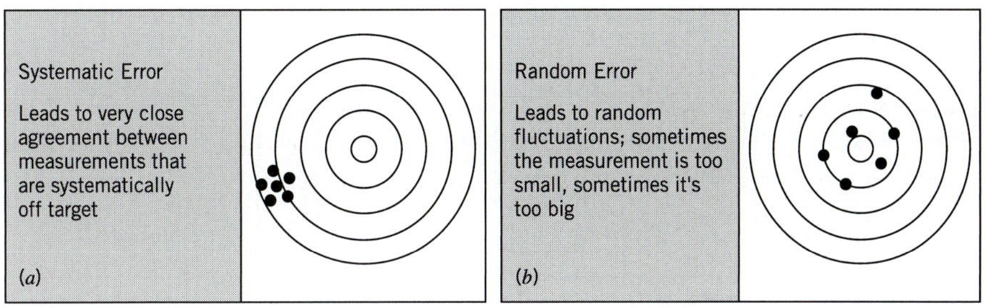

Fig. A.1 (*a*) Systematic errors give results that are systematically too small or too large. (*b*) Random errors give results that fluctuate between being too small and too large.

Exercise A.1

Which of the following would lead to systematic errors, and which would produce random errors?

(a) Using a 1-qt milk carton to measure 1-L samples of milk.

(b) Using a balance that is sensitive to ±0.1 g to obtain 250-mg samples of vitamin C.

Solution

Procedure (a) would result in a systematic error. The volume would always be too small because a quart is slightly smaller than a liter. Procedure (b) would produce a random error because the equipment used to make the measurement is not sensitive enough.

ACCURACY AND PRECISION

To most people, *accuracy* and *precision* are synonyms, words that have the same or nearly the same meaning. In the physical sciences, there is an important difference between the terms. Measurements are **accurate** when they agree with the true value of the quantity being measured. They are **precise** when individual measurements of the same quantity agree.

The difference between accuracy and precision is similar to the difference between two terms from statistics: *validity* and *reliability.* Accuracy means the same thing as validity. A measurement is accurate, or valid, only if we get the correct answer. Precision is the same as reliability. A measurement is precise, or reliable, if we get essentially the same result each time we make the measurement. Measurements are therefore precise when they are reproducible.

To understand how systematic and random errors affect accuracy and precision, let's return to the bull's-eye analogy (see Figure A.2). Systematic errors influence the accuracy, but not the precision, of a measurement. It is possible to get measurements that are consistently the same, but systematically wrong. Random errors influence both the accuracy and the precision of the measurement.

Systematic errors can be reduced by increasing the care and patience of the individual making the measurement or by improving the accuracy of the equipment used. Random errors can be reduced by averaging the results of many measurements of the same quantity because the precision of a series of measurements increases with the square root of the number of measurements.

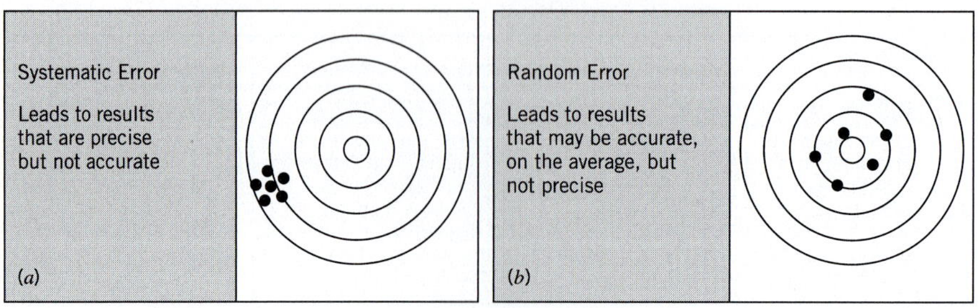

Fig. A.2 (*a*) Systematic errors affect the accuracy of a measurement. The measurement may still be precise, but the results are systematically different from the correct answer. (*b*) Random errors can affect both the accuracy and precision of a measurement.

A.3 Significant Figures

It is important to be honest when reporting a measurement, so that it doesn't appear more accurate than the equipment used to make the measurement allows. We can achieve this goal by controlling the number of digits, or **significant figures,** used to report the measurement.

Imagine what would happen if you determined the mass of an old copper penny on a postage scale and then on an analytical balance. The postage scale might give a mass of about 3 g, which means that the penny is closer to 3 g than either 2 g or 4 g. The analytical balance is inherently more sensitive; it can measure the mass of an object to the nearest ±0.001 g, in which case you might find that the penny has a mass of 2.531 g.

Postage scale	3 ± 1 g
Analytical balance	2.531 ± 0.001 g

The postage scale gave a measurement that had only one reliable digit. That measurement is therefore said to be good to only one significant figure. The analytical balance gave four significant figures (2.531). The number of significant figures in a measurement is the number of digits that are known with some degree of confidence (2, 5, and 3) plus the last digit (1), which is generally an estimate or approximation. As we improve the sensitivity of the equipment used to make a measurement, the number of significant figures increases.

At first glance, it might seem that we can determine the number of significant figures by simply counting the digits in the measurement. Unfortunately zeros represent a problem. Zeros in a number can be classified in three ways: leading zeros, trailing zeros, and zeros between two significant figures.

- Leading zeros are never significant. Leading zeros are used only to set the position of the decimal point. Consider the numbers 0.0045, 0.045, 0.45, 4.5, and 45. In each case the relative error is the same. The number is known to an accuracy of ±1 part in 45. Therefore, there are two significant figures in all five of these numbers.
- Zeros between two significant figures are always significant. In the number 3105 the zero is between 1 and 5, both of which are significant. Therefore, the zero is also significant, giving a total of four significant figures. The numbers 40.05, 0.0102, and 1706.2 have four, three, and five significant figures, respectively.
- Trailing zeros that are not needed to hold the decimal point are significant. For example 4.00 has three significant figures because the trailing zeros are not necessary to show the position of the decimal. Their function is to show that the number is known to 4.00 ± 0.01. Trailing zeros where there is no decimal point present a problem. It is often unclear how many zeros are significant. For example, does the measurement of 400 mL represent 400 ± 1 mL, 400 ± 10 mL, or 400 ± 100 mL? The only way to clearly show the number of significant figures in a measurement like this is to express the number in scientific notation. The measurement 400 mL can be written as 4.00×10^2 (three significant figures), 4.0×10^2 (two significant figures), or 4×10^2 (one significant figure) depending on the degree to which the measurement is known.

ADDITION AND SUBTRACTION WITH SIGNIFICANT FIGURES

What is the mass of a solution prepared by adding 0.507 g of salt to 150.0 g of water? If we attacked the problem without considering significant figures, we would simply add the two measurements.

$$\begin{array}{r l}
150.0 & \text{g } H_2O \\
0.507 & \text{g salt} \\
\hline
150.507 & \text{g solution}
\end{array} \qquad \text{(without using significant figures)}$$

But this answer doesn't make sense. We only know the mass of the water to the nearest tenth of a gram, so we can only know the total mass of the solution to within ± 0.1 g. Taking significant figures into account, we find that adding 0.507 g of salt to 150.0 g of water gives a solution with a mass of 150.5 g.

$$\begin{array}{r l}
150.0 & \text{g } H_2O \\
0.507 & \text{g salt} \\
\hline
150.5 & \text{g solution}
\end{array} \qquad \text{(using significant figures)}$$

Many of the calculations in this book are done by combining measurements with different degrees of accuracy and precision. The guiding principle in carrying out the calculations can be stated for addition and subtraction as follows:

When measurements are added or subtracted, the number of significant figures to the right of the decimal in the answer is determined by the measurement with the fewest digits to the right of the decimal.

MULTIPLICATION AND DIVISION WITH SIGNIFICANT FIGURES

For multiplication and division we count the total number of significant figures in each measurement, not just the number of decimal places.

When measurements are multiplied or divided, the answer can contain no more total significant figures than the measurement with the fewest total number of significant figures.

To illustrate this rule, let's calculate the cost of the copper in one of the old pennies that is pure copper. Let's assume that the penny has a mass of 2.531 g, that it is essentially pure copper, and that the price of copper is 67 cents per pound. We can start by converting from grams to pounds.

$$2.531\ \cancel{\text{g}} \times \frac{1.000\ \text{lb}}{453.6\ \cancel{\text{g}}} = 0.005580\ \text{lb}$$

We then use the price of a pound of copper to calculate the cost of the copper metal.

$$0.005580\ \cancel{\text{lb}} \times \frac{67\ \text{¢}}{1.0\ \cancel{\text{lb}}} = 0.37\ \text{¢}$$

There are four significant figures in both the mass of the penny (2.531) and the number of grams in a pound (453.6). But there are only two significant figures in the price of copper, so the final answer can only have two significant figures.

THE DIFFERENCE BETWEEN MEASUREMENTS AND DEFINITIONS

Assume that you wanted to calculate the length in inches of a piece of wood 1.245 feet long. The calculation is easy to set up.

$$1.245\ \cancel{\text{ft}} \times \frac{12\ \text{in.}}{1\ \cancel{\text{ft}}} = 14.94\ \text{in.}$$

It is not so easy to decide how many significant figures the final answer should have. The original measurement (1.245 ft) has four significant figures, but there seem to be only two significant figures in the number of inches in a foot. Thus, it might seem that the answer should contain only two significant figures.

We can clear up this confusion by remembering that some unit factors are based on definitions. For example, 1 foot is defined as exactly 12 inches. Unit factors based on definitions have an infinite number of significant figures. The answer to this problem therefore contains four significant figures.

ROUNDING OFF

When the answer to a calculation contains too many significant figures, it must be rounded off. Assume that the answer to a calculation is 1.247 and that the least accurate measurement has only three significant figures. The simplest way to round off this number would be to ignore the final digit and report the first three: 1.24. This approach has the disadvantage of introducing a systematic error into our calculations. Each time we round off, we would underestimate the value of the final answer.

There are 10 digits that can occur in the last decimal place in a calculation. One way of rounding off involves underestimating the answer for five of the digits (0, 1, 2, 3, and 4) and overestimating the answer for the other five (5, 6, 7, 8, and 9). This approach to rounding off is summarized as follows.

- If the digit is smaller than 5, drop the digit and leave the remaining number unchanged. Thus, 1.684 becomes 1.68.
- If the digit is 5 or larger, drop the digit and add 1 to the preceding digit. Thus, 1.247 becomes 1.25.

A.4 Scientific Notation

Chemists routinely work with numbers that are extremely small. (The mass of an electron, for example, is 0.000,000,000,000,000,000,000,000,000,911 g.) They also work with numbers that are extremely large. (There are 10,300,000,000,000,000,000,000 carbon atoms in a 1-carat diamond.) There isn't a calculator made that will accept either of these numbers as they are written here. Before we can use these numbers, it is necessary to convert them to **scientific notation,** that is, to convert them to a number between 1 and 10 multiplied by 10 raised to some exponent.

Before we discuss how to translate numbers into scientific notation, it may be useful to review some of the basics of exponential mathematics.

- A number raised to the zero power is equal to 1.

$$10^0 = 1$$

- A number raised to the first power is equal to itself.

$$10^1 = 10$$

- A number raised to the nth power is equal to the product of that number times itself $n - 1$ times.

$$10^5 = 10 \times 10 \times 10 \times 10 \times 10 = 100{,}000$$

- Dividing by a number raised to some exponent is the same as multiplying by that number raised to an exponent of the opposite sign.

$$\frac{Z}{10^2} = Z \times 10^{-2} \qquad \frac{Z}{10^{-3}} = Z \times 10^3$$

The following rule can be used to convert numbers into scientific notation:

> **The exponent in scientific notation is equal to the number of times the decimal point must be moved to produce a number between 1 and 10.**

According to the 1990 census, the population of Chicago was 6,070,000 $\pm$ 1000. Note that the population is known only to four significant figures because the error in the census is $\pm$1000. To convert the number to scientific notation we move the decimal point to the left six times and use the correct number of significant figures.

$$6070000 \pm 1000 = 6.070 \pm 0.001 \times 10^6$$

Atoms are very small, and therefore there are an enormous number of atoms in a small quantity of a given substance. In 1.0 gram of carbon we can calculate that there are 50,140,000,000,000,000,000,000 atoms. How can this number be reported to the correct number of significant figures in scientific notation? First we note that 1.0 gram contains only two significant figures. Thus the number of carbon atoms can be given to only two digits. The decimal point is moved to the left 22 times, and we write

$$50{,}140{,}000{,}000{,}000{,}000{,}000{,}000 = 5.0 \times 10^{22}$$

To convert numbers smaller than 1 into scientific notation, we have to move the decimal point to the right. The decimal point in 0.000985, for example, must be moved to the right four times. There are only three significant figures in this number so the number written in scientific notation can contain only three digits.

$$0.000985 = 9.85 \times 10^{-4}$$

Converting 0.000,000,000,000,000,000,000,000,000,911 g per electron into scientific notation involves moving the decimal point to the right 28 times.

$$0.000000000000000000000000000911 = 9.11 \times 10^{-28}$$

The mass of an electron was therefore listed in Table 1.3 as 9.11×10^{-28} grams.

The primary reason for converting numbers into scientific notation is to make calculations with unusually large or small numbers less cumbersome. But there is another important advantage to scientific notation. Because zeros are no longer used to set the decimal point, all of the digits in a number in scientific notation are significant, as shown by the following examples:

1.03×10^{22}	three significant figures
9.852×10^{-5}	four significant figures
2.0×10^{-23}	two significant figures

Exercise A.2

Convert the following numbers into scientific notation.
(a) 0.004694 (b) 1.98 (c) 4,679,000 ± 100

Solution

(a) 4.694×10^{-3} (b) 1.98×10^{0} (c) 4.6790×10^{6}

A.5 The Graphical Treatment of Data

If the basis of science is a natural curiosity about the world that surrounds us, an important step in doing science is trying to find patterns in the observations and measurements that result from this curiosity.

Anyone who has played with a prism, or seen a rainbow, has watched what happens when white light is split into a spectrum of different colors. The difference between the blue and red light in the spectrum is the result of differences in the frequencies and wavelengths of the light, as shown in Table A.7.

There is an obvious pattern in the data: as the wavelength of the light becomes larger, the frequency becomes smaller. But recognizing this pattern is not enough. It would be even more useful to construct a mathematical equation that fits the data. This would allow us to calculate the frequency of light of any wavelength, such as blue-green light with a wavelength of 5.00×10^{-7} m, or to calculate the wavelength of light of a known frequency, such as blue-violet light with a frequency of 7.00×10^{14} cycles per second.

The first step toward constructing an equation that fits the data involves plotting the data in different ways until we get a straight line. We might decide, for example, to plot the wavelengths on the vertical axis and the frequencies on the horizontal axis, as shown in Figure A.3. When we construct a graph, we should keep the following points in mind.

- The scales of the graph should be chosen so that the data fill as much of the available space as possible.
- It isn't necessary to include the origin (0,0) on the graph. In fact, it may be more efficient to leave off the origin, so that the data fill the available space.
- Once the scales have been chosen and labeled, the data are plotted one point at a time.

Table A.7

Characteristic Wavelengths (λ) and Frequencies (ν) of Light of Different Colors

Color	Wavelength (m)	Frequency (s^{-1})
Violet	4.100×10^{-7}	7.312×10^{14}
Blue	4.700×10^{-7}	6.379×10^{14}
Green	5.200×10^{-7}	5.765×10^{14}
Yellow	5.800×10^{-7}	5.169×10^{14}
Orange	6.000×10^{-7}	4.997×10^{14}
Red	6.500×10^{-7}	4.612×10^{14}

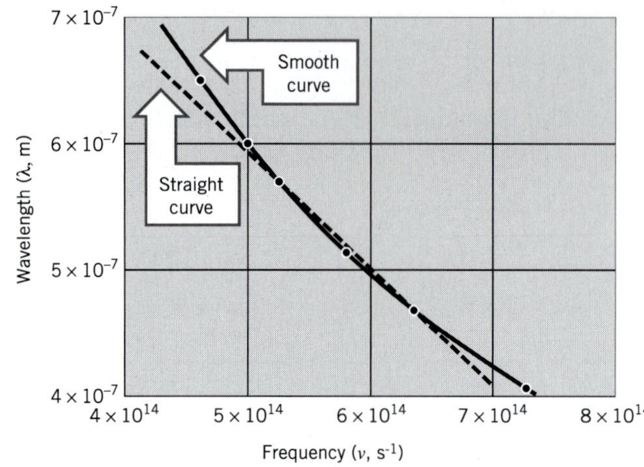

Fig. A.3 Plot of wavelength (λ) in meters versus frequency (ν) in cycles per second for light of different colors. Note that the curve is almost, but not quite, a straight line.

- A straight line or a smooth curve is then drawn through as many points as possible. Because of experimental error, the line or curve may not pass through every data point.

If this process gives a straight line, we can conclude that the quantity plotted on the vertical axis (λ) is **directly proportional** to the quantity on the horizontal axis (ν). We can then fit the graph to the equation for a straight line: $y = mx + b$.

$$\lambda = m\nu + b$$

The graph in Figure A.3 is not quite a straight line. Because the data are known to four significant figures, the deviation from a straight line is not the result of experimental error. We must therefore conclude that the wavelength and frequency of light are not directly proportional. We might therefore test whether the two sets of measurements are **inversely proportional.** In other words, we might try to fit the data to the following equation.

$$\lambda = m\left(\frac{1}{\nu}\right) + b$$

We can test this relationship by plotting the wavelengths in Table A.7 versus the inverse of the frequencies, as shown in Figure A.4.

The graph in Figure A.4 gives a beautiful straight-line relationship. We can now calculate the slope of the line (m) and the y intercept (b), as shown in Figure A.5. When doing the calculation, it is important to choose points that are *not* original data points. When this is done, the slope of the line in Figure A.4 is found to be equal to the speed of light: 2.998×10^8 meters per second.

When the graph includes the origin, the y-intercept can be determined by reading the value on the vertical axis that corresponds to a value of zero on the horizontal axis. This isn't possible for the data in Figure A.4, however, because the origin was left off the graph. Another approach to determining the intercept starts by selecting a point on the straight line. We then combine values of y and x read from the plot for this point with the slope of the line to calculate the value of the intercept. When this approach is applied to the data in Figure A.4, we find that the intercept is equal to zero. The data in Figure A.4 therefore fit the following equation.

$$\lambda = (2.998 \times 10^8 \text{ m/s})\left(\frac{1}{\nu}\right)$$

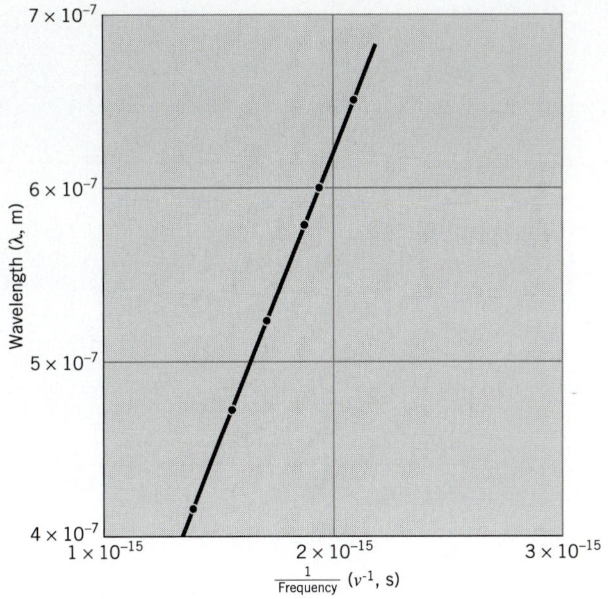

Fig. A.4 A plot of wavelength (λ) versus the inverse of frequency ($1/\nu$) for light of different colors gives a straight line, within experimental error.

Rearranging the equation, we find that the product of the frequency times the wavelength of light is equal to the speed of light.

$$\nu\lambda = 2.998 \times 10^8 \text{ m/s}$$

Exercise A.3

The following data were obtained from a study of the relationship between the volume (V) and temperature (T) of a gas at constant pressure.

Volume (mL)	273.0	277.4	282.7	287.8	293.1	298.1
Temperature (°C)	0.0	5.0	10.0	15.0	20.0	25.0

Determine the temperature at which the volume of the gas should become equal to zero.

Solution

A plot of the data gives a straight line, as shown in Figure A.6. The equation for the straight line can be written as follows.

$$T = mV + b$$

By choosing any two points on the curve, such as $T = 7.5°\text{C}$ and $T = 22.5°\text{C}$, and reading the volumes at those temperatures from the straight line that passes through the points, we can calculate the slope of the line.

$$m = \frac{\Delta T}{\Delta V} = \frac{22.5°\text{C} - 7.5°\text{C}}{295.5 \text{ mL} - 280.1 \text{ mL}} = 0.974°\text{C/mL}$$

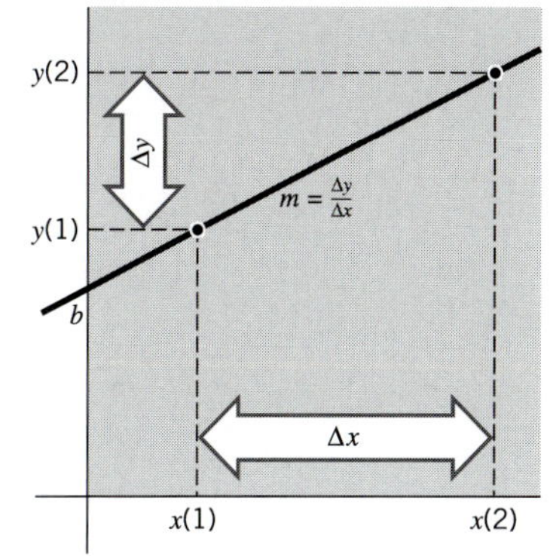

Fig. A.5 To calculate the slope (m) of a straight line, divide the distance between two points on the vertical axis by the corresponding distance between two points along the horizontal axis. The y intercept (b) can be determined by reading the value on the vertical axis that corresponds to a value of zero on the horizontal axis.

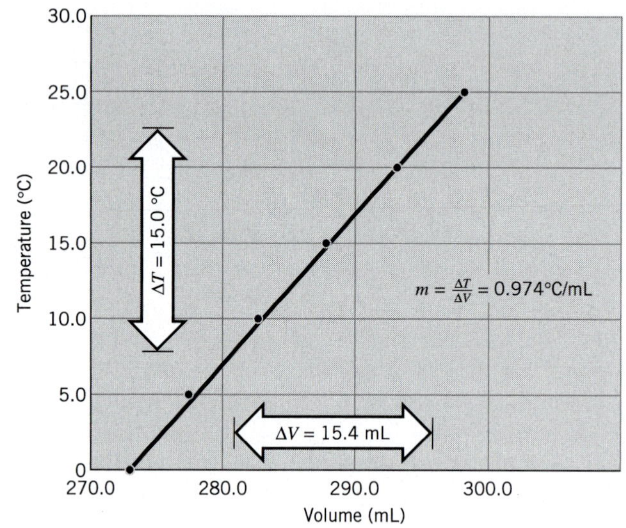

Fig. A.6 Plot of the temperature versus volume data from Exercise A.3.

The value of b for the equation can be calculated from the data for any point on the straight line, such as the point at which the temperature is 22.5°C and the volume is 295.5 mL.

$$b = T - mV = 22.5°\text{C} - \frac{0.974°\text{C}}{1\ \cancel{\text{mL}}}(295.5\ \cancel{\text{mL}}) = -265°\text{C}$$

According to these data, the volume of the gas should become zero when the gas is cooled until the temperature reaches about −265°C.

A.6 Significant Figures and Unit Conversion Worksheet

SIGNIFICANT FIGURES

One way in which a scientist can indicate the quality of a measurement or the degree to which a piece of equipment allows a measurement to be made is by the use of significant figures. Consider the measurement 32.45 g. The digits in 32.4 are known with certainty while there is uncertainty associated with the 0.05. The actual value might be 32.47 g, 32.44 g, or some other value that deviates in the second digit after the decimal. The actual value of the last reported significant figure is uncertain. All other significant figures in a measurement are assumed to be known with certainty. Thus in 32.45 there are four significant figures and the last digit is uncertain. You should use correct significant figures when recording data and reporting results in the laboratory and when working homework and test questions.

The uncertainty associated with many measurements results from estimating between the smallest divisions on a measuring device. Measurements should always be recorded to show the digits that are known with certainty (those digits corresponding to a division on the measuring device) plus one digit that is uncertain (corresponding to the estimation between the smallest divisions).

1. Figure A.7 shows a diagram of three volumes of water measured in 100 mL graduated cylinders. Record each measurement to the correct number of

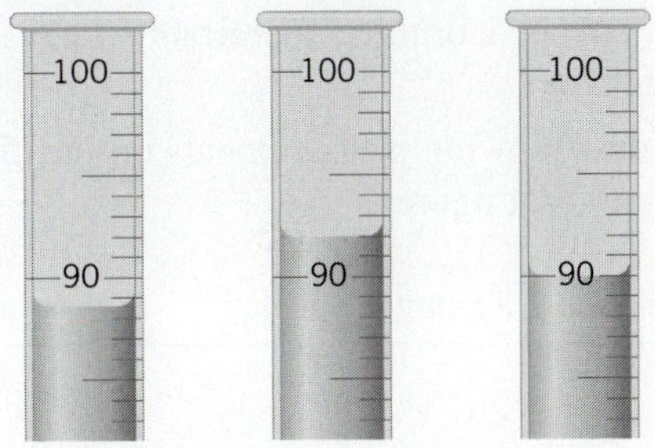

Always read from the bottom of the meniscus.

Fig. A.7 Record the volume of liquid in the graduated cylinder.

significant figures. A graduated cylinder is always read from the bottom of the meniscus.

Counting Significant Figures in a Measurement. Expressing numbers in scientific notation can help you count the number of significant figures. The exponential portion of a number expressed in scientific notation is not used in counting significant figures.

If all of the numbers are nonzero, the number of digits equals the number of significant figures.

Measurement	Number of Significant Figures
1426.5	5
32.4561	6

Leading zeros are never significant. They are used only to set the position of the decimal point.

Measurement	Scientific Notation	Significant Figures
0.369	3.69×10^{-1}	3
0.00029	2.9×10^{-4}	2
0.008957	8.957×10^{-3}	4

Zeros between significant figures are always significant.

Measurement	Scientific Notation	Significant Figures
104.56	1.0456×10^{2}	5
0.002305	2.305×10^{-3}	4

Trailing zeros that are *not* needed to set the position of the decimal point are significant. Their function is to show how accurately the measurement is known.

Measurement	Scientific Notation	Significant Figures
12.30	1.230×10^{1}	4
5.00	5.00×10^{0}	3
230.0	2.30×10^{2}	3

Trailing zeros that have no decimal point do not clearly show the number of significant figures. These measurements are best written in scientific notation. For example the measurement 700 milliliters could represent 7×10^2 mL, 7.0×10^2 mL, or 7.00×10^2 mL (i.e., one, two, or three significant figures, depending on how the measurement was made).

2. Express each of the following measurements in scientific notation and count the number of significant figures.

Measurement	Scientific Notation	Significant Figures
98.30		
104.02		
0.00285		
100.03		
0.04340		

Significant Figures in Calculations. It is important to maintain the proper number of significant figures when doing calculations. The number of digits shown in the result must correctly reflect the significant figures associated with each measurement used in the calculation.

Addition and Subtraction: In addition and subtraction the answer is restricted to the number of digits *after the decimal.* The measurement with the least number of digits *after the decimal* determines the number of digits after the decimal in the answer.

140.15	(two digits after the decimal)
34.4129	(four digits after the decimal)
2032.1	(one digit after the decimal)
2206.6629	(calculator answer)

The answer in correct significant figures is 2206.7 (limited to one digit after the decimal). Note that the last digit needed to be rounded up to 7 since the portion that was dropped (0.0629) was greater than 0.0500.

Multiplication and Division: In multiplication and division the answer can have no more total significant figures than the measurement with the least number of significant figures.

$$12.34 \times 0.0203 \times 25.673 = 6.431137846$$

Significant figures (4) (3) (5) (calculator answer)

The answer in correct significant figures is 6.43 (limited to three significant figures).

Calculations with Multiple Steps: In calculations with multiple steps it is necessary to look at the number of significant figures in the answer to each step.

3. Explain why there are only three significant figures in the answer for the following calculation.

$$\frac{(123.4 + 0.42)}{(17.48 - 2.8)} = \frac{123.8}{14.7} = 8.42$$

4. Explain why there are only two significant figures in the answer.

$$\frac{(123.4 + 0.42)}{(17.48 - 17.00)} = \frac{123.8}{0.48} = 2.6 \times 10^2$$

Measurements versus Definitions

5. Use a metric rule to find the diameter of the following circle to two significant figures (for this measurement, do not estimate between the mm marks).

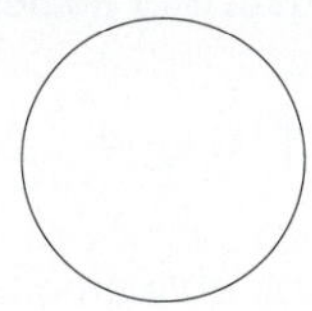

Diameter = ______ Number of significant figures = __2__

6. Now use the definition of radius (radius = diameter/2) to find the radius of the circle using the measurement you made of diameter. Maintain significant figures as described in the previous sections.

Radius = ______

Number of significant figures in numerator = ______

Number of significant figures in denominator = ______

Number of significant figures in radius = ______

7. Repeat the measurement of the diameter of the circle. This time, record the measurement to three significant figures (estimate between the millimeter marks). Now determine the radius of the circle while maintaining significant figures as described above.

Diameter = ______

Number of significant figures in numerator = ______

Number of significant figures in denominator = ______

Number of significant figures in radius = ______

Radius = ______

8. Compare the number of significant figures in the two values of the radius. Does the number of significant figures in the radii that you calculated in steps 6 and 7 differentiate between the number of significant figures in the two measurements that you made?

The purpose of significant figures is to reflect the quality of a **measurement.** The problem with the radii that you calculated is that the 2 in the denominator of the radius formula is not a measurement. It instead is a definition of radius. The 2 can be assumed to have an infinite number of significant figures. This also applies to counting numbers such as 24 people in a class. There could not be 24.5 or 23.9 people in a class. Numbers in a calculation that are not measurements (definitions and counting numbers) can be considered to have an infinite number of significant figures.

9. Now determine the radius of the circle for both measurements (steps 6 and 7), assuming that the 2 in the denominator has an infinite number of significant figures.

Measurement 1:

Radius = ______ Number of significant figures = ______

Measurement 2:

Radius = ______ Number of significant figures = ______

UNIT CONVERSIONS

When conversion factors are given (as in Table B.2 in Appendix B), you must determine the appropriate number of significant figures to use in converting a measurement to another unit.

$$1 \text{ kg} = 2.2046 \text{ lb}$$

This relationship is the result of a measurement. It therefore has a limited number of significant figures. You may assume that the number of significant figures associated with the 1 kg is equal to the number of significant figures in the pound measurement. This relationship can be determined by determining the weight in pounds of a 1.00000-kg mass using a balance or scale.

$$1.00 \text{ kg} = 2.20 \text{ lb}$$

Using a better balance or scale can give a better relationship.

$$1.0000 \text{ kg} = 2.2046 \text{ lb}$$

Some conversion factors are definitions and therefore have an infinite number of significant figures. Below are examples of two relationships between units that are based on definitions, not measurements.

$$1 \text{ ft} = 12 \text{ inches} \quad \text{or} \quad 1.0000\ldots \text{ ft} = 12.0000\ldots \text{ inches}$$
$$1 \text{ inch} = 2.54 \text{ cm (by definition)} \quad \text{or} \quad 1.000\ldots \text{ in.} = 2.54000\ldots \text{ cm}$$

To convert 14 lb into kilograms:

$$14\ lb \times \frac{1\ kg}{2.2046\ lb} = 6.4\ kg$$

To convert 14.0000 lb into kilograms:

$$14.0000\ lb \times \frac{1\ kg}{2.2046\ lb} = 6.3504\ kg$$

To convert 58 cm into inches:

$$58\ cm \times \frac{1\ inch}{2.54\ cm} = 23\ inches$$

To convert 58.0000 cm into inches:

$$58.0000\ cm \times \frac{1\ inch}{2.54\ cm} = 22.8346\ inches$$

10. Explain the difference between the number of significant figures found in each of the previous four conversions.

11. Perform the following calculations and express the result in correct significant figures.
(a) Convert 18.9 inches into centimeters
(b) Convert 47.50 kg into pounds.
(c) Convert 35.56 inches into centimeters.
(d) Convert 92.3 cm into feet.

Appendix B

TABLES

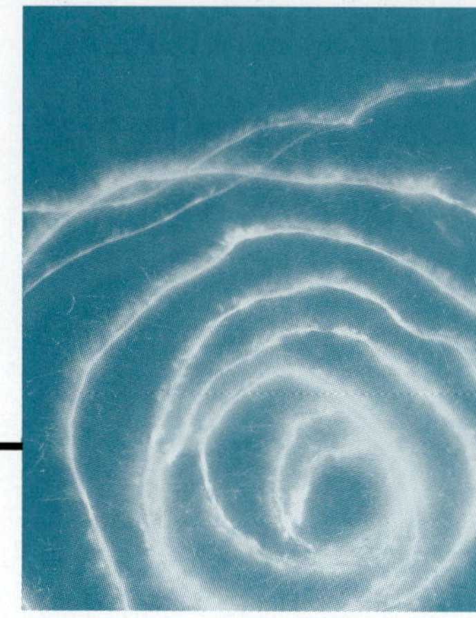

Table B.1
Values of Selected Fundamental Constants

Speed of light in a vacuum (c)	$c = 2.99792458 \times 10^8$ m/s
Charge on an electron (q_e)	$q_e = 1.60217733 \times 10^{-19}$ C
Rest mass of an electron (m_e)	$m_e = 9.109389 \times 10^{-28}$ g
	$m_e = 5.485799 \times 10^{-4}$ amu
Rest mass of a proton (m_p)	$m_p = 1.672623 \times 10^{-24}$ g
	$m_p = 1.00727647$ amu
Rest mass of a neutron (m_n)	$m_n = 1.674928 \times 10^{-24}$ g
	$m_n = 1.0086649$ amu
Faraday's constant (F)	$F = 96{,}485$ C/mol
Planck's constant (h)	$h = 6.626075 \times 10^{-34}$ J · s
Ideal gas constant (R)	$R = 0.0820568$ L · atm/mol · K
	$R = 8.31451$ J/mol · K
Atomic mass unit (amu)	1 amu $= 1.6605402 \times 10^{-24}$ g
Boltzmann's constant (k)	$k = 1.380658 \times 10^{-23}$ J/K
Avogadro's constant (N)	$N = 6.0221367 \times 10^{23}$ mol^{-1}
Rydberg constant (R_H)	$R_H = 1.0973715 \times 10^7$ m^{-1}
	$= 1.0973715 \times 10^{-2}$ nm^{-1}

Table B.2
Selected Conversion Factors

Energy	1 J = 0.2390 cal = 10^7 erg
	1 cal = 4.184 J (by definition)
	1 eV/atom = $1.6021793 \times 10^{-19}$ J/atom = 96.485 kJ/mol
Temperature	K = °C + 273.15
	°C = 5/9(°F − 32)
	°F = 9/5(°C + 32)
Pressure	1 atm = 760 mmHg (by definition) = 760 torr (by definition)
	1 atm = 101.325 kPa = 14.7 lb/in^2
Mass	1 kg = 2.2046 lb
	1 lb = 453.59 g = 0.45359 kg
	1 oz = 0.06250 lb = 28.350 g
	1 ton = 2000 lb = 907.185 kg
	1 tonne (metric) = 1000 kg = 2204.62 lb
Volume	1 mL = 0.001 L = 1 cm^3 (by definition)
	1 oz (fluid) = 0.031250 qt = 0.029573 L
	1 qt = 0.9463529 L
	1 L = 1.05672 qt
Length	1 m = 39.370 in.
	1 mi = 1.60934 km
	1 in. = 2.54 cm (by definition)

Table B.3

The Vapor Pressure of Water

Temperature (°C)	Pressure (mmHg)	Temperature (°C)	Pressure (mmHg)	Temperature (°C)	Pressure (mmHg)	Temperature (°C)	Pressure (mmHg)
0	4.6	13	11.2	26	25.2	39	52.4
1	4.9	14	12.0	27	26.7	40	55.3
2	5.3	15	12.8	28	28.3	41	58.3
3	5.7	16	13.6	29	30.0	42	61.5
4	6.1	17	14.5	30	31.8	43	64.8
5	6.5	18	15.5	31	33.7	44	68.3
6	7.0	19	16.5	32	35.7	45	71.9
7	7.5	20	17.5	33	37.7	46	75.7
8	8.0	21	18.7	34	39.9	47	79.6
9	8.6	22	19.8	35	42.2	48	83.7
10	9.2	23	21.1	36	44.6	49	88.0
11	9.8	24	22.4	37	47.1	50	92.5
12	10.5	25	23.8	38	49.7		

Table B.4
Radii of Atoms and Ions

Element	Ionic Radius (nm)	Ionic Charge	Covalent Radius (nm)	Metallic Radius (nm)
Aluminum	0.050	(+3)	0.125	0.1431
Antimony	0.245	(−3)	0.141	
	0.09	(+3)		
	0.062	(+5)		
Arsenic	0.222	(−3)	0.121	0.1248
	0.058	(+3)		
	0.047	(+5)		
Astatine	0.227	(−1)		
	0.051	(+7)		
Barium	0.135	(+2)	0.198	0.2173
Beryllium	0.031	(+2)	0.089	0.1113
Bismuth	0.213	(−3)	0.152	0.1547
	0.096	(+3)		
	0.074	(+5)		
Boron	0.020	(+3)	0.088	0.083
Bromine	0.196	(−1)	0.1142	
	0.039	(+7)		
Cadmium	0.097	(+2)	0.141	0.1489
Calcium	0.099	(+2)	0.174	0.1973
Carbon	0.260	(−4)	0.077	
	0.015	(+4)		
Cesium	0.169	(+1)	0.235	0.2654
Chlorine	0.181	(−1)	0.099	
	0.026	(+7)		
Chromium	0.064	(+3)	0.117	0.1249
	0.052	(+6)		
Cobalt	0.074	(+2)	0.116	0.1253
	0.063	(+3)		
Copper	0.096	(+1)	0.117	0.1278
	0.072	(+2)		
Fluorine	0.136	(−1)	0.064	0.0717
	0.007	(+7)		
Francium	0.176	(+1)		0.27
Gallium	0.062	(+3)	0.125	0.1221
Germanium	0.272	(−4)	0.122	0.1225
	0.053	(+4)		
Gold	0.137	(+1)	0.134	0.1442
	0.091	(+3)		
Hydrogen	0.208	(−1)	0.0371	
	10^{-6}	(+1)		
Indium	0.081	(+3)	0.150	0.1626
Iodine	0.216	(−1)	0.1333	
	0.050	(+7)		
Iron	0.076	(+2)	0.1165	0.1241
	0.064	(+3)		
Lead	0.215	(−4)	0.154	0.1750
	0.120	(+2)		
	0.084	(+4)		
Lithium	0.068	(+1)	0.123	0.152
Magnesium	0.065	(+2)	0.136	0.160

Element	Ionic Radius (nm)	Ionic Charge	Covalent Radius (nm)	Metallic Radius (nm)
Manganese	0.080	(+2)	0.117	0.124
	0.046	(+7)		
Mercury	0.127	(+1)	0.144	0.160
	0.110	(+2)		
Molybdenum	0.062	(+6)	0.129	0.1362
Nickel	0.072	(+2)	0.115	0.1246
Nitrogen	0.171	(−3)	0.070	
	0.013	(+3)		
	0.011	(+5)		
Oxygen	0.140	(−2)	0.066	
	0.176	(−1)		
Phosphorus	0.212	(−3)	0.110	0.108
	0.042	(+3)		
	0.034	(+5)		
Polonium	0.230	(−2)	0.153	0.167
	0.056	(+6)		
Potassium	0.133	(+1)	0.2025	0.2272
Radium	0.140	(+2)		0.220
Rubidium	0.148	(+1)	0.216	0.2475
Scandium	0.081	(+3)	0.144	0.1606
Selenium	0.198	(−2)	0.117	
	0.069	(+4)		
	0.042	(+6)		
Silicon	0.271	(−4)	0.117	
	0.041	(+4)		
Silver	0.126	(+1)	0.134	0.1444
Sodium	0.095	(+1)	0.157	0.186
Strontium	0.113	(+2)	0.192	0.2151
Sulfur	0.184	(−2)	0.104	
	0.037	(+4)		
	0.029	(+6)		
Tellurium	0.221	(−2)	0.137	0.1432
	0.081	(+4)		
	0.056	(+6)		
Thallium	0.095	(+3)	0.155	0.1704
Tin	0.294	(−4)	0.140	0.1405
	0.102	(+2)		
	0.071	(+4)		
Titanium	0.090	(+2)	0.132	0.1448
	0.068	(+4)		
Tungsten	0.065	(+6)	0.130	0.1370
Uranium	0.083	(+6)		0.1385
Vanadium	0.059	(+5)	0.122	0.1321
Xenon			0.209	
Zinc	0.074	(+2)	0.125	0.1332
Zirconium	0.079	(+4)	0.145	0.167

Table B.5

Ionization Energies (kJ/mol)

Atomic Number	Symbol	I	II	III	IV	V	VI	VII
1	H	1,312.0						
2	He	2,372.3	5,250.3					
3	Li	520.2	7,297.9	11,814.6				
4	Be	899.4	1,757.1	14,848.3	21,005.9			
5	B	800.6	2,427.0	3,659.6	25,025.0	32,825.7		
6	C	1,086.4	2,352.6	4,620.4	6,222.5	37,829.4	47,275.6	
7	N	1,402.3	2,856.0	4,578.0	7,474.9	9,444.7	50,370.4	64,358.0
8	O	1,313.9	3,388.2	5,300.3	7,469.1	10,989.2	13,326.1	71,332
9	F	1,681.0	3,374.1	6,050.3	8,407.5	11,022.4	15,163.6	17,867.2
10	Ne	2,080.6	3,952.2	6,122	9,370	12,177	15,238	19,998
11	Na	495.8	4,562.4	6,912	9,543	13,352	16,610	20,114
12	Mg	737.7	1,450.6	7,732.6	10,540	13,629	17,994	21,703
13	Al	577.6	1,816.6	2,744.7	11,577	14,831	18,377	23,294
14	Si	786.4	1,577.0	3,231.5	4,355.4	16,091	19,784	23,785
15	P	1,011.7	1,903.2	2,912	4,956	6,273.7	21,268	25,397
16	S	999.58	2,251	3,361	4,564	7,012	8,495.4	27,105
17	Cl	1,251.1	2,297	3,822	5,158	6,540	9,362	11,017.9
18	Ar	1,520.5	2,665.8	3,931	5,771	7,238	8,780.8	11,994.9
19	K	418.8	3,051.3	4,411	5,877	7,975	9,648.5	11,343
20	Ca	589.8	1,145.4	4,911.8	6,474	8,144	10,496	12,320
21	Sc	631	1,235	2,389	7,099	8,844	10,720	13,310
22	Ti	658	1,310	2,652.5	4,174.5	9,573	11,516	13,590
23	V	650	1,413	2,828.0	4,506.5	6,294	12,362	14,489
24	Cr	652.8	1,592	2,987	4,740	6,690	8,738	15,540
25	Mn	717.4	1,509.0	3,248.3	4,940	6,990	9,220	11,508
26	Fe	759.3	1,561	2,957.3	5,290	7,240	9,600	12,100
27	Co	758	1,646	3,232	4,950	7,670	9,840	12,400
28	Ni	736.7	1,752.9	3,393	5,300	7,280	10,400	12,800
29	Cu	745.4	1,957.9	3,553	5,330	7,710	9,940	13,400
30	Zn	906.4	1,733.2	3,832.6	5,730	7,970	10,400	12,900
31	Ga	578.8	1,979	2,963	6,200			
32	Ge	762.1	1,537.4	3,302	4,410	9,020		
33	As	947	1,797.8	2,735.4	4,837	6,043	12,300	
34	Se	940.9	2,045	2,973.7	4,143.4	6,590	7,883	14,990
35	Br	1,139.9	2,100	3,500	4,560	5,760	8,550	9,938
36	Kr	1,350.7	2,350.3	3,565	5,070	6,240	7,570	10,710
37	Rb	403.0	2,632	3,900	5,070	6,850	8,140	9,570
38	Sr	549.5	1,064.5	4,120	5,500	6,910	8,760	10,200
39	Y	616	1,181	1,980	5,960	7,430	8,970	11,200
40	Zr	660	1,267	2,218	3,313	7,870		
41	Nb	664	1,382	2,416	3,960	4,877	9,899	12,100
42	Mo	684.9	1,558	2,621	4,480	5,910	6,600	12,230
43	Tc	702	1,472	2,850				
44	Ru	711	1,617	2,747				
45	Rh	720	1,744	2,997				
46	Pd	805	1,874	3,177				
47	Ag	731.0	2,073	3,361				
48	Cd	867.7	1,631.4	3,616				
49	In	558.3	1,820.6	2,704	5,200			
50	Sn	708.6	1,411.8	2,943.0	3,930.2	6,974		
51	Sb	833.7	1,595	2,440	4,260	5,400	10,400	

Atomic Number	Symbol	I	II	III	IV	V	VI	VII
52	Te	869.2	1,790	2,698	3,609	5,668	6,820	13,200
53	I	1,008.4	1,845.8	3,200				
54	Xe	1,170.4	2,046	3,100				
55	Cs	375.7	2,440					
56	Ba	502.9	965.23					
57	La	538.1	1,067	1,850.3				
58	Ce	527.8	1,047	1,949	3,547			
59	Pr	523	1,018	2,086	3,761	5,543		
60	Nd	530	1,035	2,130	3,900			
61	Pm	535	1,052	2,150	3,970			
62	Sm	543	1,068	2,260	3,990			
63	Eu	547	1,084	2,400	4,110			
64	Gd	592	1,167	1,990	4,250			
65	Tb	564	1,112	2,110	3,840			
66	Dy	572	1,126	2,200	4,000			
67	Ho	581	1,139	2,203	4,100			
68	Er	589	1,151	2,194	4,120			
69	Tm	596	1,163	2,285	4,120			
70	Yb	603	1,175	2,415	4,216			
71	Lu	524	1,340	2,022	4,360			
72	Hf	642	1,440	2,250	3,210			
73	Ta	761						
74	W	770						
75	Re	760	1,260	2,510	3,640			
76	Os	840						
77	Ir	880						
78	Pt	870	1,791.0					
79	Au	890.1	1,980					
80	Hg	1,007.0	1,809.7	3,300				
81	Tl	589.3	1,971.0	2,878				
82	Pb	715.5	1,450.4	3,081.4	4,083	6,640		
83	Bi	703.3	1,610	2,466	4,370	5,400	8,520	
84	Po	812						
85	At							
86	Rn	1,037.0						
87	Fr							
88	Ra	509.3	979.0					
89	Ac	498.8	1,170					
90	Th	587	1,110	1,930	2,780			
91	Pa	568						
92	U	584						
93	Np	597						
94	Pu	585						
95	Am	578.2						
96	Cm	581						
97	Bk	601						
98	Cf	608						
99	Es	619						
100	Fm	627						
101	Md	635						
102	No	642						

Table B.6
Electron Affinities

Atomic Number	Symbol	Electron Affinity (kJ/mol)	Atomic Number	Symbol	Electron Affinity (kJ/mol)
1	H	72.8	37	Rb	46.89
2	He	*	38	Sr	*
3	Li	59.8	39	Y	0
4	Be	*	40	Zr	50
5	B	27	41	Nb	96
6	C	122.3	42	Mo	96
7	N	−7	43	Tc	70
8	O	141.1	44	Ru	110
9	F	328.0	45	Rh	120
10	Ne	*	46	Pd	60
11	Na	52.7	47	Ag	125.7
12	Mg	*	48	Cd	*
13	Al	45	49	In	29
14	Si	133.6	50	Sn	121
15	P	71.7	51	Sb	101
16	S	200.42	52	Te	190.15
17	Cl	348.8	53	I	295.3
18	Ar	*	54	Xe	*
19	K	48.36	55	Cs	45.49
20	Ca	*	56	Ba	*
21	Sc	*	57–71	La–Lu	50
22	Ti	20	72	Hf	*
23	V	50	73	Ta	60
24	Cr	64	74	W	60
25	Mn	*	75	Re	14
26	Fe	24	76	Os	110
27	Co	70	77	Ir	150
28	Ni	111	78	Pt	205.3
29	Cu	118.3	79	Au	222.74
30	Zn	0	80	Hg	*
31	Ga	29	81	Tl	30
32	Ge	120	82	Pb	110
33	As	77	83	Bi	110
34	Se	194.96	84	Po	180
35	Br	324.6	85	At	270
36	Kr	*	86	Rn	*
			87	Fr	44.0

*These elements have negative electron affinities. Electron affinity is the negative of the enthalpy change for the reaction $X(g) + e^- \rightarrow X^-(g)$.

Table B.7
Electronegativities[a]

Atomic Number	Element	Electro-negativity	Atomic Number	Element	Electro-negativity
1	H	2.300	41	Nb	1.25
2	He	4.157	42	Mo	1.39
3	Li	0.912	43	Tc	1.52
4	Be	1.576	44	Ru	1.66
5	B	2.051	45	Rh	1.79
6	C	2.544	46	Pd	1.91
7	N	3.066	47	Ag	1.98
8	O	3.610	48	Cd	1.52
9	F	4.193	49	In	1.656
10	Ne	4.787	50	Sn	1.824
11	Na	0.869	51	Sb	1.984
12	Mg	1.293	52	Te	2.158
13	Al	1.613	53	I	2.359
14	Si	1.916	54	Xe	2.582
15	P	2.253	55	Cs	0.66
16	S	2.589	56	Ba	0.88
17	Cl	2.869	57–71		
18	Ar	3.242	72		
19	K	0.734	73		
20	Ca	1.034	74		
21	Sc	1.15	75		
22	Ti	1.25	76		
23	V	1.37	77		
24	Cr	1.45	78		
25	Mn	1.55	79		
26	Fe	1.67	80	Hg	1.76
27	Co	1.76	81		
28	Ni	1.86	82		
29	Cu	1.83	83		
30	Zn	1.59	84		
31	Ga	1.756	85		
32	Ge	1.994	86		
33	As	2.211	87		
34	Se	2.424	88		
35	Br	2.685	89		
36	Kr	2.966	90		
37	Rb	0.706	91		
38	Sr	0.963	92		
39	Y	1.00	93–103		
40	Zr	1.12			

[a]Allen electronegativities taken from L. C. Allen, *Int. J. Quant. Chem.*, **48,** 253–277 (1993); L. C. Allen, *J. Am. Chem. Soc.*, **111,** 9003–9014 (1989).

Table B.8

Acid Dissociation Equilibrium Constants

Compound	Dissociation Reaction	K_a	pK_a
Acetic acid	$CH_3CO_2H + H_2O \rightleftharpoons CH_3CO_2^- + H_3O^+$	1.75×10^{-5}	4.757
Ammonium ion	$NH_4^+ + H_2O \rightleftharpoons NH_3 + H_3O^+$	5.6×10^{-10}	9.25
Arsenic acid	$H_3AsO_4 + H_2O \rightleftharpoons H_2AsO_4^- + H_3O^+$	6.0×10^{-3}	2.22
	$H_2AsO_4^- + H_2O \rightleftharpoons HAsO_4^{2-} + H_3O^+$	1.0×10^{-7}	7.00
	$HAsO_4^{2-} + H_2O \rightleftharpoons AsO_4^{3-} + H_3O^+$	3.0×10^{-12}	11.52
Arsenous acid	$H_3AsO_3 + H_2O \rightleftharpoons H_2AsO_3^- + H_3O^+$	6.0×10^{-10}	9.22
	$H_2AsO_3^- + H_2O \rightleftharpoons HAsO_3^{2-} + H_3O^+$	3.0×10^{-14}	13.52
Benzoic acid	$C_6H_5CO_2H + H_2O \rightleftharpoons C_6H_5CO_2^- + H_3O^+$	6.3×10^{-5}	4.20
Boric acid	$H_3BO_3 + H_2O \rightleftharpoons H_2BO_3^- + H_3O^+$	7.3×10^{-10}	9.14
Carbonic acid	$H_2CO_3 + H_2O \rightleftharpoons HCO_3^- + H_3O^+$	4.5×10^{-7}	6.35
	$HCO_3^- + H_2O \rightleftharpoons CO_3^{2-} + H_3O^+$	4.7×10^{-11}	10.33
Chloric acid	$HClO_3 + H_2O \rightleftharpoons ClO_3^- + H_3O^+$	5.0×10^{2}	−2.70
Chloroacetic acid	$ClCH_2CO_2H + H_2O \rightleftharpoons ClCH_2CO_2^- + H_3O^+$	1.4×10^{-3}	2.85
Chlorous acid	$HClO_2 + H_2O \rightleftharpoons ClO_2^- + H_3O^+$	1.1×10^{-2}	1.96
Chromic acid	$H_2CrO_4 + H_2O \rightleftharpoons HCrO_4^- + H_3O^+$	9.6	−0.98
	$HCrO_4^- + H_2O \rightleftharpoons CrO_4^{2-} + H_3O^+$	3.2×10^{-7}	6.50
Citric acid	$H_3Cit + H_2O \rightleftharpoons H_2Cit^- + H_3O^+$	7.5×10^{-4}	3.13
	$H_2Cit^- + H_2O \rightleftharpoons HCit^{2-} + H_3O^+$	1.7×10^{-5}	4.77
	$HCit^{2-} + H_2O \rightleftharpoons Cit^{3-} + H_3O^+$	4.0×10^{-7}	6.40
Dichloroacetic acid	$Cl_2CHCO_2H + H_2O \rightleftharpoons Cl_2CHCO_2^- + H_3O^+$	5.1×10^{-2}	1.29
Formic acid	$HCO_2H + H_2O \rightleftharpoons HCO_2^- + H_3O^+$	1.8×10^{-4}	3.75
Glycine	$H_3N^+CH_2CO_2H + H_2O \rightleftharpoons H_3N^+CH_2CO_2^- + H_3O^+$	4.5×10^{-3}	2.35
	$H_3N^+CH_2CO_2^- + H_2O \rightleftharpoons H_2NCH_2CO_2^- + H_3O^+$	2.5×10^{-10}	9.60
Hydrazoic acid	$HN_3 + H_2O \rightleftharpoons N_3^- + H_3O^+$	1.9×10^{-5}	4.72
Hydrobromic acid	$HBr + H_2O \rightleftharpoons Br^- + H_3O^+$	1×10^{9}	−9
Hydrochloric acid	$HCl + H_2O \rightleftharpoons Cl^- + H_3O^+$	1×10^{6}	−6
Hydrocyanic acid	$HCN + H_2O \rightleftharpoons CN^- + H_3O^+$	6×10^{-10}	9.22
Hydrofluoric acid	$HF + H_2O \rightleftharpoons F^- + H_3O^+$	7.2×10^{-4}	3.14
Hydroiodic acid	$HI + H_2O \rightleftharpoons I^- + H_3O^+$	3×10^{9}	−9.5
Hydrogen peroxide	$H_2O_2 + H_2O \rightleftharpoons HO_2^- + H_3O^+$	2.2×10^{-12}	11.66
Hydrogen selenide	$H_2Se + H_2O \rightleftharpoons HSe^- + H_3O^+$	1.0×10^{-4}	4.00
Hydrogen sulfide	$H_2S + H_2O \rightleftharpoons HS^- + H_3O^+$	1.0×10^{-7}	7.00
	$HS^- + H_2O \rightleftharpoons S^{2-} + H_3O^+$	1.3×10^{-13}	12.89
Hypobromous acid	$HOBr + H_2O \rightleftharpoons OBr^- + H_3O^+$	2.4×10^{-9}	8.62
Hypochlorous acid	$HOCl + H_2O \rightleftharpoons OCl^- + H_3O^+$	2.9×10^{-8}	7.54
Hypoiodous acid	$HOI + H_2O \rightleftharpoons OI^- + H_3O^+$	2.3×10^{-11}	10.64
Iodic acid	$HIO_3 + H_2O \rightleftharpoons IO_3^- + H_3O^+$	0.16	0.80
Nitric acid	$HNO_3 + H_2O \rightleftharpoons NO_3^- + H_3O^+$	28	−1.45
Nitrous acid	$HNO_2 + H_2O \rightleftharpoons NO_2^- + H_3O^+$	5.1×10^{-4}	3.29
Oxalic acid	$H_2C_2O_4 + H_2O \rightleftharpoons HC_2O_4^- + H_3O^+$	5.4×10^{-2}	1.27
	$HC_2O_4^- + H_2O \rightleftharpoons C_2O_4^{2-} + H_3O^+$	5.4×10^{-5}	4.27
Perchloric acid	$HOClO_3 + H_2O \rightleftharpoons ClO_4^- + H_3O^+$	1×10^{8}	−8
Periodic acid	$H_5IO_6 + H_2O \rightleftharpoons H_4IO_6^- + H_3O^+$	2.3×10^{-2}	1.64
Phenol	$C_6H_5OH + H_2O \rightleftharpoons C_6H_5O^- + H_3O^+$	1.0×10^{-10}	10.00
Phosphoric acid	$H_3PO_4 + H_2O \rightleftharpoons H_2PO_4^- + H_3O^+$	7.1×10^{-3}	2.15
	$H_2PO_4^- + H_2O \rightleftharpoons HPO_4^{2-} + H_3O^+$	6.3×10^{-8}	7.20
	$HPO_4^{2-} + H_2O \rightleftharpoons PO_4^{3-} + H_3O^+$	4.2×10^{-13}	12.38
Phosphorous acid	$H_3PO_3 + H_2O \rightleftharpoons H_2PO_3^- + H_3O^+$	1.00×10^{-2}	2.00
	$H_2PO_3^- + H_2O \rightleftharpoons HPO_3^{2-} + H_3O^+$	2.6×10^{-7}	6.59
Sulfamic acid	$H_2NSO_3H + H_2O \rightleftharpoons H_2NSO_3^- + H_3O^+$	1.03×10^{-1}	0.987

Compound	Dissociation Reaction	K_a	pK_a
Sulfuric acid	$H_2SO_4 + H_2O \rightleftharpoons HSO_4^- + H_3O^+$	1×10^3	−3
	$HSO_4^- + H_2O \rightleftharpoons SO_4^{2-} + H_3O^+$	1.2×10^{-2}	1.92
Sulfurous acid	$H_2SO_3 + H_2O \rightleftharpoons HSO_3^- + H_3O^+$	1.7×10^{-2}	1.77
	$HSO_3^- + H_2O \rightleftharpoons SO_3^{2-} + H_3O^+$	6.4×10^{-8}	7.19
Thiocyanic acid	$HSCN + H_2O \rightleftharpoons SCN^- + H_3O^+$	71	−1.85
Trichloroacetic acid	$Cl_3CCO_2H + H_2O \rightleftharpoons Cl_3CCO_2^- + H_3O^+$	0.22	0.66
Water	$H_2O + H_2O \rightleftharpoons OH^- + H_3O^+$	1.8×10^{-16}	15.75

Table B.9
Base Ionization Equilibrium Constants

Compound	Ionization Reaction	K_b	pK_b
Ammonia	$NH_3 + H_2O \rightleftharpoons NH_4^+ + OH^-$	1.8×10^{-5}	4.74
Aniline	$C_6H_5NH_2 + H_2O \rightleftharpoons C_6H_5NH_3^+ + OH^-$	4.0×10^{-10}	9.40
Butylamine	$CH_3(CH_2)_3NH_2 + H_2O \rightleftharpoons CH_3(CH_2)_3NH_3^+ + OH^-$	4.0×10^{-4}	3.40
Dimethylamine	$(CH_3)_2NH + H_2O \rightleftharpoons (CH_3)_2NH_2^+ + OH^-$	5.9×10^{-4}	3.23
Ethanolamine	$HOCH_2CH_2NH_2 + H_2O \rightleftharpoons HOCH_2CH_2NH_3^+ + OH^-$	3.3×10^{-5}	4.50
Ethylamine	$CH_3CH_2NH_2 + H_2O \rightleftharpoons CH_3CH_2NH_3^+ + OH^-$	4.4×10^{-4}	3.37
Hydrazine	$H_2NNH_2 + H_2O \rightleftharpoons H_2NNH_3^+ + OH^-$	1.2×10^{-6}	5.89
Hydroxylamine	$HONH_2 + H_2O \rightleftharpoons HONH_3^+ + OH^-$	1.1×10^{-8}	7.97
Methylamine	$CH_3NH_2 + H_2O \rightleftharpoons CH_3NH_3^+ + OH^-$	4.8×10^{-4}	3.32
Pyridine	$C_5H_5N + H_2O \rightleftharpoons C_5H_5NH^+ + OH^-$	1.7×10^{-9}	8.77
Trimethylamine	$(CH_3)_3N + H_2O \rightleftharpoons (CH_3)_3NH^+ + OH^-$	6.3×10^{-5}	4.20
Urea	$H_2NCONH_2 + H_2O \rightleftharpoons H_2NCONH_3^+ + OH^-$	1.5×10^{-14}	13.82

Table B.10

Solubility Product Equilibrium Constants

Substance	K_{sp}	Substance	K_{sp}	Substance	K_{sp}
AgBr	5.0×10^{-13}	α-CoS	4.0×10^{-21}	$MnCO_3$	1.8×10^{-11}
AgCN	1.2×10^{-16}	β-CoS	2.0×10^{-25}	$Mn(OH)_2$	2×10^{-13}
Ag_2CO_3	8.1×10^{-12}	$Cr(OH)_3$	6.3×10^{-31}	MnS	3×10^{-13}
AgOH	2.0×10^{-8}	CuBr	5.3×10^{-9}	$NiCO_3$	6.6×10^{-9}
$AgC_2H_3O_2$	4.4×10^{-3}	CuCl	1.2×10^{-6}	NiC_2O_4	4×10^{-10}
$Ag_2C_2O_4$	3.4×10^{-11}	CuCN	3.2×10^{-20}	α-NiS	3.2×10^{-19}
AgCl	1.8×10^{-10}	$CuCrO_4$	3.6×10^{-6}	β-NiS	1.0×10^{-24}
Ag_2CrO_4	1.1×10^{-12}	$CuCO_3$	1.4×10^{-10}	γ-NiS	2.0×10^{-26}
AgI	8.3×10^{-17}	$Cu(OH)_2$	2.2×10^{-20}	$PbBr_2$	4.0×10^{-5}
Ag_2S	6.3×10^{-50}	CuI	1.1×10^{-13}	$PbCO_3$	7.4×10^{-14}
AgSCN	1.0×10^{-12}	Cu_2S	2.5×10^{-48}	PbC_2O_4	4.8×10^{-10}
Ag_2SO_4	1.4×10^{-5}	CuS	6.3×10^{-36}	$PbCl_2$	1.6×10^{-5}
$Al(OH)_3$	1.3×10^{-33}	CuSCN	4.8×10^{-15}	$PbCrO_4$	2.8×10^{-13}
AuCl	2.0×10^{-13}	$FeCO_3$	3.2×10^{-11}	PbF_2	2.7×10^{-8}
$AuCl_3$	3.2×10^{-23}	$Fe_2C_2O_4$	3.2×10^{-7}	$Pb(OH)_2$	1.2×10^{-15}
AuI	1.6×10^{-23}	$Fe(OH)_2$	8.0×10^{-16}	PbI_2	7.1×10^{-9}
AuI_3	5.5×10^{-46}	$Fe(OH)_3$	4×10^{-38}	PbS	8.0×10^{-28}
$BaCO_3$	5.1×10^{-9}	FeS	6.3×10^{-18}	$PbSO_4$	1.6×10^{-8}
BaC_2O_4	2.3×10^{-8}	Hg_2Br_2	5.6×10^{-23}	SnS	1.0×10^{-25}
$BaCrO_4$	1.2×10^{-10}	$Hg_2(CN)_2$	5×10^{-40}	$Sn(OH)_2$	1.4×10^{-28}
BaF_2	1.0×10^{-6}	Hg_2CO_3	8.9×10^{-17}	$Sn(OH)_4$	1×10^{-56}
$Ba(OH)_2$	5×10^{-3}	$Hg_2(OAc)_2$	3×10^{-11}	$SrCO_3$	1.1×10^{-10}
$BaSO_4$	1.1×10^{-10}	$Hg_2C_2O_4$	2.0×10^{-13}	SrC_2O_4	1.6×10^{-7}
Bi_2S_3	1×10^{-97}	HgC_2O_4	1×10^{-7}	$SrCrO_4$	2.2×10^{-5}
$CaCO_3$	2.8×10^{-9}	Hg_2Cl_2	1.3×10^{-18}	SrF_2	2.5×10^{-9}
CaC_2O_4	4×10^{-9}	Hg_2CrO_4	2.0×10^{-9}	$SrSO_4$	3.2×10^{-7}
$CaCrO_4$	7.1×10^{-4}	Hg_2I_2	4.5×10^{-29}	TlBr	3.4×10^{-6}
CaF_2	4.0×10^{-11}	Hg_2S	1.0×10^{-47}	TlCl	1.7×10^{-4}
$Ca(OH)_2$	5.5×10^{-6}	HgS	4×10^{-53}	TlI	6.5×10^{-8}
$CdCO_3$	5.2×10^{-12}	$K_2NaCo(NO_2)_6$	2.2×10^{-11}	$Zn(CN)_2$	2.6×10^{-13}
$Cd(CN)_2$	1.0×10^{-8}	$MgCO_3$	3.5×10^{-8}	$ZnCO_3$	1.4×10^{-11}
$Cd(OH)_2$	2.5×10^{-14}	MgC_2O_4	1×10^{-8}	ZnC_2O_4	2.7×10^{-8}
CdS	8×10^{-27}	MgF_2	6.5×10^{-9}	$Zn(OH)_2$	1.2×10^{-17}
$CoCO_3$	1.4×10^{-13}	$Mg(OH)_2$	1.8×10^{-11}	α-ZnS	1.6×10^{-24}
$Co(OH)_3$	1.6×10^{-44}	$MgNH_4PO_4$	2.5×10^{-13}	β-ZnS	2.5×10^{-22}

Table B.11

Complex Formation Equilibrium Constants

Equilibrium	K_f
$Ag^+ + 2\ Br^- \rightleftharpoons AgBr_2^-$	2.1×10^7
$Ag^+ + 2\ Cl^- \rightleftharpoons AgCl_2^-$	1.1×10^5
$Ag^+ + 2\ CN^- \rightleftharpoons Ag(CN)_2^-$	1.3×10^{21}
$Ag^+ + 2\ I^- \rightleftharpoons AgI_2^-$	5.5×10^{11}
$Ag^+ + 2\ NH_3 \rightleftharpoons Ag(NH_3)_2^+$	1.1×10^7
$Ag^+ + 2\ SCN^- \rightleftharpoons Ag(SCN)_2^-$	3.7×10^7
$Ag^+ + 2\ S_2O_3^{2-} \rightleftharpoons Ag(S_2O_3)_2^{3-}$	2.9×10^{13}
$Al^{3+} + 6\ F^- \rightleftharpoons AlF_6^{3-}$	6.9×10^{19}
$Al^{3+} + 4\ OH^- \rightleftharpoons Al(OH)_4^-$	1.1×10^{33}
$Cd^{2+} + 4\ Cl^- \rightleftharpoons CdCl_4^{2-}$	6.3×10^2
$Cd^{2+} + 4\ CN^- \rightleftharpoons Cd(CN)_4^{2-}$	6.0×10^{18}
$Cd^{2+} + 4\ I^- \rightleftharpoons CdI_4^{2-}$	2.6×10^5
$Cd^{2+} + 4\ OH^- \rightleftharpoons Cd(OH)_4^{2-}$	4.2×10^8
$Cd^{2+} + 4\ NH_3 \rightleftharpoons Cd(NH_3)_6^{2+}$	1.3×10^7
$Co^{2+} + 6\ NH_3 \rightleftharpoons Co(NH_3)_6^{2+}$	1.3×10^5
$Co^{3+} + 6\ NH_3 \rightleftharpoons Co(NH_3)_6^{3+}$	2×10^{35}
$Co^{2+} + 4\ SCN^- \rightleftharpoons Co(SCN)_4^{2-}$	1×10^3
$Cr^{3+} + 4\ OH^- \rightleftharpoons Cr(OH)_4^-$	8×10^{29}
$Cu^{2+} + 4\ OH^- \rightleftharpoons Cu(OH)_4^{2-}$	3×10^{18}
$Cu^{2+} + 4\ NH_3 \rightleftharpoons Cu(NH_3)_4^{2+}$	2.1×10^{13}
$Fe^{2+} + 6\ CN^- \rightleftharpoons Fe(CN)_6^{4-}$	1×10^{35}
$Fe^{3+} + 6\ CN^- \rightleftharpoons Fe(CN)_6^{3-}$	1×10^{42}
$Fe^{3+} + SCN^- \rightleftharpoons Fe(SCN)^{2+}$	8.9×10^2
$Fe^{3+} + 2\ SCN^- \rightleftharpoons Fe(SCN)_2^+$	2.3×10^3
$Hg^{2+} + 4\ Br^- \rightleftharpoons HgBr_4^{2-}$	1×10^{21}
$Hg^{2+} + 4\ Cl^- \rightleftharpoons HgCl_4^{2-}$	1.2×10^{15}
$Hg^{2+} + 4\ CN^- \rightleftharpoons Hg(CN)_4^{2-}$	3×10^{41}
$Hg^{2-} + 4\ I^- \rightleftharpoons HgI_4^{2-}$	6.8×10^{29}
$I_2 + I^- \rightleftharpoons I_3^-$	7.8×10^2
$Ni^{2+} + 4\ CN^- \rightleftharpoons Ni(CN)_4^{2-}$	2×10^{31}
$Ni^{2+} + 6\ NH_3 \rightleftharpoons Ni(NH_3)_6^{2+}$	5.5×10^8
$Pb^{2+} + 4\ Cl^- \rightleftharpoons PbCl_4^{2-}$	4×10^1
$Pb^{2+} + 4\ I^- \rightleftharpoons PbI_4^{2-}$	3.0×10^4
$Sb^{3+} + 4\ Cl^- \rightleftharpoons SbCl_4^-$	5.2×10^4
$Sb^{3+} + 4\ OH^- \rightleftharpoons Sb(OH)_4^-$	2×10^{38}
$Sn^{2+} + 4\ Cl^- \rightleftharpoons SnCl_4^{2-}$	3.0×10^1
$Zn^{2+} + 4\ CN^- \rightleftharpoons Zn(CN)_4^{2-}$	5×10^{16}
$Zn^{2+} + 4\ OH^- \rightleftharpoons Zn(OH)_4^{2-}$	4.6×10^{17}
$Zn^{2+} + 4\ NH_3 \rightleftharpoons Zn(NH_3)_4^{2+}$	2.9×10^9

Table B.12
Standard Reduction Potentials

Half-reaction	$E°$ (*V*)
$3\ N_2 + 2\ H^+ + 2\ e^- \rightleftharpoons 2\ HN_3$	−3.1
$Li^+ + e^- \rightleftharpoons Li$	−3.045
$Rb^+ + e^- \rightleftharpoons Rb$	−2.925
$K^+ + e^- \rightleftharpoons K$	−2.924
$Cs^+ + e^- \rightleftharpoons Cs$	−2.923
$Ba^{2+} + 2\ e^- \rightleftharpoons Ba$	−2.90
$Sr^{2+} + 2\ e^- \rightleftharpoons Sr$	−2.89
$Ca^{2+} + 2\ e^- \rightleftharpoons Ca$	−2.76
$Na^+ + e^- \rightleftharpoons Na$	−2.7109
$Mg(OH)_2 + 2\ e^- \rightleftharpoons Mg + 2\ OH^-$	−2.69
$Mg^{2+} + 2\ e^- \rightleftharpoons Mg$	−2.375
$H_2 + 2\ e^- \rightleftharpoons 2\ H^-$	−2.23
$Al^{3+} + 3\ e^- \rightleftharpoons Al$ (0.1 *M* NaOH)	−1.706
$Be^{2+} + 2\ e^- \rightleftharpoons Be$	−1.70
$Ti^{2+} + 2\ e^- \rightleftharpoons Ti$	−1.63
$Zn(CN)_4{}^{2-} + 2\ e^- \rightleftharpoons Zn + 4\ CN^-$	−1.26
$Mn^{2+} + 2\ e^- \rightleftharpoons Mn$	−1.18
$Zn(NH_3)_4{}^{2+} + 2\ e^- \rightleftharpoons Zn + 4\ NH_3$	−1.04
$SO_4{}^{2-} + H_2O + 2\ e^- \rightleftharpoons SO_3{}^{2-} + 2\ OH^-$	−0.92
$Cr^{2+} + 2\ e^- \rightleftharpoons Cr$	−0.91
$TiO_2 + 4\ H^+ + 4\ e^- \rightleftharpoons Ti + 2\ H_2O$	−0.87
$2\ H_2O + 2\ e^- \rightleftharpoons H_2 + 2\ OH^-$	−0.8277
$Zn^{2+} + 2\ e^- \rightleftharpoons Zn$	−0.7628
$Cr^{3+} + 3\ e^- \rightleftharpoons Cr$	−0.74
$2\ SO_3{}^{2-} + 3\ H_2O + 4\ e^- \rightleftharpoons S_2O_3{}^{2-} + 6\ OH^-$	−0.58
$PbO + H_2O + 2\ e^- \rightleftharpoons Pb + 2\ OH^-$	−0.576
$Ga^{3+} + 3\ e^- \rightleftharpoons Ga$	−0.560
$S + 2\ e^- \rightleftharpoons S^{2-}$	−0.508
$2\ CO_2 + 2\ H^+ + 2\ e^- \rightleftharpoons H_2C_2O_4$	−0.49
$Ni(NH_3)_6{}^{2+} + 2\ e^- \rightleftharpoons Ni + 6\ NH_3$	−0.48
$Co(NH_3)_6{}^{2+} + 2\ e^- \rightleftharpoons Co + 6\ NH_3$	−0.422
$Cr^{3+} + e^- \rightleftharpoons Cr^{2+}$	−0.41
$Fe^{2+} + 2\ e^- \rightleftharpoons Fe$	−0.409
$Cd^{2+} + 2\ e^- \rightleftharpoons Cd$	−0.4026
$PbSO_4 + 2\ e^- \rightleftharpoons Pb + SO_4{}^{2-}$	−0.356
$In^{3+} + 3\ e^- \rightleftharpoons In$	−0.338
$Tl^+ + e^- \rightleftharpoons Tl$	−0.3363
$Ag(CN)_2{}^- + e^- \rightleftharpoons Ag + 2\ CN^-$	−0.31
$Co^{2+} + 2\ e^- \rightleftharpoons Co$	−0.28
$H_3PO_4 + 2\ H^+ + 2\ e^- \rightleftharpoons H_3PO_3 + H_2O$	−0.276
$Ni^{2+} + 2\ e^- \rightleftharpoons Ni$	−0.23
$2\ SO_4{}^{2-} + 4\ H^+ + 2\ e^- \rightleftharpoons S_2O_6{}^{2-} + 2\ H_2O$	−0.224
$CO_2 + 2\ H^+ + 2\ e^- \rightleftharpoons HCO_2H$	−0.20
$O_2 + 2\ H_2O + 2\ e^- \rightleftharpoons H_2O_2 + 2\ OH^-$	−0.146
$Sn^{2+} + 2\ e^- \rightleftharpoons Sn$	−0.1364
$Pb^{2+} + 2\ e^- \rightleftharpoons Pb$	−0.1263
$CrO_4{}^{2-} + 4\ H_2O + 3\ e^- \rightleftharpoons Cr(OH)_3 + 5\ OH^-$	−0.12
$WO_3 + 6\ H^+ + 6\ e^- \rightleftharpoons W + 3\ H_2O$	−0.09
$Ru^{3+} + e^- \rightleftharpoons Ru^{2+}$	−0.08
$O_2 + H_2O + 2\ e^- \rightleftharpoons HO_2{}^- + OH^-$	−0.076
$Fe^{3+} + 3\ e^- \rightleftharpoons Fe$	−0.036
$2\ H^+ + 2\ e^- \rightleftharpoons H_2$	0.0000000 . . .

Half-reaction	$E°$ (V)
$NO_3^- + H_2O + 2e^- \rightleftharpoons NO_2^- + 2OH^-$	0.01
$AgBr + e^- \rightleftharpoons Ag + Br^-$	0.0713
$S_4O_6^{2-} + 2e^- \rightleftharpoons 2S_2O_3^{2-}$	0.0895
$Sn^{4+} + 2e^- \rightleftharpoons Sn^{2+}$	0.15
$Cu^{2+} + e^- \rightleftharpoons Cu^+$	0.158
$ClO_4^- + H_2O + 2e^- \rightleftharpoons ClO_3^- + 2OH^-$	0.17
$SO_4^{2-} + 4H^+ + 2e^- \rightleftharpoons H_2SO_3 + H_2O$	0.20
$AgCl + e^- \rightleftharpoons Ag + Cl^-$	0.2223
$IO_3^- + 3H_2O + 6e^- \rightleftharpoons I^- + 6OH^-$	0.26
$Hg_2Cl_2 + 2e^- \rightleftharpoons 2Hg + 2Cl^-$	0.2682
$Cu^{2+} + 2e^- \rightleftharpoons Cu$	0.3402
$ClO_3^- + H_2O + 2e^- \rightleftharpoons ClO_2^- + 2OH^-$	0.35
$Ag(NH_3)_2^+ + e^- \rightleftharpoons Ag + 2NH_3$	0.373
$O_2 + 2H_2O + 4e^- \rightleftharpoons 4OH^-$	0.401
$H_2SO_3 + 4H^+ + 4e^- \rightleftharpoons S + 3H_2O$	0.45
$HgCl_4^{2-} + 2e^- \rightleftharpoons Hg + 4Cl^-$	0.48
$Cu^+ + e^- \rightleftharpoons Cu$	0.522
$I_3^- + 2e^- \rightleftharpoons 3I^-$	0.5338
$I_2 + 2e^- \rightleftharpoons 2I^-$	0.535
$MnO_4^- + 2H_2O + 3e^- \rightleftharpoons MnO_2 + 4OH^-$	0.588
$ClO_2^- + H_2O + 2e^- \rightleftharpoons ClO^- + 2OH^-$	0.59
$O_2 + 2H^+ + 2e^- \rightleftharpoons H_2O_2$	0.682
$Fe^{3+} + e^- \rightleftharpoons Fe^{2+}$	0.770
$Hg_2^{2+} + 2e^- \rightleftharpoons 2Hg$	0.7961
$Ag^+ + e^- \rightleftharpoons Ag$	0.7996
$Hg^{2+} + 2e^- \rightleftharpoons Hg$	0.851
$H_2O_2 + 2e^- \rightleftharpoons 2OH^-$	0.88
$ClO^- + H_2O + 2e^- \rightleftharpoons Cl^- + 2OH^-$	0.89
$2Hg^{2+} + 2e^- \rightleftharpoons Hg_2^{2+}$	0.905
$NO_3^- + 3H^+ + 2e^- \rightleftharpoons HNO_2 + H_2O$	0.94
$ClO_2 + e^- \rightleftharpoons ClO_2^-$	0.95
$NO_3^- + 4H^+ + 3e^- \rightleftharpoons NO + 2H_2O$	0.96
$Pd^{2+} + 2e^- \rightleftharpoons Pd$	0.987
$HNO_2 + H^+ + e^- \rightleftharpoons NO + H_2O$	0.99
$IO_3^- + 6H^+ + 6e^- \rightleftharpoons I^- + 3H_2O$	1.085
$Br_2(aq) + 2e^- \rightleftharpoons 2Br^-$	1.087
$Cr^{6+} + 3e^- \rightleftharpoons Cr^{3+}$	1.10
$ClO_3^- + 2H^+ + 2e^- \rightleftharpoons ClO_2^- + H_2O$	1.15
$ClO_4^- + 2H^+ + 2e^- \rightleftharpoons ClO_3^- + H_2O$	1.19
$2IO_3^- + 12H^+ + 10e^- \rightleftharpoons I_2 + 6H_2O$	1.19
$HCrO_4^- + 7H^+ + 3e^- \rightleftharpoons Cr^{3+} + 4H_2O$	1.195
$Pt^{2+} + 2e^- \rightleftharpoons Pt$	1.2
$MnO_2 + 4H^+ + 2e^- \rightleftharpoons Mn^{2+} + 2H_2O$	1.208
$O_2 + 4H^+ + 4e^- \rightleftharpoons 2H_2O$	1.229
$O_3 + H_2O + 2e^- \rightleftharpoons O_2 + 2OH^-$	1.24
$Tl^{3+} + 2e^- \rightleftharpoons Tl^+$	1.247
$ClO_2 + H^+ + e^- \rightleftharpoons HClO_2$	1.27
$2HNO_2 + 4H^+ + 4e^- \rightleftharpoons N_2O + 3H_2O$	1.27
$Au^{3+} + 2e^- \rightleftharpoons Au^+$	1.29
$Cr_2O_7^{2-} + 14H^+ + 6e^- \rightleftharpoons 2Cr^{3+} + 7H_2O$	1.33
$ClO_4^- + 8H^+ + 7e^- \rightleftharpoons \frac{1}{2}Cl_2 + 4H_2O$	1.34
$Cl_2 + 2e^- \rightleftharpoons 2Cl^-$	1.3583
$ClO_4^- + 8H^+ + 8e^- \rightleftharpoons Cl^- + 4H_2O$	1.37
$Au^{3+} + 3e^- \rightleftharpoons Au$	1.42

(*continued*)

Table B.12
Standard Reduction Potentials (continued)

Half-reaction	$E°$ (V)
$ClO_3^- + 6\,H^+ + 6\,e^- \rightleftharpoons Cl^- + 3\,H_2O$	1.45
$PbO_2 + 4\,H^+ + 2\,e^- \rightleftharpoons Pb^{2+} + 2\,H_2O$	1.467
$2\,ClO_3^- + 12\,H^+ + 10\,e^- \rightleftharpoons Cl_2 + 6\,H_2O$	1.47
$HClO + H^+ + 2\,e^- \rightleftharpoons Cl^- + H_2O$	1.49
$MnO_4^- + 8\,H^+ + 5\,e^- \rightleftharpoons Mn^{2+} + 4\,H_2O$	1.491
$HClO_2 + 3\,H^+ + 4\,e^- \rightleftharpoons Cl^- + 2\,H_2O$	1.56
$2\,NO + 2\,H^+ + 2\,e^- \rightleftharpoons N_2O + H_2O$	1.59
$2\,HClO_2 + 6\,H^+ + 6\,e^- \rightleftharpoons Cl_2 + 4\,H_2O$	1.63
$2\,HClO + 2\,H^+ + 2\,e^- \rightleftharpoons Cl_2 + 2\,H_2O$	1.63
$HClO_2 + 2\,H^+ + 2\,e^- \rightleftharpoons HClO + H_2O$	1.64
$MnO_4^- + 4\,H^+ + 3\,e^- \rightleftharpoons MnO_2 + 2\,H_2O$	1.679
$Au^+ + e^- \rightleftharpoons Au$	1.68
$PbO_2 + SO_4^{2-} + 4\,H^+ + 2\,e^- \rightleftharpoons PbSO_4 + 2\,H_2O$	1.685
$N_2O + 2\,H^+ + 2\,e^- \rightleftharpoons N_2 + H_2O$	1.77
$H_2O_2 + 2\,H^+ + 2\,e^- \rightleftharpoons 2\,H_2O$	1.776
$Co^{3+} + e^- \rightleftharpoons Co^{2+}$	1.842
$S_2O_8^{2-} + 2\,e^- \rightleftharpoons 2\,SO_4^{2-}$	2.05
$O_3(g) + 2\,H^+ + 2\,e^- \rightleftharpoons O_2(g) + H_2O$	2.07
$F_2(g) + 2\,H^+ + 2\,e^- \rightleftharpoons 2\,HF(aq)$	3.03

Table B.13

Standard-State Enthalpies, Free Energies, and Entropies of Atom Combination, 298.15 K

Substance	ΔH°_{ac} (kJ/mol$_{rxn}$)	ΔG°_{ac} (kJ/mol$_{rxn}$)	ΔS°_{ac} (J/mol$_{rxn}$ · K)
	Aluminum		
$Al(s)$	−326.4	−285.7	−136.21
$Al(g)$	0	0	0
$Al^{3+}(aq)$	−857	−771	−486.2
$Al_2O_3(s)$	−3076.0	−2848.9	−761.33
$AlCl_3(s)$	−1395.6	−1231.5	−549.46
$AlF_3(s)$	−2067.5	−1896.4	−574.36
$Al_2(SO_4)_3(s)$	−7920.1	−7166.9	−2525.9
	Barium		
$Ba(s)$	−180	−146	−107.4
$Ba(g)$	0	0	0
$Ba^{2+}(aq)$	−718	−707	−160.6
$BaO(s)$	−983	−903	−260.88
$Ba(OH)_2{\cdot}8\ H_2O(s)$	−9931.6	−8915	−3419
$BaCl_2(s)$	−1282	−1168	−376.96
$BaCl_2(aq)$	−1295	−1181	−378.04
$BaSO_4(s)$	−2929	−2673	−850.1
$Ba(NO_3)_2(s)$	−3612	−3217	−1229.4
$Ba(NO_3)_2(aq)$	−3573	−3231	−1140.7
	Beryllium		
$Be(s)$	−324.3	−286.6	−126.77
$Be(g)$	0	0	0
$Be^{2+}(aq)$	−707.1	−666.3	−266.0
$BeO(s)$	−1183.1	−1026.6	−283.18
$BeCl_2(s)$	−1058.1	−943.6	−383.99
	Bismuth		
$Bi(s)$	−207.1	−168.2	−130.31
$Bi(g)$	0	0	0
$Bi_2O_3(s)$	−1735.6	−1525.3	−705.8
$BiCl_3(s)$	−951.2	−800.2	−505.6
$BiCl_3(g)$	−837.8	−741.2	−323.79
$Bi_2S_3(s)$	−1393.7	−1191.8	−677.2
	Boron		
$B(s)$	−562.7	−518.8	−147.59
$B(g)$	0	0	0
$B_2O_3(s)$	−3145.7	−2926.4	−736.10
$B_2H_6(g)$	−2395.7	−2170.4	−763.07
$B_5H_9(l)$	−4729.7	−4251.4	−1615.45
$B_5H_9(g)$	−4699.2	−4248.2	−1523.75
$B_{10}H_{14}(s)$	−8719.3	−7841.2	−2963.92
$H_3BO_3(s)$	−3057.5	−2792.7	−891.92
$BF_3(g)$	−1936.7	−1824.9	−375.59
$BCl_3(l)$	−1354.9	−1223.2	−442.7
$B_3N_3H_6(l)$	−4953.1	−4535.4	−1408.9
$B_3N_3H_6(g)$	−4923.9	−4532.8	−1319.84

(continued)

Table B.13

Standard-State Enthalpies, Free Energies, and Entropies of Atom Combination, 298.15 K (continued)

Substance	ΔH°_{ac} (kJ/mol$_{rxn}$)	ΔG°_{ac} (kJ/mol$_{rxn}$)	ΔS°_{ac} (J/mol$_{rxn}$ · K)
	Bromine		
$Br_2(l)$	−223.768	−164.792	−197.813
$Br_2(g)$	−192.86	−161.68	−104.58
$Br(g)$	0	0	0
$HBr(g)$	−365.93	−339.09	−91.040
$HBr(aq)$	−451.08	−389.60	−207.3
$BrF(g)$	−284.72	−253.49	−104.81
$BrF_3(g)$	−604.45	−497.56	−358.75
$BrF_5(g)$	−935.73	−742.6	−648.60
	Calcium		
$Ca(s)$	−178.2	−144.3	−113.46
$Ca(g)$	0	0	0
$Ca^{2+}(aq)$	−721.0	−697.9	−208.0
$CaO(s)$	−1062.5	−980.1	−276.19
$Ca(OH)_2(s)$	−2097.9	−1912.7	−623.03
$CaCl_2(s)$	−1217.4	−1103.8	−380.7
$CaSO_4(s)$	−2887.8	−2631.3	−860.3
$CaSO_4 \cdot 2\ H_2O(s)$	−4256.7	−4383.2	−1553.8
$Ca(NO_3)_2(s)$	−3557.0	−3189.0	−1234.5
$CaCO_3(s)$	−2849.3	−2639.5	−703.2
$Ca_3(PO_4)_2(s)$	−7278.0	−6727.9	−1843.5
	Carbon		
C(graphite)	−716.682	−671.257	−152.36
C(diamond)	−714.787	−668.357	−155.719
$C(g)$	0	0	0
$CO(g)$	−1076.377	−1040.156	−121.477
$CO_2(g)$	−1608.531	−1529.078	−266.47
$COCl_2(g)$	−1428.0	−1318.9	−366.02
$CH_4(g)$	−1662.09	−1534.997	−430.684
$HCHO(g)$	−1509.72	−1412.01	−329.81
$H_2CO_3(aq)$	−2599.14	−2396.02	−683.3
$HCO_3^-(aq)$	−2373.83	−2156.47	−664.8
$CO_3^{2-}(aq)$	−2141.33	−1894.26	−698.16
$CH_3OH(l)$	−2075.11	−1882.25	−651.2
$CH_3OH(g)$	−2037.11	−1877.94	−538.19
$CCl_4(l)$	−1338.84	−1159.19	−602.49
$CCl_4(g)$	−1306.3	−1154.57	−509.04
$CHCl_3(l)$	−1433.84	−1265.20	−567.3
$CHCl_3(g)$	−1402.51	−1261.88	−472.69
$CH_2Cl_2(l)$	−1516.80	−1356.37	−540.1
$CH_2Cl_2(g)$	−1487.81	−1354.98	−447.69
$CH_3Cl(g)$	−1572.15	−1444.08	−432.9
$CS_2(l)$	−1184.59	−1082.49	−342.40
$CS_2(g)$	−1156.93	−1080.64	−255.90
$HCN(g)$	−1271.9	−1205.43	−224.33
$CH_3NO_2(l)$	−2453.77	−2214.51	−805.89
$C_2H_2(g)$	−1641.93	−1539.81	−344.68
$C_2H_4(g)$	−2251.70	−2087.35	−555.48

Substance	ΔH°_{ac} (kJ/mol$_{rxn}$)	ΔG°_{ac} (kJ/mol$_{rxn}$)	ΔS°_{ac} (J/mol$_{rxn}$ · K)
$C_2H_6(g)$	−2823.94	−2594.82	−774.87
$CH_3CHO(l)$	−2745.43	−2515.35	−775.9
$CH_3CO_2H(l)$	−3286.8	−3008.86	−937.4
$CH_3CO_2H(g)$	−3234.55	−2992.96	−814.7
$CH_3CO_2H(aq)$	−3288.06	−3015.42	−918.5
$CH_3CO_2^-(aq)$	−3070.66	−2785.03	−895.8
$CH_3CH_2OH(l)$	−3266.12	−2968.51	−1004.8
$CH_3CH_2OH(g)$	−3223.53	−2962.22	−882.82
$CH_3CH_2OH(aq)$	−3276.7	−2975.37	−1017.0
$C_6H_6(l)$	−5556.96	−5122.52	−1464.1
$C_6H_6(g)$	−5523.07	−5117.36	−1367.7
	Chlorine		
$Cl_2(g)$	−243.358	−211.360	−107.330
$Cl(g)$	0	0	0
$Cl^-(aq)$	−288.838	−236.908	−108.7
$ClO_2(g)$	−517.5	−448.6	−230.47
$Cl_2O(g)$	−412.2	−345.2	−225.24
$Cl_2O_7(l)$	−1750	—	—
$HCl(g)$	−431.64	−404.226	−93.003
$HCl(aq)$	−506.49	−440.155	−223.4
$ClF(g)$	−255.15	−223.53	−106.06
	Chromium		
$Cr(s)$	−396.6	−351.8	−150.73
$Cr(g)$	0	0	0
$CrO_3(s)$	−1733.6	—	—
$CrO_4^{2-}(aq)$	−2274.4	−2006.47	−768.51
$Cr_2O_3(s)$	−2680.4	−2456.89	−751.0
$Cr_2O_7^{2-}(aq)$	−4027.7	−3626.8	−1214.5
$(NH_4)_2Cr_2O_7$	−7030.7	—	—
$PbCrO_4(s)$	−2519.2	—	—
	Cobalt		
$Co(s)$	−424.7	−380.3	−149.475
$Co(g)$	0	0	0
$Co^{2+}(aq)$	−482.9	−434.7	−293
$Co^{3+}(aq)$	−333	−246.3	−485
$CoO(s)$	−911.8	−826.2	−287.60
$Co_3O_4(s)$	−3162	−2842	−1080.3
$Co(NH_3)_6^{3+}(aq)$	−7763.5	−6929.5	−3018
	Copper		
$Cu(s)$	−338.32	−298.58	−133.23
$Cu(g)$	0	0	0
$Cu^+(aq)$	−266.65	−248.60	−125.8
$Cu^{2+}(aq)$	−273.55	−233.09	−266.0
$CuO(s)$	−744.8	−660.0	−284.81
$Cu_2O(s)$	−1094.4	−974.9	−400.68
$CuCl_2(s)$	−807.8	−685.6	−388.71
$CuS(s)$	−670.2	−590.4	−267.7
$Cu_2S(s)$	−1304.9	−921.6	−379.7
$CuSO_4(s)$	−2385.17	−1530.44	−869
$Cu(NH_3)_4^{2+}(aq)$	−5189.4	−4671.13	−1882.5

(continued)

Table B.13

Standard-State Enthalpies, Free Energies, and Entropies of Atom Combination, 298.15 K (continued)

Substance	ΔH°_{ac} (kJ/mol$_{rxn}$)	ΔG°_{ac} (kJ/mol$_{rxn}$)	ΔS°_{ac} (J/mol$_{rxn}$ · K)
	Fluorine		
$F_2(g)$	−157.98	−123.82	−114.73
$F(g)$	0	0	0
$F^-(aq)$	−411.62	−340.70	−172.6
$HF(g)$	−567.7	−538.4	−99.688
$HF(aq)$	−616.72	−561.98	−184.8
	Hydrogen		
$H_2(g)$	−435.30	−406.494	−98.742
$H(g)$	0	0	0
$H^+(aq)$	−217.65	−203.247	−114.713
$OH^-(aq)$	−696.81	−592.222	−286.52
$H_2O(l)$	−970.30	−875.354	−320.57
$H_2O(g)$	−926.29	−866.797	−202.23
$H_2O_2(l)$	−1121.42	−990.31	−441.9
$H_2O_2(aq)$	−1124.81	−1003.99	−407.6
	Iodine		
$I_2(s)$	−213.676	−141.00	−245.447
$I_2(g)$	−151.238	−121.67	−100.89
$I(g)$	0	0	0
$HI(g)$	−298.01	−272.05	−88.910
$IF(g)$	−281.48	−250.92	−103.38
$IF_5(g)$	−1324.28	−1131.78	−646.9
$IF_7(g)$	−1603.7	−1322.17	−945.6
$ICl(g)$	−210.74	−181.64	−98.438
$IBr(g)$	−177.88	−149.21	−97.040
	Iron		
$Fe(s)$	−416.3	−370.7	−153.21
$Fe(g)$	0	0	0
$Fe^{2+}(aq)$	−505.4	−449.6	−318.2
$Fe^{3+}(aq)$	−464.8	−375.4	−496.4
$Fe_2O_3(s)$	−2404.3	−2178.8	−756.75
$Fe_3O_4(s)$	−3364.0	−3054.4	−1039.3
$Fe(OH)_2(s)$	−1918.9	−1727.2	−644
$Fe(OH)_3(s)$	−2639.8	−2372.1	−901.1
$FeCl_3(s)$	−1180.8	−1021.7	−533.8
$FeS_2(s)$	−1152.1	−1014.1	−463.2
$Fe(CO)_5(l)$	−6019.6	−5590.9	−1438.1
$Fe(CO)_5(g)$	−5979.5	−5582.9	−1330.9
	Lead		
$Pb(s)$	−195.0	−161.9	−110.56
$Pb(g)$	0	0	0
$Pb^{2+}(aq)$	−196.7	−186.3	−164.9
$PbO(s)$	−661.5	−581.5	−267.7
$PbO_2(s)$	−970.7	−842.7	−428.9
$PbCl_2(s)$	−797.8	−687.4	−369.8
$PbCl_4(l)$	−1011.0	—	—
$PbS(s)$	−574.2	−498.9	−252.0

Substance	ΔH°_{ac} (kJ/mol$_{rxn}$)	ΔG°_{ac} (kJ/mol$_{rxn}$)	ΔS°_{ac} (J/mol$_{rxn}$ · K)
$PbSO_4(s)$	−2390.4	−2140.2	−838.84
$Pb(NO_3)_2(s)$	−3087.3	—	—
$PbCO_3(s)$	−2358.3	−2153.8	−685.6
	Lithium		
$Li(s)$	−159.37	−126.66	−109.65
$Li(g)$	0	0	0
$Li^+(aq)$	−437.86	−419.97	−125.4
$LiH(s)$	−467.56	−398.26	−233.40
$LiOH(s)$	−1111.12	−1000.59	−371.74
$LiF(s)$	−854.33	−784.28	−261.87
$LiCl(s)$	−689.66	−616.71	−244.64
$LiBr(s)$	−622.48	−551.06	−239.52
$LiI(s)$	−536.62	−467.45	−232.78
$LiAlH_4(s)$	−1472.7	−1270.0	−683.42
$LiBH_4(s)$	−1401.9	−1333.4	−675.21
	Magnesium		
$Mg(s)$	−147.70	−113.10	−115.97
$Mg(g)$	0	0	0
$Mg^{2+}(aq)$	−614.55	−567.9	−286.8
$MgO(s)$	−998.57	−914.26	−282.76
$MgH_2(s)$	−658.3	−555.5	−346.99
$Mg(OH)_2(s)$	−2005.88	−1816.64	−637.01
$MgCl_2(s)$	−1032.38	−916.25	−389.43
$MgCO_3(s)$	−2707.7	−2491.7	−724.2
$MgSO_4(s)$	−2708.1	−2448.9	−869.1
	Manganese		
$Mn(s)$	−280.7	−238.5	−141.69
$Mn(g)$	0	0	0
$Mn^{2+}(aq)$	−501.5	−466.6	−247.3
$MnO(s)$	−915.1	−833.1	−275.05
$MnO_2(s)$	−1299.1	−1167.1	−442.76
$Mn_2O_3(s)$	−2267.9	−2053.3	−720.1
$Mn_3O_4(s)$	−3226.6	−2925.6	−1009.7
$KMnO_4(s)$	−2203.8	−1963.6	−806.50
$MnS(s)$	−773.7	−695.2	−263.3
	Mercury		
$Hg(l)$	−61.317	−31.820	−98.94
$Hg(g)$	0	0	0
$Hg^{2+}(aq)$	+109.8	+132.58	−207.2
$HgO(s)$	−401.32	−322.090	−265.73
$HgCl_2(s)$	−529.0	−421.8	−359.4
$Hg_2Cl_2(s)$	−631.21	−485.745	−487.8
$HgS(s)$	−398.3	−320.67	−260.4
	Nitrogen		
$N_2(g)$	−945.408	−911.26	−114.99
$N(g)$	0	0	0
$NO(g)$	−631.62	−600.81	−103.592
$NO_2(g)$	−937.86	−867.78	−235.35
$N_2O(g)$	−1112.53	−1038.79	−247.80

(continued)

Table B.13

Standard-State Enthalpies, Free Energies, and Entropies of Atom Combination, 298.15 K (continued)

Substance	ΔH°_{ac} (kJ/mol$_{rxn}$)	ΔG°_{ac} (kJ/mol$_{rxn}$)	ΔS°_{ac} (J/mol$_{rxn}$ · K)
$N_2O_3(g)$	−1609.20	−1466.99	−477.48
$N_2O_4(g)$	−1932.93	−1740.29	−646.53
$N_2O_5(g)$	−2179.91	−1954.8	−756.2
$NO_3^-(aq)$	−1425.2	−1259.56	−490.1
$NOCl(g)$	−791.84	−726.96	−217.86
$NO_2Cl(g)$	−1080.12	−970.4	−368.46
$HNO_2(aq)$	−1307.9	−1172.9	−454.5
$HNO_3(g)$	−1572.92	−1428.79	−484.80
$HNO_3(aq)$	−1645.22	−1465.32	−604.8
$NH_3(g)$	−1171.76	−1081.82	−304.99
$NH_3(aq)$	−1205.94	−1091.87	−386.1
$NH_4^+(aq)$	−1475.81	−1347.93	−498.8
$NH_4NO_3(s)$	−2929.08	−2603.31	−1097.53
$NH_4NO_3(aq)$	−2903.39	−2610.00	−988.8
$NH_4Cl(s)$	−1779.41	−1578.17	−682.7
$N_2H_4(l)$	−1765.38	−1574.91	−644.24
$N_2H_4(g)$	−1720.61	−1564.90	−526.98
$HN_3(g)$	−1341.7	−1242.0	−335.37
Oxygen			
$O_2(g)$	−498.340	−463.462	−116.972
$O(g)$	0	0	0
$O_3(g)$	−604.8	−532.0	−244.24
Phosphorus			
P(white)	−314.64	−278.25	−122.10
$P_4(g)$	−1199.65	−1088.6	−372.79
$P_2(g)$	−485.0	−452.8	−108.257
$P(g)$	0	0	0
$PH_3(g)$	−962.2	−874.6	−297.10
$P_4O_6(s)$	−4393.7	—	—
$P_4O_{10}(s)$	−6734.3	−6128.0	−2034.46
$PO_4^{3-}(aq)$	−2588.7	−2223.9	−1029
$PF_3(g)$	−1470.4	−1361.5	−366.22
$PF_5(g)$	−2305.4	—	—
$PCl_3(l)$	−999.4	−867.6	−441.7
$PCl_3(g)$	−966.7	−863.1	−347.01
$PCl_5(g)$	−1297.9	−1111.6	−624.60
$H_3PO_4(s)$	−3243.3	−2934.0	−1041.05
$H_3PO_4(aq)$	−3241.7	−2833.6	−1374
Potassium			
$K(s)$	−89.24	−60.59	−96.16
$K(g)$	0	0	0
$K^+(aq)$	−341.62	−343.86	−57.8
$KOH(s)$	−980.82	−874.65	−357.2
$KCl(s)$	−647.67	−575.41	−242.94
$KNO_3(s)$	−1804.08	−1606.27	−663.75
$K_2Cr_2O_7(s)$	−4777.4	−4328.7	−1505.9
$KMnO_4(s)$	−2203.8	−1963.6	−806.50

Substance	ΔH°_{ac} (kJ/mol$_{rxn}$)	ΔG°_{ac} (kJ/mol$_{rxn}$)	ΔS°_{ac} (J/mol$_{rxn}$ · K)
	Silicon		
$Si(s)$	−455.6	−411.3	−149.14
$Si(g)$	0	0	0
$SiO_2(s)$	−1864.9	−1731.4	−448.24
$SiH_4(g)$	−1291.9	−1167.4	−422.20
$SiF_4(g)$	−2386.5	−2231.6	−520.50
$SiCl_4(l)$	−1629.3	−1453.9	−589
$SiCl_4(g)$	−1599.3	−1451.0	−498.03
	Silver		
$Ag(s)$	−284.55	−245.65	−130.42
$Ag(g)$	0	0	0
$Ag^+(aq)$	−178.97	−168.54	−100.29
$Ag(NH_3)_2^+(aq)$	−2647.15	−2393.51	−922.6
$Ag_2O(s)$	−849.32	−734.23	−385.7
$AgCl(s)$	−533.30	−461.12	−242.0
$AgBr(s)$	−496.80	−424.95	−240.9
$AgI(s)$	−453.23	−382.34	−238.3
	Sodium		
$Na(s)$	−107.32	−76.761	−102.50
$Na(g)$	0	0	0
$Na^+(aq)$	−347.45	−338.666	−94.7
$NaH(s)$	−3811.25	−313.47	−228.409
$NaOH(s)$	−999.75	−891.233	−365.025
$NaOH(aq)$	−1044.25	−930.889	−381.4
$NaCl(s)$	−640.15	−566.579	−246.78
$NaCl(g)$	−405.65	−379.10	−89.10
$NaCl(aq)$	−636.27	−575.574	−203.4
$NaNO_3(s)$	−1795.38	−1594.58	−673.66
$Na_3PO_4(s)$	−3550.68	−3224.26	−1094.75
$Na_2SO_3(s)$	−2363.96	−2115.0	−803
$Na_2SO_4(s)$	−2877.20	−2588.86	−969.89
$Na_2CO_3(s)$	−2809.51	−2564.41	−813.70
$NaHCO_3(s)$	−2739.97	−2497.5	−808.0
$NaCH_3CO_2(s)$	−3400.78	−3099.66	−1013.2
$Na_2CrO_4(s)$	−2950.1	−2667.18	−949.53
$Na_2Cr_2O_7(s)$	−4730.6	—	—
	Sulfur		
$S_8(s)$	−2230.440	−1906.00	−1310.77
$S_8(g)$	−2128.14	−1856.37	−911.59
$S(g)$	0	0	0
$S^{2-}(aq)$	−245.7	−152.4	−182.4
$SO_2(g)$	−1073.95	−1001.906	−241.71
$SO_3(s)$	−1480.82	−1307.65	−580.3
$SO_3(l)$	−1467.36	−1307.19	−537.2
$SO_3(g)$	−1422.04	−1304.50	−394.23
$SO_4^{2-}(aq)$	−2184.76	−1909.70	−791.9
$SOCl_2(g)$	−983.8	−879.6	−349.50
$SO_2Cl_2(g)$	−1384.5	−1233.1	−508.39
$H_2S(g)$	−734.74	−678.30	−191.46
$H_2SO_3(aq)$	−2070.43	−1877.75	−648.2
$H_2SO_4(aq)$	−2620.06	−2316.20	−1021.4

(continued)

Table B.13

Standard-State Enthalpies, Free Energies, and Entropies of Atom Combination, 298.15 K (continued)

Substance	ΔH°_{ac} (kJ/mol$_{rxn}$)	ΔG°_{ac} (kJ/mol$_{rxn}$)	ΔS°_{ac} (J/mol$_{rxn}$ · K)
$SF_4(g)$	−1369.66	−1217.2	−510.81
$SF_6(g)$	−1962	−1715.0	−828.53
$SCN^-(aq)$	−1391.75	−1272.43	−334.9
	Tin		
$Sn(s)$	−302.1	−267.3	−124.35
$Sn(g)$	0	0	0
$SnO(s)$	−837.1	−755.9	−273.0
$SnO_2(s)$	−1381.1	−1250.5	−438.3
$SnCl_2(s)$	−870.6	—	—
$SnCl_4(l)$	−1300.1	−249.8	−570.7
$SnCl_4(g)$	−1260.3	−1122.2	−463.5
	Titanium		
$Ti(s)$	−469.9	−425.1	−149.6
$Ti(g)$	0	0	0
$TiO(s)$	−1238.8	−1151.8	−306.5
$TiO_2(s)$	−1913.0	−1778.1	−452.0
$TiCl_4(l)$	−1760.8	−1585.0	−588.7
$TiCl_4(g)$	−1719.8	−1574.6	−486.2
	Tungsten		
$W(s)$	−849.4	−807.1	−141.31
$W(g)$	0	0	0
$WO_3(s)$	−2439.8	−2266.4	−581.22
	Zinc		
$Zn(s)$	−130.729	−95.145	−119.35
$Zn(g)$	0	0	0
$Zn^{2+}(aq)$	−284.62	−242.21	−273.1
$ZnO(s)$	−728.18	−645.18	−278.40
$ZnCl_2(s)$	−789.14	−675.90	−379.92
$ZnS(s)$	−615.51	−534.69	−271.1
$ZnSO_4(s)$	−2389.0	−2131.8	−862.5

Table B.14
Bond Dissociation Enthalpies

Single-Bond Dissociation Enthalpies (kJ/mol)

	As	B	Br	C	Cl	F	H	I	N	O	P	S	Si
As	180		255	200	310	485	300	180		330			
B		300	370		445	645		270		525			
Br			195	270	220	240	370	180	250		270	215	330
C				350	330	490	415	210	305	360	265	270	305
Cl					240	250	431	210	190	205	330	270	400
F						160	569		280	215	500	325	600
H							435	300	390	464	325	370	320
I								150		200	180		230
N									160	165			330
O										140	370	423	464
P											210		
S												260	
Si													225

Double- and Triple-Bond Dissociation Enthalpies (kJ/mol)

Bond	kJ/mol	Bond	kJ/mol
C=C	611	C=S	477
C≡C	837	N=N	418
C=O	745	N≡N	946
C≡O	1075	N=O	594
C=N	615	O=O	498
C≡N	891	S=O	523

Table B. 15
Electron Configurations of the First 86 Elements

Atomic Number	Symbol	Electron Configuration
1	H	$1s^1$
2	He	$1s^2 = [He]$
3	Li	$[He]\ 2s^1$
4	Be	$[He]\ 2s^2$
5	B	$[He]\ 2s^2\ 2p^1$
6	C	$[He]\ 2s^2\ 2p^2$
7	N	$[He]\ 2s^2\ 2p^3$
8	O	$[He]\ 2s^2\ 2p^4$
9	F	$[He]\ 2s^2\ 2p^5$
10	Ne	$[He]\ 2s^2\ 2p^6 = [Ne]$
11	Na	$[Ne]\ 3s^1$
12	Mg	$[Ne]\ 3s^2$
13	Al	$[Ne]\ 3s^2\ 3p^1$
14	Si	$[Ne]\ 3s^2\ 3p^2$
15	P	$[Ne]\ 3s^2\ 3p^3$
16	S	$[Ne]\ 3s^2\ 3p^4$
17	Cl	$[Ne]\ 3s^2\ 3p^5$
18	Ar	$[Ne]\ 3s^2\ 3p^6 = [Ar]$
19	K	$[Ar]\ 4s^1$
20	Ca	$[Ar]\ 4s^2$
21	Sc	$[Ar]\ 4s^2\ 3d^1$
22	Ti	$[Ar]\ 4s^2\ 3d^2$
23	V	$[Ar]\ 4s^2\ 3d^3$
24	Cr	$[Ar]\ 4s^1\ 3d^5$
25	Mn	$[Ar]\ 4s^2\ 3d^5$
26	Fe	$[Ar]\ 4s^2\ 3d^6$
27	Co	$[Ar]\ 4s^2\ 3d^7$
28	Ni	$[Ar]\ 4s^2\ 3d^8$
29	Cu	$[Ar]\ 4s^1\ 3d^{10}$
30	Zn	$[Ar]\ 4s^2\ 3d^{10}$
31	Ga	$[Ar]\ 4s^2\ 3d^{10}\ 4p^1$
32	Ge	$[Ar]\ 4s^2\ 3d^{10}\ 4p^2$
33	As	$[Ar]\ 4s^2\ 3d^{10}\ 4p^3$
34	Se	$[Ar]\ 4s^2\ 3d^{10}\ 4p^4$
35	Br	$[Ar]\ 4s^2\ 3d^{10}\ 4p^5$
36	Kr	$[Ar]\ 4s^2\ 3d^{10}\ 4p^6 = [Kr]$
37	Rb	$[Kr]\ 5s^1$
38	Sr	$[Kr]\ 5s^2$
39	Y	$[Kr]\ 5s^2\ 4d^1$
40	Zr	$[Kr]\ 5s^2\ 4d^2$
41	Nb	$[Kr]\ 5s^1\ 4d^4$
42	Mo	$[Kr]\ 5s^1\ 4d^5$
43	Tc	$[Kr]\ 5s^2\ 4d^5$
44	Ru	$[Kr]\ 5s^1\ 4d^7$
45	Rh	$[Kr]\ 5s^1\ 4d^8$
46	Pd	$[Kr]\ 4d^{10}$
47	Ag	$[Kr]\ 5s^1\ 4d^{10}$
48	Cd	$[Kr]\ 5s^2\ 4d^{10}$
49	In	$[Kr]\ 5s^2\ 4d^{10}\ 5p^1$
50	Sn	$[Kr]\ 5s^2\ 4d^{10}\ 5p^2$
51	Sb	$[Kr]\ 5s^2\ 4d^{10}\ 5p^3$

Atomic Number	Symbol	Electron Configuration
52	Te	[Kr] $5s^2\ 4d^{10}\ 5p^4$
53	I	[Kr] $5s^2\ 4d^{10}\ 5p^5$
54	Xe	[Kr] $5s^2\ 4d^{10}\ 5p^6$ = [Xe]
55	Cs	[Xe] $6s^1$
56	Ba	[Xe] $6s^2$
57	La	[Xe] $6s^2\ 5d^1$
58	Ce	[Xe] $6s^2\ 4f^1\ 5d^1$
59	Pr	[Xe] $6s^2\ 4f^3$
60	Nd	[Xe] $6s^2\ 4f^4$
61	Pm	[Xe] $6s^2\ 4f^5$
62	Sm	[Xe] $6s^2\ 4f^6$
63	Eu	[Xe] $6s^2\ 4f^7$
64	Gd	[Xe] $6s^2\ 4f^7\ 5d^1$
65	Tb	[Xe] $6s^2\ 4f^9$
66	Dy	[Xe] $6s^2\ 4f^{10}$
67	Ho	[Xe] $6s^2\ 4f^{11}$
68	Er	[Xe] $6s^2\ 4f^{12}$
69	Tm	[Xe] $6s^2\ 4f^{13}$
70	Yb	[Xe] $6s^2\ 4f^{14}$
71	Lu	[Xe] $6s^2\ 4f^{14}\ 5d^1$
72	Hf	[Xe] $6s^2\ 4f^{14}\ 5d^2$
73	Ta	[Xe] $6s^2\ 4f^{14}\ 5d^3$
74	W	[Xe] $6s^2\ 4f^{14}\ 5d^4$
75	Re	[Xe] $6s^2\ 4f^{14}\ 5d^5$
76	Os	[Xe] $6s^2\ 4f^{14}\ 5d^6$
77	Ir	[Xe] $6s^2\ 4f^{14}\ 5d^7$
78	Pt	[Xe] $6s^1\ 4f^{14}\ 5d^9$
79	Au	[Xe] $6s^1\ 4f^{14}\ 5d^{10}$
80	Hg	[Xe] $6s^2\ 4f^{14}\ 5d^{10}$
81	Tl	[Xe] $6s^2\ 4f^{14}\ 5d^{10}\ 6p^1$
82	Pb	[Xe] $6s^2\ 4f^{14}\ 5d^{10}\ 6p^2$
83	Bi	[Xe] $6s^2\ 4f^{14}\ 5d^{10}\ 6p^3$
84	Po	[Xe] $6s^2\ 4f^{14}\ 5d^{10}\ 6p^4$
85	At	[Xe] $6s^2\ 4f^{14}\ 5d^{10}\ 6p^5$
86	Rn	[Xe] $6s^2 4f^{14} 5d^{10} 6p^6$ = [Rn]

Table B.16

Standard-State Enthalpy of Formation, Free Energy of Formation, and Absolute Entropy Data, 298.15 K

Substance	ΔH_f° (kJ/mol$_{rxn}$)	ΔG_f° (kJ/mol$_{rxn}$)	S° (J/mol$_{rxn}$ · K)
	Aluminum		
$Al(s)$	0	0	28.33
$Al(g)$	326.4	285.7	164.54
$Al^{3+}(aq)$	−531	−485	−321.7
$Al_2O_3(s)$	−1675.7	−1582.3	50.92
$AlCl_3(s)$	−704.2	−628.8	110.67
$AlF_3(s)$	−1504.1	−1425.0	66.44
$Al_2(SO_4)_3(s)$	−3440.84	−3099.94	239.3
	Barium		
$Ba(s)$	0	0	62.8
$Ba(g)$	180	146	170.243
$Ba^{2+}(aq)$	−537.64	−560.77	9.6
$BaO(s)$	−553.5	−525.1	70.42
$Ba(OH)_2 \cdot 8\,H_2O(s)$	−3342.2	−2792.8	427
$BaCl_2(s)$	−858.6	−810.4	123.68
$BaCl_2(aq)$	−871.95	−823.21	122.6
$BaSO_4(s)$	−1473.2	−1362.2	132.2
$Ba(NO_3)_2(s)$	−992.07	−769.59	213.8
$Ba(NO_3)_2(aq)$	−952.36	−783.28	302.5
	Beryllium		
$Be(s)$	0	0	9.50
$Be(g)$	324.3	286.6	136.269
$Be^{2+}(aq)$	−382.8	−379.73	−129.7
$BeO(s)$	−609.6	−508.3	14.14
$BeCl_2(s)$	−490.4	−445.6	82.68
	Bismuth		
$Bi(s)$	0	0	56.74
$Bi(g)$	207.1	168.2	187.05
$Bi_2O_3(s)$	−573.88	−493.7	151.5
$BiCl_3(s)$	−379.1	−315.0	177.0
$BiCl_3(g)$	−265.7	−256.0	358.85
$Bi_2S_3(s)$	−143.1	−140.6	200.4
	Boron		
$B(s)$	0	0	5.86
$B(g)$	562.7	518.8	153.45
$B_2O_3(s)$	−1272.77	−1193.65	53.97
$B_2H_6(g)$	35.6	86.7	232.11
$B_5H_9(l)$	42.68	171.82	184.22
$B_5H_9(g)$	73.2	175.0	275.92
$B_{10}H_{14}(s)$	−45.2	192.3	176.56
$H_3BO_3(s)$	−1094.33	−968.92	88.83
$BF_3(g)$	−1137.00	−1120.33	254.12
$BCl_3(l)$	−427.2	−387.4	206.3
$B_3N_3H_6(l)$	−541.00	−392.65	199.6
$B_3N_3H_6(g)$	−511.75	−390.00	288.68

Substance	ΔH_f° (kJ/mol_{rxn})	ΔG_f° (kJ/mol_{rxn})	S° ($J/mol_{rxn} \cdot K$)
	Bromine		
$Br_2(l)$	0	0	152.231
$Br_2(g)$	30.907	3.110	245.463
$Br(g)$	111.884	82.396	175.022
$HBr(g)$	−36.40	−53.45	198.695
$HBr(aq)$	−121.55	−103.96	82.4
$BrF(g)$	−93.85	−109.18	228.97
$BrF_3(g)$	−255.60	−229.43	292.53
$BrF_5(g)$	−428.9	−350.6	320.19
	Calcium		
$Ca(s)$	0	0	41.42
$Ca(g)$	178.2	144.3	154.884
$Ca^{2+}(aq)$	−542.83	−553.58	−53.1
$CaO(s)$	−635.09	−604.03	39.75
$Ca(OH)_2(s)$	−986.09	−898.49	83.39
$CaCl_2(s)$	−795.8	−748.1	104.6
$CaSO_4(s)$	−1434.11	−1321.79	106.7
$CaSO_4 \cdot 2H_2O(s)$	−2022.63	−1797.28	194.1
$Ca(NO_3)_2(s)$	−938.39	−743.07	193.3
$CaCO_3(s)$	−1206.92	−1128.79	92.9
$Ca_3(PO_4)_2$	−4120.8	−3884.7	236.0
	Carbon		
C(graphite)	0	0	5.74
C(diamond)	1.895	2.900	2.377
$C(g)$	716.682	671.257	158.096
$CO(g)$	−110.525	−137.168	197.674
$CO_2(g)$	−393.509	−394.359	213.74
$COCl_2(g)$	−218.8	−204.6	283.53
$CH_4(g)$	−74.81	−50.752	186.264
$HCHO(g)$	−108.57	−102.53	218.77
$H_2CO_3(aq)$	−699.65	−623.08	187.4
$HCO_3^-(aq)$	−691.99	−586.77	91.2
$CO_3^{2-}(aq)$	−677.14	−527.81	−56.9
$CH_3OH(l)$	−238.66	−166.27	126.8
$CH_3OH(g)$	−200.66	−161.96	239.81
$CCl_4(l)$	−135.44	−65.21	216.40
$CCl_4(g)$	−102.9	−60.59	309.85
$CHCl_3(l)$	−134.47	−73.66	201.1
$CHCl_3(g)$	−103.14	−70.34	295.71
$CH_2Cl_2(l)$	−121.46	−67.26	177.8
$CH_2Cl_2(g)$	−92.47	−65.87	270.23
$CH_3Cl(g)$	−80.84	−57.40	234.5
$CS_2(l)$	89.70	65.27	151.34
$CS_2(g)$	117.36	67.12	237.84
$HCN(g)$	135.1	124.7	201.78
$CH_3NO_2(l)$	−113.09	−14.42	171.75
$C_2H_2(g)$	226.73	209.20	200.94
$C_2H_4(g)$	52.26	68.15	219.56
$C_2H_6(g)$	−84.68	−32.82	229.60
$CH_3CHO(l)$	−192.30	−128.12	160.2
$CH_3CO_2H(l)$	−484.5	−389.9	159.8

(continued)

Table B.16

Standard-State Enthalpy of Formation, Free Energy of Formation, and Absolute Entropy Data, 298.15 K (continued)

Substance	ΔH_f° (kJ/mol_{rxn})	ΔG_f° (kJ/mol_{rxn})	S° ($J/mol_{rxn} \cdot K$)
$CH_3CO_2H(g)$	−432.25	−374.0	282.5
$CH_3CO_2H(aq)$	−485.76	−396.46	178.7
$CH_3CO_2^-(aq)$	−486.01	−369.31	86.6
$CH_3CH_2OH(l)$	−277.69	−174.78	160.7
$CH_3CH_2OH(g)$	−235.10	−168.49	282.70
$CH_3CH_2OH(aq)$	−288.3	−181.64	148.5
$C_6H_6(l)$	49.028	124.50	172.8
$C_6H_6(g)$	82.927	129.66	269.2
	Chlorine		
$Cl_2(g)$	0	0	223.066
$Cl(g)$	121.679	105.680	165.198
$Cl^-(aq)$	−167.159	−131.228	56.5
$ClO_2(g)$	102.5	120.5	256.84
$Cl_2O(g)$	80.3	97.9	266.21
$Cl_2O_7(l)$	238		
$HCl(g)$	−92.307	−95.299	186.908
$HCl(aq)$	−167.159	−131.228	56.5
$ClF(g)$	−54.48	−55.94	217.89
	Chromium		
$Cr(s)$	0	0	23.77
$Cr(g)$	396.6	351.8	174.50
$CrO_3(s)$	−589.5		
$CrO_4^{2-}(aq)$	−881.15	−727.75	50.21
$Cr_2O_3(s)$	−1139.7	−1058.1	81.2
$Cr_2O_7^{2-}(aq)$	−1490.3	−1301.1	261.9
$(NH_4)_2Cr_2O_7(s)$	−1806.7		
$PbCrO_4(s)$	−930.9		
	Cobalt		
$Co(s)$	0	0	30.04
$Co(g)$	424.7	380.3	179.515
$Co^{2+}(aq)$	−58.2	−54.4	−113
$Co^{3+}(aq)$	92	134	−305
$CoO(s)$	−237.94	−214.20	52.97
$Co_3O_4(s)$	−891	−774	102.5
$Co(NH_3)_6^{3+}(aq)$	−584.9	−157.0	146
	Copper		
$Cu(s)$	0	0	33.150
$Cu(g)$	338.32	298.58	166.38
$Cu^+(aq)$	71.67	49.98	40.6
$Cu^{2+}(aq)$	64.77	65.49	−99.6
$CuO(s)$	−157.3	−129.7	42.63
$Cu_2O(s)$	−168.6	−146.0	93.14
$CuCl_2(s)$	−220.1	−175.7	108.07
$CuS(s)$	−53.1	−53.6	66.5
$Cu_2S(s)$	−79.5	−86.2	120.9
$CuSO_4(s)$	−771.36	−66.69	109
$Cu(NH_3)_4^{2+}(aq)$	−348.5	−111.07	273.6

Substance	ΔH_f° (kJ/mol$_{rxn}$)	ΔG_f° (kJ/mol$_{rxn}$)	S° (J/mol$_{rxn}$ · K)
	Fluorine		
$F_2(g)$	0	0	202.78
$F(g)$	78.99	61.91	158.754
$F^-(aq)$	−332.63	−278.79	−13.8
$HF(g)$	−271.1	−273.2	173.779
$HF(aq)$	−320.08	−296.82	88.7
	Hydrogen		
$H_2(g)$	0	0	130.684
$H(g)$	217.65	203.247	114.713
$H^+(aq)$	0	0	0
$OH^-(aq)$	−229.994	−157.244	−10.75
$H_2O(l)$	−285.830	−237.129	69.91
$H_2O(g)$	−241.818	−228.572	188.25
$H_2O_2(l)$	−187.78	−120.35	109.6
$H_2O_2(aq)$	−191.17	−134.03	143.9
	Iodine		
$I_2(s)$	0	0	116.135
$I_2(g)$	62.438	19.327	260.69
$I(g)$	106.838	70.50	180.791
$HI(g)$	26.48	1.70	206.594
$IF(g)$	−95.65	−118.51	236.17
$IF_5(g)$	−822.49	−751.73	327.7
$IF_7(g)$	−943.9	−818.3	346.5
$ICl(g)$	17.78	−5.46	247.551
$IBr(g)$	40.84	3.69	258.773
	Iron		
$Fe(s)$	0	0	27.28
$Fe(g)$	416.3	370.7	180.49
$Fe^{2+}(aq)$	−89.1	−78.90	−137.7
$Fe^{3+}(aq)$	−48.5	−4.7	−315.9
$Fe_2O_3(s)$	−824.2	−742.2	87.40
$Fe_3O_4(s)$	−1118.4	−1015.4	146.4
$Fe(OH)_2(s)$	−569.0	−486.5	88
$Fe(OH)_3(s)$	−823.0	−696.5	106.7
$FeCl_3(s)$	−399.49	−334.00	142.3
$FeS_2(s)$	−178.2	−166.9	52.93
$Fe(CO)_5(l)$	−774.0	−705.3	338.1
$Fe(CO)_5(g)$	−733.9	−697.21	445.3
	Lead		
$Pb(s)$	0	0	64.81
$Pb(g)$	195.0	161.9	175.373
$Pb^{2+}(aq)$	−1.7	−24.43	10.5
$PbO(s)$	−217.32	−187.89	68.7
$PbO_2(s)$	−277.4	−217.33	68.6
$PbCl_2(s)$	−359.41	−314.10	136.0
$PbCl_4(l)$	−329.3		
$PbS(s)$	−100.4	−98.7	91.2
$PbSO_4(s)$	−919.94	−813.14	148.57
$Pb(NO_3)_2(s)$	−451.9		
$PbCO_3(s)$	−699.1	−625.5	131.0

(*continued*)

Table B.16

Standard-State Enthalpy of Formation, Free Energy of Formation, and Absolute Entropy Data, 298.15 K (continued)

Substance	ΔH_f° (kJ/mol$_{rxn}$)	ΔG_f° (kJ/mol$_{rxn}$)	S° (J/mol$_{rxn}$ · K)
Lithium			
Li(*s*)	0	0	29.12
Li(*g*)	159.37	126.66	138.77
Li^+ (*aq*)	−278.49	−293.31	13.4
LiH(*s*)	−90.54	−68.35	20.08
LiOH(*s*)	−484.93	−438.95	42.80
LiF(*s*)	−615.97	−587.71	35.65
LiCl(*s*)	−408.61	−384.37	59.33
LiBr(*s*)	−351.23	−342.00	74.27
LiI(*s*)	−270.41	−270.29	86.78
$LiAlH_4$(*s*)	−116.3	−44.7	78.74
$LiBH_4$(*s*)	190.8	125.0	75.86
Magnesium			
Mg(*s*)	0	0	32.68
Mg(*g*)	147.70	113.10	148.650
Mg^{2+} (*aq*)	−466.85	−454.8	−138.1
MgO(*s*)	−601.70	−569.43	26.94
MgH_2(*s*)	−75.3	−35.9	31.09
$Mg(OH)_2$(*s*)	−924.54	−833.58	63.18
$MgCl_2$(*s*)	−641.32	−591.79	89.62
$MgCO_3$(*s*)	−1095.8	−1012.1	65.7
$MgSO_4$(*s*)	−1284.9	−1170.6	91.6
Manganese			
Mn(*s*)	0	0	32.01
Mn(*g*)	280.7	238.5	173.70
Mn^{2+}(*aq*)	−220.75	−228.1	−73.6
MnO(*s*)	−385.22	−362.90	59.71
MnO_2(*s*)	−520.03	−465.14	53.05
Mn_2O_3(*s*)	−959.0	−881.1	110.5
Mn_3O_4(*s*)	−1387.8	−1283.2	155.6
$KMnO_4$(*s*)	−837.2	−737.6	171.76
MnS (*s*)	−214.2	−218.4	78.2
Mercury			
Hg(*l*)	0	0	76.02
Hg(*g*)	61.317	31.820	174.96
Hg^{2+}(*aq*)	171.1	164.40	−32.2
HgO(*s*)	−90.83	−58.539	70.29
$HgCl_2$(*s*)	−224.3	−178.6	146.0
Hg_2Cl_2(*s*)	−265.22	−210.745	192.5
HgS(*s*)	−58.2	−50.6	82.4
Nitrogen			
N_2(*g*)	0	0	191.61
N(*g*)	472.704	455.63	153.298
NO(*g*)	90.25	86.55	210.761
NO_2(*g*)	33.18	51.31	240.06
N_2O(*g*)	82.05	104.20	219.85
N_2O_3(*g*)	83.72	139.46	312.28

Substance	ΔH_f° (kJ/mol$_{rxn}$)	ΔG_f° (kJ/mol$_{rxn}$)	S° (J/mol$_{rxn}$ · K)
$N_2O_4(g)$	9.16	97.89	304.29
$N_2O_5(g)$	11.35	115.1	355.7
$NO_3^-(aq)$	−205.0	−108.74	146.4
$NOCl(g)$	51.71	66.08	261.69
$NO_2Cl(g)$	12.6	54.4	272.15
$HNO_2(aq)$	−119.2	−50.6	135.6
$HNO_3(g)$	−135.06	−74.72	266.38
$HNO_3(aq)$	−207.36	−111.25	146.4
$NH_3(g)$	−46.11	−16.45	192.45
$NH_3(aq)$	−80.29	−26.50	111.3
$NH_4^+(aq)$	−132.51	−79.31	113.4
$NH_4NO_3(s)$	−365.56	−183.87	151.08
$NH_4NO_3(aq)$	−339.87	−190.56	259.8
$NH_4Cl(s)$	−314.43	−203.87	94.6
$N_2H_4(l)$	50.63	149.34	121.21
$N_2H_4(g)$	95.40	159.35	238.47
$HN_3(g)$	294.1	328.1	238.97
Oxygen			
$O_2(g)$	0	0	205.138
$O(g)$	249.170	231.731	161.055
$O_3(g)$	142.7	163.2	238.93
Phosphorus			
P(white)	0	0	41.09
$P_4(g)$	58.91	24.4	279.98
$P_2(g)$	144.3	103.7	218.129
$P(g)$	314.64	278.25	163.193
$PH_3(g)$	5.4	13.4	210.23
$P_4O_6(s)$	−1640.1		
$P_4O_{10}(s)$	−2984.0	−2697.7	228.86
$PO_4^{3-}(aq)$	−1277.4	−1018.7	−222
$PF_3(g)$	−918.8	−897.5	273.24
$PF_5(g)$	−1595.8		
$PCl_3(l)$	−319.7	−272.3	217.1
$PCl_3(g)$	−287.0	−267.8	311.78
$PCl_5(g)$	−374.9	−305.0	364.58
$H_3PO_4(s)$	−1279.0	−1119.1	110.50
$H_3PO_4(aq)$	−1277.4	−1018.7	−222
Potassium			
$K(s)$	0	0	64.18
$K(g)$	89.24	60.59	160.336
$K^+(aq)$	−252.38	−283.27	102.5
$KOH(s)$	−424.764	−379.08	78.9
$KCl(s)$	−436.747	−409.14	82.59
$KNO_3(s)$	−494.63	−394.86	133.05
$K_2Cr_2O_7(s)$	−2061.5	−1881.8	291.2
$KMnO_4(s)$	−837.2	−737.6	171.76
Silicon			
$Si(s)$	0	0	18.83
$Si(g)$	455.6	411.3	167.97
$SiO_2(s)$	−910.94	−856.64	41.84

(continued)

Table B.16

Standard-State Enthalpy of Formation, Free Energy of Formation, and Absolute Entropy Data, 298.15 K (continued)

Substance	ΔH_f° (kJ/mol_{rxn})	ΔG_f° (kJ/mol_{rxn})	S° ($J/mol_{rxn} \cdot K$)
$SiH_4(g)$	34.3	56.9	204.62
$SiF_4(g)$	−1614.94	−1572.65	282.49
$SiCl_4(l)$	−687.0	−619.9	240.
$SiCl_4(g)$	−657.01	−616.98	330.73
	Silver		
$Ag(s)$	0	0	42.55
$Ag(g)$	284.55	245.65	172.97
$Ag^+(aq)$	105.579	77.107	72.68
$Ag(NH_3)_2{}^+(aq)$	−111.29	−17.12	245.2
$Ag_2O(s)$	−31.05	−11.20	−121.3
$AgCl(s)$	−127.068	−109.789	96.2
$AgBr(s)$	−100.37	−96.90	107.1
$AgI(s)$	−61.84	−66.19	−115.5
	Sodium		
$Na(s)$	0	0	51.21
$Na(g)$	107.32	76.761	153.712
$Na^+(aq)$	−240.13	−261.905	59.0
$NaH(s)$	−56.275	−33.46	−40.016
$NaOH(s)$	−425.609	−379.494	64.455
$NaOH(aq)$	−470.114	−419.150	48.1
$NaCl(s)$	−411.153	−384.138	72.13
$NaCl(g)$	−176.65	−196.66	229.81
$NaCl(aq)$	−407.27	−393.133	115.5
$NaNO_3(s)$	−467.85	−367.00	116.52
$Na_3PO_4(s)$	−1917.40	−1788.80	173.80
$Na_2SO_3(s)$	−1123.0	−1028.0	155
$Na_2SO_4(s)$	−1387.08	−1270.16	149.58
$Na_2CO_3(s)$	−1130.68	−1044.44	134.98
$NaHCO_3(s)$	−950.81	−851.0	101.7
$NaCH_3CO_2(s)$	−708.81	−607.18	123.0
$Na_2CrO_4(s)$	−1342.2	−1234.93	176.61
$Na_2Cr_2O_7(s)$	−1978.6		
	Sulfur		
$S_8(s)$	0	0	31.80
$S_8(g)$	102.30	49.63	430.98
$S(g)$	278.805	238.250	167.821
$S^{2-}(aq)$	33.1	85.8	−14.6
$SO_2(g)$	−296.830	−300.194	248.22
$SO_3(s)$	−454.51	−374.21	70.7
$SO_3(l)$	−441.04	−373.75	113.8
$SO_3(g)$	−395.72	−371.06	256.76
$SO_4{}^{2-}(aq)$	−909.27	−744.53	20.1
$SOCl_2(g)$	−212.5	−198.3	309.77
$SO_2Cl_2(g)$	−364.0	−320.0	311.94
$H_2S(g)$	−20.63	−33.56	205.79
$H_2SO_3(aq)$	−608.81	−537.81	232.2
$H_2SO_4(aq)$	−909.27	−744.53	20.1
$SF_4(g)$	−774.9	−731.3	292.03

Substance	ΔH_f° (kJ/mol$_{rxn}$)	ΔG_f° (kJ/mol$_{rxn}$)	S° (J/mol$_{rxn}$ · K)
$SF_6(g)$	−1209	−1105.3	291.82
$SCN^-(aq)$	76.44	92.71	144.3
	Tin		
$Sn(s)$	0	0	44.14
$Sn(g)$	302.1	267.3	168.486
$SnO(s)$	−285.8	−256.9	56.5
$SnO_2(s)$	−580.7	−519.76	52.3
$SnCl_2(s)$	−325.1		
$SnCl_4(l)$	−511.3	−440.21	258.6
$SnCl_4(g)$	−471.5	−432.2	365.8
	Titanium		
$Ti(s)$	0	0	30.63
$Ti(g)$	469.9	425.1	180.2
$TiO(s)$	−519.7	−495.0	34.8
$TiO_2(s)$ rutile	−944.8	−889.5	50.33
$TiCl_4(l)$	−804.2	−737.2	252.3
$TiCl_4(g)$	−763.2	−726.8	354.8
	Tungsten		
$W(s)$	0	0	32.64
$W(g)$	849.4	807.1	173.950
$WO_3(s)$	−842.87	−764.083	75.90
	Zinc		
$Zn(s)$	0	0	41.63
$Zn(g)$	130.729	95.145	160.984
$Zn^{2+}(aq)$	−153.89	−147.06	−112.1
$ZnO(s)$	−348.28	−318.30	43.64
$ZnCl_2(s)$	−415.05	−369.39	111.46
$ZnS(s)$	−205.98	−201.29	57.7
$ZnSO_4(s)$	−982.8	−871.5	110.5

Appendix C

ANSWERS TO SELECTED CORE PROBLEMS

Chapter 1

9. (a) diamond — element
   (b) brass — mixture
   (c) soil — mixture
   (d) glass — mixture
   (e) cotton — mixture
   (f) milk of magnesia — mixture
   (g) salt — compound
   (h) iron — element
   (i) steel — mixture

13. (a) Sb (b) Au (c) Fe (d) Hg (e) K (f) Ag (g) Sn (h) W

14. (a) sodium (b) magnesium (c) aluminum (d) silicon (e) phosphorus (f) chlorine (g) argon

15. (a) titanium (b) vanadium (c) chromium (d) manganese (e) iron (f) cobalt (g) nickel (h) copper (i) zinc

16. (a) molybdenum (b) tungsten (c) rhodium (d) iridium (e) palladium (f) platinum (g) silver (h) gold (i) mercury

22. Atoms of different elements will have different weights and chemical properties.

27. Electron, relative charge: -1
    Proton, relative charge: $+1$
    Neutron, relative charge: 0

29. Electron

30. Nucleus

33. ^{52}Cr

35. 53 protons; 53 atomic number; 74 neutrons; 53 electrons; $Z = 53$; mass number $= 127$

36. ^{20}F

37. ^{79}Se

38. 56 electrons

40. 0.92286

41. 5.96842

43. ^{16}O

47. Three: ^{16}O, ^{17}O, ^{18}O

49. 1.99265×10^{-23} g; 12.0000 amu

50. 1200.0 amu; 1300.3 amu

51. (b)

54. 56 protons; 78 neutrons; 54 electrons

56. 53 protons; 74 neutrons; 54 electrons

63. +2

65. (a) Na_2O_2 (b) $Zn_3(PO_4)_2$ (c) K_2PtCl_6

67. K_2O_2

69. Fe_3O_4

73. (a) IA (b) IVA (c) IIA (d) VIA (e) IIA (f) VIIIA (g) VIIA

75. Seven: H, Li, Na, K, Rb, Cs, Fr

79. (a) N, nonmetal (b) Sb, nonmetal (c) Sc, metal (d) Se, nonmetal
(e) Ge, semimetal (f) Sm, metal (g) Sn, metal (h) Sr, metal

87. Ca

Chapter 2

1. (b) and (d)
5. 65.4 amu
7. 24.305 amu
9. Silicon, Si
12. (a) 107.87 amu
(b) Both isotopes have 47 protons.
(c) Both isotopes have 47 electrons. ^{107}Ag has 60 neutrons and ^{109}Ag has 62 neutrons.
15. (c) is the only true answer.
17. Al
19. Chromium
21. (b), (c), and (d)
25. (a) Ni (b) Zn (c) Ge (d) Pb
27. 2:1
31. 6.0×10^{23} atoms of O; 6.0×10^{23} atoms of P; 6.0×10^{23} atoms of S
33. Mass of ^{1}H:1.674×10^{-24} g; mass of ^{12}C:1.993×10^{-23} g
35. 6.022×10^{23} atoms is 1 mol. One mole of ^{12}C weighs 12.000 g.
37. 1.27×10^{23} atoms
39. 5.26 g
43. 1.0077 g
45. 16.043 g
47. (a) False (b) True (c) False (d) False (e) True
49. (c)
51. (a) 16.043 g/mol (b) 180.155 g/mol
(c) 74.122 g/mol (d) 75.135 g/mol
53. (a) 220.056 g/mol (b) 241.859 g/mol
(c) 294.181 g/mol (d) 310.174 g/mol
55. 169.111 g/mol
57. 195.1 g/mol
61. (a) 76.470% (b) 68.421% (c) 52.000%
63. C = 47.435%; H = 2.5589%; Cl = 50.006%
65. (a) 40.044% (b) 29.439% (c) 38.763%
67. 0.2247 mol P
69. (a) 2.41×10^{23} atoms O (b) 7.5×10^{23} atoms O
(c) 1.4×10^{24} atoms O
71. SnF_2
73. (b)

75. $CuFeS_2$

77. $C_{11}H_{15}NO_2$

79. CuCl

80. No; only the empirical formula can be determined from percent composition. The molar mass must be known to determine the molecular formula.

81. $C_{40}H_{56}$

83. $C_8H_{10}N_4O_2$

87. $C_{10}H_{14}N_2$

101. (a) $2\ Pb(NO_3)_2(s) \longrightarrow 2\ PbO(s) + 4\ NO_2(g) + O_2(g)$
(b) $NH_4NO_2(s) \longrightarrow N_2(g) + 2\ H_2O(g)$
(c) $(NH_4)_2Cr_2O_7(s) \longrightarrow Cr_2O_3(s) + 4\ H_2O(g) + N_2(g)$

103. (a) $PF_3(g) + 3\ H_2O(l) \longrightarrow H_3PO_3(aq) + 3\ HF(aq)$
(b) $P_4O_{10}(s) + 6\ H_2O(l) \longrightarrow 4\ H_3PO_4(aq)$

105. 10 mol CO_2

107. 12 mol CuO

109. 2.25 mol O_2

115. 9.79 g $O_2(g)$

119. 3.73 g PH_3

125. 10.3 g HCl

127. 0.482 mol PF_5

129. 175 g Fe

131. Only (b)

133. 5.99 *M*

135. 14.8 *M*

137. 0.0814 *M*

145. 0.25 mol

151. 0.025 *M*

153. 0.10 *M*

155. 0.267 L

157. 172 mL

161. 36 mL NaI

163. 0.9808 *M* NaOH

165. 0.375 *M*

169. $2\ KClO_3(s) \longrightarrow 2\ KCl(s) + 3\ O_2(g)$

171. CrO_3

175. (d) $ZnCO_3$

177. 32.1 g Mg(*s*)

179. 80%

181. Fe_3O_4

Chapter 3

2. The nucleus

9. 6.0×10^{-7} m

11. Red
13. 11.92; radio wave
15. 5×10^{16} Hz
17. 656.3 nm: 4.568×10^{14} Hz
486.1 nm: 6.167×10^{14} Hz
434.0 nm: 6.908×10^{14} Hz
410.2 nm: 7.309×10^{14} Hz
21. Yellow
23. 3.027×10^{-19} J
25. 4.914×10^{-7} m; visible light
29. The Rutherford model of the atom does not specify where electrons might be found in the atom; the Bohr model does.
31. As the distance between the electron and nucleus increases, the force holding them together decreases according to Coulomb's law.
33. According to the Bohr model, the hydrogen atom consists of a positively charged nucleus and an electron orbiting the nucleus at a discrete radius.
35. 1312 kJ/mol
37. $n = 2$ and $n = 3$ is 3.025×10^{-19} J.
$n = 2$ and $n = 4$ is 4.085×10^{-19} J.
$n = 2$ and $n = \infty$ is 5.447×10^{-19} J.
44. Na < Li < Be < F
46. Na: +1 Mg: +2 Al: +3 Si: +4 P: +5 S: +6 Cl: +7 Ar: +8
K: +1 Ca: +2
49. Ar will have a higher ionization energy than Cl^-.
53. Because the outer shell electrons in chlorine are farther from the nucleus than the outer shell electrons of fluorine.
55. The ionization energy of F^- would be less than that of Ne.
56. Rb < Br < Kr
59. Generally the first ionization energies decrease going from top to bottom of a column of the periodic table.
65. Ca
66. H: 1 He: 2 Li: 1 Be: 2 B: 3 C: 4 N: 5 O: 6 F: 7 Ne: 8
67. In the first three rows, the core charge, number of valence electrons, and group number are all the same.
71. No
75. The relative number of electrons that are ionized at that energy.
80. Electrons are found in three different subshells: 1*s*, 2*s*, and 2*p*.
84. Mg
86. Kr
91. $n = 0, 1, 2, \ldots, \infty$; $l = 0, \ldots, n - 1$; $m_l = -l, \ldots, 0, \ldots, +l$
95. If $n = 1$, $l = 0$, $m_l = 0$, and $m_s = +1/2, -1/2$.
If $n = 2$, $l = 0$, $m_l = 0$, $m_s = +1/2, -1/2$ and $l = 1$, $m_l = -1, 0, +1$, $m_s = +1/2, -1/2$.
97. 9 orbitals in the $n = 3$ shell; 16 orbitals in the $n = 4$ shell; 25 orbitals in the $n = 5$ shell.

99. Paired electrons are electrons occupying the same orbital with opposite spin.
101. If all the electrons in an atom are spin paired, then there is no net magnetic field.
105. (b)
107. When n = 5 and l = 3, m_l = −3, −2, −1, 0, +1, +2, +3.
108. 5
109. The difference between the atomic numbers of any adjacent pair of elements in a group of the periodic table will be 8, 18, or 32 because the change between adjacent pairs corresponds to a filled shell.
111. (a)
115. As n increases, the number of subshells increases.
119. (d)
129. Scandium
131. Antimony
135. [He] $2s^2\ 2p^3$
137. $x = 2$
139. (c)
141. Six electrons
143. (d)
145. (b)
146. Si ↑↓(1*s*) ↑↓(2*s*) ↑↓ ↑↓ ↑↓(2*p*) ↑↓(3*s*) ↑ ↑ __(3*p*)
V ↑↓(1*s*) ↑↓(2*s*) ↑↓ ↑↓ ↑↓(2*p*) ↑↓(3*s*) ↑↓ ↑↓ ↑↓(3*p*)
↑↓(4*s*) ↑ ↑ ↑ __ __(3*d*)
147. Only (c) is correct.
149. (d) Fe^{3+} has five unpaired electrons.
151. Five unpaired electrons can be placed in a *d* subshell.
155. Aluminum will have the smallest radius.
162. $Fe^{3+} < Fe^{2+}$
167. (e) Se^{2-}
169. (c) Be^{2+}
173. (a) Rb^+
174. $Ba^{2+} < Cs^+ < I^-$
179. (a)
181. Mg < Be < Ne < Na < Li
183. Fe^{2+}: [Ar] $3d^6$ and Fe^{3+}: [Ar] $3d^5$
186. (a) Mg < P < Cl
(b) Se < S < O < F
(c) K < P < O
192. Nonmetals have large values for AVEE.
195. Bi < Pb < Au < Ba
197. (a) P (b) As (c) Sn (d) Ga
201. Increasing ionization energy: $O^{2-} < F^- < Ne < Na^+ < Mg^{2+}$
Increasing radius: $Mg^{2+} < Na^+ < Ne < F^- < O^{2-}$
Increasing ionization energy: Na < Mg < O < F < Ne
Increasing radius: Ne < F < O < Mg < Na

207. (a) C, 2 unpaired electrons
(b) N, 3 unpaired electrons
(c) O, 2 unpaired electrons
(d) Ne, 0 unpaired electrons
(e) F, 1 unpaired electron

215. (a) Cl
(b) Cl: $1s^2\ 2s^2\ 2p^6\ 3s^2\ 3p^5$
(c) Cl^-: $1s^2\ 2s^2\ 2p^6\ 3s^2\ 3p^6$
(d) It is easier to ionize Cl^- than Cl.
(e) AVEE (Cl) < AVEE (Ar)
(f) This element is F.
(g) 7

Chapter 4

3. (a) 8 (b) 11 (c) 5 (d) 7
5. Na^+, Mg^{2+}, Al^{3+}, and Sc^{3+} all have 8 valence electrons.
9. For main-group elements and alkaline metals the group number is the same as the number of valence electrons.
22. (a) 2 (b) 2 (c) 4 (d) 1
23. (c)
27. (a) 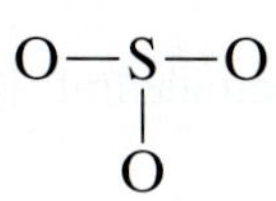 (b) O—S—O (c) O—O—O

(d) 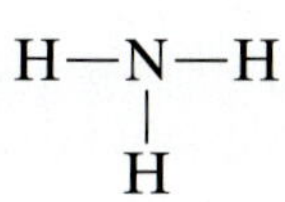 (e) 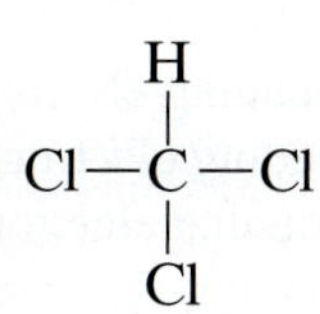

29. (a) 22 (b) 34 (c) 48 (d) 56
31. (a) 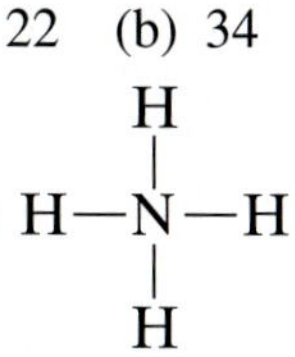 (b) 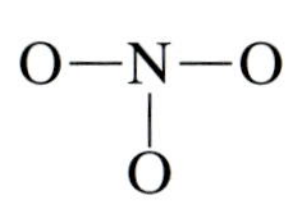 (c) 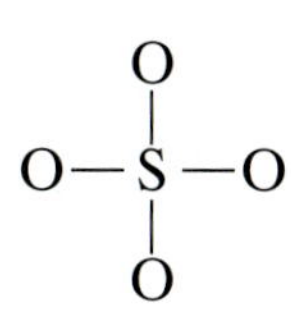

37. 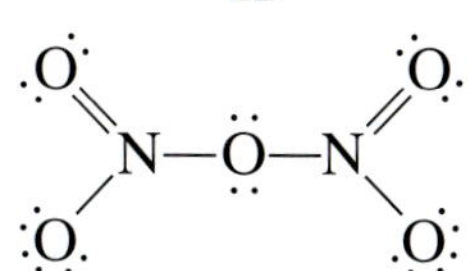

39. (a) SO_2

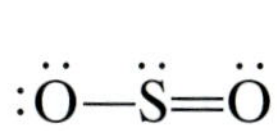

(b) SO_3

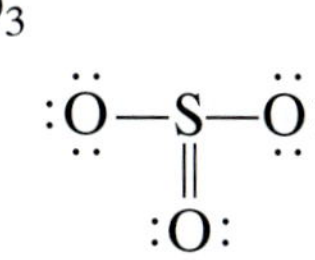

(c) $SO_3{}^{2-}$

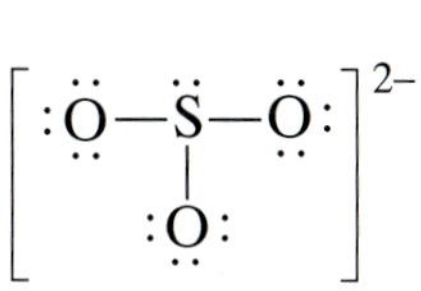

(d) $SO_4{}^{2-}$ 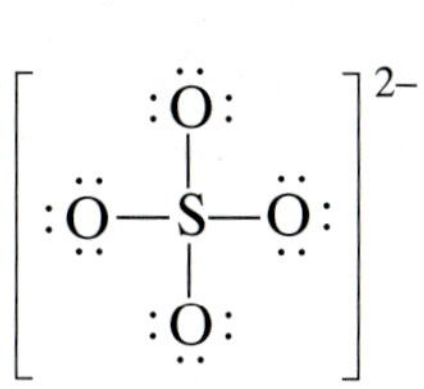

43. (a), (c), and (d)

45. 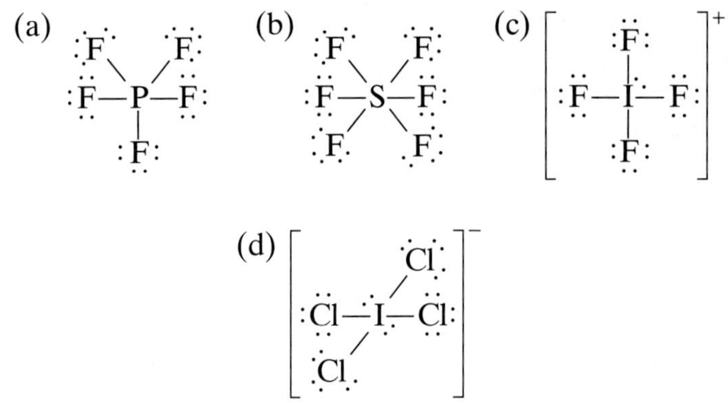

None of these structures obeys the octet rule.

47. N_2 is a triple bond, HNNH is a double bond, and H_2NNH_2 is a single bond.

51. 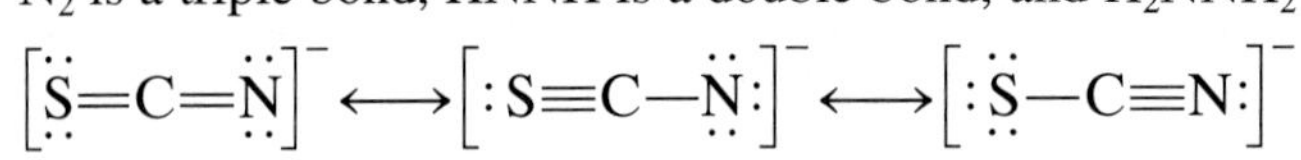

53. (b) and (c)

57. (b)

65. (a) H (b) C (c) B (d) N (e) N

67. (a) 0 (b) 0 (c) 0 (d) 0

77. (c) is the best structure.

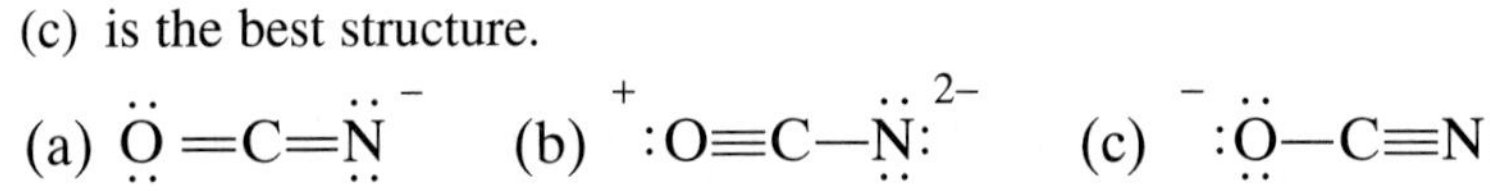

79. (a) 3 domains, trigonal planar, 120° (b) 4 domains, tetrahedral, 109.5°
(c) 6 domains, octahedral, 90°
(d) 5 domains, trigonal bipyramidal, 90° and 120°

83. (a) Three pairs of nonbonding electrons
(b) Two pairs of nonbonding electrons
(c) Eight pairs of nonbonding electrons
(d) Four pairs of nonbonding electrons
(e) One pair of nonbonding electrons

85.

Bonding Domains	Nonbonding Domains	Arrangement	Molecular Geometry
2	0	linear	linear
4	0	tetrahedral	tetrahedral
2	2	tetrahedral	bent
5	0	trigonal bipyramidal	trigonal bipyramidal
3	2	trigonal bipyramidal	T-shaped
5	1	octahedral	square pyramidal
4	2	octahedral	square planar

87. (a) tetrahedral (b) tetrahedral (c) tetrahedral (d) tetrahedral

89. (a) trigonal pyramidal (b) see-saw (c) square pyramidal
(d) octahedral

91. (a) linear (b) bent (c) trigonal planar

93. (a)

95. (c)

97. (a), (c), and (e)

99. (a), (b), and (d)

111. (a), (b), and (d)

119. (d)

121. SiF_6^{2-}

123. Group VIIA

125. (d)

127. Group VIIA

133. Structure (b) will contribute.

Chapter 4 Appendix

A-2. (a), (c), and (d) are overlapping combinations.

A-5. (a) CH_4 sp^3
(b) H_2CO sp^2
(c) HCO_2^- sp^2

A-9. A σ molecular orbital is one that results from the head-on overlap of atomic orbitals, whereas a π molecular orbital is one that results from the sideways or parallel overlap of atomic orbitals.

A-12. (a) (σ_{2s}) and (σ^*_{2s})
(b) (π_x) and (π^*_x)
(c) (π_y) and (π^*_y)
(d) (σ_p) and (σ^*_p)
(e) $(\sigma_{\sigma-p})$ and (σ^*_{s-p})

A-13. (a) 1 (b) 2 (c) 3 (d) 2 (e) 1

A-14. Calculate the bond order for the molecules. The molecule with the highest bond order is more stable.

(a) H_2^+	1 electron	$(\sigma_{1s})^1$	BO = 1/2
(b) H_2^{2-}	4 electrons	$(\sigma_{1s})^2 (\sigma^*_{1s})^2$	BO = 0
(c) H_2	2 electrons	$(\sigma_{1s})^2$	BO = 1

H_2 is more stable than H_2^+ and H_2^{2-}.

A-15. O_2 is more stable than O_2^{2-}.

A-17. The peroxide ion is not paramagnetic.

A-19. (a) diamagnetic (b) diamagnetic (c) diamagnetic
(d) paramagnetic (e) diamagnetic

Chapter 5

1. Decreasing metallic character: Na > Mg > Al > Si > P > S > Cl > Ar; increasing nonmetallic character: Na < Mg < Al < Si < P < S < Cl < Ar.

3. (d)

7. (a) K (b) Na (c) Ca (d) Ca

11. Xenon

13. AVEE will decrease from Li to Fr.

15. The second ionization energy for Li would be much higher than the first.

19. Na has a lower first ionization than Mg.

21. The atomic size increases from Be to Ra.

23. Aluminum

27. Ga: [Ar] $4s^2\ 3d^{10}\ 4p^1$
Ga$^+$: [Ar] $4s^2\ 3d^{10}$
Ga^{+2}: [Ar] $4s^1\ 3d^{10}$
Ga^{+3}: [Ar] $3d^{10}$

30. (a) C^{4-} (b) P^{3-} (c) S^{2-} (d) I^-

39. Zn: [Ar] $4s^2\ 3d^{10}$; +2

41. Ni: [Ar] $4s^2\ 3d^8$
Ni^{+2}: [Ar] $3d^8$

43. Magnesium

45. Group IIIA

50. AlN

51. Sr_3P_2

53. (a) $2\ Na(s) + F_2(g) \longrightarrow 2\ NaF(s)$
(b) $4\ Na(s) + O_2(g) \longrightarrow 2\ Na_2O(s)$
(c) $2\ Na(s) + H_2(g) \longrightarrow 2\ NaH(s)$
(d) $16\ Na(s) + S_8(s) \longrightarrow 8\ Na_2S(s)$
(e) $12\ Na(s) + P_4(s) \longrightarrow 4\ Na_3P(s)$

55. (a) ZnF_2 (b) AlF_3 (c) SnF_2 or SnF_4 (d) MgF_2 (e) BiF_3 or BiF_5

61. Coulombic forces

79. (d)

81. (c), (d), (e), and (f)

83. None of these

85. (a) and (d)

87. Using a bond type triangle as a guide:
(a) covalent (b) covalent (c) ionic (d) covalent/ionic

91. No, the compound could be metallic.

93. (a) covalent (b) semimetal (c) ionic (d) ionic (e) covalent
(f) ionic/metallic (g) covalent with some semimetal character
(h) ionic/covalent (i) covalent

97. (a) $MgZn_2$ (b) SiO_2 (c) GaAs (d) CsF (e) Hg_2Cl_2 (f) HgO

99. NO is an insulator. CdLi is a conductor.

103. $Re_2Cl_8{}^{2-}$: Re^{3+}
$Cr_2Cl_9{}^{3-}$: Cr^{3+}
$Mo_2Cl_8{}^{4-}$: Mo^{2+}

105. (a) Al^{3+} (b) Al^{3+} (c) Al^{3+}

107. (e), (g)

113. (a) 0 (b) −2 (c) −2 (d) +4 (e) +6 (f) +4 (g) +6 (h) +4
(i) +6 (j) +4 (k) +6

117. (a) $C_{CH_3} = -3$ $C_{central} = +1$ $H = +1$ $Br = -1$
(b) $C = -2$ $H = +1$ $O = -2$
(c) $C_{CH_3} = -3$ $C_{central} = +2$ $H = +1$ $O = -2$
(d) $C_{CH_3} = -3$ $C_{central} = 0$ $H = +1$ $O = -2$

119. (a) $C_{CH_3} = -3$ $C_{central} = +3$ $H = +1$ $O = -2$ $Cl = -1$
(b) $C = -2$ $H = +1$ $N = -3$
(c) $C_{CH_3} = -3$ $C_{C{=}O} = +3$ $C_{CH_2} = -1$ $O = -2$ $H = +1$

(d) $C = -3$ $H = +1$
(e) $C_{CH_2} = -2$ $C_{CH} = -1$ $C_{CH_3} = -3$ $H = +1$
(f) $C = -1$ $H = +1$

123. (a) Magnesium is oxidized. Hydrogen is reduced.
(b) Iodine is oxidized. Chlorine is reduced.
(c) Not an oxidation–reduction reaction.
(d) Sodium is oxidized. Hydrogen is reduced.

125. Carbon is oxidized. Sulfur is reduced.

129. (a) P_4S_3 (b) SiO_2 (c) CS_2 (d) CCl_4 (e) PF_5

131. (a) $SnCl_2$ (b) $Hg(NO_3)_2$ (c) SnS_2 (d) Cr_2O_3 (e) Fe_3P_2

133. (a) $Co(NO_3)_3$ (b) $Fe_2(SO_4)_3$ (c) $AuCl_3$ (d) MnO_2 (e) WCl_6

135. (a) aluminum chloride (b) sodium nitride (c) calcium phosphide
(d) lithium sulfide (e) magnesium oxide

137. (a) antimony(III) sulfide (b) tin(II) chloride (c) sulfur tetrafluoride
(d) strontium bromide (e) silicon tetrachloride

139. (a) H_2CO_3 (b) HCN (c) H_3BO_3 (d) H_3PO_3 (e) HNO_2

141. Thiosulfate ion, SSO_3^{2-} or $S_2O_3^{2-}$

143. (a) calcite, calcium carbonate (b) barite, barium sulfate

147.

	I	Br
Partial charge	+0.065	−0.065
Formal charge	0	0
Oxidation number	+1	−1

152. ZrO

153. (a) $2\ PbS + 3\ O_2 \longrightarrow 2\ PbO + 2\ SO_2$
(b) It is an oxidation–reduction reaction. The O_2 is reduced, and the S is oxidized.

155. MgO

Chapter 6

3. The temperature of a system is a measure of the average kinetic energy of all the particles in the system.

5. Water freezes at 32°F, 0°C, or 273.15K.

9. Gases, liquids, and solids

11. If both states of matter are present at the same temperature, they must both have the same average kinetic energies.

13. He, Ne, Ar, Kr, Xe are the atomic elements in the gas phase at room temperature. H_2, N_2, O_2, F_2, Cl_2 are elemental forms of the materials which are also gases at room temperature.

16. Compressibility, expandability, and volume of a gas are equal to the volume of its container.

21. $\text{Pressure} = \dfrac{\text{Force}}{\text{Area}}$

27. 29.0 psi

29. 1.2×10^{19} lb

31. The volume of the balloon will decrease.

33. 500 L
35. 2.8×10^2°C
37. 17.9%
39. They will have the same volume.
43. 18.0 L Total
47. N_2O
50. (e)
51. (a) Not always true for an ideal gas.
    (b) Always true for an ideal gas.
    (c) Not always true for an ideal gas.
55. 253 K
57. 3.13 atm
59. 3.30 g/L
61. 0.179 g/L
63. Kr
65. 1.3 L
67. 1.50 atm
69. 6.85 L
71. 12 atm
73. 2.5 atm
75. 0.452 atm
82. (a) False
    (b) False
    (c) True
    (d) True
87. (a) (i) increase (ii) remain the same (iii) remain the same (iv) decrease
    (b) remain the same
    (c) (i) increase (ii) remain the same (the changes counterbalance) (iii) decrease
    (d) The flask with He has more molecules since the pressure is higher while the temperature and volumes are the same.
89. 0.542 g Mg
91. 3.66×10^4 L CO_2
93. 2.45 L CO_2
95. (a), (b), and (d) are correct.
97. Xenon
99. The total atmospheric pressure decreases.
101. The final pressure will be three times the initial pressure.
103. (a) Kinetic energy would remain unchanged, but the frequency of collisions would decrease.
    (b) Kinetic energy would decrease as well as the frequency of collisions.
    (c) Kinetic energy and frequency of collisions would both increase.
    (d) Kinetic energy would remain unchanged, but the frequency of collisions would increase.
105. 58.1 g/mol
107. C_3H_6

Chapter 6 Appendix

A-4. In order of increasing rate of diffusion: $SF_6 < CF_2Cl_2 < Cl_2 < SO_2 < Ar$

A-5. $\frac{\text{Rate } O_2}{\text{Rate } Br_2} = \sqrt{\frac{159.8 \text{ g}}{32.00 \text{ g}}} = 2.236$

A-9. 13 s

A-12. 234.8 g/mol

A-16. Fraction empty space = 0.9987

A-19. $\frac{\text{He}}{\text{Ne}} = 1.1$ and $\frac{\text{Ar}}{\text{Ne}} = 1.2$

Chapter 7

3. Energy can be transferred from one object to another by mechanical (contact) means. Rotation of a car's tires with respect to a road surface give, an automobile kinetic energy.
6. When chemical reactions occur, some chemical bonds may be formed. Energy is released during the formation of a chemical bond because the energy of the free atoms is greater than the energy of the bonded atoms.
11. The iron atoms in the warm bar would slow down and the atoms in the cooler bar would speed up until they had the same average kinetic energy. When the average kinetic energies are the same, they have the same temperature.
12. When heat enters a system the atomic and molecular motions increase. The addition of heat to a system increases the temperature.
13. As a balloon filled with helium becomes warmer, the particles within become more agitated. This induces a greater pressure on the walls, causing the balloon to expand.
16. As the gas is heated, more energy is put into the gas molecules. Molecules with more energy will move around faster.
22. The first law of thermodynamics states that energy is conserved and it can neither be created nor destroyed.
25. State functions of a system are often associated with intensive properties of the system. Thus temperature, pressure, volume, enthalpy, and energy of an ideal gas are state functions.
27. (a), (b), (c), (d)
33. Water has the higher heat capacity.
35. $\Delta H = 891$ kJ/mol
37. (a) 131.29 kJ (b) 262.59 kJ (c) 3.93 kJ (d) 6.56 kJ
39. -1.12×10^3 kJ/mol Mg
49. In an exothermic process energy is given off in the course of the reaction: therefore the bond strengths of the products are greater than for the reactants. In an endothermic process energy is absorbed by the system to make the reaction proceed: therefore the reactants have stronger bonds than the products.
53. (a) endothermic (b) exothermic (c) exothermic

59. The reaction is endothermic.
61. $\Delta H^\circ = -1284.4\ \text{kJ/mol}_{\text{rxn}}$
63. $\Delta H^\circ = -5313.97\ \text{kJ/mol}_{\text{rxn}}$
65. $\Delta H^\circ = -410.7\ \text{kJ/mol}_{\text{rxn}}$
67. $\Delta H^\circ = -851.5\ \text{kJ/mol}_{\text{rxn}}$
69. $\Delta H^\circ = -3310.1\ \text{kJ/mol}_{\text{rxn}}$
71. P_4
75. $\Delta H^\circ = -878.22\ \text{kJ/mol}_{\text{rxn}}$
81. $\Delta H^\circ = +1.90\ \text{kJ/mol}_{\text{rxn}}$
83. $\Delta H^\circ = +180.6\ \text{kJ/mol}_{\text{rxn}}$
85. $\Delta H^\circ = -2043.96\ \text{kJ/mol}_{\text{rxn}}$
87. $F_2(g)$ and $P_4(s)$
89. $\Delta H^\circ = +49.9\ \text{kJ/mol}_{\text{rxn}}$
91. $\Delta H^\circ = -1076.87\ \text{kJ/mol}_{\text{rxn}}$
93. $\Delta H^\circ = -905.47\ \text{kJ/mol}_{\text{rxn}}$
95. $\Delta H^\circ = -91.1\ \text{kJ/mol}_{\text{rxn}}$
103. The average heat released per mole of covalent bonds broken in the hydrocarbon is 222 kJ.
107. (a) $\Delta H^\circ = -226\ \text{kJ/mol}_{\text{rxn}}$
    (b) Si-Br: 318 kJ/mol; Si-Cl: 399.8 kJ/mol; Cl-Cl: 243.358 kJ/mol; Br-Br: 192.86 kJ/mol
    (c) Si-Cl
    (d) Exothermic
109. (d)
111. Exothermic

Chapter 8

2. Intramolecular forces are stronger than intermolecular forces.
5. The level of intermolecular forces that exists between molecules of a substance will determine what phase the substance is found in at room temperature.
9. A hydrogen bond will exist only when there is a hydrogen atom that is covalently bound to an electronegative element such as N, O, or F.
10. (a) SO_2: dispersion and dipole–dipole
    (b) CH_3OH: dispersion, dipole–dipole, and hydrogen bonding
    (c) ICl_3: dispersion and dipole–dipole
    (d) SF_4: dispersion and dipole–dipole
12. $PH_3 < AsH_3 < SbH_3 < NH_3$
13. (e)
15. The dispersion forces are less in propane.
19. They have the same velocity regardless of what phase they are in.
23. Vapor pressure increases with increasing temperature.
24. Even at room temperature, some water molecules have enough energy to escape from the liquid phase.

29. The liquid and vapor phases are at equilibrium when the rate of molecules evaporating is the same as the rate of molecules condensing.
32. Water under increased pressure boils at a higher temperature.
38. (c)
41. The melting point of a substance is dependent on the intermolecular forces of the substance.
43. The atmospheric pressure is less in Denver, and water boils at a lower temperature.
47. (c)
51. High pressure and low temperature
52. Gas phase
59. It takes a large amount of energy to disrupt the hydrogen bonds in water.
61. Anesthetics and olive oil are relatively nonpolar.
63. (e)
66. Molecular solids that can form hydrogen bonds are generally soluble in water.
71. (c) is insoluble.
73. (e) is soluble in water.
75. (a) $Mg(s) + 2\ HCl(aq) \longrightarrow MgCl_2(aq) + H_2(g)$
$Mg(s) + 2\ H^+(aq) \longrightarrow Mg^{+2}(aq) + H_2(g)$
(b) $Na_2CO_3(aq) + Ca(NO_3)_2(aq) \longrightarrow 2\ NaNO_3(aq) + CaCO_3(s)$
$CO_3^{-2}(aq) + Ca^{+2}(aq) \longrightarrow CaCO_3(s)$
77. Hydrocarbons do not possess any hydrophilic groups.
79. Solubility: $NaCl > CH_3COOH > CH_3CH_2OH > CH_3CH_2CH_2CH_3$
81. The hydrogen bonds between the ethanol molecules require more energy to break than the dipole–dipole forces of the dimethyl ether.
85. (a)
89. (a) $Fe(OH)_3$ (b) no precipitate (c) $PbCl_2$
91. (a) Pentanol should have a lower boiling point than butanoic acid.
(b) Nonal should have a boiling point close to 196°C.
93. (a) 1-octanol < dibutyl ether < octane
(b) 1-octanol > dibutyl ether > octane
99. (a) Vapor pressure: $PH_3 > AsH_3 > NH_3 > SbH_3$
(b) CH_3OH is the best solvent.

Chapter 8 Appendix

A-1. Colligative properties are properties of solutions, not pure substances.
A-3. (a) Hexane and heptane would be the closest to an ideal solution.
A-6. $P_{H_2O} = 23.75$ mmHg
A-7. $P_T = 262$ mmHg
A-11. T = 53.8°C
A-13. T = 78.1°C
A-14. $k_b = 0.513$°C/m

A-17. T $=-0.191°C$

A-19. (c)

A-21. The salt lowers the melting point of ice, and highways can remain free of ice at temperatures lower than the freezing point of pure water.

A-22. The percent ionization for a 0.100 m KCl solution is 86%.

A-25. The solution would freeze at $-1.8°C$.

A-27. Hard water is water that contains a large amount of Fe^{3+}, Ca^{2+}, and Mg^{2+} ions. These ions will combine with a soap and form an insoluble and inactive precipitates. Water can be softened using an ion exchange method that replaces the "hard" ions with Na^{+} ions.

Chapter 9

1. Crystalline, amorphous, and polycrystalline
5. (d)
7. Lattice energy is the energy required to break ionic compounds into gas phase ions.
9. (a)
18. In a metallic bond the valence electrons of all the atoms in the solid are delocalized, and the metal cations are held in place by the sea of delocalized electrons.
19. In molecular, ionic, and network covalent solids the electrons are localized either on atoms or in individual bonds. In metallic bonds the electrons are delocalized.
23. (c)
25. (a), (d), (e)
34. (a) 12 (b) 12 (c) 8 (d) 6
37. Hexagonal closest packed, cubic closest packed, body-centered cubic
47. The density is mass per unit volume. If metal atoms are more closely spaced, then the metal will have a higher density since the unit cell will take up less volume.
49. 10.50 g/cm^3
51. Face-centered cubic
53. Copper
55. $N = 4$
59. 4.352×10^{-8} cm
61. $r_{Tl} + r_{I} = 0.3635$ nm
$r_{Tl} = 0.148$ nm
63. The force that holds the atoms together, the covalent bond, is much stronger than the dispersion forces that hold the molecules together.
69. (a) molecular (b) molecular (c) metallic (d) molecular/ionic (e) metallic (f) ionic (g) metallic (h) ionic (i) ionic/metallic (j) molecular (k) ionic/metallic (l) molecular/metallic (semimetal)

Chapter 10

1. Reactions that go to completion leave no unreacted limiting reagent behind. Reactions that go to equilibrium may leave an appreciable amount of unreacted starting materials.
5. The ratio of the equilibrium concentrations at 400°C is 1.27. This ratio will always have this value no matter how much material is present, so long as the system is at equilibrium at 400°C.
7. True
15. Equilibrium was initially defined as when the forward reaction appears to stop and the concentrations of the products and the reactants do no change. However, a more correct definition of equilibrium is when the rate of the forward reaction is equal to the rate of the reverse reaction.
17. (d)
19. (a) $K_c = \dfrac{[OF_2]^2}{[O_2][F_2]^2}$ (b) $K_c = \dfrac{[SO_3]^2}{[O_2][SO_2]^2}$ (c) $K_c = \dfrac{[SO_2Cl_2]^2[O_2]}{[SO_3]^2[Cl_2]^2}$
21. (a) $K_c = \dfrac{[CO_2]^2}{[CO]^2[O_2]}$ (b) $K_c = \dfrac{[CO][H_2O]}{[CO_2][H_2]}$ (c) $K_c = \dfrac{[CH_3OH]}{[CO][H_2]^2}$
23. 1.5×10^{12}
25. 5.72×10^{-6}
27. (b)
29. (c)
31. $\Delta(X) = (X) - [X]$
35. $\Delta(N_2) = 0.117$
    $\Delta(H_2) = 0.351$
37. (e)
39. $[H_2]_{eq} = 0.766\ M$ $[NH_3]_{eq} = 0.156\ M$
41. $[N_2O_4]_{eq} = 0.099\ M$ $[NO_2]_{eq} = 2.4 \times 10^{-3}\ M$
43. $[NH_3] = 0.01\ M$ $[N_2] = 0.15\ M$ $[H_2] = 0.24\ M$
45. The initial concentration of CO and H_2O is 2.18 *M*.
47. $[SO_2Cl_2]_{eq} = 0.049\ M$ $[SO_2]eq = [Cl_2]_{eq} = 8.4 \times 10^{-4}\ M$
49. 3.6%
51. $[NO_2]_{eq} = 0.098\ M$ $[NO]_{eq} = 0.0019\ M$ $[O_2]_{eq} = 9.5 \times 10^{-4}\ M$
53. $[SO_3]_{eq} \approx 0.400\ M$ $[SO_2]_{eq} = 3.7 \times 10^{-4}\ M$ $[O_2]_{eq} = 1.9 \times 10^{-4}\ M$
59. Equilibrium constants will change with temperature depending on the nature of the reaction involved.
61. Products
63. (a) The reaction will shift to the right.
    (b) The reaction will shift to the right.
    (c) The reaction will shift to the right.
65. (a) The reaction will shift to the right.
    (b) The reaction will shift to the left.
    (c) The reaction will shift to the right.
71. High temperature and low pressure

73. $K_{sp} = [Ag^+]^2[CrO_4^{2-}]$

75. $K_{sp} = 2.47 \times 10^{-9}$

77. $K_{sp} = 1.2 \times 10^{-10}$

81. 1.9×10^{-9} g AgBr

83. (d)

85. $k_r > k_f$.

87. (a) $K_{sp} = [Sr^{2+}][F^-]^2 = 8.4 \times 10^{-10}$
(b) SrF_2 will precipitate.

91. $[NOCl]_{eq} = 0.97\ M$, $[NO]_{eq} = 0.032\ M$, $[Cl_2]_{eq} = 0.016\ M$

93. (a) 3.0×10^{-2} moles
(b) $K_{sp} = [Ag^+]^2\,[SO_4^{2-}] = 1.4 \times 10^{-5}$

95. (a) $[N_2F_4]_{eq} = 1.0\ M$ $[NF_2]_{eq} = 3.5 \times 10^{-5}\ M$ (b) 1.0 M

Chapter 10 Appendix

A-2. As the reaction system approaches more closely equilibrium state, the concentrations of all components will tend to change less and less as Q_c gets closer to K_c.

A-3. The assumption of a small change in concentration is doomed to failure when the initial concentrations are very different from the equilibrium concentrations. By the fact of being "very different" the changes must become "very much larger" to accomplish the adjustments required.

A-5. (c)

Chapter 11

1. An acid is a proton donor. A base is a proton acceptor. Vinegar and lemon juice are examples of acids. Lye is an example of a base.

3. Litmus will turn red in the presence of an acid. Litmus will remain blue in the presence of a base.

5. (a) and (b)

7. $HBr(g) + H_2O(l) \longrightarrow H_3O^+(aq) + Br^-(aq)$
$H_2O(l) + NH_3(g) \longrightarrow OH^-(aq) + NH_4^+(aq)$

11. (a) and (c)

14. (a) $NH_3(aq) + H_2O(l) \longrightarrow NH_4^+(aq) + OH^-(aq)$
(b) $CH_3NH_2(aq) + H_2O(l) \longrightarrow CH_3NH_3^+(aq) + OH^-(aq)$
(c) $IO^-(aq) + H_2O(l) \longrightarrow HIO(aq) + OH^-(aq)$

15. (a) $\underset{acid}{HCl(aq)} + \underset{base}{H_2O(l)} \longrightarrow \underset{acid}{H_3O^+(aq)} + \underset{base}{Cl^-(aq)}$
(b) $\underset{base}{HCO_3^-(aq)} + \underset{acid}{H_2O(l)} \longrightarrow \underset{base}{OH^-(aq)} + \underset{acid}{H_2CO_3(aq)}$
(c) $\underset{base}{NH_3(aq)} + \underset{acid}{H_2O(l)} \longrightarrow \underset{base}{OH^-(aq)} + \underset{acid}{NH_4^+(aq)}$
(d) $\underset{base}{CaCO_3(s)} + \underset{acid}{2\ HCl(aq)} \longrightarrow Ca^{2+}(aq) + \underset{base}{2\ Cl^-(aq)} + \underset{acid}{H_2CO_3(aq)}$

17. (a) (b)

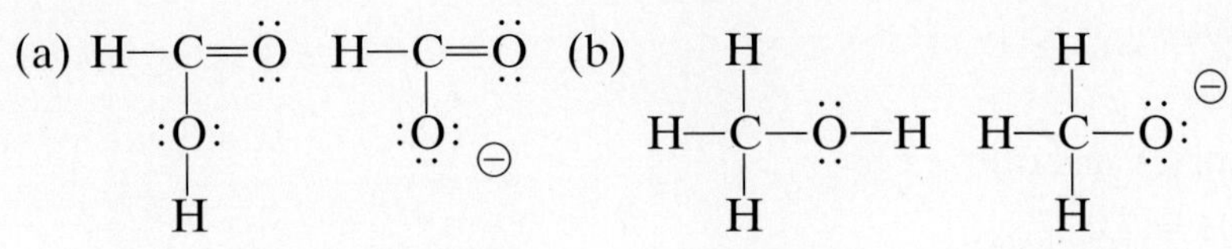

formic acid, formate ion, methanol, methoxide

21. (a) OH^- (b) H_2O (c) H_3O^+ (d) NH_3

23. True

33. $[OH^-] = 0.010\ M$

34. 6.02×10^{13} ions

35. $[H_3O^+] = 1.9 \times 10^{-4}\ M$; $[OH^-] = 5.3 \times 10^{-11}\ M$

39. (a)

41. pH = 5.8; pOH = 8.2

43. $[H_3O^+] = 1.6 \times 10^{-4}\ M$; $[OH^-] = 6.3 \times 10^{-11}\ M$

45. $[OH^-] = 1.3 \times 10^{-3}\ M$

48. The larger the value of K_a the stronger the acid.

49. (a) weak (b) weak (c) strong (d) weak (e) strong

51. (d)

53. $HOAc < HNO_2 < HF < HOClO$

55. I^-, ClO_4^-, NO_3^- will not act as bases.

59. The reaction above indicates that hydrogen bromide is a stronger acid than water.

61. This indicates that OH^- is a stronger base than HCO_2^-.

67. $AsO_4^{-3} < HAsO_4^{-2} < H_2AsO_4^- < H_3AsO_4$

69. (d)

71. $HOCH_3 < HOI < HOBr < HOCl$

73. $CH_3COOH < CH_2ClCOOH < CCl_3COOH$

75. pH = 1.2; pOH = 12.8

77. (a) pH = 0.0 (b) pH = 1.65 (c) pH = 1.44 (d) pH = 1.55

79. pH = 1.00

83. (a)

85. $[H_3O^+] = [HCO_2^-] = 4.2 \times 10^{-3}\ M$; $[HCO_2H] = 9.6 \times 10^{-2}\ M$

87. $[H_3O^+] = [PhO^-] = 1.3 \times 10^{-6}\ M$

89. $K_a = 8.1 \times 10^{-5}$

95. (c)

97. $[OH^-] = [HCO_2H] = 2.1 \times 10^{-6}\ M$; $[HCO_2^-] = 0.080\ M$

99. pH = 9.3

101. $K_b = 5.0 \times 10^{-4}$

103. pH = 11.8

105. pH = 8.6

107. NH_4Cl

109. (a) NaH_2PO_4 (b) H_2CO_3 (c) $NaHSO_4$ (d) HNO_2

111. pH = 1.72; pH = 3.1

113. pH = 9.3

117. (e)

119. (d)

121. (a), (c), and (d)

123. pH = 7.0

125. (a) $HOBr(aq) + OH^-(aq) \rightleftharpoons H_2O(l) + OBr^-(aq)$
(b) $HOAc(aq) + NH_3(aq) \rightleftharpoons OAc^-(aq) + NH_4^+(aq)$
(c) $H_3O^+(aq) + OH^-(aq) \rightleftharpoons 2\ H_2O(l)$
(d) $2\ HCOOH(aq) + Ba(OH)_2(s) \rightleftharpoons 2\ HCOO^-(aq) + 2\ H_2O(l) + Ba^{2+}(aq)$
Reactions (a), (b), and (c) will go to completion.

131. $[H_3O^+] = 0.004\ M$; $[H_3O^+] = 0.001\ M$

133. (a) False (b) True (c) True (d) False

137. $HBr < HOBrO_2 < HOBrO < NaBr < NaBrO_3 < NaBrO_2$

141. Conjugate base: (c) conjugate acid: (a)

143. (a) $Al_2O_3(s) + 6\ HCl(aq) \longrightarrow 2\ AlCl_3(aq) + 3\ H_2O(l)$
(b) $CaO(s) + H_2SO_4(aq) \longrightarrow CaSO_4(aq) + H_2O(l)$
(c) $Na_2O(s) + H_2O(l) \longrightarrow 2\ Na^+(aq) + 2\ HO^-(aq)$
(d) $MgCO_3(s) + 2\ HCl(aq) \longrightarrow H_2CO_3(aq) + MgCl_2(aq)$
(e) $3\ NaOH(aq) + H_3PO_4(aq) \longrightarrow 3\ H_2O(l) + Na_3PO_4(aq)$

147. (a) $HNO_2(aq) + H_2O(l) \longrightarrow NO_2^-(aq) + H_3O^+(aq)$
(b) pH = 1.65 (c) pH = 8.65
(d) pH = 0 (e) A

149. (a) pH = 1.0 (b) pH = 7.0 (c) $1 < pH < 7$
(d) $7 < pH < 14$ (e) pH = 14

151. $KOH < KClO_3 < KI < NH_4ClO_4 < HClO_3 < HClO_4$

153. pH = 2.87; pH = 4.57; pH = 8.72; pH = 12.0

155. (b)

Chapter 11 Appendix

A-1. $[H_2CO_3] = 0.100\ M$; $[HCO_3^-] = [H_3O^+] = 2.1 \times 10^{-4}\ M$;
$[CO_3^{2-}] = 4.7 \times 10^{-11}\ M$

A-2. $[H_2M] \cong 0.250\ M$; $[H_3O^+] = [HM^-] = 1.9 \times 10^{-3}\ M$;
$[M^{2-}] = 2.1 \times 10^{-8}\ M$

A-5. (b) and (d)

A-7. $[H_3O^+] = [HC_2O_4^-] = 0.24\ M$; $[H_2C_2O_4] = 1.01\ M$;
$[C_2O_4^{2-}] = 5.4 \times 10^{-5}\ M$

A-9. pH = 8.4

A-11. pH = 1.46

A-12. $[H_3O^+] = [H_2AsO_4^-] = 0.0745\ M$; $[H_3AsO_4] = 0.9255\ M$;
$[HAsO_4^{2-}] = 1.1 \times 10^{-7}\ M$; $[AsO_4^{3-}] = 4.4 \times 10^{-18}\ M$

A-14. (d)

A-15. pH = 1.6; pH = 4.1; pH = 10.1; pH = 12.7

A-17. Acidic

Chapter 12

3. (a) reduction (b) oxidation (c) reduction (d) oxidation
5. (a) +6 (b) +6 (c) +4
7. (c)
10. (a) Na_2S −2 (e) SF_6 +6
    (b) MgS −2 (f) $BaSO_3$ +4
    (c) CS_2 −2 (g) $NaHSO_4$ +6
    (d) H_2SO_4 +6 (h) SCl_2 +2
13. (b)
17. This is not a redox reaction.
23. Two of the carbons change from a −2 oxidation state in succinate to a −1 state in fumarate.
25. This is a redox reaction.
31. (a) $Al(s) + Cr^{3+}(aq) \rightleftharpoons Al^{3+}(aq) + Cr(s)$
    (b) $2\,Fe^{2+}(aq) + I_2(aq) \rightleftharpoons 2\,Fe^{3+}(aq) + 2\,I^-(aq)$
    (c) $Cr(s) + 3\,Fe^{3+}(aq) \rightleftharpoons Cr^{3+}(aq) + 3\,Fe^{2+}(aq)$
    (d) $Zn(s) + 2\,H^+(aq) \rightleftharpoons Zn^{2+}(aq) + H_2(g)$
33. (a) The Al is the reducing agent, while the Cr^{3+} is the oxidizing agent.
    (b) In this case the Cr^{2+} is both the oxidizing and reducing agents.
    (c) The Fe is the reducing agent, and Cr^{3+} is the oxidizing agent.
    (d) The H_2 is the reducing agent, and Cr^{3+} is the oxidizing agent.
35. Yes
41. When an atom is in the lowest, or most negative, oxidation state it cannot take in any more electrons and will therefore not be an oxidizing agent. So it can act only as a reducing agent.
43. (c)
47. (c)
49. (d)
53. (e)
55. (a) K (b) Sn (c) Ag^+
61. The sign changes.
63. (a), (b), and (c)
65. (a) and (b)
67. Copper will be deposited at the cathode and zinc will be dissolved from the anode.
69. The reaction should not occur spontaneously as written.
71. The reaction should occur spontaneously as written.
73. (b)
79. (a)
81. $E^\circ_{cell} = 0.11$ V; $E_{cell} = 0.35$ V
83. $E^\circ_{cell} = 0.809$ V
85. $E^\circ_{cell} = 0.354$; $K = 9.4 \times 10^{11}$
91. 1.1180×10^{-3} g

93. 5.4 h; 11 h; 16 h

95. 2.722×10^9 C

97. $\frac{\text{mass } O_2}{\text{mass } H_2} = \frac{8}{1}$

99. $\frac{\text{mass } Br_2}{\text{mass Al}} = 8.88$

101. (a)

103. $x = 1, y = 4$

105. +3

111. (a) This is a redox reaction. The C is oxidized, and the O is reduced.
(b) This is a redox reaction. The C is oxidized, and the Pb is reduced.
(c) There is no redox reaction.
(d) This is a redox reaction. The C's in ethylene are reduced, and the H is oxidized.

113. (b)

115. (a) Ag (b) $E^\circ_{cell} = 1.56$ V
(c) The Zn cell is the anode, and the Ag cell is the cathode.
(d) Electrons flow from Zn to Ag.
(e) $2\ Ag^+(aq) + Zn(s) \longrightarrow 2\ Ag(s) + Zn^{2+}(aq)$
(f) $K = 5.8 \times 10^{52}$

119. 6.1 g $FeCl_2$

Chapter 12 Appendix

A-3. (a) $5\ Fe^{2+}(aq) + MnO_4^-(aq) + 8\ H^+(aq) \longrightarrow 5\ Fe^{3+}(aq) + Mn^{+2}(aq) + 4\ H_2O(l)$
(b) $5\ S_2O_3^{2-}(aq) + 8\ MnO_4^-(aq) + 14\ H^+(aq) \longrightarrow 10\ SO_4^{2-}(aq) + 8\ Mn^{+2}(aq) + 7\ H_2O(l)$
(c) $5\ PbO_2(s) + 4\ H^+(aq) + 2\ Mn^{2+}(aq) \longrightarrow 5\ Pb^{2+}(aq) + 2\ H_2O(l) + 2\ MnO_4^-(aq)$
(d) $2\ MnO_4^-(aq) + 6\ H^+(aq) + 5\ SO_2(g) + 2\ H_2O(l) \longrightarrow 2\ Mn^{2+}(aq) + 5\ H_2SO_4(aq)$

A-4. $Br_2(aq) + SO_2(g) + 2\ H_2O(l) \longrightarrow H_2SO_4(aq) + 2\ HBr(aq)$

A-5. (a) $4\ Cr(s) + 12\ H^+(aq) + 3\ O_2(g) \longrightarrow 4\ Cr^{3+}(aq) + 6\ H_2O(l)$
(b) $14\ H^+(aq) + Cr_2O_7^{2-}(aq) + 6\ Fe^{2+} \longrightarrow 2\ Cr^{3+}(aq) + 6\ Fe^{3+}(aq) + 7\ H_2O(l)$
(c) $22\ H_2O(l) + 10\ Cr^{3+}(aq) + 6\ BrO_3^-(aq) \longrightarrow 10\ HCrO_4^-(aq) + 34\ H^+(aq) + 3\ Br_2(aq)$

A-7. (a) $HIO_3(aq) + 5\ HI(aq) \longrightarrow 3\ I_2(aq) + 3\ H_2O(l)$
(b) $8\ SO_2(g) + 16\ H_2S(g) \longrightarrow 3\ S_8(s) + 16\ H_2O(l)$

A-9. (a) $3\ Cu(s) + 2\ HNO_3(aq) + 6\ H^+(aq) \longrightarrow 3\ Cu^{2+}(aq) + 2\ NO(g) + 4\ H_2O(l)$
(b) $Cu(s) + 2\ HNO_3(aq) + 2\ H^+(aq) \longrightarrow Cu^{2+}(aq) + 2\ NO_2(g) + 2\ H_2O(l)$

A-11. (a) $NO(g) + MnO_4^-(aq) \longrightarrow NO_3^-(aq) + MnO_2(s)$
(b) $2\ NH_3(g) + 2\ MnO_4^-(aq) \longrightarrow N_2(g) + 2\ H_2O(l) + 2\ MnO_2(s) + 2\ OH^-(aq)$

A-13. (a) $2\ Cr(s) + 2\ OH^-(aq) + 6\ H_2O(l) \longrightarrow 2\ Cr(OH)_4^-(aq) + 3\ H_2(g)$
(b) $4\ OH^-(aq) + 2\ Cr(OH)_3(s) + 3\ H_2O_2(aq) \longrightarrow 2\ CrO_4^{2-}(aq) + 8\ H_2O(l)$

A-15. $4\ Au(s) + 2\ H_2O(l) + 8\ CN^-(aq) + O_2(g) \longrightarrow 4\ Au(CN)_2^-(aq) + 4\ OH^-(aq)$

A-17. (a) $2\ OH^-(aq) + Cl_2(g) \longrightarrow OCl^-(aq) + H_2O(l) + Cl^-(aq)$
(b) $6\ OH^-(aq) + 3\ Cl_2(g) \longrightarrow ClO_3^-(aq) + 3\ H_2O(l) + 5\ Cl^-(aq)$
(c) $3\ OH^-(aq) + 1\ P_4(s) + 3\ H_2O(l) \longrightarrow 3\ H_2PO_2^-(aq) + 1\ PH_3(g)$
(d) $2\ H_2O_2(aq) \longrightarrow O_2(g) + 2\ H_2O(l)$

A-21. When a lead acid battery is discharged, PbO_2 is reduced to $PbSO_4$. When it is charged up again, the reverse reaction is driven through electrolysis.

A-29. The $[H^+]$ is much less than 1.0 M; therefore the reduction potential will be much less than the standard reduction potential.

A-31. Rust can be better described as a hydrated oxide, $Fe_2O_3 \cdot 3\ H_2O$. It is also a mixture of oxides rather than just Fe_2O_3.

A-35. $2\ F^-(aq) \longrightarrow F_2(g) + 2\ e^-$
$Ca^{2+}(aq) + 2\ e^- \longrightarrow Ca(s)$

A-38. $E^\circ_{cell} = -4.07$ V

A-43. (c)

A-46. (b)

Chapter 13

3. Yes, the freezing of liquid water is spontaneous below 0°C but not spontaneous above 0°C.

5. Gaseous water has more entropy.

9. (d)

11. (c)

19. Gas

23. Entropy is not a driving force in the reaction.

25. Products of the reaction are more ordered than the reactants.

29. (b)

33. $\Delta H^\circ = -851.5\ \text{kJ/mol}_{rxn}$
$\Delta S^\circ = -38.58\ \text{J/mol}_{rxn} \cdot \text{K}$
The enthalpy change is so large that it is the driving force behind the reaction.

35. $\Delta H^\circ_{rxn} = -602.8\ \text{kJ/mol}_{rxn}$
$\Delta S^\circ_{rxn} = 579.7\ \text{J/mol}_{rxn} \cdot \text{K}$

39. $\Delta H^\circ = 24.9\ \text{kJ/mol}_{rxn}$
$\Delta S^\circ = 307.6\ \text{J/mol}_{rxn} \cdot \text{K}$

43. $\Delta H^\circ = -34.3\ \text{kJ/mol}_{rxn}$
$\Delta S^\circ = 75.58\ \text{J/mol}_{rxn} \cdot \text{K}$

45. (a) ΔG° will become more positive. (b) ΔG° will become more negative.
(c) ΔG° will become more negative. (d) ΔG° will become more positive.

49. −10°C $\Delta G > 0$
0°C $\Delta G = 0$
+10°C $\Delta G < 0$

51. The entropy change is positive. The enthalpy change is also positive.

55. (a) $\Delta G^\circ = -29.11$ kJ/mol$_{rxn}$ (b) $\Delta G^\circ = 326.4$ kJ/mol$_{rxn}$
(c) $\Delta G^\circ = 130.3$ kJ/mol$_{rxn}$

59. (c)

63. $P_{SO_3} = 0.471$ atm
$P_{SO_2} = 1.9 \times 10^{-2}$ atm
$P_{O_2} = 9.7 \times 10^{-3}$ atm

65. The reaction will go to the left.

67. $[N_2]_{eq} = 0.40\ M$
$[O_2]_{eq} = 0.60\ M$
$[NO]_{eq} = 3.2 \times 10^{-5}\ M$

71. When the reaction quotient, Q, equals 1.

73. $K = 2.2 \times 10^4$

75. $K = 9.7 \times 10^{33}$

77. $K = 1.9 \times 10^{-5}$

79. $\Delta H^\circ = 41.16$ kJ/mol$_{rxn}$
$\Delta S^\circ = 41.51$ J/mol$_{rxn} \cdot$ K
$\Delta G^\circ = 28.62$ kJ/mol$_{rxn}$
$K = 9.6 \times 10^{-6}$

81. 516 K

83.

T(K)	K
298	1×10^{-25}
473	9×10^{-13}
673	3×10^{-6}
873	1×10^{-2}

87. $\Delta H^\circ = -34.18$ kJ/mol$_{rxn}$
$\Delta S^\circ = -81.15$ J/mol$_{rxn} \cdot$ K

91. (a) -114.74 kJ/mol$_{rxn}$ (b) -146.54 J/mol$_{rxn} \cdot$ K (c) -71.1 kJ/mol$_{rxn}$
(d) products (e) 2.86×10^{12} (f) to the right (g) increase

93. Decrease the magnitude of ΔG°.

95. Acetic acid will be stronger.

Chapter 14

8. During the course of a reaction, the rate at which a reactant is consumed will decrease as the amount of reactant decreases.

13. time^{-1}

15. M^{-2} time^{-1}

17. (X) represents the volume of water passing through. The units of k are s^{-1}.

19. (c)

21. 0.020 M/s

23. 0.027 Ms^{-1}

29. Rate $= k \cdot (A) \cdot (B)$

33. (d)

35. (c)
37. No
41. 3.5×10^4
45. Rate $= k\,(NO)^2(Cl_2)$
47. Rate $= k\,(NO)^2\,(O_2)$
49. Rate $= k\,(CH_3I)\,(OH^-)$
53. 0.0034 *M*
55. (a) 12 s (b) 0.49 *M*
57. 24.6 s^{-1}
59. 0.36 *M*
61. 84.5%
63. 15%
65. 3.91×10^3 y
67. 4.2×10^4 y
69. 0.026 *M*
71. 1.3×10^2 h
73. The rate equation for this reaction is consistent with the mechanism as given.
79. The reaction is first order in $(Cr(NH_3)_5Cl)^{2+}$.
81. Rate $= k\,(Cr(NH_3)_5Cl^{2+})(OH^-)$
83. (d)
89. Lowering the activation barrier will increase the rate of the reaction.
93. 2.18×10^2 kJ/mol_{rxn}
95. 33.4 kJ/mol_{rxn}
97. 0.014 $M^{-1}s^{-1}$
103. The process is zero order.
105. (a) IV (b) I (c) III (d) III
113. III
115. (a) Rate $= k\,(NO)^2(Cl_2)$ (b) 180 $M^{-2}\,s^{-1}$ (c) Yes
117. (a) Second order (b) Rate $= k\,(A)^2$ (c) 0.025 $M^{-1}\,s^{-1}$
(d) The catalyst will increase the rate constant.
119. The reaction is second order. The rate constant is the slope = 0.0821 $M^{-1}\,s^{-1}$.

Chapter 15

4. Both techniques separate mixtures into their components utilizing a stationary phase and mobile phase. The stationary phase (typically a liquid) is contained in a column through which the mobile phase passes. GC is used to separate mixtures of volatile components and employs a gaseous mobile phase, while HPLC separates mixtures that are soluble in a liquid mobile phase.
6. Hexane is nonpolar and therefore will interact more strongly with the nonpolar C_{18} stationary phase than with the highly polar methanol and water mobile phase.
7. (b) and (e)

8. heptanol, butyl propyl ether, heptane; heptane, butyl propyl ether, heptanol
10. The purpose of the GC is to separate the mixture into its components. The mass spectrometer then identifies each of the components.
11. (b)
14. Quantitative (c) and (d); qualitative (a) and (b)
15. Each technique uses a different type of electromagnetic radiation. This causes different changes within the molecule and yields different information about the molecule.
17. (a) absorbance increases (b) absorbance decreases
    (c) absorbance decreases
18. (a) isobutyric acid (b) butyraldehyde
    (c) 3-pentanol (d) octane
19. Butyraldehyde
21. (c), (a)
22. 576 is the wavelength of maximum absorbance in this region of the spectrum. Using the wavelength of maximum absorbance increases the sensitivity of the measurements and increases the concentration range over which the Beer's law plot of absorbance vs. wavelength is linear.
23. 0.37
28. Elemental: (d); molecular: (a), (b), (c), and (e)
29. Ginger ale is a mixture; therefore, the first step is to separate it into its components using a chromatographic method. HPLC is the technique of choice. A chromatogram is also run of a normal ginger ale sample and the two chromatograms compared. If a component exists in the suspect ginger ale that is not in the normal sample, it should be isolated from the sample and identified using IR, NMR, or mass spectroscopy.
30. (a) GC (b) ion-selective electrode (c) atomic absorption
    (d) IR (e) IR and NMR (f) GC

Appendix D

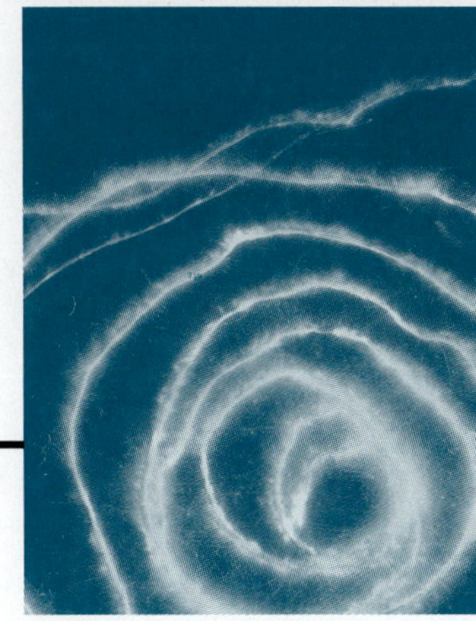

CHECKPOINT ANSWERS

Chapter 1

- (p. 7) The symbol 8 S represents 8 individual atoms of sulfur. The symbol S_8 represents 8 atoms of sulfur bonded together to form a single molecule.
- (p. 11) In 10,000 atoms of lithium you would expect to find approximately 742 ^{6}Li atoms and 9258 atoms of ^{7}Li.
- (p. 14) NaOH: Na^+ and OH^-; K_2SO_4: 2 K^+ and SO_4^{2-}; $BaSO_4$: Ba^{2+} and SO_4^{2-}; $Be_3(PO_4)_2$: 3 Be^{2+} and 2 PO_4^{3-}
- (p. 16) F: 9; Pb: 82; 24: Cr; 74: W
- (p. 18) One cm^3 of diamond is heavier than 1.0 cm^3 of coal. Two cm^3 of coal is heavier than 1.0 cm^3 of diamond.

Chapter 2

- (p. 29) K atomic weight 39.098 amu; molar mass 39.098 g/mol; U atomic mass 238.03 amu; 238.03 g/mol
- (p. 31) 2.2×10^{22} Al atoms
- (p. 35) There are two atoms of carbon in one molecule of C_2H_2; two moles of carbon atoms in one mole of C_2H_2; and 1.20×10^{24} atoms of carbon in one mole of C_2H_2.
- (p. 38) CaC_2. The molecular weight must be known to find the molecular formula.
- (p. 41) $H^+(aq) + Cl^-(aq) + Na^+(aq) + OH^-(aq) \longrightarrow Na^+(aq) + Cl^-(aq) + H_2O(l)$
- (p. 45) $x = 2$
- (p. 48) 4 moles of H_2, 2.41×10^{24} molecules of H_2, 4.82×10^{24} atoms of H.
- (p. 48) Two grams of CO correspond to 0.071 mol of CO. 0.071 mol of CO require 0.14 mol of H_2. This equals 0.28 grams of H_2.
- (p. 52) In the balanced chemical equation I_2 and Mg react in a one-to-one mole ratio. Four grams of I_2 equals 0.016 mol. Four grams of Mg equals 0.16 mol. There is ten times more Mg present than I_2. I_2 is the limiting reagent.

Chapter 3

- (p. 71) The radius of a typical atom would have to be 10,000 cm. This is equivalent to 100 meters.
- (p. 73) The frequency of microwaves covers a range of approximately 3×10^{10} to 3×10^{12} Hz.
- (p. 77) The higher the temperature of an object, the higher the frequency of electromagnetic radiation emitted by the object.
- (p. 78) Using Figure 3.5, we see that a change in energy of 1166 kJ/mol corresponds to an energy change between the first ($n = 1$) and third energy levels ($n = 3$). Electrons in the third energy level have an energy of -145.8 kJ/mol. The electrons would be in the $n = 3$ energy level.
- (p. 81) The first ionization energy of Sr would be greater than that of Rb. Sr has one more proton in its nucleus than Rb and therefore attracts its electrons more strongly than does Rb. The first ionization energy of Cl is greater than that of Br. Bromine has one more shell of electrons than Cl,

and therefore its electrons are farther from the nucleus and more easily removed.

- (p. 83) Carbon has a core charge of +4 and fluorine has a core charge of +7.
- (p. 83) The first ionization energy of sodium is much less than the first ionization energy of neon. If there were 9 electrons in the second level, we would expect these electrons to be held more tightly by the 11 protons in the nucleus of sodium than the 8 valence electrons are held by the 10 protons in the nucleus of neon. This would give sodium a larger first ionization than neon rather than the observed smaller ionization energy.
- (p. 84) Sodium has three shells whereas lithium only has two. The first ionization energy of sodium is much less than the first ionization energy of lithium. Therefore, the radius of the valence shell of sodium is larger than that of lithium.
- (p. 85) The number of peaks in a photoelectron spectrum is determined by the number of different energies that the electrons in an atom have.
- (p. 86) There is only one peak in the PES of He because both of its electrons have the same energy.
- (p. 88) Neon has 2 electrons in its $n = 1$ shell and 8 electrons in its $n = 2$ shell. Of the 8 electrons in the $n = 2$ shell, two are in the *s* subshell and six are in the *p* subshell.
- (p. 90) The energies and the relative intensities of the PES of Sc are given below:

Energy (MJ/mol)	*Relative Intensity*
433	2
48.5	2
39.2	6
5.44	2
3.24	6
0.77	1
0.63	2

Electrons are removed from the subshells of Sc in the following order: 4*s*, 3*d*, 3*p*, 3*s*, 2*p*, 2*s*, 1*s*.

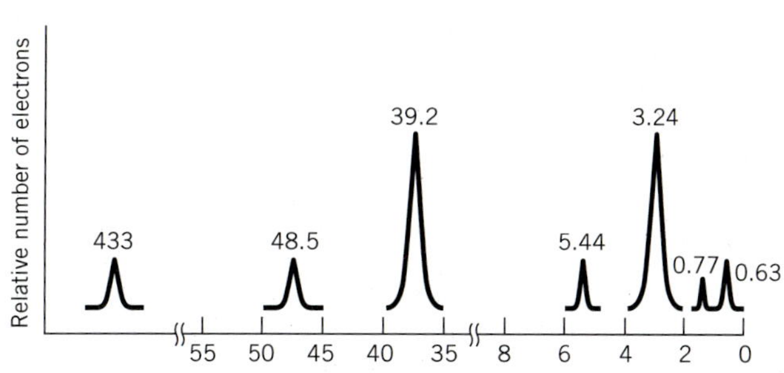

The subshells of Sc are filled with electrons in the following order: 1*s*, 2*s*, 2*p*, 3*s*, 3*p*, 4*s*, 3*d*. There is a difference in the order in which electrons are added to and taken away from Sc. Electrons are added to the lowest-energy subshell (most stable subshell closest to the nucleus) first and then to subsequent subshells in order of increasing energy until the 3*p* subshell is filled. After the $3p$ subshell is filled the $4s$ is filled followed by the lower-energy $3d$ subshell. Electrons are removed from the highest-energy subshell (subshell farthest from the nucleus) first and then from lower-energy subshells.

- (p. 92) Three quantum numbers are required to describe three dimensions in space. The numbers describe: n, size; l, shape; and m, orientation.
- (p. 95) Li, B, and N all have at least one unpaired electron and would therefore be expected to have magnetic properties. Be has no unpaired electrons and would therefore have no magnetic properties and would not be deflected in a Stern-Gerlach experiment.
- (p. 100) C: 2; N: 3; Ne: 0; F: 1
- (p. 101) A metallic radius is one-half the distance between two adjacent nuclei of two like metal atoms measured in a solid crystal lattice. A covalent radius is one-half the distance between two adjacent nuclei of two like atoms of a molecule. The molecule may be in the solid, liquid, or gaseous state.
- (p. 104) The ionic radius of Si^{4+} would be less than the ionic radius of Al^{3+}. Si^{4+} and Al^{3+} have the same number of electrons, but Si^{4+} has one more proton that pulls the electrons more closely to the nucleus. (Bottom of Table 3.9.)
- (p. 105) Ionization energies obtained from PES are for the removal of only one electron. That electron may be located in any shell or subshell. Second, third, and fourth ionization energies describe the energy to sequentially remove a second, third, and fourth electron from the outermost shells of an atom.
- (p. 106) The maximum positive charge on atoms in Group IVA is +4. For example, Sn has the electron configuration [Kr] $5s^2\ 4d^{10}\ 5p^2$. It can lose four electrons to form an Sn^{4+} ion with the electron configuration [Kr] $4d^{10}$.
- (p. 107) The difference in energy between the $2s$ and $2p$ subshells of oxygen is 1.81 MJ/mol. The difference in energy between the $2s$ and $2p$ subshells of sulfur is 1.05 MJ/mol. This trend in smaller differences between the s and p subshells would be expected to continue down the group. Therefore, Se would be expected to have a smaller difference in energy between its outer s and p subshells than does S.
- (p. 107) AVEEs increase from left to right across a period because each subsequent atom in a period has one more positive charge in the nucleus which attracts the electrons more tightly. As you move from the bottom to the top of a group, there are fewer shells containing electrons. The electrons are closer to the nucleus and therefore held tightly by the nucleus. This is the same trend followed by first ionization energy. The trend for AVEEs is the reverse of the trend followed by atomic radii.

Chapter 4

- (p. 121) An atom in Group IVA has 4 valence electrons.
- (p. 123) H_2^+ has only one electron to form the bond between the two hydrogen atoms. A covalent bond is formed by the attraction between the nucleus of one atom and the electron(s) of another atom. Because there is only one electron in H_2^+, both nuclei share the single electron. In addition, there is less electron density to shield the repulsion between the two nuclei in H_2^+. Therefore the nuclei in H_2^+ are not pulled as closely together as the nuclei in H_2.

- (p. 124) In CO_2 the nonbonding electrons are located in the nonbonding electron domains of the oxygen atoms. In CO_2 each oxygen has two nonbonding domains. There are two electrons in each nonbonding domain, but four electrons are located in each of the bonding domains between carbon and oxygen. Four electrons are required to form a double bond.
- (p. 125) Hydrogen needs only two electrons in its outer shell in order to achieve the electron configuration of He. Therefore, hydrogen is capable of forming only one bond and cannot serve as the central atom in a Lewis structure.
- (p. 125) A skeleton structure shows the relative positions of the atoms. A Lewis structure shows the positions of the electrons.
- (p. 127)

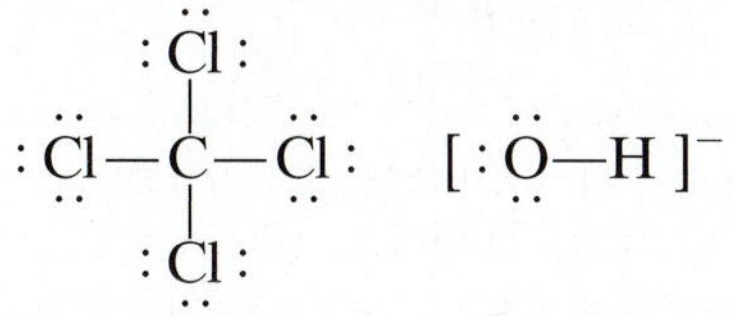

- (p. 128) The structure for BF_3 given in the checkpoint satisfies the octet for all of the atoms and therefore appears to be a good Lewis structure. However, B and F are not in the list of atoms that typically form multiple bonds. In addition, experimental evidence of the reactivity of BF_3 suggests that the Lewis structure shown in the text, with only six electrons around the B, is probably the Lewis structure that best matches the actual structure.
- (p. 129) In order for an atom to expand its octet, it must have free unoccupied orbitals with an energy close to the energy of its valence electrons (within the same shell). Nitrogen and oxygen have valence electrons in the second shell. The second shell contains one *s* orbital and three *p* orbitals. These four orbitals can contain a maximum of eight total electrons. Because there are no more subshells in the second shell, there are no more orbitals available to hold any additional electrons.
- (p. 132) The nitrogen–nitrogen bond length would be greatest in $H_2N{-}NH_2$ because single bonds are longer than multiple bonds when comparing bond lengths between the same atoms.
- (p. 133) Isomers have the same molecular formula but differ in the arrangement of the atoms. All of the atoms in the two structures shown for acetate are in the same positions. Only the location of a multiple bond differs in the two Lewis structures. This makes them resonance structures, not isomers.
- (p. 133)

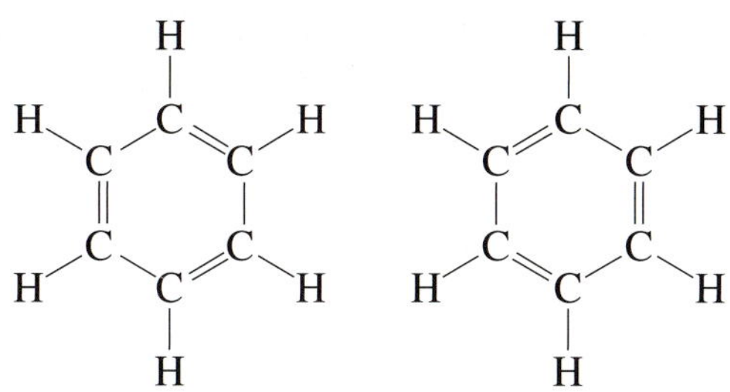

The two resonance Lewis structures of benzene have alternating single and double bonds in two different arrangements. The actual structure of benzene is an average of these two structures. Therefore, we would expect each bond in benzene to behave like a bond and a half (an average of a single bond and a double bond). The bond length of each C—C bond will be greater than the C=C double bond length of ethylene but less than the C—C single bond length of ethane.

- (p. 136) The bond in NO would be polar covalent because N and O have different electronegativities. The bond in O_2 would be pure covalent because both oxygen atoms have the same electronegativity.
- (p. 137) The most electronegative atom will have the negative partial charge. If the electronegativity of *B* is greater than that of *A*, *B* would have the negative partial charge.
- (p. 139) The formal charges are: double-bonded oxygen, 0; sulfur, +1; single-bonded oxygen, −1.
- (p. 140) The formal charge on Cl in each of the acids is as follows:

  $HClO_4$: +3
  $HClO_3$: +2
  $HClO_2$: +1
  HClO: 0

  The Cl in $HClO_4$ has the largest formal charge of +3. The hydrogen, with its positive formal charge, will be most easily removed from the $HClO_4$, which has the largest formal charge on Cl. The Cl atom in HClO has a formal charge of zero and therefore does not lose its hydrogen so easily.
- (p. 144) Each carbon in the Lewis structure of benzene has one C—C single bond, one C═C double bond, and one C—H single bond. Therefore, each carbon atom has three electron domains.
- (p. 145) A molecule with five electron domains around the central atom will have one of four possible molecular geometries: trigonal bipyramidal, seesaw, T-shaped, and linear. Which of these geometries the molecule will have is dependent on how many of the domains are bonding and how many are nonbonding.
- (p. 148) The shape of a molecule can be found using Table 4.3. The shape of a molecule with three bonding domains and one nonbonding domain around the central atom is trigonal pyramidal.
- (p. 149) The C_1—C_2—C_3 is 109°.

  The C_2—C_1—H_g is 109°.
  The H_g—C_1—O is 109°.
- (p. 149) There are three bonding electron domains around each carbon. Therefore, the bond angles would be approximately 120°.
- (p. 150) Chloroform would have a dipole moment because the polarity of the C—H bond would be different than that of the three C—Cl bonds. The polarities of the bonds would not cancel and would therefore result in a dipole moment for the molecule.
- (p. 160) N_2 has the Lewis structure: N≡N:

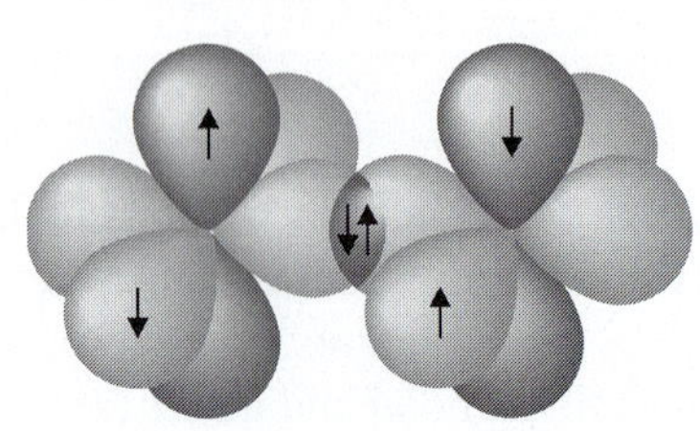

Valence bond theory describes the bonding in N_2 as one sigma and two pi bonds formed by head-to-head and side-to-side overlap, respectively, of half-filled $2p$ orbitals. The number of bonds is three in both models, but the Lewis structure depicts all the bonds as being the same.

Chapter 5

- (p. 172) Ca < K < Rb
- (p. 174) Atoms in Group IIA have two electrons in their valence shell that can be more easily lost than their inner shell electrons. Group IA elements have only one valence shell electron. Therefore, an electron must be removed from an inner shell in order to lose a second electron. These electrons are closer to the nucleus and are more difficult to remove.
- (p. 180) The most common ion of Sr is Sr^{2+}. The peroxide anion is O_2^{2-}. The formula is SrO_2.
- (p. 182)

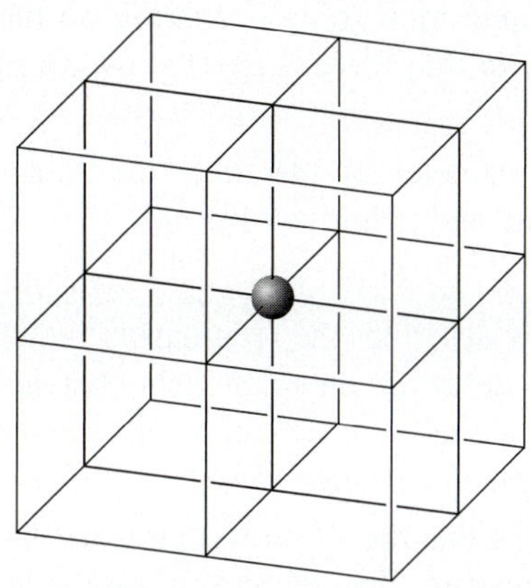

- (p. 183) The electrons in metallic bonds are delocalized and therefore free to move from one atom to another and conduct an electric current. The electrons in both covalent and ionic bonds are localized and therefore not free to move to another atom or ion.
- (p. 183) The bond in $CuAl_2$ is primarily a metallic bond, whereas NaCl has a primarily ionic bond. The electrons in the metallic bond are delocalized. An electron is transferred from sodium to chloride in the ionic bond. This electron now belongs to the chloride and is therefore localized. The ionic bond is held together by the attraction of the positive sodium ion for the negative chloride ion.
- (p. 186) The difference in electronegativity between Al and Cl is 1.26. Based only on the difference in electronegativity, we would predict that $AlCl_3$ is polar covalent. It has a difference in electronegativity similar to that between oxygen and hydrogen atoms in water.
- (p. 189) Using the bond-type triangle, we find $AlCl_3$ located directly on the dividing line between covalent and ionic bonding. We would predict that $AlCl_3$ would have approximately equal covalent and ionic character.
- (p. 192) $BaSi_2$ is primarily metallic; $BaBr_2$ is primarily ionic; $GaBr_3$ is primarily covalent. $BaBr_2$ would be the best insulator that would also stand up to high heat.
- (p. 198) Formal charge and oxidation numbers are tools that have been developed by chemists. Formal charge treats all bonds as if they were pure covalent bonds. Oxidation numbers treat all bonds as if they were pure ionic bonds. Both of these are useful tools for making predictions about chemical structures and changes that occur during reactions.
- (p. 199) The first reaction is not an oxidation–reduction reaction. None of the atoms undergo a change in oxidation state. The second reaction is an

oxidation–reduction reaction. The oxidation number of N changes from −3 in NH_3 to +2 in NO. The oxidation number of O changes from zero in O_2 to −2 in both NO and water. The third reaction is an oxidation–reduction reaction. The oxidation number for carbon in CH_3OH changes from −2 to −3, while the oxidation number of carbon in CO changes from +2 to +3.

Chapter 6

- (p. 214) If the liquid water and gaseous water are both at the same temperature, then the average kinetic energy of the water molecules should be the same.
- (p. 218) The pressure exerted on the floor by the spike heel will be greater. If the same person wears one shoe with a spike heel and the other with a normal heel, the same total force is exerted on the floor. However, in the case of the spike heel that force is exerted over a much smaller area of the floor.
- (p. 221) No, the diameter of the tube does not change the relationship between the pressure and volume of the gas.
- (p. 222) $PV \propto T$
- (p. 223) Boyle's law describes the relationship between pressure and volume and Amontons' law describes the relationship between pressure and temperature. By combining these laws, we can obtain a relationship between temperature and volume. This is Charles' law.
- (p. 230) If we assume that the 1 L of wet air and 1 L of dry air are at the same temperature and pressure, then there must be the same number of moles in both containers. The gas in the container of dry air has an average molecular weight of 29.0 g/mol. The container of wet air contains both air and water vapor. Because it contains both air and water vapor but has the same total number of moles of gas as the dry air, there are fewer moles of dry air (29 g/mol) in this container and more moles of water vapor (18 g/mol) than in the container of dry air. Therefore, the container of wet air is lighter than the container of dry air because the container of wet air has fewer heavy molecules (N_2 and O_2) and more light molecules (H_2O) of gas than the container of dry air.
- (p. 235) As the temperature of a gas increases, the molecules of gas begin to move more rapidly, resulting in their having more kinetic energy.
- (p. 238) The heavier gas molecules will be moving at a slower velocity than the lighter gas molecules. Because kinetic energy is dependent on both mass and velocity, a heavy molecule moving at a slow velocity can have the same kinetic energy as a less-massive molecule moving at a greater velocity.

Chapter 7

- (p. 256) No bonds are broken in the reactants. One bond is formed in the products. Therefore, more bonds are made than broken. This reaction will release energy because it is a bond-making process.
- (p. 257) When heat enters a balloon, the temperature of the gas particles increases. This results in an increase in the motion of the gas particles, causing them to strike the sides of the balloon more frequently and with more force.

If the balloon is elastic it will expand, increasing the volume of the gas while the pressure remains constant.

- (p. 259) According to the first law of thermodynamics, the total energy change must equal zero. This requires that the sum of the change in energy of the system and the change in energy of the surroundings equal zero. If we assume the brick to be our system and the water to be the surroundings, the brick can gain energy only if the surroundings supply that energy. Therefore, the first law does not prevent the brick from becoming hotter and the water from becoming colder. The first law requires only that what is gained by one part of the universe must be lost by another part of the universe.
- (p. 260) The temperature of the gas will increase the most for the gas in a fixed-volume piston and cylinder. All of the heat transferred to this gas is used to increase the temperature of the gas. Some of the heat transferred to the gas in the movable piston and cylinder is used to do work.
- (p. 263) While at rest on the 10th floor the crate will have only potential energy. During the fall the crate will have both potential and kinetic energy. As the crate falls its velocity increases, thus increasing its kinetic energy. As it moves closer to the ground its potential energy in relation to the ground decreases. When the crate strikes the ground it can release its energy in the form of sound and heat or by doing work on the ground (making a hole or indentation).
- (p. 265) Table sugar can provide or give off energy. This is an exothermic process, and the bond-making process must provide more energy than is required by the bond-breaking process. The products must therefore be more stable than the reactants. The sum of the bond strengths formed is larger than the corresponding sum for the bonds broken.
- (p. 267) Specific heats are based on the mass of a substance, and molar heat capacities are based on moles. One mole of iron has a much larger mass than one mole of aluminum. Given two substances with similar structures and properties, the one with the higher molecular weight will have a higher molar heat capacity and a lower specific heat compared to the substance with the lower molecular weight.
- (p. 268) If the temperature of the water bath decreases, the heat must have been absorbed as a result of the chemical reaction. The reaction must be endothermic, with a positive change in enthalpy.
- (p. 269) All of the reactions listed in Table 7.2 release or give off energy. Therefore, these are exothermic reactions and the sign of the change in enthalpy is negative.
- (p. 273) ΔH°_{373} represents the change in enthalpy for a process that occurs under standard conditions and at 373 K.
- (p. 275)

$$\mathrm{N}(g) \longrightarrow \mathrm{N}(g)$$
$$\mathrm{H}(g) \longrightarrow \mathrm{H}(g)$$
$$\mathrm{N}(g) + 3\,\mathrm{H}(g) \longrightarrow \mathrm{NH_3}(g)$$

ΔH°_{ac} for N(g) and H(g) are both zero because both are isolated atoms in the gaseous state. The ΔH°_{ac} for NH_3 is -1171.76 kJ/mol$_{rxn}$. ΔH°_{ac} for NH_3 represents the formation of NH_3 from its gaseous atoms. No bonds are broken in the reactants, but bonds are formed by the products; therefore, ΔH°_{ac} is negative.

- (p. 279)

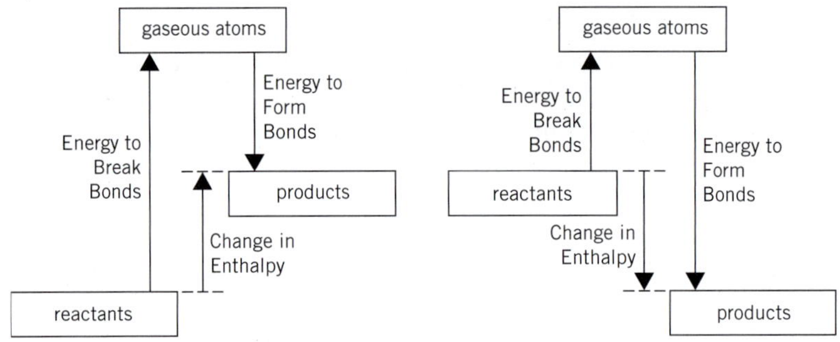

- (p. 281) Bond Breaking: $2\ CO(g)\ \Delta H^\circ = 2(1076.38\ kJ/mol_{rxn}) = 2152.76\ kJ/mol_{rxn}$

$$O_2(g)\ \Delta H^\circ = 498.34\ kJ/mol_{rxn}$$

$$\text{Total Bond Breaking:}\quad \Delta H^\circ = 2651.10\ kJ/mol_{rxn}$$

Bond Making: $2\ CO_2(g)\ \Delta H^\circ = 2(-1608.53\ kJ/mol_{rxn}) = -3217.06\ kJ/mol_{rxn}$

$$\Delta H^\circ_{rxn} = (2651.10\ kJ/mol_{rxn}) + (-3217.06\ kJ/mol_{rxn}) = -565.96\ kJ/mol_{rxn}$$

The reaction is exothermic.

- (p. 283) The enthalpy of atom combination of liquid ethanol is greater than the enthalpy of atom combination of gaseous ethanol. More energy is released in the formation of liquid ethanol than in the formation of gaseous ethanol from its gaseous atoms. The change in state from liquid ethanol to gaseous ethanol requires an input of energy. There are forces of attraction between the molecules of liquid ethanol that are not present in gaseous ethanol.
- (p. 284) The sum of the bond strengths can be estimated using the values from Exercise 7.9 and 7.10:

$$Si{-}F = 597\ kJ/mol$$
$$Si{-}O = 466\ kJ/mol$$
$$O{-}H = 463\ kJ/mol$$

$$3 \times 597\ kJ/mol + 466\ kJ/mol + 463\ kJ/mol = 2720\ kJ/mol$$

- (p. 285) When similar atoms are bonded together, the enthalpy of atom combination decreases with increasing bond length. BrF would be expected to have weaker bonds (a smaller enthalpy of atom combination) than ClF, because BrF is a larger molecule with a larger bond length.

Chapter 8

- (p. 304) As an atom becomes larger, the electrons farther from the nucleus are more easily induced into a temporary dipole. As the size of an atom increases, there is an increase in the dispersion force between atoms. Kr > Ar > Ne > He.
- (p. 307) Hydrogen bonds between NH_3 molecules are weaker than those between H_2O molecules because oxygen is more electronegative that nitrogen. This gives a larger partial charge to the H in the OH bond.
- (p. 308) (a) These three substances have very different molecular weights. We would predict that the boiling point should increase with an increase in molecular weight due to increasing dispersion forces. Therefore the boiling point of tetrabromobutane (373.6 g/mol) is greater than the boiling point of carbon tetrachloride (154 g/mol), and acetone (58 g/mol) has the lowest boiling point.

  (b) These three substances all have very similar molecular weights (142 to 144 g/mol). When comparing the boiling points of substances with similar molecular weights, it is necessary to examine their intermolecular forces. Octanoic acid is capable of hydrogen bonding and will therefore have the highest boiling point. Nonanal is polar and therefore has dipole–dipole intermolecular forces. Decane is nonpolar and will have only dispersion forces. Therefore, decane is expected to have the lowest boiling point.

- (p. 310) The phase change from liquid to gas requires more enthalpy (enthalpy of vaporization 44 kJ/mol_{rxn}) than the phase change from solid to liquid (enthalpy of fusion 6 kJ/mol_{rxn}). The solid state has the strongest intermolecular forces. Some of these forces must be overcome to melt the solid, and the remaining forces must be overcome to cause the liquid to change to a gas.
- (p. 312) If the volume of the container is decreased, the ideal gas law tells us that the pressure of the vapor in the container should increase. However, if the pressure of the vapor in the container is at its vapor pressure (the maximum pressure of vapor for a given temperature), the pressure cannot increase. Therefore, as the volume is decreased condensation of the vapor will occur, so that the pressure of the vapor remains constant at the vapor pressure. On the molecular level some of the gas phase molecules enter the liquid phase.
- (p. 324) The bond-breaking process for one mole of $BaCl_2(s)$ requires +1282 kJ/mol_{rxn}. The attractive forces that form between Ba^{2+} ions and water release 718 kJ/mol_{rxn}. The attractive forces that form between 2 Cl^- ions and water release (2) × (289 kJ/mol_{rxn}). The change in enthalpy associated with the dissolution of $BaCl_2(s)$ is −14 kJ/mol_{rxn}.
- (p. 325) The bond-breaking process for $AgCl(s)$ is +533.30 kJ/mol_{rxn}. The attractive forces that form between Ag^+ ions and water release +178.97 kJ/mol_{rxn}. The attractive forces that form between Cl^- ions and water release +288.838 kJ/mol_{rxn}. The change in enthalpy associated with the dissolution of $AgCl(s)$ is +65.49 kJ/mol_{rxn}.
- (p. 331) In order for a side chain to be hydrophilic, it must be able to interact with water. This means that the side chain should have intermolecular forces similar to the intermolecular forces of water (hydrogen bonding). The side chain on alanine contains carbon and hydrogen atoms that have similar electronegativities. This side chain is nonpolar. The side chain on cysteine is somewhat polar due to the presence of sulfur, with its two nonbonding pairs of electrons, but it is not capable of hydrogen bonding. These two side chains are therefore hydrophobic (they do not form hydrogen bonds with water). Lysine contains an NH_3 group and serine contains an OH group. Both lysine and serine are capable of forming hydrogen bond intermolecular forces. Therefore, both of these will be hydrophilic.

Chapter 9

- (p. 355) On the atomic scale, molecular solids are composed of individual molecules held together by intermolecular forces. Network covalent solids are composed of a three-dimensional array of atoms held together by covalent bonds. On the macroscopic scale, network covalent solids have high melting points and molecular solids have relatively low melting points.
- (p. 356) On the atomic scale, molecular solids are composed of individual molecules held together by intermolecular forces. Ionic solids are composed of cations and anions in a three-dimensional array held together by ionic bonds. Ionic solids have a higher melting point than molecular solids. Ionic solids conduct an electric current when melted. Molecular solids do not.

- (p. 357) The lattice energy of MgF_2 is greater than that of $MgCl_2$ because F^- is a smaller ion than Cl^-. The smaller fluoride ion is held more closely and tightly to the Mg^{2+}.
- (p. 359) Both the strength of the attraction of an atom for its valence electrons and the energy difference between valence subshells within an atom contribute to the AVEE and electronegativity. Atoms that have a relatively small attraction for their valence electrons and have small differences in energy between valence subshells have low AVEE and electronegativity. Atoms with these characteristics can lose their valence electrons to form positive cations. The lost electrons can move easily between available subshells from cation to cation. This results in metals being good conductors of electricity.
- (p. 359) Metallic solids are composed of positively charged metal cations in a three-dimensional array. The metal cations are surrounded by a sea of delocalized electrons that form metallic bonds. Ionic solids are composed of cations and anions in a three-dimensional array held together by localized electrons in ionic bonds.

  Metallic solids have a wide range of melting points compared to ionic solids, which have relatively high melting points. Metallic solids can conduct electricity in the solid state but ionic solids cannot.
- (p. 363) B_4C has an average *EN* of 2.3 and a ΔEN of 0.5. This places it in the covalent region of a bond-type triangle. MoC has an average EN of 2.0 and a ΔEN 1.1 and is in the ionic region.
- (p. 366) In Figure 9.5 each sphere touches four others. In Figure 9.10 each sphere touches six others.
- (p. 370) In order for a substance to be classified as a body-centered cubic cell, all of the positions in the unit cell must be occupied by the same type of atom or ion.
- (p. 370) When X rays strike the surface of a crystal, they are diffracted into repeating patterns. The only way that this can occur is for the X rays to remain in phase. For the diffracted X rays to remain in phase they must be diffracted by an ordered arrangement of planes of atoms. Thus, solids must consist of an ordered arrangement of particles.
- (p. 372) The three factors that contribute to the density of a metal are the atomic weight of the metal atoms, the size of the metal atoms, and the type of unit cell formed by the metal atoms.
- (p. 374) The simple cubic cell has the simplest relationship between the radii of the ions and the length of the edge of the unit cell. In a simple cubic cell the length of the edge of the cell is equal to $r + r$. In a face-centered cube $2(r + r) = a2^{1/2}$ where a is the length of the edge of the unit cell. For a body-centered cubic cell the relationship is $2(r + r) = a3^{1/2}$.

Chapter 10

- (p. 380) After the reaction has reached equilibrium, both N_2O_4 and NO_2 will be present in the container.
- (p. 382) Using Table 10.1 we see that at equilibrium the ratio of *trans*-2-butene to *cis*-2-butene is 0.559 to 0.441, which equals 1.27.
- (p. 385) An increase in the rate constant of a reaction will result in an increase in the rate of the reaction. Therefore, the rate of the reaction will increase with temperature.

- (p. 387) If the rate of the reverse reaction is greater than the rate of the forward reaction at a given moment in time, this does not mean that the reverse rate constant is greater than the forward rate constant. The rate of a reaction is dependent on both the rate constant and the concentrations of the reactants. If the rate of the reverse reaction is greater than that of the forward reaction, this may be due to a higher concentration of products (reactants for the reverse reaction) than reactants for the forward reaction.
- (p. 389) The equilibrium constant can be determined from the division of the forward rate constant by the reverse rate constant. For this reaction the equilibrium constant would be 2.
- (p. 396) The stoichiometry of the reaction is 1:1:1. For every mole of PCl_3 that reacts, one mole of PCl_5 is produced and one mole of Cl_2 is used up. Therefore, if the concentration of PCl_3 decreased by 0.96 mol/L, the concentration of Cl_2 also decreased by 0.96 mol/L, leaving only 0.04 mol/L. The concentration of PCl_5 must increase from zero to 0.96 mol/L.
- (p. 398) The equilibrium concentrations found in Exercise 10.8 give the correct $K_c(2.5)$ within ± 1 in the last significant figure: $0.71/0.29 = 2.4$.
- (p. 399) The change in concentrations that takes place during a chemical reaction must follow the stoichiometry of the balanced chemical equation.
- (p. 402) ΔC will be small when the equilibrium constant is significantly smaller than one. ΔC would be small for K values of 1.0×10^{-5} and 1.0×10^{-10}.
- (p. 406) If more P_2 is added to the system at equilibrium, the system will shift to use up the added P_2. The system will shift to the right to produce more P_4. If the pressure increases while the temperature remains constant, a reaction will shift in a direction to relieve the increase in pressure by decreasing the number of moles of gas. For this reaction the equilibrium will shift to the right to produce P_4 and use up P_2.
- (p. 412)

$$K_{sp} = [Mg^{2+}][OH^-]^2 = 1.8 \times 10^{-11}$$
$$Q = (Mg^{2+})(OH^-)^2 = (1.7 \times 10^{-4}\,M)(2.4 \times 10^{-4}\,M)^2 = 9.8 \times 10^{-12}$$

  Because $Q < K_{sp}$ these concentrations of ions may exist in solution and a precipitate will not form.
- (p. 412)

$$K_{sp} = [Li^+][F^-] = 6 \times 10^{-3}$$
$$6 \times 10^{-3} = \Delta C\,\Delta C$$
$$\Delta C = 8 \times 10^{-2}\,M = [Li^+] = [F^-]$$

Chapter 11

- (p. 429) An Arrhenius base must ionize to give OH^- when it dissolves in water. $Mg(OH)_2$ ionizes to give Mg^{2+} and 2 OH^-. It is therefore an Arrhenius base. HNO_3 and CH_3CO_2H both ionize to produce H^+ and are therefore Arrhenius acids.
- (p. 432) $H_3PO_4(aq) + H_2O(l) \rightleftharpoons H_2PO_4^-(aq) + H_3O^+(aq)$. The conjugate base is $H_2PO_4^-$. Phosphoric acid is a polyprotic acid, meaning that it can further dissociate to lose an additional proton and produce additional H_3O^+.

$$C_6H_5NH_2(aq) + H_2O(l) \rightleftharpoons C_6H_5NH_3^+(aq) + OH^-$$

  The conjugate acid of aniline is $C_6H_5NH_3^+$.
- (p. 436) The K_w will still equal 10^{-14}. The addition of acid will result in a higher overall concentration of H_3O^+ even though some of the hydronium ion from the

dissociation of water may shift back to the left. What is most significant about the shift in the equilibrium of water back to the left is the decrease in the concentration of hydroxide ion. The concentration of the hydroxide ion multiplied by the concentration of the hydronium ion will therefore equal K_w, 10^{-14}.

- (p. 440) pH is $-\log[H_3O^+]$. As the concentration of hydronium ion increases, there will be a decrease in the pH.
- (p. 443)

$$K_w = [OH^-][H_3O^+]$$

$$K_a = \frac{[OH^-][H_3O^+]}{[H_2O]}$$

The K_a expression includes H_2O as a reactant.

- (p. 444) Both solutions have a low concentration of acid. The acid with the larger K_a will be a stronger acid and will dissociate more than the acid with the small K_a. The solution with the larger K_a will produce more hydronium ion and will therefore have a smaller pH.
- (p. 446) A stronger acid produces a weaker conjugate base. HOCl is the stronger acid; therefore its conjugate base, OCl^-, will be the weaker base. OBr^- will be the stronger base.
- (p. 448) H_2S is a stronger acid than H_2O. Sulfur is a larger atom than oxygen. The sulfur–hydrogen bond is longer and therefore weaker than the oxygen–hydrogen bond. The sulfur–hydrogen bond is more easily broken than the oxygen–hydrogen bond.
- (p. 450) The only difference between the two acids (HOCl and HOI) is the oxygen bonded to Cl in one acid and to I in the other. Because Cl is more electronegative than I, it will pull electron density away from the oxygen–hydrogen bond more than I. Thus the oxygen–hydrogen bond will be weaker for HOCl than for HOI and the hydrogen will be more easily lost from HOCl. Therefore, HOCl is a stronger acid than HOI.
- (p. 452) When a strong acid is added to water, it is assumed to completely dissociate to produce hydronium ion. For a monoprotic acid the hydronium concentration will be equal to the initial concentration of strong acid.
- (p. 455) It is not possible to determine which of the two solutions has the higher pH. Solution 1 contains a stronger acid and therefore dissociation is greater than for the acid in solution 2, but the concentration of the acid in solution 2 is greater than the concentration of the acid in solution 1. It is possible that solution 2 could produce more hydronium ion because it is more concentrated even though it is a weaker acid.
- (p. 456) The 1.0 *M* solution is more acidic.
- (p. 458) A large K_b indicates a stronger base that would react with water to accept more hydrogen ions and produce more hydroxide ions. Since both bases have the same concentration, the base with the larger K_b is more basic.
- (p. 461) NO_2^- is the conjugate base of HNO_2. Therefore, $K_b = K_w/K_a$.

$$K_b = (1.0 \times 10^{-14})/(5.1 \times 10^{-4}) = 2.0 \times 10^{-11}$$

- (p. 465) If acid is added to a buffer, the pH of the buffer will decrease slightly but the decrease will not be as great as when the same amount of acid is added to water. If base is added to a buffer, the pH of the buffer will increase slightly but not as much as when the same amount of base is added to water. The resulting pH is dependent on how much acid or base is added.

- (p. 466) Lactic acid produced in the body can react with HCO_3^- by the following reaction.

$$HC_3H_5O_3(aq) + HCO_3^-(aq) \rightleftharpoons C_3H_5O_3^-(aq) + H_2CO_3(aq)$$

- (p. 471) As NaOH is slowly added to a solution of HCl, the pH will gradually increase until just enough NaOH has been added to completely react with the HCl. At this point the pH will increase rapidly for several pH units. As additional NaOH is added, the pH will resume a more gradual increase. A plot of pH vs. volume of NaOH will yield a typical titration curve.

Chapter 12

- (p. 497) Oxidation is the loss of electrons; reduction is the gain of electrons. One process cannot occur without the other.
- (p. 498) Fe = +3, Cl = −1; C = −4, H = +1; O = 0, Mn = +7, O = −2; C = +2, H = +1, O = −2, N = −3.
- (p. 504) The anode is negative and the cathode is positive.
- (p. 505) The number of atoms is balanced but the charge is not. There is a charge of +2 on the reactant side and a charge of +3 on the product side of the chemical equation. The equation is not balanced.
- (p. 507) Sn(*s*) is oxidized and nitrogen is reduced. Sn is the reducing agent and the HNO_3 is the oxidizing agent.
- (p. 512) If the reaction is reversed, the sign (+ or −) of the cell potential is changed.
- (p. 516) The standard potential corresponds to 1 *M* concentrations of both Cu^{2+} and Ni^{2+}. A concentration of 1.5 *M* Cu^{2+} on the reactant side and a concentration of 0.010 *M* Ni^{2+} on the product side will result in a greater tendency for the reaction to proceed toward the products. This will result in a greater (more positive) value for the potential (E_{cell}) under these conditions than at standard conditions.
- (p. 531) The student received no credit because the reaction was balanced in base rather than in acid.

Chapter 13

- (p. 553) Dissolving salt in water is a spontaneous process. Even though heat is absorbed during the process, the salt dissolves because the system (water and salt) becomes more disordered as the salt dissolves.
- (p. 554) The second law states that for a spontaneous process the entropy of the universe must increase. The brick cannot get hotter nor the water colder.
- (p. 555) Water freezing does not violate the second law of thermodynamics. In the process of water freezing, water is considered the system. Water becomes more ordered and therefore the surroundings must become more disordered so that the entropy of the universe can become more positive (disordered).
- (p. 557) It is possible for a process to proceed if the entropy of the system decreases as long as the entropy of the surroundings increases even more, causing the entropy of the universe to increase.
- (p. 558) At absolute zero, motion at the molecular and atomic level approaches a minimum. A substance would have minimum entropy.

- (p. 561) Like cyclopentane being converted to pentene, glucose would have greater freedom of motion (more disorder) in its open-chain form. The sign of ΔS would be positive.
- (p. 561) Both molecules consist of the same kind and number of atoms. Their reactions of atom combination would involve the same gaseous atom reactants. Because the entropy of atom combination of isobutane is more negative than the entropy of atom combination of butane, the formation of isobutane results in the loss of more freedom of motion.
- (p. 566) If ΔH for a spontaneous process, such as dissolving table salt in water, is positive, the entropy of the process must be positive enough to make the change in Gibbs free energy negative. The increase in disorder must be enough to make up for the endothermic process.
- (p. 568) No, the hardening of cement has a negative ΔS. The cement becomes more ordered. In order for this to be a spontaneous process it must therefore be exothermic (releasing heat).
- (p. 576) If $Q_p > K_p$ the reaction must proceed toward the reactants. In order for Q_p to decrease to K_p the numerator (product concentrations) of the expression must decrease and the denominator (reactant concentrations) must increase.
- (p. 583) If ΔH is > 0 (endothermic) the reactants are favored. If ΔS is > 0 the products are favored. A change in temperature will change the equilibrium constant. In the relationship among K, ΔH, and ΔS, it is the enthalpy term that is dependent on temperature. As temperature decreases, the term containing enthalpy becomes more negative and therefore K becomes smaller. Thus the equilibrium is shifted toward the reactants.

  If ΔH is > 0 (endothermic) the reactants are favored. If ΔS is < 0 the reactants are favored. These circumstances will have the same results for a decrease in temperature as in the previous question because ΔH is still > 0. Therefore, a decrease in temperature will shift the equilibrium toward the reactants.

Chapter 14

- (p. 595) The kinetics of the reaction is so slow that during a human lifetime there will be no observable change of diamond to graphite.
- (p. 597) 10.5 seconds is required for the concentration of phenolphthalein to decrease from 0.00500 to 0.00450 M. 11.8 seconds are required for the concentration of phenolphthalein to decrease from 0.00450 to 0.00400 M. As the reaction proceeds, it takes more time for the concentration to decrease by a given amount. As the reaction proceeds, the rate of reaction decreases.
- (p. 598) If a large value for dt is used, there will be a significant change in rate during the time interval. The resulting rate would then be more of an average rate during the time period rather than an instantaneous rate at some moment in time.
- (p. 598) As the reaction progresses, the concentration of phenolphthalein decreases as it reacts. The concentration term in the rate law decreases, and therefore the rate of the reaction decreases. As the reaction proceeds, phenolphthalein is used up and there is a decrease in the probability of a collision between phenolphthalein molecules and base molecules. Thus the reaction rate decreases.
- (p. 599) The rate describes how quickly the reactants are used up and products produced with respect to time. The rate constant is a proportionality

constant used in the rate law. The rate of most reactions decreases with respect to time and concentration of reactants. However, the rate constant remains constant with respect to time and concentration.

- (p. 599) The units of k will be $1/(\text{M} \cdot \text{sec}) = \text{L}/(\text{sec} \cdot \text{mol})$.
- (p. 600) The compressed air provides a constant pressure pushing on the liquid in the buret and therefore a constant rate of drainage. Without the compressed air the pressure exerted by the column of liquid at the buret tip decreases as the column of liquid decreases.
- (p. 601) The rate of disappearance of HI is twice as fast as the rate of formation of H_2:

$$\text{Rate}_{\text{HI}} = 2(\text{Rate}_{\text{hydrogen}})$$

- (p. 608) The rate of a zero-order reaction is independent of the concentration of the reactants. The rate of reaction of a first-order reaction is dependent on concentration. As a reaction proceeds, the rate of reaction of a zero-order reaction will remain constant while a first-order reaction will slow down.
- (p. 610) In trials 2 and 3 both the concentrations of NH_4^+ and NO_2^- change. Therefore there will be two unknowns (m and n) in the equation. One of the concentrations must remain constant between two trials so that its effect will cancel out of the equation. Trials 1 and 2 cannot be used to find the order with respect to NH_4^+ because the concentration of NH_4^+ does not change in these trials. However, the concentration of NO_2^- does change. Any change in the rate will therefore be due to NO_2^-, not NH_4^+.
- (p. 610) For trial 1 the following are known: $(NH_4^+) = 5.00 \times 10^{-2}$, $(NO_2^-) = 2.00 \times 10^{-2}$, $m = 1$, $n = 1$, rate $= 2.70 \times 10^{-7}$. Solving the rate law equation for the rate constant, k, we get $2.70 \times 10^{-4}\ M^{-1}\ \text{sec}^{-1}$.
- (p. 613) The rate constant for the decay of ^{15}O is less than the rate constant for ^{19}O. Therefore, the rate of decay of ^{15}O will be slower and it will have a longer half-life.
- (p. 615) For a zero-order reaction the half-life for the second reaction will be twice as large as for the first trial. For a first-order reaction the half-life will not change. The half-life of a first-order reaction is independent of concentration. For a second-order reaction, doubling the concentration will decrease the half-life by a factor of two.
- (p. 621) No, this is not true for an endothermic reaction.

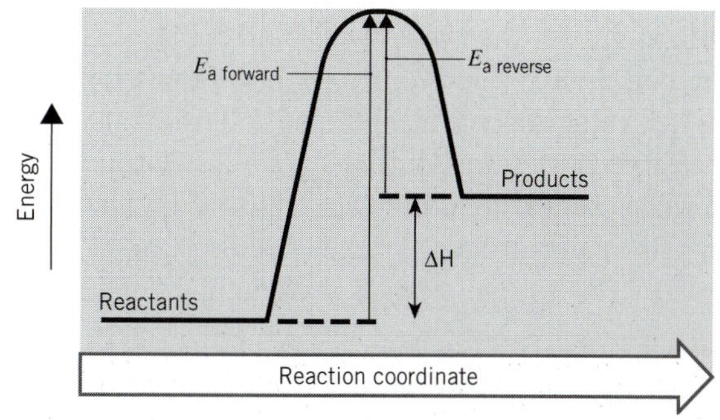

- (p. 623) ΔH is the difference between the energy of the products and the energy of reactants in both diagrams. ΔH will be the same in both diagrams because the enthalpies of products and reactants have not changed.

- (p. 623) The E_a term is a negative in the Arrhenius equation. Therefore, if E_a is large, k will be small. A small k indicates a slow reaction rate. Because a catalyst decreases E_a, the value of k will increase, leading to a faster reaction rate.
- (p. 625) Because the external pressure is less at the top of a mountain, the water will boil at a lower temperature than at sea level. Therefore, the egg will not be as hot in the boiling water. Cooking the egg to make it hard-boiled involves a chemical reaction. According to the Arrhenius equation, the rate constant for this reaction will be smaller at the lower temperature of the boiling water at the top of the mountain.

Chapter 15

- (p. 645) There is a much higher concentration of LLL in the suspect sample than in the olive oil sample. There is a much lower concentration of OOO in the suspect sample than in the good olive oil sample.
- (p. 650)

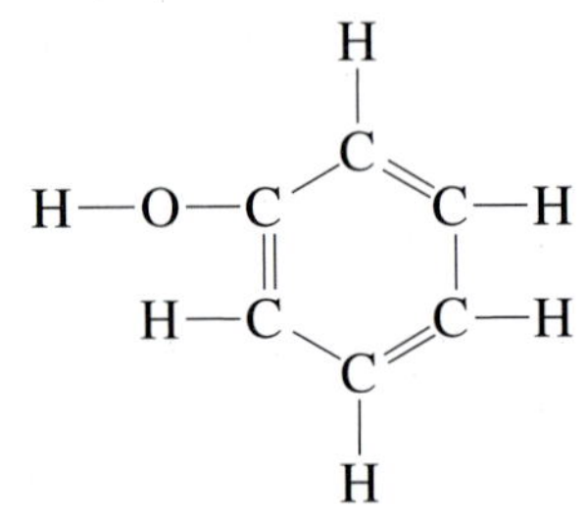

- (p. 654) The base T binds to A and G binds to C. The probe must contain the sequence AAGC.
- (p. 657) Both the second and the third electron configurations correspond to a zinc atom that has absorbed energy, and one or more electrons have moved to a higher energy level than in the ground state zinc atom. The first electron configuration represents the ground state of the zinc atom.
- (p. 660) The hemoglobin–oxygen complex has the greatest absorbance at 575 nm. Therefore, the molar absorptivity must be greatest at this wavelength.
- (p. 663) NO and NO_2 are radicals containing an odd number of electrons. The best Lewis structure for NO has a nitrogen–oxygen double bond. The best Lewis structure for NO_2 has one nitrogen–oxygen single bond and one nitrogen–oxygen double bond. However, the actual NO_2 structure will be an average of two resonance structures, which results in each nitrogen–oxygen bond behaving as a bond and a half. Therefore, the nitrogen–oxygen bond strength in NO will be greater than in NO_2. Table 15.3 shows that when comparing a carbon–carbon double bond to a carbon–carbon triple bond, the triple bond has a smaller value for wavelength and a larger value for wavenumber, and would therefore have a larger frequency of absorption. The same pattern is observed when comparing a C=O bond to a C—O bond. We would therefore expect the stronger nitrogen–oxygen double bond of NO to have a smaller wavelength of absorption, a larger wavenumber, and a larger frequency than NO_2.
- (p. 669) Acetone would show one peak and acetic acid would show two. In acetone there are six equivalent hydrogens in the same environment and therefore they absorb at the same frequency. In acetic acid the three hydrogens bonded to carbon are equivalent to one another, while the hydrogen bonded to oxygen is in a different environment and therefore absorbs at a different frequency.

Photo Credits

Chapters 1–15 Opener: Adri Berger/Stone.

Chapter 1: *Page 6:* Andy Washnik. *Page 9:* Courtesy of International Business Machines Corporation.

Chapter 2: *Page 37:* Ken Karp/Omni-Photo Communications, Inc. *Page 39:* Courtesy Bayer Corporation. *Page 51:* Bob Burch/Bruce Coleman, Inc.

Chapter 3: *Page 70:* Corbis-Bettmann.

Chapter 4: *Page 141 (top):* © Estate of Irving Geis. *Page 141 (bottom):* American Association for the Advancement of Science. *Page 169:* Ken Karp/Omni-Photo Communications, Inc.

Chapter 5: *Page 173:* Ken Karp/Omni-Photo Communications, Inc. *Page 182:* George Bodner.

Chapter 6: *Page 217:* © Ed Braveman/FPG International. *Page 217:* Ken Karp/Omni-Photo Communications, Inc.

Chapter 7: *Page 258, 268:* Andy Washnik.

Chapter 8: *Page 319:* Andy Washnik.

Chapter 9: *Page 365–367, 369:* Courtesy George Bodner.

Chapter 10: *Page 408:* Courtesy of BASF Corporation.

Chapter 11: *Page 441:* Michael Watson.

Chapter 12: *Page 547:* Ken Karp/Omni-Photo Communications.

Chapter 13: *Page 580:* Ken Karp/Omni-Photo Communications.

Chapter 14: *Page 596:* Michael Watson. *Page 613:* Douglas Burrows/Liaison Agency.

Chapter 15: *Page 654:* Courtesy Lawrence Livermore Laboratory. *Page 657:* From R. Saferstein, DNA Fingerprinting, *Chem Matters, 9,* 12 (1991). Reproduced with permission.

Index

Table of Atomic Number and Average Atomic Weight

Element	*Symbol*	*Atomic Number*	*Mass (amu)*	*Element*	*Symbol*	*Atomic Number*	*Mass (amu)*	*Element*	*Symbol*	*Atomic Number*	*Mass (amu)*	*Element*	*Symbol*	*Atomic Number*	*Mass (amu)*
Actinium	Ac	89	227.03*	Erbium	Er	68	167.26	Mercury	Hg	80	200.59	Scandium	Sc	21	44.956
Aluminum	Al	13	26.982	Europium	Eu	63	151.97	Molybdenum	Mo	42	95.94	Seaborgium	Sg	106	263.12*
Americium	Am	95	241.06*	Fermium	Fm	100	257.10*	Neodymium	Nd	60	144.24	Selenium	Se	34	78.96
Antimony	Sb	51	121.76	Fluorine	F	9	18.998	Neon	Ne	10	20.180	Silicon	Si	14	28.086
Argon	Ar	18	39.948	Francium	Fr	87	223.02*	Neptunium	Np	93	237.05*	Silver	Ag	47	107.87
Arsenic	As	33	74.922	Gadolinium	Gd	64	157.25	Nickel	Ni	28	58.693	Sodium	Na	11	22.990
Astatine	At	85	209.99*	Gallium	Ga	31	69.723	Niobium	Nb	41	92.906	Strontium	Sr	38	87.62
Barium	Ba	56	137.33	Germanium	Ge	32	72.61	Nitrogen	N	7	14.007	Sulfur	S	16	32.066
Berkelium	Bk	97	249.08*	Gold	Au	79	196.97	Nobelium	No	102	259.10*	Tantalum	Ta	73	180.95
Beryllium	Be	4	9.0122	Hafnium	Hf	72	178.49	Osmium	Os	76	190.2	Technetium	Tc	43	98.906*
Bismuth	Bi	83	208.98	Hassium	Hs	108	(265)**	Oxygen	O	8	15.999	Tellurium	Te	52	127.60
Bohrium	Bh	107	262.12*	Helium	He	2	4.0026	Palladium	Pd	46	106.42	Terbium	Tb	65	158.93
Boron	B	5	10.811	Holmium	Ho	67	164.93	Phosphorus	P	15	30.974	Thallium	Tl	81	204.38
Bromine	Br	35	79.904	Hydrogen	H	1	1.0079	Platinum	Pt	78	195.08	Thorium	Th	90	232.04
Cadmium	Cd	48	112.41	Indium	In	49	114.82	Plutonium	Pu	94	239.05*	Thulium	Tm	69	168.93
Calcium	Ca	20	40.078	Iodine	I	53	126.90	Polonium	Po	84	209.98*	Tin	Sn	50	118.71
Californium	Cf	98	252.08*	Iridium	Ir	77	192.22	Potassium	K	19	39.098	Titanium	Ti	22	47.88
Carbon	C	6	12.011	Iron	Fe	26	55.847	Praseodymium	Pr	59	140.91	Tungsten	W	74	183.85
Cerium	Ce	58	140.12	Krypton	Kr	36	83.80	Promethium	Pm	61	146.92*	Uranium	U	92	238.03
Cesium	Cs	55	132.91	Lanthanum	La	57	138.91	Protactinium	Pa	91	231.04	Vanadium	V	23	50.942
Chlorine	Cl	17	35.453	Lawrencium	Lr	103	262.11*	Radium	Ra	88	226.03*	Xenon	Xe	54	131.29
Chromium	Cr	24	51.996	Lead	Pb	82	207.2	Radon	Rn	86	222.02*	Ytterbium	Yb	70	173.04
Cobalt	Co	27	58.933	Lithium	Li	3	6.941	Rhenium	Re	75	186.21	Yttrium	Y	39	88.906
Copper	Cu	29	63.546	Lutetium	Lu	71	174.97	Rhodium	Rh	45	102.91	Zinc	Zn	30	65.39
Curium	Cm	96	244.06*	Magnesium	Mg	12	24.305	Rubidium	Rb	37	85.468	Zirconium	Zr	40	91.224
Dubnium	Db	105	262.11*	Manganese	Mn	25	54.938	Ruthenium	Ru	44	101.07				
Dysprosium	Dy	66	162.50	Meitnerium	Mt	109	(266)**	Rutherfordium	Rf	104	261.11*				
Einsteinium	Es	99	252.08*	Mendelevium	Md	101	258.10*	Samarium	Sm	62	150.36				

Source: These data are based on the definition that the mass of the ^{12}C isotope of carbon is exactly 12 amu. The atomic weight of an element is the weighted average of the masses of the isotopes of that element. Data from "Atomic Weights of the Elements," *Pure and Applied Chemistry,* **63,** 975 (1991).

*Radioactive element has no stable isotope. The mass is for one of the element's common radioisotopes.

**Mass number of the most stable isotope of this radioactive element.